PREFACE

The title of the present work is declaratory of its character. But a few words may be said as to its claims to confidence.

The mariner requires information of a specific nature respecting the countries he is interested in. For the Pacific, this information is only to be sought for in an immense mass of materials—the observations of a long series of authors on a vast variety of topics. The labour of seeking it among such a number of volumes as must be possessed to gain a complete insight into its navigation, must almost annihilate their utility in a practical sense; and this evil is also increased by the variety of languages in which many of them are composed. It is not very probable, moreover, even if it were desirable, that many of the most valuable works could be procured, as their rarity and cost have precluded them from general circulation. Thus the number of volumes exclusively applied to the description and narratives in the Pacific Ocean is above one hundred many of them of considerable bulk, and containing particulars in every department of literature. And those which less directly refer to our present subject, but still contain indispensable information, would amount to more than double that number. It is unlikely that a ship-master could either possess or properly avail himself of such a library.

It was to remedy this evil, to supply this want, that the work was undertaken; to draw up, for the use of the mariner, an hydrographical memoir, in a comprehensive and accessible form, from the mine of materials contained in the volumes which have been written on the Pacific; and in doing this, it is believed that no source of authentic information has been overlooked, but the whole range of works have been carefully referred to— a work of very considerable labour.

Although it directly refers to the navigation of the Pacific Ocean, it will be found that it offers sufficient interest to the general reader, in the varied features of the countries it describes, as it has been deemed

...e Pages 1320 ...1376 ...1335.

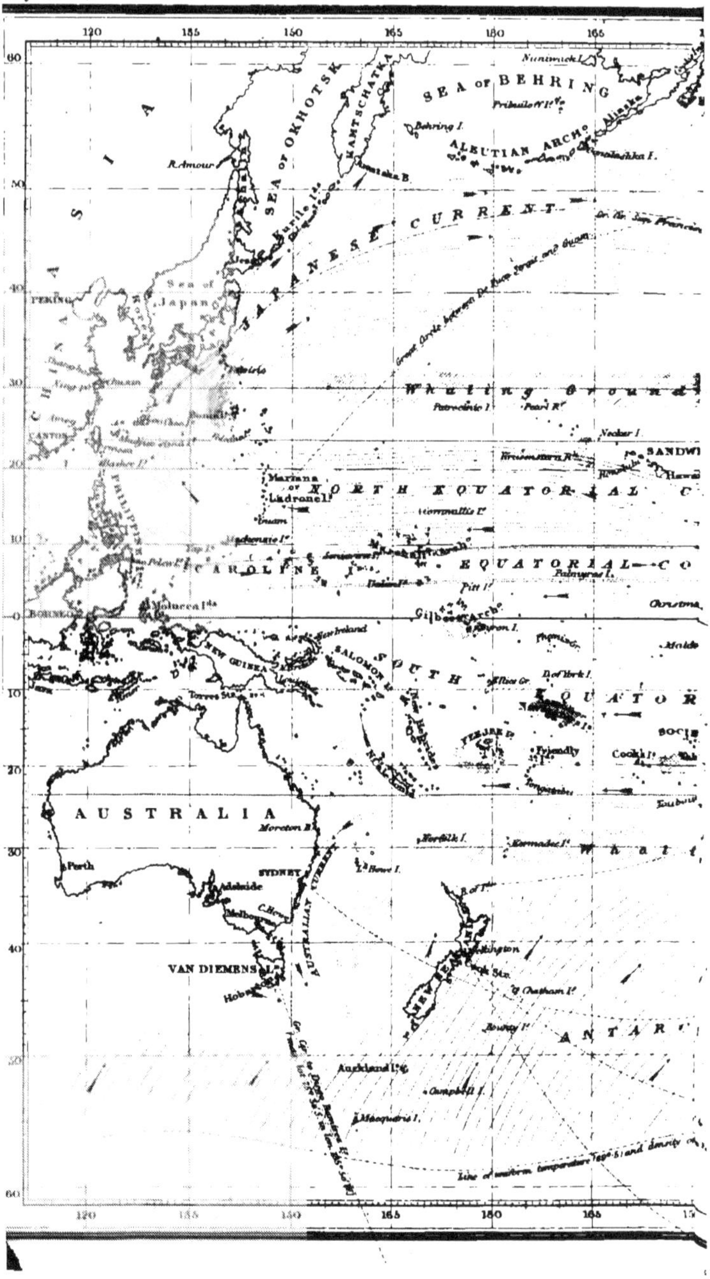

PUBLISHED BY R.H.LAURIE, 53, FLEET STREET, LONDON.

SECTION I.

INTRODUCTION.

THE existence of the great ocean to the westward of America soon became known to the European discoverers and conquerors of the New World. The Spaniards under Columbus, in 1502, had pushed their explorations along the northern coast of South America to the isthmus, and there established their power. Eleven years later, Basco Nuñez de Balboa, the Spanish governor of Darien, guided by native information, headed an expedition, among whom was Francisco Pizarro, and marched across the isthmus. He arrested their progress at the foot of a hill, from the summit of which he was told that this new sea was visible, and ascending alone, he fell on his knees and thanked heaven for having bestowed on him the honour of being the first European that beheld the sea beyond America. This was on September 25th, 1513. He subsequently took possession of it in the name of the King of Castile and Leon.*

It was called the Mar del Zur, the *South Sea*, because it was relatively so to the portion of coast from which it was first seen.

It was called *Pacific* by Fernando de Magalhaens, the first circumnavigator. This could only refer to the tropical portion, as Magalhaens sailed with great storms (*con gran tormenta*), till December, 1521, when they found themselves in lat. 32° 20′ S.

Thus the terms South Sea or Pacific Ocean are neither of them justly appropriate, seeing that it is neither more South nor more pacific than any other. But the names are recognised.

Maltè-Brun, who has been followed by many others, has proposed the name *Great Ocean*, or rather that of *Oceanica;* but as all authors who have especially written on the subject have used one or other of the former names, they ought to be adopted. Therefore that now best known, the *Pacific Ocean*, is the one to be used.

We would wish in the ensuing prefatory remarks to briefly enumerate the principal authorities upon which the present state of Pacific hydrography is based. And in so doing we ought really to commence at an era which would exclude many noble names—the leaders of that band of navigators who have added another world within the range of comparatively modern times to the knowledge of civilized men.

But as the plan of our work does not admit of mere historical detail, we must

* Vide *Gomara*, Istoria de las Indias, 1552, p. 34.

here pass over the earlier discoverers; they will be found alluded to in the subsequent pages. The reader who is interested in the progress of discovery will find the subject exhausted in the admirable work, "A Chronological History of Voyages in the South Sea," by Admiral James Burney.

We therefore commence at the period when modern science, that is, as at present understood, was first brought to bear on Pacific hydrography. This was in the famous voyage of Capt. James Cook. We shall, for convenience, enumerate the principal points in the different expeditions in a chronological order, commencing with his.

Capt. *James Cook* is justly placed at the head of English navigators, a pre-eminence which proudly overlooks all others. Perhaps there never was a narrative which attracted so much attention, or has dwelt so much in the memory of a nation, as the accounts of his voyages and his remarkable death. It is in the Pacific Ocean that his discoveries have elevated his fame, and prior to his expeditions it might be said that one-half of the oceans of the world were unknown to Europe. Cook, as is well known, was the son of an agricultural labourer, born at Marton, in Cleveland, Yorkshire, October 27, 1728. His first nautical exploits were as an apprentice on board a Newcastle collier. In 1755 he entered the navy, on board the *Eagle*, the command of which was soon taken by Sir Hugh Palliser, and this officer first recognised Cook's merits, a fact which ought not to be forgotten. Cook afterwards perpetuated his gratitude by naming the South extremity of New Zealand after him, the well-known N.E. of Cook's Strait; and also some islands in the Low Archipelago. His first public service was the survey of the River St. Lawrence, and piloting the boats to the attack of Montmorency and the heights of Abraham, at Quebec. Gaining the esteem of his superiors, as a master in the navy, he was selected, on the recommendation of his friends, Lord Colville and Sir Hugh Palliser, by Sir Edward Hawke to take the command of a scientific expedition to the South Seas; this was in 1767. The Royal Society resolved that it would be greatly to the advancement of astronomical and geographical science to send persons into some part of the South Sea, to observe the transit of the planet Venus over the sun's disc, which would happen in the year 1769, and that the islands called the Marquesas de Mendoza, or those of Rotterdam or Amsterdam, were the fittest places for that purpose. This resolution having been laid before the young king, George III., he directed that the *Endeavour* (a collier of 370 tons) should be fitted out, and the command given to Cook, then made a lieutenant.

This was while the expedition, also sent out by George III., under Wallis and Carteret was still at sea; but during the preparation Wallis returned, and then his discovery of King George's Island (Tahiti) was chosen as the best locality for the observations. Lieutenant Cook was accompanied by Mr. Green, as astronomer; and Mr. (afterwards Sir Joseph) Banks, a gentleman of fortune, and Dr. Solander, as naturalists, with proper assistants. The *Endeavour* left Plymouth August 26, 1768, and passed Cape Horn January 26th following, reaching Tahiti April 12, 1769. The transit was satisfactorily observed at the observatory on the well-known Point Venus, an account of which is given in the Philosophical Transactions, vol. lxi. part ii. p. 397, *et seq.* Quitting Tahiti, they reached Poverty Bay, New Zealand, on Friday, October 7, 1769. New Zealand at this time was little known, and most of the crew thought they had come on the *terra australis incognita*, then the great problem of geography. Having sailed all round the islands and determined their insular character, the separation between them, which he first sailed through, was called Cook's Strait; and as an evidence of his accurate observation, a dangerous rock in it is clearly described by him, but has been several times announced of late as a new discovery (page 770, part ii.). Quitting Cape Farewell, New Zealand, March 31, 1770, they reached the coast of New Holland, at Point Hicks, and Cape Howe April 19. Coasting along to the northward, he gave most of the names to the capes and bays now so familiar to the colony of New South Wales, and anchored in Botany Bay, so named from the great quantity of plants collected by Mr. Banks and Dr. Solander. Passing out from this, May 6, he saw and named Port Jackson, and then passed Broken Bay, Port Stephens, Moreton Bay, Harvey Bay, and indeed all the eastern coast of Australia, passing through Torres Strait, and thence to New Guinea, Batavia, the Cape of Good Hope, St. Helena, and arrived at Deal June 10, 1771.

The account of this interesting voyage was prepared for the press by Dr. Hawkesworth, who, by adding remarks of his own, called down some severe but unmerited remarks on the narrative. But it forms one of the most interesting memoirs in hydrography; and in this voyage, as is said by his friend and fellow-officer, King—"he discovered the Society Islands; determined the insularity of New Zealand; discovered the straits which separate the two islands, and are called after his name; and made a complete survey of both. He afterwards explored the eastern coast of New Holland, hitherto unknown; an extent of 27° of latitude, or upwards of 2,000 miles."

It certainly is marvellous that Cook should, almost within the memory of persons now living, have discovered and named those countries now so familiar to Englishmen as are formed in their populous colonies of New Zealand and Australia. Respecting the accuracy of Cook's observations, they are unimpeachable. The minute accuracy of detail is not to be found in his charts which is required in modern times, but they certainly claim for him the title of the father of modern hydrography, a character more fully sustained by the result of his second voyage, which followed immediately on the completion of the first, and obtained for him great personal honour and the rank of commander.

The *second voyage* of Commander James Cook was undertaken for a different end. King George III., his former great patron, determined on sending him to the southern hemisphere, to determine the existence or non-existence of the supposed southern continent. This was determined on soon after his return in the *Endeavour*. He was appointed to the command of the *Resolution*, November 28, 1771; and Tobias Furneaux (who had been second lieutenant with Capt. Wallis) was appointed to the *Adventure*. These two ships had been colliers, and were chosen by Cook. They were equipped and victualled in a very superior manner to the ordinary mode, and the beneficial effects of the antiscorbutics and other stores were soon manifest in the absence of scurvy throughout the voyage: it had always been a dreadful scourge in previous voyages. As a scientific corps, Mr. John Reinhold Forster, with his son, were appointed naturalists; the Board of

Longitude agreed with Mr. W. Wales and Mr. W. Bayley as astronomers, and Mr. Hodges as artist. Each of these departments was most amply illustrated by the respective parties in separate publications. Some personal pique respecting accommodation, it is said, prevented Sir Joseph Banks from again accompanying Cook. But there is little doubt but that his society must have been very irksome to Cook, placed as he was in circumstances which must have greatly interfered with the discipline of the ship.

The second expedition quitted Deptford April 9, 1772, and left the Cape of Good Hope November 22, in search of the primary object of his voyage, the southern continent, reaching lat. 67° 15′ S., lon. 40° E., January 17, 1773; but, as is well known, without meeting with any land between the meridian of the Cape of Good Hope and that of New Zealand. He entered Dusky Bay March 26. Capt. Furneaux had parted company February 7, and joined the *Resolution* in Queen Charlotte Sound, New Zealand. The ships then proceeded to Tahiti, and thence to the Tonga or Friendly Islands, called by Tasman Amsterdam and Middleburg, and again returned to Queen Charlotte Sound, in which passage they parted finally with the *Adventure*, which reached Queen Charlotte Sound after Cook had left. Here Furneaux had the sad misfortune to lose a boat's crew, ten in number, who were murdered and eaten by the natives of Queen Charlotte Sound. He then stood to the S.S.E. to lat. 56° S., and arrived off Cape Horn in little more than a month from Cape Palliser, New Zealand, and thence on to the Cape of Good Hope and England. Cook again proceeded on his search for antarctic lands to the South, seeing the first ice in lat. 62° 10′ S., lon. 172° W., December 8, 1773; and reaching his highest, lat. 71° 10′ S., lon. 106° 54′ W., January 30, 1774. He now made his way to the northward, reaching Easter Island March 11, 1774, of which he gives a detailed account. Thence he went to the Marquesas, and afterwards to Tahiti, and other of the Society Islands, passing some of the islands of the Parmento group. In the passage thence he discovered Palmerston and Savage Islands, and touched at Namuka (Namocka), in the Tonga Islands. The New Hebrides were his next point, visiting Mallicolo, Tanna, Erromanga, and other islands. His next discovery was New Caledonia, one of the largest and finest islands of the Pacific. He gives a considerable amount of detail respecting this island, and left it October 6, 1774, discovering Norfolk Island October 10, and again reached his former anchorage in Queen Charlotte Sound. He then, November 10, proceeded across the Pacific Ocean in a high latitude, supposing that his passage was the first ever made, not knowing that his consort, the *Adventure*, had preceded him. His passage was thirty-seven days to Cape Horn, and finally reached England July 29, 1775.

Beside the very important additions he made to our knowledge of the geography of the Pacific, which have never been equalled by any single voyage, this had one very marked effect on all future nautical undertakings. All the histories of early discovery by long voyages have shown at what a dreadful rate the advantages of increased knowledge were purchased, in the wholesale destruction of the crews by that frightful disease, scurvy. The fate of Behring is an example of this. But it was reserved for Cook to demonstrate that protracted voyages might be made, even extending to three and four years, in all climates, without diminishing the probability of life in any degree. His system, the basis and commencement of that now in use in all well-conducted ships, was submitted to the Royal Society, who presented him with their gold Copley medal on that occasion. This alone will place Cook among the greatest benefactors to the maritime world.

The *third voyage* of Capt. Cook was the result of the great success which attended his former expeditions, and very quickly followed. His new ships were the *Resolution* and the *Discovery*, commanded by Capt. Clerke, his former lieutenant on board the *Resolution*. Among Cook's officers were Lieutenants Gore and King, and his master was William Bligh, all distinguished afterwards. With Capt. Clerke was Lieutenant (afterwards Admiral) James Burney, the historian of the Pacific. The expedition quitted Long Reach May 29, 1776; its object was the exploration of the North Pacific, and to examine the connexion or separation of the American and Asiatic continents. They passed round the Cape of Good Hope, and anchored in Adventure Bay, Van Diemen's Land; thence they sailed to their old port, Queen Charlotte Sound, in New Zealand; thence proceeded toward the Tonga group, discovering Mangeea and Atiu on the passage, and making a more minute examination of the Friendly Islands. Thence they proceeded toward Tahiti, discovering Toobouai. They landed Omai, a chief brought to England by Cook in his former voyage, at Tahiti, and then, proceeding to the northward, made the grand discovery of the Sandwich Islands. He saw and examined all of them, and has recited many most interesting details of them. Some of these are alluded to on page 1121, part ii. From the Sandwich Islands he proceeded to the N.W. coast of America, anchoring in Nootka Sound. Thence he proceeded to the N.W., examining the coasts, and entered Prince William's Sound and Cook's River (or Inlet), reaching Ounalashka. Thence he proceeded to the northward, through Behring's Strait to the icy barrier, establishing the real character of the countries and the erroneous condition of the maps. After again touching at Ounalashka, he made away for Karakakooa Bay, in Hawaii. Here the well-known tragedy occurred. Cook, the Lono or god of the Hawaiians, was killed, and science lost one of her greatest contributors. This sad event occurred February 14, 1779 (see page 1133, part ii.). Lieutenant King then took the second command under Capt. Clerke. The ships proceeded to Awatska Bay, and again fruitlessly attempted the N.W. passage. Before the return to Kamtschatka, Capt. Clerke died, and Capts. Gore and King became the commanders. They proceeded along the Japanese coast to Macao, and thence by the Cape of Good Hope, and reached Stromness, in the Orkneys, August 22, and the Nore, October 4, 1780, the two ships having been absent four years, two months, and twenty-two days, and never having lost sight of each other but twice during their voyage. Thus concluded the most celebrated voyages of modern times, and from which dates a new era in hydrography. The account of Cook's last voyage was published under the superintendence of Dr. Douglas, Bishop of Salisbury, and is one of the most interesting works in the language.

The immediate predecessors of Cook in discovery had no mean harvest in the Pacific. The voyages referred to, as directed by King George III. in the early part of his reign, were recorded, or perhaps the narrative arranged, by Dr. Hawkesworth. The first of these was under Commodore Byron, who sailed from the Downs January 21, 1764, with the ships *Dolphin* and *Tamar*, and, passing through the Strait of Magalhaens into the Pacific, discovered the Islands Disappointment, George, Prince of Wales, Danger, Duke of York, and Byron, reaching England again May 9, 1766.

In the following August the *Dolphin* was again sent out under the command of Capt. Wallis, with the *Swallow*, under Capt. Carteret. They separated at the western entrance of the Strait of Magalhaens. Wallis discovered Whit Sunday, Queen Charlotte, Egmont, Duke of Gloucester, Duke of Cumberland, Maiatea, *Tahiti*, *Eimeo*, Howe, Scilly, Boscawen, Keppel, and Wallis Islands, and returned to England in May, 1768. Carteret pursued a different course, discovering Osnaburg, Gloucester, Queen Charlotte, Carteret, and Gower Islands, and the strait separating New Britain and New Ireland, and reached England March, 1769.

The French at this time despatched Mons. Bougainville, in November, 1766, in the frigate *La Boudeuse*, with the store-ship *L'Etoile*. He entered the Pacific by the Strait of Magalhaens, in January, 1768, and discovered several islands in the northern part of the Low Archipelago, Lanciers, Harpe, Thrum Cap, and Bow Islands. He also discovered Tahiti, and thought he was the first; then the Navigator Islands, and passed between the New Hebrides and the Louisiade, arriving in France March, 1769.

While Cook was employed in the Pacific, the Spaniards were not behind in advancing discoveries.

The expedition in 1775, in the ship *Santiago* and *Sonora*, under *Bruno de Heceta* and *Ayala*, made several discoveries on the coast of Oregon, &c. With these ships were the pilot *Antonio Maurelle* and Lieutenant *J. F. de la Bodega y Quadra*, names still retained in some places on the coast.

The discoveries of Cook had given a great impetus to the energies, commercial and exploratory, of every one at the time; and one circumstance, trivial in itself, led to great results.

During the third voyage of Cook, the sailors had procured a number of sea-otter skins on the N.W. coast of America, which sold at enormous profits at China. This fact led to important results. The voyages of *Portlock*, *Dixon*, *Meares*, *Tipping*, *Kendrick*, *Lowrie*, *Guise*, and others, and also indirectly of that of La Pérouse and Krusenstern, arose out of it.

The voyage of *La Pérouse* is one of the interesting points in maritime history. It was undertaken in order to the extension of French commerce, at the time when Cook's voyages had given so great an impetus to trade in the Pacific; and one of the first objects of the voyage of *L'Astrolabe*, under La Pérouse, and *La Boussole*, under Capt. De Langle, was to examine the N.W. coast of America, where the furs were procured. This was followed from Mount St. Elias (June 23, 1786) to Monterey, from whence they proceeded to Canton; after that to Kamtschatka, sailing into the Sea of Saghalin. Their next destination was Navigator and Friendly Islands; and lastly, Sydney, in New South Wales. After the ships quitted this port nothing more was heard of them, notwithstanding all the search and inquiry that was made; until the year 1826, some articles were found at Tucopia, which were traced to Vanikoro, where Capts. Dillon and D'Urville found undoubted evidence of the wreck of two ships, and the departure of the crews in a vessel built from the wrecks, but never more heard of. The accounts of this are given on pages 955-6, 964.

One voyage about this period, better known for its fate than the scientific results it obtained, is that of the *Bounty*, under Capt. William Bligh, who went to procure bread-fruit to convey to the West India Islands. The facts of the mutiny and escape of the mutineers to Pitcairn Island are now familiar to all. Lieutenant *Edwards*, in the *Pandora* frigate, was sent on an unsuccessful search for the mutineers in 1791. Bligh made a second but unpublished voyage to accomplish his object.

The unfortunate voyage of *D'Entrecasteaux* and *Huon-Kermadec*, in the French frigates *La Recherche* and *L'Esperance*, was for the purpose of ascertaining the fate of La Pérouse. They left Brest September, 1791, and returned in 1794. The principal additions they made to our knowledge were about Van Diemen's Land S.W. Australia, New Caledonia, Louisiade, &c. But this knowledge was purchased with the death of the two commanders, and 99 out of 219 people, forming the original crews. An account of this voyage was given to the world by Admiral Rossel, and another by M. Labillardière.

La Pérouse had examined a considerable portion of the N.W. coast of America, but much remained to be done, which was left for Vancouver.

Capt. *George Vancouver* was an officer in Cook's second voyage toward the South pole, and also under Capt. Clerk, in his third and last voyage. He was appointed to the command of the *Discovery* sloop of war, December, 1790; his consort was the *Chatham*, an armed tender, under Lieutenant Broughton. The object of his voyage was a double one. The Spaniards had made some supposed aggressions at Nootka Sound, which the English were to resent, and the old theory of a large and navigable river to the Atlantic had been revived. Vancouver was to settle both these points. If this commander had not the varied talents and enterprise of his master, Cook, his surveys at least give evidence of indefatigability. His surveys extend from the Bay of St. Francisco, lat. 30½° N., to Cape Douglas, the S.W. point of Cook's Inlet. The vast extent of a portion of the most intricate coast in the world appears to be delineated most faithfully. His object being to trace the continental continuity, the whole of the singular "canals" which penetrate the coast were minutely examined. He also determined the insularity of Vancouver Island, and the character of the dense and still unpeopled archipelagoes to the northward. As far as exact knowledge went, Vancouver may be said almost to be the discoverer of much of the coast. Besides this, his officer, Mr. Broughton, discovered the Chatham Islands, King George's Sound, South Australia, the Snares to the South of New Zealand; and a more full examination of the Sandwich Islands are a portion of the acquisitions made in this expedition, which occupied from 1792 to 1794. Vancouver's longitudes vary very considerably from those of Cook, though the details of his survey have been highly applauded. Sir Edward Belcher had for one object to reconcile these differences. Vancouver's longitudes are therefore corrected in accordance with this and other more exact determinations.

The voyage to the N.W. coast of America by Commanders *Galiano* and *Valdez*, in the Spanish ships *Sutil* and *Mexicana*, and which were unexpectedly fallen in with by Vancouver, when proceeding to the northward at the back of Vancouver Island, was the *last* voyage made by the Spaniards for discovery in the North Pacific Ocean. A meagre account of it was published by the government at Madrid, in 1802. There is, however, a valuable historical introduction prefixed to it. The book itself is superseded by the elaborate and lucid work of Vancouver. The colonization of New South Wales, in 1788, by Governor Philip, led to some discoveries in the voyages of the ships which conveyed the convicts. The ships *Scarborough* and *Charlotte*, under Capts. *Marshall* and *Gilbert*, have given two names to respective archipelagoes in the North Pacific, and also of Lieutenant Ball, in H.M.S. *Supply*.

Capt. *Don Alessandro Malaspina*, an accomplished Italian in the service of Spain, was unfortunate, not in his voyage, but in offending Godoy, the well-known Prince of the Peace, who on his return, in 1794, threw him into prison at Coruña, where he was liberated by Napoleon in 1802. His voyage was in the *Descubierta;* his consort was the *Atrevida*, commanded by Capt. *Bustamente*. They examined the N.W. coast of America, between Prince William's Sound and Cape Fairweather, and also the S.W. coast of Mexico and Central America. His journals or charts have never been published. A sketch of his voyage is given in the introduction to the voyage of Galiano and Valdez, *but his name is never once mentioned* in it. The highest laudation is there given to the officers engaged in it. What we know of it is from the charts subsequently drawn up by *Don Felipe Bauza*, who was in the expedition. To this we owe our present knowledge of the Mexican coast alluded to.

The voyage of Capt. *Etienne Marchand* to Vancouver Island, &c., in 1791, would not be worth mentioning, were it not that the account of it is preceded by a really valuable historical and critical introduction by the talented geographer, *Fleurieu*.

Some years now intervened, and the Pacific was comparatively untraversed by scientific voyagers. But circumstances arose which led to the following voyage.

Adam John Von Krusenstern is the hydrographer of the Pacific. This is a proud position, and is worthily occupied. Beside this, he was the first Russian who circumnavigated the globe in a Russian ship. This is another interesting feature, but his services to his country are not second to either of them. He was the descendant of a noble family, and in his youth, by command of his government, he served for six years (from 1793 to 1799) in the English fleet. During this period, regretting that his country, with her vast power and resources, did not participate in the advantages of the growing commerce which was being established in all parts of the world, he conceived the noble project of forming direct commercial relations between Russia, India, and China. To advance his patriotic scheme, he went to the East Indies, where he passed a year, studying its commerce, and then reached China, in order to gain some knowledge of its dangerous seas. He remained here two years (1798-99), and acquired numerous facts respecting the fur trade between China and the N.W. coast of America. On his homeward voyage he formed a plan to raise a mercantile navy to carry out his views of commerce between Russia and China. Like Columbus, he found himself neglected and misunderstood on his return, but on the accession of Alexander, his propositions were appreciated.

The expedition consisted of the *Nadiéjeda* (the *Hope*) and the *Neva*: Capt. Krusenstern commanded the first and the expedition, and the second was commanded by Capt. Lisiansky. The object of the voyage was of a varied character. It was to be one of discovery and science; also to establish some plan by which the Russian American Company could more conveniently communicate with their languishing colonies on the N.W. coast of America than by long and tedious journies overland to Okhotsk. It was also destined to convey the ambassador appointed by the Russian emperor to the court of Japan. And, finally, it was to be a voyage round the world, the first undertaken by the Russians. All these objects were triumphantly carried out by the excellent commander with but the loss of a single man, who was previously diseased!

The *Nadiéjeda* and *Neva* left Cronstadt August 7, 1803, and returned in 1806. Capt. Krusenstern's immediate sphere was Japan and its vicinity. He gave us the first true notions of its western side, and also has surveyed the eastern side of the great Peninsula or Island of Saghalin. His chart of the Japanese empire is a masterpiece. We cannot enumerate here all the points inquired into by this voyage; they will be found alluded to in various places in the work.

His consort, under Capt. Urey Lisiansky, a volunteer, but the senior officer to his commander, separated from the *Nadiéjeda* in the Pacific, and proceeded to his destination, Kodiack and the N.W. coast of America. We owe many details of the Sitka Islands, and nearly all we know of the Kodiack Archipelago, to Lisiansky.

It was during his voyages around Japan, the Sea of Okhotsk, and the Kurile Islands, Krusenstern tells us, that he conceived the idea of his great work, the Atlas of the Pacific Ocean, knowing how many perils, how many anxious moments it would have spared him and his companions, had such a work been within their reach. Busily employed in the services of his profession, it was not until 1815 that he was allowed the leisure to pursue his favourite object, and this was in consequence of impaired vision. It is difficult to speak in sufficiently high terms of this noble work, the "*Atlas de l'Ocean Pacifique*," and the accompanying "*Recueil de Memoires Hydrographiques, pour servir d'Analyse et d'Explication à l'Atlas.*" The first part of this was published at St. Petersburg, in Russian and French, at the imperial expense, in 1824, and related to the South Pacific. That for the North Pacific appeared in 1827. In 1834 a supplemental volume, and a corrected edition of the atlas appeared; and several papers subsequently appeared in the bulletin of the Imperial Academy of St. Petersburg.

In these memoirs Vice-Admiral Krusenstern had embodied everything that had previously been observed in the Pacific, and in the most masterly manner he has reconciled the discordant materials at his command, and placed its hydrography upon an entirely new basis. Up to the date of these publications, it may be said that the work had exhausted the subject. Being in the French and Russian languages, their utility was somewhat contracted for the nautical world; and moreover the chief bulk of the remarks consist of discussions on the *scientific* basis on which the charts were constructed; and the whole of the American coast was reserved for a future work which was never produced. All the determinations, except where they have been superseded by later and better observations, have been followed in the present work.

The Russians have, besides the great Krusenstern, performed no small share of the discovery and exploration of the Pacific. Capt. *Frédéric Lütke* may justly claim a high position among hydrographers. His work, the Voyage of the *Séniavine*, though but little known, is an excellent one. Capt. Lütke surveyed the inclement coasts of Asia, from the North of Behring's Strait to the extremity of Kamtschatka, and also made many observations in Behring's Sea. He has drawn up his useful memoir from many Russian sources besides his own labours. In the Caroline Archipelago, too, his explorations are very conspicuous.

The Russian language is an unfortunate barrier to most Europeans, and to this we owe our comparative ignorance of the important voyages of Capt. *Bellingshausen*, and also of Capt. *Sarytscheff*, among the archipelagoes of the Central Pacific.

The voyage of *Billings* (an Englishman who was with Cook), a very expensive one, was most unproductive. It was begun in 1785, by order of the Empress Catherine, "a secret astronomical and geographical expedition to navigate the Frozen Ocean between Asia and America." The vessels were built at Okhotsk, but were not ready till 1789, when one of them was immediately wrecked. The other only reached Mount St. Elias, stopping at Ounalashka, Kodiack, and Prince William's Sound. It then returned to Kamtschatka, and its commander abandoned it. A melancholy picture is drawn of their sufferings by *Martin Saner*, who, unfortunately for himself, was secretary. In the following year the expedition was resumed in the Sea of Behring, under Capts. *Hall* and *Sarytscheff*, but was not very successful.

Otto Von Kotzebue, the son of the celebrated author, was a cadet on board the *Nadiéjeda*, under Krusenstern, in 1803—1806. He was selected by that great hydrographer to command a vessel sent out by the munificence of Count Romanzoff, and named the *Rurick*, to endeavour to penetrate to the North of Behring's Strait, and make other explorations. It left Cronstadt July 30, 1815, and was unsuccessful in the primary object, and reached no farther than Kotzebue Sound, to the North of the West cape of America. Some discoveries in the Radack Channel, the Low Archipelago, and the Carolines, and other important services, were the results of this voyage, which lasted till August 3, 1818, when the *Rurick* anchored in the Neva.

His *second* voyage was in the Russian ship the *Predpriatie* (the *Enterprise*); and was intended to protect the Russian American Company from the smuggling then carried on by foreign traders. The ship left Cronstadt July 28, 1823, and made many important additions to our knowledge of the Low Archipelago, surveyed the Navigator Islands, the Radack Islands, Sitka, and the Ladrone Islands, returning to Cronstadt July 10, 1826.

Among those at the head of these illustrious navigators who have enriched science by their exertions, *Jules Sébastien-César Dumont D'Urville* must be placed. His first expedition, in the *Astrolabe*, left Toulon April 22, 1826. He examined parts of the coasts of New Zealand, the Tonga Islands, the Feejee Islands, the Santa Cruz Islands, the Loyalty Islands, and then the great chain of reefs extending off New Caledonia. He then passed on to New Britain and New Ireland, and the North coast of New Guinea. On his return to Hobart Town he received intelligence that Capt. Dillon had discovered the remains of La Pérouse's expedition, to the scene of the loss of which, Vanikoro, he then repaired, as related on page 965, part ii. of this work. Quitting this, he passed on to Guam and part of the Carolines, arriving at the Mauritius September 29, 1828, and at Toulon March 25, 1829. More extended examinations have since been made of many of his explorations, but at the time the expedition greatly increased our then imperfect knowledge. His second voyage, though out of its chronological order, we will notice here.

The second expedition under M. D'Urville, consisting of the *Astrolabe* and *Zelée*, the latter under the command of Capt. C. H. Jacquinot, quitted Toulon September 7, 1837, and reached the South Shetland group, where he made many additions to our knowledge; thence entering the Pacific, he visited Manga Reva, Marquesas, Society Islands, Tonga Islands, the Feejee Islands, Vanikoro, the Salomon Islands, the Ladrone Islands, and then entered the Asiatic Archipelago, and thence to Hobart Town. Quitting this, he made for the antarctic regions, and discovered portions of the supposed continent. He then again examined some portions of New Zealand, the Louisiade, thence out of the Pacific, and reached Toulon November 6, 1840.

The frightful death of the celebrated Dumont D'Urville, who, with his wife and son, were by one accident hurried into eternity, May 8, 1842, will be long remembered in France. They were travelling on one of the Paris and Versailles railways, when, in consequence of the engine failing, the whole train was overturned and burnt, together with a large number (upwards of forty) of the passengers.

To the voyage of *M. Freycinet*, with the ships *L'Uranie* and *La Physicienne*, in 1819, we owe the greater part of our knowledge of the Mariana or Ladrone Islands, and also of the Samoan group much information was acquired. This voyage was exceedingly productive in additions to natural history and science generally, and the Atlas Historique, accompanying the voyage, is really a fine work.

The voyage of Capt. *Duperrey*, the officer under Freycinet, sailed in *La Coquille* in 1822—1825, and made many additions to the hydrography of the central and western Pacific. His route was around Cape Horn, to Callao, Payta, the Low Archipelago, and Tahiti, which he reached May, 1823; thence to Port Praslin in New Ireland, New Guinea, the Moluccas, round to the West of Australia to Van Diemen's Land and Sydney; thence to the northward, through the Mulgrave Islands, &c., and examining the Caroline Archipelago, reached France *viâ* the Cape of Good Hope. The whole of the accounts of this voyage have not appeared, but Capt. Duperrey has done good service to science by his current and variation charts.

The voyage of the *Blossom*, under Capt. Frederick William Beechey, is one of the most important, in a scientific view, that we have in the Pacific. It was undertaken to afford assistance to the expeditions which had been despatched from England, under Capt. (afterwards Sir John) Parry, for the discovery of a N.W. passage through the arctic regions, and under Capt. (afterwards Sir John) Franklin, who intended descending the Mackenzie River and reaching Behring's Strait. The *Blossom* left Spithead May 19, visited Valparaiso and Pitcairn Island, minutely examined several islands in the Low Archipelago, and proceeded to Tahiti, and thence to the Sandwich Islands. The special object of the expedition was then proceeded with, and a minute survey made of the American coast, of Behring's Sea and Strait, between King's Island and Point Barrow. As is well known, the other expeditions were not met with. The *Blossom* returned to and surveyed San Francisco in California, reaching there November 6, 1826. The next point was revisiting the Sandwich Islands, and thence to Macao. The Loo-Choo Islands were next visited and surveyed, as were the Arzobispo or Bonin Islands. Petropaulovski was next visited, and then the attempt again made to pass through Behring's Strait. The return thence was made by way of San Francisco, Valparaiso, and reached Woolwich October 12, 1828.

H.M.S. *Sulphur* left England December, 1835, under the command of Capt. F. W. Beechey, accompanied by the *Starling*, Lieutenant Kellett. Capt. Beechey invalided at Valparaiso, and was at last succeeded by Capt. (afterwards Sir Edward) Belcher (who had sailed in the *Blossom*), at Panamá, February, 1837. The operations of the *Sulphur* consisted in determining the longitudes of points in dispute between Vancouver and Cook on the N.W. coast of America; examining and surveying several of the ports, and a portion of the coast of Central America and California; and returned to Panamá from her first cruise in October, 1837. After this she proceeded across the Pacific, visiting San Blas, Mazatlan, the Revilla Gigedo Islands, and the Marquesas; thence to Bow Island in the Low Archipelago, then Tahiti, the Feejee group, New Hebrides, &c.; and then to an active part in the Chinese warfare, till nearly the close of the year 1841, when she sailed for England.

The surveying voyages of H.M.S. *Adventure* and *Beagle* are certainly the most important in the present hydrography of the Pacific. The whole of the Pacific coast of South America was most accurately delineated by their means. These two ships were commissioned in 1825, Commander P. P. King to the *Adventure*, and Commander Pringle Stokes to the *Beagle*. They left Plymouth May 22, 1826, commencing operations South of the Plata in the ensuing November. In August, 1828, the sad death of Commander Stokes occurred; he was temporarily succeeded by Lieutenant Skyring, but ultimately Capt. Robert FitzRoy commanded the *Beagle*. The result of this portion of the expedition was the noble survey of the South extremity of America, with the Strait of Magalhaens, from the La Plata to Chiloe; and the excellent instructions, quoted in our work, have added more to the security of navigation than anything which preceded them. The first expedition quitted Rio for England in August, 1830.

The second expedition, under Capt. FitzRoy in the *Beagle*, was commissioned July 4, 1831. The result of this was the completion of the survey of Tierra del Fuego, and the continuation from the Gulf of Peñas northward, along the coasts of Chile and Peru, to Guayaquil. The Galapagos Islands were likewise included in this admirable survey. Leaving this coast, a chain of meridian distances was carried around the globe, the first of the kind, reaching Woolwich November 17, 1836. The directions drawn up in the survey, and given in the first part of this work from the appendix to Capt. FitzRoy's book, will say more in eulogy than we can. The labours of the accomplished naturalist, Charles Darwin, Esq., ought not to pass unnoticed here, though not connected with our subject.

The especial object of the voyage of the French frigate *La Venus*, under Capt. *Abel Du Petit Thouars*, in 1837—1839, was for the protection and encouragement of the whale-fishery, and his voyage was framed with the view of collecting all information on this head, for the purpose of increasing and establishing the French whaling interest in the Pacific. The *Venus* left Brest December 31, 1836, and returned to that port June 24, 1839. She arrived at Valparaiso April 26, after staying at Rio fifteen days; thence to Callao, from whence she made way for Oahu, in the Sandwich group, where her commander interfered in the disputes between the religious parties, as alluded to on page 1122, part ii., meeting here H.M.S. *Sulphur*, under Capt. Belcher. From Honolulu she proceeded westward, and reached Awatska Bay; thence she sailed for Monterey, arriving there October 18, 1837, with above thirty of her crew severely attacked with scurvy. Quitting Monterey she proceeded to Guadaloupe Island, which was examined, and the important position of Los Alijos Rocks (page 309, part i.) ascertained. Touching at Cape St. Lucas she reached Mazatlan, San Blas, and Acapulco, giving us some details of each of those places, and then to Valparaiso a second time, March 18, 1838. St. Ambrose and St. Felix Islands, called by the Spaniards the Unfortunate Islands, were next visited, and then Callao, surveying the Hormigas Rocks, and touching at the Galapagos. The Marquesas Islands were then surveyed. The northern part of the Low Archipelago was next traversed, and the *Venus* anchored at Papeite, at Tahiti, where, as at the Sandwich Islands, a convention was drawn up assuring protection to French ships. Passing thence toward New Zealand, they examined several islands in their track, among which were the Kermadec. reaching the Bay of Islands October 12. Sydney was the next port, which was quitted December 18, 1838, and *La Venus* entered Brest June 24, 1839. Under the excellent observations of M. U. de Tessan many great additions to the hydrography of the Pacific were made; among which may be particularized the chart of the Marquesas, and of the several ports she visited. These will be found alluded to in their respective places, and his positions deserve much confidence. The scientific portion of the voyage received due attention, and many branches of natural history were thereby much enriched. The principal object of the voyage is fulfilled in the Report to the Minister of Marine, "On the Whale Fishery in the Pacific Ocean," comprising a full account of the armament, ships, ports, instructions, crews, discipline, nature of the fishery, &c., &c., and related in the third volume of the Narrative of the Voyage.

We should not omit one author, *Von Siebold*, who visited Japan in 1823—1830, and whose noble work on Japan ought to remove the stigma sometimes attached to the Dutch of their exclusiveness, in withholding information on the subjects which they only have the power to gain. His work on "Nippon" is most certainly worthy of a nation.

The *United States' Exploring Expedition* deserves especial notice for one particular—"It was the first, and is still the only one, fitted out by national munificence for scientific objects, that has ever left the shores of the United States." Its organization appears to have been arranged under great—all but insuperable—difficulties; its result is given to the world in five goodly volumes of narrative, a series of scientific memoirs, and an hydrographical atlas—works worthy of a great nation.

The act of congress which authorized the undertaking is dated May 18, 1836, but it was not until March, 1838, that it devolved upon Lieutenant Charles Wilkes to re-arrange it and arrest it from complete failure. The ships composing the new squadron were the *Vincennes* sloop, 780 tons; the *Peacock* sloop, 650 tons; the *Porpoise*, a gun-brig, 230 tons; and two tenders, formerly New York pilot-boats—the *Seagull*, 110 tons, and the *Flying Fish*, 96 tons. Of the respective labours of these five vessels it will be too diffusive to speak, and therefore the general results of the Expedition will be enumerated.

Their first field, the coast of South America, had been amply examined and surveyed by the English Admiralty, so that little or nothing was left for the American to add to hydrography here. The central portions of the American West coast were not approached; but the next examination was that of various points on the coast of Upper California, and many particulars were gathered respecting that country which was soon, and then so unexpectedly, to be the centre of such intense interest. Capt. Beechey, however, having previously excellently surveyed its principal port, this portion is of less importance. The Columbia River was the next point which was visited by the Expedition, and a survey made of its entrance. Capt. Belcher had also surveyed it, but its changing character renders a chart of little value for any length of time. Puget Sound was next surveyed, and Lieutenant Wilkes bears ample testimony to the accuracy of our countryman, Vancouver's, delineation of that singular and interesting inlet.

It is chiefly in the islands and archipelagoes of the Pacific that the great results of the American Exploring Expedition become more apparent. The centre of their operations was the Hawaiian group. Of this they have given us a large and detailed chart, and many most interesting particulars are given in almost all branches of science, respecting the physical and social character of the various islands composing it, especially of Hawaii and its geological features.

The Low Archipelago, or Paumotu group, was also partially surveyed in its north-western groups, and the south-western range correctly placed on the charts. In these islands many discrepancies were reconciled, and a true estimate formed of their area and form. Portions of the Society Islands, Tahiti, Eimeo, Huaheinè, &c., also received their share of attention.

The Feejee group, however, is the great harvest of the Expedition. D'Urville had been here and partially surveyed it in 1827; but the Expedition made a complete survey of it in all its parts, and for the first time gave the world a correct notion of the extent and productions of this noble archipelago. The details of the customs and ferocity of its inhabitants are given with most intense interest in vol. iii. of the Narrative. One thing, however, ought to be noticed. Lieutenant

Wilkes treats the archipelago as if it had been entirely a new discovery of the American Exploring Expedition.

The Tonga or Friendly Islands were also visited, and many additions made to our knowledge. The Samoan or Navigator's group, to the northward, was also minutely and amply surveyed, and the numerous charts of these fine islands form a conspicuous feature in the Hydrographical Atlas. To the northward still, the collection of islands now named the Union group was also correctly ascertained as to its numbers, character, and position.

The Phœnix group, northward, was also surveyed; and in all these examinations the non-existence or identity of numerous doubtful islands was established, a most important fact. The islands comprising Ellice's group were also severally examined and accurately delineated. The Gilbert Archipelago, or, as it is here termed, the Tarawan or Kingsmill group, received a large share of attention; and the numerous interesting particulars of these, all singular coral islands, were collected, which, being duly arranged by the naturalist, Mr. Dana, forms a valuable addition to that branch of natural history. In the still imperfectly known Marshall Archipelago, the Radack and Ralick Channels of Kotzebue, several of the groups forming portions of it were examined and surveyed.

This embraces almost all the general result of the Expedition in the central portions of the Pacific. To this may be added the examination of the several portions of New Zealand, and the visits to Sydney. The longitudes obtained are generally dependent on the accuracy of the twenty-nine chronometers (all but two of which went well), and thus deserve all confidence.

One great point in the Expedition was the result of their antarctic cruises, in the supposed claim to priority in the discovery of the antarctic continent. We have elsewhere remarked on this subject, and to that the reader is referred.

In the ensuing pages we have endeavoured faithfully to quote from and acknowledge the remarks made in this important voyage. The Narrative has been our only source, and to this we have strictly held, without referring to the Hydrographical Atlas, which in many points varies from it.

In this brief enumeration of the services of the great navigators whose names have been quoted, we have necessarily omitted many very important particulars, but further allusion will be made to their special labours in the ensuing pages.

We have also, with the view of not unduly extending these prefatory remarks, not included many authors and observers who have contributed, in a lesser degree, to the hydrography of the Pacific; these, too, as far as their observations have been incorporated with our descriptions, have been enumerated in the body of the work. We have therefore now only to mention the names of Sir James Clark Ross, whose voyage to the South Polar regions, in company with Capt. Crozier, in the well-known ships *Erebus* and *Terror*, added the Victoria land to our geography, and has placed the magnetic phenomena of the southern hemisphere in a clear light to the world. Capt. *Kellett*, too, ought to be enrolled among the honourable band of explorers. His surveys on the western coast of Central America are invaluable; and his observations, made during his unexpected voyage through Behring's Strait, in search of the anxiously looked for expedition under Sir John Franklin, are too well known and recent to require any comment here.

Upon the basis of these scientific observers the geographical foundation of our knowledge in the Pacific rests. We need not recapitulate the subject, but in the ensuing tables the authority upon which the geodetical position and other particulars of the respective places rest is mentioned, and this will be sufficient.

GEOGRAPHICAL POSITIONS.

Although very much has been effected within a few years in the accurate determination of the relative geographical positions of the features of most countries, very much still remains to be done, more especially in the Pacific Ocean. The following tables, therefore, cannot be supposed to exhibit, in all points, that connected accuracy which is to be found in the details of the coasts nearer to Europe, which from this cause are much more frequented and important.

The immense extent, too, of this great ocean is another source of difficulty in the exact determination of the longitudes of the various points distributed over its surface. The first connected series of meridional distances around the earth was that carried by Capt. FitzRoy, in H.M.S. *Beagle*, in 1831—1836, with from fourteen to twenty-two chronometers. This was conducted with the most refined precautions, yet the entire series was 33′ of time, or 8′ of arc, in excess; a considerable portion of which discrepancy may be supposed to have arisen in the Pacific. There have been numerous chronometric observations by which the positions of a *few points* have been fixed with a considerable degree of accuracy: that is, to such a degree of refinement, that it is not probable that any but a ship or an expedition especially appointed for the purpose can be expected to improve upon. These, therefore, will serve as points of departure, by which the longitude of any other point may be gained to within much more narrow limits than by any independent astronomical observations, if good chronometers be used. It will thus be in the power of almost every navigator sailing in the Pacific, possessed of one or more of such instruments, to add very materially to hydrography, if such observations be carefully made and recorded.

The very great importance of this mode of observation—that of connecting a less-known with a well-ascertained point—has been so clearly shown of late years, and more especially by Lieutenant Raper, to whom this branch of geography is so much indebted, that it is needless to dilate on it here. It has been frequently discussed, and must be familiar to most of our readers.* Some of the leading positions will be enumerated presently, by which it will be seen how much has been done, and from this what remains to be done, in this branch of science.

Astronomical observations for the determination of longitude require to be so accurately conducted, and in such extensive series, that all observations at sea, or on board ship, can only be regarded as rude approximations to the truth, and can only be depended on to direct attention to any *great* error in the reckoning or rate of the chronometer. It must be supposed that a very large number of the isolated spots in this ocean have been determined by such means, and therefore

* See Nautical Magazine, 1839; and Raper's Navigation, 3rd ed. pp. 281, 379.

capable of every improvement. There are very few points, perhaps only one, where astronomical observations have been set up, either temporarily or permanently, which may be considered to have effected the result in an unimpeachable manner. The observations of Capt. Cook on the transit of Venus at Tahiti, the object of his first voyage, is open to doubt to the extent of several minutes of longitude. The same may be said of a great proportion of those navigators who have laboured in the same field; and it is only by a discussion of these contending results, and reconoiling their discrepancies, that anything like uniformity can be attained. Paramatta Observatory, near Sydney, New South Wales, is the only exception to these remarks: its position may be considered as determined very much within the limits required for the ordinary purposes of navigation. On the other side, on the American coast, there are numerous points, or extents of coast, which will serve as accurately determined points of departure for the navigator: and in the intermediate space there are other points more particularly to be specified, which will equally well serve the same purpose.

In all these cases, where the improvement of hydrography is concerned, the particulars of the meridian from which any new point is determined, should be accurately specified, in connexion with the new observations; as it is manifest, that any change which arises from a more definitive determination of the fundamental meridian, must equally affect that dependent on it. In the words of the great authority on this subject, Lieutenant Raper, "If navigators and hydrographers would agree to *consider*, for the time being only, certain important stations, as already established in longitude, whether really so or not, with the view of referring all the subordinate positions to them, the indistinctness which now hangs over absolute and relative positions would be forthwith cleared up. The question would be narrowed into the determination of *chronometric differences* alone, until a favourable opportunity occurred for the definitive determination of a fundamental position." There is no part of the globe to which these remarks are more espepecially applicable than in the Pacific, nor more important than in the present state of its hydrography. They are therefore now pressed upon the attention of the navigator.

For the purpose of thus rendering subordinate stations serviceable in the connexion of the various points with the *prime* meridian, Lieutenant Raper has selected *twenty* points, in various parts of the globe, which will serve as *secondary* meridians for the districts in their vicinity. Five of these, Valparaiso, San Francisco, Sandwich Islands, Otaheite, and Paramatta, refer to the Pacific; and as these have been discussed—and their accuracy is *almost* unimpugnable—as well as for the sake of uniformity, their longitudes assumed in the extensive tables in question will be followed here.

But it is by no means necessary, in the Pacific Ocean, to confine this subject to the above limits. These secondary meridians may, for the convenience of the navigator, be considerably extended: and indeed almost any well-defined position may be taken as such: but preference must be given to those which have been better determined than others; and the following enumeration of the principal points may be serviceable:—

	In time. H. M. S.	In arc. ° ′ ″	
STRAIT OF MAGALHAENS: Port Famine Observatory.	4 43 51·7	70 57 55 W.	(FitzRoy.)
VALPARAISO: Fort San Antonio	4 46 46	71 41 30 W.	(FitzRoy.)
CALLAO: Arsenal flagstaff	5 8 54	77 13 30 W.	(FitzRoy.)
PANAMA′: N.E. Bastion	5 18 4·8	79 31 12 W.	(Belcher.)
SAN FRANCISCO: Yerba Buena Cove	8 9 36	122 24 0 W.	(Beechey.)
SITKA: Arsenal Light	9 1 8·7	135 17 10 W.	
AWATSKA BAY: Petropaulovski Church	11 14 54	168 43 30 E.	(Beechey.)
SANDWICH ISLANDS: Honolulu Fort	10 31 40	157 55 0 W.	(Raper.)
LADRONE ISLANDS: Guam, Umata Bay Church	9 38 36	144 39 0 E.	(D'Urville.)
NEW GUINEA: Port Doreï	8 55 59·47	133 59 52 E.	(D'Urville.)
FEEJEE ISLANDS: Ovalau Island, Levuka Harbour, Observatory Point	11 55 30·7	178 52 41 E.	(Wilkes.)
TAHITI: Point Venus, S.E. extreme	9 7 56·7	149 29 1 W.	(Beechey.)
NEW ZEALAND: Bay of Islands, Paheha Mission	11 36 28	174 7 0 E.	(La Place, &c.)
SYDNEY: Fort Macquarie	10 4 56	151 14 0 E.	

The discussion of minor details, which might be extended to almost each item in the ensuing tables, would be both uninteresting and unnecessary. The authorities upon which the position assumed rests are quoted, and from them the confidence in the degree of accuracy they merit may be inferred; but there are numerous minor variations from the original observations, which are explained in the text, or may be found in the works quoted.

At the head of all these works stands the unrivalled production of Admiral Krusenstern. Up to the period of the cessation of his most laborious researches, it may be said that he had exhausted the subject. And although much of his great work is engrossed in those disquisitions which are not absolutely useful to the practical navigator, yet they must be referred to for the groundwork of all that has been done in the Pacific Ocean. Subsequent, and perhaps more accurate, surveys and observations, have superseded the necessity of some of his remarks; but the following pages will testify how much the service of hydrography owes to Admiral Krusenstern.

TABLE

OF

POSITIONS, TIDES, MAGNETIC VARIATIONS, &c.

PART I.

I.—THE STRAIT OF MAGALHAENS.	Lat. South.	Lon. West.	Authority.	Var. East.	H. W.	Range.	Page.
	° ′ ″	° ′ ″		° ′	H. M.	FT.	
Cape Virgins; S. E. extreme .	52 20 10	68 21 44	The surveys of				3
Dungeness Point; extremity .	52 23 50	68 23 20	H. M. S. *Adventure*	22 36	8 50	40	3
Dinero Mount; summit . .	52 19 40	68 33 30	and *Beagle*, Capts.				5
Cape Possession; middle of cliff.	52 17 0	68 56 30	P. P. King, R. N.,	22 40	8 40	40	5
Mount Aymond; summit, 1,000 ft.	52 7 10	69 32 20	F.R.S., &c.; Capt.				5
Possession Bay; western bank .	52 19 0	69 20 10	T. Stokes, R. N.;		8 19	42	5
Point Delgada; extreme . .	52 26 30	69 34 20	and Capt. R. Fitz-				7
Point Catherine; N. E. extremity	52 32 0	68 44 20	Roy, R.N., 1826—				7
Cape Orange; N. extremity .	52 27 10	69 28 10	1830; and H.M.S.	22 30	9 0	46	6
Cape Gregory; extremity . .	52 39 0	70 13 30	*Beagle*, Capt. R.	23 30	9 38	12	7
Cape San Vicente; W. extreme .	52 46 20	70 26 35	FitzRoy, R. N.,				8
Peckett Harbour; S. summit .	52 47 10	70 46 25	1831—1834.	23 29	12 0	9	9
Elizabeth Island; N.E. bluff .	52 49 10	70 37 25		23 50	0 5	7	9
Cape Negro; S.W. extreme cliff.	52 56 40	70 49 10					10
Point Geate Grande; N.W. extr.	53 0 5	70 26 55		23 0			10
Point St. Mary; extremity . .	53 21 15	70 57 55		23 26			11
Mount St. Philip; summit . .	53 36 25	71 0 10					11
Port Famine: Observatory .	53 38 15	70 57 55		23 40	0 7	6 to 7	12
" " New Observatory.		70 59 23					12
" " Point St. Anna, ex.	53 37 50	70 55 10		23 0	0 7	6	12
C. Valentyn; summit at extreme.	53 33 30	70 33 55					14
Admiralty Sound; Pt. Cook, riv.	54 17 10	70 1 53					14
" " Latitude Pt.; extr.	54 16 45	69 54 43					14
Mount Graves; summit . .	53 45 0	70 37 40					14
Port S. Antonio; Humming bd. Co.	53 54 8	70 54 18					14
Cape San Isidro	53 47 0	70 58 0		23 40	1 0	8	16
Mt. Tarn; peak at N. end, 2,602 ft.	53 45 6	71 2 20					16
Mount Vernal; summit . .	54 6 28	71 1 34					19
Nassau Island; S.E. point . .	53 50 23	71 4 40				6	17
Cape Froward; summ. of the bluff	53 53 43	71 18 25		23 20	1 0	6	17
Magdalen Sound and Cockburn and Barbara Channels.							
Mt. Boqueron; highest pinnacle.	54 10 40	70 59 54					19
Mt. Sarmiento; N.E. peak, 6,800 ft.	54 27 15	70 51 25					20
Pyramid Hill; summit . .	54 27 0	71 7 50					19
Labyrinth Isles; Jane Is. summit	54 19 10	71 0 30		28 50	0 30	6	19
Cape Turn; extremity . .	54 24 8	71 7 40		24 0	1 20	6	19
King Island; summit. . .	54 22 38	71 17 10		23 50		7	19
Prowse Islands; station . .	54 22 13	71 24 49					20
Dyneley Sound; N.E. end of S.E. Baynes Island	54 18 15	71 39 42					20
Fury Harbour; W. point . .	54 28 25	72 15 10					20
Tussack Rock	54 34 0	72 12 20				5	20
West Furies; largest rock . .	54 34 45	72 22 0		25 0	2 30	4	20
East Furies; largest rock . .	54 38 0	72 12 10					20
Mount Skyring; summit 3,000. ft.	54 24 48	72 11 30					20

d

	Lat. South.	Lon. West.	Authority.	Var. East.	H. W.	Range.	Page.
	° ′ ″	° ′ ″		° ′	H. M.	FT.	
Bynoe Island; summit . .	54 19 0	72 12 54	The surveys of H. M. S. *Adventure* and *Beagle*, Capts. P. P. King, R. N., F.R.S., &c.; Capt T. Stokes, R.N.; and Capt. R. Fitz-Roy, R.N., 1826—1830; and H.M.S. *Beagle*, Capt. R. FitzRoy, R. N., 1831—1834.				21
Bowles Island; N. summit . .	54 2 0	72 15 10		24 0	2 0	5	21
Davies Gilbert Head; N. summit	53 56 30	72 15 10					21
Cayetano Peak	53 53 4	72 9 50					22
Smyth Harbour; Mount Maxwell	53 47 10	72 15 10					22
Elvira Point; extremity . .	53 49 12	72 4 5					22
Cape Edgeworth; extremity .	53 47 3	72 9 8					22
The Strait of Magalhaens, from Cape Froward to the Pacific Ocean.							
Cape Froward; sum. of the bluff.	53 53 43	71 18 25					17
Cape Holland; S. E. extreme .	53 48 35	71 39 35		23 50	10 40	6	24
Port Gallant; Wigwam Point .	53 41 45	72 0 51		24 4	9 3	5 or 6	25
Dos Hermanas Island; summit .	53 57 45	71 25 25					25
Cordes Bay; W. outer point .	53 42 55	71 57 0					25
Cape Inglefield; islet off it. .	53 50 20	71 55 33					25
Mount Pond	53 51 45	71 56 40					26
Elvira Point; extremity . .	53 49 12	72 4 5					26
Charles Island; Wallis Mark .	53 43 57	72 5 55					27
Rupert Island; summit . .	53 42 0	72 11 52					27
Cape Crosstide; extreme . .	53 33 0	72 26 40		23 35	1 40	5	27
Jerome Channel; Jerome Point .	53 31 30	72 25 40		24 0	1 30	6	28
FitzRoy Channel; Donkin Cove.	52 45 30	71 24 37		23 40			28
Skyring Water; Dynevor Cas. sum.	52 34 30	72 32 32					28
El Morion, or St. David's Hd.; sum.	53 33 20	72 32 25		23 20		6	29
Cape Quod; extremity . .	53 32 10	72 33 35					29
Snowy Sound; extr. of islet at entr.	53 31 0	72 40 10					30
Cape Notch; extremity . .	53 25 0	72 49 5		23 40			31
Playa Parda Cove; Shelter I. sum.	53 18 45	73 1 40		23 45	1 8	6	32
Half Port Bay; point. . .	53 11 40	73 18 55		23 40	2 0	6	32
St. Anne Island; central summit.	53 6 30	73 16 40		23 30	4 0	5	33
Cape Upright; extrem. N. trend.	53 4 3	73 36 8		23 24	2 30	6	34
Cape Tamar; S. extreme . .	53 55 30	73 48 20		23 24	2 30	6	35
Point Felix; extremity . .	52 56 0	74 12 55					35
Valentine Harbour; Observn. Mt.	52 55 0	74 18 55		24 0	2 0	6	36
Cape Cortado; extremity . .	52 49 37	74 26 50		23 40			37
Cape Parker; western sum. over	52 42 0	74 14 40					37
Westminster Hall; E. summit .	52 37 18	74 24 20					37
Observation Mount; summit .	52 28 58	74 36 12		25 9	1 0	5	37
Harbour of Mercy; Bottle I. sum.	52 44 58	74 39 24		23 48	1 10	6	37
Cape Pillar; northern cliff .	52 42 50	74 43 30		23 50	1 0	6	38
Cape Victory; extremity . .	52 16 10	74 54 49					38
Evangelists, or Isles of Direction; Sugar-loaf Islet . .	52 24 18	75 6 50		24 0	1 0	5	38
II.—THE OUTER COAST OF TIERRA DEL FUEGO, FROM CATHERINE POINT TO CAPE PILLAR; INCLUDING STATEN ISLAND, CAPE HORN, &c.							
Point Catherine; N.E. extrem.	52 32 0	68 44 10					43
Cape Espiritu Santo; Nombre Head, N.E. cliff	52 39 0	68 35 0				5	43
St. Sebastian B.; Pt. Arenas, S.ex.	53 9 10	68 12 20					43
„ C. S. Sebastian; N. height	53 19 0	68 10 0		22 40	7 0	13	44
Cape Sunday; N.E. cliff . .	53 39 30	67 56 30		22 50	6 0	12	44
Cape Peñas; S.E. cliff . .	53 51 30	67 33 30		22 0	6 42	12	44
Cape San Pablo; N.E. cliff .	54 16 20	66 40 15					45
Table of Orozco; S.E. sum. 1,000ft.	54 40 40	65 59 55					45
Policarpo Point; extreme . .	54 39 0	65 39 40					45

	Lat. South.	Lon. West.	Authority.	Var. East.	H. W.	Range.	Page.
	° ′ ″	° ′ ″		° ′	H. M.	FT.	
Cape San Vicente; extreme .	54 38 40	65 14 25	The surveys of	22 50	4 30	10	45
Cape San Diego; E. extreme .	54 41 0	65 7 10	H. M. S. *Adventure*	22 50	4 30	10	45
Staten Island; Cape St. Bartholomew, S.W. cliff . . .	54 53 45	64 45 40	and *Beagle*, Capts. P. P. King, R. N.,	22 40	4 45	9	45
„ C. St. Anthony, N. extr. cliff	54 43 30	64 34 10	F.R.S., &c.; Capt.				46
„ New Year Islands; N.E. pt.	54 39 0	64 6 30	T. Stokes, R.N.;	22 30	5 30	6	46
„ Cape St. John; N. cliff .	54 42 20	63 43 55	and Capt. R. Fitz-	22 30	5 30	9	47
„ Mount Richardson; summit	54 45 50	63 51 15	Roy, R.N., 1826—				47
„ Dampier Islands; S. sum.	54 53 0	64 11 30	1830; and H.M.S.				48
Good Success Bay; N. head .	54 47 0	65 11 40	*Beagle*, Capt. R.				49
Cape Good Success; S. extreme .	54 54 40	65 21 40	FitzRoy, R. N.,				49
Campana or Bell Mount; summit	54 53 15	65 33 40	1831—1834.				51
Aguirre Bay; Kinnaird Point .	54 57 5	65 47 10		22 50	4 20	8	51
Point Jesse	55 2 45	66 22 40					51
Beagle Channel; Cape Mitchell .	54 57 30	68 14 10					52
„ „ Cape Divide .	54 59 10	69 7 20					52
„ „ Kekhlao Cape .	55 10 0	70 2 10					52
New Island; Point Waller, extr.	55 10 10	66 28 10					52
Lennox Road; Luff Island summ.	55 18 40	66 44 55		23 40	4 30	8	52
Goree Road; Guanaco Point, extr.	55 19 0	67 10 10		23 30	4 0	8	52
Terhalten Island; summit . .	55 26 15	67 1 40					52
Evout Isles; N.E. head . .	55 33 0	66 45 10					53
Barnevelt Islands; N.E. extreme	55 48 25	66 44 50		23 40	4 30	8	53
Deceit Island; C. Deceit, E. extr.	55 54 40	67 2 35					54
Wollaston Island; Cape Scoomfield	55 45 15	67 8 10					53
Herschel Isld. Mt. Herschel; sum.	55 49 45	67 19 25					53
CAPE HORN; summit . .	55 58 40	67 16 10		24 0	4 40	9	54
Hermite Island; St. Martin Cove, Observation Station . . .	55 51 20	67 34 10		24 23	4 41	9	55
„ Kater's Peak, 1742 feet .	55 51 55	67 34 0					56
„ Maxwell Island; summit .	55 47 30	67 30 55					56
„ Cape Spencer; S.E. summit	55 55 0	67 37 50		24 30	4 40	8	56
Packsaddle Island; summit .	55 23 50	68 4 30		23 50	3 30	6	57
Orange Bay; Burnt Isl. summit .	55 31 0	68 2 30		23 56			56
Point Lort; E. pitch . . .	55 40 30	67 59 10		23 30	4 30	7	56
False Cape Horn; S. extreme .	55 43 15	68 5 50		23 56	3 28	6	56
Diego Ramirez Isl.; high. summ.	56 28 30	68 42 40		24 30	4 0	6	57
St. Ildefonso Isles; highest summ.	55 52 30	69 18 40		24 10	3 20	6	58
New Yr. S.; C. Weddell, S.W. pt.	55 33 0	68 45 10					58
Henderson Isl.; M. Beaufoy, sum.	55 36 15	68 58 10					58
„ Brisbane Head; extr. summ.	55 39 0	68 57 10					58
Hope Island; central extr. summ.	55 32 30	69 40 0					59
York Minster; summit, 800 ft. .	55 24 50	70 2 40					59
Capstan Rocks; summit of largest	55 24 10	70 17 40					63
Cape Alikhoolip; S. extreme .	55 11 50	70 49 10					63
Phillips' Rocks; largest, summit .	55 14 10	70 57 10					63
Treble Island; S. summit . .	55 7 50	71 2 30		24 15	3 0	5	63
Cape Castlereagh; summit . .	54 56 0	71 28 10		24 15	2 50	4	63
Cape Desolation; S. summit .	54 45 40	71 37 20		24 30	1 40	4	63
London Id.; Horace Peaks, S. sum.	54 43 0	71 57 35					63
West Furies; largest rock . .	54 34 45	72 22 0		25 0	2 30	4	64
Tower Rocks; E. rock . .	54 36 40	73 3 0		24 34	2 30	5	65
Cape Noir Island; extreme .	54 30 0	73 5 40		25 0	2 25		64
Kempe Peaks; S. summit . .	54 23 30	72 30 20					64
Ipswich Isles; S. summit . .	54 10 30	73 20 50					65
Gloucester Cape; summit . .	54 5 18	73 29 25		24 30	1 30	5	66
Fincham Islands; summit of W.	53 44 15	73 45 40					66
Cape Tate; summit . . .	53 37 15	73 51 40					66
Landfall Isl. C. Schetky; S. pitch	53 21 40	74 12 55		24 0	2 0	5	67
„ Cape Inman; cliff summit .	53 18 30	74 19 25					67
Cape Sunday; summit . .	53 10 30	74 22 10					67
Cape Deseado; peaked sum. near.	52 55 30	74 37 40					68
Dislocation Harbour; Obs. Statn.	52 54 15	74 37 20		23 53	1 40	4	68
Judge Rocks; westernmost .	52 51 0	74 48 40		24 0	1 0		68
Apostle Rocks; W. large rocks .	52 46 15	74 48 0		23 50			68
Cape Pillar; northern cliff .	52 42 50	74 43 30		23 50	1 0	6	68

III.—THE WESTERN COAST OF PATAGONIA, FROM THE STRAIT OF MAGALHAENS TO CHILOE.	Lat. South. ° ′ ″	Lon. West. ° ′ ″	Authority.	Var. East. ° ′	H. W. H. M.	Range. FT.	Page.
The Evangelists, or Is. of Direction; Sugar-loaf	52 24 18	75 6 50	The Surveys of	24 0	1 0	5	68
Cape Victory; extremity	52 16 10	74 54 49	H. M. S. *Adventure*				70
Cape Isabel; W. extreme	51 51 50	75 13 10	and *Beagle*, Capts.				70
Cape George; bluff summit	51 37 40	75 21 10	P. P. King, R.N.,				70
Cape Santa Lucia; summit	51 30 0	75 29 10	F.R.S., &c.; Capt.	22 0			70
White Horse Islet; N. summit	51 7 50	75 14 50	T. Stokes, R. N.;				70
Cape Santiago; summit	50 42 0	75 28 10	and Capt. R. Fitz-				71
April Peak; summit	50 10 50	75 21 10	Roy, R.N., 1826—				71
Cape Three Points, or Tres Puntas; 2,000 feet	50 2 0	75 21 10	1830; and H.M.S. *Beagle*, Capt. R.				71
Port Henry; observatory	50 0 18	75 19 5	FitzRoy, R. N.,	20 50	Noon.	5	71
Cape Primero; extremity	49 50 5	75 35 40	1831—1834.	20 58			73
Mount Corso; S.W. summit	49 48 0	75 34 10					73
Cathedral Mount; summit	49 46 30	74 44 0					73
Cape Montague; W. cliff	49 7 30	75 37 10					73
Parallel Peak; summit	48 45 40	75 31 10					73
Rock of Dundee; summit	48 6 15	75 42 10					74
Cape Dyer; extremity	48 6 0	75 34 30					74
Port Sta. Barbara; N. extr. Ob. Pt.	48 2 20	75 29 30		19 10	11 45	6	74
Bynoe Islands; N. centre	47 58 0	75 23 40					75
Guaianeco Islands; northernmost islet summit	47 38 10	75 14 10					75
„ Speedwell B.; hill at N.E. pt.	47 39 30	75 10 10					76
„ Wager Isl.; E. pt. extreme	47 41 0	74 55 10					76
Ayantau Islands; summ. of largest	47 34 15	74 40 30					76
Channel's Mouth; rock of entr.	47 29 30	74 29 40					77
Xavier Island; Ignacio Beach	47 10 0	74 25 50		19 50			77
Kelly Harbour; S. point extrem.	46 59 30	74 8 40					77
Forelius Peninsula; isthmus narrowest point	46 50 0	74 41 50					78
Cirujano Island; N.E. point	46 51 10	74 21 55					78
Purcell Island; summit	46 55 20	74 39 55					78
St. Paul's Dome; summ. 2,284 ft.	46 36 16	75 13 50					79
Port Otway; Observn. spot	46 49 31	75 19 30		20 32	Noon.	6	79
CAPE TRES MONTES; extremity *	46 58 57	75 28 0	* Termination of				80
Cape Raper; rock close to	46 49 10	75 41 5	H. M. S. *Beagle's*				79
Cape Gallegos; summit	46 35 0	75 28 40	Surveys in 1830.				83
Christmas Cove; Ob. St. at S.E. ex.	46 35 0	75 34 15		20 40	12 45	5	83
The Cone; summit, 1,300 feet	46 34 10	75 31 10	The positions of				83
Rescue Point; N. summit	46 18 10	75 13 55	the various points		12 14	5	84
Hillyer Rocks; middle	46 4 0	75 14 10	in the interior				84
Cape Taytaohaohuon, or Taytao; W. extremity	45 53 20	75 8 10	Sounds, being quite unimportant for the				84
Anna Pink Bay; Patch Cove, O. S.	45 52 15	74 56 0	purposes of navi-	20 31	12 45	5	84
„ Port Refuge; Puentes I. sum.	45 51 36	74 51 35	gation, are not				84
Ynche-mo Island; S.E. summit	45 48 5	75 1 10	given in this table.	20 36	12 45	5	84
Menchuan Island; summit	45 36 0	74 56 10	The charts will be				84
Mount Iaquiliac; summ. 3,000 ft.	45 20 0	74 21 50	quite sufficient for				85
Vallenar Road; S.E. extremity; Three Finger Island	45 18 30	74 36 25	any utility.	20 48	12 18	5	85
Socorro, or Huamblin Is.; S. extr.	44 55 50	75 12 55					85
Ypun, or Narborough Island; John Point, extremity	44 40 40	74 48 40					85
Mount Mayne; summit, 2,080 feet	44 9 0	74 11 55					86
Guaytecas Island; central summit	43 52 45	74 1 10					86
Port Low; rocky islet in harbour	43 48 30	74 3 15		19 48	0 40	7	86
Queytao Islet; summit	43 43 0	73 35 40					86
Huafo I. or No Man's Land; S. ex.	43 41 50	74 46 10					87
„ summit over Weather Point	43 35 30	74 48 50		19 0			87
IV.—ISLAND OF CHILOE.							
Caucitad Rock; summit	43 30 0	73 50 30					89

	Lat. South.	Lon. West.	Authority.	Var. East.	H. W.	Range.	Page.
	° ′ ″	° ′ ″		° ′	H. M.	FT.	
San Pedro Mntn.; summ. 3,200 ft.	43 21 0	73 49 0	The survey by		11 15	9	89
San Pedro Passage; Ob. St. in cove	43 19 35	73 45 20	Capt R. FitzRoy,		12 30		89
Yanteles Mntn; S. summ. 6,725 ft.	43 30 0	72 50 30	R.N., 1831—1834.				—
Corcovado Volcano; sum. 7,510 ft.	43 11 20	72 48 40					—
Huapi Quilan Islets; S. summit .	43 29 30	74 15 0					89
Cape Quilan; S.W. extreme .	43 17 10	74 26 0					89
Cape Matalqui; W. extreme .	42 10 40	74 14 0					89
Matalqui Height, or Paps; summ.	42 10 30	74 11 10					89
Huechucucuy Head . . .	41 46 0	74 3 0					89
Corona Head; N. pitch . .	41 46 0	73 57 30					89
San Carlos Town; landing place at Mole	41 52 0	73 52 40		18 33	11 15	6	89
Point Tres Cruces; extr. pitch .	41 49 30	73 31 40			1 15	16	91
Huapilinao Head; summit . .	41 57 36	73 32 20			1 25	15	92
Lobos Head; summit . . .	42 4 0	73 27 0			12 29		92
Oscuro Port; Observation Station	42 4 0	73 29 0			1 0	20	92
Quintergen Point; summit . .	42 9 25	73 24 0					92
Chaugues Islands; N. summit .	42 15 0	73 18 0			12 28		93
Quicavi Bluff	42 15 0	73 24 0			12 51	20	93
Dalcahue; chapel . . .	42 23 0	73 40 0			12 26		93
Castro Town; easternmost part .	42 27 45	73 49 20		18 35	12 11	18	94
Yal Point; summit . . .	42 39 0	73 43 0			0 55		95
Lemuy Island; Apabon peaked hill	42 40 0	73 35 30			0 55		95
Talcan Harbour; Obs. Station .	42 47 0	72 58 0			1 3	16	95
Minchinmadom Volcano; S. sum. 8,000 feet	42 48 0	72 34 30					—
Mount Vilcun; summit . .	42 48 50	72 52 50					—
Point Sentinela; extreme . .	42 59 25	73 22 30					97
Huildad Harbour . . .	43 3 0	73 34 0		18 30	12 48	20	97
Laytec Island; S.E extreme .	43 15 5	73 36 0			12 37		97
Abtao Island; S. point . .	41 48 0	73 26 0			12 50	18	98
Calbuco Fort; E. end of island .	41 46 5	73 10 53					98
Puluqui Id. Centinela, or S. point	41 51 0	73 6 0			1 5		98

V.—THE COAST OF CHILE, FROM SAN CARLOS TO HUESO PARADO.

	Lat. South.	Lon. West.	Authority.	Var. East.	H. W.	Range.	Page.
Carelmapu Cove; Obs. Station .	41 45 0	73 45 0			0 50	10	103
Maullin, Amortajado; N. extreme	41 37 15	73 44 30					103
Point Godoy; S.W. extreme .	41 34 15	73 50 20					103
Osorno Mountain; summit . .	41 9 30	73 36 45					102
Point Coronel; S. extremity .	41 7 40	73 31 45					103
Cape Quedal; summit . . .	41 3 0	73 59 50					103
Manzano Cove; rivulet, mouth .	40 33 20	73 45 50					103
Milagro Cove; depth of . .	40 16 0	73 45 0					103
River Bueno; entrance (bar) .	40 11 0	73 44 0					103
Point Galera; W. extremity .	40 2 0	73 46 40					103
Falsa Point; summ. over (highest)	40 0 50	73 40 50					103
Valdivia; Ob. St. nr. Fort Corral	39 52 53	73 29 0		18 15	10 35	5	104
Gonzales Head; northern pitch .	39 51 15	73 30 0					104
Valdivia Town; landing place opp. church (Hospital Mole) . .	39 49 2	73 18 30		18 20	10 45		104
Chanchan Cove; islet off . .	39 26 40	73 18 30					105
River Tolten; mouth . . .	39 7 45	73 19 0					105
Cauten (or Imperial) River; mouth	38 47 40	73 26 0					105
Cauten Head Cliff; summit .	38 40 40	73 30 20					105
Mocha Island; S. summit . .	38 24 10	73 56 50					106
Cape Tirua; summit of islet off .	38 23 0	73 34 30					106
Mocha Island; Ob. St. E. side, near N. point	38 19 35	74 0 20		17 20			106
Molguilla Point; S.W. extreme .	37 48 0	73 36 0					106
Point Tucapel; extreme . .	37 42 0	73 43 0					107
River Leübu; entrance . .	37 35 45	73 42 0		17 10	10 30	5	107
Tucapel Head; summit . .	37 35 20	73 43 10					107

	Lat. South.	Lon. West.	Authority.	Var. West.	H. W.	Range.	Page.
	° ′ ″	° ′ ″		° ′	H. M.	FT.	
Carnero Head; western summit .	37 21 20	73 44 0	The survey by				107
Arauco Fort; middle . . .	37 15 0	73 23 0	Capt. R. FitzRoy,				107
Tubul River; S. head, entrance .	37 14 25	73 27 30	R.N., 1831—1834.				107
Cape Rumena; N.W. cliff, summ.	37 12 45	73 42 0					107
Laraquote River; mouth . .	37 10 30	73 14 0					108
Point Lavapie; extremity . .	37 8 50	73 38 20					108
Coloura Village; west. pitch of hill	37 2 50	73 14 0					108
Santa Maria Island; landing place	37 2 48	73 34 0		17 0	10 20	6	108
Point Coronel; W. extremity .	36 57 0	73 15 0					108
Concepcion City; mid. near to riv.	36 49 30	73 5 20					109
River Bio Bio; S. entrance point	36 48 45	73 13 0					109
Talcahuano; Fort Galves . .	36 42 0	73 10 0		16 48	10 14	5	112
Point Tumbes; N.W. cliff . .	36 37 15	73 10 20					109
Mount Neuke; summit . .	36 34 55	72 58 0					113
Coliumo Head; N. extreme .	36 31 30	73 1 15					113
Bio Bio Paps; S.W. summit .	36 6 20	73 14 40					110
Carranza Point; S.W. extreme .	35 37 20	72 42 20					113
Cape Humos; summit . .	35 22 50	72 33 0					113
Maule Church; rock . . .	35 19 40	72 29 20		16 24			113
Maule River; S. head entrance .	35 19 15	72 28 0					113
Topacalma Point; summit on ex.	34 0 50	72 5 0					114
Navidad Bay; River Rapel mouth	33 54 0	71 52 20					114
Rapel Shoal (wrongly called Topacalma)	33 51 0	71 56 30					114
Maypo River; S. entrance head .	33 39 20	71 43 15					114
White Rock Point; White Rock	33 29 0	71 46 50					114
Curaumilla Point; rock off . .	33 6 0	71 48 0					115
Valparaiso; Fort San Antonio .	33 1 53	71 41 15		15 18	9 32	5	115
Quillota; Bell; summit . .	32 57 10	71 10 20					115
Quintero Rocks; body . .	32 52 20	70 37 0					121
Quintero Point; summit . .	32 46 0	70 35 30					121
Horcon Rock; largest . .	32 41 50	70 35 30					121
Aconcagua; Mountain; summit .	32 38 30	70 0 30					102
Papudo Bay; Ob. St. landing pl.	32 30 9	71 30 45		15 12			122
Pichidanque; S.E. point of island	32 7 55	71 36 0		15 24	9 20	5	122
Conchali Bay; islet in middle .	31 53 10	71 36 0					123
Point Tablas; S.W. extremity .	31 51 45	71 37 30					123
River Chuapa; S. entrance point	31 39 30	71 38 0					123
Maytencillo Cove; N. head .	31 17 5	71 42 5					125
Talinay Mount; summit . .	30 50 45	71 41 45					123
Limari River; S. head . .	30 44 53	71 46 25					123
Lengua de Vaca; extremity .	30 13 40	71 41 30					123
Herradura de Coquimbo Port; S.W. corner	29 58 40	71 25 45		14 30	9 8	5	124
Coquimbo Port; north. islet (rock)	29 55 10	71 25 10		14 24	9 8	5	124
Arrayan Cove; S. point . .	29 42 20	71 23 45					126
Juan Soldado, Mountain; summit	29 41 30	71 20 25					126
Pajaro Islet; southern summit .	29 35 0	71 36 25					126
Yerba Buena, village; chapel .	29 34 0	71 21 50					126
Trigo Island; S. W. point . .	29 32 35	71 24 20					126
Tortoralillo; S. entrance point .	29 29 15	71 23 45		13 40			126
Chungunga Islet; summit . .	29 24 15	71 25 15					126
Toro Reef	29 21 10	71 35 25					127
Choros Islands; S.W. pt. of larg.	29 15 45	71 37 30					127
Polillao Cove; S. point extreme .	29 10 0	71 34 10					127
Chañeral Bay; S.W. point . .	29 2 40	71 33 40					127
Chañeral Island; S.W. summit .	29 1 15	71 39 5					127
Sarco Cove; middle of beach . .	28 50 0	71 32 10					128
Cape Vascuñan; islet off (rock) .	28 50 0	71 34 30					128
Huasco; Captain of Port's house .	28 27 15	71 19 0		13 37	8 30	6	128
Lobo Point; outer pitch . .	28 17 50	71 17 10					129
Herradura de Carrisal; landing pl.	28 5 45	71 15 45		13 23			129
Carrisal; middle point; S. side .	28 4 30	71 14 30					130
Matamores Cove; out. pt. on S. side	27 54 10	71 12 35					130
Pajonal Cove; S.E. corner . .	27 43 30	71 7 0		13 28			130
Salado Bay; Cachos Point; sum.	27 39 20	71 6 25					131

	Lat. South.	Lon. West.	Authority.	Var. West.	H. W.	Range.	Page
	° ′ ″	° ′ ″		° ′	H. M.	FT.	
Copiapo; landing place	27 20 0	71 1 45	The survey by Capt. R. FitzRoy, R.N., 1831—1834.	13 36	8 30	5	131
Morro of Copiapo; summit	27 9 30	71 1 45					133
Port Yngles; sandy beach in S.W. corner	27 5 20	70 56 0		13 30			134
Cabeza de Vaca; point, extreme	26 51 5	70 55 0					134
Flamenco; S.E. corner of bay	26 34 30	70 47 30		13 46	9 10	5	135
Las Animas; sum. over pt. (outer)	26 23 35	70 47 0					135
Pan de Azucar; islet, summit	26 9 15	70 47 5					136
Ballenita; islet; off Ballenita	25 45 45	70 50 40					136
Lavata; cove near S.W. point	25 39 30	70 47 15		13 30	9 20	5	136
Point San Pedro; summit	25 31 0	70 44 30					136
VI.—COASTS OF BOLIVIA AND PERU.							
Point Taltal; northern extreme	25 24 45	70 38 15					136
Hueso Parado; S. point of cove	25 24 30	70 35 15					136
Point Grande; outer summit	25 7 0	70 33 30					141
Paposo; white head	25 2 30	70 33 5		13 0	9 40	5	141
Mount Trigo; summit	24 40 0	70 36 15					141
Reyes Head; extreme pitch	24 34 30	70 39 45					141
Point Jara; summit	23 53 0	70 35 45					142
Jaron Mountain; summit	23 52 30	70 32 15					142
Moreno Mountain; summit	23 28 30	70 38 15					142
Constitucion Cove; shingle point on island	23 26 42	70 40 30		12 48	10 0	4	142
Morro Jorge; summit	23 15 10	70 39 45					142
Mexillones Hill; summit	23 6 30	70 35 0			10 32	3	143
Cobija, or la Mar; landing place	22 34 0	70 21 5		12 30	9 54	4	143
Algodon Bay; extremity of point	22 6 0	70 17 5		12 6			144
Chipana Bay	21 23 9	70 10 50		12 0			144
San Francisco Head; W. pitch	21 55 50	70 14 45					145
River Loa; mouth of	21 28 0	70 6 15					145
Point Lobo, or Blanca; out. pitch	21 5 30	70 15 45					145
Mount Carrasco; highest summit	20 58 30	70 9 45					145
Pica Pabellon; summit	20 57 40	70 14 0					145
Point Patache; extreme	20 51 5	70 18 15					145
Iquique; centre of island	20 12 30	70 14 30		12 18	8 45	5	145
Pisagua; Point Pichalo; extreme	19 36 30	70 19 0		11 30			146
Point Gorda, western low extreme	19 19 0	70 21 30					147
Point Lobos; summit	18 45 40	70 25 30					147
Arica; Mole	18 28 5	70 23 45		11 0	8 0	5	147
Sama, Mountain; highest summ.	17 58 35	70 56 15					149
Mollendo	17 0 0	71 0 0		11 5	8 0	5	151
Point Coles; extremity	17 42 0	71 26 15					150
Ylo Town; rivulet mouth	17 37 0	71 23 45		11 0	8 20	6	150
Tambo Valley; Point Mexico	17 10 50	71 52 0					151
Islay; Custom House	17 0 0	72 10 15		11 0	8 53	7	152
Quilca; Cove; W. head	16 42 20	72 31 0		10 45	8 0	6	154
Pescadores Point; S.W. extreme	16 23 50	73 20 25					156
Atico; E. cove	16 13 30	73 45 15		11 12	8 53	5	156
Point Chala; extreme	15 48 0	74 31 0					156
Lomas; flagstaff on point	15 33 15	74 54 45		10 48	8 19	5	157
San Juan; Needle Hummock	15 20 56	75 13 20		10 30	5 10	3	157
Point Beware; S.W. extreme	15 8 35	75 25 45					157
Point Nasca; summit	14 57 0	75 34 30					157
Doña Maria Table; central summ.	14 41 0	75 53 40					158
Yndependencia Bay; S. point of Santa Rosa Island	14 18 15	76 13 30		9 30	4 50	4	158
Mount Carreta; summit	14 9 50	76 20 20					158
San Gallan; Isl.; northern summ.	13 50 0	76 31 15					159
Paraca Bay; W. point N. extr.	13 48 0	76 22 15					159
Pisco; Town; middle	13 43 0	76 16 30		11 0	4 50	4	159
Point Frayles; extreme	13 1 0	76 34 50					162
Asia Rock; summit	12 48 0	76 41 55					162

	Lat. South.	Lon. West.	Authority.	Var. East.	H. W.	Range.	Page.
	° ′ ″	° ′ ″		° ′	H. M.	FT.	
Chilca Point; S.W. pitch . .	12 31 0	76 52 40	The survey by				162
Chilca Cove; Rock; summit .	12 29 20	76 52 30	Capt. R. FitzRoy,				162
Chorillos Bay			R.N., 1831—1834.		3 37	6	163
Morro Solar; summit . . .	12 11 30	77 6 15					163
CALLAO; Arsenal flagstaff .	12 4 0	77 13 30		10 0	5 47	4	165
San Lorenzo Island; N. point .	12 4 0	77 19 0		10 36			165
Hormigas Islet; largest (southern)	11 58 0	77 50 0					169
Pescador Islands; summ. of larg.	11 47 10	77 19 50					170
Chancay Head; summit . .	11 35 55	77 20 35		10 12			170
Pelado Islet; summit . . .	11 27 10	77 53 0					170
Salinas Hill; summit . . .	11 15 30	77 39 55			4 56		171
Huacho Point; extreme pitch .	11 8 45	77 40 15		9 48	4 44	3	171
Supé; W. end of village . .	10 49 45	77 47 0		9 42	4 50	3	171
Jaguay, or Gramadel Hd.; W. ex.	10 25 15	78 3 30					172
Guarmey; W. end of sandy beach	10 6 15	78 13 0		9 36	6 10	2	172
Colina Redonda; summit . .	9 38 35	78 24 20					174
Mount Mongon; western summit	9 38 15	78 21 15					174
Casma Bay; inner S. point . .	9 28 0	78 25 35		9 30			174
Samanco Bay; Cross Point . .	9 15 30	78 32 45		9 20	6 30	2	175
Ferrol Bay; Blanco Island; sum.	9 6 30	78 39 25					175
Santa; centre of projecting point	9 0 0	78 41 30		9 32			175
Chao Islet; centre . . .	8 46 30	78 49 0					176
Guanape Islands; summ. of high.	8 34 50	78 59 15					176
Truxillo; church . . .	8 7 30	79 4 0					176
Huanchaco Point; S.W. extremity	8 5 40	79 9 0		9 30			176
Macabi Islet; summit . . .	7 49 15	79 30 55					178
San Nicholas Bay . . .					5 4	3	178
Malabrigo Bay; rocks . . .	7 42 40	79 28 0		9 28	5 0	2	178
Pacasmayo Point; N.W. extreme	7 25 15	79 37 25		9 30			178
Lobos de Afuera Island; Fishing Cove on E. side . . .	6 56 45	80 43 55		9 20			180
Eten Head; summit over . .	5 56 40	79 53 50					179
Lambayeque; beach opposite .	6 46 0	79 59 30		9 10	4 0	3	179
Lobos de Tierra; central summit .	6 26 45	80 52 50					180
Point Aguja; western cliff summ.	5 55 30	81 10 0					180
Sechura Town; church . .	5 35 0	80 49 45					180
Payta, Silla (or Saddle); S. summ.	5 12 0	81 9 20					181
Payta; new end of town . .	5 5 30	81 8 15		9 0	3 20	3	181
Pariña Point; extreme . .	4 40 50	81 20 45					182
Cape Blanco; und. mid. high cliff	4 16 40	81 15 45					182
Picos Point; extreme cliff . .	3 45 10	80 47 30					182
Point Malpelo; mouth of Tumbes River	3 30 40	80 30 30		8 50	4 0	10	182
Puná Island; Consulate on Point Española	2 47 30	79 57 45	(* End of H.M.S. *Beagle's*, surveys.)	9 0	6 0	11	190
GUAYAQUIL; S. end of city* .	2 13 0	79 53 30		8 30	7 0	11	192
Point Santa Elena, N.W. extreme	2 11 10	81 1 40	Capt. H. Kellett,				198
Pelado Islet	2 3 55	80 48 45	R.N., 1836.				199
Selango Island	1 35 20	80 54 0	„				199
Callo Point	1 23 0	80 47 30	„				200
Plata Island; N.W. point . .	1 15 30	81 7 15	„				200
Cape San Lorenzo; islet off .	1 3 20	80 57 25	„	8 30			201
Monte Christo; 1,429 feet summ.	1 3 45	80 42 30	„				201
Chimborazo; Volcano . .	1 24 0	79 0 0	Humboldt.				
Port Manta	0 57 0	80 45 30	Kellett.				201
Caracas Bay; entrance . .	0 34 30	80 28 0	„				202
Cape Passado; N. extremity .	0 21 30	80 32 0	„	8 15			203
Pedernales Point; extreme .	0 4 10	80 9 0	„				203
Cape San Francisco; S.W. extr. .	0 39 45	80 9 25	„				203
Galera Point; N. extremity .	0 50 0	80 7 0	„	8 0			203
Atacames, or Tacames; mouth of river	0 57 30	79 55 0	„				204
Esmeralda River; N.E. entr. pt.	1 0 30	79 41 15	Charts.				204
Point Manglares	1 36 0	79 7 0	„				204
Tumaco	1 49 0	78 50 0	„				204
Gorgona Island, 1,296 ft. N. point	3 0 0	78 9 0	Kellett.				205

	Lat. North.	Lon. West.	Authority.	Var. East.	H. W.	Range.	Page.
	° ′ ″	° ′ ″		° ′	H. M.	Ft.	
Buenaventura Bay; town	3 53 0	76 59 0	Charts.				206
Point Charambira	4 16 0	77 30 0	„				
Cape Corrientes	5 33 0	77 29 30	„				207
Tupica Bay; C. Francisco Solano	6 37 0	77 20 0	„				208
Port Pinas	7 40 0	78 7 0	„				209
Point Gerachina	8 49 0	78 28 0	„				210
Isla del Rey; S. point	8 15 0	78 51 0	„				212
Panama'; N.E. bastion	8 56 56	79 31 12	Sir Edw. Belcher,	7 15	3 23	22 0	213
Taboga Island; vill. watering pl.	8 47 50	79 30 0	1837.				213
Otoque Islands; S. islet	8 35 25	79 34 30	Charts.				219
Point Mala	7 25 0	80 2 0	„				219
Morro de Puercos	7 13 0	80 27 0	„				219
Cape Mariato	7 12 0	80 42 0	„				219
Quibo Island; Damas B. watg. pl.	7 23 30	81 42 0	Lieut. Wood, 1848.	7 0			219
„ „ Negada, or S.E. pt.	7 13 10	81 36 0	„				219
„ „ Hermosa, or W. pt.	7 24 45	81 53 35	„				219
Hicaron or Quicara Id.; David pt.	7 10 50	81 46 18	„				219
Hicarita Island; S. point	7 6 0	81 48 0	„				219
Bahia Honda	7 45 0	81 30 0	„				221
Port Puebla Nueva, or Santiago	8 5 0	81 41 0	„				221
Montmosa Island	7 26 0	82 27 0	Charts.				221
Burica Point	8 1 0	82 59 0	„				222
VIII.—CENTRAL AMERICA, OR GUATEMALA.							
Gulf of Dulce	8 23 0	83 30 0	Charts.				226
Chiriqui Mountain, 11,266 feet	8 52 0	82 30 0	Barnett.				226
Caño Island	8 42 0	84 6 0	Charts.				226
Port Mantos	8 56 0	84 14 0	„				226
Gf. of Nicoya; Punta de Arenas H.	9 55 30	84 52 0	Sir Edw. Belcher.				226
Cape Blanco	9 34 0	85 7 0	„				227
Morro Hermoso	10 6 0	85 29 0	„				—
Cape Velas	10 13 0	85 40 0	„				228
Port Culebra; Gorda point	10 31 0	85 43 30	„				229
„ „ O.St.; head of port	10 36 55	85 33 30	„				229
„ „ Viradores Island	10 34 20	85 40 0	„				229
Point St. Elena	10 55 0	85 46 0	„				230
Salinas Bay; Salinas Island	11 2 50	85 40 45	„				230
Port San Juan; S. bluff	11 15 12	85 53 0	„				230
Cape Desolada	12 21 0	86 59 0	„				233
Realejo; Cardon Island, N. pt.	12 27 55	87 9 30	„	8 13	3 6	11	235
Fonseca, or Conchagua Gulf; Coseguina Volcano	12 58 0	87 37 0	„				240
„ „ Port La Union; Chicarene Point	13 17 5	87 42 15	„				240
„ „ Port Naguiscolo	12 59 0	87 16 0	„				242
Port Giquilisco, or Triunfo de los Libres	13 22 0	88 12 0	Spanish MS.				244
Riv. Lempa; Barra del Esp. Santo	13 21 0	88 17 0	Bauza.				245
Volcan de S. Miguel, 7,024 feet	13 30 0	88 10 0	„				245
Port Libertad; flagstaff	13 30 0	89 11 0	„				246
City of San Salvador	13 48 0	88 56 0	„				246
Port Acajutla, or Sonsonate; Point Remedios	13 30 0	89 45 0	„	10 0 (1833)	2 25	9	248
Izalco Volcano	13 48 0	89 33 0	„				249
Port of Istapa	14 0 0	90 38 0	„				249
Volcan de Agua	14 29 0	90 36 0	„				251
IX.—WEST COAST OF MEXICO.							
Tehuantepec Road, or Ventosa; Morro de Carbon	16 9 35	95 4 37	„				259

	Lat. North.	Lon. West.	Authority.	Var. East.	H. W.	Range.	Page.
	° ′ ″	° ′ ″		° ′	H. M.	FT.	
Bay of Bamba; Punta de Zipegua	16 1 0	95 28 30	Masters.	8 15			260
Morro of Ytapa, or Ayuta . .	15 56 0	95 46 0	Bauza.				260
Bay of Rosario; Morro de las Salinas	15 50 25	96 2 0	Masters.				261
Port Gustaleo; islets off . .	15 44 25	96 10 0	Sir E. Belcher.				266
Port Sacraficios; Sacraficios Id. .	15 44 0	96 19 7	Spanish MS.		3 15	6	268
Port Angeles	15 44 0	96 42 0	Bauza.				268
Alcatras Rock	15 58 0	97 30 0	"				269
Acapulco, Town of; Fort S. Diego	16 15 30	99 50 0	Sir E. Belcher.				269
Paps of Coyuca	17 6 0	100 0 0	Bauza.				273
Point Jequepa	17 20 0	101 8 0	"				273
Morro de Petatlan . . .	17 32 0	101 24 0	"				273
Port Sihuantanejo; head of port .	17 38 3	101 30 52	Capt. Kellett, 1847.				273
Paps of Tejupan . . .	18 20 0	103 18 0	Charts.				274
Colima Volcano, 12,003 feet, sum.	19 24 0	103 34 0	"				275
Manzanilla Bay; Pt. S. Francisco	19 4 0	104 26 0	"				274
Port Navidad; S.W. entrance .	19 12 0	104 48 0	"				276
Cape Corrientes; extremity .	20 25 0	105 39 0	"				276
Point Mita; extremity . .	20 46 0	105 28 0	"				276
La Corvetana Rock . . .	20 42 0	105 46 40	"				277
Tres Marias Islands; S. Juanito I.	21 44 0	106 38 0	Beechey.				277
Piedro de Mer, 130 feet . .	21 34 30	105 30 0	"				278
San Blas; arsenal . . .	21 32 20	105 16 0	"		9 41	6 or 7	278
Isabella Island	21 52 0	105 54 0	"				287
Rio Chametla, or del Rosario; W. point	22 50 0	105 58 0	"				287
Mazatlan; Creston Island extr. .	23 11 40	106 23 45	"	10 18	9 50	7	287
X.—LOWER CALIFORNIA, &c.							
Point Arboledo	23 33 0	106 48 0	Charts.				—
Culiacan Shoals; S.W. edge .	24 37 0	108 8 0	"				297
Culiacan River; S. point entrance	24 38 0	107 58 0	"				297
Point St. Ignacio . . .	25 33 0	109 1 0	"				297
Point Rosa	26 42 0	109 50 0	"				297
Lobos Marinos Island . . .	27 15 0	110 46 0	"				297
Rio Yaqui; entrance . . .	27 50 0	110 30 0	"				297
Guaymus: Morro Almagre . .	27 53 50	110 49 11	Rosamel, 1840.	12 4		3 to 11	297
Cape Naro	27 50 0	110 54 0	Charts.				298
Tetas de Cabra, or Paps . .	27 56 0	111 5 0	"				298
St. Pedro Nolasco . . .	27 56 0	111 14 0	"				299
Tiburon Island; W. point . .	28 54 0	112 26 0	"				299
Rio Colorado; mouth . . .	32 0 0	114 0 0	"				299
Angeles Island; S. point . .	29 6 0	112 52 0	"				300
Cape S. Gabriel	28 36 0	112 42 0	"				300
Cape de las Virgenes . . .	27 46 0	112 31 0	"				300
Moleje Bay; village . . .	26 52 0	112 29 0	"				300
Point Concepcion . . .	26 57 0	112 4 0	"				300
Real de Loreto	26 14 0	111 30 0	"				302
Carmen Island; E. point . .	26 10 0	111 2 0	"				302
Catalana Island; N. point . .	25 41 0	110 47 0	"				302
Espiritu Santo Island; N. end .	24 36 0	110 22 0	"				302
La Paz	24 10 0	109 45 0	"				302
Cerralbo Island; N. end . .	24 23 0	109 45 0	"				303
S. José del Cabo; mission . .	23 3 30	109 41 0	"				304
Cape San Lucas	22 52 0	109 53 0	Sir E. Belcher.				305
Mesas of Narvaez . . .	23 56 0	110 52 0	Charts.				305
Gulf of Magdalena; Observation Station; Delgada Point . .	38 24 18	112 6 21	Sir E. Belcher.	9 15	7 35	6¼	306
Cape San Lazaro, 1,300 feet .	24 44 50	112 16 0	Charts.				309
Farallones Alijos Rocks . .	24 51 0	115 47 0	Du P. Thouars.				309
Point Abreojos	26 42 0	113 34 0	Charts.				309
Asencion Island	27 8 0	114 18 0	"				
San Bartholomew, or Turtle Bay, N. head	27 39 50	114 51 20	Sir E. Belcher.				308

	Lat. North.	Lon. West.	Authority.	Var. East.	H. W.	Range.	Page.
	° ′ ″	° ′ ″		° ′	H. M.	FT.	
Cedros, or Cerros Island; S. pt. .	28 3 0	115 11 0	Charts.				309
San Benito Islands, W. I. . .	28 12 0	115 46 0	,,				309
Playa Maria Bay; Sta. Maria Pt.	28 55 0	114 31 0	Capt. Kellett.	8 44			310
St. Geronimo Island . . .	29 48 0	115 47 0	,,				310
Port San Quentin; W. pt. entr. .	30 21 30	115 56 33	Sir E. Belcher.	12 6	9 5	9 to 11	311
Point Zuniga	30 30 0	115 58 0	Vancouver.				311
Cenizas Island; N.W. point .	30 32 0	116 2 0	,,				312
Cape Colnett; S.W. point . .	30 59 0	116 15 0	,,				312
Todos los Santos Bay; Pt. Grajero	31 44 0	116 46 0	,,				312
XI.—UPPER CALIFORNIA.							
Coronados Islands; large one .	32 24 55	117 14 0	Capt. Kellett.	14 15			317
Port San Diego; E. spit of entr.	32 41 0	117 11 0	Sir E. Belcher.				317
San Juan Capistrano B.; out. rock	33 26 55	117 42 0	,,				320
San Pedro Bay	33 43 10	118 16 0	,,				320
Port Vicente	33 44 0	118 23 0	Vancouver.				321
Port Dume	34 3 0	118 45 0	,,				321
Port Conversion	34 9 0	119 9 30	,,				322
Buenaventura	34 16 0	119 13 0	,,				322
Sta. Barbara	34 24 12	119 41 0	,,		8 0	3 to 4	322
San Juan Island	32 53 0	117 44 0	,,				325
San Clemente Island; E. point .	32 46 0	118 22 0	,,				325
Santa Catalina Island . . .	32 28 0	118 38 0	,,				325
Sta. Barbara Island . . .	33 23 0	119 2 0	,,				325
San Nicolas Id.; John Begg Rock	33 22 0	119 42 3	Capt. Kellett.	14 30			325
Santa Cruz Island; W. point .	34 10 0	119 47 0	Vancouver.				326
Santa Rosa Island; N.W. point .	34 2 0	120 0 0	,,				326
Point Conception	34 31 0	120 34 0	,,				324
Point Arguello	34 38 0	120 32 0	,,				326
Rio de St. Balardo; or Geraldo .	34 44 0	120 30 0	,,				326
Point Sal	34 58 0	120 33 0	,,				327
Point Monte de Buchon . .	35 18 0	120 50 0	,,				327
Esteros Bay; Esteros Point .	35 30 0	120 56 0	,,				327
Point Pinos	36 38 30	121 55 0	,,				328
MONTEREY; Fort	36 36 25	121 53 0	Beechey.		9 42	1 to 6	328
Point Año Nuevo	36 58 0	122 8 0	,,				334
Point San Pedro	37 34 0	122 28 0	,,				—
Point Lobos	37 46 30	122 27 30	,,				335
SAN FRANCISCO, Yerba Buena Cove	37 47 20	122 24 0	,,	15 30	10 34	2 to 8	—
,, ,, Fort Point . .	37 48 20	122 27 12	Survey by Lieuts.				334
South Farallon	37 36 30	122 59 0	M'Arthur and				343
North-west Farallon . . .	37 44 0	123 7 0	Bartlett, U.S.N.,				343
Punta de los Reyes . . .	38 1 30	123 1 30	1850.				343
Point Tornales	38 14 30	123 1 30	,,				—
Bodega Head	38 18 30	123 4 0	,,				344
Fort Ross	38 33 0	123 5 30	,,				344
Blunt's Reef off Cape Mendoçino	40 27 15	124 29 0	,,				346
Cape Mendoçino; Sugar-loaf .	40 27 0	124 26 30	,,	16 30			246
False Mendoçino	40 31 0	124 25 0	,,				346
Eel River; entrance . . .	40 39 30	124 16 0	,,				—
Table Bluff	40 44 0	124 12 0	,,				—
Humboldt Harbour; entrance .	40 51 0	124 7 0	,,				—
Trinidad Bay; anchorage . .	41 5 40	124 4 0	,,				347
Trinidad City	41 6 20	124 4 0	,,				—
The Turtles, N.W. of Trinidad .	41 12 0	124 11 30	,,				—
Red-wood Creek	41 18 30	124 6 0	,,				—
Redding's Rock	41 23 0	124 6 0	,,				—
Klamath River; entrance . .	41 34 0	124 0 30	,,				354
Port St. George	41 43 0	124 3 0	,,	18 0			348
Cape St. George	41 47 0	124 6 0	,,				348
St. George's Reef, or Islets; N.W. extremity . . .	41 51 0	124 12 0	,,				348
Pelican Bay; Indian vill. anchor.	41 55 0	124 3 0	,,				348

XII.—THE COAST OF OREGON.	Lat. North.	Lon. West.	Authority.	Var. East.	H. W.	Range.	Page.
	° ′ ″	° ′ ″		° ′	H. M.	FT.	
Toatounis, or Rogue's River	42 25 30	124 20 0	Survey by Lieuts.				—
Toutounis Reef; S. extremity	42 27 30	124 27 0	M'Arthur and				—
Ewing Harbour; anchorage	42 44 0	124 20 0	Bartlett, U.S.N.,	19 0			—
Cape Orford, or Blanco	42 55 0	124 25 0	1850.				354
Orford Reef; islet above water S.W. extremity	42 49 0	124 30 30	,,				355
Coquille River	43 12 40	124 14 30	,,				353
Cape Arago	43 27 0	124 15 30	,,	20 40			—
Kowes (Cahoos) River; entrance	43 28 0	124 8 30	,,				355
Umpqua River; entrance	43 44 0	124 7 30	,,				356
Cape Perpetua; S. bluff	44 11 0	124 0 0	,,				356
,, ,, N. bluff	44 16 30	124 0 0	,,				356
Alseya River	44 39 0	123 54 30	,,				—
Three Mary's Islets (off C. Foulweather)	44 44 0	123 56 0	,,				—
Cape Foulweather	44 45 0	123 55 30	,,				356
Nekos River; entrance	44 57 0	123 51 0	,,				—
Yaquinna River	45 6 0	123 52 30	,,				—
Cape Look-out	45 23 0	123 54 0	,,				356
Killamook River	45 32 0	123 51 30	,,				—
False Killamook	45 46 30	123 57 30	,,				—
Killamook Head	45 54 0	123 57 30	,,				—
Point Adams	46 12 40	123 56 2	,,				359
Baker Bay; Curtis Point	46 16 43	124 0 7	Sir Edw. Belcher,	19 0	12 15	7½	360
Cape Disappointment	46 15 50	124 0 10	1839.				359
Fort Vancouver	45 36 53	122 39 34	U. S. Ex. Ex.				368
Cape Shoalwater	46 42 0	124 12 0	Vancouver.				370
Gray's Harbour; North point entrance	47 0 0	124 7 0	,,				370
Point Grenville	47 22 0	124 14 0	,,				372
Destruction Island	47 38 0	124 25 0	,,				372
Cape Flattery	48 9 0	124 26 0	,,				373
Cape Classet; Duncan Rock	48 24 12	124 45 45	Capt. Kellett, 1847.				374
Neeah Bay; Wyadda Island	48 22 30	124 36 45	,,	21 8			375
Klaholoh Rock, 150 feet	48 21 35	124 33 30	,,				376
Kydaka Point	48 17 20	124 22 0	,,				376
Callam Bay; Slip Point	48 16 0	124 15 10	,,				376
Pillar Point	48 13 15	124 6 0	,,				376
Crescent Bay; Tongue Point	48 10 10	123 42 20	,,				376
Freshwater Bay; Observatory Point	48 9 15	123 38 0	,,				376
Port Angelos; Ediz Hook extr.	48 8 30	123 23 50	,,				376
New Dungeness; extremity	48 11 0	123 5 30	,,				376
Port Discovery; Protection Island, S.W. point	48 7 10	122 56 0	,,	22 0	2 30	7	377
Port Townsend; Marrowstone Point	48 6 0	122 40 0	,,				379
Oak Cove, or Port Lawrence; watering place	48 1 0	122 42 40	,,				379
Admiralty Inlet; Foulweather Bluff	47 56 35	122 35 0	U.S. Ex. Ex. 1841.				379
,, ,, Port Madison; N. pt.	47 44 30	122 26 40	,,				380
,, ,, Restoration Point	47 34 40	122 27 40	,,	21 0	4 10	7 to 8	380
,, ,, Port Orchard	47 33 0	122 34 0	,,				380
Puget Sound; Narrows entrance	47 19 0	122 29 0	,,				381
,, ,, Fort Nisqually	47 6 0	122 35 0	,,		6 10	12–18	382
Hood's Canal; Port Ludlow, E. pt.	47 56 0	122 38 30	,,				383
,, ,, Head	47 27 0	122 50 0	,,				384
Possession Sd.; Penn Cove entr.	48 14 0	122 35 0	,,				385
,, Deception Pass, W. entr.	48 24 0	122 37 30	,,				386
Rosario Strait; Cypress Island, W. side	48 34 30	122 42 0	,,	22 0	2 37		386
Bellingham Bay; William Point	48 35 40	122 31 0	Vancouver.				387
Birch Bay; N. point	48 54 0	122 43 0	,,				388
Point Roberts	48 56 0	122 59 0	,,				388

XIII.—VANCOUVER ISLAND, &c.	Lat. North.	Lon. West.	Authority.	Var. East.	H. W.	Range.	Page.
	° ′ ″	° ′ ″		° ′	H. M.	FT.	
Fraser River; Point Garry	49 5 0	123 10 0	Mr. E. Simpson.				393
„ „ Fort Langley	49 8 30	122 46 0	„				393
Burrard's Canal; Point Grey	49 17 10	123 13 0	Vancouver, 1792,				394
Howe's Sound; Passage Island	49 20 0	123 16 30	1793.				395
„ „ Anvil Island	49 30 0	123 16 0	„				395
Jervis's Canal; S.E. entrance	49 35 30	123 52 0	„				396
„ „ Head	50 6 0	123 46 0	„				396
Scotch Fir Point	49 41 0	124 0 0	„				397
Favida (Feveda, Texada) Island; Point Upwood	49 28 30	123 52 0	„				397
„ „ Point Marshall	49 48 0	124 32 0	„				397
Desolation Sound; Point Sarah	50 4 30	124 48 0	„				398
Bute's Canal; vill. on N.W. pt.	50 24 0	125 5 30	„				399
Valdes Island; Cape Mudge	50 0 0	125 2 0	„				400
Discovery Passage; Menzies Bay	50 7 30	125 20 0	„				401
Point Chatham	50 19 30	125 26 0	„				401
Loughborough's Canal; head	50 54 0	125 17 0	„				402
Johnstone's Strait; Port Neville, entrance	50 28 0	126 16 0	„				402
Call's Canal; E. point	50 32 0	126 39 0	„				403
Nimpkish River; Sandy Islet off	50 35 30	127 16 0	„	20 45			403
Broughton's Archipelago; Pt. Duff	50 47 0	127 0 0	„				406
„ „ Deep Sea Bluff	50 58 0	126 43 0	„				406
Knight's Canal; head	51 1 0	126 0 0	„				406
Mount Stephens	51 1 0	126 53 0	„				407
Well's Passage; Point Bayle	50 51 0	127 18 0	„				408
Port M'Neil (coal)	50 39 0	127 20 0	„				408
Beaver Harbour; Thomas point	50 43 0	127 27 0	„				408
Goletas Channel; Port Valdes	50 54 0	128 20 0	„				408
Cape Scott	50 48 0	128 28 0	„				409
Lanz Islands; W. point	50 50 0	128 42 0	„				409
Scott Islands; W. rocks	50 51 0	129 10 0	„				409
Josef Bay; N.W. point	50 40 0	128 22 0	„				409
Woody Point	50 6 0	127 53 0	„				409
Esperanza Inlet; S. point	49 45 0	127 6 0	„				409
Nootka Sound; Friendly Cove	49 34 59	126 35 30	Sir Edw. Belcher.				410
Cape S. Estevan, or Pt. Breakers	49 24 0	126 34 0	Cook.				413
Port Cox	49 4 0	125 47 0	Meares.				414
Nitinat, or Berkeley Sound; Terron Point	48 50 0	125 24 0	Chart.				414
Bonilla Point	48 36 0	124 50 0	„				414
Port S. Juan; Observatory Rocks	48 31 30	124 27 0	Capt. Kellett.	21 0			416
Sooke Inlet; Secretary Island	48 19 0	123 42 0	„				416
Beechey Head	48 18 30	123 39 30	„				416
Race, or Rocky Islands; S.E. pt.	48 17 30	123 32 0	„				416
Pedder Bay; William Head	48 20 25	123 32 0	„				416
Albert Head	48 23 0	123 29 0	„				416
Esquimalt Harbour; Fisgard Isld.	48 25 35	123 37 28	„				417
Victoria Harbour; S.E. or Ogden Point	48 24 46	123 23 23	„	22 7			417
XIV.—COAST OF BRITISH AMERICA.							
Cape Caution	51 12 0	127 57 30	Vancouver.				422
Virgin Rocks	51 19 0	128 19 0	„				423
Pearl Rocks	51 24 0	128 7 0	„				423
Smith's Inlet; islet off entrance	51 18 0	128 1 0	„				423
Rivers' Canal; entrance	51 25 0	127 53 0	„				423
Calvert's Island; S. point	51 25 40	128 1 30	„				423
Fitzhugh's Sound; Pt. Walker	51 57 0	127 57 0	„				424
Restoration Cove	52 1 30	127 47 0	„				424
Burke's Canal; Point Edward	52 26 0	127 31 0	„				425
Fisher's Canal; Port John	52 7 0	127 59 0	„				425
Milbank Sound; Cape Swaine	52 13 0	128 30 0	„				425

	Lat. North.	Lon. West.	Authority.	Var. East.	H. W.	Range.	Page.
	° ′ ″	° ′ ″		° ′	H. M.	FT.	
Milbank Sound; Fort M'Loughlin	52 12 0	128 24 0	Vancouver.				425
Mussel Canal; Poison Cove .	52 55 0	128 10 0	,,				426
Gardner's Canal; Pt. Staniforth .	53 54 0	128 44 0	,,				427
Isle de Gil; Fisherman's Cove .	53 18 30	128 19 0	,,				427
Banks's Island; Calamity Harb.	53 11 0	129 41 0	,,				428
Canal de Principe; Pt. Stephens	53 28 0	129 48 0	,,				430
Cape Ibbetson	54 3 45	130 39 0	,,				428
Point Hunt	54 10 30	130 20 0	,,				429
Port Essington; Pt. Lambert .	54 10 20	130 2 30	,,				428
Stephen's Id.; rocks off N.W. side	54 17 0	130 48 0	,,				429
Point Maskelyne; entr. of Works Canal	54 42 0	130 24 0	,,				430
Fort Simpson	54 33 25	130 11 0	(Sir G. Simpson.)				430
Observatory Inlet; Salmon Cove	55 15 34	129 52 30	Vancouver.	25 18	1 8	16	431
Portland's Canal; head . .	55 45 0	130 5 0	,,				431
Queen Charlotte Island; Cape St. James, or S. point . . .	51 58 0	131 2 0	,,				434
,, Ibbertson's Sound . .	52 24 0	131 30 0	,,				434
,, Cape Henry . . .	52 53 0	132 25 0	,,				434
,, Cartwright's Sd.; Pt. Buck	53 10 0	132 40 0	,,				434
,, Hippah Island . . .	53 33 0	133 7 0	,,				434
,, Point Frederick . . .	53 58 0	133 6 0	,,				435
,, Cape Santa Margarita .	54 15 0	133 11 0	,,				435
,, Langara Island; N. point .	54 20 0	133 10 0	,,				435
,, Pt. Ymbisible, or Rose .	54 12 0	131 25 0	Spanish chart.				435
XV.—COAST OF RUSSIAN AMERICA.							
Cape Fox	54 45 0	130 49 0	Vancouver.				438
Cape Northumberland . .	54 52 0	131 15 0	,,				438
Behm's Canal; Point Sykes .	55 6 0	131 7 0	,,				438
,, ,, New Eddystone Rock	55 29 0	130 55 0	,,				439
Revilla-Gigedo Island; Point Whaley, N. point . . .	55 56 0	131 18 0	,,				439
Port Stewart; Islet on S. W. side	55 38 15	131 47 0	,,	28 30			440
Cape Caamano	55 29 0	131 54 0	,,				441
Duke of Clarence's Strait; Point Percy	54 55 40	131 31 0	,,				441
,, ,, Cape de Chacon	54 43 0	131 56 0	,,				442
Prince Ernest's Sound; Point le Mesurier.	55 46 0	132 13 0	,,				442
,, ,, Point Warde	56 9 0	132 0 0	,,				443
Point (and Fort) Highfield . .	56 34 0	132 22 0	,,				443
Fort Stikine (H. B. C. post) .	56 40 0						444
Point Howe	56 34 0	132 48 0	,,				444
Port Protection; Point Baker .	56 20 30	133 36 0	,,				445
Port Beauclerc; Islet off . .	56 15 0	133 48 0	,,				446
Point St. Alban	56 7 0	133 35 0	,,				446
Cape Decision	56 2 0	134 3 0	,,				447
Cape Pole	55 58 0	133 45 0	,,				447
Coronation Island; S. point .	55 52 0	134 10 0	,,				447
Cape Addington	55 27 0	133 48 0	,,				447
Cape San Bartolom . . .	55 12 30	133 36 0	,,				448
Rasa Island, or Wolf Rock . .	55 1 0	133 29 0	,,				448
San Carlos, Douglas, or Forrester's Island; S. point . .	54 48 0	133 32 0	,,				448
Cape Muzon	54 43 0	132 42 0	Quadra, 1775.				448
Port Nunez	54 43 0	132 7 0	,,				448
Christian's Sound; Port Malmesbury	56 17 0	134 11 0	Vancouver.				449
,, ,, Point Ellis .	56 31 0	134 16 0	,,				449
Prince Frederick's Sound; Point Kingsmill	56 51 0	134 22 0	,,				449
,, Pt. Camden; Pt. Macartney	57 2 0	133 58 0	,,				450

	Lat. North.	Lon. West.	Authority.	Var. East.	H. W.	Range.	Page.
	° ′ ″	° ′ ″		° ′	H. M.	FT.	
Admiralty Island; Point Gardner	57 1 0	134 32 0	Vancouver.				450
,, ,, Point Nepean	57 10 0	134 5 0	,,				450
Cape Fanshaw	57 11 0	133 25 30	,,				450
Stephens's Passage; Port Houghton, N. point	57 19 30	133 26 0	,,				452
,, Port Snettisham; Taco H.B.C. establishment	57 54 0	133 37 0	,,				453
,, Point Arden . .	58 8 0	134 10 0	,,				453
,, Point Retreat . .	58 24 0	134 59 0	,,				454
Chatham's Strait; Hood's Bay, Point Samuel	57 28 0	134 39 0	,,				454
,, ,, Point Marsden	58 7 30	134 57 0	,,				455
Lynn Canal; Point Couverden .	58 12 0	135 4 0	,,				456
,, Seduction Point .	59 2 0	135 23 0	,,				456
THE SITKA OR KING GEORGE THE THIRD'S ARCHIPELAGO.							
Cape Ommaney; Wooden's Isld.	56 10 0	134 33 0	,,				458
Port Conclusion; Ship's Cove .	56 15 0	134 33 30	,,	25 30			458
Point Augusta	58 3 30	135 1 0	,,				459
Point Adolphus	58 18 0	138 42 0	,,				460
Port Althorp; entrance . .	58 12 0	136 16 0	,,	30 0			461
Cape Cross	57 56 0	136 28 0	,,				461
Portlock's Harbour . . .	57 44 0	136 11 0	,,				463
Cape Edward	57 39 0	136 10 0	,,				463
Bay of Islands; Point Amelia .	57 17 0	135 46 0	,,				463
Cape Edgcumbe	57 2 0	135 45 0	,,				464
SITKA or Norfolk Sound; New Archangel; Arsenal, *light* .	57 2 45	135 17 10	. . . . . .				465
Point Wodehouse	56 47 0	135 41 0	Vancouver.				468
Cross Sound; Pt. Wimbledon .	58 19 0	136 15 0	,,				461
Cape Spencer	58 14 0	136 35 0	,,				460
Cape Fairweather	58 50 30	137 50 0	,,				468
Mount Fairweather	58 54 0	137 38 0	,,				468
Behring's Bay; Cape Phipps .	59 33 0	139 47 0	,,				470
,, Port Mulgrave; Pt. Turner	59 32 30	139 43 0	Sir Edw. Belcher.		12 30	9	471
,, Digges Bay; Pt. Latouche	59 51 0	139 32 0	Vancouver.				472
,, Point Manby	59 42 0	140 13 0	,,				472
Point Riou	59 54 0	141 14 0	Sir Edw. Belcher.				473
Mount St. Elias, 14,987 feet .	60 18 0	140 52 0	,,				473
Pamplona Rock	59 3 0	142 15 0	Spanish chart.				474
Cape Suckling	60 1 0	143 54 0	*Ib.*; Sir E. Belcher.				475
Kaye's Island; Cape Hamond .	59 47 0	144 28 0	Vancouver.				476
Prince William's Sound; Cape Witshed	60 29 0	145 47 30	,,				477
,, Cape Hinchinbrook . .	60 16 30	146 27 0	,,				477
,, Port Etches; Phipps Pt. .	60 21 12	146 32 0	Sir Edw. Belcher.	31 38	1 15	9½	478
,, Port Gravina; S.E. point .	60 41 0	145 18 0	,,				478
,, Snug Corner Bay . .	60 45 0	146 35 0	Vancouver.				478
,, Pt. Valdes; Pt. Freemantle	60 57 0	146 49 0	,,				479
,, Point Culross	60 44 30	147 52 0	,,				480
,, Montagu Island; S. point .	59 46 0	147 30 0	,,				482
,, Port Chalmers; peninsula .	60 16 0	146 50 0	,,	28 30	1 0	13 to	482
Cape Puget	59 55 0	148 8 0	,,			14½	480
Chiswell Islands; S. group . .	59 31 0	149 2 0	,,				483
Pie's Islands; S. extr. . .	59 19 0	149 51 0	,,				484
Point Gore	59 11 0	150 22 0	,,				484
Cook's Inlet; Cape Elizabeth .	59 9 0	151 18 0	,,				484
,, Port Chatham; watering pl.	59 14 0	151 8 0	,,	24 0	1 0	10 to	484
,, Point Bede	59 19 30	151 27 0	,,			14	485
,, Tschougatschouk Bay; Anchor Point	59 39 0	151 24 0	,,				485
,, Coulgiack Id.; Coal Bay .	60 27 0	151 31 0	,,				487
,, West Foreland	60 42 0	151 12 0	,,				487
,, North Foreland; Russ. est.	61 4 0	150 35 0	,,				488

	Lat. North.	Lon. West.	Authority.	Var. East.	H. W.	Range.	Page.
	° ′ ″	° ′ ″		° ′	H. M.	FT.	
Cook's Inlet; Ouchouganat Island, or Mount St. Augustin .	59 22 0	153 0 0	Vancouver.				486
„ Cape Douglas . . .	58 52 0	152 51 0	„				486
XVI.—KODIACK ARCHIPELAGO, ALIASKA, AND THE ALEUTIAN ARCHIPELAGO.							
KODIACK ISLAND; Greville or Tolstoy Cape	57 34 0	151 48 0	Lisiansky.				491
„ Tschinistakoy B.; Gorbun Rk.	57 41 0	151 55 0	„				492
„ St. Paul's Harbour . .	57 47 0	152 4 0	„				492
„ Igatskoy Bay; Cape Tonkoy	57 25 0	151 57 0	„				493
„ Kiluden B. Nabchmood Sett.	57 17 0	152 34 0	„				493
„ Cape Trinity . . .	56 45 0	153 33 0	Vancouver.				493
PENINSULA OF ALIASKA; Poualo B.	57 46 0	155 0 0	Wassilieff.				495
„ Wrangell Harb.; S.W. side	56 59 3	155 57 0	„	24 0			495
„ Evdokeeff Islands; S. Isld.	56 0 0	156 22 0	Golownin.				496
„ St. Stephen's Island . .	56 10 0	155 22 0	(Krusenstern.)				496
„ Tschirikoff's Id.; N.E. pt.	55 56 0	155 0 0	Vancouver.				497
„ Schumagin Ids.; Ounga N. pt.	55 42 0	160 50 0	Sarytscheff.				498
„ „ Kagay Island .	55 5 0	160 33 0	„				498
„ „ Tagh-Kiniagh Isld.	54 46 0	159 40 0	Golownin.				498
„ Sannagh, or Halibut Id.; cent.	54 27 0	162 50 0	Sarytscheff.				500
ALEUTIAN ARCHIPELAGO.							
Ounimack I. Chichaldinskoï Volc.	54 45 0	163 59 0	„				504
„ Cape Mordvinoff . .	54 51 0	164 29 0	„				504
Krenitsin Islands; Ougamook Isld.	54 17 0	164 47 0	Kotzebue.				505
Tigalga Island; centre . .	54 5 0	165 0 0	„				505
Akoun Island; N. point . .	54 22 0	165 40 0	„				505
Ounalashka Island; S.W. point	53 13 0	167 47 0	Lütke.				506
„ „ Port Illuluck	53 22 25	166 32 0	Kotzebue.	19 24	7 30	7 6in.	506
Oumnack Island; Cape Sigak .	52 50 0	168 42 0	„				509
Joann Bogosloff Island . .	53 56 20	167 58 0	Sarytscheff.				510
Younaska Island . . .	52 40 0	170 15 0	Kotzebue.				510
Amoughta Island; centre . .	52 33 0	171 4 0	Tebenkoff.				510
Segouam Island	52 22 0	172 18 0	„				511
Amlia Island; E. Cape . .	52 6 30	172 50 0	(Lütke.)				512
Atkha Island; Korovinskoï Bay, S. Cape	52 12 50	174 20 0	Inghesträm and Etoline.		1 or 2	5 or 6	513
„ Nikolskoï Village .	52 17 18	174 12 0					515
Sitkhin Island; centre . .	52 4 30	176 2 0	Stanikowitch.				516
Adakh Island; N. end . .	52 4 6	176 20 0	(Inghesträm.)				516
Kanaga Island; N. point . .	52 4 0	176 50 0	„				516
Tanaga Island; N.W. peak .	51 59 0	178 10 0	Sarytscheff.				516
Goreloy or Burnt Island . .	51 56 0	178 40 0	„				516
Amatignack Island . . .	51 5 0	178 55 0	„				517
		Lon. East.					
Semisophocnoi, or Seven Mns. I.	51 59 0	179 45 57	Inghesträm.				517
Amtschitka Island; W. point .	51 43 0	178 45 0	„				517
„ Kirilovskaïa Bay . .	51 27 1	179 9 54	„	14 5	10 0		517
Kryci or Rat Island . . .	51 45 0	179 20 0	. . . . . .				518
Kiska Island; N. point . .	52 22 0	177 50 0	. . . . . .				518
Bouldyr Island; centre . .	52 40 0	176 13 0	. . . . . .				518
Semitsch Island	53 6 0	174 0 0	. . . . . .				518
Agattou Island	52 43 0	173 37 0	. . . . . .				519
Attou Island; Tschitschagoff Bay	52 56 0	173 20 0	Etoline.	11 15	2 30	5 to 7	519
XVII.—SEA OF BEHRING, BEHRING'S STRAIT, &c.							
Point Krenitzin							524
Izenbek, or (Cte. Heiden) Bay; C. Glazenap, or Mitkoff . .	54 14 8	162 50 7	Lütke.				524

	Lat. North.	Lon. West.	Authority.	Var. East.	H. W.	Range.	Page.
	° ′ ″	° ′ ″		° ′	H. M.	FT.	
Amak or Aamak Id.; S. extreme	55 25 0	163 1 5	Stanikowitch.	21 15			524
Cape Roshnoff	55 58 0	161 0 0	"		7 30	15	525
Moller Bay; Kritakoï Id. E. pt. .	56 0 7	160 41 0	"				525
Cape Séniavine	56 23 7	160 2 7	"				526
Cape Strogonoff	56 52 0	158 51 0	"				526
Cape Menchikoff . . .	57 30 4	157 58 5	"				526
Ougatchik, or Soulima River; Cape Greig	57 43 0	157 47 2	"				527
Bristol Bay; Cape Tschitchagoff	58 17 0	157 34 0	"				527
River Nanek; Paougvigumut vill.	58 42 1	157 0 5	"				527
Chramtschenko Bay; Cape Constantine	58 29 0	158 45 0	Von Wrangel.			23–47	529
Nouchagack River; Fort Alexandroffsk	58 57 0	158 18 0	"				529
Hagemeister Island; Calm Point	58 25 0	160 55 0	Chramtschenko.				529
Cape Newenham	58 42 0	162 24 0	Cook, 1778.				530
Bay of Good News; N. pt. entr.	59 3 9	161 53 0	Etoline.	22 17	6 15	13–14	530
Kuskowine River; N.W. point .	59 50 0	162 10 0	Chramtschenko.	(1821)			530
Cape Avinoff	59 50 0	164 0 0	"				530
Nuniwack Island; N.E. extr. .	60 32 0	165 30 0	Wassilieff.				531
" " S.E. point .	60 0 0	165 3 0	"				531
Cape Vancouver . . .	60 44 0	165 0 0	Etoline.				531
Cape Romanzoff	61 51 32	166 28 0	Chramtschenko.				532
Stuart's Island; N. point . .	63 35 0	162 0 0	Cook.				533
Cape Stephens	63 33 0	162 19 0	"				533
Chaktolimout Bay; Tebenkoff Cove	63 28 30	161 52 0	Tebenkoff.	30 30		3	534
Cape Denbigh; Fort St. Michel .	64 19 0	161 10 0	Cook.				534
Cape Darby	64 21 0	163 0 0	"				535
Golovnine Bay; Stone Mole .	64 26 42	163 8 0	Tebenkoff.		6 23	3⅓	535
Aziak or Sledge Island, 642 feet .	64 31 0	166 9 0	Cook.				536
Point Rodney; northern peak .	64 42 10	166 17 50	Beechey.				537
Port Clarence; Point Spencer .	65 16 40	166 47 50	"	26 36	4 25		537
Oukivok, or King's Island, 756 ft.	64 58 49	167 57 47	Cook.				536
Cape York	65 24 10	167 19 40	Beechey.				538
Cape Prince of Wales (W. Cape of America); bluff . . .	65 33 30	167 59 10	"				539
Diomede Islands; Fairway Rock, centre	65 38 40	168 43 45	"				540
" Krusenstern Island; S. extr.	65 46 17	168 55 10	"				540
" Ratmanoff Id.; N.W. extr.	65 51 12	169 3 45	"				510
Kotzebue Sound; Cape Espenburg	66 34 56	163 36 38	"				541
" Cape Deceit . . .	66 6 20	162 40 32	"				542
" Chamisso Island; summit .	66 13 11	161 46 0	"	31 10	4 52		545
Cape Blossom	66 44 0	162 24 0	"				545
Cape Krusenstern; Low Cape .	67 8 0	163 46 0	"				546
Cape Seppings; sharp peak over	67 57 20	164 41 21	"				546
Point Hope; sandy point . .	68 19 50	166 46 24	"				547
Cape Lisburne, 849 feet . .	68 52 9	166 5 39	"				548
Cape Beaufort; coal station .	69 6 47	163 38 28	"				548
Icy Cape; village . . .	70 20 1	161 46 8	"				549
Wainwright Inlet; Cape Collie .	70 37 24	159 55 24	"				550
Point Barrow	71 23 31	156 21 30	"	41 0			551
Coast of Asia.							
Cape North, or Ir-Kaipie . .	68 55 16	179 57 0	Von Wrangel.	21 40			562
Burney, or Koliutchin Id.; S. pt.	67 27 0	175 36 0	"				562
Cape Serdze Kamen . . .	67 12 0	172 40 0	"				562
East Cape of Asia . . .	66 3 0	169 44 0	Beechey.				563
S. Lawrence B.; C. le Krleougoun	65 29 40	171 0 0	Lütke.		4 20	4	563
" Cape Pnaougoun . .	65 37 30	170 53 30	"	24 4			566
Metchigmenak Bay; entrance .	65 30 30	172 0 0	"				566
Cape Khaluetkin . . .	65 15 0	172 10 0	"				566
Cape Nygtchygan . . .	65 2 0	172 0 0	"				566
Cape Neegtchan	64 55 30	172 17 30	"				567

	Lat. North.	Lon. West.	Authority.	Var. East.	H. W.	Range.	Page.
	° ′ ″	° ′ ″		° ′	H. M.	FT.	
Arakamtchetchen I.; C. Kyghynin	64 46 0	172 7 0	Lütke.				567
Cape Mertens	64 33 15	172 20 0	„				567
Ittygran Island; Cape Postels .	64 37 0	172 21 0	„				568
Cape Tchaplin	64 24 30	172 14 0	„				569
Cape Tchoukotskoï	64 16 0	173 10 0	„				569
Port Providence; Emma Harbour	64 25 55	173 7 15	Moore, 1849.				570
Cape Spanberg	64 42 30	174 42 0	Lütke.				572
Cape Attcheun	64 46 0	175 28 0	„				572
Transfiguration Bay	64 50 0	175 25 0	„				572
Cape Behring	65 0 30	175 57 0	„	19 20			572
Gulf of St. Croix; C. Meetchken	65 28 40	178 47 0	„	21 45			574
„ Mt. Linglingaï, 1,462 ft.	65 36 30	178 17 0	„				574
		Lon. East.					
River Anadyr; mouth . . .	64 50 0	178 40 0	Charts.				576
Cape St. Thaddeus	62 42 0	179 38 0	Lütke.				576
Archangel Gabriel Bay; N. point of entrance	62 28 0	179 22 0	„				576
Cape Navarin, 2,512 feet . .	62 16 0	179 4 30	„	13 35			576
Cape Olutorskoï	59 58 0	170 28 0	„				577
Cape Govenskoï	59 50 0	166 18 0	„				577
Cape Ilpinskoï	59 48 30	165 57 0	„				577
Verkhotoursky, or Little Karaghinsky Island	59 37 30	165 43 0	„				577
Commander Islands; Behring Island, Cape Khitroff . . .	54 56 0	166 43 0	Lütke.	5 50			583
„ „ Cape Youchin	55 25 0	165 58 0	„				583
„ „ W. extremity	55 17 2	165 49 57	Beechey.				583
Medny or Copper Id.; settlement	54 47 0	168 0 0	Charts.				585
„ „ S.E. extremity .	54 32 24	168 9 0	Golownin.				585
„ „ N.W. extremity	54 52 25	167 31 0	„				585
		Lon. West.					
St. Lawrence I.; Schischmareff pt.	63 46 0	161 41 0	Schischmareff.				578
„ „ N. point . .	63 12 0	159 50 0	„				579
St. Matthew Id. (Matvoï, or Gore's Id.); Cape Upright, S.E. point	60 18 0	172 4 0	(Lütke.)				579
„ Cape Gore	60 30 0	172 50 0	„				579
„ Morjovi Island; N. point	60 44 0	172 52 0	„				579
Pribuiloff Islands; St. George's Islands, E. pt.	56 38 0	169 10 0	„				581
„ S. Paul's Id.; Sivoutchi Islet	57 5 0	169 51 0	Tchistiakoff.				582
XVIII.—KAMTSCHATKA, OKHOTSK, AND THE KURILE ISLANDS.							
		Lon. East.					
Karaghinsky Island; Cape Golenichtcheff	59 13 30	164 40 0	Lütke.				588
„ Cape Krachenninikoff	58 28 0	163 32 0	„				589
Cape Ilpinskoï	59 48 30	165 57 0	„				590
Cape Kouzmichtcheff	59 5 0	163 19 0	„				590
Karaghinskaïa Bay; mouth of the Karaga	59 8 0	162 59 0	„				590
„ „ S. point .	58 55 0	163 2 0	„				590
Cape Oukinskoï	57 58 0	162 47 0	„				591
Cape Ozernoï	57 18 0	163 14 0	„				591
River Stolbovskaïa	56 40 30	162 39 0	„				591
Cape Stolbovoï	56 40 30	163 21 0	„				591
Cape Kamtschatskoï	56 10 0	163 25 0	„				591
Klutchevskoï Volcano, 15,766 ft.	56 8 0	160 45 0	„				592
Cape Kronotskoï	54 54 0	162 13 0	„				592
Kronotskoï Volcano, 10,610 feet .	54 45 0	160 37 0	„				593
Cape Shipounsky	53 6 0	160 4 0	„				594
Villeuchinsky Peak, 7,372 feet .	52 39 43	158 20 39	Beechey.				599
Awatska Volcano, 11,500 feet .	53 20 1		„				599
AWATSKA BAY; church at Petropaulovski	53 1 0	158 43 30	„		3 30	2·2 to 6·2	598

	Lat. North.	Lon. East.	Authority.	Var. East.	H. W.	Range.	Page.
	° ′ ″	° ′ ″		° ′	H. M.	FT.	
Cape Garareab	52 21 43	158 39 8	Beechey.				600
Cape Lopatka	51 2 0	156 50 0	Lütke.				604
KURILE ISLANDS.							
Alaid Island	50 54 0	155 32 0	Krusenstern.				605
Soumshou Island; centre . .	50 46 0	156 26 0	,,				605
Poromoushir Isld.; high mountain	50 15 0	155 24 15	,,				605
,, ,, N. point .	51 0 0		Gilaeff.				605
Shirinky Island	50 10 0	154 58 0	Krusenstern.				605
Monkourushy Island; centre .	49 51 0	154 32 0	,,				605
Avos Rock	49 49 0	154 19 0	,,				605
Onnekotan Island; C. Krenitzin .	49 19 0	154 44 0	,,				606
Kharamoukotan Id.; centre peak	49 8 0	154 39 0	,,				606
Shiashkotan Island; centre .	48 52 0	154 8 0	,,				606
Tabirinkotan Island	48 44 0	153 24 0	,,				606
The Snares	48 35 0	153 44 0	,,				606
Raukoko Island; peak . .	48 16 20	153 15 0	,,				607
Mataua Island; Sarytscheff peak .	48 6 0	153 12 30	,,				607
Rashau Island	47 47 0	152 55 0	,,				607
Ushishir Island; S. point . .	47 32 40	152 38 30	Golownin.				607
Ketoy Island; S. extremity .	47 17 30	152 24 0	,,				608
Simusir Island; Prevost Peak .	47 2 50	151 52 50	,,				608
The Four Brothers; S. Torpoy Id.	46 29 15	150 33 30	,,				608
Broughton Island	46 42 30	150 28 30	,,				608
Ouroup or Staaten Island; Cape Castricum, N. point . .	46 16 0	150 22 0	,,				608
,, C. Van der Lind, S. pt.	45 39 0	149 34 0	,,				608
Itouroup Island; N.E. point .	45 38 30	149 14 0	,,				608
,, ,, C. Rikord, S. pt.	44 29 0	146 34 0	,,				609
Tschikotan or Spanberg Id.; centr.	43 53 0	146 43 30	,,				609
Kounashire Id.; St. Antony's pk.	44 31 0	145 46 0	,,				609
,, ,, Estab. in Traitor's Bay	43 44 0	144 59 30	,,				610
SEA OF OKHOTSK.							
Cape Lopatka	51 2 0	156 50 0	Lütke.				611
Bolcheretskoi	52 54 30	158 22 0	King.				611
Tigilsk	58 1 0	158 15 0	Charts.				612
Cape Outholotskoï	57 28 0	155 45 0	,,				612
Cape Bligan	59 20 0	152 50 0	,,				612
Poustaresk	61 0 0	162 30 0	,,				612
Kaminoi, at the mouth of the Pergina River	62 0 0	162 50 0	,,				613
Ingiga, or Fort Jiejiginsk . .	61 40 0	160 0 0	,,				613
Jamsk	59 29 0	153 0 0	,,				613
Taouinsk	59 56 0	148 30 0	,,				613
Okhotsk	59 20 0	143 14 0	,,				613
Jonas Island, 1,200 feet . .	56 25 30	143 16 0	Krusenstern.				615
Fort Oudskoï	54 29 0	134 58 0	Kosmin.				615
Great Shantar Island; N. point .	55 11 0	137 44 0	,,				615
,, ,, ,, Prokofieff I.	55 2 0	138 22 0	,,				615
,, ,, ,, Kossoff Id.	54 43 0	138 12 0	,,				615
Cape Linekinskoy	54 14 0	136 24 0	Charts.				616
River Tougoura; mouth . .	53 40 0		,,				616
Cape Khabaroff	53 40 0	141 22 0	,,				616
Cape Romberg	53 25 0	141 45 0	,,				616
River Amour; mouth . . .	52 30 0	140 0 0	,,				616
SAGHALIN PENINSULA.							
Cape Elizabeth	54 24 30	142 47 0	Krusenstern.	1 0			618
Cape Maria	54 17 30	142 17 45	,,				618
North Bay; Tartar colony . .	54 15 45	142 37 0	,,		2 0		619
Nadiéjeda Bay; Cape Horner .	54 10 15	142 27 34	,,				619
Cape Golovatcheff	53 30 15	141 55 0	,,				619

	Lat. North.	Lon. East.	Authority.	Var. East.	H. W.	Range.	Page.
	° ′ ″	° ′ ″		° ′	H. M.	FT.	
Cape Löwenstern	54 3 15	143 12 30	Krusenstern.				619
Cape Klokatcheff	53 46 0	143 7 0	,,				620
Cape Würst	52 57 30	143 17 30	,,				620
Shoal Point	52 32 30	143 14 30	,,				620
Downs Point	51 53 0	144 13 30	,,				620
Cape Delisle	51 0 30	143 43 0	,,				620
Cape Ratmanoff	50 48 0	143 53 15	,,				620
Cape Rimnik : . . .	50 12 30	144 5 0	,,				620
Mount Tiara	50 3 0	143 27 0	,,				620
Cape Bellingshausen . . .	49 35 0	144 25 45	,,				620
Cape Patience	48 52 0	144 46 15	,,				621
Robben Island; N.E. point .	48 36 0	144 33 0	,,				621
River Neva; mouth . . .	49 14 40	144 2 0	,,				621
Cape Soïmonoff	48 53 20	143 2 0	,,				621
Cape Dalrymple	48 21 0	142 50 0	,,				621
Cape Muloffsky	47 57 45	142 44 0	,,				622
Berniset Peak, or Mount Spanberg	47 33 0	142 20 0	,,				622
Cape Séniavine	47 16 30	142 59 30	,,				622
Cape Tonin	46 50 0	143 33 0	,,				622
Cape Löwenorn	46 23 10	143 40 0	,,				622
Cape Aniwa	46 2 20	143 30 20	,,				622
Cape Crillon	45 54 15	141 57 56	,,				623
La Dangereuse Rock . . .	45 47 15	142 8 45	,,				623
XIX.—THE JAPANESE ARCHIPELAGO.							
Island of Jesso.							
Cape Broughton; E. point .	43 38 30	146 7 30	Broughton.				628
Cape Spanberg	44 35 0	145 0 0	Japanese charts.				629
Port Atkis	43 20 0	145 30 0	,,				629
Bay of Good Hope; peaked hill .	43 0 0	144 12 0	Broughton.				630
Cape Eroen or Evoan . . .	41 59 0	142 55 0	Rikord.				630
Volcano Bay; Endermo Harbour, entrance	42 33 11	140 50 32	Broughton.	1 27½	4 30	4 to 5	630
Khakodade Point	41 43 30	141 58 0	Rikord.				631
Cape Nadiéjeda	41 25 10	141 9 30	Krusenstern.				631
Matsoumay, or Matamaï; city .	41 29 0	140 28 0	Von Siebold.				631
Cape Sineko	41 39 30	139 54 15	Krusenstern.				631
Cape Oote Nizavou . . .	42 18 10	139 46 0	,,				631
Cape Koutousoff . . .	42 38 0	139 46 0	,,				632
Cape Novosilzov . . .	43 14 30	140 25 30	,,				632
Cape Malaspina	43 42 51	141 18 30	,,				632
Mount or Peak Pallas . . .	44 0 0	141 54 0	,,				632
Cape Schischkoff	44 20 0	141 37 0	,,				632
Riosbery Island, or Pic de Langle	45 25 50	141 34 20	,,				632
Refunsbery Island; Cape Guibert	45 27 45	141 4 0	,,				632
Cape Romanzoff	45 25 50	141 34 20	,,				632
Island of Nippon.							
N.E. Cape	41 24 0	141 46 0	Chart by Ph. Fr.				633
Point King, North . . .	40 24 0	141 57 0	Von Siebold, 1840.				633
Cape De Vries	40 10 0	142 2 0	,,				—
Nanbu Harbour (King) . .	39 46 0	142 6 0	,,				—
,, ,, (De Vries) .	39 32 0	142 4 0	,,				—
Cape Gore	38 45 0	141 36 0	,,				—
Sendai Bay; Cape Nagavama .	38 23 0	141 35 0	,,				—
Nakamura	37 44 0	141 10 0	,,				—
Cape der Kennis . . .	37 2 0	141 12 0	,,				—
Minato-saki; Low Point . .	36 17 0	140 36 0	,,				—
Pitatsi Farano (Walvisch Bay) .	36 2 0	140 34 0	,,				—
Daibô-saki; Sandy Point (Zand-duirige Hoek) . .	35 40 0	140 48 0					—

	Lat. North.	Lon. East.	Authority.	Var. West.	H. W.	Range.	Page.
	° ″ ′	° ′ ″		° ′	H. M.	FT.	
Witte Hoek ; White Point . .	35 12 0	140 25 0	Chart by Ph. Fr.				—
Cape King or Firatatai . .	34 55 0	139 56 0	Von Siebold, 1840.				633
Cape Sirofana	34 54 0	139 48 0	„				—
Jedo, City of	35 42 0	139 44 30	„				633
Wodawara	35 16 0	139 6 0	„				633
Cape Nagatsuro	34 35 30	138 48 0	„				—
Barnevelds Island, Oho-sima .	34 43 0	139 24 0	„				634
Mitake or Volcano Island . .	34 4 30	139 32 0	„				634
Mikura or Unlucky Island . .	33 53 0	139 36 0	„				634
Inaniva, or Broughton's Felsen .	33 34 0	139 16 0	„				634
Fatsisio Island ; Funetsuki . .	33 6 30	139 30 30	„				634
Cape Kosan, or Omaë-saki . .	34 35 0	138 11 0	„				—
Iseno Umi Bay ; C. Isako-saki .	34 36 30	136 54 0	„				635
Toba	34 26 0	136 50 0	„				—
Idsumo-saki, or Siwono-misaki (S. point of Nippon) . .	33 26 30	135 42 0	„				636
Obosáka, Town of	34 41 0	135 22 0	„				635
Mijako, City of (capital of Japan)	35 0 0	135 40 0	„				—
Strait of De Capellen ; Simonosaki, E. point	33 58 42	130 54 0	„				636
Hamada	34 54 0	132 0 0	„				—
Uriu-saki Point	35 25 0	132 31 0	„				—
Oki Islands ; N. point . .	36 22 0	133 16 0	„				636
Dagelet Island ; Matsu-sima .	37 25 0	130 56 0	La Pérouse.				636
Argonaut Island ; Taka-sima .	37 52 0	129 50 0	„				—
Kiogami-saki Point . . .	36 49 0	135 8 0	Von Siebold.				—
Susumi-saki Point . . .	37 36 0	137 31 0	„				—
Sado Island ; N. point . . .	38 20 0	138 43 0	„				637
Awa-sima Island . . .	38 28 0	139 28 0	„				637
Tobi-sima Island . . .	39 15 0	139 41 0	„				637
Russians' Cape ; Hata-saki . .	39 59 0	139 48 0	„				637
Cape Gamaly ; Hekuri . .	40 37 0	140 2 0	„				637
Cape Greig ; Oho-saki . .	41 9 0	140 30 0	„				637
Cape Tsugar or Sangar ; Tassupi-saki	41 17 0	140 36 0	„				638
Ibid.	41 16 30	140 14 0	Krusenstern.				—
Toriwi-saki (N point of Nippon)	41 34 0	141 14 0	Von Siebold.				638
N.E. Cape ; Siria-saki . .	41 24 0	141 46 0	„				—
Tsus-sima ; Fu-tsiu, S.E. side .	34 12 0	129 18 0	„				643
Colnet Island ; Oumi-sima .	34 16 30	129 56 0	„				643
Iki Island ; Kaza-moto . .	33 52 0	129 36 0	„				—
ISLAND OF KIUSIU.							
Kokura ; Fort	33 53 30	130 50 0	„				640
Fira-to ; Town of . . .	33 21 0	129 31 30	„				645
Cape Nomo	32 35 35	129 43 0	„				644
Nagasaki ; Dezima flagstaff .	32 45 40	129 51 0	„	1 45	7 53	1 to 11½	645
Goto Islands ; N. point . .	33 16 0	129 10 0	„				643
„ „ Ohoseno-saki Point	32 33 0	128 44 0	„				643
Mesc-sima, or Linschoten Islands ; S. point	31 58 0	128 42 0	„				644
Wunzendake Volcano . .	32 44 0	130 26 0	„				642
Kosiki Id. ; Faja-saki, or S. pt. .	31 35 0	129 40 0	„				642
Symplegades ; Tsukurase . .	31 28 0	129 42 0	„				642
Cape Tcheamé, or Noma-saki .	31 24 0	130 2 0	„				642
Peak Horner ; Kaimon-ga-taki .	31 19 30	130 36 0	„				640
C. Tschitschagoff ; Satano misaki	30 56 45	130 36 30	„				640
St. Clair Island, or Kuro-sima .	30 46 0	129 55 0	„				641
Volcano Island, or Iwoga-sima	30 41 0	130 16 0	„				641
Apollos Island, or Take-sima	30 42 0	130 24 0	„				641
Seriphos, or Tanega-sima ; N. pt.	30 42 0	130 58 0	„				641
Jakûno-sima ; S. point . .	30 18 0	130 30 0	„				641
Julia Island ; Nagarabe . .	[illegible]0 26 0	130 13 0	„				641
De Zeven Zusters ; Naka-sima .	[illegible] 1 0	129 57 0	„				641
Cape Nagaeff	3[illegible] 15 0	131 11 0	„				640

	Lat. North.			Lon. East.			Authority.	Var. East.	H. W.	Range.	Page.
	°	'	"	°	'	"		° '	H. M.	FT.	
Cape D'Anville	31	28	0	131	27	0	Von Siebold.				640
Cape Cochrane	31	49	0	131	29	0	"				640
Cape Tschirikoff . . .	32	12	0	131	42	0	"				640
Tsuru-saki ; W. point of Kiusia .	32	55	0	132	10	0	"				640
Island of Sikok.											
Asi-suri-no-misaki ; S. point .	32	46	0	133	0	0	"				638
Kotsi	33	32	0	133	30	0	"				638
Murodono-saki	33	16	0	134	6	0	"				639
Tsubaki-misaki	33	50	0	134	41	0	"				639
Tok-sima	34	5	0	130	28	0	"				639

THE COLUMBIA RIVER, ETC.

During the progress of this work through the press, the increasing importance of the American acquisition of California and Oregon has led to some examination of their shores; therefore the descriptions given on pages 346—357, though nothing has been stated to impugn their accuracy, want those additional remarks acquired during the running survey made by Lieutenants W. A. Bartlett and M'Arthur. But, with the exception of the Columbia River, nothing very material requires to be added. In the Tables of Positions, pages xxxv. and xxxvi., we have given the determination of the latitudes and longitudes as stated by these officers.

The *Columbia River*, however, appears to be better known, and the following observations by Lieutenant Bartlett may be added to those on pages 366-7:—

The *South Channel* to the Columbia River is reported as a recent discovery in the early part of 1850, on page 361. It is certainly the most important entrance; as any observing seaman can cross, in or out, over this bar safely, and certainly without an hour's delay, after having once crossed in order to observe the ranges, which are well defined, and certain to lead over in good water. The depths found by the United States' schooner *Ewing*, April 19, 1850, were 16 feet at half-tide flood, deepening to 5 fathoms inside the Point of Breakers, and 6 and 7 to 9 fathoms up to Sand Island Beacon, 2 miles inside the bar. Time, from the 5 fathoms outside to Sand Island Beacon, twenty minutes. A vessel goes out from the anchorage at Sand Island into open ocean in from thirty to fifty minutes. There is abundant room for any vessel to work in or out with the wind from any point of the compass; and as the tide ebbs fair through the channel in the best water, it greatly facilitates both ingress and egress. The ranges for turning Clatsop Spit are Point Ellice, with Pillar Hill just shut in behind it, and Point Adams, in one with the highest pass of the mountains in the East C., in 8 fathoms. A vessel passes clear, either in coming in or going out. As that leading pass has no name, I propose to distinguish it as the "Ewing Pass."

A beacon has been built on Sand Island, on which is a white flag, 80 feet above the island, and 85 feet above the high-water mark. Around the base of the flag-staff is a block-house, 35 feet high and 15 feet square; it can be seen plainly, in good weather, 12 miles at sea. By bringing the beacon flag directly under the centre of the highest peak westward of Chinook Peak, and Point Adams just open South of Pillar Hill tree, a vessel will be in 12½ fathoms, in a fair way to the bar, with bar ranges on, viz., Point Adams and Pillar Hill. And the usual wind from N.W., or anywhere on the western board, is fair for crossing. Vessels cross the bar of the New Channel under all steering sails, or beat up the channel, as the case may be. It is not necessary to tack ship on the bar, in any wind. A sailing vessel can run to sea from Sand Island, or come in, in less time than she can run to Baker's Bay, after which (if in Baker's Bay) she must take her chance for wind and tide to get to sea. The anchorages at Astoria, Sandy Point (East end of Clatsop Beach), and Sand Island, are good, with abundant room for getting under weigh at any stage of the tide.

SECTION II.

CHAPTER I.

THE STRAIT OF MAGALHAENS.

1.—FROM THE EASTERN ENTRANCE TO CAPE FROWARD.

Fernando de Magalhaens (or Magalhanes), a Portuguese by birth, and a commander of reputation, offered his services, from some fancied slight, to the king of Spain, Charles V., who received his proposition favourably: this was to sail round the South extreme of America, if possible, and thus find a new route to the Molucca Islands. The expedition, consisting of five ships, left Seville (or rather San Lucar) Sept. 20, 1519, reached the American coast, and at last determined on wintering in Port San Julian; here a very serious mutiny broke out, but was quelled. They quitted the port on Oct. 18, 1520, and three days after found themselves off a cape whence a deep opening was perceived. This was on St. Ursula's day (Oct. 21), hence they called the cape De las Virgenes. The commander sent an expedition for five days to explore the opening, and then conjectured, from various evidences, that there was a passage through to the other sea. The expedition then entered the strait, and on Nov. 27, 1520 (37 days after the discovery of Cape Virgenes), they found themselves again in an open sea, and gave the name of Deseado (the Desired) to the cape, the West point of Tierra del Fuego.* The strait was soon after its discovery distinguished by a variety of names. It was called De la *Vitoria*, from that ship first discovering its eastern entrance; it was also called Streto Patagonica, from the large-footed Indians (Patagonians) first seen by Magalhaens; and it was also called the Archipelago de Cabo Deseado; but it has very properly assumed that of its discoverer. But even this has been variously spelt. The Spaniards call him Magallanes; others of his countrymen write it Fernando de Magalhanes; in Italian it has been Magaglianes; and in English, Magellan. The true form of the name, which is still a common one in Portugal, is, as re-adopted by Capt. King, and here used, Magalhaens.†

* Although Magalhaens did not live to complete the voyage, of which the discovery of the South extreme of the American continent and of the strait which now bears his name were the first and principal features, yet, from other circumstances, he must be considered as the first circumnavigator. After having proceeded northward from the West entrance of the strait, he bore away to the westward, probably passing near Tahiti, and was killed by the natives in an affray at the Island of Zebu, one of the Philippines, April 27, 1521.—*See* Herrera, dec. 3; Hakluyt, vol. iii.; Burney, vol. i. ch. 2.

† *See* Herrera, Descr. de las Indias Occ., dec. 2, lib. 2, c. 10; Peter Martyr, dec. 5, cap. 7.

It need scarcely be said, that the northern side of the strait is formed by the continent of America, the country of the Patagonians, who, however, do not show themselves in very great numbers on its shores. Interesting descriptions of them will be found in Capt. FitzRoy's Narrative.

The opposite side of the strait is Tierra del Fuego, "the Land of Fire," as it was named by Magalhaens, from his seeing many fires on its shores during the first night he approached it.* The appearance and productions of the country will be described in the course of the ensuing pages.

Of the natives of this inclement region a word may be said. Ample descriptions of their persons, manners, and customs may be found in the excellent account of the *Adventure* and *Beagle's* Voyages; also in Capt. Weddell's Narrative, pp. 148—156, &c.; and the Voyage to the Southern Seas, by Capt. Sir James Ross, R.N., 1840-1, vol. ii. pp. 303—7.

They are low in stature; their colour is of a dirty copper, or dark mahogany; their only clothing is a seal-skin, worn with the air outwards. There are no animals in the neighbourhood; and their principal food consists of mussels, limpets, and sea-eggs, which they collect and open with much dexterity; and as often as possible they procure seal's, sea-otter's, porpoise's or whale's flesh—devouring eagerly the most offensive offal: of vegetable food they only collect a few berries of the berberis, and a kind of sea-weed. Their dwellings, sometimes styled huts or wigwams, but more fitly, by old Sir John Narborough, as "arbours," consist of a few branches. They seem to possess some good feelings, and certainly appear capable of some improvement. They have few articles of traffic beyond their weapons and implements as curiosities; the seal and sea-otter skins they collect must be quite insignificant for commercial purposes. They are thievish and greedy, and have become keen traders in many parts.

In the N.E. part of Tierra del Fuego—we quote from Capt. FitzRoy—is the Yacana-kunny tribe, which may amount to 600 in number. On the shores of the Beagle Channel is a tribe formerly called Key-uhue, now probably the Tekeenica. They are the smallest, and apparently the most wretched, of the Fugians, numbering about 500 adults. To the westward, between the western part of the Beagle Channel and the Strait of Magalhaens, is a tribe now called Alikhoolip, whose numbers amount to, perhaps, about 400. About the central parts of Magalhaens' Strait is a small and very miserable horde, whose name was not known to Capt. FitzRoy. Their usual exclamation is "Pecheray! Pecheray!" whence Bougainville and others called them the Pecherais. The number of their adults is about 200. About Otway and Skyring Waters is a tribe, or fraction of a tribe, which Capt. FitzRoy is inclined to think is a branch of the Huemul people described by the Jesuit Falkner as living on both sides of the strait.

On the western coast of Patagonia, between the Strait of Magalhaens and the Chonos Archipelago, there is but one tribe, of not more than 400 adults.

Each of the tribes here mentioned speak a language differing from that of any other, though, as believed by Capt. FitzRoy, not radically differing from the aboriginal Chilian. For a more extended notice of these curious tribes, alike

* Burney, vol. i. p. 41.

insignificant to the navigator, from their numbers and their inability to administer to his wants, see Narrative of the Voyages of the *Adventure* and *Beagle*, vol. i. pp. 75—77; vol. ii. ch. 7, p. 129, &c.

The subsequent description of the Strait of Magalhaens is almost entirely derived from the Sailing Directions for East and West Patagonia, by Capt. Philip Parker King, R.N., 1832; and from the Remarks by Capt. R. FitzRoy, R.N., the Admiralty surveyors of these coasts, to whose refined observations, and admirable delineation of them, we owe our present complete knowledge of these hitherto formidable localities. In those instances where we have added anything from other sources, they are noticed; the rest must be considered as the work of the above-named officers.

The general application of these nautical descriptions will be given in a subsequent part of the work, to which the reader is referred.

CAPE VIRGINS is the northern and outer cape, forming the entrance to the strait. In approaching the strait from the eastward it is usually the first land seen, and is the best landfall, but requires caution.* It is the S.E. point of a range of steep, white cliffs, about 200 feet in height; the cape itself being 160 feet high. The cliffs extend from Cape Virgins, northward, to within 8 miles of Cape Fairweather, with only one or two breaks, in one of which, 8 miles from Cape Virgins, a boat may perhaps land, if necessary. There is good anchorage along the whole coast between Cape Virgins and the Gallegos, at from 2 to 5 miles off shore; but the bottom is rather stony, and might injure hempen cables. As Cape Virgins is approached the ground becomes more foul.

Extending nearly 4 miles to the southward of Cape Virgins is *Dungeness Point*, a low flat, formed entirely of shingle. On the opposite coast lies Cape Espiritu Santo, very similar to Cape Virgins; and Catherine Point, very like Dungeness. Cape Espiritu Santo is the N.W. end of a high range of white cliffs. Beyond it, to the westward, there is only one low short range of cliff, less than half the height of Espiritu Santo. The upper outline of this cape slopes away on both sides like the roof of a house.

Between these remarkable headlands, or rather, between the two shingle points, *lies the entrance* to the strait.

* Cape Fairweather (Cape Buen Tiempo), 52 miles to the northward of Cape Virgins, and at the North entrance of the Gallegos River, bears a very great resemblance to Cape Virgins, and also to Cape St. Vincent, on the S.W. coast of Spain. They have frequently been mistaken for each other. This error was made both in the *Adventure* and in the *Beagle*, the vessels which surveyed the coasts. Cape Fairweather was also mistaken for Cape Virgins, and the River Gallego for the entrance of the strait, by the expedition under Garcia Jofre de Loyasa, appointed by the Spanish government in 1525, and the loss of one of the ships was the consequence (Gomara, Ist. de las Indias, p. 58). In making the land this is important, as it might occasion serious consequences. There are, however, some marks by which they may be distinguished, even if the latitude should not be ascertained. In clear weather, some hills in the interior, to the S.W. of the Gallegos River, called the Friars, the Convents, and the North Hill, will be visible; in that direction from Cape Virgins is the low shore of Tierra del Fuego. In thick weather the soundings off the respective capes will be an infallible guide; for at the distance of 4 miles off Cape Fairweather no more than 4 fathoms will be found, whereas at that distance from Cape Virgins, if to the northward of it, the depth is considerable; the bottom, also, to the North of Cape Fairweather is of mud, whilst that to the North of Cape Virgins is of gravel or coarse sand; and the latter has the long low point of shingle, called Dungeness (so named by Wallis, from its resemblance to the singular promontory in the English Channel), for nearly 5 miles to the S.W.; and lastly, if the weather be clear, the distant land of Tierra del Fuego will be visible to the S.S.W.

Before approaching the land, the state of the tide should be well considered. Upon the knowledge of its movements depends the safety of the vessel and the quickness of the passage (as far as Elizabeth Island, beyond the Second Narrow), more than upon the wind or weather.

At the full and change of the moon the main stream of tide begins to run to the eastward about noon; but in the bays on each side of the channel through which the main stream passes, such as Lomas Bay, Possession Bay, &c., the times and direction of the stream vary much. This is, however, of little consequence, as they do not affect a vessel's progress through the strait. It should be borne in mind distinctly, that it is high water at nearly the same time over all the eastern entrance.

In the vicinity of Capes Virgins and Espiritu Santo, it is high water between 8 and 9 in the morning on the days of new and full moon; while the stream of flood is still running to the westward into the strait, and to the northward past Cape Virgins. Until near noon, the principal stream continues running to the westward, though the water is falling everywhere. About noon, the direction of the main stream changes; and until near 3 the water continues falling, while the stream of tide (ebb) is running to the eastward. After 3 the water again rises, the stream still running to the eastward until past 6 o'clock.

Spring tides rise from 36 to 40 feet perpendicularly; neap tides about 30 feet. In the First Narrow it is high water about 9; the water rises as above mentioned, and runs through the most confined part of the Narrows from 5 to 6 miles an hour.

Off Cape Virgins lies the *Sarmiento Bank;* this shoal has been too little considered by those who have hitherto passed that cape. After half-flood, or before half-ebb, a ship may pass Cape Virgins at any distance notless than a mile, and may cross the Sarmiento Bank without hesitation; but when the tide is low, 10 miles is not too far for a ship to keep from the cape, until it bears N.W. by W., when she should steer W.N.W., to close Dungeness. A ship might often pass over this bank without touching, even at low water, because the bottom is uneven; but there are places which no vessel, drawing more than 12 feet, could pass at a low spring-tide without injury. This bank extends about 20 miles E.S.E. from Cape Virgins, and appears to average a mile in breadth. The soundings on it are shoal and very irregular, at a less distance from the cape than 10 miles; but beyond 10 miles, as they increase they become more regular.

A *reef,* which at half-tide is hardly noticed, projects full a mile from Cape Virgins; this and Sarmiento Bank are the only dangers on the North side of the approach.

In standing eastward from Cape Virgins, the bottom is very fine brown sand, without shells or stones. When, by standing more southerly, the water shoals upon the Sarmiento Bank, the sand becomes much coarser, and is mixed with slate pebbles, or broken stones of all sizes: the sand is slaty.

This rule continues till the water deepens to 30 fathoms, or more, to the southward, when shingle only is found; and when it begins to shoal in approaching Tierra del Fuego, there is coarse dark sand, mixed with stones of various sorts, chiefly slaty. Between the shoal parts of the bank and the deep water, or from 16 to 30 fathoms, the sand is coarse, particularly near the deep water. In standing to the southward, after bringing the cape to bear West, the bottom is a very fine gray sand, until near the ridge or bank, with Cape Virgins bearing W.N.W.; with the cape in this bearing, the sand is coarser, and mixed with large and small shingle.

In crossing the Sarmiento Bank when standing to the southward, the fine dark grayish-brown sand changes to coarse slaty sand, with small stones and shingle; the stones chiefly slaty. Some casts, after crossing the ridge, were found entirely of coarse sand, while others were all shingle. As the water shoaled, the bottom was found to be coarser and more mixed.

When to the northward or eastward of the Sarmiento Bank, the lead brings up fine brown-gray sand while near the latitude of Cape Virgins; but when N.N.E. of the cape, the sand is like steel filings.

Dungeness may be passed closely, as may all the northern points as far as the middle of Possession Bay, with the exception of *Wallis Shoal,* where, at low water, there are only 2 fathoms. This shoal is of small extent, and exactly placed upon the chart.

When to the southward of Cape Virgins, *Mount Dinero* will appear; it is a sloping pointed hill, 240 feet in height. Thence to Cape Possession the land continues between 200 and 400 feet in height, rather level topped, and generally covered with grass. A few broken cliffy places show themselves near the water. Three miles East of Cape Possession there is a remarkable bare patch, serving as a fixed mark for bearings.

On the opposite, or Fuegian coast, the nearest land is very low; the hills, the tops of which are seen in the horizon, are 10 miles inland, and not, as they appear, near the water. An extensive shoal, *Lomas Bank,* projects from Catherine Point to the westward; and should be carefully avoided, as it shoals very quickly, and the tide runs near it with much strength.

CAPE POSSESSION is a bold cliffy headland on the North shore, 360 feet high, and will be seen opening round Dungeness, on the bearing of W. ¼ S. by compass; the distance between them is 20 miles: at 10 or 12 miles to the West of Dungeness, Mount Aymond* will make its appearance, bearing about W. ¼ N. by compass.

Possession Bay, which extends from Cape Possession to the entrance of the First Narrow, curves in to the northward round the cape, and is fronted by an extensive shoal, stretching off for more than 4 miles from the shore, many parts of which are dry at half-tide; on its South side the depth diminishes gradually, and offers good anchorage for vessels entering the strait to await the tide for passing the First Narrow.

There is a convenient watering place in Possession Bay, on the shore under Mount Aymond; a ship may go very near it, and lie in a safe berth out of the strength of the tide, if she crosses the Narrow Bank after half-flood, or before half-ebb.

* A hill on the North shore of Possession Bay, having near it, to the westward, four rocky summits, called "His Four Sons," or (according to the old quaint nomenclature) the "Ass's Ears," which, from a particular point of view, bear a strong resemblance to the cropped ears of a horse or an ass.—*Capt. King*, p. 12.

On the western side of the bay there are some remarkable hills of a darker green hue than others near them, called the *Direction Hills;* because, after passing Cape Possession, they afford a good mark for approaching the Narrows, which are not visible until well across the bay; by attention also to their bearings, the shoal that extends off Cape Orange may be avoided. To take up an anchorage on the bank, great attention must be paid to the soundings, which at the edge decrease suddenly; it would not be advisable to anchor in less than 10 or 12 fathoms at high water, for the tide falls 6 or 7 fathoms; but as the stream runs much weaker on approaching the edge of the bank, the nearer to it the better.

Should the distant land behind Cape Gregory be seen, which makes with a long blue level strip of land, terminating at its S.W. end with rather a bluff or precipitous fall, it is a good mark for the anchorage. The fall, or extremity, should be visible in the space between the southernmost and central of the Direction Hills. There is also a conspicuous lump on the same land, which will be seen to the northward of the northern Direction Hill; and the Ass's Ears, nearly out of sight, should be seen a little to the eastward of that part of the shore of Possession Bay where the cliffy coast commences.

To avoid the North shoals, do not get the North Direction Hill to bear more southerly than S.W. by W. *mag.*; and the mark for avoiding the reefs that extend off Cape Orange is not to get the same Direction Hill to bear more westerly than W. by S ½ S. *mag.* (for W. by S. ½ S. will just pass without the edge), until Mount Aymond bears N.W. ½ W. *mag.*, or the peak of Cape Orange South, *mag.*, when the fairway of the First Narrow will be open, bearing S.W. by S. *mag.* The North or north-western side of the First Narrow is a cliff of moderate height, and makes like a flat table-land. When abreast of Cape Orange, a S.S.W. *mag.* course must be steered. The tide sets right through; so that in drifting, which with the wind against the tide is the safest and best plan, there is no danger of being thrown upon the shoals.

Cape Orange is at the South side of the entrance to the First Narrow. There is a reef, previously mentioned, extending off it to the E.N.E. for a considerable distance. Byron struck on it, as did also the *Santa Casilda* in 1788. The *Adeona*, a sealing vessel, in 1828, also struck upon it, and was left dry; and the *Beagle*, in going to her assistance, crossed the tail of it at high water, occasionally striking the ground. Bougainville describes the position thus:—"When the hillocks, which I have named Quatre Fils d'Aymond (Ass's Ears), only offer two to sight, in form of a gate, you are opposite the said rocks."

The FIRST NARROW was called by Sarmiento, Angostura de Nuestra Señora de Esperanza. He describes it very correctly to be 3 leagues long, and less than half a league wide, with cliffy shores; the tide running strong; the depth more than 50 fathoms, sand and pebbles (callao); and on the North shore there is a beach of shingle. In this part, however, as discretion must be the best guide, it will be necessary merely to state the dangers that exist.* To

* Lieut. Simpson, who was with Com. Byron in the *Dolphin*, states, that there is a shoal in *mid-channel* at the East entrance of the First Narrow, but the bearings and distances he gives will not suffice to lay it down now; he states it to be more than 2 leagues in length, and nearly 2

the North of *Point Delgada* (meaning thin or slender) the shore is fronted by extensive shoals that dry at half-tide; these should not be approached. The South shore also, for nearly 5 miles to the West of Cape Orange, has a shoal off it, but it does not extend to a great distance from the beach; beyond this it is not safe to approach either shore within half a mile, for each is fronted by a bank that dries at low water. The western end of the Narrow on the North shore, Sarmiento's *Point Barranca* (meaning a cliff), has a considerable reef off it, upon which there is a very large quantity of kelp.

After emerging from the Narrow, the ship should be allowed to drift with the tide, the course of which is S.S.W., for at least 3 miles, before hauling up for Cape Gregory, in order to avoid the ripplings which rage furiously on each edge of the bank.

Point Barranca is a flat-topped sand-hill, the position of which being given in the chart, its bearing will indicate the situation of the ship: the point on the opposite side Sarmiento is called Point Baxa (low).

After reaching thus far, steer W.S.W. by compass, until abreast of some remarkable peaked hillocks on the North shore, where, if necessary, anchorage may be had out of the tide, in from 6 to 10 fathoms; at any part of the northern side of the bay the anchorage is good, upon a clay bottom covered with broken shells.

When through the First Narrow, where, it should be remembered, there is no anchorage, a ship may anchor, or work along either shore, but she must avoid the rocky ledge off Barranca Point, and also the dangerous shoal in the middle of the space between the First and the Second Narrows. When the water is low, there is not more than from 2 to 3 fathoms on this shoal (called the *Triton Bank*, because a ship so named struck upon it); it must therefore be carefully avoided, by repeatedly ascertaining the ship's position, by bearings of, or angles between, the western end or pitch of Gregory Range, Barranca Point, and Gap Peak. This shoal is of small extent.

It is best to anchor near the shore on account of the tide, which ripples very much all over the centre of the bay.

When abreast of the point, the land and bay to the North of Cape Gregory will be easily distinguished; the former will be seen first, and resembles an island; for the land of the bay is flat and low: but a very conspicuous hummock will also be seen half way between it and the flat table land, as soon as the land of the cape becomes visible.

The extremity of *Cape Gregory* bears from the western end of the First Narrow, S.W. ½ W. *mag.*, distant 22 miles. The anchorage is from 2 to 2½ miles to the N.N.E. of the cape, abreast of the North end of the sand hills that form the headland, and at about 1 mile from the shore, in from 13 to 15 fathoms. At low water a sand-spit extends off for one-third or nearly half a mile from the shore; close to which there are 7 fathoms water. Care should be taken not to approach too near.

---

leagues between the shoal and the South shore. The ship grounded on it as above-stated in 15 feet. It is not improbable that these banks shift, from the nature of the ground and the force of the currents.

The country in the vicinity of this anchorage seemed open, low, and covered with good pasturage. It extends 5 or 6 miles, with a gradual ascent, to the base of a range of flat-tipped land, whose summit is about 1,500 feet above the level of the sea. Not a tree was seen; a few bushes (barberry) alone interrupted the uniformity of the view. The grass appeared to have been cropped by horses or guanacoes, and was much interspersed with cranberry plants, bearing a ripe and juicy, though very insipid fruit.

At the anchorage the tide turns to the south-westward, towards the cape, for 2½ or 3 hours before it begins to run to the westward in the Second Narrow; which should be attended to, for a ship will lose much ground by weighing before an hour or more after the tide has turned.

Upon the summit of the land of the cape, four-fifths of a mile to the northward of the extreme point, is a remarkable bush; close to which the observations were made. The bush is in lat. 52° 38′ 3″ S. and lon. 70° 13′ 43″ W. The variation of the compass 23° 34′ E.

All the South shore of St. Philip's Bay is low, but gradually rising towards a range of high land which extends from Cape Espiritu Santo to Boqueron Point; anchorage may be taken, by the aid of the chart and lead, in almost any part; in some few places only the bottom is rocky. On this side the tide is felt less than on the Patagonian shore.

The country abounds with guanacoes and ostriches, and the valley, 2 miles to the westward of the cape, is frequently the abode of the Patagonian Indians; but their principal residence is upon the low land at the back of Peckett's Harbour and Quoin Hill, where guanacoes are more abundant, and the country more open. They are very friendly, and will supply guanaco meat at a small price.

At Capt. King's last interview they asked for muskets, powder, and ball, the use of which they learned from two Portuguese seamen, who left an English sealing vessel to reside with them; but these were not given, and it is to be hoped that such weapons will not be put into their hands.

The SECOND NARROW is about 10 miles long; and, with a favourable tide, which runs 5 or 6 knots, is very quickly passed. With an adverse wind a ship will easily reach an anchorage to the North of Elizabeth Island.

The North side of the Second Narrow is very shoal, and ought not to be approached, for the ground is also very foul. There are two or three very inviting bights for a ship that is caught with the tide, but it is not advisable to anchor in them; she should rather return to the anchorage off Cape Gregory.

Susannah Cove is where Sarmiento anchored in 8 fathoms, low water, half a league from the land, good bottom; but, as it was exposed to the strength of the tide, he shifted to another anchorage about half a league West of Cape Gregory, where the anchor was dropped in 8 fathoms; but the vessel tailing on the edge of the shoal in 3 fathoms, he was glad to make his escape.

The South shore of the Second Narrow, which Narborough called the *Sweepstakes Foreland*, is composed of cliffs, and is of bold approach. The projecting head in the centre is Sarmiento's St. Simon's Head; and the western end he named Cape St. Vincent, from its resemblance to that of Spain. To the southward of the eastern point of this head, *Point St. Isidro*, which is a low sandy point, is *Fish Cove*.

Three miles to the westward of Cape Gracia, the western end of the Second Narrow on the North side, is *Oazy Harbour*, so called by Narborough; it is a secure place for small vessels. The entrance is nearly 2 miles long, and too narrow for large ships, unless the weather be moderate, when they might drop in or out with the tide: the depth inside is from 3 to 10 fathoms. There is neither wood nor water to be got, and therefore no inducement to enter it.

Narborough's PECKETT'S HARBOUR is 8 miles to the West of Cape Gracia, and although very shoal, offers a good shelter, if required, for small vessels—but the space is very confined; the anchorage without is almost as safe, and much more convenient.

The entrance is between the S.W. point and the island, and is rather more than one-fifth of a mile wide. Half a mile outside, the anchorage is good in 7 fathoms: shoal ground extends for a quarter of a mile off the point.

ROYAL ROAD, the bay which is formed by Peckett's Harbour and Elizabeth Island, is extensive and well-sheltered, with an easy depth of water all over, between 5 and 7 fathoms; the nature of the bottom is clay, and offers excellent holding ground. In the centre is a patch of kelp; this is the only danger, there being 2¼ fathoms on it.

Elizabeth Island, so named by Sir Francis Drake, is a long low strip of land, lying parallel to the shores of the strait, which here take a N.N.E. direction. Its N.E. end was called by Sarmiento Point Sanisidro. Compared with the land to the southward the island is very low, no part being more than 200 or 300 feet high. It is composed of narrow ranges of hills, extending in ridges in the direction of its length, over which are strewed boulders of the various rocks which form the shingle beaches of Points St. Mary and Santa Anna to the South. In the valleys between these ridges were seen hollows that had contained fresh water, but then (April, 1827) entirely dried up, and marked by a white saline crust, indicating the quality of the soil.*

The tide is not strong to the westward of the North end of Elizabeth Island; but runs with considerable velocity in the deep channel between it and the Second Narrow. To the southward of the island the stream divides into two directions, and very soon loses its strength; one sets down the South side of the island, and the other between the Islands of Santa Martha and Magdalena. This is the flood; the ebb sets to the northward. The ebb and flow are regular; high water at the full and change, being at about 12 o'clock.

There is good anchorage, out of the strength of tide, at a mile to the North of Point San Silvestre; it is convenient for a ship to leave with the intention of passing round Elizabeth Island. This is the most difficult part of the entrance of the Strait of Magalhaens, for the tide sets across the passage with some strength.

The passage to the West of the island is clear, and without danger, by keeping in the middle part of the channel; but in passing down the South side of Elizabeth Island the shore should be kept close to, to avoid being thrown upon

* King, p. 81.

the Islands of Santa Martha and Magdalena,* as well as to clear the shoal that extends off the S.W. end of the latter island, upon which less water than 5 fathoms was not found upon any part; but the ground being irregular, and much kelp strewed about, it is not safe to trust too much to appearances.†

Capt. FitzRoy says:—In passing to the westward from the Second Narrow, it is best to keep well to the N.W. in Royal Road; avoid the shoal between Silvester Point and Peckett's Harbour; anchor in that road where convenient, or pass close round Silvester Point and along the East side of Elizabeth Island; keep clear of Santa Martha and Walker Shoal, and make for Laredo Bay.

A vessel drawing less than 16 feet may pass round to the westward of Elizabeth Island, but the eastern passage is the shortest; and if there is wind enough to ensure maintaining your position and keeping close to Elizabeth Island until past Walker Shoal, it is also far easiest. No vessel ought to pass Silvester Point without a commanding breeze, because the water there is very deep; and as the tide sets directly towards the Islands of Santa Martha and Magdalena, much inconvenience, if not danger, might be caused by the failure of the wind. The land hereabouts is low, not exceeding 200 feet in height, and without wood.

About *Cape Negro* the appearance of the land entirely changes. A low barren country gives way to hills covered with wood, increasing in height, and becoming more rocky and mountainous as you go southward (King, p. 215). The northern part is a very poor clay; whilst here, a schistose subsoil is covered by a mixture of alluvium deposited by mountain streams, and decomposed vegetable matter in great quantity (*Ibid.* p. 22).

Laredo Bay offers good anchorage in the centre and towards the North side, and particularly in the N.W. corner. Off the South point is a large patch of kelp, among which the ground is shoal and foul.

At Laredo Bay wood may be procured, and there is a fresh-water lake of a mile in diameter at about half a mile behind the beach, much frequented by wild ducks.

For the purpose of anchorage only the bay need not be entered; because a very good and secure berth may be found at from 1 to 2 miles off it, in 10 to 13 fathoms, having the S.W. extremity of Elizabeth Island on with, or a little open of, the trend of Cape Negro, which is Byron's Porpesse Point.‡

On the eastern coast, *Gente Grande Point* and the land near it are very low, and therefore dangerous. The strong tide setting along shore, and near Adventure Bridge, are additional reasons for avoiding the vicinity of this point.

Southward of Gente Grande, along shore to Cape Monmouth and Cape Boqueron, there is no danger; the water is deep, and the coast safe to approach. Cape Monmouth is a cliffy point. Cape Boqueron is the termination of the range of high land extending across the country to Cape Espiritu Santo.

* The Islands of Santa Martha and Magdalena, so named by Sarmiento (p. 254), have since been called by other names: the former St. Bartholomew, the latter St. George's, also Penguin Island.—*See Narborough's Voyage*, p. 62.

† Lieut. Simpson states that 3 fathoms was found by Byron's ship, the *Dolphin*, p. 154.

‡ Hawkesworth, vol. I. p. 35.

When to the southward of a line drawn from Laredo Bay to Gente Grande Point, the *tides* are scarcely felt; but to the northward of that line they are strong, and must be carefully guarded against during the night, or in light winds.

A vessel in mid-channel, between Gente Grande and Laredo Bay, would be set by the ebb tide, if the wind failed her, directly amongst the dangers surrounding Magdalena Island.

To the westward of the Second Narrow the tide rises about 8 feet; in the Second Narrow it rises about 12 feet; in Gregory Bay, 18 feet; and on the East side of St. Philip's Bay about 24 feet.

Between Cape Negro and Sandy Point, which is Sarmiento's Catalina Bay,* good anchorage may be had, from 1 to 2½ miles from the shore. Here the country begins to be thickly wooded, and to assume a very picturesque appearance, particularly in the vicinity of Sandy Point.

Sandy Point—Sarmiento's Cape de San Antonio de Padua—projects for more than a mile from the line of coast, and should not be passed within a mile. A shoal projects off it in an East direction (*mag.*): the mark for its South edge is a single tree, on a remarkably clear part of the country (a park-like meadow), near the shore on the South side of the point, in a line with a deep ravine in the mountain behind. One mile and a half from the point, we had no bottom with 18 fathoms.

To the southward of Sandy Point, as far as Point St. Mary, good anchorage may be had, at three-quarters of a mile from the shore, in 11 and 12 fathoms; sand and shells over clay. With the wind off shore a ship may anchor or sail along it very close to the coast, by keeping outside the kelp. The squalls off the land are very strong, sometimes so much so as to lay a ship on her broadside. These land squalls are denominated by the sealers "williwaws."

Point St. Mary, in lat. 53° 21′ 15″ and lon. 70° 57′ 55″, is 12½ miles to the South of Sandy Point, and may be known by the land trending in to the southward of it, forming *Freshwater Bay*. It has also a high bank close to the beach, with two patches bare of trees, excepting a few dead stumps. All the points to the northward are low and thickly wooded. As the bay opens, the bluff points at its *South* end become visible. There is also a remarkable round *hill* a short distance behind the centre of the bay, and a valley to the South of it, through which a river flows and falls into the bay.

It is convenient for wooding at, but from the river being blocked up by much drift timber, watering is difficult; the proximity, however, of Port Famine renders this of no material consequence.

Between Freshwater Bay and Point Santa Anna the coast is very bold, and so steep-to as to offer no anchorage, excepting in the bay that is formed by the reef off *Rocky Point;* but it is small and inconvenient to weigh from, should the wind be southerly.

"Near Rocky Point," Sir John Narborough says, "the wood shows in many places as if there were plantations; for there were several clear places in the woods, and grass growing like fenced fields in England, the woods being so even

* Sarmiento, p. 255.

by the sides of it."—(An Account of Several Voyages, &c., 1694, p. 67.) These patches are occasioned by the unusual poverty of the soil, and are covered with spongy moss.—(King, p. 24.)

Point Santa Anna will appear, on standing down near the coast, to be the termination of the land; it is a long point extending into the sea, having at the extremity a clump of trees. It bears from Cape Valentyn S.W. ½ W. *mag.* On approaching it, the distant point of Cape St. Isidro will be seen beyond it; but there can be no doubt or mistake in recognising it.

Along the whole extent of the coast, between Point Santa Anna and Elizabeth Island, the flood sets to the southward and the ebb to the northward, and it is high water about 12 o'clock at full and change. The variation is about 23° West. The strength of the tide is not great, but frequently after a southerly wind there is, in the offing, a current to the northward independent of the tide. In winter the tides occasionally rise very high, and on one occasion, in the month of June, nearly overflowed the whole of the low land on the West side.

PORT FAMINE.—The name of this harbour is a sad memento of the only colony that was ever attempted to be founded in these inclement regions. The voyage of Sir Francis Drake through this strait into the Pacific, and his successes against the Spanish colonies, induced the viceroy of Lima to send an expedition in pursuit of him. This was placed under the command of Pedro Sarmiento de Gamboa, and sailed from Peru to the strait in 1583. After encountering many difficulties in its western part, Sarmiento was so enraptured with the verdant and picturesque appearance of the shores in this portion, that he succeeded in convincing King Philip II., of Spain, of the necessity of colonizing and fortifying the strait. The undertaking was much opposed by the Duke of Alva, who said, "that if a ship carried out only anchors and cables sufficient for her security against the storms in that part of the world, she would go well laden." The result caused this remark to become proverbial. The expedition, however, left Spain in 23 ships, of which only 5 reached the entrance of the strait, and these, after several repulses, entered the strait in December, 1584, and landed 300 men under Sarmiento, between the First and Second Narrows, where they planted a city, "Jesus," which must have been near the point named in the chart N. S. de la Valle, on the North shore. Their hardships commenced with their landing; and Sarmiento set out by land with 100 men to go to Point Santa Anna (close to Port Famine). Their journey was difficult and much harassed by Indians, but at last they reached their destination, and founded the city of King Philip or San Felipe. Winter set in suddenly on them, and Sarmiento, in superintending the two colonies, was blown out of the strait, and obliged to bear up for Rio de Janeiro, and his attempts to carry supplies to the ill-fated colonists were all frustrated by the weather; to crown his misfortunes, he was captured by the English, and taken, with his three vessels, to England. The unhappy colony of Jesus then sent an expedition by land to San Felipe, but finding them equally destitute, they set out on their return, but all perished of hunger and disease. Without enumerating their subsequent disasters, it may be stated that of the two persons who were destined to be saved from the 300 individuals who formed the original detachment, one was taken off by Cavendish, and the other (who died on his passage to Europe) by Andrew Mericke, in 1589. The name of San Felipe ceased with the colony; the name of Port Famine was given to it by Cavendish; and the only memento now remaining is a thickly wooded and picturesque mountain, Mount San Felipe, on the bottom of the bay.—(Burney's Collection of Voyages, vol. ii. p. 45; King, pp. 29, 30.) Morrell states (Narrative of Four Voyages, p. 89) that the fort is still to be recognized, as also the ruins of the houses of the city.

Standing into Port Famine, pass round Point Santa Anna, if with a leading wind, at one-fifth of a mile, in 17 fathoms; but if the wind is scanty, do not get too near on account of the eddy tide, which sometimes sets towards the point. Steer in for the bottom of the bay, for the summit of Mount St. Philip, keeping it over the centre of the depth of the bay; that is, half way between the rivulet (which will be easily distinguished by a small break in the trees), and the N.W. end of the clear bank on the West side of the bay. This bank being clear of trees, and covered with grass, is very conspicuous. Keep on this course until the mouth of Sedger River is open, and upon shutting in the points of its entrance, shorten sail and anchor in 9, 8, or 7 fathoms, as convenient. The best berth, in the summer, is to anchor over towards the West side in 9 fathoms, with Cape Valentyn in a line with Point Santa Anna: but in the winter season, with N.E. winds, the best berth is more in the centre of the bay.

The strongest winds are from the S.W. It blows also hard sometimes from the South, and, occasionally, a fresh gale out of the valley, to the South of Mount St. Philip. Unless a long stay be meditated, it would be sufficient to moor with a kedge to the N.E.; the ground is excellent all over the port, being a stiff, tenacious clay. Landing may be almost always effected, excepting in easterly gales, on one side or the other. There is firewood in abundance on the beaches, and wells, containing excellent fresh water, were dug at the N.W. extremity of the clear part of Point Santa Anna, on the bank above the third, or westernmost, small shingle bay. The water of the river, as well as of the ponds, of which there are many upon the flat shore of the western side of the port, is very good for present use, but will not keep, in consequence of its flowing through an immense mass of decomposed vegetable matter; but the water of the wells drain through the ground, and not only keeps well, but is remarkably clear and well tasted.* For some time the traces of the surveying party will not fail to show the road.

Capt. King's observatory, the situation of which is indicated by the stem of a tree, 16 inches in diameter, placed upright, about 8 feet under and 3 above the ground, banked up by a mound, is in lat. 53° 38′ 12″ and lon. 70° 57′ 55″. High water at full and change at 12 o'clock; the ebb sets to the northward, and the flood to the southward; but the rise and fall is very irregular, depending entirely upon the prevalence of the winds, northerly and easterly winds causing high tides, and westerly and south-westerly low tides. The variation is about 24° 30′.

* The water in the Sedger River is excellent, and keeps well; but that from the head of the harbour is full of green slime, and will not keep.—*Capt. J. H. Smith, Naut. Mag. Apr.* 1837, p. 217.

The eastern shore of this portion of the strait is the land of Tierra del Fuego, which, southward of Point Boqueron, forms a very extensive bay, called USELESS BAY. This was formerly supposed to be the western opening of the Sebastian Channel, but the late surveys having proved the non-existence of this extension of the present Sebastian Bay, it is unimportant to navigation.

Southward of this opening in the strait, the western shore is formed by DAWSON ISLAND, which fronts Useless Bay and the deep inlet called Admiralty Sound, 46 miles long, and about 20 broad. Its northern extremity is CAPE VALENTYN. It is low, and has a small hummock near the point. Between the two points which form the cape, there is a slight incurvation of the shore, which would afford shelter to small vessels from any wind to the southward of East or West; but the water is shoal, and the beach, below high water mark, is of large stones. The coast to the S.W. is open, and perfectly unsheltered; it is backed by cliffs: the beach is of shingle, and becomes visible in passing down the opposite shore, between Sandy Point and Freshwater Bay. *Mount Graves*, 1,315 feet high, however, is seen from a much greater distance. On the western side of the island there are but two places in which vessels can anchor; viz., Lomas Bay and Port San Antonio, but both being on a lee shore, they are not to be recommended.

PORT SAN ANTONIO, which is situated about the centre of the West coast, opposite to San Nicholas Bay, has the appearance of being well sheltered, but it possesses no one advantage that is not common to almost every other harbour or cove in the strait; and for a ship, or square-rigged vessel of any kind, it is both difficult to enter, and dangerous to leave. Besides its local disadvantages, the weather in it is seldom fair, even when the day is fine elsewhere. In no part of the Strait is the vegetation so luxuriant as in this little cove; veronica, fuchsia, and other plants flourish, and shelter bees and humming-birds.—(King, p. 127.)

PORT VALDEZ is a deep inlet, South of Port San Antonio, fronting W.N.W., and not at all inviting to enter.

The GABRIEL CHANNEL separates Dawson Island from the Tierra del Fuego. It is merely a ravine of the slate formation, into which the water has found its way, and insulated the island. It extends precisely in the direction of the strata, in a remarkably straight line, with almost parallel shores. It is 25 miles long, and from half a mile to 1½ miles wide; the narrowest part being in the centre.

The South side of the Gabriel Channel is formed by a high mass of mountains, probably the most elevated land in the Tierra del Fuego. Among many of its high peaks are two more conspicuous than the rest, MOUNT SARMIENTO* and MOUNT BUCKLAND. The first is 6,800 feet high, and, rising from a broad base, terminates in two peaked summits, bearing from each other N.E. and S.W., and are about a quarter of a mile asunder. From the northward it appears very

* "This mountain was the snowy volcano (*Volcan Nevado*) of Sarmiento, with whose striking appearance that celebrated navigator seems to have been particularly impressed, so minute and excellent is his description. It is also mentioned in the account of Cordova's voyage. The peculiar shape of its summit, as seen from the North, would suggest the probability of its being a volcano, but we never observed any indication of its activity. Its volcanic form is perhaps accidental; for, seen from the westward, its summit no longer resembles a crater. From the geological character of the surrounding rocks its formation would seem to be of slate. It is in a range of mountains rising generally 2,000 or 3,000 feet above the sea: but, at the N.E. end of the range are some at least 4,000 feet high. From an attentive perusal of the voyage of Magalhaens, I have been lately led to think that this is the mountain which Magalhaens called Roldan's Bell."—*King*, pp. 26, 27.

much like the crater of a volcano; but when viewed from the westward, the two peaks are in a line, and its volcanic resemblance ceases.

It is the most remarkable mountain in the strait; but, from the state of the climate and its being clothed with perpetual snows, it is almost always enveloped in condensed vapour. During a low temperature, however, particularly with a N.E. or S.E. wind, when the sky is often cloudless, it is exposed to view, and presents a magnificent appearance.

Mount Buckland is, by estimation, about 4,000 feet high. It is a tall, obelisk-like hill, terminating in a sharp needle-point, and lifting its head above a chaotic mass of "reliquiæ diluvianæ," covered with perpetual snows, by the melting of which an enormous glacier on the leeward or north-eastern side has been gradually formed. This icy domain is 12 or 14 miles long, and extends from near the end of the Gabriel Channel to Port Waterfall, feeding, in the intermediate span, many magnificent cascades, which, for number and height, are not, perhaps, to be exceeded in an equal span of any part of the world. Within an extent of 9 or 10 miles, there are upwards of 150 waterfalls, dashing into the channel from a height of 1,500 or 2,000 feet. I have met with nothing exceeding the picturesque grandeur of this part of the strait.—(King, p. 51.)

The channel on the eastern side of Dawson Island leads to Admiralty Sound, into which there is also a passage through the Gabriel Channel. Admiralty Sound extends for 43 miles to the S.E. into the land of Tierra del Fuego; but, as it is of no service in the navigation of the Strait of Magalhaens, we shall not describe it.

Port Famine, on the western shore of the strait, has been described on page 12; from this point we again proceed with the description.

The *Sedger River*, which is fronted by a bar that dries at low water, can be entered by boats at half-tide, and is navigable for 3 or 4 miles; after which its bed is so filled up by stumps of trees, that it is difficult to penetrate farther. The water is fresh at half a mile from the entrance, but to ensure its being perfectly good, it would be better to fill the casks at low tide. (See p. 13). The low land near the mouth, as well as the beach of Port Famine, is covered with drift timber of large size.

The river was called by Sarmiento, Rio de San Juan. In Narborough's Voyage it is called Segars River. Byron describes the river, which he calls the Sedger, in glowing terms, but gives rather a more flattering account of the timber growing on its banks than it deserves.*

* Hawkesworth, vol. i. p. 38.

The enterprising Capt. Morrell tells us, in his Narrative, that he went several miles in the interior, to the W.N.W. from Port Famine. He proceeded up the Sedger River, whose banks, he says, are covered with the finest timber he ever saw; the herbage luxuriant, and the soil rich. His search for dye-woods was unsuccessful, but he found on its banks copper, lead, and iron ores; and in his opinion, if the colony, noticed on p. 12, had consisted of such men as emigrate daily from the New England States, instead of indolent Spaniards, a few years would have converted this region of Patagonia into a fruitful garden, and Philipville, or San Felipe, would have been at this moment a splendid city—a conclusion we may justly demur at. His interesting Narrative contains an account of his passages and expeditions through the Strait and in Patagonia. The authenticity of them has been doubted, but it would seem without reason; at least, *one* circumstance has been substantiated: Ichaboe (W. Africa) was visited from his description, at the suggestion of Capt. Andrew Livingston, and was converted into a *real fact* by our Liverpool merchants.—*Narrative of Four Voyages, &c.*, New York, 1832, pp. 82—96, and 155, 6.

Between this bay and Cape San Isidro (of Sarmiento—Point Shut-up of Byron) the water is too deep for anchorage, even close to the beach. The cape is the termination of the ridge, whose summit is *Mount Tarn*, the most conspicuous mountain of this part of the strait. It is 2,602 feet high by barometrical measurement. It is readily distinguished from abreast of Elizabeth Island, whence it appears to be the most projecting part of the continental shore. When viewed from the northward its shape is peaked, and during the summer it has generally some patches of snow a little below its summit; but in the winter months its sides are covered with snow for two-thirds down. From abreast, and to the southward, of Port Famine, it has rather a saddle-shaped appearance; its summit being a sharp ridge, extending very nearly for one mile, N.W. and S.E., with a precipitous descent on the N.E., and a steep slope on the S.W. sides. The highest peak near its N.E. end is in lat. 53° 45′ 6″ and lon. 71° 2′ 20″.

There is a low, but conspicuous rounded hillock covered with trees at the extremity of Cape San Isidro; and a rocky patch extends off it for 2 cables' length, with a rock at its extremity, that is awash at high water. It is covered with kelp.

Eagle Bay (Valcarcel Bay of Cordova) is about three-quarters of a mile deep; and its points 1 mile apart, bearing N.E. and S.W. The anchorage is at the head, in from 20 to 12 fathoms. There are two streams of water; but being very much impregnated with decomposed vegetable matter, cannot be preserved long.

Gun Bay, the next to the westward, although small, affords anchorage for a single vessel near the shore, at its S.W. part, in from 8 to 9 fathoms.

Neither Gun nor Indian Bays are noticed in Cordova's description of the strait, although they are quite equal to any other in the neighbourhood for stopping places.

Bouchage Bay—which is Cordova's Cantin Bay—is small, and the water very deep; except near the bottom, where anchorage may be obtained in 8 fathoms, clay. It is separated from *Bournand Bay* (Gil Bay of Cordova) by Cape Remarquable of Bougainville, which is a precipitous, round-topped, bluff projection, wooded to the summit. At 2 cables' length from the base, no bottom was found with 20 fathoms of line; but, at the distance of 50 yards, the depth was 20 fathoms. Bournand Bay is more snug and convenient than its northern neighbour, Bouchage Bay, being sheltered from the southerly winds by Nassau Island. At the S.W. end of a stony beach at the bottom, is a rivulet of good water: off which there is good anchorage in 8 fathoms stiff mud.

Bougainville Bay (Cordova's Texada Bay) forms a basin, or wet dock, in which a vessel might careen with perfect security. It is, from its small size, great depth of water, and the height of the land, rather difficult of access: which renders it almost always necessary to tow in. On entering, the anchor should be dropped in 12 fathoms, and the vessel steadied by warps to the trees, at the sides and bottom of the cove. It is completely sheltered from all winds, and an excellent place for a vessel to remain at, particularly if the object be to procure timber, which grows here to a great size, and is both readily cut down and easily embarked. A rivulet at the bottom affords a moderate supply of water; but, if more be required, the neighbouring bays will afford an abundance.*

In the passage between Nassau Island and the main, the least water is 7 fathoms, over a stiff clay bottom, gradually deepening on each side. But the winds being baffling, and the tides irregular and rippling in many parts, a vessel should not attempt it but from necessity.

Nassau Island's South extremity is Sarmiento's Point, Santa Brigida.

St. Nicholas Bay, so named by the Nodales in 1618 (but previously, by Sarmiento, Bahia de Santa Brigida y Santa Agueda, and French Bay by De Gennes), is not only of larger size than any of the bays to the South of Cape San Isidro, but is the best anchorage that exists between that cape and Cape Froward; as well from its being more easily entered and left, as from the moderate depth of water, and extent of the anchoring ground. Its points bear from each other, S.W. by W. and N.E. by E., and are distant 2 miles. Nearly in the centre is a small islet covered with trees; between which and the shore is a passage with 9 fathoms water, stiff clay. The shore is, however, fronted for its whole length by a shoal bank, which very much reduces the apparent extent of the bay. This bank stretches off to the distance of a quarter of a mile from the shore, the edge of which is steep-to, and is generally distinguished by the ripple, which, with a moderate breeze, breaks at half-tide. The *Beagle* anchored in the bay, at 3 cables' length to the N.E. of the small central islet, in 12 fathoms, pebbly bottom; but the best berth is one-quarter to one-third of a mile to the S.W. of the islet, in 10 or 11 fathoms, muddy bottom. Capt. Stokes recommends in his journal, in coming in, to keep sail upon the ship, in order to shoot into a good berth, on account of the high land of Nodales Peak becalming the sails. In taking up an anchorage, much care is necessary to avoid touching the bank. Less than 10 fathoms is not safe, but in that depth the security is perfect, and the berth very easy to leave. In passing through the strait, this bay is very useful to stop at, as well from the facility of entering and leaving it, as for its proximity to Cape Froward.

From Glascott Point the coast extends in nearly a straight line to Cape Froward, a distance of 7 miles, the land at the back continuing mountainous and woody.

CAPE FROWARD, the southern extremity of the continent of South America, is composed of dark-coloured, slaty rock, and rises abruptly from the sea. At its base is a small rock, on which Bougainville landed. Bougainville observes, that "Cape Froward has always been much dreaded by navigators." —"To double it, and gain an anchorage under Cape Holland (W. of it), cost the *Beagle* as tough à 16 hours' beat, as I have ever witnessed. We made 31 tacks, which, with the squalls, kept us constantly on the alert, and scarcely allowed the crew to have the ropes out of their hands throughout the day. But what there is to inspire a navigator with 'dread' I cannot tell, for the coast on both sides is perfectly clear, and a vessel may work from shore to shore."—(Capt. King: Narrative, &c., vol. i. p. 69.)

* It was here that M. de Bougainville cut timber for the French colony at the Falkland Islands. To sealing vessels it is known by the name of Jack's Harbour.—*King*, p. 145.

At less than a cable's length off the cape, Byron found the depth was 40 fathoms. Capt. King found, at midway between St. Nicholas Bay and Cape St. Antonio, no bottom at 256 fathoms. The hill that rises immediately above the cape was called by Sarmiento, the Morro of Santa Agueda.* Cape Froward is in lat. 53° 53′ 43″, lon. 71° 14′ 31″. The ebb tide sets to the northward, and the flood to the southward, but with very little strength. It is high water at full and change at 1, p.m.

---

2. MAGDALEN SOUND, AND THE COCKBURN AND BARBARA CHANNELS.

The *Beagle* passed into the Pacific by a route not previously used, except by sailing vessels, although it possesses many advantages over either the passage round Cape Horn, or that through the western reaches of the Strait of Magalhaens. "Mr. Low is said to be the first discoverer of it, and he certainly was the first to pass through in a ship, but I think one of the *Saxe Coburg's* boats had passed through it previously; and I much question whether Sir Francis Drake's shallop did not go by that opening into the Strait of Magalhaens in 1578."†—(FitzRoy, p. 358.)

The *Beagle* sailed from Port Famine June 9th, 1834; went down the Magdalen Channel, and anchored in a cove under Cape Turn; the following day beat to windward through the Cockburn Channel, and (leaving a small cove to be occupied by the *Adventure*) beat about all night in the limited space near the entrance of the channel. When the day at last broke on the 11th, both vessels sailed out, past Mount Skyring, as fast as sails could urge them. At sunset they were near the Tower Rocks, and with a fresh N.W. wind stood out into the Pacific.—(FitzRoy, p. 360.)

The opening of MAGDALEN SOUND was first noticed by Sarmiento.‡ Coming from the northward, it appears to be a continuation of the strait, and it is not until after passing Cape San Isidro that the true channel becomes evident. It extends in a southerly direction for 20 miles, and is bounded on either side by high and precipitous hills, particularly on the West shore. The eastern entrance of the sound, Anxious Point, is a low, narrow tongue of land, with an island off it. Opposite to it is a steep mountain, called by Sarmiento *the Vernal* (or summer-house), from a remarkable lump of rock on its summit.§

Under this mountain is *Hope Harbour;* a convenient stopping place for small vessels bound through the sound. The entrance is narrow, with kelp across it, indicating a rocky bed, on which we had not less than 7 fathoms. Inside it opens into a spacious basin, with good anchorage in 4 fathoms, sheltered from all winds, excepting the squalls off the high land, which must blow with furious violence during a south-westerly gale.

* Sarmiento, p. 218. † See Burney, vol. i. pp. 327, 368. ‡ Sarmiento, p. 220.
§ Sarmiento, p. 219; and Ultimo Viage, p. 121.

To the South of Hope Harbour, between the Vernal and Mount Boqueron, is STOKES'S INLET. It is 3 miles long, with deep water all over: there is a cove on its North side, but neither so good nor so accessible as Hope Harbour. In the entrance of the inlet are three islets (*Rees Islets*).

MOUNT BOQUERON, the extremity of which is Squally Point, is a very precipitous and lofty mountain, about 3,000 feet high, having on its summit three small but remarkably conspicuous peaks. It is the eastern head of Stokes's Inlet, and forms a part of the western shore of Magdalen Sound. The squalls that blow off this during a S.W. gale are most furious, and dangerous unless little sail be carried. On one occasion our decked sailing-boat was seven hours in passing it. The sound here is not more than 2½ miles wide. On the opposite shore, within Anxious Point, is an inlet extending to the S.E. for 2 or 3 miles, but it is narrow and unimportant.

Sholl Bay is a small bight of the coast line, 5 miles to the South of Squally Point. There is a reef off it, the position of which is pointed out by kelp.

On the opposite shore is *Keat's Sound*. It extends to the S.E. for 6 or 8 miles, and is between 4 and 5 miles wide.

In the centre of Magdalen Sound, abreast of the above opening, is a rocky islet; and, at a short distance to the southward, on the western coast, is a bay and group, called Labyrinth Islands, among which small vessels may find good anchorage.

Transition Bay is deep, and of little importance. Four miles farther, at Cape Turn, the shore trends suddenly round. Here Magdalen Sound terminates, and Cockburn Channel commences.

On the opposite shore, to the South of Keat's Sound, there are no objects worth noticing, excepting Mount Sarmiento, which has been already described, page 14, and Pyramid Hill, which was found to be 2,500 feet high.

COCKBURN CHANNEL.—The bottom of Magdalen Sound is 6 miles wide, but at Cape Turn the channel narrows to 2 miles, and in one part is not more than 1¼ of a mile wide. The South shore is much broken, and there are many sounds penetrating deeply into the land, which, in this part, according to Capt. FitzRoy's survey of Thieves Sound, is 7 miles wide. Eleven miles more to the westward, at Courtenay Sound, the width of the peninsula is not more than 3 miles.

Warp Bay, although exposed to southerly winds, is a convenient stopping place. *Stormy Bay* is a very wild, unsheltered place, unfit for any vessel to stop at. *Park Bay* is both very snug and secure, with good anchorage in 12 fathoms, sand and mud. It has the same disadvantage as Stormy Bay, in being on the lee side of the channel, and is, therefore, difficult to leave.

In working down the channel, the south side should be preferred, as it is a weather shore, and seems to be better provided with coves and harbours to anchor in.

King and FitzRoy Islands, in mid channel, are of bold approach; as are also *Kirke's Rocks*, more to the westward.

The flood tide sets to the southward, or to seaward, but was not found to run with sufficient strength to benefit or impede a vessel beating through. The rise and fall is inconsiderable, not being more than 6, or at most 8, feet at spring tides.

There are several anchorages among the *Prowse Islands*, which are very numerous, and skirt the coast for several miles. Behind them the land trends in and forms a deep sound.

The distance across the channel, between Prowse Islands and Barrow Head, is scarcely 1½ miles.

Dyneley Sound extends for more than 9 miles in a N.W. direction into the interior of Clarence Island. On the West side of its entrance is a group of islands affording several anchorages, which the chart will point out. One of them, Eliza Bay, offers shelter and security from all winds.

Melville Sound, which forms the embouchure of the Barbara and Cockburn Channels, is very extensive, and is completely filled with islands. Some of them are of large size, and all are of the most rugged and desolate character. The offing is strewed with clusters of rocks: of these, the East and West Furies are the most remarkable, as well as the most important, for the passage into the Cockburn Channel lies between them. The former are very near the land of Cape Schomberg. The West Furies bear from the Tower Rock, off Cape Noir, N.E. by E. ¼ E. *mag.* (E. ½ N.) 25 miles; and S. ½ W. *mag.* (S.S.W. ¾ W.) 11 miles from Mount Skyring. The Tussac Rocks, which are two in number, bear from the West Furies N.E. ¼ E. *mag.* (E. by N. ¾ N.) 4½ miles; and in a line between the East and West Furies, 3 miles from the latter and 2 from the former, is a rock standing alone. It bears from Mount Skyring S. by E. *mag.* (S. by W.) 12½ miles. To avoid it, in entering with a westerly wind, pass near the West Furies, and steer for the Tussac Rocks.

After passing these there are no dangers, that we know of, in the entrance of the Cockburn Channel. A reference to the chart will show everything else that need be noted.

Mount Skyring is a very prominent object. It rises to a peak nearly 3,000 feet high. It was seen from Field Bay, at the North end of the Barbara Channel; and, from its summit, Capt. FitzRoy obtained a bearing of Mount Sarmiento. Its summit is in lat. 54° 24′ 44″ and lon. 72° 11′ 32″. The variation is 25°.

The BARBARA CHANNEL diverges from the western portion of the Cockburn Channel, and pursuing a devious course to the northward, between Clarence Island on the East and the Land of Desolation on the West, enters the principal strait about 25 or 30 miles to the West of Cape Froward.

The southern entrance of the Barbara Channel is so very much occupied by islands and rocks, that no direct channel can be perceived. The chart must be referred to as the best guide for its navigation. For small vessels there is neither danger nor difficulty; there are numerous anchorages that they might reach without trouble, and which would afford perfect security.

Among *Magill's Islands* there are several coves and anchorages. Fury Harbour, on the S.E. side of the central island of Magill's group, is a very wild anchorage. From its contiguity to the East and West Furies, and the Tussac Rocks, on which seals are found, it is much frequented by sealing vessels.*

* In the winter of 1826-27, the *Prince of Saxe Coburg* sealer was wrecked in Fury Harbour, and the crew saved by the *Beagle's* boats.

Bynoe Island affords an anchorage on its N.E. side; and *Hewett Bay* is a good stopping place either for entering or quitting the channel. *Brown's Bay* is more extensive, but also affords good shelter in a small cove at the North entrance, in 8 fathoms sand, among some kelp.

North Anchorage, for a small vessel, is tolerably secure, but not to be recommended.

Between Hewett Bay and North Anchorage the channel is strewed with many rocks and shoals, some of which, although covered with kelp, only show at half-tide. Much caution is therefore necessary, and all patches of kelp should be carefully avoided.

The tide to the northward of North Anchorage, which to the southward was not of sufficient consequence to interfere with the navigation of the channels, is so much felt as to impede vessels turning to windward against it.

The country here has a more agreeable appearance, being better wooded with beech and cypress trees; but the latter are stunted, and do not attain a greater height than 15 or 18 feet. They are very serviceable for boat-hook spars, boats' masts, &c. The wood, when seasoned, works up well.

Bedford Bay is a good anchorage. It is situated on the N.W. side of the narrow part of the channel. Its depth is from 20 to 8 fathoms, good holding ground, and perfectly sheltered from the prevailing winds. At its entrance are several patches of kelp, the easternmost of which has 4 fathoms on it.

Here, as well as throughout the Barbara Channel, the flood tide sets to the southward.

Nutland Bay, having 8 and 15 fathoms over a sand and mud bottom, may be known by two small islands, Hill's Islands, which lie 1 mile N.N.E. from the anchorage.

Between Bedford and Nutland Bays, and, indeed, as far as the Shag Narrows, the channel is open, and may be navigated without impediment. There are many bays and inlets not here described or noticed that may be occupied, but almost all require to be examined. They all trend far enough into the land to afford good shelter, but in many the bottom is foul and rocky, and the water too deep for anchorage.

The western coast, being the windward shore, should of course be preferred.

Field's Bay is too exposed to southerly winds to be recommended as a stopping place, unless the wind be northerly. Nutland Bay is a more convenient place to start from with a view of passing the Narrows.

To the North of Nutland Bay is *Broderip Bay*; at the bottom, or northern part of it, are some good coves; but the most convenient of them is at the eastern extreme; it is called Dinner Cove. It extends to the North for about a furlong, and affords good anchorage in 10 fathoms, sufficiently well sheltered and distant from high land to be free from the mountain squalls or "williwaws."

Round Dinner Cove is *Icy Sound*, a deep inlet, with a glacier of considerable extent at the bottom, from which large masses of ice are constantly falling, and drifting out, occupy the waters of the inlet. The water is deep, and the anchorage not good, when there are so many better places. *Dean Harbour* is a considerable inlet trending in under the same glacier, which extends from the head of

Smyth Harbour to a considerable distance in the S.W. If of a favourable depth it might afford good anchorage.

The only navigable communication that exists between the Barbara Channel and the strait is that called the *Shag Narrows*, on the western side of Cayetano Island. The width of the opening is at least 1¾ miles; but the eastern portion is so filled with rocky islets and shoals, that the actual breadth of the only navigable part at the northern end is about 100 yards, and the widest part at the southern end scarcely half a mile: the whole length of the passage being rather less than 2 miles. It is formed on the West side by a projecting point of high land, that gradually trends round to the westward; and on the opposite side by three islands, the northernmost of which is Wet Island: on the southernmost is *Mount Woodcock*.

Between Wet Island, where the Narrows on the North side commence, and the western shore, the width is not more than 100 to 150 yards, and perhaps 300 yards long. Through this the tide sets as much as 7 miles an hour; the sides of the rocks are steep-to; so, probably, no accident can happen to a ship in passing them, notwithstanding the want of room for manœuvring. At the South end of Wet Island, the stream of tide divides; one sets to the eastward, round Wet Island, while the principal runs through the Shag Narrows; and in the same manner a part of the southern tide, which is the flood, after passing Wet Island, runs to the S.E., round the eastern side of Mount Woodcock. All the space to the eastward of Mount Woodcock is so strewed with islands and rocks, that the passage must be difficult, if not dangerous.

To avoid the danger of being thrown out of the Narrows, it is only necessary to keep the western shore on board; where there are no indentations, the tide will carry a vessel along with safety. At the North end of the Narrows, on the West side, is a shelving point, on which there are 5 fathoms: here there is an eddy, but as soon as the vessel is once within the Narrows (within Wet Island) the mid-channel may be kept. In shooting this passage it would be better to furl the sails and tow through, for if the wind be strong, the eddies and violent squalls would be very inconvenient, from their baffling, and laying the vessel upon her beam ends, which frequently happens, even though every sail be furled. It will be necessary to have a couple of boats out, ready either to tow the ship's head round, or to prevent her being thrown by the tide into the channel to the South of Wet Island.

If anchorage be desirable after passing the Narrows, there is none to be recommended until the coves between Smyth Harbour and Cape Edgeworth be reached. Of these, *Dighton Cove* is preferable. The anchorage is off the sandy beach, in 20 fathoms.

Smyth Harbour is about 4 miles deep, and half to one mile wide, surrounded by high land, and trending in a westerly direction. The water is deep, excepting in Earle Cove, on the North side, where vessels might lie, if necessary; but Capt. King thinks it a very wild place in bad weather.

The hills at the head are capped by glaciers that communicate with those at the head of Icy Sound. It seems possible that all the mountains between this and Whale Sound are entirely covered with a coating of ice.

Half a mile S.E. from Cape Edgeworth is a shoal, so thickly covered with kelp as to be easily seen in passing or approaching it; there are not more than 2 feet of water over its shoalest part.

To pass through the Barbara Channel from the North, it would be advisable to stay at Port Gallant until a favourable opportunity offers; for with a S.W. wind it would neither be safe nor practicable to pass the Shag Narrows.

The N.W. wind prevails more than any other in the western portion of the strait, in consequence of the reaches trending in that bearing. It seems to be a general rule hereabouts that the wind either blows up or down them.

Between Cape Froward and the western entrance of the strait the wind is generally from N.W., although at sea, or in the Cockburn or Barbara Channels, it may be in the South or south-western boards.

---

3. THE STRAIT OF MAGALHAENS, FROM CAPE FROWARD TO THE PACIFIC OCEAN.

Cape Froward, as before stated, is the southernmost point of the South American continent, and from this the strait takes a direction at nearly right angles to that already described. The northern side is formed by the Brunswick and Croker Peninsulas; the southern, by Clarence Island and the Land of Desolation.

A glance at the chart of this part of the strait will show the difference of geological structure in the opposite coasts. The North shore from Cape Froward to Port Gallant forms a straight line, with scarcely a projection or bight, but on the opposite side there is a succession of inlets, surrounded by precipitous mountains, which are separated by ravines. The northern shore is of slate, but the other is principally of greenstone; and its mountains, instead of running up into sharp peaks, and narrow, serrated ridges, are generally round-topped. The vegetation on both sides is almost equally abundant, but the trees on the South shore are much smaller.

There is one remarkable feature in the formation of the southern shore; although apparently very irregular and much intersected, yet the projecting points, in many portions, are situated in such an exact line, that Capt. King says, that, in taking the bearings to the S.W. of Cape Froward, the capes on a long line of coast were all in the field of the telescope at once. This is observable on much of this singular region, as may be seen by an inspection of the chart.

The northern shore of the Strait of Magalhaens, from CAPE FROWARD to Jerome Channel, a distance of 40 miles, is very slightly indented. The anchorages, therefore, are few in number, but they are of easier access, and altogether more convenient than those of the southern shore. Taking them in succession, *Snug Bay*, 5 miles N.W. of Cape Froward, is a slight indentation of the coast at the embouchure of a small rivulet; the deposits from which have thrown up a bank near the shore, on which anchorage may be had in 8 and 9 fathoms. The best anchorage is half a mile to the E.S.E. of the island, in 9 fathoms, black sand, the rivulet mouth bearing N.N.W. three-quarters of a mile. It is much exposed, being open from W.S.W. by S. to S.E. It is a convenient stopping place in fine weather.

Wood's Bay, situated under the lee of Cape Holland, is a convenient stopping place for ships, but only small vessels should anchor inside the cove. The anchorage is very good to the eastward of the river's mouth, at half a mile from the shore, in 17 and 13 fathoms water. Small vessels may enter the cove, by luffing round the kelp patches that extend off the South point of the bay, on which there is 2½ fathoms.

Entering Wood's Bay, steer for the gap, or low land behind the cape; and, as you near the South point, keep midway between it and the river's mouth; or, for a leading mark, keep a hillock, or conspicuous clump of trees at the bottom of the bay, in a line with a remarkable peak, 1 or 2 miles behind, bearing, by compass, N.W. ¾ W. Anchor in 17 fathoms, immediately that you are in a line between the two points. Small vessels may go further, into 12 fathoms. The West side of the cove may be approached pretty near, and the depth will not be less than 5 fathoms, excepting upon the 2 fathoms patch that stretches off the East point, the extent of which is sufficiently shown by the kelp; but on the eastern side the bank shoals suddenly, and must be avoided, for there are 13 fathoms close to its edge, upon which there is not more than 2 feet water. The South point of Wood's Bay is in lat. 53° 48′ 33″, and lon. 71° 39′ 33″.

Cape Holland is a bold, high, and although slightly projecting, yet a very conspicuous headland. It is precipitous, and descends to the sea in steps, plentifully covered with shrubs. It is about 14 miles to the westward of Cape Froward.

Near Cape Coventry, and in Andrew's Bay, anchorage may be had near the shore, if the weather be fine. To the westward of the former, at half a mile from the shore, there are 13 fathoms.

Cordes Bay, 4 miles to the eastward of Cape Gallant, may be known by the small bright green islet (Muscle Island) that lies in the entrance; also by a three-peaked mountain, about 1,600 or 2,000 feet high, standing detached from the surrounding hills at the bottom of the bay. The western entrance, which lies between West Point and the reef off Muscle Island, is two-thirds of a mile wide; within it is a bay 1 mile deep, but much contracted by shoals covered with kelp; between them, however, the anchorage is very good and well sheltered. The bottom is of sand, and the depth 6 and 7 fathoms. At the extremity of the bay is a large lagoon, Port San Miguel, trending in a N.E. direction for 2 miles, and two-thirds of a mile across; the entrance is both narrow and shoal, and not safe for a vessel drawing more than 6 feet. Inside the lagoon the depth is from 3 to 13 fathoms. With Fortescue Bay and Port Gallant so near, the probability is that it will never be much used; but in turning to the westward it would be better to anchor here than lose ground by returning to Wood's Bay. By entering the western channel and steering clear of the kelp, a safe and commodious anchorage may easily be reached.

Fortescue Bay is the first best anchorage to the westward of St. Nicholas Bay. It is spacious, well sheltered, easy of access, and of moderate depth. The best berth is to the S.E. of the small islet outside of Wigwam Point, in 7 or 8 fathoms.

PORT GALLANT, as a secure cove, is the best in the Strait of Magalhaens; from the stillness of its waters it is a perfect wet dock, and from its position it is invaluable. There are many coves as safe and convenient when once entered, but the prevailing steepness of the shores, as well as the great depth of water, are obstacles of serious importance. Here, however, is an exception; the bottom is even and the depth moderate; besides, Fortescue Bay, close by, is an excellent roadstead, or stopping place, to await the opportunity of entering. Having the entrance open, small vessels may sail into the port, but the channel is rather narrow. The banks on the western side, off Wigwam Point, are distinguished by the kelp. When within, the shelter is perfect; but Fortescue Bay is quite sufficiently sheltered, and much more convenient to leave. In this part of the strait, as the channel becomes narrowed by the islands, the tides are much felt. There are two good anchorages before reaching the entrance of the Jerome Channel; namely, Elizabeth Bay and York Roads, off Batchelor's River.* They are, however, only fit for stopping places. There are no anchorages among the islands that can be recommended, excepting in the strait that separates the group of Charles's Islands, in which there is security and a convenient depth. When the wind blows fresh there is a hollow sea between Charles's Islands and the North shore, which very much impedes ships beating to the westward.

The SOUTHERN SHORE of this portion of the strait is formed by the North coast of CLARENCE ISLAND, which extends from the entrance of Magdalen Sound to that of Barbara Channel. The whole of it is deeply indented by sounds penetrating in a southerly direction into the island.

Mount Vernal, mentioned on page 18, is at the West entrance of Magdalen Sound. This, with a small rocky islet called Periagua, sufficiently points out the Port of Beaubasin of Bougainville. It is a very snug place when once in, but, being on the wrong side of the strait, it possesses no advantage.

Inman Bay, Hawkin's Bay, Staples Inlet, and Sholl Harbour, are all deep inlets, surrounded by high, precipitous land.

To the westward of Greenough Peninsula is *Lyell Sound*. It is 9 miles deep, and is separated at the bottom from Sholl Harbour by a ridge of hills about 1½ miles wide.

In the entrance of Lyell Sound are two conspicuous islands, one of which is very small. They are called Dos Hermanas, and bear from Cape Froward S.S.W. 5½ miles.

Kempe Harbour, 1½ mile within the entrance, on the West side, of Lyell Sound, is rather difficult of access, but perfectly secure, and would hold six ships.

Cascade Harbour, and *Mazzaredo Bay*, are of less size, and therefore more attainable, but of the same character with Lyell Sound: viz., deep water, surrounded by high land. The former is known by the cascade which M. de Bougainville describes, from which it derives its name. On the headland that

* At three-quarters of a mile East of Batchelor's River, at half a mile from shore, is a shoal of not more than 6 feet least water; it shows itself by the weeds on it.

separates these harbours from Lyell's Sound, is a sugar-loaf hill, the position of which is well determined, in lat. 53° 57′ 32″, lon. 71° 28′ 5″.

Hidden Harbour has a narrow entrance; but, if required, offers good shelter.

San Pedro Sound is the most extensive inlet that we know in Clarence Island. It extends in a southerly direction for nearly 13 miles. There is a good, although a small, anchorage on its West side, 1½ miles within the entrance, called Murray Cove; and another close to it, which is even more sheltered.

Bell Bay (the Bahia de la Campana of Sarmiento) has a very convenient anchorage, Bradley Cove on its West side, bearing S.W. by W. from Point Taylor, the eastern head of the bay. It will be readily distinguished by a small, green, round hillock that forms its North head. The anchorage is in 17 fathoms, and the vessel hauls in, by stern-fasts or a kedge, into 9 fathoms, in perfect security. Pond Bay, to the northward, has good shelter, but it is not of such easy access; for it would be necessary to tow both into and out of it.

Mount Pond, a peaked hill over the harbour, is a conspicuous mountain, and is visible from the eastward as soon as it opens round Cape Froward. It has two summits, one of which only is visible from the eastward.

Between Cape Inglefield and Point Elvira is *St. Simon's Bay*.* It is studded with islands and rocks, and at the bottom has two communications with the Barbara Channel, separated from each other by Burgess Island; the easternmost of which, called Tom's Narrows, is the most extensive: but this, from the irregularity and force of the tides, is not to be preferred to the more direct one of the Shag Narrows, on the western side of Cayetano Island; for there is no good anchorage in St. Michael's Channel, which leads to it, and it is bounded by a steep and precipitous coast. The Gonsalez Narrows, on the West side of Burgess Island, is not more than 30 yards across; and, from the force of the tide, and the fall of the rapid, would be dangerous.

The only good anchorage in St. Simon's Bay is *Millar's Cove;* it is about 3 miles within Point Elvira, and has three rocky islets off its entrance. A conspicuous mount forms the summit of the eastern head. The anchorage is in 5 fathoms, a good bottom, and entirely sheltered. Wood and water are plentiful.

Immediately round the East head of Millar's Cove is *Port Langara*. It is rather more than a mile long, and two-thirds of a mile wide, and trends in a W.N.W. direction. The water is deep, excepting at the head, and in a cove on the North shore; in either of which there is good anchorage. This portion is opposite to Port Gallant, which is previously described.

At a short distance to the E.S.E. of Passage Point, on the North shore, is a shoal, with 2 fathoms upon it.† Elizabeth Bay has a sandy beach, and a rivulet emptying itself into it. Cordova recommends the best anchorage to be in 15 fathoms, Passage Point bearing E.S.E. distant half a mile, about 3 cables' length from the river; and to the N.W. of a bank on which there is much kelp.

Capt. FitzRoy describes the anchorage of *York Roads*, or *Batchelor's Bay*, to be good and convenient: "Half a mile off, a woody point (just to the westward of the river), bearing N. ½ E., and the mouth of the river N.E.

* Sarmiento, p. 213.

† Ultimo Viage, p. 136.

three-quarters of a mile, is a good berth; because there is plenty of room to weigh from and space to drive should the anchor drag; the bottom is good in 10 or 12 fathoms, but not in a less depth. The shore is a flat shingle beach for 2 miles, the only one in this part of the strait."

The *set and change of the tide* here are very uncertain on account of the meeting of the Jerome Channel tides with those of the strait, which occasions many ripplings; and it would require a better experience than we possess to give a correct explanation. Capt. FitzRoy says, "The tide along shore, near Batchelor's River, changed an hour later than in the offing. At Batchelor's Bay, by the beach, during the first half or one-third of the tide that ran to the S.E., the water fell; and during the latter half or two-thirds, it rose. In the offing it ran very strong." The establishment of the tide, at the entrance of the river, by an observation made by Capt. FitzRoy with the moon eight days old, would be, at full and change, at 1^h 46′. By an observation made by Capt. Stokes, two years previous, it was found to be 2^h 13′; the tide at the anchorage ran 3 knots.

Secretary Wren's Island is a small rocky islet, rising abruptly on all sides, and forming two summits. Near it are some rocks, and to the S.E. is a group of small rocks; and at a mile to the E.S.E. are two rocks above water, called Canoas.

Charles Islands, besides some smaller islets, consist of three principal islands; and in the centre there is a very good port, having good anchorage within the islets, in 13 fathoms. It has an outlet to the N.W., and one to the S.W., also a narrow point communicates with the strait to the S.E.

Opposite to Cape Gallant, on the eastern island, near its N.W. end, is a conspicuous white rock, called *Wallis's Mark*. Next to the westward in succession are *Monmouth* and *James Islands* (called by Cordova, Isla de los Infantes), then *Cordova Islet* and *Rupert Island*, and to the westward of these the island of Carlos III., so named by Cordova. The last is separated from Ulloa Peninsula by St. David's Sound, which is navigable throughout.

To the northward of *Whale Point*, the eastern extremity of Carlos III. Island, is a cove with an anchorage, in 15 fathoms, close to the shore, on a steep bank, but bad ground.

To the westward of Cape Middleton, of Narborough, is *Muscle Bay*, having deep water, and of uninviting character. Cordova describes it to be a mile wide, with unequal soundings, from 12 to 40 fathoms, stones. The bay is not to be recommended, although it appears to be well sheltered.

Choiseul Bay, and *Nash Harbour*, on the Fuegian Coast, are not in the least inviting.

Whale Sound, also on the Fuegian shore, at the back of Ulloa Peninsula, is a large inlet, trending 8 miles into the land, and terminating in a valley bounded on each side by high mountains. There is anchorage only in one place, the West side of Last Harbour; and, although this harbour appears large, the anchorage is small, and close to the shore.

St. David's Sound separates Carlos III. Island from Ulloa Peninsula. At its North end the water is deep, but where it begins to narrow, there are soundings in it.

The JEROME CHANNEL connects those singular inland lakes, the Otway and Skyring Waters, with the Strait of Magalhaens. In the old charts, this continuation is called Indian Sound, and was first explored by Capt. FitzRoy, in an open boat, in May, 1829; but the difficulties, and the short time employed, did not suffice to make a complete survey. In a subsequent voyage, however, this was remedied. The chart will give a sufficient notion of these two large inland lakes, which, as they are of no importance to the general navigator, we shall not further describe. For the particulars of the primary exploration, see Capt. King's Narrative, &c., pp. 214—236.

On the West side of the Jerome are two coves, Wood Cove and Seal Cove, that may be used with advantage by small vessels. On the eastern shore, the bights, Three Island Bay and Coronilla Cove, appeared to be commodious.

The West shore of OTWAY WATER affords several convenient anchorages. Off Point Villiers, lat. 53° 9′, at a quarter of a mile from the shore, there are from 10 to 30 fathoms; and this depth decreases in advancing more northerly. There is anchorage all across the N.E. part of the Water, in from 5 to 20 fathoms, the bottom of sandy mud.

Inglefield and Vivian Islands, at the West end of the Water, are low, but thickly wooded. An isthmus, 6 to 10 miles across, separates the Otway Water from the strait near Elizabeth Island. From an elevated station on the North side of FitzRoy Channel, this narrow neck appeared to be low and much occupied by lagoons. The South shore of Otway Water is formed by high land, with three deep openings that were not examined. *Brunswick Peninsula*, a mass of high mountainous land, is the most southern extremity of the continent.

In lat. 52° 40′ and lon. 71½° West, is the East entrance of *FitzRoy Channel;* it forms a communication between the Otway and the Skyring Waters, and takes a winding course to the N.W. for 11 miles, which is easily navigated. A strong tide running during the neaps at the rate of 5 or 6 miles an hour, in the entrance, and 2 or 3 in other parts, sets through it, six hours each way. The rise and fall, however, were scarcely distinguishable.

SKYRING WATER is 10 leagues long from East to West. Its shores are low. At the western extremity two openings were observed to wind under a high castellated-topped mountain (Dynevor Castle), which were supposed by Capt. FitzRoy to communicate with some of the sounds of the western coast. Through Euston opening, the southern one, no land was visible in the distance; but on a subsequent examination of the termination of the Ancon sin Salida of Sarmiento by Capt. Skyring, no communication was detected.

CROOKED REACH.—In the navigation of this part Wallis and Carteret suffered extreme anxiety; and no one that has read their journals would willingly run the risk of anchoring in any port or bay on its southern shore. The chart will show several inlets deep enough to induce any navigator to trust to them; and, probably, for small vessels, many sheltered nooks might be found, but they have all very deep water, and when the wind blows strong down to Long Reach, they are exposed to a heavy sea and a furious wind.

The anchorage of BORJA BAY within the Ortiz Islands (the Island Bay of Byron) is so much preferable, that it alone is to be recommended. Both Capt.

Stokes and Capt. FitzRoy speak highly of it in their journals; it is snug and well sheltered, and tolerably easy of access,* but in a gale, like its neighbours, the anchorage is much incommoded by the williwaws, which "drive the ship from one side to the other, as if she were a light chip upon the water."† Capt. FitzRoy says, "Let me recommend Borja Bay as an excellent, although small anchorage; wood and water are plentiful; under the coarse upper sand is a stiff clay, like pipe-clay. Avoid the islet off its West side as you go in or out."‡

As this is an anchorage that may be much used, Capt. Stokes's account of it is also subjoined.

"Borja Bay is situated on the northern shore of Crooked Reach, 2 miles to the eastward of Cape Quod. Its position is pointed out as well by the islet off its West point, as by its situation with respect to *El Morion*, the helmet-shaped point, previously called by the English, *St. David's Head*. The entrance to the bay is to the eastward of the largest islet, and presents no dangers; all the islets and shores of the bay may be approached to half a cable's length, even to the edge of the kelp. The only difficulty that impedes getting into the bay arises from the baffling winds and violent gusts that occasionally come off the mountains and down the deep ravines which form the surrounding coast, and the utmost vigilance must be exercised in beating in under sail to guard against their effects. The anchorage is perfectly sheltered from the prevailing winds, the westerly and south-westerly gales, and is open only to south-easterly winds, which very rarely blow here, and still more rarely with violence; and as the holding ground is good (small stones and sand) and the depth of water moderate (14 to 16 fathoms), and any fetch of sea prevented by the narrowness of the strait in this part, the greatest breadth being only 3 miles, it may be pronounced a very good and secure harbour. The best plan is to anchor with the bower, and steadied to the shore by a hawser or kedge. No surf or swell obstructs landing anywhere; good water and plenty of wood are easy to be embarked; the trees, a species of beech, are of considerable size. The shores are rocky, and the beach plentifully stocked, as indeed are all parts of the strait to the eastward, with barberries and wild celery."§

Byron anchored in Borja Bay, as did also Carteret, in the *Swallow*.‖ The former gives a plan of it, and calls it Island Bay. He attempted to anchor in it, but was prevented by the strength of the tide.¶

Capt. Stokes describes the Morion, or St. David's Head, to be a lofty granitic rock, of which the outer face is perpendicular and bare, and of a light gray colour, distinguishable from a considerable distance both from the East and the N.W., and forming an excellent leading mark to assure the navigator of his position.

Narborough thus describes *Cape Quod*:—"It is a steep-up cape, of a rocky, grayish face, and of a good height before one comes to it: it shows like a great building of a castle; it points off with a race from the other mountains, so much into the channel of the strait, that it makes shutting in against the South land, and maketh an elbow in the streight."**

* Capt. Stokes says, "Very confined, and rather difficult of access."—*King*, 154.
† FitzRoy M.S. ‡ FitzRoy M.S. § Stokes M.S.
‖ Hawkesworth, vol. i. p. 395. ¶ Hawkesworth, vol. i. p. 68. ** Narborough, p. 70.

At not a league to the eastward of Cape Quod, and a good distance from the North shore, in the fairway for working round the cape, is a rock of 9 feet least water, but shows itself by the weeds on it.

Abreast of Cape Quod, Capt. Stokes tried and found the current setting to the eastward at 1½ knots an hour.

Snowy Sound, a deep inlet, unimportant to the navigator, and not worth any person's while to enter, excepting for anchorage in a cove at about a mile, and in another at 2 miles, within its western head.

Barcelo Bay, the first to the West of Cape Quod, seems to be large and incommodious, and strewed with small islets. *Osorno Bay* follows, and, according to Cordova, has very deep water all over. Next, to the westward, is *Langara Bay*. It trends in for about a mile to the N.E., and has 10 to 12 fathoms, stony bottom. It is more sheltered than the two former bays.

Posadas Bay is, most probably, Wallis's Lion Cove. Its western point is formed by a high, rounded, and precipitous headland, resembling, in Capt. Wallis's idea, a lion's head; and although Cordova could not discover the likeness, yet it is sufficiently descriptive to point out the bay, were the anchorage worth occupying, which it is not. Wallis describes it to have deep water close to the shore; his ship was anchored in 40 fathoms.*

Arce Bay. Cordova describes it to have anchorage in from 6 to 17 fathoms, stones. It divides at the bottom into two arms, each being half a mile deep. The outer points bear from each other W.N.W. and E.S.E., half a mile across.

Flores Bay is, probably, Wallis's Good Luck Bay. Cordova describes it to be very small and exposed, with from 6 to 20 fathoms, stones and gravel. At the bottom is a rivulet of very good water.

Villena Cove has from 15 to 20 fathoms, and is very open and exposed.

Then follows *Guirior Bay*. It is large, and open to the South, and probably affords good anchorage in coves. Cordova describes it to extend for more than a league to the North, the mouth being 2 miles wide. Its West Point is Cape Notch, which will serve to recognise it.

From the above description of the bays between Capes Quod and Notch, occupying a space of 12½ miles, none seem to be convenient or very safe. The best port for shelter, for a ship, is Swallow Harbour, on the opposite shore: but small vessels may find many places that a ship dare not approach, where every convenience may be had; for if the water be too deep for anchorage, they may be secured to the shore at the bottom of the coves, where neither the swell nor the wind can reach them.

Swallow Harbour is 1½ miles to the westward of Snowy Sound. It is a better anchorage for ships than any in the neighbourhood. The plan of it is a sufficient guide, the dangers being well buoyed and pointed out by kelp. It was first used by Capt. Carteret, in the *Swallow*; and Cordova gives a short description of it.

The bay to the westward of the island is Condesa Bay. It is full of islets and rocks; and the channel behind the island, communicating with Swallow Bay, is very narrow.

* Hawkesworth, vol. i. p. 390.

At about a cable's length off the West point of the entrance of Swallow Harbour, Capt. FitzRoy saw a rock just awash. This danger should be carefully avoided.

Stuart's Bay is less than a league from Swallow Bay. Of this place Capt. Stokes makes the following remarks:—"Stuart's Bay afforded us a quiet resting place for the night, but it is by no means to be recommended as an anchorage; for though it is sufficiently sheltered from wind and sea, yet the rocks in different parts of it render the passage in or out very hazardous; every danger in it is pointed out by rock-weed, but it is so much straitened as to require the utmost vigilance."

The account in Cordova is as follows:—

"Stuart's Bay (La Bahia de Stuardo) follows Condesa Bay. It has an islet, besides several patches of kelp, an indication of the many rocks that exist. Even the best channel is narrow and tortuous; the depth from 12 to 16 fathoms, stones. At the bottom is an islet, forming two narrow channels leading into a port or basin, 2 cables' length wide; the eastern channel is the deeper, and has 15 to 20 fathoms. Inside the basin, on the East side, the depth is 6 and 9 fathoms, mud. A reef extends for half a cable's length to the westward of the South end of the islet. It would be difficult and dangerous to enter this small basin."*

Then follows a deep and extensive channel, of which we know only that it extends to the South for 5 or 6 miles, and, perhaps, is very similar in its termination to Snowy Sound. It is Sarmiento's Snowy Channel.†

At this part of the strait the breadth is about 2½ miles; but, at Cape Quod, it scarcely exceeds 1½ miles. The shores are certainly much less verdant than to the eastward of Cape Quod, but not so dismal as Cordova's account would make them appear to be; for, he says, "As soon as Cape Quod is passed, the strait assumes the most horrible appearance, having high mountains on both sides, separated by ravines entirely destitute of trees from the mid-height upwards." To Capt. King it appeared that the hills were certainly much more bare of vegetation above, but below were not deficient; the trees and shrubs, however, are of small size. For the purposes of fuel abundance of wood is to be obtained. In the winter months the hills are covered with snow, from the summit to the base; but in the month of April, when the *Adventure* passed through, no snow was visible about them.

Capt. Stokes remarks that the mountains in this part (Cape Notch) spire up into peaks of great height, connected by singularly sharp, saw-like ridges, as bare of vegetation as if they had been rendered so by the hand of art. About their bases there are generally some green patches of jungle, but upon the whole nothing can be more sterile and repulsive than this portion of the strait. This account of Capt. Stokes agrees with Cordova's; but upon examining the coves, we found them so thickly wooded with shrubs and jungle and small trees, that it was difficult to penetrate beyond a few yards from the beach.

Cape Notch, on the North side, is a projecting point of gray-coloured rock,

* Cordova, p. 147. † Sarmiento, p. 207.

about 650 feet high, having a deep cleft in its summit. It is a conspicuous headland, and cannot be mistaken.

The next place to the westward of Cape Notch that can be recommended for an anchorage is *Playa Parda Cove*, which is well sheltered, and, for chain cables, has a good bottom, being of sandy mud, strewed with stones; it is half a mile wide at the entrance, and about a quarter deep.

Playa Parda Cove is easily known by *Shelter Island*, that fronts the inlet of Playa Parda. The inlet is 1½ miles long, and half a mile broad, but with very deep water all over. By luffing round the island, a ship will fetch the anchorage in the cove; and although sail should not be reduced too soon, yet the squalls, if the weather be bad, blow down the inlet of Playa Parda with great violence. Anchor a little within, and half-way between the points of entrance, at about 1½ cables from the middle point, in 5½ and 6 fathoms.

Opposite to Playa Parda is a deep opening, which has more the appearance of a channel, leading through the Tierra del Fuego, than any opening to the West of the Barbara. It is evidently the inlet noticed by Sarmiento.* On the seaward coast there is a deep opening behind Otway Bay, which, probably, may communicate with it.

The weather here is generally so thick, that, although the distance across be only 2 to 3 miles, yet one shore is frequently concealed from the other by the mist; on which account Capt. Stokes found it impossible to form any plan of this part of the strait, on his passage through it. It is, however, a bold coast on each side, otherwise the strait would be utterly unnavigable in such weather.

Marian's Cove, 1½ miles to the West of Playa Parda, is a convenient anchorage; at the entrance it is about one-third of a mile wide, and more than half a mile deep. In entering, the West side should be kept aboard.

This cove is about midway between Cape l'Etoile and Playa Parda; and is a very advantageous place to stop at.

Opposite to Cape l'Etoile is a bay, with anchorage in 17 fathoms, in a well-sheltered situation. From Cape l'Etoile to the entrance of the Gulf of Xaultegua, the shore is straight and precipitous, and the hills are barren and rocky. On the opposite shore there are a few inlets, but the most useful one for the navigator is Half Port Bay, rather more than a league to the East of Cape Monday. It is immediately round the South side of a deep inlet. It is merely a slight indentation of the coast, but is an excellent stopping place; the anchorage is within two-thirds of a cable's length of the West point, in 16 fathoms, muddy bottom. The situation of this cove was ascertained by observation to be in lat. 53° 11′ 36″ and lon. 73° 18′ 41″ W. (or 2° 20′ 56″ West of Port Famine).

"The land on the S.W. side of the anchorage is high, and thickly wooded from its summit to the water's edge. On the eastern side it is lower, the vegetation more scanty, and the trees crooked and stunted, and pressed down to the N.E. by the prevailing winds. S.W. by W. from the anchorage is a remarkable cleft in the summit of the high land, from which a narrow strip cleared of jungle descends to the water's edge, apparently formed by the descent of a torrent or of

* Sarmiento, p. 206.

large masses of the rock.* The anchorage is well sheltered from prevailing breezes, and the holding ground is good: water and fuel are abundant."

There is an anchorage under Cape Monday for small vessels, in which Byron anchored,† and rode out a heavy gale of wind. With the exception of a shoal in midway of the entrance, on which there are 4 fathoms, it seems to offer good shelter from the prevailing winds. On the West side of Cape Monday is Cordova's Medal Bay (Puerto de la Medalla), of which a very full but florid description is given in the appendix of that voyage.‡

It has, according to the description, an island in the entrance which forms two channels, the easternmost of which is only deep enough for boats, but the western is 25 fathoms wide; it is strewed half way across with kelp; but between the kelp and the island is a good and clear passage with 6 fathoms, sandy bottom. In the kelp there are not less than 4 fathoms, and inside it the depth is 9, 8, and 7 fathoms, sandy bottom. To enter this port there are no dangers that are not visible, and those are easily avoided; they consist only of the islet in its entrance, and some patches of kelp, over which, however, there is plenty of water.

The Gulf of Xaultegua, improperly called Bulkeley's Channel, is a deep opening, trending into the land in an easterly direction for 28 miles, and approaching within 2 miles of some of the inlets on the N.W. side of Indian Sound. The entrance is about 4 miles across, but afterwards expands to a width of nearly 15 miles. At the entrance is St. Ann's Island, between which and the South point is a navigable channel, half a mile wide. St. Anne's Island is about 2 miles long, and extends in a W.N.W. and E.S.E. direction; off its N.W. end is an islet, and there is another close to its S.W. extremity.

The land forming the North side of the strait, between the Gulf of Xaultegua§ and the Jerome Channel, is called Croker Peninsula.

Should a ship be so unfortunate as to make a mistake and get into it, she must keep under weigh until she gets out again. There is no *thoroughfare*.—(FitzRoy's Journal.)

Little has been said of the *tides* in this part of the strait, and, indeed, as to their rise and fall they are really of no importance, being little more than 4 feet. It is high water, at full and change, in all parts within a few minutes of noon. The current sets constantly to the eastward with more or less strength.

Between Elizabeth Island, within the Second Narrow, and the western end of Long Reach, there is very little swell. In a heavy gale, or perhaps even a strong breeze, a short sea may be experienced in the wider part of the strait, particularly near and to the westward of Cape Froward; but nothing to be compared to the confused, breaking swell that runs in the Sea or Western Reach. It was felt by the *Beagle* when beating to the westward, immediately on reaching Cape Providence. There seems to be no danger for vessels beating through the strait hereabouts, the shore being bold-to. Byron passed a night, and a very

* More probably by the effect of a gust of wind, which to the eastward, particularly in the Gabriel Channel, is very common.

† Hawkesworth, vol. i. p. 73.

‡ Ultimo Viage, Appendix, p. 49.

§ The name of Xaultegua is from Sarmiento, who very correctly describes it.—*Sarmiento*, p. 206.

tempestuous one, here; as did also the *Beagle*, the latter not being able to find anchorage before night.

A league to the westward of Cape Monday is an inlet, which we suppose to be Sarmiento's *Puerto Angosto*. Upon its West head is a conspicuous round mount, and to the North, between the mount and a projecting point, is a confined but very snug and commodious cove for a small vessel, in 17 fathoms, at a quarter of a mile within the head.

In *Upright Bay* the anchorage, though affording excellent shelter from the prevailing winds, is bad with a southerly one; for the steepness of the bottom requiring a vessel to anchor close to the shore, sufficient scope is not left for veering cable.

Cape Upright bears due South, 5 miles from Cape Providence. It has a rocky islet a quarter of a mile off its East extremity, surrounded by kelp, which also extends for some distance from the cape towards the islet, at the end of which there are 7 fathoms. (At 3 leagues West of the cape, and some distance from the shore, is a reef.)

Cape Providence is a rugged rocky mountain, higher than the adjacent coast; it is deeply cleft at the top, and, when bearing about North, the western portion of its summit appears arched, the eastern lower and peaked. When the cape bears E. by S. *mag.*, distant about 1½ leagues, a little round rocky islet will be seen open of it, about a quarter of a point of the compass more southerly.*—(Stokes's MSS.)

The distance from Cape Providence to Cape Tamar is 9½ miles; in this space the land arches inwards, and forms a bay about 1½ leagues deep. Capt. Stokes describes the coast to the East of Cape Tamar to be formed into two large bights by the land of Cape Providence. On the western side of the latter are several islands, of which two are conspicuous; they are round and of a good height, and well wooded; at a distance their form is conical, the eastern being the lowest. Between them is a passage to two good anchorages, which Lieut. Skyring, who examined them, considered even more sheltered than Tamar Harbour.

Four miles to the eastward of Cape Tamar is *Round Island*, to the N.W. of which is a well-sheltered anchorage, but with deep water. In standing in, pass midway between Round Island and an island to the westward, which lies close to the shore, and haul round the latter to the mouth of a cove, in the entrance of which, near the South shore, there are 23 fathoms, sand. The shore to the North and N.E. of Round Island is very rocky. On the East side of the promontory of Cape Tamar, is the useful and excellent anchorage of Tamar Harbour. It is scarcely 2 miles wide, and rather more than half a mile deep. Its entrance is not exactly free from danger, but, with attention to the following directions, none need be apprehended. There is a sunken rock between a group of rocky islets, one-third over on the western side, and a patch of kelp one-third towards the eastern side of the bay. With a westerly wind it would be advisable

* There are some anchorages on the right, to the N.E. of Cape Providence, according to a plan given in Hawkesworth's Collection of Voyages, but they are too much out of the way, as well as very open and exposed to southerly winds, to be of use, or to offer any security to vessels bound through the strait.

to give the outer rock a berth of 2 cables' length to avoid this danger, on which there are only 9 feet of water, and upon which the *Beagle* struck.

An excellent leading mark for this shoal is a whitened portion of bare rock, looking like a tombstone, about one-third of the way up the green side of the mountainous land that forms the coast of the bay. This stone bears W. by N. ½ N. (by compass) from the rocks to be rounded on entering the anchorage.

The least water found among the kelp on the East side of the channel was 4½ fathoms, and near and within the edge towards the rocky islets there are 7 fathoms; so that with the lead in hand, and a look out for kelp, which should not unnecessarily be entered, there is no real danger to be apprehended.

High water at full and change takes place in Tamar Harbour at $3^h\ 5'$, and the perpendicular rise and fall are 5 feet.

The *flood tide* on this part of the northern shore of the strait sets to the eastward, and rarely exceeds half a mile an hour. At this part the strait is 7 miles wide; at Cape Philip, to the westward, the breadth increases to 5 leagues; but at Cape Parker it narrows again 4 leagues, which breadth it keeps to the end.

To the westward of Cape Tamar is *Tamar Island*. It is high, and is separated from the land of the cape by a deep channel from half to 1 mile wide. Half a mile off its S.W. end is a rock.

Between Capes Tamar and Philip, a space of 4 leagues, there is a deep bight, with two openings; the easternmost, in which are Glacier and Icy Sounds, extends to the N.E. for 10 miles from the mouth, and the westernmost is the commencement of Smyth's Channel.

Under the lee (the N.E.) of Cape Philip is *Sholl's Bay*, in which the *Beagle* anchored in 1827. Of this place Capt. Stokes writes:—"We found there an excellent anchorage in 15 fathoms. It is valuable for vessels working through the strait to the westward, inasmuch as, from the discontinuous nature of the northern shore (which here is formed into deep bays), this place will be much more easily recognised than the anchorages on the opposite coast; besides the winds hang here, in general, somewhat to the northward of West, hence a better starting place for the westward is obtained. Here, as in every anchorage on the strait, water and fuel are easily procured; but nothing more, unless we except the wild berries (*berberis, sp.*), celery, mussels, and limpets; the wild goose abounds, but its nauseous taste renders it uneatable. No inhabitants, no quadrupeds."

Of the coast of the strait on the South side, between Cape Upright and Valentine Harbour, we know very little; there are several deep bights and spacious bays, which may contain anchorage, but, in general, they are not found in the large harbours, which are mostly deep, precipitous chasms, or ravines in the rock. The smaller coves, or where the land shelves down to the sea, are more likely to afford anchorages.

In the appendix to Cordova's work are descriptions of some anchorages which it may be useful to mention here: it says, "In rounding Cape (Ildefonso) Upright we found ourselves in a bay, not very deep, 2 miles across, divided in its centre by many islets and rocks extending to the North; the outer or northernmost of which bears West from the extremity of the cape. One mile N.W. ½ N. from the northernmost islet is a round rock, which is of dangerous approach."

To the westward of this bay is another, 3 miles wide, and about as deep; the whole of it, particularly towards the eastern part, is full of islets, and at the bottom is a narrow canal trending to the S.S.E. At the western end of this bay, called by Wallis the Bay of Islands, from the number it contained, commences a third, which, with the two preceding, make the great bay, called by the Indians, according to Sarmiento, Alquilqua. It is contained between Cape Upright and a bold projecting point, 10 miles to the W.N.W., called Point Echenique. The country is there described to be poor, and the vegetation scanty.

The eastern point of the Third Bay has a string of islets extending a mile to the North; and to the S.W. are several others. On its East side is a bay called Cuaviguilgua; and a little beyond it, at the bottom of the bay, is Port Uriarte, the mouth of which is 2 cables' length across.

Port Uriarte was carefully sounded, but the bottom is generally bad and stony, with 5, 8, 14, to 18 fathoms. The harbour is surrounded by high mountains, rising vertically, and with only a few stunted trees on the shores. Its greatest extent, which is from North to South, is half a mile; the mouth is not visible until close to it: its bearing from Cape Providence is S.W. ½ S. There is no danger in entering it but what is visible; but it is not recommended as a good harbour, from the foul ground all over it. A little to the eastward also of Point Echenique is Cape Santa Casilda, a low point.

To the West of Point Echenique is a harbour 2½ miles wide, the points of entrance bearing N.W. and S.E. There is an island in the centre forming two channels, but with very deep water, no ground being found with 55 fathoms. At the bottom is a canal, trending to the S.S.W., and disappearing between the mountains. On the eastern side of the island the channel is at first a mile wide, but afterwards narrows gradually: the western channel is scarcely 2 cables' length across. The shores are high, precipitous mountains. The Indians, according to Sarmiento, call the place *Puchachailgua*.

The *Canal de la Tempestad* (or Stormy Channel), from the description, is not to be recommended. The water is very deep all over, and the place affords no security for vessels of any description. To the westward is a better harbour, which the Spanish officers thought to be Sarmiento's Port Santa Monica. It bears S.S.W. from Cape Tamar, and it is 14 miles to the westward of Cape Upright, but not more than 3 leagues according to Sarmiento's account.

Two-thirds of a mile to the westward is a point, with two islets off it; round which is *Point Churruca*, a deep and spacious bay, 2 miles wide, the points bearing E.S.E. and W.N.W., containing two ports and some coves, but with very deep water, and therefore useless, for it would be necessary to make fast to the rocks to secure a vessel.

To the westward of this is a useful cove, *Darby Cove*, in which small vessels may obtain good shelter.

From Darby Cove the coast extends to the W.N.W. for 7 miles, having in the interval several indentations, but all with deep water; at *Point Felix*, the land trends deeply in to the S.W., and forms a bay 5 miles wide and 2½ deep. At its western side is *Valentine Harbour*, in which the *Beagle* anchored, which seems to be commodious and secure, and of easy approach. On hauling round the island, there are some islets half a mile off, which must be avoided, but otherwise there seems to be no dangers.

The anchorage, as a stopping place, is in from 20 to 26 fathoms, sand, at nearly a quarter of a mile from either shore: a more sheltered situation may be obtained to the S.W.

Cape Cuevas, the extremity of an island that is close to the shore, is in lat. 52° 53′ 19″, and lon. 74° 21′ 22″.

There is plenty of wood and water in Truxillo Bay, but nobody will visit it in preference to *Tuesday Bay*, or, rather, the more convenient anchorage of Tuesday Cove, situated three-quarters of a mile South of Cape Cortado. The anchorage is in 12 to 14 fathoms. Tuesday Bay is larger, and, therefore, more exposed to the squalls; but for a ship, perhaps, might be more convenient.

On the North shore of the strait, opposite to Cape Cortado, is CAPE PARKER, a remarkable projection, with three hummocks on the summit of the high land which rises over it. To the eastward the coast trends deeply in to the North, forming a bay, the eastern head of which, Cape Philip, bears N.E. by E. 9 miles. There appeared to be several islands in the bay, and at the bottom a narrow opening, perhaps a channel, leading to the North.

On the West side of the bay the coast is indented, and affords some anchorages, but the approach is not clear. The first bay, however, to the eastward of the S.E. trend of the cape, seems to afford a good stopping place; but it is fronted by a considerable shoal, with two rocky islets; the depth is from 7 to 22 fathoms.

The land of Cape Parker will probably turn out to be an island. To the westward of it commences a range of islands, rocks, and shoals, fronting a broken coast that should never be approached but for the purpose of discovery or seal-fishery. The easternmost island is Westminster Hall,* a high, rocky island.

Sir John Narborough's Islands consist of eight or ten principal islands, and, perhaps, hundreds of smaller ones. Behind them there seemed to be a channel; and amongst them are several anchorages, but none to be recommended, especially when on the South coast there are two or three much better, much safer, and of much easier access.

It is a dangerous coast, as well from the immense number of rocks, upon which the sea breaches very high, as from the tides, which near the edge of the line of shoals set frequently in amongst them. "The coast about our unsafe anchorage was as barren and dismal looking as any part of this country, which, as Sir John Narborough says, is 'so desolate land to behold.'"—(King, 80.)

A league to the westward of Cape Cortado, on the South side, is *Skyring Harbour*. There are some islands in it, and anchorage might be obtained in 27 fathoms.

At 3½ miles from the West point of Skyring Harbour is the East head of the *Harbour of Mercy* (Puerto de la Misericordia of Sarmiento,† Separation Harbour of Wallis and Carteret), one of the best anchorages of the western part of the strait, and being only 4 miles within Cape Pillar, is very conveniently placed for

* Narborough, p. 77.

† Sarmiento, p. 182.

a ship to anchor at, to await a favourable opportunity for leaving the strait. There is no danger in entering. The depth is moderate, 12 to 14 fathoms, and the holding ground excellent, being a black clay. A ship may select her position; but the one off the first bight round the point being equally well sheltered, and much more convenient for many purposes, is the best berth.

CAPE PILLAR is 3 miles to the westward of the largest Observation Islet. "At daylight, on November 25th, 1829, we made Cape Pillar right ahead (E.N.E. by compass), distant 7 or 8 leagues. The wind became lighter, and we were set by a current to the S.W., which obliged us, in nearing the cape, to alter our course from E.N.E. to N.N.E., to avoid being carried too near the Apostle Rocks. A dangerous rock, under water, on which the sea breaks, lies half a mile more towards the North than either of the Apostles. Cape Pillar is a detached headland, and so very remarkable that no person can fail to know it easily."—(Capt. FitzRoy, Voy. *Adv.* and *Beagle*, vol. i. p. 360.)

The extremity of Cape Pillar is in lat. 52° 42′ 53″, and lon. 74° 43′ 23″; and Cape Victory in 52° 16′ 10″, and 74° 54′ 47″. These points form the western entrance of the strait.

THE EVANGELISTS, as they were named by the early Spanish navigators, but THE ISLES OF DIRECTION by Narborough, from their forming a capital leading mark for the western mouth of the strait, are a group of rocky islets, consisting of four principal ones, and some detached rocks and breakers. The islands are very rugged and barren, and suited only to afford a resting place or breeding haunt of seals and oceanic birds. There is landing on one of the islands, and anchorage round them, if necessary. The largest and highest may be seen, in tolerably clear weather, from a brig's deck, at the distance of 7 or 8 leagues. The southernmost, from its shape called the Sugar Loaf, is in lat. 52° 24′ 18″, and lon. 75° 6′ 50″. From the Sugar Loaf, the extremity of Cape Pillar bears N.W. ½ N. *mag.* 23½ miles; and from Cape Victory, according to Capt. Stokes's survey, S. by W. ¾ W. *mag.* 11 miles.

The tides here are very variable, and sometimes set to the E.N.E. towards the rocks that front Cape Victory and Sir John Narborough's Islands. The times of high water at the different points will be found in the tables.

CHAPTER II.

THE OUTER COAST OF TIERRA DEL FUEGO, FROM CATHERINE POINT TO CAPE PILLAR, INCLUDING STATEN ISLAND, CAPE HORN, ETC.

*GENERAL OBSERVATIONS.**—The Coast of Tierra del Fuego from Cape Horn to Cape Pillar is very irregular and much broken; being, in fact, composed of an immense number of islands. It is generally high, bold, and free from shoals or banks; but there are many rocks nearly level with the surface of the water, distant 2 and even 3 miles from the nearest shore, which make it very unsafe for a vessel to approach nearer than 5 miles, excepting in daylight and clear weather. The coast varies in height from 8 to 1,500 feet above the sea. Further inshore are ranges of mountains always covered with snow, whose height is from 2,000 to 4,000 feet, and in one instance (Sarmiento) 5,000 feet.

With daylight and clear weather a vessel may close the shore without risk, because the water is invariably deep, and no rock is found which is not so marked by sea-weed (or kelp, as it is generally called), that by a good look out at the mast-head, its situation is as clearly seen as if it were buoyed.

Viewing the coast at a distance, it appears high, rugged, covered with snow, and continued,—as if there were no islands. When near you see many inlets, which intersect the land in every direction, and open into large gulfs or sounds behind the seaward islands.

You now lose sight of the higher land, which is covered with snow throughout the year, and find the heights close to the sea thickly wooded towards the East, though barren on their western sides, owing to the prevailing winds. These heights are seldom covered with snow, because the sea winds and the rain melt it soon after it falls.

Opposite to the eastern valleys, where the land is covered with wood, and water is seen falling down the ravines, good anchorage is generally found. But these valleys are exposed to tremendous squalls, which come from the heights. The best of all anchorages on this coast is where you find good ground on the *western* side of high land, and are protected from the sea by low islands. It never blows so hard *against* high land as from it, but the sea on the weather side is of course too formidable, unless stopped by islets.†

* These observations are by Capt. FitzRoy, 1830.

† The lee side of high land, Capt. King remarks, is not the best for anchorage in this country. When good holding can be found to windward of a height, and low land lies to windward of you, sufficient to break the *sea*, the anchorage is much preferable; because the wind is steady, and does not blow home to the heights. Being to leeward of them is like being on the west side of Gibraltar rock when it blows a strong levanter.

Where the land is chiefly composed of sandstone or slate, anchorages abound; where of granite, it is difficult to strike soundings.

The difference between the granite and slate or sandstone hills, can be distinguished by the former being very barren and rugged, and of a gray or white appearance; whereas the latter are generally covered with vegetation, are dark coloured, and have smoother outlines. These slate or sandstone hills show few peaks, and the only rugged places are those exposed to wind or sea.

SOUNDINGS extend to 30 miles from the coast. Between 10 and 20 miles from the land the depth of water varies from 60 to 200 fathoms, the bottom almost everywhere a fine white or speckled sand. From 10 to 5 miles distant the average depth is 50 fathoms; it varies from 30 to 100, and in some places no ground with 200 fathoms of line. Less than 5 miles from the shore the soundings are very irregular indeed, generally less than 40 fathoms, but in some places deepening suddenly to 100 or more: in others a rock rises nearly to, or above, the surface of the water.

After carrying 50, 40, 30, or 20 fathoms, towards an inlet which you are desirous of entering, you will probably find the water deepen to 60 or 100 fathoms as soon as you enter the opening; and in the large sounds, behind the seaward islands, the water is considerably deeper than on the outside.*

There is a bank of soundings along the whole coast, extending from 20 to 30 miles from it, which appears to have been formed by the continued action of the sea upon the shore, wearing it away and forming a bank with its sand.

Between the islands, where there is no swell or surf worth notice, the water is deep, and the bottom very irregular.

A small ship may run among the islands in many places, and find good anchorage; but she runs into a labyrinth, from which her escape may be difficult, and, in thick weather, extremely dangerous.

FOGS are extremely rare on this coast, but thick rainy weather and strong winds prevail. The sun shows himself but little; the sky even in fine weather being generally overcast and cloudy. A clear day is a very rare occurrence.

WINDS.—Gales of wind succeed each other at short intervals, and last several days. At times the weather is fine and settled for a fortnight, but those times are few.

Westerly winds prevail during the greater part of the year. The East wind blows chiefly in the winter months, and at times very hard, but it seldom blows in summer.

Winds from the eastern quarter invariably rise light, with fine weather; they increase gradually,—the weather changes,—and at times end in a determined heavy gale. More frequently they rise to the strength of a treble-reefed topsail breeze, then die away gradually, or shift to another quarter.

* "I have heard Capt. FitzRoy remark, that, on entering any of these channels from the outer coast, it is always necessary to look out directly for anchorage; for further inland the depth soon becomes extremely great. Capt. Cook, in entering Christmas Sound, had first 37 fathoms; then 40, 60, and, immediately afterwards, no soundings with 170. This structure of the bottom, I presume, must arise from the sediment deposited near the mouths of the channels by the opposed tides and swell; and likewise from the enormous degradation of the coast rocks, caused by an ocean harassed by endless gales."—*Darwin*, pp. 266-7.

From the North the wind always begins to blow moderately, but with thicker weather and more clouds than from the eastward, and it is generally accompanied by *small rain*. Increasing in strength, it draws to the westward gradually, and blows hardest between North and north-west, with heavy clouds, thick weather, and much rain.

When the fury of the north-wester is expended, which varies from 12 to 50 hours, or even while it is blowing hard, the wind sometimes shifts suddenly into the S.W. quarter, blowing harder than before. This wind soon drives away the clouds, and in a few hours you have clear weather, but with heavy squalls passing occasionally.

In the S.W. quarter the wind hangs several days (generally speaking), blowing strong, but moderating towards its end, and granting two or three days of fine weather.

Northerly winds then begin again, generally during the summer months; but all manner of shifts and changes are experienced from North to South by the West during that season, which would hardly deserve the name of summer, were not the days so much longer, and the weather a little warmer. Rain and wind prevail much more during the long than the short days.

It should be remembered that bad weather *never* comes on *suddenly* from the eastward, neither does a S.W. or southerly gale shift suddenly to the northward. S.W. and southerly winds rise suddenly and violently, and must be well considered in choosing anchorages, and preparing for shifts of wind at sea.

The most usual weather in these latitudes is a fresh wind between N.W. and S.W., with a cloudy, overcast sky.

Much difference of opinion has prevailed as to the utility of a barometer in these latitudes. Capt. FitzRoy says, that during 12 months' constant trial of a barometer and sympiesometer (Adie's), he found their indications of the utmost value. Their variations do not of course correspond to those of middle latitudes, but they correspond to those of high northern latitudes in a remarkable manner, changing South for North (East and West remaining the same).

CURRENT.—There is a continual current setting along the S.W. coast of *Tierra del Fuego*, from the N.W. towards the S.E. as far as the Diego *Ramirez Islands*.* From their vicinity the current takes a more easterly direction, setting round Cape Horn towards Staten Island, and off to seaward to the E.S.E.

Much has been said of the strength of this current, some persons supposing that it is a serious obstacle in passing to the westward of Cape Horn, while others almost deny its existence.

It was found to run at the average rate of a mile an hour. Its strength is greater during West; less, or insensible, during easterly winds. It is strongest near the land, particularly near the projecting capes or detached islands.

* "An oar was picked up near the watering place at Desolation Harbour, on 11th Dec., 1829, and recognised by one of the men as the same which was left on a rock near Cape Pillar (in Observation Cove) by Capt. Stokes, in January, 1827. There could be no doubt of the fact, as the man's initials were on the oar: and it is a curious proof of an outset along the South side of the strait, near Cape Pillar, and of its continuation along shore."—*King*, vol. i. p. 365.

This current sets rather *from* the land, which diminishes the danger of approaching this part of the coast.

There is, in fact, much less risk in approaching this coast than is generally supposed. Being high and bold, without sandbanks or shoals, its position accurately determined, and a bank of soundings extending 20 or 30 miles from the shore, it cannot be much feared. Rocks, it is true, abound near the land, but they are very near to the shore, and out of a ship's way.

A line from headland to headland (beginning from the outermost Apostle) along the coast will clear all danger excepting the Tower Rocks, which are high above water, and steep-to.

Gales of wind from the southward, and squalls from the S.W., are preceded and foretold by heavy banks of large white clouds rising in those quarters, having hard edges, and appearing very rounded and solid.—(Cumuloni.)

Winds from the northward and north-westward are preceded and accompanied by low flying clouds, with a thickly overcast sky, in which the clouds appear to be at a great height. The sun shows dimly through them, and has a reddish appearance. For some hours, or a day, before a gale from the North or West, it is not possible to take an altitude of the sun, although he is visible; the haziness of the atmosphere in the upper regions causing his limbs to be quite indistinct. Sometimes, but very rarely, with the wind light between N.N.W. and N.N.E., you have a few days of beautiful weather. They are succeeded by gales from the southward, with much rain.

SEASONS.—It may be as well to say a few words respecting the seasons in the neighbourhood of Cape Horn, as much question has arisen respecting the propriety of making the passage round the cape in winter or in summer.*

The equinoctial months are the worst in the year, generally speaking, as in most parts of the world. Heavy gales prevail at those times, though not, perhaps, exactly at the equinoxes. In August, September, October, and November, you have the worst months in the year. Westerly winds, rain, snow, hail, and cold weather, then prevail.

December, January, and February, are the warmest months; the days are long, and you have some fine weather; but westerly winds, very strong gales at times, with much rain, prevail throughout this season, which carries with it less of summer than in almost any part of the globe.

March is stormy, and perhaps the worst month in the year with respect to violent winds, though not so rainy as the summer months.

In April, May, and June, the finest weather is experienced; and though the days shorten, it is more like summer than any other time of the year. Bad weather is found during these months, but not so much as at other times. Easterly winds are frequent, with fine, clear, settled weather. During this period there is

* "Some persons are disposed to form a premature opinion of the wind and weather to be met with in particular regions, judging only from what they may themselves have experienced. Happily, extreme cases are not often met with; but one cannot help regretting the haste with which some men (who have sailed round Cape Horn with royals set) incline to cavil at and doubt the description of Anson and other navigators, who were not only far less fortunate as to weather, but had to deal with crazy ships, inefficient crews, and unknown shores, besides hunger, thirst, and disease."—*FitzRoy*, p. 126.

some chance of obtaining a few successive and corresponding observations. To try to rate chronometers by equal altitudes would be a fruitless waste of time at other seasons. June and July are much alike, but easterly *gales* blow more during July.

The days being so short, and the weather cold, make these months very unpleasant, though they are, perhaps, the best for a ship making a passage to the westward, as the wind is much in the eastern quarter.

Capt. FitzRoy says that the summer months, December and January, are the best for making a passage from the Pacific to the Atlantic Ocean, though that passage is so short and easy that it hardly requires a choice of time. For going to the westward, he should prefer April, May, and June.

Lightning and thunder are seldom known; violent squalls come from the South and S.W., giving warning of their approach by masses of clouds. They are rendered more formidable by snow and hail of a large size.*

CATHERINE POINT is the north-eastern extremity of Tierra del Fuego; it is very low, and of shingle, precisely similar to Dungeness, upon the opposite coast of Patagonia. Between Catherine Point and Cape Orange there is a large bay, called by Sarmiento *Lomas Bay.* The land around it is very low, and the space which, in the chart, appears to be water, is chiefly occupied by extensive shoals; some of them are visible at low water.

Ten miles inshore there is a range of land from 200 to 600 feet in height, and extending from *Cape Espiritu Santo* to the westward. From Point Catherine to Cape Espiritu Santo the shore is low. Lying 2 miles northward of that cape, there is a reef with shoal water. Cape Espiritu Santo is a steep, white cliff, 190 feet high, somewhat resembling the gable end of a large but low barn. This cliff is the termination of a range of rather high land, lying nearly East and West, corresponding in height and position to the opposite range, which is terminated by Cape Virgins, but not so horizontal in outline.

The Fuegian shore has many hummocks, and does not show any extent of table-land similar to the Patagonian ranges. From Cape Espiritu Santo, cliffs from 100 to 300 feet in height extend, but with few breaks, to *Nombre Head:* the land is 300 or 400 feet high, irregularly rounded in outline, quite destitute of wood, and, excepting being rather greener, resembling the coast of Patagonia.

South-eastward from Nombre Head extends a low shingle beach, forming a spit, behind which is the large Bay of San Sebastian; an excellent anchorage as respects shelter, good bottom, and easiness of access, but without wood, or a good watering place, though water may be procured.

SAN SEBASTIAN BAY is what was formerly supposed to be the entrance of the Sebastian Channel, but the non-existence of which was not proved until the *Adventure* passed it on this side in her exploration of the coast in June, 1830. The charts of this coast had, with this exception, been tolerably correct. Capt. King says:—" Having made (from the South) what I supposed to be Cape

* See also Capt. FitzRoy's remarks, in a subsequent chapter, on the passage round Cape Horn.

Sebastian, and seeing from the mast-head a large opening to the northward of it, similar to that laid down in the chart, with low distant land yet further northward, corresponding to the shores of the 'Bahia de Nombre de Jesus,' I stood on confidently, thinking how well the chart of this coast had been laid down, and regardless of the soundings decreasing as we went on. Seeing, however, from the mast-head, what seemed to be a tide ripple, 2 or 3 miles distant, I called to the boatswain, who had been much among the tide races on this coast, to ask his opinion of it; but before he could get up aloft to me, I saw it was but very low land, almost level with the sea, and what I thought the ripple was the surf on the beach. Standing on a little further, we had but 7 fathoms water, over a bottom of dark muddy sand, with bits of black slate."* The subsequent examination of the bay expunged this channel from the charts. In February, 1828, the supposed western entrance between Capes Monmouth and Valentyn, was examined (see page 14), and also found to terminate in a deep bay, with low land, which, as it afforded neither anchorage nor shelter, nor any other advantage to the navigator, was named Useless Bay. The eastern opening was named the San Sebastian Channel by Nodales, in the year 1618; he accurately describes the bay, the low land forming the bottom of which he did not see.†

Coasting along the shingle spit, the North point of the bay, the depth is not more than 10 fathoms, but it deepens suddenly near the S.E. extremity. Within the shingle point, which is steep-to, or nearly so, the bottom is uniform, but the depth gradually decreasing.

Westward of this point, called *Arenas Point*, between it and Cape San Sebastian, there is a spacious harbour, secure from all but easterly winds, which seldom blow, and never with any strength. There is no hidden danger on the North side of the bay; the shingle is steep-to, the shores of the bay shoal gradually, the bottom is clean, and the soundings are regular. On the South side, off Cape San Sebastian, it is otherwise: a shoal rocky ledge extends under water to the north-eastward, and requires a berth of 3 miles; there is no kelp upon it.

Cape San Sebastian is a bold, cliffy headland, of a dark colour; inshore of it the land rises to near 1,000 feet above the sea, and becomes more irregularly hilly. From Cape San Sebastian a short range of cliff extends, then low land, and then another small cliff, off which there is a rock above water, about a mile off shore.

Hence to *Cape Sunday* the shore is rather low, irregularly hilly, and fronted by a shingle beach. Cape Sunday is a prominent headland, of a reddish colour, rising 250 feet above the sea; the shores near it are free from danger until near Cape Peñas, near which are some dangerous rocks.

Cape Penas is not more than 100 feet above the sea; around it, to a distance of 2 miles, there are dangerous rocks; the sea generally, if not always, breaks upon them; but they should be carefully avoided, especially at night.

* Narrative, &c., vol. i. pp. 457-8.
† Relacion del Viage, &c., por B. G. y Gonzalo Nodales, p. 59.

The bay lying to the southward of Cape Peñas appears to afford anchorage; but the appearance is deceitful, it is shallow and strewed with rocks. The hills hereabout are higher, and partially wooded, and the view of the country is pleasing.

Capes Santa Inez.Medio and *San Pablo* are high and bold; they are fronted by steep cliffs, 200 or 300 feet in height. Hence to Cape San Diego there is no outlying danger; the water is rather deep near the shore, but not so deep as to prevent a ship anchoring during westerly or southerly winds.

The *Table of Orozco* is a remarkable table-topped hill, about 1,000 feet above the sea. Between it and Cape San Diego there are three remarkable hills, called the Three Brothers, and the westernmost of these hills is very like the Table of Orozco; they are from 1,000 to 1,400 feet in height.

Continuing eastward along the coast we come to *Policarpo Cove*, which was dignified by the Spaniards with the title of Port San Policarpo, but which was found by Capt. King to be so shallow an inlet, that at its entrance, just within the heads, there was not more than a fathom of water. From the mast-head it seemed like a spacious harbour. From Policarpo Cove to Cape San Vicente the distance is 12 miles, nearly East, *true*. At 5 miles from the former is False Cove, and between them are the three hills, called the Three Brothers, before mentioned.

CAPE SAN VICENTE is a rocky point, with low bluffs above it. Between the cape and Cape San Diego is San Vicente (or Thetis) Bay, a tolerable anchorage during West or southerly winds, though the bottom is rocky in many places. Between the heads, the tides run with great strength; therefore a ship should anchor off a bluff at the West side, and within the lines of the heads, when she will have from 6 to 12 fathoms of water, over a coarse sandy bottom, mixed with patches of rock.

CAPE SAN DIEGO, the eastern extreme of Tierra del Fuego, is a long, low, projecting point. It may be approached close to. There is a *rocky ledge* projecting about 2 miles from the cape, on which are shoaler soundings than nearer the cape; 5 fathoms were found in one spot on it in the *Beagle*.

From Cape San Diego the land takes a sudden turn to the South, forming the West side of the Strait of Le Maire, which will be presently described. Off the N.E. coast of Tierra del Fuego regular soundings extend for many leagues; and good anchorage may be found near the land, on any part of the coast, during westerly winds.

STATEN ISLAND,* which was surveyed in 1828, by Lieut. E. N. Kendall, of H.M.S. *Chanticleer*, is 38 miles in extent from Cape St. John to the E.N.E., and Cape St. Bartholomew to the W.S.W. The island is described as extremely mountainous and rugged, being composed of a series of lofty, precipitous hills (2,000 feet, and some 3,000 feet, in height), clothed nearly to their snowy tops with forests of evergreen beech trees, the laurel-like winter's bark, and the holly-leaved barberry: these are all evergreens; besides, there are a host of minor

* Staten Island was so named by Schouten, Jan. 25th, 1616, in honour of the States of Holland. —*Journal du Merveilleux Voyage*, p. 18.

plants. The low ground is extremely swampy and boggy, in many parts a perfect quagmire. The writer of this description says that the cold of these regions is a fable, and at variance with truth and nature. At Cape Horn, in lat. 56° S., vegetation was in full vigour in May, or the November of their year, and snow rarely lies upon the low grounds. In fact, we have sufficient matter to elucidate the climate of the South, and to establish its comparative mildness with the North, especially if America be taken as the example. The summers of the South are by no means warm or hot, nor winters cold; but to compensate for this, it is the region of wind, storms, and rain, perpetual gales and eternal rains: never 24 hours without rain. It is the court of Eolus. The barometric pressure low, the mean being 29.32 inches,—magnetic intensity low,—the winds almost westerly, —electric phenomena extremely rare.

The CLIMATE of Staten Island is remarkably humid, and very few days can be passed there, in the course of the year, without rain; and it is rather remarkable that, however fine the weather may have been in the course of the day, some rain generally falls at night. Rain, however, is frequent there in all seasons of the year, and the sky is generally overcast. Thunder and lightning are scarcely known. The temperature may be considered as equally low, and varying little throughout the year. Frost is not very severe, nor very common in winter. The weather during the summer is cool, but still humid; and, as a general characteristic, may be considered boisterous, unsettled, wet, and dull. Vegetation lingers slowly in its summer's bloom, and is not nipped by the severity of the winter's frost.

On the shore, the weather was a few degrees warmer than on board; and at night it was colder. The most retired parts of the island were not frozen. The wind is generally from the westward, 9 days out of 10, ranging from S.S.W. to N.N.W. Gales from the S.W. prevail during the summer, and from N.W. in winter. Easterly winds are most prevalent in the winter months.*

These are the outlines of the climate, to which great attention was paid on board the *Chanticleer*, with the best possible instruments.

Off the North side of Staten Island is the group of islets called *New Year Isles*. To the S.E. by S. [*S. by E.*] from the latter is an inlet named *New Year Harbour*, about half a mile broad, and extending 3 miles to the S.W., and having the depths of 30 to 45 and 20 fathoms. A cluster of islets lie in the entrance, and the passage is on the eastern side. Capt. Morrell says,† "Here you may have any depth of water, from 30 fathoms to 5, with a bottom of mud and sand. Its shores abound with wood and fresh water. Scale-fish of various sorts may be caught with hook and line, and sea-fowls shot in several directions. Fresh green celery, in its season, can be had in any quantities, together with some berries of an agreeable flavour."

Next to New Year Harbour, at half a league to the East, is *Port Cook*, a smaller inlet, wherein the late Capt. Foster erected his observatory. It is surrounded by very high land, a mountain on its western side being 2,070 feet in

* Mr. Webster:—Voyage of the *Chanticleer*, vol. i. pp. 129-30.
† Narrative of Four Voyages, &c., p. 72.

height above the level of the sea. The entrance is very narrow, and has a depth of only 6 fathoms, but within the depth increases to 16 and 20 fathoms. Lieut. Kendall, in his Memoir on Staten Island,* states that this is decidedly the harbour most eligible for a ship in want of shelter, from the considerations of its affording good anchorage at its entrance, in not too deep water, the greater regularity of the prevailing winds, and the facility of communication with the South side of the island, by means of a low isthmus separating it from *Port Vancouver*, a shoal inlet on that side.

Cape St. John is the easternmost point of Staten Island. It is high and precipitous, and a heavy tide-rip extends from it 5 or 6 miles to seaward, setting at the rate of 6 miles an hour, to the N.N.E. with flood, and S.S.E. with the ebb: but the tide sets along-shore, both on the North and South, from East to West, from 3½ or 4 to 2 or 2½ miles an hour. Off Cape St. Bartholomew, the S.W. point of the island, the tide-rip, with flood, sets to the S.W., 5 or 6 miles an hour. This tide-rip likewise is very heavy, and extends 5 or 6 miles to seaward.

St. John's Harbour lies within the promontory of St. John, on the West. It is free from danger, surrounded by high land, and its general depths are from 25 to 20 fathoms, decreasing toward either shore. From the entrance the harbour curves in a S.W. direction to the extent of 3½ miles, but is little more than half a mile broad. The hills of the promontory, on its eastern side, are 800 or 900 feet in height, and at its head on the S.W. is a remarkable elevation, now known as *Mount Richardson*.

Lieut. Kendall has described this harbour, and says, that it may be easily recognised at a distance by Mount Richardson. On nearing it a remarkable cliff, like a painter's muller, appears on the eastern shore, which is high and steep. Allowance must be made, in steering, for the set of the tide, which at all times runs rapidly across the mouth of the harbour; it is, however, less sensible when within the headlands forming the N.W. Bay, in which, in case of necessity, or to await the turn of the tide, an anchor may be dropped in from 20 to 30 fathoms. The mouth of the harbour is wide, having 25 fathoms in the centre, with a rock standing off at some distance from the western point, to which a berth must be given. The shores, with this exception, are bold, and immediately within the western point is a small bay, where anchorage may be had in 10 fathoms. The most sheltered situation is at the head of the harbour, distant 3 miles S.S.W. from the entrance, where any depth may be chosen between 20 and 5 fathoms, with sandy bottom, and moor with an open hawse to the S.W., from whence the gusts that come from the mountains are violent. The wind, anything to the westward of W.N.W., or even N.W. outside, will be found to draw out of the harbour on nearing its head; and if at all strong, it will be impossible to beat farther, as it follows the direction imparted to it by every ravine in the hill as it passes; and therefore warping will be found the only means of advancement, taking care to have hands by a bower anchor ready to let go, and the cable stoppered at a short scope, in the event of the hawsers being carried away. A ship may readily heave down on a beach of sand at the head of the harbour.

* Appendix to Mr. Webster's Narrative, p. 258.

Wood and water are plentiful, and easily procured; celery and wild-fowl (race-horse or steamer ducks, kelp and upland geese) may also be obtained; and, in the proper season, October, a good supply of penguin's eggs may be ensured by having men in attendance at a *rookery* about a mile to the eastward of the harbour's mouth, whither they could walk along the eastern hills, from the vicinity of the ***Painter's Muller***, and remain to collect daliy the eggs as deposited, and secure them until a favourable opportunity is offered of embarking them from the foot of the cliff on which the rookery is established.

The shores of St. John's Harbour are lined with kelp, which is an excellent indication of its navigable part, the border of it being almost invariably in 8 fathoms, and that close to the shore, the depth rapidly increasing toward the centre, until near the head of the harbour, where the depth gradually decreases to the beach.

Westward of Cape Colnett, or the meridian of 64° 18′, are the small harbours named ***Port Parry*** and ***Port Hoppner***; and within New Year Isles, to the West of New Year Harbour, is another, ***Port Basil Hall.*** These are of inferior consideration, but have been described by Lieut. Kendall, as given in the Appendix to Mr. Webster's Narrative, before noticed.

To the southward of Staten Island but little amount of tide is perceptible; there is, however, a remarkable undertow, which renders it dangerous for boats to stretch across the mouths of the deep bays, as it is difficult to close again with the land, for which reason the sealers invariably follow the circuitous route of the shores.

Mr. Webster, in his copious description of the vegetable productions of Staten Island, has noticed the vast masses of sea-weed which entangle the shores. The sea teems with it, especially in the rough and open bays, while it is comparatively rare in the sequestered creeks. Did it increase in the calm harbours as upon the rougher shores, they would be choked up; and it would form an impervious mesh of cords. But it thrives best in the boisterous element; and where it would seem impossible to obtain a hold, it there grows and gathers strength to meet the storms.

The STRAIT OF LE MAIRE, between Tierra del Fuego and Staten Island, was so named from the navigator who discovered it, in 1616. It is said, in the relation of Anson's Voyage, that it is difficult to determine exactly where the strait lies, though the appearance of Tierra del Fuego be well known, without knowing also the appearance of Staten Island; and that some navigators have been deceived by three hills on Staten Island, which have been mistaken for the ***Three Brothers*** on Tierra del Fuego, and so overshot the strait. But Capt. Cook says no ship can possibly miss the strait that coasts Tierra del Fuego within sight of land, for it will then of itself be sufficiently conspicuous; and Staten Island, which forms the eastern side, will be still more manifestly distinguished, for there is no land on Tierra del Fuego like it. The Strait of Le Maire can be missed only by standing too far to the eastward, without keeping the land of Tierra del Fuego in sight; if this be done, it may be missed, however accurately the appearance of the coast of Staten Island may have been exhibited; and if this be not done, it cannot be missed, though the appearance of that coast be not known.* The entrance of the strait should not be attempted but with a fair wind and moderate weather, and upon the very beginning of the tide of flood, which happens here, at the full and change of the moon, about 1 o'clock. It is always best to keep as near to the Tierra del Fuego shore as the winds will admit. By attending to these particulars, a ship may get quite through the strait in one tide; or, at least, to the southward of *Success Bay*, into which it would be more prudent to put, if the wind should be southerly, than to attempt the weathering of Staten Island with a wind and lee current, which may endanger her being driven on that island.†

The *Cape of Good Success*, in lat. 54° 55′, is the S.E. point of the Strait of Le Maire. It is high and bluff, and some rocks lie close to it, above water.

Rather more than 2 miles north-eastward of Cape Success is a projecting headland, which at first appears to be the cape; two rocky islets show themselves close to it, and, from a distance, appear like a ship under sail. Six miles from these rocks, N. $\frac{3}{4}$ E., is *Good Success Bay*, which is visible from the northern entrance of the strait. This bay is about 2 miles wide, and extends into the land westwardly $2\frac{1}{2}$ miles. It may be easily known by a peculiar mark or feature on its southern side—a barren strip of land on the height, resembling a broad turnpike road extending into the country from the shore. This mark, which is mentioned by Cook, is still a good one for the bay, if the inbend of the land does not show it sufficiently. The anchorage is good all over it, in from 4 to 12 fathoms of water, clear ground. Here a vessel lies perfectly safe, provided she does not anchor too far in, toward the sandy beach at its head; for during S.E. gales, a heavy swell, with dangerous rollers, sets right into the bay. Elevated lands, of about 1,200 feet above the sea, surround the bay; therefore, with strong winds, it is subject to squalls, which, during westerly gales, are very violent.

The *Bay of Good Success*, or *Success Bay*, is the place within which, in the year (January) 1769, Mr. (Sir Joseph) Banks and Dr. Solander found the cold so intense, that the latter had nearly fallen a sacrifice to its severity, though in the midst of summer. Dr. Solander, who had more than once crossed the mountains which divide Sweden from Norway, well knew that extreme cold, especially when joined with fatigue, produces a torpor and sleepiness which are almost irresistible; he therefore conjured the company to keep moving, whatever pain it might cost them, and whatever relief they might be promised by an inclination to rest: "Whoever sits down," said he, "will sleep; and whoever sleeps will wake no more." The doctor, who gave this advice, was the first who yielded to the sensation which he had described; but, by exertion, he was saved: two other persons perished in the snow-storm.

* Morrell says—"Some mariners have represented it to be difficult to discover the Strait of Le Maire; but I know that any navigator who keeps the land of Tierra del Fuego in sight, cannot possibly miss or mistake the strait. The only way, therefore, that such an occurrence could take place, would be by losing sight of the land, and running too far to the eastward; which should never be done, as there is no danger that can possibly arise from keeping the western shore on board. Easterly winds are never known to blow fresh in this part of the world; and by hugging the western shore the passage to the Pacific is very much shortened."—*Narrative, &c.*, p. 73.

† Cook's First Voyage, date 16th Jan. 1769. If we may judge from the varying descriptions, we may suppose that the climate has really ameliorated since 1769. Similar remarks have been made with regard to the Falkland Islands.

Mr. Darwin says: "The harbour consists of a piece of water half surrounded by low rounded mountains of clay slate, which are covered to the water's edge by one dense, gloomy forest. A single glance at the landscape was sufficient to show me how widely different it was from anything I had ever beheld. One side of the harbour is formed by a hill about 1,500 feet high, which Capt. FitzRoy has called after Sir J. Banks, in commemoration of his disastrous excursion."—(Narrative, &c., vol. iii. p. 227; 2nd edit. pp. 205, 210.)

Although this is an excellent stopping place for vessels of any size, in which they may find wood and water, it will not answer if a vessel requires to lie steady for repairs, as a swell frequently sets in. In the winter season, when easterly winds are common, no vessel should anchor so near the head of the bay as she may in summer.

The EASTERN SIDE of the STRAIT OF LE MAIRE, already noticed, is formed by the very irregular bays and rugged capes of Staten Island: surrounding the latter are heavy tide-rips, which extend outward to a considerable distance, and render a near approach very dangerous. The *Middle Cape* lies in lat. 54° 48′ 20″, and lon. 64° 42′ 30″. This, with *Cape St. Anthony*, the N.W. cape, and *Cape St. Bartholomew*, the S.W. cape, are high, bluff promontories.

The soundings in the strait are regular near the southern entrance, 70 to 30 fathoms, over a sandy bottom; toward the North the soundings diminish, and at 2 miles from Cape St. Diego are not more than 30 fathoms, over a rocky bottom. The strait is generally clear, excepting a reef discovered by Captain E. Handfield, in passing through in H.M. sloop *Jaseur*, in 1827, which lies at about 3 miles West from the Middle Cape. It appeared to be about 1½ miles in extent, and the sea broke violently on it.

The TIDES of Good Success Bay and the Strait of Le Maire are as regular and as little to be dreaded as in any part of the world where they run with strength. They will materially assist any vessel in her passage through the strait, which is very wide, perfectly free from obstacles of any kind, and has Good Success Bay close at hand, in case wind or tide should fail. When the tide opposes the wind and swell, there is always a heavy, and, for small vessels, dangerous "race" off Cape San Diego: in one spot, where the water is more shoal than elsewhere, 5 fathoms only were subsequently found at a neap flood tide; but let it be remembered, that on another day, at the top of the springs, being the day after full moon, we passed the same spot at half-flood, with the water perfectly smooth; and, although strong eddies were seen in every direction, the vessel's steerage was but little affected by them. It is high water in Success Bay soon after 4 in the afternoon on the full and change days, and low water at 10 in the morning. The flood tide stream begins to make to the northward about an hour after low water; and the ebb to the southward about the same time after high water. The tides rise from 6 to 8 feet perpendicularly. At Cape Pillar, the turn of tide, with high water, is at noon; but along the S.W. and S.E. coast the time gradually increases to this coast. From Cape San Diego the flood tide sets North and West along the shore, from 1 knot to 3 knots each hour, as far as 20 miles along the shore, and the ebb in a contrary direction, but not so strong, except in San Vicente Bay. The flood in the Strait of Le Maire runs about 2 knots in mid-channel, more or less, according to the wind, and the ebb about 1 knot an hour. Perhaps, at times, when a strong spring tide is retarded in its progress by a northerly wind, there will be a dangerous overfall off Cape San Diego, like the bores in some parts of the world.

Capt. Wilkes says: "I cannot see why there should be any objection to the passage through the Strait of Le Maire, as it gives a vessel a much better chance of making the passage round the cape quickly. A vessel with the tide will pass through in a few hours. As for the 'race and dangerous sea,' I have fully experienced it in the *Porpoise*, on the side of Staten Island; and am well satisfied that any vessel may pass safely through it at all times, and in all weathers, or if not so disposed, may wait a few hours until the sea subsides, and the tide changes. We were only three hours in passing through."—(Narrative, vol. i. p. 107.)

VALENTYN BAY is the name applied to the inlet westward of Cape of Good Success. Good Success Bay was originally called by this name, but it is transferred in the late surveys to the present, which is unfit for vessels, being exposed to a heavy swell, and affording but bad anchorage.

Between this bay and Aguirre Bay, the next to the westward, is the *Campana* or *Bell Mountain*, 2,600 feet high, and in shape resembling a large bell. It is seen far at sea from the North as well as the South.

Cape Good Success, as before stated, is high and bluff; and the land between it and the Bell Mountain is higher than that to the westward.

Aguirre Bay is unfit for a harbour except for temporary anchorage, and Spaniard Harbour, its N.W. part, proved to be a shallow bay, full of rocks, and dangerous reefs lining the shore, and without shelter, although there is anchorage for a vessel. The country on the East side seemed level, with here and there low hills, whose eastern sides are thickly covered with wood. Some of the (beech) trees might afford topmasts or lower yards for a small ship, though perhaps of unsuitable quality.

The tide is strongly felt on this part of the coast, causing races and eddies near the projecting points. In the offing the current (or tide) sets towards the Strait of Le Maire, from 1 to 3 knots an hour, when the water is rising on the shore, and the wind westerly. While the water is falling, it runs with less strength, and with an easterly wind is not felt at all.*

Westward of *Point Kinnaird* (on which great numbers of fur-seal were seen from the *Adventure*), the southern coast of the island of Tierra del Fuego trends in nearly a due West direction, through 6 degrees of longitude; the coast in some portions being of that peculiarly straight character observed in many parts of this wild region, more particularly in the Strait of Magalhaens. South of this line the outer coast is broken into numerous islands, separated from the principal island by the Beagle Channel.

* "On May 11th (1830), we passed through a very dangerous 'tide-race' off Bell Cape. There was little or no wind, but it was scarcely possible to use our oars, so much was the water agitated; it was heaving and breaking in all directions, like water boiling in an immense cauldron. When through and again in safety, I was astonished at our fortunate escape. Looking back upon it, only a mass of breakers could be seen, which passed rapidly to the westward, and therefore led me to suppose that this race was caused by a meeting of tides, not by a strong tide passing over a rocky ledge."—*Mr. Murray, Voyage of the Adventure, &c.*, vol. i. p. 447.

A range of high mountains runs almost uninterruptedly from the Barbara Channel, in lon. 72° 20', to the Strait of Le Maire. Mount Sarmiento (p. 14) is in this range, and is 6,800 feet above the sea. Mount Darwin is the same height, and is near the point where the Beagle Channel separates into two branches, diverging to the N. and S. of West. Southward of these mountains is a succession of broken land, intersected by passages or large sounds. A boat can go from the Week Islands, S.E. of Cape Pillar, the western entrance of the Strait of Magalhaens, to the eastern entrance of the Beagle Channel, without being exposed to the outside coast, or to the sea which is there found.

The BEAGLE CHANNEL was discovered by Mr. Murray in the course of the survey of this coast, April, 1830:—"The master returned, and surprised me with the information that he had been through and far beyond Nassau Bay. He had gone very little to the northward, but a long distance to the East, having passed through a narrow passage, about one-third of a mile in width, which led him into a straight channel, averaging about 2 miles or more in width, and extending nearly East and West, as far as the eye could reach. Northward of him lay a range of mountains whose summits were covered with snow. On the South side of the channel there were mountains of considerable elevation; but, generally speaking, the shore was lower than the opposite."*

The Beagle Channel is easy of access, but is useless to a ship. Boats may profit by its straight course and smooth water. It runs 120 miles, in nearly a straight line, between snowy mountains, as above stated, and averages about $1\frac{1}{2}$ miles in width, and in general has deep water; but there are in it many islets, and rocks near them.

New Island, which lies at the South side of the entrance to the Beagle Channel, was observed at a distance by Cook. Good temporary anchorage during westerly winds may be obtained under it, or near the shore to the northward.

Lennox Island, as well as New Island, and all the coast hereabout, may be approached with confidence, using the lead, and looking out for kelp. There are no shoals, but the water is not so deep as to the West of Cape Horn, neither is the land near so high.

At the East of Lennox Island is *Lennox Harbour*, a very secure place for small vessels; but, as it is rather shallow, ships drawing more than 14 feet of water should anchor outside of the entrance, where they would be safe and in smooth water, excepting when a S.E. gale blows, with which wind they, in all probability, wish to remain at anchor. The soundings are regular in the offing, and there is anchoring ground everywhere in the vicinity. Wood and water may be obtained in any quantity. Wild fowl and fish are also to be had, but not in abundance. The easiest way of getting fish is to give bits of broken glass or buttons to the natives, who catch them in the kelp.

GOREE ROAD lies between Terhalten Island and the S.E. part of Navarin Island; or rather between Lennox Island and Navarin Island. There is good anchorage in it in 6 and 7 fathoms water over a sandy bottom.

Goree Road, according to Capt. FitzRoy, is an excellent place for ships, very

* Voyages of the *Adventure* and *Beagle*, vol. i. p. 429.

easily entered or quitted, and able to furnish wood and water with as little trouble as any harbour on the coast. It should be remarked here, that the kelp in Goree Road, as well as that which extends out from Guanaco Point, partly across the entrance to the road, does not, as far as we have been able to discover, grow upon rock, but upon loose detached stones, and need not be a subject of alarm.

South of Navarin Island is NASSAU BAY, the South side of which is formed by Wollaston Island. This bay was given in former charts under the name of St. Francis Bay; but the land having been very vaguely represented, it was found to be quite distinct from that to the North of Cape Horn Island.* It extends to the North and N.W. by the Murray Narrow into the Beagle Channel. There is nothing to lead a vessel into these openings, therefore a description of them is not necessary. They may prove useful for boats; and the charts will be sufficient guides for this purpose. Nassau Bay is very accessible and free from dangers, the only ones being some rocks or islets above water, and visible by daylight. Anchorage may be found on either coast. The compasses are very sluggish here, and might cause a serious error if not attended to.

Capt. FitzRoy says: "If bound round Cape Horn from the eastward, it might be preferable to work through Nassau Bay, and stand out from False Cape Horn, instead of making westing in the open sea, as is usually done. There are no dangers but those which are shown in the chart; the water is comparatively smooth, and an anchorage may be taken at night. For this purpose, Goree Road, or North Road, or Orange Bay may be chosen."

When it blows too hard to make any way to windward, it is at least some satisfaction, by lying quiet, to save wear and tear, and to maintain one's position, instead of being drifted to leeward, and perhaps damaged by the sea in the offing. There is less current through the bay than in the offing, near Cape Horn.

The *Evouts Isles*, consisting of one principal, with several smaller islets and rocks to the South and North, lie off the mouth of Nassau Bay. They are similar to the Barnevelts, but rather higher, and the chart is a sufficient description. Within them are the Sesambre Isles.

The *Barnevelts Isles* are to the South of Evouts. They are two low islets lying nearly North and South, covered with grass, tussac, and weeds. The largest is about half a mile long, and one-third of a mile wide; the other is about two cables' length square. Several rocks lie off the South end, both to the East and West; and one above water lies detached, towards Hermite Islands, nearly in mid-channel. There is no good landing place on the islands.

The HERMITE ISLES is the group lying southward and eastward of Navarin Island and Hardy Peninsula, and of which Cape Horn is the southernmost point. The name is given from that of Admiral Hermite, commander of the Dutch fleet, who visited the coast in 1624. The principal island is *Wollaston Island*, the northernmost, separated by Franklin Sound from Hermite and Herschel Islands, and these by St. Francis Bay from Horn Island. Wollaston Island was proved by Capt. Wilkes to consist of two islands, to the western of which he gave the name of Baily, from the Vice-President of the Royal Society. A harbour

* Voyages of the *Adventure* and *Beagle*, vol. i. p. 433.

between the two was called Sea-Gull Harbour.* Deceit Island is the eastern of the group. The passages between these islands have deep water, and are free from dangers; what few rocks there are show themselves above water, or are thickly covered with kelp.

The island next West of the Barnevelts is *Deceit Island*, the southernmost point of which is *Cape Deceit* (or Enganno), the *Mistaken Cape* of Capt. Cook. It is a rocky point, and off it are several rocks all above water; 2 miles to the S.E. is a cluster rising 30 or 40 feet above the sea. Strangers should be careful not to mistake it for Cape Horn, for such mistakes have occurred, as its name imports. It lies 11 miles S.W. by W. from the Barnevelts.

CAPE HORN is the southernmost point of the southernmost of the Hermite Islands. There is nothing very striking in the appearance of this promontory as seen from a distance; but in passing near it is more remarkable, showing high black cliffs toward the South, and is about 500 feet above the sea. Its summit is in lat. 55° 58′ 40″ S., lon. 67° 16′ 10″ W.

Capt. Morrell says: "Cape Horn may be known by a high round hill over it, which has a bold and majestic appearance, being an elevated, precipitous black rock, rising above all the adjacent land. The valleys and hill sides in the neighbourhood of the cape are covered with trees, moss, and green grass; but the summits of the hills are rough and rocky."—(p. 78). No dangers exist to the southward in approaching this part; the islands may be closed without hesitation.

Cape Horn† was but little known till the late surveys. Capt. Cook said, "In some charts Cape Horn is laid down as belonging to a small island. This was neither confirmed, nor can it be contradicted by us; for several breakers appeared on the coast, both to the East and West of it; and the hazy weather rendered every object indistinct." The surveying expedition under Capt. P. P. King landed on the cape on the 20th of April, 1830, and erected a pile of stones, 8 feet high, over a memorial of their visit.

Off the East point of Horn Island are some small rocks and breakers, and one mile to the westward of Cape Horn there are three rocks generally above water; the sea always breaks on them.

Between Horn Island and Hermite Island is St. Francis Bay, formerly much misrepresented on the charts. It is clear of obstruction, and has no other dangers than those indicated in the late surveys.

A strong *current* sets, at times, along the outer coast of the Hermite Islands, and through the Bay of St. Francis. It varies from half a knot to 2 knots an hour, according to the wind and the time of tide; and, in the bay, changes its direction with the change of tide.

The land of Hermite Island and its vicinity has a most remarkable appearance when seen from the South. Its outline is a series of peaks, following each other in regular succession, and resembling the worn teeth of an old saw. Mount Hyde is made sufficiently distinct by its rounded apex, and by being higher than

* Narrative of the U.S. Exploring Expedition, vol. i. p. 145.

† "Cape Horn was named by Le Maire and Schouten, in 1616, in honour of the town of Horne, or Hoorn, in West Friesland, on the Zuyder Zee, near Amsterdam, of which the patron was a native."—*Navig. Austral. de Le Maire: Burney*, vol. ii. p. 130.

any land near it. Kater's Peak, 1,742 feet above high water level, is also very remarkable in this view, from its conical form and very pointed summit, and from being situated at the eastern end of the island.*

WIGWAM or ST. MARTIN'S COVE, on the eastern side of Hermite Island, westward of Cape Horn, has been described by Capt. Foster, who states that it bears from the Cape W.N.W. ¾ W. [*N.W. ½ W.*] about 10 miles, and is a place of easy access with N.E., East, and S.E. winds. It is open to the East, and may be readily found by means of *Chanticleer Island*, which lies about a mile *true* East from the South head of the entrance. With westerly winds, which are adverse and prevalent, vessels should anchor off the entrance, in about 22 fathoms, and warp into the cove, where there is a convenient berth in 18 fathoms, sandy bottom, midway from either side, and about half a mile from the head of the cove. This anchorage is safe, although the gusts of wind in westerly gales, which are of frequent occurrence at all seasons of the year, rush down the sides of the mountains in various directions with impetuous violence, and may be very properly called *hurricane squalls* (williwaws). They strike the ship from aloft, and have more the effect of heeling the vessel than of bringing a strain upon the anchors, which, when once imbedded in the sandy bottom, hold remarkably well, and will cost a heavy heave in weighing.

Wood and water abound in every part of the cove, but cannot always be procured, from the steepness of the shores, and the heavy swell that sometimes sets in. The water is highly coloured by the vegetable matter through which it percolates; but no other inconvenience from its use was found than that of imparting to tea a deeper colour, and somewhat unpleasant flavour. The wood was very much twisted and stinted in growth, and did not seem fit for any other purpose than fuel.

The shores of the cove are skirted with kelp, which serves to protect the boats in landing, and amongst which fish also are to be caught with a hook and a line, abreast of the rills of fresh water that discharge themselves into the sea. From the natives was obtained a knowledge of this most valuable supply, by observing them in the act of fishing, which is ingenious: they have a line, and to the end is fastened a limpet, which the fish eagerly swallow, and not being able whilst in the water to disgorge it, are thereby drawn to the surface, and taken by the hand. In this manner they have been known to catch several dozen in the course of a few hours: but I am induced to believe that it is only in the summer months of these regions that supplies of so salutary a nature can be procured. The wild fowl that are most palatable consist of geese and race-horses, called *steamers* by the sealers. Both sorts are well tasted, and were found agreeable. They were generally seen among the kelp in the cove about daybreak, but soon afterward would depart for their daily places of resort.

At the head of the cove, and a few feet beyond the reach of high water, spring tides, abundance of celery is to be found, as also in many other places in

* The survey of this point now presents the navigator with the means of ascertaining his position to a nicety, by angles taken with a sextant between Cape Horn summit and Jerdan's Peak, or Mount Hyde and Kater's Peak; and if Jerdan's Peak and Mount Hyde be brought in a line, and an angle taken between them and Cape Horn summit, the operation will be still more simple.

the cove. During two months of the latter part of the autumnal season, a sufficiency was daily procured for the use of the ship's company, and although of not so luxuriant a growth as in December, it was, nevertheless, considered wholesome. Lat. 55° 51′ 20″, lon. 67° 34′. Variation 24° E. High water, full and change, 3^h. 50′. Rise about 8 feet. It appeared that the flood came in from the southward.

St. Joachim's Cove, to the southward of St. Martin's Cove, is more exposed than the latter, but is of easier depth. These coves are separated from each other by a steep and precipitous mass of hills of greenstone, which in many parts appears to be stratified, the dip being to the westward, at an angle of 40°. The whole surface of the hill is covered with stunted beech bushes, thickly matted, and interwoven with each other.

Temporary anchorage may be had in the small bay leading to St. Joachim's Cove, or under the South head of St. Martin's Cove, where you will find from 20 to 25 fathoms, over a clear, sandy bottom.

PORT MAXWELL is contained between Jerdan Island, Saddle Island, and a third island, forming a triangle. It has four entrances; only two of them are fit for vessels—those to the North and East; the principal one being to the North of Jerdan Island. The best berth in it is in 16 fathoms water, over a sandy bottom. This harbour is decidedly good, though it requires a little more time and trouble in the approach. It is rather out of the way, but is perfectly secure, and untroubled by mountain squalls or williwaws.

The summit of *Saddle Island* is composed of large blocks of greenstone rock, the ferruginous nature of which has a very remarkable effect upon the compass, as, indeed, is found to be the case in many parts near these islands.* This island, like the others near it, is clothed with low stunted brushwood of beech, berberis, and arbutus; and on its shores kelp-fish, a very delicate and wholesome fish, may be caught.

Cape Spencer is the southernmost point of Hermite Island. It protected the surveying vessel very well, both from wind and sea, during an anchorage there. Should a ship wish to enter St. Martin's Cove, and the wind or daylight fail her, she will find this spot a convenient stopping place. The West point of Hermite Island is low; the land at the opposite end of the island high and rugged.

FALSE CAPE HORN is a very remarkable headland. From the East or West it looks like a large horn. It is a leading mark to the best anchorage on this coast, Orange Bay. It is the S.E. point of Hardy Peninsula.†

ORANGE BAY is excellent, on the eastern side of the peninsula, and one of the few on this coast which are fit for a squadron of line-of-battle ships. Its approach from the sea is as easy as the harbour is commodious. There are 3 fathoms close to the shore, yet in no part are there more than 20; and every-

* The block of stone upon which the compass used by Capt. King was placed (and which is now in the museum of the Geological Society) had the effect of causing a deviation of 127°, and in another instance, in Port Maxwell, the poles of the needle became exactly reversed. The same was observed by Weddell, and by the surveyors at Kendall Harbour.

† Strangers should be careful not to mistake this cape for Cape Horn, for such mistakes have occurred, as the name imports. Off the cape are several rocks, all above water; and at 2 miles to the N.E. is a cluster rising 30 or 40 feet above the sea.

where there is a sandy bottom. Water is abundant, and wood grows close to the sea; wild fowl are numerous; and although shell-fish are scarce, plenty of small fish may be caught with a hook and line among the kelp, and in summer a sieve will furnish abundance. This account is also confirmed by Capt. Wilkes (*Narrative of the United States Exploring Expedition*, vol. i. p. 159).

To anchor in this bay you must pass to the eastward of the False Cape, as close as you please. Steering *N.E.* (*true*) for 4 miles, will bring you abreast of Point Lort; a bay 2 miles wide is then opened, in which you may anchor, if necessary, in 8 or 10 fathoms, over a fine sandy bottom. Some rocks above water lie at the North side. Beyond the point which forms the North side of this bay, is a small cove, with 18 fathoms water in the middle; beyond it is another cove larger, after which you open Schapenham Bay (so called by the Nassau fleet). A North course (*true*) from Point Lort will take you abreast of Orange Bay.

Schapenham Bay is 1½ miles wide; at its head is a large waterfall, marking a rocky bottom, covered with kelp. It is not recommended to anchor in this bay.

The land behind those coves that have been mentioned is high and rugged; two singular peaks show themselves, which resemble sentry boxes. Near the shore the land is low, compared with other parts of the coast.

From the heights sudden and very strong squalls blow during westerly winds. Being generally a weather shore, and regular soundings extending along it, there is no difficulty in choosing or approaching an anchorage.

Off Orange Bay anchor soundings extend to 2 miles off the land. The opening of the bay is 3 miles wide, and in that part are 18 or 20 fathoms, over fine speckled sand. Two islands, the larger having a smooth, down-like appearance, lie in the middle; behind them is the harbour, a square mile of excellent anchorage, without a single rock or shoal. The bottom everywhere is a fine speckled sand. You may go close to the shore in every part. The best watering place is a small cove at the North side, called Water Cove.

Off the North point are several small islets, which must not be approached too closely; they are, however, out of the way. Six miles N.N.W. of the outer anchorage is a curious island, like a castle or a pack-saddle. Orange Bay is somewhat open to East winds, but they seldom blow strong, and would be fair for ships bound westward. No sea can be thrown in, because of the Hermite Islands. There is no current here worthy of notice. The tide rises 6 feet; high water at half-past 3.

DIEGO RAMIREZ ISLANDS, discovered by the Nodales, in 1619, and so named after their head pilot, are a cluster of great barren rocks, 18 leagues south-westward from Cape Horn, which extend N.W. and S.E., 4 or 5 miles. The channel between is entirely clear.

There are three principal isles, and many rocks above water. The centre isle is the largest; it has tussac upon it, but neither wood nor water, and is frequented by various oceanic birds. The second in size has a shingle beach, where a boat may be hauled up in safety; and there is enough good water on the East side of the same island to supply 30 men. A furious surf breaks on the West shore, and sends a spray over the whole island. There is no sheltered anchorage for a vessel. The westernmost rock is the highest, and is surrounded by several small rocks, sufficiently elevated for birds and seals. Around the rocks the water is bold within a cable's length: and in clear weather a ship may safely run for them in the night, by keeping a good look-out. The highest point is about 200 feet above the sea. They are quite similar to the Ildefonsos: the top of a ridge of hills showing above the water, and broken through by the sea.

On the East side is a depth of 30 fathoms, with a bottom of fine green sand. The tide of flood here runs to the N.E., and apparently to the eastward, among many of the main islands.

Capt. Colnett says that, in general, the birds hereabout resemble the dun crow, common in Hampshire in the winter, and which had been seen daily from the parallel of the Falkland Islands.

St. ILDEFONSO ISLES.—These are a group of rugged islets and rocks, above water, bold-to, and within which Capt. Cook passed to the eastward, in December, 1774. They extend 5 miles in a N.W. and S.E. direction, are very narrow, and the highest and largest is about 200 feet above the level of the sea. They have been much frequented by the sealers. Their distance from the nearest point of the main is about 20 miles. The passage between them and Diego Ramirez is 35 miles wide, and entirely free from danger.

Capt. Weddell says that the largest isle is not more than a quarter of a mile long. On a N.W. or S.E. bearing, the whole appear as two islets only; but the northern one is merely a cluster of detached rocks; the southern islet is the largest and highest, and contains a quantity of tussac on its top, and sea-gull rookeries. The Isles have no beaches, and can be landed on only when the water is very smooth. Between them is a channel of a mile wide, which, being rocky, should not be used.

Bourchier Bay lies between False Cape and New Year Sound. It offers nothing inviting for ships, being a leeward bight, with rocks and islets scattered near the shore.

NEW YEAR SOUND, &c.—To the northward of the Isles of St. Ildefonso, the coast of Tierra del Fuego forms the large inlet or strait called *New Year Sound*, for a knowledge of which and its harbours the public is especially indebted to Capt. Weddell, who has given a plan of it in his useful volume. Opposite to the entrance is an extensive group of isles, of which the central and largest is *Henderson Isle;* on the East is a smaller, *Sanderson;* and on the West, *Morton*, with a number of islets. A remarkable hill, *Mount Beaufoy*, on Henderson Island, stands in lat. 55° 36′ 15″, and lon. 68° 58′. On the N.W. side of Morton Isle is a harbour named *Clearbottom Bay;* and at 5 leagues above this, on the western side of the inlet, upon the main, is another harbour, named *Indian Cove;* but the last, Capt. King says, cannot be recommended, as a vessel must go far among the islands to reach it, and when there have a bad rocky bottom, with deep water, one corner only excepted. For a description of the inhabitants hereabout, see *Weddell*, pp. 172—184.

Clearbottom Bay, on the N.W. side of Morton Island, is an anchorage which, by being close to the coast, is convenient for a vessel to touch at for wood and water. In order to gain this place from sea, bring the easternmost island of St. Ildefonso S. ½ E., and steer N. ½ W. for *Turn Point*, the western point of an islet on the parallel West of the harbour. About 1½ miles to the E.N.E. of this point is the anchorage; and, at the distance of 3 cables' length from the shore, in 22 fathoms of water, bottom of sand and clay, is the most eligible berth for anchoring.*

Ross Sound and *Trefusis Bay*, westward of New Year Sound, do not afford anchorage.

Hope Island (of Capt. Weddell) lies off the Wood Islands, and is 6 miles S.E. from York Minster. There is no good anchorage among the Wood Islands. Passages and broken land lie behind them to the northward. Off Point Nativity are two islands and an outlying rock.

CHRISTMAS SOUND.—Capt. Cook, on his return from his second voyage, December, 1774, entered here; we quote his words:—"In standing in for an opening which appeared on the East side of York Minster, we had 40, 37, 50, and 60 fathoms of water, a bottom of small stones and shells. When we had the last soundings, we were nearly in the middle between the two points that form the entrance to the inlet, which we observed to branch into two arms, both of them lying in nearly North, and disjoined by a high rocky point. We stood for the eastern branch (Christmas Sound), as being clear of islets; and, after passing a black rocky one (Black Rock), lying without the point just mentioned, we sounded, and found no bottom with a line of 170 fathoms. This was altogether unexpected, and a circumstance that would not have been regarded if the breeze had continued; but at this time it fell calm, so that it was not possible to extricate ourselves from this disagreeable situation. Two boats were hoisted out, and sent ahead to tow; but they would have availed little, had not a breeze sprung up about 8 o'clock at S.W., which put it in my power either to stand out to sea or up the inlet. Prudence seemed to point out the former; but the desire of finding a good port, and learning something of the country, getting the better of every other consideration, I resolved to stand in; and, as night was approaching, our safety depended on getting to an anchor. With this view we continued to sound, but always had an unfathomable depth.

"Hauling up under the East side of the land which divided the two arms, and seeing a small cove ahead, I sent a boat to sound; and we kept as near the shore as the flurries from the land would permit, in order to be able to get into this place, if there should be anchorage. The boat soon returned, and informed us that there were 30 and 25 fathoms of water a full cable's length from the shore; here we anchored in 30 fathoms, the bottom sand and broken shells; and carried out a kedge and hawser to steady the ship for the night.

"The morning of the 21st was calm and pleasant. After breakfast I set out with two boats to look for a more secure station. We no sooner got round or above the point under which the ship lay, than we found a cove, in which was anchorage in 30, 20, and 15 fathoms, the bottom stones and sand. At the head of the cove were a stony beach, a valley covered with wood, and a stream of

* There is a considerable tide between Morton Isle and the point next to Gold-dust Isle. The flood comes from the westward, about 1 knot, or at times 2 knots, an hour. With the ebb it is nearly slack water, or perhaps there is a slight tendency towards the West; and such appears to be the case all along this coast from Christmas Sound.—*Capt. King*, p. 421.

fresh water; so that there was everything we could expect to find in such a place, or rather more; for we shot three geese out of four that we saw, and caught some young ones, which we afterwards let go.

"After discovering and sounding this cove, I sent Lieut. Clerke, who commanded the other boat, on board, with orders to remove the ship into this place, while I proceeded farther up the inlet. I presently saw that the land we were under, which disjoined the two arms, as mentioned before, was an island, at the North end of which the two channels united. After this I hastened on board, and found everything in readiness to weigh, which was accordingly done, and all the boats sent ahead to tow the ship round the point. But, at that moment, a light breeze came in from the sea, too scant to fill our sails, so that we were obliged to drop the anchor again, from fear of falling upon the point, and to carry out a kedge to windward. That being done, we hove up the anchor, warped up to and weighed the kedge, and, proceeding round the point under our stay-sails, there anchored with the best bower in 20 fathoms, and moored with the other bower, which lay to the North, in 13 fathoms. In this position we were shut in from the sea by the point above mentioned, which was in one with the extremity of the inlet to the East.* Some islets off the next point above us covered us from the N.W., from which quarter the wind had the greatest fetch, and our distance from the shore was about one-third of a mile.

"Thus situated, we went to work, to clear a place to fill water, to cut wood, and to set up a tent for the reception of a guard, which was thought necessary, as we had already discovered that, barren as the country is, it was not without people, though we had not yet seen any. Mr. Wales also got his observatory and instruments on shore; but it was with the greatest difficulty he could find a place of sufficient stability, and clear of the mountains, which everywhere surrounded us, to set them up in; and at last he was obliged to content himself with the top of a rock not more than 9 feet over.

"Next day I sent Lieuts. Clerke and Pickersgill, accompanied by some of the other officers, to examine and draw a sketch of the channel on the other side of the island; and I went myself in another boat, accompanied by the botanists, to survey the northern parts of the Sound. In my way I landed on the point of a low isle, covered with herbage, part of which had been lately burnt; we likewise saw a hut, signs sufficient that people were in the neighbourhood. After I had taken the necessary bearings, we proceeded round the East end of Burnt Island, and over to what we judged to be the main of Tierra del Fuego, where we found a very fine harbour, encompassed by steep rocks of vast height, down which ran many limpid streams of water; and at the foot of the rocks some tufts of trees, fit for little else but fuel.

"This harbour, which I shall distinguish by the name of *Devil's Basin*, is divided, as it were, into two, an inner and an outer one; and the communication between them is by a narrow channel, 5 fathoms deep. In the outer basin I found 13 and 17 fathoms of water; and in the inner, 17 and 23 fathoms. This last is as secure a place as can be, but nothing can be more gloomy. The vast

* Adventure Cove (in which Cook anchored) is the easiest of access, but it will only hold one vessel.—*Capt. King.*

height of the savage rocks which encompass it deprived great part of it, even on this day, of the meridian sun. The outer harbour is not quite free from this inconvenience, but far more so than the other; it is also rather more commodious, and equally safe. It lies in the direction of North, 1¼ miles distant from the East end of Burnt Island. I likewise found a good anchoring place a little to the westward of this harbour, before a stream of water that comes out of a lake or large reservoir, which is continually supplied by a cascade falling into it.

"Leaving this place, we proceeded along the shore to the westward, and found other harbours, which I had not time to look into. In all of them is fresh water, and wood for fuel; but, except these little tufts of bushes, the whole country is a barren rock, doomed by nature to everlasting sterility. The low islands, and even some of the higher, which lie scattered up and down the Sound, are indeed mostly covered with shrubs and herbage, the soil of black rotten turf, evidently composed, by length of time, of decayed vegetables.

"I had an opportunity to verify what we had observed at sea, that the sea-coast is composed of a number of large and small islands, and that the numerous inlets are formed by the junction of several channels; at least, so it is here. On one of these low islands we found several huts, which had lately been inhabited; and near them was a good deal of celery, with which we loaded our boat, and returned on board at 7 o'clock in the evening. In this expedition we met with little game; one duck, three or four shags, and about that number of rails, or sea-pies, being all we got. The other boat returned on board some hours before, having found two harbours on the West side of the other channel; the one large and the other small, but both of them safe and commodious; though, by the sketch Mr. Pickersgill had taken of them, the access to both appeared rather intricate.

"Having fine pleasant weather on the 23rd, I sent Lieut. Pickersgill in the cutter to explore the East side of the Sound, and went myself in the pinnace to the West side, with an intent to go round the island under which we were at anchor (and which I shall distinguish by the name of *Shag Island*), in order to view the passage leading to the harbours Mr. Pickersgill had discovered the day before, on which I made the following observations: In coming from the sea, leave all the rocks and islands lying off and within York Minster on your larboard side; and the black rock, which lies off the South end of Shag Island, on your starboard; and, when abreast of the South end of that island, haul over for the West shore, taking care to avoid the beds of weeds you will see before you, as they always grow on rocks, some of which I have found 12 fathoms under water; but it is always best to keep clear of them. The entrance to the large harbour, or *Port Clerke*, is just to the North of some low rocks lying off a point on Shag Island. This harbour lies in *W. by S.* (*true*), 1½ miles, and has in it from 12 to 24 fathoms depth, wood and fresh water.* About a mile without, or to the southward of Port Clerke, is, or seemed to be, another, which I did not examine. It is formed by a large island, which covers it from the South and East winds.

* Capt. King says Port Clerke is a bad place for any vessel, though quite secure when in it: access is difficult, and, from its situation, it is exposed to very violent squalls. Pickersgill Cove (named by Cook), as well as Port Clerke, is unworthy of notice as an anchorage.

[It is the March Harbour of Capt. King.] Without this island, that is, between it and York Minster, the sea seemed strewed with islets, rocks, and breakers.....

"The festival which we celebrated at this place occasioned my giving it the name of Christmas Sound. The entrance, which is 3 leagues wide, is situated in lat. 55° 27′ S., lon. 70° 16′ W.; and in the direction of *N.* 37° *W.* (*true*) from the Isles of St. Ildefonso, distant 10 leagues. These isles are the best land-mark for finding the Sound. York Minster, which is the only remarkable land about it, will hardly be known by a stranger from any description that can be given of it, because it alters its appearance according to the different situations it is viewed from. Besides the black rock which lies off the end of Shag Island, there is another, about midway between this and the East shore. A copious description of this sound is unnecessary, as few would be benefited by it. Anchorage, tufts of wood, and fresh water, will be found in all the coves and harbours. I would advise no one to anchor very near that shore for the sake of having a moderate depth of water, because there I generally found a rocky bottom.

"The refreshments to be got here are precarious, as they consist chiefly of wild-fowl, and may probably never be found in such plenty as to supply the crew of a ship; and fish, so far as we can judge, is scarce. Indeed, the quantity of wild-fowl made us pay less attention to fishing. Here are, however, plenty of mussels, not very large, but well tasted; and very good celery is to be met with on several of the low islets, and where the natives have their habitations."

March Harbour was so named by Capt. King, who passed the month of March, 1830, here in the *Beagle*, and built a boat. This harbour might be useful to other vessels, its situation being well pointed out by York Minster, and affording wood and water with as little trouble as any place in which the *Beagle* had anchored. The harbour is large, with good holding ground, but there are many rocky places; and one rock under water, having only one fathom on it, marked by very thick kelp. The *Beagle* worked through the narrow passage, round Shag Island from Adventure Cove, and into the innermost corner of the harbour without using a warp; larger vessels would, of course, be more confined, and no vessel of above 500 tons should attempt to enter Christmas Sound. The *Beagle* lay moored in perfect safety here, but her chain cables became entangled with the rocks, and were not hove in without much difficulty and delay.

Waterman Island is soon known by the remarkable heights at its South part. The southernmost was named, by Capt. Cook, York Minster. "This lofty promontory, viewed from the situation we were now in, terminated in two high towers; and within them a hill shaped like a sugar-loaf. This wild rock, therefore, obtained the name of *York Minster.*" Capt. King says (1830), "I fancied that the high part of the Minster must have crumbled away since Cook saw it, as it no longer resembled 'two towers,' but had a ragged, notched summit, when seen from the westward.

"The promontory of York Minster is a black, irregularly shaped, rocky cliff, 800 feet in height, rising almost perpendicularly from the sea. It is nearly the loftiest, as well as the most projecting part of the land about Christmas Sound, which, generally speaking, is not near so high as that further West, but it is very barren. Granite is prevalent, and I could find no sandstone. Coming from the westward, we thought the height about here inconsiderable; but Cook, coming from the South Sea, called them 'high and savage.' Had he made the land nearer the Barbara Channel, where the mountains are much higher, he would have spoken still more strongly of the wild and disagreeable appearance of the coast."*

The *Capstan Rocks* are 8 miles West of York Minster, and 5 from Point May. They are above water about 20 feet. There are no other dangers to seaward of a line from York Minster to the Philip Rocks. These lie off Cape Alikhoolip, and are dangerous, though above water, because so far from shore, and so low.

Cook Bay is the large space between Waterman Island and Cape Alikhoolip. Broken land, islets, and breakers, surround and make it unfit for the approach of vessels. The entrance to the Beagle Channel (page 52) is in its N.E. part.

Treble Island, 9 miles N.W. from the Philip Rocks, is remarkable, having three peaks, visible from a considerable distance. Near it are some straggling rocks. Northward of this is Adventure Passage, clear of danger.

Gilbert Isle (or rather Islands), on the opposite side, was so named by Cook, from his master. It is nearly of the same height with the rest of the coast, and shows a surface composed of several peaked rocks, unequally high. At the north-eastern side of the eastern Gilbert Isle is Doris Cove, a safe anchorage for a small vessel; the *Beagle* lay here moored for a week.† To the West of the western island are the Nicholson Rocks.

Cape Castlereagh, the western end of the Stewart Islands, is a high and remarkable promontory. Under it is an excellent anchorage, called Stewart Harbour. Having three outlets, it may be entered or quitted with any wind, and without warping. Wood and water are as abundant as in any other Fuegian harbours. The general depth is 6 to 12 fathoms, the greatest 16 fathoms. Two rocks, just awash at high water, lie nearly in the middle; and a rock, on which the sea breaks, lies 1 mile West of the middle opening of the harbour. These are the only dangers.

Cape Desolation is the next promontory in passing along the coast. It was so named by Cook, "because near it commenced the most desolate country I ever saw." It is the South point of Basket Island. It is very remarkable, rugged, and with many peaks.

Leading north-westward from this is Brecknock Passage, which Capt. King prefers for entering or leaving the Barbara Channel (to or from the Strait of Magalhaens), rather than by passing the Fury Rocks.

London Island is one of a large group, called the *Camden Islands*. At its East end is a safe anchorage, called *Townshend Harbour*. The *Horace Peaks* point out its situation. Some rocks, on which the sea breaks violently, lie off the

* Voyages of the *Adventure* and *Beagle*, vol. i. pp. 407, 411.

† Capt. Cook says: "I have before observed that this is the most desolate coast I ever saw; it seems entirely composed of rocky mountains, without the least appearance of vegetation. These mountains terminate in horrible precipices, whose craggy summits spire up to a vast height, so that hardly anything in nature can appear with a more barren and savage aspect than the whole of this country. The inland mountains were covered with snow, but those on the sea-coast were not. We judged the former to belong to the main of Tierra del Fuego, and the latter to be islands so ranged as apparently to form a coast."

islands, and near the entrance of *Pratt Passage*. As there are no soundings in less than 50 fathoms after passing these rocks and getting into the passage, you must depend on the wind lasting to carry you into or out of the harbour. The holding ground is excellent in it; and though you have tremendous squalls off the high land to the westward, there is no fear of an anchor starting.

Between London Island and Fury Island is the entrance of the Barbara and Cockburn Channels. Rocks show themselves in every direction; the two clusters called *East* and *West Furies* being the most remarkable, and these, with the others, are much frequented for fur-seal at times.

The situation of these rocks is accurately given on the chart, but no vessel should attempt to pass them without daylight and clear weather; she must sail more by a good eye at the mast-head than by any chart.

Fury Island lies on the North side; on its South side is *Fury Harbour*, a bad place, unfit for any vessel. At its North side is a perfectly safe and snug anchorage, called North Cove, fit only for small vessels.

The entrance to the Barbara Channel is pointed out by four remarkable mountains. The *Kempe Peaks* are high and show their points. The *Fury Peaks* are high and divided. Mount Skyring, also described on page 20, is most barren, high, and has a single peak.* St. Paul's is similar to, and, in one view from near Fury Island, appears very like the dome of the cathedral whose name it bears.

Lieut. Skyring (H.M.S. *Beagle*) says: "From a summit on Bynoe Island an extensive view was obtained of the islands in Melville Sound, as well as of the entrance to the Cockburn and Barbara Channels. Such a complicated mass of islands and rocks I never before saw; to lay them all down correctly would occupy a long time. Sufficient, however, was done to take the navigator through this labyrinth; but I am well aware that very much is wanting to complete the survey."

In the Cockburn Channel the flood tide sets to seaward, but it was not found to be of consequence to a vessel in working through. The rise and fall are not more than 6, or at most 8 feet, at spring tides.

Between Kempe Island and Noir Island is the *Milky Way*, a span of sea in every part of which rocks are seen just awash with, or a few feet above, the waters; on them the sea continually breaks. It is not advisable to pass inshore of these. No chart or direction can guide any vessel here. The same observation applies to the space as far as Cape Schomberg, on London Island; daylight and a good look-out can be the only guides.

Noir Island is moderately high, about 600 feet, having a remarkable neck of land to the S.W., ended by a rock like a steeple or tower. One mile South

* Mount Skyring is on the East end of the largest of the Magill Islands. It was ascended in the course of the survey (May, 1829). Lieut. Skyring says:—"We gained the summit after three hours' hard travelling. During the last 500 feet of ascent the mountain was almost precipitous, and we had the utmost difficulty in passing the instrument from hand to hand. Its formation is remarkable, although, I believe, the same structure exists throughout the hills around. The base is a coarse granite, but this solid formation cannot be traced half the height; above is an immense heap of masses of rock, irregularly and wonderfully thrown together, many loose fragments overhanging with apparently very little hold. This station was the most commanding we had chosen during the survey. A document referring to the survey was deposited on its summit."

of this point is a sunken rock, over which the sea occasionally breaks; two other breakers are in the bight close to the point.*

Capt. Cook says: "At 3 o'clock we passed Cape Noir, which is a steep rock, of considerable height, and the S.W. point of a large island, that seemed to lie detached, a league, or a league and a half, from the main land. The land of the cape, when at a distance from it, appeared to be an island disjoined from the other; but, on a nearer approach, we found it connected by a low neck of land. At the point of the cape are two rocks, the one peaked, like a sugar loaf, the other not so high, and showing a rounder surface; and *S. by E.* (*true*), 2 leagues from the cape, are two other rocky islets."

There is an excellent roadstead under the East side of Noir Island. Several ships may lie there, secure from all winds between North and South by West, over a clear sandy bottom. Wood and water plentiful, and easily obtained. There is a cove at the South part of the island, where boats would be perfectly safe in any weather, but the entrance is too narrow for vessels of any kind. Position of Cape Noir, lat. 54° 30′, lon. 73° 5′ 40″. Variation, 25° East, the maximum of variation on this coast.

The Tower Rocks, two in number, 1½ miles apart, above alluded to by Capt. Cook, are steep-sided, high, and steep-to; a ship may pass close to them.

All the space between Noir Island and the coast is extremely dangerous for shipping, being scattered over with very numerous rocks; still there is room to go round Noir Island, and a vessel need not fear being hampered by an East wind, in the event of anchoring there.

The *Agnes Islands* and *Stokes's Bay* do not require description. No vessel ought to entangle herself in these labyrinths; in thick weather she would be in a most precarious situation.

The *Grafton Islands*, which follow next in succession, are high, and similar in character to the rest of the coast. The *Wakefield Passage*, at the back of them, has been used by a sealer, and the land beyond is broken into islets and rocks. The Grafton Islands extend about 20 miles in a N.W. direction from Isabella Island to Cape Gloucester. Between them are several anchorages, but the best is Euston Bay, between Ipswich Island and Cape Gloucester.

Hope Harbour, at the East end of the group, is one of those formerly used by the sealers.

Euston Bay is one of the best anchorages on this coast; one which can be approached and left with any wind, without risk, and in which a fleet might lie in perfect security from all but the S.E. winds, the least prevalent of any on this coast. Cape Gloucester (presently described) is a guide to it. Passing this cape, from the northward, you see a high island to the S.E., distant 7 miles; this is Ipswich Island; rounding this, you must give a good berth to the sunken rocks, 1 mile from its S.E. extremity, upon which the sea generally, but not always, breaks. After clearing them, pass close to Leading Island, and steer for

* The island itself is narrow and long, apparently the top of a ridge of mountains, and formed of sandstone (perhaps clay-slate), which accounts for the bottom near it being so good, and for the needle-like appearance of the rocks at the West end, as the sandstone, being very soft, is continually wearing away by the action of the water. It is the resort of large flocks of penguins.

the opening of *Laura Basin*, which you will see under a high, peaked mountain, and choose your berth by the eye. A large patch of kelp across the mouth of the harbour was carefully examined; the least depth found was 4 fathoms. Laura Basin has water enough for a frigate, but is better suited for a small vessel; large ships should anchor in the bay.

No place could be more convenient than this for such purposes as wooding and watering. Water casks can be filled in the boat, in perfectly smooth water, and wood cut close to the water's edge.*

Cape Gloucester is a very remarkable promontory, and cannot be mistaken. At a distance it appears to be a high, detached island; but on a nearer approach a low neck of land is seen, which connects it with Charles Island, the largest of the Grafton Islands. A rock (on which the sea breaks) lies nearly 1 mile to the N.W., and is the only danger. The cape is steep-to, and may be passed quite close.

Cape Gloucester shows a round surface of considerable height, and has much the appearance of being an island. It lies *S.S.E. ½ E. true*, distant 17 leagues from Landfall Island. The coast between forms two bays, strewed with rocky islets, rocks, and breakers. The coast appeared broken, with many islets, or rather it seemed to be composed of a number of islands. The land is very mountainous, rocky, and barren, spotted here and there with tufts of wood. From Cape Gloucester, off which lies a small rocky island, the direction of the coast is nearly *N.W. true*.—(Capt. Cook.)

Breaker Bay lies between Cape Gloucester and the Fincham Islands. "A worse place for a ship could scarcely be found; for, supposing thick weather to come on when in the depth of the bay, she would have lurking rocks and islets just awash with the water on all sides of her, and no guide to take her clear of them, for soundings would be useless, and in such weather the best chart that could be constructed could not help her. The land at the bottom of the bay appeared to be distant, and much broken. Indeed, from the Week Islands to Cape Gloucester (and thence to the Strait of Le Maire) there is an almost innumerable succession of islands and rocks, without any continued tract of land, so that channels might be found in all directions, valuable, no doubt, to Fuegians in their canoes, but not often to seamen in ships, or even to sealers; for where the natives go with their canoes, seals are never found in any numbers."

The *Fincham Islands* form the N.W. point of Breaker Bay. There are many islets and rocks near. There is no good anchorage, and the coast is very dangerous, and unfit to be approached.

Deepwater Sound runs into the land N.E. of the Fincham Islands. The *Beagle* entered here in December, 1830, and was obliged to anchor on the end of a steep-sided islet, with both anchors, one in 7, the other in 10 fathoms, on the rocks, with 40 fathoms under the stern, and no better could be found: therefore it ought to be avoided.

Cape Tate is the extremity of a mountain, the S.E. point of Otway Bay. It is rather high, and rounded at the summit; off it, to the North and West, lie the *College Rocks*. Those nearest the cape are also nearest the track of a ship

* King, vol. i. pp. 375-6.

running along the land; and half a mile West of them lies a detached and dangerous rock, under water. The sea generally breaks on it.

Otway Bay is the space comprised between the *Rice Trevor Islands*, of which *Cape Tate* is the S.W. extreme, and the Landfall Islands to the North. It is an extensive space of water, surrounded by broken land, islets, and rocks. Many of the latter are scattered about, and render it unfit for a vessel. It is probable that more than one passage exists hereabouts leading from Otway Bay to the Strait of Magalhaens, as deep inlets run in that direction as far as the eye can reach, but were not examined for want of time.

Landfall Island was so named by Capt. Cook, in December, 1774. He made the land here on his return from his second voyage. There are two principal islands, separated by a narrow channel; they are high, and towards the sea are barren; but the sides of the hills, towards the East, are thickly wooded.

Cape Schetky, the S.E. point, is a remarkable double-peaked height; some rocks, just awash, lie off it, distant 1 mile.

Cape Inman (so named in compliment to the Professor), the western point, is high, with perpendicular cliffs, and almost detached from other land; so that a vessel, knowing her latitude within 5 miles of the truth, cannot fail to make it out, if the weather be tolerably clear.

Latitude Bay, a good anchorage, though somewhat exposed to the swell thrown by heavy N.W. winds, lies on the North side of the larger island, at the East end of the opening, which separates it from the smaller. The bottom shoals gradually from 20 to 5 fathoms, over fine sand, and is sheltered from West winds and others (except North). It is remarkably easy of access, and is also easy to leave, —rather rare qualities in a Fuegian harbour. Cape Inman, being prominently situated, is a good guide to the harbour, which can be safely recommended as good anchorage for shipping. Wood and water are not to be found so close to the anchorage as in other Fuegian harbours, but they may be obtained with very little trouble, and in any quantity, by going up the passage between the islands to one of the many streams which run from the high land. There is plenty of water, also, very near the best berth, on the South side; but frequently a surf breaks on that beach. The passage is also a snug berth for a vessel of less than 12 feet draught. A vessel should not moor in less than 10 fathoms, as close to the West shore as possible, with an anchor to the eastward, in the event of the wind coming from that quarter.

The *Week Islands* lie next to the northward. At their South side is a roadstead, with good holding in 18 or 20 fathoms, coarse gravel and sand, with patches of rock. It is exposed to southerly winds, and to those from the West; therefore it is not desirable for a vessel to anchor there. Between the islands is a snug berth for a small vessel, quite secure, but difficult of access.

The eye must be the chief guide in entering most of these places; they are of one description—inlets between high land, having generally deep water, with kelp buoying the rocky places. Flaws of wind, and violent gusts off the high land, render the approach to them difficult, and, to a large ship, impracticable.

Cape Sunday is the western point of the Week Islands, and the S.W. of Graves Island: it is high and prominent. Two islets and two dangerous rocks lie off it.

Barrister Bay succeeds. It is an exposed place, full of islets, rocks, and breakers. "In sailing along the coast we passed inside of several breakers, and, I hope, noted all that lie in the offing; but we cannot be sure, for breakers on rocks which are under the surface of the sea do not always show themselves."—(Capt. FitzRoy.)

Cape Deseado is the next promontory. Three miles before arriving at it there is an opening, which probably leads into a good harbour behind a number of islands; then the coast is high and unbroken to the cape, which is the highest land hereabout, and is remarkable. A rocky islet lies 1 mile off shore.

Dislocation Harbour is 2 miles from Cape Deseado. It was so called by Capt. FitzRoy, from an accident happening here to Mr. Murray. It is a place of refuge for an embayed or distressed ship, but unfit for any other purpose; its entrance is rendered difficult to the eye, by rocks on which the sea breaks violently; and by two rocks under water, on which the sea does not *always* break. In this place water may be obtained very easily, as boats can lie in a fresh-water stream, which runs from the mountains. Wood is also plentiful. The harbour is large enough for four small vessels. The entrance is narrow, but all dangers are visible. It is much exposed to West winds and the westerly swell, which might for weeks together prevent a vessel from getting out to sea.

The situation of Dislocation Harbour is pointed out by the heights, called *Low* and *Shoulder Peaks;* they are the most remarkable on that part of the coast, and immediately over the harbour. To find the entrance, steer for the peaks; look out for the weather and lee rocks, both several feet above water, the sea breaking violently on them; and when within 4 miles of the shore, you will distinctly see the opening from the mast-head. In going in, avoid the two rocks at the entrance, and anchor in the innermost part; only a small ship can get out again without a fair wind.

To the North of Dislocation Harbour is *Chancery Point;* and hence to Cape Pillar, at the entrance to the Strait of Magalhaens, the land appears high and mountainous; southward it seems lower and more broken. Off the shore here lie many rocks, on which the sea breaks violently, besides the two clusters called the *Judges*, which lie off Chancery Point, and the *Apostle Rocks*, a little South of Cape Pillar. A dangerous rock, under water, on which the sea breaks, lies half a mile more to the North than either of the Apostles. Some of these rocks are from 5 to 50 feet above the water, but many breakers show near them, indicating an extensive reef. The outer rock is 4 miles from the land.

CAPE PILLAR is a detached headland, and so very remarkable, that no person can fail to know it easily. Close to the cape are two small rocks, called the Launches; they are not more than 3 cables' length from the shore. The cape and the shore on each side are steep-to. Off the cape, at 2 miles' distance, are 60 and 70 fathoms, fine sand.

CHAPTER III.

THE WESTERN COAST OF PATAGONIA, FROM THE STRAIT OF MAGALHAENS TO THE ISLAND OF CHILOE.

This portion of the South American coast is a continuation of the singularly broken and intersected land so characteristic of these regions. It was partially explored by Sarmiento, and also by the officers in the *Beagle*, in her surveying voyage; but even this latter is somewhat imperfect, and there are many vacancies to fill up before our knowledge of the coast is complete. This, however, is of the less consequence, inasmuch as it is likely it will be but very seldom visited, and then only by those engaged in particular pursuits.

The first notice of the intermediate coast between Tierra del Fuego and Peru was furnished by a vessel sent in 1539-40, from Seville, in Spain, by D. Gutierre de Vargas, bishop of Placentia, under Alonso de Camargo. Of the three vessels, one was wrecked in the strait, another returned to Spain without passing the strait, and one of the masts of the third was taken out and sent to Lima, where it was many years preserved as a curiosity.—(Herrera, 6. 10. 10, &c.)

The CLIMATE of the coast of Western Patagonia, described in this section, is cold, damp, and tempestuous. The reigning wind is N.W.; but if it blows hard from that quarter, the wind is very liable to shift suddenly round to the westward, and blow a heavy gale, which raises a mountainous cross sea. These westerly gales do not generally last long, but veer round to the southward, when the weather, if the barometer rises, will probably clear up. Should they, however, back round to the N.W. again, and the barometer keep low, or oscillate, the weather will, doubtless, be worse. Easterly winds are of rare occurrence; they are accompanied by fine clear weather; but westerly winds bring with them a constant fall of rain, and a quick succession of hard squalls of wind and hail.

Should a vessel be near the coast during one of these northerly gales, it would be advisable for her to make an offing as quickly as possible, to guard against the sudden shift to the westward that is almost certain to ensue. The discovery, however, of the anchorages of Port Henry, Port Santa Barbara, Port Otway, and St. Quentin's Sound, has very much reduced the dangers of the lee shore; and a refuge in either of them will always be preferable to passing a night on this coast in a gale of wind.

The barometer falls with northerly and westerly winds, but rises with southerly. It is at its minimum height with N.W. winds, and at its maximum when the wind is S.E. The temperature is rarely so low as 40°, excepting in the winter months.

At Port Otway, in the Gulf of Peñas, the maximum and minimum for 19 days, in the month of June, were 51° and 27½°.—(Capt. P. P. King, R.N.)

At the conclusion of the first chapter (page 38, and also page 68), we have given a description of the western entrance to the Strait of Magalhaens, between Cape Pillar on the South, and Cape Victory on the North. Off the entrance is the excellent leading mark formed by the Evangelists, or Isles of Direction, very clearly indicating the opening of the strait (page 38).

The following descriptions are principally drawn from those of Capts. P. P. King, Stokes, and Skyring, and Mr. Kirke, mate of H.M.S. *Beagle.*

Between CAPE VICTORY and Lord Nelson Strait the coast is very much broken, and intersected by channels leading between the islands of *Queen Adelaide's Archipelago;* on the sea coast of which, to the N.N.E. of Cape Victory, is a remarkable pyramidal hill, called *Diana Peak,* which, in clear weather, is visible to ships entering the strait. *Cape Isabel* is a steep, rocky promontory, of great height, with a peaked summit, and a sharply serrated ridge, having at its base two detached columnar masses of rock. *Beagle Island,* lying off it, is wall-sided; but, although tolerably high, is much lower than the land of the cape.

Cape Santa Lucia may be distinguished by a portion of flat table-land, about one-third of the altitude of the mountain from which it proceeds, and terminating at its outer face with a perpendicular precipice.

The coast between Capes Isabel and Lucia is dangerous to approach nearer than 10 miles; for there are within that distance many sunken rocks, on which the sea only occasionally breaks. Some of these breakers were seen to seaward at the distance of 5 or 6 miles, as Capt. Stokes proceeded along the coast. When he was off Cape Sta. Lucia, whales were very numerous.

The general aspect of this portion of the coast is similar to that of the most dreary parts of the Magalhaenic regions; bare, rugged, rocky, and mountainous, intersected by inlets, and bordered by islets, rocks, and breakers.

Cape Sta. Lucia is the westernmost point of *Cambridge Island.* Cape George, at the South end, is lower, and forms a bluff point.

According to Capt. Morrell, "the space between Cape Sta. Lucia and Cape Santiago is a numerous cluster of islands, with deep water all around them. There are many reefs and sunken rocks on the seaboard, and also among these islands, but their presence is always indicated by kelp, which, as before stated, is a good warning. They afford excellent harbours, and also furnish wood and water in abundance, and their shores are frequented by hair-seals. Timber of almost any description can be had with very little trouble, and the natives seldom visit the islands."*

The *San Blas Channel, Duck* and *Duncan Harbours,* the *Duncan Rock,* and other rocks off them, are inserted in the survey from the oral information of the master of an American schooner, and probably are very incorrectly laid down. *Augusta Island* and the *White Horse* were seen by Lieut. Skyring.

* Narrative of Four Voyages, &c., p. 98. We have inserted these descriptions from Capt. Morrell, where the information from the survey is wanting. Some of his narratives are evidently overdrawn, and we therefore submit these extracts with a caution.

Cape Santiago (or St. Jago), the South end of Madre de Dios Archipelago, is correctly placed, as are also the general direction of the coast to the northward, and the summits of the land that are particularized, viz., the opening of *West Channel, April Peak, Tower Rock,* and the bay to the North of it, and Cape Three Points, which is the South entrance of the Gulf of Trinidad.*

The southern portion of the Madre de Dios Archipelago is called by Morrell St. Martin's Island; and the strait, called West Channel in the survey, is probably that named by him Byers's Strait. He says:—"A ship of any size may pass through this strait with ease and safety, as it is clear of danger. On the North shore are two fine harbours; and there is another, which is very commodious, on the S.E. part of the Island of St. Martin, about 5 miles from Cape St. Jago. By following the eastern shore of St. Jago, which runs in a N.E. direction, this port is easily found. The entrance is plain, and the course of the channel is N.W. for about 2 miles, forming a circular basin, completely land-locked by a few small islands at its entrance. The depth of water in going in is 10 fathoms; and within the basin from 5 to 15 fathoms, mud and clay bottom. Both wood and water can be procured here with the greatest ease, and a ship may heave out with perfect safety on the West side of the basin."†

CAPE THREE POINTS, or Tres Puntas,‡ rises to a lofty rocky mountain, nearly 2,000 feet high, the summit being of peaks and sharp serrated ridges, with a detached mass of rock of pyramidal form at the base, which shuts in with the land on the bearing of *N.N.E. ¾ E. true.* The variation here is 20° 58′.

Port Henry is 3 miles to the N.E. of Cape Three Points. The shore between them is lined for nearly a league off with rocks and islets, of which several scores might be counted in the space of a square mile; but they seem to be of bold approach, and no dangers probably exist that are not above water, or are not shown by kelp.

Bound to Port Henry a vessel should keep on the South side of the gulf; for the northern part is strewed with many rocks, and seemed to be exceedingly dangerous. The soundings, also, are irregular, and the bottom is foul and rocky.

The entrance of Port Henry will be easily distinguished by its sandy beach, since it is the first that is observed on the South shore on entering the gulf. It is a small, light-coloured beach, with a lowish sandy cliff at the back, and a round, rocky, and wooded mount at its western end. The Seal Rocks, also in the offing, are a good mark; they bear *N. ¾ W. true,* 5 miles from the West point of the entrance, which is about a mile wide. The channel is bounded on each side by low rocks, lying off moderately high, round, rocky islets, that may be

* Sarmiento, p. 66. † Narrative, p. 99.

‡ The discovery of this land is thus described by Sarmiento:—"March 17, 1579. In approaching the shore we saw a great bay and gulf, which trended deeply into the land toward some snowy mountains. To the South there was a high mountain with three peaks, wherefore Pedro Sarmiento named the bay 'Golfo de la Sanctisima Trenidad.' The highest land of the three peaks was named 'Cabo de Tres Puntas ó Montes.' This island is bare of vegetation, and, at the water-side, is low and rugged, and lined with breakers; on the summit are many white, gray, and black-coloured portions of ground or rock. Six leagues to the North of Cape Tres Puntas is the opposite side of the gulf, where it forms a large high mountain, backed to the North by low land, and fronted by many islands. This high mountain, which appears to be an island from the offing, was called Cabo Primero."

approached within 1½ cables' length. The soundings are from 20 to 26 fathoms, on a sandy bottom: afterwards they decrease pretty gradually to the anchorage, which is in 9 and 10 fathoms.

When the sandy beach bears S. by E. ¾ E. *mag.*, the fairway of the entrance will be quite open; and a vessel may stand in, keeping the round mount at the western end of the sandy beach on the larboard bow, until nearly abreast of it; she may then proceed up the harbour as high as convenient, and select her berth: for the ground is quite clear of danger to the line of rock weed, which skirts the shores and islets. The depth of water is between 12 and 8 fathoms, and the bottom generally of sand and mud.

In turning in there are some patches of kelp on each side, growing upon rocks that watch at high water, which must be avoided.

As the squalls off the high land are sometimes very strong, it will be advisable for a ship to anchor as soon as possible, and warp up to her berth; which, from the smoothness of the water, may be easily effected.

The inner harbour was named *Aid Basin*. It is perfectly land-locked, and sufficiently spacious to contain a numerous squadron of the largest ships, in 20 fathoms water, over a mud bottom, and as completely sheltered from the effects of wind and sea as in wet docks. At the South side of the basin is a fresh-water lake, which discharges itself by a small stream, whence casks might be conveniently filled by means of canvas hoses, and the shores around have wood for fuel in abundance; but, from the lofty surrounding mountains, some rising almost perpendicularly to an elevation of 2,000 feet, the thick clouds with which the basin is generally overhung, and the dense exhalations that arose from it during the rare intervals of sunshine, together with the exceeding prevalence of heavy rain on this coast, this place must be disagreeable and unhealthy. Such objections do not apply to the outer harbours, for while its shores afford shelter, they do not obstruct a free circulation of air. It is sufficiently large to afford convenient and secure anchorage for five or six frigates.

It is high water at full and change within a few minutes of noon, and rises 5 feet. The stream of the tide, however, is very inconsiderable, and never exceeds half a mile an hour. The observations for latitude and longitude, &c., were made on a rock at the western side of the port. The lat. is 50° 0′ 18″, lon. 75° 19′ 5″. Variation of the compass, 20° 50′.

The Gulf of Trinidad separates Wellington Island from Madre de Dios. It is nearly 10 leagues long, and from 4 to 8 miles wide. Its South shore, or North coast of Madre de Dios, is very much broken, and, probably, contains many ports.* None of them were visited excepting for night anchorages. Under the East side of Division Island is *Port de la Morro*, which, with Point Candelaria and Port Rosario, are inserted from Sarmiento's account.†

On the northern shore are two opening-like channels: the westernmost probably communicates with the Fallos Channel; the other, Sarmiento's Brazo

* This passage throughout abounds with good harbours and excellent places of shelter.—*Morrell*, p. 101.

† Sarmiento, p. 82-3.

de Norte, or North Arm, appeared to trend under the base of the range of mountains, among which *Cathedral Mount* is a conspicuous object. From the entrance of the strait this mountain resembles the spire and roof of a church, and is visible for more than 20 leagues. Between the two openings is *Neesham Bay*, in which the *Adelaide* found a secure anchorage in 11 fathoms. There is also good anchorage for a small vessel in *Windward Bay*.

The gulf meets the *Wide Channel* at its junction with Concepcion Strait, where the channel is contracted by an island to the width of 1½ miles. There are several isles and rocks in the gulf, of which the most remarkable are the *Seal Rocks*, before mentioned; the *Van Isles*, opposite the Western Channel; and a group of numerous islands extending for a league to the southward of the land to the westward of *Neesham Bay*. On the South shore are also several isles, but they are near the coast, and are particularized in the chart. The most remarkable is *Middle Island*, which, with the reef off its S.W. end, is well described by Sarmiento.*

Opposite to Cape Three Points is Cape Primero,† the South point of the mountainous island of Mount Corso; the land of which may be seen, in clear weather, from the southward, at the distance of 10 leagues. It forms the visible northern termination of the coast line. Viewed when bearing North, or any point to the westward of North, its summit makes like a round mount rising conspicuously above the contiguous land, from which a small portion of low coast extends for 2 degrees beyond it to the westward. The land of the northern shore of the gulf makes in mountainous ridges and peaks, the average height of which Capt. Stokes estimated to be about 3,000 feet.

The harbour on the North side of Cape (or Mount) Corso is about 18 miles (13?) from its point, in the direction of N.N.E. ½ E., being a spacious bay, sheltered from all winds, and sufficiently capacious to moor 100 ships of the line. The depth of water at its entrance is 40 fathoms; but on the West and S.W. side of the bay are found from 5 to 20 fathoms, sand and mud bottom.—(Morrell, p. 99.)

The character of the land is the same as that to the southward, bare, rugged, rocky mountains, with peaks, and sharply serrated ridges.

The Island of *Mount Corso* is separated from Cape Brenton by *Spartan Passage*. For more than a league off Cape Primero are some extensive reefs: indeed the whole line of the West coast of Madre de Dios is fronted by rocks, some of which are 2 leagues from the shore. There are regular soundings in the entrance of the gulf, but the water deepens immediately after passing to the eastward of Port Henry.

Picton Opening and *Dynely Bay* very probably insulate the land that separates them, of which *Cape Montague* is the S.W. extreme. There are some rocks 8 or 10 miles off the coast to the southward; but between Cape Montague and Cape Dyer they are more numerous: several are from 8 to 10 miles off the shore; many are dry, some are awash, and others show only by the breaking of the sea.

Parallel Peak, a remarkable mountain, is at the South end of Campaña Island, and was so named by Capt. Stokes, in the *Beagle*. The South cape of

* Sarmiento, p. 86.

† *Ibid.* p. 65. It is also the Cape of Good Hope of Bulkely and Cummings' Narrative, p. 116.

the island, which is not named in the survey, was called by Morrell, Cape McIntyre. The coast here, in its general appearance, does not differ from that of Madre de Dios.

The West coast of Campaña Island did not appear on the charts previous to the survey. Capt. Stokes describes it thus:—"April 9, 1828. Throughout our run along the coast this day, we skirted a number of rocky islets, rocks, and breakers, lying off shore at the distance of 3 or 4 miles. Some of the islets were elevated several feet above the surface of the sea; others were awash; and there were breakers that only showed themselves occasionally. Along this line the surf beat very heavily; and outside, a long rolling sea prevailed, in which the ship was very uneasy.

"This line of dangers is not altogether continuous, for there is an opening about 2 miles wide abreast of Parallel Peak, to the southward of which is a bight (Dyneley Bay), where possibly a harbour may exist; but, considering the prevalence of heavy westerly gales and thick weather, if there be one, few vessels would venture to run for it; and this line must, I should think, be considered as a barrier that they ought not to pass. As seal are found on the rocks, vessels engaged in that trade might not, perhaps, be deterred by these dangers; but every other would give all this extent of coast a wide berth. We ran past the breakers at the distance of about a mile, having rocky soundings, from 20 to 23 fathoms.

"The termination of the coast line northward was a high rugged island, with a small peak at the North end. The extremity of the main land was rather a high bluff cape, whence the coast extends southward, with craggy, mountainous peaks and ridges, as far as Parallel Peak."*

Cape Dyer is in lat. 48° 6′, lon. 75° 34′ 30″.† At 5 miles W. ¼ S. from it is a rocky islet, about 45 feet high, rising like a tower from the sea, called by Bulkely and Cummings, "*The Rock of Dundee*," from its similarity "to that island in the West Indies, but not so large; it lieth about 4 leagues ‡ from the southernmost point of land out at sea."§ This rock is a good mark for *Port Santa Barbara*, from the entrance of which it bears S.W. (*S. W. by W. ¾ W. true*), distant 9 miles.

PORT SANTA BARBARA.—The land about this harbour is similar to that about Port Henry. Its shores are rocky, with some patches of sandy beach, but everywhere covered with trees, or an impervious jungle, composed of dwarfish trees and shrubs. The land, in most places, rises abruptly from the shore to mountains, some of which attain an altitude of more than 2,000 feet, and are quite bare at their summits and on their sides, except in sheltered ravines, where a thick growth of trees is found. These mountains, when we could break off specimens, were of basalt, with large masses of quartz imbedded in it: but on some parts of the shores the rocks were of very coarse granite.‖

At 1 mile to the North of the rock the depth is 23 fathoms, and gradually

* Narrative, &c., vol. i. pp. 162-3.

† It is the Cape Nixon of Morrell, p. 100.

‡ There must be a mistake here; it should probably have been 4 miles.

§ Bulkely and Cummings' Voyage to the South Seas, p. 113.

‖ Narrative, vol. i. p. 165.

decreases on approaching Port Santa Barbara; in steering for which, as soon as Cape Dyer bears South, by compass, you will be close to some rocks, which you should keep on your larboard hand. Abreast of this rock, one-eighth of a mile off, the depth will be 11 fathoms. The channel here is 1 mile wide, but gradually narrows on approaching the S.W. end of Breaksea Island; and at Wreck Point, the West head of the port, the width is about one-eighth of a mile. There are several rocks in this passage, but as the depth is from 6 to 8 fathoms, the anchor may be dropped, and the ship warped clear of them, in case of being becalmed: calms, however, are of rare occurrence here.

Breaksea Island,* more than 2 miles long, fronts the port, the heads of which are three-quarters of a mile apart. In the entrance of the port the depth is 3½ and 4 fathoms, and gradually decreases to 2½ fathoms, but at the bottom there is a basin with 6 and 8 fathoms in it.† This is a very good harbour, and from the rare opportunity of anchoring your ship in a moderate depth, is of easy access. It is also readily made out by its vicinity to the Dundee Rock, which serves to point out its position.

The West head of the port is in lat. 48° 2′ 15″, and lon. 75° 29′; variation, 19° 10′. High water takes place at full and change at 11h 45′ p.m., and rises 3 to 4 feet (*neaps*), 6 feet springs.

To the N.E. of Breaksea Island are many straggling rocks. The *Beagle* having entered the port by the western entrance, left it by threading the rocks to the eastward, in doing which she had not less than 9 fathoms.

Between the island and the mouth of the port, the depth is from 6 to 7 fathoms, good ground, which renders the entrance and exit very easy.

Flinn Sound is a deep opening to the eastward of the port; it was not examined.

Point Bynoe, with the group of islands, Bynoe Islands, extending for 2 miles off it, is the West head of the Fallos Channel, which was explored for 30 miles without offering any interesting feature.

Fallos Channel probably communicates with the sea by Dynely Bay and Picton Opening; and, beyond the latter, was supposed to communicate with the Gulf of Trinidad by the Channel to the West of Neesham Bay.

The Guaianeco Islands, 20 miles in extent, are composed of two principal islands and many smaller islets; the westernmost is called *Byron Island*, and the easternmost *Wager Island*. They are separated by Rundle Pass, called in

* "A large island, on the northern side of the harbour, is an excellent watering place, at which casks may be conveniently filled in the boats. It is also an object of great natural beauty. The hill which forms its western side rises to 700 or 800 feet, almost perpendicularly, and, when viewed from its base, in a boat, seems stupendous: it is clothed with trees, among which the light-green leaves of the winter's bark tree, and the red flowers of the fuchsia, unite their tints with darker foliage of other trees. This perpendicular part extends to the northward till it is met by the body of the mountain, which is arched into a spacious cavern, 50 yards wide and 100 feet high, whose sides are clothed with a rich growth of shrubs; and before it a cascade descends down the deep face of the mountain. On the shore were three Indian wigwams, long deserted."—*Capt. Stokes.*

† "In the afternoon we weighed anchor and warped into a berth in the inner harbour, where we moored in 3 fathoms. I found, lying just above high-water mark, half-buried in the sand, the beam of a large vessel. We immediately conjectured that it had formed part of the ill-fated *Wager*, one of Lord Anson's squadron, of whose loss the tale is so well told in the narratives of Byron and Bulkely."—*Capt. Stokes, Voyages of the Adventure and Beagle*, vol. i. p. 165.

Bulkely's Narrative, the Lagoon; * on the West side, and at the North end of it, is *Speedwell Bay*.†

Rundle Pass is only a quarter of a mile wide, but perfectly clear in the whole extent of its channel, excepting the northern entrance, where it is guarded by many detached rocks, which render the entrance to Speedwell Bay rather difficult. According to Byron's and Bulkely's Narratives, the situation of the wreck of the *Wager* is near the West end of the North side of Wager Island. Harvey Bay and Good Harbour are mentioned by Bulkely. Off the western end of Byron Island are some rocky islets; and its North coast is also very much strewed with them, even to a considerable distance from the shore.

The Guaianeco Islands are separated from the land of Wellington Island by a clear, but, in some parts, narrow passage. At its S.W. end it is contracted by rocks to 1½ miles, and at the South end of Byron Island is scarcely a mile broad; afterwards, however, it widens to 2½ and 3 miles.

Morrell says: "Among this cluster are many fine harbours; the land is low, and very fertile, clothed with heavy timber, grass, clover, &c. The islands which form the North part of the group are much frequented by hair-seals. A variety of scale and shell-fish may also be had here with more sport than labour."‡

The North part of Wellington Island is *Cape San Roman*.§ It is the West head of the Mesier Channel, which opens into Tarn Bay.

Tarn Bay is about 5 leagues wide. The *Ayautau Islands* are 4 miles from the coast, but the interval is occupied by several rocky reefs, between which, Lieut. Skyring thought, there seemed to be a "sufficiently clear passage." The pilot, Machado, in 1769, however, thought differently.||

The *Channel's Mouth* of the old chart extends in a S.E. direction for 11 miles, and then divides into two branches; it is merely deep and narrow arms of the sea, running between steep-sided ranges of mountains. The shores are rocky, and afford neither coves nor bights, nor even shelter for a boat, and are perfectly unproductive; for no seals or birds were seen, and the shores were destitute even of shell-fish.

Cape Machado, in lat. 47° 27′ 35″, lon. 74° 30′ 2″, is the North head of this opening. Two miles off it are two rocks, which Machado carefully and correctly describes, as he also does the rocks and breakers which extend off the South head for very nearly a league. The *Beagle* twice occupied an anchorage under the Hazard Isles, in the entrance, and on both occasions was detained many days from bad weather, with three anchors down. She anchored here in June, 1828, and experienced the most horrible weather. Nothing could be more dreary than the scene around; and there is no doubt that the perilous and arduous nature of the surveying service here hastened the sad termination of Capt. Stokes's existence.

Excepting this very bad and exposed anchorage, there exists none in the channel. Capt. Stokes describes it to be an extremely perilous anchorage. "The anchors," he says, "were in 23 fathoms, on a bad bottom, sand and coral. The squalls were terrifically violent. Astern, at the distance of half a cable's

* Bulkely and Cummings' Narrative, p. 106. † *Ibid.* p. 105.
‡ Narrative, &c., p. 100. § Agueros, p. 213. || *Ibid.* p. 210.

length, were rocks, and low rocky islets, upon which a furious surf raged, and on which the ship must have been inevitably driven, if the anchors, of which three were down, had started."*

Between Channel's Mouth and Jesuit Sound the coast is more unbroken and low than usual. In lat. 47° 17′ are some reefs, which project 2 miles to sea; behind them there was an appearance of a bight, which may afford anchorage.

Jesuit Sound, like Channel's Mouth, is quite unfit to be entered by any ship.

Separated by *Cheap Channel* from the main is *Xavier Island*,† the Montrose Island of Byron's Narrative.‡ It is 11½ miles long, and 4 wide, and is very high, and thickly wooded with lofty trees. The only two anchorages which the island affords are noticed and named by Machado—the northern one, *Port Xavier;* the southern, *Ignacio Bay*.§ The former is by much the better place, being secure from prevailing winds, with 17 fathoms at 800 yards from the shore. The South end of the bay is a sandy beach, backed by tall beech trees. The shore to the South of Xavier Bay, for the first 4 or 5 miles, consists of a high, steep, clay cliff, with a narrow stony beach at its base, backed by mountains of 1,200 or 1,400 feet high, and covered by large and straight-stemmed trees. The remainder of the coast, to Ignacio Bay, is low, and slightly wooded with stunted trees; and its whole extent is lashed with a furious surf, that totally prevents boats from landing.

Ignacio Bay affords anchorage in 9 fathoms. The western coast of the island is lined by reefs extending 2 miles off, upon which the sea breaks high.

KELLY HARBOUR is situated at the bottom of the N.E. corner of the Gulf of Peñas, in the bay formed between the land of St. Estevan Gulf and Xavier Island. It trends inwards in an easterly direction for 8 miles. The land about the harbour is high, rugged, and rocky, but by no means destitute of verdure. In the interior are lofty peaked and craggy ranges of snow-covered mountains. The points of the entrances are 2 miles asunder, are thickly wooded, and low, compared with the adjacent land; their magnetic bearing is N.E. ½ E. and S.W. ½ W. Between them is a channel of from 35 to 40 fathoms deep, over a mud bottom, without danger, to a cable's length of the rocky islets that fringe the shore for a quarter of a mile off. On approaching the harbour the remarkable muddied appearance of the water is rather startling: but the discoloration proceeds only from the freshes of the river, and the streams produced from a very extensive glacier that occupies many miles of the country to the North. The course in is E.S.E. by compass, until in a line between the inner North point, and an inlet on the South shore that is fronted by five or six wooded islets. Then haul up along the larboard side of the harbour, as close to the shore and as far as you please, to an anchorage. The best berth is when the two points of entrance are locked in with each other, and within 1½ cables of the sandy spit that extends off the western end of a high and thickly-wooded island. The ground is excellent, and so tenacious, that it was with difficulty that the *Beagle* lifted her anchors. Shelter, wood, and water, however, are the only advantages offered by the harbour.

* Voyages of the *Adventure* and *Beagle*, vol. i. p. 170.

† Agueros, pp. 209, 231. ‡ Byron's Narrative, pp. 73, 94, 95. § Agueros, l. c.

For knowing Kelly Harbour the glacier is a capital leading mark. It is a large field of ice, lying on the low part of the coast, about 2 miles to the northward of the harbour. The water at the anchorage, at half-tide, was perfectly fresh, but was too muddy to be fit for use.

St. Estevan Gulf.—The entrance of this gulf, which is situated 9 miles North of the N.E. end of Xavier Island, is 4 miles wide. *Forelius Peninsula*, on the western side, is a narrow tongue of land nearly 5 leagues long. The eastern side of the gulf is a long sandy beach, curving round to the N.W., towards the entrance of the River San Tadeo, between which and Cirujano Island, forming the South (or rather the West) point of entrance, the width is less than 5 miles; and at a league farther to the westward, it is not more than 3½ miles across. Here, in the centre, there is a small islet called *Dead-tree Island.*

Beyond this is *St. Quentin's Sound*, 10 miles deep;* and, at its N.W. corner, *Aldunate Inlet* extends in for about 8 miles. St. Quentin's Sound terminates in continuous, low land, with patches of sandy beach, over which, among other lofty mountains, the Dome of St. Paul's is seen. The shores are thickly wooded with shapely and well-grown trees; the land near the beach, for the most part, is low, rising into mountainous peaks; a little distance in the interior of which, some are 1,500 feet high, but they are not craggy.

St. Estevan Gulf is one of the best harbours of the coast, being easy of access. The best anchorage is at about 2 miles above Dead-tree Island, in from 4 to 6 fathoms, sandy bottom. This will be at 2 miles from either shore, but the berth is perfectly land-locked; if necessary, anchorage may be taken up much nearer to it; and as in all parts of the Sound there is anchorage depth, with a muddy or sandy bottom, the advantages offered to shipping would be of great consequence in parts of the world more frequented than the Gulf of Peñas. Whales were numerous, and seals were also seen.†

Cirujano Island, above mentioned, is that on which the surgeon of the *Wager* was buried. The island is separated from the extremity of Forelius Peninsula by a strait, 1 mile to three-quarters of a mile wide.

The mouth of the *River San Tadeo* is easily distinguished on entering the gulf, by the sand hills on each side of its entrance, and the bearing of the East trend of Cirujano Island, *S.W. ½ S. true* (by compass S. by W. ¾ W.). A sandy beach extends to the East and West of it for many miles; the land is low and marshy, and covered with stumps of dead trees. It has a bar entrance, much of which must be nearly dry at spring tides.

Purcell Island is separated from the land of *Forelius Peninsula* by a good channel, 2 miles wide; it is moderately high and thickly wooded, and about 6 miles in circuit. About mid-channel, and nearly abreast of the East end of the island, is a rock only a few feet above the water.‡ The channel to the South of the rock is from 18 to 22 fathoms deep, and the bottom sandy.

Upon the peninsula, opposite the West end of Purcell Island, is an isthmus of low, sandy land, scarcely a mile wide; the one over which, it may be

* Agueros, p. 209. † Voyages of the *Adventure* and *Beagle*, vol. i. p. 174-5.
‡ Byron's Narrative, pp. 149—156; and Agueros, pp. 200, 220, 244.

inferred from the Narrative,* the canoes in which Byron and his companions were embarked were carried.

The *Beagle* anchored in Bad Bay after dark, in 8 fathoms, sandy bottom. Of this place, Capt. Stokes remarks: "At daylight, we found that we had anchored in a small bay about half a mile off a shingle beach, on which, as well as on every part of the shore, a furious surf raged that effectually prevented our landing to get chronometer sights. The mouth of this bay is N.E. ½ E. (*mag.*) 9 leagues from Cape Tres Montes, which, in clear weather, may be seen from its mouth. Like all this shore of the gulf, it is completely open to the S.W., and a heavy rolling sea."†

To the westward, between Bad Bay and the land of Cape Tres Montes, is an extensive bight, 16 miles wide, and about 12 deep. The centre is occupied by a group of islands called *Marine Islands*,‡ upon which the *Sugar Loaf*, a mountain 1,840 feet high, is very conspicuous. It was seen from the *Wager* the day before her wreck.§ Upon the main, 5¾ miles N. 15° E. from the Sugar Loaf, is another equally remarkable mountain, called the Dome of St. Paul's, 2,284 feet high.

NEUMAN INLET, at the N.E. corner of this gulf, extends for 17 miles into the land, where it terminates; but it is of no use, as the water is too deep for anchorage. It is the resort of large numbers of hair-seals. At the N.W. corner is *Hoppner Sound*, about 5 miles in extent. At its S.W. end is a deep inlet, extending 7 miles to the S.W., and reaching to within 2 miles of the sea coast, from which it is separated by an isthmus of low and thickly-wooded land. Capt. Stokes walked across it to the sea-beach, from whence he saw Cape Raper.‖ The *Beagle* anchored at the bottom of Hoppner Sound, off the mouth of the inlet. The mouth of the sound is very much blocked up by the Marine Islands; but the southern channel, although narrow, has plenty of water. On the S.W. side of the Marine Islands is *Holloway Sound*, in which is Port Otway, an inlet extending for 5 miles into the land in a S.W. direction.

The entrance of PORT OTWAY is on the West side of Holloway Sound, about 14 or 15 miles distant from Cape Tres Montes, and may be readily known by its being the first opening after passing the cape. Off the mouth are the Entrance Isles, among which is the Logan Rock, having a strong resemblance to the celebrated rock whose name it bears. It is broad and flat at the top, and decreases to its base, which is very small, and connected to the rock upon which it seems to rest. Immediately within the entrance on the West shore is a sandy beach, over which a rivulet discharges itself into the bay. Here anchorage may be had in 9 or 10 fathoms. It is by far the most convenient one the port affords, and contains anchorage all over it, but the depth is generally inconveniently great, from 20 to 30 fathoms.¶

* Byron's Narrative, pp. 119, 120. † Voyages of the *Adv.* and *Beagle*, vol. i. p. 174.

‡ It was here that four marines voluntarily remained on shore during Byron's perilous boat voyage, after the wreck of the *Wager*.—*Byron's Narrative*, p. 85.

§ Bulkely and Cummings, p. 15. ‖ Voyages of the *Adv.* and *Beagle*, vol. i. p. 170-1.

¶ Capt. Stokes says, in his journal:—"Among the advantages which this admirable port presents to shipping, a capital one seems to be the rich growth of stout and shapely timber, with which its shores, even down to the margin of the sea, are closely furnished, and from which a

CAPE TRES MONTES is a bold and remarkable headland, rising from the sea to the height of 2,000 feet. It lies in lat. 46° 58′ 57″, and lon. 75° 28′, and is the South extremity of the Peninsula of Tres Montes.

Capt. Stokes says: "At sunset Cape Tres Montes bore N.W. (*N.* 25° *W.*) distant 18 miles. In this point of view the cape makes very high and bold; to the eastward of it land was seen uninterruptedly as far as the eye could reach. We stood in shore next morning, and were then at a loss to know, precisely, which was the cape. The highest mountain was the southern projection, and has been marked on the chart as Cape Tres Montes; but none of the heights, from any point in which we saw them, ever appeared as 'three mounts.' The land, though mountainous, seemed more wooded, and had a less rugged outline than that we had been hitherto coasting, since leaving the Strait of Magalhaens."

This was the northern termination of the *Beagle's* surveys in 1830.

Tides.—High water, at most parts of this coast, takes place within half an hour on either side of noon. The stream is inconsiderable, and the rise and fall rarely more than 6 feet.

The *variation of the compass*, at the western entrance of the strait, is 23¾°; at Port Henry, 21°; at Port Santa Barbara, 19°; at Xavier Island, 20°; and at Port Otway, 20½°.

THE INTERIOR SOUNDS AND CHANNELS.—The western coast, between the Strait of Magalhaens and the Gulf of Peñas, is formed by a succession of islands of considerable extent, the largest of which, Wellington Island, occupies a length of coast of 138 miles. It is separated from the main by the Mesier and Wide Channels; * and from Madre de Dios by the Gulf of Trinidad. Madre de Dios, which is probably composed of several islands, has for its inner or eastern boundary the Concepcion Strait.

Hanover Island has the Sarmiento and Estevan Channels on its eastern side, and on the South is separated from Queen Adelaide Archipelago by Lord Nelson Strait, which communicates by Smyth Channel with the Strait of Magalhaens.

Smyth Channel commences in the strait at Beaufort Bay, on the eastern side of Cape Philip; E. by N. *true*, 5½ miles from which are the Fairway Isles; and, at a little more than 6 miles from the cape, on the West shore, is the anchorage of Deep Harbour, the entrance of which is a quarter of a mile wide. The anchorage is about half a mile within the head.

Good's Bay, the next anchorage, is better than the last, the depth being from 20 to 25 fathoms. It is convenient for vessels going to the northward, but when bound in the opposite direction, North Anchorage will be better, from the depth being less.

Opposite to Cape Colworth is Clapperton Inlet, beyond which is a considerable tract of low country,—a rare sight in these regions. Two miles further, on the

---

frigate of the largest size might obtain spars enough to replace a topmast, topsail-yard, or even a lower-yard. On each side of the harbour we found coves so perfectly sheltered, and with such inexhaustible supplies of fresh water and fuel, that we lamented their not being in a part of the world where such advantages would benefit navigation."—*Voyages of the Adventure and Beagle*, vol. I. p. 170.

* Brazo Ancho of Sarmiento, p. 99.

eastern side, is *Hose Harbour*, suitable for a small vessel; and, on the opposite shore, is *Retreat Bay*, fronted by low rocky islets.

Onwards, the channel is clear as far as *Oake Bay*, where the depth is 9 fathoms: but the anchorage is better among the *Otter Islands*, the depth being 6 and 7 fathoms, and the ground clean.

The channel for the next 8 miles becomes more strewed with islands and rocks, and has much shoal water off every low point. The coast, also, is very low on the eastern shore, as far as the base of *Mount Burney*, which is 5,800 feet high, and covered with perpetual snow.

The best channel is on the East side of the Otter Islands, and between the Summer Isles and Long Island, for which the chart and a good look out for kelp will be sufficient guides.

Fortune Bay is at the S.E. extremity of, apparently, an island in the entrance of a deep channel, named the *Cutler Channel*, after an American captain who gave the surveyors much valuable information. At the bottom of the bay is a thickly-wooded valley, with a fresh-water stream.

A league to the North of Point Palmer, on the opposite shore, is Isthmus Bay, affording excellent anchorage. Five miles North of Point Palmer is *Welcome Bay*, also affording an excellent place to anchor in, with moderate depth.

In *Sandy Bay*, on the East side of the channel, and off Inlet Bay, on the opposite shore, there are good anchorages: both have a moderate depth, and are sheltered from the prevailing winds, which generally are north westerly.

In latitude 52° 1′ is *Victory Passage*,* separating *Zach Peninsula* from Hunter Island, and communicating with *Union Sound*, which leads to the Ancon Sin Salida of Sarmiento.†

At the South extremity of *Piazza Island* is *Hamper Bay*, with anchorage in from 7 to 15 fathoms. *Rocky Cove* is not to be recommended, and *Narrow Creek* seems confined.

Hence to the mouth of the channel, which again widens here to 5 miles, and in which, during strong N.W. winds, the sea runs heavy, we know of no anchorage; but a small vessel in want will, doubtless, find many, by sending her boat in search. The *Adelaide* anchored among the Diana Islands, and in Montague Bay, having passed through Heywood Passage. The northern point of Piazza Island is Sarmiento's West Point (Punta del Oeste‡), and a league to the South is his Punta de Mas-al-Oeste, or Point more West. Lieut. Skyring concludes the journal of his survey of Smyth Channel with the following remarks:—

"So generally, indeed, do the northerly winds prevail, that it would be troublesome even for working vessels to make a passage to the northward; but it is a safe channel for small craft at any time. The tides are regular; the rise and fall at the southern entrance are 8 and 9 feet, but at the northern only 5 and 6. The flood tide always sets to the northward, and the strength of the stream is from half to 1½ miles an hour; so that a vessel is not so likely to be detained here for any length of time, as she would be in the Strait of Magalhaens, where there is

* Sarmiento, p. 139. † *Ibid.* p. 142. ‡ *Ibid.* p. 143.

little or no assistance felt from westerly tides. The channel, besides, is comparatively free from sea, and the winds are not so tempestuous."

As the sounds within Smyth Channel will never be used for any purpose of navigation, little need be said in a work destined solely for the use of shipping frequenting the coast. The chart will be sufficient to refer to for every purpose of curiosity or information. They possess many anchorages for small vessels, affording both shelter and security.

Sarmiento,* on his third boat voyage to discover a passage through the land into the Strait of Magalhaens, gives a detailed account of his proceedings. All his descriptions are so good, that we had no hesitation in assigning positions to those places he mentions, to all of which his names have been appended.

The Canal of the Mountains, nearly 40 miles long, is bounded on each side by the high, snow-capped Cordillera, the western side being by very much the higher land, and having a glacier of 20 miles in extent, running parallel with the canal.

Worsley Bay and *Sound* extend 15 miles into the land.

Last Hope Inlet is 40 miles in length. Its mouth is 3½ miles wide, but at 8 miles the breadth is contracted by islands† to less than a mile.

Obstruction Sound extends for 30 miles in a South by East direction, and then for 15 miles more to the W.S.W., where it terminates. It is separated from the bottom of Skyring Water by a ridge of hills, perhaps 12 miles across.

Sarmiento Channel, communicating between the East side of Piazza Island and Staines Peninsula, continues to the northward of the mouth of Peel Inlet, where it joins the San Estevan Channel, from which it is separated by the islands of Vancouver and Esperanza: between these is a passage nearly a league wide, but strewed with islands.

Relief Harbour, at the South end of Vancouver Island, is a convenient anchorage; but the best hereabouts is *Puerto Bueno*, first noticed by Sarmiento.‡

Peel Inlet extends in for 7 leagues, communicating with Pitt Channel, and insulating Chatham Island, which is separated from the North end of Hanover Island by a continuation of the Sarmiento and San Estevan Channels, of which the principal feature is the *Guia Narrows*.§

The N.W. coast of Chatham Island has many bights and coves fronted by islands, but the coast is too exposed to the sea and prevailing winds to offer much convenient or even secure shelter.

Concepcion Strait separates Madre de Dios and its island to the southward from the main land. It commences at Cape Santiago, in lat. 50¾°, and joins the Wide Channel, or Brazo Ancho of Sarmiento, in 50° 5′. On the West side (the eastern coast of Madre de Dios) are several convenient anchorages, all of which, being on the weather shore, afford secure anchorage: but the squalls off the high land are not less felt than in other parts.

St. Andrew's Sound is 4 leagues wide; but the mouth is much occupied by the

* Sarmiento, p. 129, *et seq.*

† These islets were covered with black-necked swans, and the sound generally is well stocked with birds.

‡ Sarmiento, p. 133.

§ So called after Sarmiento's boat. It was by this route he passed down to the examination of his Ancon Sin Salida; he describes it as a narrow, 300 paces wide.—*Sarmiento*, p. 130.

Canning Isles. The principal entrance of St. Andrew's Sound is to the North of Chatham Island. It is 5 miles wide, and, at 6 leagues within, divides into two arms; the northern one is 5 or 6 leagues long, and terminates; but the southern channel, which is Pitt Channel, trends behind Chatham Island, and communicates, as before mentioned, with Peel Inlet.

At Point Brazo Ancho the Gulf of Trinidad commences, and the Concepcion Strait terminates; for its continuation to the N.E. bears the name of Wide Channel, which is 40 miles long, and from 1¾ to 3½ miles broad.

At Saumarez Island it joins the Mesier Channel, and to the N.E. communicates with Sir George Eyre Sound, which is 40 miles long, and with an average breadth of 4 miles. Near the entrance on the East side was found a large rookery of seals, and another, 13 miles farther up, on the same side, in lat. 48° 21′.

The southern end of the *Mesier Channel*, for nearly 10 leagues, is named *Indian Reach.* It is narrow, and has many islets, but the water is deep. Then follows English Narrows, 12 miles long, and from half to 1½ miles wide; but many parts are contracted by islands to 400 yards. The passage lies on the West side of the channel, to the westward of all the islands.

From the North end of the Narrows to the outlet of the Mesier, at Tarn Bay, in the Gulf of Peñas, a distance of 75 miles, the channel is quite open and free from all impediment.

Every bight in the Wide and Mesier Channels offers an anchorage, and almost any may be entered with safety. On all occasions the weather shore should be preferred, and a shelving coast is generally fronted by shoaler soundings, and more likely to afford moderate depth of water than the steep-sided coasts; for in the great depth of water alone consists the difficulty of navigating these channels.

Throughout the whole space between the Strait of Magalhaens and the Gulf of Peñas, there is abundance of wood and water, fish, shell-fish, celery, and birds.

We again resume the description of the outer coasts.

The coast northward of Cape Tres Montes is lofty and weather-beaten. It is remarkable for the bold outline of its hills, and the thick covering of forest, even on the most precipitous flanks. There is no outlying danger off it. The water is deep, and the land from 2,000 to 4,000 feet high.

Cape Gallegos is a bold promontory, barren to seaward, and rising abruptly from the water.

San Andres Bay is 28 miles round the coast to the northward of Cape Tres Montes. *Cove Creek*, in this bay, is narrow, and may be easily recognised by its proximity to a singular cone, 1,300 feet high, an unfailing land mark.* It is a place difficult to be got out of, and not to be recommended unless in distress. There is also anchorage in Christmas Cove in the S.W. part of the bay. *Useless Cove*, on the northern side of the bay, is well named. Pringle Point is the North

* "This hill is even more perfectly conical than the famous Sugar-loaf, at Rio de Janeiro. The next day after anchoring, I succeeded in reaching the summit of this hill. It was a laborious undertaking, for the sides were so steep, that, in some parts, it was necessary to use the trees as ladders. There were also several extensive brakes of the fuchsia, covered with its beautiful drooping flowers, but very difficult to crawl through."—*Darwin*, p. 343.

point of San Andres Bay, and hence to Rescue Point the land is considerably less high. Stewart Bay and Cliff Cove seemed to promise anchorage.

Port San Estevan is 20 miles N. by E. of San Andres Bay It is sheltered by Rescue Point,* in lat. 46° 18′ S., and lon. 75° 14′ N., and terminates in a fresh-water river, or rather mountain stream. Dark Hill is an excellent mark for it.

The *Hellyer Rocks* are a very dangerous patch, having soundings around them, and lying 6 miles from the nearest land, in lat. 46° 4′, 14 miles *N. by W. ¼ W. North* (*true*) from Rescue Point.

Cape Taytaohaohuon, or *Taytao*, is in lat. 43° 53′ 20″ S., and lon. 75° 8′ 10″ W. It is one of the most remarkable promontories on this coast, and forms the S.W. point of the land encircling the Chonos Archipelago. It is a high, bold promontory, and its neighbourhood is an unprofitable wilderness of rocky mountains, woody and swampy valleys, islands and rocks in profusion, and inlets or arms of the sea penetrating in every direction. It makes like a large island, pointed at the summit, and is near 3,000 feet high, rugged, barren, and steep. Several rocks above water lie around it; none, however, a mile off shore. Northward of the cape is Anna Pink Bay, so named by the surveyors from the narrative in Anson's Voyage. Within a cove of this bay the *Anna Pink*, one of Anson's squadron, employed as a victualler, took refuge from westerly gales. She anchored under Ynche-mo Island, but drove from thence across the bay, and subsequently brought up in Port Refuge.

Port Refuge is a safe but out-of-the-way place. It is described in Anson's Voyage (chap. 3) in very glowing colours; but it may be remarked, that those who discovered it were here saved from destruction.†

Canaveral Cove, West of Port Refuge, though small, is very convenient for refitting, or executing any repairs. Patch Cove ‡ is so small as to be unfit for vessels of any size exceeding 200 tons.

* Upon Rescue Point the *Beagle* found five seamen, who had lived in these solitudes for thirteen months. They had deserted from an American whaler, in Oct., 1833, near Cape Tres Montes, and in landing had destroyed their boat. They then endeavoured to penetrate northward on foot, but found, to their dismay, that there were so many arms of the sea to pass round, and that it was so difficult to walk, or rather climb, along the rocky shores, that they abandoned the idea, and remained stationary at Port San Estevan, after losing one of their companions. They lived, for thirteen months, only upon seals' flesh, shell-fish, and wild celery; yet these five men, when received on board the *Beagle*, were in better condition, as to healthy fleshiness, colour, and actual health, than any five individuals belonging to the ship. During the whole period they had only once seen eight vessels sailing together to the northward, until their most fortunate and unhoped-for rescue by the *Beagle*, Dec. 28, 1834. These remarks are quoted as an example to any who may have the dreadful misfortune to be cast away on this coast.—*Voyages of the Adventure and Beagle*, vol. ii. p. 370-1; and vol. iii. p. 344.

† See also Morrell's Narrative, p. 159.

‡ "We anchored (Dec. 30, 1834) in a snug little cove (Patch Cove, in Anna Pink Bay), at the foot of some high hills, near the northern extremity of Tres Montes. After breakfast the next morning, a party ascended one of these mountains, which had an altitude of 2,400 feet. The scenery was remarkable. The chief part of the range was composed of grand, solid, abrupt masses of granite, which appeared as if they had been coeval with the beginning of the world. The granite is capped with slaty gneiss, and this, in the lapse of ages, has been worn into strange finger-shaped points. These two formations, thus differing in their outlines, agree in being almost destitute of vegetation. This barrenness had, to our eyes, a still stranger appearance from our having been so long accustomed to the sight of an universal forest of dark-green trees. The complicated and lofty ranges bore a noble aspect of durability, equally profitless, however, to man and to all other animals."—*Mr. Darwin*, p. 345.

North-east of Ynche-mo, about 6 miles distant, are the *Inchin** or *San Fernando Islands*, and next to them are the *Tenquehuen*, *Menchuan*, and *Puyo Islands*, among which no doubt there are many good anchorages, and abundance of fresh water, wood, wild herbs, and fish.

As a general rule it may be observed that there are no sand banks, little or no currents, and, generally speaking, no hidden dangers on the West coast of South America, between the Strait of Magalhaens and Chiloe. Rocks under water are either buoyed by kelp, or are distinctly visible to an eye aloft, if the sea does not break on them so as to show their position exactly to an eye on deck. The western extremity of Menchuan Island is low, and has several rocks near it, therefore a good berth should be allowed in passing.

The inner coasts hereabouts, those of *Skyring*, *Clements*, *Garrido*, and *Isquiliac Islands*, are high, rugged, and barren, ranging to about 3,000 feet above the sea. In the middle of Darwin Bay, that large bight between Tenquehuen and Vallenar Islands, is a detached and somewhat dangerous islet, named *Analao*.

The wide inlet called *Aguea*, or *Darwin Channel*, leads to the interior sounds behind the Chonos Archipelago, in which harbours are as numerous as islands.

Within the *Vallenar Islands*, well pointed out by the mountain of Isquiliac, which is 3,000 feet high, very rugged, and triply peaked, is an excellent road-stead, easy of access and easy to leave. The best anchorage is in about 12 fathoms water, near the little islet which lies off the S.E. end of Three Finger Island. The *Beagle* lay there quietly during a heavy S.W. gale.

North-westward, about 30 miles from the Vallenar Islands, is the island of Huamblin,† under which there is good anchorage.

SOCORRO or HUAMBLIN ISLAND lies off the Chonos Archipelago. Its South extreme is in lat. 44° 55′ 50″ S., lon. 75° 12′ 45″ W. It is a comparatively low and level island, about 300 or 400 feet high, except one hill, which is about 700 feet. When made from the offing, it is considerably detached from those which seem like Tierra del Fuego, being a range of irregular mountains and hills, forming apparently a continuous coast.

The three outlying and neighbouring islands of Huamblin or Socorro, Ypun or Narborough, and Huafo or No-man's Island, are thickly wooded, and, as before stated, rather level, compared with their neighbours, and not exceeding 800 feet in height. There are few, if any others, like them in the Chonos Archipelago; almost all the rest, however portions of some may resemble them, being mountainous, and very like those of Tierra del Fuego and the West coast of Patagonia, beyond 47° South. The vegetation is therefore much more luxuriant, and there is a slight difference on it, consequent, probably, upon a milder climate; there are some

* This island was noticed by Anson, from the circumstance of the *Anna Pink* having been driven into the bay. He considered that it would be a good position for a settlement. But in the Spanish expedition, undertaken subsequent to this voyage for colonization, it was thought that it might be safely abandoned, holding out no inducement whatever for settlers. The English, when here, seized one of the boats belonging to the Indians of the neighbouring continent, who came here to fish; and upon asking how they called that island, were answered, "*Inchin*," "it is ours;"—from which mistake arises the name now given to the island.—*Sir Woodbine Parish*, *Journal of the Geographical Society*, vol. iv. p. 184.

† "Huamblin—if, as I suppose it, a corruption of Huampelen—means 'on watch,' 'posted as a sentinel.' Ipun means 'swept off,' or 'swept away;' Lemu means 'wood;'—names singularly applicable to each of those islands respectively."—*Capt. FitzRoy*.

productions, such as canes and potatoes, &c., found, which do not grow near the Strait of Magalhaens. The formation of the three islands is a soft sandstone, which can be cut with a knife as easily as a cake of chocolate.

Abundance of vegetables and live stock might be raised on Huamblin and Ypun for the supply of shipping. Easy places to approach, or leave, they both are, and the rocks which lie around their more exposed points are all distinguished by the sea always breaking on them, and may therefore be easily avoided. These two islands are valuable even now, and likely to become more so.

The whole of the surface of the Chonos Archipelago, of which little was correctly known prior to the *Beagle's* survey, is a succession of high and considerable islands, so near one another, that, from the offing, they make like a solid unbroken coast.

Adventure Bay, to the eastward of Huamblin, is bounded by dangerous outlying rocks. *Paz* and *Liebre Islands*, in the middle of the bay, are remarkable from their height and conical form, but they afford no shelter. Under Ypun or Narborough Island, however, there is good anchorage in 12 to 16 fathoms, over clay and sand: and *Scotchwell Harbour*, at the S.E. part of the island, is not only a valuable place of refuge, but a perfectly secure and agreeable place for wooding, watering, or refitting.* On these islands were a considerable number of seals.

The *Island of Ypun* resembles Huamblin in its character, and therefore differing totally from the rest of the neighbouring islands, which are high, rugged, and generally barren to seaward; while these are comparatively low, level, and fertile. Scotchwell Harbour should be approached from the northward, because, although the passage South of it has been examined, and appeared to have no hidden dangers, it is narrow, and there may be undiscovered rocks.

The cluster of islands between Narborough and the Guaytecas offer no anchorages so easy of access to a stranger as those previously mentioned.

At the North end of the Chonos Archipelago, among the group of islands called *Guaytecas*,† will be found *Port Low*, an excellent harbour; in approaching it a good berth must be given to the numerous rocks that lie along the N. and N.W. shores of those islands, and allowance made for the stream of tide, which is felt off Huacanec Island, and causes a race off Chaylaime Point.

In running for Port Low from the westward, the Guaytecas Islands appear in a hummocky ridge, at the N.E. point of which there is a remarkable flat-topped island. This table looks like the N.E. point of a large island, of which the S.W. part diminishes into low land. When seen from a considerable distance, about 20 miles, the flat island, summit-knobbed hill, and hummocky ridge are still conspicuous. This hummocky ridge appears to be the middle of the group of islands. On the left, looking to the S.E., is a single-knobbed hill inland, which

* "On the shore, near Scotchwell Harbour, was a large bed of strawberries, like those that grow in English woods; and there was a sweet-scented pea, besides abundance of other vegetable produce, both herbage and wood, and plenty of water."—*Mr. Stokes, Voyages of the Adventure and Beagle*, vol. ii. p. 374.

† "About the Guaytecas (or Huaytecas) Islands quantities of excellent oysters were found, quite as good as any sold in London. No quadrupeds were seen, except nutrea and otters, which were numerous."—*Voyages of the Adventure and Beagle*, vol. ii. p. 375.

looks higher and insulated. As far again to the left, is the flat-topped island already mentioned, beyond which there appears to be an opening; the low land to the westward makes like many islands.*

Port Low has the usual supplies: water of excellent quality, wood, fish, shell-fish, including oysters, and wild herbs. Of late years, potatoes have been planted by otter-hunting and sealing parties from Chiloe, therefore a small supply may be looked for. This is a port in which a number of large ships might lie conveniently, it being one of the best harbours on the coast.

Westward of the Guaytecas Islands, distant about 20 miles, is HUAFO or NO-MAN'S ISLAND, a large island, but without a harbour, except for boats; the highest part is the N.W. head (Weather Point), 800 feet above the sea. Reefs extend 3 miles seaward to the North and West. It is highest at the N.W. end, low in the middle, and high again at the S.E. extremity; it is well wooded, and formerly had many sheep on it, while the aborigines lived there in peace.†

CHAPTER IV.

THE ISLAND OF CHILOE.

THE Island of Chiloe is the northernmost of that vast chain of islands which fronts the American continent from lat. 42° southward to Cape Horn. From its situation, it is of great consequence, and, under its former domination, was considered as the key to the king of Spain's possessions in the Pacific. The policy of that government was to conceal everything relating to it, so that up to a recent period but little was known of it. Although much wilful ignorance has been attributed to the Spaniards regarding their possessions during the last century, it must not be supposed that they were as much so as has been stated. Alarm for their own interests did occasionally rouse them, and the voyages of our own great navigators stimulated them to exertions which have remained unknown to the rest of the world. Thus, Lord Anson's voyage, of which the account appeared in 1748, led to an expedition, which, among other results, made a careful examination of Chiloe and the Chonos Archipelago. The same results attended the publication of Cook's Voyages.—(See Sir Woodbine Parish, in the Journal of the Royal Geographical Society, vol. iv. 182—191.) Should it

* "From Port Low we saw a notable mountain, one of the Cordillera of the Andes, having three points upon a small flat top, about 8,000 feet above the sea. I called it the Trident at the time, but afterwards learned that there are four peaks (one of which was hid by another from our point of view), and that it is called by the aborigines Meli-moyu, which in the Huilli-che language signifies four points. Three other remarkable mountains, active volcanoes, are visible from the northern Huaytecas Islands, as well as from Chiloe: the Corcobado (hump-backed); Yanteles (or Yanchiñu, which means 'having a shivering and unnatural heat'); and Minchenmadom, which in the Huilli-che language means 'under a firebrand;'—names so expressive and appropriate, as to put to shame much of our own nomenclature."—*Capt. FitzRoy.*

† Voyages of the *Adventure* and *Beagle*, vol. ii. p. 377.

emerge (as it apparently is doing) from the darkness and indolence which, in common with other Spanish colonies, characterise its inhabitants, it will become of very great importance in the future history of the South American continent.

It is about 100 miles in length, by about 35 in breadth, and within, or to the East of it, is a numerous archipelago of islands, 63 in number, of which 36 are inhabited. The Island of Chiloe itself, or Isla Grande, is hilly, but not mountainous, and covered with one great forest, particularly cypress, and "alerse," a variety of it; these affording a large article of export, in the form of planks. The interior of the island is not known, and the only road is an artificial one between S. Carlos and Castro. The other articles produced are potatoes, wheat, barley, and hams, for which it is famed. Poultry may be had in abundance.*

The province of Chiloe is one of the eight divisions of the Chilian republic, and extends on the main land as far as the South bank of the Maullin, and southward to Tres Montes, but there is nothing to govern on the Chonos Archipelago. Its name is significant of the origin of its people—"Chili-hue," which means "farther," or "new," or "the end of" Chili. Hence, Chilóe, or Chiloé, as it is sometimes written and pronounced, into which it has been corrupted. It was first discovered by the Spaniards, in 1558, when it was thickly populated by Chonos Indians, or Huyhuen-che. The first Spanish settlement was made in 1566, when the city of Castro was founded, during the government of Don José Garcia de Castro in Peru, from which it takes its name of Castro. Chacao was then made its seaport, but was afterwards abandoned. It continued under Spanish domination until the overthrow of that power in Chile, and was the last possession held by Spain upon the continent of South America. The principal portion of the natives are now Huilli-che, nominally Christians, but painfully ignorant of Christianity. The language used is almost exclusively Spanish.

The island is divided into 10 districts or partidos; the principal population, which would appear to be considerable for the districts occupied, are centred at S. Carlos, and in Castro, Quinchao, and Lemuy. In 1832, the census gave a total of 43,832 for the population of the province.

The *climate* of Chiloe is considered by those who live in other parts of Chili to be rigorous, cold, and damp: certainly there is much reason for such an opinion, particularly in the winter months, when it almost always rains, and the wind, with little cessation, blows hard from North to N.W., and by the West to South; but, notwithstanding the great quantity of rain that falls, the evaporation is great, and it cannot therefore be called unhealthy; indeed, from experience, it is considered quite otherwise.—(Capt. King, R.N.)

There is a marked difference of climate between the East and West sides of Chiloe, as to the quantity of rain and wind. A proportion of both appears to be arrested (as it were) on the windward side of the heights, so that the neighbourhood of Castro, and the islands in the Gulf of Ancud, enjoy much finer weather than is met with about San Carlos. There is an idea prevalent in Chiloe, that after a

* There is a good deal of *coal* in Chiloe (as in the island of Lemuy), but of an inferior description, like that of Concepcion. It is not true coal—lignite would be a more appropriate term. However, it burns readily. Seals are now rare, and whales are fast diminishing in numbers.—*FitzRoy*, p. 382. See hereafter on this subject.

great eruption of Osorno, in particular, or indeed of any of the neighbouring volcanoes, fine weather is sure to follow.—(Capt. FitzRoy.)

The coasts abound in shell-fish, which afford an inexhaustible supply to the inhabitants, and are obtained with the greatest facility. Money is almost unknown, therefore all transactions are by barter. Of the various ports and places we shall remark in the subsequent descriptions.*

Off *Olleta Point*, which is the southernmost point of Chiloe, and to the southward of the mountainous island of San Pedro, are the *Caduhuapi* and *Canoitad Rocks*, dangerous in the night or during calms. The Canoitad Rocks are 4½ miles from the nearest land, and the tide stream sets towards them.

HUAMBLIN or SAN PEDRO ISLAND makes at a distance like a rounded lumpy hill; when near, it appears to be wooded to the very summit, though 3,200 feet in height. At the entrance of its harbour a white rock, near the N.E. head, may be noted as a mark; care should be taken to avoid the three-fathoms bank, extending two-thirds across the entrance of the harbour, if the tide is low. There are nearly 2 fathoms rise at springs in this place.

CAPE QUILAN is rounded and woody; there are cliffs in its vicinity of a light yellowish colour, about 300 feet in height.

From Cape Quilan to *Pirulil Head*, a similar character of coast line continues; there is no kind of anchorage, scarcely can even a whale-boat find a place of shelter where she could be hauled ashore.

Cucao Bay is bounded by a low beach, always lashed by a heavy surf. Cucao heights are remarkable, as being the highest and most level high lands in the island: they are wooded to their summits, and in height from 2,000 to 3,000 feet.†

Cape Matalqui is remarkable; the heights over it rise about 2,000 feet, and make from seaward in three summits. Off all this coast from Cape Quilan northward, there are no outlying or hidden dangers. Between Cocotue and Caucahuapi Heads, a low isthmus joins the peninsula of Lacuy to the rest of the main island.

Caucahuapi, *Guabun*, and *Huechucucuy Headlands*, are bold, cliffy promontories, needing scarcely any remarks. The latter is a high, steep, bare bluff; these three headlands are the first seen when making the land near the port of San Carlos. In approaching that excellent harbour, the *Huapacho Shoal* must have a good berth; it lies 1½ miles West of Corona Head, which is a light-coloured cliffy head, bare at the top, and broken at the seaward extremity; in the night, more especially, this shoal should be guarded against, the land behind being a low sandy beach, not then distinguishable.

PORT SAN CARLOS.—The low extreme of *Corona Head* is sometimes called Tenuy Point; probably it extended farther seaward, and was more

* See Account of Chiloe, by Capt. Blanckley, R.N., 1834, Journal of the Royal Geographical Society, 1834, pp. 344—361; Voyages of the *Adventure* and *Beagle*, vol. i. p. 269, *et seq.*; vol. ii. p. 378, *et seq.*; vol. iii. p. 333, &c.; Morrell's Voyages, p. 161; and Agüeros' History of the Province of Chiloe.

† "The district of Cucao is the only inhabited part on the whole West coast of Chiloe. Its Indian population are very much secluded from the rest of Chiloe, and have scarcely any kind of commerce, except sometimes in a little oil, which they get from seal-blubber."—*Darwin*, 2nd edit. p. 295.

remarkable formerly. The stream of tide is strong hereabouts, and must be allowed for, according to the direction. Ships should steer for Corona Head, keeping it to the southward of East, on account of the Huapacho Shoal, and thence along shore at less than half a mile distant round Huapilacuy Point and Aguy Point (on which there was a battery in the time of the Spaniards, just above the point), to an anchorage near Arena Point, under Baracura Heights. Between Aguy Point and Baracura Head, on which there was also a small battery, a ship should not close nearer than half a mile, as a shoal called *Pechucura* extends to nearly that distance, half-way between the two points.

On no account ought a vessel to get near the islets of Carelmapu or Doña Sebastiana, for there the tide runs strongly, and with dangerous eddies. The *Yngles Bank* must also be particularly avoided; it is a very dangerous shoal, over which the tide runs with great strength: the shallowest spot found by Capt. FitzRoy, at slack water, after vainly trying to stem the tide at half-ebb, in a fast-pulling whale-boat, had not a fathom of water, the bottom sand, or hard *tosca*, or sandstone. A long lance forced into the bank penetrated about 2 feet into sand generally, but in one small spot it was stopped by a hard substance which bent the lance, and turned its point. This substance felt like wood rather than rock. If not sand, sandstone, or tosca, it might have been a piece of a wreck.

The best anchorage for a large ship is with Aguy Point bearing North, and the extremity of Arena Point bearing S.W. Trading vessels anchor off the town of San Carlos in 4 fathoms water, with the town bearing about East, but it is an exposed position, and insecure.

The town of *San Carlos* is built on two rising grounds, and in the valley that separates them; through which a rivulet runs into the bay, at a mole which affords sufficient protection to the boats and periaguas frequenting the port. The houses, which are all of wood, are generally small, and have but little comfort. The plaza, or square, without which no town in Chili of the least importance is to be found, is situated on a flat piece of ground at the summit of the southern hill, and commands an extensive view. It is about 180 yards square, and has a flagstaff in the centre. On the North side there is a strong, well-built, stone storehouse, and opposite to it is the church, also built of stone. On the side next the sea is the yntendente's residence, a low range of wooden buildings, erected without regard to taste, convenience, or comfort; and opposite to this are two or three dwellings, very little superior to common huts or ranchos. Within the last few years, however, some substantial stone buildings have been erected by the wealthy people in the town, an example which is likely to be followed.*

The people of Ancud, the local name of San Carlos, formerly so simple and artless, have gradually changed this character, and become corrupt and degenerate; they have acquired the vices consequent upon a frequent intercourse with the whale fishers, and other ships which frequent the port.

Off the East point of *Cochinos Islet* a shoal extends about a mile; on it are 2 fathoms water. N.N.W., 1 mile off *Point Mutico*, there is a patch of rocks; and all the bottom thereabouts is very irregular; patches of kelp are seen

* King, pp. 274-5.

frequently, but they seem to be attached to large stones as well as to rocks.* *Pecheura Point* is about 3 miles from the entrance to the Narrows: a rocky patch runs about a quarter of a mile off it, which forms the termination of a bank extending from the shore between it and Matico Point.

Nearly a mile to the westward of Punoun Point there is a rock awash at low neap tides, and another rock lies just to the westward: they are called the *Periagua Rocks*. The rise of tide here is about 11 feet at springs, and 7 feet at ordinary neaps; the soundings are very irregular. During the strength of the flood tide there is a heavy tide rip off Punoun Point, caused by the strong stream running over so very irregular a bottom.

Punoun Point is low, with a sandy beach: it is the S.W. limit of the Chacao Narrows, a dangerous passage, and seldom used except by the large coasting boats called *periaguas*.

On the opposite coast, that of the mainland, are *Carelmapu* and *Chocoy Heads*; steep cliffs, in front of which runs a powerful stream of tide.

Excepting the great strength of tide which may prevent a ship from being under command, the only danger to be encountered in the Chacao Narrows is a rock in the mid-channel, at the narrowest part; this rock, called by the natives *Petucura*, is awash at half-tide. The stream runs very strongly over and past it, during ebb, as well as flood tide.

San Gallan Point, on the Chiloe shore, is steep, with a remarkable clump of bushes on its summit, which is about 500 feet high. The North shore opposite is low, except near Coronel Point, where there are cliffs, about 100 feet in height. Behind these cliffs the land rises to about 20 feet, and is thickly wooded.

Between San Gallan Point, and *Santa Teresa Point*, the distance is just 1 mile: it is the narrowest part of the passage from shore to shore, and half a mile farther eastward the rock Petucura divides the channel into two narrow passages, either of which may be used.

Another rock, more dangerous to large ships than Petucura, lies E. ½ S. from it, distant half a mile. On this rock, called the *Seluian*, there are 2 fathoms at low water. Round this and the other rock there is deep water, except to the eastward, in which direction a rocky ridge extends a quarter of a mile.

Between Coronel Point and Tres Cruces Point there is deep water, about 50 fathoms; but in Chacao Bay there is excellent anchorage in about 10 fathoms half a mile North of Chacao Head.

* "The water of the bay is clear and good; only round the little island of Cochinos, and along the harbour, it is covered with a quantity of sea-moss, which renders landing difficult. It frequently happens that commanders of ships, wishing to go on board to make sail during the night, get out of the right course, and instead of going to the ship steer to Cochinos, and get into the moss, when their boats stick fast, till returning daylight enables them to work their way out.

"The poor inhabitants boil this sea-moss, and eat it. It is very salt and slimy, and is difficult of digestion. Among the people of Chiloe this sea-moss occupies an important place in surgery. When a leg or an arm is broken, after bringing the bone into its proper position, a broad layer of the moss is bound round the fractured limb. In drying, the slime causes it to adhere to the skin, and thus it forms a fast bandage, which cannot be ruffled or shifted. After the lapse of a few weeks, when the bones have become firmly united, the bandage is loosened by being bathed in tepid water, and it is then easily removed. The Indians of Chiloe were acquainted long before the French surgeons with the use of the paste bandage."—*Dr. Von Tschudi*, pp. 11, 12. We give this quotation, as it may afford a useful hint to commanders in an emergency, when surgical assistance cannot be very acquirable.

When the Spaniards first settled in Chiloe, their head-quarters were at Chacao, and their vessels anchored in this bay.* The state of the tide, and having sufficient wind to keep a vessel under command, are the principal points to consider when about to pass Chacao Narrows. A temporary anchorage may be had on the South side of the Narrows, between Punoun and San Gallan Points, by getting close to the shore; and as the tide, strong as it is, sets to each side of, rather than towards, the Petucura Rocks, the passage of these Narrows is not so formidable as it appears to the Chiloe boatmen.

About a mile South of Tres Cruces there is a stony point, after passing which the tide is scarcely felt; and in the Bay Manao there is no stream of tide: near that bay the tides usually meet. The nature of the tides around Chiloe will be hereafter described.

Chilen Bluff is a low shingle point, with a remarkable tree on its extremity. About half a mile in shore the land rises suddenly to about 150 feet. Off the N.E. point of *Huapilinao Head*, to the S.E. of it, there is a reef of rocks extending above a mile from the point. About 4 miles from Huapilinao is the small village of *Lliuco*. *Queniao Point*, which follows, is a low stony point, with a remarkable single tree on its extremity. Shoal water extends nearly a mile off the point.

Lobos Head, on the island of *Caucahue*, is a steep bluff, above 250 feet high; behind it the land falls suddenly, and is very low for a short distance, after which it rises again.

About 1½ miles from Queniao Point there is a sandy spit, with 2 fathoms of water on it about a quarter of a mile from the shore, when it deepens suddenly to 8 and 12 fathoms near the shingle spit, which forms the small but valuable harbour called Oscuro Cove. This cove may become of great use, as the tide rises in it to about 20 feet: the water is deep close to the shore, and always perfectly still. The entrance is about 3 cables wide, the point of the spit steep-to; the length of the cove is three-quarters of a mile, and its breadth 3 cables: there are 7 fathoms of water within 50 yards of low-water mark, and from 12 to 16 in the middle, over a bottom of mud and sand. The West side of the entrance is a rocky point, with stones lying off it half a cable's length. A vessel should keep close to the other side.

In this cove any ship might be laid ashore, hove down, or thoroughly repaired, with perfect safety and great ease. I am not aware that there is any similar place on the West coast of South America: the flood tide here runs to the northward, and strongly at spring tides.

Chogon, a bluff point about 200 feet in height, lies a long mile to the southward of Quintergen Point, which is low and stony, with a shoal spit of about a quarter of a mile in length. Between them lies Caucahue Strait, and in the entrance there is no bottom with 50 fathoms.

Between the Chaugues Islands and Quicavi Bluff, there are some reefs, and a

* As before stated, this was formerly the principal port in the island; but many vessels having been lost, owing to the dangerous currents and rocks in the strait, the Spanish government burnt the church, and thus arbitrarily compelled the greater number of the inhabitants to migrate to S. Carlos.

tide race. Pulmun Bank, 4 miles East of Chogon, is shown always by its breakers.

The flood tide sets close round Tenoun Point, which has a reef extending half a mile off it, and then across the channel towards the Chaugues Islands; the ebb tide sets to the S.W. close round the point, and at the commencement of the springs, at the rate of 2 knots.

About a mile to the southward of Quicavi Bluff, lies the *Laguna of Quicavi*, which is an excellent place for boats, and when inside, they can lie afloat at low water, but it cannot be entered until the tide has flowed some time.

Eastward of this cove, and between the points before mentioned, at the distance of 3 miles, lie the group of islands called the *Chaugues;* they are four in number, and separated by a channel, running nearly North and South, and 1½ miles wide in its narrowest parts, in which there are from 48 fathoms to no bottom with 55 fathoms. The western island is the highest, being about 350 feet high, and forms a ridge East and West; the N.E. island has a round hill upon it, nearly as high as the former, but the other parts are much lower: there are some cleared patches, but they appeared thinly inhabited by Indians.

The *Island of Linlin*, which lies 4 miles to the south-westward of Tenoun Point, is low in the centre, gradually rising to a round hill terminated by a bluff, both to the northward and southward.

To the southward of Linlin stands the smaller island of *Linna*.

The channel narrows gradually to the westward, as far as the N.W. point of Quinchao; it then turns suddenly to the S.W., and is not more than a mile wide. On the Chiloe shore there is a small village called *Dalcahue*, with its saw-mills; the best water in the channel runs close to the shore of Quinchao, and the deepest water is 4 fathoms. The tide runs through it about 4 knots at springs.

The channel opens out to the southward into a broad bay on each shore. On the Chiloe side lies the small cove and village of *Relan:* in the entrance of the cove there are 18 fathoms. The ebb tide sets very strongly across Relan Reef to the S.E. towards the channel between the islands of Lemuy and Chelin. Between Lemuy and the main the tide was scarcely perceptible; what little was found appeared to set to the eastward; but the springs rise 18 feet.

From Relan Reef to the entrance of Castro Inlet, the channel is from 2 to 3 miles wide; the East entrance point of the inlet is low and stony, but a vessel may pass at a quarter of a mile off it in 12 fathoms: the western side of the entrance is formed by Lintinao Islet, to the southward of which lies the small harbour of Quinched, in which a vessel bound to Castro might wait for a favourable opportunity to go up, in case she found the winds baffling in the two first reaches; this is generally the case with northerly winds, however strong, and no anchorage can be found in either reach until too near the shore for safety.

The village of *Quinched* is about 3 miles to the westward of the harbour; the country is well cultivated and thickly inhabited for about 3 or 4 miles on either side of Castro Inlet, and the houses are numerous, and surrounded with apple trees.

On the outer point of Lintinao Island, which is joined to Chiloe by a sandy spit that dries at low water, a stony point runs off about 1 cable to the eastward, but the South side of the point is steep-to.

At half a mile above the second reach of Castro Inlet, the eastern shore may be approached within half a cable, but the other side is flat and shallow for nearly half a mile from the beach, and shoals too suddenly for a vessel to go by the lead; in working up or down, a vessel should keep the former aboard, not going farther across than two-thirds the breadth of the channel.

The eastern shore is composed of steep-wooded slopes, rising to about 150 feet above the sea; the western rises gradually from the beach, forming several level steps, which increase in height to 400 or 500 feet: behind them, at a distance of 5 miles from the beach, there is a range of hills nearly level, about 1,000 feet high and thickly wooded. Two miles below Castro, on the Chiloe shore, there is a small cove where vessels might anchor if necessary; but there are 20 fathoms between the points, and it shoals suddenly a little inside of them. The point of Castro is a level piece of land, about 100 feet above the sea, running out between the small harbour to the northward, and the River Gamboa to the southward; it terminates in a low shingle point, which is steep-to on its North side, but to the southward of it a flat commences, which follows the western shore all down that reach of the inlet. The small harbour to the northward of the point is half a mile in length and one-third of a mile wide; between the points there are 7 fathoms, but it shoals gradually to 3, about a quarter of a mile further in; the best anchorage is nearest to the South point, as the North side is shoal for about a cable's length off. In running for the harbour, a vessel should keep the eastern shore aboard till she is abreast of it, when she may stand across, and will thus avoid the shoal to the southward of Castro Point, which extends half a mile off.

CASTRO stands near the outer part of the level point, and consists of two or three short streets of wooden houses; there are two churches, one of which, belonging to the Jesuits, has been a handsome building, but is fast falling to decay, and shored up on all sides; the other also appears to have been well built, but is now nearly in ruins: altogether Castro has been much neglected, and the people are poor.* Between San Carlos and Castro is an artificial road. The road itself is a curious affair; it consists in its whole length, with the exception of very few parts, of great logs of wood, which are either broad and laid longitudinally, or narrow and placed transversely, the longitudinal logs fastened down by transverse poles pegged on each side into the earth, rendering a fall rather dangerous.—(Mr. Darwin.)

The tide at springs does not run above $1\frac{1}{2}$ knots in the strongest part, and at neaps it is felt very little.

Opposite to the entrance of Castro Inlet, on the North shore of Lemuy, is *Poqueldon*, the principal village on the island. The landing is bad; the tide

* Castro, the ancient capital of Chiloe, is now a most forlorn and deserted place. The usual quadrangular arrangement of Spanish towns could be traced, but the streets and plaza were coated with fine green turf, on which sheep were browsing. The church, which stands in the middle, is entirely built of plank, and has a picturesque and venerable appearance. The poverty of the place may be conceived from the fact that, although containing some hundreds of inhabitants, one of our party was unable anywhere to purchase either a pound of sugar or an ordinary knife. No individual possessed either a watch or a clock; and an old man, who was supposed to have a good idea of time, was employed to strike the church bell by guess.—*Darwin*, p. 339. Dec. 1834.

at high water flows close up to the trees; and at low water the shores are very muddy.

After passing Quinched, the land is only cultivated in patches, the rest being thickly wooded.

Off the *Point of Yal*, on the Chiloe shore, there are two small low shingle islands; they are connected by a spit, which is covered at high water.

A mile to the S.E. of Yal Point there is a bluff head, which forms the North point of *Yal Bay;* and a little in-shore of the point there is a remarkable flat mound covered with trees. Half a mile inside the bluff lies the entrance of the small harbour of Yal.

There is no anchorage in the bay until within a quarter of a mile of the head; it is not a fit place for vessels to anchor in unless obliged to do so.

The S.W. extreme of *Lemuy Island*, called *Detif Point*, terminates in a perpendicular cliff, about 150 feet high, surmounted by a round hill 250 feet above the sea.

About a league to the N.E. of Detif Point the same headland throws out *Apabon Point*, with a reef extending to the eastward 3½ miles; near its outer edge there is a rock always dry, and at low water the reef uncovers for about a quarter of a mile on each side of it. No vessel should attempt to cross this reef, although there are 9 feet at low water, between it and the shore, because the tide sets over it strongly and irregularly.

Between the S.E. point of Quinchao Island and Apabon Point lie the islands of *Chelin* and *Quehuy;* the N.E. extreme of the latter is called Imeldeb, and is detached from it by a narrow isthmus. Off this point, for a mile to the S.E., there is a shingle bank that dries at low water, on which a French ship struck, and which very considerably narrows the channel between Imeldeb and Chaulinec.

Besides *Chaulinec*, there are two smaller islands lying to the eastward of Quinchao Point, called *Alau* and *Apiau;* and reefs extend off the North end of both of them, from the latter as far as 2½ miles. At the S.W. end of Alau, close to the entrance of the channel, between it and Chaulinec, there is a small harbour or cove.

South-eastward of Linlin and Linna, and midway between Quinchao and the Chaugues Islands, lie *Meulin, Quenac, Cahuache* and *Tenquelil Isles;* on Cahuache there is a round hill, 250 feet high, which commands a good view of the neighbouring islands.*

Tiquia Reef, which lies from 2 to 3 miles East of Cahuache, is about a league in length, N.W. and S.E., half a mile broad, and dries at low water. To the S.E. of Chaulinec Island, lie the Desertores.

The largest of these islands is called *Talcan;* it is 9 miles long and 4 broad, and has a deep inlet on its S.E. side; just outside the entrance, a bay is formed between the points, in which lie several patches of kelp: and about half a mile beyond the line of the points, there is a reef of rocks which dry at low water; a small channel leads into the bay to the northward of them, which is visited for fishing in the season.

* This neighbourhood is the most cultivated part of the Archipelago; for a broad strip of land on the coast of the main island, as well as on many of the adjoining ones, is almost completely cleared.

Many scattered rocks lie off the S.W. and southern part of the island to the distance of a mile, and off its North point a shoal extends as far as 1½ miles, with from 4 to 6 fathoms on it. Two miles from the point there is a rock about 10 feet above the sea, on which many seals rest. Vessels seeking anchorage among these islands should be cautious in approaching them in consequence of the rocks before named.

The smaller islands, *Chulin*, *Chiut*, *Nihuel*, *Ymerquiña*, and *Nayahue*, do not afford any shelter for vessels, except off the North end of the latter, which is divided into two islands by a narrow channel, with from 2 to 10 fathoms in it, but quite useless except for boats. Some rocks lie half a mile off the S.E. points.

On the main land, abreast of the S.E. point of Talcan, there is a remarkable sugar-loaf hill, which rises direct from the water's edge, and is thickly wooded to the summit; to the southward of it there is a deep inlet, with an islet at the mouth of it.

Returning again to the main island of Chiloe, *Ahoni Point* lies opposite to Detif Point. *Lelbun Point* lies about 4 miles from Ahoni; and abreast of it the shoal widens to nearly 1½ miles, and is covered with patches of kelp. The ebb tide sets to the S.E. about 2 knots at springs.

Aytay Point is low and rocky, and about 3 miles to the southward of Lelbun Point. Some rocks of a reef which runs out from it dry about 2 miles from the shore.

Quelan Point is a long, narrow strip of land, very low, and covered with trees. Three miles to the eastward of the point is the small island of Acuy.

The *channel* between Quelan Point and the island of Tranque is about a mile wide; and the ebb sets through it to the westward about 2 knots at neap tides. After rounding the point of Quelan, by keeping along the inside of the spit it will lead to the small harbour or cove of Quelan, the entrance to which is about half a mile wide; but the shores on either side should not be approached within a cable's length, at which distance there are 3 fathoms, and 13 in mid-channel.

The cove is about three-quarters of a mile long, and the same broad, but the West side is shallow for a quarter of a mile; the edge of the shoal is in a line with the shingle point, on the eastern side of the entrance; but in every other part of the cove there is good anchorage in from 5 to 8 fathoms, with 3 fathoms a cable's length off the beach. In its N.W. corner there is a narrow creek, but fit only for boats; and there are three or four houses, with patches of clear land around them. The inhabitants were Indians, but the surrounding country seemed thinly peopled. To the westward the land rises suddenly to about 200 feet, and is thickly wooded.

On the North shore, about 1 mile to the westward of the cove, there is a small bay with an island off it, which affords anchorage for a vessel in from 10 to 13 fathoms; the island may be approached within a cable's length, when it shoals suddenly from 10 to 3 fathoms.

A ridge of hills runs through the island of Tranque, from N.W. to S.E.; they are about 300 feet high in the highest part, which is nearest the N.W. end; from thence Tranque Island slopes gradually towards the S.E., and terminates in a low point called *Centinela;* the North shore slopes gradually, and is well wooded; the island appears thinly inhabited.

In the Chiloe shore, abreast of the N.W. end of Tranque Island, there is a deep inlet, called *Compu*, and a little to the eastward of it a smaller one, neither of which were examined. The *flood tide* runs close round the points, and then strikes across toward the North shore, outside the small island, within which there is very little tide; in the narrow channel it runs at least 4 knots at neap tides, sweeping round the rocky points.

Off the entrance to the S.W. channel, between Tranque Island and Chiloe, there is a small island called *Chaulin*. About 1 mile from Cuello Point on the Chiloe shore, in the direction of the West point of Chaulin, there is a stony reef, extending in a N.W. and S.E. direction about half a mile.

Five miles S.E. by S. of Cuello Point lies the *Inlet of Huildad;* its entrance is only 150 yards wide, but is wider within. In the outer harbour there is good anchorage in from 5 to 9 fathoms: the shores are steep-to, except along the bend behind the shingle spit, which is shoal for about a cable and a half from the beach.

The *tide* at the entrance runs on the ebb at springs nearly 4 knots, but inside it slackens considerably: should a vessel wait in Huildad for a change of wind or weather, the outer harbour would be the best, as N.W. gales blow very heavy down the upper harbour, while in the outer one a vessel would be sheltered from every wind.

On the South shore stands the church, with three or four houses round it, the remainder (there are about twenty in all) are scattered along the sides of the harbour.

To the southward of Huildad, between it and Chayhuao Point, a shoal extends above a mile from the shore; it is nearly covered with kelp; the tide at the outer edge of it runs about 1½ knots at springs: the shoal terminates in a long stony reef, which runs off Chayhuao Point to the S.E. There is a channel between the South end of it and the N.E. side of Caylin Island. The reef commences half a mile inside the outer point, and deepens suddenly to 7 and 12 fathoms; at a quarter of a mile inside the reef on the Chiloe shore there are 27 fathoms within half a cable of the beach. The flood tide sets to the East in the channel across the reef at least 3 knots at springs: after passing the reef it meets the outside tide coming from the southward. Between Chayhuao Point and San Pedro passage there is a deep bay, fronted by the islands Caylin, Laytec, and Colita, with the small cove of Yalad to the N.W. of the latter.

The former of these islands is 5 miles long N.W. and S.E., and about a league broad; the North shore is steep-to.

Caylin is called here "El Fin de la Christiandad," the termination of South American christendom. Here is an Indian village, the southernmost place at which provisions can be procured. The village has about forty houses, containing about 250 inhabitants, who were glad to supply Capt. FitzRoy's party with sheep and poultry in exchange for tobacco and handkerchiefs; they seemed anxious to know when the king of Spain would retake the islands.

Laytec Island is 2 leagues N.W. and S.E., and about a league in breadth;

it is separated from Caylin by a channel 2 miles across; off its S.E. extreme there are a few rocks, but no danger appeared beyond half a mile.

The *Island of Colita* is low and thickly wooded, about 4 miles long and 1½ miles broad; the channel between it and Chiloe is very narrow, and apparently not fit for a ship. Between Colita and Laytec Islands the passage is 1½ miles broad. The tide sets about 1 knot through the channel North of the islands.

We return now to the Chacao Narrows; two leagues E. of which lies the *Island of Abtao*: it is 2½ miles long and 1 mile broad; the N.W. point is the highest, and ends in a bluff, 80 feet high; off which a stony flat runs two cables' length: close to the flat there are 12 fathoms, and a quarter of a mile to the N.E. 30 fathoms; the shoal from the main runs off nearly a mile. From the S.E. end of Abtao a shoal extends a mile and a quarter, with 5 fathoms near the extremity.

N.E. of Abtao lies the small island of *Carva*, a round hummock, about 200 yards long, surrounded by a bed of shingle; a shoal extends a mile off its S.E. end. Two miles East of Carva lies the N.W. edge of the *bank of Lami*, always dry in several places; the North side is about 2 miles long, and runs parallel to the shore, at the distance of about 1½ miles; in mid-channel there are 35 fathoms. The passage between the islands of Quenu and Calbuco is about three-quarters of a mile wide, with 21 fathoms in mid-channel; the points of both islands are low. A rocky flat runs off the *Point of Quenu*, but it does not obstruct the channel.

The town of CALBUCO (or El Fuerte), situated near the N.E. end of the *Island of Quenu*, on a steep slope, is about one-third of the size of San Carlos, and superior to any of the other settlements: the church is a large wooden building, but not equal to either of those at Castro; and the land about Castro is better cleared and cultivated. The beach off El Fuerte dries at low water, about a cable's length, and close outside there are 6 fathoms, and a very little farther 17 fathoms near it; the channel then deepens to 24 fathoms. The best anchorage is abreast of the town, about a third of a mile distant, and in from 20 to 22 fathoms, muddy bottom.

The *Island of Puluqui* is thickly wooded; on the East side the patches of clear land are very few, but on the West side, where the land is lower and swampy, they are more numerous.

Centinela Point, the southern extremity of Puluqui, is a low shingle point, thickly wooded; the high land rises about 200 yards in shore, and a flat extends a cable's length from the point: it runs nearly East and West for 3 miles, and then turns to the N.W.; after rounding this point about a mile to the northward there is a small cove, the entrance of which is very narrow, and too shallow for a boat after half-tide, but inside it is about half a mile across, with 8 fathoms in one part.

The *Island of Chidhuapi*, to the westward of Puluqui Island, is low, and nearly all cultivated. The *Island of Tabon* is composed of a number of detached hummocks of land joined together by low shingle spits, some of which are overflowed at high water; it is entirely clear, except the apple trees round the houses; the highest part does not exceed 150 feet. Half a mile to the N.E. of its western extremity, a stony reef runs to the northward, in the direction of the banks of Lami, and is dry at low water three-quarters of a mile from the shore. The channel between it and the South end of Lami bank is about three-quarters of a mile wide, and at two cables' length from the end of the reef there are 7 fathoms. Another reef runs off more to the westward, and to the distance of a mile.

Off Aulen Point, on the main shore, lies the *Island of Cullin*, about a league East and West, and 2 miles North and South; between it and the S.E. point of Puluqui Island is the entrance to *Reloncavi Sound*, which extends 20 miles to the northward, and about 12 miles across, from East to West.

A mile and a half to the northward of Cullin is the *shoal of San Jose*, but there is a clear space of a league between it and Puluqui Island.

Huar Island lies on the West side of the sound, and to the S.E. of it are two shoal patches, *Pucari* and *Rosario;* the eastern side of the latter lies 3 miles from the island. As far as information could be obtained, there is no bottom with 120 fathoms throughout the sound, except in the neighbourhood of the islands and shoals; anchorage may be found under both the former, and doubtless along the shores on either side, according to the prevailing wind. In the entrance between Cullin and Puluqui Islands, there is no bottom with 60 fathoms.

There is a deep inlet on the eastern side of Reloncavi Sound, by way of which and the river Raleon, through Todos Santos Lake, and up the Peulla, a communication was formerly kept up with the Spanish missionaries' settlement on an island in the great lake of Nahuelhuapi; but this mission was abandoned towards the close of the last century.

The *Volcano of Osorno*, or *Purraruque*, or *Hueñauca*, is 7,550 feet above the sea level, and is 26 miles to the N.E. of the head of Reloncavi Sound. This mountain is most striking in form. It is not only quite conical from the base to the summit, but it is so sharply pointed that its appearance is very artificial. When seen from the sea, at a distance of 90 or 100 miles, the whole of the cone, 6,000 feet in height at least, and covered with snow, stands out in the boldest relief from among ranges of inferior mountains. The apex of this cone being very acute, and the cone itself regularly formed, it bears a resemblance to a gigantic glass house, which similitude is not a little increased by the column of smoke so frequently seen ascending. It is one of the indications of what is the actual physical condition of the country; and its eruptions and actions are intimately connected with those tremendous convulsions which we shall have to record in the description of Valdivia, Concepcion, and other places to the northward.

TIDES.—The tide wave from the ocean sets against Chiloe, looking at the whole island and its vicinity from the westward. The body of water impelled round the South end of the large island drives the waters of the Corcovado Gulf northward into those of the Gulf of Ancud, at the N.W. point of which they meet the stream impelled through the Narrows of Chacao. Very little stream is felt in the gulfs, but there is a considerable rise and fall, from 10 to 20 feet, and more or less stream along shore and among the islands.

The tides on the East coast of Chiloe are very irregular, being much influenced by the winds. The time of high water at Castro, and other places, is earlier in going to the southward, yet at Huildad, which is more than 30 miles South of Castro, it was high water three-quarters of an hour later than at Castro; but at the time it was blowing a heavy N.W. gale at Huildad. The average time of high water in the North part of the Archipelago is probably about 1 o'clock, on full and change days, which decreases gradually to about 12^h 15′ near the South end. It appears to be never regular, as it was found to vary half an hour in two following tides.

The rise was also very irregular, as the tides often rose higher when they were taking off. The night tides were always higher than the day.*

In Port Oscuro the rise and fall at one time at dead neap tides were 18 feet, and the next springs it only rose 16; by the marks on the shore the rise and fall at some high tides had been above 24 feet. The greatest rise and fall are at this place, and it is the best for heaving down in the gulf, or for cutting docks, if they should ever be required. The only other place that would answer well for that purpose is the outer part of Huildad Inlet, on the West side of which there are 9 fathoms close to the shore, and the coast is composed of rock, which would answer better than the sand and shingle of Port Oscuro; but the rise and fall are only 15 feet at spring tides, which would be too small for large ships. Port Oscuro may, therefore, be considered preferable.

CHAPTER V.

THE COAST OF CHILE, FROM SAN CARLOS TO HUESO PARADO.

The republic of Chile extends from the ridge of the Andes to the Pacific, from the Island of Chiloe to about lat. 25° 25′ S., or at Point Taltal; but its limits are not exactly defined. It is thus, including Chiloe, about 1,100 miles in length, with an average breadth of 110 and 120 miles; area, including Chiloe, perhaps 130,000 square miles. Its population, variously estimated at from 600,000 to 1,500,000, is probably about 1,200,000. It is divided into nine provinces, including Araucania, which is still independent of the others. Santiago is the capital; Valparaiso the principal port.

There is a marked difference in the appearance and climate of the northern

* "Round Chiloe the flood tide streams run both ways from the S.W., and meet in the N.W. part of Ancud Gulf. The times of syzygial high water, in all the Archipelago, vary only from noon to an hour and a half after noon. In December and January (1834–5), our boat expedition found that the night tides were always higher than those of the day, and the inhabitants said that was always the case in the summer. In the months of July and August, 1829, the day tides were higher than the night, I am quite certain; and an old Biscayan, resident near Point Arena, told me that they were always so in winter; hence we may conclude that they are regularly higher at that time of year."—*Capt. FitzRoy, Voyages of the Adventure and Beagle*, vol. ii. p. 388.

and southern parts of Chile. In the South, vegetation is abundant and luxuriant; in the North, the sea-coast has an irreclaimable barren appearance, very repulsive to an eye accustomed to woodland scenery. Chiloe, as before stated, is exposed to an excessive amount of rain. About Valdivia the climate is similar, and must always be an obstacle to cultivation. Northward of Valdivia, towards Concepcion, is one of the finest countries in the world, in a very healthy climate.*

Capt. Charles Wilkes says:—"The climate of Chile is justly celebrated throughout the world; and that of Santiago is deemed delightful, even in Chile; the temperature is usually between 60° and 75°. Notwithstanding this, it has its faults. It is extremely arid, and were it not for its mountain streams, which afford irrigation, the country would be a barren waste for two-thirds of the year. Rains fall only during the winter months (June to September), and after they have occurred, the whole country is decked with flowers. The rains often last several days, are excessively heavy, and, during their continuance, the rivers become impassable torrents. The temperature, near the coast, does not descend below 58°. The mean temperature, deduced from the register kept at Valparaiso, gave 63°. At Santiago, the climate is drier and colder, but snow rarely falls. On the ascent of Cordilleras, the aridity increases with the cold. The snow was found much in the same state as at Tierra del Fuego, lying in patches about the summits. Even the high peak of Tupongati was bare in places, and to judge from appearances, it seldom rains in the highest regions of the Cordilleras, to which cause may be imputed the absence of glaciers."—May, 1839.†

In the southern part, the surface is not formed by a series of table heights (as in the North) reaching from the sea to the Cordilleras; but it is a broad expansion of the mountainous Andes, which spreads forth its ramifications from the central longitudinal ridge towards the sea, diminishing continually, but irregularly, till they reach the ocean.‡

The Andes, which form so important a feature in the physical condition of South America, commence in the South part of continental Chile, the connected chain which extends northward to the farthest extreme of the continent. One of the southernmost peaks in this part is the volcano of Osorno, mentioned on page 99. The range southward of this forms a series of detached peaks along the East side of the Gulf of Ancud, and may be traced southward, at a minor elevation, to Cape Horn, which is its South termination, varying here from 5,000 to 9,000 feet high.

North of this, the Chilean Andes attain a mean elevation of 13,000 or 14,000 feet, rising with an extremely sharp ascent from the plain below.

The principal point of interest to the mariner respecting these mountains is their aspect from the offing. Capt. FitzRoy says:—"There is an effect in these lofty mountains, which seem to rise abruptly, amost from the ocean, that charms one for a time. Just before sunrise is generally the most favourable moment for enjoying an unclouded view of the Andes in all their towering grandeur; for scarcely have his beams shot between their highest pinnacles into the westward

* Geographical Journal, vol. vi. p. 319.
† Narrative of the United States Exploring Expedition, vol. i. p. 184.
‡ Miers's Travels in Chile, &c., vol. i. p. 378.

valleys, when clouds of vapour rise from every quarter, and during the rest of the day, with few exceptions, obscure the distant heights."*

The principal peaks are Osorno, Villarica, 16,000 feet; Antuco, at which is the southernmost Chilean pass; Chillan, Tupongati or Tupungato,† Aconcagua,‡ Limari, and numerous others, most of which are volcanoes more or less in activity. Their appearances, when observable from seaward, are noticed hereafter. In the ensuing chapter, a fuller description, though brief, will be found of this great chain of mountains. From the volcanic nature of the Andes, the whole of this region is liable to earthquakes. Some notice of their devastations are given in subsequent pages; and it may be observed, that through their agency great alterations may be effected, not merely in the actual condition of the harbours and coast, but in the state of the inhabitants.

The rivers of Chile, as will be evident, are unimportant. In the middle and southern provinces they are sufficiently numerous. The North part of the country is scarcely watered by any; and from the Maypu to Atacamá, a distance of 1,000 geographic miles, all the streams and rivers together would not make so considerable a body of water as that with which the Rhone enters the Lake of Geneva, or as that of the Thames at Staines.§ They are quite useless for navigation, but are serviceable for the purposes of irrigation.

From this cause, the southern provinces are those devoted to agricultural industry, cattle breeding and the raising of grain being the chief employments. In the North part mining is the most important commercial pursuit, and for which Chile is best known. The mines are principally in the province of Coquimbo; and at Copiapó the principal part of the copper and copper ore is exported. In 1830, this department contained 103 mines—75 copper, 24 silver, and 3 gold. The copper mines were worked at Checó, by an English company, producing the largest proportion, but silver exceeds it in amount.

Prior to the Spanish conquest, Chile belonged to the Peruvian incas. In 1535 Pizarro sent Almagro to invade the country; and in 1540 he sent Valdivia; the latter conquered most of it, except its South part, held to this day independent by the Araucanian Indians. Chile was divided by the Spaniards into 360 portions, which were given to as many individuals; some of these estates still remain in their great extent. In 1810, the Chilenos revolted from Spain; from 1814 to 1817 it was kept by the royalist forces; but in the latter year the victory of Maypu, gained by San Martin, permanently secured the independence of Chile.‖

Chile is almost the only Spanish republic of South America which is improving; and the late vigorous measures of the government promise continued prosperity.

* Narrative of the *Adventure* and *Beagle*, vol. ii. p. 481.—"During the winter months, both in northern Chile and Peru, a uniform stratum of clouds hangs (at no great height) over the Pacific. From the mountains we had a very striking view of the great white and brilliant field, which sent arms up the valleys, having islands and promontories, in the same manner as the sea now intersects the Chonos Archipelago, or the West coast of Tierra del Fuego."—*Darwin*, p. 427.

† Narrative of the United States Exploring Expedition, vol. i. p. 189.

‡ Aconcagua appears to be the highest in this portion of the Andes, being 23,200 feet. See Journal of the Geographical Society, vol. vii. p. 143.

§ Schmidtmeyer's Travels, p. 28.

‖ An interesting account of these proceedings will be found in Capt. Basil Hall's Extracts, &c., part 1, chap. 3, &c.; and the subsequent political history in the Narrative of the United States Exploring Expedition, vol. i. chap. 11.

THE COASTS of the southern part of Chile, comprising the island of Chiloe, have been described in the preceding chapter.

Continuing along the coast from Chiloe, northwards, the islets of Doña Sebastiana and Carelmapu require another notice, in order that their vicinity may be widely avoided. The tide sets strongly at times in races near them; and when there is a swell from seaward with an ebb tide running, the short high sea northwestward of these islets is very straining to a ship, as well as dangerous to boats or even to small vessels. Corona Head should always be closed, but Doña Sebastiana avoided. There is water for any ship between Sebastiana and Chocoy Head, avoiding the sandbank half a mile from the East point of the island, as well as to the eastward of the Carelmapu rocks, but should not be attempted. The Carelmapu islets should not be approached to the westward within a league, and it will be but prudent to give them a berth of more than 4 miles.

Westward from Doña Sebastiana a *sandbank* extends 4 or 5 miles, and over it there is considerably disturbed water, rippling and swelling during a calm, but breaking in short high seas during a gale. About 3 miles westward of the island are 6 fathoms at low water on this ridge, and at 2 miles about 4 fathoms. This ridge extends westward, in a line with Chocoy Head and Doña Sebastiana: by some it is called the Achilles Bank.

Maullin Inlet is a shallow but wild place, exposed to a heavy breaking sea, and unfit for vessels. It is only remarkable as being the division between the province of Chiloe and the country of the Araucanian Indians.

From *Godoy Point* the coast trends N.W. 8 miles to *Quillahua Point*, thence N.N.W. 17 miles to *Estaquillas Point*, and 9 miles beyond this to *Cape Quedal*, a projecting and bold promontory: under a height which is very conspicuous (a part of the range called Parga Cuesta) is a point called Capitanes.

Most of the projecting points on the coast between Godoy and Galera Points, in lat. 40°, have many detached rocks about them, all close to the shore, and the greater part above water. The land is high and bold, without any outlying danger; but at the same time without a safe anchorage between San Carlos and the port of Valdivia. Soundings extend some miles into the offing, though the water is deep. At 2 miles to the westward of this shore there are usually about 40 fathoms water; at 3 miles about 60; and at 5 miles from 70 to 90 fathoms, over a soft sandy and muddy bottom.

Point San Antonio is a high, bold headland, dark coloured, and partly wooded; the land hereabouts ranges from 1,000 to more than 2,000 feet in height.

Manzano Cove, lat. 40° 33′, and *Milagro Cove*, lat. 40° 16′, may afford temporary shelter for small coasters. The *River Bueno* is navigable within, and flows through a valuable tract of country, but there is a bar at its mouth which excludes all but the smallest craft.

Punta de la Galera is a point of land with a low hill on it, backed by the remarkable heights called Valdivia Hills, three in number, very conspicuous, pointed at their summits, and about 1,500 feet in height. Two miles and a half N.N.E. from Galera Point is Falsa Point, a low projection, with rocks half a mile off it, but above water; it is in a line with the ridge of Valdivia Hills, which are excellent marks for this part of the coast.

VALDIVIA.—From Falsa Point the shore trends north-eastward 13 miles to *Gonzales Head*, a woody, bluff cliff, immediately behind which is the port of Valdivia. N.E. 2¾ miles from Gonzales Head, is *Punta del Molino*, or Mill Point, off which some rocks lie about 3 cables' length. Mill Point is rather steep, and covered with wood; between these is the entrance to Valdivia, a port apparently spacious and really secure, but the portion affording sheltered anchorage for large ships somewhat confined.

On the second point from Gonzales Head stands a battery, *Fort San Carlos*, which may be passed close; on the opposite shore, nearly East of San Carlos, is *Niebla Castle*, off which there are 3 fathoms at 2 cables' length, 5 fathoms at 3. In mid-channel there are about 7 fathoms water, from which the depth gradually increases seaward. *Amargos Point*, which is rather less than a mile from Niebla Castle, is low, and has a small battery on it, close to which there is deep water. At 3 cables' length South of Amargos Point, and a cable's length off shore, is a rock, which is awash, and should be carefully avoided. About 1 mile to the southward of Amargos Point, at the further side of a well-sheltered cove, 3 or 4 cables square, is the Corral Fort.

In a line from Corral Fort to *Piojo Point*, and midway between them, there is a sandbank that increases gradually;* this bank, the Manzera Shoal, is dangerous to a stranger, because there is but little to indicate its situation in the appearance of the water, which is usually discoloured during the ebb tide by that brought down the river. This bank, which extends nearly across to Corral Fort, and has shallow water on it, detracts materially from the goodness of Valdivia Harbour.

From an island 300 feet high, called *Manzera*, to the south-eastward of this bank, there are three river-like inlets, extending southward, south-eastward, and to the north-eastward: the latter is a river, but winding and full of banks, and navigable only for small vessels assisted by a local pilot. About 9 miles up this river, on the South bank, is the town of Valdivia. Water is too plentiful, the climate being almost as rainy as that of Chiloe. Provisions are cheap, but not abundant. The best anchorage is in the cove near Fort Corral, and the best watering place is in that cove.

"The town of Valdivia, formerly dignified with the appellation of city, disappointed our party extremely. It proved to be no more than a straggling village of wooden houses, and the only building even partially constructed of stone was a church. The town is situated on the low banks of the river Calla-calla, and is so completely buried in a wood of apple trees, that the streets are merely paths in an orchard. Around the port are high hills, completely covered with wood,† and they attract clouds so much, that almost as great a quantity of rain falls there as on the western shores of Chiloe. Several rivers empty themselves at this one mouth, which is the only opening among hills that form a barrier between the ocean and an extensive tract of champaign country, 'Las Llanos,' reaching to the

* In evidence of this increase the excellent plan drawn up by Don Jose Moraleda, in 1788, does not show this shoal, but gives a depth of 20 or 30 feet over the space. In other respects the plan seems to be as minute in its detail as the recent survey.

† "I was informed that there is *coal* in many places about Valdivia; but I did not see any."—*FitzRoy*, p. 401. In a subsequent part of this work we shall give some notices of the existence of coal on the shores of the Pacific.

Cordillera of the Andes.* The quantity of matter brought down by this outlet is deteriorating the harbour as before mentioned."

Since the period of the *Beagle's* visit, just cited, February, 1835, the town has been laid in ruins by an awful earthquake, November 7, 1837. While the surveying expedition was here, one of very great intensity was felt, but not so great as the later one. A most interesting account of this terrible phenomenon, which took place February 20, 1835, is given by Capt. FitzRoy and Mr. Darwin—(Voyages of the *Adventure* and *Beagle*, vol. ii. pp. 402—418, and vol. iii. pp. 368—381).

VALDIVIA was founded by Don Pedro de Baldivia, or Valdivia, in 1551. Eight years afterwards the Araucanians defeated the Spanish troops, and destroyed the town. In 1643 it was taken by the Dutch, who were soon compelled to abandon it. The fortresses erected by the Dutch, afterwards strengthened by the Spaniards, appear at the present day as if of importance, but are in reality almost in ruins. The authority of the republic of Chile is limited to a small space around the town, the country beyond being possessed by the unsubdued Araucanians.

Although the plan of the port and river made by the *Beagle's* officers was exactly correct in 1835, it ought not to be trusted either for the river banks, or for the limits of the Manzera shoal, for more than a few years. The land about Valdivia ranges to 1,000 feet in height, and everywhere wooded.

Bonifacio Head is about 8 miles North of Gonzales Head; it is bold, and has deep water near: 2 miles off it are 20 fathoms. Thence the coast trends North, about 20 miles to *Chanchan Cove*, at the mouth of the River Mehuin, a tolerably good anchorage for coasters in summer only.

At *Cocale Head*, 8 miles North of Chanchan Point, the coast changes its character, becoming low and sandy, with occasional cliffs; the high lands, which to the southward of this point bordered the ocean, here retreat 5 or 6 miles, leaving a level and apparently fertile country, as far as abreast of Mocha Island. This piece of coast lies N.W. by N., and extends nearly 60 miles. Off its whole extent there are comparatively shoal soundings, 10 fathoms at 2 miles' distance, 20 at 4 miles, everywhere a sandy bottom: it is therefore dangerous to approach at night without the lead going; and a heavy surf breaks everywhere, even in fine weather.

The rivers *Tolten* and *Cauten*,† though said to have been navigable formerly, are scarcely distinguishable at 2 miles' distance from the shore; and are both closed by bars. Ranges of cliffs extend for several miles at a time along this shore, and on their level summits may often be seen troops of the unconquered Araucanian Indians riding lance in hand, watching the passing ships.‡ The summits of the Andes are visible for a great distance northward and southward, whenever the weather is clear.

About 7 miles N.W. of the Cauten stands *Cauten Head*, in latitude 38° 40′, a bold, cliffy headland, about 300 feet in height, with 20 fathoms 2 miles off shore, and apparently steep-to. From thence cliffs, more or less broken, extend 10

* FitzRoy, pp. 397-8; Darwin, pp. 363; Morrell, pp. 168—170.

† The city of Imperial was founded by Valdivia in 1550, and stood on the Cauten River. It has been obliterated by the successful Araucanians, and near its site now dwell the remarkable Boroa tribe.—*FitzRoy*, p. 402.

‡ FitzRoy, p. 401.

miles to a point bearing East from Mocha, distant 20 miles: 8 miles N.N.W. from this point is Cape Tirua, the point of the main land nearest to Mocha.

Cape Tirua has a small islet close to it, and in a little bay just to the northward is the mouth of the River Tirua, whence a communication used to be kept up by the Indians of the main land with those who lived on Mocha Island, by means of rafts (Balsas) and large canoes. The channel between Mocha and the main land is perfectly free from danger; the depth varying regularly from 10 to 20 fathoms, over a sandy bottom. The tide runs about a knot during springs, the flood to the northward. There is no sheltered anchorage on the part of the coast that has been described; but 9 miles north of Cape Tirua there is a cove which may afford temporary protection, and possibly a good landing place.

MOCHA ISLAND, a prominent landmark for navigators, is a lofty, hilly island, about 7 miles long and 3 miles broad, its summit 1,250 feet above the sea. Its South extreme is abrupt, but the North end descends gradually into a long, low point. In clear weather it may be seen at 30, 40, or even 50 miles' distance; but soundings are no guide in its neighbourhood. They are irregular, and indeed not to be got, except very near the land. It should not be approached too closely, however, on its North, West, or South sides, as dangerous rocks lie off it, those from the South end extending 3 miles out. During the flood tide these rocks are particularly dangerous, as it sets toward them from the south-westward. Sometimes the ebb stream is scarcely felt for days together, and then the flood stream has the effect and appearance of a continual northerly current.

Previous to the eighteenth century it was inhabited by Araucanian Indians, but they were driven away by the Spaniards, and since that time a few stray animals have been the only permanent tenants; the anchorages are indifferent, one on the N.E., the other near the S.E. point, called by the Spaniards *Anegadiza.* The landing is bad, and there are now no supplies to be obtained except wood, and with considerable difficulty water, but of excellent quality. The anchorage near Anegadiza is good in northers, in front of the first little hills, in 6 or 7 fathoms sand: the other, at the *English creek,* in 13 to 20 fathoms, over a sandy bottom; nearer the shore is rocky. Were there any adequate object in view, a good landing place might easily be made, and there is abundant space on the island for growing vegetable produce, as well as pasturing animals.*

From *Cape Tirua* to *Tucapel Point* is a wild, exposed coast, totally unfit to be approached: it is incessantly lashed by the S.W. swell, and has no kind of shelter.

At the N.W. end of a long low beach, on which there is always a heavy surf, is MOLGUILLA POINT, on which H.M. ship *Challenger* was wrecked in 1835.† Eight miles N.W. of Molguilla Point is Point Tucapel, a low, projecting rocky point, flat-topped, and dark coloured. The interior country hereabouts is very

* Mocha afforded water and fresh provision to Mr. Wafer and Capt. Davis, in 1686. There were then some Spanish Indians on the island, and they killed a great number of guanacoes. See also Colnett's Voyage, 1798, p. 30.

† H.M. frigate *Challenger*, Capt. Seymour, was proceeding from Rio de Janeiro towards Talcahuana, to resume her station on the West coast of America. She was going N. by E., with strong wind from W.N.W. during the day, when, at 8 p.m., she hauled to the wind for the night. There was a thick haze around; and, about 10 o'clock, April 18th, 1835, she struck on the rocks on Molguilla Point, and became immoveably fixed. Her head swung to seaward, to which circumstance the safety of the crew is owing. The mizen-mast only was cut away, and the

fertile and beautiful. Hill and dale, woodland and pasture, are everywhere interspersed, while numerous streams plentifully irrigate the soil.

TUCAPEL HEAD is a high, bold hill, 7 miles N.N.W. of Tucapel Point. Between Tucapel Head and Millon Point, a rocky projecting point, 2 miles farther North, there is a cove, into which the *River Leübu* runs. Coasters may find shelter there if the wind does not blow strong from N.W., but it has no defence from that quarter. Boats can enter Leübu at half-tide, when there is not much swell on the bar. In former days there was a settlement called Tucapel Viejo at the mouth of this river.

Carnero Bay is a wild, exposed bight, unfit for shipping, but *Yannas Cove*, at its northern end, affords anchorage for coasting vessels of small size. *Carnero Head* is a cliffy bluff. From thence to *Cape Rumena* and *Lavapie Point*, the shore is bold and cliffy, and backed by high land, well wooded: it is a deep-water shore. Cape Rumena is recommended by Capt. Beechey as a landfall in going to Concepcion.

Santa Maria Island is comparatively low and dangerous, on account of numerous outlying rocks. It has a cliffy coast, and somewhat irregular currents. Between it and Lavapie Point there are two particularly dangerous rocks under water, on which the sea does not always break; one is 1½ miles North of the East side of that point, the other is half a mile North of the same point, around which there are several other rocks; but the sea always breaks on them. The outer one above mentioned, called *Hector Rock*, requires especial care, as it lies so near mid-channel, and exactly in the track that most vessels would incline to take. Except some rocks half a mile South and S.W. of Cochinos Point, the southern extremity of Santa Maria, there is no danger in this passage to Arauco Bay.

There is tolerably good anchorage in *Luco Bay*, but not quite sheltered from N.N.W., and liable to heavy squalls off the heights over Cape Rumena when it blows strong from the south-westward: there are 5 fathoms water over good ground.

For 3 or 4 miles on each side of the River Tubul the coast is steep and cliffy, with high down-like hills. *Tubul River* was formerly capable of receiving vessels of considerable burthen—vessels of 200 tons could pass up nearly a mile; but the earthquake of 1835 raised its bar so much as to prevent access to more than boats; but that the bar will remain is unlikely. The neighbouring country is very beautiful and fertile.*

---

anchors were not cast out. With the exception of two of the crew, all got to shore, and much of the equipment was saved. After remaining fortified against the Indians, who, however, behaved in a friendly manner, they all got safely to Concepcion. While encamped on the sandy beach, they got beef and mutton, water, and abundance of most excellent potatoes. Of the cause of her loss it is difficult to conjecture, except on the score of a current to the South and East. She was out of her reckoning more than 40 miles; and such a current is not to be expected here, for the general set of the water is to the northward, excepting near the land. Perhaps the effects of the earthquake had not subsided, though it occurred four months before. It was not so considered on land. The fate of the *Challenger* may afford a useful lesson of caution to the mariner on this now barren coast.—*FitzRoy*, p. 451; *Nautical Magazine*, 1835, pp. 789—796.

* In the river Capt. FitzRoy saw the remains of the *Hersilia* whaler, captured by the pirate Benavides. He was a most remarkable character, a native of Concepcion, taken prisoner at the battle of Maypu, in 1818, and for his crimes sentenced to be shot; but, though terribly wounded, had the fortitude to feign death, and escaped. He then entered the Chilian army, and afterwards became pirate. The particulars of this singular man are given by Capt. Basil Hall, in his Extracts from a Journal, &c., vol. i. chap. 22.

Off the N.W. point of the long cliff West of Tubul River, and 1 mile from the land, is a rock called *El Frayle;* but the sea always breaks on it, unless the water is unusually still. In southerly winds there is good anchorage throughout Arauco Bay, but, except in Luco Bay, it is everywhere exposed to northerly winds and sea.

Laraquete Beach extends 10 miles to the N.E. from Tubul Cliffs ; and 2 miles off it are from 8 to 10 fathoms water, over a sandy bottom. The *River Carampangue* is not navigable at its mouth, though deep and rather wide 2 miles inland : its exit is choked by sandbanks.

ARAUCO, famous in Spanish song and history, is simply a small collection of huts, covering a space of about 2 acres, and scarcely defended from an enemy by a low wall or mound of earth. It stands upon a flat piece of ground, at the foot of the Colocao heights, a range of steep, though low hills, rising about 600 feet above the sea. In the sixteenth century, Arauco was surrounded by a fosse, a strong palisade, and a substantial wall, the work of the Spaniards. This was the first place assaulted by the Indians, after their grand union against the Spaniards, at the end of the sixteenth century. It was surrounded by the hostile Indians, who at first unsuccessfully attacked the fortress; but the Spaniards, seeing that they must be overpowered, escaped in the dead of the night. Thus began the famed insurrection which caused the destruction of seven towns, and drove every Spaniard from Araucania. S.E. of Santa Maria Island, there is a tolerable roadstead, with from 4 to 8 fathoms of water, over good ground : but the only place now sheltered is quite close to the South point of the above island. Formerly there was good anchorage between this point and Delicada Point, but the earthquake of 1835 raised the land nearly 1½ fathoms ; so that where there was a depth of 5 fathoms in 1834, the *Beagle* found only 3½.

From the River Laraquete to *Coronel Point*, the coast runs N. by W. ½ W., high and bold, free from outlying dangers, and affording temporary anchorage for small vessels, or at the least shelter for boats, in three or four coves ; and at the mouth of the little *River Chivilingo* affords shelter for small craft, except during S.W. gales. *Fuerto Viego Cove*, immediately South of Colcura, is equally exposed to both the S.W. and N.W. The little cove just to the northward of Colcura, called *Lotilla*, is the best of the three, but it also is open to the S.W.

In passing round Santa Maria, to the eastward, a wide berth must be given to the shoal which now runs off towards the S.E. ; it is not prudent to go a cable's length to the northward of a line drawn E.S.E. ¼ E. from the southern point of the island, until 3 miles eastward of that point, where there are but 4 fathoms at low water. From thence the shoal turns to the northward round Delicada Point, off which the water deepens suddenly to 10 and 20 fathoms : there is anchorage to the N.E. side of the island during southerly winds. Water is good and abundant, there is also plenty of wood and vegetables, but little else at present.

Off the N.W. end of the island are many rocks, one of which, the *Dormido*, lies 3 miles off shore, and the *Vogelborg* 4 miles.* They are sometimes undistinguish-

* *John Renwick Rock.* A statement—upon the veracity of which considerable doubt has been thrown—of the wreck of the *John Renwick*, on a rock off Santa Maria, was forwarded to the Lords of the Admiralty by her commander, Mr. John H. Bell. He states that in the night of July 4, 1848, his vessel struck on a rock, in lat. 37° S., lon. 74° 44′ 30″ W., and became a total

able by breakers; and it is not safe to pass between them and the island, neither is it prudent to approach the western side of Santa Maria nearer than a league.

The entrance of the great *River Bio Bio* is not accessible, on account of its sandbanks, and the S.W. swell: its situation, together with that of Port San Vicente and Concepcion Bay, is well pointed out by the remarkable pointed hills, about 800 feet high, called the Paps of Bio Bio: there is no danger near them except rocks close to the shore. *Port San Vicente* is an exposed bad anchorage.

Close off the *Heights of Tumbes*, the western promontory of the fine bay of Concepcion, there are a few straggling rocks, some under, some above water, near Lobo Point, Pan de Azucar, and Tumbes Point; this piece of coast trends North 6 miles from Port San Vicente. N.W. from Tumbes Point there is a rock above water, called Quebra Olla, or Break-pot Rock; between it and the point it is not prudent to pass.

CONCEPCION BAY.—Between Tumbes and Loberia Head, 6¼ miles to the N.E., lies the entrance to the bay of Concepcion, the finest port on this coast; being about 6 miles deep and 4 miles wide, having anchorage ground everywhere, and abundant space, well sheltered.

Quiriquina Island, lying North and South, 3 miles long by nearly 1 mile wide, gives shelter from northerly winds: and near Arena Point, at its S.E. extreme, is a good place for ships to anchor temporarily.

Near the principal anchorage off Talcahuana are the Belen, the Choros, and the Manzano Banks, but their positions are clearly shown on the chart. On the Belen there is generally a *beacon* (or *red buoy*).

About 1½ miles W.S.W. from Lirquen Point, at the S.E. part of the bay, Capt. Beechey found a rock, or rocky shoal, with only 15 feet on it. The *Beagle's* boats searched for it in every direction near the place indicated by him, but could not succeed in finding less water than 9 fathoms. Nevertheless such authority as that of Capt. Beechey is not to be doubted, and ships should avoid that part of the bay, till the exact situation of this danger is decided: it is not at all necessary to stand over so far towards the East shore when working up to Talcahuana.

Coal is abundant near Concepcion, and has been noticed by almost every writer; it will be therefore needless to cite authorities. Up the river Aldarien, between Talcahuana and Old Penco, are the coal works of Dr. Mackay. The coal near the surface, which is what has hitherto only been used, is similar in appearance to English cannel. There are some objections to it; but still its abundance and facility of working are of very great importance to this part of the country.*

The following are the directions given by Capt. Beechey, R.N., and will be well understood from the foregoing by Capt. FitzRoy:—

During the summer months southerly winds prevail along this coast, and occasion a strong current to the northward. It is advisable, therefore, to make the land well to the southward of the port, unless certain of reaching it before night.

---

wreck in 24 minutes. The position was ascertained by several bearings, taken in a small boat. If it really exists, it is a formidable danger, bearing N.W. 12 miles from Point Lavapie, and W.S.W. 6 miles from the North end of Santa Maria Island. Those well acquainted with the locality disbelieve the statement, and consider that it was the Dormido Rocks on which the wreck took place. It is necessary, however, to mention it here.

* As before stated, we shall give some remarks upon coal in a subsequent page.

Punta Rumena appears to me to be a preferable landfall to that of St. Mary's Island, which has been recommended, as it may be seen considerably further, and has no danger lying off it. But should the latter be preferred, it may be known by its contrast to the main land, in having a flat surface and perpendicular cliffs, as well as by a remarkable peaked rock off its N.W. extremity.* If the port cannot be reached before dark, it would be advisable to bring to the wind, between Saint Mary's and the Paps of Bio Bio, as there will almost always be found a southerly wind in the morning to proceed with. In doing this, take care of the Dormido Bank, lying off the N.W. end of St. Mary's. Having daylight to proceed by, close the land near the Paps of Bio Bio; and, keeping 1½ miles from shore, keep along the coast of Talcahuana Peninsula.

Should the Paps of Bio Bio be clouded, the land about them may still be known by the opening into St. Mary's Bay, and by the land receding in the direction of the Bio Bio River, as well as by high rocks lying off the points. The capes of St. Vincent's Bay, on both sides, are high, and terminate abruptly, and the South one has a large rock lying some distance off it. The northern cape is tabled, and has a small tuft of trees near its edge. Table land extends from here to Quebra Ollas. The Paps, viewed from the westward, appear like an island; the wide opening of the Bio Bio being seen to the southward, and St. Vincent's Bay to the northward. The high rocks off the capes, at the foot of the Paps, are an additional distinguishing mark; and, when near enough, the rock of Quebra Ollas will be seen lying off the N.W. end of the peninsula. About one-third of the way between Quebra Ollas and St. Vincent's Bay there is a large rock, called the Sugar Loaf. All this coast is bold, and may be sailed along at 1¼ miles' distance. Quebra Ollas Rock lies the farthest off shore, and is distant exactly 1¼ miles from the cliff; it may be rounded at a quarter of a mile distance if necessary, but nothing can go within it.

Having passed Quebra Ollas, steer to the eastward, in order to round Pajaros Ninos as closely as possible, and immediately haul to the wind (supposing it from southward) for a long beat up to the anchorage. There are two passages into Concepcion, but the eastern is the only one in use. On the eastern shore of this channel there is no hidden danger until near Punta Para and Lirquen, when care must be taken of the Para Reef, the Penco Shoal, and the Roguan Flat. When near the two latter, and when the southern head of St. Vincent's Bay comes open with Talcahuana Head, it will be time to go round; and it is not advisable *at any time* to open the *northern* cape of St. Vincent's Bay, distinguished by a tuft of trees upon it, with Talcahuana Head. These two landmarks a little open, and the pointed rock at the South extremity of Quiriquina a little open with Point Garzos, the N.E. extremity of the peninsula, will put you on a 2¼ fathom shoal. There is a safe channel all round this shoal; but ships can have no necessity for going to the southward or eastward of it.

On the Quiriquina side of the channel avoid the Aloe Shoal (situated one-sixth of a mile off the first bluff to the northward of the low sandy point), by keeping

* This rock bears *S. 53° 8′ W. true*, from the Look-out Hill, Talcahuana, and is 24′ 48″ W. of it. Its latitude is 32° 58′ 10″ S., as found by Mr. Foster.

the N.W. bluff of Espinosa Ridge open a sail's breadth (5°) with Talcahuana Head,* and do not stand into the bay between the Belen Bank and Frônton Reef (off the South end of Quiriquina); yet, as there are no good cross marks for the shoal, a stranger had better not run the risk, particularly as there will be found ample space to work between this line and the Para Reef. When the hut on the Look-out Hill is over the N.W. extremity of Talcahuana village, and the Fort San Joao bears W. by S. ½ S., the Belen is past,† and the anchorage may be safely approached by a proper attention to the lead. Be careful to avoid drifting down upon the Belen, either in bringing up in squally weather or in casting; and remember, that on approaching it the soundings are no guide, as it has 8 fathoms close to it. There is no passage for ships inside the shoal, except in case of urgent necessity, There is no good landmark for the channel.

Men of war anchor in 6 or 8 fathoms: Fort St. Augustine, *S. 45° W. true;* Fort Galvez, *N. 57° W. true;* Talcahuana Head, *S. 7° 39′ W. true.* Merchant vessels usually go quite close in shore, between the Shag Rock, a flat rock near the anchorage, and Fort Galvez, and anchor in 3 or 4 fathoms; in doing this, until the Shag Rock is passed, keep a *red mark*, which will be seen upon a hill South of Espinosa Ridge, open with Talcahuana Head. A good berth will be found in 3 fathoms, mud, close off the town, the *eastern* slope of Espinosa Hill in one with Talcahuana Head. At Talcahuana, moor open hawse to the north-eastward; but many think this unnecessary, as the holding ground is so excellent, that it is sufficient to steady the ship with a stream.

Should it happen by any accident that ships, after having passed Quebra Ollas, should not be able to weather Pajaros Ninos (supposing the wind to be from the northward), or should be set upon the northern shore of Talcahuana Peninsula, off which lie scattered rocks, they may run through the channel between Quiriquina and the peninsula. In doing this it is safest to keep close over on the island side, but not in less than 7 fathoms water. On the opposite shore a reef extends, eastward of the Buey Rock, to the distance of 700 or 800 yards from the foot of the cliffs; the mark for clearing it is Fort St. Augustine *open* with all the capes of Talcahuana Peninsula: but this danger will generally show itself, except the water be particularly smooth, as there is a small rock near its outer edge, which dries at half-tide.‡

Having passed the Buey Rock, haul a little to the westward, to avoid a reef off the S.W. extremity of Quiriquina, and be careful not to stand into either of the sandy bays of Quiriquina, between this point and the range of cliffs to the northward of it, or towards the peninsula, so as to bring the Buey Rock to bear eastward of *North, true,* until you have advanced full half a mile to the southward, when the lead will serve as a guide. If it be found necessary to anchor, haul into Tumbes Bay on the peninsula, and bring up in 7 or 8 fathoms mud.

* These two remarkable bluffs are situated to the left of Talcahuana, Espinosa being furthest inland.

† This mark, it must be remembered, carries you well clear of the Belen, and in bringing them on, take care not to shoot too far over Talcahuana Head, or to shoal the water on that side to less than 5 fathoms.

‡ The narrowest distance between this rock and the reef on Quiriquina sides is exactly half a mile.

This is the northernmost bay, and may be known by several huts and a large storehouse. When through, give the South and S.W. points of Quiriquina a berth of half a mile; and, having passed them, steer over towards Lirquen, until the two heads (Espinosa and Talcahuana) are open; then pursue the directions before given.

If vessels put into Concepcion for supplies, the anchorage of Talcahuana is unquestionably the best, on account of being near the town; but if wood and water only be required, or if it be for the purpose of avoiding bad weather from the northward, &c., the anchorage under the sandy point of Quiriquina will be found very convenient; it is in many respects better sheltered than Talcahuana, particularly from the northerly, north-westerly, and north-easterly winds. The depth is 12 fathoms, the bottom of a blue clay, and the marks for the anchorage South point of Fronton *S.* 76° 20′ *W. true;* Punta Arena, *N.* 45° *E. true;* one-sixth of a mile off shore; the sandy point being shut in with Point Darca, and the South end of Quiriquina in one with a hut which will be seen in a sandy bay in the peninsula. On rounding the sandy point (Punta de Arena), which may be done quite close, clear all up, and the ship will shoot into a good berth. Wood may be procured at this island at a cheaper rate than at Talcahuana, and several streams of water empty themselves into the bay, northward of the point.

The common supplies of Talcahuana are wood, fresh beef, live stock, flour, and a bad sort of coal. We found stock of all kinds dear, and paid the following prices: for a bullock, twenty-nine dollars; sheep, three dollars; fowls, three reals each, or four-and-a-half dollars a dozen; nine dollars per ton for coal, although we dug it for ourselves.

It is high-water, full and change, at Talcahuana, at 3^h 20′; and the tide rises 6 feet 7 inches; but this is influenced by the winds.*

Earthquakes are very common in this region, and must considerably affect the general prosperity of the country. We have before alluded to the awful earthquake of February 20th, 1835. Its effects were particularly ruinous to Concepcion. It is largely and excellently described by Capt. FitzRoy, and also by Mr. Darwin. "At 10 in the morning of the 20th of February, very large flights of sea-fowl were noticed passing over the city of Concepcion, from the sea-coast towards the interior. At 40 minutes past 11, a shock of an earthquake was felt, slightly at first, but increasing rapidly. During the first half minute, many persons remained in their houses; but then the convulsive movements were so strong, that the alarm became general, and they all rushed into open places for safety. The horrid motion increased; people could hardly stand; buildings waved and tottered: suddenly an awful, overpowering shock caused universal destruction; and, in less than six seconds, the city was in ruins. The stunning noise of falling houses; the horrible cracking of the earth, which opened and shut rapidly and repeatedly in numerous places; the desperate heart-rending outcries of the people; the stifling heat; the blinding, smothering clouds of dust; the utter helplessness and confusion; and the extreme horror and alarm, can neither be described nor fully imagined." About half an hour after the shock,—the sea having retired so much that vessels which had been lying in 7 fathoms water were aground, and every

* Beechey's Voyage to the Pacific, part 2, appendix, pp. 643-5.

rock and shoal in the bay were visible,—an enormous wave was seen forcing its way through the western passage, which separates Quiriquina Island from the main land. This terrific swell swept the steep shores of everything moveable within 30 feet (vertically) from high-water mark, and then rushed back again in a torrent, which carried everything within its reach out to sea, leaving the vessels again aground. A second wave, and then a third, apparently larger than either of the two former, completed the ruin. Earth and water trembled; and exhaustion appeared to follow these mighty efforts.

This earthquake was felt at all places between Chiloe and Copiapo, between Juan Fernandez and Mendoza, an area of 700 by 400 miles: and Mr. Darwin says, "We can scarcely avoid the conclusion, however fearful it may be, that a vast lake of melted matter, of an area nearly doubling in extent that of the Black Sea, is spread out beneath a mere crust of solid land." One of the permanent effects of the earthquake has been to raise the level of the land. It has been shown that the island of Sta. Maria, off Arauco Bay, was raised 9 feet; and it is almost certain, that there has been an uplifting of the bottom of Concepcion Bay, to the amount of 4 fathoms, since the famous convulsion of 1751. Other notices on this subject will be found elsewhere.—(See also Voyages of the *Adventure* and *Beagle*, vol ii. pp. 402—418; iii. 368—381.)

The highest land in the vicinity is *Mount Neuke*, 1,790 feet in height, and 5 miles eastward of Loberia Head. Four miles to the northward of this head, which is dark coloured, and has several straggling rocks close about it, at Cullen Point, the coast again trends short round East and then South, so as to form the small *Bay of Coliumo*, where coasters may anchor in security, but there is not much shelter for large ships during northerly winds. It has always been the scene of smuggling transactions. The best anchorage is close under the height over Coliumo Head, where *Rare Cove* offers good landing for boats, and a convenient watering place.

From Coliumo Bay, 16 miles North, to *Boquita Point*, and thence 40 miles farther, in a similar direction, to *Carranza Point*, the coast assumes an unbroken line, without any place for shipping. It is a deep-water shore; the land rises to a considerable height, and is partially wooded. *Fox Bay* does not deserve the name.

Among the rocks at Carranza Point boats find shelter occasionally; it is a projecting and rather low part of the coast, and therefore to be avoided: for about 10 miles on each side of the Point there is a sandy or shingle beach.

Seventeen miles N. by E. nearly, from Carranza Point, is *Cape Humos*, a remarkable headland projecting westward, and higher than any other land near that part of the coast; it is bold-to, and there are no outlying dangers in the vicinity.

Four miles N.N.E. from Cape Humos there is a remarkable rock, called *the Church* from its appearance, 1 mile N.E. of which is the entrance of the *River Maule*. There is no mistaking the entrance, for on the South side the land is high and the shore rocky; while on the North side a long, low, sandy beach extends beyond eye-sight. Not far from Church Rock a remarkable bare space of gray sand may be seen on the side of a hill, but generally the heights between Cape Humos and the Maule are covered with vegetation and partially wooded:

the highest hills in the vicinity range from 1,000 to 1,300 feet; those actually on the coast between Humos and the river, from 500 to 900 feet.

But for the bar, which shuts up the River Maule, there would be a thriving trade to this place; and notwithstanding this disadvantage, the little town of *Constitucion*, on the South bank of the river, 1½ miles from its mouth, may flourish hereafter, by the help of small steamers, and some engineering assistance at the bar. A most productive country surrounds it, abounding with internal and external wealth, and a fine river communicates with the interior, and is navigable far inland; besides which, the best pass through the Andes (discovered in 1805) is not far from the latitude of the Maule, being nearly level, and even fit for waggons, the only pass of such a description between the Isthmus of Darien and Patagonia.*

A ship may anchor, in fine weather, in from 10 to 15 fathoms, sandy ground, from 2 to 3 miles N.W. of the Church Rock: there is no hidden danger, but an extensive sandbank North of the river shelves out to seaward, and should have a wide berth. Behind this sand, evidently formed by the detritus brought down the river, there is a flat, several miles in extent; this flat in front of the high ground reaches to within 5 miles of a very remarkable valley, called the Falsa Maule, from its having been taken for the place of that river.

Thence (lat. 35° 6′) the coast trends North to *Lora Point*, and thence nearly N. by W. to *Topacalma Point*, a distance of 55 miles, without an anchorage or any outlying danger; the shore is high and bold, and there is deep water everywhere.

Bucalemo Head is a bold cliff 200 feet high; and 2 miles West of this head lies the *Rapel Shoal*, sometimes but erroneously called *Topacalma Shoal*. This shoal extends near a mile, and has three rocks above water, on which the sea breaks in all weathers. North from Bucalemo Head is *Toro Point*, close off which are a few rocks. The coast trends north-eastward from Toro Point, forming a bay as far as the River Maypu, which disembogues at a point 10 miles from Toro point; a bar extends across its mouth, and stretches for nearly 2 miles to the northward, parallel to the shore. Three miles North of the Maypu is San Antonio Cove, a small place affording indifferent shelter to a few coasters, and immediately under a pointed hill. Two miles North of this hill there is a diminutive cove called *La Bodega;* but large boats frequent it occasionally.

Cartagena beach, 5 miles from the Maypu, is quite exposed to S.W. winds. *Tres Cruces Point* is low and rocky: and N.W. 5 miles from it is *White Rock Point*, so called from the remarkable appearance of the white rock, a good landmark.

Behind *Algarroba Point*, 3¼ miles North of White Rock, there is a cove, where small coasters find temporary shelter during southerly winds. About Algarroba Point the coast is cliffy, but the cliffs are dark-coloured; the land in the neighbourhood is high and rather barren, of a dark colour, generally a brownish hue. In the distance the Andes, stretching from North to South, show their majestic height, and appear much nearer than they are in reality.

Gallo Point is a steep cliff, 7 miles North of Algarroba Point; between them

* FitzRoy, p. 425.

are two sandy bights, divided by a rocky point. At the corner of the northern bight, called *Tunquen*, and close under Gallo Point, a boat might find shelter in a northerly wind, but there is no place for a sailing-vessel. Steep cliffs extend 6 miles North of it. *Quintay Cove* affords no good anchorage. From thence the steep cliffs extend 3 miles to *Curauma Head*, a remarkable promontory, and one that demands special notice, because it is generally the first land made out distinctly by ships approaching Valparaiso from the southward. The head itself is a high cliff; above it the land rises steeply to the two high ranges of Curauma, the higher one being 1,830 feet above the sea, and 2 miles inland N.E. of the head.

Usually, when first made out from seaward, the high part of the range of Curauma appears directly over the Head, and, if tolerably clear weather, the Campana (Bell) de Quillota is seen in the distance, 6,200 feet high. If the Andes are also visible, *Aconcagua* will at a glance be distinguishable by its superior height of 23,200 feet.

Projecting 4 miles W.N.W. from the heights of Curauma, the high land over Curauma, is the well-known point *Curaumilla*, often, but incorrectly, called *Coroumilla*: this point, low by comparison with the neighbouring land, though not so really, is rugged and rocky; two or three islets lie close off it. From Curaumilla Point, the N.W. extreme of the land forming Valparaiso Bay bears N.E. by N., distant 7 miles; between them is a deep angular indentation of the coast, bordered by scattered rocks on the West side, and steep cliffs on the East.

VALPARAISO is now one of the principal commercial places on the western coast of South America. It has been described by numerous travellers, whose accounts will afford a history of its progress. It lies in the southern part of the bay, behind Piedra Branca, or Angeles Point, on which stands the lighthouse.

One of the most graphic, and at the same time recent, travellers who have portrayed it is Dr. Von Tschudi, from the translation of whose work we transcribe the following :—

"The impression produced by the approach to Valparaiso on persons who see land for the first time after a sea voyage of several months' duration, must be very different from that felt by those who anchor in the port after a passage of a few days from the luxuriantly verdant shores of the islands lying to the South. Certainly none of our ship's company would have been disposed to give the name of 'Vale of Paradise' to the sterile, monotonous coast which lay outstretched before us.

"The town of Valparaiso looks as if built on terraces at the foot of the range of hills (which continue rising in undulating outlines, and extend into the interior of the country, when they unite with the great chain of the Andes). Northward it stretches out on the level sea shore, in a long double row of houses, called the Almendral (almond grove); towards the South it rises in the direction of the hills. Two clefts or chasms (quebradas) divide this part of the town into three separate parts, consisting of low, shabby houses. These three districts have been named by the sailors after the English sea-terms, 'Fore-top,' 'Main-top,' and

'Mizen-top.' The numerous quebradas, which all intersect the ground in a parallel direction, are surrounded by poor looking houses. The wretched, narrow streets running along these quebradas are, in winter, and especially at night, exceedingly dangerous, Valparaiso being very badly lighted. It sometimes happens that people fall over the edges of the chasms and are killed, accidents which not unfrequently occur to the drunken people who infest these quarters of the town.

"Viewed from the sea, Valparaiso has rather a pleasing aspect; and some neat detached houses, built on little levels artificially made on the declivities of the hills, have a very picturesque appearance.

"The scenery in the immediate background is gloomy; but, in the distance, the summit of the volcano, Aconcagua, which is 23,000 feet above the level of the sea, and which, on fine evenings, is gilded by the rays of the setting sun, imparts a peculiar charm to the landscape.

"The bay is protected by three small forts. The southernmost, situated between the lighthouse and the town, has five guns. The second, which is somewhat larger, called 'El Castillo de San Antonio,' is in the southern inlet of the bay. Though the most strongly fortified of the three, it is, in reality, a mere plaything. In the northern part of the town, on a little hillock, stands the third fort, called 'El Castillo del Rosario,' which is furnished with six pieces of cannon. The churches of Valparaiso are exceedingly plain and simple, undistinguished either for architecture or internal decoration.

"The custom-house is especially worthy of mention. It is a beautiful and spacious building, and, from its situation on the Muele (Mole), is an object which attracts the attention of all who arrive at Valparaiso. In the neighbourhood of the custom-house is the exchange. It is a plain building, and contains a large and elegant reading-room, in which may always be found the principal European newspapers. In this reading-room there is also an excellent telescope by Dollond, which is a source of amusement, by affording a view of the comical scenes sometimes enacted on board the ships in the port.

"The taverns and hotels are very indifferent. The best are kept by Frenchmen, though even these are incommodious and expensive. The apartments, which scarcely contain the necessary articles of furniture, are dirty and often infested with rats. In these houses, however, the table is tolerably well provided, for there is no want of good meat and vegetables in the market. The second-rate taverns are far beneath the very worst in the towns of Europe. —(pp. 21—4.)

"The mole in front of the custom-house is exceedingly dangerous; in stormy north winds it is impossible to pass along it. From the shore a sort of wooden jetty stretches into the sea to the distance of about 60 paces. This has been sometimes damaged and destroyed by the sea. The harbour-masters' and men-of-wars' boats go on the right side, those of merchantmen on the left. Day and night, custom-house officers perambulate the port to prevent smuggling, which, however, is but partially successful. The police of Valparaiso is as good as in any part of South America. Among the most remarkable objects is the moveable prison, a number of large covered waggons.

"Valparaiso is yearly increasing in extent and in the number of its inhabitants; but the town makes little improvement in beauty. That quarter which is built along the quebradas is certainly susceptible of no improvement, owing to its unfavourable locality, and it is only the newly-built houses on the heights that impart to the town anything like a pleasing aspect. My visits to Valparaiso did not produce a very favourable impression upon me. The exclusively mercantile occupations of the inhabitants, together with the poverty of the adjacent country, leave little to interest the attention of a mere transient visitor."*

Capt. Wilkes, of the United States Exploring Expedition, says—

"I have had some opportunity of knowing Valparaiso, and contrasting its present state with that of 1821 and 1822. It was then a mere village, composed, with but few exceptions, of straggling ranchos. It has now the appearance of a thickly settled town, with a population of 30,000, five times the number it had then.

"They are about bringing water from one of the neighbouring springs on the hill, which, if the supply is sufficient, will give the town many comforts.

"It was difficult to realize the improvement and change that had taken place in the habits of the people, and the advancement in civil order and civilization. On my former visit, there was no sort of order, regulation, or good government. Robbery, murder, and vices of all kinds, were openly committed. The exercise of arbitrary military power alone existed. Not only with the natives, but among foreigners, gambling and knavery of the lowest order, and all the demoralising effects that accompany them, prevailed.

"I myself saw on my former visit several dead bodies exposed in the public squares, victims of the *cuchillo*. This was the result of a night's debauch, and the *fracas* attendant upon it. No other punishment awaited the culprits than the remorse of their own conscience.

"Now, Valparaiso, and indeed all Chile, shows a great change for the better; order reigns throughout; crime is rarely heard of, and never goes unpunished; good order and decorum prevail outwardly everywhere: that engine of good government, an active and efficient police, has been established. It is admirably regulated, and brought fully into action, not only for the protection of life and property, but in adding to the comforts of the inhabitants."†

The LIGHTHOUSE of Valparaiso stands on Piedra Blanca or Angeles Point, about 3 cables' length from the Baja Rocks, and 1¼ miles from the custom-house of Valparaiso. It stands 311 feet above the sea, and was first lighted on the 1st of August, 1838.

The tower, which is constructed of wood, and painted white, is 21 feet square at its base, and at the height of 56 feet, or the foot of the lantern, it is 10 feet square. The cylindric iron lantern is 12 feet high, and the light (which is *fixed*) may be seen in fine weather at the distance of 23 miles.

To vessels coming from the southward the lighthouse will first appear behind a bluff point. When it opens from this point it will bear N.E., and may be then

* Travels in Peru during the Years 1838-42, by Dr. J. J. von Tschudi, translated by Thomasina Ross, London, 1847, pp. 21—33.

† Narrative of the United States Exploring Expedition, vol. i. pp. 166-7.

steered for, on which course a vessel will clear Point Curaumilla, and its dangers, as far as Piedra Blanca, or the point of Valparaiso Bay. To this point the berth of a mile should be given until the lighthouse is brought to bear South, when all its dangers will have been cleared.

The rocks called the *Baja* lie E.N.E. ¼ E. from the above point of the bay: they are always above water, and are about 55 yards from North to South, and 27 from East to West. After passing the Baja rocks the vessel may freely enter the bay, and anchor in from 12 to 30 fathoms sand and mud.

Vessels bound to Valparaiso should make the coast in about 33° 20′ S. lat., during ten months of the year, as the wind prevails then to the southward; but it should be observed that, even in fine weather, the mountains will be seen before any part of the coast can be distinguished, so as to enable a vessel to make the port. Among those the volcano of Aconcagua is conspicuous, from its great elevation, and from its summit being almost always covered with snow. The highest or western part of it has an irregular outline marked by several peaks, but the S.E. part is entirely plain and even. When the summit of the volcano (which is 30 leagues from Valparaiso) bears N.E. by E. ½ E., it will be on the line of bearing of the lighthouse.

There is, at 9 leagues from Valparaiso, another remarkable height, called the *Bell of Quillota*. The middle part of its broken summit is called the Bell, and when it bears E.N.E. it will be on the line of bearing of the lighthouse. As these two peaks are the first seen in making the land, they serve well to direct a vessel to the lighthouse.

Vessels which make the coast to the southward of 33° 20′ S., and run along shore within 5 or 6 leagues, will not see the lighthouse till Curaumilla Point bears E. by N.

Coming from the northward, and having made Quintero Point, which is 18 miles N. by W. from the lighthouse, care must be taken not to approach too near to the coast in the night, as some sunken rocks lie S.S.E. 4 miles from the point. There is a channel between them, but it is too dangerous to be attempted without a skilful pilot.

Bearings (by compass) and distances from the lighthouse:—Quintero Point, N. ½ E.; Concon Point, N.N.E.; Volcano of Aconcagua, N.E. by E. ½ E; Bell of Quillota, N. 66° 45′ E.; the Baja, always above water, E.N.E. ¼ E.

The *Look-out*, from whence signals* are made, is 3,940 yards inland from the lighthouse, and 1,072 feet above the sea.†

The following remarks on making the land about Valparaiso are by Mr. George Peacock, late Marine Superintendent of the Pacific Steam Navigation Company, and appeared in the Nautical Magazine for June, 1847:—

"All vessels bound to Valparaiso should endeavour to make the land about *Curaumilla Point*, which lies 7 miles S.W. of Valparaiso lighthouse; and by no

* It was stated in the Valparaiso Mercury that Mr. Mouat had established a time-ball at his Observatory in the N.E. angle of the Castle of St. Joseph, for the purpose of enabling vessels to rate their chronometers.—*Nautical Magazine*, 1843, p. 768.

† These directions are from the Hydrographer's office, Admiralty, February 21st, 1839, and are a corrected version of those given by the captain of the port, Senr. Paul Delano, June 1st, 1838; they are also very similar to those by Capt. Vancouver, 1795, (vol. iii. p. 456,) and others.

means approach the coast in the neighbourhood of the *Rapel* or *Topacalma Reefs*, which lie 15 leagues S. ¾ E. of *Curaumilla Point*, and 7 leagues N. ¾ E. of *Topacalma Point*, as the heavy S.W. swell sets right down upon this highly dangerous part of the coast, as well as the prevailing current, which sometimes runs upwards of a knot an hour round *Topacalma Point* towards the reefs; and in thick weather, on approaching the land at night, the greatest attention should be paid to the *deep-sea-lead, which ought to be kept ready on deck for immediate use*, as soundings may be obtained at from 2 to 6, and even in some places 12 miles off the land, which is not generally known.

"I have had soundings in the parallel of the *Rapel* or *Topacalma Reefs* in 94 fathoms (coarse sand); 14 miles West of *Bucalemo Head*, and 6 miles off *Curaumilla Point*, you will strike soundings in 100 fathoms, also off Valparaiso Point, where the bottom is muddy. I have on several occasions taken a steamer into Valparaiso and *Talcahuana* in a thick fog, by close attention to the *deep-sea-lead*; and I believe many of the vessels which have been wrecked near Valparaiso, from time to time, might have been saved, had the officer in charge of the deck made use of *this highly necessary precaution*; for nearly in every case I have heard of, they had got too close in before they were aware of it, in light winds at night, or in thick weather by day, and when the land was suddenly seen, or the breakers heard, it was too late; for the heavy S.W. swell hurried them on to instant destruction.

"A revolving light upon the hummock of *Curaumilla Point* would be very desirable, for the light on Valparaiso Point is a miserable affair, and strangers are apt to be led astray by fires on shore. The lighthouse would also be a conspicuous object in closing the land, to make it out in thick hazy weather.

"South of *Curaumilla Point*, about 8 leagues, and from 4 to 5 miles off *White Rock Point*, a sunken rock *is said* to exist, but I think its existence is very doubtful; nevertheless, it would be advisable not to come in with the land to the southward of the parallel of 33° 15′ S. in the summer, nor 33° S. during the winter months, *i.e.*, in June, July, and August. I have known the current in these months set to the southward a mile an hour at intervals, and northerly gales are very prevalent during this season of the year."

In entering the Bay of Valparaiso with southerly winds, care must be taken to reef in time, for however moderate and steady the southerly winds may be in the offing, squalls blow from the high land into the bay, which are not to be disregarded. When it is blowing fresh outside from the southward, so as to require one reef in the topsails on a wind, probably treble-reefed topsails without the mainsail will be quite sail enough in the bay; when it is blowing strong in the offing from the same quarter, close-reefed topsails, over reefed courses, or over reefed foresail only, will be quite as much sail as can be carried. Should a ship find it blowing too hard to work up to an anchorage, she had better stand out, and remain under easy sail off Angeles Point till it moderates, which it does generally in a few hours.

In the event of a ship approaching with a northerly wind, likely to blow strong, she should keep an offing till the wind has shifted to the westward of N.W., which it always does after some hours of strong northerly winds: the best anchorage is close off Fort San Antonio, or in the S.W. corner of the bay; but, occupied as that part always is, a ship must take as good a berth near that part as she can find. During summer the closer in shore the better; and during winter, on the West side of the bay, outside other vessels, if it can be managed, so as to be safe from their driving during a northerly gale, which sends a heavy sea into the bay.

A norther, as it is called, often passes over without doing damage, but at intervals the effects are most disastrous, and all the ill-secured or ill-placed vessels are driven ashore.* Some prefer riding near the shore, on account of the undertow, but in such a position you risk having vessels driven upon you, besides feeling the sea very much. In the summer, southerly gales blow in furious squalls off the heights. Clear weather and a high glass presage strong southerly winds: cloudy weather, with a low glass, and distant land, such as the hill over Papudo, called Gobernador, or Cerro Verde, and the heights over the little port of Pichidanque, called La Silla, or on the coast northwards, being remarkably visible, are sure indications of northerly winds.†

Capt. Basil Hall, who was here in December, 1820, says:—"The climate was generally agreeable in the daytime, the thermometer ranged from 62° to 64°; and, at night, from 54° to 62°; between half-past 10 and 3 in the day, however, it was sometimes unpleasantly hot. Whenever the morning broke with a perfectly clear sky overhead, and the sun rose unconcealed by haze; and when, also, the horizon in the offing was broken into a tremulous or tumbling line, as it is called, a very hard southerly wind is sure to set in about 10 o'clock, blowing directly over the high ridge of hills encircling the town. The gusts, forced into eddies and whirlwinds, bore the sand in pyramids along the streets, drove it into the houses, and sometimes even reached the ships, covering everything with dust. About sunset those very troublesome winds gradually died away, and were succeeded by a calm, which lasted during the night. From sunrise till the hour the gale commenced, there never was a breath of wind; or, if the surface of the bay was occasionally ruffled, it was here and there by those little transient puffs, which seamen distinguish by the name of cats'-paws.

* Capt. Wilkes observes: "The northers are greatly dreaded, although, I think, without much cause. One of them, and the last of any force, I had myself experienced in June, 1822 (whilst in command of a merchant vessel). In it eighteen sail of vessels were lost. But since that time vessels are much better provided with cables and anchors, and what proved a disastrous storm then would now scarcely be felt. I do not deem the bay so dangerous as it has the name of being. The great difficulty of the port is its confined space, and, in the event of a gale, the sea that sets in is so heavy, that vessels are liable to come in contact with each other, and to be more or less injured. The port is too limited in extent to accommodate the trade that is carried on in it. Various schemes and improvements are talked of, but none that are feasible. The depth of water opposes an almost insuperable obstacle to its improvement by piers. The enterprise of the government, and of the inhabitants of Valparaiso, is, I am well satisfied, equal to any undertaking that is practicable.

"From the best accounts I am satisfied that the harbour is filling up, from the wash of the hills. Although this may seem but a small amount of deposition, yet after a lapse of sixteen years, the change was quite perceptible to me, and the oldest residents confirmed the fact. The anchorage of the vessels has changed, and what before was thought an extremely dangerous situation, is now considered the best in the event of bad weather. The sea is to be feared rather than the wind, for the latter seldom blows home, because the land immediately behind the city rises in abrupt hills, to the height of from 800 to 1,500 and 2,000 feet."—Vol. i. p. 166.

† FitzRoy, Narrative, &c., vol. ii. p. 426.

"On the other hand, when the morning broke with clouds, and the atmosphere was filled with haze, a moderate breeze generally followed during the day, sometimes from one quarter, sometimes from another; and on such occasions we were always spared the annoyance of the southerly gales.

"These varieties take place only in summer. During the winter months, that is, when the sun is to the northward of the equator, the weather is very unsettled. Hard northerly gales blow for days together, accompanied by heavy rains, and a high swell, which, rolling in from the ocean, renders the anchorage unsafe for shipping, and by raising a vast surf on the beach, cuts off all communication between the shore and the vessels at anchor. These gales, however, are not frequent. At that season the air is cold and damp, so that the inhabitants are glad to have fires in their houses."*

The N.E. side of Valparaiso Bay is formed by alternate beaches and rocky points, as far as that of Concon, behind which there is a cove, where boats can land in moderate weather. Three miles N.N.W. from that point are the *Concon Rocks*, always above water. These rocks should have a wide berth given to them during light winds, as there is usually a swell and a northerly current setting towards them from the southward.

Quintero Bay is roomy, and during southerly winds sheltered; it is quite open to the N.W. *Liles Point*, the West extreme of Quintero Bay, may be passed close. This bay affords spacious and good anchorage in the summer months, some even prefer it to Valparaiso: the best anchorage is in 13 fathoms, half a mile East of Liles Point. Some shelter during northerly winds, and fresh water, when the season is not very dry, may be found at the N.E. corner of the bay, under Ventanilla Point. There is a little shoal or rocky patch on the West side, nearly 2 cables' length off shore, and 4 cables from the junction of the cliff and sandy beach at the S.W. corner; this shoal, called Tortuga, does not watch, and requires caution in approaching the shore closely. The land between this bay and Concon is rather high and rugged; and all this coast has rather a barren and weather-beaten aspect, here and there only any trees being visible. During the winter and spring alone is there verdure near the sea-coast.

N. by W. 4 miles from Liles Point, and 1¾ miles West of Horcon Head, are the *Quintero Rocks*, above water, but low, straggling, and dangerous; they are of a dark colour, and spread over half a mile of space. *Horcon Head* has a remarkable hole in the extreme point of the cliff: the cliffs are dark coloured, about 80 or 100 feet high, and the land immediately behind them, though higher, is level. Inland are considerable heights, and in the distance the Cordillera of the Andes.

E.N.E., 1 mile from Horcon Head, there is a landing place between projecting rocks; and good water and plenty of fish may be procured, as well as fire-wood, and fresh provisions in small quantities. The roadstead is good during southerly winds, that is, in effect, during nine months out of the twelve; and there are 10 to 15 fathoms of water half a mile North of the landing place, over a clean

* Extracts from a Journal, &c., by Capt. Basil Hall, R.N., chap. 1.

sandy bottom. This bay was somewhat unaccountably omitted in all the Spanish charts of this coast.

Papudo Port is 13 miles from Horcon; between them there is no anchorage, the shore is steep, and free from outlying dangers. The high pointed hill over Papudo, called *Gobernador* or *Cerro Verde*, 1,020 feet in height, is an unfailing landmark for this small open bay. *Zapallar Point*, at the West extreme of this bay, is low, and must have a berth of nearly half a mile. It is safe during nine months of the year, but quite the reverse during the other three. There is a fresh-water stream close to the landing place: wood and small quantities of fresh provisions may be obtained, but not cheaply.

Five miles to the northward of Papudo, behind a low rocky point, is the mouth of the *River Ligua*, not navigable, nor affording anchorage for any but the smallest craft. *Point la Cruz de la Ballena* and *Muelles Point* are steep and bold-to. The trend of the coast from Ligua River to Ballena Point is W.N.W. for 5 miles, then it trends about 5 miles to the northward, and finally West 4 miles to Muelles Point, which is low, dark-coloured, and rocky. The shore round Muelles Bay is sandy, with low rocky points, backed, as all the coast is, by high land.

From Muelles Point to the western point of *Herradura* or *Pichidanque Bay*, the broken, dark-coloured, rocky shore runs nearly N. by W. ½ W. for 8 miles. The high saddle-topped hill of *Santa Ynez*, overlooking Pichidanque, is an excellent mark: it is 2,000 feet in height, and only 2 miles from the harbour. The best anchorage here is close to the little island of *Locos*, on its East side, in about 5 fathoms water. Care must be taken to avoid a rocky patch, called the *Casualidad*, very dangerous at low water for vessels drawing above 10 feet water, as there is neither ripple nor weed upon it in fine weather, though it breaks when a swell sets in rather heavily. This rock, on which there is a depth of 15 feet at low water, is in a line between the North end of Locos Island and a gully at the N.E. part of the harbour, through which a river runs from the neighbouring village of Quilimari, distant 4 cables' length from the islet. *The tide* rises 5 feet at springs. On full and change days it is high water at 9 o'clock. Pichidanque is used occasionally for loading copper ore, or for smuggling affairs; there are only a few fishermen's huts near the harbour, but at the village of Quilimari, behind the nearest hills, supplies can be obtained. It is not a good place for watering.*

In sailing along the coast, near Pichidanque, care should be taken to avoid a few outlying rocks, which may be seen by day close to Salinas Point, and those which lie half a mile off shore, 3 miles N. by W. ½ W. from Locos Islet. Nine miles N. by W. ¾ W. from the same islet is Ballena Point, dark-coloured, broken, and with an islet close to it: it is the extreme point of a ridge of land, extending southward.

From Ballena Point the coast trends N. by W. ½ W. to *Penitente Point*, a low rocky point; between which and *Tabla*, a projecting and dangerous point

* "Quilimari to Conchali. The country became more and more barren. In the valleys there was scarcely sufficient water for any irrigation; and the intermediate land was quite bare, not supporting even goats. In the spring, after the winter showers, a thin pasture rapidly springs up, and the cattle are then driven down from the Cordillera to graze for a short time."—*Darwin*, p. 417

4 miles to the N.W., lies *Conchali Bay*, an exposed roadstead, seldom used but by smugglers; the anchorage and landing are both bad, except in one little cove at the North part of the bay. There is an outlying rock awash, about 1½ miles South of Tabla Point, and N.N.W. ½ W., nearly 3 miles from Penitente Point; there are also two islets, and some rocks above water in the bay. It is a wild place, exposed to much swell.

All the coast, except a few corners, is steep, high, and barren, but picturesque in the outline.

North of Tabla Point there is an indentation of the coast, at the N.E. corner of which, 5 miles from Tabla Point, is a cove called *Chigua Loco*: thence to the *River Chuapa* is a nearly straight piece of cliffy coast, extending N.N.W. 7 miles; and thence to Maytencillo is 22 miles in a similar direction without a break.

*Maytencillo** is a little cove, fit only for Balsas; at certain times a boat may land, but there are many hidden rocks. Its situation is pointed out by a large triangular space of white sand, having an artificial appearance, on the face of the steep cliffs which here line the coast; this mark is made by the sand that is drifted by the eddy winds against the North side of the cove.

From that cove the coast extends in an unbroken line 33 miles N.N.W. to the next opening, which is that of the *River Limari*; the opening here looks large from seaward, but it is inaccessible. The coast near Limari is steep and rocky. For 10 miles North of Maytencillo the coast is composed of blue rocky cliffs about 150 feet high; the land above the cliffs rises to between 300 and 400 feet, and then about 3 miles further in-shore the range of hills runs from 3,000 to 5,000 feet in height.

Mount Talinay is a remarkable hill, 2,300 feet high; it is 3 miles from the coast, and 7 miles southward of the river; it is thickly wooded on the top, but the sides are quite bare.

About 14 miles northward of Limari there is a small bay, with a sandy beach in the North corner, but a heavy surf. From this bay to the northward the coast is rocky and broken; and about 8 or 9 miles further we come to a small rocky peninsula, with a high sharp rock rising from its centre, and a small deep cove South of it, without landing. This cove is called the *Tortoral de Lengua de Vaca*.

The *Lengua de Vaca* is a very low rocky point, rising gradually in-shore to a round hummock about a mile to the southward of the point. There are rocks nearly awash about a cable's length from the point, and at 2 cables' length distant there are but 5 feet. After rounding the Lengua, the coast turns short to the S.E., into *Tongoy Bay*, and is rocky and steep for about 2 miles from the point, where there are 15 fathoms about half a mile from the shore. About 3 miles from the point a long sandy beach commences, which extends the whole length of that large bay as far as the *Peninsula of Tongoy*: the South part of the beach is called *Playa de Tanque*, and the eastern side of the bay *Playa de Tongoy*. Off the S.W. end of the beach near Tanque there is anchorage

* The remarks on the coast of Northern Chile, which follow, are the work of Mr. Sulivan, Lieut. R.N., and were originally given in the Appendix to the Voyages of the *Beagle*, by Capt. FitzRoy, pp. 209—231. They have been repeated, with emendations, in the Sailing Directions for South America, part ii.

about half a mile from the shore, in from 6 to 7 fathoms; the bottom is a soft muddy sand in some places, but in others it is hard. With a southerly wind the bay is smooth, and the landing good, but a heavy sea sets in with a northerly breeze. This anchorage was once frequented by American and other whalers. The village, which is called the *Rincon de Tanque*, consists of about a dozen ranchos. The only water to be had is brackish; and about 2½ miles to the E.N.E., where there is good water, the landing is generally very bad, besides which the water is some distance from the beach.

All the way from Tanque to the peninsula of Tongoy there is anchorage in any part of the bay within 2 miles of the shore, in from 7 to 10 fathoms, sandy bottom. There is also good anchorage with a northerly wind, for small vessels, to the S.W. of the peninsula, abreast of the small village on the point, with the Lengua bearing W.N.W., in 4 fathoms, sandy bottom, with clay underneath; but no vessel, however small, should go into less than 4 fathoms, as the sea breaks inside of that depth, when blowing hard from the northward. Even large vessels might find a little shelter there with the wind to the northward of N.W. With a strong south-westerly breeze the sea across the bay would render any vessel unable to remain at anchor in this berth. There is a small bay on the North side of the peninsula, which is completely sheltered from southerly winds. In the S.W. corner of this bay there is a small creek, into which, when smooth, boats can go: it runs about a mile inland, and near its head there is fresh water, for which the whalers sometimes send their boats. The village of Tongoy consists of half a dozen small houses, built on a high point on the South side of the peninsula. To the northward of Huanaquero Hill there is a deep bay, well sheltered from southerly and westerly winds, but open to the northward; between this and Port Herradura there is no place fit for a vessel.

From Huanaquero Point it is 13 miles to the narrow entrance of *Herradura de Coquimbo*, a small land-locked harbour, separated from Coquimbo Bay by an isthmus of about a mile in breadth. Vessels, however, of any size, may freely enter with a leading wind, by keeping the southern shore on board, in order to avoid a rock off Miedo Point; and when in may anchor in any depth they please, on a bottom of sand covering very tenacious marly clay. In the S.W. angle they will find perfect shelter from all winds, and the water so smooth, that they may carry on any repairs with the utmost security. The *Beagle* lay there some weeks refitting, her crew encamped on the beach; but found neither fresh water nor wood.

COQUIMBO.—If the lead be kept briskly going, when approaching either the eastern shore or the bottom of Coquimbo Bay, the chart will be a sufficient guide, as the water shoals gradually towards the beach, which is low and sandy. It is necessary, in going in, to give *Pajaros Niños Islets and Rocks* a berth in case of falling calm, lest you should be obliged to anchor, for the ground near them is rocky; and for those reasons vessels are advised to pass outside of them.

The western shore of the bay is high and bold, particularly at its northern end, off which lies an insulated rock, the *Pelicanos*, having 4½ fathoms within a boat's length of it. On the point there is a platform with two guns, and a hut that answers the purpose of a guardhouse, but they are scarcely visible, having much the appearance of the rock on which they stand. About one-third of a mile from this point to the southward there are 9 fathoms at half a cable's length off the rocky beach, and then another point projects a little from the line of coast, having also two guns mounted on an open platform, and a shed amongst the rocks, more visible than those on the outer point.

The usual anchorage for strangers is in 8 fathoms, with the extreme North point of the western shore N.W. ½ W., the church at Serena (or town of Coquimbo) N.E., and the houses near the landing place S.W. ¼ W. The best anchorage is in 6 fathoms, in the S.W. angle of the bay, and the holding ground excellent; but a swell usually rolls in and produces such a surf along the beach that landing is difficult, except in a few sheltered spots. The winds at Coquimbo are in general moderate and southerly, or chiefly off shore during the greatest part of the year, and are interrupted for short intervals only in winter by strong breezes from the N.W. In short, the weather is so uniformly fine, the climate so charming, and the atmosphere so clear, as to have given to the city the name of La Serena. In approaching this port vessels must guard against being swept to the northward by the prevailing swell current and wind, which almost always come from the south-west. The land is remarkable, and easily recognised by the views in the chart; and *Signal Hill*, being upwards of 500 feet, can be easily made out at a moderate offing.

The Town of Coquimbo, or La Serena, is clean, and tolerably well laid out. The streets are straight, and intersect each other at right angles, like other Spanish towns. The houses, to each of which is attached a garden, are shaded by myrtle trees. The town, and its grounds, are supplied with water by canals, cut from the river on its North side; and, by its irrigation, increases the fertility of the place. The houses are mostly of sun-dried bricks, and only one story in height; so built in consequence of the earthquakes, to which all Chile is subject. There are several churches and other public buildings. The vicinity abounds in mines of silver and copper. Much copper and copper ore are exported, as also are chinchilla skins, &c.* Mr. Darwin says that "the town is remarkable for nothing but its extreme quietness;" but all travellers unite in lauding the kindness and hospitality of the inhabitants.†

This is a much frequented port, though one great inconvenience attends it, which is, that the fresh water is not good and difficult to be procured, the watering place being at a lagoon on the eastern side of the bay; wood is also

* Miers; American Encyclopædia; Morrell's Narrative, &c., p. 114; Capt. Basil Hall's Extracts, &c., chap. 26.

† One of the most singular features of the neighbourhood is up the valley, and is thus described by Capt. Basil Hall:—"On the 18th November, 1822, our friendly host accompanied one of the officers of the *Conway* and me in a ride of about 25 miles up the valley of Coquimbo; during which the most remarkable thing we saw was a distinct series of what are called parallel roads, or shelves, lying in horizontal planes along both sides of the valley. They are so disposed as to present exact counterparts of one another, at the same level on opposite sides of the valley; being formed entirely of loose materials, principally water-worn rounded stones, from the size of a nut to that of a man's head. Each of these roads, or shelves (of which there are three distinctly characterized sets), resembles a shingle beach; and there is every indication of the stones having been deposited at the margin of a lake, which has filled the valley up to those beds." Without pursuing Capt. B. Hall's notices of these singular formations, and which were fully described by him, we may state that Mr. Lyell concluded from the account that they must have been formed by the sea, during the gradual rising of the land. Mr. Darwin, who examined these ancient sea margins, coincides with Mr. Lyell's conclusions:—"In every case it must be remembered, that the

scarce, and far from the anchorage. Plenty of fish may be caught with the seine; and fresh provisions are cheap and plentiful. There is no landing at the town of Coquimbo, in consequence of the heavy surf, except on Balsas; but its distance from the landing place under Signal Hill is only 6 or 7 miles, where horses and conveyances are readily obtained, and when there the kindest hospitality more than makes up for the distance. It contains about 7,000 inhabitants.*

Teatinos is a bold rugged point, the land behind it rising in ridges, which gradually become higher, as they recede from the coast to *Cobre Hill*, which is 6,400 feet high. The point, which makes the North extremity of the bay in coming from the northward, is a low rocky point, called *Poroto*; about 4 miles to the northward of which is the port of *Arrayan*, or *Juan Soldado*, but it does not deserve any name, it being merely a small exposed bight. A little to the northward of Cobre Hill is another mountain in the same range, called Juan Soldado, 3,900 feet high; its northern side is steep, and at its foot lies the small unsheltered bay of Osorno, which is about half a mile long, but it would not afford any shelter for the smallest vessel: about half a mile to the northward of the bay there is a hamlet, consisting of a few small houses, called Yerba Buena.

The *Pajaros* are two low rocky islets, lying about 12 miles from the coast. A little to the northward of Yerba Buena, a small hamlet, there is a small island, called *Tilgo*, separated from the shore by a channel about a cable's length broad, but it is only fit for boats. The island, except when very close, appears to be only a projecting point; there is a large white rock on its West point.

About 3 miles to the northward of *Tilgo Island* is the port of *Tortoralillo*, which is formed by a small bay facing the North, with three small islands off the West point. In coming from the southward, the best entrance for small vessels is between the southernmost island and the point, where there is a channel about a cable's length wide, with from 8 to 12 fathoms water; the dry rock off the point on the main land should not be approached nearer than half a cable, as a sunken rock lies nearly that distance from it. There is no channel between the islets, as the space is blocked by breakers. A vessel may anchor about half a mile from any part of the beach, in from 6 to 8 fathoms, sandy bottom; the landing is not good; the best is on the rocks near the entrance, but nothing could be embarked from thence; the East end of the beach is the best for that purpose. From the land to the northward running so far westward, it is not likely that a heavy sea would be caused by a northerly gale.

Temblador is a small cove in the N.E. side of Tortoralillo, but the landing there is worse than on the other beach, and it is not so well sheltered. About 4½ miles to the northward of Tortoralillo lies the small island of *Chungunga*, at about a mile from the shore, and it is a good mark for knowing the little port of

---

successive cliffs do not mark so many distinct elevations; but, on the contrary, periods of comparative repose during the gradual, and perhaps scarcely sensible, rise of the land." The celebrated parallel roads of Glen Roy, in the West of Scotland, which tradition has made the work of Fingal, are of a precisely analagous nature, and have excited a great deal of research and curiosity. —See Capt. Basil Hall's Extracts from a Journal, &c., vol. ii. last pages; Mr. Darwin's Voyage of the *Beagle*, vol. ii. p. 423; Lyell's Principles of Geology, 5th edit., vol. iv. p. 18; and Mr. Darwin's Account of Glen Roy, in the Philosophical Transactions.

* FitzRoy, p. 427.

Chungunga: there is a rocky point abreast of it; and a little way in-shore is a remarkable Saddle Hill, with a nipple in the middle, which to a person coming from the southward appears as the end of the high range that runs thence to the eastward of Tortoralillo, and is from 2,000 to 3,000 feet high.

A little to the northward of Chungunga Point there is a large white sand-patch, which is seen distinctly from the westward; it is at the South end of the Choros Beach, which runs for 7 or 8 miles to the N.W. to Point Choros; a heavy surf always breaks upon it. Off *Choros Point* there are three islands; the inner one is low, and so nearly joins the shore, that nothing but a boat can pass. About 2 miles West of this island there is another small island, and between them the channel is clear of danger. To the S.W. of the latter, about a mile, lies the largest of the Choros Islands; it is about 2 miles long, the top is very much broken, and the S.W. end resembles a castle; there is a small pyramid off the South point, and rocks break about a quarter of a mile from the shore. The channel between the two outer islands is clear of danger; but about half a mile to the westward of the northern island there is a rock nearly awash. Five miles to the south-eastward of the southern Choros Island there is the very dangerous *Reef of Toro*, only a little above the water.

Carrisal Point is low and rocky, about 7 miles to the northward of Choros Point, with a remarkable round hummock; to the southward of it is the smae *Cove of Polillao*, where there is shelter for small vessels, but the landing is bad; there are two small rocky islets off the South point of the cove. To the northward of Carrisal Point is the bay of the same name, but it is not fit for sea-going vessels:* in the bay a heavy surf breaks about half a mile from the shore. The North side of the bay is formed by a rocky point, with outlying rocks and breakers about a quarter of a mile off all sides of it; there is a landing place in the bay near the S.E. corner, where the rocky coast joins the beach, but in bad weather the surf breaks outside of it.

Nearly one mile to the northward of the North point of Carrisal Bay is the *Port of Chañeral*; it is well sheltered from northerly and southerly winds, but the swell sets in heavily from the S.W., which makes the landing bad; the best landing is in a small cove on the North side near the beach; there is also a landing place on the South side of the bay, but it is bad when there is any swell. On the beach, in the bight of the port, there is always too much surf to land, except after very fine weather. About 4 miles to the westward lies the *Island of Chañeral*; it is nearly level, except on the South end, near which there is a remarkable mound, with a nipple in its centre. There are rocks nearly half a mile from the South point of the island, and one about the same distance off the N.W. point. On the North side there is a small cove, where boats can land with the wind from the southward, and there is anchorage close off it, but the

* "June 3rd, 1835:—Yerba Buena to Carizal. During the first part of the day we crossed a mountainous rocky desert; and afterwards a deep sandy plain, scattered over with broken marine shells. There was very little water, and that little saline; hence the few streamlets were bordered on each side with white incrustations, amongst which succulent, salt-loving plants grew. The whole country, from the coast to the Cordillera, is desert and uninhabited. At Carizal there were a few cottages, some brackish water, and a trace of cultivation; but it was with difficulty that we purchased a little corn and straw for our horses."—*Darwin, Journal, &c.*, p. 426.

water is deep; an American sealing schooner was lost there a few years ago from a norther coming on while she was at anchor.

The land round Chañeral Bay is low, with ridges of low hills rising from the points; their tops are very rugged and rocky, and the land is sandy and very barren. A range of high hills will be seen several miles from the shore, but between them and the coast there are several smaller hills springing out of the low land. The village of Chañeral is about 3 miles from the port, and is said to consist of about twenty houses; there are none near the port.* The people that came off to the *Beagle* said that the only vessel that had ever been here was a small schooner, called the *Constitucion*, which had taken a cargo of copper to Huasco. There was a large quantity of copper ready to be embarked.

From the North point of Chañeral Bay to *Leones Point*, off which are several rocks and reefs, it is about 3½ miles N.W. by N.; the coast between is low, and falling back forms a small bay. Leones Point has several rocks and reefs extending from it to the distance of a mile; there is also a reef, which projects nearly a mile from the shore, a little to the northward of Chañeral Bay. From Leones Point the coast projects N. by W. ¼ W. 4¾ miles to *Pajuras Point*, and from thence about North 4 miles to *Cape Vascuñan;* this cape has a small rocky islet off it about 2 cables' length from the shore. The land in-shore rises gradually to a low ridge about half a mile from the sea; the high range is about 3 miles in-shore.

From Cape Vascuñan the coast runs in to the north-eastward, forming a small bay, open to the northward, but well sheltered from southerly winds; there is anchorage in from 8 to 12 fathoms about one-third of a mile from the shore, but the landing is bad. To the eastward lies the *Bay of Sarco*, in which there is also shelter from southerly winds. To the northward of Sarco the high land comes close to the coast; the sides of the hills are covered with yellow sand; the summits are rocky, and the whole coast has a miserable, barren appearance. To the northward of the deep gully, about 4 miles, there is a projecting rocky point at the foot of a high range of hills, with a very remarkable black sharp peak near its termination; the coast to the northward of this runs nearly North and South, and is very rocky for about 8 miles, when it turns to the westward, forming a deep bay, in the N.E. corner of which is a small beach called Tongoy. To the northward of the bay a high range projects towards *Alcalde Point*, the extreme point of the bay, which is nearly 8 miles to the southward of Guasco: the point is very rocky, with small detached rocks close to it: in-shore it rises a little, and there are several small rocky lumps peeping out of the sand, one of which from the southward shows very distinctly; it is higher than the rest, and forms a sharp peak, a little in-shore of which the land rises suddenly to the break of the high range. Eight miles to the northward of Alcalde Point is the point forming the port of *Guasco*, or *Huasco;* it is low and rugged, with several small islands between it and Port Guasco, one only of which is of any size, and it is separated from the shore by a very narrow channel, so as to appear from seaward

* "The valley of Chañeral is the most fertile one between Guasco and Coquimbo. It is very narrow, and produces so little pasture that we could not purchase any for our horses."—*Darwin*, p. 427.

to be the point of the mainland; it is covered with low rugged rocks, one of which, on its North side, is much higher than the rest, and shows distinctly coming from the southward, but from the northward it is mixed with the other rocks behind it: to the S.W. of the island there are several other small rocky islets, which appear as two small islands when seen from a distance. A little in-shore of the extreme point there is a short range of low hills, forming four rugged peaks, which show very distinctly from the southward and westward; the land falls again inside of them for a short distance, and then rises suddenly to a high range, running East and West, and directly to the southward, of the anchorage. The top of the range forms three round summits, the easternmost of which, being 1,900 feet, is a little higher, and the middle one a little lower, than the other; they are all called the *Cerro del Guasco.*

Guasco Port.—Nearly 3 miles to the N.E. of the anchorage, there is another range of hills about 1,400 feet high, on the South slope of which there is a sharp peak, from which it slopes to the valley that conveys the river. The river is small, and a heavy surf breaks outside of it; the water, however, is excellent. The anchorage is much exposed to northerly winds, and a heavy sea then rolls in; but a mischievous norther does not occur more than once in two or three years, The village consists of about a dozen small houses scattered among the rocks, on the point dividing the old and new ports. The country round presents a more barren and miserable appearance than any part even of this desolate coast.*

Lobo Point, about 10 miles to the northward of Guasco, is rugged, with several small hummocks on it; to the southward of this there are many small sandy beaches, with rocky points between, but a tremendous surf breaks on them, allowing no shelter even for boats. A little in-shore of the point there are two low hills, and within them the land rises suddenly to a range about 1,000 feet high.

About 11 miles to the northward of Lobo Point is another rugged point, with several sharp peaks on it, and half a mile to the northward lies the small *Bay of Herradura,* which can hardly be distinguished till quite close. Between it and Herradura, which is distinguished from other Herraduras by the additional name of *de Carrisal,* there are breakers a quarter of a mile from the shore. Off Herradura Point there is a patch of low rocks, which, in coming from the southward, appears to extend right across the mouth of the bay; but the entrance faces the N.W., and lies between a low patch of rocks and a small islet to the N.E. of it, and there is no danger within half a cable of either of them. The bay curves in about three-quarters of a mile to the eastward of the islet, and is sheltered from both northerly and southerly winds, but with a strong northerly breeze a swell rolls in round the islet; it is rather small for large vessels, and they would not be able to lie at single anchor in the inner part of the cove, but there is room enough to moor across it, about a quarter of a mile above the islet, in 4 fathoms, fine sand. In this

* At Guasco there is a similar natural feature to that described as existing in the valley of Coquimbo, the ancient sea margins. "At Guasco the phenomenon of the parallel terraces is very strikingly seen: no less than seven perfectly level but unequally broad plains, ascending by steps, occur on one or both sides of the valley. So remarkable is the contrast of the successive horizontal lines, corresponding on each side with the irregular outline of the surrounding mountains, that it attracts the attention of even those who feel no interest regarding the causes which have modelled the surface of the land."—*Darwin, Journal, &c.,* p. 423, and page 125 *ante.*

place an American ship, the *Nile*, of 420 tons, was moored during a northerly gale, which blew very heavily, and she was perfectly sheltered; the landing is better than in any place between it and Coquimbo, but there is a very serious inconvenience in the want of water. There is a small lagoon about a mile from this place, in the valley at the head of Carrisal Cove, but it is worse than brackish, yet the *Peones*, who work at shipping the ore, make use of it. A deep valley runs in from the head of the cove, separating the high ranges of hills, and is a good mark to know the place.

Carrisal is a small cove, about a mile to the N.E. of Herradura, well sheltered from southerly winds; but as it is so close to Herradura, which is so much superior, it is not likely to be of much use. To the northward of Carrisal the coast is bold and rugged, with outlying rocks a cable's length off most of the points. About 7 miles to the northward there is a high point, with a round hummock on it, and several rugged hummocks a little in-shore. To the northward of this there is a cove sheltered from the southward, where small vessels may anchor, but it is not fit for large vessels: there is another cove similar to it about a mile farther to the northward. A little to the northward of the second cove there is a high rocky point, which is the termination of the high part of the coast. To the northward of the point there is a small port, which, from the natives, appears to be Matamores; it is well sheltered from southerly winds, and the landing is good. In the inner part of it, a vessel, not drawing more than 10 or 12 feet, might moor, sheltered from northerly winds, in 3 or 4 fathoms, but with a northerly wind there would be a heavy swell. There is anchorage farther out under the point in from 8 to 10 fathoms, but a vessel should not go nearer the shore there than 8 fathoms, as the bottom inside of that depth is rocky.

About 2 miles to the northward of Matamores, and 10 miles from Carrisal, we come to the low rocky *Point of Totoral*, a little to the northward of which there is a small deep bay, at the mouth of the valley Totoral Baxo.

About 6 miles to the northward of Totoral there is a remarkable rocky point, with a detached *white rock* off it, and a hump with a nipple on it a little in-shore. About 1½ miles to the northward of this lies the small *Cove of Pajonal*, which, in coming from the southward, may be easily known by the above nipple, and by a small island, with a square-topped hillock in its centre, off the point to the northward of the cove. A range of hills, higher than any near this, rises directly from the North side of the cove; and in the valley, about a mile from it, there is a range of small and very rugged hills, rising out of the low land.

The anchorage is better sheltered from southerly winds than any to the southward, except Herradura, and there ought not to be much swell, as the point and island to the northward project considerably to the westward. The southerly swell rolls into the mouth of the cove, but along the South shore it is smooth, and the landing pretty good: there is a dangerous breaker about a quarter of a mile W. by S. of the South extreme point, which only shows when there is much swell; the best anchorage is about halfway up the cove, near the South shore, in 5 fathoms: near the head it is shallow. There was a cargo of copper ore ready to be shipped here, but no vessel had yet been in the cove: there is no water within 2 miles, and there it is very bad indeed.

At 4 miles to the N. of Pajonal, and 1½ from the beforementioned square-topped island, *Cachos Point* appears with an island, and several rocks: both these islands may be passed within half a mile, but there is no passage inside of them. At that point the coast turns to the eastward, forming the spacious *Bay of Salado*, and close round the point the large *Cove of Chasco*, which, at a distance, looks very inviting, but a mile from its head there are only 3 fathoms, with rocks all round, some watching and others sunk, which, from the bay being well sheltered from the southward, do not show. A mile to the northward of these rocks there is another recess, which may be called *Middle Bay*, and which is quite clear of danger, and in the South corner a small cove; there is good anchorage in 7 fathoms, well sheltered from southerly winds, but very open to northerly: the water is perfectly smooth, with a southerly wind, and no swell could ever reach it unless it blew from the northward. *Salado Point* is a steep rocky point, with a cluster of steep rocky islets off it. To the northward of this point the coast is rocky and broken, with rocks a short distance from the shore for about 4 miles; then a rugged point, with a high, sharp-topped hill a little in-shore, which, from the southward, shows a double peak. Directly to the northward of this point there is a deep rocky bay, called the *Baranquilla de Copiapó*, with a small cove close to the point where the *Beagle* anchored in 5 fathoms, but half a cable off shore on either side; it is not fit for a vessel. The bay is partly sheltered from northerly winds, but a northerly swell rolls in, and it does not appear to be a proper place for a vessel to enter. There is no fresh water nearer than the river of Copiapó, which is about 12 miles off.

From Baranquilla to *Dallas Point* the coast is rocky and broken, without any place sufficient to shelter the smallest vessels. Dallas Point is a black rocky point, with a hummock on its extreme, which, coming from the southward, appears to be an island: the land rises to a range of low sandy hills, with rocky summits.

The *Caxa Chica* is a small sharp-topped rock, and is the only one of the reefs that shows above water. The patch near Dallas Point was awash; and the channel between it and the Dallas Point appears to be wide enough for any vessel, though the reef off the point projects so far as to show in a high sea a breaker above a quarter of a mile out; but at a quarter of a mile farther there were 11 fathoms. When the swell is not high, the breakers off the point would not show: they appeared to be detached from the reef which joins the point.

COPIAPO' is a very bad port, the swell rolls in heavily, and the landing is worse than in any port to the southward: it may easily be known by the Morro to the northward, which is a very remarkable hill, nearly level at the top, but near the eastern extreme there are two small hummocks: the East fall is very steep; the end of another range of hills shows to the northward. To the S.W., apparently forming part of the same range, stands another hill, the West side of which forms a steep bluff: in coming from the southward these hills will be seen in clear weather before the land about the port can be made out.

The island to the northward of Copiapó Bay, called *Isla Grande*, is very remarkable, having a small nipple on each extremity; that on the eastern end is the highest; and just to the westward of the middle of the island there is another small round nipple.

The idea some persons have of Copiapó being a difficult place to make is rather unfounded.*

"The chief dangers to be avoided in entering the harbour of Copiapó are the Caxa Grande and Caxa Chica shoals, and between these and Dallas Point several other small but dangerous patches of rock, on one of which the *Anacachi*, a Chilian brig, was recently wrecked. It lies about half a mile N.W. ½ W. from the Caxa Chica, and carries only 10 feet at low water.

"The *Caxa Grande*, the northernmost of the two first mentioned, is a bed of rocks under water, about three-quarters of a mile long and one-third of a mile broad, and lying nearly in a North and South direction; its situation is apparent from the heavy breakers on it, whenever a swell sets into the bay. The Caxa Chica is a small rocky shoal, having in its centre one large rock always above water: it lies South of the Caxa Grande, with a passage between them of nearly a mile in breadth, though appearing much less, from the rollers which extend sometimes across it on the Caxa Grande side. In going through this passage, the Caxa Chica should be given a convenient berth of from one to two cables' length, but unless the wind is steady, and to be depended on, it should not be taken on any account. The flagstaff above the town of Copiapó, bearing E. ½ N., leads through the passage.

"The passage between Dallas Point and the southern shoals should never be taken; as, should the wind fail, which, when so near the high cliffs in this vicinity is a common occurrence, a ship would be placed in a very dangerous position.

"The obvious and best passage is to the northward of the Caxa Grande; and to avoid those rocks, when coming from the southward, bring Isla Grande to bear N.E., and steer for it on that bearing till the northern end of the sandstone rocks, to the northward of the town of Copiapó, bears at least E. by S.; then haul in for that mark, and when the flagstaff above the town of Copiapó bears S.E. ½ E., steer towards it, and anchor where convenient. Should the flagstaff, which is small, not be quickly seen, a large house in the town, remarkable from its bright green roof, which is of copper, and always visible when the flagstaff is not brought on the same bearing, will be an equally good mark.

"Coming from the northward, vessels will most probably have to work in; in which case the shore may be approached to half a mile, and Isla Grande to within that distance, and when approaching the Caxa Grande, stand no nearer to it, or any of the shoals, than to bring the western extreme of Isla Grande to bear N.N.W., or the bluff part of Dallas Point to bear S.S.E. Should the wind be from the northward, the flagstaff on a S.E. ½ S. bearing will lead up to the anchorage, in from 12 to 6 fathoms. A large scope of cable should always be given in this road, and it might be prudent to drop another anchor under foot, as the rollers often set in with very little warning, and the bottom is bad holding ground. The soundings are very regular from 12 fathoms to 3, close up to the beach, but the bottom is chiefly a hard yellow sand, with occasional patches of yellow sandstone rock. Several vessels have been driven on shore here from their anchors by the rollers suddenly setting in.

* See Capt. Basil Hall's Extracts from a Journal, &c., vol. ii. chap. 29. The new surveys will remove all difficulties.

"The in-shore anchorage for a large vessel is in 5 fathoms, with the following bearings, viz., the Caxa Chica, W.S.W.; West extreme of Isla Grande, N.W. ¾ N.; the jetty (or landing place), S. ¼ E.; and the flagstaff over the town, S.S.W. ½ W.

"When the Morro de Copiapó, which is so high as to be seen 30 or 35 miles in clear weather, is open of Isla Grande, you are well to the westward of all the dangers off Copiapó."*

Capt. Basil Hall says of the journey to the town of Copiapó, 18 leagues in the interior:—"The valley was three or four miles across, and bore every appearance of having been, at some former period, the channel of a mighty river, now shrunk into a scanty rivulet, flowing almost unseen amongst dwarf willows, stunted shrubs, and long, rank grass. The soil was completely covered, at every part of the valley, by a layer, several inches thick, of a white powder, since ascertained, by analysis, to be sulphate of soda, or Glauber salts. It looked like snow on the ground, an appearance it still retained, even when made into roads and beat down. The dust thrown up by the horses' feet almost choked us; and the day being dreadfully hot, our thirst became excessive, so that we hailed with delight the sight of a stream; but, alas! the water proved to be brine, being contaminated by passing through the salt soil."—(Extracts from a Journal, &c., vol. ii. chap. 29.)

The town itself is so subject to earthquakes, that their constant effects are everywhere visible. In chap. 31, Capt. Hall gives an interesting account of the Chilian mining system.†

The channel between Isla Grande and the main is clear of danger in the middle, but such a heavy swell rolls through it that it is scarcely fit for any vessel. Off the North end of the island there is a reef projecting under water two cables to the eastward: but at a cable's distance from that reef there are 8 fathoms. The main land abreast of the island appeared to have no danger off its points: and the rocks to the southward of it are inside the line of the northern points. The swell in the channel was by far the worst the *Beagle* had experienced on this coast. To the northward of the island there are several small rocks, one of which is high, but there is no danger within a quarter of a mile of them.

Medio Point, on the main to the northward of the island, is very rocky: on the S.W. point there are two rugged hummocks, and several rocks and islets close to the shore, but no danger outside of them. From this to Morro Point the shore is steep and cliffy, with remarkable patches of white rock on the cliffs to the South of the point, which is steep, with rugged lumps on its summit. The *Morro* rises suddenly a little in-shore.

On rounding Morro Point a deep bay opens to the S.E.: there are several

* Directions by Mr. J. W. R. Jenkins, master R.N.—*Naut. Mag., April*, 1846, pp. 177-8.

† It is thought by Dr. Meyen that no part of America is more subject to earthquakes than that around Copiapó, and assigns as the reason for this peculiarity the entire want of volcanoes on the adjacent range of the Andes; as it appears that no volcano is found between that of Coquimbo in 30° S. lat. and that of Atacama, nearly 8° further North. Hence the countries lying between these parallels on the Pacific are continually agitated by earthquakes. It would also seem that the Andes in this space rarely rise to the snow line; for the small rivers which descend from them bring down all the year round nearly the same volume of water; which, between the parallel of 23° and 30°, could not happen if the mountains were covered a considerable part of the year with snow. It is also confirmed by the great number of passes which here traverse the range; for in the department of Copiapó alone there exist five mountain passes, distant from each other about 20 leagues; and many others might be opened without great expense.—*Reise um die Erde in den jahren* 1830, 1831, *und* 1832, *von Dr. F. J. F. Meyen, Berlin*, 1834.

small rocky patches in it, and at the North end of the long sandy beach there is a piece of rocky coast, off the North extreme point of which there is a small island. The entrance to *Port Yngles* is to the northward of this point, round the peninsula of Caldereta, off which there is a rock awash at high water about a cable's length, but it always shows; after passing this rock the land is steep-to, and may be approached within a cable's length. The harbour inside forms several coves, in the first of which, on the starboard hand going in, there is anchorage for small vessels, but the bottom is stony and bad: there is a low island to the S.E. of this cove, to the eastward of which is the best anchorage with southerly winds. About halfway between it and a projecting rocky point on the East shore, small vessels may go much closer into the cove to the S.E. of the island, where the landing is very good. The bay in the N.E. corner is well sheltered from northerly winds, and no sea could ever get up in it, but the landing is not good; the best there is at a rocky point at the South end of the N.E. beach, where there is a small cove among the rocks perfectly smooth. The N.E. cove is by far the best in the harbour, but it has no fresh water. The South cove is too shoal for any vessel to go higher up than abreast of the projecting rocky point on the East shore, where she would have 4 and 5 fathoms in mid-channel: the bottom is hard sand, and may be seen in 12 fathoms water, which makes it appear shallow. In the entrance there are 18 fathoms close to the shore on both sides.

Port Caldera is close to the northward of Port Yngles, and is directly round a point with a small island off it: it is a fine bay, pretty well sheltered, but more open than Port Yngles, and the landing not so good. There is water near the beach on the East side, but it is very salt; and it appears wonderful how they can make use of it, but they have no other nearer than Copiapó. The land is entirely covered with loose sand, except a few rocks on the points: the bottom of the bay is low, but the hills rise a little inland, and the ranges become higher as they recede from the coast. The first hill to the eastward is a very remarkable sharp-topped hill, the sides of which are covered with sand, with two low paps to the eastward of it. They have sometimes strong northers here, which throw a good deal of sea into the South corner of the bay; but in the N.E. corner, which they call the Calderillo, it is always smooth. There are fish to be got in the bay, but only with a net; and in none of the ports visited by the *Beagle* were any caught alongside. Near the outer points of these two ports rock fish are to be caught, but there is always a heavy swell in such places.

The *Cabeza de Vaca* is a remarkable point, about 12 miles to the northward of Caldera: it has two small hummocks near its extreme; inside of them the land is nearly level for some distance, and then rises to several low hills, which form the extremity of a long range. The coast between Caldera and the Cabeza forms several small bays, with rocky points between them, off all of which there are rocks at a short distance: there is no danger within a quarter of a mile of this point. To the northward of it there is a small rocky bay called *Tortoralillo*.

To the northward of this the coast is steep and rocky for 3 or 4 miles, with a high range of hills running close to the shore; then, a small cove called *Obispito*, with a white rock on its South point: and to the northward of this the land is low and very rocky, with breakers about a quarter of a mile from the shore.

About 2 miles from that cove there is a point with a small white islet off it: to the northward of which the coast trends to the eastward, and forms the small cove of *Obispo*, in which the *Beagle* anchored, but it is not fit for any vessel. There is a very high sandhill there, with a stony summit. A little in-shore of the cove, and to the northward, a higher range of stony hills runs close to the coast for about 7 miles, where it terminates in low rugged hills a little in-shore of a brown rugged point with a white patch on its extremity, which is an islet, though it does not show as one from the sea. To the northward of this lies the fine bay of *Flamenco*; it is a very good port, well sheltered from southerly winds, and better from the northward, as the point projects far enough to prevent a heavy sea getting up. The landing is good in the S.E. corner of the bay, either on the rocks, or on the beach of a small cove in the middle of a patch of rocks a little more to the northward, where there are a few huts, in which two brothers with their families were living: their chief employment was catching and salting fish, called *congre*, and drying them to supply Copiapó; in one day they had caught 400. They appeared to live in a miserable way, in huts made of seal and guanaco skins, much worse than a Patagonian *toldo*: the only water they had to drink was half salt, and some distance from the shore. They sometimes get guanacoes that they run down with dogs, of which they have a great number. *Flamenco* may be known by the white mark on Patch Point. Eleven miles North of Flamenco is Las Animas, a rocky bay, and 4 miles N.E. of it is the Bay of Chañeral, neither of which appears to be fit for ships.

In the bay to the northward of Flamenco, in which *Las Animas* was said to be, Capt. FitzRoy could see no place fit even for a boat to land: the whole bay is rocky, with a few little patches of sand, and a heavy surf always breaking on the shore. The North point of this bay is low; but a little in-shore there is a high range of hills, the outside of which is very steep; and to the northward of this point there is a small rocky bay, which appears to answer better to the description of Las Animas than the other: it did not, however, appear to be a fit place for vessels, and the landing was bad. To the northward there is a much deeper bay, which, from the description, must be Chañeral: the South side of it is rocky, with small coves, but the landing appeared to be bad; the East and North shores are low and sandy, and a heavy surf was breaking on the beach. The North point of Chañeral Bay is low and rocky, with a high range a little in-shore. To the northward of this point the hills and coast are both composed of brown and red rocks, with a few bushes on the summits of some of the hills: the sandy appearance that the hills have to the southward ceases, and the prospect is, if possible, more barren.

Nearly 9 miles to the northward of the bay of Chañeral stands *Sugar Loaf Island*, about half a mile from the shore. In coming from the southward there is a similarly shaped hill on the main, a little to the southward of the island, for which it may be mistaken; but the island is not so high, and the summit is sharper. Between Sugar Loaf Island and Chañeral the coast is rocky, and affords no shelter: but there is a small bay to the southward of the island, which might afford some shelter from northerly winds, though with southerly it would be exposed, and the landing is very bad. In the middle of the passage, between the island and the main, there are 5 fathoms in the shallowest part: the water in its northern end is smooth, and a vessel might anchor off the point of the island, sheltered from southerly winds, in 6 or 7 fathoms; but outside of 8 fathoms, it deepens suddenly to 13 and 20 fathoms about half a mile from the island. There is a small bay on the main to the northward of the channel, where a vessel might apparently be sheltered from southerly winds.

About 19 miles to the northward of Sugar Loaf Island there is a projecting point, with some small rocky islets off it, which was supposed to be *Ballena Point*, from the description given at Port Caldera. Between the point and Sugar Loaf Island the coast falls back a little, and is rocky, with a high range of hills running close to the shore. A little to the northward of Ballena Point there is a small bay, with a rocky islet about half a mile off the South point of it; the top of the islet is white, and answers the description given of a port called *Ballenita*, but it is not worthy of the name of a port; it is very rocky, with two or three small patches of sandy beaches, on which a heavy surf was breaking: the hills come close to the water, and have a rugged appearance. A little to the northward of this there is another bay, which seems to be *Lavata:* the South point has several low rugged prongs from it, and in-shore the hills rise very steeply: there is a small cove, with excellent landing, directly behind the South point, in which the *Beagle* anchored; and there was a still better looking port inside, but it was far from the outer line of coast, and her time would not allow of more than a hasty glance. The outer cove, in which she anchored, appeared to afford good shelter from southerly winds, and the water was very smooth.

A little to the northward of Lavata there is a point which, till close, appears to be an island; but it is joined to the shore by a low shingle spit; and several rocky islets that lie scattered off the point are named the *Tortolas*.

Nearly 3½ miles to the northward of them comes the *Point of San Pedro*, very rugged, and with a high round hummock a little way in-shore. To the eastward of this point there is a deep bay, in which it was expected to find the town of Paposo, according to its position given in the old charts, but there was no appearance of any houses or inhabitants. The bay is very rocky, and does not afford good anchorage; several rocks lie off San Pedro Point, and inside of it there is a reef projecting half a mile from the shore: in the bottom of the bay there are several small white islets, and two or three small sandy coves, none of which are large enough to afford shelter for a vessel. This is the bay of *Ysla Blanca*, and is bounded to the northward by Taltal Point.

About 3 miles from Taltal Point there is a white islet, with some rugged hummocks upon it; and a little way in-shore there is a hill of much lighter colour than any in the neighbourhood.

To the northward of this there is a deep bay, in which it was thought that Paposo would certainly be found; and, as the *Beagle* was becalmed, a boat was sent to search for it. On landing at the point, she saw a smoke on the East side of the bay, and, on pulling there, two fishermen stated that the place was called Hueso Parado, and that Paposo was round another point about 17 miles to the northward. On inquiring for water they brought some, which was better than that tasted in some other places to the southward, but still it was scarcely fit for use; they said it was as good as that at Paposo, and they thought both good. In the South corner of the bay there appeared to be anchorage for vessels, and the landing good, but very open to northerly winds. No vessel had ever been there in the recollection of those men, neither had they heard of any.

On this uninhabited coast it was impossible to ascertain the exact spot where the republic of Chile terminates, and that of Bolivia begins; but according to the best information, afterwards obtained, the line of limitation comes down to the shore in the bight of Hueso Parado, or between it and the point of San Pedro.

TIDES.—The only place at which the time of high water was satisfactorily determined was at Guasco,* where it is $8^h\ 30'$ at full and change; the rise 4 feet at neap tides, and at springs about 2 feet more. From the swell on all this coast it is very difficult to get the time of high water at all near the truth: the rise and fall appeared to be 5 or 6 feet in all parts of it.

The only perceptible *current* that was experienced was in the channel between Sugar Loaf Island and the main, where there was a very slight stream setting to the northward, but not more than a quarter of a mile an hour; and this was after a fresh breeze from the southward for several days. It is said, however, by coasters, that there is usually a set towards the North, of about half a mile an hour.

CHAPTER VI.

THE COASTS OF BOLIVIA AND PERU, FROM HUESO PARADO TO THE RIVER TUMBES.

THE coast described in this chapter was, during the Spanish domination in South America, that of Upper and Lower Peru; but when these states threw off the yoke in 1824, they became separate republics, the former receiving, in 1825, the name of Bolivia, from the Liberator, General Bolivar.†

The sea-coast of Bolivia is a desert. It is seldom visited by European shipping, except at one point, the miserable town of Cobija, or Puerto la Mar. The whole of this district is called the Desert of Atacamá. Towards its North part there are some fertile valleys, but the greater part is covered with dark brown or black moveable sand, absolutely arid, uninhabitable and uninhabited. The River Loa, whose mouth is in lat. 21° 28′ S., is the boundary between the two republics.

The republic of Peru extends on the Pacific to the Rio Tumbes, in lat. 3° 31′ S. The name Peru was given to the country in an early part of the Spanish dominion.‡

* The tide was very carefully observed in a cove where there was no swell, yet from the small rise the exact time could not be taken; the water remaining at the same level above half an hour.

† The republic of Bolivia embraces those provinces of Alto Peru which, under the dominion of Spain, formed the Presidency or Audiencia of Charcas.

‡ An expedition in 1524, under Pizarro and Almagro, being in want of provisions, marched three days along the banks of a river named *Biru*, in the territory of a cazique, whose name was Biruqueta. From this river, or from the cazique, originated the name given to the country now called *Peru*, which was not then so called by the inhabitants, nor by any other name before the entrance of the Spaniards. Garcilaso de Vega derives the name from one of the natives, who

The precise extent and population of these republics are but vaguely estimated. Bolivia may have an area of 320,000 square miles, and its population from 500,000 to 1,500,000. Peru has an area of 500,000 square miles, and a population of 1,700,000, but all these computations are mere estimates. Among the people the native races greatly predominate, but intermixed occasionally with Spaniards and with negroes.

The coast of Peru is sandy, bare, and scorched, and the inland country, called the Valles, extends to the slopes of the Andes. This part of the state has little wood, and includes but small districts fit for culture; sandy and stony deserts prevail.

The great feature of this country is, as farther South, the great chain of the Andes. There are numerous descriptions of them; foremost among which must be placed that given by the illustrious Von Humboldt.* Another interesting account is given by Mr. J. B. Pentland.† Many narratives relative to their general features will be found interspersed throughout many writers; among others is the Narrative of the U. S. Exploring Expedition.‡ The following extract is from the excellent work of Dr. Von Tschudi, who resided here from 1838 to 1842:—

"The two great mountain chains of the western regions of South America are not sufficiently distinguished by their respective names; the terms Andes and Cordillera being used indiscriminately for either of them, which confusion of names is done even in Peru. Nevertheless, a strict distinction ought to be observed;—the western chain should properly be called the Cordillera, and the eastern chain the Andes. The latter name is derived from the Quichua word Antasuyu; *anta* signifying metal generally, but especially copper, and *suyu* a district; the meaning of Antasuyu, therefore, is metal district. In common *parlance*, the word *suyu* is dropped, and the termination *a* in *anta*, was converted into *is*. Hence the word *antis*, which is employed by all old writers and geographers; and even now is in common use among the Indian population of southern Peru. The Spaniards have corrupted this Quichua word into Andes.§

"The old inhabitants of Peru, who dwelt chiefly along the base of the eastern chain, where they worked the mines, consequently gave the name Antis, or Andes, to that chain. But the Spaniards, advancing to the western mountains, gave them the name of *Cordillera*, the usual Spanish term for any mountain chain. These two great mountain chains stand, in respect to height, in an inverse relation to each other; that is to say, the greater the elevation of the Cordillera, the more considerable is the depression of the Andes. The medium height of the Cordillera, which is of the most importance in the present work, is in South Peru about 15,000 feet above the sea, but with particular points here and there which rise to a greater height. The Andes are here about 17,000 feet. In Central Peru the Cordillera is the highest.

---

was taken prisoner to the South of the Bay of Panama in the time of Nuñez de Balboa.—*Herrera*, dec. 3, lib. vi. c. 16.

* Relation Historique, tome iii.

† Journ. R. Geog. Soc., vol. v. 1835, p. 70.

‡ Vol. i. pp. 253—278.

§ Some derive the word Andes from the people called Antis, who dwelt at the foot of these chains of mountains. A province in the department of Cuzco, which was probably the chief settlement of that nation, still bears the name of Antas. See also Von Humboldt, Relation Historique, tome iii. pp. 194-5, where these points are fully considered. The name of *Chile* as spelt by the Spaniards, or Chili as originally, is said to be derived from the Indian word *quauhchilli*, capsicum-seeds. See Life in Mexico, by Madame C. de la B. (Calderon de la Barca), 1843, p. 384.

"The Cordillera is more wild and rugged, its ridge broader, and its summits less pyramidical than those of the Andes, which terminate in slender, sharp peaks, like needles. The Cordillera descends in terraces to the level heights, and is moreover the ridge of division, or watershed, between the waters which flow to the Atlantic and those which reach the Pacific. All the waters of the eastern declivity of the Cordillera work their way through the Andes to the Atlantic, while there is not a single instance of the Cordillera being intersected by a river; a fact more remarkable, because in Bolivia and southern Peru it is the lower chain."*

Some of the more prominent peaks visible from sea are noticed hereafter.

The rivers which flow into the Pacific are chiefly used for irrigation; none of them are navigable, except the Rio Piura, which is so for some months, as far as the town of Piura. The principal productions for which Peru is known to the rest of the world are from its mines. It is not to be doubted that immense supplies of gold and silver have been produced from the Montaña, or mountain district, but it was probably much overrated. Humboldt calculates that the annual produce of gold and silver at the beginning of the present century was about £1,248,000, but, since the anarchy consequent upon the revolutionary struggles, many of the mines are abandoned, and the produce reduced, perhaps to one-half, in 1849. Those of the Cerro de Pasco, lying in the interior mountains N.E. of Lima, are the richest mines in South America, and formerly produced silver to the value of £1,800,000 annually. The quicksilver mines of Huancabelica, E.S.E. of Lima, were one of the richest in the world, and at one time were unwrought, but a private company has since been formed to work them to a considerable extent.†

Other metals are also produced, and coal mines are met with in various parts. Among the more recent articles of export must be named nitrate of soda (or saltpetre, as it is sometimes but erroneously called). This is collected on the Valles, and used in European agriculture as a fertilizer, opening a great trade in a comparatively unnoticed article. More important than this, however, are the stores of *guano* which abound on some of the detached islets on the coast. The principal point hitherto frequented for this, which is worked by a company under the auspices of the Peruvian government, is the great Chincha Island, near Pisco.‡ The islands on which it occurs are mentioned incidentally hereafter in Capt. FitzRoy's descriptions.

For the last ten years the Peruvian trade has been better understood. The demand and the means of payment have been more accurately ascertained, and a

* Travels in Peru during the Years 1838-42, by Dr. J. J. Von Tschudi, translated by Thomasina Ross, 1847.

† Smith's Peru, vol. ii. p. 24.

‡ *Guano*, or *huano*, is stated by some to be a Spanish word, but by others Peruvian: there is no doubt of its origin from the latter; it is a corruption from the Quichua word *huana*—animal dung (Von Tschudi), and it refers to one of the most wonderful agents of fertility ever discovered. It was first introduced into England by M. Barhoillet and Mr. Bland, of the firm of Myers, Bland, and Co., of Valparaiso, who sent several cargoes in 1839-40, but had great difficulty in introducing it to agriculturists. Its value soon became apparent; when these gentlemen, in connexion with another English firm, obtained an exclusive privilege of shipping it from the Peruvian coasts for a term of years. The Ichabo, or African *guano*, was, as is well known, introduced through the instigation of Capt. Andrew Livingston, of Liverpool, a gentleman well known in the nautical world. For an account of the Peruvian *guano*, and its application, see Dr. Von Tschudi's work, p. 239, *et seq.*, before quoted. See also Naut. Mag., Nov. 1845, p. 621, *et seq.*

healthy and increasing commerce has been carried on as far as the state of the country and the fluctuations which are inseparable from distant traffic would permit. The commerce of Peru will not bear a comparison with that of Chile, and while the former has been diminishing, the latter has been rapidly increasing. A portion of the supplies which were formerly sent to Peru direct, are now obtained in Chile, and sent to their destination in coasting vessels. This change has been brought about by the unwise policy pursued by the various Peruvian rulers, in imposing heavy transit duties. This is also in part to be attributed to the advantageous situation of Valparaiso, where purchasers are always to be found for articles for the leeward coast. There is little doubt in the minds of those who are most competent to judge, that Valparaiso must become the principal mart of foreign commerce on the western coast of America; unless, indeed, that San Francisco in the North should become so, as may not be improbable.

The foreign trade of Peru is principally carried on by the English, Americans, and French. Of late years many German and Spanish vessels have been sent thither, and occasionally some of the Mediterranean flags are seen on the coast.*

On the coast of Peru each bay or landing place has its own peculiarly constructed vessel, adapted for the surf it has to go through. Thus at Malabrigo, the fishermen have what they call "caballitos," bunches of reed tied together, and turned up at the bow like a Chilian balsa, but much higher. These are so light, that they are thrown from the top of the surf to the beach, when the people jump off, and carry them to their huts. But the most important and best known of these contrivances is the *balsa* (raft, Spanish), which is formed of seal skins sewed together and inflated; two of these bags, about 8 feet in length, are fastened together at one end for a prow, and completed by small pieces of wood covered with matting sewed across. It is paddled with a piece of wood with a blade at each end. It is difficult at times to launch, but will land three passengers, besides the steersman, at any time, with great facility.

In Lima the seasons are usually distinguished as spring, summer, autumn, and winter, but the usual division of the aborigines, into wet and dry, are the true distinctions.

"In May the mornings become damp and hazy; and, from the beginning to the latter end of June, more or less drizzly. In October, again, the rains, which even in the months of July and August are seldom heavier than a Scotch mist, cannot be said to be altogether over, as the days are still more or less wet, or occasionally there may be seen to fall a light passing shower; the evenings and mornings being damp and foggy.

"In November and December, when the dry season may be reckoned to have set in, the weather, except for an interval at noon, is for the most part cool, bracing, and delightful: and April, too, is in this respect an agreeable month; at the latter end of which, the natives of the capital, being so exceedingly sensitive as to feel a difference of only two or three degrees betwixt the temperature of two succeeding days like an entire change of climate, are admonished, by a disagreeable change in their sensations, to protect themselves by warm apparel

* Wilkes' Narrative of the U. S. Exploring Expedition, vol. i. p. 303.

against the chills arising from an occasional N.W., or from the influence of the common S.W., wind.

"Throughout summer the wind blows almost uniformly, and in gentle breezes, from the South; but the prevailing wind for nine months in the year is the S.W., which, as it mingles with the warmer air along the arid coasts of Peru, tends to moderate the temperature of the atmosphere, and to produce the fog and '*garua*,' or thick Scotch mists, of which we have taken notice. During the dry seasons on the coast, the rains are experienced in the interior of the country and lofty range of the high table-lands; especially in the months of January, February, and March, when the rain that falls inland is often very heavy, and, on the most elevated regions, it is not unfrequently alternated with snow and hail. Thus, the dry season of the coast is the wet in the sierra, or mountain land, and *vice versâ;* and by merely ascending nigher to the sierra, or descending close to the sea, without any appreciable shifting of latitude, the favoured Peruvians may enjoy, by the short migration of a few leagues, a perpetual summer or an endless winter; if that, indeed, should be called winter which is the season of natural growth and herbage." *

Earthquakes are more common, perhaps, in Peru than in any other country. Every traveller tells us of his experience of some shocks of greater or less violence. To enumerate them, therefore, would be a very prolix work. The great shock of 1746, described by Juan and Ulloa, destroyed Callao, and greatly injured Lima; those of 1687, 1806, and 1828, were remarkable also for their violence. These convulsions have effected very evident changes in some parts of the sea-coast and harbours, and some of their traces have been much speculated on.

In a subsequent part of this volume we shall give some lengthened observations on the *winds* and seasons of the coasts. The reader is therefore referred to them for more full information on these heads.

The survey of this coast, made under Capt. FitzRoy's orders by Mr. Usborne, R.N., has superseded all preceding charts. Mr. Usborne's directions are given in the subsequent pages, as they will be found in Capt. FitzRoy's appendix. To these have been added some observations made by M. Lartigue in the French frigate *La Clorinde*, Capt. Le Baron Mackau, in 1822-3; other information, from the Narrative of the U.S. Exploring Expedition, &c., will be found as quoted.

At San Pedro Point, or rather at *Taltal Point*, near Hueso Parado, the coast of Bolivia is supposed to commence; and from thence it sweeps round the *Bay of Nuestra Señora* to *Grande Point*, a distance of 17 miles. This point, when seen from the S.W., appears high and rounded, terminating in a low, rugged spit, with several hummocks on it, and surrounded by rocks and breakers to the distance of a quarter of a mile. N. by W. ½ W. 9½ miles from it, lies *Rincon Point*, along with a large white rock, and between these two points, in the latitude of 25° 20′ S., the village of Paposo was at length discovered. It is a miserable place, containing about 200 inhabitants, under an *Alcalde;* the huts are scattered, and difficult to distinguish, from their being the same colour as the hills behind them. Vessels touch there occasionally for dried fish and copper

* Smith's Peru.

ore; the former plentiful, but the latter scarce. The mines lie in a S.E. direction, 7 or 8 leagues distant; but are very little worked. Wood and water may be obtained on reasonable terms; the water is brought from wells 2 miles off, but owing to the swell which constantly sets in on the coast, is difficult to embark. Vessels bound for this place should run in on a parallel of 25° 5′, and when at the distance of 2 or 3 leagues, the white rock off Rincon Point will appear, and shortly after the low white head of Paposo. The course should be immediately shaped for the latter; for, with that bearing S.S.E., distant half a mile, they should anchor in from 14 to 20 fathoms, sand and broken shells. Should the weather be clear (which is seldom the case) a round hill, higher than the surrounding ones, and immediately over the village, is also a good guide.

N.N.W. from Grande Point, at the distance of 23 miles, is *Plata Point*, similar to it in every respect, and terminating in a low spit, off which lie several small rocks, forming a bay on the northern side, with from 17 to 7 fathoms water, rocky, uneven ground.

From this point to Jara Head, which lies N. ¾ W. 53 miles, the coast runs in nearly a direct line; a steep, rocky shore, surmounted with hills from 2,000 to 2,500 feet high, and without any visible shelter, even for a boat.

Jara Head is a steep rock, with a rounded summit, and has on its northern side a snug cove for small craft; it is visited occasionally by sealing vessels, who leave their boats to seal in the vicinity. Water is left with them; and for fuel they use kelp, which grows there in great quantity, as neither of these necessaries of life are to be had within 25 leagues on either side.

Nearly 4 miles N. ½ E. from the Head, the large *Bay of Moreno*, or *La Playa Brava*, commences, the intermediate land being high and rocky, with a black rock lying off it; and N.N.W. ¼ W. 22 miles from the Head, the S.W. point of Moreno Peninsula, sloping gradually from the summit of *Mount Moreno*, and two nipples terminating in, from whence its name of *Tetos Point*.

Mount Moreno, formerly called *Monte Jorge*, is the most conspicuous object on this part of the coast; its summit is 4,160 feet above the level of the sea, inclined on its southern side, but to the northward ending abruptly over the barren plain from which it rises. It is of a light brown (*moreno*) colour, without the slightest sign of vegetation, and split by a deep ravine on its western side.

Immediately under Mount Moreno lies *Constitucion Road*, a small but snug anchorage, formed by the main land on one side and by *Forsyth Island* on the other. Here a vessel might haul in to the land, and careen, without being exposed to the heavy rolling swell which sets into most of the ports on this coast. The landing is excellent. Neither wood nor water are to be found in this neighbourhood, therefore provision must be made accordingly.

N. by W. 12 miles from this harbour, *Mount Jorgino*, a steep bluff, terminates the range of table-land which runs in a line from Mount Moreno; on the northern side of this headland lies the *Bay of Herradura de Mexillones*, a narrow inlet, running in to the eastward, but without affording any shelter.

North 9 miles from thence, *Low Point* is surrounded with sunken rocks; and 5 miles N.N.E. ½ E. of it is *Leading Bluff*, a very remarkable headland, which with the hill of Mexillones, a few miles farther South, is an excellent guide for the port of Cobija. The Bluff being about 1,000 feet high, and facing the North, is entirely covered with *guano*, which gives it the appearance of a chalky cliff. There is an islet about half a mile to the N.W. of the Bluff, and attached to it by a reef; but there is no danger of any description outside it. *Mount Mexillones* is 2,650 feet high; it has the appearance of a cone with the top cut off, and stands conspicuously above the surrounding heights. In clear weather this is undoubtedly the best of the two marks; but as the tops of the hills on the coast of Bolivia are frequently covered with heavy clouds, Leading Bluff is a surer mark, for it cannot be mistaken; for, besides its chalky appearance, it is the northern extremity of the peninsula, and the land falls back many miles to the eastward of it.

Round this head is the spacious BAY OF MEXILLONES, 8 miles across, but of little use, as neither wood nor water is to be obtained. The shore is steep-to; but there is anchorage on the western side, 2 miles inside the Bluff.

COBIJA, or LA MAR.—From Mexillones the coast runs nearly North and South, without anything worthy of remark, as far as the Bay of Cobija, or Puerto La Mar, which lies N. by E. 30 miles from Leading Bluff. It is the only port of the Bolivian republic, and contains about 1,400 inhabitants. Vessels call occasionally to take in copper ore and cotton; but the trade was very small in 1835, after the recent revolution in Peru.

Good water is very scarce: an occasional rill (caused by the condensed fog) runs down a ravine to the northward of the town, but so small that a musket-barrel is sufficient to convey it to the reservoir of the inhabitants: there are wells, but the water from them is very brackish, and will not keep in casks. Fresh meat may be procured at a high price; but fruit and vegetables, even for their own consumption, are brought from Valparaiso, a distance of 700 miles. They have a mud-built fort of five or six guns on the summit of the Point; and it is the only fortification noticed about the place.

If coming from the southward toward this bay, after having passed Leading Bluff (which should always be made), it would be advisable to shape a course so as to close the land 2 or 3 leagues to southward of the port, and then coast along until two white-topped islets, off *False Point*, are seen; a mile and a quarter to the northward of them is the port. On the slope of Cobija Point there is a *white stone*, which shows very plainly in relief against the black rocks in the background: a white flag is usually hoisted at the fort when a vessel appears in the offing, which is also a good guide. In going in there is no danger; the point is steep-to, and may be rounded at a cable's length, and the anchorage is good in 8 or 9 fathoms, sand and shells. In the bay there are a number of straggling rocks, but they are all well pointed out by kelp. Landing at all times is indifferent, and owing to the heavy swell, it requires some skill to wind the boats in through the narrow channel formed by rocks on each side. Two miles about N.N.E. is *Copper Cove*, a convenient place for taking in the ore; and where there is anchorage in 12 fathoms, a short distance from the shore.

It is high water, at the full and change, at $9^h\ 54'$, and the tide rises 4 feet.

After passing the North point of Cobija Bay, off which lie a number of straggling rocks at a short distance, the coast takes about a N. ½ W. direction: generally shallow sandy bays with rocky points, and hills from 2,000 to 3,000 feet high close to the coast, but no anchorage or place fit for shipping, until you reach *Algodon Bay*, 28 miles from Cobija.

This bay is small, and the water deep: the *Beagle* anchored a quarter of a mile from the shore, in 11 fathoms, sand and broken shells, over a rocky bottom. The only use of this anchorage is to send from it for water to the *Gully of Mamilla*, 7 miles to the northward. The spring there is a mile and a half from the beach; and the usual method of bringing it is in bladders made of seal-skin, holding 7 or 8 gallons each, with which most of the coasters are provided, the only vessels that profit by a knowledge of these places.

Algodon Bay may be distinguished by a gully leading down to it, and by that of Mamilla to the northward, which has two paps on the heights over its North side. There is also a white islet off Algodon Point.

About North, 10 miles from Algodon Bay, there is a projecting cape, called in the Spanish chart *San Francisco*, but known more generally by the name of *Paquiqui;* on its North side, and near its extremity, there is a large bed of *guano*, which is so much used on this coast for manure, as to be a regular trade.

N. ½ W. 16 miles from Cape Paquiqui lies *Arena Point*, a low, sandy point, with rocky outline: between the two is a small fishing village, near a remarkable hummock. Anchorage may be obtained under Point Arena, in 10 fathoms, fine sandy bottom.

N. ¾ E. 12 miles from Arena Point, come the *Gully* and *River of* Loa, which forms the boundary line between Bolivia and Peru. It is the principal river on this part of the coast; but its water is extremely bad, in consequence of running through a bed of saltpetre, as well as from the hills surrounding it containing copper ore. It is said that the ashes of a volcano also fall into it, which add greatly to its unwholesomeness; but, bad as it is, the people residing on its banks have no other. At *Chacansi*, in the interior, the water is tolerably good. In the summer season it is about 15 feet broad and a foot deep, and runs with considerable strength to within a quarter of a mile of the sea, where it spreads, and flows over, or filters through the beach; but does not make even a swatchway, or throw up any banks, ever so small. A chapel on the North bank, half a mile from the sea, is the only remains of a once populous village. People from the interior visit it occasionally for *guano*, which is in abundance.

The best distinguishing mark for the Loa is the gully through which it runs; and that may easily be known from its being in the deepest part of the bay formed by Arena Point on the South and Lobo Point on the North; as well as from the hills on the South side being nearly level, while those on the North are much higher and irregular.

There is good anchorage, but rather exposed to the sea breeze, with the chapel bearing North, half a mile from the shore, in from 8 to 12 fathoms, muddy bottom; and landing may be effected under *Chileno Point;* but the best anchorage here is in the *Bay of Chipana*, 6 miles N.N.W. ½ W. from Loa River (the land projecting between) and a snug cove for landing, near the tail of the point; but at the full and change a heavy swell sets in, and a boat would scarcely be able to land with goods at those times.

For the Bay of Chipana, after making the land in the latitude of the Loa, a large white double patch may be seen on the side of a hill near the beach, and a similar one a little to the northward; on discovering these marks (which are visible 3 or 4 leagues) a course should be shaped directly for the southern point, where lies the anchorage in 7 fathoms, sand and broken shells, sheltered by low level ground. No danger need be feared in anchoring; for, though the land is low, it may be approached within half a mile, in from 10 to 6 fathoms. The anchorage inside the long kelp-covered reef might perhaps be preferred; but the landing is not so good there.

N.N.W. ½ W. of this bay, at the distance of 18 miles, is *Lobo* or *Blanca Point*, high and bold, and on its extremity there are several hillocks. In the interval there is a small fishing village, called *Chomache*, under a point, with a long reef, on the outer part of which a cluster of rocks show themselves a few feet above the water. The people of this village get water from the Loa, a passage requiring, on a balsa, four days or more.

N.N.W. 14 miles from Lobo Point, is the low, rugged, projecting *Point of Patache*, with an inlet a quarter of a mile in the offing, and all clear outside. Halfway between these two points is the *Pabellon* (tent shape) *of Pica*, a remarkable hillock, said to be all *guano*; its appearance being in strong contrast with the barren, sunburnt brown of the surrounding hills. This is also a place of resort for the guano vessels, as they find pretty good anchorage close to the northward of the Pabellon.

East, a little southerly, a few miles in-shore of this, is a bell-shaped mountain, named *Carrasco*, 5,520 feet high.

From Patache Point to Grueso Point, N. ¾ W. 28 miles, the coast is low and rocky, the termination of a long range of table-land, called the *Heights of Oyarvide*, or *the Barrancas*, from its cliffy appearance: it has innumerable rocks and shoals off it, and should not be approached on any account within a league, for the frequent calms and heavy swell peculiar to this coast render it unsafe for nearer approach.

Grueso Point, at the North end of the Barrancas, is low but cliffy, with three white patches on its northern side, round which lies the *Bay of Cheuranatta.*

N. ¼ W. 11 miles from that point, come the anchorage and town of Iquique.

"IQUIQUE contains about 1,000 inhabitants, and stands on a little plain of land at the foot of a great wall of rocks, 2,000 feet in height, which line forms the coast. The whole is utterly desert. A light shower of rain falls only once in very many years: and hence the ravines are filled with detritus, and the mountain sides covered by hills of fine white sand, even 1,000 feet high. During this season (July) of the year, a heavy bank of clouds, extending parallel to the ocean, seldom rises above the wall of rocks on the coast. The aspect of the place was most gloomy. The little port, with its few vessels, and a small group of wretched houses, seemed overwhelmed and out of all proportion with the rest of the scene.

"The inhabitants live like persons on board a ship, every necessary coming from a distance. Water is brought in boats from Pisagua, about 40 miles to the northward, and is sold at the rate of 9 reals (4*s*. 6*d*.) an 18 gallon cask: I bought a wine bottle full for threepence. In like manner firewood, and of course every article of food, is imported. Very few animals can be maintained in such a place. On the ensuing morning I hired, with difficulty, at the price of £4 sterling, two mules and a guide, to take me to the saltpetre (or nitrate of soda) works, 14 leagues' distance. These are the present support of Iquique. During one year the value of £100,000 sterling was exported to France and England. Formerly there were two exceedingly rich silver mines (Huantacayhua) in this neighbourhood; but they now produce very little."*

There are no imports. All the property belongs to merchants in Lima, where the vessels are chartered, and have only to call here and take in their cargoes.

Vessels bound for this place should run in on the parallel of Grueso Point, until the white patches on that point are discerned, when a course should be shaped for the northern of three large sandhills: stand boldly in on this course till the church steeple appears, when, or shortly after, the town and low island will be seen, under which is the anchorage; care must be taken in rounding this island to give it a good berth, *a reef* extending off it to the westward, to the distance of 2 cables' length.

The anchorage is good in 11 fathoms, with Piedras Point bearing N. by W.; the outer point of the island, S.W. by W.; and the church steeple, S. by E. ½ E.

Vessels have attempted the crooked passage between the island and the main by mistake, and thereby got into danger, from which they were extricated with some difficulty; it is only fit for boats or very small vessels.

Landing is bad, and the approach to the shore hazardous, owing to the number of blind breakers with which it abounds; and at the full and change of the moon the heavy swell sets in. Balsas are employed to bring the cargoes to launches at anchor outside the danger, as is the case in most of the ports on this coast.

Off *Piedras Point* there is a cluster of rocks; and N. by W. 18 miles from it the small low black *Island of Mexillon*, with a white rock lying off it. It may be known by the Gully of Aurora, a little to the southward, and a road, apparently well trodden, on the side of the hills, leading to the mines. And N. by W. ¾ W. 33 miles from Piedras Point is *Pichalo Point*, a projecting ridge at right angles to the general trend of the coast, with a number of hummocks on it. Round to the northward of this point is the village and roadstead of *Pisagua*: this, as well as Mexillon, is connected with Iquique in the saltpetre trade, and is resorted to by vessels for that article. In rounding the point, a sunken rock lies about half a cable's length off, and should be looked out for, as it is necessary to hug the land closely, in order to ensure fetching the anchorage off the village; for baffling winds are frequent, and may throw you near the shore, but that does not signify, as the water is smooth and the shore steep-to. The best anchorage is with the extreme of Pisagua Point N. ¾ W., and Pichalo Point W. ¾ S., a quarter of a mile off the village, in 8 fathoms; by which you will avoid a rock with 4 feet water on it, lying off the sandy cove at the distance of 2 cables.

North of this, 2½ miles, the *River of Pisagua* makes a conspicuous break in the shore; and its water supplies all the neighbouring inhabitants. For a few months during the winter season, when this river attains the greatest strength, it appears to be about 10 feet in width, but even then has not sufficient force to make an exit for itself into the sea; like the Loa to the South, it merely filters

* Mr. Darwin, pp. 442-3.

through the beach, or is lost in the parched up soil around. During nine months of the year no water is found in its bed; though a scanty supply may always be had from the wells dug near it, yet no vessel should trust to renewing her stock at this place, for, besides its unwholesomeness, the difficulty and expense attending embarkation would be very great.

From Pichalo Point to *Gorda Point*, 18 miles, the coast is in low broken cliffs, with a few scattered rocks off it, and ranges of high hills near. Gorda Point is a low jutting prong, where a long line of cliff, several hundred feet high, commences, and continues, with only two breaks or interruptions (the *quebradas* of the Spaniards), as far northwards as Arica.

These breaks in the cliffs, or gullies, as they are called by the sailors, are remarkable, and very useful in making Arica from the southward. The first is the *Quebrada de Camarones*, which lies 7 miles North of Gorda Point, and is about a mile in width, lying at right angles to the coast, with a stream of water running down it, and a quantity of brushwood on its banks; it forms a slight sandy bay, but not sufficient to shelter a vessel from the heavy swell.

The *Gully*, or *Quebrada, of Victor* is the other; it lies 29 miles to the northward of Camarones, and 16 miles to the southward of Arica; it is about three-quarters of a mile in width, and from a high bold point, called *Cape Lobos*, jutting out to the south-westward, forms a tolerably good anchorage for small vessels: it traverses the country in a similar manner to that of Camarones, and has likewise a small stream passing through, with verdure on its banks. Vessels bound to Arica should endeavour to make this gully or ravine, and when within three or four leagues of it they will see Arica Head, which appears as a steep bluff, with a round hill in-shore, called *Monte Gordo*. Upon nearer approach the *Island of Alacran* will be observed, joined to the head by a reef of rocks. To the northward of this island, and round the head, is the port and town of Arica, which is the seaport of Tacna.

ARICA.—We give the description of this part of the coast as detailed by M. Lartigue, and then the observations of Mr. Usborne :—

"At 2 miles to the South of the Morro of Arica is a cove where the coast is formed of gravel or shingle, but it cannot be entered on account of the surf. The Morro of Arica is perpendicular on the West side, and falls rapidly towards the East. Coming from the South, it may be seen at 12 leagues off; at this distance it seems detached from the land. Coming from the West, the Morro of Arica appears confounded with the high lands which lie to the East of it, and cannot be distinguished more than 8 leagues off. Its great whiteness indicates it. At the Morro of Arica the direction of the coast changes to the N.E. for 2 miles, and then runs N.W. ¼ W., and is lined with great stones as far as the river of Juan Diaz, or Juan de Dios.

"The coast and the high lands both change their appearance at Arica; from hence to Juan Diaz it is low, the land rising in a gentle slope to a plateau, interspersed with trees, called the *Plain of Arica*. The portion of this plain above the city is cut through with a gorge called the *Valley of Sapa*, thickly set with trees.

"The summits of the interior mountains have an appearance of verdure, which increases as you advance to the N.W. These summits are sometimes hidden by clouds, but they are less thick than to the South of Arica. The differences which exist between the appearance of the coast on either side of Arica make it easy to be recognised, either coming from the South or N.W. Notwithstanding, the land should always be made between the Quebrada Camarones and the Morro, because, as is stated in the description of the prevalent winds,* there is almost always found in the afternoon, South of Arica, a fresh breeze from the S.S.W., by the aid of which you may proceed. You may run for the Morro as soon as it is distinguished; when within 4 leagues you will see the Guano Island, which care must be taken not to approach on the South side.

"The *best anchorage* at Arica is to the North or N.N.E. of this island, not too near the land, for inside the proper anchorage the small spaces where the bottom is good are embarrassed by coarse sand, rocks, and lost anchors.

"There is at all times in this road a heavy swell coming from S.W. or S.S.W.; as the breeze is in general moderate, vessels generally make fast with a single heavy anchor, and keep the head to the sea by another. Landing can only be effected at one point, which is to starboard of the flagstaff; to reach it, leave to starboard the rocky flat nearest the shore, and pass it at an oar's length. This flat uncovers only at one-third tide. When the tide is low, small boats only can be beached, and then only with difficulty. There is then but very little water at the landing place, and it is only by the lift of the sea that you can reach the shore; at least, in very fine weather, it would be dangerous even for the smallest boats to go ashore on any other part of the coast.

"For watering you must anchor the long-boat outside, a small boat must take the casks, when filled, to it. The spring is not very abundant, and is nearly level with the surface; and you must clear it out to bale."†

Of late Arica has been the seat of civil war, from which it has severely suffered. It was in contemplation, in the latter end of 1836, to make it the port of the Bolivian territory; and, had that taken place, it would perhaps have become next in importance to the harbour of Callao, the principal port of Peru; its present exports are bark, cotton, and wool, for which is received in return merchandise, chiefly British. Fresh provisions and vegetables, with all kinds of tropical fruit, may be had in abundance, and upon reasonable terms; the water also is excellent, and may be obtained with little difficulty, as a mole is built out into the sea, which enables boats to lie quietly while loading and discharging; the only inconvenience is having to carry or roll it through the town. Fever and ague are said to be prevalent; this in all probability arises from the bad situation which has been chosen for the town, the high head to the southward excluding the benefit of the refreshing sea breeze, which generally sets in about noon.‡ In entering this place there is no danger whatever; the low island may be rounded at a cable's distance in 7 or 8 fathoms, and anchorage chosen where convenient.

The *Western Cordillera* of the Andes between Cobija and Arica attains a very great elevation, and offers several snow-capped peaks well known to navigators.

The most southern group of these peaks consist of four majestic *nevados*, known

* See the chapter upon the winds at the latter part of this volume.
† Description de la Côte du Perou, 1827, pp. 25, 26.
‡ Capt. FitzRoy.

to the aboriginal inhabitants of the neighbouring provinces of the interior by the names of Gualatieri or Sehama, Chungara, Parinacota, and Anaclache. *Gualatieri*, or *Sehama*, which Mr. Pentland thinks the most elevated of the four, is in the form of the most regular truncated cone, enveloped to its base in perpetual snow. Masses of ashes and vapour are seen to issue from its summit at intervals, so as to leave no doubt of its being a volcano in activity. Its elevation is estimated at 22,000 feet.

North of *Gualatieri* rise two magnificent nevados, which, owing to their similarity of form, and their contiguity to each other, are known to the Creole population by the name of *Melizzos* or *Twins*, whilst they are called *Chungara* and *Parinacota* by the Indian population. The most southern of these nevados forms a very perfect truncated cone, whilst the most northern rather resembles a dome or bell (campana). There is little doubt but that both are of igneous origin, and that Chungara possesses an active crater at its summit, still in activity.

The *Nevado of Anaclache* is certainly less elevated than the three preceding, and perhaps does not exceed 18,500 feet. It forms a ragged ridge, in the direction of the Cordillera, of considerable length.

Still farther North several snow-capped peaks rise at the back of Arica. The centre of this group may be fixed where the Gualillas Pass, a col or passage of the western Cordillera, which attains an elevation of 14,830 feet, is crossed by the great commercial road from Arica to La Paz, and the interior of Bolivia. The *Nevado of Chipicani*, which is about the mean elevation of this snow-capped group, is 16,998 feet high, and consists of a broken down crater, with an active solfatara in its centre.*

From Arica the coast takes a sudden turn to the westward, as far as the river Juan de Dios or Juan Diaz; it is a low sandy beach, with regular soundings: from this river it gradually becomes more rocky, and increases in height till it reaches the *Point* and *Morro of Sama*, 3,890 feet high. This is the highest and most conspicuous land near the sea about this part of the coast, and, at a distance, appears from its boldness to project beyond the neighbouring coast line. On its western side there is a cove formed by Sama Point, bearing N.W. by W. ¼ W. 45 miles from Arica Head, where coasting vessels occasionally anchor for *guano*: and there are three or four miserable-looking huts, the residence of those who collect the guano. It would be quite impossible to land there, except in a balsa, and even then with difficulty. Should a vessel be drifted down there by baffling winds and heavy swell, which has been the case, she should endeavour to pass the head (as a number of rocks surround it) about a mile to the westward; and there anchorage may be obtained in 15 fathoms.

N.W. by N. ½ W. 9 miles from Sama Point, is a low rocky point, called *Tyke*, and between those points issues the small *River Cumba*, with low cliffs on each side: like most of the rivers on the coast, it has not strength to make an outlet, but is lost in the shingle beach at the foot of the before-mentioned cliffs. Regular soundings, which continue gradually increasing as far as Coles Point, may be obtained at the distance of 2 miles, in from 15 to 20 fathoms.

* Journal of the Royal Geographical Society, vol. v., 1835, p. 72, and Meyen, Reise um die Erde, vol. i.

W.N.W., at the distance of 31 miles from Sama Point, is *Coles Point;* the shore between is alternately sandy beach, with low cliff, and moderately high table-land a short distance from the coast. It is doubtful if a landing could be effected anywhere between Arica and Coles Point, as a high swell sets directly on, and appears to break with universal violence.

Coles Point is very remarkable; it is a low sandy spit, running out from an abrupt termination of a line of table-land. Near its extremity there is a cluster of small hummocks; and at a distance it appears like one island. Off the point, to the S.W., there is a cluster of rocks or islets, but no hidden dangers. The rebound of the sea beating against both sides of the point causes a ripple, and much froth, which leads one to suspect a reef in the vicinity.

YLO.—N.E. 5½ miles from this point is the village and roadstead of Ylo. This is a poor place, containing about three hundred inhabitants, under the local governor and captain of the port. But little trade is carried on, and that chiefly in *guano:* a mine of copper has been lately discovered, which may add to its importance. The inhabitants have full occupation in collecting the necessaries of life, and do not care, therefore, to trouble themselves about luxuries. *Water* is scarce, and wood is brought from the interior, so that it is not, on any account, a suitable place for shipping.*

The best anchorage is off the village of *Pacocha* (1½ miles South of the town), in 12 or 13 fathoms, and the best landing is in *Huano Creek:* but bad indeed is the best, and care must be taken lest the boat be swamped, or hurled with violence against the rocks.

In going into Ylo, the shore should not be approached nearer than half a mile (as many sharp rocks and blind breakers exist), until three small rocks, called the Brothers, which are always visible under the Table End, bear East, when the village of Pacocha may be steered for, and anchorage taken abreast of it, as convenient.

English Cove affords the best landing, but boats are forbidden that cove, to prevent the contraband trade carried on there.

"The Morro Sama is the point which ought to be made in order to reach the anchorage of Ylo, lying behind Point Coles. When you are 4 leagues to the West of the Morro Sama you will see that point, and also the plateau of rocks, which is half a mile distant from it to the S.W.

"You must then steer so as to pass 1¾ miles outside these rocks; and as the part of the coast which is between Point Coles and the anchorage is bestrewed with rocks, some never uncovering, and only breaking at times, you should not near the land before the village of Ylo bears N.E. ½ E.

"The *anchorage* is sheltered from the swell on the South side by Point Coles; a hill, which appears sandy, commences at this point, at a mile from its most projecting point; it then runs S.W. and N.E., and joins the high mountains of the interior. The village of Ylo is near the shore, at the opening of a valley formed on the South side by this sandy height; and on the North, by an elevation also sandy, and of the same height, but connected with the interior mountains. The

* See Capt. B. Hall's Expedition, &c., vol. i. chap. xiii.

best anchorage is in 15 fathoms, fine sandy bottom, N.N.W. ½ W. of some houses (abandoned) which are on the coast; the village church bearing N. 64° East. The swell is less heavy farther to the North; the S.E. and S.S.E. breezes freshen almost always in the afternoon. You must moor fore and aft. You can land near the deserted houses. It is dangerous to approach the sandy beach, which extends to the North. The landing is probably very difficult in winter; in fine weather, in summer, it is not without difficulties.

"The *anchorage* of Ylo is the best on the coast. In the months of August and September the river is very low, but even then water may be procured. As this place only communicates with the town of Moguegua, it is not much frequented." —(M. Lartigue, pp. 29, 30).

The *Cordillera of the Andes*, behind Ylo, has not been very accurately examined, its being frequently clouded prevents observations. It is traversed by a road leading from Ylo and Moguegua to the interior of Bolivia, along which the merchandise of the sea-coast is carried. One of the peaks of this part of the Cordillera, probably an elevated cone-shaped nevado, was called by Dr. Meyen the *Volcan Viejo*, who gives its height as between 19,000 and 20,000 feet.* It may be mentioned, however, that some of the names as given by the Indians were not always correct; a remark which holds good in respect to the other mountains of the Cordillera.

From Ylo the coast trends to the north-westward, with a cliffy outline, from 200 to 400 feet in height, and with one or two coves, useful only to small coasters, as far as the *Valley of Tambo*, which is of considerable extent, and may be easily distinguished by its fertile appearance, contrasting strongly with the barren and desolate cliffs on either side; those to the eastward maintaining their regularity for several miles, while on the other they are broken, and from their near approach of the hills the aspect is bolder.

The point off this valley is called *Mexico;* it bears N.W. ¾ W. 40 miles from Coles Point, and E.S.E. ½ E. 21 miles from Ilay Point; it is low, and covered with brushwood to the water's edge, and projects considerably beyond the general trend of the coast. At the distance of 2 miles to the southward, soundings may be obtained in 10 fathoms, muddy bottom; from that depth, in the same direction, it increases to 20 fathoms; but on each side of the bank there are 50 fathoms. W.N.W. ½ W. 21 miles from Mexico Point, is *Ilay Point*, and between the two, 5 miles from the latter, the *Cove of Mollendo*, once the port of Arequipa; but of late years the bottom has been so much altered, that it is only capable of affording shelter to a boat or very small vessel; † in consequence of which it has been thrown into disuse, and the Bay of Ilay now receives the vessels that bring goods into the Arequipa market.

"The town of Mollendo, which is one of the seaports of the great city of Arequipa, 60 miles inland, consists of forty or fifty huts, built of reed mats, without any coating of mud, as the climate requires no exclusion of air. Each hut is surrounded by a deep shady verandah, and covered by a flat cane roof.

* Meyen, Reise um die Erde, vol. i.

† M. Lartigue gives ample directions for Mollendo, but it is presumed they are now unnecessary for the above reason.

There are no windows, and, of course, no chimneys; and the doors, like the walls, are constructed of basket-work. The original ground, with all its inequalities, forms the floors;—in short, a more primitive town was never built. The inhabitants of this wide seaport were very kind to us, and remarkably gentle in their manners." *

ILAY, the port of Arequipa,† is formed by a few straggling islets and by Flat Rock Point, which extends to the N.W.: it is capable of containing 20 or 25 sail. The town is built on the West side of a gradually declining hill, sloping towards the anchorage, and is said to contain 1,500 inhabitants, chiefly employed by the merchants of Arequipa. As in all the seaports of Peru, a governor and a captain of the port are the authorities; and it is also the residence of a British vice-consul. Trade was in a more flourishing condition here, even during the civil war, than at any place that was visited by the *Beagle;* there were generally four or five, and often double that number of vessels, discharging or taking in cargoes. The principal exports were wool, bark, and spice, in exchange for which British merchandise was chiefly coveted.

A *vigia* or *look-out house* was established here in 1837, as a beacon for vessels making the port of Ilay, to avoid their passing and getting to leeward of it, which has often been the case. The vigia is painted white, with a flagstaff, on which a flag is hoisted when any vessel is discerned in the offing. It stands upon an eminence, bearing E. by S. 1 mile from the anchorage, and may be clearly perceived by vessels coming from the southward and eastward, as far as the point of Tambo.

In case any vessels should be off the port at nightfall, a *light* will be exhibited from the peak of the flagstaff, as a guide to the entrance of the harbour, for which a charge of two dollars each vessel is made.—(T. Crompton, H.B.M. Consul, Feb. 20th, 1837.)

Ilay being the seaport of the second city of Peru, is much frequented by British merchant vessels, and the following directions will assist them in making it. Vessels have frequently been in sight, to the westward of the port, yet from the set of the current—half a knot, and at the full and change often as much as one knot, to the westward—have been prevented from anchoring for several days.

This, no doubt, has been partly owing to the hitherto inaccurate position assigned to it, and from a proper reluctance to expose a vessel on an imperfectly known coast to be baffled and drifted about by light and variable airs, in addition to a heavy swell continually rolling directly towards the shore.

Coming from the southward, the land abreast of Tambo should be made, and a certainty of that place ascertained, which, according to the state of the weather, may be seen from the distance of 3 to 6 leagues: the course should then be shaped toward a gap in the mountain to the westward, with a defined sharp-topped hill in the near range, a short distance from it. Through this gap lies the road which leads to Arequipa, and which winds along the foot of the hill from Ilay.

* Capt. Basil Hall, Extracts, &c., vol. i. chap. 13.

† AREQUIPA, the second city of Peru, was founded by Pizarro's orders, in 1536. It is a tolerably well built and trading town, standing 7,797 feet above the sea, and a few miles from the volcano of Arequipa, which is 18,300 feet high. The houses are low and massive, on account of the frequent earthquakes from which it has at times greatly suffered. Its population is about 30,000.

As the coast is approached, the foot of the hills will be seen to be covered with white ashes.* This peculiarity commences a little to the westward of Tambo, and continues as far as Cornejo Point, and when within 3 leagues, Ilay Point and the White Islets forming the bay will be plainly observed, and should be steered for.

Care must be taken in closing *Ilay Point*, as a rock, barely covered, lies a quarter of a mile to the southward of the cluster of islands off the point. It is the custom to go to the westward of the White Islands; but, with a commanding breeze, it would be better to run between the third outer and next island, which enables a vessel to choose her berth at once; for the wind heads on passing the outer island, and obliges a vessel to bring up and use warps, or endanger her being thrown by the swell too near the main shore. The best anchorage is just within Flat Rock Point, off the landing place, in 10 or 12 fathoms. A hawser is necessary to keep the bow to the swell, to prevent rolling heavily, even in the most sheltered part. Vessels from the eastward should close the land about Tambo, and observe the same directions.

If coming from the westward, run in on the parallel of 17° 5′, which will lead about a league to the southward of Ilay Point; and if the longitude cannot be trusted, Cornejo Point, being the most remarkable land, and easily seen from that parallel, should be recognised in passing. It lies N.W. by W. ½ W. 14 miles from Ilay Point, and is about 200 feet high, with the appearance of a fort of two tiers of guns, and perfectly white; the adjacent coast to the West is dark, and forms a bay; and on the East there are low black cliffs, with ashes on the top, extending halfway up the hills. If the weather be clear, the Valley of Quilca may be seen, which is the first green spot West of Tambo. Cornejo Point, however, must be searched for, and, when abreast of it, Ilay Point will be seen, topping to the eastward, like two islands, off a sloping point.† The sharp hill before named in the near range will also be seen, if favourable weather; and shortly after the town will appear like black spots in strong relief against the white ground, when a course may be shaped for the anchorage under the White Islets, as before. Landing at Ilay is far from good; a sort of mole, composed of a few planks, with a swinging ladder attached to it, enables a boat generally, with a little manage-

* Near Ilay the land is in several places covered with a whitish powder or dust, which lies many inches thick in hollows or sheltered places, but is not found abundantly in localities exposed to wind. Much difference of opinion has arisen about this powder. People who live there say it was thrown out of a volcano near Arequipa, a great many years ago; other persons assert that it is not a volcanic production, and appertains to, or had its origin, where it is found. My own idea was, before I heard anything of the controversy, that there could be no doubt of its having fallen upon the ground within some hundred years, for it was drifted like snow, and where any quantity lay together had become consolidated, about as much as flour, which has got damp in a damaged barrel.

In one of the old voyages there is a passage which seems to throw some light upon this subject: "As they (of Van Noort's ship) sailed near Arequipa, they had a dry fog, or rather the air was obscured by a white sandy dust, with which their clothes and the ship's rigging became entirely covered. These fogs the Spaniards called 'arenales.'"—*Voyage of Van Noort in* 1660, *from Burney*, vol. ii. p. 223. See Extracts from a Journal, &c., by Capt. B. Hall, vol. i. p. 43.

† M. Lartigue says: "Off Point Cornejo is a large islet, of which the lowest part is nearest the land, but it does not seem to be detached, except when it bears to N.W. or S.E. When it bears East it is confounded with the high lands in the interior. Its reddish tint, which differs from the colour of the surrounding lands, will also serve to distinguish it. The point is thus called by the inhabitants Point Colorado (red), and De los Hornillos (the little ovens)."

ment, to get on shore in safety; but often, at the full of the moon, vessels are detained three days or more, without being able to land or take in cargo. Fresh provisions may be had on reasonable terms; but neither wood nor water can be depended on. There are no fortifications of any description.

The coast between Ilay and Cornejo Point is an irregular black cliff, from 50 to 200 feet high, bounded by scattered rocks, to the distance of a cable's length; about 3 or 4 miles from Ilay is a cove, called *Mollendito*, the residence of a few fishermen; there is a similar cove, Santano Cove, very plainly seen, a little to the eastward of Cornejo Point. Westward of that point the coast retires and forms a shallow bay, in which are three small coves—Aranta, Guata, and Noratos.

"NORATOS COVE is a mile North of Point Cornejo; the swell is not felt in it, and there is water enough for a ship of the line near the land. A ship who wants to re-stow her cargo, or in need of repairs, cannot find a more commodious basin than the Caleta Noratos; but she must be furnished with everything, the place offering nothing. To enter, range along the islet at Point Cornejo, and very near to the land, by the assistance of the breezes which blow in the afternoon, anchoring in 25 fathoms, fine sand, at one-fifth of a mile from the entrance; then tow the vessel into the cove, and moor to the land.

"CALETA LA GUATA is half a mile North of Noratos. The swell is not more felt than in the latter, but there is less water. In the gorge at the bottom of La Guata there is a well of water, which may be of some utility. Fish was also taken in the cove.

"ARANTA COVE, which is about 5 miles from Cornejo Point, and 4 miles S.E. of Quilca, is the place where vessels discharge their cargoes at the time when the Quilca River overflows, and frequently stops the communication between Quilca and the city of Arequipa. The custom-house of Quilca is then established at Aranta, and remains there during the months of February and March. You anchor opposite the cove, very near the land. The swell is not heavy in the cove, and you may land easily, but outside the swell trends directly on to the shore; you must, therefore, wait for a fresh breeze to get away with."*

QUILCA.—N.W. ¾ W. of Cornejo Point, 13 miles distant, are the Valley and River of Quilca, off which vessels occasionally anchor, under the Seal Rock, lying to the S.E. of Quilca Point. This anchorage is much exposed; but landing is good in the cove westward of the valley. Watering is sometimes attempted, by filling at the river and rafting off, but must always be attended with much difficulty and danger. The valley is about three-quarters of a mile in width, and, differing from the others, which are level, runs down the side of the hill: from the regularity of the cliffs by which it is bounded, it has almost the appearance of a work of art.

"When proceeding for the anchorage of Quilca, you must first make Cornejo Point. It is difficult to distinguish this point, as before-mentioned, but its reddish tint, and the difference and height of the parts North and South will serve to make it out. Hence to Quilca it is 13 miles, which is not difficult to reach from Cornejo Point.

* M. Lartigue, 1821, pp. 35, 36.

"The ANCHORAGE is to S. ¾ W. from the little church of Quilca, in 16 fathoms water, on fine gray sand, at the point where a moderately high rock, near the Quilca Point, bears N. 70° E. There are 110 fathoms at 2 cables' length to West of this anchorage; there are some small spaces to the North, where the bottom is good, but they are surrounded by gravel and rocks; still further North the bottom is very unequal. It seems that when the lead shows mud, it is but a very thin covering over rocks; for many vessels, which have anchored on a muddy bottom, have had their cables chafed. When the river overflows the water is very muddy, and this covers the bottom in some parts.

"The best anchorage is on fine sand; it may also be taken to the S.E. of the port above spoken of, but it is too far from the land. You must moor N.N.E. and S.S.W., having a heavy anchor in the latter direction, because the current at times runs very strongly to S.E. This anchorage is on the edge of the flat lying against this coast, formed probably by the matter brought down by the Quilca River, and, therefore, very liable to change, and increase. The anchorage is very uneasy; even in the middle of summer the swell was so strong as to occasion great inconvenience." *

AREQUIPA, a populous city, is the capital of southern Peru, and has about 30,000 inhabitants. It is about 45 miles up the river from Quilca; it is tolerably well built, and has some trade. The city is 7,797 feet above the sea, and towering over it rise three snow-capped mountains, nearly of equal height, viz., Pichu-Pichu, the volcano of Arequipa, or Guagua-Putina, and Chacani. The first and third of these mountains form two elongated serrated ridges, whilst the second presents a very regular volcanic cone, truncated at its summit, and rising to an elevation of 18,300 feet above the Pacific. This volcano has a deep crater, from which ashes and vapour are constantly seen to rise. The summits are usually covered with snow, but at times, after very warm summers, it disappears.†

W. ½ N. from Quilca, at the distance of 6 leagues, is the *Valley of Camana*: the coast between is nearly straight, with alternate sandy beach and low broken cliff, the termination of the barren hills immediately above. The valley is from 2 to 3 miles broad (M. Lartigue says 5 or 6 miles) near the sea, and apparently well cultivated: the village stands about a mile from the beach, but, being small and surrounded with thick brushwood, is scarcely perceptible from seaward.

On approaching from the eastward, MONTE CAMANA, a remarkable cliff, resembling a fort, will be seen near the sea; this is an excellent guide till the valley becomes open. There is anchorage in 10 or 12 fathoms muddy bottom, due South, about a mile; but landing would be dangerous.

M. Lartigue says:—"The coast, between Quilca and Camana, is formed by high and perpendicular rocks, as far as 1 mile to the N.W. of Quilca; there it changes its appearance, and begins to form undulations, of which the landslips, of a whitish colour, are spread out as far as the sand beach which lines the shore, and show like semicircular spots, appearing like a long range of arches. You first see these arches separated from each other, but as you approach they become less defined. They terminate by running one into the other near the Valley of

* M. Lartigue, pp. 37, 38.
† Mr. Pentland, Journal of the Royal Geographical Society, vol. v. 1835, pp. 74, 84.

Camana, where they have lost their first regularity. The beach which lines this singular looking coast is very narrow, and cannot be seen at low water but at a short distance. You follow this beach to go from Quilca to the Valley of Camana, a distance of 15 miles. The Valley of Camana is the largest of those to the S.E., and may be seen at a great distance when brought to N.E.; it is, like the others, lined with trees, and surrounded by whitish arid mountains. The *Clorinde* anchored in 14 fathoms, muddy bottom, 2 miles S.S.E. from the middle of the valley."

W.N.W. ½ W. 23 miles, is the *Valley of Ocoña*, the next remarkable place; it is smaller and less conspicuous than the former, but similar in other respects. An islet lies at its southern extreme, and several rocks near the end of the cliff, on its eastern side.

M. Lartigue says:—"The Valley of Ocoña is very narrow, but fertile, and you must be very near to distinguish it. In general, vessels ought not to anchor opposite to the Valleys of Camana and Ocoña, only when the breezes are light, and when there is danger of being carried on to the coast by the swell."

W. by N. 12 miles, is a projecting bluff point called *Pescadores*: it has a cove on its eastern side, surrounded by islets; and off the point, at the distance of three quarters of a mile in a southerly direction, lies a rock barely covered. To the westward of the point there is a bay, but no anchorage; and the coast then runs in a direct line W. ¾ N. 26 miles, as far as *Atico Point*, a rugged peninsula, with a number of irregular hillocks on it, and barely connected with the coast by a sandy isthmus. At a distance it appears like an island, the isthmus not being visible far off: there is tolerable anchorage in Atico Road, in 19 or 20 fathoms on its western side, and excellent landing in a snug cove at the inner end of the peninsula. By keeping a cable's length off shore, no danger need be feared in running into this road. The *Valley of Atico* lies a league and a half to the eastward, where there are about 30 houses, scattered among trees, which grow to the height of some 20 feet. From this point the coast continues its westerly direction (low and broken cliff, with hills immediately above) to the foot of *Capa Point*; it then forms a curve towards Chala Point; and in these two intervals several sandy coves were observed, but none that appeared serviceable for shipping.

Chala Point bears from Atico Point W.N.W. ¼ W., distant 17 leagues; it is a high rocky point, the termination of the *Morro*, or mount of that name. This mount shows very prominently, and has several summits: on the East side there is a valley separating it from another but lower hill, with two remarkable paps, and on the West it slôpes suddenly to a sandy plain: the nearest range of hills to the westward are considerably in-shore, making Morro Chala still more conspicuous.

N.W. by W. ¾ W. 18 miles from Chala Point, *Chavini Point* appears like a rock on the beach: between them there is a sandy beach, with little green hillocks and sandhills, and two rivulets, running from the *Valleys of Atequipa* and *Lomas*; these valleys are seen at a considerable distance.

Half a mile to the westward of Chavini there is a small white islet and a cluster of rocks level with the water's edge; hence to the roadstead of Lomas a sandy beach continues, with regular soundings off it, at 2 miles from the shore.

Lomas Point projects at right angles to the general trend of the coast, and, like Atico, is all but an island; it may easily be distinguished, although low, from the adjacent coast by its marked difference in colour, being a black rock.

Lomas Road is the port of Acari, and affords good anchorage in from 5 to 15 fathoms, and tolerable landing; it is the residence of a few fishermen, and used as a bathing place by the inhabitants of Acari, which, from the information obtained, is a populous town several leagues inland. All supplies, even water, are brought here by those who visit it; the fishermen have a well of brackish water scarcely fit for use. Boats occasionally call here for otters, which are plentiful at particular seasons.

W.N.W. ¼ W. 23 miles from Lomas Point, we come to *Port San Juan*, and 8 miles farther, that of *San Nicolas*. The former is exceedingly good, and offers a fit place for a vessel to undergo any repairs, or to heave down in case of necessity, without being inconvenienced by a swell; but all materials must be brought, as well as water and fuel, none being found there.

The shore is composed of irregular broken cliffs, and the head of the bay is a sandy plain; still the harbour is good, indeed much better than any other on the S.W. coast of Peru, and might be an excellent place to run for, if in distress. It may be distinguished by *Morro Acari*, a remarkable sugar-loaf hill, rising very steeply from the cliff, on the North side of the bay; and 3 leagues to the eastward, a short distance from the coast, a high bluff head forms the termination of a range of table-land, and is well called *Direction Bluff*. Between this bluff and the harbour the land is low and level, with few exceptions, and has a number of rocks lying off it to the distance of half a mile.

S.W. three-quarters of a mile from *Steep Point* (the southern point of *Port San Juan*), lies a small black rock, always visible, with a reef of rocks extending a quarter of a mile to the northward; and nearly 2 miles to the S.E. there is an islet that shows distinctly. A passage may exist between that reef and the point, but prudence would forbid its being attempted; the safest plan is to pass to the southward, giving it a berth of a cable's length, and not close the shore until well within Juan Point, off which lies a sunken rock. Then haul the wind and work up to the anchorage at the head of the bay, and come to in any depth from 5 to 15 fathoms, muddy bottom. In working up, the northern shore may be approached boldly; it is steep-to, and has no outlying dangers.

The *Harbour of San Nicolas* lies N.W. ¼ N. 8 miles from San Juan, is quite as commodious and free from danger as the latter, but the landing is not so good.

Harmless Point may be rounded within a cable: there are a number of scattered rocks to the southward of it, but, as they all appear, there is no danger to be feared. There are no inhabitants at either of these ports, so that vessels wanting repairs may proceed uninterruptedly with their employment.

N.W. by W. ¼ W. 8½ miles from Harmless Point, is *Beware Point*, high and cliffy, with a number of small rocks and blind breakers in its immediate vicinity. From this point the coast is alternately cliffs and small sandy bays, for 14 miles to Nasca Point, round which lies Caballas.

Nasca Point may be readily distinguished; it has a bluff head of a dark brown colour, 1,020 feet in height, with two sharp-topped hummocks moderately high at its foot: the coast to the westward falls back to the distance of 2 miles, and is composed of white sandhills: in the depth of this bight is *Caballas Road*, a rocky shallow spot, that should only be known to be avoided.

N.W. by W. ¾ W. 28 miles from Nasca Point, is *Santa Maria Point*, and the rock called *Ynfiernillo*. The point is low and rugged, surrounded by rocks and breakers. At the distance of a league and a half inland, to the eastward, is a remarkable flat-topped hill, called the *Table of Doña Maria;* this hill may be seen in clear weather at a considerable distance from seaward, and from its height and peculiar shape is a good mark for this part of the coast.

The *Ynfiernillo Rock* lies due West from the northern end of Santa Maria, at the distance of a mile; it is about 50 feet high, quite black, and in the form of a sugar-loaf; no dangers exist near it, and there are 54 fathoms at 2 miles' distance. Between this rock and Caballas Road the coast to a short distance West of the small *River Yca* is a sandy beach, with ranges of moderately high sandhills. From thence to the Ynfiernillo it is rocky, with grassy cliffs immediately over it, and some small white rocks lying off.

N.N.W. ½ W. 10 miles from Santa Maria, is *Azua Point*, a high bluff with a low rocky point off it; between there is a sandy beach, interrupted by rocky projections, and a small stream running from the hills.

N.W. by W. from Azua Point, and at the distance of 21 miles, is the *Dardo Head*, forming the northern entrance to the Bay of Yndependencia.

YNDEPENDENCIA BAY.—This extensive bay, which is 15 miles in length in a N.W. and S.E. direction, and 3½ miles broad, was, till lately, unknown, or at least unnoticed: no mention was made of it in the Royal Spanish charts, and it was not till the year 1825 that the hydrographer at Lima became aware of its existence; and then only by an accidental discovery. The *Dardo* and *Trujillana*, two vessels that were conveying troops to Pisco, ran in, mistaking it for that place, and were wrecked; and many of the people on board perished. It has *two entrances:* the southern, called *Serrate*, which takes its name from the master of the vessel by whom it was discovered, is formed by the islets of *Santa Rosa* on the North, and *Quemado Point* on the South: it is three-quarters of a mile wide, and free from danger. The northern, or *Trujillana entrance*, is named after one of those unfortunate ships, and is formed by *Carretas Head* on the North, and to the southward by *Dardo Head*, so called after her consort; it is 4¾ miles in width, and clear in all parts. The bay is bounded on the West by the islands of *Vieja* and of *Santa Rosa*, and on the East by the main land, which is moderately high, cliffy, and broken by a sandy beach, at the end of which is a small fishing village called *Tungo*. The people of this village are residents of *Yca*, the principal town in the province, which is about 14 leagues distant;* they come here occasionally to fish, and remain a few days, bringing with them all their supplies, even to water, as that necessary of life is not to be obtained in the neighbourhood. There is anchorage in any part of this spacious bay; the bottom is quite regular, about 20 fathoms all over, excepting off the shingle spit on the N.E. side of Vieja Island,

* YCA is moderately large and very agreeably situated. The vine is almost the only object of industry, and flourishes with astonishing facility. The fruit is chiefly employed in making brandy, which, from being shipped at Pisco, is called Aguardiente de Pisco. The vale of Yca supplies all Peru, and much of Chile, with this liquor. A superior and much dearer brandy is also made from the muscatel grape, and called Aguardiente de Italia.—*Von Tschudi*, p. 234.

where a bank runs off that spit to the northward, on which there are 5 and 6 fathoms: this is the best place to anchor, for, on the weather shore near Quemado Point, it blows strong with sudden gusts off the high land, and great difficulty would be found in landing; whereas at the spit you are not annoyed by the wind, and there is a snug cove or basin within it, where boats may land or lie in safety at any time.

Approaching this part of the coast from seaward, it may be distinguished by the three clusters of hills, Quemado, Vieja Island, and Carretas; they are nearly of the same height, and at equal distances from one another. The S.W. sides of Mount Carretas and the Island of Vieja are steep dark cliffs; but Mount Quemado slopes gradually to the water's edge, and is of a much lighter colour. At the southern extremity of Vieja there is a remarkable black lump of land, in the shape of a sugar-loaf; off which lies the white level island of *Santa Rosa*, the S.W. side of which is studded with rocks and breakers, but there is no danger a mile from the shore.

N.W. ¾ N. 6¼ leagues from Carretas Head, is the *Boqueron of Pisco*, or the entrance to that bay; the shore between them forms a deep angular bay, with the Island of Zarate near its centre. The Boqueron is formed by the main land on the East and the *Island of San Gallan* on the West; this island is 2½ miles long, in a N.N.W. and S.S.E. direction, and 1 mile in breadth; it is high, with a bold cliffy outline. There is a deep valley dividing the hills, which, when seen from the S.W., gives it the appearance of a saddle, the South end terminating abruptly, while its northern end slopes more gradually and carries several peaks. Off this point there are some detached rocks, the northernmost of which has the appearance of a nine-pin, and shows distinctly.

S. ½ E., at the distance of a mile from San Gallan, lies the *Pinero Rock*, which is much in the way of vessels bound to Pisco from the southward; it is just level with the water's edge, and in fine weather can always be seen; but when it blows hard and the weather tide is running, there is such a confused cross sea, that the whole space is covered with foam, rendering it difficult to distinguish the rock; at such a time the shore should be kept well aboard on either side, and when in a line between the South point of the island and the white rock off *Huacas Point*, you will be within the rock, and may steer for Paraca Point; on rounding which the Bay of Pisco will open.

PISCO.—This extensive bay, formed by the *Peninsula of Paracas* on the South, and the Ballista and Chincha Islands on the West, is the principal port of the province of Yca. The town of Pisco is built on the East side, about a mile from the sea; and is said to contain 3,000 inhabitants, who derive considerable profit from a spirit they distil, known by the name of Pisco or Italia, great quantities of which are annually exported to different parts of the coast: sugar is also an article of trade, but the pisco is the staple commodity. Refreshments may be obtained on reasonable terms: wood is scarce: excellent water may be had at the head of Pisco Bay, under the cluster of trees, 2 miles South from the fishing village of Paraca: the landing there is very good, and the wells are near the beach.

The *best anchorage* off the town is with the church open of the road, bearing

E.N.E. ¾ E., in 4 fathoms, muddy bottom, three-quarters of a mile from the shore. A heavy surf beats on the beach with rollers to the distance of a quarter of a mile off, rendering it dangerous to land in ship's boats; launches built for the purpose are used in loading and discharging vessels; but at times even these cannot stand it, and all communication is cut off for two or three days together.

There may be said to be four entrances to this capacious bay: the Boqueron, already mentioned; between San Gallan and the Ballista Islands; between those and the Chincha Islands; and the northern entrance between them and the main; all of which, from appearances, may be safely used; but of those between the islands time would not allow a full examination, and, therefore, there may be dangers that were unseen.

In coming from the southward, after passing Paraca Point, a course may be shaped rather outside of Blanca Island, in order to give a berth to some doubtful ground about a mile to the northward of the peninsula, in 4 fathoms, and then towards the church of Pisco, which will lead directly to the anchorage. Abreast of the island you will have 12 fathoms, muddy bottom; and from this depth it decreases gradually to the anchorage.

In coming from the northward it is all plain sailing; after passing the Chincha Islands, stand in boldly to the anchorage; the water shoals quicker on this side Blanca Island, but there is no danger whatever. Vessels having to ballast here, should work up and anchor under Shingle Point; they can lie close to the shore, and boats may land with expedition.

In coming from seaward, this part of the coast may easily be known by the Island of San Gallan, and the high Peninsula of Paracas at the back of it, which make like large islands, the land on each side being considerably lower and falling back to the eastward, so as not to be visible at a moderate distance. As the shore is approached, the Chincha and Ballista Islands will be seen; which will confirm the position, there being no other islands lying off the coast near this parallel.*

The CHINCHA ISLANDS have become one of the principal points of commercial interest on the South American coast. The guano upon them, which exceeds greatly in abundance that once on the well-known island of Ichaboe, similarly situated in respect to the African coast, is now the object of a very considerable shipping trade, under the government of a commercial company, as previously stated.† They were not very perfectly surveyed by the *Beagle*, but are now so well known, that there will be no difficulty in making or approaching them. The principal danger discovered is a small sunken rock, named the Peacock Rock, having 5 feet at low water on it, and 5 fathoms all round it. It is conical, and the size of a small boat, lying about half a cable's length off the N.E. point of the island.

* Capt. Livingston, when off Pisco, August 26, 1824, in lat. 13° 56′ 28″ S., lon. 78° 30′ W., found the water much discoloured, indeed quite green, and the temperature of it fallen to 60°. He prepared for sounding, but not liking to delay the vessel, then going 7 knots, did not do so. Perhaps he passed over a bank not laid down in the charts, and, according to his opinion, it was very probable that it was so.

† "Masters of vessels are cautioned against throwing overboard ballast at the loading anchorage: any master or masters permitting ballast to be thrown overboard, in less than 12 fathoms of water, will be fined in the penalty of 100 dollars."—"*Comercio*," *Lima*, *Sept*. 2, 1845.

"I have now laid down a buoy upon it for the Guano Company," says Mr. Peacock, in the Nautical Magazine, March, 1847, p. 119, "in 5 fathoms, a boat's length from the rock North of it; it carries a staff and red vane 15 feet high, and is very conspicuous. It would not be prudent to pass inside of the buoy, as another rock lies some 60 yards S.W. of it, although the channel between it and the nearest point is 150 yards wide, with 5, 4, and 3 fathoms in it. Should the buoy be gone at any time, the marks to clear the rock on hauling round the N.E. point, coming from Pisco roads, are, the easternmost Salmadina islet kept open of the point until the N.W. pinnacle ship rock comes open of the Mangara cliff, to which cliff the ships lash, and receive the *mangaras* (canvas hoses) down their hatchways." *

From Pisco the coast runs in a northerly direction, a low sandy beach with regular soundings off it, till you reach the River Chincha; and from thence a clay cliffy coast, which continues as far as the River Cañete. From this river to Frayle Point a beautiful and fertile valley fringes the shore, and to the north-eastward of Frayle Point stands the town of Cerro Azul. This valley produces rum, sugar, and *chancaca*, a sort of treacle, for which it is resorted to by coasters.

Cerro Azul,† or the Port of Cañete, or Canyete, is an open bay, in which landing at all times is very precarious: but the nature of the coast affords great facility for constructing a breakwater, which would render this bay more deserving of the name of port. In its present state they continue to embark sugar, which is produced in tolerable quantity in the fertile *valleys of Canyete*. These I overlooked from my station on the summit of Cerro Azul, or about 300 feet above the sea level. The town or village consists of one house, one church or chapel,

* "From an approximate calculation I made some two years ago of the quantity of *guano* on this little group from actual survey, there is sufficient, at fifty thousand tons exported per annum, to last upwards of a thousand years! so that there need be no fear of exhausting the supply, as there are several other spots on the coast of Peru and Bolivia that would yield equal quantities. This substance is upwards of 100 feet deep in the centre of the northern island, gradually decreasing in thickness at the edges, and resting on a granite formation; the other two islands appear to contain a still larger quantity than this one, and notwithstanding the many thousands of tons that have been carried away from this (the only island yet worked), the quarry, in comparison with the unwrought *guano*, is only a fractional part. The Indians had worked this island long before the Spanish conquest, and it has given constant employment to a number of coasting vessels since the sixteenth century, besides the no small quantity exported to Europe and the United States within the last five years. Birds' eggs, the interior filled with native sal ammoniac, are frequently dug out at great depths (I have had two in my own possession), and also the bones, beaks, and claws of the various birds which have frequented the islands, such as pelicans, boobies, cormorants, and a bird called *potayunka* (Indian), a kind of petrel, which burrows in the *guano*, and is so numerous that it is generally believed this class has chiefly contributed to the formation of this invaluable manure, called by the Indians '*guano*' or '*huano*,' signifying the dung or excrement of animals, and which name the Spaniards afterwards retained, and the vessels employed in this trade are hence called *guaneros*.

"The Chincha group may be reckoned amongst the hundred and ninety-nine wonders of the world, and in walking over the island, the mind can hardly conceive the mass to be the excrement of birds, unless one goes into a simple calculation, when it will be found that a million of birds, in three thousand years, are more than sufficient to make this deposit, for here it has never been known to rain in the memory of man! The surrounding sea is literally alive with fish at all seasons, and, after all, a million of birds distributed over 7 square miles of surface (which is about the superficial contents of the Chinchas) is by no means overrated. In fact, we are quite at liberty, in such a calculation, to take as many birds as would actually cover the ground; for every one who has made a voyage to the tropics, must have seen the innumerable flocks of sea-fowl that hover over rocks and islands, away from a civilized coast."—*George Peacock.*

† The following observations are by Capt. Sir Edw. Belcher, in H.M.S. *Sulphur*, August, 1838, vol. i. pp. 199—201.

and a few huts arranged on three sides of a square, the fourth open to the sea, with other straggling huts, amounting altogether to about twenty.

Cerro Azul is a high, bluff, insulated clump, projecting into the sea, and at a short distance might be mistaken for an island. Its predominant colour is yellowish red.

There are no objects of interest between this and the *Asia Islands*, which are distant 17 miles N.W. ½ N., and are merely a patch of high rocks projecting about 2 miles to seaward from a very flat sandy beach, having a channel carrying 4 fathoms, but well studded with rocks, which, by daylight, are easily avoided. *Asia Peak* is situated in lat. 12° 47′ S., lon. 76° 34′ W., and its island is about half a mile long by a quarter broad, having no vegetation. There is good landing in a very snug bay, on its eastern side, where a seal fishery has apparently been carried on at times.

Between Cerro Azul and Asia Island the coast is dangerous, and landing generally impracticable, but the lead will always afford timely warning. A little to the northward of Asia Island is a deep bay, but neither here nor at any point, until reaching Chilca, could we find landing; although we were informed that this could be effected at the River Mala. We did not see the river, nor anything like one. It was possibly screened by the surf.

Chilca Point, 20 miles N.W. from Asia Islands, forms a sharp elbow in the land, making a very deep bay, in which a small town was noticed. It is about 300 feet high in its highest part, has several rises on it, and terminates in a steep cliff, with a small flat rock close off its pitch. A remarkable peak, called *Devil's Peak*, rises about 300 feet perpendicularly, and forms the eastern limits. Northerly from Chilca Point, 3 miles, lies the *Port of Chilca*, formed by a large island, which enables vessels of small draught to lie in a complete dock, land-locked, the outer harbour having good anchorage in 10 to 14 fathoms.

A small village of huts, with a chapel, is situated on the eastern beach of the inner harbour, and is apparently merely the resort of fishermen. . . . . . The whole soil is so entirely impregnated with salt, that every stone has an incrustation of pure white crystalline salt on it; and in many cases I noticed that it cemented the stones together to a thickness of 4 inches solid salt. This, of course, is of great importance to the fishery, but a sad drawback to the seamen who may seek for water in this neighbourhood. A road runs through the valley of Chilca to the town in the bay before mentioned, where bright green tints afford assurance of fertility.

Chilca is a snug cove, but very confined; anchorage is good in any part of it, and landing tolerable; there is a small village at the head of the bay, but no information could be obtained from the inhabitants about Chilca, for they deserted their huts on the arrival of Capt. FitzRoy.

From Chilca the coast forms a bend to the valley of Lurin, off which are the Pachacamac Islands. The northern is the largest, half a mile in length, and about a cable's length broad; *San Francisco* is the most remarkable, being quite like a sugar-loaf, perfectly rounded at the top: the others are mere rocks, and not visible at any distance. At the northern end of these islands lies a small reef even with the water's edge; the group runs nearly parallel to the coast, in a

N.W. and S.E. direction, and is about a league in extent. There is no danger on their outer side, but towards the shore the water is shoal, which causes a long swell to break there heavily.

This island is one of the most interesting places on the Peruvian coast. At the time of the Spanish invasion it was connected with the main land, and formed a promontory. Possibly it may have been detached from the coast during the great earthquake of 1586.* On the summit of a hill on the island is the Temple of Pachacamac, or Castle, as it is called by the Indians, a square edifice of three terraces, built of rock, and covered with sun-dried bricks. This ruin of the ancient Peruvian times resembles the view given by Humboldt of the pyramid of Cholula. The remains of the town occupy some lower ground, a quarter of a mile to the northward. Many interesting antiquities have been procured here, evidences of the domestic life of the "children of the sun."

North of the Pachacamac Islands the *River Lurin* brings its small stream from the interior, but without sufficient force to make its way into the sea; the valley, however, which it waters, appeared fertile and well cultivated when seen from the offing. From thence to the Morro Solar is a sandy beach, with moderately high land a short distance from the sea. The MORRO SOLAR is a remarkable cluster of hills, standing on a sandy plain; when seen from the southward it has the appearance of an island in the shape of a quoin, sloping to the westward, and falling very abruptly in-shore; on its sea face, however, it terminates in a steep cliff, with a sandy bay on each side.

Off *Solar Point*† there is an insignificant islet, with some rocks lying about it, and off *Chorilla Point* a reef of rocks projects about 2 cables' length; round this reef, on the North side of the Morro, lies the town and road of *Chorillo.* The former is built on a cliff, at the foot of one of the slopes of that mountain.

CHORILLO, 3 leagues to the South of Lima, is the favourite watering place, and frequented during the sultry months by gambling parties and persons of rank and fashion from town. It is a small village of fishermen, constructed of cane and mud. The Indian owners of the shades, and of some houses or *ranchos,* let them to the bathers during the bathing season; and some persons either take these for a term of years, or construct light houses for themselves, which they fit up tastefully, and pass the summer months in them in the midst of gaiety and mirth. Chorillo is sheltered from the south-western blast by the elevated promontory, the *Morro Solar,* which rises like a gigantic *guaca* overlooking the numerous monuments or Pagan temples of this name, which are scattered over the naturally rich, but now, in a great measure, waste and desolate plain, that extends from Lima to Chorillo.

During the raw, damp, and foggy months of July and August in Lima, Chorillo enjoys a clear sky and a genial air. The south-westers, laden with heavy clouds,

* Von Tschudi, p. 45.—See Narrative of the U. S. Exploring Expedition, vol. i. pp. 279—81.

† "Lachira Bay, under the point of Morro Solar, having been named as the rendezvous for British shipping should the blockade of Callao be maintained, became my next point of interest. Its character may be summed up in a few words. The bay is open, landing bad (if practicable), anchorage untenable and even dangerous; in proof of which we left there the fluke of our anchor."—*Capt. Sir Edw. Belcher.*

spend their strength on the Morro Solar (on which burst the only thunder-storm witnessed by the Limenians in the memory of any one now living), and divide into two currents; the one pursues the direction of the village of Miraflores, and the other, the hacienda of San Juan, leaving Chorillo clear and serene between. Thus protected, Chorillo does not experience the chilly mists of winter; and it is the great hospital of convalescence for aguish, asthmatic, dysenteric, rheumatic, and various other sorts of invalids, from the capital during the misty season.

When political revolutions rendered the road of Callao a dangerous berth, the vessels all resorted to Chorillo Bay, though in every other respect an unfit place for anchoring, as the bottom is a hard sand, with patches of stones and clay mixed together, called tosca; and the heavy swell that sets round the point, causing almost a roller, brings a vessel up to her anchor and throws her back again with a sudden jerk, which endangers dragging the anchor or snapping the cable.

Capt. Livingston says:—"This bay has very foul ground. A reef lies off from the S.W. point of the bay (N.W. point of the Morro Solar); from the break it seems to extend 1½ or 2 cables' length from the point, and a good berth ought to be given to it. At the same point there is a small detached rock, which cannot be distinguished from the main land until you have shut in all the land to the southward. When once you can see out to sea betwixt this rock and the point with the two spires, turrets, or steeples, of the church of Chorillo in one, you will be in the best anchorage for moderate-sized vessels, in 4½ fathoms, but in this situation you should be sure of having good ground-tackle, as, in the event of parting a cable, you have very little room left to work in.

"In sounding in this bay it is advisable to do it with a boat, and to drag the lead after the boat, as thus you more readily discover the heads of rock.

"Excellent water may be procured at Chorillo, and with as much or more facility than at Huacho. The landing at Chorillo with boats is generally bad, often dangerous, and sometimes impracticable. There is generally a heavy swell in this bay, and vessels, if not moored with a stream cable out to the northward, roll exceedingly. Outside the point, and about N.W. by N. from it, there is a patch of mud, but I do not know its exact position. This is the best place for line-of-battle ships to anchor."*

Vessels that must anchor there ought not to shut Solar Point in with the next, or Codo Point: by keeping those points open they will ride in 8 or 9 fathoms, and not have so much swell as there is further in. The landing in the bay is very bad; canoes built purposely, and dexterously managed, are the usual means of communication: for though, no doubt, there are times when a ship's boat may land without danger, yet very seldom without the crew being thoroughly drenched.

From Chorillo the coast runs in a steady sweep, with cliffs diminishing in height, till it reaches the Point of Callao, which is a shingle bank, stretching out towards the Island of San Lorenzo, and which with it forms the extensive and commodious Bay of Callao.

* Observations by Capt. Andrew Livingston, of the brig *Jane*, a gentleman to whom the nautical world is indebted.—July, 1825.

CALLAO AND LIMA.

Callao is the port of Lima, the Peruvian capital. The harbour is protected from the Pacific by the Island of San Lorenzo; the town standing on a sandy peninsula, which projects beyond the general line of coast towards the middle of the island.

The Bay of Callao forms a fine harbour. The climate and prevalent winds from the southern quarters render it so. Its northern side is entirely exposed, but there is no danger to be apprehended from that quarter. San Lorenzo keeps off all swell from the shore immediately around the town, but a few miles to the northward the surf breaks heavily on the beach, and effectually prevents all landing.

The Island of San Lorenzo lies in a N.W. by W. and S.E. by E. direction; and off its S.E. end is the Isla del Fronton, extending a mile farther.

Capt. Wilkes says:—"We remained at San Lorenzo ten days (July, 1839), during which time its three highest points were measured with barometers at the same time. The result gave 896 for the southern, 920 for the middle, and 1,284, for the northern summit. Upon the latter the clouds generally rest, and it is the only place on the island where vegetation is enabled to exist. The others are all barren sandy hills. It is said that the only plant which has been cultivated is the potato, and that only on the North peak. This becomes possible then from the moisture of the clouds, and their shielding it from the hot sun.

"The geological structure of the island is principally composed of limestone, clay, and slate. It presents a beautiful stratification. Gypsum is found in some places between the strata, and crystals of selenite are met with in one or two localities. Quantities of shell-fish are found on the shore, and the waters abound with excellent fish.

"The burying-ground is the only object of interest here. The graves are covered with white shells, and a white board, on which is inscribed the name, &c. They appear to be mostly of Englishmen and Americans." *

The small strait separating San Lorenzo from the land will perhaps afford some insight into those remarkable geological changes which have occurred in the coast and land of Peru. Mr. Darwin considers that this part of the land has risen 85 feet since it had human inhabitants. He founds this opinion upon the facts observed on the shores of Lorenzo, the ledges and the deposits of shells, on one of which, at that height above the sea, he found some bits of cotton thread, plaited rush, and the head of an Indian corn stalk. There are statements that the land *sunk* at different periods, among others in 1746, when the great earthquake swallowed up the city of Callao; the ruins upon the tongue of land from the fortress, supposed by Darwin to be of this city, are those of Callao, destroyed in 1630. Subsequently to this it must have been upheaved, for boys used to throw stones over to the island. At present the distance is nearly 2 English miles. Another proof of sinking is the shoal between the coast and San Lorenzo, called the Comotal, which in early times was cultivated, particularly with *comote*

* Narrative of the United States Exploring Expedition, vol. i. p. 231.

(sweet potatoes), hence its name. This occurred perhaps in 1687 or 1630.—(See Darwin, 2nd edit., pp. 43, 44.)

May not both these phenomena, upheaving and subsiding, have occurred at different times? The plan of Ulloa (1742) does not make the distance between the main land and San Lorenzo very different from the recent surveys.*

"The principal street of Callao runs parallel with the bay. There are a few tolerably well-built two-story houses on the main street, which is paved. These houses are built of adobes, and have flat roofs, which is no inconvenience here, in consequence of the absence of heavy rains. The interior of the houses is of the commonest kind of work. The partition walls are built of cane, closely laced together. The houses of the common people are of one story, and about 10 feet high; some of them have a grated window, but most of them only a doorway and one room. Others are seen that hardly deserve the name of houses, being nothing more than mud walls, with holes covered with a mat, and the same overhead.

"The outskirts of Callao deserve mentioning only for their excessive filth; and were it not for the fine climate it would be the hotbed of pestilence. One feels glad to escape from this neighbourhood."—(Wilkes.)

"The market, though there is nothing else remarkable about it, exhibits many of the peculiar customs of the country. It is held in a square of about one and a half acres. The stands for selling meat are placed indiscriminately, or without order. Beef is sold for from 4 to 6 cents the pound, is cut in the direction of its fibre, and looks filthy. It is killed on the commons, and the hide, head, and horns, are left for the buzzards and dogs. The rest is brought to market on the backs of donkeys. Chickens are cut up to suit purchasers. Fish and vegetables are abundant, and of good kinds; and good fruit may be had if bespoken. In this case it is brought from Lima. Everything confirms, on landing, the truth of the geographical adage, 'In Peru it never rains.' It appears everywhere dusty and parched up.

"The tide at Callao is small, generally of 3 and 4 feet rise. The temperature of the water during our stay was 60°; of the air from 57° to 63°. Since my visit to Callao in 1821, it had much altered, and for the better, notwithstanding the vicissitudes it has gone through since that time. A fine mole has been erected, surrounded by an iron railing. On it is a guardhouse, with soldiers lounging about, and some two or three on guard.

"The mole affords every convenience for landing from small boats. The streets of Callao have been made much wider, and the town has a more decent appearance. Water is conducted from the canal to the mole, and a railway takes the goods to the fortress, which is now converted into a depôt. This place, the seaport of Lima, must be one of the great resorts of shipping, not only for its safety, but for the convenience of providing supplies. The best idea of its trade will be formed by the number of vessels that frequent it. I have understood that there is generally about the same number as we found in port, namely, forty-two, nine of which were ships of war: five American, two French,

* Dr. Von Tschudi, pp. 43, 44. See likewise Darwin, p. 451; and also second edition.

one Chilian, and thirty-five Peruvian, merchantmen, large and small. The castle of Callao has become celebrated in history, and has long been the key of Peru.* Whichever party had it in possession were considered as the possessors of the country. It is now converted into a better use, viz., that of a custom-house, and it is nearly dismantled."†

The city of Lima stands on a plain, in a valley formed during the gradual retreat of the sea. It is distant 7 miles from Callao, and, according to Capt. Wilkes' measurement, Mr. Bartlett's (the United States' consul) house is 420 feet above the sea; but, from the slope being very gradual, the road appears absolutely level, so that, when at Lima, it is difficult to believe that one has ascended some hundred feet. Humboldt has remarked on this singularly deceptive case, as has almost every other traveller. Steep barren hills rise like islands from the plain, which is divided by straight mud-walls into large green fields, having only a few willows here and there, and an occasional clump of oranges and bananas.‡

LIMA, the capital of Peru, was founded by Pizarro, Jan. 15th, 1535; he is buried in the cathedral of this "City of the Kings," as he named it. Its present name is derived from the river which flows through it, the *Rimac* of the Peruvians, softened into its European form by the Spaniards.

The houses are tolerably built of adobes, or sun-dried bricks, canes, and wood; they are low, in order to stand the shocks of earthquakes, being seldom above two stories, with small balconies to the second floor, with generally an archway from the street, and with a strong door leading to a court within. The lower or ground floor is commonly used as store-rooms and stables, and all kinds of rubbish are stowed away on the tops. The staircase is generally spacious and handsome, and the apartments of the lodgers often adorned with common fresco paintings. For the climate these houses are, however, sufficiently well adapted. The cathedral, the palaces of government and of the archbishop, the university, several colleges, and some churches, are the most remarkable edifices. The population is estimated at about 70,000. There are several unimportant manufactures carried on, and its trade in foreign merchandise, and its exports of the produce of the mines, and of the interior, are through the nearly adjacent port of Callao.

It will be unnecessary to give a more detailed description of this important but faded city. The best and most recent account of it is that of Dr. Von Tschudi in 1842. His description and statistics are more complete than most of his

* Prior to its destruction, the collection of batteries, known under the name of the Castle of Callao, had an imposing appearance. One of the scenes for which it is best known to Englishmen is the exploit of Lord Cochrane, who cut out the *Esmeralda* Spanish frigate, by means of fourteen boats, on the night of Nov. 5th, 1820, from under its guns, thus destroying the Spanish naval power in the Pacific, and giving a great impetus to the success of the Chilians against the Spanish domination. The president of the Peruvian republic sold all its beautiful brass guns but five in July, 1835, during the time the *Beagle* was there, assigning as a reason that he had no officer to whom he could entrust the command of so important a fortress; he himself having obtained his presidentship by a successful rebellion while in command of it. Full particulars of the war of liberation and the events of Callao are given by Capt. Basil Hall, in his Extracts from a Journal, &c., part 1, chap. 3.

† Wilkes' Exploring Expedition, vol. i. p. 233.

‡ Mr. Darwin, p. 449.

predecessors. The most authentic and complete account of it, prior to the great earthquake of 1746, is given by Ulloa, Voyage de *l'Amerique*, vol. i. pp. 422—66. See also Capt. Wilkes, U.S.E., before quoted; Dr. Meyen, Reise um die Erde, vol. xi. pp. 56—64 (this work, though excellent, has not been translated); Poeppig; Scarlet; Basil Hall, &c., &c.

The Island of San Lorenzo is 1,050 feet at its highest part, 4½ miles long in a N.W. and S.E. direction, and 1 mile broad. Off its S.E. end lies a small but bold-looking island, called Fronton; and to the S.W. are the Palominos Rocks. Its northern point, or Cape San Lorenzo, is clear, and round it is the usual passage to the anchorage at Callao. In rounding it, however, do not close the land nearer than half a mile, for within that distance there are light baffling airs, caused by the eddy winds round the island, by getting among which you would be more delayed than if you gave the island a good berth, and should have to make an additional tack to fetch the anchorage.

There are no dangers in working in, except the long spit that stretches off from Callao Point towards San Lorenzo Island; part of it, however, just shows at the water's edge, and the sea breaks violently along its ridge. Callao Point is very low, and consists of a bank of small round stones, as far nearly as the battery of San Rafael.

Should there be occasion to work to windward to fetch the anchorage, the above shoal with another rock, said to lie off the Galera Point, of the Island of San Lorenzo, are so far to the southward that you need scarcely apprehend borrowing on them. Run or work up close to the shipping, and anchor in from 7 to 5 fathoms; with the pier-head bearing about S.E. and San Lorenzo W. by S. Although the above mark is given for the most convenient anchorage, yet ships may lie with the greatest safety in any part of the bay, and in any depth of water, on clear ground and gradual soundings from 20 to 3½ fathoms up to the molehead and landing place.

This is the obvious route to Callao; but there is another which, with common precaution, may be used to great advantage, by vessels coming from the southward; and passing through the Boqueron Channel between the Island of San Lorenzo and Callao Point. After making Fronton Island, steer so as to keep its southern end about a point open on the port bow; continue on this course until Callao Castle is seen, which has two martello towers on it, and stands on the inner part of the shingle bank that forms the point: then steer for that castle till Horadada Island (which has a hole through it) comes in one with the middle of the southern sandy bay of the Morro Solar, bearing about E.S.E. with these marks in one, and therefore steering about W.N.W.: for the furthest point of Lorenzo that can be seen, you will be clear of all danger; and when the western martello tower in the castle comes in one with the northern part of Callao Point, you may haul gradually round to the northward till that tower opens clear of the breakers on the spit, when a direct course may be shaped for the anchorage; taking care not to come nearer the sand called the Whale's Back than 6 fathoms. There is no regular tide in this passage, yet a little drain always felt, sometimes to the N.W., and at others the contrary: should the stream be adverse, and it fall calm while in the channel, there is good anchorage in 8 or 9 fathoms with the leading marks in one.

These marks will also lead clear of the bank that extends three-quarters of a mile to the northward of Fronton Island; and as soon as the rock between Fronton and San Lorenzo bears S. by W., the Fronton Shoal will have been passed, and San Lorenzo may be approached as above directed.

H.M.S. *Collingwood* came through with the above marks, in not less than 5½ fathoms on the port side, and 5¾ on the starboard.

Directions *by M. Lartigue.*—"In proceeding from any of the ports on the South coast of Peru, it is best to take advantage of the breezes to get off the land as quickly as possible, and keep at 30 leagues from the coast; it will be unnecessary to close again until abreast of the Isle of San Lorenzo, for which you can proceed direct if you are sure of your longitude. If not, steer a little East of this island to make the Isle of San Gallan, which is visible 12 leagues off. As soon as this is made out, run for San Lorenzo. The currents are generally to the northward, following the direction of the coast; so that, whether you proceed directly for San Lorenzo or first make San Gallan, you must always allow for a difference of 10 miles at least in your reckoning.

"During the night, and a part of the morning, the weather is often foggy in the neighbourhood of Lima; the land near the coast cannot be seen but at a short distance; it is then possible to mistake the Morro Solar for San Lorenzo, which is 12 miles S.E. ½ E. from it, and appears like an island from a distance, or when the neighbouring land is hidden by the haze. There is to the N.W. of the Morro Solar several rocks, some of which are at a distance from the land, and, therefore, it must not be approached; it is prudent, in foggy weather, to keep board and board to the South until the breeze freshens, for then the haze never fails to be dissipated.

"The Island of San Lorenzo ought not to be neared too closely on the South; but you may range near to the West point of the island, so as to reach the anchorage on that tack. If you should be becalmed at the mouth of the bay, you must anchor to the North of San Lorenzo, so as to avoid being carried by the current to leeward of the port."

Supplies of all sorts may be obtained for shipping; fresh provisions as well as vegetables, with an abundance of fruit: watering is also extremely convenient, a well-constructed mole being run out into the sea, at which boats can lie, and fill from the pipes that project from its side; wood is the scarcest article and very dear, so that vessels likely to remain at this port should husband their fuel accordingly.

HORMIGAS.—Due West from the North end of San Lorenzo, at the distance of 31 miles, lie a small cluster of rocks called the *Hormigas de Afuera;* the largest is about three-quarters of a mile in circumference, 25 feet high, and covered with *guano;* no sign of vegetation was observed: it is merely a resting place for birds and seals; landing might be effected, if requisite, on its North side, but with difficulty. Being somewhat in the way of vessels bound to Callao from the northward, and of those leaving that port for the West, care should be taken not to approach too closely, for fear of being overtaken by dense fog, so frequent on the Peruvian coast, while in their neighbourhood. The water is deep close-to all round, and no warning would be given by the lead.*

The Cordillera of the ANDES, on this part of the Peruvian coast, approach it within 60 or 70 miles. Of their elevation we have but few measurements. One singular feature of this part of the range is, that it seems to be rather a continuation of the eastern range, with which it unites at the back of Arequipa, than the western chain of the Bolivian Andes. Several of the peaks to the South rise above the limits of perpetual snow, but the elevation of none of these have been determined. The best known are the Toldo de la Nieve, S.E. of Lima, from which it is seen: the Altunchagua, about 10° South lat.; and the Nevado de Guaylillas, above Truxillo, or 7° 50′ South. The passes over the Cordillera are of very great elevation. The best known is that from Lima to Tarma and Cuzco; it rises on the principal chain at the Portacuelo de Tucto to 15,760 feet above the sea. Farther North are the passes to the silver district of the Cerro de Pasco; the westernmost, the Alto de Taicabamba, is at 15,135 feet. Other passes cross the chain to the northward, but we are comparatively unacquainted with the particulars.

From Callao the coast is a sandy beach, lying in a northerly direction, until it reaches *Pancha Point;* it there becomes higher and cliffy, and maintains this character as far as *Mulatas Point*, round which is the little *Bay of Ancon*.

To the West and S.W. of Ancon lie the *Pescador Islands*, the outer and largest of which bears N.N.W. ¾ W. from Callao Castle, and at the distance of 18 miles. There is no danger among these islands; they are steep-to, with from 20 to 30 fathoms near them.

N.W. by N. from Mulatas Point, 12 miles distant, is the *Bay of Chancay* and river of that name; this bay may be known by the bluff head that forms the point, and has three hills on it, in an easterly direction; it is a confined place, and fit only for small coasters. From Chancay, the coast runs in a more westerly direction, as far as Salinas Point, a shingle beach, with a few broken, cliffy points; the hills are near the coast, and from 400 to 500 feet high.

The *Point of Salinas* is 27 miles N.W. by W. ¼ W. from Chancay Head. It is 5 miles in length, in a North and South direction; off its southern face there is a reef of rocks, a quarter of a mile from the shore; and at its northern angle, called *Las Bajas*, an islet at a cable's distance; two coves, between these points, are fit only for boats. There is a remarkable round hill, called Salinas, at a short distance from the coast, and further in-shore, a level, sandy plain; at the South side of which plain lie the Salinas, or salt-ponds, that give the headland its name. These ponds are visited occasionally by people from Huacho.

Off Salinas Point, in a S.W. direction, lie the HUAURA ISLANDS, the largest of which is called *Mazorque*. It is 200 feet in height, three-quarters of a mile long, and quite white: sealers occasionally frequent this island, as there is a landing place on its North side.

The next in size is *Pelado;* it lies S.W. ½ W. 6½ miles from Mazorque, is about 150 feet high, and apparently quite round; and between these two islands a safe

* See also remarks by Mr. Babb, R.N., Naut. Mag., 1853, p. 248.

passage exists, and may be used without fear in working up to Callao. Between Mazorque and Salinas stand several other islands, which, from their appearance, may be approached without danger; but as no advantage could be gained, it would not be prudent to risk going between them. Vessels, in working up, do sometimes pass between the inner one and the point; but what they gain thereby does not appear, for when the current sets to the southward, it runs equally as strong between Mazorque and Pelado as it does nearer the shore.

Round Bajas Point is the *Bay of Salinas*, of large dimensions, and affording roomy anchorage. From thence the coast is moderately high and cliffy, without any break, until you reach the *Bay of Huacho*, which lies round a bluff head and is small, but the anchorage is good, in 5 fathoms, just within the two rocks off the northern part of the head. The town is built about a mile from the coast, in the midst of a fertile plain, and in coming from seaward has a pleasant appearance; it is not a place of much trade, but whale ships find it useful for watering and refreshing their crews. Fresh provisions, vegetables, and fruit are abundant, and on reasonable terms; wood is also plentiful, and a stream of fresh water runs down the side of the cliff into the sea. Landing is tolerably good: yet rafting seems to be the best method of watering.

In coming from seaward, the best distinguishing marks for this place are the *Beagle Mountains*, three in number in the near range, and each of which has two separate peaks. They lie directly over the bay, and, on closing the land, the round hill near Bajas Point as well as the island of Don Martin, to the northward, will be seen; about midway between them is the Bay of Huacho, under a light brown cliff, the top of which is covered with brushwood. To the southward the coast is a dark rocky cliff.

N.N.W. $\frac{1}{2}$ W. 3$\frac{2}{3}$ miles from Huacho, lies the *Bay of Carquin*, scarcely as large as Huacho, and apparently shoal and useless to shipping: off Carquin Head, which is a steep cliff, with a sharp-topped hill over it, there are some rocks above water, and an islet a short mile distant. N.N.W. $\frac{3}{4}$ W. 3 miles from the islet near Carquin Head, stands the Island of Don Martin, and round to the northward of the point, abreast of it, is the Bay of Begueta, no place for a vessel.

From this bay the coast is moderately high, with sandy outline, all the way to Atahuanqui Point, distant 8 miles N.N.W. $\frac{1}{2}$ W. This is a steep point, with two mounds on it, and is partly white on its South side; there is a small bay on its North side, fit only for boats. Between this point and the South part of Point Thomas, the coast forms a sandy bay, low and shrubby; with the town of *Supé* about a mile from the sea.

Point Thomas is similar in appearance to Atahuanqui, without the white on the South side. To the northward of this point there is a snug little bay, capable of containing four or five sail; it is called the *Bay of Supé*, and is the port of that place and of Barranca.

There is a fishing village at the South end of the bay, which is used by the inhabitants of Barranca during the bathing season. Hitberto it had been a forbidden port by the government; in consequence of which it is little known, and has had few opportunities of exchanging its produce for the goods of other countries. Very little information could be gained there, as to the size of the neighbouring towns or the number of inhabitants they contain, but from their appearance it was thought they might be of considerable extent. These places produce chiefly sugar and corn, cargoes of which are taken in the various little vessels that trade along the coast. Refreshments may be obtained; but water is scarce, the greater part of which is brought from Supé town for the use of the inhabitants of the village.

The best anchorage is in 4 fathoms, with Point Thomas shut in by Patillo Point, about a cable's length from the rocks off that point, and rather more than a quarter of a mile from the village. Good anchorage may be obtained further out in 6 or 7 fathoms, though but little sheltered from the swell. In entering, no danger need be apprehended; Point Thomas is bold, with regular soundings, from 10 to 15 fathoms, three-quarters of a mile off it. Off Patillo Point, though there are a few rocks, yet there is no necessity of hugging the shore very closely, as you can always fetch the anchorage, by keeping at a moderate distance when standing in.

To recognise this port, the best guide at a distance is *Mount Usborne*, the highest and most remarkable mountain in the second range; it bears from the anchorage N.E. ¾ E.; it has something of the shape of a bell, and has three distinct rises on its summit—the highest at the North end. On that side it shows very distinctly, there being no other hills within a considerable distance. On approaching the coast, the *Island of Don Martin* to the southward, and to the northward *Mount Darwin*, and *Horca Hill* near the beach, with a steep cliffy side to it facing the sea, with apparently an islet off it, will be seen nearly 4 leagues. The harbour itself has a white rock off its northern point, and cannot be mistaken, for there is no other like it near this part of the coast.

From Supé the coast is a clay cliff, about 100 feet in height, to the distance of 1¾ leagues; it then becomes low and covered with brushwood, to the foot of Horca Hill; here it again becomes hilly near the sea, with alternate rocky points and small sandy bays, which continue for the distance of 6 leagues, to *Jaguey Point* and the bay called *Gramadel*. This is a wild-looking place, with a heavy swell rolling in; but it is visited occasionally for the hair-seal, with which it abounds: there is anchorage in 6 or 7 fathoms, sandy bottom, with the bluff that forms the bay bearing S.S.E. about half a mile from the shore; landing is scarcely practicable.

The coast maintains its rocky character, with deep water off it, as far as the *Bufadero*, a high steep cliff, with a hill having two paps on it, a little in-shore. From this bluff a rocky cliff from 200 to 300 feet high, with a more level country, extends as far as Legarto Head, round which is the *Port of Guarmey*.

GUARMEY.—In comparison with other places, this may be considered a tolerable harbour, having good anchorage everywhere, in from 3½ to 10 fathoms, over a fine sandy bottom.

Firewood is the principal commodity, for which it is the best and cheapest place on the whole coast. Vessels of considerable burden touch here for that article, which they carry up to Callao, and derive great profit from its sale. There are also some saltpetre works, established by a Frenchman, but little business is done in that line. The town lies in a north-easterly direction, about 2 miles from the anchorage, but is hidden by the surrounding trees, which grow to the height of 30 feet. It has only one street, and cannot contain more than 500 or 600 inhabitants. At the anchorage there is a small house used for transacting business, but no other building, which is unusual, as at most of these places a small village has been established near the sea. Large stacks of wood are piled up on the beach, ready for embarkation.

Fresh provisions, vegetables, and fruit, are plentiful and moderate; but water is not to be depended on. It is true there is a river, and for several months after March a plentiful supply may be obtained, but in the summer season great drought is sometimes experienced. At the time the *Beagle* was there a whale ship put in to supply her wants, and had to remain several days, waiting for the water to come down from the mountains.

Legarto Head is a steep cliff, with the land falling immediately inside it, and rising again to about the same height. In sailing in, after having passed the Head, a small white islet will be seen in the middle of the bay; steer for it, that you may not border on the southern shore, for there are many straggling rocks running off the points; and when sufficiently far to the northward to shape a mid-channel course between this *Harbour Islet* and the point opposite it, to the southward, do so, and it will lead to the anchorage. In standing in, in this direction, the water shoals gradually to the beach, but the southern shore must on no account be approached nearer than a quarter of a mile.

The BEST ANCHORAGE is in 4 fathoms, with Harbour Islet bearing N.N.W. ¼ W., and the ruins of a fort on a hill in-shore E. ½ N. about a quarter of a mile from the landing place on the beach. This landing place does not seem to be so good as at a steep rock on the outer side of the bluff, where the sandy beach commences; but probably it is the most convenient for loading boats.

TIDES.—The rise and fall of tides are very irregular, and the time of high water uncertain; but, generally speaking, 3 feet may be considered about the extent to which it ranges. The sea-breeze sets in so strongly occasionally, that it is difficult for boats to pull against it: this is particularly the case under the high land, whence it comes in sudden gusts and squalls.

In coming from seaward, the best way to make this port is to stand in on a parallel of 10° 6′, and when within a few leagues of the coast, a sharp-peaked hill, with a large white mark on it, will be seen standing alone a little North of the port: the break in the hills through which the river runs is high and cliffy on each side. The land is also much lower to the northward of Legarto Head, and there is a large white islet at the North end of Guarmey Bay.

N.W. by N. 7½ miles from the white islet at the North end of Guarmey Bay is *Culebras Point*, a level projecting point, similar in appearance to Legarto Head, when seen from the northward; the intervening coast is a mass of broken cliffs and innumerable detached rocks, with moderately high land near the shore.

On the North side of Culebras Point there is anchorage off the valley of that name. From that point the coast is rocky, with small sandy bays, and some rocks lying off it, for three-quarters of a mile; there is also a white cliffy islet, *Cornejos Islet*, 5 miles to the northward of Culebras, from whence the coast takes a bend inwards, forming a bay, and then out to *Mogoncilla Point*. A straight shore, of 10 miles in length, then leads towards the *Colina Redonda*, a point with two hummocks on it, and when seen from the southward, appearing like an island. On its North side is the *Caleta*, or *Cove*, but only fit for boats, and immediately over it the Cerro Mongon.

The *Cerro* or *Mount Mongon* is the highest and most conspicuous object on this part of the coast: when seen from the westward it has a rounded appearance, though with rather a sharp summit; but from the southward it shows as a long hill with a peak at each end. It is said there is a lake of fresh water on the range between those peaks, aud that its valleys abound with deer; but the truth of this depends on report only, as the examination of the officers did not extend so far.

From Mongon a range of hills run parallel to the coast, which is high and rocky, with some white islets lying off it, as far as *Casma*, where they terminate in *Calvario Point*, a steep rocky bluff that forms the southern head of that port.

The *Bay of Casma* is a snug anchorage, something in the form of a horse-shoe; between the entrance points it is a mile and three-quarters in a N.W. and S.E. direction, and a mile and a half deep from the outer part of the cheeks, with regular soundings from 15 to 3 fathoms near the beach.

The best anchorage is with the inner part of the South Cheek bearing about S.S.E. a quarter of a mile off shore, in 7 fathoms water: by not going farther in you escape, in a great measure, the sudden gusts of winds that at times come down the valley with great violence. Capt. Ferguson, of H.M.S. *Mersey*, mentions a rock with 9 feet water on it, on the South side of the bay, half a mile from the shore, that sometimes breaks.

This place seemed quite deserted; the only indications of its having been visited were a few stacks of wood piled up on the beach.

The best distinguishing mark for Casma is the sandy beach in the bay, with the sandhills in-shore of it contrasting strongly with the hard dark rocks of which the heads at the entrance are formed; there is also a small black islet lying a little to the westward of the North Cheek.

Capt. Andrew Livingston says: "Casma lies but a short distance to the northward of the Mountain of Mongon, which is the first high hill near the shore to the northward of Lima. Mongon Hill is long from North to South, brown-coloured, and pretty regular at the top.

"The entrance of Casma is about a mile wide; and about two-thirds over from the S.W. point to the N.W. one, lies a rock about 12 feet above water; it seems bold-to. The Admiralty directions are full enough about Casma, but Capt. Browne, of the *Tartar*, seems to have estimated it too highly. Many human bones were seen near Casma, and even skulls with long hair on.

"It is excessively squally, even in fine weather, at Casma. No person can be too cautious in carrying sail when entering Casma."

From hence the coast takes rather a more westerly direction, but continues bold and rocky.

N.W. ¼ N. 14 miles from Casma is the great *Bay of Samanco*, or *Guambacho;* and midway between them the shore recedes into a deep bight, with the two islands, in front, of *Tortuga* and *Viuda*.

The BAY of SAMANCO is the most extensive on the coast of Peru, to the northward of Callao; being 6 miles in length, in a N.W. and S.E. direction, and 3 miles wide: the entrance is 2 miles across, between Samanco Head on the South and Seal Island on the North, and there are regular soundings all over the bay.

At the S.E. corner, in a sandy bay, stands a small village (the residence of some fishermen) at the termination of the River Nepeña. This river, like most others on the coast, has not sufficient strength to force a passage for itself through the beach, but terminates in a lagoon within a few yards of the sea.

The town of *Guambacho* is about a league distant, at the eastern extremity of the valley. And *Nepeña*, which is the principal town, lies to the N.E., about 5 leagues off. There is very little trade at this place; small coasting vessels from Payta sometimes call here with a mixed cargo, and they get in exchange sugar and a little grain.

Refreshment may be obtained from the neighbouring towns, but wood is scarce. The water of the river is brackish and unfit for use, but there are wells on the left bank, a short distance from the huts. When taken on board, this water is not good; but, contrary to the general rule, after it has been some time confined on board, it becomes wholesome and pleasant tasted.

When at a distance, the best mark to distinguish this bay is *Mount Division*, a hill with three sharp peaks, rising from the peninsula between Samanco and the Bay of Ferrol. There is also a bell-shaped hill on the South side of the bay that shows very distinctly. *Mount Tortuga*, a short distance inland to the eastward, will also be seen: it is higher, and similar in appearance to the Bell Mount.

Samanco Head is a steep bluff, with some rocks lying off it to a cable's length; on opening the bay, Leading Bluff will be seen, a large mass of rock on the sandy beach at the N.E. side, that looks like an island. In going in, give Samanco Head a berth in passing; you may then stand in as close as convenient to the weather shore, and anchor off the village in 4, 5, or 6 fathoms, sandy bottom: when rounding the inner points, take care of your small spars; for the wind comes off the Bell Mount in sudden and variable puffs.

Three leagues from Samanco, the Bay of Ferrol opens, nearly equal in size to Samanco, and separated from it by a low sandy isthmus; it is an excellent place for a vessel to careen, being entirely free from the swell that sets into most of these ports. On its N.E. side is the Indian village of *Chimbote*, where, it is said, that refreshment of any kind might be had, but no water. The entrance is clear; but there is a reef of rocks off Blanca Island, a mile and a half to the northward, which must be avoided.

N.W. ½ N. 6 miles from the entrance of Ferrol stands *Santa Island*, about a mile and a half in length, lying N.N.E. and S.S.W., and of a very white colour: just without it are two sharp-pointed rocks, 20 feet above the sea. Two miles N.N.E. from the island, *Santa Head* forms the South side of the bay of that name; which, although small, is a tolerable port; the best anchorage is in 4 or 5 fathoms, with the extreme of the Head bearing S.W. Fresh provisions and vegetables may be obtained on moderate terms. It is also a tolerable place for watering.

The town lies about 2 miles East from the anchorage, and the mouth of the river a mile and a half North of it. This is the largest and most rapid river on the coast of Peru: from Santa Head it is seen to wind its way along the valley with several islets interrupting its course; but at its termination it branches off and becomes shallow, with only sufficient strength to make a narrow outlet through the sandy beach that forms the coast-line: a heavy and dangerous surf lies off it, so that no boat could approach with any degree of safety.

This part of the coast may be known by the wide-spreading valley through which the river runs, bounded on each side by ranges of sharp-topped hills; and as you approach, Santa Island will be plainly seen, with the Head of the same name: there is also a small but remarkable white island, called *Corcovado*, to the N.W. of the harbour. No danger exists in entering; the soundings are regular for some distance outside, and you may anchor anywe re between the islands and the main, in a moderate depth of water, but of course exposed to the swell.

N.W. ½ N. 5 leagues from Santa lie the *Chao Islands*, one mile and three-quarters off the *Point* and *Hill* of that name. The largest is a mile in circumference, about 120 feet high, and, like most of these islands, quite white.

Between Santa and Chao the coast is a low sandy beach, which continues and forms a shallow bay, as far as the *Hill of Guañape*, with moderately high land a few miles in-shore.

The *Hill of Guañape* is about 300 feet high, rather sharp at its summit, and when seen from the southward appears like an island; on its North side there is a small cove, with tolerable landing just inside the rock that lies off the point. S. by W. from this point, between 6 and 7 miles from the coast, lie the *Guañape Islands*, with a safe passage between them and the shore; they may be said to be two, with some islets and rocks lying about them; the southern is the highest and most conspicuous.

From the Hill of Guañape the coast continues a sandy beach with regular soundings, and ranges of high sharp-topped hills, about 2 leagues from the sea, until you near the little Hill of Carretas, which is on the beach, with the *Morro de Garita* overlooking it. Here commences the *Valley of Chimu*, about the middle of which stands the *City of Truxillo*, and 5 miles farther North, the *Village* and *Road of Huanchaco*. This is a bad place for shipping, and seems to have been badly chosen: for the North side of Carretas Hill would be a better place for landing and embarking goods: and might be further improved by sinking some small craft laden with stones, plenty of which the hill would afford.

The *Road of Huanchaco* is on the North side of a few rocks that run out from a cliffy projection; sheltering the beach in a slight degree, but affording no protection to shipping. The village is under the cliff, and not distinguishable till to the northward of the point; but the church, which is on the rising ground, shows very distinctly, and is a good guide when near the coast.

The usual anchorage is with the church and a tree that stands in the village in one, bearing about East, a mile and a quarter from the shore, in 7 fathoms dark sand and mud. Vessels often have to weigh, or slip, and stand off, owing to the heavy swell that sets in: it is also customary to sight the anchor once in twenty-four hours, to prevent its being imbedded so firmly as to require much time to weigh it when required.

Capt. Livingston thus describes it:—"The best anchorage is about 1½ or 2 miles off shore, with the belfry of the church open, or in one with the only palm tree in the village of Huanchaco. The church is very remarkable, and the village lies between it and the sea, low down, and, the houses being the same colour as the ground behind, it is rather difficult to perceive it at any considerable distance."

Landing cannot be effected in ships' boats; there are launches constructed for the purpose, manned by Indians of the village, who are skilful in their management; they come off on every arrival, and will land you safely, for which they charge six dollars, equal to £1 4*s.* sterling: it is to be remembered that no more is charged for a cargo of goods, the risk of the surf being that for which you pay.*

Fresh provisions may be had from *Truxillo*, but watering is out of the question.† That city is said to contain 4,000 inhabitants.‡ It was founded in 1535 by Pizarro, who gave it the name of his native city in old Spain. The houses are low, in consequence of the earthquakes, and in its neighbourhood are the ruins of several ancient Peruvian monuments. Rice is the principal production of the valley; and it is for that article and spice that vessels call here.

* "Landing here is always bad, and often impracticable. No stranger ought to attempt it without having *cholos* (natives) in the boat; and it generally is advisable to employ the launches kept at the place; several persons were drowned about the time I was there. The Indians will, however, come off to ships in the very worst weather, on what they call *caballitos*, which are merely bundles of a kind of triangular bulrush, called *totora*. The *caballito* is generally formed of two bundles, but some large ones are of three bundles of *totora*. Some carry two men, and others only one. Their paddles are about 6 feet long, and are merely the half of a split bamboo. With a heavy swell the Indians are frequently washed off the *caballitos*, but never fail to regain them. They are admirable swimmers, the children seem almost amphibious. They do not swim like persons in England, but paddle with their arms like a dog with his fore feet in swimming. The Indians do not generally sit astride on the *caballitos*, but generally with both their legs straight out on the top."—*Capt. Livingston.*

† "At Huanchaco good fresh water can seldom be procured except at times (such as when I was there) when a fleet of transports arrives, and then a small branch of some river is turned so as to come down to Huanchaco; but the water is very muddy."—*Ibid.*

‡ "Truxillo lies about 8 or 9 miles to the south-eastward of the village of Huanchaco. On the road are very extensive ruins, said to be those of an Indian town called Shimbo, or Chimbo, as near as I could understand. The ruins cover a prodigious space of ground; I should think 4 or 5 leagues in circumference. There are many large tumuli; out of one of them a family of the name of Toledo dug so much wealth, that the fifth, paid to the King of Spain, amounted to 500,000 dollars. All these tumuli seem built of a kind of brick baked in the sun. Some rounded water-worn stones are also intermixed in the large one's interior, and in some of the smaller ones I observed cane-reeds had been used.

"An Indian is said to have given Toledo the information relative to the treasure in the tumulus, or *huako* (pronounced whakko), as it is called, and was to have showed him another, in which there was much more treasure, but the Indian died, and the knowledge died with him. There are, however, many large tumuli still unopened, and I have no doubt but immense treasures are concealed in them. The Indians frequently dig up earthenware vessels, of much the same appearance as the ancient Etruscan ware, and some of their shapes even elegant, others very grotesque.

"The city of Truxillo is walled with a kind of mud wall. There are some good houses. The streets are regular. The population is variously stated at from 8,000 to 12,000. I suppose that 8,000 is nearest the mark. There are several large churches, and a tolerable theatre. I attended it twice, and both the actors and actresses acquitted themselves much better than I expected. To the S.W. of Truxillo lies the village of Moché, the church of which is also conspicuous from the sea, but not so much so as that of Huanchaco. Moché is but a small straggling village. In running down the coast from the southward, and pretty close in-shore (say 4 miles off), the church at Moché will be seen before that at Huanchaco; so persons ought to take care not to mistake Moché for Huanchaco. The ground about these places does not seem so carefully cultivated as about Huacho, though there seems to be no great difficulty in getting water to irrigate the fields."—*Ibid.*, *August*, 1824.

If bound for this road you should stand in on a parallel of 8° 7′ (which is a mile to windward), and you will see *Mount Campana*, a bell-shaped mount, standing alone, about 2 leagues to the northward, and *Huanchaco Peak*, which is very sharp, and the first hill in the range on the North side of the valley. Shortly after the church will come in sight, and the shipping in the road.

"The winds on this coast," says Capt. Livingston, "are almost always from the southward, though there are instances to the contrary. The current, also, generally (though not always) sets to the northward. The swell is reckoned always to be heaviest at the full and change of the moon, but, during the seven weeks we lay in Huanchaco Roads, I thought the heaviest swell was generally at the quarters of the moon.

"The ground in Huanchaco Roads is very foul, and there are few vessels which do not lose one or more anchors. A vessel should always come to with a very light anchor, and a long scope of chain. We came to with an anchor of only 4½ cwt., but we lost it. One of 4 cwt. is quite heavy enough for a vessel of 200 tons.

"In beating to windward from Huanchaco to Callao it is advisable to stand off shore about 14 hours, and in shore about 10 hours, on account of the land trending so much to the eastward. If possible, it is always best to be pretty close in with the shore at sunset, as the wind frequently draws more off the shore about that time. It is also highly improper to stand far off shore, and any person not well informed is apt to get into this error, as the vessel generally comes up with her larboard tacks on board as you stand farther off shore, even sometimes lying up South. This tempts many to stand off shore, never considering that the trend of the shore to the eastward, and the loss they must sustain by standing in-shore with their starboard tacks on board, when they cannot lie higher than N.E., more than balances the seeming advantage of lying so high on the other tack, and they also lose all chance of advantage from the land-breeze at night when they stand so far off shore."

The coast is cliffy for a few miles to the northward of Huanchaco; the low sandy soil with bushes on it then commences, with regular soundings off it, and continues as far as *Malabrigo Road*. This bay, although bad, is considerably preferable to Huanchaco; it is formed by a cluster of hills, projecting beyond the general trend of the coast, which at a distance appears like an island; there is a fishing village at the S.E. side, but no trade is carried on. The town of *Paysan* lies some leagues to the S.E., and by the account they gave of it at Malabrigo, must be of considerable extent. The best anchorage here is with the village bearing about E.S.E. three-quarters of a mile from the shore, in 4 fathoms sandy bottom: landing is bad.

The small island of *Macabi* lies S. ¼ W. 6½ miles from Malabrigo, with a safe channel of 10 fathoms between it and the main.

N.W. by N. 20 miles from Malabrigo is *Pacasmayo Road*; the coast is low and cliffy, with a sandy beach at the foot of the cliff, and soundings of 10 fathoms 2 miles off shore. Pacasmayo is a tolerably good roadstead, under a projecting sandy point, with a flat running off it, to the distance of a quarter of a mile. The best anchorage is with the point bearing about S. by E. and the village East;

you will there have 5 fathoms sand and mud: there is no danger in standing in; the soundings are regular, shoaling gradually towards the shore. Landing is difficult: such launches are used as at Huanchaco. The principal export is rice, which is brought from the town of *San Pedro de Yoco*, 2 leagues inland. Fresh provisions may also be obtained from the same place; wood and water may be had at the village on the beach, which is principally inhabited by Indians employed by the merchants of San Pedro.

To distinguish this road from seaward, the best guide is to stand in on a parallel of 7° 25′ to 30′, and when within 6 leagues, the Hill of Malabrigo will be seen, like an island sloping gradually on each side; and a little to the northward, *Arcana Hill*, rugged with sharp peaks. As you approach, the low yellow cliffs will appear (those North of the road the highest), on the summit of which, on the North side of the point, there is a dark square building that shows very distinctly. The best mark for the anchorage is the shipping when any are there. From this road the coast continues low, with broken cliffs, 11 leagues N.W. ¾ N. as far as *Eten Point*, which is a double hill (the southern one the highest), with a steep cliff facing the sea. The North side of this cliff is white, and shows very conspicuously.

LAMBAYEQUE ROAD.—N.W. ½ N., a little more than 4 leagues, is the road of Lambayeque, the worst anchorage on the coast of Peru. There is a small village on the rising ground, with a white church; off which vessels anchor in 5 fathoms, 1½ miles from the shore. The bottom is a hard sand, and bad holding ground; it is always necessary to have two anchors ready, for the heavy swell that sets on this beach renders it almost impossible to bring up with one, particularly after the sea-breeze sets in.

Rice is the chief commodity for which vessels touch here: the only method of discharging or taking in a cargo (or in fact landing at all) is by means of the *balsa*. This is a raft of nine logs of the cabbage-palm, secured together by lashings, with a platform raised about 2 feet, on which the goods are placed. They have a large lug-sail which is used in landing; the wind being along the shore enables them to run through the surf and on the beach with ease and safety, and it seldom happens that any damage is sustained by their peculiar mode of proceeding. Supplies of fresh provisions, fruit, and vegetables may be obtained, but neither wood nor water.

"A vessel bound to this place from the southward should make the Hill of Eten, the highest land about here near the coast, and distant from the town about 6 leagues. The coast off the town, and to the N.W. of it, is very low, and should be approached with caution, allowance being made for the current, which sets to the N.W. sometimes 1½ miles per hour. Vessels by not attending to this particular have been drifted to leeward of the place, and have lost three or four days in beating up again. Having made the Hill of Eten, a vessel may stand in for the anchorage. Care must be taken to keep the lead going. The *Alert's* anchorage was about 4 miles off the shore, in 7 fathoms, with the Hill of Eten S.E. ¾ E. Landing can only be effected safely in balsas, and no boat can cross the bar." *

* Observation by Mr. Babb, M. H.M.S. *Alert*.

The coast continues low and sandy, similar in appearance to that of Lambayeque, to the distance of 25 leagues W.N.W.: an extensive range of table-land of considerable height, with broken rocky points, then commences, and continues to Point Aguja, or the Needle.

LOBOS DE AFUERA.—Fifteen leagues from Lambayeque, in a W.S.W. direction, lies a small group of islands called Lobos de Afuera. These islands are 3 miles in length North and South, 1½ miles broad, and about 100 feet high, of a mixed brown and white colour. They may be seen several leagues, are quite barren, and afford neither wood nor water. There is a cove on the North side formed by the two principal islands, but with deep water and rocky bottom; within this cove there are some little nooks, in which a small vessel might careen without being much interrupted by the swell.

These islands are resorted to by fishermen from Lambayeque on their balsas; they carry all their necessaries with them, and remain about a month salting fish, which fetches a high price at Lambayeque. There is no danger round the islands, at the distance of a mile; and regular soundings will be found between them and the shore, from 50 fathoms abreast of the islands.

LOBOS DE TIERRA.—N.N.W. ¼ W. 10 leagues from Lobos de Afuera, lies the Island of Lobos de Tierra, nearly 2 leagues in length, North and South, and a little more than 2 miles wide; when seen from seaward it has a similar appearance to the former islands, and many rocks and blind breakers lie round it, particularly to the westward. There is tolerable anchorage on the N.E. side, in 11 or 12 fathoms, sand and broken shells. A safe passage is said to exist between this island and the main, which is distant 10 miles; but as no advantage can be gained by using it, it was not thoroughly examined.

Aguja Point is long and level, terminating in a steep bluff 150 feet high, and has a finger-rock (Aguja or Needle Rock) a short distance off it, with several detached rocks round the point.

Three miles and a half N. by E. ½ E. of this is *Nonura Point*, and 5 miles farther in a N.E. by N. direction is *Pisura Point*, the South point of the Bay of Sechura; between Aguja Point and Pisura Point there are two small bays, where anchorage might be obtained if required. The land about this part of the coast is much higher, and has deeper water off it than either N. or S., and may be known by its regularity and table-top.

The BAY OF SECHURA is 12 leagues in length, from Pisura Point to Foca Island, bearing N.N.W., and is 5 leagues deep; on the S.E. side the coast shows low sandhills; but as it curves round to the northward it becomes cliffy and considerably higher.

Near the centre of the bay is the entrance to the River Pisura, and the *town of Sechura* is situated on its banks. This town is inhabited chiefly by Indians, who carry on a considerable trade in salt, which they take to Payta on their balsas, and sell to the shipping. The river is small, but of sufficient size to admit the balsas when laden. There is anchorage anywhere off the river, in from 12 to 5 fathoms, coarse sand; the latter depth being better than a mile from the shore. This place may easily be distinguished by Sechura church, which has two high steeples, and shows conspicuously above the surrounding sandhills; one of these steeples has a considerable inclination to the northward, which, at a distance, gives it more the appearance of a tree than of a stone building. From Foca Point the coast is cliffy, about 120 feet high, and continues so as far as Payta Point, which is 3 leagues distant N. ¼ E.; between these two, 1½ miles from the coast, is a cluster of hills called the Saddle of Payta, thus described by Capt. Basil Hall:—"The *Silla* or *Saddle of Payta* is sufficiently remarkable; it is high and peaked, forming three clusters of peaks joined together at the base, the middle being the highest; the two northern ones are of a dark brown colour; the southern is the lowest, and of a lighter brown. These peaks rise out of a level plain, and are an excellent guide to vessels bound for the Port of Payta from the southward."

PORT OF PAYTA.—A few leagues to the northward, as already mentioned, is Payta Point, round which is the port of that name; and it is, without exception, the best open port on the coast. A considerable trade is carried on. Vessels of all nations touch there for cargoes, principally cotton, bark, hides, and drugs; in return for which they bring the manufactures of their several countries. In 1835 upwards of 40,000 tons of shipping anchored at this port. Communication with Europe (*vid* Panama) is more expeditious than from any of the other ports.

The TOWN is built on the slope and at the foot of the hill, on the S.E. side of the bay; at a distance it is scarcely visible, the houses being of the same colour with the surrounding cliff. It is said to contain 5,000 inhabitants, and is the seaport of the Province of Piura, the population of which is estimated at 75,000 souls.

The city of *San Miguel de Piura** stands on the banks of the River Piura, in an easterly direction from Payta, and between 9 and 10 leagues distant. Fresh provisions may be had at Payta on reasonable terms, but neither wood nor water, except at a high price, the latter being brought from Colan (a distance of 4 miles) for the inhabitants of the place.† At the time of this survey hopes were entertained of a supply of water from the West side of the bay: an American having commenced boring with an apparatus proper for the purpose.

"The heat is always considerable at Payta; and as no rain falls, the houses are slightly constructed, of an open sort of basket-work, through which the air blows freely at all times; the roofs, which are high and peaked, are thatched with leaves; some of the walls are plastered with mud; but, generally speaking, they are left open. . . . . The extraordinary economy of water arose, as they told us, from there not being a drop to be got nearer than 3 or 4 leagues off; and as the supply, even at this distance, was precarious, water at Payta was

* "In 1531 the first town built by the Spaniards in Peru was founded a short distance to the South of the city of Tumbes, and named San Miguel de Tangurala; but the site being unhealthy, it was shifted, and the name changed to San Miguel de Piura."—*Burney*, vol. i. p. 164.

† "Payta is an excellent position for supplies of cattle and vegetable or table necessaries; but, unfortunately, does not abound in wood or water, for both of which *payment* must be made, and that exorbitant.

"We were fortunate in obtaining here some excellent cordage, which is rather scarce on this coast, very probably that exchanged by some of the whale ships which frequently touch here for supplies of stock, and more particularly the sweet potato, which is an excellent antiscorbutic."—*Capt. Sir Edw. Belcher, Sept.* 1838—*Voy. of the Sulphur*, vol. i. p. 207.

not only a necessary of life, but, as in a ship on a long voyage, was considered a luxury." *

There is no danger in entering this excellent port: after rounding the outer point with a signal station on its ridge, you will open False Bay: this must be passed, as the true bay is round Inner Point. That point ought not to be hugged closely, for there are some rocks at the distance of a cable's length, and the wind baffles often. After rounding Inner Point you may anchor where convenient, in quiet still water, with from 4 to 7 fathoms, over a muddy bottom. The landing place is at the mole about the centre of the town.

N.W. ¼ N. 9 leagues from Payta, PARINA POINT rises to a bluff about 80 feet high, with a reef out to the distance of half a mile on its West side; between this point and Payta the coast is low and sandy, with table-land of a moderate height at a short distance from the beach, and the *Mountain of Amatape* (3,000 to 4,000 feet high) 5 leagues in the interior.

After rounding Parina Point (which is the western extreme of South America) the coast trends abruptly to the northward, and becomes higher and more cliffy in approaching Talara Point.

CAPE BLANCO is high and bold (apparently the corner of a long range of table-land), sloping gradually toward the sea; near the extremity of the cape there are two sharp hillocks; and midway between them and the commencement of the table-land is another rise with a sharp top. There are some rocks that show themselves about a quarter of a mile off, but no danger exists without that distance. From Cape Blanco the general trend of the coast is more easterly, in nearly a direct line to Malpelo Point, which is 21 leagues distant. N.E. by N. 7½ leagues from the former is *Sal Point*, a brown cliff, 120 feet high; along the coast lies a sandy beach, with high cliffs as far as the *Valley of Mancora*, where it is low, with brushwood near the sea; the hills being at a distance inland.

Northward of Sal Point the coast is cliffy, to about midway between it and Picos Point; it then becomes lower and similar to Mancora.

Picos Point is a sloping bluff, with a sandy beach outside of it, and another very similar point a little to the northward: behind there is a cluster of hills with sharp peaks; from whence arises probably the name given by the Spaniards to the point. From Picos Point the coast is a sandy beach, with a mixture of hill and cliff of a light brown colour, and well wooded. There are several small bays between it and Malpelo Point, which bears N.E. ½ N. 7¼ leagues distant.

MALPELO POINT forms the southern side of the entrance of Guayaquil River, and may be readily known by the marked difference between it and the coast to the southward. It is low, and covered with bushes, and a short distance in-shore there is a clump of bushes more conspicuous than the rest, which shows plainly on approaching. At the extremity of the point the River Tumbes issues, and a reef extends to the distance of a quarter of a mile. This place is frequented by whalers for fresh water, which is found a mile from the entrance, where they fill their boats from alongside; great care is necessary in crossing the bar, as a dangerous surf beats over it, and renders that operation at all times difficult.

* Capt. Basil Hall, part ii. chap. 35.

The Rio Tumbes is, in some measure, classic ground; for here, in 1526, Pizarro landed with his Spanish army destined to conquer Peru. According to the Spanish accounts there was a temple of the sun, an inca's palace, and other edifices, at the town of Tumbes, the remains of which are now nowhere to be seen.

The entrance to the river may be distinguished by a hut on the port hand going in, which is perceived immediately on rounding the point. About 2 leagues up the river stood the old town of Tumbes, now scarcely more than a few huts, and barely sufficient to supply the whalers with fruit and vegetables. You may anchor anywhere off the point in 6 or 7 fathoms. This river is the boundary between Peru and the State of Ecuador.*

The following remarks are made by Capt. Pipon, R.N.:—"On the 4th of July, 1814, we anchored in the Bay of Tumbes, in 6½ fathoms, soft clay, and good holding ground; Point Malpelo bearing S.W., a reef extending without the Point S.W. by W., the Island of St. Clara N. ½ W. From this anchorage it was impossible to discover the entrance of the River Tumbes; we sounded from the ship to the shore, in every direction, and found the soundings very regular, from 6 to 5½, and 5 and 2 fathoms, within 2 cables' length of the beach, which is a fine sand, with considerable surf on it. We found here a commodious inlet for wooding, the entrance about one cable wide, and only 5 feet water; this was at the beginning of the flood tide; there is a bar across it, indeed it is dry at low water in many places, so that we could only pass it with the boats during the flood; this, however, was not attended with any material inconvenience, for, having once got the boats into the inlet before low water, the people were employed filling them with fuel during the time the bar was not passable.

"We shifted our berth off the mouth of this inlet, for the convenience of the boats passing and repassing, and anchored in 5½ fathoms.

"This inlet is extremely well calculated to admit boats any time before low water; it is, however, best to enter it with a flowing tide. Here wood may be procured in great abundance, and it being very spacious, any number of boats may wood at the same time. The boats lay close to the beach of the different little islands scattered within the inlet, without the least surf or swell, so that they load with great facility. There are no inhabitants about this inlet; we found, by the shore, the tide rose in general about 6 feet. The mosquitoes are extremely troublesome, and alligators very numerous. With our seine a great quantity of very fine fish were caught. After examining the shores, we at length discovered the entrance into the River Tumbes; a bar, with a violent and dangerous surf, lies at the mouth of it, so that the utmost caution must be used in entering it. It lays very near to Malpelo Point, where we first discovered a reef with breakers on it. In this river only is fresh water to be found, and I would recommend boats employed on this service to enter it together, and to keep a good look out, and steer tolerably close to the point on the larboard hand; entering, avoid going near the breakers. On this point, however, you will discover huts, the residence of pilots and fishermen, who will not only point

* This is the termination of the excellent nautical descriptions given in the appendix to Capt. FitzRoy's volume of the "Narrative of the Surveying Voyages of the *Adventure* and *Beagle*;" they are the result of that survey, and will be fully appreciated by the mariner.

out the course you should pursue, but will willingly embark in your boats, and direct you how to avoid all dangers. After proceeding about 1 mile up the river, you will find the water perfectly sweet and good; and as you take it from alongside your boats, watering is here very expeditiously effected. Wooding would be difficult here, the mosquitoes being more troublesome than it is possible to describe; alligators of a large kind are also very numerous, (some small ones we shot,) also large guanas; but the woods are so thick as to be impenetrable, besides being swampy and muddy; I would therefore recommend this river for watering, and the before-mentioned inlet for wooding.

"The weather during our stay here was very fine; and it is pleasing to remark, that although the duty the people were employed upon was uncommonly severe from the great heat, and notwithstanding the ground on which the wood was cut was swampy, yet I did not perceive that any of my men felt any ill effects from it, owing, perhaps, to the precaution taken (as recommended by my surgeon) to serve out a small quantity of wine, previous to their leaving the ship; it may certainly have been the means of preventing sickness. There are no forts of any description here, and wooding and watering might easily be effected at all times, even were you at war with the natives. The country around appears an almost impenetrable wood; though we discovered many cultivated spots up the river, consisting of plantains, bananas, Indian corn, sweet potatoes, &c. To these plantations the workmen repair in canoes along the river, for I do not imagine the woods in this neighbourhood are passable even to the Indians or slaves.

"The town of Tumbes does not even merit the appellation of a village; it lays about 7 miles up the river, and is situated on a level plain, surrounded by a wood. It is composed of a few miserable huts, and the inhabitants appear to exist in a very wretched state. A governor resided here, who was extremely polite, and offered to procure us any refreshment the place afforded."

The prevailing winds on the shores of Peru blow from S.S.E. to S.W.; seldom stronger than a fresh breeze, and often in certain parts of the coast scarcely sufficient to enable shipping to make a passage from one port to another. This is especially the case in the district between Cobija and Callao.

Sometimes during the summer, for three or four successive days, there is not a breath of wind; the sky beautifully clear, and with a nearly vertical sun.

CHAPTER VII.

THE COAST OF COLOMBIA, BETWEEN GUAYAQUIL AND PANAMA.

THE country whose western coast, with the exception of Guayaquil and Panamá, is imperfectly described in the ensuing chapter, is one of the most important of all the South American territories; but at the same time it is one in which the vast capabilities it possesses have been least tested, and of which we are in many points most ignorant.

The Pacific coast is now the limit of the two separate and independent republics of Ecuador and New Granada, which have thus existed since 1831.

The territory called under the collective name of Colombia was the first portion of the new continent discovered by Columbus from the Atlantic side in 1498; hence its name. The Spaniards found more difficulty in establishing their sway over it than in any other portion of America; but eventually, by the middle of the sixteenth century, both the territories now known as Venezuela and New Granada were subjected to their dominion, and erected into captaincies, governors, and Spanish viceroys.

There are few portions of the coasts of this world which are so commercially unimportant as the portion embraced between Guayaquil and Panamá. None of its ports are resorted to by Europeans, or for European commerce; and indeed up to the present time the whole district has remained a complete *terra incognita;* though beyond all question in future ages the fine country which it bounds on the West must become of great importance. At present the few Indian and mixed breed families at the different ports accessible from the ocean constitute the sole links between it and the civilized world. Under these circumstances our present ignorance of the nautical condition of its coasts is less to be deplored, as many years must elapse before it can become a question of necessity that instruction should be attainable for so desert a district.

The fine surveys of our English Admiralty, conducted under the superintendence of Capt. H. Kellett and Lieut. Wood, have made us acquainted with its actual present condition; but there is no published description, and the portion between Esmeraldas and Panamá remains yet unpublished, the survey having been but recently brought home by Lieut. Wood, while these sheets are at press.

In the absence of more modern accounts we have drawn from the interesting and unvarnished statements of the great patriarch of nautical description, William Dampier, who was on this coast on a buccaneering expedition in 1684-5. As his veracity and accuracy have never been impeached, and his fidelity and minuteness ever acknowledged, we shall make no apology for giving descriptions of these parts from so antiquated a source, convinced as we are that they will be found to be borne out by present circumstances. The other detached notices which we have collected will be quoted when given, and we less regret their paucity from the nautical unimportance of the coast.

ECUADOR, which is the southernmost republic of the present political division of Colombia, is divided from Peru by the Rio Tumbes: on the North it is separated from that of New Granada by the River Mora, in lat. 1° 45′ N. Its eastern limits are comparatively undefined and unimportant to our present subject. Prior to the present political state it formed a portion of the vice-royalty of New Granada; but upon the discontent and rebellion consequent upon the French invasion of Spain in 1808, it separated with that state from the Spanish rule in 1811. In 1819 these states coalesced, and were declared to found one republic of Colombia; but political feeling was very far from settled, and led to fresh warfare: in 1822 the royalists were defeated in Ecuador by General Sucre, while General Bolivar was victorious on the same side, in other parts. In 1823 Ecuador adopted the convention of Cucuta, and remained an integral portion of Colombia until November, 1831, when the territory separated into the present three independent republics before named.

According to the census of 1827, the population of Ecuador amounted to about 492,000, exclusive of the Indians of the eastern plains. Subsequently it was estimated as follows :—

DEPARTMENTS.	Area in Square Miles.	Inhabitants.	Inhabitants to a Square Mile.
	Number.	Number.	Number.
Chimborazo, or Ecuador	190,000	190,000	1
Guayaquil	25,000	150,000	nearly 5
Assuay	105,000	210,000	2
Total	320,000	550,000	

NEW GRANADA, which occupies the remainder of the Pacific littoral to the boundary with the states of Central America or Guatemala, is in many of its interior parts but very little known.* Prior to the recent (but as yet unpublished) English Admiralty surveys, before alluded to, the charts of this coast were most deplorable.

According to a census published in 1827, the whole population of New Granada amounted to 1,270,000 inhabitants. The number was some time after estimated at 1,360,000, distributed among the five provinces as follows :—

PROVINCES.	Area in Square Miles.	Inhabitants.	Inhabitants to a Square Mile.
	Number.	Number.	Number.
Istmo	25,000	100,000	4
Magdalena	50,900	250,000	5
Boyaca	83,000	450,000	5½
Cundinamarca	152,800	370,000	less than 2
Cauca	68,300	190,000	3
Total	380,000	1,360,000	

* Bogotá, or, as it was formerly called, Santa Fé de Bogotá, is the capital of New Granada; it was founded by Gonzalo Ximenes de Quesada, who built 12 huts here in 1538; ten years after-

The countries are rich in almost every tropical production, and stately timber is met with in perfection in almost all parts. The different features of the coast vary very much; the interior of the country exhibits some of the most magnificent natural features in the ridges of the Andes, which extend along the western part of the continent parallel to, and, in many parts, within sight of, the coasts.

The ANDES of Ecuador increase in magnitude and elevation in advancing northward from the Peruvian boundary. The portion between 5¼° and 3½° S. lat. forms the great mountain knot of Loxa, but which, however, does not rise into the limits of perpetual snow. Here it separates into two principal parallel ridges, which enclose the valley of Cuença, which extends from 3° 15′ to 2° 30′ S., and is about 7,800 feet above the sea. The mountains of Assuay, which form the North boundary of the Valley of Cuença, and the western extreme of which approaches Guayaquil, rise to the elevation of 15,500 feet, and some of the peaks are above the line of perpetual snow. This transverse ridge is narrow, occupying only about 3 minutes of latitude (2° 30′ to 2° 27′). North of this are the valleys of Alausi and Ambato, extending to 40′ S., and are about 7,920 feet above the sea. The summits of the ranges on the East and West, which enclose them similar to those to the southward, are of great elevation, and on the western range stands the famous Chimborazo, 21,420 feet in height, and its peak covered with perpetual snow. It is a very conspicuous object from Guayaquil and the shores of the Pacific about Cape San Francisco. This majestic mountain, which has been so vividly described by Humboldt, was, from his measurements, considered as the highest summit of the Andes. But the more recent measurements of Mr. Pentland and others of the Bolivian Andes, have shown that several peaks, as the Nevado of Zorata and the Illimanni, rise from 3,000 to 4,000 feet higher. Chimborazo has every appearance of being an extinguished volcano.

Like the Valley of Cuença to the South, that to the North of the Assuay is bounded by a narrow transverse ridge—the Alto de Chisinche, but it hardly rises 300 feet above the adjacent level ground. But at its extremities, or rather at its junction with the eastern and western ranges, rise two very high peaks; the one to the East is the terrible volcano of Cotopaxi, 18,880 feet, that to the West the peak of Yliniza, 17,376 feet above the sea. The eruptions of Cotopaxi have been more frequent and destructive than those of any volcano in South America. It is stated that in 1744 the roaring of the volcano was heard at Honda, near Bogotá, a distance of 200 leagues. In 1758 the flames shot up 2,700 feet above the crater. On April 4th, 1768, the vast quantity of ashes discharged from the crater made it as dark as night, until 3 P.M., at the towns of Hambato and Tacunga. In January, 1803, the volcano having been quiet for 20 years previously, an eruption was preceded by the sudden melting of all the snow on its summit; and Humboldt heard, day and night, at Guayaquil, the roaring of the volcano, like repeated discharges of artillery, the distance being 52 leagues.

---

wards it was made the seat of an *audiencia réal*, and, in 1561, a metropolitan see. From its elevation its appearance is at first imposing; but dirt, narrow but straight and regular streets, and low houses, do not keep up the illusion. The number of inhabitants is from 30,000 to 40,000, a scanty population for its area. Lat. 4° 37′ North, lon. 74° 10′ West.

The Valley of Quito, still bounded East and West by parallel ranges, is elevated 9,600 feet above the sea level. On the East range is Antisana, 19,136 feet, and Cayambe Urcu, 19,648 feet above the sea. On the West range are Pichincha, 15,936 feet, at the foot of which is the city of Quito,* and the Cotocache, 16,448 feet above the sea.

The three narrow longitudinal valleys of Cuença, Hambato and Alausi, and Quito, extending 240 miles in length, from 3° 15′ South to 21′ North, form the most populous and the richest portion of the republic of Ecuador, and, from its great elevation (the barometer stands at 21·33 inches), is of a totally different character to what its latitude would indicate, and everything proclaims the industry of a nation of mountaineers. To the North of this, the western portion of the great chain is called the Andes de los Pastos, crowned with several high summits and volcanoes, as those of Cumbal, Chiles, and Pasto. North of this the westernmost range is called the Cordillera of Sindagua, which is traversed in about 1° 20′ North by the Rio de los Patias, falling into the Pacific, but rising in the intermediate valley South of the Alto de Robles.

The countries lying on both declivities, and at the foot of the Andes of Ecuador, are very thinly inhabited, and almost entirely by aboriginal nations, unacquainted with civilization or commerce, which is confined to the elevated valleys between the ranges. In this very imperfect sketch of this important feature, we have chiefly followed Humboldt, in whose fine works the reader will find ample details.

The Andes of New Granada are separated into three chains, which divide at the mountain knot at Socboboni and the Alto of Robles, near Popayan.

The western chain, called also the Cordillera de Chocó and the coast range, separates the provinces of Popayan and Antioquia in the East from those of Barbacoas, Raposo, and Chocó, on the West. In general it is but moderately elevated, compared with the eastern and central ranges, but offers great difficulties to the communications between the Valley of Cauca and the coast region. The terrible roads which traverse it are those of the passes of Chisquio (to the East of the Rio de Micay); that of Anchicaya de las Juntas; of S. Augustin, opposite Cartago; of Charmi; and of Urras.†

It is on its western slope that the famous auriferous and platiniferous region lies, which has for ages past thrown into commerce more than 13,000 marcs of gold per annum. This alluvial zone is 12 or 14 leagues broad, and attains its maximum richness between latitudes 2° and 6°; it sensibly diminishes in value

* The city of Quito, the capital of Ecuador, is in a valley 9,543 feet above the sea. Eleven snow-capped mountains are in view from it. The volcano of Pichincha is the nearest. It is in some parts regularly built, and has some handsome buildings; as the president's palace, formerly that of the Spanish viceroy, that of the archbishop, the cathedral, &c. As a place of education it stands high in South America. It has a good university and several colleges. Its inhabitants are variously estimated at from 40,000 to 70,000. Earthquakes are frequent, and the climate is a perpetual spring. Its exports are principally corn and agricultural produce to Guayaquil, through which it receives European manufactures. In 1736 the French and Spanish astronomers measured a degree of the meridian on a plain about 4 leagues North of Quito, which is admirably recorded in the excellent work, the *Viage a la America Meridionale*, by Don J. and D. A. de Ulloa, liv. v. chaps. 4, 5. Quito is in lat. 13′ 27″ S., lon. 78° 10′ 15″ W. For descriptions, see Stevenson's South America, chap. 11, p. 279—325, &c. There are roads which lead from most of the chief ports over the Andes to Quito, and there is one to Guayaquil, a portion of which is extremely steep.

† Semanario de Bogotá, vol. i. p. 32.

North and South of this, and disappears entirely between 1¼° and the equator. The auriferous region fills the basin of the Cauca, as well as the ravines and plains to the East of the Cordillera of Chocó. It rises sometimes to 4,000 feet above the sea, and descends to less than 250 feet. Platinum (and this geognostic fact is worthy of attention) has up to this hour been found *only to the West* of the Cordillera of Chocó, not to the East, notwithstanding the geological and mineralogical similarity of either side of the mountains.* From the Alto des Robles, which separates the plateau of Almaguera from the basin of the Cauca, the western range then forms, in the Cerros de Carpintaria, to the East of the Rio San Juan de Micay, the continuation of the Cordillera of Sindagua, broken by the Rio Patias; then it becomes lower towards the North, between Cali and Las Juntas de Dagua, being from 5,000 to 6,000 feet high, and sends out considerable counterforts (lat. 4½° to 5°) towards the sources of the Calima, the Tamana, and the Andagueda Rivers. The first two of these auriferous rivers are affluents of the Rio San Juan del Chocó, and the last, into the Atrato, which falls into the Mexican Gulf. This enlargement of the western range forms the mountainous parts of Chocó, and it is here, between the Tado and Zitara, called also S. Francisco de Quibdó, that is, the Isthmus of Raspadura, celebrated as being the site of the first navigable communication between the two oceans, hereafter more particularly alluded to.

The foregoing description of this portion of the mountains, though of less consequence to the navigator, yet important for its mineral riches, has been taken from Humboldt.†

The RIVER TUMBES, as stated on a previous page, forms the boundary between the republics of Peru and Ecuador. Malpelo Point, at the mouth of this river, with Salinas Point, at the S.W. end of the Island of Puna, may be considered as the limits of the southern and principal branch of the entrance to the Guayaquil River.

GUAYAQUIL.

Guayaquil is the most important port of this section of South America, being the entrepôt of the rich valleys of Ecuador, and the chief outlet of all the produce of the republic.

The estuary, which is much embarrassed with shoals, and is extensive, is (or rather *was*), however, buoyed, and means are generally adopted to avoid all inconveniences arising from the difficulties of navigation; so that a vessel may now proceed up the river to the city, a distance of 80 miles from the outer entrance, with tolerable facility; though above Puná the depth at low water prevents any great draught being carried beyond, except at the top of spring tides.

* "Platina is found abundantly on Chocó and on the coast of the Pacific as far to the southward as Barbacoas; it accompanies the gold, and is obtained in the washings of that metal, from which it is afterwards separated. When the republic was established, the exportation was prohibited, and the government was to purchase it all in its crude state, at four to eight dollars a pound, and to refine it, and then sell it at four dollars an ounce. But, from the facility of smuggling it out to Jamaica, chiefly down the river Atrato, it was determined by the government to export it when purified, and marked with its arms, at six dollars an ounce.—*Present State of Colombia*, p. 309. See also Cochrane, vol. ii. p. 421, for an account of the method of procuring it.

† Relation Historique, tome iii., p. 204.

The LIGHTHOUSE on the ISLAND of SANTA CLARA or AMORTAJADA, is the best mark for making the river, and is excellently situated for this purpose, lying as it does quite outside of all the points of the river. It is situated in lat. 3° 10′ S., lon. 80° 26′ W. The island itself is so remarkable, that it cannot well be mistaken; it is high, and on many bearings assumes the appearance of a gigantic shrouded corpse, which it exactly resembles when the centre bears W. ½ S. Thence comes the name of "Amortajada" or "Muerto," given to it by the Spaniards.

The lighthouse is erected on the breast of the island, about one-third from the head, showing a FIXED LIGHT, about 230 feet above the level of the sea; having ten argand lamps, with burnished silvered parabolic reflectors, the light from which is visible in every direction, except from N. ½ W. to N. by E.; and should a vessel approach too near the island, in a southerly direction, it will be shut in by the edge of the cliff. Its most brilliant face extends from W.N.W. to E.N.E. by the South, in which directions it will be seen, in clear weather, from 5 to 6 leagues off. It was begun to be erected under the superintendence of Mr. George Peacock, of the Pacific Steam Navigation Company, November 20th, 1841, and was first lighted on December 1st, in the same year.*

The SOUTH COAST of the river N.E. of Malpelo Point recedes so as to form a shallow bay, called Tumbes Bay, the points of which bear N.E. and S.W. from each other, 16 miles apart. The north-easternmost of these points is that of the Tembleque Islands, forming a portion of the low land at one of the mouths of the Tumbes. To the N.E. of this are some extensive shoals, called the *Payana Shoals*, dry at low water for 2½ miles from the shore.

A black buoy with staff and ball was laid on the *Payana Spit*, at the same time as the erection of the lighthouse. It lies in 4½ fathoms at low water, with Point Tembleque bearing S.S.W. ¾ W.; a bluff of trees on an islet, and Chupador Inlet, S.S.E.; extreme point of a sandy point, E. by S.; lighthouse on Santa Clara, W. by N. ¾ N., 10 miles; starboard point of Chupador entrance, S. by E., about 2 cables' length from the nearest breaker at low water, and about 2½ miles from the nearest land. The water is deep, 15 to 18 fathoms immediately outside this buoy, and the whole space is clear between it and the lighthouse.

At 12 miles E.N.E. from this buoy is the West point of the entrance to *Jambeli Creek*, which runs to the South, and has a good depth of water for its breadth. Above this part the depth of the main river becomes irregular, and has much shoal water, though there is a deeper channel over on the West side.

The ISLAND of PUNA, which forms the N.W. side of this part of the river, is about 28 miles long, N.N.E. and S.S.W., and 12 miles broad. Its S.W. point is called *Point Salinas*, and is in lat. 3° 3′ S., lon. 80° 16′ W. Shoals extend from it for 5 miles toward Amortajada Island, and northward along the western coast to a greater distance; but as they lie out of the general track of shipping, and have, moreover, not been amply examined, they will not be approached

* A recent statement, however, has been made, that the light is at times not shown for several nights together, from the keeper deserting it to procure provisions, and that the *buoys here described have disappeared*. Under these circumstances we cannot but caution all approaching and depending on their existence. (March, 1850.)—*Melange Hydrographique, Depôt de la Marine*, 1849-50.

unnecessarily. The lighthouse will be an excellent guide by night. At 9 miles E. by N. ¾ N. of Point Salinas, is *Arena Point* (which has been sometimes called Salinas Point). The shore in the interval is fronted by shoals, which reach 2 or 3 miles off the land. At Arena Point the coast assumes a more northerly direction, and extends 16 miles to the foot of *Mala Hill;* a direction mark, 3½ miles farther, in a E.N.E. direction, is *Española Point*, from whence the coast trends to N.E. and North, 4 miles to Mandinga Point, and the town of Puna at the N.E. extremity of the island, and above which a vessel cannot proceed without the aid of a pilot.

The estuary here varies from about 14 miles to 6 miles in width. The space between the island and the main, though containing some good channels, is of irregular depth, and has much shoal water on its eastern side.

The principal shoal is called the *Baja de Mala*, and consists of a chain of banks of different depths, extending from off Arenas Point to Española Point. The channels through are only fit for small vessels.

On its South extreme a *white buoy* was placed in 1841. It lies in 4 fathoms water, with the termination of the trees on Point Arenas bearing W. by S. ¼ S.; Peak of Cerro de las Animas, W. ½ N.; Cerro de Mala, N. ¼ E.; Point Puna Vieja, N.W. by N., and the lighthouse on Santa Clara, S.W. by W. 25 miles.

Capt. Peacock, who established the buoys and lighthouse, says: "After a great deal of trouble, time, and patience, we had the good fortune to find the shoal of Punta Arenas, and laid down the white buoy off its S.E. extremity. This is a very dangerous shoal, and I should strongly recommend that, whenever an opportunity occurs, a balsa should be moored on the spot the buoy now occupies, carrying a light elevated 50 or 60 feet above the level of the sea, and the buoy removed to the middle of the Mala shoal. I should likewise recommend that a black buoy should, as soon as possible, be placed on the spit of the shoal lying off the mouth of Balao River; one also on the shoal lying 1½ miles East of Punta Mandinga (or Puna Bluff), which is sometimes dry; and another black buoy close to the West side of the sunken rock lying off Punta Piedra, which would complete the navigation of this noble river, and render it safe and secure. Whenever the trade of the port shall warrant it, a revolving light on Punta Mandinga (or Puna Bluff) would be very serviceable to lead vessels up to Puna in the night."—Guayaquil, November 26, 1841; from the Lima "*Comercio*," December 18.

The *North buoy* of the Baja de Mala is *white*, and is 15½ miles N.N.E. of that just described. It lies in 4 fathoms sand, with Mr. Cope's (the English consul) summer house on the hill just shut in with the sandy bluff of Punta Española, bearing W. by N.; the West point of Mondragon Island a ship's length open of Puna Bluff, bearing N. by W.; and Cerro de Mala, W. ¾ S. The white buoys must be left on the port hand, and the black on the starboard in sailing up the river.*

* "At 2 miles East of Puna is a shoal, shown in Capt. Kellett's survey, lying in the route of vessels going to Puna for a pilot. It is 4 miles above the North buoy of the Baja de Mala. The *Adele*, Capt. Game, struck on this bank of hard sand. It was not laid down in the Spanish chart, and is about half a mile in circumference. The shoalest part has from 1 to 5 feet of water, and bears from Punta Mandinga E. by S. ¾ S., and from Punta Española N.E. by E. per compass. Distant from the nearest land from 1¼ to 1¾ miles."—*Nautical Magazine, February*, 1843, p. 134.

The SAILING DIRECTIONS given by Mr. Peacock, in connexion with the announcement of the placing of the lighthouse and buoys, are as follow; but we refer the reader to the note on p. 190 respecting these buoys and light:—

"From a fair distance off the South point of the island (of Sta. Clara) to the South buoy of the Baja de Mala, the course is N.E. by E. From this buoy steer N.E. by N. 10 miles, and then N.N.E. 5 miles, to the North buoy of the Baja de Mala, white, taking care not to come into less than 4 fathoms water. From this buoy you may keep right for Cape Mandinga, or Puna Bluff, N. by W. East from this bluff, 1½ miles, lies a dangerous shoal, which is sometimes dry at low water. Near the bluff the water is deep. From the North buoy of the Baja de Mala a direct course may likewise be shaped to Punta Española, and *vice versâ*, and it points out the entrance of both channels."

The CITY of GUAYAQUIL is the only port of the republic of Ecuador, and is, therefore, its chief point of interest to the mariner. It is the seaport of Quito, Catacunga, Hambato, Riobamba, &c., &c., and indeed of all the rich valleys between the Andes. It is built on the West bank of the river, and extends about a mile in a straight line, at 50 yards from the water. It is divided into the old and new towns, the former inhabited by the poorer classes. It is tolerably well laid out, and the houses are chiefly built of wood; from this cause it has frequently suffered greatly from fire. The private houses are mostly tiled, and are furnished with arcades. A promenade runs the whole length of the town between the houses and the river. It has some good edifices, as the custom-house, &c.; but from its being on a dead level, and intersected with many creeks, the drainage is bad, and the streets are sometimes so swampy, as to be almost impassable. The town, nevertheless, has a good appearance at a distance. One great article of commerce here is cocoa, which is shipped in large quantities for Spain and the United States. The French possess most of the retail trade. The population may be about 20,000.

The river is about 1½ miles in width, very rapid and muddy, the banks of slimy mud, dotted in every direction with alligators. The scenery is very like that of the rivers on the coast of Africa, and almost as productive of fever. About 10 miles below Guayaquil the river is not more than half a mile wide, and the banks dense mangrove swamps. All breezes are excluded, and the air is insufferably hot, even in the "cool season."

The water for the use of the town is brought from a considerable distance up the river, in earthen jars; from a hundred to one hundred and fifty of which are packed together in a *balsa*, formed of logs of a very light wood, lashed together with vine, and floated down. The water opposite to the town is fresh at the last of ebb, but is considered as unfit for drinking, passing, as it does, through a mass of poisonous mangroves. Plenty of large timber and firewood are also brought in the same way. Fresh beef and various kinds of fruit are likewise in abundance, and of course cocoa, which is the staple commodity of the place. The mosquitoes are so troublesome, that the ships lying opposite to the town are obliged to send their crews on shore at night.

On the opposite side of the river is a dry dock, where several ships of a superior construction have been built.

The position of the arsenal at Guayaquil is lat. 2° 12′ 25″ S., and the longitude, as adopted by Lieut. Raper, 79° 52′ 40″ W.—(See Naut. Mag., 1839, p. 757.)

It is high water on full and change at 7^h; springs rise 7 feet.*

DIRECTIONS.—The following remarks on the navigation of the river were made during the passage up and down by Capt. Basil Hall, in H.M.S. *Conway*, between the 23rd and 31st of December, 1821. Taken along with the chart of the river, and every possible precaution used, they may enable a stranger to proceed up the river, if very urgent service were to require it, and that no pilot could be procured; otherwise it would always be much preferable for a stranger to take a pilot at Punta de Arena, or at the town of Puna, on the island of that name, off the entrance of the river.

Having made the Island of Santa Clara or Amortajado, pass between 3 and 4 miles to the southward of it, and steer a mid-channel course to the eastward. The island may be approached much nearer, but there are said to be shoals lying off it on all sides. By night the lighthouse will be an indication of its place.

The passage into the river to the northward of Santa Clara is not recommended, in consequence of the numerous shoals between it and the coast to the north-eastward.

In all parts of the river with light winds, and the ebb tide making, it would be well to anchor, as it sets to the southward and westward in this part of the river at the rate of $1\frac{1}{4}$ miles per hour, as we found. The flood, which did not run so strong here, sets to the northward and eastward at the rate of three-quarters of a mile per hour.

As we advanced to the northward and eastward we gave Point Arena, on the S.E. angle of the Island of Puna, a good berth on passing, and stood more to the eastward on the opposite side of the river, which is clear of shoals, and may be approached with safety to 4 fathoms at the distance of 2 or 3 miles.

This stretch over is advisable before hauling up to the northward, to avoid a small bank, that lies a little to the northward, but at some distance to the eastward, of Point Arena, on which there is (or was) the white buoy previously described. This point, which is a pilot station, does not appear sandy at the distance of 2 or 3 miles, as its name implies, but, in common with the rest of the coast, woody, and this renders it, as well as other places, difficult to be known on many bearings. The pilot gave us a mark by which it might be known, when a vessel was in the direction of the before-mentioned shoal. It was as follows:—When the mouth of a small river to the northward and eastward of Point Arena, and near Puna Vieja, was no longer open; that is to say, when the two points forming the river's mouth, as it appeared, came in one, and bearing at that time N.W. by W. by compass, Hill of Mala also on the Island of Puna, at this time bearing N. $\frac{3}{4}$ W., and the Boco del Galao W. $\frac{1}{2}$ S., we had $6\frac{1}{2}$ fathoms. This mark is principally useful in coming from the northward, or when coming from the southward, and having opened the Hill of Mala, situated on the N.E. side of the Island of Puna, bring it to bear N. $\frac{3}{4}$ W. before steering to the northward;

* For a good account of Guayaquil, see Stevenson's Peru, vol. ii. chap. 7. Capt. Basil Hall also describes the state of the political relations, &c., at the time of his visit in 1821—Extracts, &c., vol. ii. chapters 35—37; and of the buccaneer's attempts, see Dampier, vol. i. p. 154, *et seq.*

it might be approached nearer, but there are no marks sufficiently conspicuous by which one could with propriety go closer. The Hill of Mala is well characterized in the chart, and is seen at some distance off, making like a moderately high island, when about 5 or 6 leagues to the southward of it.

Being to the northward of this shoal, keep the mid-channel, or rather to the eastern side of the river; and in advancing to the northward the *Point Mandingo* will be seen: it is a bold bluff, forming the extreme N.E. point of the Island of Puna; it is easily known, and must not be brought to bear to the eastward of North by compass; but, when bearing N. by W., steer direct for it, when it is flood tide, as otherwise you will be set to the N.N.E. on the bank, off the South end of the *Island of Mondragon*.

By not bringing the Point Mandingo to the eastward of North the dangerous bank of Mala will be avoided, the northern extreme of which lies off shore nearly a mile, and due East of the Hill of Mala; from thence it stretches to 6 or 7 miles in a S. by W. direction, maintaining a distance of about 4 miles from the island. The land nearest the North end of this shoal forms an ill-defined point, but it may be known by the land trending suddenly off to the westward. You will, however, be just clear of the North end of this shoal when the Hill of Mala bears W. ½ S. by compass; with this bearing of the hill, and a hut on the beach W. by N., Point Mandingo N. by E., we anchored in 7 fathoms muddy bottom; here the ebb tide sets S.W. ¾ W. at the rate of 2 miles an hour; a few miles to the southward of this we found the flood setting N.N.E. 1½ miles per hour.

The coast of the Island of Puna to the northward of the bank of Mala, as far as Point Mandingo, is steep-to, except off *Point Centinela*, where a small spit of sand runs out to the eastward about one-third of a mile, on which there are only 2½ fathoms; in other parts, close to the beach, 4 fathoms.

Point Mandingo might be rounded at the distance of one-third of a mile.

The anchorage off the town of Puna is in 6 fathoms, with the town bearing South about three-quarters of a mile. The *Conway* anchored with the extremes of the town from S. by W. ¼ W. to S.S.W.; Point Mandingo, S. by E. ¼ E.; extreme western point of Puna, W. by N.: here we observed the flood tide set to the N.W. ½ W. at the rate of 2½ miles per hour. The ebb sets to the S.S.E. in this part of the river.

It may not be unnecessary to remark, that hitherto the soundings laid down in the chart were of little or no service to us; they are extremely irregular, and very few of them.* We had, when in near mid-channel, 5, 4¾, 6, but most commonly 5½ fathoms; and we were informed by the pilot that, near the bank of Mala, the water deepens from its usual depth in mid-channel to 7 and 8 fathoms, and immediately afterwards shoals on the bank to 2 fathoms.

From Puna to Guayaquil, Dec. 24th, p.m.—Having got a pilot on board from Puna, we weighed with the flood tide, and steered for the centre of *Green Island*, bearing N.W. ¼ W.; and when Point Mandingo bore S. 30° E., the western extreme of Green Island W. by N. ½ N., and the extreme point of Mondragon E. by N. ½ N., we hauled up for the eastern extreme point of Green

* This of course does not apply to the recent and minute surveys by Capt. Kellett.

Island, bearing N.N.W., having 4¾ fathoms, and when the western extreme of Green Island bore W. ¾ N., we steered N. ½ W. until Point Mandingo bore S.W. ½ S.; western end of Green Island, W. ¾ S.; Point *Little Chupador*, N. by E. ¼ E.: here we had 5¾ fathoms, and steered North along the banks of Green Island, to which we gave a berth of one-third of a mile on passing, and had 5 fathoms when the North end of Green Island was just on with the south-western end of the Island of *Maguinana* bearing W. ¾ S.: we then steered for Point Little Chupador N.N.E., and gave it a berth of a quarter of a mile on passing. Between Point Little Chupador and the North end of Green Island lies a sandbank, and directly opposite to Point Little Chupador, on the Island of Mondragon, a sandbank begins to run out to the southward and westward; its south-western extreme point lies with Point Mandingo S. by E., South end of Mondragon, S.S.E. ¾ E., and the North end of Green Island, S.W. by W.; we had 3½ fathoms on it at high water. Therefore, between Green Island and Chupador, be careful in not bringing Point Mandingo to the southward of S. by E.

Being past Point Little Chupador, we kept the western bank of the river close on board, to avoid the sandbanks lying off the Island of Mondragon, and we had 5 fathoms soft mud when a small sort of look-out house bore West.

To pass the sunken rock called the *Baja* is considered one of the difficulties of this river navigation; it lies in the direction of the Island of Mondragon, when bearing E.S.E. by compass, and both the ebb and flood tides set upon it; at half tide there is a ripple, which points out its situation, and on still nights the water roars over it so as to be heard at a sufficient distance, by which the danger may be avoided: between the Baja and the western shore we had 3½ fathoms at low water; we had also the same soundings between Point Piedra and the Baja.

Point Piedra lies between 2 and 3 miles to the northward of the Baja, and may be easily known, there being a few houses on it, and the wood cleared away; to this point we gave a berth on passing: the river is deep hereabouts; close to the northward of the point we had 9 fathoms when nearly low water.

We kept on Point Piedra side of the river for about 2 miles further to the northward, and then steered N.E. across, to avoid the bank that lies to the southward, at some distance from the Island of Sona (part of this bank we saw dry at half tide), as well as to avoid those extending to the northward from the small islands lying off the North end of the Island of Matorillos. Here it may be observed, that the tide of ebb separates itself into two currents, the one running S. by W., along the western shore of the river, and the other S.E. by E., between the Island of Matorillos and the main land to the eastward: therefore, when coming down the river, it will be advisable to haul over to the western shore, before the S.E. by E. current is felt.

In going up, after crossing the river, and having hold of the eastern shore, we kept it very close on board, and as we advanced to the northward we saw a red-tiled house, and a small sort of out-house near it, which is to the northward of a clump of trees that we passed very close to; and when this house bore East, we were just to the northward of the spit that runs out to the eastward of the Island of Sona; then keeping nearer to the eastern shore than to the island side of the river, in steering to the northward we saw a round shaped hill to the N.N.W., having a building on its summit; that hill is the northern boundary of the town of Guayaquil, and when brought to bear N.N.W. $\frac{1}{4}$ W. by compass, and the centre of the eastern branch of the river (which is divided by the Island of Santay) N. $\frac{3}{4}$ W., we steered across, keeping nearer to the western side of the two, as a bank lies at a little distance from the South end of the Island of Santay, where there are two or three houses.

When these houses bore N.E. by E. $\frac{1}{2}$ E., Point Gordo S. by E. $\frac{1}{4}$ E., and the hill to the northward of the town of Guayaquil N.N.W., we had 5 fathoms at high water, and then kept the western shore, or Guayaquil side of the river close on board, to avoid the banks lying off the Island of Santay and the wreck of a French vessel. Our anchorage was off the southern end of the town, in 4$\frac{1}{2}$ fathoms at low water, soft muddy bottom, with the steeples of the custom-house N.N.W. $\frac{1}{4}$ W.; North end of the Island of Santay, N.E. by E. $\frac{3}{4}$ E.; wreck off the North end of ditto, N.E. $\frac{3}{4}$ E.; hill to the northward of the town, having a building on the summit, N. $\frac{3}{4}$ E. Here the rise was 9 feet, and the greatest strength of the flood setting to the northward was 3$\frac{1}{2}$ miles per hour, and the ebb to the southward at the rate of 3 miles per hour, spring tides.

It may be necessary to state that all the bearings in the foregoing directions of the river are by compass.

The *following directions* are from a manuscript drawn up by an Italian, an experienced trader on the coast, and communicated by Capt. J. W. Monteath, of the brig *Mary Brade*. It will be seen that they were written prior to the establishment of the lighthouse and the buoys.

In going to Guayaquil you ought to make the land about Cape Blanco, which is easily known by its white colour. Having made the cape, sail along the coast 2 leagues off, though you may approach to half a mile. Sailing on you will see the Point of Los Picos, which is very remarkable, being full of small peaks, one larger than the rest. Sailing on a few miles farther you will see a low point full of trees; this is Point Malpelo, from which the shoals of Payana begin, extending about 2 miles off, and are very dangerous; the sea breaks on them at low water. Do not approach this point to less than 10 fathoms, and 5 miles off, with a leading wind, which you generally have on going in; therefore it is best to keep in mid-channel, from 18 to 22 fathoms, good ground for anchorage.

The Muerto (or Amortajada) may be seen from the mast-head at 7 leagues off. It is tolerably large; at the first sight it appears like three hummocks; the head of it is rather perpendicular, and sloping toward the North part of the island. It has a reef on each end; that toward the Island of Puna is about a league in length. You may approach any part of the Muerto to within a mile, except the ends of it. In general you will see the Muerto and Puna at the same time. The latter island (Puna) is of a good height on the South side. When the Muerto bears West, 6 miles off, you ought to steer for mid-channel, and rather nearest the main land; and when Punta de Arena on the Puna bears W. by S. you ought, if not before, to go nearer the main in 6 fathoms, but if with foul wind, approach the main to 5 and even 4 fathoms; and about towards the Puna take care to deepen to not more than 7$\frac{1}{2}$ fathoms, for in 8 you may find yourself on the banks before you are aware of it, for the banks are steep-to.

With a fair wind keep in 5½ and 6 fathoms, and when the low point of the Island of Puna bears N.W. by N., sail for it, taking care to keep the point open. When abreast of the point you may approach to half a mile, but keep a mile from it, and sail coasting the island till you see the houses, which are large and white. At the foot of them is a creek or river. You cannot fail seeing the houses, and when abreast of them, 1 mile off, let go the anchor in 9 fathoms mud. Should the pilot not come to you, you must go to the houses to get him, for you must not proceed any further without, as 5 miles beyond there is a flat extending all across the river, on which there is no more than 18 feet at spring tides, but in general from 12 to 16 feet.

At Guayaquil you moor with two bowers close to the town, so close that when your vessel swings you are not more than the length of your vessel from the shore.

Capt. Basil Hall describes the descent of the Guayaquil River by means of kedging, which, as it is very interesting from its connexion, we transcribe from his Extracts from a Journal, &c., part ii., chap. 38.

The *operation of kedging* is a device to produce a relative motion between the ship and the water, in order to bring the directing power of the rudder into action. This object is accomplished by allowing the anchor to trail along, instead of being lifted entirely off the ground, as in the first supposition. It is known practically, that the degree of firmness with which an anchor holds the ground depends, within certain limits, upon its remoteness from the ship. When the anchor lies on the ground immediately under the ship's bows, and the cable is vertical, it has little or no hold; but when there is much cable out, the anchor fixes itself in the bottom, and cannot without difficulty be dragged out of its place. In the operation of kedging, the cable is hove, or drawn in, till nearly in an upright position; this immediately loosens the hold of the anchor, which then begins to trail along the ground, by the action of the tide pressing against the ship. If the anchor ceases altogether to hold, the vessel will, of course, move entirely along with the tide, and the rudder will become useless. However, if the anchor be not quite lifted off the ground, but be merely allowed to drag along, it is evident that the ship, thus clogged, will accompany the tide reluctantly, and the stream will in part run past her; and thus a relative motion between the vessel and the water being produced, a steering power will be communicated to the rudder.

In our case, the tide was running 3 miles an hour; and had the anchor been lifted wholly off the ground, we must have been borne past the shore exactly at that rate; but by allowing it to drag along the ground, a friction was produced, by which the ship was retarded 1 mile an hour; and she was therefore actually carried down the stream at the rate of only 2 miles, while the remaining 1 mile of tide ran past, and allowed of her being steered: so that, in point of fact, the ship became as much under the command of the rudder as if she had been under sail, and going at the rate of 1 mile an hour through the water.

This power of steering enabled the pilot to thread his way, stern foremost, amongst the shoals, and to avoid the angles of the sandbanks; for, by turning the ship's head one way or the other, the tide was made to act obliquely on the opposite bow, and thus she was easily made to cross over from bank to bank, in a zig-zag direction. It sometimes happened that, with every care, the pilot found himself caught by some eddy of the tide, which threatened to carry him on a shoal; when this took place, a few fathoms of the cable were permitted to run out, which in an instant allowed the anchor to fix itself in the ground, and consequently the ship became motionless. By now placing the rudder in the proper position, the tide was soon made to act on one bow; the ship was sheered over, as it is called, clear of the danger; and the cable being again drawn in, the anchor dragged along as before. The operation of kedging, as may be conceived, requires the most constant vigilance, and is full of interest, though rather a slow mode of proceeding; for it cost us all that night, and the whole of the next day and night, to retrace the ground which we formerly had gone over in 10 hours.

POINT SANTA ELENA is 68 miles N.W. ½ N. from the Island of Santa Clara at the entrance of Guayaquil River. Its N.W. extremity is in lat. 2° 11′ 10″ S., and lon. 81° 1′ 40″ W. Of the intervening coast we have no account; and here commences the survey continued to the northward by Capt. H. Kellett in 1836. It forms the southern side of Santa Elena Bay, and is thus described by Dampier. The allusion to the singular bituminous spring is still correct, and might be turned probably to some useful purpose.

"Point Santa Elena is pretty high, flat, and even at the top, overgrown with a great many thistles, but no sort of tree: at a distance it appears like an island, because the land within it is very low.

"This point strikes out next into the sea, making a pretty large bay on the North side. A mile within the point, on a sandy bay, close by the sea, there is a poor small Indian village, called Sancta Hellena: the land about it is low, sandy, and barren; there are no trees nor grass growing near it. There is no fresh water at this place, nor near it; therefore the Indians are obliged to fetch all their water from the River Colanche, which is in the bottom of the bay, about 4 leagues from it. Not far from this town on the bay, close by the sea, about 5 paces from high-water mark, there is a sort of bituminous matter boils out of a little hole in the earth; it is like thin tar; the Spaniards call it algatrane. By much boiling it becomes hard like pitch. It is frequently used by the Spaniards instead of pitch; and the Indians that inhabit here save it in jars. It boils up most at high water; and then the Indians are ready to receive it. These Indians are fishermen, and go out to sea on bark logs. Their chief subsistence is maize, most of which they get from ships that come hither for algatrane. There is good anchoring to leeward of the point right against the village, but on the West side of the point it is deep water, and no anchoring."*

The Bay of Sta. Elena, thus alluded to by Dampier, is a miserable spot, though the anchorage is good. Capt. Pipon anchored July 9th, 1814, in 6¾ fathoms, sandy bottom, about 2 miles off shore. It was found on examining the shore that firewood might be procured for ships in want of that article; but no fresh

* Dampier, vol. i. pp. 133-4.

water, and from what intelligence could be obtained from the inhabitants of the miserable village at the bottom of the bay, there is no fresh water within 5 miles of the anchorage, all they have to make use of being brought on rafts from that distance. No refreshment is to be procured here; the poor inhabitants living chiefly on the fish caught in the bay. They have some saltpans in the neighbourhood.

At 20 miles N.N.E. ¼ E. *mag.* from Sta. Elena Point is the little *Islet of* PELADO, in lat. 2° 3′ 55″, lon. 80° 48′ 45″ W., according to Capt. Kellett's, R.N., survey. Morrell says that he found nothing on it but birds and hair-seals. It lies 3 miles N.W. off *Ayanqui Point*, the channel between being clear, but a shoal extends nearly a mile off the point.

The coast hence is clear, and trends in a general N.N.W. direction for 24 miles to Selango Island; but at 10 miles North of Ayanqui Point is *Montanita Point*, off which are some rocks, and 8 miles further is Jampa Point, also rocks around it. At 2 miles N.W. of the latter, with a clear passage between, are the *Ahorcados* (hanging islets), a detached bank of rocks and islets.

SELANGO ISLAND is in lat. 1° 35′ 20″ South, lon. 80° 54′ 0″ West. Capt. Pipon, in H.M.S. *Tagus*, on the 12th July, anchored off this island, opposite a fine sandy beach, in 25 fathoms water, about 1 mile off shore, and began their operation of wooding; finding, however, that they ran rather too far distant from the rivulet, they shifted their berth, and anchored nearer the main land, for the convenience of watering, in 19 fathoms, blue sand; the bearings when at anchor were, the Island of Plata, N.W. ¼ N. 7 or 8 leagues; the N.W. point of a small island, N.W. by W. ¼ W.; and a rocky point on the main, N.E. The channel between the island and the main consists of a ledge of rocks across, many of them above water. It was found that in this anchorage wood was procured with greater facility and plenty than at Selango Island; and by penetrating a little into the woods, spars of large dimensions and various sizes were found. You may anchor here in between 15 and 23 fathoms water, and will not then be more than half a mile off shore.

Bamboos of large dimensions were also found here; and such an abundance of excellent fish was caught with the seine as is almost incredible; one in particular, whose name was not known, of a reddish hue, and large scales, resembling much in flavour the red mullet, though considerably larger. Here the ships' companies were fully and very pleasantly occupied; and although in all the operations of wooding, watering, and hauling the seine, they had to toil in general against a heavy surf, yet their labours were invariably well repaid. The greatest surf prevailed with a rising tide; but at low water the casks could always be rafted off with tolerable facility.

There were two huts erected by the rivulet, the residence of a few poor fishermen. These people had little to dispose of, though they were civil and willing to supply the wants as far as their abilities would permit, and had Capt. Pipon been inclined to remain there a few days longer, they offered to furnish him with live cattle and vegetables. Plantains, of which there were several plantations in the neighbourhood, with a few lemons and Seville oranges, were the only productions they saw there, and the country around an entire forest.

This will be found a most convenient place to supply ships cruising in these seas with those indispensable necessaries, for here you need be under no apprehension of disturbance or molestation.*

Callo Point is 14 miles N. by E. ½ E. *mag.* from Selango Island; and a mile to the North of it is *Callo Island*, a small island only the resort of birds and seals, having no channel inside between it and the coast. Hence to Cape San Lorenzo the distance is 21 miles. Off the coast, 14 miles distant, is Plata Island. The soundings between, as indeed they are all hereabout, are tolerably regular, increasing from 5 to 30 fathoms.

PLATA ISLE was so named by the Spaniards, from Sir Francis Drake dividing his plunder at it.† Its N.W. point, according to Capt. Kellett's survey, is in lat. 1° 15′ 30″ S., lon. 81° 7′ 15″ W. It is of moderate height, and of a verdant shaggy appearance, from the large bushes, or low trees, that cover it. Its length is about 3 miles: and the western side is an entire cliff of an inaccessible appearance. A few small islets appear off the South end of it. The watering and anchoring places are said to be on the eastern side, in a small sandy bay, half a mile from the shore, in 18 or 20 fathoms water.

This island was a favourite place of resort with the buccaneers, it being most conveniently situated to watch the Plata fleets to and from Lima; but all traders, either to or from the coast of Mexico, or between Panama and the coast of Peru, make the coast a little to the northward of it. If we may believe the buccaneers, this island has plenty of water and turtle, and abounded with goats, till the Spaniards destroyed them.‡

Dampier has described it:—"It is bounded with high steep cliffs, clear round only at one place on the East side. The top of it is flat and even, the soil dry and sandy; the trees it produceth are but small bodied, low, and grow thin, and there are only three or four sorts of trees, all unknown to us. I observed that they were much overgrown with long moss. There is good grass, especially in the beginning of the year. There is no water on this island but at one place on the East side, close by the sea; there it drills slowly down from the rocks, where it may be received into vessels. There were plenty of goats, but they are now all destroyed. There is no other sort of land animal that I did ever see; there are plenty of boobies and men-of-war birds. The anchoring place is on the East side, near the middle of the island, close by the shore, within 2 cables' length of the sandy bay. There is about 18 or 20 fathom good fast oazy ground, and smooth water; for the S.E. point of the island shelters from the South winds which constantly blow there. From the S.E. point there strikes out a small shoal a quarter of a mile into the sea, where there is commonly a great rippling or working of short waves during all the flood. The tide runs pretty strong, the flood to the South, and the ebb to the North. There is good landing on the sandy bay against the anchoring place, from whence you may go up into the island, and at no place besides. There are two or three high, steep, small rocks at the S.E. point, not a cable's length from the

* Remarks by Capt. Pipon, R.N., in Capt. Basil Hall's Memoir, pp. 74—77.

† This can scarcely be correct, as we are told that Sir Francis Drake sailed for [illegible] land with his prize, and then sailed westward.—See page 203 hereafter.

‡ Colnett, pp. 62-3.

island, and another much bigger at the N.E. end; it is deep water all round but at the anchoring place and at the shore at the S.E. point. At this island are plenty of those small sea-turtle I have spoken of." *

Capt. Pipon, R.N., says: "On approaching this island we found the soundings very irregular, having only 16 fathoms as we rounded the southern part, at a good distance, and then deepening to 34 and 35, as we hauled in for the little sandy bay. We anchored in 35 fathoms abreast of this bay, about 1 mile off shore. The marks when at anchor were: the South point (off which is a reef), S.S.W.; the extreme point, W. ½ N.; the sandy beach, S.W. by ½ W.; Cape San Lorenzo, N. by E. ¾ E. It is, however, advisable to anchor more within the bay, in 19 and 20 fathoms; you will not then be above 2 or 3 cables' length off shore.

"Wood may be procured here, but not very abundant. Water was not to be found. By report of a fisherman, whose vessel was hauled up in the little sandy bay opposite the anchorage, after rain a small quantity might be obtained; but, whilst we were here, we found everything completely dried up. Fish may be caught here in great abundance, but *not* turtle, as mentioned by Colnett. Considering that no water is to be procured here, it is by no means a desirable place to touch at; I would, in preference, recommend anchoring off the Island of Selango, where abundance of firewood may be cut, and fresh water procured from a considerable rivulet that empties itself into the sea from the main land."

Cape San Lorenzo is in lat. 1° 3′ 20″ S., lon. 80° 57′ 0″ W. It is the West extreme of a projection of the continent, the general line of which, in a N.N.E. direction, is indicated by some hills, of which Monte Christo, 14 miles inland of the cape, and 1,429 feet high, is the principal. The cape itself is a small tongue, off which, for half a mile further, some small islets and rocks extend.

The bank of soundings off the coast, which preserves a generally uniform line from Plata Island, is much narrower off the cape, and does not reach further off than about 3 miles, and the depth itself immediately off the cape is much greater than in other parts, irregular from 30 to 70 fathoms.

From Cape San Lorenzo the coast, which is of moderate height, trends to the eastward, and 14 miles along it is the little Port of Manta. Off the intermediate points, reefs and rocks extend a short distance, and the port in question lies to the S.E. of one of these projections.

"MANTA is a small Indian village on the main. It stands so advantageously to be seen, being built on a small ascent, that it makes a very fair prospect to the sea, yet but a few poor scattered Indian houses. There is a very fine church with a great deal of carved work. It was formerly a habitation for Spaniards, but they are all removed from hence now. The land about it is dry and sandy, bearing only a few shrubby trees. These Indians plant no manner of grain or root, but are supplied from other places, and commonly keep a stock of provisions to relieve ships that want, for this is the first settlement that ships can touch at which come from Panamá bound to Lima or any other port in Peru. The land being dry and sandy, is not fit to produce crops of maize; which is the reason they plant none. There is a spring of good water between the village and the seas.

* Dampier, vol. i. pp. 132-3.

"On the back of the town, a pretty way up in the country, there is a very high mountain,* towering up like a sugar-loaf. It is a very good sea-mark, for there is none like it on all the coast. The body of this mountain bears due South from Manta. About a mile and a half from the shore, right against the village, there is a rock which is very dangerous, because it never appears above water; neither doth the sea break on it, because there is seldom any great sea; yet it is now so well known that all ships bound to this place do easily avoid it. A mile within this rock there is good anchoring, in 6, 8, or 10 fathom water, good hard sand and clear ground. And a mile from the road on the West side, there is a shoal running out a mile into the sea. From Manta to Cape Sta. Lorenzo the land is plain and even, of an indifferent height."†

At 27½ miles to the N.E. of Manta is Caracas Bay.

Caracas Bay.—The entrance is in lat. 0° 34′ 30″ S., lon. 80° 28′ 0″ W. At 2 miles off its entrance is a rocky reef nearly a mile in extent.

"In entering this bay, strict attention must be paid to the lead, as there are many shoals to the North and in front of the entrance; and there are some also on the S.W. side of the bay. The water being generally smooth here, these dangers seldom show themselves on the surface, and therefore render the greater caution necessary. If it be the navigator's wish to anchor near the mouth of the river, he will approach it on the S.W. side, where he may anchor within half a mile of it, between two banks that are nearly dry at low water. The western bank will completely shelter him from seaward, and he will have 4 fathoms of water at low tide, with sufficient room for four or five other ships to lie in his company with perfect safety.

"The country on both sides of Caracas Bay and river is the most beautiful that can possibly be imagined. The soil is rich and fertile, producing in great abundance cocoa, coffee, rice, Indian corn, tobacco, and a great variety of excellent fruits. All kinds of vegetables are plentiful, as are also honey and wax. This is one of the best places on the coast to procure a cargo of cocoa, as you may depend on its being of the very best quality that grows in this country; whereas, if you go to Guayaquil, you are liable to be imposed upon by adulterations. The best coffee and wax may likewise be had at this place, and at a much lower rate than at Guayaquil.

"Among the animal productions of this country are cattle, horses, sheep, goats, hogs, and poultry, in abundance. The forests are well tenanted with a great variety of wild animals, including a multitude of birds of very beautiful plumage. The usual temperature of the atmosphere being warm and moist, brings into existence innumerable swarms of insects and animals of a noxious kind. But the period of their existence is not very protracted, as the S.W. winds, which generally prevail from May to December inclusive, destroy them in great numbers. In the height of the wet season, alligators and other disagreeable reptiles spread themselves over the country, and become very troublesome to the natives; but in the fair weather season they cause very little annoyance.

* 1,429 feet, according to Capt. Kellett.

† Dampier, vol. i. p. 136.

"The S.W. winds, just alluded to, commence blowing about noon, and continue until daylight the next morning. During those months of the year in which these winds prevail the atmosphere is very clear, and it is seldom or never known to rain; but from January to the last of April the heat is very oppressive, accompanied with frequent and heavy falls of rain, with tremendous thunder-storms, and very sharp lightning."*

Cape Passado is 13 miles North of Caracas Bay, and is in lat. 0° 21′ 30″ S., and lon. 80° 32′ W.

"Cape Passao (Passado)," says Dampier, "runs out into the sea with a high round point, which seems to be divided in the midst. It is bold against the sea, but within land and on both sides it is full of short trees. The land in the country is very high and mountainous, and it appears to be very woody. Between Cape Passao and Cape San Francisco the land by the sea is full of small points, making as many little sandy bays between them; and is of an indifferent height, covered with trees of divers sorts; so that, sailing by this coast, you see nothing but a vast grove or wood; which is so much the more pleasant, because the trees are of several forms, both in respect to their growth and colour."†

From hence to *Pedernales Point* (Shingle Point), in lat. 0° 4′ 10″ North, the distance is 35 miles, and the same distance, nearly *true* North from the latter, is Cape San Francisco. Between these points there is no place of commercial importance. Before the whole of the coast a shoal extends, from 1 to 3 miles off, to the depth of 5 fathoms and under, so that it is advisable to keep 2 leagues off the land, by which all danger will be avoided, and a depth of 10 to 30 fathoms found between Point Pedernales and Cape Passado.

North of Pedernales Point the coast runs in nearly a *true* North direction, and is for the most part low and unhealthy. At 18 miles from it the *Cogimies Shoals* extend from thence to 5 miles from the land, and up to Mangles Point, in lat. 27′ 40″ N. Twelve miles further North than the latter is Cape San Francisco, with which it forms an open bay.

"CAPE SAN FRANCISCO is a high bluff, clothed with tall great trees. Passing by this point, coming from the North, you will see a small low point, which you might suppose to be the cape, but you are then past it, and presently afterwards it appears with three points. The land in the country within this cape is very high, and the mountains commonly appear very black. The sea winds are here at South, and the land winds at S.S.E."‡

It was off Cape San Francisco that Sir Francis Drake captured the Spanish ship *Cacafuego*, March 1st, 1579, and steering from the land all day and night, they lay by their prize, taking out her cargo. The treasure found consisted of 13 chests of rials of plate, 80 lb. of gold, 26 tons of uncoined silver, and a quantity of jewels and precious stones. The whole was valued at 360,000 pesos, each

* Morrell's Narrative, &c., pp. 188—190. We have before quoted this author, and with a caution as to the veracity of his *personal* knowledge of the parts his work professes to describe. It is evidently in many points derived from other sources. In the paucity of information on this coast we extract the above, but for its minute accuracy we must not guarantee.

† Dampier, vol. i. pp. 162-3. ‡ *Ibid.* vol. i. p. 132.

nearly equal to 8*s*. English. The uncoined silver was worth upwards of £200,000. After this the unfortunate ship was allowed to proceed on her voyage to Panamá.*

At 12 miles around the coast to the northward is Galera Point. The intervening coast is bold-to, and at the latter point it assumes a more easterly direction.

ATACAMES, or TACAMES, a small seaport town, is about 13 miles from Point Galera. The mouth of the small river on the East side of which it lies, is, according to Capt. Kellett's survey, in lat. 0° 57′ 30″ N., lon. 79° 55′ W.

Capt. Morrell, who states that he visited it in September, 1823, gives the following description of it:—

"At Atacames vessels will find good anchorage and safe shelter a little to the eastward of a rock that lies on the West side of the bay, about two cables' length from the shore, rising nearly 75 feet above the level of the sea.

"The best watering place is in a small river on the West side of the bay, at the mouth of which, on the last of the ebb, water-casks may be filled not more than three-fourths of a mile from the ship. This is also the best place to cut wood, which may be procured in any quantity at the mouth of this river. The water taken from this stream is of an excellent quality for long voyages, no other having ever, to my knowledge, kept sweet and pure so long.

"The town of Atacames is small, containing about 500 inhabitants, the construction of whose habitations is somewhat singular, but well adapted to the climate and other localities. They are built similar to those of New Guinea, being elevated on posts about 10 feet from the ground, and consisting of only one story. On the posts or stakes driven into the earth, which support the building, the floor is laid, above which most of the materials are bamboos. The roof is thatched with a kind of long grass that is common in this country. Each house has one door only, which is entered by means of a ladder, the latter being hauled up into the house every night, when the family is about retiring to rest, to prevent their being disturbed by wild animals, with which this part of the country abounds.

"The soil is very fertile, and yields two crops a year; so that vegetables and fruit are always abundant at Atacames. The temperature is like that of Guayaquil; and accordingly it produces the same kind of fruit, grain, and vegetables, some of them in greater perfection on account of its more elevated situation. It likewise produces, in great abundance, vanillas, balsams, achiote, copal, cocoa, sarsaparilla, and indigo. Considerable quantities of wax are made here; and the forests of the country afford a great variety of trees of large size and lofty height, fit for naval and domestic purposes, including many rare and valuable woods. They likewise procure a considerable quantity of gold dust from the streams of the mountains, beside many valuable minerals. Notwithstanding the ample resources of this place, however, it has hitherto been very little frequented by nautical adventurers, either for trade or refreshment."†

* See Hakluyt, vol. iii. p. 747, "The Famous Voyage," &c. The quantity of treasure taken in these marauding excursions, at the different towns along the coasts of Peru, &c., seems almost incredible, yet the relations given by Burney and other historians of the buccaneers are generally consistent with each other, and afford collateral proof of each other's veracity.

† Narrative, &c., pp. 122-3. See also Account of Colombia, vol. i. p. 327; and Capt. Woodes Rogers' Voyage, 1708, in Burney.

At 14 miles N.E. of Atacames is the entrance of the *Esmeralda River*. The coast is nearly straight, and off *Gorda Point*, 9 miles from Atacames, there are some rocks. A shoal spit extends due North from Atacames to the distance of 11 miles, having on its outer point 10 fathoms of water, and only 4 fathoms at 7 miles. In proceeding, therefore, along this coast, this circumstance should be remembered.*

The town of Esmeraldas is on the left bank of the river, 9 miles from its mouth, the course being *South*, true. Hereabouts was the famous emerald mine, which has long been supposed to be lost.†

Northward of this our information becomes very scanty; we know only the names of the points as laid down in the Spanish charts, which give a very imperfect notion of the reality. The boundary between the two republics of Ecuador and Nueva Granada is at the river Mira, which falls into the Pacific West of Tumaco.

TUMACO is a port of which we know nothing, except its existence. In a decree given at Bogotá, March 22, 1844, Tumaco is declared a free port until 1861; what that year may see at Tumaco it would be hazardous to say. The decree states, "that every class of vessels, national and foreign, can freely enter and leave the free ports (Buenaventura is the other) without paying port dues, or import or other national duties." By another article of this decree, this freedom only applies to merchandise to be consumed in the Island of Tumaco; that which leaves these ports pays national duties.‡

Gallo, according to Dampier, is a small island, moderately high, and covered with very good timber. It lies in a wide bay, about 3 leagues from the mouth of the Tumaco, and 4½ leagues from the Indian village Tumaco. There is a spring of good water at the N.E. end, where there is good landing. The roadstead is before this bay, and there is secure and good riding in 6 and 7 fathoms water; here also a ship may careen. All around this island the water is shoal, but there is a channel to enter by, in which is not less than 4 fathoms water, entering with the flood tide and quitting with the ebb, sounding all the way.

"Tumaco (Tomaco) is a large river, that takes its name from an Indian village so called. It is shoal at the mouth, yet barks may enter. The small village of Tumaco (Tomaco) is seated not far from the mouth of the river. The land between Tumaco and the River St. Jago, a distance of 5 leagues, is low and full of creeks, so that canoes may pass within land from thence into the Tumaco River."§

The RIVER ST. JAGO, according to Dampier, is large, and navigable some leagues up. At 20 miles from the sea it separates into two branches, making an island 12 miles wide against the sea. Both branches are very deep; but the mouth of the narrowest is so shoal, that, at low water, even canoes cannot enter. Above the island it is 3 miles wide, with a strong current.‖

GORGONA ISLAND.—Proceeding to the N.E., the next place we have any

* This spit, and the irregularity in the depth off this part of the coast, compared with that to the southward, are remarkable. Perhaps they may be occasioned by the action of the *currents*, which, as we shall show hereafter, are very strong and peculiar in the vicinity.

† Account of Colombia, vol. i. p. 327. ‡ Commercial Tariffs, part xix. 1847, p. 318.

§ Dampier, vol. i. p. 169. ‖ *Ibid.* vol. i. p. 164.

notice of is the Island of Gorgona, which, according to the recent survey, is 5 miles long, N.N.E. and S.S.W., and 1,296 feet in height. Its North point is in lat. 3° N., lon. 78° 9′ W. The following is Dampier's account of it:—

"(Gorgona) Gorgonia is an uninhabited island. It is pretty high, and very remarkable, by reason of two saddles, or risings, or fallings, at the top. It is about 2 leagues long, and a league broad; and it is 4 leagues from the main. At the West end is another small island. The land against the anchoring place is low; there is a small sandy bay, and good landing. The soil, or mould, of it is black and deep in the low ground; but, on the side of the high land, it is a kind of red clay. This island is very well clothed with large trees, of several sorts, that are flourishing and green all the year. It is very well watered with small brooks that issue from the high land. Here are a great many little black monkeys, some Indian conies, and a few snakes, which are all the land animals that I know here. It is reported of this island that it rains on every day in the year, more or less; but that I can disprove. However, it is a very wet coast, and it rains abundantly here all the year long. There are but few fair days: for there is little difference in the seasons of the year between the wet and the dry; only, in that season which should be the dry time, the rains are less frequent and more moderate than in the wet season, for then it pours as if out of a sieve. It is deep water, and no anchoring anywhere along this island only at the West side. The tide riseth and falleth 7 or 8 feet up and down. Here are a great many periwinkles and mussels to be had at low water. Then the monkeys come down by the sea-side, and catch them; digging them out of their shells with their claws. Here are also pearl oysters in great plenty."*

BUENAVENTURA BAY is the next place we have any notice of, and appears to be the most important port in this part of the republic, and is to remain a free port until 1879!

Considering the importance and beauty of its situation, San Buenaventura ought to be a considerable town: an active commerce should animate its port; a rich and industrious population fill its streets; lastly, it should be frequented by numerous vessels. Nothing of all this is to be seen. A dozen huts, inhabited by negroes and mulattos, a barrack with eleven soldiers, a battery of three pieces of cannon, the residence of the governor, built, like the custom-house, of straw and bamboo, on a small island called Kascakral, covered with grass, brambles, mud, serpents, and toads: such is San Buenaventura; yet the commerce carried on is not without importance, though chiefly very common articles; for instance, salt, onions, and garlic. These, in general, are the only cargoes brought from Payta. Straw hats and hammocks are brought from Xipixapa, and salt meat, causing dysentery, from Costa Rica. The exportations consist of rum, sugar, and tobacco. This unwholesome place suffers a continual scarcity of provisions; it is difficult to procure green bananas, or bread made of maize, and cheese. Fowls cost a piaster apiece, and can hardly be obtained even at that price; fish is scarce, and said to be injurious to health.

* Dampier, vol. i. pp. 172-3. A similar account is given in Woodes Rogers' Voyage, in Burney's Collection.

San Buenaventura, therefore, is, at present, a village of no importance, but may rapidly increase, if, conformably to a plan which has recently been suggested, it be removed to the N.E. of its present site. The place where it is proposed to make the new port being rather elevated, is consequently drier.

To the advantages which all the ports of the Pacific possess, the bay of San Buenaventura joins a considerable extent and depth of water. The bottom is excellent, and ships of war can enter and remain at anchor without danger. The entrance is to the W.S.W. of Kascakral, whereas the mouth of the Dagua* is to the S.E. of the same point. This is not the only river which empties itself into it. The port depends on the imperfectly known district of Chocó.†

The Dagua was descended by Mollien, but with some difficulties. High up it is rapid and precipitous. As it approaches its mouth it has attained its level, and flows sluggishly through low and marshy banks, but without a bar at its mouth, into the bay.

The *River Chinquiquira* falls into the Bay of Cascajal, in San Buenaventura. The Cascajal may be ascended in boats for four or five days' journey from its mouth; sometimes the current is very strong, and running over a rocky bottom, it forms rapids and cascades. An account of the ascent to the gold washings, with the productions, the picturesque appearance, and the scanty inhabitants of the river banks, is given by Capt. Gabriel Lafond, Bull. de la Société de Géographie, March, 1839, pp. 121—143.

Malpelo Island, in lat. 4° N., lon. 81° 32′ W., lies off Buenaventura Bay, but as it scarcely belongs to the coast, though it serves as a useful landmark, being 1,200 feet high, we shall describe it among the islands hereafter.

CAPE CORRIENTES, in lat. 5° 33′ N., lon. 77° 29′ 30″ W., is the next point we know anything about.

Cape Corrientes, according to Dampier, is high, bluff land, with three or four small hillocks on the top, and appears, at a distance, like an island. Here, also, in indication of its name, he found a strong current setting to the North; and the day after he passed the cape, going towards Panamá, he saw a small white island, which he chased, supposing it to be a sail, till he discovered his error on approaching it.‡

TUPICA lies to the northward of this. It has been the subject of very great uncertainty in its geographical situation, and even in its name. It is called in some works "Cupica;" but the Indian pronunciation, obtained during the late Admiralty surveys, is as given, "Tupica."

This uncertainty as to its position will appear less strange if it is remembered that on the whole of the South coast of the Isthmus of Panamá, and the littoral between Capes Charambira and San Francisco, is never sailed along within sight of land by seamen provided with accurate instruments. Tupica is a port of the little-known province of Biruquete, that the charts of the Deposito Hydrografico of Madrid place between Darien and Chocó de Norte. The province takes its

* The ancient *Provincia*, or rather Partivo del Rasposo, comprehends the unhealthy district extending from the Rio Dagua, or San Buenaventura, to the Rio Iscuandè, the southern limit of Chocó Proper.

† Voyage dans la Colombie, par *G. Mollien*, 1824, p. 313.

‡ Vol. i. p. 174.

name from that of a cacique named Biru, or Biruqueta, who reigned in the country around the Gulf of San Miguel, and who fought as an ally of the Spaniards. —(Herrera, dec., vol. ii. p. 8.)*

The environs of the Bay of Tupica abound with excellent timber, fit to be carried to Lima; but this advantage, which in other places would be no ordinary one, is here of no importance, seeing that it only shares it with so many other equally neglected ports.

This bay has been of some celebrity, in consequence of its being one of the points proposed for the junction of the Atlantic and Pacific oceans, as mentioned below† respecting the San Juan. The junction was to be effected by means of the Niapippi River, which flows into the Atrato just below Citera, according to Capt. Cochrane, and not 180 miles below, as it is laid down in the best maps. The Atrato is navigable, and flows into the Gulf of Darien in the Caribbean sea.

The Niapippi is partly navigable, but the navigation is very dangerous, and unfitted for commerce; according to the information given to Capt. Cochrane by Major Alvarez, a Colombian officer, as to forming a canal or railroad, it is *impossible*. Major Alvarez, who travelled over it to Panamá, found the Niapippi shallow, rapid, and rocky; that the land carriage to Tupica was over three sets of hills; and that he could perceive no possibility of making a communication between this river and the Pacific. It must, therefore, be inferred, that the Baron Humboldt (who did not visit this spot himself) was misinformed on the subject.‡

The following is the general account of the matter as given by Humboldt in his Relation Historique, tome i.:—

"A monk of great activity, padre of a village near Novita, employed his parishioners to dig a small canal on the Quebrada de la Raspadura, which is a branch of the San Juan; by means of which, when the rains are abundant, canoes laden with cacao *pass from sea to sea*. This interior communication has existed since 1788, unknown in Europe. The small Canal of Raspadura unites, on the coasts of the two oceans, two points 75 leagues distant from one another."—(Humboldt.)

This communication can never become of great utility, from its distance, and the brief season of the year in which it is practicable. M. Gogueneche, a very intelligent Biscayan pilot, is said to have been the first who turned the attention of the government to this part for communicating between the oceans. An

* See Account of Colombia, vol. i. The allusion to Biru, or Biruqueta, has been before noticed, as giving the name to the region now called "Peru."—See page 137, *ante*.

† The communication between the Pacific and the Gulf of Mexico might be formed by the junction of the upper parts of the rivers San Juan and Atrato which falls into the Gulf of Darien.

The junction would be between the Tambo of San Pablo, on the San Juan, and the Tambo of Citera, on the Atrato. Capt. C. S. Cochrane thus speaks of it:—"After an hour's travelling I came to the rising ground that divides this stream from the one on the Citera side. I particularly inspected it, and found the distance from one stream to the other to be about 400 yards, and the height of the ground necessary to be cut through about 70 feet; but after digging a very few feet you come to the solid rock, which would make the undertaking expensive; besides which, it would be necessary to deepen each stream for about a league, so that I think the least cost would be 500,000 dollars to make a good communication from the Atrato to the San Juan."—*Cochrane*, vol. ii. p. 431; but it never can become serviceable for reasons stated above.

‡ Cochrane, vol. ii. p. 440.

Englishman, also, is among the competitors for augmenting this plan. We shall give the whole in a subsequent chapter.

PORT PINAS is the next point to the northward, and of it we find in Dampier the following :—

"Puerto Pinas is named from the numerous pine trees growing there. The land is tolerably high, rising gently as it runs into the country. Near the sea it is all covered with high woods. The land which bounds the harbour is low in the middle, but high and rocky on both sides. At the mouth of the harbour there are two small islands, or rather barren rocks. In the Spanish pilot books this is recommended for a good harbour; but it lies open to the S.W. winds, which frequently blow here in the wet season." Besides this, he says that within the islands it is of small extent, and has a narrow entrance. He did not enter with his ship, but sent his boats, and they found a good stream of water, but they could not fill them, on account of the great swell which rolled into the harbour.—(March, 1685.)*

THE BAY OF PANAMÁ,

according to the imperfect charts hitherto published, is about 100 miles wide, between the Gulf of San Miguel on the East, and Point Mala on the West. The city of Panamá, now again rising into importance, is at the head of this extensive space, 60 or 70 miles within the arms of its opening. For the greater part of the bay we have only the Spanish charts, but the immediate vicinity of the city has been surveyed and published by the English Admiralty; of this part, therefore, we possess now an accurate knowledge—of other portions less so. As in the parts we have just described, Dampier is our great authority. We extract the description of the climate given in his Collection :—

"The weather is much the same here as in other places in the torrid zone in this latitude, but inclining rather to the wet extreme. The season of rains begins in April or May; and during the months of June, July, and August, the rains are very violent. It is very hot, also, about this time, wherever the sun breaks out of a cloud; for the air is then very sultry, because then usually there are no breezes to fan and cool it, but it is all glowing hot. About September the rains begin to abate, but 'tis November or December, and, it may be, part of January, ere they are quite gone; so that 'tis a very wet country, and has rains for two-thirds, if not three-quarters, of a year. Their first coming is after the manner of our sudden April showers, or hasty thunder-showers, once in a day at first; after this, two or three in a day; at length a shower almost every hour, and frequently accompanied by violent thunder and lightning, during which time the air has often a faint sulphureous smell, when pent up among the woods.

"After this variable weather for about four or six weeks, there will be settled continued rains of several days and nights, without thunder and lightning, but exceeding vehement, considering the length of them; yet at intervals between these, even in the wettest of the season, there will be several fair days, intermixed with only tornadoes or thunder-showers, and that sometimes for a week together.

* Dampier, vol. i. p. 198.

These thunder-showers cause, usually, a sensible wind, by the clouds pressing the atmosphere, which is very refreshing, and moderates the heat. But then this wind shaking the trees of this continued forest, their dropping is as troublesome as the rain itself. When the shower is over, you shall hear a great way together the croaking of frogs and toads, the humming of moskitos or gnats, and the hissing or shrieking of snakes and other insects, loud and unpleasant, some like the quacking of ducks. The moskitos chiefly infest the low swampy or mangrove lands, near the rivers or seas; but, however, this country is not so pestered with that uneasy vermin as many other of the warm countries are. When the rains fall among the woods, they make a hollow or rattling sound; but the floods caused by them often bear down the trees. These will often barricade and dam up the river, till 'tis cleared by another flood that shall set the trees all afloat again. Sometimes, also, the floods run over a broad plain, and for the time make it all like one great lake. The coolest time here is about our Christmas, when the fair weather is coming on."*

POINT GARACHINA, which may be taken as the S.E. limit of Panamá Bay, Dampier says is tolerably high land, rocky, and destitute of trees, but within land it is woody. It is surrounded by rocks seaward; within the point by the sea, at low water, there are abundance of oysters and mussels. The tide rises here 8 or 9 feet; the flood sets N.N.E., the ebb S.S.W.†

The Gulf of San Miguel occupies the S.E. side of the Bay of Panamá, and its entrance lies between Point Garachina on the South and Point San Lorenzo, or Point Brava, as it is named in the chart. This cape is unimportant, and has shoal water extending for several miles to the South of it. It is the outlet of several rivers, the principal of which are the Sambo, the Sta. Maria, and the Congo.

The *Sambo* (or *Sambu*) *River* seems to be considerable, according to Dampier's account. It falls into the gulf a few miles to the East of Point Garachina, on the South side. We copy his further account.

"Between the mouths of these two rivers, on either side, the gulf runs in towards the land somewhat narrower; and makes five or six small islands, which are clothed with great trees, green and flourishing all the year, and good channels between the islands. Beyond which, further in still, the shore on each side closes so near, with two points of mangrove land, as to make a narrow or strait scarce half a mile wide. This serves as a mouth or entrance to the inner part of the gulf, which is a deep bay, 2 or 3 leagues over every way; and about the East end thereof are the mouths of several rivers, the chief of which is that of Santa Maria. There are many outlets or creeks besides this narrow place, but none navigable besides that. This is the way that the privateers have generally taken, as the nearest between the North and South seas. The River Santa Maria is the largest of all the rivers of this gulf. It is navigable 8 or 9 leagues up, for so high the tide flows. Beyond that place the river is divided into many branches, which are only fit for canoes. The tide rises and falls in this river about 18 feet."‡

In Wafer's description of the Isthmus is also an account of this gulf. The

* From Mr. Lionel Wafer's description of the Isthmus of America, given in Dampier's Collection, vol. iii. pp. 314-15.

† Dampier, vol. i. pp. 174, 198.

‡ Dampier, vol. i. p. 194. "About 6 leagues from the river's mouth, on the South side of it, the Spaniards, about 20 years ago, upon the first discovery of the gold mines here, built the town

country all about here is woody and low, and very unhealthy, the rivers being so oazy that the stinking mud infects the air. But the little village of *Scuchadero* lies on the right side of the River of Santa Maria, near the mouth of it; it is seated on a fast rising ground, open to the Gulf of St. Michael, and admitting fresh breezes from the sea, so that it is pretty healthy, and serves as a place of refreshment for the mines, and has a fine rivulet of very sweet water, whereas those rivers are brackish for a considerable way up the country.

"Between Scuchadero and Cape San Lorenzo, which makes the North side of the Gulf of St. Michael, the River of Congo falls into the gulf; which river is made up of many rivulets that fall from the neighbouring hills, and join into one stream. The mouth of it is muddy, and bare for a great way at low water, unless just in the depth of the channel; and it affords little entertainment for shipping. But further in the river is deep enough; so that ships coming in at high water might find it a very good harbour, if they had any business here. The gulf itself hath several islands in it; and up and down, in and about them, there is in many places good, very good, riding, for the most part in oazy ground. The islands, also, especially those toward the mouth, make a good shelter; and the gulf hath room enough for a multitude of ships. The sides are everywhere surrounded with mangroves, growing in swampy land.

"North of this gulf is a small creek, and the land between these is partly such mangrove land as the other, and partly sandy bays. From thence the land runs further on North, but gently bending to the West: and this coast also is much such a mixture of mangrove land and sandy bay quite to the River Cheapo; and in many places there are sholes for a mile or half a mile off at sea. In several parts of this coast, at about 5 or 6 miles' distance from the shore, there are small hills, and the whole country is covered with woods. I know but one river worth observing between Congo and Cheapo; yet there are many creeks and outlets; but no fresh water that I know of, in any part of this coast, in the dry season; for the stagnancies and declivities of the ground, and the very droppings of the trees, in the wet season, afford water enough.

"*Cheapo* is a considerable river, but has no good entering into it for sholes. Its course is long, rising near the North Sea, and pretty far towards the East. About this river the country something changes its face, being savannah on the West side, though the East side is woodland as the other.

"Between the River of Cheapo and Panamá, further West, are three rivers of no great consequence, lying open to the sea. The land between is low, even land, most of it dry, and covered here and there by the sea, with short bushes."—(Dampier's Collection, vol. iii. pp. 309—311.)

Chepillo, or Chepelio, is the pleasantest island in the Bay of Panamá. It is

---

of Santa Maria. This town was taken by Captains Coxon, Harris, and Sharp, at their entrance into these seas, it being then but newly built."—*Dampier*. In the third volume of his Collection, which as far as modern observation has followed the narratives may be implicitly relied on for accuracy, there is an account of an expedition in 1702, to these same mines of Santa Maria, related by Nathaniel Davis. This, as well as the others, is very interesting, and evidently faithful. See Dampier's Collection, 1729, vol. iii. p. 461, to the end. Dampier mentions, also, the great quantity of gold extracted from the mines, by means of slaves, under the Spaniards. After these buccaneering expeditions these towns were deserted.

8 leagues from Panamá, and a league off shore. The island is about a mile long and nearly as broad. It is low on the North side, and rises by a gentle ascent toward the South. It was planted with excellent fruits, the centre growing good plantains. The anchorage is on the North side, and is good about half a mile from shore. There was a well close by the sea in this neighbourhood. The island lies off the mouth of the River Cheapo.*

In the eastern part of the bay is a group of islands, called sometimes the *Islas del Rey*, or *King's Islands*, and also the *Pearl Islands*; but Dampier says, "I cannot imagine wherefore they are called so, for I did never see one pearl oyster about them, nor any pearl oyster-shells; but on the other oysters I have made many a meal there."

The Ylas del Rey (now the Islands of Columbia) cover about 400 square miles, and comprise numerous islets, and probably 30 or 40 fishing villages. At Casalla, Capt. Sir Edward Belcher witnessed the pearl fishing carried on at these islands in full activity. The quantity of pearls estimated at the season is about two gallons. Capt. Belcher describes the mode and state of the diving trade he witnessed in October, 1838.

The principal of these is Isla del Rey, which is about 17 miles long by 10 in its greatest breadth. To the West of this are *San José* and *Pedro Gonzales*, and to the northward of it are a group of several, of which *Pachea*, or *Pacheca* (Pacheque), is the northernmost; it is small, and 33 miles S.E. of Panamá. On the South side of it there are two or three small islands, leaving a very narrow, but clear, channel between them and Pacheque. Dampier says, the southernmost of the group is called St. Paul's; but the chart does not notice this. He says, at the S.E. end of the islands, about a league from St. Paul's, there is a good place for ships to careen or haul ashore. It is surrounded with the land, and has a good, deep channel on the North side to enter by. The tide rises here about 10 feet.

The islands are low and woody, and the soil fertile, but not much cultivated. Between them the channels are narrow and deep, but only fit for boats to pass. They are pleasant in appearance from their flourishing verdure.

Between the South end of Isla del Rey and Point Garachina is *Galera Island*, small, low, and barren; S.W. from it extends a shoal for some miles. Six miles S.E. of this island is the *San José Bank*, according to the chart, having 32 feet of water on it, and is 15 miles W.N.W. of Point Garachina.

The channel between these islands and the main on the East is clear and deep, having good anchorage all the way.†

PANAMÁ.

This place is much more familiar in Europe than its own present importance would warrant; this arises from its being the nearest point of approach between the two great oceans, and therefore it has been more minutely examined than other more commercial parts, for the purpose of effecting some means of communication between the Gulf of Mexico and the Pacific, which scheme seems

* Dampier, vol. i. pp. 202—204. † Dampier, vol. i. pp. 175—177.

to be in some prospect of realization since the great accession of interests to the country of California to the N.W. In former times, during the Spanish domination, it was a place of great commerce and strength. This arose from the policy pursued by that government in restricting the transit of merchandise from and to Mexico and Peru and Europe to this port; and down to 1824 Panamá was the highway between Spain and her colonies on the Pacific. But the downfall of the Spanish power, combined with the trade opened by British ships passing around Cape Horn, has been the destruction of the prosperity of Panamá.

The Isthmus between Panamá and Chagres was surveyed by Mr. J. A. Lloyd, F.R.S., in 1827 and 1828; and from his description we extract the following:—

"The site of Panamá has been once changed. Where the old city stood, which is about 3 miles East of the present situation, was already, when the Spaniards first reached it in 1515,* occupied by an Indian population, attracted to it by the abundance of fish on the coast, and who are said to have named it 'Panamá' from this circumstance, the word signifying much fish. They were, however, speedily dispossessed, and even so early as in 1521, the title and privileges of a city were conferred on the Spanish town by the emperor Charles the Fifth. In the year 1670 it was sacked and reduced to ashes by the buccaneer Morgan; and it was only after this built where it now stands.

"Its present position is in lat. 8° 57′ N., lon. 79° 31′ W., on a tongue of land, shaped nearly like a spear-head, extending a considerable distance out to sea, and gradually swelling towards the middle. Its harbour is protected by a number of islands, a little way from the main land, some of which are of considerable size, and highly cultivated. There is good anchorage under them all, and supplies of ordinary kinds, including excellent water, may be obtained from most of them.

"The plan of this city is not strictly regular; but the principal streets extend across the peninsula from sea to sea; and a current of air is thus preserved, and more cleanliness than is usually found in the Spanish-American towns. The fortifications are also irregular, and not strong, though the walls are high, the bastions having been constructed from time to time as the menaces of pirates, or other enemies, have suggested. The buildings are of stone, generally most substantial, and the larger with courts or patios. The style is the old Spanish. Of public edifices there are a beautiful cathedral; four convents (now nearly deserted), belonging respectively to the Dominican, Augustin, Franciscan, and Mercenarios monks; a nunnery of Sta. Clara; a college *de la Campania*, and also the walls of another (the Jesuits' college), which was begun on a magnificent scale, but was never finished, and is now falling into ruins."†

Capt. Basil Hall, who visited it a few years previous to Mr. Lloyd, says that the finest ruin at Panamá is that of the Jesuits' college, a large and beautiful edifice, which, however, was never finished; yet the melancholy interest which it inspires is rather augmented than diminished by that circumstance; for it

* "Nata, on the West side of the Bay of Panamá, was the first town built by the Spaniards on the coast of the South Sea. It was founded in 1517. The following year they established themselves at Panamá."—*Herrera, Historia de las Indias Occidentales*, dec. 2, lib. iv. chap. 1.

† Journal of the Royal Geographical Society, vol. i. pp. 85-6.

reminds us not only of the destruction of the great order which founded it, but also of the total decay of Spanish taste and wealth which accompanied that event. The college is a large quadrangular building, which had been carried to the height of two stories, and was probably to have been surmounted by a third. . . . . In a field a little beyond the square, on the side opposite the college, stand the remains of a church and convent, which is reached, not without difficulty, by wading, breast high, through a field of weeds and flowers, which in this climate shoot up with wonderful quickness. . . . . In some districts of the town of Panamá whole streets are allowed to fall into neglect, grass has grown over most parts of the pavement, and even the military works are crumbling fast to decay. Everything, in short, tells the same lamentable story of former splendour and of present poverty.*

Although this has been written for twenty years, yet the remarks hold good to the time just previous to the California attractions, which, necessarily, have occasioned great changes in its position. A writer in July, 1848, says:—

"Panamá is indeed a city of ruins, and a feeling of melancholy is excited on contrasting it with what it once was. It stands on an area of about 12 acres, entirely surrounded by fortifications, the sea wall of which is in the best state, but even that in many places fallen away.† Only in one part on the land side to the N.W. have I seen any symptoms of repairing. The ditch and walls luxuriant with weeds, and grass not uncommon in the streets. On the S.E. bastion there are a few beautiful brass guns, three only mounted, but the carriages in such a state that I doubt much their standing a couple of rounds. Should a war ever take place (which God forbid!) Panamá would certainly be a place worth having (even if only in trust for the present possessors), from its being the key to the Pacific. The country around is beautiful, and capable of producing everything: that is, as far as I am capable of judging, and comparing the soil with other countries well known as fertile.

"On the morning of the 24th July, 1848, I saw a sight which certainly raised feelings of pride at being an Englishman, and knowing that all before me was going to my home. Before the house of the British Consulate were upwards of one hundred and twenty mules loading with upwards of half a million of treasure, which had been landed the day before from the steamer just arrived from the ports between Valparaiso and this. It is to be shipped at Chagres on board the Royal Mail steamer for England, which I certainly think the best and quickest mode of transmitting specie, instead of by Cape Horn; and when steamboats from the northward are running more will go this way. The quantity has been gradually increasing, commencing about 16 months ago with 16,000 dollars.

"As far as I have seen and can learn from others, the arrangements are perfect, and now only want the road between Cruces and Panamá widened and

* Extracts from a Journal, &c., part ii. chap. 40.

† "There is every facility for erecting a substantial pier, and improving the inner anchorage, which must follow the arrival of the steamers, unless they still submit to the miserable landing at the seaport gate, which is as filthy as it is inconvenient. Panamá affords the usual supplies which are to be obtained in these tropical regions, and at moderate prices; but vessels wishing to procure water, bullocks, &c., can obtain them more readily at the Island of Taboga."—*Sir E. Belcher, Voyage of the Sulphur*, vol. i. pp. 23-4.

repaired. At present it is in a terrible state; but still nothing to prevent a good mule travelling with a load of specie, which is packed in boxes or bars in hide or canvas, two of which are a load, when each weigh from 120 to 140 lb.

"At present, voyaging from Chagres to Cruces is performed in canoes. The large ones for one person can be made very comfortable, and in the dry season the voyage can be performed in a day and a half; in the wet season of course longer, the downward current being very strong.

"The population of Panamá, with its suburbs, is about 6,000. What is here greatly wanted is capital, backed up with English energy; and I certainly think the Royal Mail West India Company will reap great benefit by what they are about to do respecting the repair of the Cruces road; not forgetting that thanks and praise are due to those who first advocated such a thing. We may then expect to see Panamá in a different state from what it is now: all the ruins of immense convents turned into stores, with plenty throughout the land."*

As might be expected, the California movement has effected a change in the fallen fortunes of Panamá. There have been some comfortable inns established, as is also the case at Cruces and Chagres on the Gulf of Mexico, called into existence by the transit of the emigrants from the United States and Europe to San Francisco. Between New York and Chagres, at the commencement of 1850, three different lines of first-class steamers were constantly plying. From this it may be judged that a great influx of prosperity had reached this deserted city, and that many of the old inhabitants, who for many years have been very poor, now deem themselves rich. Consequently Panamá is changing its appearance very much. The Americans have been buying and leasing property for various purposes, and rents have increased ten-fold. It was believed that the seat of government would be removed to Panamá, which would give it much additional importance. Large improvements were going on at the Island of Taboga, opposite the city, where the steamers lie. A steam ferry to the island was about to be established, and American commercial houses were beginning business there on a large scale.

From Panamá, also, to San Francisco there were three lines of steamers plying regularly.

All this sounds very differently from the old accounts. How far this sudden prosperity may have stability, or whether it be but ephemeral, time must unfold.†

Immediately about Panamá, East along the coast, and N.W. from it, the land is low and flat, but West and N.E. the mountains approach it closely; and from a hill called Cerro Ancon, about a mile West from the city, and 540 feet high, an excellent bird's-eye view is obtained of the whole adjoining country, including the city, the islands in the bay, the neighbouring plantations, the mountains of Veragua, the Pearl Islands, the flat country towards Chagres, the elevated chain between Porto-Bello and Panamá, the Rio Grande, the low land along the coast towards the Pacora and Cheapo, Panamá Vieja, &c., all which come successively under review, and together constitute a landscape beyond measure beautiful.‡

* Naut. Mag., Nov. 1848, pp. 609-10. † See Daily News, March 6, 1850.
‡ Mr. Lloyd, Journal of the Royal Geographical Society, vol. i. p. 86.

Panamá is celebrated for its gold chains. They are very well known for their beauty and peculiarly neat workmanship. One weighing about an ounce costs £4 8*s.*; and silver chains of the same kind of manufacture are to be had very reasonably. The Panamá hat, which is a favourite wear in the country, is brought from Guayaquil, and costs from two to twenty dollars.*

At 9 miles South by compass from the city is the little island of *Taboga*. It is about 2 miles long N.W. and S.E.; and off its S.E. end is a smaller island, *Urava*, and a rock, *Terupa;* the former separated from Taboga by a narrow, shallow, rocky channel.

Taboguilla lies 1⅓ miles N.E. of Taboga, and is about 1 mile long. Off its South point is a rock called the *Farallon;* and about midway between it and Urava, that is, three-quarters of a mile from each, is a *sunken* rock, having 8 to 14 fathoms close around it. Otherwise the channel between the islands is clear, with a depth of 15 to 20 fathoms. These islands are important to Panamá, as they are the gardens of the town, and supply it plentifully with fruit and vegetables.

On the N.E. side of Taboga is a village, opposite which vessels visiting Panamá usually anchor, to procure water and supplies, which are not readily to be got at the city. "The anchorage is in a snug cove, opposite to a romantic little village, the huts of which, built of wattled canes, are so completely hid by the screen of trees which skirts the beach, that they can scarcely be seen from the anchoring place, though not 200 yards off; but the walls of a neat whitewashed church, built on a grassy knoll, rise above the cocoa-nut trees, and disclose the situation of the village. The stream, from which vessels fill their water casks, is nearly as invisible as the houses; the whole island, indeed, is so thickly wooded with shrubs, tamarind, plantain, and cocoa-nut trees, and thick grass, that nothing can at first be discovered but a solid mass of brilliant foliage."

There is a watering place on each side of the village, but which are not seen until you land; that on the S.E. side is the largest stream, and at night runs clear and constant, but in the day time it is liable to be contaminated by washing operations; the other, to the northward, though smaller, lies more out of the village.

"As the days were intolerably hot, I determined to water the ship by night; and she was accordingly moored as close to the shore as possible. The sea in this corner of the cove being quite smooth, the boats rowed to and fro all night with perfect ease; and the moon being only one day short of the full, afforded ample light to work by. The casks were rolled along a path, to the side of a natural basin, which received the stream as it leaped over the edge of a rock, closely shrouded by creepers and flowers interlaid into one another, and forming a canopy over the pool, from which our people lifted out the water with buckets. This spot was only lighted by a few chance rays of the moon, which found their way through the broken screen of cocoa-nut leaves, and chequered the ground here and there. Through a long avenue in the woods, we could just discover the village, with many groups of the inhabitants sleeping before their doors, on mats

* Mr. Webster, Narrative of the Voyage of the *Chanticleer*, vol. ii. p. 166.

spread in the moonlight. The scene was tranquil and beautiful, and in the highest degree characteristic of the climate and country.

"Here we procured various kinds of tropical fruits, together with a supply of yams, fowls, and vegetables, and fodder for the oxen.

"The town of Panamá is supplied from these and the neighbouring islands with the last-mentioned articles."—(Capt. Basil Hall.)

The coast of the main land to the North of Taboga, or to the S.W. of Panamá, cannot be approached to within 1¼ miles, on account of the shoal water which extends that distance off. On the edge of these shallow soundings are several small islands. Those nearest to the city are *Perico*, *Ileñao*, and *Culebra*, which are connected with each other, and San Jose outside of them. These are from 2 to 3 miles from the city. S.W. of them are the *Pulberia Reefs*, and *Changarni Islet* upon them; and still farther, in the same direction, are the Islets of *Tortola* and *Tortolita*. Within these islands a vessel ought not to entangle herself.

At 1½ miles nearly East of the city is a large bank, nearly dry at low water. It is of some extent, and should be carefully avoided.* To the South of it is the anchorage.

The *Baja de Afuera*, according to the observations of M. Fisquet, in the French corvette *La Danaïde*, in 1843, is 2¼ miles E. by S. from the East point of the fortification of Panamá. It has only 12 feet water on it, and is surrounded by a depth of 4½ to 5 fathoms.

These are the principal features of the environs and city of Panamá, which may, at some future day, assume its former importance, though it must necessarily be upon a very different basis.†

The notices relative to the proposed communications to be effected by railroad or canal between the two oceans, will be given in a subsequent chapter, as stated with respect to the Atrato and San Juan rivers (page 208), when these considerations will be united under one head.

To Capt. Basil Hall's Directions we are indebted for the following instructions for Panamá. They are extracted from Lieut. Foster's Memoir:—

"It is recommended when going to Panamá, either from the northward or

* "H.M.S. *Actæon*, in working into the Bay of Panamá, December 26, 1840, grounded on this bank, called the *Condado*, not laid down in the charts at that time, with the following bearings:—Panamá cathedral, W. by S.; the outer Penco Island, S.S.W. ½ W. The ship took the ground from 4½ fathoms, payed off before the wind, and in ten minutes was again afloat. The upper part of Panamá Bay, off the town, has very shoal water; and as there is no good survey yet made of it, a measure so highly desirable in these days of steam navigation, vessels are recommended not to make too free in standing in-shore."—*Naut. Mag.*, Nov. 1841, p. 781. Of course these remarks are now obviated by the excellent survey made by Sir Edw. Belcher, in 1837.

† In the Colombian congress of 1823, an exclusive privilege was granted to Messrs. Rundell, Bridge, and Rundell, the well-known goldsmiths of London, to fish for pearl oysters, *with machinery*, on certain parts of the Colombian coast. There were three places: one about the mouths of the Orinoco and Cumana, including Margarita; a second, on the coast of the Rio Hacha; and the third was in the Bay of Panamá. The stipulations were, ceding one-fifth of the pearls procured to the government, and, at the expiration of ten years, their monopoly and the machinery were to revert to the government. They consequently formed a company, called the Colombian Pearl Fishery Association, which despatched one ship to the Pacific, provided with diving bells; but their machinery did not answer the expectation. See Present State of Colombia, 1827, p. 324; and Colombia in its Present State, by Col. Francis Hall, 1824.

southward, to get hold of the eastern shore, and work up along it, as the prevailing wind is from the N.N.W. in the day time, North and N.N.E. in the night, which enables a vessel to make a long stretch to the W.N.W. in smooth water, whereas it would be a dead beat to windward between the Isla del Rey and the western shore, independently of the drain of current out and a short rough sea.

"In the Spanish surveys of this bay, there is a bank extending 8 miles to the S.E. from the Island of Galera, on which the least water marked is 8 Spanish fathoms, and we were informed that there was not less.

"When working up this side, it may be advisable not to stand in-shore on the main side to less than 6 fathoms, as there is a bank laid down off it at the distance of 4 miles, but we were told that the water gradually shoaled to it.

"If the wind proves light, and the small islands lying to the northward of the Isla del Rey have not been seen, and night coming on, it would be well to anchor, and not to run in in the night time, unless it were moonlight, before being satisfied on this head.

"Anchorage of H.M.S. *Conway* in 3½ fathoms, muddy bottom, with the cathedral N. ½ W.; tower of old Panamá Castle, N. by E. ½ E.; a detached rock lying off the Island of Penco, S.S.E. ¼ E. (by compass).

"The flood tide appears to set to the W.N.W., and the ebb to the S.S.E.; but the current was so slight, that the direction of either tide is uncertain. The rise, as got by means of the lead alongside of the ship, appeared to be between 11 and 12 feet, and the time of high water about two hours.

"Supplies of fresh beef and live oxen are obtained here in plenty; from the inconvenient distance that ships lie off the town, watering is rendered somewhat troublesome, and it is seldom done, since such facilities are found at the Island of Taboga, which lies between 8 and 9 miles to the southward, and directly in the way out from this anchorage."

For the next we are favoured by a communication from Mr. Jeffery, R.N.:—

"At the bottom of the bay the land is rather high and irregular; the town lies on a low point under a high bluff, and in making it the first objects to be seen are two white spires. To the right, for a considerable distance, the land is low, and covered with trees; the low land should not be approached nearer than 6 miles. To the southward is a considerable number of islands, they are also covered with trees. Sailing up the bay with a fair wind, keep the islands close on board; and after making the town keep well to the southward, as there are two rocks, one bearing E. by S. (*true*), 3 miles from the cathedral, and the other E. ½ S., 2½ miles from the town. The difficulty of obtaining water is very great, it being on the Island of Taboga, 8 miles from the town; but you may anchor close in under the island in 8 fathoms, abreast a small village, nearly hid in cocoa-nut trees, with a small church to the right; here is an excellent stream, and you can obtain almost every tropical fruit, and plenty of yams. The flood tide sets right into the bay, and very strong; the ebb sets from the town, towards Ileñao, and out between Taboga and the Isla del Rey; at neap it rises 10 or 12 feet, and considerably more at springs."

According to the observations of Capt. Sir Edw. Belcher, in H.M.S. *Sulphur*, the N.E. bastion of Panamá is in lat. 8° 56′ 56″ N., lon. 79° 31′ 12″ W. The variation, in 1837, was 7° 15′ E. High water at full and change at 3ʰ 23′; greatest rise 22 feet.

Otoque, an island in the West part of Panamá Bay, is, or was, inhabited, and produced some provisions, as hogs, fowls, and plantains. It lies 14 miles South of the more important island of Taboga.

The S.W. point of Panamá Bay is *Point Mala*, in lat. 7° 25′, lon. 80° 2′; we have no particulars of it. The land here forms a peninsula, the West point of which is Cape Mariato. Between them is the *Morro de Puercos*, a high round hill. According to Dampier (vol. i. p. 211), there are on this coast many rivers and creeks, but none so large as those on the South side of the bay. It is a coast that is partly mountainous, partly low land, and very thick woods bordering on the sea; but a few leagues within land it consists mostly of savannahs, which are stocked with bulls and cows. The rivers on this side are not wholly destitute of gold, though not so rich as the rivers on the other side of the bay. The coast is but thinly inhabited, for except the rivers that lead up to the towns of Nata and Lavelia, I know of no other settlement between Panamá and Puebla Nueva.

Montijo Bay and the Islands of *Cebaco* lie between Punta Marrato and Quibo Island, but we have no particulars of them.

QUIBO ISLAND has been surveyed by Lieut.-Com. J. Wood, R.N. (1848). From his plan it appears that the island has the form of an irregular crescent, 19 miles in length, North to South, or from lat. 7° 13′ 15″ N. to 7° 32′ 15″ N. Its breadth is from 5 to 8 miles on an average; the convexity facing the West.

The principal anchorage, *Damas Bay*, is on the S.E. side; on its southern part rocky shoals extend for nearly a mile, and off the mouth of the little river San Juan, which falls into the bottom of the bay, there are some sandy flats. The watering place is in the North part of the bay, which is about 7 miles wide between its outer points.

Off its S.W. side are the smaller islands of *Hicaron* (Quicara of Bauza and Malaspina, and all others) and *Hicarita*; together they are about 5 miles in length, North and South. Hicaron is irregular in its surface, the highest hill, on the East side, is 830 feet high. Its N.E. point is called *David Point*; and off its N.W. point some rocks and rocky shoals extend for three-quarters of a mile in all directions, and should be avoided. Colnett says that the smallest island is entirely covered with cocoa trees; and the larger one bears an equal appearance of leafy verdure, but very few of the trees which produce it are of the cocoa kind.

The channel between Hicaron and Quibo, 4 miles in breadth, is clear, the depth being irregular, from 6 and 7 to 20 fathoms; but to the eastward, off the South end of Quibo, are some outlying rocks, which must be very dangerous to ships attempting to pass between the two islands. The most to be avoided is the *Hill Rock*, a very small detached knoll of only 6 feet, with 9 to 15 fathoms close around it. It bears about *true* East 5½ miles from David Point, and 2 miles South from Barca Island, a small islet half a mile off the shore of Quibo.

To the eastward of this latter, the shoal water extends for some distance off shore, as far as and around *Negada Point*, the S.E. point of Quibo.

Off the N.E. point are several smaller islands, the largest of which, *Rancheria*, is 1½ miles long, and the same distance off the shore of Quibo. At 4½ miles beyond this, or to the N.E., are the Islands of *Afuera* and *Afuerita*,* the channel between being quite clear, from 40 to 70 fathoms depth. All the West coast of the island, round to Hicaron Island, appears bold-to, and free from any outlying danger.

In former times, when the system of reprisals against the Spaniards was so vigorously pursued by the buccaneers and ships of war sent by England, as related by Dampier, Woodes Rogers, Lord Anson, and others, Quibo was a point of very considerable importance, as affording means of shelter, and also water, near to the principal field of action against the Spanish galleons. In the account of Commodore Anson's voyage, the whole island is described to be of a very moderate height, excepting one part of it (near the N.E. end), and its surface covered with a continual wood, which preserves its verdure all the year round. Tigers, deer, venomous snakes, monkeys, and iguanas exist upon the island, a statement repeated by Capt. Colnett (1794). In the surrounding sea, alligators, sharks, sea-snakes, and the gigantic ray abound. Pearl oysters, which attracted the pearl fishers from Panamá, were also to be gathered from the surrounding rocks, and the huts of these men and the heaps of shells still existed at Colnett's visit. On the N.E. part of the island, Anson describes a cascade of very great natural beauty, a river of clear water, about 40 yards wide, rolling down a rocky declivity of near 150 in length.

Capt. Colnett, who was here, as before stated, in February, 1794, anchored in Damas Bay (Port de Dames) in 19 fathoms, the North point of the bay in a line with the North point of Cebaco Isle, bearing N.N.E., the watering place N.W., and the South point of Quibo S.E. by S. He says:—"Quibo is the most commodious place for cruisers of any I had seen in these seas, as all parts of it furnish plenty of wood and water. The rivulet from whence we collected our stock was about 12 feet in breadth, and we might have got timber for any purpose for which it could have been wanted. There are trees of the cedar kind a sufficient size to form masts of a ship of the first rate, and of the quality which the Spaniards, in their dockyards, use for every purpose of ship building, making masts, &c. A vessel may lay so near the shore as to haul off its water; but the time of anchoring must be considered, as the flats run off a long way, and it is possible to be deceived in the distance. The high water, by my calculation, is at half-past three o'clock: at full and change the flood comes from the North, and returns the same way, flowing 7 hours, and ebbing 5, and the perpendicular rise of the tide 2 fathoms.

"It would not be advisable for men-of-war and armed vessels, acting upon the offensive or defensive, to anchor far in, as the wind throughout the day blows fresh from the eastward, and right on shore, so that an enemy would have a very great advantage over ships in such a situation. There is good anchorage throughout the bay, at 5 or 6 miles' distance, 33 and 35 fathoms, with a mud bottom, and firm holding ground.

* These islands are called Canales and Cantarras, by Dampier, vol. i. p. 213. He says that Rancheria has plenty of Palma Maria trees on it; which wood, on account of its toughness, length, and twisted grain, is greatly esteemed for making masts.

"The most commanding look-out is at the top of Quicara; we saw it over Quibo (which is low and flat) whilst we lay at anchor; and is, I presume, the remarkable mountain which Lord Anson mistook for part of Quibo, as mentioned in his voyage. Indeed, a good look-out on the top of this island may be necessary for many obvious reasons, as it commands the whole coast and bay."*

To the northward of Quibo, on the main land, is the harbour of *Bahia Honda*, in lat. 7° 45′ N., lon. 1° 30′ W., and 24 miles farther to the N.N.W. is *Port Puebla Nueva*, in lat. 8° 5′ N., and lon. 81° 41′ W.†

PUEBLA NUEVA, or SANTIAGO.—The river takes its name from a small village situated on the River Santiago, where the Spaniards probably first appointed the seat of government. The port is formed by a neck or island about 3 miles in length, which affords good anchorage for vessels of any class. Three larger streams discharge themselves into the main basin, at the western end of the island, where the apparent great entrance is situated, but so studded with rocks and shoals as to be unnavigable for anything larger than boats. It is, in fact, an extensive archipelago, as most of the region towards the Chirique territory will be found on a future examination.

A plan was made, which will prove interesting to those who may visit this port for refuge or refit, but water cannot be procured in any quantity; it may probably be found by digging wells. The natives were timid, and little disposed to trade. Their principal article is sarsaparilla, said to be of superior quality from this place. The stream runs fresh at some miles up. Sugar cane of good quality was offered, and tortoiseshell may be procured in the season.‡

BAHIA HONDA is a most capacious, safe, and convenient harbour, completely land-locked, and perfectly adapted for refit, heaving out, &c., there being no tide or current. Water was in abundance at the beach, and nothing wanting but a town and civilization to render it a favourite resort; timber of every kind, and the best abundance. The islands at its entrance are beautifully adapted for defence, with but trivial labour.§

Montuosa Island lies 50 miles S.W. of Puebla Nueva. "It rises to a considerable height, and is 5 or 6 miles in circumference, its summit covered with trees; the greater part are those which bear the cocoa-nut, which gives it a very pleasant appearance; but islets and breakers extend off its East and West ends, to the distance of 3 or 4 miles. The bottom is rocky on the South side, as is the shore near the sea. There is a beach of sand behind some little creeks that run in between the rocks, which makes a safe landing for boats. Here we went on shore, and got a quantity of cocoa-nuts, with a few birds. The Spaniards or Indians had lately been there to fish on the reef for pearls, and had left great heaps of oyster shells. There were a great plenty of parrots, doves, and iguanas; and it is probable that other refreshments might be obtained, of which we are

* Colnett's Voyage, pp. 131—137.

† Dampier, when at Quibo, June, 1685, sent 150 men to take Puebla Nueva, to get provisions. "It was in going to take this town that Capt. Sawkins was killed, in 1680, who was succeeded by Sharp. Our men took the town with much ease, although there was more strength of men than when Capt. Sawkins was killed."—*Dampier*, vol. i. p. 213. However, they got no provisions.

‡ Sir Edward Belcher's Voyage of the *Sulphur*, vol. i. pp. 240-50.

§ Voyage of the *Sulphur*, vol. i. p. 251.

ignorant. At all events, it may be useful to whalers or cruisers, by offering a place where the sick may be landed and cocoa-nuts procured, whose milk will supply the want of water."*

Between Montuosa and Burica Point, the S.E. limit of the Gulf of Dulce, are the *Ladrone Islands* (Zedzones of Colnett), consisting of small barren rocks. Besides these, in the offing there are several islands on the coast between Point Mala and Point Burica, which are all covered with trees. The coast, also, in this interval, is high and covered with trees, and is tolerably well delineated in the charts, excepting a considerable error in the longitudes.†

CHAPTER VIII.

THE COAST OF CENTRAL AMERICA, OR GUATEMALA.

The country whose southern coast is described in this chapter, includes that long, narrow, and irregular tract which forms the junction between the northern and southern continents of America. In a political sense, the divisions between the states on either side of it are, to the South, the River Escudo de Veragua, which falls into the Caribbean Sea, opposite the island of the same name,‡ separating it from the republic of New Granada, lat. 9° N., lon. 81° 20′ W.; and on the N.W., from that of Mexico by the Rio Sintalapa, falling into the Pacific in lon. 93° 20′ W.

This territory, including an area of from 120,000 to 196,000 square miles, is divided into the five republican states of Guatemala, Honduras, Nicaragua, San Salvador, and Costa Rica. The Federal District, which is common to them, is a circle round the capital, San Salvador, 20 miles in diameter, with a further extension of 10 miles to the South, so as to include its port, the roadstead of Libertad, on the Pacific.

Guatemala, according to Don Domingo Juarros, derives its name from the word Quanhtemali (which in the Mexican language means a decayed log of wood), because the Mexican Indians who accompanied Alvarado found, near the court of the kings of Kachiquel, an old worm-eaten tree, and gave this name to the capital, and afterwards it was applied by the Spaniards to the country. Some have derived it from U-hate-z-mal-ha, which, in the Tzendal language, means "a mountain that throws out water," doubtless alluding to the Volcan de Agua, near the city.

The jurisdiction of the royal chancery of Guatemala extended from the bar of the River Parredon, in the province of Soconusco, to the mouth of the Boruca, in Costa Rica.§

* Colnett, p. 180. † Mr. Jeffery, R.N. ‡ Galindo, Jour. R. Geo. Soc., vol. vi. p. 131.
§ Juarros, Lieut. Bailey's Translation, p. 7.

The N.E. coast of Guatemala, that is, the West India part, was discovered by Columbus in 1502. The greater portion of it was usurped by the Spaniards by 1524, and it was erected into a captain-generalship by the Emperor Charles V. in 1527. From the fact of its being only a minor state, its expenditure was on a less magnificent scale, and consequently comparative benefit accrued to the people. On the overthrow of the Spanish power, Guatemala became independent in 1821, and was subsequently incorporated with Mexico; but when Iturbide fell, it separated, and declared its independence on July 1st, 1823, adopting a constitution drawn up for it by Mr. Livingston, the U.S. statesman. Affairs were, however, far from settled, and much internal commotion continued; but all Spanish influence was thrown over at Omoa, September 12th, 1832. Notwithstanding its very great geographical importance, the resources of the country have hitherto been very imperfectly developed: and, according to a consular statement for 1843-44, "The foreign import trade of the state of Nicaragua, and of the republics in general, has been greatly affected by the continuance of the internal commotions; nevertheless, owing to the great competition of speculators, the consumption of foreign manufactures increased. Indigo and cochineal suffered, in consequence of drought and civil war. Agriculture was making some progress in Costa Rica, and its superior coffee was becoming an article of export to some extent. Activity, also, was visible in the mining districts of Costa Rica, aided by English exertions."*

Of the interior, as of its Pacific sea-coasts, our knowledge is very scanty and unsatisfactory; and there are few parts of the globe accessible to our ships and commerce of which we are so ignorant. Some detached portions of its southern side have been accurately surveyed by Sir Edward Belcher, in H.M.S. *Blossom*, and we extract many isolated facts, bearing upon the subject of this present work, from different authors and travellers; but we possess no continuous nautical description, nor does it seem possible, at present, to draw up one.†

Mountains.—In describing the general physical features of the country, these naturally become the first in order, influencing as they do the rest of its surface.

The ensuing description is that by Colonel Galindo, in the Geographical Journal, vol. vi. pp. 122-3.—The elevated range (a continuance of the Andes) in Central America has no determined name, and is in many parts without a visible existence. It commences in Costa Rica, at a distance from the Pacific of about one-fourth of the whole breadth of the isthmus, and, at the beginning of this course, separates this state from Veragua; in Nicaragua it inclines close to the

* Commercial Tariffs, &c., 1847, vol. xix. p. 330.

† The principal authorities for any description of Central America are the works of Padre Thomas Gage, an English friar, 1632, an excellent and interesting work; that of Don Domingo Juarros, a native of Guatemala, in 1780, which has been translated by Lieut. Bailey, R.M., 1823; Thomson's Visit to Guatemala in 1825, gives an excellent account of much of the interior; Reise naar Guatemala, 1829, by J. Haefkens, and a work by the same author, Centraal Amerika, 1832, both useful; Narratives, &c., by Mr. Roberts, chiefly on the Atlantic side; a paper in the Geographical Journal, vol. vi., 1836, on Costarrica, by Colonel Don Juan Galindo (an Englishman); Diccionario de las Indias Occidentales, by Col. Don A. de Alcedo; L'Isthme de Panamá, &c., by M. Michel Chevallier; and also the important work of Capt. Sir Edward Belcher, the Voyage of the *Sulphur*. Upon its antiquities and general information, the works of Dupaix, Waldeck, Kingsborough, Rouchaud, and Dumartray, may be consulted. One of the most interesting is that by Mr. Stephens, who describes the ruins in Yucatan, but who passed through portions of the other republics.

borders of the Pacific, leaving the lakes on the East; in Honduras it returns towards the Atlantic, leaving the whole state of Salvador on the South; traversing Guatemala, the new city and Chimaltenango stand on the top of the ridge, which now becomes more elevated as it approaches Mexico, and, branching into various groups, forms, in the western part of the state, that region which is demonstrated the highlands. The population on the Pacific side of the chain is much greater in proportion to its extent than on the Atlantic slope.

The chain is apparently interrupted in its course through Central America by the transversal valleys containing the Lake of Nicaragua and the plain of Comayagua, but still the elevation between the two oceans is considerable, and will be more dwelt upon when we describe the proposed canals, which would render Central America of very great importance in the commercial world, should they ever be carried into execution.

The *Lakes* of Nicaragua and of Leon, or Managua, are among the most important features of the country. On the Pacific side, the rivers which are met with rarely have their sources above 60 miles from the sea. The Lempa is the principal, but is not navigable. The next in size is the Rio Choluteca, falling into the Bay of Conchagua.

Although not eminently possessed of good harbours, yet it is still superior to Mexico in this respect. The principal on the Pacific coast are, Realejo, Calderas, La Union, Libertad, Acajutla, and Istapá.

Volcanic phenomena are frequent, and their devastating effects have been, at times, very severe. The principal volcanoes now, or recently, in activity, are those of Coseguina, Isalco, de Agua, and de Fuego, and many others; of these the Volcan de Agua is the loftiest, being differently stated as 14,895 or 12,620 feet above the Pacific.

The *productions* of Central America, before alluded to, are important. The Tisingal gold mines, near the Chirique Lagoon, on the Atlantic side of Costa Rica, have afforded as much riches as those of Potosi; but the vegetable productions are of greater importance than the mineral. Of cultivated articles, cocoa, indigo, coffee, sugar, and cotton, are the most prominent. These crops vary with the height of the country. At a lower elevation than 3,000 feet, indigo, cocoa, sugar, and cotton are grown. Cocoa is chiefly grown along the shores of the Pacific, and that of Soconusco was esteemed by the Spaniards to be the best furnished by their American possessions. Indigo is grown in the Federal District; but is general throughout the country. Cochineal, or the nopal cactus, is cultivated between the heights of 3,000 and 5,000 feet, particularly in the neighbourhood of Guatemala. Of native woods, &c., abundance is produced, but principally refer to West India trade.*

The TRADE on the coast of Central America, which is almost exclusively British, is increasing rapidly. At Puntas Arenas, in the Gulf of Nicoya, excellent *coffee* is exported, giving occupation to between fourteen and twenty vessels, and is the best coffee in the Pacific; at Realejo, *dye-wood* (Brasil-wood), &c.; at La

* "This country is so pleasing to the eye, and abounding in all things necessary, that the Spaniards call it Mahomet's Paradise."—*Gage*, 1650, p. 165. An interesting account of the author's twelve years' residence in it will be found in his curious and valuable book.

Union (Gulf of Fonseca), *indigo*, &c.; at Sonsonate, *indigo*; and at Istapá, *cochineal*.

Provisions of every description are very cheap, and the inhabitants obliging; the European residents exceedingly so.*

BRAZIL-WOOD grows in several localities in the vicinity of the great lakes. That of most importance to the Pacific is the district from the Gulf of Papagayo to Point San Andres, between San Juan and Point Desolada.

It is also cut on the N.E. side of the lakes Nicaragua and Managua. There is also a tract N.E. of the latter, and a more extensive district on the borders of the Mosquito territory. The produce of these districts is carried in large canoes down the Rio San Juan to the town of San Juan de Nicaragua, a journey generally performed in four or five days.

Climate.—The whole of Central America is situated between the tropics; but the temperature and salubrity of its climate are as variable as are the diversities of its abrupt elevations, mountains, plateaux, ravines, sands, low districts, lakes, and forests.

It freezes sometimes during the night on the highest part to the table lands, in November, December, and January. At the city of Guatemala, situated in the mean height of the table land (4,961 feet above the sea), the dry season begins towards the close of the month of October, and lasts till the end of May: during which time only a few showers occasionally fall. In the beginning of June thunder-storms become frequent, and are followed by heavy rains. From six o'clock in the morning till three or four o'clock in the afternoon, the sky is generally without clouds, and the air clear and refreshing. About the middle of October the North winds blow and the rains cease. The absence of either the windy or rainy seasons is accompanied by thunder; and, it is said, with slight shocks of earthquake. In March and April the thermometer sometimes rises to 86°. It generally ranges between 74° and 82° in the middle of the day. In December and January, when the North winds sometimes blow with great force, the thermometer varies between 68° and 72°. During the summer heat it rises at about seven o'clock in the morning to between 60° and 67°, and in the evening at the same hour, to 67° and 68°; in winter it falls in the morning to 60° and 58°, and sometimes even to 56°, but in the evening only to between 60° and 64°. Towards the end of the dry season the trees shed their leaves, and in many places vegetation appears suspended. The region in which the capital stands is considered healthy; *goîtres* are frequent in the high and mountain districts, especially among the mixed races.

On the sea-coast of the Pacific, the seasons correspond with those of the table lands, but the temperature is much hotter. It is said that the Pacific shores are healthy, although they are almost entirely covered with woods. This salubrity is, however, not without exceptional districts.

On the coast, during the fine season, which commences in November and ends in May, the land and sea breezes blow alternately, with a clear sky and but little rain; strong winds rarely occur during this period, except at the fall and

* Note by Capt. Worth, R.N., 1847.

change of the moon, when occasionally a strong breeze from the northward may be experienced.

In the rainy season, May to November, heavy rains, calms, light variable breezes, with a close sultry atmosphere, heavy squalls, with thunder and lightning, and not unfrequently strong gales from the S.W., are prevalent.

"On the coast of Central America the *currents* are variable, but almost always setting to the S.E., sometimes rather strong. The land wind never blows far off shore, and, except in the harbours, is not certain; the sea breeze is seldom felt, but there are of course exceptions, as when we left Realejo and Punta Arenas, we experienced on both occasions a remarkably strong land breeze.

"The coast has soundings with the hand lead from Cape Blanco to Istapá, and from the information I could gain from trading vessels, and from our own experience, it appears to be perfectly clear, affording anchorage nearly everywhere."—(Capt. Worth, R.N., 1848.)

Population.—The inhabitants of Central America comprise three classes: whites, or creoles of Spanish race; mestizos, or the offspring of whites and Indians; and aboriginal natives. There are but few negroes or Zamboes. In the department of Guatemala the Indian inhabitants are said to constitute the great majority of the people; in Costa Rica those of European race predominate; and in the three other departments, the mestizos, mixed with a few mulattoes, prevail. Haefkens estimates the whole population at 1,500,000, which he distributed as follows, viz.:—of European races, 125,000; mixed races, 500,000; Indians, 875,000: total, 1,500,000. But it is doubtful whether any approximate estimate can be formed.

The GULF of DULCE is the first place that may be noticed proceeding westward from Point Burica, but we know nothing of it. In the background of this part the *Mountain of Chiriqui* rises to the height of 11,266 feet.

Of *Caño Island* and *Port Mantos* we have no account. Dampier says (vol. i. p. 215), that the coast is all low land, overgrown with thick woods, and but few inhabitants near the shore.

The GULF of NICOYA is the next place to be noticed.* It is of considerable commercial importance, because it contains the port known by the name of *Punta de Arenas*, which is the chief port for the coffee trade of this country. This name, and the port itself, have been but little known to the rest of the world, and its locality has been the object of great embarrassment to shipmasters coming here.

Mr. Stephens visited Nicoya during a portion of his travels in these regions. Leaving Guatemala, and sailing along the coast, they passed the volcanoes of San Salvador, San Vincente, San Miguel, Tolega, Momotombo, Managua, Nindiri, Nasaya, and Nicaragua, forming an uninterrupted chain.

Mr. Stephens remarks: "This coast has well been described as bristling with volcanic cones. For two days we lay with sails flapping in sight of Cape Blanco,

* Very soon after the discovery of the South Sea, the Spaniards in pursuit of plunder increased their knowledge of the coast of this sea; for in 1516 Hernan Ponce de Leon sailed from Panamá and discovered the port, which he named San Lucar, but afterwards Nicoya, from the cacique then dominant over this portion of the country.—*Admiral Burney's Collection*, vol. i. p. 10.

the upper headland of the Gulf of Nicoya. On the afternoon of the 31st we entered the gulf. In a line with a point of the cape was an island of rock, with high, bare, and precipitous sides, and the top covered with verdure. It was about sunset; for nearly an hour the sky and sea seemed blazing with the reflection of the departing luminary, and the islands of rock seemed like a fortress with turrets. It was a glorious farewell view. I passed my last night on the Pacific, with the highlands of the Gulf of Nicoya close around us.

"Early in the morning we had the tide in our favour, and very soon leaving the main body of the gulf, turned off to the right, and entered a beautiful little cove, forming the harbour of Caldera. In front was the range of mountains of Aguacate; on the left the old port of Puntas Arenas; and on the right the volcano of San Pablo. On the shore was a long low house, set upon piles, with a tile roof, and near it were three or four thatched huts and two canoes. We anchored in front of the houses, and apparently without exciting the attention of a soul on shore."

He says that, "All the ports of Central America on the Pacific are unhealthy—but this was considered deadly. I had entered, without apprehension, cities where this plague was raging, but here, as I looked ashore, there was a death-like stillness that was startling."

From Caldera the country inland is level, rich, and uncultivated, with here and there a wretched *hacienda*, the owners of which live in the towns. Herds are stationed on the estates, from time to time, to gather and number the cattle, which roam wild in the woods.

Cape Blanco is the outermost point of this gulf, and it is thus described by Dampier:—"Cape Blanco is so called from two white rocks lying off it. When we are off at sea right against the cape, they appear as part of the cape; but being near the shore, either to the eastward or westward of the cape, they appear like two ships under sail at first view, but coming nearer, they are like two high towers; they being small, high, steep on all sides; and they are about half a mile from the cape. This cape is about the height of Beachy Head in England, on the coast of Sussex. It is a full point, with steep rocks to the sea. The top of it is flat and even for about a mile: then it gradually falls away on each side with a gentle descent. It appears very pleasant, being covered with great lofty trees.

"From the cape on the N.W. side the land runs in N.E. for about 4 leagues, making a small bay called by the Spaniards *Caldera*. A league within Cape Blanco, on the N.W. side of it, and at the entrance of this bay, there is a brook of very good water running into the sea. Here the land is very low, making a saddling between two small hills. It is very rich land, producing large tall trees of many sorts; the mould is black and deep, which I have always taken notice of to be a fat soil. About a mile from this brook, towards the N.E., the wood land terminates. Here the savannah land begins, and runs some leagues into the country, making many small hills and dales. These savannahs are not altogether clear of trees, but are here and there sprinkled with small groves, which render them very delightful. The grass which grows here is very kindly, thick and long; I have seen none better in the West Indies.

"Toward the bottom of the bay, the land by the sea is low and full of mangroves, but farther in the country the land is high and mountainous. The mountains are part wood land and part savannah. The trees in those woods are but small and short; and the mountain savannahs are clothed with but indifferent grass. From the bottom of this bay it is but 14 or 15 leagues to the Lake of Nicaragua, on the North sea-coast. The way between is somewhat mountainous, but most savannah."*

Cape Blanco, according to the survey, is in lat. 9° 34′ N., lon. 85° 7′ W. Capt. Sir Edw. Belcher says, in his account of the Voyage of the *Sulphur* :—†

"March 30, 1837, passed close to the island termed Cape Blanco at its western point. Here we found ourselves obstructed by a point off which the breakers and rocky ledges above water extended a considerable distance to seaward. The soundings were regular, from 25 to 11 and 8½ fathoms, hard sand, in which latter depth we tacked successively within 1½ miles of the shore surf, and an outer roller about half a mile from us on the last tack.....

"Cape Blanco still in sight. A short distance to the westward observed a sandy, sloping bluff, off which a shelf, apparently composed of sand, with conical studded rocks, extended a considerable distance seaward. On a sandy islet near the bluff, two very remarkable ears jutting up."

The Gulf of Nicoya is frequently called by the name of Punta Arenas, which is only one of its ports, as before stated.

Punta Arenas was formerly the port of this gulf in the state of Costa Rica; but interested parties whose property lay near to Calderas, about 5 miles southerly, on the eastern side of the gulf, managed to have the port or custom-house officers, &c., drafted thither. It is very unhealthy, almost fatal, to all new residents; and the highest authorities take care to excuse residence.

Firewood, water, cedar timber, bullocks, and oysters, are to be obtained; the latter on the banks, dry at low water, above Venado, on the western shore; bullocks, either at Arenas, Calderas, or Verugate, on the western shore; water at San Lucas, or better and more easily at Herradura Bay, whence the casks are rolled into a small lake *at the beach*, and vessels may ride safely close to the shore, by veering the whole cable with a warp to the beach. Wood may be cut anywhere by the crew, or more easily purchased at Calderas or Punta Arenas.‡

From Cape Blanco the coast trends, in a north-westerly direction, to *Cape Velas*, in lat. 10° 13′ N. lon. 85° 40′ W.§ It is so called from the rock being sometimes mistaken for a sail.‖

* Dampier, vol. i. pp. 111—113. Father Gage, in his Survey of the Spanish West Indies, says that the Indians in his time (*circa* 1645), "were all like slaves to the Spaniards, and employed by them to make a kind of thread called *pita*, which is a very rich commodity in Spain, especially of that colour wherewith it is dyed in these parts of *Nicoya*, which is purple, for which the Indians are here much charged to work about the sea-shore, and there to find certain shells, wherewith they make this purple dye. There are also shells for other colours, not known to be so plentifully in any other place as here."—P. 437.

† Vol. i. pp. 25-6.

‡ Sir Edward Belcher (January, 1839), vol. i. pp. 248-9.

§ "We lay along a deep bay, and passed some very remarkable rocks or rocky islands, white with green tops, the Port of Matapala bearing S.S.E. Between that and these rocky islands a number of small, high, white rocks shot up, resembling vessels under sail; bearing E., E. by N., and E.N.E., a little bay extending landwards, and called, as I suppose, from these little rocks, 'Puerto Velas.'"—*G. U. Skinner, Esq.*

‖ Sir Edward Belcher, vol. i. p. 185.

The BIGHT of PAPAGAYO may be said to commence here, extending to the northward to Realejo; being but a slight curve in the general line of coast. It is scarcely worthy of the name, but becomes more familiar from the fact of the peculiar winds experienced off it called by the name, which are elsewhere described. Suffice it here to state, that the papagayo is a strong wind, blowing from N.E. to E. by N., with a bright, clear sky overhead, and a glaring sun, with a dense atmosphere.

At Cape Velas Capt. Belcher lost the papagayo; "therefore," he says, "the limits may be included in a line drawn from Cape Desolado to Point Velas, and it is rather a curious phenomenon that the strength of this breeze seldom ranges so far as this chord, but seems to prefer a curve at a distance of 15 or 20 miles from the land."*

GORDA POINT,† according to the chart, lies 18 miles northward of Cape Velas, in lat. 10° 31′ N., lon. 85° 43½′ W. At this point the coast turns abruptly to the E.N.E., towards Port Culebra.

"Off Point Gorda are several high rocks, the two largest, which are close together, are about 2½ or 3 miles from the land, the others lay principally more to the North and N.E.; they are *all* high, and the smaller ones have very much the appearance of upright tombstones; others again, at first sight, appear like a ship under canvas."—(Mr. E. P. Brumell.)

PORT CULEBRA was surveyed by Capt. Sir Edward Belcher in 1838. The spot at which he observed, at the head of the port, he places in lat. 10° 36′ 55″ N., lon. 85° 33′ 30″ W.; variation 7° 5′ 54″ E. The entrance to the port is between the *North* and *South Viradores*, some detached cliffy islands, 1¾ miles apart. Between the South Viradores and *Cacique Point*, to the N.W. of which they lie, there is a channel of 5 to 10 fathoms; but, as a rocky reef runs off a quarter of a mile to the West of the point, and some detached rocks lie South of the Viradores, it should not be used.

COCOS BAY lies to the southward of the South Viradores, and between Cacique Point and *Miga Point*, bearing S.W. by S. from the former; the distance is about 1¼ miles. These points are both rocky cliffs, surmounted by hills. Cocos Bay may be about a mile in depth within the line of opening. The bottom is formed by a sandy beach, off the South part of which a line of rocks runs North about a quarter of a mile, and another small rock lies in its eastern part. It lies entirely open to the N.W. *Sesga Point* lies a mile and a half S.W. from Miga Point, the west extreme of the bay, and midway between is a cliffy islet. To the eastward of Cacique Point is a similar bay to Cocos Bay, having about the same width to *Buena Point*, which forms the South Point of Port Culebra, which extends nearly 4 miles within the two entrance points, Buena and Mala, a mile asunder, and is about two miles wide, the depth even, 6 to 18 fathoms, and anchorage everywhere.

The following is Capt. Sir E. Belcher's description of the port:—

"Port Culebra is certainly magnificent; and from information by the natives,

* Sir Edward Belcher, vol. i. p. 185.

† Point Catalina (of Bauza); from the disjointed portions or islands, it might have caused that of Murciellagos to be mistaken for it.—*Belcher*, vol. i. p. 185.

it is connected with Salinas, and thence to Nicaragua, Granada, &c. If any railroad is contemplated in this quarter, it ought to enter at the Bay of Salinas, which would render these two ports important. (Capt. Sir E. Belcher adds, that he has little doubt but that it will become the chief port of Nicaragua.)

"Water fit for consumption was not found at the beach, but may be obtained a short distance up the creek, which a boat may enter at high water. If wells were dug, doubtless it would be found at the N.W. side, as the surrounding country is mountainous. Another symptom in favour of this is the thickly-wooded sides and summits, as well as bright green spots of vegetation throughout the bay.

"Brazil-wood is very abundant; mahogany and cedar were observed near the beach, but as they have been employed cutting the Brazil, probably all the cedar and mahogany, easily attainable, has been taken. Timber, in great variety, abounded. In the bay, where H.M.S. *Starling* was at anchor, there was a large village, where the natives were anxious to dispose of their productions, consisting of fruit, stock, cattle, &c."*

POINT ST. ELENA is 23 miles North of Gorda Point, and is in lat. 10° 55′, lon. 85° 46′.

Point St. Elena is a remarkable cape, and to the South of it are the ***Murciellagos*** or ***Bat Islands***, eight in number, almost forming two distinct harbours, the smaller islands making a crescent by the South, one large island protecting the East, and another of similar size forming the line of separation. Capt. Belcher anchored in the inner or eastern harbour, and completed his water at a very convenient position, in 32 fathoms, with a hawser fast to the shore.

The springs are numerous, and there are tolerable rivulets; but only that they watered at (between the centre point and the main) is safe to approach, by reason of the constant surf. The gulf squalls, even in this sheltered position, come down the gullies with great force, and impeded the work as well as endangered the boats. The geological character of the cape and islands is a schistose serpentine, containing balls of noble serpentine.

TOMAS or ST. ELENA BAY is immediately to the northward of St. Elena Point, and is separated from SALINAS BAY by a promontory, of which *Descarte Point* is the West extreme. The N.W. point of Salinas Bay is *Cape Natan*, which is just 12 miles due North (*true*) of St. Elena Point. In Tomas Bay are some detached islets and rocks. These have been named the *Vagares Rocks*, *Juanilla* and *Despensa Islands*. To the S.E. of the first is a small inner bay, which, from the survey, seems to afford anchorage, sheltered from the S.W. winds.

Salinas Island, in Salinas Bay, is placed, by Sir Edward Belcher, in lat. 11° 2′ 50″ N., lon. 85° 40′ 45″ W.

PORT SAN JUAN DEL SUR.—This portion of the coast is interesting, on account of its proximity to the navigable Lake of Nicaragua; but it is for this reason only, as with the exception of the Port of San Juan, called del Sur, to distinguish it from the other San Juan in this state, at the mouth of the River San Juan de Nicaragua, in the Caribbean Sea, it scarcely possesses any harbour or foreign trade, except in dye-wood.

* Vol. i. pp. 180—183.

The South bluff of Port San Juan is in lat. 11° 15′ 12″, lon. 85° 53′. The proposed communication with the Atlantic, by Mr. Baily,* was to terminate here—a canal, 15¾ miles in length, cut across the narrow tract, separating this port from the lake, which, with the Rio San Juan, would form the navigable connexion.

At the present moment, however, a treaty is pending between the United States and England, respecting Central America, which promises fairly to open some rivalry, if carried into effect. It guarantees the protection and free navigation of the projected canal through the lakes, and the neutrality of the country through which it may pass, as also of the sea within a reasonable distance of either termination to the route; both powers to pledge themselves not to exercise any jurisdiction whatever in any part of the territory, and to protect the operations of the company who, under the authority of the state of Nicaragua, shall construct and maintain this ship-canal.

The construction of the canal has been, it is said, undertaken by a commercial company of New York; and, if it is carried through, must prove immensely beneficial and valuable to the country and its projectors.

"The effect of this treaty must be to render the uncultivated and revolutionary states of Central America a prosperous and fertile country, while between the two contracting powers it cannot but prove a bond of peace and a union of interests, the beneficial effects of which will be felt throughout the habitable world."†

This may take place ere long, but it has been so frequently stated that the works were to be commenced at one or other of the proposed points, that, until something more definitive than this takes place, but little speculation can be made as to its completion.

Mr. Stephens says: "Our encampment was about the centre of the harbour, which was the finest I saw in the Pacific. It is not large, but beautifully protected, being almost in the form of the letter U. The arms are high and parallel, running nearly North and South, and terminating in high perpendicular bluffs. As I afterwards learned from Mr. Baily, the water is deep, and under either bluff, according to the wind, vessels of the largest class can ride with perfect safety. Supposing this to be correct, there is but one objection to this harbour, which I derive from Capt. D'Yriaste, with whom I made the voyage from Zonzonate to Caldera. He has been nine years navigating the coast of the Pacific, from Peru to the Gulf of California, and has made valuable notes, which he intends publishing in France, and he told me that during the summer months, from November to May, the strong North winds which sweep over the Lake of Nicaragua, pass with such violence through the Gulf of Papagayo, that during the prevalence of these winds it is almost impossible for a vessel to enter the Port of San Juan. Whether this is true to the extent that Capt. D'Yriaste supposes, and, if true, how far steam-

* Mr. Baily is a British officer, and was employed by the government of Central America to make a survey of this canal route, and had completed all except the survey of an unimportant part of the Rio San Juan (the outlet of the lake into the Caribbean Sea) when the revolution broke out. This not only put a stop to the survey, but annihilated the prospect of remuneration for Mr. Baily's arduous services. In the subsequent notice of these communications Mr. Baily's project will be more fully noticed.

† Times, May 19, 1850.

tugs would answer to bring vessels in against such a wind, is for others to determine. But at the moment there seemed more palpable difficulties.

"The harbour was perfectly desolate, for years not a vessel had entered it; primeval trees grew around it, for miles there was not a habitation; I have walked the shore alone. Since Mr. Baily left, not a person had visited it; and probably the only thing that keeps it alive, even in memory, is the theorising of scientific men, or the occasional visit of some Nicaragua fisherman, who, too lazy to work, seeks his food in the sea. It seemed preposterous to consider it the focus of a great commercial enterprise; to imagine that a city was to rise up out of the forest, the desolate harbour to be filled with ships, and become a great portal for the thoroughfare of nations. But the scene was magnificent. The sun was setting, and the high western headland threw a deep shade over the water. It was, perhaps, the last time in my life that I should see the Pacific, and in spite of fever and ague tendencies, I bathed once more in the great ocean.

"At 7 o'clock we started, recrossed the stream, at which we had procured water, and returned to the first station of Mr. Baily. It was on the river San Juan, 1½ miles from the sea. The river here had sufficient depth of water for large vessels, and from this point Mr. Baily commenced his survey to the Lake of Nicaragua.

"My guide cleared a path for me with his machete; and working our way across the plain, we entered a valley, which ran in the great ravine called Quebrada Grande, between the mountain ranges of Zebadea and El Platina.

"Up to this place manifestly there could be no difficulty in cutting a canal; beyond the line of survey follows the small stream of El Cacao for another league, when it crossed the mountain; but there was such a rank growth of young trees, that it was impossible to continue without sending men forward to clear the way. We therefore left the line of the canal, and crossing the valley to the right, reached the foot of the mountain over which the road to Nicaragua passes."—(Incidents of Travel, &c.)

Port Nacascolo, Naguiscolo, or *Playa Hermosa*, lies almost adjoining to, and to the N.W. of, Port San Juan, which it somewhat resembles, and, like it, is only the resort of a few natives occasionally. There is no village or town near it, and it never has been resorted to for general European commerce. In its S.E. portion is a sort of canal, excavated for a short distance to facilitate and shorten the transit of the local trade to the town and lake of Nicaragua, to which there is a road or pathway through the forest.

Northward of San Juan del Sur the coast trends nearly straight in a due N.W. direction. As was stated in a former page, the district on the coast produces dye-wood, or Brazil-wood, for which its ports are much frequented. From information received by Capt. Eden, H.M.S. *Conway*, in 1835, the coast between Brito and San Andres was then much resorted to by vessels to load that article. The landing at some of the places is rather difficult; but the anchorage is perfectly safe, particularly from November till May. The winds are then constantly from the N.E., though they sometimes blow very strong: but the sea breezes during those months never reach the coast. Between November, 1834, and May, 1835, about 60,000 quintals of Brazil-wood were embarked on board

British vessels, and they continued their loading during the whole of the winter months.*

Of the ports on this part of the coast we know little more than their names, for on the recent edition of the Admiralty charts there is not one of them inserted.

Brito is the first point North of San Juan. Its distance from it is not indicated.

Then follows *Mogote*, an open anchorage; next *Casares*, off the mouth of a river between some reefs. This, by the road, is 7 leagues from the town of Ximotepe, and which is 12 leagues from Managua on the lake. Three leagues further along the coast is the road of *Masapa*, and 5½ leagues further is the anchorage of *Masachapa*, to the southward of *Point San Andres*. Here the Brazil-wood district terminates.†

Capt. Sir Edward Belcher, in passing along to the North, began to experience gusts from the Lake of Managua, *no high land intervening in its course*, causing him to go under treble-reefed topsails, &c.‡

The next point is *Cape Desolada*, a most appropriate name; it seems almost in mockery that one or two stunted shrubs are allowed to stand on its summit.§ Mahogany and cedar grow in the vicinity of the cape, and to the North of it is *Tamarindo*.

The LAKE of NICARAGUA (or Grenada) is a fine sheet of water, and, according to Mr. Baily's account of it, is 90 miles long, its greatest breadth is 40, and the mean 20 miles. The depth of water is variable, being in some places close to the shore, and in others half a mile from it, 2 fathoms, increasing gradually to 8, 10, 12, and 15 fathoms, the bottom usually mud (according to Capt. A. G——, quoted in a work on this subject by Prince Louis Napoleon, his soundings gave a depth of 45 fathoms in the centre). The level of the lake is 128 feet 3 inches above that of the Pacific Ocean at low water, spring tides.

This basin is the receptacle of the waters from a tract of country 6 to 10 leagues in breadth on each side of it, thrown in by numerous streams and rivers, none of them navigable except the River Frio, having its source far away in the mountains of Costa Rica, which discharges into the lake a large quantity of water near the spot where the River San Juan flows out of it. The embouchure is 200 yards wide, and nearly 2 fathoms deep. There are several islands and groups of islets in different parts of the lake, but none of them embarrass the navigation, nor is this anywhere incommoded by shoals or banks, other than the shallow water in shore; and even this is but very trifling, or rather it is no impediment at all to the craft at present in use, the practice being to keep the shore close aboard, for the purpose of choosing convenient stopping places at the close of day, as they scarcely ever continue their voyage during the night.

The largest islands on the lake are Omotepe, Madera, and Zapatera. Taken together, the first two of these islands are 12 miles long, and have gigantic volcanoes on them. Zapatera is almost triangular, and 5 miles long. Sonate, Solentiname, and Zapote, are smaller, and uninhabited, but some of them, and the last in particular, are capable of cultivation.

* Communicated by G. U. Skinner, Esq.
† G. U. Skinner, Esq.
‡ Sir E. Belcher, vol. i. p. 180.
§ *Ibid.* vol. i. p. 27.

Near the town of Grenada there is the best anchorage for ships of the largest dimensions.

The Lake of Nicaragua is connected with that of Leon by means of the river Panaloya (or Tipitapa), navigable for the boats employed in that country for 12 miles, as far as the place called Pasquiel, where the inhabitants go to cut and bring away Brazilian timber. The 4 miles which remain between that place and the Lake of Leon are not navigable by any kind of boat, whatever may be its construction, because, beyond Pasqueil, the channel is obstructed by a vein of rocks, which, when the river is swollen, are covered with water; but in the dry season, the water sinks so low that it can only escape through gradually diminishing fissures in the rocks. At a distance of a mile beyond this first vein of rocks, we find another more solid, which, crossing the river at right angles, forms a cascade of 13 feet descent.

The River Tipitapa, which discharges itself into the Lake of Nicaragua, is the only outlet for the Lake Leon. The lands bordering this river are somewhat low, but fertile, having excellent pasturage; as at Chontales, they are divided into grazing and breeding farms. All this country, covered with Brazilian timber, is scantily inhabited. The only village is that of Tipitapa, situated near the above-mentioned waterfall. It contains a small church, and about 100 cottages. The river is crossed by a wooden bridge.

The Lake of Leon or Managua is from 32 to 35 miles long, and 16 miles at its greatest width. It receives from the circumjacent lands, chiefly from the eastern coast, a number of small streams. According to Mr. Lawrence, of H.M.S. *Thunderer*, it is not so deep as that of Nicaragua; but, according to Capt. A. G——, it is still deeper.

The Lake of Managua is 28 feet 3 inches above that of Nicaragua; and, according to M. Garella, the difference between *high* water in the Pacific and *low* water in the Atlantic is 19½ feet. In the proposition for making use of these lakes, it is stated that the ground is perfectly level between the head and Realejo, one of the best ports on the coast; but the distance is 60 miles, and to Mr. Stephens the difficulties seemed to be insuperable. Capt. Sir Edward Belcher is of opinion that there is no insurmountable obstacle to connecting the Lake of Managua with the navigable stream, the Estero Real, falling into the Gulf of Fonseca, as is hereafter mentioned.

The principal noticeable points on the shores of the Lake of Nicaragua are the city of Nicaragua and the Omotepeque Volcano, 5,040 feet above the sea. Mr. Stephens says it reminded him of Mount Etna, rising, like the pride of Sicily, from the water's edge, a smooth unbroken cone to the above height.

Cape Desolada, on the Pacific, is nearly 100 miles along the coast, N.W. of the Port of San Juan. The city of Leon lies inland from Cape Desolada.

Leon is the capital of the state of Nicaragua; it was formerly a place of importance, with a population of 32,000 souls, but has been since greatly reduced by anarchy and other distracting circumstances. It is situated on a plain about 40 miles from Realejo, 10 from the sea, and 15 from the Lake of Managua. It has a university, cathedral, and 8 large churches, and other public institutions. It carries on some trade through Realejo. The houses are described by Mr. Roberts as very similar to those of Guatemala, none being above two stories high. The population in 1820 was about 14,000.*

The *Plain of Leon* is bounded on the Pacific side by a low ridge, and on the right by high mountains, part of the chain of the Cordilleras. Mr. Stephens says:—

"Before us at a great distance, rising above the level of the plain, we saw the spires of the cathedral of Leon. This magnificent plain, which in richness of soil is not surpassed by any land in the world, lay as desolate as when the Spaniards first traversed it. The dry season was near its close; for four months there had been no rain, and the dust hung around us in thick clouds, hot and fine as the sands of Egypt. Leon had an appearance of old and aristocratic respectability, which no other city in Central America possessed. The houses were large, and many of the fronts were full of stucco ornaments; the plaza was spacious, and the squares of the churches and the churches themselves magnificent. It was under Spain a bishop's see, and distinguished for the costliness of its churches and convents, its seats of learning, and its men of science, to the time of its revolution.

"In walking through its streets I saw palaces in which nobles had lived dismantled and roofless, and occupied by half-starved wretches, pictures of misery and want, and on one side an immense field of ruins covering half the city."

REALEJO is the next place in proceeding north-westward, and is one of the most important ports on the coast, and has in consequence been more frequently visited and described. It has been, moreover, minutely surveyed by Capt. Sir Edward Belcher, in the *Sulphur*, in 1838.

The port is formed by the three islands of *Castaňon*, separating the Estero Doña Paula† from the Pacific on the South, *Cardon Island* in front of it, and forming two entrances, and the larger island of *Asseradores* (Sawyers) to the northward.

Sir Edward Belcher thus concisely describes the harbour:—

"Cardon, at the mouth of the Port of Realejo, is situated in lat. 12° 27′ 55″ N., and lon. 87° 9′ 30″ W. It has two entrances, both of which are safe, under proper precaution, in all weather. The depth varies from 2 to 7 fathoms, and safe anchorage extends for several miles; the rise and fall of tide 11 feet, full and change 3^h 6′. Docks or slips, therefore, may easily be constructed, and timber is readily to be procured of any dimensions; wood, water, and immediate necessaries and luxuries, are plentiful and cheap. The village of Realejo is about 9 miles from the sea, and its population is about 1,000 souls.

* The city of Leon is lauded by Father Gage in his interesting work as the pleasantest place in all America, and calls it the "Paradise of the Indies." Dampier was here in 1685, and his men marched up to it to take it, and they set it on fire, but did not procure much plunder.—The way to it, he says, is plain and even, through a champion (champagne) country, of long grassy savannahs, and spots of high woods. About 5 miles from the landing place there is a sugar work, 3 miles further there is another, and 2 miles beyond that there is a fine river to ford, which is not very deep, besides which there is no water all the way till you come to an Indian town, which is 2 miles before you come to the city, and from thence it is a pleasant straight sandy way to Leon.—*Dampier*, vol. i. p. 218.

† Capt. Sir E. Belcher states, in his appendix, that the Estero (or Creek) of Doña Paula takes a course toward the city of Leon, and is navigable to within 3 leagues of that city. It has been suggested to carry a railroad from Leon to the Lake of Nicaragua. As to any canal into the Pacific, unless behind Monotombo Telica and Viejo Range into the Estero Real, Capt. Belcher saw little feasibility in the scheme.

The principal occupation of the working males is on the water, loading and unloading vessels. It has a custom-house and officers under a collector, comptroller, and captain of the port;"* (and, it may now be added, an English vice-consul.)

The Island of Cardon is of volcanic origin, and the beach contains so much iron that the sand, which probably is washed up, caused the magnetic needle to vibrate 21° from zero. On the West end they found a mark, probably left by the *Conway* a few years back.†

The town was founded by some of the companions of Don P. Alvarado, during his descent on the territory. Its situation, near the sea, exposed it to the depredations of the buccaneers; and in consequence the inhabitants returned inland, and founded the city of Leon; but even this was not free from their attacks. Juarros, the historian of his country, full of the *amor patriæ*, gives a very lofty description of its capabilities and excellence. This might all, perhaps, be realized but for the fundamental objections of climate and political discord, which must for years prevent a portion of its real merits being elicited.‡

The present village of Realejo (for the name of town cannot be applied to such a collection of hovels) contains one main street about 200 yards in length, with three or four openings leading to the isolated cottages in the back lanes of huts.

With the exception of the houses occupied by the commandant, our vice-consul, Mr. Foster, administrador of customs, and one or two others, there is not a decent house in the place. The ruins of a well-constructed church attest its former respectability; but the place is little more now than a collection of huts.

The inhabitants generally present a most unhealthy appearance, and there is scarcely a cottage without some diseased or sickly-hued person to be seen.§ About a mile below the town the ruins of an old but well-built fort are yet to be traced. Vessels of 100 tons have grounded at the pier of Realejo custom-house, but above that they would be left dry at low water.

Realejo is the only port after quitting Panamá where British residents can be found, or supplies conveniently obtained. Water of the finest quality is to be had from a powerful stream, into which the boat can be brought, and the casks filled, by baling, alongside of a small wall raised to cause a higher level. Here the women resort to wash, but, by a due notice to the alcalde, this is prevented. A guide is necessary on the first visit, after entering the creek which leads to it, and which should only be entered at half-flood; it is necessary to pole the remainder, the channel not having sufficient width for oars.

* Voyage of the *Sulphur*, vol. ii. p. 307.

† Pearl oysters are found near the South of Cardon; but few pearls, however, are found in them, and the search has been found very unprofitable.—*G. U. Skinner, Esq.*

‡ This port, if a settlement were established on the Islands of Asseradores, Cardon, or Castañon, would probably be more frequented; but the position where vessels usually anchor (within Cardon) to Realejo, is a sad drawback to vessels touching merely for supplies. Rum is also too cheap, and too great a temptation to seamen. Supplies of poultry, fruit, bullocks, grain, &c., are, however, very reasonable, and of very superior quality; turkeys are said to attain an incredible weight; they still, however, justly maintain a very high reputation.

At the period of our visit a young American had imported machinery for a cotton mill, but the success of it is doubtful.—*Sir Edward Belcher*, 1838.

§ "This is a very sickly place, and I believe hath need enough of an hospital; for it is seated so nigh the creeks and swamps that it is never free from a noisome smell. The land about it is a strong, yellow clay; yet where the town stands seems to be sand."—*Dampier*, vol. i. p. 221.

The water from the well on the Island of Asseradores is good;* but Sir. E. Belcher says, "I have a great objection to water infiltrated through marine sand and decayed vegetable matter, and consider the chances of sickness one step removed by obtaining it from a running stream."

The mountains in the neighbourhood of Realejo are magnificent, particularly to a spectator at 12 or 15 miles off shore.†

The northern channel, or entrance to the port, lying around the N.W. end of Cardon Island, according to the survey, has a depth of 6 to 10 fathoms. The North end of the island appears to be bold-to. The N.W. point of the island is called *Ponente Point*, and is a detached rock. The N.E. point is *Cardon Head*, and is 30 feet high. The channel passes round close against this point, and all over towards the South end of Asseradores Island the water is very shallow, and a large patch, the Sawyer Bank, is nearly awash.

The mark given for taking this channel, called the *Cardon Channel*, is: run toward the entrance, with the North point of Cardon and the South point of Asseradores touching, when they will bear *East, true*, or E. ¾ S. by compass, and then haul close round Cardon Head, as the current sets direct on Sawyer Bank.

Cardon Island is three-quarters of a mile in length, N.W. and S.E. Shoal water extends some distance off its seaward face. Its southern point is *Cape Austro*, surrounded by a shoal. S.S.E. of this is *Castañon Bluff*, the western point of Castañon Island. These two points are a quarter of a mile apart, and the channel between, which has a depth of 5 or 6 fathoms, is the *Barra Falsa*.

The mark for entering it is a *vigia* on a hill inland, or about 5 miles within the entrance, kept between the two points (Cape Austro and Castañon Bluff), and bearing about N.N.E. This mark kept on leads into the port.

The town of Realejo is up the channel which runs at the back of Asseradores Island, which is 8 or 9 feet deep.

It is high water at Realejo, on full and change, at 3^h 6′; springs rise 11 feet; variation, 8° 13′ E., 1838.

The following are the remarks made by Mr. P. C. Allan, R.N., which will prove very serviceable:—"Vessels bound to Realejo from the southward should (passing about 20 miles to the eastward of Isle Cocos) steer to make the land to the eastward of the port during the period between November and May, as the winds prevail from the N.E., and sometimes blow with great violence out of the Gulf of Papagayo, causing a current to set along shore to the N.W.

"A range of mountains in the interior may be seen at the distance of 60 miles; the most remarkable of them is the Volcan de Viejo, the highest part of which, bearing N.E. by N., is the leading mark to the anchorage. The shore, for some distance on each side of the entrance, is low and woody. Cardon Island, which is on the right side of the entrance, is rather higher, and its western end is a brown rocky cliff. The wooden tower, or look-out house, which is situated 5 or 6 miles inland, may be seen rising above the trees.

* "We established our tide-gauge on the Island of Asseradores, although directly open to seaward through Barra Falsa, and we were fortunate to find a good well of water close to the beach."—*Sir E. Belcher*, vol. i. p. 28.

† Voyage of the *Sulphur*, vol. i. p. 31.

"In coming from the southward, and running along the land, ships must avoid a rocky reef, which lies about 7 miles E.S.E. of the anchorage off Realejo, on which H.M.S. *Conway* struck. This reef was examined by the boats of that ship.

"The two rocks that are above water (the one 8, and the other 5 feet high) are distant from the beach rather more than three-quarters of a mile. The ground between these rocks, and 1½ miles to seaward of them, and probably more, is very uneven. The rock on which the *Conway* struck lies S.S.W. ½ W. ¼ of a mile from the N.W. or highest of the two rocks. In passing this reef give the rocks above water a berth of 2 miles.

"The flood tide comes from the N.W. The tides are irregular; one day during our stay here it was low water 16 hours."*

Mr. E. P. Brumell also says:—"In steering for Port Realejo from the southward, after passing the parallel of Port Culebra, keep the land well aboard (during the papagayos) as there is generally a strong offset. The land is low in front, without anything to make it remarkable at 8 or 10 miles off; however, there are most excellent marks inland, in the event of not getting observations. A very high peak inland, rising evenly and gradually to a fine point, bearing E.N.E., will place you to the southward of the port, and another high mountain, El Viejo, bearing N.E. ½ N. by compass, will place you right off the port. This mountain is very remarkable, there being none other bearing the least resemblance to it, and in fact none in the immediate vicinity."

Capt. Worth, of H.M.S. *Calypso*, visited Realejo in 1847, and the following are the remarks made from observation during his stay:—

"The breeze of Papagayo is always strong, and is felt 40 or 50 miles off shore, strongest nearest the land. Coming from Puntas Arenas to Realejo, although this breeze does not usually allow a ship to lay up for it at first, it will, as you proceed northward, draw more off shore, and lead directly up for the El Viejo mountain, the leading mark for Realejo.

"El Viejo is an extinct volcano, and a remarkable landmark, being the westernmost of a number of conical mountains; one of which, Momotombo, is an active volcano, and almost constantly smoking, and having the appearance of a slice cut of its top, slanting to the eastward. Westward of El Viejo the land is low, with a sandy beach, thickly covered with wood; with very clear weather, and approaching El Viejo from the southward, you will see the volcano of San Miguel (extinct), making like a round island to the westward.

"Having made out El Viejo, a N.E. by N. course leads directly up to the anchorage of the entrance of Realejo Harbour; 7 or 8 miles from the beach is 19 fathoms, shallowing quickly, but regularly, to 12 fathoms, at about 4 or 5 miles from the shore. The land to the westward is a continuous low beach, wooded close down; carrying the eye along this beach to the eastward will be observed a break, the land protruding further into the sea; the beach white, and the surf heavier, having detached trees upon it, with low abrupt cliffs, also an opening, which is the South entrance into Realejo. The island to the westward of this opening (Castañon Island) has few trees upon it, and is abrupt at the West end;

* Nautical Magazine, Feb., 1836, p. 70.

to the westward of it is Cardon Island, which has many detached trees upon its West end, of larger size than upon the apparent main land (Asseradores Island), and is higher and more abrupt at this end than Castañon; there is also a large green tree jutting out from it to the westward, and a higher rock close to it, which can be seen as soon as the island itself.

"The face of Cardon Island is of a reddish-brown colour, occasioned by burning the bamboo, which grows thickly upon it. Having made out Cardon Island, the best anchorage is with its N.W. end bearing East 1¼ miles off El Viejo, N.E. by N., in 6½ and 7 fathoms sand and mud.

"No ship should anchor to the northward of the West end of Cardon bearing E. by S., as the soundings shoal very quickly from 6 to 5 and 4½ fathoms, the rise being about 11 feet, and always a swell, sometimes very heavy, the ship rolling deeply; nor is the bottom so good when nearer the beach.

"At the commencement of the dry season, the land-winds frequently last for several days together, blowing very fresh and preventing ships from entering the harbour; but as the season advances, the land and sea-breezes become more regular, but never strictly so. Sir E. Belcher's plan of Realejo is very correct, though there did not appear to me quite so much water near the East end of the North side of Cardon as laid down in it; the entrance is narrow, making it necessary to pass close to the tree on the West end of Cardon Island. A pilot takes vessels in and out, but requires to be watched, as he is not a sailor, and is old and incapable. When leaving the harbour, you should have a commanding land-wind, as the ebb sets directly across the South entrance, and very strong.

"The *Calypso* was awkwardly situated; the land-winds failing us, we drifted down towards Cardon Island, and were obliged to anchor and warp against the tide into deeper water; the anchor was let go in 3¾ fathoms, but it soon shoaled to less than 3 fathoms.

"The proper anchorage, which is nearer the Island of Asseradores than the opposite shore, is confined; the bottom, soft mud; consequently, should a vessel take the ground, no damage would occur; a large frigate would find considerable difficulty in swinging here.

"Realejo is healthy during the dry season (November to April), but subject to fever and ague during the rainy season, which commences in May. Water can be had, but not very good, and is very dear; the natives fill the casks out of the river at low water, a little above Realejo; wood is cheap, 4 dollars per thousand pieces; fresh beef is cheap, purchased by the bullock; the stock cheap and very good: fowls, 1½ reals; sucking pigs, 2 reals; ducks, 3 reals; lard, 1 real per pound; rice, 2 dollars a quintal; and vegetables scarce.

"This is the best place for stock on the whole coast: great advantages are derived from its being the residence of the vice-consul, Mr. Foster, whose kindness and attention are very great. Washing is also comparatively very cheap—1 dollar the dozen. Fire-wood may be cut in any quantity, and good."

Dampier, who was on this coast in September, 1685, says:—"We had very bad weather as we sailed along this coast; seldom a day passed but we had one or two violent tornadoes, and with them very frightful flashes of lightning and claps of thunder: I never did meet with the like before nor since. These tornadoes commonly come out of the N.E. The wind did not last long, but blew very fierce for the time. When the tornadoes were over, we had the wind at West, sometimes at W.S.W. and S.W., and sometimes to the North of West, as far as N.W."

The GULF of FONSECA, or CONCHAGUA, is about 40 miles N.W. of Realejo. Its S.E. point is that of Coseguina, upon which stands the celebrated volcano of Coseguina. It is 3,800 feet high, and is in lat. 12° 58′ N., lon. 87° 37′ W. A view of it gives the best description of its distant appearance. It was anciently called Quisiguina, and stands, as before stated, on the S.E. point of the entrance. The verge of the crater, which is half a mile in diameter, is elevated about 3,800 feet above the mean level; thence the interior walls fall perpendicularly to a depth of about 200 feet, when the bottom of the crater becomes flattish, with a small transparent lake in its centre. One of the most remarkable volcanic eruptions on record occurred from it. It commenced on the 20th of January, 1835, and its first evidence was, as seen at 60 miles' distance, an immense column of smoke and flame emitted from the crater. At 9 A.M. a very heavy shock of an earthquake was felt; the night following five shocks; and during the 21st several shocks, accompanied with the noise resembling distant thunder, or "*retumbo*," as the Spaniards call it. On the 22nd the ground was covered with fire, ashes, or sand, darkness and the roar of the volcano prevailing. On the 23rd, the fall of ashes and noise increased till it became darker than the darkest night, and continued so till 3 P.M., when it cleared a little; everything covered thickly with the volcanic dust, the noise, and odour of sulphur, being overpowering.* Its devastating effects were continued for many weeks after.

In proof of its tremendous effects the eruption shook all the windows and doors in the city of Guatemala, which is between 240 and 250 statute miles distant, most forcibly. This was occasioned, not by the earthquake, but by the explosions transmitted through the air; this was on January 23, 1835. But the distance to which the thunder of the volcano was heard, and the dust felt, was very much greater than this. According to the official account, these were both felt and heard at Ciudad Real de Chiapas, a distance of 420 geographic miles. It occasioned very great alarm at Tonala and other parts in Soconusco, 450 miles to the N.W.; and on the coast about Merida, in Yucatan, 800 miles off. Those to the southward were not slight evidences, but the air was darkened and the noises terrific, and the sulphureous vapours most suffocating. The flocks perished from the pastures being destroyed by the dust, and great sickness ensued among all from the water becoming tainted from the same cause.†

In the *Jamaica Watchman* (January 29, 1835), too, it is announced that all the ships about that island were covered with the fine volcanic dust, which continued to fall for some days, covering everything.

G. U. Skinner, Esq., left the city of Guatemala and Istapa in the middle of March, passing large banks of floating pumice during the passage to Conchagua. When at many leagues' distance, they were almost suffocated by the sulphureous vapour and the volcanic dust, which obscured the sky and settled on everything, causing most violent burning pains in the eyes.

* Voy. of the *Sulphur*, vol. i. pp. 242-3. † Boletin Oficial, No. 78, p. 726, Mar. 7, 1835.

Dampier's account of the gulf appears to be so graphic that we insert it here in preference to others of later date :—"The Gulf of Amapalla (or Conchagua, or Fonseca) is a great arm of the sea running 8 or 10 leagues into the country. It is bounded on the South side of its entrance with Point Casivina (Coseguina), and on the N.W. side with St. Michael's Mount (Volcan de San Miguel). Both these places are very remarkable. Point Casivina (Coseguina) is a high round point, which at sea appears like an island, because the land within it is very low. St. Michael's Mount is a very high peaked hill, not very steep; the land at the foot of it on the S.E. side is low and even for at least a mile. From this low land the Gulf of Amapalla enters on that side. Between this low land and Point Casivina (Coseguina) there are two considerable high islands. The southernmost is called Mangera, the other is called Amapalla, and they are two miles asunder.

"Mangera is a high round island, about 2 leagues in compass, appearing like a tall grove. It is environed with rocks all round, only a small cove or sandy bay on the N.E. side. The mould and soil of this island is black, but not deep; it is mixed with stones, yet very productive of large timber trees. In the middle of the island there is an Indian town, and a fair Spanish church. There is a path from the town to the sandy bay, but the way is steep and rocky. At this sandy bay there are always 10 or 12 canoes lie hauled up to dry, except when they are in use. Amapalla is a larger island than Mangera. The soil is much the same. There are two towns on it, about 2 miles asunder, one on the North side, the other on the East side: that on the East side is not above a mile from the sea; it stands on a plain on the top of a hill; the path to it is so steep and rocky, that a few men might keep down a great number only with stones. There is a very fair church, standing in the midst of the town. The other town is not so big, yet it has a handsome church.

"There are a great many more islands in this bay, but none inhabited as these. There is one pretty large island belonging to a nunnery, as the Indians told us; this was stocked with bulls and cows. They are all low islands, except Amapalla and Mangera. There are two channels to come into this gulf; one between Point Casivina and Mangera, the other between Mangera and Amapalla: the latter is the best. The riding place is on the East side of Amapalla, right against a spot of low ground; for all the island, except this one place, is high land. Running in farther, ships may anchor near the main on the N.E. side of Amapalla. This is the place most frequented by the Spaniards. It is called the Port of Martin Lopez. This gulf, or lake, runs in some leagues beyond all the islands, but it is shoal water, and not capable of ships."*

PORT LA UNION is an inner harbour or bay of the Gulf of Fonseca; it lies on the North side, around Chicarene Point, extending 8 or 9 miles inland, but the upper and N.E. sides are shallow and uncovered at low water. This contracts the limits of the port within much smaller dimensions. On these flats oysters are very abundant.†

San Carlos, better known by the name of Conchagua, is situated on the South side of the port. The site is badly chosen, as the difficulty in landing is

* Dampier's Collection, vol. I. 121—124.

† Mr. E. P. Brumell.

at all times great, and at low water nearly impossible; during strong northerly winds the communication is frequently cut off for days, independent of unsafe holding ground for shipping. Near Chicarene this might have been entirely avoided.

The port is entirely landlocked ; in fact, a complete inland sea.

The actual town or village of Conchagua, from which this port derives its name, is situated about 3 miles up the Amapala Mountain, or extinct volcano, immediately over San Carlos. The selection of this spot is said to have originated in the piracies committed on this race of Indians by the buccaneers. They were then located on the Islands of Conchaguita and Manguera. They then fled to this secluded spot of Conchagua, which is destitute of water, that necessary of life being daily carried up in calabashes.

The volcano of Amapala, situated immediately above the port, has been extinct beyond tradition. It has no evident remains of volcanic agency, and may owe its appellation to its conical form. The height was determined to be 3,800 feet above the mean tide level.

Deer are said to abound on Manguera, Conchaguita, and Tiger Islands. Rabbits and squirrels, with the addition of jackals, may complete the list. Bullocks and other stock are easily procured at San Carlos, and prices moderate.*

Port La Union forms only a small portion of this extensive inlet. The entrance to it is formed by Chicarene (or Chiquirin) Point on the West, and Sacate Island on the East. There are very strong ripplings between these points, but no danger. South of this is Perez, and also Conchaguita and Manguera (or Miauguera) before mentioned. Between the last and *Punta Arenas*, the North point of the promontory on which Coseguina stands, are the islets called the Farallones Blancos. Into the head of Port La Union the rivers Siriano† (or Estero Jirausa) and Guascoma fall. Off the S.E. point of the port, on the sandflat, stands *Conejo Islet.* S.W. of this are *Garrobo, Violin*, and *Esposicion Islands*, and to the S.W. of this again is *Tigre Island.* The East and northern portions of the gulf receive the rivers Nacoame, Estero Hermoso, Chuloteca, and some others. The S.E. portion of the gulf is extensive, and the Estero Real falls into its head, and in this part is Naguiscolo Port.

Mr. Stephens visited Conchagua. From Lake Managua he started for the Port of Naguiscolo, 7 leagues distant, through a forest. He overtook the bungo men, nearly naked, moving in single file, with the pilot at their head, and each carrying on his back an open network, containing tortillas and provisions for the

* Voyage of the *Sulphur*, vol 1. pp. 234—236, 244.

† The *Sirano*, Siriano, or San Miguel River, has been one of the sites thought of for forming the communication between the two oceans, to be connected with the Bay of Honduras through the transverse valley, the Llanura de Comayagua, watered on the Atlantic side by the Jagua, and on the Pacific by the *Sirano*, as before mentioned, and both of which are navigable, but how far, or how long, our knowledge will not determine. This scheme, therefore, is very desultory.—*L'Isthme de Panamá, par M. Chevallier*, p. 72.

The town of San Miguel is situated on a plain at the base of the volcano, which is 7,024 feet high, and suddenly springs on this side to its apex; and is surrounded on its other sides by ranges of 500 to 600 feet above its level, entirely excluding it from any prospect beyond their outlines. There is nothing in the city itself which calls for remark, and its consequence arises principally from the fairs held here for the purpose of transacting the indigo trade. The fair at the period of Capt. Belcher's visit is the principal (November 23, 1838), and had a large quantity of cattle, horses, sheep, &c.—*Belcher*, vol. 1. p. 228.

voyage. When he arrived at the port he found only a single hut, at which a woman was washing corn.

"In front was a large muddy plain, through the centre of which ran a straight cut called a canal, with an embankment on one side dry, the mud baked hard and bleached by the sun. In this ditch lay several bungoes high and dry, adding to the ugliness of the picture.

"The bungo in which we started was about 40 feet long, dug out of the trunk of a guanacaste tree, about 5 feet wide and nearly as deep, with the bottom round, and a *toldo*, or awning, round like the top of a market-waggon, made of matting and bull's hides, covered 10 feet of the stern. Beyond were six seats across the sides of the bungo for the oarsmen. The whole front was necessary for the men, and in reality I had only the part occupied by the awning, where, with the mules as tenants in common, there were too many of us.

"The sun was scorching, and under the awning the heat was insufferable. Following the coast, at 11 o'clock they were opposite the volcano of Coseguina, a long dark mountain promontory, with another ridge running below it, and then an extensive plain covered with lava to the sea.

"The wind died away, and the boatmen, after plying a little while with the oars, again let fall the big stone and went to sleep. Outside the awning the heat of the sun was withering, under it the closeness was suffocating, and my poor mules had had no water since their embarkation. Fortunately, before they got tired we had a breeze, and at about four o'clock in the afternoon the big stone was dropped in the harbour of *La Union*, in front of the town. One ship was lying at anchor, a whaler from Chili, which had put in in distress and been condemned."*

The Estero Real appears to be of considerable importance, as Capt. Sir E. Belcher took the *Starling* up it for 30 miles from its mouth, and might easily have gone further, but the prevailing strong winds rendered it too toilsome a journey at this period; he considered that it might be ascended much higher—the natives say, 60 miles—by vessels drawing 10 feet, but steamers would be absolutely necessary to tow against the prevalent breezes.

According to Capt. Belcher's opinion, this unquestionably is the most advantageous line for a canal; for, by its approaching thus the Lake of Managua, the entire lake communication might readily be effected.

Mr. G. U. Skinner says that there is considerable traffic carried on by means of bungoes, or large canoes, and that the distance from the embarcaderà to the Conchagua is 65 miles.

Leaving Realejo for Conchagua,† with the land-wind, it is advisable to steer obliquely off shore to meet the sea-breeze, which takes a ship to Coseguina Point, on the East side of the Gulf of Conchagua. Sailing in the morning, you will generally reach this point late in the evening, or the next day after noon.

The Gulf of Conchagua will be well understood by reference to Sir E. Belcher's plan; there appears to be no dangers. The *Calypso* worked in and out twice,

* Incidents of Travel, &c.
† The ensuing directions are by Capt. Worth, R N., H.M.S. *Calypso*, 1847.

tacking at less than two cables' length from the shore and islands.* The tides are very strong; full 2 knots at full and change, which take place at 3 P.M. About the change of the moon the land-wind blows strongly during the night and a greater part of the day; you can, however, see it coming by the foam on the water.

As in coming from Realejo you generally arrive here in the evening, it is advisable to anchor when the land-wind comes off; for, should you be driven off the coast, it will take all the next day to reach the islands, and to arrive at the proper anchorage, the sea-breeze being weak after such strong land-winds.

In the *Calypso*, on our first going to Conchagua, the land-wind came down with such force that the gulf appeared to break across; and we anchored in 18 fathoms, and worked up the next day to Chicarene Bay.

This gulf contains the best and most easily obtained water on the coast; it is a stream running down the mountains, clear and sweet, into the bay called *La Playeta de Chicarene*, which is just to the southward of the Chicarene Point: you can anchor close in. The best way to water is by rafting; the water in the bay being quite smooth, you can pull well to windward, and alongside to the eddy, and then across the tides to the ship. We filled 26 times a day, although badly off for boats, having only a 28-foot pinnace. There is a surf on this beach, sometimes heavy, but seldom enough to prevent landing. Merchant vessels anchor so close, as to be able to hand their casks off with long lines. When we first anchored in this point we tried, through ignorance, to water in the Playa de Chicarene, but we found the surf so heavy, the water so bad, and such great difficulty from the tides, that we could not get more than 18 tons a day, and that after great labour to the men, and much damage to the casks and boats.

Wood is not so cheap here as at Realejo; beef about the same price, also washing; but stock is dearer, and difficult to get in any quantity: turtle is plentiful, about 2 reals for one weighing 50 lb. Sir Edward Belcher's plan clearly shows the only dangers in entering Port La Union; they are visible at low water, the only difficulty is the very great strength of tide; it is quite a sluice round Chicarene Point. H.M.S. ship *Dublin* lay here for a few months.

Of the coast westward of Conchagua we have no very recent connected account, with the exception of the different ports, or rather roadsteads, which constitute the remainder of the maritime ports of the republic. The coast, as delineated on the chart, is taken from the survey of Don Alexandro Malespina, in 1794, as drawn up subsequently for the Spanish government by Don Felipe Bauza, F.R.S. In the introduction to the next chapter (p. 255) we have noticed the unfortunate voyage of this commander, and have there stated the reasons why we have not a more complete description of the tract he explored.

PORT GIQUILISCO, or *del Triunfo de los Libres*, according to Colonel Don Juan Galindo (an Englishman in the service of the republic), is about 24 miles

* In entering the Conchagua from the westward, bring the Island of Tigre to bear exactly between the Islands of Conchaguita and Manguera, to avoid the rocks off Point Candadillo. Tigre is a high conical hill. It is thus quite safe entering to an anchorage, even at night, if this island be seen, which is seldom not the case at all seasons.—*G. U. Skinner, Esq.*

beyond *Point Candadillo*, the N.W. entrance point of Conchagua; or, as stated in the M.S. plan, in lat. 13° 22′ N., lon. 88° 12′ W., nearly.

In 1798, the Royal Consulate of Guatemala ordered Don Vincente Rodriguez del Camino to survey it. He states that it was then named the *Bay of San Salvador de Jiquilisco* (a species of wood), *anciently* called the Bay of Fonseca. May this account for the double name applied to Conchagua to the S.E.? The name given to it by Colonel Galindo evidently has reference to one of those "triumphs," so common and so ephemeral in these distracted countries. We have therefore preferred the old name.

According to the plan of Don V. del Camino, the anchorage is good and well sheltered. Like Tehuantepec, the coast of the main land appears to be fronted by a long narrow island, perhaps formed by the tremendous surf raised by the prevalent winds.

According to Colonel Galindo's M.S., the port is formed by Arenas Island, off the South point of which, on the East, a sandy bar runs for some distance, and breaks on its extreme. A similar bar extends from the West point, and a beacon and flag is (or *was*) placed on each, the entrance lying between them. Good water is stated to be obtained at the town, or Pueblo, up the port.

The RIVER LEMPA enters the ocean immediately to the West of this port. It is the largest river of Guatemala, and has a great volume of water, but is only navigable for boats.

In passing along the coast of this river the navigator should be very cautious of his distance, for it is stated that a long bar or flat runs off, as indeed might be anticipated, from the magnitude and character of the river. This flat, which reaches the larger part of a mile, is called the *Barra del Espiritu Santo*, and on it the *Lucretia*, a brig drawing 12 feet, was wrecked, February 18, 1847. This was in consequence of the ignorance of the existence of such a projection.

In Father Gage's work, before quoted, he mentions the Lempa:—"This river is privileged in this manner, that if a man commit any heinous crime, or murther, on this side of Guatemala and San Salvador, or on the other side of St. Miguel or Nicaragua, if he can flie to get over this river, he is free as long as he liveth on the other side, and no justice, on that side whither he is escaped, can question or trouble him for the murther committed. So likewise for debts, he cannot be arrested." *

Above this port and the mouth of the river the *Volcan de San Miguel* rises. It is 7,024 feet high, and is a very conspicuous object in the offing, and will serve well to point out the locality.

PORT LIBERTAD is about 50 miles to the westward of the Lempa. It was visited by H.M.S. *Sulphur*, and we copy her commander's observations on it.

One would naturally expect from this title that something pretending to a bay, or deep indentation at least, would have warranted the appellation. But a straight sandy beach, between two slightly projecting ledges of rock about a mile asunder, forms the *plaza* of Libertad: it is *law* and *interest* only that have made it a *port*.

* A Survey of the Spanish West Indies, by Thomas Gage, p. 417.

At times the bay is smooth, but the substratum at the beach being of large smooth boulders of compact basalt, the instant the surf rises they are freed from their sandy covering, and a dangerous *moving* strong bottom left, on which the boat grounded. We were informed that it is generally violent for three or four days, at full and change, which corresponded to the time of our visits.

The village contains about twelve huts, with a family of about six in each. There is also a long government building constructed of adobes, in which the tackle of the bungoes used for landing cargoes is usually stored, and a cabin for the commandant at its further extremity, served for parlour, bedroom, kitchen, &c. The only pet birds were fighting cocks, perched under the chairs, or probably tethered in the corners. Cock-fighting is a complete passion in Spanish America.

This is all that can be hoped for at Libertad.

The rollers which set in on this beach curl and break at times in 4 or 5 fathoms, at least a quarter of a mile off. Those within, which are the most dangerous, are caused by the offset or efflux.

The sand beach is composed chiefly of magnetic iron sand, the dried superstratum, about one inch in thickness, caking in flakes free from admixture.

The anchorage is uneasy, and I should think unsafe, and should be avoided near the full moon. Sudden rollers come in, which are apt to snap chain cables, unless with a long range.

Poultry, bullocks, &c., are to be obtained, but compared with those of San Salvador or Realejo, the prices are exorbitant. Bullocks can only be embarked in one of these bungoes.*

A boat belonging to H.M.S. *Sulphur* was capsized on the surf in the Bay of Libertad, when her coxswain was drowned. Extreme apathy was shown by the authorities towards the crew, who were not able to effect a communication with the ship for several days. This renders it by no means desirable for vessels to visit the bay, more especially as no refreshments, or even bullocks, can be advantageously obtained.—(Naut. Mag., Sept. 1837, p. 609.)

San Salvador, the capital of the republic, is to the N.E. of Libertad.

Capt. Sir Edward Belcher visited this city in April, 1837, going thither from Realejo, the road being through a very mountainous tract, and in parts little better than a goat path.

The town is very prettily situated on a level plain, or amphitheatre, from which several lofty mountains rise, that of the Volcan de San Salvador being the most conspicuous.† The streets are broad, and very clean for a foreign town: the

* Voyage of the *Sulphur*, vol. i. p. 39.

† When we first saw the mountain of Guatemala, we were, by judgment, 25 leagues' distance from it: as we came nearer the land it appeared higher and plainer; yet we saw no fire, but a little smoke proceeding from it. The land by the sea was of a good height, yet but low in comparison with that in the country. The sea, for about 8 or 10 leagues from the shore, was full of floating trees or drift wood, as it is called (of which I have seen a great deal, but nowhere so much as here), and pumice stones floating, which probably are thrown out of the burning mountains, and washed down to the shore by the rains, which are very violent and frequent in this country.

The Volcan of Guatemala is a very high mountain, with two peaks or heads, appearing like two sugar-loaves. It often belches forth flames of fire and smoke from between the two heads; and this, as the Spaniards do report, happens chiefly in tempestuous weather.—*Dampier*, vol. i. pp. 225—230.

houses have very projecting eaves; they are substantial, although lightly constructed, and of one story only, in consequence of the liability to frequent shocks of earthquakes. They have internal courts, and appear to possess convenience, space, and comfort. All are well supplied with water by aqueducts; have a good market, every necessary being cheap and abundant; and nothing is wanting to their comfort but society and strictly enforced order. The want of this latter, I am informed, is a sad drawback; and it never can be attained under the present laws, habits, &c. One of these habits, arising from their new system of *independence*, is entering your house, and seating themselves without invitation; any opposition might be attended by unpleasant results, even to assassination.*

PORT ACAJUTLA, or SONSONATE ROADS, is the next attainable point beyond Libertad.

The principal town of this port is Sonsonate,† which is situated about 15 miles inland. There is also a small village on the coast which gives its name to the port; it consists of about thirty habitations of various descriptions, most of them of the meanest order; they are constructed of bamboo open work at the sides, and the top is rudely thatched of palm leaves, which latter is, however, made impervious to the heavy rains that fall almost perpendicular in the wet season. There is also still remaining the ruins of an old Spanish fort, in which is situated the dwelling-house of the governor. This officer performs all the official duties of captain of the port, administrator, &c. Mr. Thompson estimates the population at about 130 individuals.

The port consists of an open bay, of which Point Remedios is the eastern boundary. There is anchorage all over it at a prudent distance from the shore in from 7 to 15 fathoms water; the bottom appears to be of sand, with here and there a patch of mud. Large vessels should not anchor in less than 12 fathoms.

The surf breaks heavily on the beach, which renders landing in ships' boats almost impracticable. The usual mode of effecting this object is in large canoes or bungoes, which belong to merchants residing at Sonsonate, and are kept for the purpose of discharging cargoes. There is generally one of these kept afloat, moored just without the surf in the N.E. corner of the bay, near where the village is situated: persons desirous of landing usually pull in in their own boats, transfer themselves, with a portion of their crew, into the bungo, and haul in through the surf to the beach by a line fast to the shore for that purpose. To get on shore dry, they will then require to be carried out through the receding surf, which is about 1 foot or 18 inches deep. This contrivance, called at Istapa the *anda-nivel*, is described more fully on page 250. There is another landing place, which is practicable only in the finest weather, on the rocks about 1 mile South of the

* Sir Edw. Belcher, vol. i. pp. 34-5.

† Sonsonate, or Zonzonate, derives its name from the Rio Grande, formed by almost innumerable springs of water, to which the name of *Zezontlatl* is given, a Mexican word meaning 400 springs, corrupted to Zonzonate.

"Santissima Trinidad de Zonzonate is situated on the Rio Grande. It is a pleasant town, although the climate is very hot. Each of three monastic orders have (had) a convent here. The church is very spacious, besides which there are three oratories. On the opposite side of the river it has a suburb called the Barrio del Angel, on which there is a chapel. The communication between the town and the suburb is by means of a stone bridge. In the vicinity are three small Indian villages, &c."—*Don Dom. Juarros: Translation by Mr. Baily*, p. 28.

village towards Point Remedios; but great care and judgment are required in attempting it.

Point Remedios has a reef off it extending in a south-westerly direction nearly 3 miles in fine weather. This reef scarcely shows itself, therefore more caution is necessary in rounding it. Vessels of a light draught have frequently passed safely over the outer part of it unknowingly, whereas several others, less fortunate, have been brought up by detached rocks, and a total wreck has ensued. The point is long and low, thickly wooded, and from the eastward easily recognised. Its extreme is in lat. 13° 30′ N., and lon. 89° 45′ W., of Greenwich; variation, 10° easterly.

Beef, poultry, vegetables, and fruit are plentiful and cheap. Water is plentiful on shore, but the difficulty of getting it off through the surf is very great; however, if much wanted, it may be had with a little extra labour and perseverance.

The following observations on this port, and on approaching it, are by Capt. Worth, of H.M.S. *Calypso*, in 1847:—

"Acajutla, or Sonsonate Roads, although not much known, is safe, the oldest inhabitant remembering only one wreck; the reef off Remedios Point breaks, the sea setting directly into the anchorage. Here the salt water is very injurious to the cables and copper; although at anchor not more than a fortnight, the cable and anchor were completely covered with small shell-fish, as also the boom boats; this remark is applicable to all the ports we visited on this coast, though not so much as at this place.

"The passages to the westward are uncertain as to time, the land and sea breezes being so very unsettled.

"The land-breeze always blows (if ever interrupted, only for a short period) at all the ports we visited, except Conchagua, and is nearly always sufficient to take a ship to sea. As a rule, I should recommend on leaving any port that you stand off shore, always bearing in mind that the sea-breeze is from South to S.W. There is a current always setting to the S.E.

"Between Conchagua and Acajutla the passage is very tedious, being never less than two days, and sometimes five, and even longer, the land-breeze being not to be depended on, and the sea-breeze often very light, although at times the sea-breeze will blow very fresh indeed. The best plan, after leaving Conchagua, is to stand rather off shore, so as to make a long leg off with the sea-breeze.

"The coast is quite clear, there being anchorage nearly all along it.

"We found the sea-breeze seldom set in before noon, and often later, and a continual set to the S.E. The leading mark for Acajutla is the Isalco Volcano, which smokes, and frequently sends up large jets of fire. On a N.E. by N. bearing it leads to the anchorage off Acajutla.

"This anchorage is difficult for a stranger to find; the best plan is to take notice of the several volcanoes on the coast, after leaving Conchagua; viz., San Miguel, San Vincent, and San Salvador. The land is a low beach, the soundings decreasing gradually to 10 fathoms, at 3 or 4 miles off shore, until the volcanoes are past, when it becomes tolerably high, and has 25 fathoms at a little distance from the beach, particularly in the bight to the S.E. of Point Remedios, where in that depth the surf can be heard quite distinctly. Point

Remedios, which runs out from this moderately high land, is low, and thickly wooded, appears to stretch a long way into the sea, and has several black rocks, one nearly a solid square, lying just off it; these rocks are the inside part of a reef extending 3 miles in a S.W. direction, on which the sea breaks heavily at times. On the S.E. side of this point the beach is clear, having no rocks upon it; but on the N.E. side it is broken by rocks and clumps of trees, dividing it into a number of small sandy bays. Should you not be close in, you cannot make the point out, as it appears part of the low, thickly-wooded land that stretches from the before-mentioned moderately high coast to the westward of Istapa.

"The Volcano of Isalco is decidedly the surest mark, if it can be made out; but, as it does not smoke constantly, and is situated on the side of, and is lower than, the mountains behind (Sierra Madre), it is very difficult to find.

"The Madre, and mountains to the westward of Isalco, are very high; it may be known by the table-land top, in which it differs from the others, they being conical, on approaching that form. The Isalco is a conical volcano, apparently on the East side of the Madre, and the crater is about one-fourth down from the table land."—(Capt. Worth, R.N.)

From the paucity of our knowledge of the detail of this coast following to the westward, we extract the following from the work of Juarros, which, though not nautical, may be interesting.

The *Province of Escuintla* stretches 80 leagues in length. The Port of Istapa is in it. The climate is hot, yet there are a few spots where it is temperate; and in a few others it is even cold. It carries on very little trade, confined chiefly to fish, artificial salt, maize, plantains, and other fruit, that are carried to the market of Guatemala. Of the numerous rivers which water the district, the most distinguished is the *Michatoyat*, which flows out of the Lake Amatitlan; after a course of a few leagues it has a fall or cascade, the largest in the kingdom, called the Falls of San Pedro Martyr (from being near a small village of that name), presenting one of the most agreeable points of view in the country; this river discharges into the Pacific, and forms the *Barra de Michatoyat*. The *Rio de los Esclavos* (Slave River) attracts notice from the bridge, built over it in 1592, by the city of Guatemala; it is by far the finest and best constructed in the country, 128 yards long and 18 broad.* The *River Guacalat* rises in the province of Chimaltenango, passes by the site of Old Guatemala, where it is called the Magdalena, and joined by the River Pensativo. Receiving so many tributary streams it becomes navigable, and finally disembogues into the Pacific, where it forms the Port of Istapa, celebrated for being the place where Pedro de Alvarado equipped his squadron in the year 1534.

The PORT of ISTAPA, at the mouth of the River Michatoyat (or Michitoya),

* "The *Rio de los Esclavos*, which falls into the sea West of Acajutla, is described by Stephens as a wild and majestic river. He crossed the bridge over it, erected under the Spanish dominion, and is the greatest structure of that period. The village beyond was a mere collection of huts, standing in a magnificent situation near the river, and above which mountains rise covered to their summits with pines. Every predatory or fighting expedition between Guatemala and San Salvador passed through this miserable village. Twice within his route Morazan's army was so straitened for provisions, and pressed by fear of pursuit, that huts were torn down for firewood, and bullocks slain and eaten half raw in the street, without bread or tortillas. After leaving this village, the country was covered with lava."—*Incidents of Travel.*

is the outlet of the Lake of Amatitlan, and is said to be navigable from the Falls of San Pedro Martyr, 70 miles from its mouth; but there were no boats upon it, and its banks are still in a wilderness state. The crossing place was at the old mouth of the river.

The Port of Istapa is an open roadstead, without bay, headland, rock, or reef, or any mark whatever to distinguish it from the adjacent shores. "There is no light at night, and vessels at sea take their bearings from the great volcanoes of the Antigua, more than 60 miles inland. A buoy was anchored outside of the breakers, with a cable attached, and under the sheds were three large launches for embarking and disembarking the cargoes of the few vessels which resort to this place." At the time of Mr. Stephens' visit, a ship from Bordeaux lay off, more than a mile from the shore. Her boat had some time before landed the supercargo and passengers, since which she had had no communication with the land. Behind the sand-bar were a few Indian huts, and Indians nearly naked. Generally the sea is, as its name imports, pacific, and the waves roll calmly to the shore; but in the smoothest times there is a breaker, and to pass this, as a part of the fixtures of the port, an anchor is dropped outside with a buoy attached, and a long cable passing from the buoy is secured on the shore.* It was from this place that Alvarado fitted out his armament, and embarked with his followers to dispute with Pizarro the riches of Peru. Around the base of the volcano *de Agua* are cultivated fields, and a belt of forest and verdure extends to the top. Opposite there is another volcano, with its slopes wooded with magnificent trees. Between the two there is a convent of Dominican friars, and a beautiful valley, in which there are hot springs, smoking for more than a mile along the road, near which the *nopals*, or cochineal plantations, commence. On both sides are high clay walls, and Mr. Stephens says these *nopals* are more extensive than those of Antigua, and more valuable, as, though only 25 miles from it, the climate is so different that they produce two crops in each season.

MOUNTAINS.—Capt. Basil Hall makes the following observations on the coast in his Extracts from a Journal:—"The watering of the ship was completed in the course of the day, after which we tripped our anchor, and made all sail out of the bay, on our course to Acapulco, which lies on the S.W. coast of Mexico,

* The discharge and loading of vessels lying off Istapa, which is not effected in the easiest manner through such a tremendous surf as lashes the shore, is done by means of what is called here the "*anda-nivel*," Anglicè, "guess-warp," above alluded to.

This contrivance consists of a cable made fast to a strong post on the shore, the outer end of which is secured by an anchor some distance outside the surf. Within this anchor, which is marked by a buoy, but still sufficiently clear of the broken water, is another buoy attached to the warp, by means of which the strong launch employed is siezed to it, or casts it off. The launch having been brought to the warp buoy, the warp is thrown into rowlocks, one on the bow, the other on the stern of the launch. These are then bolted in with a pin to prevent their slipping off, and secured by a stopper, wormed round it near the bolts. The *bogas*, or watermen, watch the heaves of the sea, which, singularly enough, are always heaviest in threes, and, when the heaviest wave approaches, the pilot gives the signal, the lashings which secure the bow and stern are slipped, and at the same moment all hands haul in the warp; while running on the tremendous wave she is propelled with immense rapidity, and is usually driven on to the beach with the succeeding wave, when generally 40 or 50 Indians, with the fall of an "*aparejo*," or treble-purchase line, which is hooked to a ring on the launch's stern-post, and secured to the post, haul her high and dry with the next wave. It sometimes happens that the *practico*, or bowman, does not take the right sea, and then a larger breaks over her, swamping the launch, or damaging the cargo, or perhaps losing it.—*G. U. Skinner, Esq.*

at the distance of 1,500 miles from Panamá. There are two ways of making this passage, one by going out to sea far from the land; the other, by creeping, as it is called, alongshore. I preferred the latter method, as the most certain, and as one which gave an opportunity of seeing the country, and of making occasional observations on remarkable points of the Andes, the great chain of which stretches along the S.W. coast of Mexico, precisely in the manner it does along the West shore of South America.

"On the 23rd of February, 18 days after leaving Panamá, when we had reached a point a little to the northward of Guatemala, we discovered two magnificent conical-shaped mountains towering above the clouds. So great was their altitude, that we kept them in sight for several days, and by making observations upon them at different stations, we were enabled to compute their distances, and, in a rough manner, their elevation also. On the 23rd, the western peak was distant 88 miles, and on the 24th, 105. The height deduced from the first day's observations was 14,196 feet; and by the second day's, 15,110; the mean, being 14,653, is probably within 1,000 feet of the truth; being somewhat more than 2,000 feet higher than the Peak of Teneriffe. The height of the eastern mountain, by the first day's observations, was 14,409 feet, and by the second it was 15,382, the mean being 14,895. How far they may have preserved their peaked shape lower down, we do not know, nor can we say anything of the lower ranges from whence they took their rise, since our distance was so great, that the curvature of the earth hid from our view not only their bases, but a considerable portion of their whole altitude. On the first day 5,273 feet were concealed; and on the second day, no less than 7,730 feet of these mountains, together with the whole of the coast ridge, were actually sunk below the horizon. Owing to the great distance, it was only at a certain hour of the day that these mountains could be seen at all. They came first in sight about 40 minutes before the sun rose, and remained visible for about 30 minutes after it was above the horizon. On first coming in sight, their outline was sharp and clear, but it became less so as the light increased. There was something very striking in the majestic way in which they gradually made their appearance, as the night yielded to the dawn, and in the mysterious manner in which they slowly melted away, and at length vanished from our view in the broad daylight."*

The following remarks on *Istapa* are by Mr. H. Thompson, master of H.M.S. *Talbot*:—"We experienced some difficulty in finding this place, owing to the hidden position of the village, the similarity of the coast, and the want of instructions to guide us; therefore I conceive the following simple description will assist a stranger in hunting it out.

"The whole of this country is remarkable for its mountainous ranges, which may be seen in clear weather from a great distance seaward, many of their lofty peaks and volcanoes serving admirably as beacons to guide strangers to the various little ports and roadsteads situated on its coast, which otherwise would not be easily found. Such is the case when bound to the roadstead off the village of Istapa. There are visible from the vicinity of this roadstead, to many

* Extracts from a Journal, &c., part ii. chap. 41.

miles seaward, four conspicuous mountains, which are situated as follow: commencing with the easternmost one, which is the volcano of Pacayo; next West of this is the water volcano (Volcan de Agua) of Guatemala; then the fire volcano (Volcan de Fuego) of Guatemala, and the last and westernmost is the volcano of Tajumulco. The first and last of these volcanoes are of a moderate height, and flattened or scooped out at the top; but the two middle ones, which are the volcanoes of Guatemala, are considerably higher, and much more peaked at their tops. The easternmost one of the two last-mentioned is the water volcano; it has but one peak, which, at some periods of the year, is slightly snow-capped, and from the holes and crevices near its summit ice is procured the whole year round for the luxurious inhabitants of Guatemala. The fire volcano is to the westward of the last-mentioned, and appears to have two peaked summits, which open and close according to their bearing. From the roadstead it has the appearance of one mountain with a deep notch in its summit. The upper part of this mountain has a whitish appearance, which might be mistaken for snow; but I am informed that it is caused by the action of fire; smoke is constantly emitted from it, and may be seen from the sea in clear weather. From the anchorage at Istapa the true bearings of these four mountains are as follow: viz., volcano of Pacayo, N. 22° E.; water volcano of Guatemala, N. 5° E.; fire volcano of Guatemala, N. 8° W.; and the volcano of Tajumulco, N. 28° W. The thatched roof of a large hut, in the village of Istapa, which was just visible over the high white beach, then bore N. 17° E., distant about 2¼ miles, and the depth of water was about 17½ fathoms. The above bearings of either of the volcanoes of Guatemala nearly on will guide a vessel to within a few miles of the anchorage, and sufficiently near to make out the thatched roof of the above-mentioned hut, which is the only object that marks the spot, the remaining small huts, which constitute the village, being hidden behind the beach. There is also a small flagstaff close to the largest hut; but, unless the flag be flying, it is difficult to distinguish it, in consequence of its being mixed up with the trunks of trees that stand behind it. The entrance to the river is choked up by the sea-beach, through which it has not strength enough to force itself. The village consists of about fifteen huts, which afford shelter to about forty or fifty inhabitants, who occasionally find employment in discharging merchandise from the very few vessels that call here.

"In the bad season I should imagine this a very unsafe place to anchor at, owing to its being entirely exposed to the ocean swell, which, with the southerly winds, is exceedingly heavy. Landing is only practicable in the finest weather.

"No supplies whatever are to be had, except a few plantains." *

Of the coast of Guatemala to the W.N.W. of Istapa we have no particular account. From Malaspina's survey there does not appear to be any port, and the ocean swell must set on it with more than ordinary force. To the sailor, then, it is unimportant. The republic extends to the boundary of that of Mexico, the River Sintalapa (or Tilapa), a distance of 180 miles from Istapa. It forms the southern side of the province of Suchiltepeques, described by Juarros, and is the eastern portion of the extensive bay called the Gulf of Tehuantepec.

* Nautical Magazine, 1848.

The *Province of Suchiltepeques*, says Juarros, is bounded on the West by the Mexican province Soconusco, and extends along the Pacific 32 leagues. It was much more populous formerly than in Juarros' time. The climate is warm, but less so than Soconusco. The province is watered by 16 rivers; of these the Samala, which discharges itself into the sea, under the name of the Xicalapa, is the most important. It is fertile from its situation, and abundance of water; the chief article of commerce is cocoa, so excellent in quality as to be preferred by many to that of Soconusco. The cultivation of this valuable commodity has materially decreased since the Province of Caraccas has been the great mart for it. The people of Suchiltepeques also trade in cotton and *sapuyul* (the kernel of the *sapote*, a fruit 6 inches in length, which is used by the Indians in making chocolate). This province was subdued by Pedro de Alvarado in 1524.

CHAPTER IX.

THE WEST COAST OF MEXICO, BETWEEN TEHUANTEPEC AND MAZATLAN.

The coast described in the present chapter may be said to be that of the South extreme of the North American continent, and is the southern sea-board of the provinces of Chiapa, Oaxaca, Puebla, Mexico, Valladolid, and Guadalajara, portions of the Mexican republic.

Our geographical knowledge of the republic, generally, is very incomplete and unsatisfactory. On the Pacific shores, if the commercial importance of its few ports were at all commensurate with the natural riches of the districts of which they might be the outlets, navigators would be much embarrassed by the deficiency of our charts and descriptions. But as few points are now visited for any purposes of trade, we have tolerably accurate and recent descriptions of those ports; and in the ensuing pages it is hoped that there will be found ample notices to allow the ship-master to approach them with confidence and safety.

Mexico, as is well known, has been the scene of constant intestine warfare and change for many years past; and to this evil must be added the very great ignorance of the great mass of the people, for it is estimated by Mr. Branz Mayer,* that only *one in a hundred* of the entire population are able to read and write. This is a startling fact, as he remarks, in a republic, the basis of whose safety is the capacity of the people for an intellectual self-government.

The total population of the Mexican republic, estimated by the last-named author, in 1842, is 7,015,509, but this is inclusive of California, Nuevo Mexico,

* "Mexico," by Branz Mayer, Secretary to the United States' Legation to that country, in 1841 and 1842.

and Yucatan, which have since been separated from it. It is needless here to give the particulars of this, other works must be consulted.

The original inhabitants of Mexico are believed to be the Toltecans, a tribe of Indians from the Rocky Mountains, who migrated to the vicinity of the capital. They were displaced by the Aztecans, who came from the present territory of Upper California, founding their city of Mexico, thus called after their god of war, *Mexitli*, a term afterwards applied to the whole country. The historian Don Ygnacio Cuevas, however, says that Mexico, in the Indian language, means "below this," alluding to the tradition of the population buried beneath the Pedregal in the city of Mexico.

When Nunez de Balboa first landed on its shores (in the Mexican Gulf), Montezuma I. was emperor, and had extended the Aztec dominions to the Pacific. The conquest of his kingdom by Cortes is well known, and Mexico became a vice-royalty to Spain; and, with powers almost as absolute as that of the parent monarchy, Mexico was scarcely known to Europe, except by its issue of the precious metals. When Charles VI. of Spain abdicated in 1808, the royal authority here received a shock from which it never recovered; for an open insurrection broke out in 1810, and a national congress assembled in 1813, one of the earliest acts of which was a declaration of the independence of Mexico. Subsequently, the history is one of a sanguinary guerilla warfare, until, in 1821, Iturbide was made emperor, under the title of Augustine I. He soon abdicated and retired, but returning, he was apprehended and executed. The government was then modelled on a similar constitution to that of the United States; but the original party divisions have remained, though under different names, and are each perpetually in the ascendency, or are subdued by some other faction. M. Chevallier says:—"I have been two months in Mexico, and I have already witnessed five attempts at revolution;" and this might be perhaps stated of any similar period. Some very interesting details of the state of Mexican society may be found in Madame Calderon de la Barca's work, describing such scenes during her residence there in 1840-42.* To these disturbances, and consequent insecurity of property, must be attributed the embarrassments of commercial enterprise, the enactments of the law, and the long train of evils which lie so heavy on this fine country, and its otherwise, in many respects, good population. With the late events connected with the campaign carried on against them by the United States in 1847, which led to the final separation of Upper California from the United Mexican States, most are familiar. The chief warfare was carried on on the Atlantic side, from the ports of Vera Cruz, &c.; but the system of blockading and attack was also applied to the Pacific ports, and Mazatlan, Guaymas, San Diego, &c., were made the scene of an unequal warfare with the United States' Pacific fleet. An interesting account of this portion of the Mexican history, more particularly connected with the subject of the present volume, is to be found in a recent work with a singular title, "Los Gringos," by Lieut. Wise, of the U.S. Navy,† who was employed in the surveying squadron.

* Life in Mexico, by Madame C——n de la B——a, Jan. 1st. 1840-42.

† "Los Gringos;" or, an Inside View of Mexico and Guatemala, with Wanderings in Peru, Chili, and Polynesia. By Lieut. West, U.S.N. London, 1849. "Los Gringos" is an epithet (and

The country of Mexico, especially that part on the Pacific, is divided by the natives into *tierras calientes*, or hot regions; the *tierras templadas*, or temperate regions; and the *tierras frias*, or cold regions: the first including those beneath the elevation of 2,000 feet; the latter tract occupies the most important part of Mexico, and, in fact, is that vast plateau on which Mexico stands. On the low lands of the coasts the heat, during part of the year, is insupportable, even by the natives, and thus the town of San Blas becomes annually depopulated for a season.

The Cordillera or mountain chain, which, in the southern Andes, is a well-marked line of lofty ranges, and less distinctly so in the North and throughout Guatemala, in Mexico divides to two somewhat indistinct branches, following either coast. That to the South is irregular, and in some parts but little known. At the head of the Gulf of Tehuantepec it is about 60 miles of the coast, on an average, but leaves many valleys of slight elevation between the detached portions of it. To the mariner, most of them are unimportant, unless we mention the volcano of Colima, which becomes an excellent landmark for that portion of the coast. Most of the peaks are volcanic, some in activity; and the usual volcanic phenomena of eruptions and earthquakes are frequent, the latter particularly so; and many severe visitations of this sort are upon record. At Acapulco this becomes a serious bar to its permanent prosperity.

Of our hydrographical knowledge of the Pacific coast a word may be said. With some detached portions we are intimately acquainted, through the excellent surveys made in 1837-38, by Capt. Sir Edward Belcher, of the English navy, when on this coast in H.M.S. *Sulphur*. These points of Guatulco, Acapulco, San Blas, Chamatla, and Mazatlan, will be found described hereafter. In the voyage of the French frigate, *Venus*, commanded by Du Petit Thouars, who was on the coast at the same time with Sir Edward Belcher, we find some information: and to M. Tessan, his hydrographical engineer, we owe some of the graphic information we possess. By Capt. Beechey, who was here in H.M.S. *Blossom*, the nautical world is informed of the exact nature of the islands and ports near the Gulf of California. For the remainder of the coast, the Spanish charts of the Madrid Hydrographic Office furnish the details. In 1790 the Spanish government despatched an expedition, under Don Alexandro Malespina, for the exploration of these shores, a task which we must suppose was completely performed; but the publication of his journal, which was looked for by the learned of Europe at that period, was frustrated by Malespina, a little time after his return to Cadiz, being arrested by order of government, and thrown into the prison of Buen Retiro, and afterwards transferred to one of the strong castles of Coruna. In this captivity El Padro Gil, a man of great learning and merit, also shared, and all papers and drawings belonging to, and collected by, the expedition, were seized and suppressed by the government. Of the cause of this little is known. The disturbed state of Spain, in reference to her overgrown and ill-attached colonies, might have led to suspicions against Malespina when in the country. Suffice it to say, that the charts, resulting from the survey, were subsequently

---

rather a reproachful one) used in Mexico and California, to designate the descendants of the Anglo-Saxon race.

published, as drawn up by Don Felipe Bauzá, F.R.S., from his observations while accompanying the expedition.*

Climate.—The following outline of the climate, weather, &c., by Commodore C. B. Hamilton, will be found useful.

The West Coast of Mexico is considered highly dangerous in the bad season, namely, from June to 5th of November, and all the vessels obliged to remain in the neighbourhood lie up either in the secure harbour of Guaymas or at Pichilique, in the bay of La Paz, both in the Gulf of California.

The hurricanes that occasionally visit this coast are so much dreaded, that in the months of July, August, September, and October, the ports are deserted, and trade ceases.

I believe the *Frolic* is the first vessel of any nation, whether man-of-war or merchant ship, that ever remained the whole bad season on the coast, and that off the two most dangerous ports, viz., San Blas and Mazatlan. I shall, therefore, give all the information I can relative to the bad season.

The hurricane so much dreaded on this coast is called the Cordonazo de San Francisco, a name given by the Spaniards, on account of the hurricane prevailing about the time of San Francisco's day, the 4th of October; the word *cordonazo* signifying a heavy lash with a rope or whip; but, from my own experience, and all I can learn, these cordonazos may be expected any time from the middle of June to the 5th of November. The worst ones that have been experienced of late years have occurred on the 1st of November, although the weather usually clears up about the 20th of October, and sometimes even sooner; and as soon as the weather does begin to clear up, a ship may, with common precautions, venture into the anchorages again, for this reason, the weather will give ample warning of a coming hurricane, whereas, in the previous four months, before the weather has cleared up, the circumstance that adds to the dangers of this coast is, that owing to the threatening appearance of the sky every evening, and the violent thunder storms and squalls at night, accompanied by heavy rain and lightning, the wind veering about, you are at first led to believe that the hurricane is coming every night, and latterly you see it is utterly hopeless to foresee the coming of it, as every night appearances were as bad as they could be; the barometer here being of little or no use, and a tremendous sea occasionally setting in. Thus the remaining off this coast during the hurricane season will cause great anxiety.

The squalls and gales usually commence about S.E., and quickly fly round to the southward and S.W.; you have generally time to get to sea when it commences at S.E.; but, as I have before shown, you must go to sea every night, if you can, if you would be free from the dangers of the cordonazos coming on. But a tremendous swell frequently sets in whilst the weather is in this threatening state, and the wind still light, which makes it impossible to get out. Moreover, if our boats happened to be out, and on shore when the swell came, it was impossible to hoist them in, and for this reason we have frequently been obliged to send our

* It will be unnecessary to quote further authority here for our work. Those who feel interested will find much information in the great work of Humboldt; Ward's Mexico in 1827, an excellent and copious work; Poinsett's Notes on Mexico in 1824; Bullock's Mexico; Alcedo's Diccionario de las Indias Occidentales; Gilliam's Sketches of his Travels in 1844, are interesting, &c., &c.

boats from the ship, with their crews, to be hauled up on shore, and remain there until the swell went down, that I might be ready to slip and go to sea.

It appears that the cordonazos come on an average once in 6 or 8 years, and we experienced none during our stay, although we had a gale on the night of the 21st of September. I was fortunately under way, and had plenty of room when it came on, having stood out to sea on the evening of the 19th, on account of the weather being bad, and fearing the full of the moon on the 20th.

It commenced about 9h 20′ P.M. from the S.E., flying round to S.W., heavy rain, thunder, and lightning, with a very heavy sea, reducing us to close-reefed main-topsail, and fore-staysail, washing away a boat, and obliging us to batten down. The squalls come on very suddenly, the prevailing winds being in the bad season S.E. to S. and S.W., and the heavy swell usually before and after the full and change of the moon. The swell is such as is seen in the Bay of Biscay in a heavy gale, and, unfortunately, usually sets into the bays before the wind comes.

I therefore think, that a ship caught at anchor off San Blas, or Mazatlan, by a cordonazo, would have small chance of escape, especially off the former, as she would either go on shore or go down at her anchors; to slip and stand out the instant it commences from S.E., is her best course.

The range of the thermometer for June was 77° to 86°; July, 80° to 87°; August, 81° to 89°; September, 83° to 92°; October, 83° to 90°. The barometer appeared to be of little service, usually remaining 30 inches; seldom varying above a tenth, except during a heavy squall, when it *rose considerably*.

After the 4th of November the coasting and other vessels again make their appearance on the West coast of Mexico.

San Blas is very sickly during the bad season. Guaymas is healthy, although the thermometer stands there at the astonishing height of 106° in July, August, and September; and, owing to the extreme dryness of the atmosphere, ships receive much injury by the wood opening. Furniture, apparently well seasoned, there cracks and falls to pieces.

On this coast there are some immense fish, of the ray species. I caught two of them, and with difficulty hoisted one on board; it measured 19 feet in breadth across the back, the mouth was 3 feet 5 inches wide, and the flesh was 3 feet 6 inches in depth in the centre. I had no means of ascertaining the weight, but found I could not lift it with the yard tackles and 60 men, it requiring 130 men with the heaviest purchases in the ship to hoist it in. These fish are common on the West coast of Mexico and Gulf of California, where they are more dreaded by the pearl divers than sharks or any other fish.*

Of the *eastern part of the Gulf of Tehuantepec* we know very little more than is shown in the chart from Malaspina's survey. The country is all volcanic inland, and, from the proximity of the mountains to the sea, there is no stream of sufficient strength to penetrate across the beach. The following is an extract from Juarros's work.

The *district of Soconusco*, or the Province of Cheapa, extends along the Pacific for 58 leagues, from the River Tilapa (or Sintalapa), which separates

* Nautical Magazine, Sept. 1840, pp. 407-8.

Mexico from Guatemala, to the plains of Tonalá, adjoining to Tehuantepec. The climate is extremely hot; the country level, pleasant, and fertile. It is watered by fifteen rivers: though very fruitful, very little of the land is under cultivation. The principal articles of commerce now carried on, are cocoa (the most esteemed of any in the kingdom), and fish, caught in the rivers and on eight fishing banks on the coast. There is some salt manufactured. In proportion as the products of the earth in this beautiful country are numerous, the abundance of wild beasts and reptiles is so great, as to render it intolerable and almost uninhabitable. This was the first province in the kingdom that Pedro de Alvarado conquered in 1524.*

Dampier sailed along the coast of the Gulf of Tehuantepec to the westward, in September, 1685, seeking for some refreshment for his men, for which Capt. Townley intended to "romage the country." After leaving Guatemala, he says: "We ran along by a tract of very high land, which came from the eastward, more within land than we could see; after we fell in with it, it bore us company for about 10 leagues, and ended with a pretty gentle descent towards the West.

"There we had a perfect view of a pleasant low country, which seemed to be rich in pasturage for cattle. It was plentifully furnished with groves of green trees, mixt among the grassy savannahs. Here the land was fenced from the sea with high sandy hills; for the waves all along this coast run high, and beat against the shoar very boisterously, making the land wholly unapproachable in boats or canoes. So we coasted still along by this low land 8 or 9 leagues farther, keeping close to the shoar, for fear of missing Capt. Townley. We lay by in the night, and in the day made an easy sail.

"The 2nd day of October (1685), Capt. Townley came aboard; he had coasted along shoar in his canoes, seeking for an entrance, but found none. At last, being out of hopes to find any bay, creek, or river, into which he might safely enter, he put ashoar on a sandy bay, but overset all his canoes, had one man drowned, and several lost their arms, &c.

"Upon his return we presently made sail, coasting still to westward, having the wind at E.N.E., fair weather and a fresh gale. We kept within 2 miles of the shoar, sounding all the way, and found, at 6 miles' distance from land, 19 fathoms; at 8 miles' distance, 21 fathoms, gross sand. We saw no opening, nor sign of any place to land at."—(Dampier, vol. i. pp. 231-2.) This was up to the head of the gulf.

The following are the remarks of Sir Edw. Belcher:—"From Morro Ayuca, on the western side, I shaped a direct course across the Gulf of Tehuantepec, expecting to meet with some of the gusts which are assigned to that region. In this we were entirely disappointed, although a fresh breeze favoured us for a short period. On approaching the eastern shore, near the Amilpas range, I was surprised, when at a considerable distance from the land, to strike soundings in 68 fathoms, which continued to decrease, very regularly, until ten at night, when we changed our course off shore in 11½ fathoms, without perceiving land, or hearing the 'surf sound,' which generally can be detected at night at 7, or even 10 miles.

* Translation by Mr. Baily, p. 20.

"Light baffling airs prevented our making much progress, but the tedium was in some measure dissipated by splendid views of these volcanic ranges. At one view no less than twelve conspicuous volcanic cones were visible, as far as the sea horizon was available, we endeavouring to fix their positions, by anchoring daily before noon. Our draughtsmen attempted to delineate them, but no effort of the pencil could convey an adequate idea of such magnificence. Far as the eye could reach to the N.E. numerous cones of extinct volcanoes were readily traced, as friends of yesterday; while to the westward we could barely trace, through the tropical haze, those with which to-morrow would bring us more intimately in connexion. Although apparently *overlooking* us, the nearest cone was at least 60 miles distant." *

TEHUANTEPEC is a city which has attracted some attention in Europe, from the surveys made of the Isthmus by General Don Juan de Orbegoso and Don Tadeo Ortiz in 1825, and Don José Garay and Signor Gaetano Moro in 1842-43, for the same purpose, of forming a communication between the Pacific and the Bay of Campeché, by means of a canal and the rivers falling into the lagoon of Tehuantepec, and the River Goazacoalcos, which runs into the Gulf of Campeché. It will be unnecessary here to give any detail of these operations, or of the proposed schemes; they will be noticed elsewhere. It would appear that there is a very great difficulty in the southern approaches to such a canal, on account of the shallowness of the lakes and the decreasing depth of the entrances of the Tehuantepec Lagoons.

Since the end of the sixteenth century Tehuantepec has been but very little frequented: the sea retires daily from its shores, and the anchorage deteriorates every year. The sand brought by the Chimalapa increases the height and extent of the sandy bars lying at the exit of the channel from the first lagoon into the second, and from this into the sea, so that Tehuantepec is no longer capable of admitting ships larger than ordinary brigs.†

The following are the observations of General Orbegoso, made during his exploration, by order of the Mexican government, in 1825:—"Between the base of the Cordillera and the ocean is a place which separates them from the lagoons, which, like an immense bay, communicates with the Pacific.

"This *llano*, or plain, consists of a shifting soil, formed by the detritus of the slate composing the adjacent hills, a species of rock, which appears from time to time in crossing it towards the lakes, and even on the coasts, where it forms the islands and capes.

"From the Cordillera to the lagoons the plain occupies a space of about 6 leagues. Those of the lagoons most inland may be about 4 leagues broad; and from its mouth, called the *Barra de Santa Teresa*, to the point where the two discharge themselves into the ocean, called the *Bocca Barra*, may be 3 leagues. This second bay, or inner lagoon, extends to the westward in the form of a marshy lake, to the extent of 9 leagues, under the name of *Tilema*; and to the East, to the *Barra de Tenola* (*Tonala*), to about 30 leagues.

* Voyage of the *Sulphur*, vol. i. p. 167.
† M. Chevallier, L'Isthme de Panamá, &c., p. 67.

"There is but little depth in either of these; that outside has not more than 16 feet (Castilian) in the centre, in the line of the canoe navigation. The bar communicating with the ocean has not been sounded; but, by observing the motion of the waves, it may be judged that, with a slight wind off the land, and in the season when the bad weather is not prevalent, there is not more than 6 feet mean water; but this may be raised to 2 feet more by the tide."*

The sandy tongue of land dividing the interior lagoon of the Tilema marsh and that dividing this from the ocean, is formed by the waters brought down by the rivers coming from the Sierra Madre, particularly the Chicapa and the Juchitan.

If the small harbour lying to the West of the mouth of the Tehuantepec, and which the floods and the want of a vessel prevented my examining, was more fit than it is to receive large vessels, it would be very easy to open a passage to the inner lagoon at Tilema, through the tongue of land separating them, by means of a short canal extended to the mouth of the Tehuantepec to the South of the hills of Huilotepec. The small port of which we have been speaking was the point at which Cortez prepared, and went to sea in, the first vessel which explored the coasts of the South Sea. The names here have varied at different times, in consequence of the changes which have occurred in the mouth of the rivers. The Rio de Tehuantepec sometimes discharges itself into the Tilema marshes below Huilotepec. It is eighteen years since it left this exit, and again ran into the sea. Some years afterwards a small portion of its waters again took the direction we have before mentioned. This will prove its small current, except during the season of the rains.†

The present roadstead of Tehuantepec, according to the Admiralty chart, is at Ventosa or Tehuantepec Road. It is protected on the West by the *Morro de Carbon*, a hill occupying the extreme point, and on which stands a vigia. N.E. from this a well on the beach is marked in the plan. The depth appears to decrease regularly from 9 to 3 fathoms, within a cable's length of the shore, on a sandy bottom. The Morro de Carbon is placed in lat. 16° 9′ 35″ N., lon. 95° 4′ 37″ W.

From Ventosa to the Morro de Ayuta of Malaspina's chart, the distance is 45 miles. In the interval are marked *Port Salinas, Estrete Island*, and *Port Don Diego*, but we have not been able to meet with any particulars of them.

The BAY OF BAMBA, does not appear on the chart. The following description is by Capt. Masters:—

"*Punta de Zipegua* is in lat. 16° 1′ N., lon. 95° 28′ 30″ W.; variation of the compass, 8¼° E. From this point to the Morro de Ystapa the coast runs about W.N.W. by compass. Between these points are several bluff headlands. They do not project far out from the general line of coast, and afford no shelter. Punta de Zipegua forms the eastern part of what is called the Bay of Bamba, and is a very remarkable headland. From the westward it shows itself with a bold dark cliff to the sea about 400 feet high. It projects out from the western line of coast nearly a mile, and forms a kind of double head. A short distance within the outer bluff is a peaked hill, with the appearance of a light coloured sandstone.

* Bulletin de la Société de Geog., tome xiv., 1830, pp. 16-17. † *Ibid.*

It is quite bare of vegetation. Further inland, between 1 and 2 miles, the ground rises higher in small hummocks. A few of them are quite bare, and others have a small quantity of stunted trees and bushes scattered over them. The head which forms the West side of the Bay of Bamba is not so high, nor does it rise so suddenly from the sea as Punta de Zipegua. It is also covered with bushes. The eastern side of Punta de Zipegua is covered with bushes and stunted trees; the sand only showing through the soil in very few places. When abreast of it, and off shore from 2 to 8 miles, the current was running to windward, W.S.W., from 2½ to 3 miles per hour. About N.E., by compass, from the Punta de Zipegua, and distant from 4 to 5 miles, is a high reef of rocks, called *Piedra de Zipegua*, or *Machaguista*, in the chart, Island of Eschevan. Its greatest elevation is from 60 to 70 feet; its length is about a third of a mile, running in an E.S.E. and W.N.W. direction. It is said there are no dangers near it but what can be seen. Between it and the main, from which it is about 4 miles distant, in a N.W. direction, is good anchorage; the best is close to the reef. The pearl oysters are plentiful near this reef; they are caught by the divers in the rainy season; the general line of coast from Punta de Zipegua toward Tehuantepec runs about N.E. by N., easterly.

"Shortly after I landed in the Bay of Bamba, the land-proprietor came down on horseback. He believed there was some Brazil-wood at a place called Rosario (in my instructions it was called San Francis de Aguatulco), and that Rosario was several leagues nearer Aguatulco. He said we were the only vessel larger than a canoe that had been on this part of the coast for a great number of years. No vessel had ever loaded hereabouts. The beach, or *Playa de Bamba*, is about 5 miles long, and must be very bad to land on with a fresh sea-breeze. There was more surf on it when we landed than was very agreeable. The boat was half filled, although the wind was blowing along the coast."

The *Morro de Santiago de Ystapa*, Mr. Masters says, is called the Morro de Ayuta in the chart. Near it is the entrance of the small river of Ayuta, the stream that runs by Huamilulu (hereafter alluded to) and Ystapa. There is a bar runs across the entrance of it. The canoes land on the beach in preference to going over it, as this is attended with danger.

A few miles to the westward is the *Morro de la Laguna*, near which is a large lake, from which the headland takes its name.

The BAY OF ROSARIO.—We do not find this name on the chart, but the following description, and directions for it, will be interesting. They are by Capt. Peter Masters, of Liverpool, who we presume to be the same gentleman to whom the West India navigators are indebted for his descriptions of the Tabasco River, &c., in the Bay of Campeché.

The West side of the Bay of Rosario is formed by the Morro de las Salinas de Rosario, and is in lat. 15° 50′ 25″ N., lon. 96° 2′ W., by four sets of lunars taken East and West of the moon. It projects about a mile beyond the line of coast. On its western side is a beach 4 or 5 miles in length to the next head. When abreast of Morro de las Salinas it appears like an island with two large rocks abreast of its eastern and western part; but the whole is connected to the main. What appears to be the eastern rock, is a broken rocky head, about 160 feet high. The western is about half the elevation. Both these heads terminate with a broken cliff; the tops of them are bare, and of grayish colour; the lower part is quite black, caused by the sea washing against them. Between these heads is a small sandy bay, which is at the foot of the Morro, and rises gradually from the beach to the top of the hill, and is about 180 to 200 feet high. It has a few straggling bushes on it, but its general appearance is very barren. The beach of Rosario is 10 miles long from Morro de las Salinas to Morro de la Laguna Grande, which is its eastern extremity. About half the distance between the Morros is a rock on the beach, about 40 feet high, and nearly the same diameter at spring tides. The water flows round it.

During the time of our lying in the Bay of Rosario, which was from the 12th of February to the 1st of April, we had three smart northers. These came on at the full and change of the moon. At this time the surf runs very heavy on the beach. Our boat was capsized several times whilst we lay here, in landing and coming off. At times the sea broke very heavily in all parts of the bay, that is, on the beach. I was caught on shore, a few days after arriving here, during the first norther, which came on suddenly with a parching hot wind. A cross, confused sea hove in from the South and N.E. The wind must have blown strong out in the gulf, from the same direction, and though it blew very heavily for three days, with the wind at times to the westward of North, the sea kept up until some time after the norther had ceased blowing. This is not generally the case, for a strong norther (and in particular if it veers to N.N.W.) beats the sea down, at which time landing is attended with little or no risk, which was the case when we had the last two northers. I was informed (and judging from appearances I think correctly) that very often when the wind is in North, or N.N.W., close in shore, it is N.E. in the offing, which makes it impossible to land on the coast. I remarked whilst lying here, at the full and change of the moon, when no norther was blowing, that although the surf ran so high that no boat could land, the vessel lay without any motion. We were moored less than 300 fathoms from the shore. The surf appeared not to be caused by a swell rolling in, and agitating the sea at the surface, but to rise from below, and without any apparent cause, as we had light winds and fine weather the most of the time we lay here. On another occasion I was caught on shore with a boat's crew for three days. In attempting to get off to the ship, the boat was capsized and stove. It was then, and had been for a week previous, nearly a calm. The heavy ground swell invariably hove in from the S.S.W. We fortunately escaped from this beach without losing any of our people, which was more than I expected, having had three laid up at different times, who were saved from being drowned by a mere chance.

Had our cargo been shipped off from the western part of the bay, which is well sheltered from the south-western swell by the Morro de las Salinas, we should have been loaded in one-third the time. It is intended, by the parties who are cutting the wood, to take the next cargo there. The place where our cargo was piled was at the most exposed part of the beach. It was 3½ miles to the eastward of the Morro de las Salinas, and was open to every swell. Our only hope of getting the cargo off was by mooring our long boat outside the surf, and having a messenger passed from her to the shore. A capstan was fixed on the beach, which hove off the wood as it was made fast to the messenger from the long boat, and we had two small canoes which brought it to the ship.

In the event of a vessel coming here, or any other part of the coast, to load, (excepting at the port of Acapulco,) the great expense of getting the cargo off should be considered. The whole of our cargo was little less than 90 tons, equal to about 235 tons measurement of 40 feet. It cost the shipper upwards of 8,000 dollars, and if the labour of getting the cargo off had fallen on the ship it would have cost a great deal more; for, independent of the deficiency of system and energy on the part of the people who sell the wood, working people cannot always be got, even by themselves, who are well known, so that a stranger would stand but a poor chance of getting any work done, the Indians are so idle.

The nearest habitation to the part of the bay which we loaded at was the Hacienda de Rosario, about 2 leagues off; but it scarcely deserved the name of a Rancho, yet this was where my consignees lived. In consequence of a little chapel being built there, dedicated to the Virgin, in which mass is said once or twice a year, it has the title of Hacienda.

The land is poor, and the soil is very thin, at a trifling elevation above the coast, and also very rocky. The low land is well covered with trees, and in places appears to be fertile; as there had been no rain everything was parched up. The Nicaragua, or Brazil-wood, does not grow on land which is fit for cultivation; it is mostly found where the soil is very thin, and where the ground is rocky, and at an elevation not more than 400 feet above the level of the sea; at least such was the case here.

About 5 leagues, in a north-easterly direction from the Hacienda de Rosario, is the Indian town of Huamilulu, from which we got a few supplies.

The Indians of Tehuantepec are a better looking race than those of Huamilulu, their features nearly approaching the European, and almost as fair: their language is very different from that of Huamilulu; it is quite harmonious. With regard to the liquid sounds, it resembles the Spanish, though decidedly it is a language quite different. The dress of the Indians of Tehuantepec is similar to the people of Huamilulu; but, at the town of Tehuantepec, I was informed, their dress is often very expensive, on account of so many ornaments that are used. As I saw none of their gala suits, I can say nothing about it. Although the distance from Huamilulu to Tehuantepec is not more than 15 leagues, and there are villages a great deal nearer, yet I think there cannot be the least doubt but that they are of different origin. The manner in which the Huamilulu Indian would pronounce the word "Tehuantepec," for instance, is this: each syllable is pronounced in a middling uncouth manner, until the "p" is on the lip, in order to pronounce the last syllable: then he finishes with a sound as near as possible like a person who has the hiccough; whereas the Tehuantepec Indian finishes the "e" of "pec" as if it were a "pee."

In the woods about Rosario I saw a number of very fine deer. It is said there are also a large quantity of tigers and wild boars, and if all that is said of them be true, they are very fierce, but I rather think the contrary, as no person that I spoke to on the subject remembered any of their acquaintances being attacked by them. For my own part I saw none. The woods also abound with parrots, parroquets, macaws, and a number of other birds of most beautiful plumage. There were more doves than any other species. A few squirrels are caught occasionally. On the beach we saw a few snipes, curlews, and plovers; also an immense number of zapalates and pelicans. The people who were working about the cargo caught a number of iguanas, in one of which there were 54 eggs, as large as those of pigeons. The price of a live ox is from 5 to 6 dollars; sheep are very scarce; indeed I saw neither sheep nor goats whilst we lay here. Fowls are 3 dollars per dozen; turkeys are from 6 to 8 reals each; rice and sugar (panache) are moderately reasonable. This is sent here from Tehuantepec and the neighbourhood. Plantains and tamarinds are also cheap, and other things equally reasonable. The difficulty is to get it from the towns; although I employed a person to procure what trifling supplies we were in want of, yet we came away short of many things: the constant excuse was, when desired to get the articles which had been ordered, that they would be at the beach in a day or two.

In the bay there is an immense quantity of fine large fish; we tried every method, without success, to catch them with hook and line. We caught a few messes with grains, but the most of them were caught by going in the boats after dark, and having a large torchlight in the bows of her, striking them as they came towards the light. A small seine would pay well for itself here, as there is plenty of salt to be had to cure them, from a lake about a mile from the beach, which is nearly dry. In April it had a crust of excellent salt over it, about 6 inches thick, and has a very pretty appearance. There are two guards stationed here in the dry season, to prevent its being made use of, as it would injure the government. Salt is a government monopoly; but, nevertheless, the salt of Rosario is sold by the guards themselves. When the rain sets in the salt dissolves; the sea also breaks into it during heavy gales of wind, and fills the lake up, which again evaporates in the dry season.

There was an immense number of rays in the bay, commonly called by sailors devil-fish: these fish are very dangerous to the pearl divers. We struck several of them, but did not succeed in getting any on board. In coming from the shore one day, a very large one came under the fore part of the boat, and attempted to clasp her with its fins; it could not have been less than 8 feet across, as the fins were above the gunwale of the boat, and lifted her bows more than a foot out of the water. They make tremendous leaps at times, and allow the boat to approach quite near, so that there is no difficulty in striking them. The ray we had alongside weighed not less than 600 lb. They are said to be very good eating; their shape is very like the skate, excepting that their tails are smaller. We saw several lobsters in the western part of the bay; they were most plentiful near the Morro de las Salinas. We tried several methods to catch some, but without success.

In addition to what has already been said about this part of the coast, it can be known by the low land at the back of the beach of Rosario. This runs in from 1 to 2½ leagues before there is much rise in it, and is thickly covered with trees. From North to N.W. of Morro de las Salinas, nearly 2 leagues from the shore, the rising ground is formed by a number of small barren hillocks. From our anchorage where we loaded at, the following bearings were taken, lying in 9½ fathoms water, sandy bottom. There are two large patches of a whitish appearance, the farthest range of the Cordilleras, the eastern is also the lowest, and bore N. 59½° W. The appearance cannot be seen, unless from a little to the westward of Morro de las Salinas. This has every appearance of being a waterfall, and rises from the other patch in a N.W. direction at about an angle of 45°. It issues from a small valley in the Cerro del Chonga. The highest point of this range has but a small elevation above it, and is covered with trees. The waterfall inclines towards the South, and can be seen for several hundred feet descending before it is lost sight of amidst the forest below. Cerro de Zadan bore N. 89° W., and the extreme bluff of Morro de las Salinas, S. 36° W., 3½ miles. The eastern point well within the bearings, and Punta de la Laguna Grande, N. 71° E. 6 to 7 miles, and rock on the beach (already mentioned as 40 feet high), N. 65° E., and the galena or shed, under which the cargo was piled, N. 26° W. half a mile; bearings by compass.

At the western part of the bay are four palm trees close to the beach. The distance from the Morro de las Salinas is about half a mile, and between these trees and the Morro is a larger cluster of palms. Between these two clusters is at all times the best place to land, as a boat can beach here with comparative safety, when at every other part of the bay the sea runs very heavy. At the neaps we found the place quite smooth, with the exception of a sea heaving in at about every 10 or 15 minutes; but it causes no risk to a boat provided she is kept end on.

At the south-western part of the beach, and where a small pathway leads to cross the Morro de Salinas, close to the sea-side, in the cliff of a rock, is a small spring of excellent water. We always found it clear and cool, even at noon; my consignee said we could fill the ship's stock of water from it with dispatch, but I soon found out that he knew nothing about it. The quantity that could be filled in a day did not exceed 30 gallons, and after having landed all our water casks we had to re-ship them, through a great deal of surf, and land them at the galena abreast the ship. We filled our water at a well about a mile from the beach, but the supply was very limited, it being the only well that had water in it up to the day of our sailing. We did not complete our stock.

A captain of a ship should trust to no promises when he comes here, either with regard to supplies or anything else, no matter by whom made; and, as water and fuel are indispensable articles, the filling the one, and cutting the other, should be immediately commenced on their arrival by some of the crew. It is useless to employ Indians to work for the ship, (that is, on shore,) the greatest part of them will neither be led nor driven. On board they answer better (that is, a few of them) to haul the wood about in the hold. I found the promises of Indians, and, as they called themselves, "*gente decente y Civilizado*," on a par.*

From the Bay of Rosario to the Island of Tangolatangola there are several small headlands, which do not project much beyond the general line of coast, with the exception of Morro de las Salinas de Rosario. Most of them have a steep

* Nautical Magazine, February, 1840, pp. 73—82.

cliff facing the sea, with fine sandy beaches between them; at the back of which are scattered a few small trees and bushes, the land rising in very irregular shaped hills towards the Cordilleras. Abreast of the beaches, between the heads, the anchorage is quite clear, and when in from 9 to 12 fathoms water the distance off shore is about a mile, with sandy bottom.

The *Island of Tangolatangola* is not shown on the charts. It is, however, mentioned by Dampier.

"At the small high island of Tangola there is good anchorage. The island is indifferently well furnished with wood and water, and lies about a league from the shoar. The main against the island is pretty high champion savannah land, by the sea; but 2 or 3 leagues within the land it is higher, and very woody." *

Capt. Masters describes it thus:—"The Island Tangolatangola is E.N.E. 3 miles from Guatulco, and is separated from the main by a channel a quarter of a mile wide. This makes from the westward as a part of the main land; the outer part of it is quite bluff, or rather a cliff of a brownish stone, the strata of which is horizontal, and has the same geological appearance as the land on the main nearest it to the N.E., and of the same height, namely, about 150 feet. Within the island, and round the western side, is the entrance of the Bay of Tangolatangola; it runs in about N.E. 2 miles. At the bottom of the bay is a fine sandy beach; the anchorage is said to be very good in it, but not equal to Guatulco; its entrance is more than a mile across, and continues nearly the same to the bottom."

The *River Capalita*, both according to Dampier and Malaspina's chart, must fall into the sea hereabouts. Dampier says that it is rapid and deep near its mouth.

PORT GUATULCO lies next along the coast, and is a very secure harbour. According to Sir Edward Belcher's survey of it in 1838, some islets that lie off its mouth are in lat. 15° 44′ 25″, and lon. 96° 10′ W. Dampier's clear and graphic account of it is as follows:—"Guatulco is one of the best ports in all this kingdom of Mexico. Near a mile from the mouth of the harbour, on the East side, there is a little island close by the shoar; and on the West side of the mouth of the harbour there is a great hollow rock, which, by the continual working of the sea in and out, makes a great noise, which may be heard a great way. Every surge that comes in forceth the water out of a little hole on its top, as out of a pipe, from whence it flies out just like the blowing of a whale; to which the Spaniards also liken it. They call this rock and spout 'the buffadore' (*bufadero*, Spanish, a roarer), upon what account I know not. Even in the calmest seasons the sea beats in them, making the water spout at the hole, so that this is always a good mark to find the harbour by.†

* Dampier's Collection, vol. i. p. 232.

† This description will also exactly apply to another of these singular phenomena, the *Souffleur* (French, blower), at the South point of the Mauritius. Here the water is driven up with enormous force to the height of 120 to 140 feet above the waves, and may be heard a long distance. They are also seen, too, at times, around the bases of icebergs, and there was one, the Devil's Trumpet, on the coast of Cornwall. Other instances, less striking, might be adduced of these singularities, which are well worthy of a seaman's attention, showing, as they do, the power the waves exert, which, to raise such a column of water as above-mentioned, must be from 3 to 5 tons per square foot.

"The harbour runs in N.W., but the West side of the harbour is best to ride in for small ships, for there you may ride land-locked, whereas anywhere else you are open to the S.W. winds, which often blow here. There is good clean ground anywhere, and good gradual soundings from 16 to 6 fathoms: it is bounded by a smooth, sandy shore, very good to land at, and at the bottom of the harbour there is a fine brook of fresh water running into the sea. Here formerly stood a small Spanish town, or village, which was taken by Sir Francis Drake; but now there is nothing remaining of it besides a little chapel standing among the trees, about 200 paces from the sea.* The land appears in small short ridges, parallel to the shore and to each other, the innermost still gradually higher than that nearer the shore; and they are all clothed with very high flourishing trees, that it is extraordinarily pleasant and delightful to behold at a distance; I have nowhere seen anything like it."

Dampier, when at this place, went about 14 miles inland from the port, when they reached a small Indian village, where there was a great quantity of "*vinellos*," vanilla, drying in the sun in course of preparation. This valuable article is only met with in this immediate neighbourhood, and Dampier relates all he could gather of its preparation.†

The following remarks, by Capt. Masters, will complete the description:— "Santa Cruz, Port of Aguatulco (*Guatulco*), is very difficult to make; it is situated in a small bay about half a mile wide at its entrance, and runs in to the northward upwards of one mile and a half. At the bottom of the bay is a sandy beach; on its eastern part two huts are built, which cannot be seen unless close in-shore. E.S.E. three-quarters of a mile from the eastern point which forms the bay, is the *Piedra Blanca*. This is a reef of rocks extending East and West about a quarter of a mile. The western part of the reef is nearly 40 feet high; for about one-third of its length it is of the same elevation; the remaining two-thirds to the eastward is low, in places level with the water. When abreast of it and off shore a few miles, it appears to be a part of the coast. Although it is called Piedra Blanca, it is a dark, irregular shaped reef of rocks.

"The anchorage in Guatulco is said to be good. It is well sheltered from all winds, except between East and S.E. by S.; but, as the strongest winds blow from the northward, except in the rainy season, it may be considered a very safe port. It is the only place that can be considered a harbour to the eastward of Acapulco, and even in the rainy season, I was informed that a vessel might lay there in perfect safety. The depth of water in the bay is from 7 to 9 fathoms, with a clear bottom.

"To the westward, half a mile from the head which forms the western part of the bay or harbour of Guatulco, is a bluff point, or head, under which is a good leading mark for knowing the harbour. There is a cave in one of the rocks, level with the water, and close in-shore, and every swell that heaves in throws a quantity of water into the cave; and as the cave has a small aperture in the upper

* Guatulco seems to have been an unfortunate place during the buccaneering expeditions against the Spaniards, for Sir Francis Drake sacked the place in 1574, and it was burnt in 1587 by Sir Thomas Cavendish, among other places. The reader will find many notices of these and similar incidents in Admiral Burney's Collection.

† Dampier, vol. i. pp. 233-4.

part of it, the water flies up resembling the spout of a large whale. It has often been taken for one by strangers. We were also deceived by its appearance. In the night or foggy weather when it is calm, or blowing a light breeze, the sound can be heard some distance like a whale blowing. This place is called the *Bufadero*.

"When about 5 miles off the shore from the Bufadero, the western extreme point of land has a broken rocky appearance, and is not so high as the land adjoining. When about 2 leagues off shore from the Bufadero, another cape further to the westward can be seen. Its extreme point is rather low, but rises gradually inland to a moderate elevation.

"To the westward of Santa Cruz are two bluff heads, which, when abreast of them, might be taken for islands. The first is about 3 miles from the port, the other is 2 miles further to the westward, and has a white sandy beach on its western side. On the eastern side, also, of the eastern head is a small sandy beach, from which to the Bufadero the coast is rocky. The land which crowns this part of the coast is covered with stunted trees and brushwood. N. 8¼° W. (by compass), between 4 and 5 leagues, is the *Cerro Zadan*. Its top is bell-shaped, and it has a ridge on its N.E. side, connecting it with the higher range of the Cordilleras. The Cerro Zadan is elevated above the sea rather more than 6,000 feet. The mountains further inland a few leagues cannot be much short of 10,000 feet high, as they can be seen over the Cerro Zadan.

"The Port of Aguatulco is so bad to make, that vessels have been upwards of a fortnight in searching for it. It was by the greatest chance possible we had not passed it, although we were not a mile and a half from the shore. The two huts which were on the beach can scarcely be distinguished from the trees near which they are built."

The coast beyond Guatulco trends a little to the South of West, for 20 or 30 leagues. At about a league West of Guatulco is a small green island called *Sacraficios*, about half a mile long, and half a mile off the land. There appears to be a fine bay to the West of the island, but it is full of rocks. The best anchorage is between the island and the main, where the depth is about 5 or 6 fathoms. The tide runs strongly, rising and falling 5 or 6 feet.

The land winds are here at North, and the sea breezes generally W.S.W., sometimes at S.W. with an easterly current.

Westward of Sacraficios the shore is all formed of sandy bays, the country tolerably high and wooded, with an enormous swell tumbling in on the shore.

Port Angeles is a broad, open bay, with two or three rocks on the West side. There is good anchorage all over the bay in 30, 20, or 12 fathoms water, but you must lie open to all winds, except the land winds, until you get into 12 or 13 fathoms; you will then be sheltered from the W.S.W., which are the common trade winds. The tide rises about 5 feet; the flood setting to N.E., and the ebb to S.W. The landing in the bay is bad, behind a few rocks; the swell is always very great. The land bounding the harbour is tolerably high, the earth sandy and yellow, in some places red. It is partly wooded, partly savannahs.*

* Dampier, vol. i. pp. 230-40.

Eighteen miles westward of Port Angeles is a small rocky island, half a mile off shore. The coast is all small hills and valleys, and a great sea falls upon the shore.*

Near the *Alcatras Rock* (in lon. 97° 30′ W., Malaspina), the land is moderately high and wooded; farther within land it is mountainous. Five or six miles to the West of the Alcatras are seven or eight white cliffs by the sea, which are very remarkable, because there are none so white nor so close together on all the coast. There is a dangerous shoal lying S. by W. from these cliffs, 4 or 5 miles off at sea. Two leagues to the West of these cliffs there is a tolerably large river, which forms a small island at its mouth. The eastern channel is shallow and sandy, but the western channel is deep enough for canoes to enter."†

ACAPULCO.

This celebrated seaport has sadly fallen from the high position it held among the places of commercial importance in the world. It owed all its prosperity to the system pursued by the Spanish colonial policy, and, when that power became annihilated in the new world, Acapulco descended, not to its level as a harbour, for it is one of the finest in the world, but to that of the capability of the surrounding country in supporting it. This, as is well known, is very limited, and the foreign trade that it has across the Pacific is of very minor importance. Its principal support is in the commerce carried on between Callao, Guayaquil, &c.

In addition to the changes in its external relations, it has some very serious drawbacks to any permanent prosperity. The climate is extremely hot, and pernicious to European constitutions. This is increased by the proximity of a marshy tract to the East of the town. During the dry season this marsh dries up, and occasions the death of great quantities of small fish, whose decay under a tropical sun produces no ordinary amount of pestilential vapours to be diffused, a fruitful source of the putrid bilious fevers so prevalent here and in the vicinity.

Being entirely surrounded with high mountains, the sun has intense power, and the usual breezes are in a measure intercepted. To remedy this, an artificial cut was made through the chain of rocks which surrounded the town; this has caused a freer circulation of air.

It has been well surveyed, and the plan from the united observations of Sir Edw. Belcher and M. de Tessan, the Hydrographical Engineer to the expedition of Admiral Du Petit Thouars, in the *Venus*, will give a perfect idea of the port.

It consists principally of one extensive basin, in an angle of which, on the N.W. side, stands the town. At its head are some whitish rocks, the Piedras Brancas, useful as marks in entering. There are two entrances, formed by Roqueta or Grifo Island. That to the North of it is called the Boca Chica or Little Entrance, and is narrow. The principal entrance is between the South point of Grifo (Siclata Point) and the S.E. point (Bruja Point) of the harbour, and is above 1½ miles in width. It is quite clear.

Eastward from Bruja Point is Port Marques, extending 2¼ miles E.S.E. Its S.W. point is named *Diamante Point*. This bay, too, is quite clear.

* Dampier, vol. i. pp. 239-40. † *Ibid.* p. 242.

"Acapulco is familiar," says Capt. Basil Hall, "to the memory of most people, from its being the port whence the rich Spanish galleons of former days took their departure to spread the wealth of the western over the eastern world. It is celebrated also as Anson's delightful voyage, and occupies a conspicuous place in the very interesting accounts of the buccaneers: to a sailor, therefore, it is classic groud in every sense. I cannot express the universal professional admiration excited by this celebrated port, which is, moreover, the very *beau idéal* of a harbour. It is easy of access; very capacious; the water not too deep; the holding ground quite free from hidden dangers, and as secure as the basin in the centre of Portsmouth dockyard. From the interior of the harbour the sea cannot be discovered; and a stranger coming to the spot by land would imagine he was looking over a sequestered mountain lake." *

EARTHQUAKES are a great scourge to Acapulco, and must prevent its ever becoming a substantial town; at present it is poor and mean. When Capt. Basil Hall came here there were about thirty houses, with a suburb of huts built of reeds, in open work, to admit the air. The town is commanded by the "extensive and formidable castle of San Carlos.

Besides the earthquakes, the heaviest of which occur between March and June, the rainy season is also another great drawback, and is felt here severely. It commences about the middle or end of July, and continues until the end of October. Owing to the immediate vicinity of a very lofty chain overlooking the town (one of 2,790 feet), the fall is heavy, and almost incessant. It has been asserted that, in 1837, the rain gauge frequently indicated 28 inches in 24 hours. During this period the inhabitants are compelled to use every precaution to keep their houses dry, particularly under foot; a neglect of this is supposed to produce fever. The heat during this period is excessively oppressive, especially in May, when the temperature seldom falls below 98°. Water then becomes scarce, and, towards the end of the dry season, the ponds run dry, and wells are their only resource.—(Sir E. Belcher.)

Capt. Sir Edw. Belcher says (January 12, 1838): "We made the high paps of Coyuca, to the westward of Acapulco; but I cannot persuade myself that they are good landmarks for making the port. In the offing they may be useful, if *not obscured*.

"Acapulco may be approached from the southward or westward, by keeping the western cone open of the land, which will lead up to the Boca Chica entrance, or until Acapulco port is so close under the lee, that no further marks are necessary. There is not *any hidden* danger in the entrance to Acapulco. Keep a moderate distance from either shore; 5 fathoms will be found alongside all the rocks, and 25 to 30 in mid-channel. Round Point Grifo *sharp*, rather than stand over to San Lorenzo, as the wind, generally westerly, heads on that shore. If working, tack when the rocks on the South point of Tower Bay show in the *gap*.

"The two best berths are off the rocks alluded to: that outside is preferable; but in either case let the outer rock bear W.S.W. or W.N.W., so that a hawser

* Extracts from a Journal, &c., vol. II. p. 1.

fast to the rock may keep your broadside to land or sea breezes, and prevent a foul anchor.

"The harbour of Acapulco has long been reckoned, for its size, one of the most complete in the world. It affords sheltered, land-locked anchorage of 16 fathoms and under, in a surface of one mile square; which, allowing for moorings, would, at half a cable's range, or one cable asunder, accommodate 100 sail of vessels, even of the line. The bottom is sandy at its surface, but clayey beneath, and holds well.

"It would naturally be inferred that, surrounded on its North and East sides by mountains ranging from 2,000 to 2,700 feet, and by others of 300 to 500 feet on the West, the breeze would scarcely be felt, and the heat be intolerable. This is confined to the town limits: at our observatory (Capt. Belcher's), and the port, San Carlos, we enjoyed a constant breeze.

"In all harbours there may be objectionable berths, but in that of Acapulco, if care be taken to keep in the line of what I have designated 'West gap,' or neck of the peninsula open of the South point of the town bay, both land and sea breezes will be felt in their full strength, and free from causes which would heat them before entering the port, the neck being but a few feet above the sea level.

"Water of good quality was found at several points between the fort and Obispo rock; but the two best streams are between the fort and San Lorenzo."*

The following remarks are by Capt. Basil Hall, as given in his Hydrographical Memoir:—"The entrance into Acapulco lies about midway between the East and West extremes of a high portion of the coast, which stands forward in a very prominent way to the southward of the rest of the coast; the centre part is the highest, probably about 3,000 feet above the sea.

"Both ends run off to bluff points: the eastern one is called Point Marques, and is distinguishable by its presenting a set of steep white cliffs; it is succeeded on the eastern side by a long line of white sandy beach, backed by a lower range of country, which reaches to the foot of the hills.

"The entrance of the harbour, when bearing North, has Point Marques on the East, and a small promontory on the West, not unlike each other, and both having white cliffs.

"The entrance may also be distinguished by a remarkable white rock, which lies nearly abreast of the middle part of the white beach at the bottom of the Bay of Acapulco; this may be seen with ease at the distance of 3 or 4 leagues, when it bears on any point of the compass between N. ½ W. and N.N.E. ½ E. by compass; in other cases it is shut in, either by the land near Port Marques or by the Island of Grifo, off the entrance.

"Port Marques lies close round the point of that name, and is not very distinguishable till it be approached within a couple of leagues.

"There are no dangers in Acapulco harbour, except one shoal nearer the shore than any ship would think of going; there can be no difficulty in making out the situation of this harbour, when it is understood that its latitude is 16° 51′ N.,

* Voyage of the *Sulphur*, vol. i. p. 143.

and that it is pointed out by a lofty promontory, which maintains its height and abruptness to the very sea, without any low land; this high land is covered with trees or shrubs, and everywhere presents a green surface, except where it meets the sea, and then its face is laid bare, and shows only naked white or gray cliffs of granite, not of a massy character, but splintered in all directions.

"The anchorage is abreast of the town in the western corner of the bay, near two white rocks, to one of which a hawser may be made fast, and the ship canted to the sea breeze. There is a remarkable high land considerably to the eastward, and much further inland than the promontory of Acapulco, having a long tabular top, which rises considerably above the neighbouring peaks. But there is no difficulty in distinguishing this promontory, when coming along shore from the eastward, as it is the first high coast land which reaches to the sea, and terminating a line of low white beach. On the western side, the coast land is high, and offers such a variety of forms and heights, that it may not be easy to distinguish the high land of Acapulco, or rather a stranger might perhaps mistake some other part of the coast for it.

"The paps of Coyuca are the marks generally pointed out as affording the means of distinguishing the land; they lie some leagues to the westward or W.N.W. of the promontory of Acapulco, and might be better described, I think, as a castle or fort-like mountain, than as paps. There is first a very abrupt precipice facing the West, with a surface somewhat tabular, but not quite level; the top being nearly equal, in horizontal length, to what the cliff is in abrupt height; then there is a nick or gully which is succeeded by a flat peak, not very unlike a pap. After this there is a long hog-backed ridge, with an irregular peaked termination at the eastern end. The land between this and Acapulco sinks considerably, and though it still maintains a tolerable height, the promontory is always sufficiently conspicuous.

"When its extreme South point bears about East, and indeed when it bears considerably to the northward of East, there is no high land to be seen beyond it to the eastward.

"At the distance of 7 or 8 leagues the land about Point Marques makes like an island."

Mr. Jeffery, R.N., has also favoured us with the ensuing observations on approaching Acapulco:—

"On the 5th December, 1833, made the land near Acapulco, to the eastward of the harbour. It is very high and conspicuous, and known by the name of the tabular mountain. It slopes a little to the westward, and the eastern end is rather the highest. Being about North 20 miles, Point Marques, which also appears tabular, will be seen under the East part of it. From Point Marques the coast stretches nearly as far as the western part of the tabular mountain, where it again forms a point; as this is a very remarkable part of the coast, it can seldom be mistaken for the East end of the above high land; the land hollows out a little, and runs up to a peak; it again hollows out and forms another peak. These are not what are called the Paps of Coyuca, for the paps lay to the westward of the harbour. From this last peak it runs rather low in with a few hillocks, until it is lost to the view. When near the harbour, and having Point

Marques bearing N.E. ½ E., and the paps N. ¾ W., the entrance to the harbour will bear N.E. ½ N.; as you near the shore Point Marques will make like an island, with a sandy cove inside of it. Towards Acapulco, the Island of Grifo, which forms the West side of the harbour, will appear like a part of the coast, and not unlike the land of Point Marques. To the westward of Grifo is another sandy beach. If the weather be clear, you will see the continuation of the land to the eastward, running out like a low point, covered with trees. When the entrance to Acapulco bears N. ½ E. 10 or 12 miles, you will see Point Marques and the Island of Grifo making like each other, and both presenting white cliffs; the harbour is directly between the two, and in the centre you will see a large gray rock (*Piedras Blancos*); the paps at this time will be seen to the westward of the harbour; they are two small hillocks on very high land, the westernmost one appears somewhat higher and sharper than the other.

The coast to the West of Acapulco is low, and formed by what is called the beaches of Coyuca. The swell sets so strongly upon this long, shallow, sandy bay to the West, that it is impossible to get near it in a boat or canoe; yet it is good clean ground, and good anchorage a mile or two from the shore. The land near the sea is low, and moderately fertile, producing spreading palms and other trees, which grow in clumps all along this bay. The land-wind here is generally from N.E., and the sea-breezes from S.W.

The country inland is full of small, peaked, barren hills, making as many small valleys, which appear flourishing and green.*

The Paps of Coyuca, serviceable as a mark for approaching Acapulco, have been before noticed. They are remarkable, and may be readily distinguished, as Capt. Belcher says, "when not obscured." According to the Spanish chart, they are in lat. 17° 6′ N., lon. 100° 0′ W.

Having arrived at *Point Jequepa* (or Tequepa), the coast trends rather more northerly, and at about 20 leagues is the *Morro de Petatlan*, a high mountain, which may be known by the islands which surround it.

"The hill of Petaplan (Petatlan) is a round point stretching out into the sea, appearing at a distance like an island. A little to the West of this hill are several round rocks."† Dampier anchored on the N.W. side of the hill, passing inside of these rocks, between them and the round point, where he had 11 fathoms water.

Between this point and several white islands, is the small port of Siguantanejo, or Sihuantanejo.

Port Sihuantanejo has been surveyed by Capt. Kellett, R.N., 1847, and from his plan it would appear to be an excellent harbour, but open to the S.W. There is no hidden danger in going, and the entrance is sufficiently marked by the bold coast to require no directions. The position of the observations at the head of the port, Capt. Kellett places in lat. 17° 38′ 3″ N., lon. 101° 30′ 52″ W.

Dampier passed along this coast, and says:—"About 2 leagues West from Petaplan is Chequetan (Sihuantanejo?). At 1½ miles from the shore is a small bay, and within it is a very good harbour, where ships may careen; there is also a small river of fresh water, and sufficient wood."

* Dampier, vol. i. p. 247.

† *Ibid.* vol. i. p. 248.

He landed at a place he calls Estapa, a league to the West of Chequetan, and taking a mulatto woman for his guide, his companions plundered the unfortunate Acapulco carrier, as related in vol. i. p. 250.

To the West of Estapa, Dampier says the land is high, and full of ragged hills; and West from these ragged hills the land makes many pleasant and fertile valleys among the mountains.

All this coast is lined with villages and salt-works (salines), worked by the inhabitants. The approach to the coast is clear, but there is no safe anchorage, and there is not a single important river. That of *Sacatula* or *Zacatula*, which comes from the Volcan de Jorullo, as also the *Rios Camuta* and *Coalcaman*, are not navigable. The bays of Tejupan and of Santiago, lying to the South and East of the promontory called the *Paps of Tejupan*, can only be considered as open roadsteads.

The Paps of Tejupan (Thelupan, Dampier) is a very remarkable hill, which, towering above the rest of the hills around, is divided in the top and makes two small parts.*

The coast then runs more northerly towards the entrance of Manzanilla Bay, and in the interval are the mouths of the rivers *Coaguanaja*, *Apisa*, and *Armeria*.

MANZANILLA.—This harbour was visited by Sir Edward Belcher. The situation of it is in lat. 19° 3′ 30″ N., lon. 104° 16′ W. The ensuing description is that given by M. Duflot de Mofras, who was secretary to the French Legation to Mexico.

The harbour of Manzanilla, or Salagua, is far better than those of San Blas and Mazatlan. It has four excellent anchorages, and ships of every tonnage may anchor here safely at all times.

In making the harbour of Manzanilla the latitude must be reached in the offing, and then steer for the land, having as a landmark the double peak of the Volcano of Colima. When near the harbour, the entrance of which is wide, it will be seen that it is divided into two bays by the *Punta de la Audiencia*, which runs to the South. The bay to the East is called Manzanilla; that to the West, Santiago; and in this is the best watering place. When the wind is from the South, the preferable anchorage is in the eastern bay, which may be reached from the entrance by bearing to N.E. ½ E., and anchor in 7 or 8 fathoms, in front of the *San Pedrito Rock*. You may also enter the western bay by steering N.E. ½ N., passing the rocks called *Los Frayles*, which surround the second point, *Juluapan*, and drop anchor behind the hill in 5 or 6 fathoms, at a short distance from shore. To fetch the anchorage in Santiago, run North, a little East or West, to avoid the *Estrada Rock*, lying at the extremity of the *Punta de la Audiencia*, which, as has been before mentioned, is right in front of the entrance. The tide occurs every 24 hours, the flood in the morning and the ebb in the evening. It rises about 7 feet, and the current runs to the South. At Salagua wood and water are very abundant, and bullocks very cheap. Vanilla, tortoise (with good shell), good pearls, &c., may be had.

Manzanilla has been thrown open to foreign ships, and received several rich

* Dampier, vol. i. p. 251.

cargoes from Europe; but in 1836, the jealousy of the merchants of Tepic and San Blas caused it to be closed, as well as that of Mazatlan, which was opened later to external commerce.

The advantageous situation of Salagua would give it superiority over other ports in supplying the provinces of Colima, Michoacan, and Jalisco; and above all, it would export more quickly, and at less cost, the merchandise from Guadalajura and the celebrated fair of San Juan del Rio.

Manzanilla is about 20 leagues' distance from the city of COLIMA, the capital of the territory of that name. Except frequent earthquakes and goitres, with which the inhabitants are affected, there is nothing remarkable in the city, which contains about 30,000 or 40,000 inhabitants, wholly occupied in agriculture and commerce.*

At 8 leagues to the E.N.E. is the *Volcan de Colima*, the westernmost of the Mexican group. Its entire height is 12,003 feet; it is in activity, and sulphureous vapours, cinders, and stones are emitted; but it has not discharged any lava for a long period. The diameter of its crater is 500 feet, and its mouth is perpendicular. The flanks of the mountain are barren and cliffy. The sulphur on it is of a bad quality. At a league North of the volcano there is an extinct crater, which exceeds the former in elevation by 710 feet, and the height of which above the port is 12,713 feet. Its summit is covered with snow, and it may be seen at very great distances at sea, and offers, when the sky is clear, an excellent point of recognition for navigators approaching Manzanilla.

The valley in which Colima is situated seems to be formed of volcanic products and decomposed lava. The vegetation consists of palms, aloes, and superb orange trees. Above the usual level these tropical plants are replaced by forests of sombre pines.

The next remarks are those of Sir Edward Belcher:—

"The bay is small but safe, the anchorage is good, the water brackish. There are no houses; men and families living exposed under the trees: and had not the *Leonora* (an English bark, then at anchor) been there, it is probable that we should not have seen a soul."—(1838.)

While at Manzanilla, Sir Edward Belcher was furnished with the following information respecting it by a friend at Colima:—

"This port has a good anchorage, and is well protected against the southerly winds prevalent during the rainy season; but, on account of a very considerable lake of stagnant water in its immediate neighbourhood, is very unhealthy during the summer. Infested by myriads of mosquitoes and sand flies, even in the dry season, it is nearly impossible to reside there.

"This port has been open to foreign countries for several years, but has not been able to make much progress. The port itself has not a single house, and the first adjacent town is Colima, formerly the capital of the territory bearing the same name, now embodied with the department of Michoacan.

"Colima, it is true, is a large town, of considerable consumption, containing about 38,000 inhabitants; but the distance from the port (30 leagues, or 18

* Exploration de l'Oregon, &c., par M. Duflot de Mofras, vol. I. p. 146.

hours' travel), and the difficulty of communication, the roads being passable in the dry season only, naturally augment the expenses on any mercantile transaction to such a degree that it scarcely pays, as any cargo which could be introduced would be merely to supply the district of Colima. Such drawbacks, added to the detention, deter vessels from touching at Manzanilla.

"Another cause which must draw the maritime trade from Colima and Manzanilla is the preferable market at the capital of Guadalaxara, for its produce of maize, coffee, cocoa, indigo, &c.; and as these articles are not eligible for exportation, on account of the high cost prices, the foreign merchant could only deal in cash payments; whilst Guadalaxara, which is generally overstocked with goods, viâ Tampico on the East and San Blas on the West, can supp'y Colima with the necessary merchandise by barter.

"The articles saleable at Colima are linens, cotton goods, woollens, and a little hardware; but, as already stated, in small quantities, calculated perhaps for the consumption of about 10,000 or 15,000 souls." *

N.W. of Manzanilla Bay the land is moderately high, full of ragged points, which at a distance appear like islands. The country is very woody, but the trees are not very large. There are a good many harbours, continues Dampier, "between Sallagua (Manzanilla) and Cape Corrientes, but we passed by them all. As we drew near the cape, the land by the sea appeared of an indifferent heighth, full of white cliffs; but in the country the land is high and barren, and full of sharp peeked hills, unpleasant to the sight. To the West of this ragged land is a chain of mountains running parallel with the shore; they end on the West with a gentle descent; but on the East side they keep their heighth, ending with a high steep mountain, which hath three small sharp peeked tops, somewhat resembling a crown, and therefore called by the Spaniards Coronada, the Crown Land." †

CAPE CORRIENTES lies in lat. 20° 25′ N., lon. 105° 39′ W. Between this and Manzanilla there are the three little-frequented anchorages of *Guatlan*, *Navidad*, and *Tamatlan*.

"The cape is a bold and well-characterised promontory, jutting far into the sea, with a tolerably straight sky line, broken here and there by ravines and small peaks; it is everywhere clad with underwood to the top, and has the appearance of being a safe, bold shore; sandy beaches were noticed at different places, but in general the cliffs appear to be washed by the sea; we had no bottom with 64 fathoms of line, at the distance of 4 or 5 miles from the shore."—(Capt. B. Hall.)

Beyond Cape Corrientes, which all ships from the South should make in going to San Blas, is the great *Bay of Ameca* and the *Valle de Banderas*, 12 or 15 leagues in extent, where foreign ships sometimes take in the Brazil-wood, with which the coast abounds.

In front, and a little to the South, of *Point Mita*, which forms the northern limit of this bay, lie, nearly on the same parallel, the three small islands *Las Marietas*, and a fourth, to the West, called *La Corvetana*. This group being only a degree distant from that of the Tres Marias, presently described, should be carefully made, in order to prevent any error.

* Voyage of the *Sulphur*, vol. i. pp. 42-3.

† Dampier, vol. i. pp. 255-6.

The coast beyond Point Mita retires a little for the space of 20 leagues, in which is met with, after *Point Tecusitan*, the anchorages of *Chacala* and *Matanchel*, to the South of the small *Cape Los Custodios*, which marks the South entrance to San Blas.

The TRES MARIAS lie 70 miles to the north-westward of Cape Corrientes. They were discovered by Mendoza in 1532, and often served, in after years, as a refuge to pirates and the buccaneers who scoured these coasts. Dampier was here in 1686, and he says:—"I had been for a long time sick of a dropsy, a distemper whereof, as I said before, many of our men died: so here I was laid and covered all but my head in the hot sand; I endured it near half an hour, and then was taken out and laid to sweat in a tent. I did sweat exceedingly while I was in the sand, and I do believe it did me much good, for I grew well soon after."—(Vol. i. p. 276.) A very singular sand bath.

They were surveyed by Capt. Beechey, whose directions, in connexion with those for San Blas, are given below. Sir Edw. Belcher also visited them in 1838, and he says:—

"There is nothing inviting in either of the Marias. In the rainy season water may flow, but from what I witnessed of the channels through which it must pass, they should be well cleansed by floods before it would be fit for consumption. What remained in the natural tanks was sulphureous and brackish, although far above the influence of the sea, and formed by a strong infusion of decayed leaves.

"By tracks observed, turtle appeared to have visited the island lately, but none were seen or taken. Wood is plentiful, particularly a species of lignum vitæ. Cedar, of the coarse species, used for canoes, we met with, but none of the fine grain.

"Firs are probably in the mountains, as I found a cone in one of the water-courses. The other trees are similar to those found at San Blas.

"The soil is chiefly composed of a sandy mud, similar to that discharged from volcanoes, and which in some cases assists in forming an amygdaloidal stratum, of which the cliffs and water-courses, especially on the northern island, are chiefly composed.

"On George's Island the water-courses were of this nature, with large boulders of greenstone. On the eastern point a small delta occurs, which has coral sand for its substratum, skimmed over with a covering of mud and soil, on which rank grass luxuriates.

"Fish appear to be numerous, particularly sharks; and the dead shells on the beach, including almost every known species in these seas, hold out a prospect of employment for the conchologist.

"But the capricious character of the ocean about these islands renders visits at any time hazardous, as a few moments may imprison the naturalist for weeks. Ten years since, nearly to a day, I found landing on any part of these shores impracticable, although the weather previously had been fine.

"Here Vancouver tried ineffectually for water, and I was induced, by the assertion of a master of a vessel belonging to San Blas, 'that wells were sunk and good water conveniently to be had,' to make this examination. It is not improbable that if wells were sunk water could be obtained; but is the result worth the trouble or risk?"*

Capt. Beechey's directions are as follow:—

The Tres Marias, situated 1° 15′ W. of San Blas, consist of three large islands, steep and rocky to the westward, and sloping to the eastward, with long sandy spits. Off the S.E. extremity of Prince George's Island (the centre of the group) we found that the soundings decreased rapidly from 75 fathoms to 17, and that after that depth they were more regular. Two miles from the shore we found 10 and 12 fathoms, bad holding ground. There is nothing to make it desirable for a vessel to anchor at these islands. Upon Prince George's Island there is said to be water of a bad description; but the landing is in general very hazardous.

There are passages between each of these islands. The northern channel requires no particular directions;† that to the southward of Prince George's Island is the widest and best; but care must be taken of a reef lying one-third of a mile off its S.W. point, and of a shoal extending 1½ miles off its S.E. extremity. I did not stand close to the South Maria, but could perceive that there were breakers extending full three-quarters of a mile off its S.E. extremity; and I was informed at San Blas that some reefs also extended from 2 to 4 miles off its south-western point. There is an islet off the N.W. part of this island, apparently bold on all sides; but I cannot say how closely it may be approached.

From the South channel Piedro de Mer bears N. 76° E., *true*, about 45 miles. It is advisable to steer to windward of this course, in order that, as the winds during the period at which it is proper to frequent this coast blow from the northward, the ship may be well to windward.

The Piedro de Mer is a white rock, about 130 feet high, and 140 yards in length, with 12 fathoms all round it, and bears from Mount San Juan N. 77° W., 30 miles.‡

Having made Piedro de Mer, pass close to the southward of it, and, unless the weather is thick, you will see a similarly shaped rock, named Piedro de Tierra, for which you should steer, taking care not to go to the northward of a line of bearing between the two, as there is a shoal which stretches to the southward from the main land. The course will be S. 79° E., *true*, and the distance between these two rocks is very nearly 10 miles.

SAN BLAS.—To bring up in the road of San Blas, round the Piedro de Tierra at a cable's length distance, and anchor in 5 fathoms, with the low rocky point of the harbour bearing N. ½. E., and the two Piedros in one. This road is very

* Belcher, vol. i. pp. 138-9.

† The channel between the two North islands (Tres Marias) appears to be quite safe, and in the narrowest part has from 16 to 24 fathoms water; but the ground in other places is very steep, and at 2 miles' distance from the shore there is no bottom with 100 fathoms. When the wind is from the northward it is calm in this channel, and a current sometimes sets to the southward, which renders it advisable, on leaving the channel, to take advantage of the eddy winds which intervene between the calm and the true breeze to keep to the northward, to avoid being set down upon St. George's Island.—*Beechey*, vol. ii. p. 584.

‡ The mountains above San Blas may be seen towering above the vapour which hangs over every habitable part of the land near it. The highest of these, St. Juan, 6,230 feet above the sea, by trigonometrical measurement, is the best guide to the road of San Blas, as it may be seen at a great distance, and is seldom obscured by the fogs, while the low lands are almost always so.—*Ibid.* vol. ii. p. 584.

much exposed to winds from S.S.W. to N.N.W., and ships should always be prepared for sea, unless it be in those months in which the northerly winds are settled. Should the wind veer to the westward, and a gale from that quarter apprehended, no time should be lost in slipping and endeavouring to get an offing, as a vessel at anchor is deeply embayed, and the holding ground is very bad. In case of necessity, a vessel may cast to westward, and stand between the Piedro de Tierra and the Fort Bluff, in order to make a tack to the westward of the rock, after which it will not be necessary again to stand to the northward of a line connecting the two Piedros.

The road of San Blas should not be frequented between the months of May and December, as during that period the coast is visited by storms from the southward and westward, attended by heavy rains and thunder and lightning. It is besides the sickly season, and the inhabitants having all migrated to Tepic, no business whatever is transacted at the port.

It is high water at San Blas at 9^h 41′ full and change; rise, between 6 and 7 feet, spring tides.*

The next remarks and directions are those given in Capt. Basil Hall's Hydrographical Memoir:—

Having passed about 8 or 10 leagues to the southward of the group called the Tres Marias Islands (the westernmost of which lies 54 miles West of San Blas by chronometers), steer a N.N.E. course, until Piedra Blanca comes in sight, when it would be advisable to steer directly for it, and pass about a league or two to the southward of it; from thence you will see Piedra Blanca de Tierra, which points out the harbour; for this you may steer direct, taking care not to go in-shore of the line of bearing of the two rocks, as a sandbank lies off the coast about halfway between them, to a considerable distance on the pitch of which there is only 2½ fathoms.

Vessels provided with chain cables should use them here, as the barnacles will soon destroy the hempen ones, and the boats should be frequently out of water, and their bottoms scrubbed, as the worm is very destructive at this place; a mixture of tallow and lime, laid on thick, is described as being the only preservative of boats' bottoms.

Supplies of vegetables may be procured from the market, which is held on Saturday evenings and Sunday mornings; the watering place is at a well dug near the end or commencement of a rope walk, on the beach; the water is somewhat brackish, but improves by keeping.

Plenty of fish are to be caught at the anchorage, and oysters are found clustered to the roots of the trees on the banks of the river.

Early in June the rainy season commences with great violence, like the monsoons of the East Indies, and continues for six months, during which period there prevail violent squalls, a heavy rolling sea at the anchorage, thunder and dangerous lightning, and almost constant rain. The inhabitants, at this season, retire to the neighbouring town of Tepic, and to other parts of the country, not only to avoid the bad weather, but the ardent fevers which are prevalent.

* Beechey, vol. ii. Appendix, pp. 660-1.

In the evenings and mornings the air is so filled with mosquitoes and sand flies, that those periods, which otherwise would be the best to work in, are not the most suitable for communicating with the shore.

There is always reason, too, to apprehend the effect of marsh miasma, in the mornings and evenings, at a spot so surrounded with swamps. Our boats passed backwards and forwards in the heat of the day, and a constant communication was kept up with the town, but little or no sickness ensued; it is true some cases of fever occurred, and it is worthy of remark, that it fell most severely upon those who had gone up the river in a boat. There is no doubt, however, that this place is extremely unhealthy at times, and every precaution, especially against exposure at night, would be at all times of importance.

Nearly all sorts of provisions* and stores may be procured at San Blas, either in the town, or by having them sent from Tepic. The spirits of the country are bad and dear. Vegetables are only to be procured in the town at the market; they are of various kinds, principally pumpkins, cabbages, and onions.

Live oxen are in plenty and reasonable; the only difficulty likely to arise in completing a ship with salt provisions is owing to the extreme heat of the weather.

The following is by Capt. Masters, of Liverpool:—"In the rainy season, when the wind blows strong from the southward, a heavy swell sets in at San Blas, and as there is nothing to protect the anchorage, it must be felt very severely; but I never heard of any damage having been done to the shipping in consequence.

"There is some advantage in a vessel lying outside in the roads during the rainy season, for there the crews have purer air to breathe; and probably it might be more healthy than that of the port, besides being partially clear of mosquitoes, and other tormentors of the same cast, which are very numerous. The crew also are easier kept on board. But if the ship has to discharge her own cargo, the expenses of doing it will be considered; and if her long-boat is too small for that purpose, the launch, which must be hired, will most likely be manned by the crew; so that they are more liable to become ill, by being more exposed to the rain and sun, than if a vessel was in the port; and in the next place, their meals would not be very regular: they would also get spirits by some means, whoever was in charge of the boat or launch (providing that he should be even disposed to prevent it), whenever she went on shore.

"There are 13 feet water on the bar of San Blas, in the shallowest part of the entrance, and very seldom less even in the neaps. By giving the point which forms the harbour a berth of 15 or 20 fathoms, you will avoid a large stone, which is awash at low water, and is about 8 fathoms from the dry part of the rocks

* The foregoing remarks upon the supplies are by Mr. Inderwick, purser of H.M.S. *Conway*. The following mode was adopted in curing the beef:—The oxen were killed early in the morning, and the meat was immediately cut from the bones, except from some of the ribs; the meat was cut thin and wiped as dry as could be done; this occupied but a few minutes, when the men commenced rubbing salt well into the meat while warm, which being done, the whole was put on a stage erected in the open air, from whence the sun was excluded; planks of wood were laid upon it, on which were placed all the weights that could be put, for the purpose of pressing out the blood, &c.; in this state it remained till the following morning, when it was examined, and appearing good, was again well rubbed with salt, and the weights again put on; it remained so for three days, when it was weighed and put into casks, and, on getting them on board, they were filled with strong pickle, for making which, a quantity of salt and saltpetre had been procured.

or breakwater. As soon as you are so far in, that the innermost or eastern part of the breakwater is in a line with the other part of it inside, which runs to the N.N.E., it may be approached to within 10 or 15 fathoms, and by keeping well off from the low sandy point which is on the starboard hand as you warp up the harbour, you will have the deepest water. But as the sea sometimes in the rainy season (although but seldom), breaks over the breakwater which forms the harbour, it would be best to moor close under the high part of the land on which the old ruins of a fort stand, with the ship's head up the river, and a bower laid off to the eastward, and an anchor from the starboard quarter, from the larboard side to the shore, either by taking an anchor out or making fast to the rocks. It would be next to impossible that any accident could happen to the ship: the cargo can also be discharged with despatch, and immediately under the eye of the master or mate, as the place where the cargo is landed would be about 100 fathoms from the ship. The ship's long-boat would do more inside than two launches if she was outside; and besides, when the sea is heavy in the roads, the discharging of the cargo could go on. As there are no established pilots here, it would be advisable to engage a person to point out where the stone lays; the captain of the port is the best to apply to, and if he does not come off himself, would most likely recommend a person. Should any vessel be inside, it is customary to apply to the captain for information; and although there is not much danger in getting in, it is still satisfactory to have a person on board who has even entered but once. Of course a Mexican would be preferred by being more accustomed to the place. The town of San Blas is built on a hill, with a cliff on its western side; from the *commandancia* (or governor's house) and the custom-house is an excellent view both by land and sea. On the approach of any vessel a flag is hoisted on the southern flagstaff, which is that of the *commandancia*; the northern one is that of the custom-house.

"When the Spaniards were in possession of San Blas, the harbour was kept clear of sand and mud, which has since collected and filled up the eastern part of the harbour; and it is also very much shoaler on the bar. I was informed by people who lived here before the revolution, that they had seen ships of war inside mounting sixty guns, and upwards of a dozen others of smaller size, at one time; but now, at half tide, a boat can hardly get up to the arsenal, as the bight nearest the town is almost filled up with mud. The shipping as well as the port was neglected during the revolution. One fine frigate, with her guns and stores on board, was allowed to sink at her moorings; and even after the Mexicans had taken possession of the place, others shared a similar fate; and now the vessels which the Mexicans have purchased are rotting and going to pieces from the same neglect. San Blas and La Playa (the small village at the port) were stirring places under the Spaniards. They had generally 600 men employed at the different works; and even after the Mexicans had gained their independence, the place was kept in repair and work carried on for a short time, but on a small scale to what it had been. At one time there were from 60 to 100 men employed in the arsenal.

"During the rainy season, the town of San Blas is nearly deserted; not above two or three families remain, and these not from a matter of choice, as all who can afford it remove off to Tepic, or to a village between Tepic and San Blas. Fevers are said to be more prevalent at San Blas than at La Playa, the village at the port: mosquitoes and sand flies are not felt at the town, excepting in the rainy season. The land to the westward is low, thickly wooded, and swampy; the effluvia rising from it in the rainy season must be anything but healthy or agreeable; and people in Mazatlan have as much dread of San Blas as people in England have of Sierra Leone.

"Tepic, the principal town near San Blas, is 18 leagues from it. It is well elevated, but is not entirely free from fevers in the rainy season. The authorities (in the rainy season) of every department of the government all start off from San Blas, as it is only considered a temporary residence at any time. The charge of affairs is left in the hands of some inferior officer who has not the means to move, and on the arrival of a vessel, a courier is despatched to Tepic, when they return, transact their business, and then go off again. But as all the merchants reside at Tepic there is no time lost by it. At La Playa, the people remain the year through; and being used to the climate they do not think anything of it. As their chief dependence for a livelihood is on the shipping, it is their interest to remain. There are several shops at San Blas, but the greatest part of their stock appears to be a quantity of bottles filled with different coloured liquors.

"The town of San Blas is built on a small hill, about 450 feet above the sea; it boasts of very few houses that do not want repairs; one I particularly remarked, in which a family was then living, that had trees of a small size growing through the windows, doing considerable damage to the walls; in some parts of it the roof had fallen in, and bushes had grown up; but this is, with a few exceptions, the general state of the houses at San Blas. The church, which is rather large, is built of stone (as are the houses), but is now completely in ruins. Mass is performed in a small place within the walls. As some scaffolding is up, there must have been some intention to repair it; but it has been put off such a length of time for want of hands, that there appears to be no chance of it. The custom-house, government warehouses, *commandancia*, and what were dwelling houses for the principal officers of the different departments, form a square, and is the nearest part of the town. These buildings are all going fast to decay: the three first bespeak the declining trade, and misery; the last, in addition, appears as if war and pestilence had passed over the place with all its horrors: I never saw a place with a more wretched appearance. The population, I think, could not amount to 300 in the healthy season; but on Saturday, which is market day, the place has more life in it. The market is very well supplied with fruit, poultry, and vegetables, moderately cheap; and good fowls are 2 to 2½ reals each, plantains 2½ reals per head, and other things in proportion; the sallads and cabbage are excellent; and bananas, oranges, plantains, passado, green peas, potatoes, sweet potatoes, onions, with various other fruits and vegetables, can be purchased here: the beef, for a warm climate, is not bad, and can be had daily at La Playa. Ducks and turkeys can also be procured. The greatest part of these things come from the neighbourhood of Tepic. Whatever is purchased at La Playa, that is, vegetables or stock, is a hundred per cent. above the Saturday market price; and many things cannot be had. Fish is also as plentiful as on all parts of the coast; and there are likewise excellent oysters here. Turtle at this time of the year were rather scarce, and as they are not made use of by the inhabitants, they are never caught, unless a ship-of-war is in the roads, or a bargain is struck beforehand. I could only get two small ones. Fruit is better and cheaper here than at Mazatlan.

"The launch hire in San Blas is 10 dollars per day; of which the patron gets one, and the crew half a dollar each trip, to be paid by the ship; this is when she lies in the roads."*

The following recent remarks, which were given (anonymously) in the Nautical Magazine for March, 1849 (pp. 140-1), give a picture of the *present* condition of San Blas:—

"Care must be taken in standing in for the land not to get to leeward of San Blas, as there is a strong southerly current along the coast, especially off Cape Corrientes.† If possible, keep San Blas on an E.N.E. bearing. The Tres Marias Islands, off the port of San Blas, are convenient points for making; and here a master could leave his vessel in perfect safety to water while he communicated with his consignees, or got his overland letters from his owners at home. There is a safe mid-channel course between the middle and southern islands: we brought a saddle-shaped hill on the main, a little South of San Blas, one point open of the South island, and steered by compass N.E. by E.

"The two Pedro Brancos, that of Del Mar and De la Tierra, are excellent marks for the roadstead; a good anchorage for vessels awaiting orders (for which purpose San Blas is now almost alone visited, except by English men-of-war, and Yankee clippers for smuggling purposes) will be found with Pedro Branco del Mar, N. 70° W., De la Tierra, N. 43° W., and village in the Estero, N. 26° W.

"Since the days of Hall and Beechey, the town of San Blas has very much changed. Its population of 20,000 have dwindled to 3,000, and their unwholesome appearance fully accounts for the decrease of residents; and nearly all its trade has been transferred to its rival—Mazatlan.

"The large town of Tepic, in the interior, with a small factory, owned by an English merchant, causes a small demand for European luxuries, and a cargo or two of cotton; which petty trade is carried on during the six healthy months in the year. A great deal of smuggling is carried on from the neighbourhood of this port, the extensive bay to the southward affording great facilities to the men-of-war's boats in that employment.

"The town is built on the landward slope of a steep hill, almost perpendicular to seaward, and its crest crowned by the ruins of a custom-house; but this being about three-quarters of a mile distant from the beach, a large assemblage of huts has been formed at the landing place in the Estero del Arsenal, for the convenience of supplying the shipping: the occupants being for the most part grog venders, fishermen, and an agent to the harbour-master.

* Nautical Magazine, Nov. 1837, pp. 712—716.

† Capt. Beechey came to San Blas from the northward, and on approaching it found himself more to leeward than he was aware, in consequence of the current setting out of the gulf. To save time he passed between the two northernmost islands, and in doing so was becalmed for several hours, fully verifying the old proverb that the longest way round is often the shortest way home.—*Beechey*, vol. ii. p. 584.

"In the Estero del Arsenal, small craft of less than 10 feet draught will find convenient anchorage, means of heaving down, &c. The watering place is at least 3 miles distant from the above anchorage, and to assist the boats in this heavy work, it would always be advisable to shift the vessel into such a position that they might make a fair wind off and on whilst the daily sea-breeze blows.

"The watering place is at the northern extremity of a large open bay South of San Blas: the beach is shoal, and the casks have to be rolled three or four hundred yards through the jungle to a stream of water. This stream during the spring tides is liable to be found brackish, but even then we succeeded in obtaining supplies, by immersing the empty cask with the bung in such a position that only the fresh water (which of course would be on the surface) could enter.

"By rigging triangles with spars in such a position that the boats could go under them to load, we succeeded in embarking daily 32 tons of water.

"Many useful and ornamental woods are to be procured on shore, for the mere trouble of cutting, especially lignum vitæ. Fresh beef we found good in quality. Game moderately plentiful; oysters good and abundant; vegetables scarce and expensive. The climate may be summed up by the word *execrable*."

Mr. Jeffery, R.N., thus describes his passage from Acapulco to San Blas, which will form a fitting supplement to the preceding:—

"On the 8th of December, 1833, we weighed with a light sea-breeze, which enabled us to work out of the harbour: towards the afternoon the wind increased from the westward, and continued to blow fresh for 30 hours, which is considered unusual on this part of the coast; we were near 60 or 70 miles off the land, and for the next 10 days had nothing but calms and light variable airs alternately, and the weather excessively hot and sultry; at last we got in with the land, about 100 miles west of Acapulco, and on the 21st were off Point Tejupan. The only remarkable objects on this part of the coast are the Paps of Tejupan: they are two sharp hills on very high land. We now found the good effects of being close in shore, and I think it advisable for vessels making this passage to keep as near the land as possible, for in the day time we had a regular breeze from the westward; it fell light about sunset, and then about 10 or 11 o'clock the land-breeze would come from the N.E. The advantage of keeping in-shore can seldom be doubted when we consider that vessels are frequently 40 or 50, and sometimes 60, days making this passage, through not keeping near the land. Between Tejupan and Cape Corrientes we found a current setting along the land to the westward from 12 to 15 miles a day. On the 28th, when within 30 miles of the harbour, we had very heavy rain, and thick weather; we arrived on the 29th, and made the passage in 21 days. After rounding Corrientes, if the weather is clear, you will see the saddle mountain near San Blas; and to the N.W. of it, another high mountain, with a remarkable peak at its N.W. extremity; if you look at it with a glass it will appear split in two, but to the eye it will appear as one. This is so very remarkable that it ought never to be mistaken, as there is nothing like it on this part of the coast. When the above peaked mountain bears N.E., the anchorage at San Blas will bear N.E. also.

"The watering place at San Blas is in a bay to the eastward of the roads; you have to land the casks and roll them about 300 yards through the woods to a river. At high water it is rather brackish, but at all other times the water is excellent. Wood is very plentiful."

San Blas has an unhappy celebrity for the heat and pestilential nature of its climate. The following extracts from Capt. Basil Hall's Journal will amply explain its nature:—

"We experienced a great difference in the climate of San Blas and that of Tepic, especially at night. At both places it was disagreeably hot during the day; but at Tepic, which stands on an elevated plain, the thermometer fell 15° or 20° at night, whereas at San Blas, which is close to the sea, there was much less variation of temperature. Throughout the day it was generally, in the coolest part of the shade, 90°, sometimes, for several hours, 95°. The reflection from the walls and from the ground made the air in the open streets often much hotter, and I have several times seen it above 100°. The highest temperature, however, in a shaded spot was 95°. At night the thermometer stood generally between 80° and 85°. Between 10 and 11 o'clock in the morning the sea-breeze set in. None but those who have felt the bodily and mental exhaustion caused by the hot nights and sultry mornings of low latitudes can form a just conception of the delicious refreshment of this wind. For some time before it actually reaches the spot its approach is felt, and joyfully hailed by people who, a few minutes before, appeared quite subdued by the heat, but who now acquire sudden animation and revival of faculties; a circumstance which strangers, who have not learned to discover the approach of the sea-breeze, are at a loss to account for. When it has fairly set in, the climate in the shade is delightful, but in the sun it is scarcely ever supportable at San Blas. Between 3 and 4 o'clock the sea-breeze generally dies away; it rarely leaves till 5. The oppression during the interval of calm, which succeeds between this and the coming of the land-wind, baffles all description. The flat-roofed houses resemble ovens, from having been all day exposed to the sun, and as it is many hours before they part with their heat, the inhabitants are sadly baked before the land-wind comes to their relief.

"During the morning the thorough draught of air, even when the sun is blazing fiercely in the sky, keeps the rooms tolerably cool; but, when the breeze is gone, they become quite suffocating. The evil is heightened most seriously by clouds of mosquitoes, and, what are still more tormenting, of sand flies, insects so diminutive, as scarcely to be distinguished till the eye is directed to the spot they settle upon by the pain of their formidable puncture. San Blas, as mentioned before, is built on the top of a rock, standing in a level, swampy, and wooded plain. During ordinary tides in the dry season, this plain is kept in a half-dried steaming state; but at spring tide a considerable portion of it is overflowed. The effect of this inundation is to dislodge myriads of mosquitoes and sand flies, and other insects, which had been increasing and multiplying on the surface of the mud during the low tides. These animals, on being disturbed, fly to the first resting place they can find; and the unhappy town of San Blas, being the only conspicuous object in the neighbourhood, is fairly enveloped, at the full and change of the moon, in a cloud of insects, producing a perfect plague, the extent of which, if properly described, would scarcely be credited by the inhabitants of a cold climate."—(Capt. Basil Hall, Extracts, &c., vol. ii. chap. 49.)

"The fine season lasts from December to May inclusive. During that interval the sky is always clear, no rain falls, land and sea-breezes prevail; and, as there is no sickness, the town is crowded with inhabitants. From June to November a very different order of things takes place: the heat is greatly increased, the sky becomes overcast, the sea and land-breezes no longer blow, but in their stead hard storms sweep along the coast, and excessive rains deluge the country, with occasional violent squalls of wind, accompanied by thunder and lightning. During this period, San Blas is rendered uninhabitable, in consequence of the sickness, and of the violence of the rain, which not only drenches the whole town, but by flooding the surrounding country, renders the rock on which the town is built literally an island. The whole rainy season, indeed, is sickly, but more especially so towards the end, when the rain becomes less violent and less frequent, while the intense heat acts with mischievous effect on the saturated soil, and raises an atmosphere of malaria such as the most seasoned native cannot breathe with impunity.

"This being invariably the state of the climate, nearly all the inhabitants abandon the town as soon as the rainy season approaches, that is, by the end of May."—(Capt. Basil Hall, Extracts, &c., vol. ii. chap. 50.)

The whole of the coast about San Blas, Mazatlan, and Guaymas, is perfectly clear, and may be approached to within a short distance.

The year is divided into the two seasons described above. It must be remarked that the change occurs gradually, and its period varies. During the dry season the weather is constantly fine. The winds blow regularly during the day from N.W. and W., following the direction of the coast, and then give place during the night to a slight land-breeze or calms. The rainy season, which commences in June, is then indicated by calms and slight showers of rain; as the season advances, the showers become heavier, and, instead of beginning at night, they do so in the afternoon, and terminate by violent storms, accompanied with very dangerous lightning and thunder, the fierce winds blowing from all points of the compass. The weather keeps of this nature until the end of September, and it sometimes occurs that the season terminates by a terrible hurricane, which generally occurs from the 1st to the 5th of October, the festival of St. Francis. These hurricanes, which always blow from S.E. to S.W., are of short duration, but they are so violent, and raise such a heavy sea, that nothing can resist them. They are called in the country the *Cordonazo de San Francisco* (the lash with St. Francis's cord or belt). A vessel surprised when at anchor ought to slip her cables, or cut the moorings, and make sail. At the approach of the cordonazo, she ought to gain an offing, or if obliged to keep in the road, to moor at such a distance off shore that she can easily get under sail on the first intimations of its approach. These observations are not applicable to roadsteads entirely open, but such ought to be avoided during the months of September and October. Sometimes the cordonazo occurs later than St. Francis's day: thus, on the 1st of November, 1839, twelve ships, who thought it had passed, were surprised in the Port of Mazatlan, and the greater part were lost, and all perished. On the 1st November, 1840, three vessels were lost on the Road of San Blas, and several people were drowned, without it being possible to render them any assistance.

A phenomena has been observed on the N.W. coast of Mexico, and in the Gulf of California, known in meteorology under the title of the *inversion of the trade winds.* In reality, this wind, almost constantly blowing from the N.E. in the Atlantic, in the parts North of the equator, is here supplanted by one from the S.W., and even by winds directly from the West. This inversion, which only prevails in the Vermilion Sea, is not experienced on the Californian coast, washed by the Pacific beyond the latitude of 23° North.*

At 20 leagues N.W. of San Blas, in front of the mouth of the Rio San Pedro, lies the little isle Isabella.

Isabella Island is of moderate height, and nearly barren; the herbage and grasses are scarcely to be distinguished. Neither water nor wood are to be got from it. The beach is lined with rocks, with the exception of a small sandy cove, open to the West, where boats may be hauled up on the shore. This island is only frequented by sealers.†

In lat. 22° 25′ N. the small hills of Bagona are seen; and anchorage may be found in 8 fathoms near the N.W. point, sheltered from the N.E. winds. The mouth of the *Rio Bagona* is designated under the name of the *Boca de Teacapan.* At 8 leagues further North, the *Hillocks of Chametla* are seen. The West point of the *Rio Chametla,* or *Del Rosario,* is in lat. 22° 50′ N. It was in the small port formed by its continuation that Hernan de Cortes embarked, April 15th, 1535, on his voyage to discover California. A mile outside the depth is 8 and 9 fathoms.

On the coast several large farms are seen after leaving San Blas. These are the haciendas, *Del Mar, San Andres, Santa Cruz, Teacapan,* and *Del Palmito.* Bullocks may be bought at them for 8 piastres, and some vegetables. The water of all the rivers is good, and fine wood very abundant.‡

MAZATLAN lies 40 leagues from San Blas, and has supplanted the latter in its commerce, and consequent importance. It was surveyed in 1827 by Capt. Beechey, who places the extreme bluff of Creston Island, lying off it, in lat. 23° 11′ 40″ N., lon. 106° 23′ 45″ W., variation 10° 18′ E.

The harbour of Mazatlan is entirely open to the winds, which are most dangerous in the rainy season. It is formed by a bay, in the centre of which stands the town, but small vessels can only approach it. Larger ships anchor to the southward, under the lee of the *Island of Creston,* a small but very high island, forming the North side of the road. It is of a roundish form, and green at the top, but is perpendicular seaward, so that it appears only as a white cliff. Creston is separated from another island by a narrow channel, and this last from the main land by a cable's length. In approaching Mazatlan, Creston will be first made out, seemingly detached from the coast; to the North of it are two islets, called *Islas De los Pajaros* and *De los Venados,* which at a distance will appear like two patches on the coast, and will also seem to make out the anchorage, for it is the only point on this part of the coast where there is a group of islands. The anchorage now used is that to the South of Creston, but the islets form between them

* These remarks are by M. Duflot de Mofras, Exploration de l'Oregon, vol. i. pp. 169—172.
† Du Petit Thouars, tome ii. p. 184.
‡ M. Duflot de Mofras.

and the main land another road, formerly used by the Spaniards, which is much preferable in the rainy season. It is sheltered from the South and S.W. winds, which blow heavily at times, and gives it the advantage of getting away between the islets, or between them and the coast: but as the prevalent N.W. winds of the dry season blow right into it, and raise such a heavy sea on the beach, that landing goods, &c., is rendered a very difficult task, the anchorage South of Creston, where these inconveniences are not felt, is preferred.

The port of Mazatlan has been opened to foreign commerce for some years. Under the Spanish dominion it was unknown, but on the proclamation of the Mexican independence it was placed on a different footing; and Capt. Sir E. Belcher says, that between his former visit in 1827 and that in 1839, it had increased from a village to a town, and, of course, had also increased in bad characters. It was dangerous to be out at night unarmed or alone.*

The official name which was applied to it by the Mexican government is *La Villa de los Costillos*. Its population amounts to about 8,000 during the rainy season, but rises to 10,000 or 12,000 at the dry season, or when the vessels arrive, for at that time the merchants of the provinces of Chihuahua, Jalisco, Sonora, Sinaloa, and Durango, come here to transact their business.

The town of Mazatlan is open on all sides, having neither fortifications nor batteries, a few *outré* soldiers forming the entire garrison.

Ships ought to get their water in the peninsula which forms the South side of the road; everywhere else it is brackish. Although Mazatlan is less unhealthy than San Blas, severe fevers are common during the rainy season, and as there is no hospital in the town, commanders should be strict in not allowing their men to run into any excesses, which are highly dangerous.

Mazatlan is the only port in this part of America, North of Guayaquil, where a ship can procure provisions completely. A bullock costs (or rather did cost) 8 or 12 piastres. Flour from Guaymas, which is excellent, is from 12 to 14 piastres for two arrobes (300 lb. French). Sail-cloth, pitch, tar, cordage, chains, anchors, and timber (partly from wrecks) may be procured in the stores.

At 10 leagues to eastward of the port, on the road leading to San Blas and Tepic, and 3 leagues from the sea, is the old Presidio of Mazatlan. This village is scarcely anything but ruins; for since the removal of the trade to the port, it has lost all its military importance. There are no vestiges of the old fortification, and the fine barracks built by the Spaniards only serve now to shelter a few cavalry soldiers; the population is about 500. The Rio de Mazatlan, which runs near the Presidio, and falls into the harbour, is not navigable.†

Capt. Beechey's directions are as follow:—

"The anchorage at Mazatlan, at the mouth of the Gulf of California, in the event of a gale from the south-westward, is more unsafe than that at San Blas, as it is necessary to anchor so close to the shore, that there is not room to cast and make a tack. Merchant vessels moor here with the determination of riding out the weather, and for this purpose go well into the bay. Very few accidents,

* Voyage of the *Sulphur*, vol. i. p. 342.
† Duflot de Mofras, vol. i. pp. 172—178; Sir E. Belcher, vol. i. p. 342.

however, have occurred, either here or at San Blas, as it scarcely ever blows from the quarter to which these roads are open, between May and December.

"There is no danger whatever on the coast between Piedro de Mer and Mazatlan; the land is a sure guide. The Island of Isabella is steep, and has no danger at the distance of a quarter of a mile. It is a small island, about a mile in length, with two remarkable needle rocks near the shore, to the eastward of it.

"Beating up along the coast of Sonora, some low hills, of which two or three are shaped like cones, will be seen upon the sea-shore. The first of these is about 9 leagues South of Mazatlan, and within view of the Island of Creston, which forms the Port of Mazatlan. A current sets to the southward along this coast, at the rate of 18 or 20 miles a day.

"Having approached the coast, about the lat. of 23° 11′ N., Creston and some other steep rocky islands will be seen. Creston is the highest of these, and may be further known by two small islands to the northward of it, having a white, chalky appearance. Steer for Creston, and pass between it and a small rock to the southward, and when inside the bluff, luff up and anchor immediately, in about 7½ fathoms, the small rock about S. 17° E., and the bluff W. by S. Both this bluff and the rock may be passed within a quarter of a cable's length; the rock has from 12 to 15 fathoms within 30 yards of it in every direction. It is, however, advisable to keep at a little distance from the bluff, to escape the eddy winds. After having passed it, be careful not to shoot much to the northward of the before-mentioned bearing (W. by S.), as the water shoals suddenly, or to reach so far to the eastward as to open the *West* tangent of the *peninsula*, with the *eastern* point of a low rocky island S.W. of it, as that will be near a dangerous rock (*the Blossom Rock*), nearly in the centre of the anchorage, with only 11 feet water upon it at low spring tides, and with deep water all round it. I moored a buoy upon it; but, should this be washed away, its situation may be known by the eastern extreme of the before-mentioned low rocky island (between which and Battery Peak there is a channel for small vessels) being in one with a *wedge-shaped protuberance* on the *western* hillock of the *northern island* (about 3 miles North of Creston), and the N.W. extremity of the high rocky island to the *eastward* of the anchorage being a little open with a *rock off the mouth of the river* in the N.E. The South tangent of this island will also be open a little (4°) with a dark *tabled hill* on the second range of mountains in the East. These directions will, I think, be quite intelligible on the spot.

"The winds at Mazatlan generally blow fresh from the N.W. in the evening; the sea-breeze springs up about 10 in the forenoon, and lasts until 2 o'clock in the morning.

"It is high water at this place at 9ʰ 50′, full and change; rise, 7 feet, spring tides."*

To these may be added the following extract from the remarks by the French Admiral Du Petit Thouars:—

In approaching Mazatlan there is no difficulty. The latitude of the port must be gained, or rather a few minutes to the North of it, on account of the

* Beechey, Appendix, pp. 661-2.

currents which generally run strong out of the Gulf of California, and which would thus send a ship to leeward of the port. In clear weather the land may be seen at 40 or 50 miles off, and, if you are on the parallel of Mazatlan, the first which will be seen will appear in the N.E., the N.E. ¼ E., or E.N.E., according to the distance. Nearing the land it will be seen successively extending towards the South as far as E., and even as E.S.E. From the offing nothing can be made out. The land on the shores is generally very low, and, in the first instance, only the interior high land will be seen; and it is only when you are within 18 or 20 miles of the coast, for example, that its different points can be distinguished. The islands of *Creston*, *Venado*, and *Pajaros* will be made out; the first seen is Creston. As soon as all uncertainties of the position of Mazatlan cease, steer either for the anchorage of Venado or that of Creston.

In proceeding for Venado, with the ordinary winds of the season, steer for the S.W. point of Venado Island, carefully looking out for, and avoiding, a small rock awash, called the *Laxa*, which lies about 200 yards from this point, and does not always break: after having passed this rock, which may be done within a hundred yards, bear more to the northward to enter the bay, and bring the S.W. point of Venado to bear West, or even W.S.W., according as you intend to anchor more or less within the bay.*

The subsequent description and directions are by Capt. Masters, who we have before quoted. It must be premised, however, that his names differ in some few points from that of the recent Admiralty chart. We have therefore altered them to agree with that, leaving the originals within parentheses.

Mazatlan is a port very easily made. It is formed by a cluster of islands; to the southward of them is a long line of beach, with low land, thickly covered with trees, running several miles in before it reaches the foot of the mountains, and continues the same as far to the southward as the North side of the bar of Tecapan, where the land is high.

The Port of Mazatlan, at its entrance, is formed by the Island of Creston on its western, and Ciervo Island on the eastern side. From the sea the former has nearly a regular ascent, the length of the island lying from East to West, where it terminates in an abrupt precipice, and is covered with small trees. It has from 8 to 10 fathoms water to within a few fathoms of it. The Island of Ciervo has a very similar appearance, and is about half the height of Creston, being partially covered with trees. These islands can be seen several miles before the land at the back of the town makes its appearance. The outer rock is situated well outside the roadstead, and forms nearly an equilateral triangle with the islands of El Creston and Ciervo; it is about 8 feet high, and nearly the same in breadth, and from 7 to 8 fathoms long from North to South; there are 5 fathoms water close to it.

In the excellent plan of the harbour of Mazatlan the soundings in general are very correct; but the stranger, in coming to an anchor in the night time, should not attempt to pass within the line from the outer part of El Creston to El Ciervo, but anchor outside in from 9 to 12 fathoms, where he will find sand

* Du Petit Thouars, Voyage du *Venus*, vol. ii. pp. 175–6.

and mud. Within the port is a long sand, which extends out from the bottom of it, a great part of which is dry at low water, and is shoal for some distance to the S.E., extending nearly as far as the Island of Ciervo, with a boat channel between it and the island. The inner anchorage is to the westward of this sand. It is said that the bank is increasing, and that the port has filled very much within a few years past.

In the summer season large vessels anchor between the two islands at about a third the distance from Creston to Ciervo, and moor East and West. The depth of water is from 7 to 9 fathoms, sand and muddy bottom. A vessel drawing 12 feet water might go inside to the minor anchorage without the least risk; but, as the pilot has launches to get employed, he cannot be persuaded to take a vessel in even drawing 9 feet.

North of the Island of Creston, and between it and the main land, is the Island of Azada (Gomez), which is low, and is separated from Creston by a narrow boat channel. From about the middle of Azada a bar extends to the eastward across the port nearly to the sand-bank already mentioned. The pilot informed me that there were patches of shoal water on it when the water was low, not having more than 6 feet on them, which might be the case; but where I sounded, there was not less than 12 feet. Inside the bar it deepens to 4 or 5 fathoms, and close up to the town, abreast of the custom-house, at low water, there are from 2½ to 3 fathoms, with a sandy bottom.

The custom-house is built near the inner point of sand and rock to the northward of what was called a castle. It was merely a platform, with a few guns mounted on it, but it is now dismantled.

When the wind blows strong from the N.W. there is a short chop of a sea heaves in between the Island of Gomez and Point Pala (Calandare), although the distance they are apart is short, but by anchoring, as already mentioned, opposite Creston, most of it is avoided.

In the rainy season it is very unsafe to lay inside, as gales come on from the southward, which bring in a heavy sea. Vessels of all sizes anchor in this season in the outer roads, between the islands and the outer rocks, from which they can be got under way, and stand clear of the coast.

To the northward of the present Port of Mazatlan, about 5 miles, is the N.W. Port of Mazatlan. It is a very fine bay, and well sheltered from the N.W. winds by Venado and Pajaros Islands. It was in the southern part of this bay that vessels formerly discharged their cargoes, but the present port being more secure, it was established in its stead.

Watering is attended with great risk at all times in this place, especially at full and change, the boats having to cross the heavy surf of the bar formed between a long spit which runs down the centre of the river, and a bank joining it from the South shore. Several boats and lives are annually lost here. In pulling in care should be taken to cross the surf pretty close to the middle ground, and, when through the first rollers, to pull over to the South shore, and keep it on board up to the watering place. In coming out no casks ought to be allowed in the head sheets, everything depending upon the buoyancy of the boat; inattention to this point caused the loss of two lives, to my own knowledge.

The water is procured from a number of wells dug by seamen, on a low alluvial island, formed on a quicksand in the bed of the river: none of them are consequently more than 10 feet deep. The water is by no means sweet, being merely sea water, which undergoes a partial purification in filtering through the soil.

Supplies of all sorts come from the neighbourhood of San Blas; and as the bullocks are driven that long distance, and on arrival they are immediately killed, from the want of grass, the beef is necessarily lean and bad. Pork, fish, and oysters are, however, plentiful; vegetables are scarce. The river abounds in turtle of excellent quality; wood of various descriptions, principally hard, was plentiful, and at a short distance oak and cedar might be obtained.

CHAPTER X.

THE GULF AND PENINSULA OF LOWER CALIFORNIA.

When Cortes had conquered Mexico, more, perhaps, by the strength of his genius than by the superiority of European arms, the Spanish emperor (Charles V.), satisfied with the treasures which the new acquisition opened to him, and also too prudent to trust the successful general with greater power, granted him great titles of honour, but limited his authority to the idle command of troops reduced to inaction, and forbad him to enter into any new military expedition on the continent. Thus confined, his ardent spirit entered into new projects of extending the knowledge of the hitherto all but unknown countries in the West. He had previously caused the N.W. coast to be examined, in order to discover whether some strait might not afford a free passage to navigation from the old world. In one of these expeditions California was first discovered, by Fortun Ximenes, the pilot of the *La Concepcion*, a vessel sent by Cortes to explore the coast to the N.W. of Acapulco. Ximenes having murdered the captain (Diego Bezerro), in November, 1533, to avoid the consequences sailed to the N.W., and anchored in a port which is probably on the Gulf of California, and was attacked and slain by the Indians on shore. This appears to have been the first discovery of California (not then known by this name), though the Spaniards had received some imperfect intimations of it from the natives of Colima.

Several other expeditions despatched by Cortes also met with as ill success as this; and, mortified with reverses to which he was not accustomed, he ordered a fresh armament, in 1537, which he put under the command of Don Francisco de Ulloa, whose reputation had been acquired by long services. But, tired of inactivity and of intrusting to others the execution of his projects, he took the command in person, and, after encountering a long series of fatigues and dangers, he finally discovered the great peninsula of California, and he entered the gulf, which has sometimes been called by his name. Thus was California really made known to the civilized world. It has been objected that Cortes returned without doing anything new, but this is disparaging too much, and the merit of discovery is certainly partly due to him.*

The Gulf of California was first formally explored by Francisco de Ulloa, by direction of Corte . He had under him three vessels, which sailed from Acapulco in July, 1539. He entered the gulf, and sailed all around its shores, and anchored probably in Port Guaymas, and took formal possession of the country; proceeding northward, they came to the low and sandy head of the gulf. They then returned along the shore of the peninsula, examining its features, and subsequently (January 11, 1540) discovered and named the Isle de los Cedros. His farthest North was Cabo del Enganno (in 30½° N.?). This was the last expedition of discovery in which Cortes was concerned.—(Burney, p. 209.)

The name of California is of uncertain origin. It is not known to have been used by the natives in any part of the country. It has been conjectured that it is derived from the heat of the weather experienced here by Cortes, but of this we are not told. It appears at first to have been only applied to a bay. "Cortes went to discover other lands, and met with California, which is a bay." †

By some it is thought that it is derived from a custom prevalent throughout California, of the Indians shutting themselves in ovens until they perspire profusely. It is not improbable that the custom appeared so singular to Cortes, that he gave it the name in consequence. It has also been considered as a corruption of *colofon*, the Spanish term for resin; the pine trees yielding it being in such profusion here: and this might be changed into californo, a more familiar term in the Catalonian dialect, signifying a hot oven.‡

The peninsula is sometimes called Old or Vieja California, in contradistinction to Nueva or New California, the *later* discovery, hence the appellation. California *Baja*, or *Lower*, is also a term of some antiquity; and the distinction of Baja and Alta, Lower and Upper, seems to be that now more generally recognized, since the United States have denominated their portion of the territory from its *higher* latitude.

The Spaniards do not appear to have made any great advances towards occupying the territory thus explored, though there is no doubt that they were quietly gaining ground. During the seventeenth and eighteenth centuries, large quantities of pearls were procured by the Spanish adventurers, and these pearl fisheries in the Gulf of California increased in fame. Violent means, attended with frequent loss of life to the unhappy Indians, were resorted to, to carry on the fishery. It was customary with the Spaniards to kidnap and employ by force, as divers, all the inhabitants of the coasts and islands of the gulf they could lay their hands on. When the Jesuits had established the missions in the prosecution of their labours to the improvement of the condition of the Indians, they obtained a prohibition of these practices against those under their protection, and then they were brought from the continent opposite the gulf; but all were obliged to have

* See *Herrera*, dec. viii. bk. 8, ch. 9, 10; dec. viii. bk. 6, ch. 14; *P. M. Venegas*, Noticias de la California, Madrid, 1757, p. 124; *Lorenzana*, Hist., p. 322; *Robertson's* History of America, bk. v.; and *Reinhold Forster*, Voyages and Discoveries made in the North, Translation, London, 1776, 4to. p. 448; *Burney's* History of Voyages in the Pacific, vol. i.; the Introduction to the Journal of *Juliano* and *Valdez*; *Flurien's* Introduction to Marchand's Voyage; and *Greenhow's* History of Oregon and California.

† Bernal Diaz, Ist. and Conq. Mexico.

‡ Beechey, vol. i. pp. 284-5.

the viceroy's licence to fish for pearls. At the period of its greatest prosperity, about 600 to 800 divers were employed; the fishery being carried on in small vessels of 15 to 30 tons burden. The oysters, immediately upon being taken, were separated into five portions, two for the armador or owner, two for the diver, and one for the king, whose fifth, at the commencement of the seventeenth century, often produced no less than 12,000 dollars per annum for every bark employed. In the early part of the present century, a different distribution took place, and in 1825 sixteen or eighteen small vessels were employed. In 1831 four tolerably large vessels, from Mexico, with 180 divers, together obtained pearls to the value of £2,660.*

After the establishment of the missions in the northern Californian territory, under the Franciscan order, the Dominicans, subsequently to 1769, founded sixteen missions, of which Loreto was the chief, and considered as the capital of the peninsula; and all these missions were protected from the attacks of hostile tribes, by the guard of five soldiers at each, a force, as a garrison, almost unexampled, and yet it was sufficient. Upon the destruction of the missions (upon which subject more will be said in the succeeding chapter), this state of affairs was naturally very much altered, and, with the decline of these institutions, of course all other civilized relations also fell. Of the present condition of the missions, and their tenants, little can be said here. The attractions of Upper California must have withdrawn all or most of the former state of society and prosperity from its less fortunate and less rich neighbour. Therefore, of its population now we can give no idea. Mr. Forbes reckoned it at 14,000 or 15,000, in 1839.

The general feature of the peninsula consists of a chain of rugged mountains, which traverses it from one end to the other, the greatest elevation not exceeding 5,000 feet, but very uneven as to its general height. The country itself consists usually of groups of bare rocks, divided by ravines and hills, and intersected with tracts of a sandy soil, nearly as unproductive as the rocks themselves. There are but very few streams, as must be evident from the fact of the mountainous ridge which occupies nearly the whole of the surface; but this again leads to the formation of torrents in the wet season, which wash away all portions of earth, which otherwise might become fertile and productive. In some few places there are small valleys not subject to these drawbacks, which exhibit great fertility. From these causes Lower California is one of the most barren and unattractive regions of the temperate zone. There are some tolerable harbours, but, from the foregoing reasons, they are ineligible for the site of large towns. It is said to be rich in minerals, and near its South extremity there is said to be some rich argentiferous lead mines,† but these and others have been comparatively

* The most valuable pearls in the possession of the court of Spain were found in the gulf in 1615 and 1665, in the expeditions of Juan Yturbi and Bernal de Pinadero. During the stay of the Visitador Galvez in California, in 1768 and 1769, a private soldier in the *presidio* of Loreto, *Juan Ocio*, was made rich in a short time by pearl fishing on the coast of Ceralvo. Since that period the numbers of pearls of California brought annually to market were almost reduced to nothing. The Indians and negroes who followed the severe occupation of divers have been frequently drowned, and often devoured by sharks. The divers have always been poorly paid by the whites. —*Diccionario de las Indias Occidentales, by Col. Don Ant. de Alcedo*, 1768-69. *Translation, by G. A. Thompson*, 1812.

† Vancouver, vol. iii. p. 408.

little worked. Timber, and indeed vegetation itself, is very scarce, by far the largest portion of the surface being incapable of producing a single blade of corn. In some of the sheltered valleys fruits suitable to a warm climate have been successfully cultivated, and wine has been made and exported. Cattle are rather more numerous than would have been supposed, as they feed on the leaves of the musquito tree, a species of acacia. Some observations of the climate of the gulf are given presently; but for the land it may be said to be excessively hot and dry: unlike Mexico, the rains, except in the most southerly parts of the peninsula, occur during the winter months; summer rains scarcely occur North of Loreto. Near lat. 26° violent earthquakes are not uncommon.

In the subsequent descriptions the works of Capt. Sir Edward Belcher, of H.M.S. *Blossom*, who surveyed several of its ports; that of Du Petit Thouars; of M. Duflot de Mofras (who was formerly attached to the Madrid Embassy, and then to the Mexican Legation from France in 1840, and who was subsequently sent to examine the capabilities of the whole coast for commerce and navigation); and the voyage of Capt. George Vancouver in 1793, have been consulted. *

The GULF OF CALIFORNIA may be considered to terminate on the South at Mazatlan and Cape St. Lucas. These points have been geographically fixed by modern science, and Guaymas also, farther North, has been determined. Dr. Coulter, and Lieut.-Colonel Emory, U.S.E., made some observations on the Rio Colorado at its head, and the result of these determinations has been to remove the Gulf farther to the East than had been previously exhibited on the charts, thus giving the peninsula a greater breadth than it had been supposed to have; but this cannot be considered as finally determined.

This gulf was designated by the first Spanish navigators under the name of the Red or Vermilion Sea (*Mar Rojo, Mar Vermejo*), on account of the colour of its waters, and for its resemblance to the Red Sea of Arabia. The learned Jesuit missionaries traversed it entirely, calling it *Seno*, or *Mar Lauretaneo*, Gulf or Sea of Loreto, in honour of the Virgin, their protector and patroness, and it has been called the *Sea of Cortes*, from the great general.

The length of the Gulf is about three hundred leagues; its greatest breadth is 60 leagues at its entrance; but throughout its extent the distance from one side to the other does not vary but from 25 to 40 leagues. Beyond the thirty-first parallel, its breadth rapidly diminishes to the Rio Colorado, which falls into it at its head.

It has been often remarked, that a singular phenomenon occurs here, which science does not explain, and of which we possess but few examples. It is that of rain falling when the atmosphere is quite clear, and the sky perfectly serene. The savant Humboldt and Capt. Beechey have related the fact; the first having testified as to its occurring inland, the second in the open sea.

The tides are felt throughout the Gulf of Cortes; their height varies with the direction and force of the wind, and the configuration of the coasts: thus it is

* For accounts of the country and its nature, from whence we have drawn the previous particulars, see Mr. Alex. Forbes's California, 1839; Dr. Coulter's Account of California; and also the Journal of the Royal Geographical Society, vol. v. part i. 1835. The older authors who treat on this country have been before noticed.

7 feet at Mazatlan, the road of which is open, and at Guaymas, the port of which is full of islets, and sheltered from the wind, it does not exceed 5½ feet.

Independently of a great number of fish, of a variety of species, there are two species of immense shark found in the Gulf (*el tiburon*, and *la tintorera*), which often seize the pearl fishers. Whales are also met with in considerable numbers, but, up to the present time, no whaler has pursued them. On the islands are numerous seals and sea-calves. The pearl fishery is, or rather was, also followed, as stated in the introductory remarks.

The two shores of the Vermilion Sea run parallel with each other toward the N.W.; they are very low and full of salt marshes, tenanted by alligators, reptiles, and insects. The general aspect of the country is horrible; the imagination cannot conceive anything more naked, more desolate. There is an entire want of water and vegetation; there are only mangroves, and some thorny plants, such as the *cactus*, magueys (aloes), or acacias, to be seen. Orange trees or palms are rarely met with, and one must proceed some leagues into the country to find vegetable mould. The shore is formed by sand and lands quite unfit for cultivation.

At the entrance of the Gulf, on the eastern side, the summits of the *Sierra Madre* may be seen in the distance; these separate the provinces of Jalisco, Sinaloa, and Sonora, and those of Nuevo Mexico, Chihuahua, and Durango. The coast of Lower California presents, without interruption, a series of rugged peaks of volcanic origin, and without any vegetation. This mountain chain, which comes down from the North, and extends throughout the whole peninsula, gradually decreases in elevation as it approaches Cape San Lucas.

The eastern side of the Gulf is comprised in the Mexican provinces of Sinaloa and Sonora, separated by the Rio del Fuerte. Their principal riches consisted in their gold and silver mines, of course now eclipsed by Upper California. There are some considerable cities in the two departments. The chief are in Sinaloa, Culiacan, the residence of the governor, &c., population, 5,500. Rosario, above Mazatlan, has or had the quarters of the troops. Up to 1839 Arispe was the capital of Sonora, in which year it was carried 40 leagues to the South, on account of the incursions of the Apache Indians. It is now at the ancient mission of S. José de los Ures. The port of Guaymas concentrates all the maritime affairs of Sonora, and Hermosillo is the centre of its commerce and riches. In the North part of the state large quantities of gold have been found.*

Mr. Jeffery, R.N., who was here in January, 1834, makes the following observations in his Journal:—In the Gulf of California two winds are prevalent during the year. The N.W. from October until May, and the S.E. from May until October. During the former of these winds fresh breezes and fine weather will prevail, and a vessel making a passage up the gulf should keep the western shore on board, and she will find a little current in her favour, while on the eastern shore it will be against her: when the latter wind prevails you get nothing but heavy rains, oppressive heat, and sultry weather, and the reverse must be observed with respect to the currents. N.B. The above is a copy from an old Spanish manuscript, and we proved it correct when we made the passage in 1834."

* Duflot de Mofras, vol. i. ch. vi. p. 201.

Of the coast between Mazatlan and Guaymas our information is exceedingly defective. The charts, too, are very incorrect. At 80 miles N.N.W., *true*, from Mazatlan is the mouth of the *Piastla River*, the several points between are indicated on the chart. Thirty-five miles further on is the entrance of the *River Culiacan*, which runs to the southward of the mountains of the same name. At 7 leagues to the North of this is the mouth of the *Tamazula River*. The town of Culiacan is situated near the union of the Umaya and Tamazula Rivers, and is the capital of Sinaloa. Tamazula, on the river, is a declining town. According to the charts the coast here is rocky, and continues so to the mouth of the *Sinaloa* (or *Cinaloa*) *River*, 50 miles to the N.W. This last is 10 miles from *Point Ignacio*.

San Ignacio Rock lies 3 miles South of the point; more to the North is the *Rio Santa Maria de Aome*. The mouth of the *Rio del Fuerte* is shallow, and you must anchor a mile to the North or South of it to find a depth of 5 or 6 fathoms. The *Rio Mayo* has anchorage in 6 fathoms at 5 leagues to the N.W. Very near the land is seen the small barren island of *Lobos Marinos*, and at 4 leagues to the South of the entrance of Guaymas, the *Rio Yaqui*, the banks of which are inhabited by a tribe of Indians of the same name.

Though none of these rivers are navigable, the entrances of some of them will admit coasters.

GUAYMAS.—The harbour of Guaymas is recognised from the offing by a mountain surmounted by two peaks, which are called *Las Tetas de Cabra*, from their supposed resemblance to the teats of a goat. When this is made out, run along the coast, leaving it a little to larboard, and the Island of *Pajaros*, which forms the East side of the entrance, will soon be seen. Then steer so as to leave it to starboard, entering the channel between it and the land, and the town and harbour will soon be discovered. The entrance of the harbour once doubled, two islands are seen in the inner part of the bay, and you pass between these to reach the anchorage, near or off the land, according to the draught of water. Vessels under 100 tons make fast to the landing place, and those drawing 12 to 15 feet anchor a quarter of a mile off, in 3 or 4 fathoms. Men-of-war, corvettes, and frigates, ought to cast anchor outside these islands in 6 or 7 fathoms. This harbour, which would hold a considerable number of vessels, is very safe in all seasons; the bottom is good holding ground, and it is sheltered from all winds, and forms a large basin, surrounded with islands, which prevent any heavy swell reaching it. The bank lying in front of the entrance is the only danger to shipping, but it is easily avoided in leading winds, by keeping along the land. Should a vessel be obliged to beat in, she should be careful not to touch this rock.

The town of Guaymas has about 5,000 inhabitants in the fine season, and, during the rainy, 2,000 return to the small towns in the interior. The port possesses neither fortifications nor garrison.* The low price and excellent quality of the flour at Guaymas offer considerable advantages to shipping

* On October 20th, 1847, the Port of Guaymas was bombarded and captured by the American frigate *Congress* and sloop-of-war *Portsmouth*. About 500 shot and shells were thrown into the town. One English resident was killed, and some houses were burned. This was a portion of the campaign against the Mexican republic, which resulted in the entire separation of California from its former masters.

requiring such. Bullocks are worth about 12 piastres. Vegetables are scarce and dear, and the water is so bad in the harbour that they send boats to bring it from the Rio Yaqui and from 4 leagues to the southward.

Guaymas is surrounded by high mountains, which make it extremely hot in the rainy season. The same fevers are prevalent here as at San Blas and Mazatlan.*

The following remarks on Guaymas are by Lieut.-Com. S. O. Woolridge, H.M.S. *Spy*:—"I arrived at Guaymas on July 21st, 1847, in four days from Mazatlan. During this passage we experienced *strong currents running to the N.W.*,† from a mile to a mile and a half an hour. They were much influenced by the wind, which, from the 19th to the 21st, was south-easterly and southerly. Current also runs with more force on the eastern shore, which side we kept.

"Cape Naro can be easily distinguished by the Tetas or Paps, which resemble the teats of a goat; they are to the northward. The Island of St. Pedro Nolasco is just visible from the deck to the N.W. The land on the Yaqui shore is high and peaked; keeping this broad on your starboard bow, steer to the northward of a deep bay, where the land breaks off, and you will soon perceive the Island of Pajaros, which is at the entrance, or facing Guaymas. The water is deep all along the Island of Pajaros; that is to say, 4 fathoms, so close as to throw a biscuit on shore.

"A large ship will have to anchor soon after passing Pajaros; that is, abreast the Morro, in 5 fathoms. A small ship, and those of *Carysfort's* class, can anchor inside the Isles of Ardilla and Almagro, in 4 and 3½ fathoms, just inside them; and in 3 fathoms, as far in as the point off the town. You may go close to either of the Isles Ardilla or Almagro, in 3 and 3½ fathoms.

"Fresh beef and vegetables are to be obtained here, but the price depends greatly on the season of the year. In August, when the *Spy* was there, it was a bad time of the year, being the hottest season, the thermometer averaging 98° in the shade, when the country is very dry, and there is no herbage for cattle, which makes it difficult to obtain, and higher in price. The *Spy* paid 8 dollars a quintal for it, and the same for vegetables; but in October I am told it is much lower. There is also very good flour to be obtained here between July and March, that is, all the end and beginning of the year; and with little difficulty (chiefly depending on time) very good biscuit can be made. In August this article is also scarce, and dear in comparison, because the new batch of flour is just coming in, and the difficulty of transportation from the interior is very great, owing to there being no herbage for the mules. I contracted for 640 quintals of biscuit for H.M.S. *Constance* and *Spy*, at 10 dollars the quintal, to be delivered at the contractor's house, and bags found by the purchaser. Flour at this time was 16 dollars the carga, or 300 lb. But I am told in September and October it will fall to 9 dollars the carga, when biscuit will be proportionably cheaper. Water is very difficult to be got; it is to be obtained by sending about 4 miles for it, or it can be purchased; but, owing to its having to be brought in on

* Duflot de Mofras, vol. I. pp. 183-4.

† This will demonstrate that these currents are greatly, if not entirely, dependent on the wind, for we are told by other commanders that there is a *great outset* from the gulf, which must be guarded against in crossing the entrance.

mules, or in carts, the price is very high. I wanted 12 tons, which I found could not be obtained for less than 30 dollars, which would be nearly 10*s*. a ton." *

Mr. Jeffery, R.N., makes the following remarks :—"This is a very blind harbour to make, the entrance being entirely hid by the Island of Paxaros (or Pajaros) which is quite barren, and of a light gray colour, and about one mile long. After rounding Paxaros, steer right for the Island of Morro, which is on a spit of land, and will be the first island you open, bearing about N.W. from the West end of Paxaros; when you open the two islands off the point abreast of Paxaros, you may steer right between them, and anchor in 3½ or 4 fathoms, good holding ground, and perfectly landlocked. Fresh water can be obtained by drawing it out of wells, wood can be cut, fresh provisions are plentiful."

Proceeding northwards along the coast of Sonora, at 4 leagues from Guaymas is an excellent harbour, called *Puerto Escondido*, the small islands of *San Pedro Nolasco*, *La Tortuga*, *San Pedro*, and in lat. 29° the *Isla del Tiburon*, inhabited by Seris Indians, who have some huts on the continent. This island is 10 leagues in length, and is the only one in the gulf which is inhabited. It forms with the coast a narrow and dangerous channel (*el Canal Peligroso*), which is terminated by the islet *De los Patos*. All this part of the province is barren, and only a few miserable Tépocas Indians are to be met with, and the old *Misione de Caborca*, lying 22 leagues in the interior, on the banks of a small river, in lat. 31° N.

To the North of the *Rio de la Concepcion de Caborca* is the small bay of Santa Sabina and the Island of Santa Inez, the Rio de Santa Clara, and the watering place, *Los Tres Ojitos*. As far as the *Rio Colorado*, at the head of the gulf, the coast is barren, and very low. The wind perpetually raises clouds of the fine sand which composes it.

RIO COLORADO.—The river called the Rio Colorado de l'Occident, to distinguish it from that which falls into the Gulf of Mexico, rises in the Rocky Mountains, about lat. 41° N. The French Canadians named its upper portion the Riviére Espagnole. Its length is about 300 miles, and during the fine season it is very shallow. During the rainy season, and after the melting of the snows, it overflows its banks, and inundates the flat country through which it flows. Its mouth, at the head of the gulf, is nearly 2 leagues in width, and is divided into three channels, by two small islands called the *Islas de los Tres Reyes*. The tide rises here 20 or 25 feet, causing violent currents, which run at the rate of 10 or 12 miles an hour. The depth near the entrance of the river is very small, and to enter it the Californian coast must be kept close in order to find the passage, which is very narrow, and often is not more than 5 or 6 feet in depth. The bed of the river is filled with banks, which are left dry at low water.

At 8 leagues above its mouth the Colorado receives the *Rio Gila*, coming from the East, after it is increased in volume by the *Rio de la Ascencion*, formed again by the junction of the Rios Verde and Salado. All these rivers rise in the branches of the Sierra Madre, have but little depth, and overrun their banks during the rains. The boundary between Upper and Lower California is now, as under the old order of things in Mexico, from the mouth of the Colorado to the

* Nautical Magazine, Dec. 1848, pp. 637-8.

South of the mission of San Diego on the Pacific. The Rio Gila forms the eastern continuation of this boundary between the new American province of Upper California and the Republican States of Mexico.

In re-descending the gulf on its western shore, from North to South, we pass the marshes which extend to *Cape San Buenaventura*, the watering places of *San Felipe de Jesus*, *San Fermen*, *Santa Ysabel*, *La Visitacion*, *San Estanislao*, the Bay of *San Luiz de Gonzaga*, *San Juan y San Pablo*, *Los Remedios*, the *Bay de los Angeles*, *San Rafael*, *Capes San Miguel* and *San Juan Bautista*, the Islet of *San Barnabé*, *Cape Trinidad*, *Santa Anna Islet*, and the *Cape de las Virgenes*, which is the last extinct volcano in Lower California. According to the accounts of the Jesuits, it was still in activity in 1746. There is much sulphur in the neighbourhood of the crater.

In lat. 29° the Island of *del Angele de la Guardia*, which is long and narrow, forms with the coast the *Canal de Ballenas*, where a great number of whales have been met with. Abreast of the island, at 9 leagues from the shore, is the mission of *San Francisco de Borja*. To the South, the islands of *Sal si Puedes* (get out if you can), *Las Animas*, and *San Lorenzo*, present a very dangerous passage. To the South of the Cape de las Virgenes are seen the Bay of *Santa Agueda*, *Galapagos Island*, the Cape and Island of *San Marcos*, which, with the islands *Tortuguitas* and Cape San Miguel, form the *Bay of Molejé*.

Opposite the Island of San Marcos, but 6 leagues inland, there exists still the mission of San Ignacio. That of Santa Rosalia lies half a league from the sea, on the banks of the *Rio Molejé*. This point is easy to be found. In approaching it a small hill will be seen, called the *Sombrerito*, having the form of a hat. The bay is shallow, and boats of 15 or 20 tons can only enter it. Some pearls are found, and on the banks of the river some fruits and grain are, or have been, produced.

MOLEJE.—This is an extensive bay, which runs to the S.S.W., between the above river and Point Concepcion on the East. The following observations on it are by Lieut.-Com. S. O. Woolridge :—" Point Concepcion is difficult to make out, when you have about a dozen of the same kind within a few miles of each other. However, the best marks I can give are some table land, which is very remarkable, and is rather to the right of Molejé village. Keep this about two points on your starboard bow, and you may stand in until you discover some sandy islets, which are off a point called Punta Ynes. When you are East and West with them, you will be distant from them about 3 miles. After passing these islets, then steer South and S.S.W., until you make out the Pyramid Rock, spoken of by Capt. Hamilton. This rock is called Sombrerito, or Little Hat. I think it bad to call it Pyramid Rock, as there is a point which, in standing in, may be easily mistaken for it, resembling also a pyramid; but the rock is a pyramid fixed on a round pedestal like a fort. Another good way of making out this place is, when the wind is fair, to keep Tortuga Island, about 20 miles distant, bearing about N.W., and steer in S.E. till you make out the sandy islets, and proceed as above. There is a passage between the islets and the main land for small vessels, but, though very inviting, should not be attempted. I tried it, but getting into 2½ fathoms, I put about as quick as possible. My anchorage marks in Molejé Bay were as follow, in 5 fathoms :—Point Concepcion,

N. 84° E.; Tortuga Isle, N. 4° W.; Lobos Isle, N. 2° E.; Sombrerito, S. 67° W. (Pyramid Rock of Capt. Hamilton); Equipalito, S. 22° W. (Rock on South side of entrance to the river); Punta San Ynes, N. 10° W.

"This is very close in, but I wished to facilitate the watering; about half a mile further to the northward, in 8 fathoms, is a very good berth. In going into the bay after making out the Sombrerito, if you wish to go close in, take care not to bring the Sombrerito at all on your starboard bow; that is, do not open the mouth of the river, as by sounding I discovered a rock with only one fathom on it; it is on a sandbank with 3 fathoms all round it, about three-quarters of a mile from the shore; but the rock itself has only one fathom. It lies with the entrance of the river open, directly between the Sombrerito and Equipalito rocks, distant from half to one mile off shore. I am surprised Capt. Hamilton has not mentioned it, for I must have gone very close to it in rounding my vessel to. The report of the facility of watering is very delusive and uncertain.

"In the first place I cannot think it possible to water out of the river, as it is salt for at least 2 or 2½ miles, and a great portion of the time boats could not possibly get up so far. I was there, fortunately, when the moon was nearly full, and the water was only low between eleven at night and four in the morning, so that I was enabled to water about eighteen hours out of twenty-four, and though I had but one small boat (23-feet cutter), I managed to get 12 tons in two days. She had to go 1½ miles up the river, to the house of Joseph Padras, and the casks were rolled about 100 yards to a small stream in his garden; with a force pump it might be obtained without running the casks. The water is delicious to drink at the stream, but it is so very low, and our water, after being a day or two on board, became so black, and smelt so strong of decayed vegetable matter, that though it improved by keeping, it served chiefly for cooking and washing.

"In going up the river in boats keep close to the Sombrerito, and keep the starboard shore on board till you are a mile, or one and a quarter up the river, when you will encounter a sandbank in the centre of the river, and must keep over on the port shore to clear it. Abreast this sandbank is the Rancho of José Padra. At the time I visited this place, in August, owing to the dryness of the season, and the want of fodder, there was no beef or vegetables to be procured. The bay is open to the N.E. winds, and there is no shelter from the sea, which rolls in heavily; but with all other winds I think any man-of-war could get out, if she did not leave it too late, till the sea was too heavy, as there is plenty of room for beating.

"The passage from Guaymas to Molejé Bay can be easily done in 20 hours. I left Molejé again on the 14th, and arrived at Guaymas on the 16th, being 36 hours. On the 26th of August I sailed from Guaymas for Mazatlan, where I arrived on the 3rd of September, in eight days. This, at this season of the year, is considered very fair, as south-easterly winds and calms prevail. I kept over by advice on the western shore, and passed inside of Catalon Island; but I think the more you can keep in mid-channel the better. We experienced little or no currents, but the wind was very light and the weather fine all the way. On the 30th and 31st of August we had an easterly current about 14′ per diem."

From Molejé Bay to Loreto there is always, near the land, from 20 to 30 fathoms, and the coast offers good anchorages: the Points of *Santa Teresa*, *Punta Colorada*, those of *Púlpito de San Juan*, the bay of that name, that of *Mercenarios*, *Point Maglares*, and *San Bruno Cove*.

At 3 leagues to the North of Loreto the little island of *Coronados* offers shelter from the N.E. Near the mission there is a depth of 4 fathoms, and under the lee of the *Island del Carmen* 13 to 16 fathoms.

The anchorage at Loreto is pointed out by the church and a clump of palm trees, and it may be distinguished at a distance by a very lofty peak, surrounded by smaller hills.

This mountain, called *El Cerro de la Giganta*, is the highest in Lower California. Its height above the sea level is 4,560 feet, according to trigonometrical measurement; it is of volcanic formation, as is all the rest of the chain which runs through the peninsula. The anchorage of Loreto is open to the winds from North, N.W., and S.W. When they blow very strongly, the ship must get under way to escape being driven on shore. If she is of small draught, she may make for Puerto Escondido, 14 leagues to the South.

The mission of *Real de Loreto*, opposite Carmen Island, was the capital of Lower California; but it is so much decayed that the authorities were transferred to the *Real de San Antonio*. The presidio, the mission, and the church, are fallen to ruins. These buildings, very substantially built by the Jesuits, were intended to serve, in case of attack, as an asylum to the inhabitants. They are surrounded by a thick wall, which turns the waters of a torrent which comes from the mountains; and which, several times, washed away the houses and the vegetable earth. The presidio was defended by some small artillery, but never used. The church, for a long time after its decay, contained many pictures, silver vessels, and jewels of considerable value, which, though left quite open, were considered safe from spoliation. At Loreto there are some gardens, but water is generally scarce; and that from the wells is brackish and unwholesome.

Under the Spanish government a courier left Guaymas every month, and went to Loreto, from whence the letters were despatched to the different missions; but this is long since abolished.

To the South of the Isle Carmen are the Islands *Catalana*, three leagues in length, *Monserrate*, *De los Danzantes*, the *Pearl Banks*, &c., of which the chart gives the best guide. The only points visited by shipping are the harbour of *La Paz* and *San José del Cabo*.

La Paz, where Cortes landed, May 3rd, 1535, is in lat. 24° 10′, lon. 109° 45′. In coming here vessels anchor in *Pichilingue Bay*, to the East of the Island of *San Juan Nepomuceno*, in from 5 to 9 fathoms, and at 2 leagues' distance from the houses, passing at equal distances from the Island of Espiritu Santo and Point San Lorenzo; only small vessels can approach near to the houses. The population of La Paz consists of about 400 people, the greater part descended from foreign seamen. This is the most commercial port of Lower California. Vessels from San Blas, Mazatlan, and Guaymas, often come to purchase shell at from 16 to 18 piastres the quintal, and mother-of-pearl shells at 6 piastres the hundredweight.*

* The Port of La Paz was named at the time of its discovery Bahia de Santa Cruz; afterwards

A vessel anxious to keep on the coast of Mexico, or in its neighbourhood, during the bad season, cannot do better than run over to the Bay of La Paz on the West shore of the Gulf of California, and but little to the North of Mazatlan. This splendid harbour is formed by the main land of Lower California on the starboard hand going in, and a long chain of islands with shallow passages between on the port hand. The most eastern island is Espiritu Santo.

North end of Espiritu Santo, lat. 24° 36′ N., and lon. 110° 22′ W., with a large rock due North of it 5 miles.

In approaching this bay from Mazatlan the Island of Cerrabo will be first made, high and mountainous; North end lat. 24° 23′ N., lon. of South end 109° 45′ W.; from it Espiritu Santo will be seen bearing about W. by N. The bay is at least 30 miles deep, and for the first 20 miles a deep bold shore on either hand, no bottom with 20 fathoms close to the islands. Large vessels anchor under the Island of San Juan de Nepomuceno; but small ones anchor within half a mile of the village of La Paz: fish, water, turtle, cheese, and fruits are to be obtained here; and cattle also in the wet season when pasturage is to be found on the coast. Snakes are very numerous and venomous.

A knowledge of the tides or currents in the neighbourhood of this port would be very serviceable; it has been much frequented by the Americans during their operations against Mexico. A vessel bound to California could only have one object in making the Mexican coast, *en route*, namely, that of communicating with her owners by overland despatch through Mexico, and as that is a possible occurrence the few following notes will serve for general guidance.

A vessel making the passage northward from San Blas, had better make an in-shore track until she reaches the latitude of or sights Cape Lucas, the southern promontory of South California, as she will there get the true wind, which blows almost without intermission along the line of coast from the northward. A West, or may be a *South* of West, course will only be first made good, but as an offing is obtained, the wind will be found to veer a little to the eastward. However, it will always be the object to make headway, and get out of the tropic without any reference to the longitude, as a strong N.W. wind will soon in 25° or 28° N. run off the distance, provided you have sufficient northing.

The attempt to beat up in-shore amounts to perfect folly, if it does not deserve a worse name, a strong current accompanying the wind; and the latter must be taken into consideration when running in for your port with westerly winds.*

At La Paz, the Rio Yaqui, and Guaymas, eight or ten small vessels, of 20 to 40 tons, are fitted out for the pearl fishing. The divers are all Yaquis Indians. The fishing begins in May, and ends in October. The principal pearl-banks in the gulf are in the Bay of La Paz and near Loreto, the S.W. point of the Isle Carmen, Puerto Escondido, Los Coronados, Los Danzantes, San Bruno, and S. Marcos Islands.†

---

it was called under the name of Porto del Marques del Valle (Cortes), and lastly that by which it is now known. It was during his stay in this bay that Cortes received the unpleasant news of his disgrace, brought on by the jealousies which his great services raised, and that, not less painful, the news of the arrival of the first Mexican viceroy, sent to replace him in the command.—*Du Petit Thouars*, vol. ii. p. 161.

* Nautical Magazine, March, 1849, pp. 143-4.

† Duflot de Mofras.

The Bay of *San José del Cabo,** open from the South to N.N.E., is very dangerous, when, in bad weather, the winds prevail from this quarter; the bottom being bad holding ground, and the anchorage so near the shore, that there is no chance of getting off with the wind on shore. It follows, therefore, that this bay ought not to be frequented, except from the end of November until May, the season when winds from West and N.W., and fine weather, prevail throughout the gulf, as well as on all the Mexican coast. It is at this period that it is visited by the whalers and merchantmen trading to Lower California; but these latter prefer the port of La Paz, before described, which is more secure.

The Bay of San José, apart from its insecurity, a fact proved lately by the loss of several whale ships, is a good point for refreshment. Wood is easily procured, and at a low price. Water also is got from a small rivulet which falls into the sea to the North of the anchorage. Bullocks, sheep, and vegetables, are to be had also in abundance.†

The *Bay of San Lucas* offers safe anchorage and shelter from westerly winds, but is exposed to a very heavy and most dangerous sea from the S.W. The soundings are very irregular, and the anchorage, by reason of its great depth in the centre, is completely a lee shore.

At the village, consisting of four houses at the time of Capt. Belcher's visit, inhabited by two Americans and some Californians, water, wood, cattle, cheese, oranges, and pumpkins, were obtained. The water, which is procured from wells, is sweet when drawn, and very bright, but is impregnated with muriate of soda and nitre, which pervade the soil. It consequently soon putrefies on board. Cattle were fine, varying in price from 5 to 8 dollars; and ships-of-war calling or passing *en route* for San Blas or Mazatlan, will do well to take their bullocks here, as the Mexican beef is very inferior, and does not afford so much nourishment even as the salt provision now supplied to Her Majesty's service. Wood is about the same price as at San Blas, The cheese is good, at times excellent, and may be procured at any age; it is the refuse of this market, and at treble price of that which is met with at Mazatlan or San Blas.

The country about Cape San Lucas is mountainous, and probably granitic. The plains, as well as the hills, are very abundant in cacti (or Indian fig).

The navigator has no hidden dangers to fear; all are above water. After rounding the Frayles from the westward, he may safely stand for the houses, dropping his anchor in 15 fathoms. The bad season is supposed to commence in June, and terminates on the 1st of November.

A word of caution in anchoring here may be given. The *Sulphur*, on shortening sail in coming to, had 10 fathoms, and immediately after they had no bottom with 88 fathoms, just as they were about letting go the anchor. This shows the necessity of keeping the lead on the bottom before letting go an anchor, or you may lose it.‡

* At the mission of San José in the southern part of the peninsula, the learned Abbé Chappe d'Auteroche died, whom the Royal Academy of Sciences sent to observe the transit of Venus in 1769. He accurately determined the position of Cape San Lucas, which served as a landfall and departure for the ships for China and Europe.

† Du Petit Thouars, vol. ii. pp. 161-2.

‡ Voyage of the *Sulphur*, vol. i. pp. 338-9.

In Mr. Bennett's whaling voyage there is this notice of the bay :—

"Hove-to off the mouth of the bay, between Cape St. Lucas and Cape Palmo, the southern extremity of the Isthmus of California, where is a small grazing settlement, which supplied us with excellent beef, poultry, and cheese. The land about Cape St. Lucas, which forms the S.W. extreme, is bold, rugged, and mountainous; the low lands appear flat and sandy. The settlement on the shores of the bay is a little to the N.E. of the cape, and the bay affords a fair roadstead with 17 fathoms water at half a mile from the beach ; but open to the S.E. gales, which are very severe. The tide here is regular, with a rise and fall of 5 or 6 feet. Shipping may here procure fuel, water, and provisions ; a bullock costs from 5 to 10 dollars. The residents number about thirty, and the whole farm belongs to one person. Their commerce is confined to the English and American South Sea men, who visit the bay for supplies. Close in with the land I obtained a species of sea-weed exactly resembling the sargasso or gulf-weed of the Atlantic."*

CAPE SAN LUCAS, the southern extremity of the Peninsula, is in lat. 22° 52′ N., lon. 109° 53′ W. It gradually, though not very regularly, descends from the very broken and uneven range of mountains which extend from the N.W., and terminates in its South extremity in a hummock of low, or very moderately elevated, land, of a rocky, sterile appearance.

The coast between Cape San Lucas and the Island of Santa Margarita, a distance of 130 miles, is in many parts composed of steep, white, rocky cliffs; the country rising with a very broken and uneven surface to the ridge of stupendous mountains previously mentioned, and which are visible at a great distance into the ocean. The shores jut out into small projecting points that terminate in abrupt cliffs, and having less elevated land behind them, gives them at first the appearance of being detached islands; but on a nearer approach this does not seem to be the case. The general face of the country is not very inviting, being destitute of trees and other vegetable productions.

After having doubled Cape San Lucas, and proceeding along the coast, at a short league from it, you find the mission of *Todos los Santos*, which, at the period of M. Duflot de Mofras' work, still contained a few Indians. There is an anchorage, with a small rivulet where water can be procured, and also provisions could be had.

At the parallel of 24°, the mountains form a promontory, surmounted by three peaks, the truncated summits of which resemble tables, and which are, from this reason, called *Las Mesas de Narvaez*.

From this point the coast runs nearly N.E. to the large island of Santa Margarita, forming the South entrance of the immense bay of La Magdalena.

The GULF of MAGDALENA is an extensive inland sea, whose existence was scarcely suspected in Europe prior to the visit of Capt. Sir E. Belcher, in 1839, when he minutely surveyed it, and the fine chart recently published is the result of that survey. Nearly at the same time the French frigate *La Venus*, under Du Petit Thouars, also surveyed its shores, and his plan, much less elaborate

* This is also quoted in the Journal of the Geographical Society, vol. vii. chap. 2, pp. 228-9.

than that of our English hydrographers, forms a portion of the atlas accompanying the voyage.

It may properly be said to consist of two extensive bays; Almejas Bay to the S.E., and Magdalena Bay, the principal, to the N.W.

Santa Margarita Island, which forms the seaward face of the greater part of these two bays, is about 22 miles in length, by 2½ in average breadth. In its centre it is so low, that at a distance it might be taken for two separate islands, the northern and southern portions being high land.

Cape Tosco is its south-east point, and is bold-to. Four miles N.N.W. from it is Mount Santa Margarita, about 2,000 feet in height. To the East of it is the southern and intricate entrance to the two bays, through the *Rehusa Channel*, formed by the low sandy extremity of *Cresciente Island*, called *Sta. Marina Point*. The tide runs very strongly through this narrow channel, which is much embarrassed by sandy patches nearly and quite awash. No instructions can therefore be given for this, and will be scarcely ever required.

Almejas Bay is about 12 miles in extent to the N.W., and is divided from the principal bay by a narrow but clear and deep channel, formed by a sandy projection from the main land on the N.E. The extremity of this is called *Lengua* (Tongue) *Point*, and connected with it is an extensive low island, named *Mangrove Island*.

Southward of this sandy tract, the Island of Santa Margarita is divided into two separate tracts of mountainous land by a low sandy neck. From the chart, these points would seem to be the effect of currents, probably in combination with the geological changes hereafter noticed.

The North side of Santa Margarita Island, from this strait, trends nearly West, *true*, to its N.W. point, *Cape Redondo*, off which, to the distance of a quarter of a mile, some sunken rocks extend. About midway between its extremities, on the seaward face, is *Pequena Bay*, lying against the low tract previously mentioned; *Cape Judas* forms its N.W. extremity. Two species of tortoises are found on the island, one of which is very good eating, but the shell worthless; the other, on the contrary, is unfit for food, but the shell is excellent and valuable.

The entrance to Magdalena Bay lies between Cape Redondo and *Entrada Point*, the southern point of the San Lazaro peninsula: like the opposite side, some sunken rocks lie off it. *Mount Isabel*, 1,270 feet high, lies 3¼ miles N.W. of it. The width of the entrance is 2½ miles, and the depth between the head 15 to 20 fathoms. At the foot of Mount Isabel, within the bay, a fresh-water marsh is marked on the chart. This is 3 miles within the entrance point. Five miles farther on, on the N.E. face of the peninsula, and near to where the high land sinks to the long sandy neck connected with it northward, is the anchorage under Delgada Point; here Capt. Belcher established his observatory. This is in lat. 38° 24′ 18″ N., lon. 112° 6′ 21″ N. This was at the foot of a hill 600 feet in height, according to M. de Tessan's chart.

Northward of this the bay becomes very shallow, but has some deeper channels extending northward, separated by extensive shoals, which have received the names of *Du Petit Thouars*, *Tessan*, and *La Venus*, the French surveyors and their ship. As there can be no inducement whatever for entering them, no further notice here is necessary.

In the account of his exploration, Capt. Sir Edward Belcher gives the following description of his progress :—

I was fully prepared to have found, as the name imported, an extensive bay, but on entering the heads, which are about 2 miles asunder, no land could be discerned from the deck, from N.W. to N.E. or East; and even after entering, it was quite a problem, in this new sea, where to seek for anchorage, our depths at first, even near the shore, ranging from 17 to 30 fathoms. However, as the prevailing winds appeared to be westerly, I determined on beating to windward, in which it eventually proved I was correct. About 4 P.M. we reached a very convenient berth in 10 fathoms, with a very sheltered position for our observatory. Preparations were immediately made for the examination of this extensive sea, or what I shall in future term the Gulf of Magdalena.

It is probable that this part of the coast formerly presented three detached islands, viz., St. Lazarus range, Magdalena range, and Margarita range, with one unnamed sand island, and numerous sand islets. It is not improbable that its estuary meets those from La Paz, forming this portion of Southern California into one immense archipelago.

The first part of our expedition led us up the northern branch of what held out some prospect of a fresh-water river, particularly as frequent marks of cattle were noticed. In the prosecution of this part of our survey, we noticed that the St. Lazarus range is only connected by a very narrow belt of sand between the two bays, and that the summits of some of these sand hills were covered in a most extraordinary manner by piles of fragile shells, which resembled those found recently in the gulf. At elevations of 50 and 60 feet, these minute and fragile shells were found *perfect*; but on the beaches, either seaward or within, not a shell was visible. This is the more extraordinary, as these sandy wastes are constantly in motion, and drowning everything else, and yet these shells are always exposed. On digging beneath them to erect marks, no beds of shells occurred, nothing but plain sand. It was further remarkable, that they appeared to be collected in families, principally arca, venus, cardium, and murex; when ostrea appeared they were by themselves.

The cliffs throughout the gulf abound in organic remains, and I cannot but believe that the same cause has produced the above unaccountable phenomena, which I witnessed throughout a range of at least 90 miles.

Having explored the westernmost estuary, about 17 miles North of our observatory, until no end appeared of its intricacies, I resolved on attempting a second, which afforded a wider entrance, about 4 miles beyond the last, and it still offered ample scope for employment, the advance boats being at that moment in 4 fathoms, and distant heads in view; but finding no hope of reaching fresh water, I determined on adhering to its main outlines.

By November 9th (1839), we had reached the East end of the first gulf, and found the channel or strait connecting them not more than a quarter of a mile wide. I was sanguine in expecting that we should discover a safe channel out by the East end of Margarita, but I found that our boats, and, upon emergency, the *Starling*, might have passed out, but it was far too difficult and doubtful for the ship.

We had frequently seen, indistinctly, the outlines of very high mountains to the eastward, distant 50 or 60 miles; but on this day I could detect breaks which indicated water courses between them, and could plainly follow out yellow breaks of cliffs as far as the eye could trace inland. I have not the slightest doubt that these estuaries flow past them, and probably to the very base of the most distant mountain, even into the Gulf of California. As I am informed that there are no fresh streams in the district of La Paz, and that similar esteros run westerly from that neighbourhood, it is not improbable that they meet. Although the solution of this question may not be commercially important, it is highly interesting in a geographical point of view.

After all the time expended (eighteen days) on this immense sheet of water, it will naturally be inquired, what advantages does the port offer? The reply is, at the present moment, shelter; and from several water courses, which were nearly dry at the time of our visit, it is evident that very powerful streams scour the valleys in the winter season, which, in this region, is reckoned between May and October.

Fuel can be easily obtained in the estuary (mangrove).

As a port for refit after any disaster it is also very convenient; and for this purpose either our northern or southern observatory bays may be selected. The latter would afford better shelter; but the former is certainly more convenient, the access to it being entirely free from shoals.

In war it would be a most eligible rendezvous, particularly if watching the coasts of Mexico or California, as no one could prevent the formation of an establishment without adequate naval force, and the nature of the country itself would not maintain an opposing party.

The Island of Margarita would afford an excellent site for a deposit for naval stores. Martello towers on the heads of entrance would completely command it, and, excepting on the outside, no force could be landed.

Water would doubtless flow into wells, of which we had proof in spots where the wild beasts had scraped holes; but from some (no doubt removeable) causes, it was intensely bitter. There is nothing in the geological constitution of the hill to render it so.

The ranges of hills composing the three suites of mountains vary from 1,500 to 2,000 feet.*

It is high water, full and change, in Magdalena Bay, at 7^h 35′; rise, 6 feet 3 inches; variation, 9° 15′ E.—(1838.)

From Entrada Point, the entrance to Magdalena Bay, to *Cape Corso*, which is at the North end of the elevated part of the peninsula, the distance is 9 miles, the breadth of the peninsula being here, on an average, about 2 miles.

From Cape Corso to the S.E. point of Cape Lazaro, the distance is 7½ miles, and the outer coast is formed by the long, narrow, sandy neck which separates the ocean from the entrance channels in the North part of Magdalena Bay, before

* Sir Edw. Belcher, vol. I. p. 331—336.

noticed. This recedes 4 or 5 miles from the general line of the coast, and forms under Cape Lazaro, that is, to the southward of it, the *Bay of Santa Maria*.

Cape San Lazaro is in lat. 24° 44′ 50″, and is 1,300 feet in height; hence the coast trends to the East and northward, and is steep-to as far as the small bay (*Pequena Bay*) formed by Point Santo Domingo, in about lat. 26°.

The ALIJOS ROCKS, or *Farallones Alijos*, a cluster of four high, detached, and remarkable rocks, lie off 140 miles from this part of the coast. They are in lat. 24° 51′, and lon. 115° 47′. This position was accurately determined by Admiral Du Petit Thouars, who saw them for the second time. They will be again noticed hereafter, but, from their dangerous character, they must be alluded to here.

The coast beyond Point Santo Domingo trends to the N.W. for nearly 20 leagues, and then turns abruptly for 10 leagues to the S.E., when it forms *Point Abreojos* (open your eyes), the extremity of which is surrounded with dangerous rocks. The point is in lat. 26° 42′ N., lon. 113° 34′ W.

The two small islands of *La Asuncion* and *San Roque* lie on the coast to the N.W. The coast is clear thence for 80 miles to the northward.

SAN BARTHOLOMEW BAY, or Turtle Bay, as it is also called by the whalers, lies to the South of Point S. Eugenio. The bay is formed by a high range of loose cliffs on the North and a fine gravelly bay on the East, and a coarse sandy tongue connects a high peninsula, or island at high water, in its centre, forming a third southern bay. From this peninsula rocks extend northerly, partly under water, jutting into the heart of the bay, and forming a safe land-locked position, having 5 fathoms within.

The sheltered position where the whalers resort to cooper is within a range of reefs which divides the bay, from seaward, into two parts. The anchorage taken by the *Sulphur* was in 7 fathoms, sheltered from all but S.W. winds; but bad holding ground. The surrounding land is high and mountainous.—(Belcher, vol. i. p. 330.)

The place of observation on the northern head of the bay is in lat. 27° 40′ N., lon. 114° 51′ 20″ W.; var. 10° 46′ E.—(1839.)

In proceeding along the coast, the *Point San Eugenio*, which appears to form but one promontory with Natividad Island, lying off its western extremity, should be avoided.

The Island of Natividad lies S. ¼ E., distant 14 miles from the peaked mountain on Cerros Island. It appeared to Vancouver to be about 4 miles long in a N.E. and S.W. direction, and, like the latter, presented a barren and dreary aspect. It is moderately elevated, and lies off (Point S. Eugenio) the S.W. point of the large Bay of S. Sebastian Vizcaino.

Here, as at Cape San Lazaro, the whalers come, passing by night between the island and the main land, and anchoring against the point, which is very low. The passage between the island and this point is 7 or 8 miles wide, and has a depth of from 17 to 25 fathoms. In approaching from the South, a tolerably high mountain, called the *Morro Hermoso*, shows itself to the south-eastward of the point. In coming from the North the islands of Cerros and San Benito will be a sufficient guide.

To the East of Point San Eugenio, the coast recedes considerably for the space of a degree of latitude, forming the great *Bay of Sebastian Vizcaino*,* in which, in lat. 28° 56′, is the small bay Del Pescado Blanco, formed by a point 5 or 6 miles in extent, which projects to the southward.

CERROS or CEDROS ISLAND forms the most western side of the Bay of S. Sebastian Vizcaino. In the Spanish charts it is represented to be about 10 leagues long. Vancouver, who approached it from the northward, says:—"The shores of the Island of Cerros wore an uneven, broken appearance, though on a nearer view they seemed to be all connected. The southern part, which is the highest, is occupied by the base of a very remarkable and lofty peaked mountain, that descends in a peculiar rugged manner, and by projecting into the sea, forms the S.W. end of the island into a low, craggy point; this, as we passed at the distance of 5 or 6 leagues, seemed, like the other part of the island, to be destitute of trees, and nearly so of other vegetable productions."†

The Island of San Benito lies W. by N. ¾ N., 20 miles from the peaked mountain on Cerros Island. It is a small island, and has some islets and rocks about it.

Playa Maria Bay.—Northward of Pescado Blanco Bay, in lat. 28° 55′ N., lon. 114° 31′ W., is *Santa Maria Point*, to the East of which is Playa Maria Bay, the limits of which may be taken as Black Point, about 6 miles S.E. by compass from Santa Maria Point. The bay is open, but clear; and inland, 2 miles from the head of it, is a hill called the *Nipple*, 1,132 feet high; and in the North part of the bay, on the coast, is what Capt. Kellett calls *Station Peak*, 256 feet high. The variation line in 1847 was 8° 44′ E.

We have no particulars of the coast northward of this, further than what the Spanish charts afford us, until we come to *St. Geronimo Island*, whose position is determined as lat. 29° 48′, lon. 115° 47′. The next place is Port San Quentin.

PORT SAN QUENTIN is the name applied or chosen by Sir Edward Belcher for the harbour, which, under several Spanish and English charts, is called the Bay of San Francisco. This is much preferable, as the triple repetition of this name on this coast, applied by the Spaniards in honour of their patron saint, peculiarly so regarded by the navigators in the Pacific, has led to confusion. Capt. Sir Edward Belcher's plan does not notice former synonymes, but we take the following extracts as applying to this spot:—

Cape San Quentin (or Virgenes), the western point of the Port San Quentin, is a long, low, projecting point of land. From this it takes a course of about N.N.W., for 8 miles, to Point Zuniga of Capt. Vancouver. This portion of the coast consists of five remarkable hummocks, nearly of equal height and size, moderately elevated, with two smaller ones close to the water side, the whole rising from a tract of very low and nearly level land, forming a very projecting

* Sebastian Vizcaino, from whom this name is derived, was charged by the Spanish viceroy of Mexico, Don Gaspar de Zuniga, Count de Monterey, to survey these coasts. He set sail on this commission from Acapulco, May 5th, 1602, with four vessels, and among others, discovered and named the Port of Monterey, which has remained the capital of this country up to recent times.

† Vol. iii. p. 342.

promontory. This was named, by Vancouver, Point Five Hummocks, who says that it is as conspicuous and remarkable as any projecting point the land affords.*

In coming down the coast from the northward, he had taken it for a series of detached islands.

The following is by Capt. Morrell, the American commander, whose work we have before quoted, with a *caution*. His observations appear to agree with the Admiralty survey, therefore they are here inserted.

It is difficult to obtain fresh water or wood at this place. There are many kinds of scale-fish at the head of this bay, which may be caught with a small seine in great abundance; among them are very large mullets, which average three pounds a piece.

Vessels intending to enter this port must steer for Cape San Quentin, the S.W. point of the bay, which opens to the South; and when you are within 2 miles of its southern extremity, steer N.N.E. until the point bears W.N.W., when you may haul N.N.W.; or if the wind is out of the bay, you may make short tacks, taking care not to stretch under the East shore in less than 5 fathoms of water, from which it shallows very suddenly. In approaching the West shore, you may stand within a cable's length of the beach, after the point bears West, and choose your anchorage in from 7 to 3 fathoms, muddy bottom. There is a sand and rocky bank, running off the S.W. point, in a S.S.W. direction, with 4 feet water on it at low ebb. The tide rises here about 9 feet on the neap, and 11 feet on the spring tides.†

The following are Sir Edward Belcher's own observations on the port, which he has surveyed:—

"Port San Quentin does not afford anything equal to San Diego, but it is more secure when within, and might afford fresh water.

"The sandy point on the West side of the entrance is situated in lat. 30° 22′ N., lon. 115° 56′ 33″ W.; variation, 12° 6′ E.; high water, full and change, 9^h 5′, rise 9 feet.

"The whole coast is dreary, being either sand hills or volcanic mountains, five of which, very remarkably placed, caused one of the early navigators to call it the Bay of Five Hills. It is the Bay of the Virgins of the former, and Port San Quentin of the later Spanish surveyors.‡ As it appears engraved under the latter, on an extensive scale (which misled us and caused our touching), I have preferred that name for it.

"The Island and Paps of Las Virgenes (Cenizas?) are situated to seaward, about 2 miles from what has been termed Observatory Peak in our plan."§

Point Zuniga, according to Vancouver's chart (he passed near to it), is the North extreme of what he called Five Hummock Point, and is 6 or 7 miles N.W. by N. of Cape San Quentin.

* Capt. Vancouver, vol. ii. pp. 482-3. "The S.E. point of San Quentin (Bay of S. Francisco) has a remarkable mount of white barren sand behind it."—*Ibid.*

† Morrell's Narrative, &c., pp. 199, 200.

‡ The North promontory is called, by Vancouver, Point Five Hummocks, as before mentioned. In his chart, the bay to the North of Point Zuniga, the North end of the promontory, is called the Bay de los Virgenes.

§ Sir Edward Belcher, vol. i. p. 329.

Cenizas Island (or S. Hilario Island; Virgenes of Sir E. Belcher ?) lies off Point Zuniga about 3 miles. It is 4 miles in circuit, of a triangular form; its western side is formed of steep high cliffs, but its N.E. and S.E. sides terminate in low sandy land, extending toward the continent.* Capt. Morrell says it is of volcanic origin, and is entirely barren; the rocks have been melted into a complete lava, and the low land is covered with pumice stone. There is a reef lying off the N.E. end of the island about 2 miles, and another off the N.W. part, at nearly the same distance. He here found about 800 sea-elephants, and on the North and West sides 400 sea-leopards, the former very tame, the latter extremely wild and difficulty to approach.

CAPE COLNETT is about 30 miles northward of Cenizas Island. The interval forms the large bay of *San Ramon*, or *De los Virgenes*. Cape Colnett is very remarkable from its shape and appearance, as likewise by its forming a bay on its N.W. and another on its S.E. side. It was thus called by Vancouver, who says (vol. ii. p. 481):—"This promontory bore a very singular character as we passed; the cliffs already described as composing it are, about the middle, between their summit and the water side, divided horizontally, nearly into two equal parts, and formed of different materials; the lower part seemed to consist of sand or clay, of a very smooth surface, and light colour; the upper part was evidently of a rocky substance, with a very uneven surface, and of a dark colour: this seemed to be again divided into narrow columns by vertical strata. These apparent divisions, as well horizontally as vertically, existed with great uniformity all round the promontory."†

From Cape Colnett to Point Grajero, the distance, according to Vancouver, is 50 miles. He was prevented from minutely examining this portion of the coast from the effects of extensive conflagrations of the grass and shrubs on the land, the smoke and heat of which, brought by an E.N.E. wind, enveloped the vessel in an uncomfortable manner.

At 3 leagues S. by E. of Point Grajero, lie a cluster of detached rocks (Solitarios Island), about half a league from a small projecting point, that forms a bay or cove on each side of it.

At *Point Grajero* (of Vancouver) the coast takes a sharp turn to S.E., forming the *Bay of Todos los Santos*. Off the cape some rocky islets and rocks extend N.W. ½ W., a league distant.

The coast northward of Todos los Santos Bay consists of high, steep, rocky cliffs, rising abruptly from the sea, and composing a craggy, mountainous country, extending in a N.W. by N. direction for about 10 leagues to *Point San Miguel*, when it assumes a more northerly direction, or N. by W. ¾ W., for 6 leagues. The shores still continue to be of steep rocky cliffs, which in general rise, though

* Vancouver.

† M. Duflot de Mofras (whose work, however, is not to be depended on for geographical detail) says, that at Cape Colnett, and also at Todos los Santos Bay and Point Grajero, salt is to be procured. He says, that at Port San Quentin, the American ships have visited the Salinas (which are unworked) to procure this necessary article, but, for want of hands, they prefer bringing it from the United States to cure the beef which is got on the Californian coast. The Hudson's Bay Company, too, have examined these ports for the same purpose. At Point Grajero, in the dry season, great quantities of crystallized salt may be taken.

not very abruptly, to a very hilly country, remarkable for three conspicuous mountains, entirely detached from one another, rising in quick ascent at a little distance from the shore, on a nearly plain and even surface. The northernmost of these presented the appearance of a table in all directions from the ocean; the middle one terminated in a sharp peak; and the southernmost in an irregular form. The centre one of these remarkable mountains lies from Port San Diego, S.E. by S., distant 9 leagues, and, at a distance, may serve to point out that port. They are called the Tables, or *Mesas de Juan Gomez*; and hereabout the boundary between Upper and Lower California, and the United States and Mexican territories reaches the coast.*

CHAPTER XI.

THE COAST OF UPPER CALIFORNIA.

In the commencement of the preceding chapter, some remarks are given which may be taken in connexion with the present subject. Until a very recent period, the countries respectively denominated Upper and Lower, or New and Old, California, were under one dominion, subject to the same laws and under the influence of the same social system; intimately connected with each other morally, and having the same origin in a political sense, their histories may be considered as identical. When, however, a new order of things became established, and Upper California was ceded to the dominion of the United States, while the Lower peninsular remained an appendage to Mexico, a wonderful change took place in the importance of these two territories, not so much arising from the change of masters as from the great event of the present century—the discovery of the gold produce of the basins of the rivers falling into the Bay of San Francisco.

From the very recent date of this circumstance, which placed the country, from being one of the most unimportant, commercially, at the head of all the territories bordering on this side of the Pacific, and the consequent vast influx of population, the actual state of the places, hitherto so insignificant, must have changed so greatly, that no former accounts of them can hold good at the present time. This fresh distribution of its population, the consequent annihilation of trade in one part, and the growth of importance in others before unknown or disregarded, not to mention the numerous points of interest which time alone can bring to our knowledge or establish on a firm basis, must manifestly leave us at a very great disadvantage in describing the actual present condition of Upper California.

It is with some diffidence, therefore, that we enter upon the task of drawing into one view the account of the ports and towns of this territory. But, as the natural features of each place cannot change, much must be absolutely as correct

* Vancouver, vol. ii. p. 27.

for one age as for another. The indulgence of the reader is therefore requested for that imperfection which must necessarily be included in the present chapter. The accounts hitherto received have been so vague, so contradictory, except in one point—the riches of the country—that no certain description can be gathered from them. They have, therefore, been comparatively disregarded.

The first discovery or exploration of this part of the American coast was made in 1540, under the orders of Don Antonio de Mendoça, the viceroy of Mexico, who despatched Francisco Vasques Coronado by land, and Francisco Alarçon by sea, in search of the supposed Strait of Anian, which, it was said, communicated with the Atlantic Ocean. Alarçon reached the 36th parallel (South of Monterey), but was then forced to return, from sickness among his crew, without making any particular discovery. The same project was resumed in 1542 by Rodrigues de Cabrillo, a Portuguese in the Spanish service. He reached the latitude of 44°, but the extreme cold and sickness caused him then to return. His principal discovery was a projecting point in lat. 40½°, to which he gave the name Cape Mendoçino, by which it is still known.

The Spaniards seem to have forgotten these discoveries, till, in 1578, Sir Francis Drake, whose exploits are no less celebrated than his voyages, passed the Strait of Magalhaens, then scarcely known, and traversed the great ocean from South to North, reaching the N.W. coast of America in lat. 48° North, ravaging in his way the Spanish possessions in the South, as has been previously alluded to. No Spanish navigator had reached thus far. He then coasted the shore downwards, and discovered the harbour now bearing his name to the N.W. of San Francisco, where he made some stay. He here formally took possession of the country in the name of Elizabeth, Queen of England, imposing the name of New Albion on it.* He gave this name for two reasons: first, because from the nature of the rocks and shoals with which the coast is skirted it presents the same aspect as England; and the second, because it was reasonable that it should bear the name of the country of the first navigator who landed there.

The expeditions of Drake, of Cavendish in 1587, and of Van Noort in 1598, gave rise to considerable jealousy with the Spaniards; they determined to colonize these coasts. Accordingly, Don Gaspar de Zuniga, Count de Monterey, the viceroy of Mexico, despatched Sebastian Viscaino, in 1602, whose most northern important discovery was the harbour, which, in honour of the viceroy, he named the Puerto de Monterey. Whether any subsequent attempts were made to improve this knowledge, the historians are silent. On June 25th, 1767, the Emperor Charles III. abolished the Society of Jesus (Jesuits) in Lower California, and gave the administration of their missions, which they certainly had conducted with wisdom and success, to the Dominican monks, and their property to the Franciscan order. Sixteen of the monks of this latter fraternity landed at Loreto, in Lower California, in April, 1768; and the Visitador, Don Josef de Galvez, arrived there in July of the same year, bearing a royal order to found an establishment either at Monterey or at San Diego. He, with the priests, determined on establishing one at each extremity of the upper province, that is, the

* "The World Encompassed," by Fletcher. London, 4to, 1653, p. 64, *et seq.*

presidios and missions of San Carlos de Monterey, and the same at San Diego, in such a way as to protect all the country, adding, as an intermediate point, the mission of San Buenaventura.

In the establishing these missions, for which Don Vincente Vila set sail in January, 1769, the vessels met with the greatest difficulties from adverse winds; but they were overcome. All this, however, did not make known to the explorers the existence of the finest harbour of all, that of San Francisco, which was subsequently discovered by a *land expedition*, in 1770.

When the Franciscan establishments were completed in Upper California, they were separated into the districts, or jurisdictions, of Monterey, San Francisco, Santa Barbara, and San Diego, in each of which was a *presidio*, or military fort, in which the military governor and guard, for the protection of the missions, dwelt. Apart from these, the missions themselves, entirely under the superintendence of the priests, drew around them their Indian converts, who laboured for the good of the community. Of these missions there were twenty-one at the time of their downfall, and the population attached to their influence, as estimated by Alcedo, was, in 1790, 7,748 souls; in 1801, 13,668; and in 1802, 15,562. In 1831, the population of the four presidios, including all under them, was estimated as follows:—

	Population in 1831.	Population of Chief Towns.
San Francisco	6,328	371
Monterey	4,143	708
Santa Barbara	5,293	613
San Diego	7,261	1,575

Or about 23,000 souls, of which only about 5,000 were free settlers.

In the course of their usefulness, these missions acquired great possessions, at least in natural riches, in flocks, and the fruits of cultivation; and, as a natural consequence, this led to their destruction from the rapacity of those in power. There has been very great difference of opinion as to the actual merit of these missions, one party endeavouring to asperse them by every slander, while another lauds them to the utmost. It must be very unfair to apply any European standard to their influence, and from the evidence of the majority of writers, it must be confessed that they were of inestimable benefit to the country and people among whom they were founded. Commodore Wilkes, who might be considered as *nationally* opposed to the system, thus speaks of them:—"Fortunately for the country, the padres and rulers of the missions were men well adapted for their calling—good managers, sincere Christians; they exerted a salutary influence over all in any way connected with them, practising at the same time the proper virtues of their calling, in order more effectually to inculcate them upon others. These reverend men were all old Spaniards, and greatly attached to their king and country. When the revolution broke out they declined taking the oath to the new government; many, in consequence, left their missions, and some of the others have since died." *

When the revolution in 1823 occurred, which separated California from Old

* Narrative of the United States Exploring Expedition, vol. v. p. 162.

Spain, a fresh order of things was established, and the country was deprived of their religious establishments, upon which its whole prosperity was dependent, and their properties placed in the hands of administradores, whose sole aim was to make the most of their offices; the consequent ruin of these once flourishing communities naturally followed, and nothing can be more deplorable than the subsequent accounts of their downward progress. Politically, the country went on with varied prosperity until 1836, when, about the beginning of November in that year, a dispute arose between Colonel Chico, the governor, and Alvarado, the inspector of customs, who was very popular. Alvarado raised the standard of revolt at San Juan, some leagues from Monterey, upon which place he afterwards advanced. Gutierez, the second in command of the province, believing the presidio to be impregnable, shut himself up with his troops in it; he was formally summoned by Alvarado and his two hundred men, twenty-five of whom were American hunters, to surrender. While they were deliberating within the presidio upon the message, one of their hunter adversaries, impatient at the delay, and moreover with sufficient courage to fire one of the wretched honeycombed guns at their command, fired an eighteen-pound ball, the only shot in this bloodless civil war, which struck the roof of the room in which the council were deliberating. This brought a speedy surrender. Gutierez and his party were transhipped in the British brig *Clementine* to Cape San Lucas, in the lower province. Under Alvarado the ruin of the missions was complete, and society altogether remained in a very disordered state.

The warfare between the United States and the republic of Mexico has been before alluded to.

The result of that campaign was the cession by the latter of the territories of California and New Mexico to the United States government. The exchange of ratification of this treaty took place on May 30th, 1848, and within one year from that date the respective governments undertook to mark the boundary from San Diego to the mouth of the Rio Bravo del Norte. The exact limits of this boundary have not yet been published.

Among the settlers who had introduced themselves into Upper California, after the revolution, was Captain Suter, by birth a Swiss, and who had been a lieutenant in the Swiss guards during the time of Charles X. of France. He had obtained a large grant on the Sacramento River. He fixed his abode and fortification at the head of the tide on the Sacramento, calling it New Helvetia. He constructed a water-mill here, and after one of the freshets to which the stream is liable, on examining the earth brought down by the waters, some particles of gold were picked up, the discovery being almost purely accidental. Further search was made, and an immense quantity of the precious metal, no doubt, was collected. Ere long this fact acquired publicity, and immediately almost the entire male population of California flocked to the gold region, being the pioneers for the vast influx from every part of the Pacific and neighbouring countries, increasing the population one hundredfold in the course of a very few months. We cannot trace out here the progress of the gold movement, which received its first impetus in 1848; it is fresh in the memory of all, and this very singular chapter in the world's history will have many illustrations in other places.

The province of Upper California has formed a government, similar in its model to that of many of the United States; but much time must necessarily elapse before society can solidify from such incongruous elements as have been called together.

The features of the coast are sufficiently described in the succeeding pages for the navigator, and all that is necessary of the country itself is there alluded to for the present work; it will be needless, therefore, now to enter more into detail.

In CLIMATE, California varies as much, if not even more, than in natural features and soil. On the coast range, it has as high a mean temperature in winter as in summer. The latter is, in fact, the coldest part of the year, owing to the constant prevalence of N.W. winds, which blow with the regularity of a monsoon, and are exceedingly damp, cold, and uncomfortable, rendering fire often necessary for comfort in Midsummer. This is, however, but seldom resorted to, and many persons have stated that they have suffered more from cold at Monterey than in places of a much higher latitude. The climate 30 miles from the coast undergoes a great change, and in no part of the world is there to be found a finer or more equable one than in the valley of San Juan. It more resembles that of Andalusia, in Spain, than any other, and none can be more salubrious. The cold winds of the coast have become warmed, and have lost their force and violence, though they retain their freshness and purity. This strip of country is that in which the far-famed missions have been established, and the accounts of these have led many to believe that the whole of Upper California was well adapted for agricultural purposes. This is not the case, for the small district already pointed out is the only-section of country where these advantages are to be found. This valley is of no great extent; it extends beyond the pueblo of San Juan, or to the eastward of Monterey, being about 20 miles long by 12 wide.*

"There was a curious anomaly observed in the movements of the barometer and sympiesometer during our stay at San Francisco (December, 1826); the former *rose* with the winds which brought bad weather, and *fell* with those which restored serenity to the sky. The maximum height was 30·46, the minimum 29·98, and the mean 30·209."—(Beechey, vol. ii. p. 397).

The division between the upper and lower province of California meets the Pacific somewhere to the southward of San Diego, which is thus the southernmost port in the new state. Off this point lie the Coronados.

The CORONADOS consist of two islets and three rocks, situated in a South direction, 4 or 5 leagues from Point de la Loma, occupying the space of 5 miles, and lying N.W. by N. and S.E. by S., from each other. The southernmost, which, in point of magnitude, is equal to all the rest collectively taken, is about a quarter of a mile broad and two miles long, and is a good mark to point out the Port of San Diego, which, however, is otherwise sufficiently conspicuous, and not easily to be mistaken.†

Capt. Kellett places the North end of the chief island in lat. 32° 24′ 55″ N., lon. 117° 14′ W.; variation, 14° 15′ E.—(1847.)

PORT SAN DIEGO, according to Morrell, is as fine a bay for vessels under

* Wilkes' Narrative, vol. v. pp. 155-6.

† Vancouver, vol. ii. p. 474.

300 tons as was ever formed for mariners. It was first discovered by Sebastian Viscaino in 1603, and now, under the new domination of the Californian territory, is the southernmost port belonging to the United States.

The mission from whence the port takes its name is not within sight of the sea nor of the port; it is situated in a valley within the view of, and about 2 miles distant from, the presidio, to the N.E. The entrance to the port runs to the eastward of Point Loma, between it and the *Barros de Zuniga*, a shoal stretching in a parallel direction from the opposite point of the opening. From the South extremity of the Barros de Zuniga to the Point Loma, the soundings do not exceed (according to Vancouver) more than 4 fathoms at high water, and form a narrow bar from the shore to the shoal, gradually deepening as well on the inside as on the outside of the bar, with increase in mid-channel from 5 fathoms close to the shore, to 10 fathoms between the two low points that form the entrance to the port. The western of these points is the *Punta de Guiranas*, a low spit of land projecting from the high steep cliffs, the eastern shore of Point Loma. The opposite point is also very low, but not a spit of land.

From the presidio, south-eastward, the eastern side of the port is bounded by high land as far as its head, a distance of 11 or 12 miles. On the S.W. side it is separated from the open bay, before-mentioned, by a narrow tract of land covered with bushes.

Point Loma, the western point of this bay, is the southern extremity of a remarkable range of elevated land that commences from the South side of an inlet on the outer coast, 7 miles to the northward, called *Puerto Falso*; and, at a distance, has the appearance of being insular, which effect is produced by the low country that connects it with the other mountains. The top of this tract of land seems to terminate in a ridge, so perfectly and uniformly sharp as apparently to render walking very inconvenient. The fact, however, is not so; but, when viewed from sea, it has that singular appearance. It descends in very steep rocky cliffs to the water side, from whence a bed of growing weeds extends into the ocean half a league or 2 miles.

The sharp ridge of land is connected with other mountains by an isthmus or tract of very low land, which in the rainy season is flooded, and at high spring tides makes the sharp land forming the West and N.W. side of the port an island. The presidio is on the continental side of this low sandy isthmus.

The peninsula bears a very different appearance when seen from the port, from that before described as observed from the ocean. It descends with an uneven surface, and some bushes grow on it, but no trees of a large size.

The channel between Point Loma on the Zuniga Shoal is the only navigable passage for shipping; that to the north-eastward of the shoal does not anywhere exceed half a mile in width, which, with its shallow depth of water, renders it ineligible, excepting for boats or vessels of very small draught. The port, however, affords excellent anchorage, and is capable of containing a great number of vessels; but the difficulty, nay, almost impossibility, of procuring wood and water under its present circumstances, reduces its value as a port of accommodation.*

* Vancouver, vol. ii. p. 473.

"The kelp at this port is rather a nuisance to vessels drawing less than 20 to 24 feet, as it leads you at least 2 miles out of your way to clear its tongue, having nothing under 3 fathoms over its whole bank. To those accustomed to the ground it is well; but to us lead-going gentry, who are compelled to stick rigidly to the laws we lay down, it would have been quite *en regle* to round this kelp, even by going ten miles further out of our way. It is *very* doubtful if the *Sulphur* would have done so if any breeze had helped her.

"The position on which our observations were made is the eastern spit of entrance, which was found to be in lat. 32° 41′ N., lon. 117° 11′ W.

"The Port of San Diego, *for shelter*, deserves all the commendation that previous navigators have bestowed on it; and with good ground-tackle a vessel may be perfectly land-locked. The holding ground is stubborn, but in heavy, southerly gales, I am informed that anchors 'come home,' owing to the immense volume of kelp driven into the harbour. It has been stated to me by an old sailor in this region, that he has seen the whole bank of *fucus giganteus* (which comprises a tongue of 3 miles in length by a quarter broad), forced by a southerly gale into the port. This, coming across the bows, either causes the cable to part, or brings the anchor home. No vessel, however, has suffered from this cause. The chief drawback is the want of fresh water, which, even at the presidio, 3 miles from the port, is very indifferent. This is strange; for I am perfectly satisfied that if proper precautions were observed in digging wells, the bight of the peninsula must furnish water. This, however, to be maintained sweet must be constantly worked, and occasionally as dry as it can be reduced.

"The mission is situated up a valley about 7 miles from the presidio. They have here not only the finest water, but a river or torrent flows from thence to the presidio during the rains, which in the dry season loses itself in the sand about half way. The soil is very loose, chiefly volcanic sand and mud, mixed with fine pumice and scoriæ, which on the flats between the elevated ridges, when the rain has carried off the lighter particles, presents the appearance of finely gravelled terraces. Several varieties of *cacti*, particularly of the Turk's Head variety, were abominably abundant, very much to the discomfort of our ascending parties in the prosecution of the survey.

"Since the missions have been taken away from the padres, and placed under the administradores, they have fallen into decay and ruin; and it is not improbable that the whole country will, ere long, fall back into the hands of the Indians, or find other rulers. They even now attack it. The garden formed in former days has fallen entirely into decay, and instead of *thousands* of cattle and horses to take care of, not twenty four-footed animals remain.

"The trade of the port consists entirely of hides and tallow; but not, as formerly, from the missions, for they have long been fleeced. It has now become a complete speculation. It is necessary that one of the parties should reside on the spot, probably marrying into some influential family (*i.e.*, in hides and tallow), to ensure a constant supply for the vessels when they arrive. It is dangerous for them to quit their post, as some more enterprising character might offer higher prices, and carry off their cargo.

"But little wine is made since the virtual death of the mission, and that little of very inferior quality. I believe that at the neighbouring mission of San Luis Rey the principal wines are made, as well as a very pure spirit, resembling Italia, whisky, or the *pisco* of Peru.

"We found tolerable sport between daylight and the breakfast hour in killing rabbits, hares, ducks, and cordoneces (quail of California); the seine afforded also a plentiful supply of excellent fish."—(Sir Edward Belcher.)

The tides were found by Vancouver to run in general about 2 knots (though faster at spring tides), six hours each way. High water nine hours after the moon passes the meridian. Variation, December, 1793, 11°. E.; dip, 59° 13'.

The shores between San Diego and San Juan Capistrano, a distance of 57 miles, are in general straight, and entirely compact. The face of the country here assumes a more uniform appearance than that to the North, and rises from the sea-coast, which chiefly consist of sandy beaches or low cliffs, with a gradual ascent. It is broken into some chasms and valleys, where a few small trees and shrubs in two or three places are seen to vegetate.*

SAN JUAN CAPISTRANO, a Spanish establishment founded in 1776, is erected close to the water-side, in a small sandy cove, near the centre of which is a small detached rock, and another lying off its North point.

This mission is very pleasantly situated in a grove of trees, whose luxuriant and diversified foliage, when contrasted with the adjacent shores, gave it a most romantic appearance; having the ocean in front, and being bounded on its other sides by rugged dreary mountains, where the vegetation was not sufficient to hide the naked rocks, of which the country in this point of view seems to be principally composed.†

The bay, or rather the outer half tide rock, on which Sir E. Belcher observed, is in lat. 33° 26' 55" N., lon. 117° 42' W. It has a high cliffy head to the N.W., but terminates in low sandy beaches to the southward. The anchorage is foul under 5 fathoms, is unprotected, and the landing bad.

"The mission is situated in a fruitful-looking sheltered valley, said to abound in garden luxuries, country wines, and very pretty damsels, whence the favourite appellation *Juanitas*. I suppose, therefore, that they all assume this name. As many call here apparently, to my view, at risk of anchor and cable, I was induced to ask the master of a vessel who called upon me what brought him here? 'It is only visited for stock, fruit, or vegetables,' was his dry reply."—(Sir Edward Belcher, vol. i. p. 324.)

At 21 miles from San Juan is *Point Lasuen*, the eastern point of the Bay of San Pedro. Behind this bay the once flourishing mission of *San Gabriel*, or *Pueblo de los Angelos*, "the country-town of the angels," was formed in 1781.

SAN PEDRO.—The Bay of San Pedro, which is situated in latitude 33° 43' North, lon. 118° 16' West, is open to the S.W., but tolerably sheltered from the N.W. Inside of the small island in the bay is a very snug creek, but only accessible to small craft, by reason of a rocky bar, having only 5 feet on it at low water springs.

The only house near the bay is supplied with water from some miles inland;

* Vancouver, vol. ii. p. 468.

† *Ibid.* vol. ii. p. 467.

and Capt. Sir E. Belcher was informed, that at times the inhabitants are in great distress. It is only maintained for the convenience of trading with vessels which touch here for the purchase of hides and tallow.

The cliffs on the western side of the bay, which form the beach line, are very steep, about 50 feet perpendicular, descending from an elevated range, about 500 feet above the sea. They are composed of a loose mud, mixed with lumps of a chalky substance, enclosing organic remains, sometimes running into chert or chalcedony. It apparently results from organic action, and the frequent shocks of earthquakes have left the ground full of deep fissures. This chalky or pipeclay substance also occurs at Santa Barbara.

Vessels occasionally anchor within the 5 fathoms or kelp line, but are always prepared to warp out. This is a kind of inner bar. The kelp doubtless prevents much surf, and renders it more convenient to vessels discharging, but the kelp, during heavy gales, is generally washed up.*

The opposite point of the bay is *Point Fermin*, and W. by N. ¾ N. near 10 miles from this is *Point Vicente*, composed of steep barren cliffs, and forming the north-western extremity of a conspicuous promontory, extending to Point Fermin. From this latter, the western shores of the bay take a northerly direction, and constitute a projecting promontory between two bays, the shores of which terminate on all sides in steep cliffs of a light-yellowish colour. These extend along the bay about a league, and have a small island lying off their North extremity; the northern and eastern sides of the bay are composed of a low country, terminating in alternate low white cliffs and sandy beaches. On this low, extensive tract some small trees and shrubs are produced, but the interior country, consisting of rugged, lofty mountains, presents a dreary and sterile appearance.

Point Dume (off which lie two or three small rocks) and Point Vicente lie from each other N.W. ½ W. and S.E. ½ E. 26 miles asunder. Between these lies an extensive open bay, the N.W. side of which was observed by Vancouver to be composed chiefly of steep barren cliffs; the North and eastern shores terminate in low, sandy beaches, rising with a gradual ascent until they reach the base of a mountainous country, which has the appearance of being rugged and barren.†

* Sir Edward Belcher, vol. i. pp. 323-4. Sir George Simpson, who visited it in his overland journey round the world in 1841-42, says:—"San Pedro is an open bay, which has no better claim to the character of a harbour than almost any other point on the coast, being exposed to both the prevailing winds, and being destitute of everything in the shape of a house, or even of a shed. Its only recommendation is, that it affords access to the Pueblo of Nuestra Señora, about 18 miles distant, which contains a population of 1,500 souls, and is the noted abode of the lowest drunkards and gamblers of the country. This den of thieves is situated, as one may expect from its being almost twice as populous as the two other pueblos taken together, in one of the loveliest and most fertile districts of California; and being, therefore, one of the best marts in the province for hides and tallow, it induces vessels to brave all the inconveniences and dangers of the open and exposed Bay of San Pedro."—Vol. i. p. 402.

† At the bottom of this bay is a bitumen spring. In reference to this, when Vancouver anchored in a small bay 60 miles to the north-westward, he says:—"The surface of the sea, which was perfectly smooth and tranquil, was covered with a thick slimy substance, which when separated or distributed by any little agitation, became very luminous, whilst the light breeze that came principally from the shore brought with it a very strong smell of burning tar, or of some such resinous substance. The next morning the sea had the appearance of dissolved tar floating upon its surface, which covered the ocean in all directions within the limits of our view, and indicated that in this neighbourhood it was not subject to much agitation." This singular fact, which might

POINT CONVERSION is 20 miles westward from Point Dume, the coast between having a steep and rugged character. Toward the N.W., in the direction of the mission of San Buenaventura, the shores are low and flat, producing some small trees and shrubs. At 8 miles from Point Conversion is the mission above mentioned.

BUENAVENTURA was established in 1782. The establishment was about three-quarters of a mile from the sea-side, and Vancouver speaks greatly in its praise. "The coast immediately opposite, and to the northward of our anchorage, chiefly consisted of high deep cliffs, indented with some small sandy coves. The general face of the country was mountainous, rugged, barren, and dreary; but toward the mission, a margin of low land extended from the base of the mountains, some of which are of great height, and at a remote distance from the ocean; and being relieved by a few trees in the neighbourhood of the establishment, gave this part of the country a less unpleasing appearance." He here procured abundance of vegetables, roots, fowls, and sheep from the mission.

In consequence of the general serenity of the weather almost throughout the year, the roadstead may be considered as a tolerably good one, and anchorage may be had nearer the shore, in the vicinity of the mission; that occupied by Vancouver being 2 miles from the nearest shore, and 3 miles W.S.W. of the landing place near the mission.*

But neither situation is so commodious as at Santa Barbara to the westward, being much more exposed to the S.E. winds and ocean swell, which frequently render the communication with the shore very unpleasant.

SANTA BARBARA.—The mission of Santa Barbara is 20 miles from that of San Buenaventura, and as far as within 2 or 3 leagues from the former the coast rises with a steep ascent in rocky cliffs that chiefly compose it. Vancouver says the establishment is in a small bay, and has the appearance of a far more civilized place than any other of the Spanish establishments to the northward had exhibited. The buildings appeared to be regular and well constructed, the walls clean and white, and the roofs of the houses were covered with a bright red tile. The presidio was nearest to the sea-shore, and just showed itself above a grove of small trees, producing, with the building, a very picturesque effect. He speaks in very high terms of the Spanish missionary priests, who furnished him with ample supplies of refreshments. This was in 1792. It is now sadly changed. Wood was procured from the holly-leaved oak, growing at some distance from the sea-shore. The water, which was not of the best quality, was in wells close to the sea-shore. They were very dirty, and their contents brackish; but a few yards further than where the wells had been made, a most excellent, though not abundant, spring of

---

be turned to profitable account, has also been noticed by others:—"Off this part of the coast (near Santa Barbara) to the westward," Capt. Sir Edward Belcher says, "we experienced a very extraordinary sensation, as if the ship was on fire, and after very close investigation attributed it to a scent from the shore, it being much more sensible on deck than below, and the land breeze confirming this, it occurred to me that it might arise from naphtha on the surface."—*Voyage of the Sulphur*, vol. i. p. 320.

* A short distance to the southward of Buenaventura the coast spits out in a low sandy point, off which the water shoals suddenly to 7 fathoms. There is no danger if the lead be kept going. Sir E. Belcher noticed a lagoon over the sands.—*Ibid.*, vol. i. p. 322.

very fine water was discovered. There was also a large well of excellent water at the presidio.

To sail into the bay, or more properly speaking the roadstead, of Santa Barbara, requires but few directions, as it is open, and without any kind of interruption whatever; the soundings on approaching it are regular, from 15 to 3 fathoms; the former from half a league to 2 miles, the latter within 1½ cables from the shore. Weeds were seen growing about the roadstead in many places; but so far as was examined, which was only in the vicinity of the anchorage, they did not appear to indicate shallower water, or a bottom of a different nature.

The shores of the roadstead are for the most part low, and terminate in sandy beaches, to which, however, its western point is rather an exception, being a steep cliff, moderately elevated. To this point Vancouver gave the name of *Point Felipe*, after the commandant of Santa Barbara.

The interior country, a few miles only from the water side, is composed of rugged barren mountains, which he was informed rise in five distinct ridges behind and above each other, a great distance inland towards the E.N.E.

The tide, though showing no visible stream, regularly ebbed and flowed every six hours; the rise and fall, as nearly as could be estimated, seemed to be about 3 or 4 feet, and it is high water about eight hours after the moon passes the meridian.

Santa Barbara was visited by Sir Edward Belcher, who makes the following remarks on it:—"At sunset we were unable to discover the bay, and could barely distinguish a long, low, yellow line spitting to the southward, and terminating abruptly. This low land proved to be the high yellow cliffs of the western head, at least 50 feet above the sea.

"The customary guide in approaching the coast is the 'kelp line,' which generally floats over 5 to 7 fathoms. So long as a vessel can keep on its verge there is no danger; this is the general opinion of those who have navigated this coast during their lives, and our observation has tended to confirm it. I know, however, that less than *two fathoms* have been found *within* it, barely at its edge. It is the *fucus giganteus*, and sufficiently strong to impede the steerage if it takes the rudder.

"This mission at Santa Barbara is situated on an elevation of about 200 feet, gradually ascending, and is about 3 miles from the sea. The town is within a few hundred yards of the beach, on which the landing is at all times doubtful. The bay is protected from northerly and westerly winds, which prevail from November until March, and the swell is in some measure broken by the islands of Santa Cruz, Santa Rosa, and San Miguel, to the westward. In March the south-westers blow with fury, which is contrary to the seasons southward of Cape San Lucas. Even during the fine weather months, vessels are always prepared to slip when the winds veer to the S.E., from which point it blows with great violence, but soon expends itself."*—(Voyage of the *Sulphur*, vol. i. pp. 321-2.)

* "To the southerly winds which prevail during the winter, every point of the bay is a lee shore; so that when the push comes, the vessels in the port have no other choice than that of making the best of their way past Point Conception into the open ocean, and there remain until the storm has blown over.

"During the season of the south-easters, the surf is sometimes so heavy as to prevent boats

Sir George Simpson, in his Narrative of his Journey round the World, says:—"Santa Barbara is somewhat larger than Monterey, containing about 900 inhabitants, while the one is just as much a maze without a plan as the other. Here, however, anything of the nature of resemblance ends. Santa Barbara, in most respects, being to Monterey what the parlour is to the kitchen.

"The site of the town has doubtless been fixed by the position of the port, if port it can be called. The bay, as the shore of the main land may perhaps be termed, is exposed at every point to the worst winds of the worst season of the year, and, to crown all, the bottom is not to be trusted in the hour of trial, being hard sand, covered with sea-weed. But the port, such as it was, had been selected for want of a better, while the superiority of the climate, which was at once drier than that of San Francisco or Monterey, and cooler than that of San Pedro and San Diego, rendered the neighbourhood the favourite retreat of the more respectable functionaries, civil and military, of the province. Hence, among all the settlements as distinguished from the rascally pueblos, Santa Barbara possesses the double advantage of being both the oldest and the most aristocratic.

"The houses are not only well finished at first, but are throughout kept in good order, and the whitewashed adobes and the painted balconies and verandahs form a pleasing contrast with the overshadowing roofs blackened *by means of bitumen*,* the produce of a neighbouring spring. Compared with the slovenly habitations of San Francisco and Monterey, the houses of Santa Barbara are built and maintained at an additional cost, the greater that it is at the expense of the most expensive articles in this country—the time of hired labourers and mechanics. Nor is the superiority of the inhabitants less striking than that of their houses.

"The church, which is one of the finest in the country, is large and well proportioned, and is a good monument of the skill and indefatigability of its founders. The mission is also well supplied by water brought down from a range of rocky hills, at the distance of 5 or 6 miles, by the labour of the priests and their converts. The garden, a walled enclosure of 5 or 6 acres, produces an endless variety of vegetables and excellent fruits; but since 1836 it has been running to decay and ruin." Sir George Simpson speaks in the highest terms of the kindness and hospitality of the inhabitants.†

For 20 miles from Santa Barbara the coast runs to the northward of West, the country rising to mountains of different heights. In the vicinity of the shores, which are composed of low cliffs or sandy beaches, are produced some stunted trees and grovelling shrubs. From the distance previously mentioned the coast takes a westerly direction to Point Conception, at which the coast assumes a new direction to the northward.

POINT CONCEPTION is rendered very remarkable by its differing much in form from the points to the northward. It appears to stretch out into the ocean from an extensive tract of low land, and to terminate like a wedge, with

---

from landing, to say nothing of their grounding on the sand, and being entangled in the sea-weed. In summer, however, the surf is less dangerous, while the shallows are said to be deepened by the banking up of the sand on the beach in the absence of the seaward gales."—*Sir George Simpson.*

* See note on page 321.

† Vol. i. chap. 9.

its large end falling perpendicularly into the sea, which breaks against it with great violence.

Point Conception thus terminating the belt of coast, which is affected, more particularly in the mornings, by the N.W. fogs, by its sudden turn places all below it in the same position as the interior. It is therefore for this reason that San Francisco and Monterey on the one hand, and Santa Barbara, San Pedro, and San Diego on the other, are respectively classed as the windward and leeward ports.—(See Sir George Simpson, vol. i. p. 373.)

SANTA BARBARA CHANNEL.—Point Conception is the N.W. point of the Canal of Santa Barbara, which is formed by the main land and a range of islands, which we have not yet mentioned, lying off the coast between San Diego and Point Conception.* These we shall describe in their order, commencing with the southernmost.

SAN JUAN is the nearest to San Diego, from which it is but seldom visible, being about 24 miles off, and remained undiscovered until seen by Martinez a few years before Vancouver's visit in 1793. It is low and flat, and of small dimensions.

SAN CLEMENTE is one of the largest of the group, and lies about 25 miles to the W.N.W. of San Juan.

SANTA CATALINA.—To the North of this is Santa Catalina, another of the larger islands. "We ran along the north-western face of Santa Catalina, on which there is a small but very snug cove or harbour, quite defended from all winds. In this harbour we anchored the vessel, and found a bay on the opposite side of the island, from which the head of the harbour is divided by a low peninsula of not more than a quarter of a mile broad; high hills rising on either hand, directly from the coast line. We had some difficulty in getting to sea the next day from the calm and flaws caused by the high land; which, with the great depth of water close to the entrance, from 40 to 60 fathoms, and the swell constantly running, must always prove a great drawback to its use as a harbour.

"Goats in thousands were met with; but we could find no water in either of these bays. There are said to be some wells in a bay on the southern end, which we did not meet with; but most of the gulleys must contain pools of water, else so many thousand head of goats could not exist."—(October 7th, 1848.)†

SANTA BARBARA is a mere islet, 1½ miles in diameter, with a large rock off its southern end, and lies abreast of the north-western end of Santa Catalina, from which it is distant 19 or 20 miles.

SAN NICOLAS lies about 20 miles S.W. of Santa Barbara, and off its western side is the John Begg reef, on which the ship of that name, commanded by Capt. Lincoln, struck, and was nearly lost, 20th September, 1824. The shoal consists of foul ground covered with kelp. The island has been surveyed by Capt. H. Kellett, R.N., in 1847. It is about 7 miles long, East and West by compass, by

* "Excepting the very light and baffling winds that prevailed, there were neither currents nor any other obstruction, so far as our examination went, to interrupt its navigation; which, to those who may have occasion only to pass through it, will be found neither difficult nor unpleasant."—*Vancouver*, vol. ii. p. 466.

† Nautical Magazine, May, 1849.

3 miles in breadth. Off its East and West ends breakers extend; the former to the distance of 1¾ miles.

The *John Begg Rock* lies 7½ miles W.N.W. from its West point. The rock itself is about 30 or 40 feet high, and is in lat. 33° 22′ N., lon. 119° 42′ 3″ W.; variation, 14° 30′ E.

The northern group consists of three islands lying nearly East and West: San Miguel, the western and smallest, and also the north-westernmost of the group; Santa Rosa, and Santa Cruz, the eastern and largest.

Off the eastern end of the last is a group of three rocky islets, called by the natives *Enneeapah*, or *Anacapes*, the westernmost of which is the largest and highest. They are composed of rugged rocks, nearly destitute of wood and verdure. The westernmost, already stated to be the largest, is about a league in length from North to South, and about 2 miles in breadth. The easternmost of them is 2 miles in circuit, and lies East by North 3 miles from the above.

The ISLAND of SANTA CRUZ is, perhaps, the largest of the whole group, which all wear the same barren appearance. "We found good shelter from a strong north-westerly breeze in a small bay abreast of the Enneeapah Islets. There is said to be also good anchorage in a large bay on the northern face of the island, which the boisterous state of the weather did not permit us to look at. Here we found good water, but in small quantities, as also on Santa Rosa."—(October, 1848.)

SANTA ROSA is the middle island; on the eastern side is a bay which affords excellent shelter from gales from the western quarter. Both cattle and horses were seen on the island.—(1848.)

All these islands have furnished abundance of sea-otter to the vessels employed in the fur and sealing trade.

At Point Conception, as before stated, the coast assumes a new direction, trending to the northward. Besides this change in the direction of the coast, there is also as great a change in the climate; to the northward fogs and mists prevail during the early part of the day for three-fourths of the year.

This haziness renders the coast somewhat dangerous to approach; but the smell of the asphaltum or bitumen, which we have referred to as having been observed by Vancouver, and has also been noticed by others at long intervals, indicating it to be of a permanent nature in this vicinity, may indicate the proximity of danger at times.

POINT ARGUELLO lies N.W. ½ W. 10 miles from Point Conception. It is a high, steep, rocky, projecting point, rising very abruptly in rugged, craggy cliffs. Near it are two or three detached rocks, lying close to the shore. The coast between this and Point Conception falls a little back to the eastward. The intermediate shores and interior country are destitute of wood, and nearly so of every other vegetable production.

At 6 miles North by East is a rivulet, which appeared to Vancouver to be the largest flow of water into the ocean he had seen, excepting the Columbia river; but the breakers that extended across its entrance seemed to preclude the possibility of its being navigable for boats. In the Spanish charts it was called *Rio de St. Balardo* (or *Rio San Geraldo*).

POINT SAL lies N. ¼ W. 19 miles from Point Arguello, and is a high, steep, rocky cliff, projecting from the low shore. The interior country to the northward of it, similar to that to the south, consists of lofty barren mountains; the intermediate land descends gradually from their base, interspersed with eminences and valleys, and terminating on the coast in sandy beaches, or low white cliffs.

At 20 miles N.W. ½ N. from Point Sal, is the Point named in the Spanish charts the *Monte de Buchon*. Off it are some rocks and breakers one mile distant. Inland from the bay formed by these two points were the missions of Guadaloupe and S. Luis Obispo; the latter formed in the year 1772. At about 8 leagues off Point Buchon, it was said that an island had been discovered by the Spaniards, but we know nothing more of it.

ESTEROS BAY is formed by a point to the northward of Point Buchon and Esteros Point, N.N.W. ¼ W. 13 miles from it. The southern point is formed by steep cliffs, falling perpendicularly into the ocean; and about it are some detached rocks. From the line of the two outer points, the shores of the bay fall back about 5 miles; they appeared to be much exposed, and not to afford any shelter. They seem to be composed of a sandy beach, stretching from a margin of low land, extending from the rugged mountains forming the interior of the country. Four small streams appear to enter the bay, and on the coast in the middle of the bay is a high conical hill, flat at the top, apparently an island.

At Point Esteros the character of the country changes greatly. To the southward of it the shores are abrupt, and the country barren and uninviting. To the northward it is wooded, and north-westward from the point the mountains are at a distance from the coast, the intermediate country being apparently a plain, rising with a very gradual ascent for the space of 4 leagues along the coast, and is tolerably well wooded, even close down to the shore; some of the trees being very large, and for the greater part distributed in detached clumps, produce a very pleasing effect. This alteration in the features of the country may, perhaps, be attributed to the change in the climate and the prevalence of mists and fogs, as previously mentioned occurring to the northward of Point Conception.

At 6 leagues northward of Point Esteros is a low projecting point, off which lie two or three rugged detached rocks. To the northward of this the coast continues in a N.W. ½ N. direction; the country, composed of valleys and mountains, gradually descending towards the sea-shore, which consists of alternate rocks and sandy beaches, having some dwarf trees, and the surface interspersed with a few dull verdant spots.

Immediately in the rear are the *Santa Lucia Hills*, a part of the Coast Range, extending as far as Punto Pinos, the southern point of the Bay of Monterey.*

* On the opposite or eastern slope of the ridge is the valley of Salinas, through which the Rio Buenaventura flows. The hills are rendered much more fertile by their exposure to the fogs and mists of the coast, which supply them plentifully with moisture, and this is seen running in many rills down the hill sides.

The valley of Salinas is 50 miles in length, and has an average width of 6 or 7 miles; the valley descends to the N.W., and at its lower end is contracted by the hills through which the river passes, a low and well-wooded bottom being formed on each side; the whole of it is well drained, and admirably adapted for stock farms; it may be called an open country covered with grass; the tops of the hills are covered with oaks, pines, and cedars.

The river having passed through a narrow range of hills, the valley again opens and now

The plains and neighbouring mountains are well covered with large timber, and here the olive and other fruits of this region grow in perfection; on the hills the California cedar (pale colorado) is found of large size.

Still further to the northward the coast continues nearly straight and compact; the mountains form one uninterrupted, though rather uneven, ridge, with chasms and gullies on their sides; the whole, to all appearance, nearly destitute of vegetation. This continues to a small high rocky lump of land, lying about half a mile from the shore, and 4 leagues S. by E. of the North point of the Bay of Carmelo.

The BAY of CARMELO is a small, open, and exposed situation, containing some detached rocks; and having a rocky bottom, is a very improper place for anchorage. Into this bay flows the River Carmelo, and at a distance from the sea, it is said to abound with a variety of excellent fish. From hence the exterior coast takes a N.N.W. $\frac{1}{2}$ W. direction for 4 miles to Point Pinos, a low projecting point of land, covered with trees, chiefly the stone-pine, and forming the South point of the Bay of Monterey.

POINT PINOS is 100 miles N.W. by N. of San Luis Obispo; the Santa Lucia range still continues to rise until it reaches the altitude of 2,700 feet in the rear of Point Pinos; they are thickly wooded, and generally have, from the trees being for the most part pines, a sombre appearance.

MONTEREY.

Monterey, from its capabilities in affording supplies, was fast rising into importance, and was generally resorted to by vessels visiting the coast. We will first give Vancouver's description of it in December 1792:—

"This famous bay is situated between Point Pinos and Point Año Nuevo, lying from each other N.N.W. $\frac{1}{2}$ W. and S.S.E. $\frac{1}{2}$ E. 22 miles apart. Between these points this spacious but very open bay is formed by the coast falling back from the line of the two points nearly 4 leagues. The only part of it that is at all eligible for anchoring is near its South extremity, about a league south-eastward of Point Pinos, when the shores form a sort of cove, that affords clear, good riding, with tolerable shelter for a few vessels. These, for their necessary protection, must lie at no very great distance from the S.W. shore, where, either at night or in the morning, the prevailing wind from the land admits the vessels sailing out of the bay, which otherwise would be a tedious tack, by the opposition of the winds along the coast, which generally blow between the N.W. and N.N.W. To these points of the compass this anchorage is wholly exposed; but as the oceanic swell is broken by the land of Point Pinos, and as these winds, which prevail only in the day time, seldom blow stronger than a moderate gale, the anchorage is rendered tolerably safe and convenient. . . . . The soundings are

---

receives the name of La Soledad, which is 20 miles wide, and extends to the Bay of Monterey. The land on either side rises into undulating hills, and from these into mountains, some 2,000 feet high. The valley of La Soledad is considered very fertile, the plains affording large areas of arable land, while the hills are covered with grass and groves of oak, and the mountains with trees of higher growth.

regular, from 30 to 4 fathoms, the bottom a mixture of sand and mud, and the shores are sufficiently steep for all the purposes of navigation, without shoals or other impediments. Near Point Año Nuevo are some small rocks, detached from the coast at a very little distance; the shores of Point Pinos are also rocky, and have some detached rocks lying at a small distance from them, but which do not extend so far into the ocean as to be dangerous. The rocky shores of Point Pinos terminate just to the South of the anchoring place, where a fine sandy beach commences, extending all round the bay as far as Cape Año Nuevo. In a direction N.E. ½ N., 4 leagues from Point Pinos, is what the Spaniards call Monterey River, which, like the River Carmelo flowing by San Carlos de Monterey, is no more than a very shallow brook of fresh water that empties itself into that part of the bay. Near Point Año Nuevo is another of these *rivers*, something less than the other, in whose neighbourhood the mission of Santa Cruz was placed." *

The shores of this bay, and indeed of the whole of the coast near Point Pinos, are armed with rocks of granite, upon which the sea breaks furiously; and as there is no anchorage near them on account of the great depth of water, it is dangerous to approach the coast in light or variable winds. Fortunately some immense beds of sea-weed (*fucus pyriformis*) lie off the coast, and are so impenetrable that they are said to have saved several vessels which were driven into them by the swell during calm and foggy weather.—(Beechey, vol. ii. p. 408.)

The old Spanish presidio, first established in 1770, is about three-quarters of a mile to the southward of the spot where the sandy beach previously mentioned commences. The mission of San Carlos de Monterey is about a league from the presidio in a southerly direction. As it is certain that the new order of things now existing in California will totally change the character of this and other places, we will refer the reader to the descriptions given by various authors.

The first extract is from the work of the unfortunate La Pérouse:—

"The Bay of Monterey, formed by the New Year Point to the North, and that of Cypees on the South, is 8 leagues wide in that direction, and is nearly 6 in depth towards the East, in which quarter the land is low and sandy. The sea rolls on to the foot of the sandy downs, with which the shore is lined, with such force that La Pérouse states that he heard the noise at above a league distant.

"The land on the North and South of this bay is high and covered with trees, and ships intending to enter should follow the South side; and, after having doubled the Point Pinos, which projects to the North, they will see the presidio, and they can then anchor in 9 fathoms inside, and rather inland of this point, which will shelter them from the winds from seaward.

"The Spanish vessels intending to make a long stay at Monterey take a position nearer the shore, at 1 or 2 cables' length distance, in 5½ fathoms, and make fast to an anchor which they bury in the sandy beach; they have thus nothing to fear but the southerly winds, which sometimes come rather strong, but are not exposed to any danger when they come off the land.

"We found bottom throughout the bay, and we anchored at 4 leagues from

* Vancouver, vol. ii. pp. 41-2.

the land in 52 fathoms muddy bottom. But the swell is very heavy here, and anchoring should not be attempted except to wait for daylight or until it clears.

"It is high water at new and full moon at 1^h 30′, and it rises 7 feet (French), and as the bay is very open, the current in it is almost insensible; I have never seen it run half a knot. (But see Capt. Beechey's remarks on page 322.)

"I cannot express the number of whales with which we were surrounded, nor yet their familiarity; they blew every half minute within half pistol shot of our frigate, and occasioned a very strong smell in the air. We had not known of this effect from whales, but the inhabitants told us that the water they spouted was impregnated with this unpleasant odour, and that it extended to considerable distance.*

"The Bay of Monterey is enveloped in almost eternal fogs, which circumstance renders approach difficult; without this phenomena nothing could be easier than making the harbour; there is not a hidden rock which extends a cable's length from the beach; and if the haze is too thick, you can readily anchor until it clears, when the Spanish establishment is readily seen, situated in an angle formed by the South and East coasts.†

"The sea was covered with pelicans; it seems that these birds never go farther than 5 or 6 leagues from the land, and seamen, when they meet with them in the fog, may be assured that they are at most within that distance of the land. We saw them for the first time in the Bay of Monterey; but I had been previously informed that they were very common on all the Californian coast. The Spaniards call them *alkatræ* (algatras).‡

The following is Sir George Simpson's account of it in 1842:—

Though infinitely inferior, as a port, to San Francisco and Diego, yet Monterey, from its central position, has always been the seat of government. It was, however, only after the revolution of 1836, that it could be compared with the other settlements in point of commercial importance, having suddenly expanded from a few houses into a population of about seven hundred souls.

The town occupies a very pretty plain, which slopes towards the North, and terminates to the southward in a tolerably lofty ridge. It is a mere collection of buildings scattered as loosely on the surface as if they were so many bullocks at pasture; so that the most expert surveyor could not possibly classify them, even into crooked streets. What a curious directory of circumlocutions a Monterey directory would be! The dwellings, some of which attain the dignity of a second story, are all built of adobes, being sheltered on every side from the sun by

* Many whales were also seen by Sir George Simpson, in 1842, the bay being a favourite resort of that animal; the shark, the cod, and the sardine also abound; the last is sometimes thrown in millions on to the beach by westerly gales.

"The whale has been known to burst among his human persecutors with the report of a cannon, and almost to suffocate them with the stench."—*Sir George Simpson*, vol. i. pp. 364, 401.

† The mission of Monterey, the principal place of the two Californias, was founded June 3, 1770; but prior to this, and subsequent to its discovery by Sebastian Vizcaino, in 1602, (see note, page 310) the galleons, on their return from Manilla, sometimes touched here for refreshment after their long voyages. La Pérouse, vol. ii. chapters 11 and 12, contains a lengthened account of the Californian missions and their Indian inhabitants. La Pérouse speaks in the highest terms of the missionaries, and the excellence of their dominion over these simple people.

‡ "We were apprised of our approach to the coast of California by some large white pelicans which were fishing a few miles to the westward of Point Pinos."—*Beechey*, vol. ii. p. 582.

overhanging eaves, while, towards the rainy quarter of the S.E., they enjoy the additional protection of boughs of trees, resting like so many ladders on the roof. In order to resist the action of the elements, the walls, as at the mission of San Francisco, are remarkably thick; though this peculiarity is here partly intended to guard against the shocks of earthquakes, which are here so frequent that a hundred and twenty of them were felt during two successive months of the last summer. This average, however, of two earthquakes a day, is not so frightful as it looks, the shocks being seldom severe, and often so slight, according to Basil Hall's experience in South America, as to escape the notice of the uninitiated stranger. Externally the habitations have a cheerless aspect, in consequence of the paucity of windows, which are almost unattainable luxuries.

As to public buildings, this capital of a province may, with a stretch of charity, be allowed to possess four. First is the church, part of which is going to decay, while another part is not yet finished; next comes the castle, consisting of a small house surrounded by a low wall, all of adobes. It commands the town and anchorage, if a garrison of five soldiers and a battery of eight or ten rusty and honeycombed guns can be said to command anything.* Third in order is the guardhouse, a paltry mud hut, without windows. Fourth, and last, stands the government house, which is, or rather promises to be, a small range of decent offices; for though it has been building for five years, it is not yet finished.

The neighbourhood of the town is pleasantly diversified with hills, and offers abundance of timber. The soil, though light and sandy, is certainly capable of cultivation; and yet there is neither field nor garden to be seen.

Monterey is badly supplied with water, which in consequence of the extraordinary drought of last year, lately brought a dollar a pipe. The small stream which runs through the town is generally dry in summer, the very season when its water is most wanted.

The mission of San Carlos is about 4 miles from Monterey, lying near the sea on the Carmelo. The intervening country is very picturesque, presenting a succession of grassy slopes, with a sufficient sprinkling of timber to relieve the monotony, with the snow-capped mountains in the distant interior. This country is covered with a very considerable number of cattle, who rove very much at their own will.

Near the mission, when Sir George Simpson visited it, there was a very distinct rent in the earth of a mile or so in length, and of 30 or 40 feet in depth, the result of one of the recent earthquakes. The mission itself, in addition to the hand of the spoiler, has also had this same subterranean enemy to encounter; for the beautiful church, which, as usual, superstition has wrested from rapacity, has had one side pretty severely shattered by a recent shock. With the exception of the church, the immense ranges of buildings were all a heap of ruins. Of the 700 converts residing here according to Humboldt, in 1802, not one remained;

* The presidio, or fortress, though much better than that of San Francisco, at the time of Beechey's visit, was quite useless as a place of defence; and its strength may be judged of from its having been taken by a small party of seamen, who landed from a Buenos Ayrean pirate in 1819, destroyed the greater part of the guns, and pillaged and burnt the town."—*Beechey*, vol. ii. p. 408.

and the only living tenants of the establishment were a man and his wife, whose single duty was to take care of a church that had no priest.*

DIRECTIONS.—The bay was examined by Capt. Beechey, R.N., whose directions for it follow:—

The anchorage at Monterey is at the South extremity of a deep bay formed between Punta Año Nuevo and Punta Pinos. This bay is about 7 leagues across, and open in every part, except that frequented by shipping, where it is shut in by Point Pinos. Ships should not enter this bay in light winds, in any other part than that used as an anchorage, as there is generally a heavy swell from the westward, and deep water close to the shore.

It is impossible to mistake Point Pinos if the weather be at all clear, as its aspect is very different to that of any other part of the bay to the northward. It is a long, sloping, rocky projection, surmounted by pine trees, from which it takes its name; whereas the coast line of the bay is all sandy beach. There is no danger in approaching Point Pinos, except that which may ensue from a heavy swell almost always setting upon the Point, and from light winds near the shore, as the water is too deep for anchorage. With a breeze from the southward, Point Pinos should be passed as close as possible, a quarter of a mile will not be too near; and that shore should be hugged in order to fetch the anchorage. In case of having to make a tack, take care of a shoal at the S.E. angle of the bay, which may be known by a great quantity of sea-weed upon it; there is no other danger. This shoal has 3½ and 4 fathoms upon its outer edge, and 7 fathoms near it. With wind steer boldly towards the sandy beach, at the head of the bay, and anchor about one-sixth of a mile off shore in 9 fathoms; the fort upon the hill near the beach bearing W.S.W., and moor with the best bower to the E.N.E.

This anchorage, though apparently unsafe, is said to be very secure, and that the only danger is from violent gusts of wind from the S.E. The north-westerly winds, though they prevail upon the coast, and send a heavy swell into the bay, do not blow home upon the shore; and when they are at all fresh, they occasion a strong effect in the bay. This, I believe, is also the case at Callao and at Valparaiso, to which this anchorage bears a great resemblance.

There is no good water to be had at Monterey, and ships in want of that necessary supply must either proceed to San Francisco, or procure a permit from the governor,† and obtain it at Santa Cruz, or some of the missions to the southward.

By the mean of many observations on the tides at this place, it is high water, full and change, at 9^h 42′. The rise at spring tides is about 6 feet, and 1 foot 2 inches at neaps. There is very little current at the anchorage.‡

The ordinary wind at this place is from the S.W. and W.S.W. in the morning; towards ten o'clock it veers to the West and W.N.W., from which quarter it freshens till three or four o'clock, afterwards decreasing, and finally becomes calm, which lasts until midnight, when light airs come off from the land,

* Journey round the World, vol. i. pp. 370-1.

† This remark, as well as others, it will be seen, cannot apply to the present condition of the Californian ports.

‡ Beechey's Voyage to the Pacific, vol. ii., Appendix, p. 655.

which continue until daylight. In November there are frequent short gales from the S.E., which blow from off the high land, rushing down in violent squalls. The most dangerous gales are from the North and West, on which side the bay is completely open; the sea sets in very heavy, and is more to be apprehended than the wind. Fogs generally prevail in the morning to seaward; these, however, do not extend into the bay, and when the wind from the N.W. sets in they are generally dissipated.

The following useful remarks will form a fitting supplement to the foregoing:—

A vessel bound to Monterey ought to make the high land of Santa Cruz on the northern extremity of the bay, and then shape a course for the anchorage. A man-of-war in 1846 was nearly cast away by standing for Point Pinos, the southern extreme, and being set into the dangerous Bay of San Carmelo in one of the thick and sudden fogs peculiar to this coast.

The anchorage of Monterey is in a small elbow formed on the northern side of Point Pinos. With N.W. winds there is sufficient scope from the Santa Cruz shore for a considerable sea to get up; but, with good ground tackling, a vessel need not hesitate to ride out any gale: the effect from the shore, which is steep-to, and the kelp, materially increasing her chances of safety. The shore should not be approached nearer than 10 fathoms, and the vessel moored, open hawse, to the N.W.

Bearings from our anchorage were as follow:—Point Santa Cruz, N. 45° W.; land of Cape Pinos, N. 62° W.; the pier head, S. 22° W.

Watering at Monterey is attended with much trouble and inconvenience. The Americans promised to improve it by carrying the water to some point near the landing place; but the present gold mania has most likely destroyed their projects.

Supplies in the shape of beef and mutton are plentiful, good, and were cheap. The climate all the year round allows of the most successful salting. Vegetables grow in profusion at Carmelita, and potatoes, though small, are good; cheese is to be got in any quantity.

Wood for planks or spars is plentiful, though not of the first quality; but from the little port of Santa Cruz a superior quality of pine is obtainable, it growing to a very large size. Santa Cruz boasts of a ship-building establishment, and many of the schooners trading on the coast have been launched from the yard of an enterprising Frenchman residing there.

From the pine-forest about Monterey we cut three top-gallant masts, the trees requiring little trimming, as the following dimensions will show: diameter of tree at base, 3 feet; whole height of tree, 110 feet; height of first branches from the ground, 70 feet. These top-gallant masts, when fidded, stood tolerably well, and were only set aside in consequence of getting very superior spars from Puget Sound. American oak grows at Monterey in vast profusion, and affords excellent crooked timber for knees, &c., and is, moreover, first-rate firewood.*

The sportsman will be well repaid in the neighbourhood of the town; indeed the whole country is one great game preserve, from elk to quail.

* Remarks, &c., H.M.S. *Collingwood*, Nautical Magazine, March, 1849, p. 144.

From Point Año Nuevo to the entrance of San Francisco the distance is 53 miles. The coast for the whole distance is uninviting in appearance; the San Bruno hills gradually decrease in altitude, and become sandy and barren without any symptoms of cultivation.

SAN FRANCISCO.

The Port of San Francisco was discovered as late as the year 1770, and that, too, not by the obvious mode of such an expedition, but by one overland. In 1767, when the Jesuits were replaced by the Franciscans, the viceroy of Mexico, the Marquis de Croix, finding that England and France were taking an interest in these countries, as evidenced by the expeditions of Cook and Bougainville, and that Russia, too, was steadily progressing from the North, proposed to the ecclesiastics the colonization of this territory. Accordingly it was divided into districts, and missions were planned for San Diego and Monterey, the only two ports then known to exist in the upper province, as recited in the introduction to this chapter. Three vessels were despatched from San Diego, but were eminently unfortunate, from the fact that the N.W. or opposing winds blow during three-fourths of the year. Under these circumstances the remainder of the distance was undertaken by land, and though the explorers did not recognise what is now Monterey, they made the far more valuable discovery of the inland sea to which the name of the patron saint of their order, and of sailors in general, was given.

The mission was founded, and its progress had but comparatively little to do with the external world, and this little commercial importance was all but annihilated by the revolution in 1836.

It is now wonderfully altered in its importance, and bids fair to become the metropolis of the Pacific, when but two years since a few miserable cabins alone indicated the presence of civilized man.

Whales are seen around the neighbourhood of San Francisco. But the Californian whale is not so accommodating an animal as elsewhere. As soon as he sees the boats approaching him, he turns upon them, and becomes the aggressor. This has driven the whalers away from the battle in this region, and thus, while the species multiply on the coast of California, they, on the contrary, are disappearing in the parts where they do not act on the defensive. At the present time, the Russian whale is found in shoals in the distant seas of Japan and Okhotsk, and even in these parts of such difficult access, they do not attempt to shelter themselves against their pursuers.*

The Harbour of San Francisco has been minutely surveyed by Capt. Beechey, in H.M.S. *Blossom*, in 1828. The very excellent plan, on a large scale, published by the British Admiralty, gives so accurate an idea of its physical character that it is itself its best elucidator.

It has only two drawbacks; that of a narrow entrance in an unsheltered line of coast, where fogs are both sudden and dense; and the sudden manner in which the rollers set in on the bar at the mouth.

The harbour consists of the narrow entrance 5 or 6 miles in length, opening into

* M. Dillon, Revue des Deux Mondes, *January*, 1850.

a vast inland sea, extending to the southward about 30 miles, and to the northward towards the Bay of San Pablo about 8 miles. This entrance, which has been compared with the Goulet of Brest, is sufficiently narrow for the forts, which it is at present under consideration to construct on either side, to cross their fires, and command the entrance. The depth in it is very irregular; but the shoalest spots have more than 26 feet.

The outer points of the narrow entrance are *Point Lobos* on the South, and *Point Boneta* on the North, about 2 miles asunder; and 2½ miles N.N.E. of the South point is a projection on the South shore, on the extremity of which stands an old Spanish fort, at the base of which an extempore jetty has been lately formed. In front of the entrance, within the harbour, *Alcatraces Island* seems evidently destined as the site of a battery; a new element of power and security for a harbour which already possesses so many. Two and three-quarter miles East of this is *Yerba Buena Island*, and midway between them is the *Blossom Rock*, mentioned in the directions hereafter. A mile and a half northward of Alcatraces is Angeles Island, the largest in the bay, lying before Sausalito Bay, or Whaler's Harbour.

From the harbour itself having nearly the same general direction as the coast, it forms two peninsulas which the entrance separates. Along the southern of these a mountainous ridge runs, called the *Sierra San Bruno*, the principal point of which is called the *Blue Mountain*, 1,087 feet high, to the S.E. of which are two paps. The principal hill on the northern side is the *Table Hill*, 2,569 feet in height, but which is 7 miles N.N.W. of the entrance. Above the old fort, on the South side, is the presidio; and 2½ miles S.E. of this is, or was, the mission of San Francisco, once the only mark of civilization in this part of the harbour.

To the East of the harbour, immediately before the entrance, is the *Sierra Diavolo*, a mount 3,770 feet in height, to the northward of which is the mouth of the Sacramento and San Juan rivers.

SAN FRANCISCO, or YERBA BUENA, as it is otherwise called, lies to the right of the entrance, a little beyond the old Spanish fort. It is now (at the end of 1849) a city of 50,000 souls, which promises to become, in a few years, the capital of the Pacific. At the present moment there are more than 340 merchant ships at anchor near the city, without reckoning a very considerable number of brigs and schooners. All, without exception, have lost their crews, and with many of them the captains themselves have deserted. An American corvette, carrying the flag of Commodore Jones, is the only guardian of this mass of property.

At San Francisco, where 15 months before you only found half a dozen rude huts, may now be seen an exchange, a theatre, churches for every sect, and a large number of houses of good appearance. Some are built of stone, but the largest portion of wood or adobes (sun-dried earth).* The fronts of the houses are whitened, or painted, and the streets are well laid out. On both sides of the city

* La Californie, dans les Derniers Mois de 1849, par M. Patruc Dillon, French Consul at the Sandwich Isles, October, 1849; *Revue des Deux Mondes*, January, 2nd part, 1850, pp. 193—519. This paper gives a very good and interesting account of the state of society in California at the present time.

following the beach, lines of tents extend beyond the reach of sight. These serve as places where a large part of the emigrants rest before they start for the mines.

The temporary nature of the construction of the principal part of the town led to the destruction by fire of above one-half of it at the end of 1849. It was soon after reinstated.

Of the state of society in San Francisco at the present time, perhaps the least said of it the best. One very important consideration with the shipmaster visiting the harbour is the desertion of the crews; the evidence of which is the fleet now lying there. But since September, 1849, in consequence of the lawless state of society, a police has been formed, which, though composed of only fifteen men, still is (or was) very effective. They undertook to get all deserters back for three ounces of gold per man.

The BAY of SAN PABLO, which lies to the North of the principal portion of the harbour, is of a form nearly circular, and 10 miles in diameter; many small streams enter it on all sides from the neighbouring hills. On the East side of this bay the Sacramento River empties into it through the Straits of Karquines. The land is high, and the sandstone rock on each side of the straits resembles that seen about the Straits of De Fuca, to the northward. The hills are thickly covered with wild oats, which were ripe at the period of Commodore Wilkes's brief visit in August, 1841, and the landscape had that peculiar golden hue before remarked. The contrast of this with the dark foliage of the scattered oaks, heightens the effect, which, although peculiar, is not unpleasing to the sight. The trees all have an inclination towards the S.E., showing the prevalence and violence of the bleak N.W. winds, producing on them a gnarled and mountain character. This feature is general throughout the coast of California, and gives the trees a singular appearance, the flat tops having the air of being cut or trimmed after the manner of box trees. The tops are bent to one side, and the larger branches hidden by the numerous twigs that compose the mass. The only place where a similar character was observed to be impressed upon the foliage was at Tierra del Fuego.*

The STRAITS of KARQUINES connect the Bay of San Pablo with Suisun Bay and the mouth of the Sacramento. One of the towns resulting from the influx of population is rising here, on the North side of the straits—the town of Benicia, beautifully situated, and possessing greater commercial advantages than any other part of the Bay of San Francisco. Ships of the largest class can lie close to the shore, and some steamers are already building there. At the period of the last intelligence, there were 40 or 50 houses here; but as San Francisco had got the start, of course it carried the sway as regards prosperity and numbers of immigrants, though doubtless this must rapidly increase in importance. After winding about 5 leagues to the East and S.E., Suisun Bay extends from it to the N.E. about 20 miles, but its upper portion is shoal. In the lower, or S.E. part of the bay, the Sacramento enters, and at this juncture is the city of Montezuma, a single and deserted hut. Fifteen miles above this, on the South

* Wilkes, vol. v. p. 177.

side, is Suisun, a beautiful site, a city without inhabitants. Proceeding up the Sacramento, a couple of log houses on the eastern bank represents the town of Webster; and 10 or 15 miles higher is Sutersville, so called after the original possessor of the soil, mentioned in the introductory part of this chapter.

SACRAMENTO CITY is 3 miles above this, at the junction of the American Fork with the Sacramento at the head of the tide, and, as it was the focus of the gold finders, rose to considerable extent as regards population, but not substantially, for disregarding the natural character of the situation, so well applied by the native Indians, and described by more than one author, the whole of it, with the exception of *one* house, raised on piles, was washed away by the river floods at the end of 1849. Sir Edward Belcher's remarks on this are given below.

The situation has been known as Suter's Fort, and was called by the original proprietor New Helvetia; it is placed by Commodore Wilkes in lat. 38° 33′ 45″, lon. 121° 20′ 5″ W.

Capt. Sir Edw. Belcher, when here in October and November, 1837, ascended the Rio Sacramento for 150 miles to Point Victoria, in lat. 38° 46′ 47″ N., lon. 0° 47′ 41·5″ E. of the observatory on Yerba Buena. He says:—"Throughout the whole extent, from Elk Station to the Sacramento Mouth, the country is one immense flat, bounded in the distance, N.W., by Sierras Diavolo; West, Sierras Bolbones; and E.N.E. to E.S.E. by the Sierras Nievadas, from whence no doubt this river springs, and rises in proportion to the rains and thaws. Our course lay between banks, varying from 20 to 30 feet above the river level; apparently, from its strata, of differently composed clay and loose earth, produced by some great alluvial deposit. Sand did occur at times, but not a rock or pebble varied the sameness of the banks.

"The banks of the River Sacramento were belted with trees, particularly the plane (*platanus occidentalis*), which were of immense size. Some of the oaks, too, which were seen disposed in clumps on the immense park-like extent on each side, like a sea of grass, were abundant, and of remarkable bulk. All the trees, &c., on the banks afforded unequivocal proofs of the force of the flood streams, and indications of a rise of 10 feet were recent and evident. During the rainy season, commencing in the middle of November, and ending about the end of February, the river overflows, and ascent is then impossible. At times these floods, with the annual rains and heavy falls of snow, cause all the plains to be as one immense sea, leaving only a few scattered eminences, which art or nature have produced, as so many islets or spots of refuge. The Indians are thus frequently obliged to seek shelter at these "Rancherías," which they raise about 15 feet above the level of the plain, and none exceeding 100 yards in diameter. The fearful ravages of disease and want upon these unfortunate people, thus confined to such narrow bounds, were at many places evident, and some of these mounds were but the tombs of the whole tribes who resorted to and formed them."*

On approaching the coast in the neighbourhood of San Francisco, the country has by no means an inviting aspect. To the North it rises in a lofty range,

* Voyage of the *Sulphur*, vol. i. pp. 123-5.

whose highest point is known as the Table Hill, and forms an iron bound coast from Punta de los Reyes to the mouth of the harbour.

To the South, there is an extended sandy beach, behind which rise the sand hills of San Bruno, to a moderate height. There are no symptoms of cultivation, nor is the land on either side fit for it; for in the former direction it is mountainous, in the latter sandy, and in both barren. The entrance to the harbour is striking; bold and rocky shores confine the rush of the tide, through a narrow passage into a large estuary. In this, several islands and rocks lie scattered around; some of the islands are clothed with vegetation to their very tops, others are barren and covered with guano, having an immense number of sea-fowls hovering over, around, and alighting upon them. The distant shores of the bay extend North and South far beyond the visible horizon, exhibiting one of the most spacious, and at the same time safest, ports in the world. To the East rises a lofty inland range, known by the name of La Sierra, brilliant with all the beautiful tints that the atmosphere in this climate produces.*

DIRECTIONS.—The following account of the harbour, with the necessary instructions for vessels entering or leaving, are by Capt. Beechey, R.N., who surveyed it in 1828:—

The harbour of San Francisco, for the perfect security it affords to vessels of any burden, and the supplies of fresh beef and vegetables, wood and fresh water, may vie with any port on the N.W. coast of America. It is not, however, without its disadvantages, of which the difficulty of landing at low water, and the remoteness of the watering place from the only anchorage which I could recommend, are the greatest.

Ships bound to San Francisco from the northward and westward, should endeavour to make Punta de los Reyes, a bold, conspicuous headland, without any danger laying off it sufficiently far to endanger a ship. In clear weather, when running for the land, before the latitude is known, or the Punta can be distinguished, its situation may be known by a table-hill terminating the range that passes at the back of Bodega. This hill, in one with the Punta de los Reyes, bears East by compass. If ships are not too far off, they will see, at the same time, San Bruno, two hills to the southward of San Francisco, having the appearance of islands; and from the mast-head, if the weather be very clear, the South Farallon will in all probability be seen. Punta de los Reyes, when viewed from the West or S.W., has also the appearance of an island, being connected by low land to the two hills eastward. It is of moderate height, and as it stands at the angle formed by the coast line, cannot be mistaken. Soundings may be had off this coast in depths varying with the latitude. In the parallel of the Farallones, they extend a greater distance from the main land, in consequence of these islands lying beyond the general outline of the coast.

The Farallones are two clusters of rocks, which, in consequence of the shoals about them, are extremely dangerous to vessels approaching San Francisco in foggy weather. The southern cluster, of which, in clear weather, one of the islands may be seen from the mast-head 8 or 9 leagues, is the largest and highest,

* Wilkes's Narrative, vol. v. pp. 151-2.

and lies exactly S. 3° E. *true*, 18 miles from the Punta de los Reyes. The small cluster of rocks lies to the N.W., and still further in that direction there are breakers, but I do not know how far they extended from the rocks above water. In a thick foggy night, we struck soundings in 25 fathoms, stiff clay, near them; and, on standing off, carried regular soundings to 32 fathoms, after which they deepened rapidly.

Coming from the southward, or when inside the Farallones, the position of the entrance to San Francisco may be known by the land receding considerably between the table-hill, already mentioned, and San Bruno hill, which at a distance appear to terminate the ridge extending from Santa Cruz to the northward. The land to the northward or southward of these two hills has nothing remarkable about it to a stranger; it is, generally speaking, sufficiently high to be seen 13 to 15 leagues, and inland is covered with wood.

About 8½ miles from the fort, at the entrance of San Francisco, there is a bar of sand, extending in a S. by E. direction, across the mouth of the harbour. The soundings, on approaching it, gradually decrease to 4½ and 6 fathoms, low water spring tide, depending upon the situation of the ship, and as regularly increase on the opposite side, to no bottom with the hand-leads. In crossing the bar, it is well to give the northern shore a good berth, and bring the small white island, Alcatrasses, in one with the fort, or South bluff, if it can be conveniently done, as they may then ensure 6 fathoms; but if ships get to the northward, so as to bring the South bluff in one with the Island of Yerba Buena, they will find but 4½, which is little enough with the heavy sea which sometimes rolls over the bar; besides, the sea will sometimes break heavily in that depth, and endanger small vessels. To the northward of this bearing the water is more shallow. Approaching the entrance, the Island of Alcatrasses may be opened with the fort; and the best directions are to keep in mid-channel, or on the *weather side*. On the South shore the dangers are above water, and it is only necessary to avoid being set into the bay between the fort and Point Lobos. If necessary, ships may go inside, or to the southward of the *One Mile Rock*, but it is advisable to avoid doing so, if possible. On approaching it, guard against the tide, which sets strong from the outer point toward it, and in a line for the fort. Off Punta Boneta there is a dangerous reef, on which the sea breaks very heavily; it lies S.W. from the point, and no ship should approach it nearer than to bring the fort in one with Yerba Buena Island.

In the entrance it is particularly necessary to attend to the sails, in consequence of the eddy tides and the flaws of wind that come off the land. The boats should also be ready for lowering down on the instant, as the entrance is very narrow, and the tides, running strong and in eddies, are apt to sweep a ship over upon one side or the other, and the water is in general too deep for anchorage; besides, the wind may fail when most required. The strongest tides and the deepest water lie over on the North shore. Should a ship be swept into the sandy bay West of the fort, she will find good anchorage on a sandy bottom, in 10 and 15 fathoms, out of the tide; or in the event of meeting the ebb at the entrance, she might haul in, and there await the change. There is no danger off the fort at a greater distance than 100 yards.

As soon as a ship passes the fort, she enters a large sheet of water, in which are several islands, two rocks above water and one under, exceedingly dangerous to shipping, of which I shall speak hereafter. One branch of the harbour extends in a S.E. by S. direction, exactly 30 miles, between two ridges of hills, one of which extends along the coast towards the Bay of Monterey, and the other from San Pablo, close at the back of San José, to San Juan Bautista, when it unites with the former. This arm terminates in several little winding creeks, leading up to the missions of Santa Clara and San José. The other great branch takes a northerly direction, passes the Puntas San Pablo and San Pedro, opens out into a spacious basin, about 10 miles in width, and then conveying to a second strait, again expands, and is connected with three rivers, one of which is said to take its rise in the Rocky Mountains, near the source of the River Columbia.

As a general rule in San Francisco, the deepest water will be found where the tide is the strongest; and out of the current there is always a difficulty in landing at low water. All the bays, except such as are swept by the tide, have a muddy flat, extending nearly from point to point, great part of which is dry at low water, and occasions the before-mentioned difficulty of landing; and the north-eastern shore, from Punta San Pablo to the Rio Calavaros, beyond San José, is so flat that light boats only can approach it at high water. In low tides it dries some hundred yards off shore, and has only one fathom water at an average distance of one mile and a half. The northern side of the great basin beyond San Pablo is of the same nature.

After passing the fort a ship may work up for the anchorage without apprehension, attending to the lead and the tide. The only hidden danger is a rock (the *Blossom Rock*), with one fathom on it at low water spring tides, which lies between Alcatrasses and Yerba Buena Islands. It has 7 fathoms alongside it; the lead, therefore, gives no warning. The marks when on it are the North end of Yerba Buena Island in one with two trees (nearly the last of the straggling ones) South of Palos Colorados, a wood of pines situated on the top of the hills over San Antonio, too conspicuous to be overlooked: the left hand, or S.E. corner of the presidio, just open with the first cape to the westward of it; Sausalito Point open a quarter of a point with the North end of Alcatrasses; and the Island of Molate in one with Punta de San Pedro. When to the eastward of Alcatrasses, and working to the S.E., or indeed to the westward, it is better not to stand toward this rock nearer than to bring the Table Peak in one with the North end of Alcatrasses Island, or to shut in Sausalito Point with the South extreme of it. The position of the rock may generally be known by a ripple; but this is not always the case.

There are no other directions necessary in working for Yerba Buena Cove, which I recommend as an anchorage to all vessels intending to remain at San Francisco.

In the navigation of the harbour much advantage may be derived from a knowledge of the tides. It must be remembered that there are two extensive branches of water lying nearly at right angles with each other. The ebbs from these unite in the centre of the bay, and occasion ripplings and eddies, and other irregularities of the stream, sometimes dangerous to boats.* The anchorage at Yerba Buena Cove is free from these annoyances, and the passage up to it is nearly so after passing the presidio. The ebb begins to make first from the Santa Clara arm, and runs down the South shore a full hour before the flood has done about Yerba Buena and Angel Island; and the flood, in its return, makes also first along the same shore, forcing the ebb over the Yerba Buena side, where it unites with the ebb from the North arm.

The flood first strikes over from the Lime Rock, and passing the Island of Alcatrasses, where it diverges, one part goes quietly to Santa Clara; the other, sweeping over the sunken rock, and round the East end of Angel Island, unites with a rapid stream through the narrow channel formed by Angel Island and the main, and both rush to the northward, through the Estrecho de San Pablo, to restore the equilibrium of the basin beyond, the small rocks of Pedro Branco and the Alcatrasses Island lying in the strength of the stream.

The mean of eighty observations gave the time of high water, full and change, at Yerba Buena anchorage, 10^h 52′.

The tide, at springs, rises	..	7 ft. 10 in., sometimes 8 ft. 3 in.
,, neap ,,	..	1 ft. 10 in.
Average of ebb at spring tides		2 kn. 0 fm., at neap, 1 kn. 0 fm.
,, flood ,,	..	1 kn. 0 fm., ,, 0 kn. 6 fm.
Duration of flood	..	5 h. 25 m.

At Sausalito, the mean of seventeen observations gave the time of high water, full and change, 9^h 51′.

Rise, full and change	..	..	..	6 ft. 0 in.
Neap	..	..	..	2 ft. 6 in.
Duration of flood ..	..	..	..	4 h. 53 m.

On quitting San Francisco, the direction of the wind in the offing should be considered. If it blow from the S.W. there would be some difficulty in getting out of the bay to the southward of Punta de los Reyes. The residents assert that an easterly wind in the harbour does not extend far beyond the entrance; and that a ship would, in consequence, be becalmed on the bar, and perhaps exposed to a heavy swell, or she might be swept back again, and obliged to anchor in an exposed situation. Northerly winds appear to be most generally approved, as they are more steady, and of longer duration, than any other; they may, indeed, be said to be the trade-wind on the coast. With them it is advisable to keep the North shore on board, as the strength of the ebb tide takes that side, and as on the opposite shore, near the One Mile Rock, the tide sets rather *upon* the

* "We passed the presidio and fort under the influence of a strong ebb tide, which, after rounding the southern side of the entrance, rushes to the southward at the rate of 6 knots an hour. In the very direction of the current there lay some rocks; and, as the wind failed us just at the point, the vessel, which no longer had any way upon her, was turned toward them like a log. The anchor was dropped, with 30 fathoms of chain, but dragged till we were within a few yards of the object of our fears; and when at last it did hold, it was raised so as barely to touch the bottom, that by thus counteracting, in some degree, the action of the tide, it might enable the ship to obey her helm. By this operation of kedging, we steered clear of the rocks, when the wind freshened sufficiently for us to stand off the shore, which was not above a cable's length distant. Luckily the rocks in question show all their dangers above water; for there is a depth of 7 fathoms around each of them, so that a vessel is sometimes carried in safety between them."—*Sir George Simpson*, Journey, &c., vol. i. p. 342.

land. In case of necessity, a ship can anchor to the eastward of the One Mile Rock; but to the S.W. of the rock the ground is very uneven. The wind generally fails in the entrance, or takes a direction in or out. From the fairway steer S.W. ½ W., and you will carry 7 fathoms over the bar, half ebb, spring tide. This I judge to be good course, in and out, with a fair wind. I would avoid, by every endeavour, the chance of falling into the sandy bay to the southward of Lobos Point, and also closing with the shore to the N.W. of the Punta Boneta.*

As a supplement to these directions by Capt. Beechey, the following remarks may be useful:—

"If the Farallones are not made, and the position of the harbour not very certain, some difficulty may be experienced in discovering the entrance, particularly from the northward. It may, however, be known by a long sandy strip of land just to the southward of the entrance, which has the appearance of a hay field, and also not far from this shore is a remarkable rock, having an arch in it.

"To the northward of the entrance are three or four rocks, close in shore, very white on their tops, and at nearly equal distances from each other.

"It is to be hoped that these remarks, however meagre, may be of service in hazy or foggy weather, and when close to the land, in assisting the stranger to find out the entrance to this magnificent port." †

The next, which may be taken in connexion with those by Capt. Beechey, are derived chiefly from Mr. Richardson, captain of the port (November, 1846), and also an experienced pilot of the harbour.

"Ships coming in from the South Farallones should run in on a N.E. by E. ½ E. course, and bring Point Lobos on the same bearing (N.E. by E ½ E.) in order to cross the bar in 6½ fathoms, and to keep as nearly mid-channel as possible, there being a bank of 4 fathoms on the South shore, outside, which has generally a heavy swell on it. There is a similar bank also on the North shore, extending at least 5 miles out.

"Between these two banks there is anchorage in 10, 12, and 15 fathoms, as you draw in. After getting inside, and having passed the fort, you can anchor anywhere in as far as the Alcatrasses, there being no hidden danger.

"In going for Sausalito with a light wind and ebb tide, it will be very advisable to steer directly for Angel Island, as the tide sets strong against Sausalito Bay, and tends to heave the ship into deep water.

"A ship leaving Sausalito should avoid being set into Lime Rock Bay, by standing over towards the fort point, and from the fort point stand across to the northern shore to keep out of the eddy current in the S.E. bay, outside the fort.

"The ebb makes on each shore at least two hours before it sets out in the stream, and therefore a ship should not leave the anchorage until the tide had fallen a foot by the shore. These remarks apply chiefly to vessels leaving with a foul wind.

"If the wind be fair, and of sufficient strength to render the ship perfectly under command, she can then start at the last of the flood.

"The ebb tide makes from Yerba Buena Bay, across towards Lime Rock, thence into Mile Rock Bay (so that ships going out have not unfrequently been

* Voyage of the *Blossom*, part ii. appendix, pp. 651-654.
† R. M. I., in Nautical Magazine, March, 1849, p. 154.

set between Mile Rock and the main), and from that bay it runs to the N.W. round Point Lobos.

"Outside the fort point the ebb sets to the N.W., round Point Boneta, and the flood runs to the S.E."

THE COAST, from the entrance of Port San Francisco, trends in a N.W. direction, and rises in abrupt cliffs, with very unequal surfaces, presenting a most dreary and barren aspect. A few scattered trees grow on the more elevated land, with some patches of dwarf shrubs in the valleys. The rest of the country presents either a surface of naked rocks, or a covering of very little verdure. At about 12 miles from the harbour, it suddenly falls, and forms a low sandy projecting point, off which some breakers extend nearly 2 miles to the E.S.E.

PORT SIR FRANCIS DRAKE.—At 8 leagues N. 62° W. from San Francisco is Point de los Reyes, on the eastern side of which is the bay in which, according to the Spaniards, Sir Francis Drake anchored.* The eastern side of the Bay of Sir Francis Drake is composed of white cliffs, as is the coast between it and Point de los Reyes, though the latter are lower. The bay extends a little distance to the northward, is entirely open, and much exposed to the South and S.E. winds. Vancouver says:—"However safe Sir Francis Drake might then have found it, yet at this season of the year (November, 1792) it promises us little shelter or security."†

The white cliffs, on the eastern side of this bay, in all probability induced Sir Francis Drake to bestow upon this country the name of New Albion. The bay is considered by Capt. Beechey as too exposed to authorize the conjecture of Vancouver, that it is the same in which Sir Francis refitted his vessel.‡

POINT DE LOS REYES, about a league westward of the Bay of Sir Francis Drake, is one of the most conspicuous promontories South of the Straits of De Fuca. It cannot easily be mistaken. When seen from the North or South, at the distance of 5 or 6 leagues, it appears insular, owing to its projecting into the sea, and the land behind it being less high than usual near the coast; but the interior country preserves a more lofty appearance, although these mountains extend in a direction further from the coast than those to the northward. Point de los Reyes, as before stated, stretches like a peninsula to the southward into the ocean, when its highest part terminates in steep cliffs, moderately elevated, and nearly perpendicular to the sea, which beats against them with great violence.

The FARALLONES are nothing but a cluster of rocky islets, destitute of vegetation. They have been alluded to by Capt. Beechey on page 338. The northernmost, which is the largest, is about 2 miles in circumference, of an oblong shape, lying E.N.E. and W.S.W. On each end is a hill, rising about

* The harbour of San Francisco (Sir Francis Drake) was sought for by Don J. de Ayala, who commanded a Spanish expedition in 1775, and who also discovered the Puerto de Trinidad (lat. 41° 4′), hereafter mentioned. Maurelle was the pilot. There is a confusion of circumstances here. The Hon. Daines Barrington, who published the accounts of these expeditions, reproaches the Spaniards with having thought religion interested in suppressing the name of the *brave heretic*, who, by his exploits, had exceeded those of the Spanish navigators, and unblushingly substituting the name of the ensign *La Bodega* for that of St. Francis, or even of Sir Francis Drake. But these objections are futile, as will be seen. Sir Francis Drake's harbour is distinct from La Bodega, and the name of the saint was imposed without any reference to the English admiral.—*Miscellanies, by the Hon. Daines Barrington*, 1781.

† Vol. i. p. 430.

‡ Beechey, vol. i. p. 343.

300 feet, and declining to a valley in the centre of the island, forming the appearance, when viewed from the North or South, of a saddle. Many years previous this place was the resort of numerous fur seal, but the Russians, who had established themselves on them, had made such havoc that there was scarcely a breed left. This island is of volcanic origin; most of the rocks have evidently been once in a state of fusion, and the low land is covered with pumice stone. Aquatic birds, in considerable variety, resort hither for the purposes of incubation. There is no fresh water on them.*

PORT BODEGA.—At 4½ leagues to the northward of Point de los Reyes is the western point of the entrance of Port Bodega, which is about 90 miles from San Francisco. Its South point is formed by steep, rocky cliffs, with some detached rocks lying near it. The port is both small and inconvenient, and cannot be entered, except by vessels of a light draught of water; the anchorage outside is rocky and dangerous. The following is by Sir Edward Belcher.

Bodega is an extensive bay, almost joining (by a creek) the Port of Sir Francis Drake at its southern end, which is very shallow. On the northern side of the bay, at a small creek or estuary (nearly dry at low water springs), stand two Russian buildings; one a storehouse, the floor of which was filled with grain and a few marine stores, and the other the residence of those left in charge, amounting, perhaps, to three men, their wives, and children.

The anchorage is within a rocky islet, with a reef and bank extending about three-quarters of a mile, which is thickly covered by *fucus giganteus*. The bottom is coarse sand, with some patches of clay, but bad holding ground. Here, however, it is customary for the Russians (who have excellent ground-tackle) to ride out the S.W. gales, inasmuch as the heavy swell which immediately tumbles in, or generally precedes, prevents any moderate sailing vessel from making any head, and the sea-room is but scant. I am informed that the Russians have experienced several losses here, but no lives.

At the houses excellent water, carefully conducted by spouts, for the convenience of hose, which allow of filling without removing the casks; and although we found the runs small, yet, being steady and continuous, they afforded employment for two boats. I am satisfied, also, that it is as good, and more expeditiously obtained, than at Sausalito, San Francisco, where it is necessary to fill from wells, and injure the boats embarking it.†

ROSS.—The Russian presidio of Ross (Little Russia) is about 30 miles to the northward of the bay or port of Bodega, on land elevated about 100 feet above the sea, the outline of which is cliffy, with alternate rocky and gravelly margins, rendering landing, excepting in fine weather, nearly impracticable. The anchorage off it is bad, by reason of the beds of rock above and below water, and the constant liability to fogs rendering it unsafe to break ground unless with a fair wind.

The hills above it, which command the presidio, are sparingly clothed with fir trees. The main government establishment, or fort, as such enclosures are termed in these countries, consisted of a large square, formed, as usual, with

* Morrell's Narrative, p. 209.

† Voyage of the *Sulphur*, vol. i. pp. 316-7.

towers at two angles, commanding the sides, having but one entrance by large folding gates, facing the sea. The governor's house faced this gate at the back of the officers' houses and chapel on either side. On the N.W. were situated the stables for cattle, a granary, &c., and to the southward is a deep ravine, forming part of the bay; three large buildings, occupied as factories and storehouses for boats, &c. On the slope of the hill were about twenty huts for the Kodiac Indians employed in the service. Besides this presidio and Bodega, the Russians had a small rancho, perhaps the subject of some dispute as to property.*—(Sir Edward Belcher.)

Northward of Port Bodega the land forms a bay, in which appear three small openings, that to the South being Bodega. The North point of the bay is formed of low steep cliffs, and, when seen from the South, has the appearance of an island, but is firmly connected with the main land.†

The land to the northward is high, steep to the sea, and has a rude and barren aspect. At 45 miles from the point previously mentioned is *Point Barra de Arena*. It forms a conspicuous mark on the coast; the shores to the North of it take a direction of N. by W. Its northern side is composed of black rugged rocks, on which the sea breaks with great violence; to the South of it the coast trends S.E. by S.; its southern side is composed of low sandy or clayey cliffs, remarkably white, though interspersed with streaks of a dull green colour; the country above it rises with a gentle ascent, is chequered with copses of forest trees and clear ground, which gives it the appearance of being in a high state of cultivation.

The coast North of Point Barra de Arena stretches about N. by W. ½ W. (*true*) for 57 miles, where *Point Delgado* forms something of a projection, extending about 2 leagues to the westward. At a distance the land appears much broken, but is compact; the irregularity of its surface occasions this: it rises abruptly in low sandy cliffs from a connected beach which uniformly composes the sea-shore. The interior country appears to be nearly an uninterrupted forest, but

* The property of the Russian establishment at Ross and Bodega had (August, 1841) just been transferred to Capt. Suter, of New Helvetia, at San Francisco, for the consideration of 30,000 dollars. In the purchase was included all the stock, houses, arms, utensils, and cattle belonging to the establishment. It was understood that this post was abandoned, by order of the Russian government, the Russian Company no longer having any necessity to hold it to procure supplies, as they are now to be furnished under a contract with the Hudson Bay Company; and by giving it up they avoid many heavy expenses.

Bodega was first established by the Russians in 1812, under a permission of the then governor of Monterey to erect a few small huts for salting their beef. A small number of men were left to superintend this business, which in a few years increased until the place became of such importance in the eyes of the Spanish authorities, that on the Russians attempting to establish themselves at San Francisco (on the Island of Yerba Buena), they were ordered to leave the country. This they refused to do, and, having become too strong to be removed by the Spanish force, they had been suffered to remain undisturbed until the time of our visit.

From what I understood from the officers who had been in charge of Bodega, it had been a very considerable expense to the Russian American Company to fortify it, and the disposal of the whole, on almost any terms, must have been advantageous. Capt. Suter had commenced removing the stock, and transporting the guns, &c., to his establishment.

The buildings at the two posts numbered from fifty to sixty, and they frequently contained a population of four or five hundred souls. Since the breaking up of the establishment, the majority of the Russians returned to Sitka, the rest have remained in the employ of the present owner.—*Commodore Wilkes, Narrative, &c.* vol. v. 179-80.

† Off this point Vancouver anchored on a rocky bottom, S.W. by W., 2 miles from it, November 12, 1792. His buoy ropes were cut by the rocks.

toward the sea-side are a pleasing variety of open spaces. Vancouver says:— "We sailed at the distance of 4 or 5 miles from the shore, which still continued compact, with two or three small rocky islets lying near it. As we proceeded, a distant view was obtained of the inland country, which was composed of very lofty, rugged mountains, extending in a ridge nearly parallel to the direction of the coast. These were in general destitute of wood, and the more elevated parts were covered with snow."*

CAPE MENDOCINO is formed by two high promontories, about 10 miles apart; the southernmost, which is the highest when seen either from the North or South, much resembles Dunnose. Off the cape lie some rocky islets and sunken rocks, near a league from the shore. The southernmost of these from the northernmost promontory lies S.W. by W. ½ W. (*true*), about a league distant; and within it are two rocky islets, in shape much resembling haycocks. The northernmost of them lies N. ¼ W., distant 5 or 6 miles, nearly of the same shape and size with the other, to which it is apparently connected by a ledge of rocks, whose outermost part lies from the above promontory N.W. ½ N., about 2 leagues distant, having a smaller islet about midway between them. On some parts of this ledge the sea constantly breaks with great violence; on others, at intervals only. The whole of this cape, though by no means a very projecting headland, is doubtless very remarkable, from being the highest on this part of the coast of California.† The mountains at its back are considerably elevated, and form altogether a high steep mass, which does not break into perpendicular cliffs, but is composed of various hills that rise abruptly, and are divided by many deep chasms.

From Cape Mendoçino the coast takes a direction of N. by E. ½ E.; beyond the islets the shores are straight and compact, not affording the smallest shelter; and, although rising gradually from the water's edge to a moderate height only, but agreeably varied, yet the distant interior country is composed of mountains of great elevation. This part of the coast is generally defended by a sandy beach, but, about lat. 41°, it becomes of a very different description: the shores are composed of rocky precipices, with numberless small rocks and rocky islets extending about a mile into the sea. The most projecting part was named by Vancouver *Rocky Point.*

TRINIDAD BAY is the first shelter North of Cape Mendoçino, and lies just south-eastward of Rocky Point.

"Trinidad Bay,‡ or Porto de la Trinidad, can in no respect be considered as a safe retreat for ships, not even the station occupied by the Spanish explorers in 1775, which I conceived to be close up in the N.N.W. part of the bay, between the main and a detached rock lying from the headland that forms the N.W. point of the bay, E. by W. ½ N. (*true*), about half a mile distant. There two or

* Vol. i. p. 426.

† This cape, as previously mentioned (page 314), was discovered by Rodrigues de Cabrillo, a Portuguese in the service of Spain, in 1542. He was sent by Antonio de Mendoça, the Spanish viceroy, and hence its name.

‡ Trinidad Bay was first explored by the Spanish expedition sent by Senr. Quadra, in 1775. Don F. Maurelle was the pilot, and from his description (given in the Annual Register for 1781, by the Hon. Daines Barrington) it would appear to be an eligible place for shipping; but Vancouver's account of it, as given above, will modify such an opinion.

three vessels, moored head and stern, may lie in 6 and 7 fathoms water, sandy bottom. The point above mentioned will then bear by compass S.W., and the rocks lying off the S.E. point of the bay S.E. ½ E. Between these points of the compass it is still exposed to the whole fury and violence of those winds which, on our return to the southward the preceding autumn, blew incessantly in storms; and, when we approached the shores, were always observed to take the direction of the particular part of the coast we were near. Under these circumstances, even that anchorage, though the most sheltered one the place affords, will be found to be greatly exposed to the violence of these southern blasts, which not only prevail during most part of the winter seasons, but continued to blow very hard in the course of the preceding summer. Should a vessel part cables, or be driven from the anchorage, she must instantly be thrown on the rocks that lie close under her stern, when little else than inevitable destruction is to be expected. The points of Trinidad Bay lie S.E. ¾ E. and N.W. ¾ W., about 2 miles asunder. From this line of direction the rocks that line the shore are nowhere more than half a mile distant. The round barren rocky islet, made white by the dung of sea-fowl, lies from the N.W. point of the bay S. by W., distant three-quarters of a mile: this is steep-to, and has 8 or 9 fathoms water all round it, and admits of a clear channel from 9 to 6 fathoms deep, close to the above point. The soundings in the bay are regular from 9 to 5 fathoms, clear sandy bottom, perhaps not very good holding ground.

The tides appeared to rise and fall about 5 feet, but they were so very irregular that no positive information could be gained of their motion." *

There is a small sunken rock, between two islets, which escaped Vancouver's observation (or rather Mr. Whidbey's, who made the survey). Plenty of good water may be had at this place. The anchorage is situated in lat. 41° 4′ N., lon. 123° 55′ W.—(Capt. Wilkes.†)

A rock, a high round lump, about half a mile in circuit, apparently steep-to, lies N. by W. 13 miles from Rocky Point, and about half a league from shore.

Thirty miles North of Trinidad Bay is the mouth of *Smith's River*,‡ on the bar of which there are 3 fathoms water. The land on the North is high and abrupt, but on the South a narrow neck of dry sand projects with a perpendicular rock on the extreme point. The current sweeps out of the river with great velocity, causing heavy breakers on the bar, which at most times prevent vessels from entering. The soundings outside are regular, and there are three sunken rocks about a mile distant from the sandy point.—(Wilkes.)

Mount Shaste is a magnificent peak in the rear of the entrance to Smith's River. Its summit gives evidence of its having been an active volcano. The snow lies in patches on the sides and part of the peak of this mountain, which is

* Vancouver, vol. ii. pp. 240—246.

† It is much to be regretted that portions of the really fine work of Capt. Wilkes and his coadjutors should be deficient in that accuracy of detail which is expected to be found in such works. The reader will search in vain for anything relating to the correct features of this coast in the *apparently* elaborate map of the Oregon Territory, forming part of the atlas of the Exploring Expedition.

‡ When abreast of Rocky Point, the colour of the sea suddenly changed from the oceanic hue to a very light river-coloured water, extending as far ahead (northward) as could be discerned. This gave us reason to suppose some considerable river was in the neighbourhood. This continued to lat. 41° 36′.—*Vancouver*, vol. i. p. 201.

stated to be 14,390 feet above the sea; though Lieut. Emmons, of the United States' Exploring Expedition did not think it was so high.

POINT ST. GEORGE, in lat. 41° 46', lon. 124° 25', is very conspicuous; and the very dangerous cluster of rocks extending from thence were also named by Vancouver the *Dragon Rocks*. The outwardmost of these lies from Point St. George N.W. ¾ W. 3 leagues distant. The rocks above water are four in number, with many sunken ones, and numerous breakers stretching from the outermost, or southward of Point St. George. This point forms a bay on each side; that upon the North side (*Pelican Bay*) is perfectly open to the N.W., yet apparently sheltered from the W.S.W. and southerly winds by the Dragon Rocks; the soundings are regular from 35 to 45 fathoms, black sand and muddy bottom.* There are several small rocks close to the shore, near which a stream discharges itself: from it water can be procured in boats.—(Capt. Wilkes.)

North of this the face of the country may be considered mountainous, and bounded by innumerable rocky islets. Capt. Morrell, who was on this coast in June, 1825, says:—"On these islands or keys I expected to find fur-seals; whereas I found them all manned with Russians standing ready with their rifles to shoot every seal or sea-otter that showed his head above water."

The boundary between Upper California and the Oregon Territory, the 42° parallel, strikes the coast to the northward of Point St. George; and to the northward of this latitude the Klamet River debouches, after traversing the boundary more than once in its course from the Cascade Range.

CHAPTER XII.

THE COAST OF OREGON, FROM CAPE BLANCO TO ADMIRALTY INLET.

THE more important section of the Oregon Territory now forms an integral portion of the United States of America, the sovereignty having been acknowledged by the treaty of July 15th, 1846, between that power and Great Britain, and is that country described in the present chapter, lying between the parallels of 42° and 49° North latitude, the former separating it from the newly-acquired region of California, described in the preceding pages.

The actual right of possession of the Oregon Territory has been the theme of long and angry discussion, and notwithstanding the cession of the claim by Britain to its present owners, it must ever be acknowledged that their right

* Vancouver, vol. i. p. 202.

by the usual laws of sovereignty was indefeasible. We cannot give here even an outline of the controversies and disputes which preceded the cession of this territory to the United States, nor can we enter upon the justice of the claim acceded to.

That Oregon, or this part of the West coast of America, was known to the ancient Chinese, is endeavoured to be proved from the concurrent testimony of several of their early writers. *M. de Guignes* states that these navigators, having followed the Asiatic coast towards the North as far as Kamtschaka, which they called *Ta Han*, crossed the ocean in an easterly direction, and, at the distance of 20,000 lis, or about 2,000 miles, arrived nearly under the same parallel, at a country which they called Fou Sang, being, according to them, the land where the sun rises.*

The country was described as rich in copper and silver, but had *no iron*. Its first discoverer, as stated in the history of the two empires in China, which resulted from the destruction of the Tsin dynasty in A.D. 420, was *Hoeï-chin*, a Buddhist priest, in about A.D. 499. Some curious particulars of the country are given, such as its fruits, its spotted deer, its cattle with great horns, &c., which are difficult to reconcile with the present state of the countries of N.W. America. From the subsequent dissertations which have ensued, it certainly would appear in some degree probable that the Chinese reached as far South as San Francisco, or at least the Columbia River, at this early age: and it seems not impossible but that they have left some traces of having done so, in the singular ruins of Mexico, or still more singular sculptures of Yucatan. On the other hand, the deduction was denied by M. Klaproth, Von Humboldt, and others.†

The N.W. coast of America, in this part, was first made known to Europe by Sir Francis Drake, in his voyage in 1578, before mentioned. He reached the lat. of 48° North, and coasted southward to the harbour now bearing his name. It was next seen by the Spaniards.

We shall not dwell here upon the much-disputed accounts of De Fuca's voyage in 1592, which will be mentioned elsewhere, nor of Martin de Aguilla in 1603, nor of Admiral Bartolomeo de Fuente, or de Fonte, in 1640. All these have been denied the merit of truth, but there certainly would appear some reason for believing a portion of the first-named narrative. The next or really authentic account of any voyage to this part of the coast is from the Spanish authorities. Ensign Juan Perez sailed from San Blas in the year 1774, and after encountering storms, made the land on July 16th, in lat. 54°, the South point of which was named Cape Santa Margarita, the land being what is now called Queen Charlotte's Island, and the cape, Cape North. He then made Nootka Sound, which he called Port San Lorenzo.‡

The next, in March, 1775, was under Capt. Bruno Heceta, under whom was

* See Acad. des Inscriptions et Belles Letters, tome xxviii., 1757, Recherches sur les Navigations des Chinois du Côte de l'Amerique, &c.

† See L'Amerique sous le nom de pays de Fou Sang, by *M. de Paravey*; Annales de la Philosophe Chretienne, Paris, Feb. 1844; and Nouvelles Preuves, &c., by *M. de Paravey*, in the same work, 3me serie, No. 20, tome xv. p. 449, *et seq*.

‡ The account of this expedition was not published till 1802, in the Introduction to the Journal of the Voyage of the *Sutil* and *Mexicana*, by order of the king of Spain.

Perez. Don J. de Ayala has been frequently named as chief of the expedition, but was at first in command of the second vessel, the *Sonora*. They made Port Trinidad, North of Cape Mendoçino, where they left a cross, which Vancouver found in 1793. They went northward, but being separated by a storm, Heceta returned, and saw the opening of the mouth of the Columbia, which he called the Ensenada de Ascencion; it was also called Heceta's Inlet in some subsequent Spanish charts. The other vessel, under Bodega and Maurelle, made for the North, making the land about King George III.'s Archipelago, and landed on Point Remedios, called the Bay of Islands by Cook, three years afterwards. They then proceeded southward. Such are the first Spanish voyages.

In 1776, Capt. Cook left Plymouth on his last voyage, and after discovering the Sandwich Islands, in January, 1778, he made the coast 200 miles North of Cape Mendoçino, proceeding northwards.

From some notices given respecting the fur to be procured on the coast during Cook's voyage, some vessels fitted out for this region from China, the first of which was under Capt. James Hanna; but in 1787, Capt. Berkeley discovered an inlet in 48° 30'. In 1788, Capts. Duncan and Colnett were on the coast; and Duncan, running down the coast from the North, anchored on the South coast of a strait off a village called *Claasit*, or *Claaset*, in 48° 30'.

In 1788 Capt. Meares entered this strait, and communicated his discovery to Capt. Gray, of the *Columbia*, in which vessel the latter discovered the great river now known by the name. This fact is recorded by Vancouver, to whose expedition, in 1792, we owe almost the whole of our present knowledge of the coasts. In 1839, the mouth of the Columbia was surveyed by Capt. Sir Edward Belcher. The American Exploring Expedition also examined the Columbia River in 1841, and part of the Straits of de Fuca, afterwards more elaborately completed by Capt. Kellett, in 1847.

The sovereignty of the territory in question has, until recent times, been the subject of continual dispute. In the subsequent description of Nootka Sound, which was at first believed to form a portion of the main land, and consequently included in the possession of this region, it will be seen that the question of this right was nearly involving England and Spain in a war, most extensive preparations having been made for it on both sides. But the convention of the Escurial, signed between these powers in 1790, stipulated that subjects of either nation should not be molested in their possession, so that it was open to both nations. In 1818 the Americans concluded a treaty with Spain, in which the latter ceded all her right North of lat. 42°, in North America, to the United States. In the same year it was stipulated between England and the United States that the country should be open to both nations for ten years. In 1827 this convention was indefinitely extended, with the proviso that it might be terminated by either party giving twelve months' notice to the other.

The Hudson's Bay Company, during this period, having established settlements, derived almost the sole benefit from its productions; but in 1846 it was determined that the right of sole possession should be divided. It was against the claims of England; and the treaty before alluded to stipulated that all South of 42° should belong exclusively to the United States, the navigation of the Columbia River to be open to both parties, and indemnification granted to settlers of either nation on either side of the boundary.*

The origin of the name "Oregon" is involved in some obscurity. It is, perhaps, first found in some travels in the interior of North America, in 1766-68, by Jonathan Carver, published in London in 1778. He does not state his authority for calling the river by the name of Origan, or Oregon; and it has been supposed by some that it was an invention of his own. It has also been stated that it is from the Spaniards, from the "oregano," or wild marjoram, (*origanum, Lat.*) said to grow on its banks.

The native names appear to be very incapable of being rendered into European orthography. Their pronunciation is so very imperfect, that it is almost impossible to arrive at any satisfactory conclusion as to the real names, and each voyager has represented the same word in very different forms, so that an absolute standard must not be expected. Sir George Simpson gives some amusing instances of the imperfection of their powers of speech.

The whole territory has been divided into three belts of country by Capt. Wilkes, and as that officer's means of information have been great as to the new United States' territory, we quote his words:—†

The three divisions are the *western*, between the Pacific Ocean and Cascade Mountains; the *middle* section, between the Cascade Range and the Blue Mountains; and the *eastern*, between the Blue and the Rocky Mountains.

The first, or *western* section, is much the smallest, but by far the most valuable and important, comprising as it does the greater proportion of the arable land, or that fit for cultivation, and enjoying a climate every way suited to the productions that are the objects of man's labour for his sustenance. It extends from De Fuca's Straits to the parallel of 42° North, a distance of 400 miles, with an average width of 120 miles. It is well covered with timber, from the slope of the mountains to the sea, with many fertile valleys and prairies that are well watered; though with the exception of the Columbia and Willamette, the rivers are not navigable. The inlets on the South leading from De Fuca's Straits offer many fine harbours which will be described hereafter. The principal value of this section is its adaptation for agricultural purposes. The coast range of mountains gives it a rough appearance, particularly that portion of it which forms the peninsula of Cape Flattery. Mount Olympus is the highest peak, and rises to the elevation of 8,197 feet. With the exception of the Columbia, all the rivers which flow through this section have their sources in the Cascade Mountains. Those to the South flow directly towards the sea; those of the middle run both North and South, and are tributary to the Columbia; while the more northern ones discharge themselves into Puget's Sound and Admiralty Inlet. They are all

* The long controversy is contained generally in the History, &c., of the Oregon Territory, by *John Dunn*, 1844; the History of Oregon, &c., by *Robert Greenhow*, London, 1844, a work disfigured by vituperation and invective against *all* previous authors, and therefore unworthy of its semi-official character; the Oregon Question, and other Remarks, by *Thomas Falconer*, London, 1845; *J. C. Fremont's* Expedition to Oregon and California, London, 1846; the Oregon Territory, by *Nicolay*; the Oregon Question Examined, by *Dr. Travers Twiss*, London, 1846, an excellent work; and others.

† See Western America, Philadelphia, 1849, chap. vii.

rapid and small streams, having great fall, and many of them offer mill sites of some extent.

The most fruitful portions of this section are the valleys through which the Willamette and Umpqua flow, particularly the latter.

The Columbia River and its valley is by far the most interesting and important part of Oregon, not only on account of the variety of soil, productions, and climate, but also from its being the great and only line of communication between the sea-coast and the interior. The river is 750 miles long; that portion in the western section is 120 miles in length, and from 3 to 5 miles in width; it is navigable as far as the Cascades, during its lowest stages, for vessels not drawing more than 12 feet water. The tides rise and fall above Vancouver 80 miles from its mouth, but they cause no change of current beyond Oak Point; during the freshets the Columbia rises at Vancouver 19 feet above the low-water mark.

The Cowlitz discharges itself from the North, having its source in the Cascade Mountains, and is only navigable for boats during the spring and fall months. The lands on its banks (which are high) are fertile, and here a settlement has been made by the Hudson's Bay Company. The Willamette enters from the South, 30 miles above the Cowlitz.

The valley of the Columbia, as high as the Cascades, is divided into high and low prairies; the latter are not suitable for cultivation, on account of being overflown by the annual freshets, but they are admirably adapted for grazing-lands. The soil of the upper or higher prairie is light and gravelly; it is well covered with pines, arbutus, oaks, ash, and maples; and the hills that border it are generally volcanic. In passing up the river several low islands are met with; the soil on them is of the same character as the lower prairie; they are bordered by a thick growth of trees, cottonwood, ash, &c.

Vancouver is situated on the North bank of the river, and was the head-quarters of the Hudson's Bay Company: here they have large dairies, and 6 miles above are their grist and saw-mills.

That portion of the western section of Oregon North of the Columbia which lies between it and Puget's Sound is watered by several streams, some of which flow into the Columbia on the South, others into the Pacific on the West, and others into Puget's Sound on the North. These all rise in the spurs of the Cascade Range, and drain this part of the country. The land between the Cowlitz and the Chicaylis or Chickeeles Rivers is an extensive prairie, known as the Cammas Plains.

The country from the seaboard to the Cowlitz is covered with a dense forest of spruce pine and hemlock. The soil is a brown or black vegetable earth, with a substratum of clay. The patches of alluvial land bordering the Chickeeles River are fertile, and of some extent, studded with white-oaks, and would yield good crops of wheat; they are excellent sites for farms, having an abundance of fine water, and but a short distance from water communication.

The country in the neighbourhood of Puget's Sound presents an inviting aspect, and with the exception of some bluffs, is undulating, and covered with trees of the species spoken of above. The soil of this forest-land is a thin brown stratum of sandy vegetable earth, the subsoil of clay and gravel; the latter having the appearance of being water-worn. These are succeeded by the tract of prairie lands in the vicinity of Nisqually, which are valuable as pasture lands for flocks of sheep and dairy cows. These prairies have a very extensive range in a S.E. direction, and connect with the valley of the Cowlitz on the South towards the Cascade Mountains, intersected by strips of forests. Within this district are numerous ponds or lakes, surrounded by rich meadow land, furnishing luxuriant crops of nourishing herbage. No part of Oregon is better adapted for dairy purposes than this; and wheat, rye, barley, oats, &c., come to perfection.

The peninsula of Cape Flattery, North of the Chickeeles, between Puget's Sound and the Pacific, is rough and mountainous, and covered with a dense forest. The principal trees are hemlock, spruce, and arbor vitæ. The high ridges which jut in all directions from Mount Olympus leave but little space for tillage, except along the western side of Hood's Canal. Little, however, is known of the interior of this portion.

The Indian inhabitants, who are scattered in numerous tribes throughout the territory, are, it is supposed, rapidly decreasing in numbers, from their dissipated lives and their rude treatment of diseases. They acted in many instances as the allies and hunters for the white men settled among them, but on more than one occasion have proved the treachery of the wild man's character. Thus, on the 29th November 1847, the Presbyterian mission in the Wallawalla valley was destroyed by the Cayuse tribe, and Dr. Whiteman and thirteen others were killed, and sixty-one taken prisoners, but were released by the praiseworthy efforts of Peter Skene Ogden, Esq., chief factor of the Hudson's Bay Company. The occurrence led to the severe battles in January and February following, in which the Indians were completely routed. Many other instances of their treachery and cruelty might be adduced, but this will suffice.

On the coast between Cape Flattery and Cape Look-out, and up the Columbia to the first rapids, the singular custom prevails among the Indians of compressing the skulls of the infants. Thus the heads of all acquire a remarkable deformity, but which does not appear to affect their intellectual capacity. The process and its results are described by Mr. Hinds, in the Voyage of the *Sulphur*, vol. i. pp. 306—311, and by Dr. Scouler, in the Zool. Jour., vol. iv. p. 304, *et seq.*

The total number within the territory is very difficult to be estimated, but it has been guessed at about 13,000.

Fish is very abundant in these regions, especially salmon; and in the subsequent pages some accounts are given of enormous quantities being found in the fresher water at the heads of the extensive and singular inlets which penetrate the coast to the northward. The salmon is of several varieties, and in the spawning season they ascend the Columbia and other rivers for 600 or 800 miles above the mouth. A singular fact occurs in this migration: one variety ascends the Cowlitz, another the Columbia, another the Willamette, &c., &c., that which is peculiar in one stream never being found in the other. When they are taken in the upper parts of the rivers they have their tails and fins nearly worn off with the effect of their long and difficult ascent of these rapid streams, and are almost unfit for food.

The climate of western Oregon is mild, having neither the extremes of heat during the summer, nor of cold during winter; this is probably owing to the prevalence of the S.W. winds, and the mists which they bring with them from the ocean. The winters are short, lasting from December to February, and may be termed open. Snow seldom falls, and when it does, lasts but a few days. Frosts are, however, early, occurring in the latter part of August, which is accounted for by the proximity of the snowy peaks of the Cascade Range, a mountain or easterly wind invariably causing a great fall in the temperature. These winds are not frequent; and during the summer of 1841 they were noted but a few times. The wet season lasts from November till March; but the rains are not heavy, though frequent. The climate during winter is not unlike that of England; and as to temperature, is equally mild with that of 10° lower latitude on our eastern coast. The fruit trees blossom early in April.

The weather at Nisqually, between the middle of May and the middle of July, may be inferred from the following statement of winds, as observed by the American Exploring Expedition :—

Dir.	No. of Days.	Dir.	No. of Days.
South-west	21	North-west	2
South	16	Calm	8
North	19	East	2

The mean temperature during the same period was 67°, maximum 98°, minimum 39°; the barometer 30.04 in.

From June to September, at Vancouver, the mean temperature was 66°, maximum 87°, minimum 51°; out of 106 days, 76 were fair, 19 cloudy, and 11 rainy.

The *Klamet* or *Too-too-tut-na River* has a very narrow entrance, with a low shingle beach on either side, by which it may be known. There are 2 fathoms water on the bar, and inside from 4 to 5 for a quarter of a mile; the flats then begin, and the overfalls extend for several miles; the tide rises 6 feet. Wood and water may be had in any quantity. The mouth is situated in lat. 42° 26′ N., and lon. 124° 11′ W..

CAPE BLANCO or ORFORD is the extremity of a low, projecting part of land, and forms a very conspicuous point, bearing the same appearance whether approached from North or South. It is covered with wood as low down as the surf will permit it to grow. The space between the woods and the wash of the sea seems composed of black craggy rocks, and it may be seen, from the mast-head, at the distance of 7 or 8 leagues.

Capt. Vancouver, whose descriptions are given above, was the first to examine this coast with any minuteness; and of the coast between Point St. George and Cape Orford, he says :—

"Northward (of Point St. George) the shores are composed of high, steep precipices and deep chasms, falling abruptly into the sea. The inland mountains were much elevated, and appeared, by the help of our glasses, to be tolerably well clothed with a variety of trees, the generality of which were of the pine tribe; yet amongst them were observed some spreading trees of considerable magnitude. Although some of these mountains appeared quite barren, they were destitute of snow; but on those at the back of Cape Mendoçino, which were further to the South, and apparently inferior in point of height, some small patches of snow were noticed. The shores were still bounded by innumerable rocky islets, and in the course of the afternoon we passed a cluster of them, with several sunken rocks in their vicinity, lying a league from the land, which, by falling a little back to the eastward, forms a shallow bay, into which we steered. As the breeze which had been so favourable to our pursuit since the preceding Sunday died away, and as a tide or current set us fast on shore, we were under the necessity of coming to an anchor in 39 fathoms of water, black sand and mud. A remarkable black rock, the nearest shore, bore N.E. by E. ¾ E. 3½ miles; a remarkably high black cliff, resembling the gable end of a house, N. 1° E.; the northernmost extremity of the main land, which is formed by a low land projecting from the high rocky coast a considerable way into the sea, and terminating in a wedge-like, low, perpendicular cliff, N.N.W. ½ W. This I distinguished by the name of Cape Orford, in honour of my much-respected friend, the noble Earl (George) of that title; off it lie several rocky islets."*

The name of Cape Blanco, by which this promontory is known, is derived from Martin de Aguilar, who commanded one of the vessels under Vizcaino, in 1602. Aguilar was separated from the squadron by a storm at Cape Mendoçino, and discovered a second point, January 19th, 1603, which he named Cape Blanco, beyond which the coast began to decline, and near it a safe and navigable inlet or large river. The Cape of Aguilar is not well fixed as to position, but has by some been taken for Cape Gregory of Capt. Cook, to the northward.†

North of Cape Blanco or Orford the coast takes a direction about N. by E. ½ E., and the rocky islets which abound so much to the southward of it cease to exist about a league to the northward of the cape, and in their stead an almost straight sandy beach presents itself, with land behind, gradually rising to a moderate height, near the coast. The interior is considerably elevated, much diversified, and generally well wooded, though with intervals of clear spots, which gives it some resemblance to a country in an advanced state of cultivation.

Cape Gregory is to the northward of Cape Blanco or Orford, and was so named by Capt. Cook, who, thwarted by winds and bad weather, had occasional glimpses of the coast; there is consequently some difficulty in assigning the position given by him. Vancouver places a cape more to the South, which he takes for Cape Gregory. This cape, though not so projecting a point as Cape Orford, is nevertheless a conspicuous one, particularly when seen from the North, being formed by a round hill on high perpendicular cliffs, some of which are white, a considerable height above the level of the sea. Above these cliffs it is tolerably well wooded, and is connected with the main land by land considerably lower.

Between Cape Orford and Cape Gregory are some small rivers: the *Sequalchin*, *Cotamyts*, *Coquils*, and *Cahoos*. Most of these, as well as those between it and the Columbia, can be entered with boats in fine weather, and fresh water easily obtained.—(Capt. Wilkes.)

* Vancouver, vol. i. p. 203.

† Torquemada, Monarquia Indiana, lib. v. chap. 45 & 55.

About a league North of the pitch of Cape Gregory, the rocky cliffs composing it terminate, and a compact, white, sandy beach commences, which extends along the coast 8 leagues, without forming any visible projecting point or headland.

UMPQUA RIVER.—To the northward of Cape Gregory is the mouth of the Umpqua River. The entrance is between two sand spits; the northern one is 1¼ miles long; on the southern spit is a small rock. There are 2 fathoms water on the bar which lies outside of the spits; between them 6 and 7, and inside 11 fathoms. The river, for the distance of 10 miles, admits vessels not drawing more than 12 feet.

In the Narrative of the American Exploring Expedition, however, Capt. Wilkes says that the Umpqua has only 9 feet water on the bar.

A party, detached from the American Exploring Expedition, travelled overland from the Columbia to San Francisco (September and October, 1841). They crossed the Umpqua at the fort, a few miles from the mouth of the river.

"*Fort Umpqua* was, like all those built in this country, enclosed by a tall line of pickets, with bastions at diagonal corners; it is about 200 feet square, and is situated more than 150 yards from the river, upon an extensive plain; it is garrisoned by five men, two women, and nine dogs, and contains a dwelling for the superintendent, as well as storehouses, and some smaller buildings for the officers' and servants' apartments."

The Umpqua country yields a considerable supply of furs, principally of beaver, most of which are of a small size.*

CAPE PERPETUA was so named by Capt. Cook because he discovered it on the day distinguished by that name in the calendar (March 7, 1776), as was the case with Cape St. Gregory, to the southward. It is a high, rocky bluff, nearly perpendicular to the sea, which breaks against it with immense violence. For some distance South of it a straight and compact coast, composed of steep, craggy, rocky cliffs, replaces the sandy beaches and low shores still farther South.

CAPE FOULWEATHER is a conspicuous promontory, singular in its appearance. A high, round, bluff point projects abruptly into the sea; a remarkable table-hill is situated to the North, and a lower round bluff to the South of it. Cape Foulweather is in lat. 44° 49′ N., lon. 124° 18′ W.

From Cape Foulweather the coast takes a direction a little to the eastward of North, is nearly a straight and compact shore, considerably elevated, and in general steep to the sea. The face of the country is much chequered, and in some places covered with a pleasing verdure, in others occupied by barren rocks and sand; but in none very thickly wooded.

CAPE LOOK-OUT forms only a projecting point, yet it is remarkable for the four rocks which lie off it, one of which is perforated, as described by Mr. Meares,† and is about 40 miles North of Cape Foulweather, which is most likely the same cape as that so named by Capt. Cook, on March 7, 1778, from the very bad weather he soon after met with.‡

"From Cape Look-out, which is in lat. 45° 32′, the coast takes a direction of

* Wilkes's Narrative of the Exploring Expedition, vol. v. pp. 225-6.

† Vancouver, vol. i. p. 200.

‡ Third Voyage, vol. ii. p. 258.

about N. ¾ W., and is pleasingly diversified with eminences and small hills near the sea-shore, in which are some shallow, sandy bays, with a few detached rocks lying about a mile from the land. The more inland country is considerably elevated: the mountains stretch towards the sea, and at a distance appear to form many inlets and projecting points; but the sandy beach that continued along the coast renders it a compact shore, now and then interrupted by perpendicular rocky cliffs, on which the surf violently breaks. This mountainous inland country extends about 10 leagues North from Cape Look-out, when it descends suddenly to a moderate height; and, had it been destitute of its timber, which seemed to be of considerable magnitude, and to compose an entire forest, it might be deemed low land. Noon brought us up with a very conspicuous point of land, composed of a cluster of hummocks, moderately high, and projecting into the sea from the low land before mentioned. These hummocks are barren, and steep near the sea, but their tops thinly covered with wood. On the South side of this promontory was the appearance of an inlet, or small river, the land behind not indicating it to be of any great extent; nor did it seem accessible for vessels of our burden, as the breakers extended from the above point 2 or 3 miles out into the ocean, until they joined those on the beach nearly 4 leagues further South. On reference to Mr. Meares's description of the coast South of this promontory, I was at first induced to believe it to be Cape Shoalwater; but on ascertaining its latitude, I presumed it to be that which he calls Cape Disappointment, and the opening to the South of it Deception Bay."*

Thus says Vancouver, honestly, what his impression was on passing this important point, which is no other than the mouth of the great Columbia River. Later navigators, with the knowledge of the existence of the real character of this Deception Bay, have expressed surprise that Vancouver should have been in any doubt about it. But on his return to the southward, he entered and explored the river nearly to its navigable extent, being the first who had done so.

COLUMBIA RIVER.

This river, by far the most considerable of any that enter the sea on this side of the Pacific, is the principal feature of the territory it waters. If its capabilities were at all commensurate with its magnitude, it would really become an important point in the commercial history of the Pacific. It has some insuperable obstacles to its ever becoming of any great service to the country it drains.

The entrance to the Columbia is impracticable for two-thirds of the year. It cannot be entered at night, and in the day only at particular times of the tide and direction of the wind. Unlike all known ports, it requires both the tide and wind to be contrary to ensure any degree of safety.

Vessels frequently lie for several weeks in Baker's Bay, inside the entrance, during the winter, for fine weather to get out, for which a fair wind and smooth water are indispensable. The difficulties of ingress must also be greater, inasmuch as a vessel in the open ocean cannot watch her opportunity so conveniently as

* Vancouver, vol. i. pp. 209-10.

when at anchor in Baker's Bay; and the hazard would be still greater were it not that the openness of the coasts, and the prevalent gales, lesson the hazard of a lee-shore.

"But these obstructions, in proportion as they lessen the value of the river, enhanced at the same time the merit of the man who first surmounted them—a merit which cannot be denied to the judgment, and perseverance, and courage, of Capt. Gray, of Boston. Whether or not Capt. Gray's achievement is entitled to rank as a discovery, the question is one which a bare sense of justice, without regard to political consequences, requires to be decided by facts alone. *First*, in 1775, Heceta, a Spaniard, discovered the opening between Cape Disappointment on the North, and Point Adams on the South,—a discovery the more worthy of notice, inasmuch as such an opening can hardly be observed excepting when approached from the westward; and being induced partly by the appearance of the land, and partly by native traditions as to a great river of the West, he filled the gap by a guess with his Rio de San Roque. Secondly, in 1788, Meares, an Englishman, sailing under Portuguese colours, approached the opening in question into 7 fathoms water, but pronounced the Rio de San Roque to be a fable, being neither able to enter it nor discover any symptoms of its existence. Thirdly, Gray, though after an effort of nine days he failed to effect an entrance, was yet convinced of the existence of a great river by the colour and current of the water. Fourthly, in April, 1792, Vancouver, while he fell short of Gray's conviction, then unknown to him, correctly decided that the river, if it existed, was a very intricate one, and not a safe navigable harbour for vessels of the burden of his ship. Fifthly, in May, 1792, Gray, returning expressly to complete his discovery of the previous year, entered the river, finding the channel very narrow, and not navigable more than 15 miles upwards, even for his *Columbia*, of 220 tons. According to this summary statement of incontrovertible facts, the inquiry resolves itself into three points—the discovery of the opening by Heceta, the discovery of the river by Gray on his first visit, and the discovery of a practicable entrance by the same individual revisiting the spot for the avowed purpose of confirming and maintaining his previous belief. Gray thus discovered *one* point in a country, which, as a whole, other nations had discovered, so that the pretensions of America had been already forestalled by Spain and England." *

The Columbia River possesses but few advantages as a port; the difficulties and dangers of its entrance, which have been manifest to all who have come hither since its first discovery, have not been exaggerated, and, until very efficient means be at command for towing ships over the bar into the quieter water within, it will be almost impossible for a ship to enter in safety. One feature which renders it still more difficult for a sailing vessel to cross the bar is, that a good and commanding breeze within or without the mouth often falls to a calm when the breakers are reached. The shifting of the shoals forming the bar and entrance, which they do apparently very considerably at short intervals, increases the difficulty, from the impossibility of having any established mark for the guidance of the commander that will be good for a lengthened period.

* Sir George Simpson, Journey Round the World, vol. i. pp. 260-1.— See page 351.

In addition to this, the cross-tides, and their great velocity, increase the difficulties. The heavy swell of the Pacific, and the influence of an under current, add their embarrassments, and all these become greater from the distance of the leading marks of the channel, and their indistinctness when the weather will permit entrance. It is necessary to use them, because the compass bearings are of little or no use. The land near it is well marked, and this is some little advantage. A *new channel entrance* to the southward has lately been discovered, which may be free from some of these objections, as hereafter noticed.

CAPE DISAPPOINTMENT, in lat. 46° 16′ N., lon. 124° 5′, is the North point of the entrance. It is formed of high, steep precipices, with several lofty spruce and pine trees on its summit. POINT ADAMS, the southern point, is 4¾ miles S.E. from it. It is low and sandy, but is covered with high trees. From each of these points a sand spit runs out; that from Point Adams for about 4 miles in a W. by S. direction, and that from Cape Disappointment 2½ miles to the S.S.W., being thus at nearly right angles to each other. Their outer points, about a mile apart, are connected by the outer bar, on which the least depth is generally 28 feet, but which, when there is any swell on, or in bad weather, forms with the spits on each side one continued line of formidable breakers. On the spits the depth varies from 1 to 3 fathoms, with some spots awash, but from their character are constantly varying. It would seem that at present the entrance channel is straightening its course, that is, the South end of the North spit is wearing away, and the channel shifting accordingly. It is this part which is the point of greatest danger, as it has to be closely approached, and here most of the wrecks have occurred.

Between Cape Disappointment and Point Adams is a *middle bank*, which leaves but a narrow channel between the points. It is extensive, and through a great part of its extent has but very little water over it; in the centre, a portion is dry, forming a sandy island. It is formed by the trees and timber which have floated down the river, and which become lodged here, forming a nucleus, around which the debris brought down by the stream increase. From the West side of this Middle Bank a spit runs out to the S.W. and joins the South spit off Point Adams. Over this bar or spit is the track to the South or Queen's Channel, as given in Capt. Belcher's chart, the depth on it being 3 fathoms.*

* This middle bank and the spit were quite omitted in the sketch chart given to Vancouver by Mr. Gray, who first entered the river, and on the spit in question the *Chatham* grounded. Commodore Wilkes says:—"There is little doubt that the spits are undergoing constant change, and are both increasing. This is corroborated by those who have had the most experience. In the memory of many, Cape Disappointment has been worn away some hundred feet by the sea and the strong currents that run by it. The middle sands, which lie within the two spits, and occupy a great extent of the bay, are subject to still greater changes. In the course of two months a large portion of what was dry sand was washed away. The sea usually breaks on the western edge of these sands. Two vessels have been wrecked here within the last year; the bark *Vancouver*, one of the Hudson's Bay Company's vessels, and the whale ship *Maine*, both totally lost.

"After passing the spits, the old channel leads to Baker's Bay, the usual anchorage for vessels awaiting an opportunity for departure. This bay is by no means well sheltered during the stormy months, being exposed to the S.E. The New or Clatsop Channel leads directly to Point Adams. Changes in both the channels are said to have taken place since the surveys made by the Exploring Expedition. The alteration in the former is caused by the accumulation of sand about the wreck of the *Peacock*; that in the latter by the greater deposit of sands from the failure of the common spring freshets. I am led to believe that neither is of the extent reported, and I feel satisfied that

On Point Adams some missionaries are established. Capt. Wilkes visited these pioneers in the wilderness. He says:—"We landed, and after walking a mile, came to the mission, when we had the pleasure of seeing Mr. and Mrs. Frost. Mr. Frost gave us a kind welcome at his new dwelling, which I understood him to say, had been built with his own hands. His wife appeared cheerful and happy, and made herself quite agreeable. The house is a frame one, of one story, and contains three rooms; it is situated in a young spruce and pine grove, which is thought to be the most healthy situation here." Two other houses were being built here.

In walking on the sand hills, and about Point Adams, Capt. Wilkes says that he had never in all his life seen so many snakes as were on the beach, where they were apparently feeding at low water.*

Clatsop village, near the mission, consists of a few rough Indian lodges, constructed of boards, or rather large hewn planks, the interior resembling a miserably constructed ship's cabin, with bunks, &c.; the only light is admitted from above, near the ridge and gable end. Around the whole is a palisade made of thick planks, fifteen feet long, with one end set in the ground.

"On the Clatsop beach we saw a great number of dead fish. Mr. Birnie, informed me that they were thrown up in great numbers during the autumn, and were supposed to be killed by a kind of worm generated in their stomachs."†

BAKER'S BAY is to the eastward of Cape Disappointment, and is so named after Vancouver's officer. It affords good anchorage under the promontory forming the cape.

The North shore of the river here forms a large bay, terminated on the East by *Chenoke* (or Chinook) *Point* and village, five miles and a half from Cape Disappointment, but its eastern point is not very advantageous for anchorage, on account of the confused sea, which reaches it at times over the bar.

"We found the tide exceedingly strong, and having some apprehensions that the boats might lose their way, I thought it better for us to make for the Chinook shore, and follow it until we reached the cape. It may seem strange that this precaution should be taken, but it is necessary at all times, even in clear weather, for the tide is frequently so strong, that it cannot be stemmed by oars; and too much caution cannot be observed in passing across the bay. As little frequented as it is, many accidents have occurred to boats and canoes, by their being swept by the tide into the breakers on the bar, where all hands have perished. The Indians are very cautious, and it is only at certain times of the tide that they will attempt to make the passage."‡

From Chenoke Point the North shore takes a more easterly and then N.E. direction, extending to the bay, which terminated Mr. Gray's first exploration,

---

the Clatsop Channel will be kept open by the action of the current of the river, and ought to improve in depth. Before any degree of reliance can be placed on these reports, an accurate survey of the river ought to be made, which will serve to show the changes that have occurred since that made by the Exploring Expedition; and a comparison of the two surveys will point out the causes that may be at work to effect it, and the probable remedies that may be used to prevent the change, or arrest its progress."

* Capt. Wilkes, Narrative of the United States' Exploring Expedition, vol. iv. p. 322.

† *Ibid* vol. iv. p. 323. ‡ *Ibid.* vol. v. pp. 113-14.

and hence termed, by Vancouver, Gray's Bay. Beyond this is another bay, though unapproachable for shoal water, and may be taken as the N.E. extreme of the estuary or broad mouth of this river. From the East point of Gray's Bay to this part, the shore is nearly straight and compact, and tends in a E. ½ N. direction.

Beyond this again, where the river contracts to a narrower stream, is a remarkable rock, called the *Pillar Rock*. It is 25 feet high, and only 10 feet square at the top. It is composed of conglomerate, or pudding-stone, and is fast crumbling to pieces. The ascent is very difficult. The Indians call it Taluaptea, after the name of a chief, who in bygone days lived at the falls of the Columbia, and who, having incurred the displeasure of their spirit, called Talapos, was turned into a rock, and placed where he would be washed by the waters of the great river. About this rock the water is very shallow, with overfalls from 2½ to 6 fathoms.

Returning to the entrance on the South side of the estuary, at about 4 miles East by South of the land of Point Adams, is Point George, forming the East limit of *Young's Bay*. The entrance of *Young's River*, so named by Vancouver after Sir George Young, is about 1½ miles E.S.E. of the point; from its banks a low meadow, interspersed with scattered trees and shrubs, extends to the more elevated land. This is of easy ascent, and is agreeably variegated with clumps and copses of pine, maple, alder, birch, poplar, and several other trees, besides a considerable number of shrubs, greatly diversifying the landscape by their several tints. The marshy edges of the river afford shelter to wild geese, which fly about in very large flocks; ducks, too, are in abundance, as are a variety of large brown crane. From a sort of bar across its entrance it is not easily navigable, though the depth is above 2½ fathoms.

ASTORIA, which lies to the eastward of Point George, on the southern shore of the river, has been much celebrated; but its fame depends mainly upon its historian, that delightful writer, Washington Irving, who has told all the world of its progress and fortunes. It has sunk from the scene of revelry and hospitality to a neglected collection of hovels. Though the site is still known by its original appellation, yet that was superseded for *Fort George* when it was taken possession of by the British.*

Astoria is very beautifully situated. It is 11 miles from Cape Disappointment in a direct line. From it there is a fine view of that high promontory and the ocean bounding it on the West; the Chinook Hills and Point Elliu, with its rugged peak, on the North; Tongue Point and Katalamet Range on the East; and a high background, bristling with lofty pines, to the South. The ground rises from the river gradually to the top of a ridge 500 feet in elevation. This was originally covered with a thick forest of pines; that part reclaimed by the first occupants is again growing up in brushwood.†

* Wilkes, vol. iv. p. 321.

† For full accounts of this settlement of Astoria see Adventures on the Columbia River, by Ross Cox, London, 1831; Relation d'un Voyage à la Côte Nord-Ouest de l'Amerique Septentrionale dans les Années 1810-14, par Gabriel Franchère, Montreal, 1820; and also Astoria, or Anecdotes of an Enterprise beyond the Rocky Mountains, by Washington Irving, first published at Philadelphia, 1836.

It has been stated elsewhere that the American ships, the *Columbia* and the *Lady Washington*, prosecuting the fur trade on this coast in 1787-88, first closely examined its details. During the first of those years the North-West Company of Montreal was established, to prevent collisions with the Hudson's Bay Company, and to reorganize the fur trade on a larger and more secure system. But this rivalry too soon gave rise to violent outbreaks; but each company confined themselves within different chartered limits.

In 1809 Mr. John Jacob Astor, a German, who had emigrated from his native country in 1783, engaged in the fur trade, and in that year obtained a charter from the New York legislature for organizing a Pacific Fur Company, *all the capital of which belonged to himself*. His plan was to establish posts on the coast of the Pacific, on the Columbia, &c., &c. For the execution of this project, two expeditions were sent out, one by sea and one by land. In September, 1810, the ship *Tonquin* left New York, and in March, 1811, founded the post of Astoria.* In the ensuing summer the necessary buildings were constructed. In the July following, a detachment from the North-West Company arrived here, after traversing the continent, for the purpose of taking possession of the mouth of the Columbia. The trade of the settlement was arrested by the destruction of the *Tonquin* and her whole crew by the Indians, near Nootka Sound, in the spring of 1812. In 1813 the news of the war between the United States and England reached Astoria, and M'Dougall and Ross Cox, the managers, immediately quitted the service of the American Company, and entered that of the North-West Company, the whole of the profits being transferred to the latter for the sum of 40,000 (or 58,000) dollars. While this transfer was in progress, a British ship-of-war hove in sight, and the post was surrendered, and the name of it was changed to Fort George. By the treaty of Ghent, of 1814, between Great Britain and the United States, the post was given up to the Americans, which was carried into effect in October, 1818, but its importance had then vanished. In conclusion, respecting the rival companies it may be stated that, after numerous outbreaks, a battle was fought between the rival traders on June 19th, 1816, when Lord Selkirk's Highland settlers on the Red River were routed, and their governor, Mr. Temple, and several others, were killed. In consequence of these fatalities, arrangements were made, in 1819-20, which led to the union of the two interests in that of the present Hudson's Bay Company.

Astoria has now very little to boast of. Since it was given up as a post, it has been little regarded, and half a dozen log houses, as many sheds, and a pigsty or two, and these going to decay, are all that are there. There is but one field under cultivation. This is a very different picture from the former accounts under the North-West Company, when it had its gardens, forts, and banqueting halls tenanted by as jovial a set as were ever met together. Since the Hudson's Bay Company have removed their operations to Vancouver, neglect and consequent ruin has followed. There is not more room than for about a dozen vessels at Astoria, so that it would be difficult to accommodate an extensive trade.

* One of the first establishments on the banks of the Columbia was by Capt. Smith, of Boston, in 1810. He built a house, &c., on the South bank of the river, but abandoned it before the close of the year.

The peninsula upon which Astoria stands forms a bay with Young's River, where the intrepid travellers, Lewis and Clark, wintered. The position of their huts (long since gone to decay) is still pointed out.

The primeval forest in the rear of Astoria is well worth seeing. The soil in which this timber grows is rich and fertile, but the obstacles to the agriculturist are almost insuperable. One of the largest trees was 39 feet 6 inches in circumference 8 feet above the ground, and had a bark 11 inches thick. The height could not be ascertained, but it was thought to be upwards of 250 feet, and the tree was perfectly straight.*

These trees, though of gigantic size, are of very little use as timber. It may be supposed that they are produced from the cones brought down from the upper parts of the course of the rivers, and being drifted on to the shores where the alluvial soil is so very much richer than that of the parent ground, that the trees are thus forced into a most unnaturally vigorous growth, their texture deteriorating accordingly.

Eastward of Astoria is a remarkable projection from the southern shore, called *Tongue Point*, forming a peninsula; and to the North of this point is a channel, which has been designated the *Tongue Point Channel*. This leads over to the northern shore, and then turns to the eastward on the northern side.

Tongue Point is considered to be the best position for a fortification to defend the channel up the river. It is a high bluff of trap rock, covered with trees of large dimensions; the top had been cleared and taken possession of by Mr. Birnie, prior to 1841; he erected a hut and planted a patch of potatoes. The hut was inhabited for a year by a Sandwich Islander and his wife. It is rather a rough spot for cultivation, but the end of occupancy was answered by it. There is a small portage on Tongue Point, which canoes often use in bad weather, to avoid accidents that might occur in the rough seas that make in the channel that passes round it.

At the time of Vancouver's visit (December, 1792), a range of five small, low, sandy islets, partly covered with wood, extended about 5 miles to the eastward, the largest of which was the easternmost, and forms the S.E. side of the actual entrance to the river. The land on this side is low and marshy.

The space, whose shores are thus imperfectly described, is from 3 to 7 miles wide, and very intricate to navigate, on account of the shoals which nearly occupy its whole extent. These shift very materially, and that, too, in very short periods of time, so that no established directions can here be given.

In here citing the necessary instructions to vessels about to enter, we should incur some amount of responsibility were they not accompanied by a caution of their necessary inevitable uncertainty. The shipmaster, therefore, who would enter unassisted by a proper pilot, must be quite prepared to exhaust all a seaman's energies in the formidable undertaking. And he will the more readily comprehend the necessity of this, when it is said that nearly every vessel whose visit is recorded gives an account of her grounding on some part of the passage;

* Capt. Wilkes, vol. v. p. 116.

and the *Peacock*, one of the United States' Exploring Expedition, was totally lost on the terrible bar.

The first that we will give are those of Mr. Broughton, who surveyed the river under Vancouver's orders, in December, 1792, and, as his exploration (of the upper part of its course) was the first ever made, it will be perhaps of some service now to compare the changes which have taken place since that period. While giving these remarks, however, it must be understood that they are not extracted as being proper *directions* for the present day, but rather as matter of curiosity.

"The discovery of this river, we were given to understand, is claimed by the Spaniards, who call it Entrada de Ceta (Entrada de Heceta), after the commander of the vessel, who is said to be its first discoverer, but who never entered it: he places it in 46° North latitude. It is the same opening that Mr. Gray stated to us in the spring (1792) he had been nine days off the former year, but could not get in, in consequence of the outsetting current; that in the course of the late summer, he had, however, entered the river, or rather the sound, and named it after the ship he then commanded. The extent Mr. Gray became acquainted with on that occasion, is no further than what I have called Gray's Bay, not more than 15 miles from Cape Disappointment, though according to Mr. Gray's sketch it measures 36 miles. By his calculation its entrance lies in lat. 46° 10′, lon. 237° 18′, differing materially in these respects from our observations.

"The entrance, as already stated, lies between the breakers extending from Cape Disappointment on the North side, and those on the South side from Point Adams, over a sort of bar, or more properly speaking, over an extensive flat, on which was found no less depth of water than 4½ fathoms. The best leading mark is to bring the Tongue Point, which looks like an island near the southern shore, to bear by compass about E. by N., and then steer for it; this was observed in the passages of the *Chatham* in and out, though on the latter occasion circumstances were too unpleasant to allow of great precision.

"From the information and experience derived by this visit, it appears to be highly advisable, that no vessel should attempt entering this port but when the water is perfectly smooth; a passage may then be effected with safety, but ought even then to be undertaken with caution: bordering on the breakers off Point Adams, and keeping the Tongue Point well open with Chenoke or Village Point, will avoid the spit bank, and give a clear channel up to Chenoke; but, in case of failure in the wind or tide, it will then be advisable to anchor in Baker's Bay, bringing its entrance to bear North, and keeping close round the cape breakers, where the depth of water is from 11 to 9 and 6 fathoms, close to the cape shore. Within the cape are three rocky islets in the bay, the middle one being the largest; just on with the cape is the line of direction going in or out; leading along the southern side of the spit bank in deep water, and near this islet, bringing the cape to bear between S. and S.E., is good anchorage in 5 fathoms water. The greatest rise and fall of the tide in this bay observed by Mr. Baker were 12 feet; high water at full and change at half-past one o'clock. Mr. Manby's observations on board the *Chatham* confirmed those of Mr. Baker, as to the time of high water; but the rise and fall of the tide with him did not exceed 6 feet, and the greatest strength of the tide was about 4 knots.

"This bay, besides affording good and secure anchorage, is convenient for procuring wood and water; and by keeping upon good terms with the natives, who seemed much inclined to be friendly, a supply of fish and other refreshments may easily be obtained. The heavy and confused swell that in bad weather constantly rolls in from the sea over its shallow entrance, and breaks in 3 fathoms water, renders the space between Baker's Bay and Chenoke Point a very indifferent roadstead. Cape Disappointment is formed by high, steep precipices, covered with coarse grass, the sides and tops of the hills with pine trees. Point Adams, being the S.E. point of entrance, is low and sandy, from whence the country rises with a gradual ascent, and produces pine and other trees."*

These notes may be well followed by those of *Commodore Wilkes*, as given in his sketch of Western America, 1849, pp. 74-5.

It is safest to enter on the ebb tide, with the usual N.W. wind, which sets in about 10 or 11 o'clock, A. M., during the summer months. The entrance should never be attempted on a flood tide and N.W. wind, unless the Clatsop Channel is followed, and the sea is smooth.

After making Cape Disappointment, which is easily distinguishable by the dark hummocks and tall pines, trimmed up, with the exception of their tops, you may lead in for it on a N.E. bearing, if to the southward; if to the northward, you may run in until you have that bearing on. A hummock or saddle-hill to the northward, on with the outer part of the cape-land, will give you notice that you are on the bar, in 4½ or 5 fathoms water: in ordinary weather, the outer line of the North spit is readily perceived by the rollers breaking; the inner line is always perceptible. When Young's Point is open with dead trees on Point Adams, you will be to the northward of the end of the North spit, and may run down along it until those two points are on range; then haul in for Point Ellice, or the green patch on Chinook (or Chenoke) Hill, if intending to take the channel by the cape. When *Leading-in Cliff* is well open with the inner point of the cape, haul up for the latter, and steer in: you will then have doubled close round the North breaker, in 7 fathoms water; and it is better to keep the North spit aboard if the wind is not so scant as to oblige you to beat up for Cape Disappointment: on opening Green Point you must go about; it is not safe to go nearer the middle sand. On ordinary occasions there will be scarcely ever a necessity to tack; the ebb-tide on your lee bow will keep you sufficiently to windward.

The cape will be required to be passed close aboard, in order to avoid the *sand-spit* making off from the *middle sands* towards the cape: the two outer bluffs of the cape, in range, will strike it. After you have passed this range, you may steer into Baker's Bay, and, having passed an opening in the wood on the cape, you may anchor in from 7 to 10 fathoms. In passing the cape, care must be taken not to be becalmed by it; if this should happen, the only resource is to down anchor at once, and wait a favourable tide. The current will be found

* Vancouver, vol. ii. pp. 74—76.

very strong. It sometimes runs from 5 to 6 knots an hour—a perfect mill-race —and no boat can make way against it when at its strength.

If desirous to proceed to Astoria, and one of the native pilots is not to be had, the only precaution necessary in proceeding up is to keep the small islet in the *cove* of the cape open until you have the *dead trees* nearly S.S.E. (compass), and then steer over for them, as it will be probably young flood. It is necessary to keep the starboard or sand island side of the channel; and if near high water, this island, in running up, must be kept open on the starboard bow; otherwise, the approach to it would be too near for safety. On reaching the Clatsop Channel, steer up for Young's Point, keeping in 5 or 6 fathoms water. The sand shoals on either side are very bold. When abreast of Astoria, moor with an ebb and flood anchor, with open hawse, to the northward and westward.

If the intention be to take the Clatsop Channel, the same directions are to be observed in passing the North spit. When the Leading-in Cliff is open, instead of hauling up for the cape, steer direct for the Clatsop village on Point Adams, which will take you into fair channel way; the breakers on each side will be visible: keep in the middle, and steer up for Young's Point, following the directions as before given.

In coming out, the state of the bar may be distinctly seen from the top of the cape, but due allowance must be made for the distance. The surf beating on the cape is a good guide; if there is much of it the swell will be very heavy and sharp between the North and South spits, if it does not actually break: the best time is with a N.W. wind, and about half ebb; you will then have tide enough to carry you to sea.

It is always dangerous to drop anchor in the channel between the cape and the end of the North spit; if it is done, it should only be in case of absolute necessity, and not a moment is to be lost when possible to proceed out or in. If the ship gets off with only the loss of an anchor, she may consider herself fortunate. The sea-breeze or N.W. and westerly winds blow at times very fresh; a sure indication of them is a thick hazy bank in the west, to seaward.

In entering the river, the following cautions must be attended to:—1. The entrance should never be attempted when the passage between the North and South spits is not well defined by breakers; it is equally dangerous, whether it be concealed by the sea's breaking all the way across, or so smooth as not to show any break.

2. The wind generally fails, or falls light, in the passage between the North and South spits, if it blows but a moderate breeze, and leaves a vessel at the mercy of a strong tide and heavy swell.

3. The best time to enter and depart is after half ebb and before quarter flood; the tide then runs direct through the channels, and is confined to them. With the prevailing westerly winds, for those intending to take the North channel, the best time to enter is after half ebb, though the wind may be scant; yet the ebb tide acting on the lee bow, will enable the vessel to keep to windward, and avoid the spit on the middle sands.

A SOUTH PASSAGE through the bar has been discovered in the early part of the present year (1850). Capt. White had sounded it through, and found *four fathoms* at half-tide at the *shoalest* part, which is only about two ships' length. Through the rest the water is very deep. The certainty of this passage into the Columbia is of very great consequence, as it is much shorter than the channel by Cape Disappointment. It is nearly straight, and avoids all the delays incident to getting into Baker's Bay. Vessels can moreover enter with winds which do not permit them to come in by the old channel. It was added, that it is Capt. White's intention, as soon as a buoy could be put down, to bring vessels in by this route.—*Oregon Spectator, March 7th*, 1850.

The Columbia River, above the broader portion of its course, just described, continues towards the N.E., the two points of its entrance being low and marshy; but, after advancing about two leagues, the land becomes high and rocky on both sides, the river turning from a N.E. to a S.E. direction, at Katalamet Point.

At about 4 miles above this point is the western point of *Puget Island*, well wooded, and about a league and a half in length, dividing the stream into two branches, and affording a good passage on either side of it, that on the North side being the deepest, having 10 to 12 fathoms water. It continues in the same direction about 5 miles further, when it takes a turn to the N.N.E., for a league to Oak Point on the South shore. At Oak Point the river turns nearly at right angles, taking its course along a barrier of trap rocks, which here meets on the West side, and which rises 800 feet perpendicularly above its surface. On the opposite side of the river is one of the remarkable prairies of the country, covered with tall waving grass, and studded with many oaks, from which the point takes its name. What adds additional interest and beauty to the scene is Mount St. Helen's, about 40 miles off in the eastern quarter, which may be seen from the sea when 80 miles distant. Its height was ascertained by Capt. Wilkes to be 9,550 feet.

From Oak Point to Walker's Island the distance is about 6 miles. It is small and wooded; and was named by Vancouver after his surgeon. Seven miles and a half above Walker's Island is the mouth of the Cowlitz River, entering the Columbia on the North shore.

The Cowlitz River, which here enters the Columbia as a broad stream, 35 miles below Vancouver, rises in the Cascade Range near Mount Rainier. It is very tortuous, and flows between very high banks until near the Columbia. At high water, during the spring and fall of the year, the river may be used for boating, and only at these times, for during the other seasons, at a few miles from its mouth, there is not water enough to float a boat, and it is, moreover, filled with rapids. It is not navigable for barges for more than three months in the year. When navigable, supplies are sent by it from Vancouver, and the grain, &c., returned in large flat barges.

"On this river it was reported that *coal* of a good quality existed, but I examined all the places that indicated it, and found only lignite. This exists in several places, but the largest quantity lies above the East Fork; several specimens of it were obtained." *

Mount Coffin, a high conical hill, is near the mouth of the Cowlitz, 710 feet

* Wilkes's Narrative, &c., vol. iv. p. 318.

great economy. Everything may be had within the fort; they have an extensive apothecary's shop, a bakery, a blacksmith's and cooper's shops, trade offices for buying, others for selling, others again for keeping accounts and transacting business; shops for retail, where English manufactured articles may be purchased at as low a price, if not cheaper, than in the United States, consisting of cotton and woollen goods, ready-made clothing, ship-chandlery, earthen and iron ware, and fancy articles; in short, everything of every kind and description, including all sorts of groceries, at an advance of 80 per cent. on the London prime cost. This is the established price at Vancouver; but at the other ports it is 100 per cent., to cover the extra expenses of transportation. All these articles are of good quality, and suitable for the servants, settlers, and visitors. Of the quantity on hand, some idea may be formed from the fact that all the posts West of the Rocky Mountains get their annual supplies from this depôt.

In concluding the description of this large river, which for our purpose requires less consideration on account of its comparative unimportance in a nautical view, it must be remembered that, since the cession of the territory, and the dominion of the United States' government necessarily carrying different interests to the previously established order of affairs, the Hudson's Bay Company's post will probably be gradually relinquished as such, and the seat of its trade transferred to their new territory of Vancouver Island.

SHOALWATER BAY lies on the North side of Cape Disappointment. It is a deep indentation on the coast, but, from its exposure to the N.W., can be of little use for the protection of vessels. It is surrounded by a low sand beach.

Forty miles to the North of the Columbia is Gray's Harbour, at the mouth of the Chickeeles or Chicaylis River.

GRAY'S HARBOUR was surveyed by Mr. Whidbey, in H.M.S. *Dædalus*, under Vancouver. His survey is sufficient for all nautical purposes. It was also visited by a party from the American Exploring Expedition. The North point of the entrance was named *Brown Point*, and the southern *Point Hanson*. They lie S. by E. and N. by W. (*true*), 2¼ miles apart. These points are composed of low sand-hills, and from them project two spits of sand for upwards of 3 miles beyond the line of the two points; and from the termination of these reefs the bar stretches across from point to point, on which, at high neap tides, there is only 20 feet water. On the northern sand spit is *Eld's Island*, with several hillocks, which appear as one from seaward, &c. The channel is narrow though deep, and is 2 miles in length. This harbour is easy to enter, the wind being usually fair, and it is only necessary to keep clear of the breakers on either side; but the same cause makes it very difficult to depart from.

Four miles and a quarter E.N.E. from Point Brown lies *Point New*; and between them the northern shore forms a deep bay, nearly 1¼ leagues deep, and occupied by shoals and overfalls, and these extend along the North shore to the mouth of the river. The South side is also lined with shoals, some of which dry at low water, and on which are lodged large quantities of dead trees and logs of drift timber. To the southward of Point New are two shoals: the easternmost, which is the largest, nearly connects the two shoal banks, having only a narrow passage to the North, and another to the South of it. The mark given by

Vancouver for entering is, two small red cliffy islets lying to the N.W. of Point New; the outermost of these, having the resemblance of a flower-pot, in a line with Point Brown led over the centre of the bar.

The land about the bay is low, with the exception of *Brackenridge Bluff*, on the North, and *Stearn's* on the South. These are both covered with pines; the latter bearing S. ½ E., and on with the South side of Eld's Island, will lead into and through the channel.—(Capt. Wilkes.)

The *River Chickeeles*, or *Chicaylis* according to Sir George Simpson, before entering into the harbour, increases in width to several hundred feet, and is navigable for vessels drawing 12 feet water 8 miles above its mouth. The harbour is only suitable for vessels of from 100 to 200 tons; and there are places where such vessels may find security between the mud shoals some distance within the capes.

The tides here are irregular, and influenced by the winds and weather: the time of high water at full and change was found to be 11^h 30′. Mr. Whidbey says, that it is high water about 50′ *after* the moon passes the meridian, and the rise about 10 feet. Fogs prevail very frequently during the summer season.

This port appears to be of little importance in its present state, as it affords but two or three situations where the boats could approach sufficiently near to effect a landing; the most commodious place was at Point Brown, another near Point Hanson, and one in a cove or creek to the S.E. of that point. The shallowness of the water on the bar also renders it by no means a desirable port. To pass this is impracticable, unless near high water, even with vessels of a very moderate size, and then it should be attempted with the utmost caution, since Mr. Whidbey had great reason to believe that it is a shifting bar, there being a very apparent difference in the channel on their arrival and at their departure, when it seemed to have become much wider, but less deep. A dry sand-bank, which lay near their anchorage the first evening, on the North side of the channel, was now entirely washed away by the violence of the sea, which had incessantly broken upon the shoals and bar.

Wood and water are at too great a distance to be easily procured, particularly the latter, which is found in small springs only, running through the sand near Point Hanson, at the distance of a mile from the landing place, over a heavy sand. The surrounding shores are low, and apparently swampy, with salt marshes: the soil is a thin mixture of red and white sand, over a bed of stones and pebbles. At a small distance from the water-side, the country is covered with wood, principally pines of an inferior stunted growth.

The vessels procured a most abundant supply of excellent fish and wild fowl; the productions of Gray's Harbour being similar to those found in and about Columbia River. Salmon, sturgeon, and other fish, were plentifully obtained from the natives; and geese, ducks, and other wild fowl, shot by themselves in such numbers as sometimes to serve the whole of their crews. The best sporting ground in Gray's Harbour was found to be on its South side. The number of Indians found here was about 100; they did not differ from those in other parts, and were then very friendly.*

* See Vancouver, vol. i. p. 410; vol. ii. pp. 70—84; Wilkes's Narrative of the American Exploring Expedition, vol. v. pp. 131-2; and Wilkes's Western America, p. 77.

POINT GRENVILLE, lat. 47° 22′, lon. 124° 14′, was so named by Vancouver, and projects from a straight and compact shore. Lying off the point are three small rocky islets, one of which, like that at Point Look-out, is perforated.

From hence to the North the coast increases regularly in height, and the inland country, behind low land bordering on the sea-shore, acquires a considerable degree of elevation. The shores are composed of low cliffs, rising perpendicularly from a beach of sand or small stones, with many detached rocks of various romantic forms lying at the distance of about a mile, with regular soundings, between 16 and 19 fathoms, soft sandy bottom.*

The coast affords no harbour or shelter except for very small vessels, and may therefore be accounted particularly dangerous, both on account of its outlying rocks, and also from the fact that the current sets in upon the coast, which was ascertained by Capt. Wilkes, whose vessels were unexpectedly set far inwards of their reckoning.

In addition to this several accidents have occurred near Point Grenville, both to English and Russian vessels; and a boat's crew belonging to the latter was inhumanly massacred by the Indians. It was also near this spot that the very remarkable occurrence of the wreck of a Japanese junk happened, in the year 1833. The officers of the Hudson's Bay Company were apprised of this by the receipt of a drawing, on a piece of China paper, of three shipwrecked persons, &c., expressive of the fact. They were rescued, and sent to England, and finally to China.†

As a precaution, the soundings will indicate the proximity of the coast; and safety may be ensured by not approaching the coast in less than 70 fathoms.

DESTRUCTION ISLAND is in lat. 47° 37′, and is much the largest detached land on this part of the coast from the southward. It was so named by Mr. Barclay. It is about a league in circuit, low, and nearly flat on the top, presenting a very barren aspect, producing only one or two dwarf trees at each end. Some breakers extend from its North point.

* Vancouver, vol. i. p. 212.

† *Wreck of the Japanese junk.*—The following account of this singular fact is given by Sir Edward Belcher, in his Voyage of the *Sulphur*. It is also mentioned in Washington Irving's Astoria.

"We received from the officers of the Hudson's Bay Establishment several articles of Japanese china, which had been washed ashore from a Japanese junk, wrecked near Cape Flattery. Mr. Birnie knew little of the details of the event; but, in the appendix to Washington Irving's Rocky Mountains, vol. i. p. 240, is the following account of it, in a letter from Capt. Wyeth:—'In the winter of 1833, a Japanese junk was wrecked on the N.W. coast, in the neighbourhood of Queen Charlotte's Island, and all but two of her crew, then much reduced by starvation and disease, during a long drift across the Pacific, were killed by the natives. The two fell into the hands of the Hudson's Bay Company, and were sent to England. I saw them on my arrival at Vancouver, in 1834.' Mr. Birnie states that it was at Cape Flattery, and not as above; and on this point his local knowledge makes him the best judge. 'There were,' he says, 'two men and a boy purchased from the natives. As soon as it was known that some shipwrecked people were slaved among the natives, the Hudson's Bay Company sent their vessel *Lana*, Capt. M'Neil, to obtain them by barter; and there was some trouble in redeeming the boy. They were subsequently sent to England, and then home, but their countrymen refused to receive them.' Further my informant could not acquaint me."

By the winter of 1833, Capt. Wyeth means, probably, the commencement of that year, as will presently appear more likely. There had been many people on board the junk, but distress had greatly thinned them; and several dead bodies had been headed up in casks. About the same time another Japanese junk was wrecked on the island of Oaher, Sandwich Islands. The particulars of this occurrence will be found noted in the account of those islands hereafter.

A current is found setting along this coast to the northward. Vancouver was so drifted at the rate of 10 or 12 miles every day, after leaving Cape Orford; he found it to be at times 1½ knots an hour.

Advancing along the coast, it increases in height, with numberless detached rocky islets, amongst which are many sunken rocks, extending in some places a league from the shore.

The *Flattery Rocks* are 10 miles South of the cape of the same name, and stand 4 or 5 miles from the land; they are from 50 to 100 feet high, black, and pillar shaped.

CAPE FLATTERY is placed on the recent charts in lat. 48° 9′ N., lon. 124° 46′ W. Capt. Cook, who so named a point hereabouts, makes it in 48° 15′ N., but he may have been drifted beyond his reckoning by the current above alluded to.—"At this time (7 P.M., Sunday, March 22, 1778) we were in 48 fathoms water, and about 4 leagues from the land, which extended from North to South, E. ½ E., and a small round hill, which had the appearance of being an island, bore N. ¾ E., distant 6 or 7 leagues, as I guessed; it appears to be of a tolerable height, and was just to be seen from the deck. Between this island, or rock, and the northern extreme of the land, there appeared to be a small opening, which flattered us with the hopes of finding an harbour. These hopes lessened as we drew nearer, and at last we had some reason to think that the opening was closed by low land. On this account, I called the point of land to the North of it, *Cape Flattery*. It lies in lat. 48° 15′ N., lon. 235° 3′ E. There is a round hill of a moderate height over it, and all the land upon this part of the coast is of a moderate and pretty equal height, well covered with wood, and had a very fertile and pleasant appearance. It is in this very latitude where we now were, that geographers have placed the pretended Strait of Juan de Fuca. But we saw nothing like it, nor is there the least probability that ever any such thing existed."*

The great commander was obliged to stand off the coast for the night, but a hard gale obliged him to get an offing, and he did not make the land again until he had reached the parallel of Nootka Sound. This singular fatality prevented him from having the credit of being the discoverer of that vast extent of inland navigation, soon after explored, but then believed to be apocryphal.

The name of Cape Flattery has been considered as appertaining to the S.W. cape of the entrance of De Fuca Strait, but he would have hardly thus passed the opening. This latter cape is now called Cape Classet.

DE FUCA STRAIT, separating Vancouver Island from the continent on the South, runs eastward from Cape Classet. Of its discovery we shall speak in the ensuing chapter. It is 95 miles in length, and has an average width of 11 miles; at the entrance, 8 miles in width, there are not any dangers, and it may be very safely navigated throughout. Altogether it forms one of the finest channels in the world, leading to a vast collection of fine harbours and safe anchorages, covering an area of 2,000 square miles, the whole unsurpassed by any tract of similar extent in the whole world. The country possesses the advantages neces-

* Cook's Third Voyage, vol. ii. pp. 262-3.

sary for the support of a great commercial people, and may, at some future period, act a prominent part in the world's history. At present, it is of comparatively small importance, and its new relations have not had time to solidify.

As before stated, we reserve further particulars respecting its discovery and exploration, till it is considered in connexion with Vancouver Island.

CAPE CLASSET is the S.W. point of the entrance of the Strait of Juan de Fuca. It has been supposed to be the same as that named by Capt. Cook Cape Flattery, a title now restricted to the cape to the southward; but, from the name of one of the chiefs in the neighbourhood, it has been called Classet by Vancouver, though he afterwards inclined to Cook's original name. In the new survey it appears under the former. It is moderately high, and on an E. by S. bearing, distant 12 miles, appears like an island high in the centre, with a gradual rounding slope from N.W. to S.E., and is thickly wooded with pine and cedar trees of gigantic size. At a little distance S.W. from the foot of the cape, and just within the confines of the beach, is a rock in the shape of a pillar, about 400(?) feet high, and 60 in circumference; the upper part of it has a slight bend or inclination to seaward.* These columnar rocks are very numerous just hereabout; and De Fuca, the discoverer, remarked one in particular, which may be that here adverted to. Capt. Wilkes has given a sketch of it in his work.

Tatouche Islands, a group of several detached masses, half a mile in extent, lie half a mile N.W. of the cape. They are flat on the top, having a verdant and fertile appearance, the sides almost perpendicular. Near them are some white barren rocks, some sunken ones, and some rocky islets of curious and romantic shapes. The rock in question is difficult to make out among the thousands of every variety of form about it. Tatouche Islands leave a passage between the capes, but there are some sunken rocks in it, which must render it very dangerous. "On the East side is a cove, which (nearly) divides the island into two parts; the upper part of the cliff, in the centre of the cove, had the appearance of having been separated by art, for the protection or convenience of the village there situated, and has a communication from cliff to cliff, above the houses of the village, by a bridge or causeway, over which the inhabitants were seen passing and repassing."†

The DUNCAN ROCK is in lat. 48° 24′ 12″ N., lon. 124° 45′ 45″, according to Capt. Kellett's survey in 1847. It lies three-quarters of a mile N.N.W. of the Tatouche Islands, and 1½ miles N.W. of Cape Classet. It is a dark, rugged rock, from 8 to 10 feet above water, at low spring tides, and generally at three quarters flood it is wholly covered. The surf breaks over it with great violence, and a quarter of a mile *North, true,* of it, is a dangerous shoal of 3 fathoms, with deep water all round, over which the water whirls and breaks very strongly; the tide trends here to the southward of West, and runs at the rate of 3 to 5 miles an hour, influenced by the wind and weather. The rock was so named by Vancouver after Mr. Duncan, who made an excellent sketch of the entrance.

The *Village of Classet*, which appeared to Vancouver to be extensive and populous, is about 2 miles East of the cape, but the anchorage off it being much

* Remarks by Mr. R. J. Gibbon, R.N.

† Vancouver, vol. i. pp. 217, 417.

exposed, he did not stay; and at 3½ miles, according to the Admiralty survey, nearly *East* (*true*) from Cape Classet, is *Koikla Point*, the West point of Neeah Bay.

NEEAH BAY, or Scarborough Harbour of Wilkes, is but a small indentation in the coast, which is partly sheltered on the N.E.by *Neeah* or *Wyadda Island*. This is the position where the Spaniards attempted to establish themselves in 1792, which they called Port Nunez Gaona, and the remains of the old fort can still be seen, and some bricks were found that were supposed to have belonged to it. Water is to be obtained here in some quantity, and a small vessel would have no difficulty in being supplied. It offers a tolerably safe and convenient anchorage, though exposed to N.W. gales:* yet, by anchoring well in, which a small vessel may do, protection even from these gales may be had. The shores of the island are lined with kelp, particularly to the westward of it, one-fourth of a mile off the island, and within the kelp is a detached rock.

According to Capt. Kellett, R.N., a point near the South end of Wyadda or Neeah Island is in lat. 48° 22′ 30″ N., lon. 124° 36′ 45″ W.; but, according to Commodore Wilkes, the North point of the island is in lat. 48° 24′ 40″, nearly 2 miles further North. Variation, 21° 8′ 14″ E.

When the American Expedition was there, they had a great many visitors from the Classet tribe.—"This tribe of Indians is one of the most numerous on the coast that I had an opportunity of seeing, and seems the most intelligent. These Indians wore small pieces of an iridescent mussel-shell attached to the cartilage of their nose, which was, in some, of the size of a ten cent piece, and triangular in shape. It is generally kept in motion by their breathing. They had seldom any clothing, excepting a blanket; but a few, who have contrived to make friends with the visitors, have obtained some old clothes; while others seem to be in the pay of the Hudson's Bay Company. The principal articles of trade are, tobacco, powder (paulalee), and leaden balls. These are preferred to most other merchandise, although more can be obtained for spirits than for any other article. This shows very conclusively, to my mind, the sort of trade that was carried on when the Boston ships entered into rivalry with the North-West Company for the purchase of furs." †

These Indians take some whales in the season, August and September: this they do by a seal-skin buoy attached to the harpoon, each buoy being differently painted; the booty is divided among the separate owners of the attached buoys.

On the East point of the bay, called *Mee-na Point* on the plan, is a village of this name. Between the point and the island there is a passage, though it must be used with caution.

We have no particular description of the South shore to the eastward. It follows nearly a straight line to the West by compass. It is composed of perpendicular sandy cliffs, that run back into high and rugged peaks, and is covered with a forest of various species of pines, that rises almost to the highest points of the range of mountains. The highest points themselves are covered with snow;

* Capt. Wilkes says that some of his sailors were fired at with ball by the Indians, and cautions visitors accordingly.—Vol. iv. p. 488.
† Narrative, vol. iv. p. 487.

and among them Mount Olympus is conspicuous, rising to an altitude of upwards of 8,000 feet.

At 1¾ miles westward of Mee-na Point, passing *Scarborough Point*, is *Klaholoh Rock*, 150 feet high, within which is the *Okho River*, beyond which there is nothing marked on the coast till *Kydaka Point* is reached, 19 miles farther; *Callan Bay*, between *Sekou Point* and *Slip Point*, 2 miles apart, is 3 miles farther West; 7½ miles beyond is *Pillar Point*, to the southward of which is *Canel River* and *Ketsoth Village*. For 15 miles the shore is more or less fronted with kelp, and at that distance is *Crescent Bay*, off the West point of which, some distance from the shore, is a rock which breaks at low water. *Freshwater Bay* is 3 miles farther, and is about the same distance between its points, *Observatory Point* on the West, and *Angelos Point* on the East; at the latter the *River Elwha* enters the strait. *Puerto de los Angelos*, of the Spanish charts, is 7 miles beyond this. It is formed, like Dungeness Bay to the eastward, by a long, narrow spit, similar to New Dungeness at the latter. Its extreme is *Ediz Hook*, and is 12 miles S.W. from New Dungeness. About midway between them, in the bight of the shallow bay formed by the coast, is a sunken rock, with 11 feet least water, lying half a mile off shore.

NEW DUNGENESS forms a safe roadstead, and lies 80 miles from Cape Flattery, E. by S., *true;* the trend of the strait being E. by S. and W. by N., nearly. The point of New Dungeness is well adapted for the position of a light-house:* it projects into the strait, and would be seen a long distance, both up and down; the water close to the point is deep; a vessel may approach to within a quarter of a mile, and after turning it, safe and secure anchorage may be had in from 10 to 15 fathoms water; it is extensive enough to accommodate a very large fleet. An abundance of wood, water, and fine fish, may be obtained there.

Budd's Harbour lies adjoining it, and is connected with the roadstead of New Dungeness by a narrow channel, which has a depth of 2½ fathoms, and may be easily deepened if necessary; it is a fine and very capacious harbour, being 4 miles long and 1¾ miles wide, and perfectly secure at all times for repairs.

The following remarks are by Mr. R. J. Gibbon, R.N., of H.M.S. *Modeste*:—

"Being abreast of the Race Islands, the S.E. projection of Vancouver Island, and bound for New Dungeness, steer S.E. ½ E. to S.E. by E., depending on the direction of wind and set of the tide, and having reached within sight of the Ness, being the eastern end of a remarkable belt of pine trees, which extends the whole length of the southern side of Dungeness Bay, bear a little to the southward of S.S.E., which will clear you of a long shelving spit of sand, extending about three-quarters of a mile North, from the extreme point of Dungeness Spit, over

* Vancouver found on the low land of New Dungeness, about Port Townsend and elsewhere, a number of very tall straight poles, erected perpendicularly, and with much regularity. They were like flagstaffs, or beacons, and supported by spurs. On a low projecting point near Port Townsend were two of these singular memorials, rudely carved, and on the top of each was a human skull, recently placed. The intentions of these were quite incomprehensible.—*Vancouver*, vol. i. pp. 225—234. Capt. Wilkes states that these poles still remain as thus described, and that the Indians informed him that they were used for taking wild-fowl. Nets were suspended on them, and the birds frightened towards them at night by means of fires; this causes them to fly against the nets, by which they are thrown upon the ground and then readily killed.—*American Exploring Expedition*, vol. iv. p. 298.

which the tide runs very strong, and produces a dangerous race for boats. Run on that bearing until within the bay, and when you have brought the extreme point of Dungeness Spit to bear from N. by W. to N. ½ W., you may then haul up for the head of the bay, and anchor in any depth, from 7 to 4 fathoms, good tough holding ground.

"The best position for a small vessel to anchor in is about a quarter of a mile from the North side of the bay, with the extreme point of the sandy spit bearing from N.N.E. to N.E. in from 5 to 4 fathoms water.

"There is excellent water to be had from a small river in the S.W. part of the bay, where you can enter your boats at half tide and fill the casks in the boat. Potatoes, salmon, and a great variety of fish, are brought by Indians to barter for clothes. Whale and salmon oils are also offered in small quantities.

"The Indians here are not unlike those at Cape Classet; the men have a wild and savage aspect, are of middle stature, and somewhat robust, their complexion tawny, their only covering a small plaited grass fringed apron round their loins. The women present a soft and mild appearance, and not of an uninteresting complexion; they have a pleasing tone of articulating words of their language, and were it not for the deformity of the upper part of their heads, which are flattened when infants, they would be symmetrically formed. They are blessed with a bewitching risibility of expression, which is seldom displayed with more becoming modesty by an enlightened race. Their dress is simply a fringed belt, made from the fibres of the inner part of bark of the white cedar, worn round the loins.

"The climate, I think, is in no way insalubrious at any season of the year; the distant country is exceedingly mountainous, many of the highest peaks (particularly Mount Baker) are capped with perpetual snow; whilst there are many extensive plains, clothed with exuberant pasture, where numerous herds of elk and deer are found. The dense forests of pine, cedar, oak, yew, maple, poplar, ash, willow, alder, elder, and hazel, are the abiding places of the bear, panther, wolf, fox, racoon, lynx, and squirrel; whilst the lakes and rivulets abound with wild-fowl (swans, geese, ducks, seals, &c.), and in season the swamps and marshes afford cover for snipe and plover.

"It is in the lakes of the high land where the sagacious beaver builds its huge dam with such adroitness and skill, so, as would seem, to surpass the genius of man himself by its instinct." *

PORT DISCOVERY, 7 miles to the S.E. of New Dungeness, is very easy of access, and a well-protected harbour; but the depth of water, and the high, precipitous banks, will almost preclude its being made the seat of a settlement. The anchorage is close to the shore, in 27 fathoms water.† The appearance of the country here is almost as enchantingly beautiful as the most elegantly finished pleasure grounds in Europe. The watering place at which Vancouver encamped

* Nautical Magazine, 1848, pp. 226-7.

† Vancouver refitted in this port, in May, 1792. He encamped near the North side of the brook or watering place on the West side, and found it very convenient for his purpose.—Vol. i. pp. 229-30. "The description of Vancouver," says Capt. Wilkes, "is so exactly applicable to the present state of this port, that it is difficult to believe that almost half a century has elapsed since it was written."—Vol. iv. p. 298.

is on the western shore, 5 miles from the entrance. The stream of water appears to have its source at some distance from its outfall through one of those low spits of sand which constitute most of the projecting points on the shores of the inlet. They usually acquire a form somewhat circular, though irregular; and in general are nearly steep-to, extending from the cliffy woodland country, from 100 to 600 feet towards the water's edge, and are composed of a loose sandy soil. The shores of the harbour are supplied with large quantities of shell-fish. The Indians also bring abundance of venison, geese, fish, &c.

The name of Port Discovery was given by Vancouver; it is 7 miles long, 1½ miles average width, and its points, which terminate in low sandy projections, interlock each other. *Protection Island* covers it completely to the North. Had this singular production of nature been designed by most able engineers, it could not have been placed more happily for the protection of the port, not only from the N.W. winds, to the violence of which it would otherwise be greatly exposed, but against all attempts of an enemy, when properly fortified; and hence Vancouver called it Protection Island.—(Vol. i. p. 228.)

The Indians who dwell here are of the Clalam tribe. They occupy a few miserable lodges on one of the points, and are a most filthy race, so much so that the appearance of their lodges is absolutely disgusting.

There are few places where the variety and beauty of the flowers are so great as they are here; the general character of the soil around this harbour is a thin, black, vegetable mould, with a substratum of sand and gravel. The trees grow so closely, that in some places the woods are almost impenetrable. The timber consists principally of pine, fir, and spruce. Of the latter there are two species, one of which resembles the hemlock-spruce of the United States; it has a very tall growth, and puts out but few, and those small, lateral branches. Some maple trees grow in the open ground and on the banks, but they are too small to be of any service to the settler.

Port Discovery has its outer points 1¾ miles asunder, bearing from each other S.W. by W. ½ W. and N.E. by E. ½ E. from its entrance; the port first takes a direction S.S.E. about 8 miles, and then terminates S.W. by W. about a league farther. If it lies under any disadvantage it is its great depth of water; in which respect, however, we found no inconvenience, as the bottom was exceedingly good holding ground, and free from rocks. Towards the upper part of the harbour it is of less depth; but I saw no situation more eligible than that in which the vessel rode, off the first low sandy point on the western shore, about 4½ miles within the entrance. Here our wooding, watering, brewing, and all other operations were carried on with the utmost facility and convenience. The shores of Protection Island form on its South side, which is about 2 miles long, a most excellent roadstead, and a channel into Port Discovery near 2 miles wide on either side, without any interruption. The country in the neighbourhood of this port may generally be considered of a moderate height, although bounded on the West side by mountains covered with snow, to which the land from the water's edge rises in a pleasing diversity by hills of gradual ascent. The snow on these hills probably dissolves as the summer advances, for pine trees were produced on their very summits. On the sea-shore the land generally terminated in low sandy cliffs, though in some spaces of considerable extent it ran nearly level from high-water mark.*

It is high water in Port Discovery at 2h 30′; the rise 7 feet.

The S.W. point of Protection Island is in lat. 48° 7′ 10″, lon. 122° 56′.

PORT TOWNSEND lies at the entrance of Admiralty Inlet; it is a fine sheet of water, 3½ miles in length, and 1¾ in width. On the West side is an extensive table-land, free from wood, which would be a good site for a town. This bay is free from dangers, and is well protected in the direction from which stormy winds blow. It has anchorage of a convenient depth, and there is abundance of fresh water to be had; the best anchorage is on the North side. The soil in this place is a light sandy loam, and appears to be very productive; it was covered with wild flowers and strawberries, in blossom, in May. Vancouver, its first explorer, says:—"It proved to be a very safe and more capacious harbour than Port Discovery, and rendered more pleasant by the high land being at a greater distance from the water-side. Its soundings also give it a further advantage, being very regular from side to side, from 10 to 20 fathoms' depth of water, good holding ground. To this port I gave the name of Port Townsend, in honour of the noble marquis of that name."—(Vancouver, vol. i. p. 234.) From this place, Mount Baker is distinctly seen to the N.E., and adds a beautiful feature to the landscape when its conical peak is illuminated by the setting sun.

The N.E. point of Port Townsend is a high steep cliff formed of an indurated clay, resembling fuller's earth, and hence called by Vancouver *Marrowstone Point*. East of this cliff the shore is extended about a quarter of a mile by one of those sandy projecting points so frequently met with. The country to the East of this, between it and the eastern snowy range, preserves the same luxuriant appearance. At its North extreme is Mount Baker, bearing N.N.E. by compass, and its southern, the round snowy mountain, called by Vancouver after his friend Admiral Rainier, S.E. ½ S. Mount Rainier was ascertained by the American Exploring Expedition to be 12,330 feet high. Marrowstone Point is the extremity of a narrow island which forms the eastern boundary of the smaller area of Port Townsend. At its South end is Oak Cove or Bay, so called by Vancouver from the fact of some of those trees having been discovered there. It is called Port Lawrence in Wilkes's chart.

OAK COVE, or *Port Lawrence*, is just at the junction of Admiralty Inlet and Hood's Canal; it is a convenient anchorage, and is separated from one of the arms of Port Townsend by a narrow strip of land.

Just above this the inlet divides into two principal arms, that to the eastward the most extensive, called Admiralty Inlet, and the western Hood's Canal. The point at which these separate is a high perpendicular point, called *Foulweather Bluff*, the North extreme of what appears to be a long low island, but is not so.

ADMIRALTY INLET is a collection of very singular and labyrinthian channels terminating at Puget Sound and its various branches, which reach the latitude of 47° 2′. Some portions of their shores are remarkably fertile and beautiful, and it would seem to be almost the only eligible part for civilization in the ceded

* Vancouver, vol. i. p. 248.

Oregon Territory. The settlements formed by its original settlers, emanating from the Hudson's Bay Company, still remain under the same dominion, though the government is in the hands of the United States.

The whole of these inlets have been accurately and amply explored and surveyed, for the first time by Vancouver, in May, 1792, and most of the names were applied by him. Hood's Canal was thus named after Lord Hood. The American Exploring Expedition, under Capt. Wilkes, also surveyed the assemblage of inland waters, adding the names not given by Vancouver, chiefly those of the officers of the Expedition.

The shores of all these inlets and bays are remarkably bold; so much so, that in many places a ship's side would strike the shore before the keel would touch the ground.

Passing the entrance to Hood's Canal, and up Admiralty Inlet, there are several anchorages where a vessel may await tide, in beating up; such as *Pilot's* and *Apple-Tree Coves.*

Port Madison is the first harbour, and affords every possible convenience for shipping; it is on the West side of the inlet, and communicates on the South, by a ship channel, with Port Orchard.

Seven miles South of Port Madison is *Restoration Point*, at the northern side of the entrance to Port Orchard.

During Vancouver's stay here, off Restoration Point, he observed that the tides were materially affected by the direction or force of the winds, not only in respect to their rise and fall, but as to the time of high water. The former seldom exceeded 7 or 8 feet, and the latter generally took place about $4^h\ 10'$ after the moon passed the meridian. The variation was 19° 36′ E. in 1793. Capt. Kellett found it to be 21° E. in 1847.

PORT ORCHARD is one of the most extensive and beautiful of the many fine harbours on these inland waters, and is perfectly protected from the winds. The only danger is a reef of rocks, nearly in the middle of the entrance from Admiralty Inlet. It includes three arms, the most northern of which, though entered by a narrow channel, 200 yards in width, is from a half to one and a half miles in width, and extends for a distance of 6 miles. The water is deep enough for the largest class of vessels, with a bold shore and good anchorage.*

"The sheet of water is very extensive, and is surrounded by a large growth of trees, with here and there a small prairie, covered by a verdant green sward, and with its honeysuckles and roses just in bloom, resembling a well-kept lawn. The soil is superior to that of most places around the sound, and is capable of yielding almost any kind of production. The woods seemed alive with squirrels, while tracks on shore and through the forest showed that the larger class of animals also were in the habit of frequenting them."†

The country that surrounds this harbour varies in its elevation; in some places

* "The scenery of this portion of Admiralty Inlet resembles strongly parts of the Hudson River (New York), particularly those about Poughkeepsie and above that place. The distant highlands, though much more lofty, reminded us of the Kaatskills. There were but few lodges of Indians seen on our way up; and the whole line of shore has the appearance of never having been disturbed by man."—*Wilkes*, vol. iv. p. 304.

† Wilkes, vol. iv. p. 479.

the shores are a low level land, in others of a moderate height, falling in steep low cliffs on the sandy beach, which in most places binds the shores. It produces some small rivulets of water, is thickly wooded with trees, mostly of the pine tribe, and with some variety of shrubs. This harbour, after the gentleman who discovered it, obtained the name of Port Orchard. The best passage into it is found by steering from Restoration Point for the South point of the cove, which is easily distinguished, lying from the former S. 62° W., at the distance of about 2½ miles, then hauling to the N.W. into the cove, keeping on the larboard or S.W. shore, and passing between it and the rocks in the cove; in this channel the depth of water is from 9 to 15 fathoms, gradually decreasing to 5 fathoms in the entrance into the port. There is also another passage round to the North of these rocks, in which there are 7 fathoms water; this is narrow, and by no means so commodious to navigate as the southern channel.*

The shores are covered with a large growth of trees, with here and there a small prairie; the soil is superior to that of most places around the sound, and is capable of yielding almost any production of the temperate zone.

VASHON'S ISLAND lies in Admiralty Inlet, above Port Orchard, and there is a ship channel on both sides of it; the best one is on the West: the two again unite just before entering the narrows leading into Puget Sound.

Commencement Bay lies at the bottom of Admiralty Inlet, on the East channel; it affords good temporary anchorage, and an excellent supply of wood and water can be obtained. There is a small stream emptying into it, called, by the Indians, *Puyallup*. This river forms a delta; none of the branches of which are large enough for the entrance of a boat. Capt. Wilkes's party found Indians on almost all the points, in May, 1841, the same filthy creatures before described.

The NARROWS, which connect Admiralty Inlet with Puget Sound, are a mile in width, and 4½ miles long; the tide here runs with great velocity, causing many whirlpools and eddies, through which a ship is carried with great rapidity, the danger appearing to be imminent. The banks rise nearly perpendicular, and are composed of sandstone: a great variety of shrubs grow along their base. This narrow pass seems as if intended by nature to afford every means for the defence of Puget Sound. *Point Defiance*, on the East, commands all the approaches to it.

PUGET SOUND is a most singular termination to Admiralty Inlet. The fertility of its shores, and its fine climate, may possibly make it of considerable importance in future years. It received the name of Mr. Puget, the officer under Vancouver's expedition, who originally surveyed it.†

Puget Sound may be described as a collection of inlets, covering an area of 15 miles square, the only entrance to which is through the Narrows, which, if strongly fortified, would bid defiance to any attack, and guard its entrance against any force.

The inlets, in the order in which they come from the entrance, have received the names of *Carr's*, *Case's*, *Hammersley's*, *Totten's*, *Eld's*, *Budd's*, and *Hen-*

* Vancouver, vol. i. p. 265. † *Ibid.* vol. i. p. 275.

derson's, from the officers of the United States' Exploring Expedition; they are united by passages, which form several islands and peninsulas. All these inlets are safe, commodious, and capacious harbours, well supplied with water, and the land around them fertile. On many of the islands and peninsulas are to be found slate and sandstone, which, though soft and friable in some places where it has been exposed on the surface, will be found suitable for building purposes.

The southern part of the Sound is the site of the farming operations of the Puget Sound Company, a branch of, or connected with, the Hudson's Bay Company. According to the charter of the latter company, they are precluded from entering into any operations not immediately connected with its primary object, but the great want of some such establishment on this side of the American continent led to the formation of the Puget Sound Company, by the officers and servants of the Hudson's Bay Company. The capital is £500,000, subscribed entirely among themselves, and the portion of this amount paid up was invested in the extensive establishments now alluded to. From this quarter the provisions, &c., are supplied to all the ports and stations of the Hudson's Bay Company on the western side of the American continent; the Russian establishments to the northward were also supplied largely with grain, butter, cheese, &c. Nisqually, as it is called by Europeans, but by the natives 'Squally, is the head-quarters.

NISQUALLY is 9 miles distant from the Narrows. Here the anchorage is very much contracted, in consequence of the rapid shelving of the bank, that soon drops off into deep water, and only a few vessels can be accommodated. The shore rises abruptly to a height of 200 feet, and on the top of the ascent is an extended plain, on which Fort Nisqually is built. On the hill side is a well-constructed road, of easy ascent. Fort Nisqually, with its outbuildings and enclosures, stands back about half a mile from the edge of the table-land. It is constructed of pickets, enclosing a space about 200 feet square, with a bastion at each corner. Within this enclosure are the agent's stores, and about half a dozen houses, built of logs and roofed with bark. Its locality is badly chosen, on account of the difficulty of obtaining water, which has to be brought a distance of nearly a mile.

In the garden at Nisqually, on the 12th of May, peas were a foot high; strawberries and gooseberries in full bloom, and some of the former nearly ripe, with salad that had gone to seed, 3 feet high and very thrifty.

The hill at Nisqually is an insuperable objection to the place ever becoming a deposit for merchandise, as it would very much increase the labour and expense of transportation. Water, however, can be obtained for vessels with great ease from a small stream that flows in abreast of the anchorage. The harbour is also exposed to the S.W. winds.

The spring tides were found to rise and fall 18 feet; neaps 12. High water at full and change, at 6^h 10′, P.M. During the whole of the stay of Capt. Wilkes's party, there was found to be a great discrepancy between the day and night tides, the latter not rising so high as the former by 2 feet.

The country in this vicinity is thought to be remarkably healthy; the winter is represented to be mild, and but of short duration. The greatest difference, or range of temperature, was found to be 55°, the lowest 37°; and the mean, during the same period, 63° 87′; the barometer standing at 29·970 in.*

The climate is propitious, while the seasons are remarkably regular. Between the beginning of April and the end of September there is a continuance of dry weather, generally warm, and often hot. The mercury rose, in 1841, at Nisqually, to 107° in the shade. March and October are unsettled and showery, and during the four months of winter there is almost constant rain, while the temperature is so mild, that the cattle and sheep not only remain out of doors, but even find fresh grass for themselves from day to day.†

Better sites than Nisqually for the location of a town are to be found in this neighbourhood. There is one, in particular, just within *Kitson Island*, about 2¼ miles North of the Nisqually anchorage, where the shore has a considerable indentation, and, although the water is deep, vessels would be partially protected from the S.W., S.E., and N.W. winds, which blow with great violence, and also from any sea. Water can be obtained with as much facility, and the hill is not so precipitous.

CASE'S INLET extends to within 2 miles of the waters of Hood's Canal. Between these there lies *Kellim* or *Kellmso Pond*. The communication might be easily made between them.

MOUNT ST. HELEN'S was first seen by Vancouver when in the southern parts of Admiralty Inlet. Like Mount Rainier, it seemed covered with perpetual snow. He called it by this name in honour of our ambassador at the court of Madrid.‡

In former times they have been quiet, but during and since 1845 the craters on the top of Mount Rainier and Mount St. Helen's have been in activity.§

HOOD'S CANAL diverges to the West of Admiralty Inlet at Foulweather Bluff, as previously stated, and was so named by Vancouver after Lord Hood. It extends for a distance of 40 miles in a S.S.W. direction, and then turns to the N.E. Its usual width is about 1½ miles.

Port Ludlow, open to the North, is opposite Foulweather Bluff, and affords anchorage; 3½ miles further up the inlet on the same side is what Vancouver at first took to be a high round island, but afterwards found it to be connected with the main by a low sandy neck, almost entirely occupied by a salt-water swamp. On the opposite side of the canal is *Port Gamble*, 2½ miles in depth to the southward, having a narrow entrance between the kelp. The canal here assumes a direction to S.W. ½ S. for 14 miles, and is half a league wide; and S.W. of the above island is *Suquamish Harbour*, with a dry sand bank off it. In the small harbour formed by the first-mentioned island or peninsula a few small streams of fresh water descended, with which the country did not seem to abound, for the few natives seen by Vancouver brought some with them in small square boxes, which they could not be tempted to part with. The extremity of Hood's Canal was not found by the American Exploring Expedition to terminate where Vancouver's survey ends, but takes a short turn to North and East for 10 miles, approaching Puget Sound within 2½ miles, the intervening country being rough and hilly. At its South extreme also there is a large inlet called *Black Creek*,

* Wilkes, vol. iv. p. 416.
† Sir George Simpson, Journey, &c., vol. i. p. 179.
‡ Vancouver, vol. i. pp. 421-2.
§ Capt. Wilkes.

by which the Indians communicate with the Chickeeles and Columbia rivers. The water in the centre of the canal is too deep for anchorage, but there are several good harbours.—(Wilkes, vol. iv. p. 411.)

The shores of the canal in the reach South from this part exhibit by no means the luxuriant appearance of the country to the northward, being nearly destitute of open verdant spots, and alternately composed of sandy or rocky cliffs, falling abruptly into the sea, or terminating on a beach, whilst in some places the even land extended from the water side, with little or no elevation. The low projecting points cause the coast to be somewhat indented with small bays, where, near the shore, Vancouver had soundings from 5 to 12 fathoms; but in the middle of the canal no bottom could be reached with 110 fathoms of line. This latter circumstance must have arisen from the effect of the strong current, because in Wilkes's survey the depth nowhere exceeds 65 fathoms.

Hazel Point, so named from the number of those trees on it, is the S.E. end of a peninsula which separates the two branches of the inlet. That to the West runs up North for 6 miles, separating *Dahap* and *Colseed Inlets*, running 3 or 4 miles farther. The South point of this peninsula is *Tokutki Point;* South of it is *Hahamish Harbour*, formed on the West by Seabeck Island, and to the West of it is *Syclopish Rock*. Beyond this the canal runs 20 miles to the end of its S.W. direction, which was as far as Vancouver explored.

POSSESSION SOUND may be considered as the southern entrance to the channel, separating Whidbey Island from the main land. Its eastern shore is compact, forming a deep bay, into which the *Sinahomis River* falls, but this river was not observed by Mr. Broughton, despatched by Vancouver. Seven miles within the entrance from the South is a high round island. It was observed by Vancouver that the tide or current constantly set outwards here.

At 2 miles N.W. from this round island is *Point Alan*, the South extremity of *Caamano Island*, which lies between Whidbey Island and the main. Point Alan is the end of a high narrow strip of land, which separates Port Susan on the East from Port Gardner on the West of Caamano Island. *Port Susan* extends about 11 miles north-westward, and is terminated by a line of kelp fronting a tract of swampy land, through which a rivulet extends which forms the island to the West. The land farther back is more elevated, and covered with a growth of timber similar to that in other parts. It was in the upper part of this inlet that Vancouver's ship, the *Chatham*, ran aground, but was soon got off.*

Eastward of Alan Point, on the main land, is a small bay, before which Vancouver anchored. There were two excellent streams flowing into it, but they were so nearly on a level with the sea, that it became necessary to procure the water at low tide, or at some distance up the brook, which latter was easily effected, as the boats were admitted to where the fresh water fell from the elevated land. They also took some fish with the seine.

Port Gardner (so named by Vancouver after Vice-Admiral Sir Alan Gardner) is the western arm of the continuation of Possession Sound, its western shore being formed by Whidbey Island. This shore was found by Vancouver to be well

* Vancouver, vol. i. p. 282.

peopled by Indians, who were very friendly. At about 14 miles from Point Alan the branch which runs thus far about N.N.W. assumes the directions of about West and N.E.

Penn's Cove is the termination of the western branch, and is a very commodious and excellent harbour, with regular soundings from 10 to 20 fathoms, good holding ground. Its western extent, in lat. 48° 14′ (according to the United States' survey) is not more than a league from the eastern shore of the main inlet within the strait. On each point of the harbour Vancouver found a deserted village, in one of which were several sepulchres, formed exactly like a sentry-box.

"The surrounding country, for several miles in most points of view, presented a delightful prospect, consisting chiefly of spacious meadows elegantly adorned with clumps of trees, amongst which the oak bore a very considerable proportion, in size from 4 to 6 feet in circumference. In these beautiful pastures, bordering on expansive sheets of water, the deer were seen playing about in great numbers. Nature had here provided the well-stocked park, and wanted only the assistance of art to constitute that desirable assemblage of surface which is so much sought in other countries, and only to be acquired by an immoderate expense in manual labour. The soil principally consisted of rich black vegetable mould, lying on a sandy or clayey substratum; the grass, of excellent quality, grew to the height of 3 feet, and the ferns, in the sandy soils which occupied the clear spots, were nearly twice as high. The country in the vicinity of this branch of the sea is, according to Mr. Whidbey's representation, the finest we had yet met with, notwithstanding the very pleasing appearance of many others; its natural productions were luxuriant in the highest degree, and it was by no means ill supplied with streams of fresh water. The number of its inhabitants was estimated at 600, which I should suppose exceeded the total of all the natives we had before seen." *

The main channel to the N.E. has no northern outlet. It leads to a branch whose general direction is N.W. From the eastern shore of this branch a shallow flat of sand, on which are some rocky islets and rocks, runs out until within half a mile of the western shore, forming a narrow channel, navigable for about 3 leagues. The depth in its entrance is about 20 fathoms, but gradually decreases to 4 fathoms in advancing northward, and the sand bank, continuing with great regularity, makes it about half a mile wide to lat. 48° 24′, where it ceases to be navigable for vessels of any burden, in consequence of the rocks and over falls, from 3 to 20 fathoms deep, and a very irregular and disagreeable tide.

In the bay just to westward of the South entrance point of Possession Sound, that is, of Whidbey Island, there is a shoal lying a little distance from the shore; it shows itself above water, and is discoverable by the soundings gradually decreasing to 10, 7, and 5 fathoms, and cannot be considered as any material impediment to the navigation of the bay.

WHIDBEY ISLAND is well fitted for settlement and cultivation. The soil is good, the timber excellent, and there are several open plains, which

* Vancouver, vol. i. pp. 287-8.

have been prepared by nature for the plough. It is about 38 miles in length, of irregular figure and breadth.

The S.W. coast of Whidbey Island continues in an irregular N.W. direction to *Partridge Point*, which forms the N.E. point of the entrance of Admiralty Inlet. It is formed by a high white sandy cliff, having one of the verdant lawns on either side of it. According to the Admiralty chart, drawn up from the survey by the United States' Exploring Expedition, it is in lat. 48° 12′ 30″ N., lon. 122° 45′ W.

"Passing at the distance of about a mile from this point, we very suddenly came on a small space of 10 fathoms water, but immediately again increased our depth to 20 and 30 fathoms. After advancing a few miles along the eastern shore of the gulf, we found no effect from either the ebb or flood tide." *

Bonilla (or Smith) *Island* lies 6¾ miles N.W. of Partridge Point, and 4½ from the nearest land. It is low and sandy, forming at its West end a low cliff, above which some dwarf trees are produced. Some rocks lie on its western side, nearly three-quarters of a mile of its shores, and its eastern part is formed by a very narrow low spit of land, over which the tide nearly flows. From this the remarkably lofty and snowy peak of Mount Baker bears N. 63° E., and that of Mount Rainier S. 27° E. Two other very lofty round snowy mountains are also seen to the southward of these. They appear to be covered with perpetual snow, as low down as they can be seen, and seem as if they rise from an extensive plain of low country.

Deception Passage, which runs into Port Gardner, to the North of Whidbey Island, is a very narrow and intricate channel, which, for a considerable distance, is not 40 yards in width, and abounds with rocks above and beneath the surface of the water. These impediments, in addition to the great rapidity and irregularity of the tide, render the passage navigable only for boats or vessels of very small burden.

To the northward of this the Strait of St. Juan de Fuca is limited by a collection of islands which separates it from that explored by Vancouver, and named by him the Gulf of Georgia. The North side of Deception Passage has been proved to be in reality an island by the United States' Exploring Expedition, and named *Fidalyo Island*, separated from the main land on the East by a tract of low land intersected by a narrow stream. The country here assumes a very different aspect from that seen to the southward. The shores are here composed of steep rugged rocks, whose surface varies exceedingly as to height, and exhibits little more than the barren rock, which in some places produces some herbage of a dull colour, with some few dwarf trees.

ROSARIO STRAIT forms the connection, according to the later charts, between the Strait of De Fuca and the Gulf of Georgia, running northward between Fidalgo Island and that next westward, now named *Lopez Island*. Off the S.E. point of this island, which is the entrance of this strait, and which is low and rocky, there is a very dangerous sunken rock visible only at low tide; and 2½ miles to the northward is a very unsafe cluster of small rocks, some constantly

* Vancouver, vol. i. p. 291.

and others visible only near low water. The strait varies from 5 to 3 miles in width. On its eastern side, that is, against Fidalgo Island, are Alan and Burrows's Islands, off the South end of which are some detached rocks.

CYPRESS ISLAND lies in front of the opening, some 7 miles within it. It is about 4½ miles in length, and on its western side is *Strawberry Bay*, so named by Vancouver from the great quantity of very excellent strawberries found there when Mr. Broughton first visited it.

This bay is situated on the West side of the island, which, producing an abundance of upright cypress, obtained the name of Cypress Island. The bay is of small extent, and not very deep. When at anchor in 16 fathoms, fine sandy bottom, its South point bears S.E. ½ S., a small islet forming nearly the North point of the bay, round which is a clear good passage West; and the bottom of the bay East, at the distance of about three-quarters of a mile. This situation, though very commodious in respect to the shore, is greatly exposed to the winds and sea in a S.S.E. direction.

The bay affords good and secure anchorage, though somewhat exposed, as before stated; yet, in fair weather, wood and water may be easily procured. Cypress Island is principally composed of high rocky mountains, and steep perpendicular cliffs, which in the centre of Strawberry Bay fall a little back, and the space between the foot of the mountains and the sea-side is occupied by low marshy land, through which are several runs of most excellent water, that find their way into the bay oozing through the beach. It is situated in lat. 48° 34½′, lon. 122° 42′ W.

The variation of the compass, as found by Vancouver in 1792, was 19° 5′ E., while that by Capt. Kellett, R.N., 1847, is 22° E. By Vancouver's observations the rise and fall of the tide were inconsiderable, though the stream was rapid; the ebb came from the East, and it was high water 2^h 37′ after the moon had passed the meridian.

Guemes Island lies to the eastward of Cypress Island and North of Fidalgo Island, and to the eastward of these the main land forms Padilla Bay.

BELLINGHAM BAY is separated from Padilla Bay to the South by a long, narrow peninsula, of which William Point is the West extreme. There are a number of channels leading into it through the cluster of islands before alluded to; and the bay itself extends about 12 miles North and South. It everywhere affords good and secure anchorage. Opposite to its North point of entrance the shores are high and rocky, with some detached rocks lying off it. Here is a brook of most excellent water. To the North and South of these rocky cliffs, the shores are less elevated, especially to the northward, where some beautiful verdant lawns are seen. The land generally is inconvenient for communicating with, on account of a shallow flat of sand or mud which extends a considerable distance off the land.

It may be stated that in Valdez and Galiano's charts, this bay appears in two portions, the northernmost being named *Gaston Bay*. This part is separated from the gulf by a long, narrow peninsula, terminating in Point Francis; an inlet lying in the middle of the bay is called *Puerto del Socorro*, and the southern part of the bay is called *Padilla Bay*, an appellation confirmed in the recent charts to that still farther South.

BIRCH BAY is 10 miles to the northward of Port Francis. The S.E. part of this bay is formed by nearly perpendicular rocky cliffs, from whence the higher woodland country retires for a considerable distance to the north-eastward, leaving an extensive space of low land between it and the sea, separated from the high ground by a rivulet of fresh water, that discharges itself at the bottom or North extremity of the bay. On the low land very luxuriant grass was produced, with wild rose, gooseberry, and other bushes in abundance.

Here Vancouver's vessels anchored, and erected an observatory for ascertaining its position, &c. The variation of the compass there was 19° 30′ E. The tides were found to be very inconsiderable, though not particularly observed. Nothing further occurred at this station worthy of notice, with the exception of an observation which had been repeatedly made, that, in proportion as they advanced northward, the forests were composed of an infinitely less variety of trees, and their growth was less luxuriant. Those most commonly seen were pines of different sorts, the *arbor vitæ*, the oriental arbutus, and, probably, some species of cypress. On the islands some few small oaks were seen, with the Virginian juniper; and at this place the Weymouth pine, Canadian elder, and black birch, which latter grows in such abundance, that it obtained the name of Birch Bay.

The coast to the northward of this forms two open bays; the southernmost, which is the smallest, has two small rocks lying off its South point; it extends in a circular form to the eastward, with a shoal of sand projecting some distance from its shores. This bay affords good anchorage in from 7 to 10 fathoms of water; the other is much larger, and extends to the northward; but the shoals attached to each, and particularly those of the latter, prevent reaching them within 4 or 5 miles of their heads.

The point which forms the West extreme of these bays was named by Vancouver *Point Roberts*. It is the S.W. extremity of a very low, narrow peninsula. Its highest part is to the S.E., formed by high white sand cliffs, falling perpendicularly into the sea, from whence a shoal extends to the distance of half a mile round it, joining those of the larger bay; while its S.W. extremity, not more than a mile in an East and West direction from the former, is one of those low, projecting, sandy points, with 10 to 7 fathoms water within a few yards of it.

It is in the bay to the East of Point Roberts that the boundary between the Oregon Territory and that belonging to Great Britain reaches the coast, the parallel of 49° North latitude, as defined by the treaty of 1846.

The country to the north of this, being the territory of the Hudson's Bay Company, will be described with Vancouver Island in the next chapter.

CHAPTER XIII.

VANCOUVER ISLAND.

Prior to 1789 the outer coast of this island was supposed to be that of the American continent, but in that year its insular character was established.

The first intimation in Europe of the existence of the channel which separates Vancouver Island from the continent was in the observations prefixed by Capt. Meares to the Narrative of his Voyages.* In the chart accompanying that work there is a sketch of the track of the American sloop *Washington*, in the autumn of 1789, which is that through the inland navigation presently described. The name of the commander is not given, but it was naturally supposed that Capt. Gray, of the *Columbia*, previously mentioned, was the person. In the angry discussion which ensued between Meares and Dixon† relative to the remarks of the former, it is stated that Mr. Kendrick was the commander of the *Washington*, who perhaps took it after it had been quitted by Gray. Therefore Kendrick, in the sloop *Washington*, must be taken as the person who really made known the real character of the territory in question, after the formerly discredited voyage of its discoverer, De Fuca, in 1592.

Vancouver reached the coast in March, 1792, and, after navigating through the strait to the eastward, he applied the names Quadra and Vancouver Island to it; the first name in compliment to the Spanish commandant at Nootka Sound, from whom he received much politeness during the negotiations relative to the restoration of the tract of country claimed by Great Britain, as mentioned hereafter. Lately, however, the first name has been dropped, and that of the surveyor only retained for it, which is certainly preferable, as the Spanish governor took no part in its geographical *advancement*.

It remained in the same state, untenanted by Europeans during the subsequent years, until recent events have given it a new existence. It was only visited, at regular intervals, on its north-eastern side, by the officers and agents of the Hudson's Bay Company for the trade in furs and other commodities collected by the natives, and at Nootka Sound, on its S.E. side, for refreshments, &c., by the vessels in the North Pacific.

When, however, the treaty which severed the Oregon Territory from Great Britain, and the thriving settlements established therein by the Hudson's Bay Company became surrounded by a foreign power and interest, Vancouver Island assumed a fresh importance to English commerce, and may hereafter occupy a conspicuous position in the affairs of the North Pacific Ocean.

In the previous chapter we have described some flourishing settlements estab-

* Narrative of the Voyages of Capt. John Meares, &c., 4to, London, 1789.
† Remarks on the Voyage of John Meares, by George Dixon, Commander of the *Queen Charlotte*, London, 1790; and Answer to Mr. George Dixon, &c., by John Meares, London, 1791.

lished by the Hudson's Bay Company, either officially or by their servants in a private character. The principal of these are the Forts Vancouver and Nisqually, the latter the property of the Puget Sound Company. In consideration of the great advancement of colonization by the company, and their claims upon our government in consequence of the above-named cession, by which the sovereignty of the territory in question has changed (subject, however, to the purchase of the actual property so ceded), the British crown has granted the dominion of Vancouver Island to the Hudson's Bay Company.

This grant, which is by letters patent under the great seal, and dated January 13th, 1849, gives the absolute lordship and proprietorship of the whole island, with its mines, &c., to the Hudson's Bay Company, at the yearly rent of seven shillings, with the condition that, upon the expiration of the company's exclusive privilege of trading with the Indians, the government may re-purchase all improvements effected by the company.

The administration is vested in a governor and council, appointed by the crown, with discretionary powers; the first governor being Richard Blanchard, Esq.*

With these powers the company have announced their intention of colonizing the island, upon which the terms of the grant depend, and will sell to colonists tracts of land at the sum of one pound per acre, those taking large quantities of land being under obligation to take out a proportionate number of individuals for its cultivation. One-eighth of the entire territory is to be reserved for church and school purposes, or roads and public works.

In the infancy of this scheme, of course it is premature to speak of its results, but in the neighbouring districts they have been very successful.

Vancouver Island extends from lat. 48° 19′ to 50° 53′ N., and between lon. 123° 17′ and 128° 28′ W., being about in the latitude of Belgium and the North of France. It is 240 geographical miles in length from N.W. to S.E., and with a breadth varying from 50 to 65 miles.

Its N.E. shores are much more entire than its seaward face, which appears to be broken into sounds, in the same manner as the continent in its neighbourhood. With its inner shores we are tolerably well acquainted from the examinations and descriptions of Vancouver; with its S.E. side less so, being dependent in a great degree upon the Spanish surveys, which have been published without any descriptive account. The Strait of De Fuca, which separates its southern end

* The most material provisions of the commission and instructions to the governor for the government of the colony are as follow:—

The governor is appointed by the crown, with a council of seven members, likewise so appointed.

The governor is authorized to call assemblies, to be elected by the inhabitants holding 20 acres of freehold land.

For this purpose, it is left to the discretion of the governor to fix the number of representatives; and to divide the island into electoral districts if he shall think such division necessary.

The governor has the usual powers of proroguing or dissolving such assembly.

Laws will be passed by the governor, council, and assembly.

The legislature, thus constituted, will have full power to impose taxes and to regulate the affairs of the island, and to modify its institutions, subject to the usual control of the crown.

The crown has already power, under 1st and 2nd George IV., c. 66, to appoint courts of justice and justices of the peace in the Indian territories, of which Vancouver Island forms a part; but as the jurisdiction of such courts is, by the 12th section of that act, limited in civil cases to causes not involving more than £200 in value, and in criminal cases to such as are not capital or transportable (all of which must be tried in Canada), it is intended to extend the jurisdiction created by the existing act by the entire removal of those restrictions.

from the United States' Oregon Territory, has been minutely surveyed by Capt. Kellett, R.N., in 1847. The principal portion of our knowledge is derived from the first-named authority, and of it Sir George Simpson remarks:—"We found Vancouver's chart so minute and accurate, that, amid all our difficulties, we never had to struggle with such as mere science could be expected to overcome; and in justice to both our own navigator and to one of his successors in the same path, I ought to mention that Commodore Wilkes, after a comparatively tedious survey from the mouth of the Columbia to that of Frazer River, admitted that he had required to make but few and inconsiderable corrections."* The ensuing descriptions, therefore, derived from different sources, have very various degrees of accuracy and authenticity; but the discredit, if any be necessary, must rest on the respective authors. With the details and resources of the island itself we are as yet very ignorant, but it would appear to hold out great encouragement for enterprise, commercial and agricultural.

Of its interior nothing has as yet been described. Its shores possess the excellent harbours subsequently enumerated. In the northern part of the island coal is very abundant, and promises to be of great importance. The only river at present known, as being in any way navigable, is that of Nimkis, to the southward of Port M'Niel, on the N.E. end of the island.

The northern end of Vancouver Island would be an excellent position for the collecting and curing of salmon, which being incredibly numerous in these waters, might easily be rendered one of the most important articles of trade in this country. The neighbouring Newetrees (or Neweetg), a brave and friendly tribe, would be valuable auxiliaries, not only in aiding the essential operations of the establishment, but also in furnishing supplies of venison.†

Victoria, at the S.E. extremity, is the chief point of interest at present, and here the speculations of the company have commenced, their dairy and corn growing establishments increasing daily in importance. It is probable that ere long the interests will be removed from Vancouver on the Columbia, and Nisqually at Puget Sound will be removed hither.

The *natives*, who have been described at length by Cook, have been understated in former enumerations. The following is a list of the tribes according to the last census, allowing a decrease of one-fifth for the effect of the late mortality amongst them from the measles, influenza, &c., which have made great havoc.

Tribes.			
1	Songes	700	inhabiting the S.E. part of the island.
2	Sanetch ..	500	,, N.E. 60 miles, N.W. of Mount Douglas.
3	Kawitchin..	1763	,, country N.W. of Sanetch territory to the entrance of Johnson's Straits.
4	Uchulta ..	1000	Kawitchin country in ditto.
5	Nimkis	500	Uchulta country ,,
6	Quaquiolts .	1500	Nimkis country ,,
7	Neweetg ..	500	At N.W. entrance of Johnson's Straits.
8	Quacktoe ..	1000	Woody part N.W. coast of the island.

* Sir George Simpson's Journey round the World, vol. i. pp. 185-6. † *Ibid.* vol. i. pp. 234-5.

Tribes.			
9	Nootka....	1600	Of that name on the West coast.
10	Nitinat	1200	,, ,,
11	Klay-quoit..	1100	,, ,,
12	Soke	100	East point of San Juan to the Songes territory.
	Total..	11,463	

Along the whole coast the savages generally live well. They have both shell-fish and other fish in great variety, with berries, sea-weed, and venison. Of the finny race, salmon is the best and most abundant, while at certain seasons the ullachan, very closely resembling the sardine in richness and delicacy, is taken with great ease in some localities. This fish yields an extraordinary quantity of very fine oil, which being highly prized by the natives, is a great article of trade with the Indians of the interior, and also of such parts of the coast as do not furnish the luxury in question. This oil is used as a sauce at all their meals, if snapping at any hour of the day or night can be called a meal, with fish, with sea-weed, with berries, with roots, with venison, &c.; nor is it less available for the toilet than for culinary purposes. It is made to supply the place of soap and water, smearing the face or any other part of the body that is deemed worthy of ablution, which, when well scrubbed with a mop of sedge, looks as clean as possible. In addition to this essential business of purifying and polishing, the oil of the ullachan does duty as bears' grease for the hair, and some of the young damsels, when fresh from their unctuous labours, must be admitted to shine considerably in society.

These "savages," natives of this region, have become tolerably acquainted with European trade and visitors, through the periodic circuits made by the Hudson's Bay Company's officers. They are very treacherous to each other still, as they were formerly to whites, but as they have found that they cannot do so with impunity, they do not now molest such, though what any great temptation might induce, it would be hazardous to say.

White shells, "hiaquays," found only on the West side of Vancouver Island, are used as small change all along the coast and in many parts of the interior, thus practically corresponding to the cowry of the East. They also form them into fanciful ornaments.

The name of Wakash nation, which has been applied to the natives in some charts, was so used by Cook. The word "Wakash," was very frequently used by them at Nootka Sound, and seemed to express approbation or friendship. Hence he proposed the appellation Wakashians.*

CLIMATE.—The winter is generally very stormy, with heavy rain in the months of November and December, the S.E. wind then prevails. There is some frost about the low land in the beginning of January, which is seldom of long duration, and never interrupts the agricultural operations.

Early in February vegetation begins to advance, and about the commencement of March everything assumes the beautiful hue of spring. April and May bring in alternate warm showers of rain and sunshine, and the heat becomes extremely

* Vol. II. p. 337.

oppressive in the months of June and July. In August and the beginning of September vegetation dries up from the drought of summer, and is then easily ignited, which is generally done by the natives when passing along the coast in their canoes. The weather being then very foggy, still, and close, the atmosphere becomes so much darkened by the fog and smoke combined, that the sun occasionally appears to be of a deep red colour even at noonday. In the month of October the rainy season sets in, the soil being then moist and the weather not very cold, the grass grows vigorously. The pasturage for the cattle is then better than during the two preceding months.

Fogs are very embarrassing from their density and duration. Sir George Simpson says that Mr. Finlayson was, in the year 1837, held a prisoner for a fortnight within a few miles of his home, by a fog worthy of keeping Christmas in London. This was in Johnston's Straits.—(Vol. i. p. 187.) These occur also to the southward.

The fur bearing animals generally hunted on the island are beaver, both black and grizzly, racoon, minks, land-otter, &c., and the sea-otter is hunted about Nitinat and Scott's Islands. The elk and deer are said to be abundant in the interior of the island. The fish generally taken by the natives in the vicinity are as follows, viz., halibut, flounders, skate, rock cod, sardine, salmon, trout, and several varieties of the herring species.

In the ensuing description of Vancouver Island and the adjoining continental shore, we resume the accounts from the point at which the preceding chapter concluded, proceeding along the *inner* side of the *island*, by the Gulf of Georgia, to the northward, and then to the southward on the *outer* or seaward face, terminating at De Fuca's Strait and the settlement of Victoria.

FRAZER RIVER debouches 7 miles to the northward of Point Roberts, described on page 388, but its mouth was passed unsuspected by Vancouver. He found his progress along the eastern shore much impeded by a shoal extending W. by N. 7 or 8 miles from Point Roberts, from whence it stretches N.W. by W., about 5 or 6 miles farther. It is at this time that the entrance to the channel of the river is to be found.

According to the Admiralty chart, executed from a drawing by Mr. Emilius Simpson, in the Hudson's Bay Company's schooner *Cadborough*, in 1827, this entrance was about half a mile in width, with a depth of 5 or 6 fathoms in it, abruptly commencing from 20 and 50 fathoms outside. This depth, with one or two exceptions of 3 and 4 fathoms, may be carried up to Fort Langley.

Frazer River had never been wholly descended by whites previously to 1828, when, in order to explore the navigation all the way to the sea, Sir George Simpson started from Stuart's Lake with three canoes. He found the stream hardly practicable even for any craft, excepting that, for the first 25 miles from its mouth, it might receive large vessels. This river, therefore, is of little or no use to England as a channel of communication with the interior; and in fact the trade of New Caledonia, the very country which it drains, is carried overland to Okanagan, and thence down the Columbia.*

* Sir George Simpson, Journey Round the World, in 1841-42, vol. i. p. 182. Sir George Simpson, in 1842, found a large camp of about 1,000 Indians, inhabitants of Vancouver Island, who

POINT GREY,* a low bluff point, is 7 leagues northward from Point Roberts. Vancouver's remarks upon the mouth of Frazer River are as follow:—"The intermediate space (between Points Roberts and Grey) is occupied by very low land, apparently a swampy flat, that retires several miles, before the country rises to meet the rugged snowy mountains, which we found still continuing in a direction nearly along the coast. This low flat being very much inundated, and extending behind Point Roberts to join the low land in the bay to the eastward of that point, gives its high land, when seen at a distance, the appearance of an island; this, however, is not the case, notwithstanding there are two openings between this point and Point Grey. These can only be navigable for canoes, as the shoal continues along the coast to the distance of 7 or 8 miles from the shore, on which were lodged, and especially before these openings, logs of wood and stumps of trees innumerable."†

The bank to the North of the river was named *Sturgeon Bank* by Vancouver, from the circumstance of his having purchased from the natives some excellent fish of that kind, weighing from 14 to 200 pounds each.

A most excellent leading mark was observed by him for carrying along its western extremity in 6 fathoms; this is, Passage and Anvil Islands, in Howe's Sound to the northward, in one. It deepens from this suddenly to the westward, and in many places to the eastward, shoaling as suddenly to 3, 2, and 1 fathom.

The N.E. coast of Vancouver Island is laid down from the surveys of Galiano and Valdes. Vancouver only touched here and there upon it, his object being to trace the connexion of the continental coast, in order to set at rest the question then much controverted, of an entrance to a great inland navigation. The West coast of the Gulf of Georgia in this part is composed of steep rugged rocks.

BURRARD'S CANAL, the Brazo de Florida Blanca of Galiano and Valdes, runs to the eastward from Point Grey. Its North point is *Point Atkinson*, lying North, about a league distant from the former. The southern side of the canal or inlet is of moderate height, and though rocky, well covered with trees of large growth, principally of the pine tribe. On the northern side the rugged snowy barrier, whose base is now nearly approached, rises very abruptly, and is only protected from the wash of the sea by a very narrow border of low land. At about a league within its entrance is an island, which nearly closes up the passage, but forming a passage of from 10 to 7 fathoms deep, not more than a cable's length in width. There is apparently a similar passage South of it, with a smaller island lying before it. It extends eastward about 6 miles farther, to a small opening leading to the northward, with two little islands before it; it runs nearly North for 3 leagues, and is very narrow, terminating in a small rivulet.

Vancouver met in the middle part of the inlet with a party of about 50 Indians, who were very friendly.‡

HOWE'S SOUND was named by Vancouver after the Admiral Earl Howe, and was called Brazo de Carmelo, by Galiano and Valdes. It extends to the

---

periodically cross the gulf to Frazer River for the purpose of fishing. They assisted in getting wood, water, &c., for the steamer he was in.—Vol. i. p. 183.

* Point Langara of Galiano and Valdes. † Valcouver, vol. i. p. 300.

‡ Vancouver, vol. i. pp. 300—302, 3n7.

northward of Point Atkinson, which is its S.E. point of entrance, Point Gower being its western limit. Between these is an extensive group of islands of various sizes. The shores of these, like the adjacent coast, are composed principally of rocks rising perpendicularly from an unfathomable sea; they are tolerably well covered with trees, chiefly of the pine tribe, though few are of a luxuriant growth.

The entrance between Point Atkinson and the island West of it is about 3 miles wide; and nearly in the centre between these two points is a low rocky island producing some trees, to which Vancouver gave the name of *Passage Island.* The channels on either side of it appear to be good.

"Quitting Point Atkinson, and proceeding up the sound, we passed on the western shore some small detached rocks, with some sunken ones amongst them, that extend about 2 miles, but are not so far from the shore as to impede the navigation of the sound. The low fertile country we had been accustomed to see to the southward, though lately with some interruption, no longer existed; its place was now occupied by the base of the stupendous snowy barrier, thinly wooded, and rising from the sea abruptly to the clouds, from whose frigid summit the dissolving snow or foaming torrents rushed down the sides and chasms of its rugged surface, exhibiting altogether a sublime though gloomy spectacle, which animated nature seemed to have deserted. Not a bird nor living creature was to be seen, and the roaring of the falling cataracts, in every direction, precluded their being heard, had any been in our neighbourhood."

At 4½ miles above Passage Island, on the eastern shore, is an island lying opposite to an opening on the western, which separates the islands, and opens into a numerous assemblage of rocky islands and rocks. At 5 miles N. by E. from the North point of the channel, is an island, which, from the shape of the mountain composing it, was called by Vancouver *Anvil Island,* and has been previously alluded to as forming with Passage Island a leading mark to clear West of the Sturgeon Bank. Between Anvil Island and the point above mentioned are three white rocky islets, lying about a mile from the western shore. The width of this branch of the sound is about a league, but northward of Anvil Island it soon narrows to half that breadth; taking a direction to the N.N.E. as far as lat. 49° 39′, when it terminates in a round basin, encompassed on every side by the dreary country already described. At its head, and on every part of its eastern shore, a narrow margin of low land runs from the foot of the barrier mountains to the water-side, which produces a few dwarf pine trees, with some little variety of underwood. When Vancouver, whose is the only account we have, visited this place in June, 1792, he found the water of the sound in the upper part nearly fresh, and in colour a few shades darker than milk; this was attributed to the melting of the snow, and its water passing rapidly over a chalky surface, which appeared probable by the white aspect of some of the chasms that seemed formerly to have been the course of waterfalls, but were then dried up. When at anchor in a little cove they were visited by a party of natives.

The starboard or continental shore of the Gulf of Georgia takes a direction about W.N.W. from Point Gower, and affords a more pleasing appearance than the shores of Howe's Sound. This part of the coast is of a moderate height for some distance inland, and it frequently juts out into low sandy projecting points.

The country in general produces forest trees in great abundance, of some variety and magnitude; the pine is the most common, and the woods are little encumbered with bushes or trees of inferior growth.

For about 5 leagues along the coast there are some rocks and rocky islets, when an island of about 2 leagues in circuit is reached, with another about half that size to the westward of it, and a smaller island between them. From the North point of this island, which forms a channel with the main about half a mile wide, and is situated in lat. 49° 28½′, the coast of the continent takes a direction for about 8 miles N. 30° W., and is composed of a rugged, rocky shore, with many detached rocks lying at a little distance, to the entrance of Jervis's Canal.

JERVIS'S CANAL or INLET, named by Vancouver after Admiral Sir John Jervis, is the Brazo de Mazarredo of Galiano and Valdes. Its entrance, in lat. 49° 35½′ N., is about half a mile wide, and is navigable, leading amongst a cluster of rocks and rocky islets, lying just in front of its entrance. About 2 miles to the S.S.E. is an island about a mile long, with a very dreary, uncomfortable cove at its South point. This branch of the inlet winds towards the N.N.E. for about 9 miles, where it meets the western branch of the entrance formed by a large island. Hence it continues in an irregular course to the northward. At the above junction are a few rocky islets, which are steep, as no soundings with the hand line could be gained, nor can the bottom in the channel southward be reached with 100 fathoms of line, though the shores are not a mile asunder. The shores are of a moderate height within a few miles of the lower part of the inlet, and are principally composed of craggy rocks, in the chasms of which a soil of decayed vegetables has been found, producing some few pine trees of inferior dwarf growth, with a considerable quantity of bushes and underwood. From the point above mentioned, the inlet assumes a N.W. direction, with an increased width for 2 leagues, when it proceeds nearly N.E. to a point in lat. 50° 1′, when it assumes a N.W. by W. direction, without any contraction in its width, for 3 leagues, and terminates as usual in swampy low land, producing a few maples and pines, in lat. 50° 6′. The surrounding country in the upper portion of this arm has an equally dreary appearance with that about Howe's Sound, but the rugged surface of the mountains is infinitely less productive. A few detached dwarf pine trees, with some berry and other small bushes, are the only signs of vegetation. The cataracts here rushed from the rugged snowy mountains in greater number, and with more impetuosity, than in Howe's Sound, yet the colour of the water was not changed, though in some of the gullies there was the same chalky aspect.

Through the small space of low land which extended from the head of the inlet to the base of the mountains there flowed three small streams of fresh water, apparently originating in one source in the N.W. corner of the bay formed by the head of the inlet; in which point of view there was an extensive valley, that took a northerly direction, uninterrupted as far as could be seen, and was by far the deepest chasm seen by Vancouver in the descending ridge of the snowy barrier, without any appearance of elevated land rising behind it. In all these arms of the sea a visible, and sometimes a material, rise and fall of the tide were observed, but a constant drain down to seaward. A few civil native Indians were also communicated with.

The entrance to the inlet, as before mentioned, is formed by an island of about 3 leagues in length, with several small islets about it. The N.W. point of the entrance was named *Scotch Fir Point*, from its producing the first trees seen of that species.

The *Gulf of Georgia*, at this point, is divided into two channels by the island named Favida (Vancouver, vol. i. p. 313), Feveda (*ibid.* vol. i. p. 318), or Texada, according to the charts.

It was in this part that Vancouver met, to his great surprise and mortification, with the two Spanish surveying vessels which had preceded him. These were the brig *Sutil*, under Don D. Galiano, and the schooner *Mexicana*, under Don C. Valdes, detached from the commission under Malaspina, from whom, however, he met with the most polite and friendly attention.

The two straits have been named Malaspina Strait to the northward of the island, and Rosario Strait to the southward. The former is much the narrowest, and is that traversed by Vancouver. For the representation of the other we are indebted to the labours of the above-mentioned Spanish commanders, and also for nearly all the line of the N.E. coast of Vancouver Island. Their observations, however, did not extend beyond the exterior shores, as the extensive arms and inlets Vancouver examined, he states had not claimed their attention.*

FAVIDA ISLAND is 32 miles in length. Its S.E. point, named *Point Upwood*, situated in lat. 49° 28½′, is chiefly composed of one lofty mountain, visible at the distance of 20 leagues and upwards, and is very narrow, appearing from the lower parts of the gulf like a small lofty island. The island is very narrow, and its northern shores, nearly straight and compact, are principally formed of rocky substances of different sorts, amongst which slate is in abundance; and the trees which it produces are of infinitely more luxuriant growth than those on the opposite shore. Its N.W. extreme is *Point Marshall*, in lat. 49° 48′.†

Sir Geo. Simpson, who traversed this strait in the Hudson's Bay Company's steamer with Capt. M'Neill in 1842, says:—"We anchored in the snug harbour of the Island of Favida, to take in wood and water. Capt. M'Neill preferred halting here on account of the superiority of the fuel, which was both close in the grain and resinous; and he stated that a cord of it was almost as durable as two cords of any other growth. For this singular fact there must be a reason, which may be expected to lurk rather in the soil than in the climate; and, whether or not the two peculiarities be respectively cause and effect, the isle in question is almost entirely composed of limestone, which, if it exist elsewhere on the coast, is found only in very small quantities."—(Journey, &c., vol. i. p. 184.)

From Point Marshall, N.W. by W., about a league distant, lies an island of

* Vancouver, vol. i. p. 313.

† Beware Harbour, at the North end of Favida Island, is said to be famous for the abundance of its herring spawn. The natives collect it by means of pines laid on the beach at low water, where, after the next flood has retired, they are found to be covered with the substance in question to the thickness of an inch. When dry it is rubbed off for use, being washed in fresh water before eating. Sir George Simpson found snug anchorage in the harbour during bad weather.—Vol. i. pp. 239-40.

moderate height, with a smaller one to the S.W. of it named *Harwood's Island*, (Concha of Galiano and Valdes). Between it and Point Marshall are some rocky islands and sunken rocks.

The MALASPINA STRAIT is about 5 miles wide off Scotch Fir Point, and from this point the shores approximate to the westward until they come within 2 miles of each other at its western end; and are, as well on the island as on the continental side, nearly straight, perfectly compact, and rise gradually, particularly on the continental shore, from a beach of sand and small stones, to a height that must be considered rather elevated land, well clothed with wood, but without any signs of being inhabited. From hence the continental shore takes a N.W. direction, and opposite Marshall Island is a small brook, probably of fresh water. In advancing, the shores put on a very dreary aspect, chiefly composed of rugged rocks, thickly wooded with small dwarf pine trees. The islands, however, which here close up the breadth of the gulf, have a more pleasing and fertile appearance.

SAVARY'S ISLAND is the first of these; it is about 2 leagues in length, East and West, and half a league broad. Its N.E. point, in lat. 49° 57¼', forms a passage with *Hurtado Point* on the continental shore, and on its South side are numberless sunken rocks, nearly half a league from its shores, visible probably only at low water.

The *N.E. coast of Vancouver Island* from that part opposite to Point Marshall is compact, rising in a gentle ascent from the sea-shore to the inland mountains (some of which were covered with snow), wearing a pleasant, fertile appearance. This continues to the narrow channel formed by the peninsula of Point Mudge.

DESOLATION SOUND.—This tract, which may be considered as the head of the Gulf of Georgia, is, as before stated, occupied by a numerous archipelago of islands, of very various areas.

Point Sarah, which is the limit of the N.W. direction of the continental shore, is in lat. 50° 4½'; to the southward there are numerous islets and rocks. It is also the point where the coast turns to the eastward, and forms an arm, called by the Spanish surveyors before alluded to, and who here formed a portion of the surveying party with Vancouver, the *Brazo de Malaspina*. It leads in a S.E. direction, almost parallel with, and 2 or 3 miles from, the northern shore of the gulf to the distance of about 3 leagues, with a smaller branch near the middle, extending about a league from its northern shore to the N.N.E.

From the mouth of this inlet the continental shore continues in an easterly and N.E. direction, and for 2 leagues is much indented; and several small islands and rocks lie near it to the lat. of 50° 10′ N., where they disappear, and the coast takes a winding course N.W. and West to a point bearing N.W. by N., about 2 leagues from the former point, and forming the East point of the *Bruzo de Toba*; the entrance of which, about half a league wide, has two small islets in it. This arm extends in an irregular N.E. direction to the latitude of 50° 22′, where it terminates in shallow water and a little low land, through which flow two small rivulets. In these rivulets and on the shoal parts several weirs were erected. Along the shores of the upper part of this arm, which are mostly composed of high, steep, barren rocks, were several fences formed by thin laths, stuck either in the ground or in chinks of the rocks, with others placed on them in different directions. Ranges of these were fixed in various ways, and were imagined to be intended for drying fish, but this was very uncertain.

The country presents one desolate, rude, and inhospitable aspect, and was nearly destitute of inhabitants. An extensive deserted Indian village was found by Vancouver's party on a rock, whose perpendicular cliffs were nearly inaccessible on every side, and connected to the main by a narrow neck of land, about the centre of which grew a tree, from whose branches planks were laid to the rock, forming by this means a communication that could easily be removed, to prevent their being molested by their internal unfriendly neighbours; and protected in front, which was presented to the sea, by a platform, which, with much labour and ingenuity, had been constructed on a level with their houses, and overhung and guarded the rock.

"Whilst examining these abandoned dwellings, and admiring their rude ingenuity, our gentlemen were assailed by an unexpected enemy, whose legions made so furious an attack upon each of their persons, that, unable to vanquish their foes, or to sustain the conflict, they rushed up to their necks in water. This expedient, however, proved ineffectual; nor was it until after all their clothes were boiled, that they were disengaged from an immense hoard of fleas, which they had disturbed by examining too minutely the filthy garments and apparel of the late inhabitants."* This incident will afford a hint to any other visitor to such locality.

BUTE'S CANAL or INLET lies 13 or 14 miles to the westward of the previous arm. Its entrance is about 2 miles wide, and leads in an irregular northern direction to the lat. of 50° 52′, or a distance of 12 leagues, when, in the usual manner, it is terminated by a small tract of low land, from whence a shallow bank stretches into the arm, which rapidly increases from 2 to 50, 70, and 100 fathoms in depth, and then becomes unfathomable.

Behind this low small spot of ground, the mountains rise very abruptly, divided by two deep valleys, from whence issued streams of fresh water, though not sufficiently copious to admit boats. In these valleys, and on the low plains, pine trees grow to a tolerable size; the few seen on the mountains are of very stinted growth. High, steep, barren rocks, capped with snow (in June), formed the sides of the canal, the water of which, at its head, was nearly fresh, and of a pale colour.

On a point, in lat. 50° 24′, near the West side of the entrance, Vancouver's party found an Indian village, situated on the face of a steep rock, containing about 150 natives, who were particularly civil and friendly to his people.

To the West of the point on which this village stood, a channel, with a very narrow opening, stretches to the westward, and through which so strong a current flowed that they could not row against it, but were towed from the rocky shores by their Indian friends. Having passed these narrows the channel widens, and the rapidity of the tide decreased.

We will now advert to the islands which occupy the whole of the interval of the coast thus described. Those to the southward are numerous, and are mostly of moderate height from the sea, tolerably well wooded, and the shores not

* Vancouver, vol. i. p. 325.

wholly composed of rugged rocks, affording some small bays bounded by sandy beaches. At the time of Vancouver's visit, in June, numberless whales enjoying the season were playing about the ship in all directions, as were also several seals.

It would be almost impossible to give a definite description of each island composing this archipelago, even if it could prove useful; the chart must fill up all vacancies in the verbal accounts. The first island of magnitude is *Mary Island*, whose S.E. extreme, *Point Mary*, lies N. 72° W. about half a league distant from Point Sarah. Its northern part is named *Cortes Island* on the Spanish chart. To the N.E. of this island is *Redonda Island*, which exceeds it in size, and forms the channel in front of the entrance to the Toba Inlet. Vancouver anchored between these two islands. He says:—"Our situation presented as gloomy and dismal an aspect as nature could well be supposed to exhibit, had she not been aided a little by vegetation, which, though dull and uninteresting, screened from our sight the dreary rocks and precipices that compose these desolate shores, especially on the northern side (that is, the S.W. shore of Redonda Island); as the opposite shore, though extremely rude and mountainous, possessed a space of nearly level land, stretching from the water side, on which some different sorts of the pine tribe, arbor vitæ, maple, and the oriental arbutus seemed to grow with some vigour, and in a better soil.

"Our residence here was truly forlorn; an awful silence pervaded the gloomy forests, whilst animated nature seemed to have deserted the neighbouring country, whose soil afforded only a few small onions, some samphire, and here and there bushes bearing a scanty crop of indifferent berries: nor was the sea more favourable to our wants; the steep rocky shore prevented the use of the seine, and not a fish at the bottom could be prevailed on to take the hook."* This character caused Vancouver to apply the name of Desolation Sound to their locality.

Stuart's Island lies in the entrance of the Bute Inlet, off the point on which stands the village previously mentioned. It is nearly a round island, 3 or 4 leagues in circuit.

VALDES ISLAND is by far the largest of all the islands in the vicinity, and forms the continuation with Vancouver Island of the Gulf of Georgia, called Discovery Passage.

Point Mudge, its S.E. extremity, is the termination of a long and narrow peninsula, which Valdes Island here forms. It is in lat. 50° 0′ N., lon. 125° 2′ W. On Point Mudge Vancouver found a very large village of natives, many of whom visited the party on their passing and repassing by it, who uniformly conducted themselves with the greatest civility and respect. On the western shore, immediately without the entrance of the inlet, they found a rivulet of excellent fresh water. The passage up the inlet is perfectly free from danger, and affords good anchorage. Round Point Mudge, at the distance of about half a mile, is a ledge of sunken rocks; these are, however, easily avoided by the weeds which they produce. The village stands nearly at the summit of a steep sandy cliff or precipice, about 100 feet high, and nearly or quite perpendicular. The houses were built similar to those at Nootka, and were considered capable of containing

* Vancouver, vol. i. pp. 331-2.

300 persons. About 70 canoes also were lying on the beach at the foot of the cliff. Sir George Simpson says, that in 1842 there were three villages of Comoucs opposite Point Mudge.

DISCOVERY PASSAGE, so named after Vancouver's sloop, runs N.W. from Point Mudge to Point Chatham, in lat. 50° 19½'. About 4 leagues up the inlet is Menzies Bay, on the West side. From Point Mudge to this bay the channel is nearly straight; the western shore is compact; the eastern arm has some rocky islets and rocks lying near it; it is about half a league wide. In turning up Vancouver found not the smallest obstruction, as the shores are sufficiently bold for vessels to stand as close to them as inclination may direct.

Menzies Bay, on the western side, affords commodious anchorage; there is also another but smaller situation on the opposite side of the arm. Nearly in the centre of Menzies Bay is a shallow bank of sand, with a navigable passage all around it. The ships were stationed between this bank and the North side of the bay, near a small Indian village, whose inhabitants had but little to dispose of, though they were civil and friendly. The latitude was observed at 50° 7' 30".

Immediately above this part the passage contracts to a short half mile, by the projecting land that forms the North side of these two bays, and by an island on the eastern shore (navigable round for boats only), which projects so far as to reduce the channel to nearly one-half its width. The tide setting to the southward through this confined passage, rushes with such immense impetuosity as to produce the appearance of falls considerably high, though not the least obstruction of either rock or sand, so far as was examined, appeared to exist. The returning tide to the North, though very rapid, does not run with such violence; this was estimated to move at the rate of 4 or 5 miles, the other at 7 or 8 miles an hour. They seemed regular in their quarterly change, but the visible rise and fall of the shore in this situation was so inconsiderable as to allow merely to distinguish the ebb from the flood tide.

To give an idea of the strength of the current in these straits during strong winds, Sir George Simpson states, "that being driven through with the speed of a race, no bottom at times could be found with two deep sea lead lines fastened together, even when the actual depth did not exceed 30 or 40 fathoms."* He also says, that in passing a village near the narrowest part of Johnstone's Straits, that they "were greatly impeded by deep whirlpools and a short sea, which were said generally to mark these narrows, and to be caused by the collision of current or tides flowing round the opposite ends of Vancouver Island from the open ocean."

From these narrow parts the passage gradually increases its width to a mile and half a league, and communicates with Johnstone's Straits, in nearly the same N.N.W. direction, about 4 leagues farther, without any visible obstruction or impediment to the navigation. The eastern shore, like that to the northward, is much broken; the western shore is firm, and affords some small bays, in which there is good anchorage.

Point Chatham, the N.W. termination of Discovery Passage, and named after that vessel's consort, is in lat. 50° 19½', and is rendered conspicuous by the

* Journey Round the World, vol. i. p. 230.

confluence of three channels, two of which take their respective directions to the westward and south-eastward toward the ocean, as also by a small bay on each side of it, by three rocky islets close to the South of it, and by some rocks, over which the sea breaks, to the North of it.

The channel to the northward is *Nodales Channel*, and leads from the branch noticed on page 399, as being the point where Vancouver met with such strong tides. Its N.W. side is formed by *Thurlow Island*, which is narrow, and about 8 leagues long, East and West.

In pursuing the continental coast to the westward of the entrance to Bute's Inlet, two inlets leading to the N.W., each about a league in extent, will be passed; the easternmost, according to the Spanish chart, leading to a basin, called the *Estero Basin*. The channel formed to the southward by Thurlow Island is called by the same authority, *Cardero Canal*; it is very intricate, and unfit for shipping, in consequence of the irregular direction and rapidity of the tides, and to the great depth of water.

LOUGHBOROUGH'S CANAL, which runs to the northward, at the western end of Cardero Canal, is one of those extensive inlets so characteristic of this region. It extends for nearly 12 leagues inland, though in many parts not more than a mile wide, between steep and nearly perpendicular mountains, from whose summits the dissolving snow descended their rugged sides in many beautiful cascades. An islet and some rocks lie off its East point of entrance; and in lat. 50° 46′ it appears, in ascending it, to terminate, but this arises from two interlocking points, as the inlet itself proceeds N.E. by N., about 3 leagues farther, terminating in the usual low land. The flood tide here varies about four hours from what occurs in the Gulf of Georgia, an evidence that it comes from the North end of Vancouver Island.

Thurlow's Island, previously mentioned, lies before the entrance to this inlet. On the South side of the island Vancouver stopped in a bay, in lat. 50° 23′, which afforded good anchorage; wood and water, too, were easily procured.

Hardwicke's Island lies to the westward of Thurlow's Island, and is about 4 leagues in extent. The channel formed by it and the main is intricate, containing many sunken rocks and rocky islets, occasioning great irregularity in the tides, which are here extremely violent.

PORT NEVILLE, the Brazo de Cardenas of the Spanish surveyors, is an inlet on the North shore, about 7 miles West of the West end of Hardwicke's Island. Off its West point lies a small island; its entrance is about half a mile wide, but with no more than 4 fathoms water in mid-channel, from whence it extends about 8 miles in a direction N. 75° E., the depth increasing to 5, 6, and 7 fathoms, affording good anchorage about two-thirds of the way up; beyond which limit, like all the rest of these canals, it terminates in shallow water. The country in it is more pleasant than that seen from Johnstone's Straits, and the soil at the upper part is composed of black mould and sand, producing pine trees of large dimensions. A run of water was seen at its head, but inaccessible from the sea, on account of shoals. A deserted village was seen on the northern shore, possibly proving the existence of the only undiscovered requisite, fresh water, to constitute Port Neville a very snug and commodious port.

The southern side of Johnstone's Straits, opposite to Port Neville, is nearly straight and unbroken, rising abruptly from the sea to mountains of great height. Some few signs of inhabitants were seen by Vancouver's party on this side, but not the least indication of such to the westward of the Narrows.

CALL'S CANAL or CREEK, the Brazo de Retamal of Galiano and Valdez, commences on the North shore, at 15 miles to the westward of Port Neville. It extends irregularly, like the rest, in a N.E. direction, to lat. 50° 42½′, terminating in shoaler water at its head. To the N.W. of it is the collection of islands to which the collective name of Broughton's Archipelago was given by Vancouver.

At 23 miles westward of the entrance of Call's Canal, and on the South side, is the native village, under the chief Cheslakees, at which Vancouver anchored in July, 1792. This collection of timber houses, thirty-four in number, constructed in the same manner as those at Nootka Sound, but rather less filthy, was pleasantly situated on a sloping hill, above the banks of a fine fresh-water rivulet, discharging itself into a small creek or cove. It was exposed to a southern aspect, whilst higher hills behind, covered with lofty pines, sheltered it completely from the northern winds.

A small sandy island, lying to the eastward of the village, affords, between it and the land on which the town is situated, a small but very commodious anchorage. This is not, however, to be approached by the passage to the South of the island, that being navigable only for very small vessels. To the South of the village the valley extended, apparently to a considerably distance, in a south-westerly direction; through it the very fine stream of fresh water (*Nimpkish River*) emptied itself into the sea, and from the many weirs that were seen in it, it was unquestionably well stocked with fish, though none were offered for sale. It is the only stream as yet known to be navigable for any distance inland. All the people of the village were very friendly, and seemed as familiar with fire-arms as if they had been accustomed to them from their earliest infancy. The sandy island is placed by Vancouver in lat. 50° 35½′ N., lon. (corrected) 127° 16′ W. Variation (July, 1792) 20° 45′ E.

The shores on each side of the channel materially decrease in height toward this part. That to the northward is much broken, and is composed of islands, whilst that to the southward continues compact and entire. The islands to the North are generally formed by low land near the shore, rising to a moderate height, well wooded, and on them the smoke of several fires was observed. This circumstance, together with the number of inhabitants on the southern shore, and the many canoes that were seen passing and repassing, evidently bespoke this country to be infinitely more populous than the shores of the Gulf of Georgia.

JOHNSTONE'S STRAITS.—The channel we have been describing may be said to terminate here; its eastern limit being Point Chatham, mentioned on page 401. Discovery Passage, extending thence to Point Mudge, is not particularized in Vancouver's original survey. The following observations are made by that excellent observer upon the entrance passage:—

"The length of coast from Point Mudge to this station, about 32 leagues, forms a channel which, though narrow, is fair and navigable; manifested by the adverse winds obliging us to beat to windward every foot of the channel, and to form a complete traverse from shore to shore through its whole extent, without meeting the least obstruction from rocks or shoals. The great depth of water, not only here, but that which is generally found washing the shores of this very broken and divided country, must ever be considered as a very peculiar circumstance, and a great inconvenience to its navigation. We, however, found sufficient number of stopping places to answer all our purposes, and, in general, without going far out of our way. In coming from the westward, through Johnstone's Straits, the best channel into the Gulf of Georgia, in thick weather, might, though not easily, be mistaken. Such an error, however, may be avoided by keeping the southern shore (that is, of Vancouver Island) close on board, which is compact, and so steep that it may be passed within a few yards with the greatest safety; indeed, I have every reason to believe the whole of the passage to be equally void of dangers that do not evidently show themselves. The height of the land that composes these shores and the interior country has been already stated to decrease as we proceeded westward. The land on the southern side, which is an extensive island, appeared the most elevated, composed of very lofty mountains, whose summits, not very irregular, were still in some places covered with snow. The northern side, for a considerable distance, seemed less elevated, and the entire forest that covered its surface might have favoured the belief of great fertility, had we not known that pine trees innumerable are produced from the fissures and chasms of the most barren rocks, of which we had good reason to suppose the whole of the country before us was composed. Its low appearance may possibly be occasioned by its being much divided by water, as we evidently saw through an opening, about 4 miles only to the westward of that appointed for our rendezvous, a much greater space so occupied than that which comprehended these straits. Our general view to the northward was, however, bounded by a mountainous country, irregular in the height of its eminences, and some of them capped with snow. The retired hills of the most eastern part of the straits were, as we passed, so obscured by the high, steep, rocky cliffs of the shores, that we were unable to describe them with any precision. As the elevation of the northern shore decreased, I was in expectation of seeing a continuation of that lofty and connected range of snowy mountains, which I have repeatedly had reason to consider as the insurmountable barrier to any extensive inland navigation. Herein I was disappointed, as this lofty structure either decreases in its vast degree of elevation, or it extends in a more inland direction.

" The residence of all the natives we had seen, since our departure from Point Mudge, was uniformly on the shores of this extensive island (Vancouver Island), forming the southern side of Johnstone's Straits, which seems not only to be as well inhabited as could be expected in this uncultivated country, but infinitely more so than we had reason to believe the southern parts of New Georgia. This fact established, it must be considered as singularly remarkable that, on the coast of the opposite or continental shore, we did not discover even a vestige of human existence, excepting the deserted villages! This circumstance, though it countenances the idea of all the original inhabitants of the interior country having migrated, fallen by conquest, or been destroyed by disease, still leaves us unable to adduce any particular reason as the cause of this evident depopulation. The width of the passage, scarcely anywhere exceeding 2 miles, can hardly have induced the inhabitants of the northern side to quit their dwellings for a residence on the opposite shore, merely for the purpose of being that small distance nearer to the commerce of the sea-coast. On regarding the aspect of the two situations, and on reflecting that the winter season, under this parallel, must be inclement and severe, it appears reasonable to suppose that any human beings, not restrained in fixing their abode, would not hesitate to choose the very opposite side to that which is here preferred, where, in general, their habitations front a bleak northern aspect, with mountains rising so perpendicularly behind them, that if they do not totally, they must in a great measure exclude the cheering rays of the sun for some months of the year. The northern side labours not under this disadvantage, and enjoying the genial warmth denied to the other, at certain seasons, most probably possesses the requisites necessary to their present mode of life, at least in an equal degree, especially as this country has in no instance received the advantages of cultivation. This would appear to be the situation of choice, the other of necessity; for the same source of subsistence, which is evidently the sea, affords equal supplies to the inhabitants of either shore. And that there was a time when they resided on both, is clearly proved by their deserted habitations yet in existence on the northern shore."*

Sir George Simpson, who visited the northern part of this most extraordinary course of inland navigation in the world, in the Hudson's Bay Company's steamer, thus traversed it twice. He concludes:—"According to the whole tenor of my journal, this labyrinth of waters is peculiarly adapted for the powers of steam. In the case of a sailing vessel, our delays and dangers would have been tripled and quadrupled, a circumstance which raised my estimate of Vancouver's skill and perseverance at every step of my progress. But, independently of physical advantages, steam, as I have already mentioned, may be said to exert an almost superstitious influence over the savages; besides acting without intermission on their fears, it has, in a great measure, subdued their very love of robbery and violence. In a word, it has inspired the red man with a new opinion—new, not in degree, but in kind—of the superiority of his white brother."†

BROUGHTON'S ARCHIPELAGO, the cluster of islands to the northward of Johnstone's Straits, were so named from Mr. Broughton, Vancouver's officer, who first explored them. The chart, the result of that survey, gives the best idea of their relative configuration. The separate islands have not received distinct names, and of course are commercially unimportant at present.

From the mouth of Call's Canal, page 403, the North shore of the straits is formed by an island about 7 miles in length, having between it and that next West of it a narrow and rocky channel. This island is about 6 miles in length, and the western channel, though very unpleasant, on account of the many rocks in it, is infinitely less dangerous than that to the eastward of the island, which is by no means advisable for ships to attempt.

From this entrance to Point Duff, the distance is about 12 miles, nearly North;

* Vancouver, vol. i. pp. 349—351. † Journey round World, vol. i. pp. 240-1.

and to the South and East of the latter is an opening (unexplored) which leads to Knight's Canal.

KNIGHT'S CANAL or INLET, the Brazo de Vernacci of Galiano and Valdes, runs a little northward of East, to the latitude of 50° 45', when its width increases to near a league, taking an irregular northerly direction, to its final termination in lat. 51° 1', lon. 126° 0'. The shores of it, like all the rest, are formed by high, stupendous mountains, rising almost perpendicularly from the water's edge. The dissolving snow on their summits produced many cataracts, that fell with great impetuosity down their barren, rugged sides. The fresh water that thus descended gave a pale white hue to the canal, rendering its contents entirely fresh at its head, and drinkable for 20 miles below it. This dreary region was not, however, destitute of inhabitants, as a village was discovered a few miles from its upper extremity. At the point where Knight's Canal turns to the northward, the head of Call's Canal approaches the southern shore to within half a mile, forming the western part into a peninsula.

At 15 miles due West from this narrow isthmus there is a wide passage leading from the North side of Knight's Canal, with two islands lying in its entrance. The passage winds to the northward and westward for about 5 leagues, having on the starboard or continental shore two smaller inlets, named by the Spanish surveyors the *Brazo de Balda* and *Baldinat*.

Deep Sea Bluff, the N.W. extreme of this passage, and so named from its appearance and situation, is in lat. 50° 52', lon. 126° 43'. The channel itself is in some parts full of innumerable rocks and rocky islets, and in it the tides are very violent. The channel to the S.W. from Deep Sea Bluff separates two of the larger islands of Broughton's Archipelago, that to the northward being called Broughton Island in the recent Admiralty chart. It is about 12 miles in length, and was named *Fife's Passage* or *Sound*. Like all the rest, it is very deep, and the tides very irregular; on some days they were very rapid, on others scarcely perceptible; the uncertainty of the rise and fall, the time of high water, and other fluctuations and irregularities, being attributed to the influence of the winds, and the peculiar position as regards the open ocean.

Point Duff is the S.E. point of Fife's Passage. It is in lat. 50° 48'; a small rocky islet lies off it, covered with shrubs. *Point Gordon*, the opposite point of entrance, bears W. ¾ N. from Point Duff, and off it are several white, flat, barren rocks, lying a little distance from the shore.

The land within this western entrance is not very high, composed of rugged rocks steep to the sea, in the chasms and chinks of which a great number of stunted or dwarf pines are produced. Some few natives also visit this part.

Between Deep Sea Bluff and the East point of Broughton Island is an opening leading to a series of channels, which insulate the latter. The first, leading to the N.E., bears first to the N.E. and then to the West, forming a narrow isthmus. Vancouver found a convenient station for procuring wood and water, the only supplies this dreary region affords, near the southern part of this arm.

The shore of the main across this small opening from Deep Sea Bluff takes a N.W. ½ W. direction for about 4 miles: then extends N.N.E. about a league to a point where the arm takes an easterly course from an island and several rocky islets, forming passages for boats only to the eastward, while to the westward of the island, between it and *Point Philip*, a channel of about a mile in width. From this point it leads N.E. and E. for 6 leagues, to lat. 51° 0′ N., lon. 126° 25′ W., when it terminates in a similar way to the many before described. Its shores, about a mile apart, are composed of steep, high, craggy mountains, whose summits are capped with perpetual snow; the lower cliffs, though apparently destitute of soil, produce many pine trees, that seem to draw all their nourishment out of the solid rock. The water near 4 leagues from its upper end, in August, 1792, was of a very light chalky colour, and nearly fresh. From its shores two small branches extend, one winding about 4 miles to the S.E. and S.W.; the other about a league to the N.N.W.

MOUNT STEPHENS, conspicuous for its irregular form and its elevation above the rest of the hills in its neighbourhood, is in lat. 51° 1′, lon. 126° 53′. It will serve as an excellent guide to the entrance of the various channels with which this country abounds. It stands on the N.W. side of the inlet just described, and to the northward of Point Philip.

From Point Philip, the continental shore takes a W. by N. direction for about 2 leagues, forming an irregular channel with the North shore of Broughton's Island. At this point it becomes divided into several channels; that to the northward takes a direction E. by N. for near 2 leagues, terminating, as usual, at the base of Mount Stephens. This narrow and intricate passage is only admissible for boats, and appears to be a chasm caused by some natural convulsion. This idea originated in its differing materially in one particular from all the canals previously examined, namely—in its having regular soundings, not exceeding the depth of 13 fathoms, although its shores, like all those of the bottomless canals, were formed by perpendicular cliffs, from their snowy summits to the water's edge. The stupendous mountains on each side of this chasm prevent a due circulation of the air, and exclude the rays of the sun, rendering it a most uncomfortable place.

Westward of the entrance of this passage there is an excessively dangerous channel, so caused by the innumerable rocks with which it is bestrewed, and the irregularity of the tides, which runs first to the West and then to the South. To the westward of this again, an arm leads for 16 miles to the West. There are many rocky islets and sunken rocks in it, rendering it dangerous even for boats. Near its termination is a very narrow opening on its northern shore, winding towards the E.N.E., replete with overfalls and sunken rocks, and ending in a cascade, similar to many others hereabouts. These are perfectly salt, and seem to owe their origin to the tidal waters, which in general rise 17 feet, and, at high water, render these falls imperceptible, by the rocky bar being 4 or 6 feet beneath the surface. Within a few yards of one of these cascades, a considerable stream of *warm* fresh water was discovered.

Vancouver found a small Indian village on a rocky islet near this part. It was nearly all occupied, and rendered almost inaccessible by platforms, &c., similar, to those elsewhere described.

From the mouth of the principal arm, and of that previously described, a channel leads S.W. into the main channel, separating Vancouver Island from the continent. It was named by Vancouver *Well's Passage*, and is about 8 miles long. The S.W. point, in lat. 50° 51′, lon. 127° 18′, was named *Point Boyles*. Off it lie some islets, and hence the continental shore runs to the north-westward for 11 leagues to Cape Caution, which may be taken as the limit of Queen Charlotte Sound, the N.W. entrance of the very singular and extensive inland navigation we have been describing.

PORT M'NEIL, a few miles to the N.W. of the mouth of the Nimpkish or Nimkish River, previously described, is a point of some importance, as it is here that the principal *coal* deposit has been worked.*

The harbour has been frequented by the Hudson's Bay Company's steam-vessels, for trading in otter skins with the Quakeolths, in which these natives prove themselves most expert bargain makers. The name of the harbour, we presume, is derived from the commander of the company's vessel, *Beaver*. Of the nature of the harbour we have no particulars; from the chart, it appears to be separated from the anchorage, where the coal is procured, by a peninsula projecting to the eastward.

Thomas Point is 4½ miles from this anchorage, and is the S.E. point of *Beaver Harbour*, a bay fronted on the N.E. by a group of islands, having a sunken rock on its northern part. It is separated from *Hardy Bay* on the West by a peninsula.

The GOLETAS CHANNEL of Galiano and Valdes commences about this point, and forms an exit from the inland navigation to the southward of Queen Charlotte Sound, being separated from it by a range of one large and several smaller islands, to which the names of the Spanish surveyors, *Galiano* and *Valdes Islands* have been given. Vancouver did not examine this portion. To this survey,

* The coal of Vancouver Island is but a new feature in the importance of this territory. The following extract from the *Times*, Jan. 1848, will explain its origin:—

"On the North and East side of Vancouver Island, a recently discovered river debouches into Johnstone's Straits, near the mouth of which large seams of coal crop out on the surface of the soil. At the point, the trading steamer of the Hudson's Bay Company navigating the straits of Juan de Fuca obtains ready and plentiful supplies, which are put on board by the Indians at a mere nominal price. Mr. Dunn, who was a trader and interpreter in the Hudson's Bay Company's steamer *Beaver*, gives an interesting account of the discovery of this coal. He states:—'The cause of the discovery (of the coal) was as curious as the discovery itself was important. Some of the natives at Fort M'Loughlin having, on coming to the fort to traffic, observed coal burning in the furnace of the blacksmiths, in their natural spirit of curiosity made several inquiries about it; they were told that it was the best kind of fuel, and that it was brought over the great salt lake six months' journey. They looked surprised, and in spite of their habitual gravity, laughed, and capered about. The servants of the fort were surprised at their unusual antics, and inquired the cause. The Indians explained, saying that they had changed in a great measure their opinion of the white men, whom they thought endowed by the Great Spirit with the power of effecting great and useful objects, as it was evident they were not then influenced by his wisdom in bringing fuel such a vast distance, and at so much cost. They then pointed out where it could be found, of the richest quality, close to the surface, rising in hillocks, and requiring very little labour to dig it out. This intelligence having been reported at Fort Vancouver, we received instructions to make the necessary inquiries and exploration. Mr. Finlaison and part of the crew went on shore, and after some inquiries, and a small distribution of rewards, found from the natives that the original account—given at Fort M'Loughlin—was true. The coal turned out to be of excellent quality, running in extensive fields, and even in clumpy mounds, and most easily worked, all along that part of the country. The natives were anxious that we should employ them to work the coal. To this we consented, and agreed to give them a certain sum for each large box. The natives being so numerous and labour so cheap, for us to attempt to work the coal would have been madness.' It is earnestly to be hoped that this rich and valuable deposit may ere long be brought within the reach of the fast increasing number of our steamers on the West coast of America and the Pacific."

then, we are indebted for our charts. The eastern extremity of the range has been named the *Cordon Islands*, and commence off the N.W. point of Hardy Bay. Four miles W.N.W. from this point, on the South side of the channel, that is, on the coast of Vancouver Island, is *Port Mier*, and 14 miles farther in the same direction, the intervening coast being straight and compact, is *Port Shucartie*, a good harbour, of which we have no particulars. Opposite to this, on the South side of Galiano Island, is *Port Valdes*, and a short distance to the westward is the western entrance of the Goletas Channel, the N.W. point of which is *Mexicana Point*.

From the South point, the coast of Vancouver Island runs S.W. by W. ½ W. about 11 miles to Cape Scott.

CAPE SCOTT terminates in a low hummock, joined to the main land by a narrow isthmus, and forms, with the islands which lie from it W. ¾ N., a clear navigable channel, about 3½ miles wide. There are a few breakers at a small distance from the cape, in a direction from it S.S.E. ½ E. 7 miles. About 7 miles to the S.E. of the cape, Vancouver passed an opening on the exterior coast, with two small islets lying off its North point of entrance. This appeared clear, and promised to afford very good shelter. Cape Scott forms the West point of Vancouver Island.

Off Cape Scott is a group, consisting of three small and almost barren islands, with many small rocks and breakers about them. West of the westernmost of them, a ledge of rocks extends nearly 2 miles, and South of it is another, a league distant. The eastern or inner of these islands is called *Cox Island*, the outer or principal are the *Lanz Islands* of the Spanish survey.

The SCOTT ISLANDS, which lie 16 miles West of the Lanz Islands, are surrounded by rocks for a considerable distance. The larger island is the *Triangulo* of the Spaniards, and the group the *Is. de Bereford* of the same.

JOSEY BAY is the opening mentioned as being seen by Vancouver. From what is published of it, it consists of two bays, one trending to the W.N.W., and the other N.N.W.; the latter is the *Sea Otter Harbour* of Capt. Hanna, as inserted in Meares's Narrative.

The coast trends hence irregularly for 35 miles to *Port Brooks*, whose southern point is the Woody Point of Capt. Cook, most likely the Cape Santa Clara of Ensign Juan Perez, who preceded him, in 1744.

WOODY POINT projects considerably to the S.W., and is high land. At the time of Cook's visit, in March, 1778, the summit of the high mountains, which filled the interior, were covered with snow; but the valleys between them, and the grounds on the sea-coast, high as well as low, were covered to a considerable breadth with high, straight trees, that formed a beautiful prospect, as of one vast forest. With Point Breaker, the land apparently formed a large bay, which Cook called Hope Bay, hoping from the appearance of the land to find in it a good harbour.

At 32 miles from Woody Point is *Esperanza Inlet*, which, entering the coast, proceeds first to the E. by N. for 7 leagues, and then to the South for 5½ leagues; the latter arm being named the *Tasis Canal* when it enters Nootka Sound, thus forming the land on which Friendly Cove is situated into a large island, called

Nootka Island. In the northern branch of these inlets, the arms or canals called *Espinosa* and *Zeballos Arms*, and at the junction of the two, *Mocuina Basin*, run to the northward. The entrance of Esperanza Inlet, in which *Catala Island* lies, and is also much embarrassed with rocks and shoals, is about 23 miles from the N.W. point of Nootka Sound. On the S.W. face of Nootka Island, *Ferrer Cove* is marked, and a rocky shoal, 2½ miles off the land, lies 9 miles West (mag.) from Mocuina Point.

NOOTKA SOUND.—This celebrated place was, until 1789, supposed to be on the continent of America, but the discovery of its insular character deprived it of the great importance previously attached to it.

It was named by Capt. Cook, who came hither in his last and disastrous voyage, *King George's Sound*, but he afterwards found that it was called *Nootka* by the natives, and hence by that name it has ever since been known. It is stated, however, that no word more nearly resembling Nootka than *Yukuatl*, or *Yucuat*, the name applied to Friendly Cove, has been since found. But this circumstance must not have undue importance attached to it; the native pronunciation is so exceedingly imperfect, that different individuals express themselves by very different sounds for the same word.

It communicates with the Pacific by two openings, the southern one of which is probably the Port San Lorenzo in which the Spanish navigator, Ensign Juan Perez, was, with his ship, the *Santiago*, August 10th, 1774 (St. Lawrence's day). He had been despatched by the Mexican viceroy to survey and take possession of these lands.*

Cook had heard that a Spanish expedition had been hither, and the foregoing fact was confirmed by the former purchasing from the natives two silver table-spoons, apparently of Spanish make, one of which was worn as an ornament, and had probably been stolen from the *Santiago*.

From the accounts contained in Cook's voyage of the furs procured here, several vessels were fitted out from the East Indies to take advantage of the enormous profits they afforded. The earliest of these appears to be that of James Hanna, an Englishman, under Portuguese colours, who reached Nootka in April, 1785; he repeated the voyage in 1786, but then he had to compete with others, Capts. Lowrie and Guise, in two small vessels from Bombay, and Capts. Meares and Tipping from Calcutta, all under the East India Company.†

It was in the prosecution of this trade that Capt. Meares stated that he had purchased a tract for the erection of a house and factory in Nootka Sound, which subsequently led to very serious results. The Spaniards had claimed, by right of a papal bull, dated 1493, the whole of these countries, and also by the subsequent right by discovery; but they had not erected any fort, or in any other way taken possession of it. In consequence of the orders, or the misunderstanding of them, from the Spanish government, the *Iphigenia* was taken

* Of this voyage no account appeared until 1802, when a short notice of it was given in the Introduction to the Journal of the *Sutil* and *Mexicana*. A more perfect notice from some Spanish MSS. will be found in Greenhow's History of Oregon, &c., pp. 114—116.

† The accounts of these fur trades between 1785 and 1789 will be found in the Narrative of the Voyage of the Ship *Queen Charlotte*, by her Captain, John Dixon, or rather by her Supercargo, Beresford; the Narrative of the Voyage of the Ship *King George*, by Capt. Nathaniel Portlock; and the Narrative, before quoted, of the Voyages of Capt. John Meares.

possession of by Estevan Martinez, who had accompanied Perez in 1774, and who had been sent hither to assert and maintain this claim. The *Argonaut*, under Capt. Colnett, was subsequently seized, and the captain sent a prisoner to San Blas, he suffering from delirium or insanity in consequence. From this and other matters the respective governments warmly took up the cause, and the consequences were some warlike preparations, which cost England three millions sterling, but which were quieted by the *Nootka convention* of October, 1790, by which the South Seas were opened to British enterprise. The preparation of these fleets, however, was not without some results; they did good service afterwards, as Lord Howe's victory was gained by them four years later.

Vancouver was subsequently sent to recover possession of these lands, but did not then succeed as he intended, the Spaniards having erected a fort here, and the possession still remained undecided. In 1818 Spain concluded a treaty with America, in which she ceded all her possessions South of lat. 42° N. to the former power, leaving the northern part as it had been. This convention was renewed in 1827, and the whole question was definitively settled by the Oregon treaty of 1846.

The first who more minutely examined this important inlet was our great circumnavigator, Cook. The following is extracted from his account of it:—

"On my arrival in this inlet, I had honoured it with the name of King George's Sound; but I afterwards found that it is called Nootka by the natives. The entrance is situated in the East corner of Hope Bay, in the latitude of 49° 33′ N., and in the longitude of 233° 12′ E. The East coast of that bay, all the way from Breaker's Point to the entrance of the Sound, is covered with a chain of sunken rocks, that seemed to extend some distance from the shore; and, near the Sound, are some islands and rocks above water.

"We entered this sound between two rocky points, that lie E., S.E., and W.N.W. from each other, distant between 3 and 4 miles. Within these points the Sound widens considerably, and extends into the northward 4 leagues at least, exclusive of the several branches towards its bottom, the termination of which we had not an opportunity to ascertain. But, from the circumstance of finding that the water freshened when our boats crossed their entrance, it is probable that they had almost reached its utmost limits. And this probability is increased, by the hills that bounded it toward the land being covered with thick snow, when those toward the sea, or where we lay, had not a speck remaining on them, though in general they were much higher.* In the middle of the sound are a number of islands of various sizes. . . . . The depth of water in the middle of the sound, and even close home to some parts of the shore, is from 47 to 90 fathoms, and perhaps more. The harbours and anchoring places within its circuit are numerous; but we had no time to survey them. The cove in which our ships lay is on the East side of the sound, and on the East side of the largest of the islands. It is covered from the sea, but has little else to recommend it, being exposed to the S.E. winds, which we found to blow with great violence; and the devastation they make sometimes was apparent in many places.

* The arms in question extend farther than Cook supposed; for the two in the N.W. part join the Tasis Canal before mentioned, as connected with Esperanza Inlet, together forming Nootka Island.

"The land bordering upon the sea-coast is of middling height and level, but within the sound it rises almost everywhere into steep hills, which agree in their general formation, ending in round or blunted tops, with some sharp, though not very prominent, ridges on their sides. Some of these hills may be reckoned high, while others of them are of very moderate height; but even the highest are entirely covered to their tops with the thickest woods, as well as every flat part toward the sea. There are sometimes spots upon the sides of some of the hills, which are bare; but they are few in comparison of the whole, though they sufficiently point out the general rocky disposition of the hills. Properly speaking they have no soil upon them, except a kind of compost, produced from rotten mosses and trees, of the depth of 2 feet or more. Their formations are therefore to be considered as nothing more than stupendous rocks of a whitish or gray cast, when they have been exposed to the weather; but, being broken, they appeared to be of a bluish-gray colour, like that universal sort which were found at Kerguelen's Land. The rocky shores are a continued mass of this: and the little coves in the sound have beaches composed of fragments of it, with a few other pebbles. All these coves are furnished with a great quantity of fallen wood lying on them, which is carried in by the tide, and with rills of fresh water, sufficient for the use of a ship, which seem to be supplied entirely from the rains and fogs that hover about the tops of the hills; for few springs can be expected in so rocky a country, and the fresh water found farther up the sound most probably arose from the melting of the snow, there being no reason to suspect that any large river falls into the sound, either from strangers coming down it, or from any other circumstance. The water of these rills is perfectly clear, and dissolves soap easily. The weather during our stay (March 29th to April 26th, 1778) corresponded pretty nearly with that which we had experienced off the coast. That is, when the wind was anywhere between North and West, the weather was fine and clear; but, if to the southward of West, hazy with rain. The climate, as far as we had any experience of it, is infinitely milder than that on the East coast of America under the same parallel of latitude. The mercury in the thermometer never, even in the night, fell lower than 42°; and very often in the day it rose to 60°. No such thing as frost was perceived in any of the low ground; on the contrary, vegetation had made a considerable progress; for I met with grass that was already above a foot long." *

FRIENDLY COVE, which lies to the N.E. of the entrance of the sound, or on the S.E. point of Nootka Island, is its most important position. It has been surveyed by Capt. Sir Edward Belcher in 1839. The village stood on the side of a rising ground, which has a tolerably steep ascent from the beach to the verge of the wood, in which space it is situated. A view of this is given in Vancouver's narrative, and Sir Edward Belcher's plan is the best guide for it.

From Friendly Cove the western coast of the sound runs nearly in a true North direction, and for about 3 miles is covered with small islands, which are so situated as to form several convenient harbours, having various depths of water from 30 to 7 fathoms, with a good bottom. To the West of the northernmost of

* Cook's Third Voyage, vol. ii. pp. 288—291.

these islands is the entrance to a land-locked lake, called in the Spanish chart *Boca del Infierno*; and 2½ miles to the North of this is *Marvinas Bay*, a stream of good water flowing into its southern part. Two leagues within the sound, on this West side, there runs in an arm in a N.N.W. direction; and 2 miles farther is another, nearly in the same direction, with a tolerably large island before it. Capt. Cook had not time to examine either of these arms, but inferred that they did not extend far inland, as the water was no more than brackish at their entrances. A mile above the second arm there were the remains of a village, and behind it a plain of a few acres in extent, covered with the largest pine trees that Capt. Cook had ever seen. This was thought the more remarkable, as the elevated ground, on most other parts of the West side of the sound, was rather naked. As before mentioned, these two arms join and form the Tasis Canal, leading to the northward.

East of this an arm, called *Hapana Reach*, or *Tlupana Arm*, in the Spanish chart, extends to the N.N.E., apparently to no great distance, but really terminates in two basins 8 miles distant, to the South of which arm is the large island, on the S.W. extremity of which is the cove where Capt. Cook refitted his ship in 1778. On the West side of this island are many smaller ones, of which we have no further particulars than their existence.

RESOLUTION COVE, on the S.E. side of the large island previously alluded to as being the place where Capt. Cook refitted his vessel during March and April, 1778, is 4 miles within the entrance. It would seem to be well adapted for repairs, as he took out his main and foremast here, and replacing the latter with a new one cut in the immediate vicinity. In the second volume of the account of his last voyage there are given very ample details of the people, their manners, appearance, disposition, and resources, which are familiar to most. He calculated, of course at a venture, the number of people occupying the two villages, the only inhabited parts of the sound, to be about 2,000.

From the observations made by Cook while here, he fixed the time of high water at about 12ʰ 20′ on the days of new and full moon. The perpendicular rise and fall, 8 ft. 9 in.; that is, for the day tides, and those which happen two or three days after the full and new moon. The night tides at this time rise near 2 feet higher, of which they had many instances on the removal of timber during the night, which had been drawn up above the reach of the day tides. There was no decision arrived at as to the direction of the flood tide, whether it fell into the sound from the N.W., S.W., or S.E.; but the latter is scarcely probable, as the S.E. gales diminished the rise of tide.

CAPE SAN ESTEVAN, so named by Ensign Juan Perez, is the *Point Breakers* of Capt. Cook. It forms the S.E. extreme of the land South of Nootka Sound, and the South point of the supposed great bay named Hope Bay by Capt. Cook. It is low, and off it extend a great many sunken rocks, from which circumstance Capt. Cook applied the name.

CLAYOQUOT SOUND.—To the south-eastward of this is the extensive range of inlets called *Clayoquot* by the natives. We know nothing of it but the configuration given in the Spanish charts, except one harbour in its southern part, a few miles within the outer coast, named Port Cox by Meares.

PORT COX, so named by Meares (June, 1788), in the district he describes as under the domination of a chief called Wicananish, lies within some islands, and a bar of 3½ to 4 fathoms separates the inner harbour from the roadstead.

This roadstead bore the wildest appearance that can be imagined, and was defended from the sea by several small islets and reefs, which nearly connected them. The port, we observed, was situated about 2 miles from the anchoring ground we occupied, the entrance of which did not appear more than 2 cables' length in breadth. The village of Wicananish stands on a rising bank near the sea, and is backed by the woods.

The harbour of Wicananish affords very secure shelter, with good anchorage, both in the roads and the inner port. An archipelago of islands seems to extend from Nootka or King George's Sound to this place, and still farther to the southward. The channels between these islands are innumerable; but the necessary occupations of the ship would not allow us time to examine them: as far, however, as our observation extended, we are disposed to believe that there is no good channel for ships but that which we entered, and which is an exceedingly good one.

These islands are thick with wood, with but very few clear spots, at least that we could discern. The soil is rich, producing wild berries and other fruit in great abundance. The timber is of uncommon size as well as beauty, and applicable to any purpose. We saw frequently groves, almost every tree of which was fit for masts of any dimensions. Among a great variety of other trees we observed the red oak, the larch, the cedar, and black and white spruce fir.

In proceeding to the southward, we saw numerous villages on the shore, from whence we were visited by canoes, filled with people, who in their persons and manners very much resembled those of Port Cox. The chart by Maurelle convinced us that he had either never seen this part of the coast, or that he had purposely misrepresented it.*

This account of what may be, at some future time, an important place, is necessarily very imperfect, for the want of proper charts and the pursuits of trade prevented Capt. Meares from giving any clear account of this port.

NITINAT SOUND, according to the Spaniards; *Berkeley Sound*, according to the fur traders, lies 30 miles to the S.E. of Port Cox. It is an extensive inlet, filled with islands, the entrance being 4 leagues in breadth, and extending nearly the same distance inland. The northern channel, under Terron Point, is called the *Cayuela Entrance*. Its S.E. point is Carrasco Point in the Spanish charts. At 10 miles S.E. of the latter is Bonilla Point, the entrance of the famous Strait of Juan de Fuca.

STRAIT OF JUAN DE FUCA.

In some former pages we have described the southern shore of this extensive arm of the sea, which for many years was the subject of much speculation and inquiry; and almost all the expeditions which were dispatched to the N.W.

* Voyage of Capt. John Meares, pp. 137, 148, 152.

coast of America had for one principal object the discovery of this supposed channel, leading to a vast inland sea and to the Atlantic Ocean. And it is very singular, that in so many cases—as in that of Capt. Cook, to mention one—it should, by fortuitous circumstances, have been so overlooked, or its broad and clear entrance passed unperceived, and that the relations of Dé Fuca and De Fonte were considered so long as entirely fabulous.

The first of these accounts we shall give the outline of below. The second was circulated in Europe in the beginning of the eighteenth century, and was an account of an expedition performed in 1640 by one Admiral Bartolomeo de Fuente, or de Fonte, according as his origin is Spanish or Portuguese. It first appeared in April and June, 1708, in a periodical work called, "Memoirs for the Curious," and is a strange account of an extensive inland navigation performed on the N.W. coast. Its greater part consists of romantic falsehoods, but later discoveries give some colour to the reality of the expedition.* The earlier narrative of De Fuca is more definite; it is in substance as follows:—

Juan de Fuca was a mariner and pilot in the service of Spain. He was a Greek, of the Island of Cephalonia, and his real name was Apostolos Valerianos. In 1592 he was dispatched from Acapulco by the viceroy of Mexico, to explore the coast to the northward, in order to find a communication between the Great Ocean and the Atlantic Ocean, which had been then, as now, a great desideratum with European navigators. He entered a strait between 47° and 48°, and continued sailing in it for 20 days. In some places the land extended to the N.E., in others to the N.W., and the passage, which became much wider than its opening, contained several islands. The inhabitants of the shores were clothed in skins, and the country appeared to him to be very fertile, and, it was stated, abounded with gold, silver, and pearls. *He thus reached the Atlantic Ocean*, and for two years after his return he warmly solicited the reward for so great a discovery.† Such is the substance of De Fuca's narrative. It was garnished with many fictions, and consequently was not believed; but later times have proved it substantially correct, except its reaching the Atlantic, which De Fuca evidently only supposed it would. In 1787, Capt. Berkley, and in 1788, Capt. Duncan, and also Capt. Meares, all visited this entrance; and in 1789, the American sloop *Washington*, Capt. Kendrick, entered the strait, and rounded the archipelago ‡ to the N.N.E., for a distance of 160 leagues. The assertions of De Fuca were thus revived, and some disquisition ensued.§

This navigation was subsequently minutely surveyed by Vancouver, as described in the preceding pages, which will afford all other information necessary. The shores of De Fuca Strait were surveyed by Capt. Henry Kellett, R.N., in 1847, and from his charts we gain most of the subsequent particulars.

The *shores on the South side* are composed, generally, of low sandy cliffs,

* Those who are curious in such matters may consult the excellent History of Voyages and Discoveries made in the North, by Dr. Reinhold Forster. See, also, Burney's Collection; the Introduction to Fleurieu's Voyage of Marchand; and Greenhow's History of Oregon, &c.

† Purchas, Collection of Voyages, vol. iii. pp. 849—852.

‡ Kendrick never returned to Europe. He was killed, in 1793, at Karakakooa Bay, by a ball accidentally fired from a British vessel, while saluting him.

§ See Voyages made in the Years 1788 and 1789, &c., by John Meares, 4to, London, 1790.

from the top of which the land takes a farther gentle ascent, and is entirely covered with trees, chiefly of the pine tribe, reaching to a range of high craggy mountains.

The *northern shore* is not quite so high; it rises more gradually from the sea-side to the tops of the mountains, which have the appearance of a compact range, infinitely more uniform, and much less covered with snow (in April and May, 1792), than those on the southern shore.

PORT SAN JUAN, the first inlet within the entrance, is 13 miles from Bonilla Point, the coast between being very straight and compact. Its entrance, between *Owen Point* and *Island* on the West, and San Juan Point on the East, is 1½ miles in width, and the harbour extends about 3¾ miles in a N.N.W. direction, being 1 or 1½ miles in breadth. At its head, the River Gordon enters on the North side, but is shallow; and Cooper Inlet on the South, also shallow. The only danger in it, apparently, is a rock, *awash*, which lies about half a mile N.E. by N. of Owen Island, and a quarter of a mile off shore. It has 4 or 5 fathoms within it. A quarter-tide rock also lies a short distance from the North shore, at about a mile from Owen Point.

The *Observatory Rocks*, off its S.E. point, are in lat. 48° 31′ 30″, lon. 124° 27′. Variation, 21° E., 1847.

From this point the North shore continues the same uniform direction of E.S.E., *true*, for 20 miles, scarcely varied by two small rivers, *Sombriò* and *Jordan*, in the interval. At that distance the coast becomes somewhat more irregular, and continues so for 10 miles, to Sooke Bay and Inlet.

SOOKE INLET is a tortuous channel, 3½ miles in length, leading to a land-locked basin of 2 miles in length by 1½ broad.

On the N.W. side of this entrance the depth is shallow, and several spits of shingle project from it, one of which, having Whiffin Island on its extremity, narrows the entrance to a few yards. The channel throughout keeps on the S.E. side, and is from 3 to 9 fathoms in depth. The eastern point of its entrance is named *Company Point;* and *Secretary Island*, a small, bold, rocky islet, is a mile to the S.E. of it. The latter is in lat. 48° 19′ W., lon. 123° 42′, and has deep water inside it.

Beechey Head is 2 miles eastward from Secretary Island, and is the South extreme of the peninsula formed by Sooke Inlet and Becher Bay.

Becher Bay is open to the South, about 1½ miles in the opening, and a mile in depth. Some islands, named *Frazer* and *Wolfe Islands*, lie in it.

Between *Smyth Head*, its eastern point, and *Cape Church*, a mile to the W.S.W., lie several islands; and from the latter cape, *Christopher Point*, the southernmost point of Vancouver Island, bears E.N.E. a mile distant. Off the coast between this and *Cape Calver* lies Bentinck Island; and a mile to the S.E. of which are the *Race Islands*, a group of rocky islets, nine or ten in number, and bold-to. They are sometimes called the *Rocky Islands*.

Pedder Bay, a narrow inlet, extends for 2 miles in a W.N.W. direction, being shallow at its upper end. Its N.E. limit is a peninsula, the East end of which is *William Head*, separating it from *Parry Bay*, the North limit of which is *Albert Head*, 3½ miles distant from it.

ROYAL BAY, which occupies the S.E. extremity of Vancouver Island, lies to the N.E. of Albert Head. It contains the important settlement of Victoria, in the harbour of the same name, and Esquimalt Harbour to the West of it.

ESQUIMALT HARBOUR, which lies in the North part of Royal Bay, is entered between *Fisgard Island*, off *Rodd Point*, on the West, and *Duntze Head* on the East, the width being about a quarter of a mile. Shoal water extends about a hundred yards off Fisgard Island; Duntze Head itself is bold-to, but to the South of it are some islets, off which some sunken rocks extend 150 yards, the outermost being S. by E., *mag.*, one-third of a mile from Duntze Head. They are named the *Srogg Rocks* in the Admiralty chart. In the entrance the depth is 8 and 9 fathoms, inside the depth gradually decreases. When within the entrance, the harbour widens considerably to the eastward, where it forms Village Bay, quite clear and deep in all parts. Northward the bay extends for 2 miles above the entrance, but at about three-quarters of a mile there is a sunken rock, with 10 feet water on it, which lies in the middle of the channel, about West, by compass, from the *Inskip Islands*, which contract the harbour at this part.

VICTORIA HARBOUR, the original native name of which is *Camosack*, or Cammusan, though not the best, is the most important of all on Vancouver Island, and lies 2 miles to the West of Esquimalt. Its entrance, nearly half a mile in width, between *M'Loughlin Point* on the West, and *Ogden Point* on the East, runs to the northward for a quarter of a mile, when it turns at a right angle to the eastward. In the centre of the first branch of the entrance is a shoal spot of 12 feet. At the point where the harbour takes an easterly direction is a sort of bar, and above this the depth is irregular, but the shoalest spots are all buoyed according to the Admiralty plan. The plan will furnish a better guide than any written description.

An important shoal, the *Brotchy Ledge*, with one spot of 7 feet only on its southern end, lies half a mile off shore, to the S.E. of the entrance, that is, about three-quarters of a mile S.S.E. of the outer entrance. Two beacons, one on the shore, the other on the summit of a mount, and bearing N.E. ½ E., when in one mark its exact line of direction; great care must, therefore, be taken of this.

VICTORIA, the settlement of the Hudson's Bay Company, lies at the head of the first branch of the harbour, on the eastern side.

Above Victoria, the harbour extends 1½ miles in a N.W. direction, but its upper part is shallow, and it is nearly connected with Esquimalt Harbour by a kind of natural canal, which at some future period may be turned to good advantage.

Fort Victoria is situated on the southern end of Vancouver Island, in the small harbour of Cammusan, the entrance to which is rather intricate. The fort is a square enclosure of 100 yards, surrounded by cedar pickets, 20 feet in height, having octagonal bastions, each containing six 6-pounder iron guns, at the N.E. and S.W. angles; the buildings are made of squared timber, eight in number, forming three sides of an oblong. This fort has lately been established; it is badly situated with regard to water and position, which latter has been chosen for its agricultural advantages only. About 3 miles distant, and nearly connected by a small inlet, is the Esquimalt or Squimal Harbour, which is very commodious, and accessible at all times, offering a much better position, and having also the advantage of a supply of water in the vicinity.

This is the best built of the company's forts; it requires loop-holing, and a platform or gallery to enable men to fire over the pickets; a ditch might be cut round it, but the rock appears on the surface in many places.*

The site of Victoria was chosen by Mr. Chief Factor Douglas, after a careful survey of the South end of the island, in 1842. Where the settlement is founded there is a range of plains nearly 6 miles square, containing a great extent of valuable tillage and pasture land, abundance of timber around, and water power for flour or saw mills on the canal of Camosack, or Cammusan.

Mr. Douglas, on deciding on the situation, says:—"Camosack is a pleasant and convenient site for the establishment, within 50 yards of the anchorage, on the border of a large tract of clear land, which extends eastward to Point Gonzalo, at the S.E. extremity of the island, and about 6 miles interiorly, being the most picturesque, and decidedly the most valuable, part of the island that we had the good fortune to discover.

"More than two-thirds of this section consists of prairie land, and may be converted either to purposes of tillage or pasture, for which I have seen no part of the Indian country better adapted; the rest of it, with the exception of the power of water, is covered with valuable oak and pine timber. I observed, generally speaking, but two marked varieties of soil on the prairies; that of the best land is a dark vegetable mould, varying from 9 to 14 inches deep, overlaying a substrate of grayish clayey loam, which produces the rankest growth of native plants that I have seen in America. The other variety is of inferior value, and to judge from the less vigorous appearance of the vegetation upon it, naturally more unproductive. Both kinds, however, produce abundance of grass, and several varieties of red clover grow on the rich moist bottoms. In two places, particularly, we saw several acres of clover growing with a luxuriance and compactness more resembling the close sward of a well-managed lea than the product of an uncultivated waste.

"Being pretty well assured of the capabilities of the soil, as respects the purposes of agriculture, the climate being also mild and pleasant, we ought to be able to grow every kind of grain raised in England. On this point, however, we cannot speak confidently until we have tried the experiment and tested the climate, as there may exist local influences, destructive of the husbandman's hopes, which cannot be discovered by other means. As, for instance, it is well known that the damp fogs which daily spread over the shores of Upper California blight the crops, and greatly deteriorate the wheat grown near the sea-coast in that country. I am not aware that any such effect is felt in the temperate climate of Britain, nearly corresponding in its insular situation and geographical position with Vancouver Island; and I hope the latter will also enjoy an exception from an evil at once disastrous and irremediable. We are certain that potatoes thrive and grow to a large size, as the Indians have many small fields in

* Report from *Lieut. Vavasour*, R.E., to Capt. Holloway, March 1st, 1846.

cultivation, which appear to repay the labour bestowed upon them, and I hope that other crops will do as well. The canal of Camosack is nearly 6 miles long, and its banks are well wooded throughout."

The results of the Hudson's Bay Company's farming at Vancouver Island have answered, it is understood, the most sanguine expectations.

The S.E. entrance point of the harbour, *Ogden Point*, is in lat. 48° 24′ 46″ N., lon. 123° 23′ 23″ W. Variation, 22° 7′ E., 1847.

The following brief directions, by Mr. *G. J. Gibbon*, master of H.M.S. *Modeste*, will form a fitting conclusion to the foregoing:—

"I should recommend all strangers, when bound into the straits, after they have made both capes, to get into a mid-channel course, steering about E. ¾ S. until the centre of Classet Island bears anything to the westward of S.S.W., then an E. ¾ N. course will take you up off Rocky Point (which is the most south-eastward projection of the Island of Vancouver), off which are many rocks above water. The largest of them is about the size of the hull of a ship of about 200 tons, and is the southernmost of the whole group, and sufficiently to the northward to clear the shelving ground off False and New Dungeness. Should you be bound for Port Victoria, and from the direction of the wind have to borrow on the North shore, be sure to give those rocks a berth of at least half a mile, for the tide here runs very strong and irregular. From the outer or southern rock, the mouth of Victoria bears N.N.E. ½ E., about 11¼ miles distant. It has a dangerous rock lying off it, covered with only 11 feet of water, and bears from the south-eastern point of the harbour about South, 500 fathoms. I knew of no good anchorage without the harbour, the land being steep-to, and bottom rocky. Her Majesty's sloop *Modeste*, in September, 1844, anchored in 35 fathoms about 1 mile S. ¾ E. from the mouth of the harbour, and N.N.E. ½ E. from Rocky Point."

Of the extremity of Vancouver Island, beyond Victoria, we have no particulars but the charts. Between the island and Rosario Strait, described on a former page, is a group of islands of various dimensions, some of which were surveyed by the United States' Exploring Expedition, and others by Capt. Kellett. The principal are Lopez Island, *Orcas Island*, and *San Juan Island*. Between the East extreme of Vancouver Island, which has numerous islets against it, and the last-named island, is the *Haro Strait* (not Arro), named after one of the early explorers. It does not appear that this has been examined throughout to the northward, though doubtless such a communication does exist.

We have thus described the circuit of Vancouver Island as far as the imperfect nature of the materials at command will allow. The coast in continuation is that with which this chapter commences.

CHAPTER XIV.

COAST OF BRITISH AMERICA, FROM QUEEN CHARLOTTE SOUND TO THE PORTLAND CANAL.

The little frequented coast described in the following pages forms part of the territory of the Hudson's Bay Company, and is the western seaboard of the districts formerly known under the names of *New Hanover* and *New Cornwall*. The whole of it is fronted by an immense collection of islands of all dimensions and forms. Within these is a correspondingly extensive series of channels and arms of the sea, forming a most complete chain of inland navigation, which may be pursued for many degrees to the northward, without interruption or exposure to the oceanic swell. It is true that in many parts these canals are too narrow to be very advantageous for sailing vessels to work through, and are therefore more adapted for steam navigation than by other modes; yet the great depth of water, in most parts all but unfathomable, and the boldness of the shores, make this disadvantage of less importance. One feature adverted to by Vancouver, who has excellently surveyed part of this inland navigation, and which is remarkable, is, that caution ought to be used in passing close to some of the projecting points, for he found that, notwithstanding the perpendicularity of the cliffs composing the shores, that a *shelf* would sometimes project *under* water, from the general line of the upper portion, a fact which, if neglected, might lead to serious consequences.

Besides these channels the continent is penetrated with numerous and peculiar canals, whose characteristics, greatly similar in all cases, will be best gathered from the ensuing descriptions. There are no rivers, or at least none of importance have been discovered. They are mere torrents, fed in summer by the melting of the snow, and in the winter by the untiring deluges of this dismal climate. The Babine, the Nass, and the Stikine, are the only ones within the territories visited by the Hudson's Bay Company that may be ascended to any distance, and even these only with considerable difficulty and danger.

The primary discovery of the country has been before adverted to, and many of the remarks are as applicable to this as to other portions; but it is to the zeal and perseverance of two intelligent men, who traded hither under the licence of the South Sea Company. These were Capts. Portlock and Dixon; they made the principal discoveries on the coast subsequent to Capt. Cook's visit in his last voyage. Their narratives are, nevertheless, too diffuse, and abound too much with personal narrative and minutiæ to afford much general information on the country they visited.

It is to the excellent surveys of Vancouver, in 1792, that we owe the greater part of the knowledge we possess of the inland navigation and nautical information of this country, and in the subsequent pages it must be considered that all portions not otherwise noticed have been derived from the narrative of his voyage.

Besides Portlock, Dixon, and Vancouver, the Spaniards have surveyed a portion, and their charts fill up the vacancies left by the other; but of their surveys we have no verbal description to be of service. In the voyages of Meares, previously adverted to, there are also some notices of visitors to this coast during the origin of the fur trade, as before explained, as arising out of the notice afforded by Capt. Cook. Then Capt. Ingraham visited the S.E. side of Queen Charlotte Island; and Capt. Gray, who discovered the Columbia River, first explored it in the *Washington*, which name he applied to it.

With this extensive island, second only in extent to Vancouver Island, we are very imperfectly acquainted, though it is believed to possess many advantages, and to be a fine island, destined, perhaps, at no very distant period, to become of considerable importance in the trade and operations in the Pacific.

As before stated, the territory is under the government of the Hudson's Bay Company; and is consequently subject to their laws and trade. But little, if any, foreign commerce is therefore carried on. The whole productions of the country, chiefly furs and peltry, are procured from the natives at certain points, and perhaps at stated intervals, by the periodic visits of the company's officers. Their posts are liable to change, and therefore cannot now be definitely described, and are of less importance, as the local knowledge of these affairs must be so much more complete than any we can procure in Europe. These imperfect notes must be considered as rather intended to give a vague notion of the country to strangers than as specific instructions for visitors.

Of the climate or other productions of the territory we have no information, no resident or traveller, except Sir Alexander Mackenzie, and more recently Sir George Simpson, having given anything to the world respecting it.

The *Indian* inhabitants are of very different character, habits, and manners, to those of the American races to the East of the Rocky Mountains, with which Europeans are more familiar from numerous travellers' descriptions. Several causes contribute to produce this remarkable variety. On the East side of the mountains the buffalo is the great source of provision to the Indian tribes. They are therefore hunters, dependent on their skill and activity for subsistence. This animal has never penetrated to the West side of these mountains; at the same time, the great rivers rising on their West side abound with salmon, almost to their source. The inland tribes living chiefly on the margins of these streams live on salmon during the summer, and prepare great quantities of the same fish for their winter supply. By thus obtaining their subsistence by fishing, they are more sedentary than those East of the mountains, and, from their more settled mode of life, have made considerable progress in the rude arts of the savage. They are more accustomed to continuous labour, and show great aptitude for agriculture. Westerly winds prevail on the Pacific coasts throughout the greater part of the year, and render the climate extremely moist and mild; the natives then go about, even during winter, with very slight clothing. The custom of flattening the head, mentioned on page 353, so common among the southern tribes, appears to be unknown to the tribes North of Vancouver Island. But it is replaced by one equally singular and disgusting. This is, the labret or lip-piece, worn after they arrive at maturity. An incision is made under the lower lip, in which a piece of wood or bone is inserted, and the deformity extended by this means to the utmost limits. In the South parts this custom is going out of fashion, from European influence. In the North it flourishes unimpaired. We cannot enter into an enumeration or description of the numerous tribes and families they are subdivided into.*

One feature of the social system of the natives of this coast is, the prevalence of slavery. Fully one-third of the large population are slaves, or "thralls" as they are termed, of the most helpless and abject description. Though some of these poor creatures are prisoners taken in war, yet most of them have been born in their present condition. They are as much the property of their masters as so many dogs, with the distinction that they are the subjects of an infinitely greater amount of cruelty and atrocity. Every indignity and privation is heaped on them. But this is nothing when compared with the purely wanton atrocity to which these most helpless and pitiable children of the human race are subjected. They are beaten, lacerated, and maimed; the mutilations of fingers or toes, the splitting of noses, the scooping out of eyes, being ordinary occurrences. They are butchered, without the excuse or the excitement of a gladiatorial combat, to make holidays; and, as if to carry persecution beyond the point at which the wicked are said to cease from troubling, their corpses are often cast into the sea, to be washed out and in by the tide. To show how diabolically ingenious the masters are in the work of murder, six slaves, on the occasion of a late merry-making at Sitka, were placed in a row, with their throats over a sharp ridge of a rock, while a pole, loaded with a chuckling demon at either end, ground away at the backs of their necks till life was extinct. What a proof of the degrading influence of oppression, that men should submit in life to treatment from which the black bondmen of Cuba or Brazil would be glad to escape by suicide! †

QUEEN CHARLOTTE SOUND was so named by Capt. S. Wedgborough, of the *Experiment*, in 1786. Capes Scott and Caution form its western limits, and it is the northern outlet of the chain of inlets which insulate Vancouver Island. Here there is an interruption to the continuity of the inland navigation, which extends from De Fuca's Strait to Cross Sound. The navigation along this coast is one of considerable danger, on account of the prevalence of fogs, and the presence of the Virgin and Pearl Rocks. The tidal currents, too, are violent and irregular.

CAPE CAUTION, which forms the N.E. limit of Queen Charlotte Sound, was so named by Vancouver on his second visit, from the dangerous navigation in its vicinity. It makes a conspicuous cape, terminating in rugged, rocky, low hummocks, that produce some dwarf pine, and other small trees and shrubs. Off the cape are some very dangerous breakers, consisting apparently of three distinct patches, occupying the space of a league. Their eastern part bears from Cape Caution W. by N. ½ N., distant about 5 miles; but the rocks that lie off the shore to the northward of the cape reduce the width of the channel between them and the breakers to about a league, in which there does not appear any obstruction which is not sufficiently conspicuous to be avoided.

* See *Dr. Scouler* on the North-West American Tribes, combining information from Dr. Tolmi, of the H.B.C.S., Journal of the Royal Geographical Society, vol. xi. 1841, pp. 215—250.

† Sir George Simpson's Journey Round the World, vol. i. pp. 242-3.

The VIRGIN and PEARL ROCKS are two very dangerous clusters off the entrance to Smith's Inlet. They were discovered and named by Mr. Hanna, in 1786. The Virgin Rocks lie W. by N. ½ N. 13 miles from the South point of Smith's Inlet; and the Pearl Rocks, N.W. ¾ N. 8 miles from the same point. They lie in a line, W.S.W. from the South extreme of Calvert's Island, 11 and 4 miles distant respectively.

SMITH'S INLET.—The entrance lies about 7 miles North of Cape Caution, the intervening coast being bestrewed with rocks and islets. The entrance into it is nearly closed by rocky islets, some producing shrubs and small trees, others none; with innumerable rocks, as well above as beneath the sea, rendering it a very intricate and dangerous navigation for shipping. About 3 leagues within the entrance the rocks and islets cease to exist, and the inlet contracts to about half a mile in general width. The shores are both formed of high rocky precipices covered with wood.

RIVER'S CANAL is about a league to the northward of the North point of Smith's Inlet. The entrance to it appears less dangerous than the latter. It has, however, on its southern side many rocky islets and rocks, but none were discovered beneath the water level. By keeping on the North side of the entrance, which is 1½ miles across, a fair navigable passage is found, about half a mile wide, between the North shore and the rocky islets that lie off its southern side. The land about its mouth is of moderate height, but towards its head, where it branches off in different directions, the shores are composed of high, steep, rocky mountains, and, like Smith's Inlet and many other of the canals in the neighbourhood, there was no bottom found in the middle with 80 fathoms of line; though in the bays found in most of these canals anchorage may, in all probability, be procured.

From River's Canal a channel diverges towards the South end of Calvert's Island. It is very narrow and intricate, leading through an immensity of rocks and islets to Point Addenbrooke, in Fitzhugh's Sound.

CALVERT'S ISLAND forms the exterior coast northward of River's Canal, and within it is *Fitzhugh's Sound;* the former was discovered and named by Mr. Duncan; the latter by Mr. Hanna. Off the South point of the island are two small islets. The eastern side of the island forms a steep, bold shore, rising abruptly from the sea to a great height, composed of rock, and, like the eastern shore, entirely covered with pine trees.

SAFETY COVE (*Port Safety* of Mr. Duncan?) is 2 leagues North of the South extreme of Calvert's Island, on the West shore of Fitzhugh's Sound. It terminates in a small beach, near which is a stream of excellent water, and an abundance of wood. The depth is, however, rather great; 17 to 30 fathoms. A small rock and two rocky islets lie off its North point of entrance, which is about a quarter of a mile wide. It is the first place that affords safe and convenient anchorage on the western shore within Fitzhugh's Sound. Vancouver found it a comfortable retreat, in August, 1792. High water at the time the moon passes the meridian; rise and fall about 10 feet.

At 13 miles northward of Safety Cove is the passage which insulates Calvert Island. Vancouver places its N.E. point in lat. 51° 45′. South of this point lies a sunken rock, which, though near the shore, is dangerous, being visible at low tide only by the surf which breaks on it. From this point the passage extends S.W. by W. ½ W., about 7 miles. Its northern shore is composed of rocky islets and rocks, with some scattered rocks off its northern shore. Between these rocks is a passage, generally 1 or 2 miles wide, but rendered unpleasant by the want of soundings, the depth being beyond 150 fathoms.

FITZHUGH'S SOUND extends for 26 miles from Safety Cove, in nearly a *true* North direction to *Point Walker*, where it separates into two arms. The easternmost was named by Vancouver after Edmund Burke. Its S.E. point is *Point Edmund*, about 2 miles E.S.E. from Point Walker. There are some rocks off the points, but the channel is fair. The sides of the canal are composed of compact, stupendous mountains, and nearly perpendicular rocky cliffs, producing pine trees to a considerable height above the shores, and then barren, or nearly so, to their lofty summits, which were mostly covered with snow in August.

RESTORATION COVE is about 8 miles within the entrance to the eastern branch of Burke's Canal. It has a fine sandy beach, through which flows an excellent stream of water. The breadth of the cove at its entrance, in a North and South direction, is about 1½ miles, and its depth about three-quarters of a mile. The soundings, though deep, are regular, from 60 fathoms at the entrance, to 5 and 10 fathoms close to the shore. The land on the opposite side of the arm is 2½ miles distant. The tide rises and falls 14 feet, those in the night 1 foot higher than the day tides; the flood comes from the South, and it is high water at the time the moon passes the meridian. Variation, 19° 15′ E.

BURKE'S CANAL extends to the N.E., maintaining the same breadth. Its north-western side is formed by a large island, named by Vancouver *King's Island.* After continuing 11 miles in an easterly direction from the N.E. point of King's Island, the canal separates into two arms to the N.E. and S.E. at Point Menzies.

These arms were named by Vancouver *Bentinck's Arms.* The width of that to the S.E. in general a little exceeds a mile, and the country exactly resembles that contiguous to the branches, which have been so repeatedly described. On the eastern side of this canal, near the head of a small rivulet, a native house, of singular construction, was observed by Mr. Johnstone, of Vancouver's party, and some Indians led them to a village different to any they had seen.

Sir Alexander Mackenzie reached the Pacific after his long, arduous, and perilous journey across the continent at this point a month after Vancouver's party had left. He came to a village of twenty-six large houses, where Mr. Johnstone had come on June 1st, 1793, as above noticed. He coasted along King's Island, and learned that Macubah (as the natives termed Vancouver) had been there with his large canoe. He commenced his return July 22, 1793.*

"This is as desolate, inhospitable a country as the most melancholy creature could be desirous of inhabiting. The eagle, crow, and raven, that occasionally had borne us company in our lonely researches, visited not these dreary shores. The common shell-fish, such as mussels, clams, and cockles, and the nettle samphire and other coarse vegetables, that had been so essential to our health and

* Mackenzie's Travels, p. 342, *et seq.*; and Vancouver, vol. ii. pp. 273-4.

maintenance in all our former excursions (in the southward), were hardly found anywhere to exist; scarcely any signs of human beings were found in the country, which appears to be devoted entirely to the amphibious race; seals and sea-otters, particularly the latter, were seen in great numbers."*

The N.E. point of King's Island was named *Point Edward.* Opposite to it is the entrance of *Dean's Canal*, which penetrates many miles in a N.E. and North direction, terminating in low marshy land. In this inlet neither ebb nor flood occasion any visible stream.

CASCADE CANAL is to the N.W. of Point Edward. Its shores are bounded by precipices more lofty than any hereabouts; and from the summits of the mountains, particularly on the N.E. shore, are some extremely grand and tremendous cascades. The canal which forms the N.W. side of King's Island runs S.W. to Fisher's Canal.

FISHER'S CANAL separates the southernmost of the *Princess Royal Islands* from the main land.

The PRINCESS ROYAL ISLANDS form a portion of that immense archipelago which here fronts the American continent. On its western side it is uneven, rocky, and of moderate height. The eastern shore rises more abruptly, and bounded behind with lofty snowy mountains.

PORT JOHN is 10 miles nearly North of the entrance to Burke's Canal, before described. It is on the western side of King's Island, and forms a good harbour. Its North point of entrance bears N. by E. 2 miles from its South point. Before its entrance are two small islands, and towards its northern shore are some rocks.

To the North of Port John is the entrance to the canal previously mentioned, which passes round the N.W. side of King's Island toward Cascade and Dean's Canals. Fisher's Canal continues its northern course for 12 miles above Port John, its shores being comparatively of moderate height. Its surface, covered with wood, is very uneven. From this point it takes a westerly course to Milbank Sound, and forms the North limit of the southern Princess Royal Island. The course is first about W. by N. ½ N. a league, the shores being low and rocky, with many detached rocks lying off them. The channel then takes a more southerly course, and, although there are many rocks and breakers in it, they are all sufficiently conspicuous to be avoided in fair weather.

MILBANK SOUND is an opening between the Princess Royal Islands, in lat. 52° 13′. Its S.E. point is Cape Swaine, so named after the third lieutenant of the *Discovery*, Vancouver's vessel. Milbank Sound was discovered and named by Mr. Duncan Its N.W. point is *Point Day*, off which lie several very barren rocky islets. The southern side of the channel or sound, being entirely covered with trees and with low shores, is very pleasant in appearance, but the northern sides are a rude, confused mass of low, rugged cliffs, bounded by innumerable rocky islets and rocks.

FORT M'LOUGHLIN, distant a few miles from Milbank Sound, is (or was) one of the Hudson's Bay Company's posts. "This very neat establishment was planned, in 1837, by Mr. Finlayson, of Red River, who left the place in an

* Vancouver, vol. i. p. 374.

unfinished state to Mr. Manson, who, in his turn, had certainly made the most of the capabilities of the situation. The site must originally have been one of the most rugged spots imaginable; a mere rock, in fact, as uneven as the adjacent waters in a tempest; while its soil, buried as it was in its crevices, served only to encumber the surface with a heavy growth of timber. Besides blasting and levelling, Mr. Manson, without the aid of horse or ox, had introduced several thousand loads of gravel, while, by his judicious contrivances in the way of fortification, he had rendered the place capable of holding out, with a garrison of twenty men, against all the natives of the coast. Mr. Manson's successor, Mr. Charles Ross, had made considerable additions to the garden, which, when Sir George Simpson visited it in 1842, was about three acres in extent, with a soil principally formed of sea-weed, and produced cabbages, potatoes, turnips, carrots, and other vegetables.

"In the neighbourhood of the fort was a village of about 500 Bollabollas, who spoke a dialect of the Quakeolth language. Here Sir George Simpson first saw that disgusting and singular ornament of the fair sex, the lip-piece. The fashion, however, is now wearing out, from respect to the opinion of the whites."*

Several inlets or arms run up to the northward from the canal leading from Fisher's Canal into Milbank Sound. They are very similar in character, and need no particular description. The principal arm out of Milbank Sound is the westernmost, and runs in a general northerly direction for 30 miles, when it divides, one portion continuing to the northward, and a wider branch extending eastward to *Carter's Bay* and *Mussel Canal.* These two last derive their names from one of Vancouver's party having died from the effects of poisonous mussels collected in *Poison Cove*, lat. 52° 55', lon. 128° 1' W. The whole party who partook of them were seized with a numbness about their faces and extremities, which soon extended to their whole bodies, accompanied with sickness and giddiness. This may serve as a caution here, though generally the shell-fish is wholesome.

From Carter's Bay the principal inlet continues its northward course, and 5 miles beyond the junction is an opening, apparently communicating with the sea, running southward on the western side of the channel; 13 and 17 miles farther on are two openings on the opposite or eastern, which extend but a short distance inland. Vancouver found scarcely any inhabitants here. The tides rose 15 feet, and it was high water 10^{h} 15' after the moon passed the meridian. Continuing northward, but bearing more to the westward, the canal still skirts the western shore of the Princess Royal Island, as far as its North extreme. Near this, on the eastern shore, is a commodious cove, where Vancouver anchored. This little bay is formed by a stony beach, through which a considerable run of water falls into the sea. The South point of it is a rocky lump, covered with trees, which becomes an island at high water. The anchorage was in 46 fathoms 1½ cables from this lump, the nearest shore. A league northward is a small inlet, where a *hot spring* was discovered.

In sailing among the rocky precipices which compose the shores of the

* Journey Round the World, vol. i. p. 204.

channels hitherto described, it is not always safe to make too free with them in sailing by; for they are frequently found to jut out a few yards at or a little below low-water mark; and if a vessel should ground on any of those projecting points about high water, she would, on the falling tide, if heeling from the shore, be in a very dangerous situation.

To the North of this anchorage before mentioned, the channel continues between the main land and Hawkesbury's Island. The shores are like the rest described, partly composed of lofty steep mountains rising nearly perpendicularly from the sea, and covered from the water side to their summits with pines and forest trees. The other parts, equally well wooded, are less elevated, and terminate in sandy beaches with projecting points, forming several small bays and coves. It takes an irregular northerly direction for about 15 miles, when it turns eastward to *Point Staniforth*, placed by Vancouver in lat. 53° 54′, lon. (corrected) 128° 33′ W., before reaching which Mr. Whidbey, who explored it, observed more drift wood than on any part of the coast.

GARDNER'S CANAL runs 45 miles in an irregular course to the eastward; its upper part passing through a country that is almost an entirely barren waste, nearly destitute of wood and verdure, presenting to the eye one rude mass of almost naked rocks, rising into lofty mountains, whose towering summits, seeming to overhang their bases, gives them a tremendous appearance. The whole is covered with perpetual ice and snow, and many waterfalls descend in every direction in the summer.

Northward from Point Staniforth the principal branch of the inlet continues 17 miles to *Point Hopkins*, on the eastern shore, which is up to this nearly straight and compact, moderately elevated and well covered with wood. Northward of Point Hopkins the inlet continues to the lat. of 54° 4′, where it is terminated by a border of low land, differing from the generality of these arms, from the abrupt mountains on either side, continuing parallel some leagues beyond the head, forming a narrow valley covered with tall forest trees. The Salmon River falls into an inlet running to the East at 7 miles above Point Hopkins.

The North point of HAWKESBURY'S ISLAND is opposite Point Hopkins. Its South point is *Point Cumming*, in lat. 53° 18½′. It is thus about 33 miles in length, and from 3 to 11 miles broad. The continent to the westward forms a point, which extends to about the same latitude as the South point of Hawkesbury's Island, and having one of the numerous arms or canals dividing them.

The North point of the *Isle de Gil* is opposite Point Cumming, and the South opening to the inlet last mentioned. There is anchorage at this North point: it is in a bay on the N.E. part of the island, about 2 miles from its northern extremity on the western shore of the islet. Here Vancouver anchored in 40 fathoms, stones, shells, and sandy bottom, mooring with a hawser to the shore, the outer points of the bay bearing by compass N.W. to S.E. by E., distant a cable's length from the shore. The shore affords abundance of berries and Labrador tea. Fish may also be caught, which, in these regions, is a very scarce commodity, and hence, from his success in this, Vancouver called it *Fisherman's Cove*. In this cove are two considerable runs of fresh water, and wood may easily be procured in abundance. Lat. 53° 18½′, lon. 128° 57′.

There is also anchorage directly to the S.W. of the North point of the Isle de Gil, or Ysla de Gil, which was so named by Senr. Caamano, in 1792. It is about 5 leagues long, North and South, and 5 miles broad; of a moderate though uneven height, composed chiefly of rocky materials, covered with inferior pine trees, and having to the North and N.W. of it much broken and divided land.

The *Isle de la Campania*, to the westward of it, has a conspicuous ridge of mountains, and, when seen from eastward, with a remarkable peak, nearly in the centre, considerably above the rest. Their summits are naked rocks, without the least appearance of verdure.

Nepean Sound is to the northward of these islands, and that of San Estevan, which is the outermost. The general character of these islands differs little from that of the surrounding region. That on the sea-coast is somewhat less mountainous, chiefly covered with wood, and less encumbered with snow than the summits of the interior.

To the north-westward of these lie the extensive islands forming Pitt's Archipelago and Banks's Island, separated from the continent by Grenville's Canal, and from each other by the Canal de Principe; the first so named by Vancouver, on his exploration in 1793, and the latter by Senr. Caamano, who first navigated it.

GRENVILLE'S CANAL.—The southern entrance to Grenville's Canal is opposite to Fisherman's Cove, the North point of the Isle de Gil. Its direction is N.W. ½ N., and is nearly straight for 14 miles on this bearing to a small harbour, or rather cove, on its eastern shore. For 2 miles within it Mr. Whidbey (July, 1793) found the sea abounding in sea-otters, who sported about the boats. At this part it is not more than half a mile wide, with straight and compact shores on each side. The shores of the arm beyond this are mountainous on the East or continental side, and low and rocky on the opposite; both producing pine trees, interspersed with bare patches. From the small cove above mentioned, which has a sandy beach at its head, and a lagoon of water behind it, the arm continues in the same direction for 4 miles farther, from an island off the N.W. point of the cove. It then stretches N.N.W. ½ W., about 8 miles to the South point of an opening on the eastern or continental shore, about a mile wide; its opposite point of entrance lying North. At this point the width of the main arm increases to nearly half a league. Off the South part are many rocks above and below the surface. This inlet divides into three short arms, terminating in the usual way.

N.W. 3½ miles from this is a small cove on the East shore; and 10 miles farther, N.N.W. ¾ W., is a bay about a mile wide and 2 deep, in a N.E. direction, with many islets and sunken rocks in it. The continental shore between these last is lined with innumerable rocks and islets, nor is the middle of the channel free from these obstructions. Seven miles to the north-westward of this, on the opposite or western shore, is an extensive opening, running to S.S.W., apparently dividing the land. Through this opening the ebb-tide sets with a very rapid stream, no part of which apparently enters the passage to the N.E.

To the northward of this opening is a high island, about 7 miles long; and 10 miles North of it is *Point Lambert*, on the continental shore. To the N.E. of this point is *Port Essington*, an extensive sound, surrounded by a moderately elevated country, particularly on the N.W.; but, to the North and East, the view is bounded by lofty barren mountains, covered with perpetual snow. The entrance to Port Essington is narrowed by a shoal against Point Lambert, forming a rounding spit, of 3 to 6 feet water, with many dead trees in it, causing the channel to be on the North side. The tide rushes in furiously, the flood 4, and the ebb 5 knots, the water being perfectly fresh at low tide. Many sea-otters were seen.

Opposite to the entrance to Port Essington are some islands, forming the North side of the opening into Chatham's Sound, to one of which the name of *Raspberry Island* was given by Vancouver, from the quantity of excellent raspberries he found here. The passage through is 2 miles long and about a mile wide between the islands, but mostly occupied by shoals, which contract it to a very narrow channel close on the southern side. To the N.W. of this channel Chatham's Sound is interspersed in most directions with small islands, rocks, and shoals; one in particular, an extensive sandbank, bears N.W. by W. a league from the opening. To the S.W. of this bank are some islands, on which the compass is strongly affected, Mr. Whidbey finding a difference of 13° in the direction of the needle.

Point Hunt is very conspicuous, and forms the North point of Pitt's Archipelago. It is in lat. 54° 10½′, and bears West 3 miles from the above-mentioned islands. From Point Hunt the shores of the land take an irregular direction of S.W. ½ W. to *Point Pearce*, the intermediate space bounded by innumerable rocks and other impediments. Westward of this again, the shore falls back considerably, forming a deep bay, with several small openings running south-eastward. Cape Ibbetson is the western point of this bay, and the N.W. of Pitt's Archipelago. It is a very conspicuous, projecting land.

Stephen's Island lies to the N.W. of these points, and is about 4 leagues long. Between Cape Ibbetson and its S.W. point is a cluster of rocky islets and sunken rocks, which thus lie in the opening seaward of the channel between Stephen's Island and Pitt's Archipelago. Northward of Stephen's Island is an extensive and intricate cluster of islets and rocks, forming a complete labyrinth to navigators, but on its eastern side is a very commodious anchoring place, in lat. 54° 18′, lon. 130° 41′, where Vancouver stayed in company with three ships in search of furs, &c., under the command of Mr. Brown. The group extends W.N.W., a league and a half from the North side of Stephen's Island, and occupying a space of 2 miles in width. To the westward of this group, at the distance of 2 or 3 miles, lies a low detached rock, with some breakers near it; there are other lurking rocks lying about the same distance from Stephen's Island.

The land, which is separated from the continent by Grenville's Canal, and which we have been describing, although it was not traversed in the extent of 20 leagues, was still believed to consist of several islands, and therefore received the name of Pitt's Archipelago, after the celebrated statesman.

The CANAL DE PRINCIPE, between the archipelago and Banks's Island, extends from the North point of entrance into Nepean Sound to the North point of Banks's Island, first N.W. ¼ N. to the South point of *Puerto de Canaveral*, and thence N.W. by W. ½ W. to its N.W. point, in all about 14 leagues. The southern shore is nearly straight and compact, without soundings; the northern shore is much broken, bounded by many rocks and islets, and affording soundings in several places. On the S.W. side the acclivity is the greatest; but both sides of the canal may be considered as elevated land, and are entirely covered with pine trees. The shores abounded with a great number of very shy sea-otters.

PORT STEPHENS is 18 miles from the South end of the channel on the eastern shore. It was so named by Capt. Duncan, in the *Princess Royal*, in 1788. It is a small opening, the entrance of which is obstructed by many islets and rocks presenting no very tempting appearance as a port.

PORT DE CANAVERAL (of Senr. Caamano) is also on the eastern shore. Its entrance, 4½ miles wide, seems to be free from obstruction. Off its S.E. point is a small round island.

The channel between Pitt's Archipelago and Banks's Island and Queen Charlotte Island appears to be of irregular depth. To the S.W. of the North point of Banks's Island, 20 miles, is a bank of sand and shells in 23 to 25 fathoms, suddenly rising from 30 fathoms, mud, on each side. This bank apparently extends towards Point Ibbetson.

CHATHAM'S SOUND lies between Dundas and Stephen's Islands and the main land. The southern entrances have been before described. *Brown's Passage* enters the sound between the islands to the North of Stephen's Island and Dundas Island. This latter, in a N.N.W. direction, is 15 miles long and 5 broad, East and West. The eastern shores of the sound are low, and somewhat indented with small bays, and bounded by a reef of rocks at the distance of a quarter of a mile off shore. The interior country is snowy mountains. The shores and islands in the sound produce large numbers of pine trees. In the northern part of the sound are two clusters of rocks with breakers around, one S.S.W. ½ W. 8 miles, and the southernmost S.W. by S. 10¼ miles from Point Maskelyne. By daylight they are easily avoided, but by night or in fogs they must be very dangerous.

POINT MASKELYNE, so named after the astronomer, forms the S.E. point of the entrance to Observatory Inlet and Portland Canal. Off it lie two rocky islets, and to the South of it a rocky island close to the shore. The opposite or N.W. point is *Point Wales*.

WORKS CANAL.—Immediately East of Point Maskelyne is the entrance to a branch which takes a S.E. direction for 32 miles; its head approaching within about half a mile of the N.E. part of Port Essington, thus forming the land into a peninsula. Its S.W. shores are nearly straight and compact; its general width from 1½ to 2 miles, excepting near the entrance. An arm diverges from its N.E. shore, at 23 miles within the entrance, and trends in a general N.E. direction, but is made into a tortuous channel by a remarkably steep, rocky precipice, which at high water becomes an island. It had formerly been appropriated to the residence of a very numerous tribe of Indians.

"FORT SIMPSON, one of the Hudson's Bay Company's establishments, was originally formed at the mouth of the Nass River, but had been removed to this peninsula, washed on three sides by Chatham Sound, Port Essington, and Works Canal. It is the resort of a vast number of Indians, amounting in all to about

14,000, of various tribes. All these visitors are turbulent and fierce. Their broils, which are invariably attended with bloodshed, generally arise from the most trivial causes; such, for instance, as gambling quarrels, or the neglect of points of etiquette. Here the lip-piece is in more general use than on any other part of the coast, but is clearly going out of vogue; for it was far more common among the ancient dames than among the young women. The anchorage is in lat. 54° 33′ 25″, lon. 130° 18′."—(Sir George Simpson, vol. i. p. 207.)

At the mouth of Works Canal, N.E. of Point Maskelyne, is an island which divides the entrance into two channels; in the rear of this is a short arm called *Nass Bay*, and farther N.E. is one more extensive; neither require particular notice.

OBSERVATORY INLET.—The principal inlet runs in a N.E. ½ N. direction; and at 21 miles above Points Maskelyne and Wales the Portland Canal diverges from the principal one at Point Ramsden. Off this point are some dangerous rocks, only visible at low water, and opposite to it is a deep bay, with very shallow water all around it, except in the N.E. part, where a branch enters, bringing down muddy water, which is distinguished flowing down the principal arm. Beyond this bay to the N.E. the inlet is in general about half a league wide. The shores on both sides are straight and compact; a counter tide, or strong under-tow, is felt here, which very much embarrasses a vessel.

Salmon Cove is 20 miles above Point Ramsden, and on the western shore of Observatory Inlet. It affords good anchorage, and every convenience. Here Vancouver's vessel remained for some time, in July, 1793; and here he placed his observatory, from which circumstance the name of the inlet is derived. A very great abundance of salmon were taken here, up a very fine run of fresh water that flows into the cove; but they were small, insipid, of a very inferior kind, partaking in no degree of the flavour of European salmon.

The latitude of the observatory was deduced as 55° 15′ 34″, lon. 131° 3′ 30″; variation, 25° 18′ E., dip. 75° 54½′. High water at 1h 8′ after the moon passes the meridian, and the tide generally rose about 16 feet.

Beyond Salmon Cove the inlet extends 5 leagues in a North direction, when the western arm terminates, and the eastern arm extends the same distance, and forms the mouth of the River Simpson. The head of Observatory Inlet is much indented with small bays and coves, and abounding in some places with sunken rocks.

PORTLAND'S CANAL (so named from the noble family of Bentinck) diverges from Point Ramsden, in a N. by W. ½ W. direction, for about 5 miles; thence it bears in a more northerly direction 5 leagues farther, and then trends a little to the eastward of North, terminating in low, marshy land, in lat. 55° 45′, 70 miles from its entrance in Chatham's Sound. The shores of this inlet are nearly straight, and in general little more than a mile asunder, composed mostly of high, rocky cliffs, covered with pine trees to a considerable height; but the interior country is a compact body of high, barren mountains, covered with snow. As the surveying party ascended, salmon in abundance were leaping in all directions. Seals and sea-otters were also seen in great numbers, even where the water was nearly fresh, which was the case upwards of 20 miles from its termination.

The northern shore of the inlet, between Point Wales and Point Ramsden, is formed by several islands, behind which a channel runs parallel with the direction of the main inlet. This gradually decreases in width south-westward, continuing 13 miles from its N.E. entrance to an opening to the S.E. into the main channel. Pursuing the same direction, it enters much broken land, intersected by arms, forming an island about 10 miles in circuit, to the N.E. of which is an arm running in a N.E. direction, ending in low, steep, rocky shores. The shores form many little bays and coves, abounding with islets and rocks. An immense number of sea-otters, and some few seals, were here seen by Vancouver.

At 7 miles a little to the North of West from Point Wales, is the S.W. extremity of an island, from whence an arm extends in a N. ¾ W. direction, terminating in a fresh-water brook, in lat. 54° 56′, lon. 130° 40′. Its shores are nearly straight and compact. A league within the entrance, on the eastern side, are three small bays or coves, with four or five islets before them.

The S.W. shore, composing the entrance to the above inlet, is much indented with small bays, and bounded by innumerable rocks, and, from opposite the three small bays or coves, it trends to the S.W. to Cape Fox, so named from the statesman. About 2½ miles S.E. from this point is the outermost of a cluster of rocks and islands, extending nearly in a S.W. and N.E. direction, about half a league. There is a channel between them and the cape. About half a league westward from Cape Fox is a very commodious and well-sheltered little cove.

The Portland Canal, which may be considered to terminate here, is the boundary between the Russian and British possessions on the North American continent, as stated in the introduction to this chapter. The territory we have just described is only frequented by the Hudson's Bay Company's officers in their steam-vessel for the purposes of occasional trade with the natives. The continental shore to the northward belongs to Russia, but is leased by the Russian-American Company to the Hudson's Bay Company for trading purposes, as subsequently explained. One of the most important features of this region, as yet unknown and undeveloped, is the extensive island or archipelago, named Queen Charlotte Island, which is imperfectly described as follows.

QUEEN CHARLOTTE ISLAND.

This land was discovered nearly at the same period by the navigators of two nations. La Pérouse made the outer coast on August 10th, 1786, and followed it from South to North, for 50 leagues, in the ensuing ten days. Capt. Lowrie, in the *Snow*, Capt. Cook, and Capt. Guise, in the *Experiment*, sailed from Nootka on July 27th in the same year, and made the land in question soon afterwards, though the day itself is not now known. Thus the honour of discovery belongs to both the English and French. The name by which it is now known is derived from the vessel in which Capt. Dixon made it in the year following, but only assumed that it was an island from conjecture, as it was not proved to be such till Capt. Douglas, in the *Iphigenia*, sailed through the strait which divides it from the continent of America. It has also been called *Washington Island*, by Ingraham. Dixon's Channel, which runs in between Queen Charlotte Island and the Prince of Wales's Archipelago, to the North of it, was discovered, perhaps, by Ensign Juan Perez, in 1774. It was next seen by Dixon, on July

1st, 1786, though he himself acknowledges that Capt. Douglas was the first who sailed through it. He then sailed nearly round the island, afterwards repairing to Nootka. The eastern coast of the island was also examined and traded on by Capt. Duncan, in the *Princess Royal*, in 1787; after doing so he proceeded to the eastward to some other islands, which he named the Princess Royal Islands (which have been before described), but which have been supposed to be identical with the archipelago of San Lazaro of De Fonta, previously considered to be apocryphal. A part of the features of this latter were examined by Duncan, and he anchored in *nineteen* of its harbours, not without being frequently exposed to the danger of losing his vessel, but he was indemnified by an ample trade in furs. These are the principal of those early traders who have made us acquainted with the existence and a part of the natural features of this large and fine island. It is included in the British possessions, and up to the present time we are very ignorant of its actual character and resources. The following accounts are, therefore, necessarily very imperfect.

Dixon, or rather Beresford, whose letters form part of the account of the voyage, says:—"There is every reason to suppose, not only from the number of inlets we met with on coasting along the shore, but from meeting the same inhabitants on the opposite sides of the coast, that it is not one continued land, but rather forms a group of islands. The land, in some places, is considerably elevated, but not mountainous, and is totally covered with pines, which in many places afford a pleasing contrast to the snow that perpetually covers the higher grounds.

"The weather, whilst we were cruising here, was generally mild and temperate (August, 1787), the mean of the thermometer 54°. The whole time we coasted along from Cloak Bay to Cape St. James, the wind was generally steady at N.W. and W.N.W., but no sooner had we doubled the cape, and got to the N.E. side of the land, than we fell in with light variable winds and intervening calms.

"The number of people we saw during the whole of our traffic was about eight hundred and fifty; and if we suppose an equal number to be left on shore, it will amount to 1,700 inhabitants, which I have reason to think will be found the extreme number of people inhabiting these islands, including women and children. The great plenty of furs we met with here sufficiently indicated that these people have had no intercourse whatever with any civilized nation; and I doubt not but we may justly claim the honour of adding these islands to the geography of this part of the coast. The ornaments seen amongst them were very few; and it is probable that their knives and spears have been obtained by war rather than traffic, as there seems to be a universal variance amongst the various tribes; however, be all this as it may, they undoubtedly approach much nearer to a state of savage brutality than any Indians we have seen on the coast.

"The women distort the under lip in the same manner with those at Norfolk Sound* (Sitka), but with this difference, that here this wooden ornament (labret)

* See Dixon's Voyage, pp. 224-5.

seems to be worn by either sex indiscriminately, whereas at Norfolk Sound it is confined to those of superior rank."

Queen Charlotte Island, according to the running survey made of its outer coast by Vancouver in 1794, which must be taken as nearly correct, is about 160 miles in length in a N.N.W. and S.S.E. direction. Its greatest breadth, at the North extremity, is about 60 miles, from which it gradually diminishes towards Cape St. James, its South extreme.

CAPE ST. JAMES is in lat. 51° 58′ N., lon. 131° 2′ W. From the cape some rocks and rocky islets extend between the directions of S.S.E. and S.E. by S., at the distance of about a league; though Mr. Gray, in the *Columbia*, informed Vancouver that he had struck and received some material damage upon a sunken rock, which he represented as lying at a much greater distance, though nearly in the same line of direction. The cape was so named by Mr. Dixon from the common circumstance of the saint's day on which it was first seen. About it the land is very moderately elevated; but, like that on the northern part of the island, it rises gradually to rugged and uneven mountains, which occupy the centre of the country, descending towards its extremities to a less height, and is of a more uniform appearance.

IBBERTSON'S SOUND, an inlet running to the northward, is placed 30 miles to the northward of Cape St. James in Dixon's chart, but he gives no particulars of it. The weather was foggy here during his cruise off the coast, but he had many interviews with the natives all along.

CAPE HENRY, which is 24 leagues from Cape St. James, is in lat. 52° 53′ N., lon. 132° 25′. It is a conspicuous projecting cape, and forms the South point of a deep bay or sound, the shores of which are apparently much broken, to which Vancouver gave the name of *Englefield Bay*, after his much-esteemed friend, Sir Henry Englefield. Its North point of entrance, lying from Cape Henry N. 27° W., at the distance of 7 leagues, was named *Point Buck*; which also forms the South point of entrance into a sound falling deep back to the eastward, named by Vancouver *Cartwright's Sound*. Its North point of entrance, *Point Hunter*, lies from Point Buck N. 25° W., distant 10 miles, and a little within this line of direction is an island near the northern shore.

RENNELL'S SOUND, so named by Capt. Dixon, appears to be very extensive, and takes an easterly direction to the northward of Point Hunter. Its entrance, according to the observations of Vancouver, is in lat. 53° 28′ N., lon. 132° 49′ W. The land appears much broken, and the coast composed of steep mountainous precipices, divided from each other by water. These gradually decrease in height toward the North extremity of the island, to which point the shore is more or less lined with scattered islets and rocks, at a small distance from the land.

HIPPAH ISLAND lies N. 32° W. 15½ leagues from Point Hunter, and forms the northern limit of Rennell's Sound. It is a high, steep, cliffy hill, ending in a low projecting point, to the N.E. of which lie some breakers, though at no great distance. The island was thus named by Dixon,* from its being inhabited

* Voyage, p. 206.

by a tribe of Indians who had fortified themselves precisely in the manner of a *hippah* (e-pah) of the New Zealanders. It is in lat. 53° 33′, lon. 133° 7′.

POINT FREDERICK lies N. 17° W. 26 miles from Hippah Island, and is the West extremity of a projecting land, appearing like two islands; it is about 22 miles S. 14° W. from Point North, the N.W. extreme of Queen Charlotte Island: between it and Hippah Island is *Clonard Bay*, but we have no particulars of it.

The N.W. point of the principal portion of these islands is named in the Spanish charts *Cape Florida Blanca*, and near to it is *Cape Santa Margarita.*

CLOAK BAY lies to the southward of North Island. It was so named by Dixon, from the number of fur cloaks (principally of sea-otter skins) purchased from the natives here. Dixon says, "There appeared to be an excellent harbour, well land-locked, about a league ahead; we found soundings from 10 to 25 fathoms water, over a rocky bottom; but unluckily could not reach it, from the contrary wind and strong tide."—(P. 200.)

LANGARA ISLAND forms the N.W. extremity of the group; and North Point, its outer end, is in lat. 54° 20′, lon. 133° 11′. The coast here turns to the eastward, continuing so, irregularly, for about 65 miles, to *Point Ymbisible*, or *Rose;* having in the interval *Port Masaredo, M'Intyre's Bay*, and *Port Estrada.* That these exist, and probably many other places of shelter, is the whole amount of our knowledge.

Of the *eastern* side of the islands we have only the representations of Mr. Duncan, and other early navigators. At 35 miles from the N.E. point an extensive opening is marked, called *Trollope's River*. In some of the charts, as in that of Meares, this opening is made to communicate with Rennell's Sound on the western shore, thus separating the land into two larger islands at least. This is more than probable, from what has been said in the former part of the description, and is exactly analogous to the formation of the Sitka Archipelago, the assumption of which was formed from similar facts, that the same individuals of the native tribes were seen by the traders, Dixon and others, on both sides of the group at different times. On the coast, southward of this, there is a settlement called *Skidegats.*

At 4 leagues from the South extremity of Cape St. James is *Rose's Harbour*, or *Bay de Lujan* of the Spaniards; it appears to be embarrassed with rocks, on its South side especially.

DIXON'S CHANNEL is the strait separating the Queen Charlotte Archipelago from the islands fronting the continental shore. As stated previously, its northern entrance was first seen by Perez in 1774, and therefore ought perhaps to be called by his name. Capt. Douglas, in the *Iphigenia*, also, was the first who passed through it, and he too has some claim to its designation. But Capt. Dixon, who was the principal officer in these expeditions, has the priority of discovery, except that of the Spaniards, as above mentioned.

The following observations on it are by Capt. Douglas, its first explorer:—"Ships which arrive early on the coast, where they must expect to meet with heavy gales of wind, will find it to their advantage to make the South end of Queen Charlotte Island, and to enter the straits in the lat. of 52°, and lon. of

130° 30′, when they will find shelter either in the island or in the continent. It may also be added, that as ships which are returning from the North at a late period of the season are liable to be blown off the coast, it would be advisable for them to make Douglas Island and enter the straits in the lat. of 54° 30′, where they will find good anchorage, as well as inhabitants, on the North side of the island. On the continent they will have also the advantage of Port Meares and Sea Otter Sound, besides several other bays which have not yet been explored, between 56° and 54° North latitude.*

CHAPTER XV.

COAST OF RUSSIAN AMERICA, FROM PORTLAND CANAL TO COOK'S INLET.

The whole of the American coast North of lat. 55° 40′, which is the South point of the actual territory of the Russians according to the international treaty, is under the colonization of the Russian-American Company. The lat. of 56° would intersect Prince of Wales Island, which it was necessary to include. This division passes down the Portland Canal, as described in the previous chapter.

The Russian-American Company was established under charter from the Emperor Paul, July 8th, 1799; and the extensive territory in question was granted to them to occupy and bring under the dominion of Russia. The Russian Company and the Hudson's Bay Company were thus brought into collision, and the latter experienced considerable loss in their endeavours to prevent this extension of Russian power. But in justice to Russia it must be said, that no country had a better claim to the territory; for as early as 1741, Behring, and his companion Tschirikoff, had touched on the continent in the lat. of 59° and 56° respectively; the former seeing much of the intervening countries, too, on his return; and by 1763 many other adventurers had penetrated eastward as far as Kodiak—and it must be remembered that no other nation claims to have penetrated farther North than lat. 53°. In addition to this, Russia had as gradually improved her knowledge by possession as these discoveries advanced, and this, too, not from any jealousy of other powers interfering, as was the case between Spain, England, and France, to the South. Thus the settlement at Kodiak was formed four years before our countryman Meares purchased, or said he did so, his tract of land in Nootka Sound, and Sitka was founded ten or twelve years before Astoria was.

Notwithstanding this, the Hudson's Bay Company expended considerable sums in the establishment of trading posts on the large River Stikine, in lat. 56° 20′. The Russians resented by force this procedure of the company, although

* Douglas, in Meares, p. 332.

England claimed the privilege of navigating the rivers flowing from the interior of the continent to the Pacific, across the line of boundary established under the treaty of 1825. The British government required redress for this infraction of the treaty; and after negotiation between the two governments and the two chartered companies, it was agreed, in 1839, that from the 1st June, 1840, the Hudson's Bay Company should enjoy for ten years the exclusive use of the continent assigned to Russia by Mr. Canning in 1825, and extending from 54° 40′ N. to Cape Spenser, near 58° W., in consideration of the annual payment of 2,000 otter skins to the Russian-American Company.

The boundary between the Russian and English possessions was fixed by the convention agreed to by the respective powers, February 28th, 1825. In the appendix to this volume will be found the treaty at length. By its articles the trade is open to both nations in the Pacific; that subjects of other powers shall not land without permission at the establishments of either respectively. The boundary is fixed as commencing in lat. 54° 40′ N., between lon. 131° and 133° W., running northward along the Portland Canal as far as the parallel of 58° N.; then north-westward along the summits of the mountains, parallel with the coast, to the meridian of 141° W.; always provided that this line shall not exceed the distance of 30 miles from the coast; that no establishments shall be formed by either party within the limits of that claimed by the other; that all streams or rivers in the Russian territory shall be open for navigation to the British, either from the ocean or the land; that the trade at Sitka (except in spirituous liquors) should be open for ten years; that vessels taking shelter, from distress, shall pay the same dues as national vessels, unless she disposes of any of her cargo.

The charter of the Russian-American Company, granted in 1799, was renewed in 1839, when they had thirty-six hunting and fishing establishments.

Sitka, or New Archangel, founded in 1805, is their chief post, and here all the business of the company is centred. Subordinate to it there is a smaller establishment of a similar kind at Aliaska, which supplies one post in Bristol Bay, and three posts in Cook's Inlet, all connected with minor stations in the interior. Another station in Norton Sound has its own inland dependencies.

The whole of the territory is divided into six agencies, each controlled by the governor-general. The inhabitants of the Kurile and Aleutian Islands, and those of the large island of Kodiak, are regarded as the immediate subjects of the Russian Company, in whose service every man between eighteen and fifty may be required to pass at least three years.

The natives of the country adjacent to Cook's Inlet and Prince William's Sound also pay a tax to the company, in furs and skins. The other aborigines in the Russian territory are not allowed to trade with any people but those of the Russian Company.

In 1836 the number of Russians in the territory of the company was 730; of native subjects, 1,442 creoles; and about 11,000 aborigines of the Kurile, Aleutian, and Kodiak Islands.

At the time of Sir George Simpson's visit to Sitka in 1842, the returns of the trade, he says, were nearly as follows:—10,000 fur seals; 1,000 sea-otters;

12,000 beavers; 2,500 land-otters, foxes, martens, &c.; and 20,000 sea-horse teeth.

The character of the country, and its trade, &c., will be gathered from the previous remarks and the subsequent descriptions. These are not always perfect. A portion of the interior sounds were explored and surveyed by Vancouver, doubtless with his usual accuracy, but of course this occurred before it was colonized by the Russian Company. Its chief settlement, Sitka, has been visited by many navigators since that time, and indeed is the only point of interest to the world in general. Of detailed description we possess but little: this want is, however, of but little consequence, though for the sake of rendering their description of the entire shores of the Pacific in some degree complete, the ensuing details are given to afford an insight into the character of the country.

The PORTLAND CANAL, forming the boundary, has been described before, pages 431, 2. *Cape Fox* forms the N.W. point of the approaches to it.

From this the coast takes a rounding direction N.W. by W. 4 miles, and then N. by W. ½ W. near 7 miles farther, to a projecting point called *Foggy Cape*, the coast being very rocky and dangerous.

North of Foggy Cape is a large bay filled with a labyrinth of small islands, rocks, and shoals, the north-westernmost and largest being N. by W. ¾ W. nearly a league distant.

Cape Fox on the East, and Cape Northumberland on the West, bearing E. by S. and W. by N. 5 leagues apart, form the southern entrance to the Canal de Revilla Gigedo of Senr. Caamano, hereafter noticed.

Four miles to the northward of the island above mentioned is the entrance to the BOCA DE QUADRA, which is almost rendered inaccessible by islets and rocks. The inlet first takes a direction of N.E. ½ E., to a point 7 miles within the entrance, whence the shores become less elevated, and the inlet takes a S.S.E. direction, and is here about 2 miles in width for a distance of 4 miles from the point. On the South shore above this are three inconsiderable rocky inlets, from the N.E. of which the main branch, about three-quarters of a mile wide, takes a direction of N.N.E. ½ E. for 4½ leagues to its head, in lat. 55° 9′, a small border of low land, through which flow two rivulets. The sides of this canal are nearly straight, firm, and compact, composed of high, steep, rocky cliffs, covered with wood.

Near the entrance, in the Canal de Revilla Gigedo, is an islet called by Vancouver Slate Islet, a prodigious mass of this stone differing from any other about here. N. ¾ W. 4 miles from this is Point Sykes, and N.W. ½ W. 5 miles, is Point Alava, between which points is the entrance to Behm's Canal.

BEHM'S CANAL (so named after Major Behm) is one of those extensive and singular arms which abound on this forbidding and inhospitable coast. It runs northward for 55 miles, then westwardly and southwardly, encircling the large island of Revilla Gigedo, and this is separated on the S.W. by the strait of the same name from the Island de Gravina.

From *Point Sykes*, the S.E. point of the entrance, the South shore runs N.E. ¾ N. 10 miles to *Point Nelson*, the inlet being from 2 to 4 miles wide. Eastward of Point Nelson an inlet takes an East and N.E. direction for 10 miles, terminating in the usual manner. The surrounding country consists of a huge mass of steep, barren, rocky mountains, destitute of soil, the summits covered with perpetual snow: the shores are nearly perpendicular cliffs rising from the water's edge. The N.E. point of this inlet is *Point Trollope*, 4½ miles from Point Nelson. In this part of the canal are several islands, which separate it into different navigable channels. Northward of Point Trollope are two long narrow islands on the East side, forming a narrow channel 7 miles long inside of them. Off the N.W. point of the northernmost of these, bearing N.N.W. nearly a league distant, is a very remarkable rock, named by Vancouver the *New Eddystone*, from its resemblance to the celebrated lighthouse and rock. Its circumference at its base is about 50 yards, standing perpendicularly on a surface of fine dark-coloured sand. Its surface is uneven, and its diameter regularly decreases to a few feet at its apex. Its height was found to be above 250 feet, lat. 55° 29′. Except the bed of sand on which it stands, and a ledge of rocks to the North, visible only at low tide, the surrounding depths are unfathomable.

On the East shore of the canal, a league above the New Eddystone, is an unimportant arm, 2 leagues in depth, terminating in two coves, and winding between an immense body of high, barren, snowy mountains. The coast beyond this is straight and compact, trending N. ½ W. 9 miles to *Walker's Cove*, an inlet extending 2 leagues E.N.E. through rocky, barren precipices to a marshy termination.

The main inlet extends in a N.W. by N. direction from Walker's Cove. The water is of a very light colour, not very salt, and the interior country, on the Island of Revilla Gigedo, rises into rugged mountains, little inferior in height to those on the eastern side; there is a cluster of rocks a mile in extent on the East shore, to the northward of the cove and the shores of the canal, which are nowhere more than 2 miles asunder, and afford some small bays and coves. Proceeding northward, the canal takes a more westerly direction to *Fitzgibbon Point* on the East side, in lat. 55° 56′; and the opposite point on the island is called *Point Whaley*.

Burrough's Bay extends N.E. from Point Fitzgibbon about 2 leagues, where it is terminated by low land, through which *three or four small rivulets* appear to flow over a bank of mud stretching from the head of the arm, and reaching from side to side, on which was lodged a quantity of drift wood. When Vancouver was here, August 11th, 1793, he found the water perfectly fresh, and the whole surface of the bay strewed over with salmon, either dead or in the last stages of existence. They were all small, of one sort, and called by him hunch-backed salmon, from a sort of excrescence rising along the backs of the male fish. The mouths of both fish were formed into a sort of hook, resembling the upper mandible of a hawk; they had little of the colour and none of the flavour of salmon, and were very indifferent and insipid food. In all parts of the inlet, particularly in the arms, and in every run of fresh water, vast numbers of these fish were seen, but all in a sickly condition. If any just conclusion could be drawn from the immense numbers found dead, not only in the water, but lodged on the shores below high-water mark, it would seem that their death takes place immediately after spawning, for the purpose of which they ascend these inlets.

From the mouth of Burrough's Bay, the main inlet takes an irregular S.W. ¾ W. direction, to a point on the North shore, 4 miles from Point Whaley, named *Point Lees*. Here the channel decreases in width to less than a mile, and the water gradually assumes a darker colour. Beyond Point Lees the North shore of the principal channel is formed by *Bell's Island*, which is about 2 leagues long in a N.E. and S.W. direction. Behind this island is a channel with steep rocky shores, covered with pine trees, of irregular width, in some places not more than a quarter of a mile, on the North side of which are these unimportant arms.

The point on the South shore, beyond the West point of Bell's Island, is in lat. 55° 50′, lon. 130° 41′ (Vancouver, vol. ii. p. 357); and here the channel turns sharp to the South, and widens in that direction. On the opposite side of the canal is an inlet extending in a N.W. ½ W. direction about 4 miles, containing several sunken rocks; and on a bay on the N.E. shore the remains of a considerable Indian village were found, overrun with shrubs, among which a small fruited crab was in great abundance.

South of this inlet is a large bay, terminating in a sandy beach nearly all round, its shores being very moderately elevated and thickly wooded. Off its S.E. point is an island, but no channel inside it. The interior country is not very high, particularly westward, where a low wooded country extends as far as the eye can reach.

PORT STEWART, named after one of the mates of Vancouver's ship, is to the southward of this. Its South point of entrance is in lat. 55° 38′ 15″ N., lon. 131° 47′ W.; var., 28° 30′ E. (1793). Here Vancouver remained with his vessel in August and September, 1793. He found it a small but convenient bay, secured, by several islets before it, from the wind in all directions. Great plenty of excellent water was found close at hand; the shores of moderate height, and covered with pine trees, berry bushes, and other shrubs.

It is formed, as before stated, by a bay in the land, having several islets and rocks lying before it: within these, from the South point of its entrance, it takes a course of N.N.W. ½ W., about half a league in length and three-quarters of a mile in breadth. In this space it affords good and secure anchorage, from 4 to 18 fathoms water, good holding ground. The communication with the shore is easy, and wood and water may be conveniently procured in the greatest abundance. Towards its head are two very snug coves or basins, one of which is a continuation of the port, the other formed by an indent in the land; the soundings are from 6 to 9 fathoms, admitting of a navigable though narrow channel into them. There are passages in several directions between the islets lying before the harbour, but they are not very safe, in consequence of several rocks between and about their shores, visible only at low tide. The best passage into Port Stewart is between the southernmost isle and the main land; this is perfectly free from any obstruction, with soundings from 4 fathoms at the side to 11 fathoms in the middle.

The eastern shore of the canal, southward of the point where it assumes a southerly direction, is much broken and intersected with arms; and opposite to Port Stewart is a cove near which Vancouver was attacked by the Indians, in which two of his men were severely wounded; hence he called it *Traitor's Cove*, and a point to the South on which he landed, in lat. 55° 37′, *Escape Point*.

Cape Caamano is the South point of the peninsula, dividing the arm from Clarence Strait. It is in lat. 55° 29′, lon. 131° 54′. It was so called after the Spanish commander who first delineated (though imperfectly) these shores.

On the opposite side of the channel, the westernmost point of the island of Revilla Gigedo, is called *Point Higgins*, after the then president of Chile, Senr. Higgins de Vallenar, and this latter name is applied to the North point of the Island Gravina, S. ¾ W. 2 miles from Point Higgins. From *Point Vallenar* lies a ledge of rocks, parts of which are only visible at low tide; this ledge nearly joins on to two small islands off the point.

Betton's Island lies to the northward of Point Higgins, against the eastern coast. On its N.W. side are several dangerous rocks, lying half a mile from its shore; and between it and the eastern shore are several smaller islands.

The CANAL de REVILLA GIGEDO, of Señor Caamano, separates, as before stated, the Island, or rather Islands, of Gravina from the island of its name and the main land. It runs south-westward from between Points Higgins and Vallenar to between Foggy and Northumberland Capes, described previously. It was not explored by Vancouver, and is apparently badly represented in the Spanish chart.

DUKE of CLARENCE'S STRAIT separates the Prince of Wales's Archipelago on the West, from the islands we have been describing on the South, and from the Duke of York's and other islands northward, and is probably the opening distinguished in Caamano's chart as the "Estrecho del Almirante Fuentes, y Entrada de Nostra Senr. del Carmin."*

CAPE NORTHUMBERLAND is the southernmost point of the Islands de Gravina. Off Cape Northumberland are several clusters of rocks, the bearings of the principal of which, from a tolerably high round island lying South from the cape, are as follow: the outermost to the N.W., N.W. by W., 3½ miles; the south-westernmost, W.S.W. 4½ miles; the southernmost, which are the most distant, South, 6½ miles; and the south-easternmost, S.E. ¼ E., 5 miles distant; within some of these the intermediate spaces are occupied by an immense number of rocks and breakers. The southernmost is a round lump of barren rock, always above water, with some breakers a short distance from its S.E. side. N.E. ¼ N. 4½ miles from this lies the south-easternmost of these rocks; it is low, flat, and double, always above water, but has much broken ground in its neighbourhood. The south-westernmost of them bear from the South rock N.W. 5½ miles; they are two small rocks, with much broken ground North and N.E. of them. Between these and the eastern shore lie many dangerous rocks and breakers; but to the northward of the South rock, and between it and these two latter clusters, there did not appear any dangers.

Point Percy lies N.W. by W. ¾ W. 9 miles rom Cape Northumberland. It is the western extremity of a long, narrow cluster of low islands, extending about

* The Strait of De Fonta. See Fleurieu's Introduction to Marchand, &c.

5 miles in an E.N.E. direction, nearly uniting to the eastern shore, which is much broken North and South of them. Between this point and Cape Northumberland are several clusters of dangerous rocks, lying in all directions, a considerable distance from shore, with very irregular soundings, from 4 to 3 fathoms water; but in the day time they are sufficiently indicated by the weeds growing on them.

Point Davison bears N. $\frac{1}{2}$ E. 4 miles from Point Percy, and is in lat. 55° $0\frac{1}{3}'$. The coast then runs North towards an opening about 2 miles wide, appearing to divide the Island of Gravina. In it are innumerable rocks and rocky islets. Northward of this the shores trend N.N.W. 5 miles, and then about N. by W. 6 leagues to Point Vallenar; they are nearly straight and compact, with a few rocks, extending from the projecting points. The shores of the Islands of Gravina are of moderate height, and covered with wood.

The southern entrance to the Duke of Clarence's Strait lies, as before stated, between Cape Northumberland on the East and *Cape de Chacon* on the West. This latter cape is the S.W. point of the Prince of Wales's Archipelago, and bears W.S.W. from the former, 8 or 9 leagues off, lat. 54° 43′, lon. 131° 56′.

The first considerable opening on the western shore of the strait, North of Cape de Chacon, is *Moira Sound;* a smaller one is just to the South of it. It takes a south-westerly direction, and appears to be divided into several branches, with some islands lying before its entrance.

From this sound the western shore takes a direction nearly North, and forms some bays; the largest of these, situated in lat. 55° 8′, has, in and before it, several smaller islets; the outermost is by far the largest; and as it in many points of view resembled a wedge, it was called *Wedge Island;* off its South point lies a ledge of dangerous rocks.

The land in the neighbourhood of Moira Sound is high, and rather steep to the sea; but beyond Wedge Island the straight and compact shores are more moderately elevated, and the interior country is composed of lofty, though uneven mountains, producing an almost impenetrable forest of pine trees, from the waterside nearly to their summits.

Nine miles North of Wedge Island is a projecting point, in lat. 55° $16\frac{1}{3}'$, and to the West of this is *Cholmondeley Sound,* which extends to the southward, divided into several branches. A small island lies to the N.W. of the entrance.

Point Grindall bears from Cape Caamano S.W. by W. 4 or 5 miles distant. This point projects from the main land to the westward, with some rocks and breakers extending about a mile from it; 4 or 5 miles to the S.E. of it is a small bay, with some islets and rocks lying off it.

The strait up to this part varies in width, from $1\frac{1}{4}$ to $2\frac{1}{4}$ leagues; and, with the exception of the dangers immediately adjacent to the shores, is open and clear throughout.

From Cape Caamano to Point Le Mesurier the coast first bears N.W. by W. near 2 leagues, and then N.N.W. Halfway between these points is a small island, with a passage between it and the eastern shore. *Point Le Mesurier* projects from the main land to the westward, and has some islets and rocks extending about a mile from it. Opposite to Point Le Mesurier is *Point Onslow,*

N.N.W. ¾ W. 5½ miles distant, and between these points is the entrance of an inlet, nearly as extensive as the one it enters, named PRINCE ERNEST'S SOUND (after the Duke of Cumberland, the present King of Hanover). Point Onslow is the South extreme of the island or islands forming the Duke of York's Archipelago, and the above-named sound encircles it in a similar manner to Behm's Channel around the Island of Revilla Gigedo.

The continental shore from Point Le Mesurier trends N.N.E. ½ E. for about 4 leagues, indented with bays of different capacity, and some scattered rocks and islets along its shores. The opposite shores then incline more to the eastward from this point; and to the northward of it is the South point of an island extending N.N.W. ¼ W. 5 miles in front of a bay on the East shore, on which are some islets and rocks. The western shore of the island is very much broken, and has some islands off it, but it allows a tolerably good channel inside it. N. by W. ¼ W. from the above bay, 2½ leagues, brings you to *Point Warde*, in lat. 56° 9′. The western shore is irregular in its direction, and much broken; opposite the island it is 6 miles across, but here its shores are moderately elevated, and covered with the usual productions, and approach within a mile of each other. From Point Warde the coast takes a sharp turn N.E. by E. ¼ E. 4 miles, to a point where the channel divides into two branches; the easternmost extends eastward about 3 leagues, terminating in the usual way, and named *Bradfield Canal*. The main branch extends in a N.N.W. direction, 3 leagues to a point in lat. 56° 20′. This branch is here not more than three-quarters of a mile broad, with a small island and two islets in its entrance. Here it again divides into two branches, but the N.N.E. one is insignificant; the main channel to the West, before which lie several rocks and small islets, is not more than a quarter of a mile wide, extending irregularly to the N.W. and S.W., forming a passage about a league long to *Point Madan*, where the channel is more spacious, and again takes two directions, one to the S.S.W., through a broken insulated region; the other stretching to the N.N.W. ½ W., nearly 2 miles wide. In this direction it proceeds about 16 miles to a very conspicuous point, in lat. 56° 34′, named *Point Highfield*, where the channel again appears to divide into two branches to the N.N.W. and West. On Point Highfield the Hudson's Bay Company once had a fort.

The apparent opening to the northward of Point Highfield is entirely closed by a shoal extending across it from *Point Rothsay* on the East or continental shore, and *Point Blaquiere* on the opposite side, on the edge of which are only 6 and 9 feet water. To the South of this shoal, and in its immediate vicinity, are four small islands and two or three islets; one of the former upon the shoal, and the others, at the distance of 1½ leagues from Point Highfield, extend to the West and S.W. of it. This shoal is very steep-to, and, by its connexion with the adjoining land, it may be said to make the latter form a portion of the continent.

Just to the northward of this the RIVER STIKINE debouches on the eastern shore, and near its mouth is the fort of the same name.

FORT STIKINE was originally founded by the Russian-American Company, and had been recently (1842) transferred to the Hudson's Bay Company, on a lease of ten years, together with the right of hunting and trading in the continental territories of that association as far up as Cross Sound. The establishment, of which the site had not been well selected, was situated on a peninsula barely large enough for the necessary buildings; while the tide, by overflowing the isthmus at high water, rendered any artificial extension of the premises almost impracticable; and the slime, that was periodically deposited by the receding sea, was aided by the putridity and filth of the native villages in the neighbourhood in oppressing the atmosphere with a most nauseous perfume. The harbour, moreover, was so narrow, that a vessel of a hundred tons, instead of swinging at anchor, was under the necessity of mooring stem and stern; and the supply of fresh water was brought by a wooden aqueduct, which the savages might at any time destroy, from a stream about 200 yards distant.

The Stikine or Pelly's River empties itself into the ocean by two channels respectively 4 and 10 miles distant from the fort. One of them is navigable for canoes; while the other, though only in the season of high water, can be ascended by the steamer about 30 miles.

The establishment is frequented by the Secatquonays, who occupy the main land about the mouths of the river, and also the neighbouring islands, and amount to about 3,000 souls. About 4,000 or 5,000 people are, in all, dependent on Fort Stikine for supplies. Most of these Indians make trading excursions into the interior, in order to obtain furs. Their grand emporium is a village, 60 miles distant from Dease's Lake, and 150 from the sea, and thither they resort three or four times a year.*

The North shore of the principal arm now takes a direction of S.W. ½ S. for 14 miles from the Stikine River to *Point Howe*. The shores are indented with small bays, with some small islets; the opposite, or South shore, is about a league distant; and, to the westward of *Point Craig*, lying from Point Hood S.E. by E. 2 leagues, the shore appears firm and compact; to the East of it, it is much broken and divided. From Point Howe the shore rounds in a westerly direction to *Point Alexander*. This point is the easternmost of the entrance to Duncan's Canal, which stretches irregularly North and N.W. to its termination in a shallow bay, bounded to the North by a low sandy flat, in lat. 56° 58′. The entrance is formed into two channels by an island; the easternmost is a narrow arm, 6 miles long, with a rock nearly in the centre of the entrance; and, from the point where it diverges, a narrow arm extends 4½ miles N.N.E., to a low place producing very long grass. The channel passes through broken land in a S.W. direction, with only 3 fathoms, to Point Hood, in lat. 56° 44′. Here it communicates with the more spacious western branch, about 2 miles wide, leading South on the western side of the island before mentioned. Above it Duncan's Canal stretches irregularly, having on it several islets and shallow bays, the latter principally on the S.W. shore. *Point Mitchell* forms the S.W. point of the canal, and is opposite the opening of the southern branch of the Duke of Clarence's Strait, the description of which we will resume from the point where Prince Ernest's Sound diverges from it.

Point Onslow, as before mentioned, is the North point of the entrance of Prince

* Sir George Simpson, vol. i. pp. 209–10.

Ernest's Sound; and, from this to *Point Stanhope*, the next projection on the eastern shore of the channel, the distance is 15 miles. The interval forms a bay, the shores of which appear much broken, and has some rocky islets near it. The coast then extends N. ¾ W., about 10 miles, to *Point Harrington*. Three miles and a half southward of Point Harrington is a small island, on the North side of which is an anchorage, close under the shores of Duke of York's Island. It is tolerably well sheltered from the South and S.E. winds, but the soundings are irregular, and the bottom in parts is rocky.

Point Nesbitt is in lat. 56° 15′, and bears from Point Harrington N.W. about 2 leagues, the interval forming the opening to an inlet bearing to the N.E., which possibly communicates with the channel to the East of the islands. Off Point Harrington, and nearly in mid-channel, is a cluster of low rocks; and also off Point Nesbitt, extending southward from the point, is a ledge: these seem very dangerous, as most of them are only visible at low water. The western shore of the strait is moderately elevated, of an uneven surface, and very much divided by water. The soundings in this part are very irregular, from 10 to 30 fathoms, and, in some places, rocky bottom.

Bushy Island, which lies in the channel to the northward of Point Nesbitt, is about 2 miles long, having from its shores, on both sides, some detached rocks, but admitting between it and the eastern shore a navigable channel. From the N.W. side of this island lies also a chain of small islets, extending northward to the entrance of this opening, which is between *Point Macnamara* on the East, and *Point Colpoys* on the West; this bears West, 1½ leagues from the former. Here the channel enters from the north-eastward, as before described, and bears to the westward and S.S.W. to the ocean. *Point Mitchell*, on the S.W. side of the entrance to Duncan's Canal, is the point on the North shore opposite to Point Colpoys, and is 8 miles distant. The northern shore of this branch of the strait extends a little to the southward of West to *Point Barrie*, a distance of 18 miles. In that space are innumerable rocks; and nearly midway between the two points there is a large bay, about 4 miles wide at the entrance, and the same depth; there are two or three islets and many rocks in it.

The southern shore, which forms the North coast of the Prince of Wales's Archipelago, and the distance between Point Colpoys and Point Baker, its East and West extreme, is 17 miles. Just to the south-westward of Point Baker is an excellent harbour, Port Protection, which was a haven which afforded Vancouver an asylum when he little expected it, amidst impending dangers, in September, 1793.

PORT PROTECTION will be most readily found by attending to the following directions. It is situated at the N.W. extremity of the Prince of Wales's Archipelago; its southern extreme comprises the base of a very remarkable barren peaked mountain, named Mount Calder. This is conspicuous in many points of view, not from its elevation when compared to the mountains on the neighbouring continent, but from its height above the rest of the country in its immediate vicinity, and from its being visible in various directions at a great distance. Point Baker, in lat. 56° 20′ 30″, lon. 133° 36′, on an islet close to the shore, forms the N.E. point of entrance, from whence the opposite point lies S.S.W. ½ W. three-quarters of a mile distant; the channel is good, and free to enter, yet there is one lurking rock, visible only at low tide, lying in a direction from Point Baker S. by E. ½ E. 3 cables' length distant. It is indicated by weeds, and is clear all around. There is also an irregular bank North of Point Baker, with from 15 to 32 fathoms; this, with the meeting of the tides around the Prince of Wales's Archipelago, causes an agitation, or race, especially at the flood tide, but there is no danger; the depth is very great.

The harbour takes a general direction from its entrance S.E. by S. ½ S. for 2½ miles, and its navigable extent is from 5 to 3 cables' length in width, beyond which it terminates in small shallow coves. The depth is rather irregular, from 30 to 50 fathoms: the shores are in most places steep and rocky, and are covered with an impenetrable forest of pine and other trees. They afford several streams of fresh water; some fish and fruit were found, as also wild-fowl. The tides appear to be irregular, but come from the South, and it is high water 7ʰ 40′ after the moon passes the meridian.

Points Baker and Barrie form, as before stated, the western extremes of the branch of the Duke of Clarence's Strait, which trends East and West. Westward of this the strait takes a southerly direction to the Pacific, and the western shore of this portion is formed by the southern end of an island which is singularly intersected by deep bays and inlets, and the shores of which are bestrewed with innumerable rocks. Although this is an island, yet to the seaman it cannot be considered so, because the narrow channel separating it from the main land North of Point Barrie is so full of rocks and dangers that it certainly is not navigable.

The western shore of the strait is distant from Point Barrie in a West direction, but between is *Conclusion Island*, about 3½ miles long, N.W. and S.E., with some rocks off its shores, and lying in a large bay full of an infinite number of rocks, very dangerous even for boats; consequently it is unimportant. Between Point Baker and Conclusion Island, distant from the former 4 miles, is a smaller island, low, and about a mile long North and South, with a ledge of very dangerous rocks extending from its South point. Off the western shore of the strait, abreast of this latter island, is another small island, 1½ miles long, and having two smaller ones lying off its South point.

From hence the coast takes an irregular direction about S. by E., to a point in lat. 56° 17′, forming the N.E. point of entrance into Port Beauclerc.

Port Beauclerc is of easy access and egress, free from every obstruction but such as are sufficiently evident to be avoided. The opposite point of entrance lies West 2 miles distant; it extends N.W. 4½ miles, and S.W. 2 miles, from the points of entrance. Nearly in the middle is a small island and some rocky islets, and a rocky islet with some rocks before its entrance, lying S.S.E. ½ E., one mile from the N.E. point of the entrance. The surrounding shores are in general moderately elevated, well covered with wood, and water is very easily to be procured, as the communication with the land is sufficiently commodious.

Point Amelius lies S.E. by S. ½ S. about a league from the entrance of Port Beauclerc, and South of it the coast forms a bay about a league to the westward, and thence it takes a more southerly direction, about 7 miles, to *Point St. Albans*, which is a low rocky point, in lat. 56° 7′, lon. 133° 55′.—(Vancouver, vol. ii.

p. 414.) Off this portion of the coast, islets, rocks, and breakers extend about a league. About 3 miles North of the point is a snug boat cove.

AFFLECK'S CANAL extends to the N. by W. 15 miles, immediately to the westward of Point St. Albans. Its eastern shore has rocks off it for the first league and a half, and then becomes straight and compact to its termination in some low land, through which flow some streams of fresh water. The western shore, which is from half a mile to 2 miles from that opposite, is indented with three large bays in its southern part. The eastern sides of the canal are mountainous, but not so steep as the more interior country. The western side is moderately elevated, of uneven surface, and is covered with dwarf, pine, and other trees.

CAPE DECISION is a very conspicuous promontory, extending in a South direction into the ocean, and forms the southernmost point of the island we have been describing; it is in lat. 56° 2′, lon. 134° 3′. Southward of the cape are some islands; the largest, *Coronation Island*, is about 7 leagues in circuit. From the N.E. point of this island, which bears S. by E. 4 miles from Cape Decision, is a range of rocky islets extending to the North, within half a league from the main land; the space between them and the cape appearing free from interruption.

CAPE POLE is the promontory on the western shore of the Prince of Wales's Archipelago which forms, with Cape Decision, the entrance to the Duke of Clarence's Strait. They bear W. by S. ½ S., and E. by N. ½ N., 11 miles asunder. Off Cape Pole is *Warren's Island*; it is high, and between it and the cape many lurking rocks were observed. To the southward of it also are three clusters of very dangerous rocks, the first lying from its S.W. point S. by E. ½ E. 3½ miles distant; the second, South, 6 miles, and a small islet lying from them S.E. at half a league distant; and the third cluster lies off the S.E. point of the island, which from its N.W. point lies S.E. ¾ E. 4 miles, from whence those rocks lie in a direction S.E. by S. ½ S. about 4 miles distant. Nearly in mid-channel, between Warren's and Coronation Islands, there was no bottom at 120 fathoms.

Although the navigation of Clarence's Strait may be free from interruption, yet it ought not to be prosecuted without much circumspection.

Of the coast of the Prince of Wales's Archipelago we know but very little, and that little is chiefly comprised in the Spanish charts before alluded to, and a portion of it in the atlas accompanying the voyage of La Pérouse.

CAPE ADDINGTON, which appears to be the next most remarkable promontory to the South of Cape Pole, was so named by Vancouver, after the Speaker of the House of Commons. It is very conspicuous, and, according to Vancouver, is in lat. 55° 27′, lon. 133° 48′.

PORT BUCARELI, a very extensive inland sea, lies at the back of and to the southward of Cape Addington. It was discovered by Ayala and Quadra, the two Spanish navigators, of whom mention has been made previously. They anchored here on August 16th, 1775, and named it Puerto del Baylio Bucareli, in honour of the Mexican viceroy. It seems also to be the same as Sea Otter Sound of Meares. It is formed by several islands, but the account of the Spaniards' visit, as given by Maurelle, the pilot, is so vague that no description can be drawn from it. They here took possession, in the name of his Catholic Majesty, of all the country they saw, and all they did not see. They saw no inhabitants, though the remains of a destroyed hut, and some paths here and there, indicated that it was, or had been, inhabited. The unfortunate La Pérouse also explored it, and a large plan, No. 26 of the atlas of his voyage, is the result of that examination.

CAPE SAN BARTOLOM is the S.W. point of the entrance of this inland sea, and is in lat. 55° 12½′, lon. 133° 36′, and, according to the charts, is the South extremity of a long narrow peninsula, extending in a southerly direction, with some islets off it; it is probably the Cape Barnett of Meares.* From hence the S.W. coast of the Archipelago extends to the S.S.E. and East, to Cape Chacon, at the entrance of Clarence's Strait, described on page 441. At the South end of the island is the extensive bay called Port Cordova in the Spanish charts, and is the Port Meares of that commander.†

The RASA ISLE, or the WOLF ROCK, lying off the mouth of Port Bucareli, is one of the most dangerous impediments to navigation on the exterior coast, and from these circumstances it obtained from Vancouver its latter name. It is a very low, flat, rocky islet, surrounded by rocks and breakers that extend some distance from it: it lies 14 miles S. 21° E. from Cape St. Bartolom, 12 miles from the nearest point of the contiguous shore, and 3 leagues N. 11° E. from San Carlos Island. It was seen by the Spaniards in 1775, who called it Rasa, or low. By Capt. Douglas it was called Forrester's Island in 1786.

SAN CARLOS is a small high island; its South point is in lat. 54° 48′, lon. 133° 32′. The channel between it and the Wolf Rock appears to be free from interruption, and was passed safely by Capt. Douglas. It was discovered by Ayala and Quadra, in August, 1775, and by them named San Carlos Island. It is called *Douglas Island* by Meares and others, and *Forrester's Island* by Vancouver, but its real name must be that first applied by the Spaniards. It was seen by Douglas, in the *Iphigenia*, August 13th, 1788. He says it is very high, covered with verdure, and visible 16 or 17 leagues off.

We now return to the northward.

Between Cape Decision and Cape Ommaney, which latter is in lat. 56° 10′, lon. 134° 33½′, and 16 miles distant from the former, is Christian's Sound, and this forms the southern entrance to a very extensive inland navigation, extending

* Cape Adamson is high bluff land, in lat. 55° 28′ N., lon. 226° 21′ E. Cape Barnett is in lat. 55° 39′, lon. 226° 4′. It is low towards the sea, but rises within to a considerable height. Between these points is the mouth of a large bay. "Having run up a considerable way into the bay, they entered the mouth of a straight passage, not more than half a mile across from shore to shore, steering North. A number of whales within showed that there was plenty of water. When the ship anchored in 17 fathoms, sandy bottom, about half a mile from the shore, she was entirely land-locked, except at the entrance; and her situation was named Sea Otter Harbour, from the very great number of those animals seen there."—*Meares*, p. 326.

† Port Meares, "the western point of land which forms the bay when the ship was at anchor in 23 fathoms, sand and shells, bore E.S.E., and the eastern point E.N.E. So that a ship lying there is only exposed to four points of the compass; that is, between E.S.E. and E.N.E. She will be land-locked every other way, about a mile from the western shore. The latitude observed was 54° 51′ N., and lon. from several lunars, 227° 54′ E.

"In Port Meares there are two large arms, or branches of the sea; the one turns N.N.E. and the other about N.N.W., which Capt. Douglas supposes to have a communication with Sea Otter Sound."—*Meares*, p. 329.

through upwards of 3° of latitude, separating a series of large islands from the continent of America. The principal of these are Chatham's Strait, leading immediately from Christian's Sound to the northward; Prince Frederick's Sound, diverging eastward from it; and Stephens's Passage, which branches northward out of the latter. These principal arms insulate the Sitka Islands (or King George the Third's Archipelago), Admiralty Island, and numerous subordinate islands, which will be described in due order, commencing with the continental shores.

CHRISTIAN'S SOUND is the passage between Cape Ommaney and Cape Decision. It is noticed again hereafter. From Cape Decision the coast trends N.N.W. ¾ W. 3 leagues, and then N. by W. the same distance, to the North point of *Port Malmesbury*. This is about 2 leagues deep N.E. and then S.S.E., and has some islets and rocks on it; notwithstanding which it affords very excellent shelter in from 17 to 34 and 12 fathoms water, and is conveniently situated towards the ocean. The North point, called *Point Harris*, is rendered very remarkable by its being a projecting point, on which is a single hill, appearing from many points of view like an island, with an islet and some rocks extending near a mile to the S.W. of it. Seven miles farther northward is the South point of a large bay full of innumerable islets and rocks, with a great number of very small branches in various directions. Its N.W. point of entrance, *Point Ellis*, is in lat. 56° 31′, lon. 134° 15′. This also forms the S.E. point of another small inlet, equally intricate, and as much incommoded with islets and rocks. *Point Sullivan*, which is the next point in the main inlet to the northward, is in lat. 56° 38′, and East of this, also, is an inlet full of rocks and islets. From Point Sullivan the shores to the northward are less rocky, and become firm and compact, taking a direction of N. ¾ W. 13 miles, to Point Kingsmill, which is conspicuous.

Point Kingsmill is the S.W. point of Prince Frederick's Sound, the opposite point of entrance being Point Gardner, the S.W. extreme of Admiralty Island. This sound extends to the N.E. and East.

PRINCE FREDERICK'S SOUND.—From Point Kingsmill to Point Cornwallis the bearing and distance are N.E. ½ E. 6½ miles, the space between being occupied by two bays, each taking a south-easterly direction, a mile or a mile and a half wide, and 4 or 5 miles deep, and containing many islets and dangerous rocks. To the eastward of Point Cornwallis is *Port Camden*, the West shore of which trends first E.S.E. 9 miles, and then S.S E. ½ E. for 7½ miles farther. From this a branch about half a league wide runs in a S.S.W. direction, 8 miles, to within 2 miles of the head of the inlet N.E. of Point Ellis, previously mentioned. The shores of this branch of Port Camden are pretty free from islets and rocks, but those to the N.W. of it are lined with them, and render the approaching of it extremely dangerous. From the point whence this branch diverges to the South, another extends to the E.S.E., and then southward as far as the northern part of Clarence's Strait, having Point Barrie and Conclusion Island at its South extreme. But it is perfectly unnavigable for shipping, in consequence of the numerous rocks, islets, and shoals which extend throughout it. It serves to insulate the land we have been describing.

The peninsula, which is connected with the more eastern land by the last-mentioned narrow isthmus, is by no means so high or mountainous as the land composing the adjacent countries on the opposite or north-eastern side of the sound, which at no great distance consists of very lofty, rugged, dreary, barren mountains, covered with ice and snow; but the land composing the peninsula is chiefly of moderate height, and producing a noble forest of large and stately pine trees, of clean and straight growth, amongst which were a few berry bushes and some alders. The shores along the bays are in general low, and the bays and arms abounded with a greater number of salmon and sea-otters than was observed by Vancouver's party on any other part of the coast; and as they were found in the greatest abundance at the heads of those places, it was inferred that salmon and other small fish form a large proportion of the food of the sea-otters, which are thus induced to frequent these inland channels, to which, at this season of the year (August), such fishes resort.

Point Macartney forms the N.E. point of Port Camden. It is a large, rounding, though not lofty promontory, in which are several small open bays, and near it several detached rocks. From hence the shore trends N. by E. ½ E. about a league, where the width of the sound is about 7 miles across in a N.W. direction, to Point Nepean. From this station N.N.E. ½ E. 4½ miles distant, lies a small island with patches of rock from this point reaching nearly to its shores.

The promontory still takes a rounding direction about E.N.E. 5 miles farther, from whence the southern shore of the sound extends E. by S. ½ S. 17 miles, to the West point of a small cove, the only opening in the shore from Point Macartney; but off the little projecting points between this cove and that point, are detached rocks lying at no great distance from the shore. The cove extends S.E. by S. about a league, forming a narrow isthmus, 2 miles across, from the head of Duncan's Canal (p. 444), another striking instance of the extraordinary insular state of this region.

POINT GARDNER, as before mentioned, forms the N.W. point of the entrance to Prince Frederick's Sound. Off it, in a S.S.E. direction, lie some rocks and a small island; the former at the distance of three-quarters of a mile, and the latter at that of 3 miles. The coast hence rounds irregularly to Point Townsend, a distance of 9 miles; off the projecting points are some rocks. Six and a half miles E.N.E. from this is *Point Nepean*, situated in lat. 57° 10′, lon. 134° 5′. It is a high, steep, bluff, rocky point, and off it lies a ledge of rocks about half a mile. From this the coast takes a more northerly direction, or N.E. ½ N. 10½ miles, to *Point Pybus;* the coast between is much indented with small bays, and vast numbers of islets and rocks both above and beneath the water. It is in general but moderately elevated; and although it is composed of a rocky substance, produces a very fine forest, chiefly of pine timbers. Northward of this is a large channel, called Stephens's Passage; Prince Frederick's Sound continues to the eastward and south-eastward.

Point Fanshaw, which is the point of the main land opposite, and forming the angle at which the two channels diverge, is low and projecting, but very conspicuous; in lat. 57° 11′, lon. 133° 25½′. The branch is here 8 miles wide, and its northern shore takes a course E.S.E., 16 miles, to a low, narrow point of land 2½ miles long, and half a mile broad, stretching to the South, called *Point Vandeput.* Here the breadth of the branch decreases to 3½ miles in a South direction, to a steep bluff point; from this part the branch takes a more southerly course. South of Point Vandeput a shoal extends about a mile, and on its East side a small bay is formed, from whence the eastern shore trends S.E. by S. 7 miles to another point, off which a shoal extends about three-fourths of a mile. The shore here is a small extent of flat land, lying immediately before the lofty mountains, which rise abruptly to a prodigious height immediately behind the border. A few miles to the South of this margin the mountains extended to the water-side, when a part of them presented an uncommonly awful appearance, rising with an inclination towards the water to a vast height, loaded with an immense quantity of ice and snow, and overhanging their base, which seemed insufficient to bear the ponderous fabric it sustained, and rendered the view of the passage beneath it horribly magnificent.

At a short distance to the South of this the head of the inlet appears closed by a beach extending all round the head of it. At high water this becomes a shallow bank, with an island on it. Beyond it is the mouth of the Stikine River. Thus at low water the land to the West, of which Cape Decision is the S.W. point, becomes a portion of the continent; at high water it is insulated, and, by means of this channel, an inland navigation, for canoes and boats, is found from the southern extremity of Admiralty Inlet, in lat. 47° 3′, to the North extremity of Lynn Canal, in lat. 59° 12′, lon. 135° 37′. The southern end of the shallow portion of Prince Frederick's Sound enters the arm of the Duke of Clarence's Strait between Point Blaquiere and Rothsay, described on page 443. Prince Frederick's Sound was so named by Vancouver after the late Duke of York, on whose birthday his three years' survey of this desolate coast was here brought to a conclusion.

It was observed by Mr. Whidbey, during Vancouver's survey, that in no one instance, during his researches, either in the several branches of Prince William's Sound, in those extending from Cross Sound, or in the numerous branches about Admiralty Island, did he find any immense bodies of ice on the islands; all those which he had seen on shore were in the gullies or valleys of the connected chain of lofty mountains so frequently mentioned, and which chiefly constitute the continental shore from Cook's Inlet to Prince Frederick's Sound; though, in different places, these mountains are at different distances from the sea-side. He likewise observes, that all the islands, or groups of islands, were land of a moderate height, when compared with the stupendous mountains which compose the continental boundary, and were still seen to continue in a S.E. direction from this shallow passage, whilst the land to the westward of the passage assumed a more moderate height, was free from snow, and produced a forest of lofty pine trees.*

STEPHENS'S PASSAGE, which is over 95 miles in length, opens into Prince Frederick's Sound, between Point Pybus and Cape Fanshaw, which are 16 miles asunder; but it should be remembered that its north-western end is rocky, intricate, and very dangerous for shipping in the entrance into Lynn Canal, as hereafter shown.

N. by E. 6¾ miles from Cape Fanshaw is *Port Houghton*. There are many rocks on the shore between. The South point of the harbour is Point Walpole,

* Vancouver, vol. iii. pp. 282-3.

near which are some islets and sunken rocks: its North point is Point Hobart, N. by W. a league from the other, and from which extends a bank of sand, a little distance from the shore, but leaving a clear passage between it and the islets into the port. It extends E.S.E. 5 or 6 miles, and is bounded by lofty mountains, forming the shores of a snug harbour, with soundings of 10 to 6 fathoms a considerable distance from the shore, sand and muddy bottom. From Point Hobart to *Point Windham* the bearing and distance are N.N.W. ¼ W. 12½ miles. Between are several islets, in various directions. Opposite to Point Windham is *Point Hugh*, and here perhaps it may be considered that Stephens's Passage more properly begins.

Beyond Point Hugh is *Point Gambier*, bearing S.S.W. ½ W. 5 miles from it; and this latter is N.E. ½ N. 6 miles from Point Pybus, previously mentioned. Between the two former points is the entrance to *Seymour's Canal*, which extends N.W. by N. 29 miles from Point Hugh to its head, in lat. 57° 51′. At its entrance it is from 2 miles to 3 miles wide, which gradually increases towards its head to 2 leagues. At its termination is a small brook of fresh water. South of this the centre of the inlet is occupied by two islands, together about 8 miles long, having a great number of islets on their N.E. sides. The adjacent country is moderately high, and covered with timber of large growth, excepting towards *Point Hugh*, which is a lofty rocky promontory, from whence extends a ledge of rocks, on which the sea breaks with considerable force.

This point forms the South extreme of a long, narrow strip of land, dividing Seymour's Canal from Stephens's Passage. The S.W. coast of this passage, which is here about 5 miles in breadth, is nearly straight, compact, and free from rocks or other interruptions up to a high round island lying in the middle of the channel, in lat. 58° 1′, from which the western shore extends N. by W. 8 miles to Point Arden, where the branch divides into three arms, the principal one directed to the westward.

The eastern shore of the passage, up to this part, is composed of a compact range of stupendous mountains, chiefly barren, and covered with ice and snow, but affording some inlets. From Point Windham, on the South, to Point Astley, 13 miles to the North, the shores are very rocky, and contain many small, open coves. The latter is the South point of a deep bay, about 4 miles wide, named *Holkham Bay*, on which are three small islands, to the westernmost of which a shallow bank extends from each side of the bay. Much floating ice was seen within the islands. From *Point Coke*, its North point, in a direction S.W. ¼ S. 2½ miles, are two small rocky islets, nearly in the middle of the branch; and the eastern shore trends from it N.W. ½ W. 9 or 10 miles to *Point Anmer*, the South point of *Port Snettisham*. This harbour first extends about a league from its entrance in a N.E. direction, where on each side the shores form an extensive cove, terminated by a sandy beach, with a fine stream of fresh water. On the N.W. side of the entrance (*Point Styleman*) is a small cove, on which there is also a run of water, with an islet lying before it. The shores are high and steep, and produce very few trees. From the latter point, which is in lat. 57° 53′, lon. 133° 49′, the eastern shore trends about N.W. 12 miles towards the high round island before mentioned. This part is much indented with small bays.

TACO, an Hudson's Bay Company's establishment, in Port Snettisham, is on a little harbour almost land-locked by mountains, being partially exposed only to the S.E. One of the hills, near the fort, terminates in the form of a canoe, which serves as a barometer. A shroud of fog indicates rain; but the clear vision of the canoe itself is a sign of fair weather.

The fort, though it was only a year old, was yet very complete, with good houses, lofty pickets, and strong bastions. The establishment was maintained chiefly on the flesh of the cheveril, which is very fat, and has an excellent flavour. Some of these deer weigh as much as one hundred and fifty pounds each, and they are so numerous that Taco has this year (1842) sent to market twelve hundred of their skins, being the handsome average of a deer a week for every inmate of the place, the post being conducted by Dr. Kennedy, with an assistant and twenty-two men.

But extravagance in eating venison is here a very lucrative business; for the hide, after paying freight and charges, yields in London a profit on the prime cost of the whole animal.

Seven tribes, three of them living on islands, and four on the main land, visit Taco. They muster about 4,000 souls, and are delighted to have the English settle among them, and on this ground are jealous of other Indians. The bighorn, sheep, and mountain goat are very numerous in the neighbourhood.

The River Taco, falling into the gulf to which it gives its name, according to Mr. Douglas, who ascended it about 35 miles, pursues a serpentine course between stupendous mountains, which, with the exception of a few points of alluvial soil, rise abruptly from the water's edge. In spite of the rapidity of the current, the savages of the coast ascend it 100 miles in canoes, and thence trudge away on foot the same distance to an inland mart, where they drive a profitable business, as middlemen, with the neighbouring tribes. Besides facilitating this traffic, the establishment of the fort has done much to extinguish a traffic of a very different tendency. Though some of the skins previously found their way from this neighbourhood to Sitka and Stikine, yet most of them used to be devoted to the purchasing of slaves from the Indians of Kygarnie and Hood's Bay.*

Opposite to *Point Arden*, on the West shore, is the mouth of the arm leading to the N.E. from Stephens's Passage. Its West point of entrance is *Point Salisbury*, and it extends about N. by E. 13 miles, when the shores spread to East and West, and form a basin about a league broad and 2 leagues across, N.W. and S.E., with a small island lying nearly at its N.E. extremity. From the shores of this basin a compact body of ice extended some distance nearly all round at the time of Vancouver's visit; and the adjacent region is composed of a closely united continuation of the lofty range of frozen mountains, whose sides, almost perpendicular, were formed entirely of rock, excepting close to the water-side, where a few scattered dwarf pine trees find sufficient soil to vegetate on; above these the mountains were wrapped in undissolving ice and snow. From the rugged gullies in their sides were projected immense bodies of ice, that

* Sir George Simpson, vol. i. pp. 214—216.

reached perpendicularly to the surface of the water in the basin, which admitted of no landing place for the boats, but exhibited as dreary and inhospitable an aspect as the imagination can possibly suggest. The rise and fall of tide here were very considerable, appearing to be upwards of 18 feet.

From Point Arden the principal inlet takes a general course of W. ¾ N., and is about a league in width. About 5 leagues along the South shore is *Point Young*, forming the East point of a cove, with an island and rock in its entrance, and another at the bottom of the cove. Here the width of the arm is decreased to half a league, and the South shore stretches N.W. ½ N. 7 miles to another cove with an islet lying near it. North from this cove, 1½ leagues distant, is the West point of *Douglas Island*, so named after the Bishop of Salisbury, and forms the North side of this portion of the passage. It is about 20 miles long, and 6 broad in the middle, narrowing towards each end, and separated from the continent by a narrow channel, rendered impassable from its being filled with ice; an evidence of the partial nature of this phenomenon, for the other arms were free from this obstruction.

To the N.W. of this part is a rocky and intricate portion of the passage, very dangerous for the navigation of shipping, so that the communication between it and the large channel to the North and West of it impeded.

The channel beyond the N.W. point of Douglas's Island is divided into two branches by a very narrow island about 4½ miles long, and half a mile broad. The passage on its N.E. side may be considered as next to impassable for shipping, by the rocks and islets at the S.E. end of it. The other channel is equally unsafe and intricate, from the same cause.

Opposite the North end of the above island is *Point Retreat*, the northernmost point of Admiralty Island; it is in lat. 58° 24′, lon. 134° 59′. About a league southward from Point Retreat, in the southern channel, is a deep cove, *Barlow's Cove*, which, with the narrow island lying before it, forms a very snug harbour, of good access by the passage round to the North of Point Retreat, as the rocky part of the channel lies to the S.E. of it. To the West of Point Retreat and Admiralty Island is that extensive branch named by Vancouver after the nobleman, Chatham's Strait, to the South opening of which we will now return.

CHATHAM'S STRAIT.—Cape Decision, the S.E. point of entrance, has been before described. It was so named by Vancouver, from his having so far decided that the great openings stated to exist by De Fonte, De Fuca, and others, did not exist—a conclusion he was scarcely warranted in making, inasmuch as a more careful attention to their narratives show some truth, although much alloyed with the fabulous.

The other point of entrance is Cape Ommaney, the South extremity of the Sitka Archipelago; it was so named by Capt. Colnett. Off it lies a rock called *Wooden's Rock*, from one of Vancouver's men having been drowned here.

The opening between Capes Ommaney and Decision was named by Colnett Christian's Sound, and off the opening is a group of small rocky islets, a league in extent, called the Hazy Islands. They lie S. ¾ E. 16 leagues from Cape Ommaney, S.W. by W. ¼ W. from Cape Decision, and 3 leagues West from Coronation Island, which is the nearest land to them.

The eastern shore of the strait, from its southern point to the entrance of Prince Frederick's Sound, has been before described (p. 449). Point Gardner, the North point of its entrance, is also the southern extremity of Admiralty Island. From this point the eastern shore of the strait runs about N. ¾ W. 22 miles to *Hood's Bay*, which is about 1½ leagues across to *Point Samuel*, and has some islands nearly in its centre. In the intermediate distance are two smaller bays, off the points of each of which islets and rocks lie at a little distance.

Point Parker is 9 miles N. ¾ W. from Point Samuel. The coast is indented into several small bays; the shores are low, and much divided by water. In one of these inlets Vancouver's party found some cultivation of a species of tobacco, the only instance they had met with among the North-West American Indians. Beyond this, still following the same direction for 30 miles, is *Point Marsden*. The land is very moderately elevated, covered with fine timber, chiefly of the pine kind, and terminating at the water-side with alternate steep rocky cliffs and small sandy bays, with a few detached rocks and islets lying near it. Hence to Point Retreat, which is the North extremity of Admiralty Island before mentioned, the distance is about 16 miles, the coast being nearly in the same direction and of the same character as that more to the South.

ADMIRALTY ISLAND.—The shores of Admiralty Island, which have thus been described, are about 60 leagues in circuit. With the exception of its N.W. and S.E. parts, they are very bold, affording many convenient bays likely to admit of safe anchorage, with fine streams of fresh water flowing into them, and presenting an aspect very different from that of the adjacent continent, as the island in general is moderately elevated, and produces an uninterrupted forest of very fine timber trees, chiefly of the pine tribe; whilst the shores of the continent, bounded by a continuation of those lofty frozen mountains which extend south-eastward from Mount Fairweather, rise abruptly from the water-side, covered with perpetual snow, whilst their sides are broken into deep ravines or valleys, filled with immense mountains of ice; notwithstanding that the island seems to be composed of a rocky substance covered with little soil, and that chiefly consisting of vegetables in an imperfect state of dissolution, yet it produces timber which was considered by Mr. Whidbey to be superior to any he had before noticed on this side of America. The ocean hereabouts, too, encroaches most rapidly on the low land. The stumps of trees, in various stages of decay, still standing erect, are to be found below high-water mark, and many of the low shores, now covered with the sea, produced, at no very distant period, tall and stately timber.

LYNN CANAL.—In lat. 58° 35′ a point projects from the West shore of Lynn Canal, which bears N.N.W. from Point Retreat and N. by W. from Point Couverden, the extremity of a peninsula separating the canal from Chatham Strait. Both sides of the arm are bounded by lofty, stupendous mountains, covered with perpetual ice and snow, whilst the shores in the neighbourhood appear to be composed of cliffs of very fine slate, interspersed with beaches of paving-stone; the channel continues to be about 5 miles wide, and the western shore straight and compact. In lat. 58° 54′ is a small islet about 2 miles from the West shore. Another islet lies to the North, between it and the South point of an island 5 miles long and 1 broad, lying along the western shore, and forming a channel about a mile wide, having at its southern entrance shoals that extend nearly from side to side. Beyond this the arm diverges into two branches, the West one terminating in its navigable part in lat. 59° 12′. At its head, according to Lisiansky's chart, is the native village of *Chilkat*. There are some islets and rocks in mid-channel, and above these the water is perfectly fresh. Above the shoal limiting the navigation, the arm extends half a league, and through a small opening a rapid stream of fresh water rushes over the shoal. The eastern side of this portion of the arm is low and indented into small bays and coves extending S.E. ½ S. 4½ leagues to *Seduction Point*, from off which lies a range of small islands about 4 miles in a South direction. The southernmost is a flat barren rock, but on the other trees were produced. This peninsula is a narrow strip of low land 1 or 2 miles across, separating the western from the eastern arm, which extends N. by W. ¾ W. about 11 miles, and thence winds in a westerly direction about 3 miles farther, where it terminates in low land, formed immediately at the foot of high stupendous mountains, broken into deep gullies, and loaded with perpetual ice and snow. The eastern shore of the inlet trends in a compact manner to Point St. Mary's, in lat. 58° 43½′, forming the North point of a bay called *Berner's Bay*, about 4 miles across in a S.S.E. direction, and about 5 miles deep to the N.N.E. From its South part, Point Bridget, the continental shore takes a direction S.S.E. ½ E., and at 18 miles lies a small island, with some rocks and islets about it. Beyond this navigation is difficult, even for boats, being incommoded with numberless islets and rocks. Beyond these islets, to the westward, lies a larger island, and from the shore of the main land a shallow bank extends nearly half a league. The large island is about 6 miles long and 2 broad, and opposite to its S.W. point is Point Retreat, the West side of the entrance to Stephens's Passage. This entrance, which is almost impassable for shipping from the rocks and islets lying within it, has been before described, page 454.

Lynn Canal was first made known and surveyed by Vancouver's party, under Mr. Whidbey, in July, 1794, and was named by the commander after his native town in Norfolk.

It receives a river, which the Indians ascend about 50 miles to a valley running towards Mount Fairweather, and containing a large lake, which pours its waters into the open ocean at Admiralty Bay. The natives of this valley are called the Copper Indians, from the abundance of virgin copper in the neighbourhood.*

THE SITKA OR KING GEORGE THE THIRD'S ARCHIPELAGO.

The land forming this collection of islands was first discovered by Alexoi Tschirikow, the second in command of the expedition under the unfortunate Behring, in 1741. This was their third voyage, and they were separated by a storm soon after they had set out on their voyage. Tschirikow directed his course

* Sir George Simpson, vol. i. p. 217.

to the East, from the parallel of 48°, and, towards the middle of July, made the land of America, between the fifty-fifth and fifty-sixth parallels; but others place his landfall in 58°. The coast which he found was steep, barren, guarded by rocks, and without a single island that could afford shelter. He anchored off the coast, and detached his long-boat, with orders to put on shore wherever she could land. Several days elapsed without her reappearing; he despatched his other boat to gain tidings of her, but the latter no doubt experienced the same fate as the former, and it is unknown what became of either. Some canoes, manned by native Americans, presented themselves a few days after, to reconnoitre the ship; but they durst not approach her, and there remained on board no boat of any sort that could be detached to join or pursue them, and prevail on them to come to the ship, where they would have been detained for hostages. Tschirikow, despairing to see again the men whom he had sent on shore, resolved to quit the coast, and accordingly returned to Kamtschatka.* These discoveries became known to France and Europe from the fact of Delisle de la Croyère, one of the brothers of the French savans, and Dr. Steller, the naturalist, having accompanied Tschirikow. Such was the first authentic discovery of North-West America, which arose out of the original plans projected by Peter the Great, and subsequently carried into effect by the Empress Catharine.

The land in question, like Vancouver Island and others to the southward, was then supposed to form part of the American continent; and it was not until Vancouver's expedition that Chatham's Strait was discovered, and thus showed the real nature of the land on the Pacific. Vancouver, too, as will be seen from the preceding remarks, did not very minutely examine the western shore of the strait to which he gave the name of his vessel, but justly inferred that it was penetrated by one or more channels leading to the open ocean, from the fact of some of the natives being found in the strait who belonged to the other side of the islands.

Capt. Urey Lisiansky, of the Russian navy, examined the group in 1805; and, by his survey, it appears that it consists of four principal islands, viz.: Jacobi, Crooze, Baranoff, and Chichagoff.

BARANOFF ISLAND is the southernmost, and is about 85 miles in length, by about 20 miles on its maximum breadth. On its West side is Sitka or Norfolk Sound, the principal place of resort in these seas, and the situation of the head-quarters of the Imperial Russian Company.

CROOZE ISLAND, the South extremity of which is formed by Cape and Mount Edgcumbe, lies before Sitka Sound. It was named so by Capt. Lisiansky, after the Russian Admiral. It is separated from Baranoff Island by Neva Channel. It is about 7 leagues in length. At its North extreme is the Bay of Islands, but which leads to the strait, separating the two principal islands of the group called by Lisiansky *Pagoobnoy* or *Pernicious Strait*. It joins Chatham's Sound, is deep, and derives its name from a party of Aleutians having been poisoned there some years previously by eating mussels, an accident, it will be remembered, which occurred to one of Vancouver's party, as mentioned on page 426. Crooze Island is called Pitt Island in La Pérouse's and some other early charts.

* See Müller's Discoveries of the Russians, vol. i. pp. 41—43.

CHICHAGOFF ISLAND is the next and northernmost large island. It is divided from Baranoff Island by the strait previously mentioned, and extends from it to Cross Sound, which separates it from the continent to the northward.

JACOBI ISLAND, the fourth of those described by Lisiansky, lies at the N.W. extremity of Chichagoff Island. The passage separating them was not explored by Lisiansky, but a vessel belonging to the Russian Company was said to have passed through it, and to have found a sufficient depth of water.

The Sitka Islands are plentifully supplied with wood, chiefly of pine, larch, and cedar, and abundance of wild berries are found. The rivers, during summer, are full of excellent fish. Herrings swarm in the sound every spring; fine cod may be caught, and halibut, of great weight, with the hook and line only. There are few land animals, but a great quantity of almost every species of amphibious ones. The birds are not so numerous as at the Kodiack islands to the northward.

The climate of these islands is considered to be capable of producing barley, oats, and European fruits and vegetables; but the Russian establishment is in a great measure dependent on the produce of the more southern settlements of the Hudson's Bay Company at Victoria, Vancouver, and Puget's Sound. The summer is warm, and extends to the end of August: the winter differs from the Russian autumn, only that there are frequent falls of snow.

Of the population it is difficult now to speak: they are an uncouth race, similar to those previously mentioned. They are brave, but extremely cruel; an instance of this has been related. The labret, or lip-piece, among the women, is one of the greatest characteristics.

CAPE OMMANEY, in lat. 56° 9′, lon. 134° 34′, is the South extreme of the Archipelago; off it lies the *Wooden's Rock*: they have been alluded to before on page 454, as forming the S.W. entrance point of Chatham's Strait.

PORT CONCLUSION.—Two leagues N. by E. ½ E. from Cape Ommaney, which has been before described, the South point of the Sitka Archipelago, is the southern point of the entrance to Port Conclusion, so called because it was here that Vancouver's vessels awaited the conclusion of the survey of this coast in August, 1794, having left England on the 1st of April, 1791, for that service. The North point of Port Conclusion bears from the southern N. ¾ W. 2 miles distant. The depth of water in mid-channel, between these points, is 75 fathoms, but decreases to 8 or 10 close to the shores, without rocks or sands, excepting near the points, which are sufficiently evident to be avoided. S.S.W. about half a mile from the North point of entrance is a most excellent and snug basin, *Port Armstrong*, about a mile long, and a third of a mile wide; but its entrance is by a very narrow channel, half a mile in length, in a W.S.W. direction, with some islets and rocks lying off its South points; these are steep, nearly close to them, as are the shores on both sides, which vary from a sixth to a twelfth of a mile asunder, with a clear navigable passage from 8 to 12 fathoms deep in the middle, and 5 fathoms on the sides. The soundings are tolerably regular in the basin, from 30 in the middle to 10 fathoms close to the shores. Immediately within its North point is a fine sandy beach, and an excellent run of water, as is the case also at its head, with a third sandy beach just within its South point of entrance. In the vicinity of these beaches, especially the first and third, is a small extent of low land; but the other parts of the shores are composed of steep rugged cliffs, on all sides surrounded by a thick forest of pine trees, which grow with more vigour than those in other parts of the harbour.

From its entrance to its head the port extends about a league S.W. by S., free from any interruption, although it is inconvenient from its great depth of water. Near the southern side of the entrance lies a small islet and some rocks, but these are entirely out of the way of its navigation. The soundings cannot be considered as very regular, yet in general they are good; in some places it is stony, in others sand and mud: but in the cove, where the vessels were at anchor, a mile and a quarter within the South point on the other side, the bottom is rocky. The head of this cove approaches within the fourth of a mile of the head of another cove, whose entrance on the outside is about 2 miles to the South of the South point of the harbour.

In the entrance of that cove the depth is 7 fathoms; weeds were seen growing across it, and to the northward of it is a small islet, and some rocks. The surrounding shores are generally steep and rocky, and covered with wood to the water's edge; but on the sides of the adjacent hills are some spots clear of trees, and chiefly occupied by a damp, moorish soil, with several pools of water. The surface produces some berry bushes, but the fruit, in August, was not ripe.

The western shore of Chatham's Strait was not minutely surveyed by Vancouver's party. It follows a nearly straight direction from Port Conclusion, about N. by W. for 105 miles to *Point Augusta*, in lat. 58° 3½′, lon. 135°, preserving a nearly parallel direction to the opposite side, which varies from 5 to 9 miles distant.

This extensive arm, as far as was ascertained, is without danger, and probably affords many places of refuge. The flood tide, although of short duration, not running more than two hours, was regularly observed to come from the South.* To the northward of Point Augusta the western shore diverges more to the westward, while the eastern side beyond Port Marsden still preserves the same direction. Into the opening thus formed a peninsula of the main land projects to the southward, thus dividing it into two channels; the principal being to the N.W., while that which continues on in a North or West course is called Lynn Canal.

POINT COUVERDEN, the South extremity of the above peninsula, is in lat. 58° 12′, lon. 135° 4′. It was so named after the seat of Vancouver's ancestors in Holland. The continental shore in this neighbourhood constitutes a narrow border of low land, well wooded with stately trees, chiefly of the pine tribe, behind which extends a continuation of the lofty snowy mountains. About 2 miles North of Point Couverden is one small island and three rocky islets, one of which lies nearly in mid-channel. Beyond this the western shore of the arm is

* Mr. Whidbey considered that Chatham's Strait was likely to be one of the most profitable places for procuring the skins of the sea-otter on the whole coast, not only from the abundance observed in the possession of the natives, but from the immense numbers of these animals seen about the shores in all directions. Here the sea-otters were in such plenty that it was easily in the power of the natives to procure as many as they chose to be at the trouble of taking. They were also of extremely fine quality.—(Vancouver, vol. iii. p. 264.) This opinion has been verified by the success of the Russian Company since established here.

firm and compact, indented with a few coves, and some islets and rocks lying near it. The eastern shore, described on page 455, presents a broken appearance.

From Point Couverden the continental shore takes a somewhat irregular direction, N.W. ½ W. 7 leagues, to a part of which lies a low and nearly round island, about 2 leagues in circuit. It is moderately elevated, its shores pleasant and easy of access, and well stocked with timber, mostly of the pine tribe. About a league to the eastward of it lie some islets. To the North and West of this the shores of the continent form two large open bays, terminated by compact, solid mountains of ice, rising perpendicularly from the water's edge, and bounded to the North by a continuation of the united lofty frozen mountains that extend eastward from Mount Fairweather. These bays were filled with large quantities of broken ice, set in motion and drifting to the southward by a northerly wind, and obliged Mr. Whidbey's boats (July, 1794) to take shelter round the N.E. part of an island which lies W. ½ S. from the low round island above mentioned, distant from it 3 leagues. This island is about 7 miles long, N.E. and S.W., and 3 miles broad. On its North side is a channel 2 to 3 miles wide, between it and the continental shore. The N.W. point of this channel is Point Dundas, in lat. 58° 21′, lon. 135° 55′.

To the westward of this point is a branch extending to the North and N.W. At about 2 leagues up it the channel is nearly stopped by shoals, rocky islets, and rocks, 4 miles beyond which it is finally closed, being in most places greatly encumbered with ice. On its N.E. side are some shoals, which extend to within half a mile of its S.W. side. The entrance, which is about 2 miles wide between Points Dundas and Wimbledon, has, in mid-channel, only 18 fathoms water. About the entrance the soundings are regular, of a moderate depth, and afford good and secure anchorage; but in the summer season (or in July) vessels would be much inconvenienced by the immense quantities of floating ice.

CAPE SPENCER, the North point of the entrance of Cross Sound, on the Pacific Ocean, is a very conspicuous, high, bluff promontory. Off it extend some rocks for about half a league. It is in lat. 58° 14′, lon. 136° 35′, and bears from Point Wimbledon S.W. ¾ W. distant 11 miles. To the N.W. of Point Wimbledon the main land terminates in steep, rugged, rocky cliffs, off which, at a little distance, are three small rocky islands. The bay, extending to the N.W. between these points, was occupied by an icy barrier, and a great quantity of floating ice (in July, 1794), rendering the navigation across it very difficult and tedious. The head of the bay, which decreases to 3 miles in width, is formed by a range of lofty mountains, connected with which is an immense body of compact perpendicular ice, extending from shore to shore. In the upper part of the eastern side the shores are composed of a border of low land, which, on high tides, is overflown, and becomes broken into islands.

The southern shore of this portion of the strait is of a more broken character than the northern. We have described it as far as Point Augusta, lying opposite to Point Marsden and Couverden. From this to *Point Sophia* is N. by W. ½ W. 17 miles; the coast composed chiefly of rocky cliffs, with islets and detached rocks lying at some distance from the shore, which is compact, not very high, but well covered with wood. The latter point is at the N.E. of the entrance of

Port Frederick, the entrance of which is about a league wide, East and West, winding to the southward, and apparently much divided by water. From the West side of this sound the shore takes a more northerly direction, with some islets near it, to a point which is the north extreme of King George's Archipelago, in lat. 58° 18′. Hence the coast takes an irregular course, W. by S. 17½ miles, to *Point Lavinia*, containing many open bays. On the opposite shore of the sound is *Point Wimbledon*, just mentioned, bearing N. by W. 6 miles from it. Between these points is a group of one low and two high rocky islands, with some rocks and islets about them.

PORT ALTHORP is to the westward of Point Lavinia, which extends 11 miles to the southward of it. Its South point of entrance is Point Lucan. From Point Lucan, in a direction about N.W., lies a narrow, high island, about 2½ miles in length; and between its S.E. point and Point Lucan there are two small islets, which render that passage not so commodious for sailing in and out of the port as that to the North of the island, between it and the western part of a cluster of three small islands, which extend about 2 miles from the eastern side of the port. This channel is clear, free from danger, and is about 1½ miles in width, with a tolerably snug cove, in which Vancouver anchored, just within its N.W. point of entrance. The high narrow island affords great protection to the northern part of this port, which, opposite to that island, is about 2½ miles wide; but nearly in the middle of the harbour, and opposite the South point of the island, are some detached rocks; and at *Point Lucan*, which is situated from Vancouver's ships' cove, S.S.E. 4½ miles distant, the width of the harbour is 2 miles, from whence it extends S.E. ¾ S. about 2 leagues, and terminates in a basin that affords good and secure anchorage, the best passage into which is on the eastern shore. The cove in which Vancouver's ships were stationed afforded good anchorage also, but it was not so well sheltered as the basin, nor was there any fresh water in it that could be easily procured, a disadvantage that can readily be done away with by resorting to a stream of excellent water close at hand, on the eastern shore, where the casks may be filled in the boat. The surrounding country is chiefly composed of a rugged rocky substance, covered with a forest consisting principally of pine trees; and where the steep acclivities do not forbid their growth, they extend down to the water's edge.

CROSS SOUND was discovered by Capt. Cook, in his last voyage, on Sunday, May 3rd, 1778, and was named by him from the day marked in the calendar. Its existence was denied by some after its original discovery, but the survey of it proves that Cook's description is much more accurate than from the transitory, distant view he had of it might have been reasonably expected. Its eastern limits may be placed at Points Lavinia and Wimbledon, which have been previously described. From seaward it appears to branch into many openings, the largest seemingly to the northward; this is the large bay described between Cape Spencer and Point Wimbledon. Its southern shore, from Point Lucan to Point Bingham, which is opposite to Cape Spencer, trends S.W. by W. ½ W. 10 miles. Between these points an opening takes a S.E. by S. direction for some distance, and probably affords some shelter and secure anchorage. *Point Bingham*, which lies S. by E. ½ E. 10 miles from Cape Spencer, affords a bold entrance into the sound, without rock, shoal, or any permanent obstacle. The group of rocky islands, noticed as existing to the South of Point Wimbledon, form a kind of termination to Cross Sound, and almost separate the ocean from Chatham's Strait, to the eastward of it; but on either side of these islands there are two narrow channels; the northernmost, being the widest, is near a mile across, the southernmost is about half that width; both of which are free from rocks, shoals, or any other impediment, excepting the large masses of floating ice, which, at the time of Vancouver's survey (July, 1794), rendered them very dangerous to navigate, although in the summer season and in the winter they are most probably entirely closed or impassable.

Every part of Cross Sound appears to be free from any rock, shoal, or permanent obstacle; and if it does possess any navigable objection, it is the unfathomable depth of it, which everywhere exists, excepting very near the shores, along which, in many places, are detached rocks; these, however, lie out of the way of its navigation, and are sufficiently conspicuous to be avoided.

The unfortunate La Pérouse touched on this part of the coast, previous to his departure for the West, in 1786. He makes the following remarks upon it:—

At Cross Sound the high mountains covered with snow terminate, the peaks of which are 8,000 or 9,000 feet high. The country bordering on the sea, S.E. of Cross Sound, although elevated 5,000 or 6,000 feet, is covered with trees to the summit, and the chain of primitive mountains seems to penetrate farther into the continent. *Mount Crillon*, almost as elevated as Mount Fairweather, is to the North of Cross Sound, in the same way that Mount Fairweather is to the North of the Baie des Français; they will serve to point out the ports they are near to. The one may be readily mistaken for the other, in coming from the South, if the latitude should not be correct within 15′. Otherwise, from all points, Mount Fairweather appears accompanied by two mountains, less elevated; and Mount Crillon, more isolated, has its peak inclined towards the South.*

CAPE CROSS, which was considered by Cook as forming the S.E. point of entrance to the sound, is not precisely so, but lies about 7 miles South of *Point Bingham*, which forms the true S.E. point. The interior part is a low, rocky land, free from any danger.

From Cape Cross the coast takes a direction of S. 31° E., about 7 leagues, to another promontory, to which Vancouver gave the name of *Cape Edward*, and off which lies a cluster of small islets and rocks. The coast between these capes is much broken, and has several openings in it that appear likely to afford shelter; but the vast number of rocks and small islets, some producing trees, and others entirely barren, that extend to the distance of 3 or 4 miles from the shore, will render the entering of such harbours unpleasant and dangerous, until a more competent knowledge of their several situations may be better acquired.† That

* Voyage de La Pérouse, vol. ii. p. 219.

† "All the land next the sea, beginning about 8 leagues to the S.E. of Cross Cape, and trending to within 10 leagues of Cape Edgcumbe, seems to be composed of low woody islands, among which there appear several places of good shelter. The inland country forms into a number of peaked hills, some well wooded, and others quite bare."—*Portlock*, p. 257.

which appeared to Vancouver to be the easiest of access, lies about 2 leagues to the northward of Cape Edward, and as it is in lat. 57° 44′, he was led to conclude that this opening was Portlock's Harbour.

PORTLOCK'S HARBOUR.—"On drawing near the opening, and about 2 miles from the shore to the N.W. of it, we had 20 and 25 fathoms water over a muddy bottom, and just within the entrance were some high, barren rocks. On getting into the entrance of the passage, which is about a mile across, we deepened the water to 30 fathoms, sandy bottom, the barren rocks just mentioned (and Hogan's Island) forming the South side; the northern side is Hill's Island, low land, forming itself into several small bays, from whose points are breakers at no great distance. About half a mile within the barren rocks we had 30 fathoms over a rocky bottom, which depth and bottom we carried at least a mile farther, steering N.E. by E., which is nearly the course into the harbour. The passage so far is nearly a mile across, with bold rocky shores on each side. Presently afterwards we shoaled the water to 10 fathoms, being then in the narrowest part of the channel, which in that situation is not more than half a mile across, having to the North some bold rocks, and to the South a bluff point of land; to the East of which, a small distance from shore, are some rocks which just show themselves above water. Immediately on passing these rocks we deepened the water very quickly to 30 and 40 fathoms, and a most spacious and excellent harbour opened itself to our view, bending to the N.W. and S.E., and running deep into the northward, with a number of small islands scattered about. We ran up towards the N.W. part of the harbour, and after passing a small island near the North shore, covered with trees, we anchored in 31 fathoms, mud, entirely land-locked; the rocks lying in the inner part of the passage, just shut in with the small island already mentioned, and bearing South 3 or 4 miles distant."*

The adjacent country abounds with white cedar, and excellent spruce may be made. Wooding and watering are performed with facility, and some salmon may also be caught.

Goulding's Harbour is a branch of Portlock Harbour, extending from its N.W. part. It runs in a zig-zag direction, between North and N.E., about 5 miles to the head of it from the island on the entrance. It appears navigable for at least 4 miles up, for vessels of any size, and there are a number of small islands covered with trees scattered about in various parts of it. At the head of it Portlock saw some Indians' houses, whose filthiness he describes.†

From Cape Edward the coast takes a direction about S. 30° E. to a very conspicuous opening, named by Capt. Cook the *Bay of Islands;* he rightly considered that it was the entrance to a channel which separated the land, on which Mount Edgcumbe is situated, from the adjacent shores. It is also the entrance to the channel separating the two principal islands of the Sitka Archipelago. Near the land forming the southern part of the Bay of Islands are several small islets, and its South point was named by Vancouver *Point Amelia.*

* Portlock, pp. 257-8. † *Ibid.* p. 270.

From Point Amelia the coast extends S. 5° E., 16 miles, to Cape Edgcumbe, having nearly in the middle of that space an opening, with two small islets lying before it, and presenting the appearance of a good harbour, called by Vancouver *Port Mary*. The other parts of the coast are indented with small open bays.

This is the western face of the island, named Crooze Island by the Russian, and Pitt Island by the early English navigators. The following notice of the channel which insulates it is given by Capt. Portlock :—

"Between Cape Edgcumbe and Portlock Harbour, Portlock's long-boat fell in with a strait about a league wide at the entrance, and running in about East, or E.S.E., with bold shores and good anchorage. Soon after getting in (the southern and eastern point in lat. 57° 30′, and the northern and western in 57° 36,′) they stood up between South and S.E. near 4 leagues, the strait being about 3 leagues across, with several islands in it. They kept along under the southern shore, and after getting up the passage about 4 leagues, they found it not more than half a league across : two leagues higher it became narrow and shoal, drying at S.W. above 2 miles. After passing this it grows wider still, trending away to N.E., the depth increasing gradually: they soon came into a large sound, where they saw a great number of whales, and also Mount Edgcumbe, and some islands lying to the S.E. of Cape Edgcumbe. On getting round the North point of this passage, through which they had come thus far, there was an opening running up in a North direction, and branching in several ways; in the entrance of the main opening were several small islands. They still continued to steer on to S.E. for a passage about 1½ miles across, made by the North part of an island just under Mount Edgcumbe (Pitt or Crooze Island) and the main, through which they saw the sea. After getting through this passage, they steered among a cluster of islands, lying near the shore, to the North of Cape Edgcumbe." *

CAPE EDGCUMBE, the Cabo del Engano of the Spanish charts of Maurelle, is low land, covered with trees, which projects considerably into the sea, lat. 57° 2′, lon. 35° 45′.†

MOUNT EDGCUMBE, which stands on the South end of the island, inland of the cape of the same name, is the Mount San Jacinto of the Spanish charts. It was estimated by Lisiansky, who ascended it, to be about 8,000 feet in height. The side toward the sea is steep, and was covered with snow (in July, 1805); that towards the bay (to the southward) is smooth, and of gradual ascent, and overgrown with woods to within 1½ miles of the top. This upper space exhibits a few patches of verdure, but is in general covered with stones of different colours. On the summit is a basin, or crater, about 2 miles in circumference, and 40 fathoms deep, the surface covered with snow. To judge from the appearance of the top of this mountain, it may be concluded that it was formerly much higher; but the eruptions having ceased, that time has crumbled to pieces the highest points, and filled up the abyss out of which the materials forming the exterior mountain were vomited. Many years must have

* Portlock, p. 274—279.

† Voyage de La Pérouse, tome ii. p. 221.

elapsed since this volcano was in action, as several sorts of the ejected lava are turning to earth.*

SITKA or NORFOLK SOUND.—The first of these two names is that of the natives, who call themselves Sitka-hans. It is probably the same as that called by the Spaniards, *Baya de Guadalupa.*† It is also called *Tchinkitânay Bay* by Marchand and other authors. The name of Norfolk Sound is that applied to it by Dixon, whose industry first made known its real character; he anchored, probably, in the first cove round Cape Edgcumbe, and did not penetrate to the eastward, to where the present Russian establishment is. The charts and descriptions of this period are so imperfect that they would probably rather tend to mislead than instruct.‡ By referring to these authors, ample descriptions of the natives, their habits and manners, will be found. These, however, are probably much changed since their contact with Europeans.

The coast of Sitka Bay is intersected by many steep rocks, and the neighbouring waters thickly sprinkled with little rocky islands overgrown with wood, which are a protection against the storms, and present a strong wall of defence against the waves. A bold, enterprising man, of the name of Baranoff, long superintended the company's establishment. Peculiarly adapted by nature for the task of contending with a wild people, he seemed to find a pleasure in the occupation. Although the conquest of the Sitkaens (Sitka-hans), or Kalushes, was not so easily achieved as that of the more timid Aleutians and Kodiacks, he finally accomplished it. A warlike, courageous, and cruel race, provided with fire-arms by the ships of the North American United States in exchange for otter-skins, they maintained an obstinate struggle against the invaders. But Baranoff at length obtained a decisive superiority over them. What he could not obtain by presents he took by force; and, in spite of all opposition, succeeded in founding the settlement on this island. He built some dwelling houses, made an intrenchment, and having, in his own opinion, appeased the Kalushes by profuse presents, confided the new conquest to a small number of Russians and Aleutians. For a short time matters went on prosperously, when suddenly the garrison left by Baranoff, believing itself in perfect safety, was attacked by great numbers of Kalushes, who entered the intrenchments without opposition, and murdered all they met there with circumstances of atrocious cruelty. A few Aleutians only escaped in their little baidars (or air-tight canoes) to Kodiack, where they brought the news of the destruction of Sitka. This took place in 1804, at the period that Admiral Krusenstern made his voyage round the world, and his second ship, the *Neva*, was bound for the colony. Baranoff took advantage of this, and with three armed vessels he accompanied the *Neva* to Sitka. The Kalushes retired at his approach to their fortifications, and attempted to maintain a siege, but the guns from the ships soon caused a speedy surrender. They were allowed to retire unmolested, but they stole away secretly on a dark night, after murdering all of their party who might have been an encumbrance to them. Baranoff thus

* Lisiansky, pp. 227—229.

† See Voyage of Don Juan de Ayala; Barrington's Miscellanies, Lond. 1781.

‡ See Dixon's Voyage, p. 183, *et seq.*; Marchand's Voyage, by Fleurieu, Eng. Trans. vol. i. p. 291, *et seq.*

became nominally possessed of the island, but in reality of a hill forming a natural fortification, and formerly inhabited by a Kalush chief called Katelan. The savages continued to make aggressions on the Russians, who could not venture beyond the range of the fortification, which Baranoff had rendered perfectly safe from attack. The necessary houses sprung up around it, and the settlement, under the name of New Archangel, became the capital of the Russian-American possessions. Such is the history of this isolated but interesting place.*

NEW ARCHANGEL, the Russian establishment, stands on the N.W. point of a bay on the eastern side of the Sound. The arsenal, on which a *light* is occasionally shown, is in lat. 57° 2′ 45″, lon. 135° 17′ 10″ W.

The following extracts will give a sufficient idea of the place. We have no sailing directions that may be trusted.

"The Sound of Sitka is formed by Mount Cape Edgcumbe to the North, and Point Wodehouse to the South, 14 miles apart; the former landmark is easily distinguished, being a dome-shaped mountain, evidently an old volcano, capped with snow, which lines its red sides in stripes from its apex down to its centre.

"The settlement of New Archangel is about 14 miles up the sound in a north-easterly direction, and cannot be approached closer than within 6 to 8 miles by a stranger, who should hoist his ensign and fire two guns. This we were aware of, and fired eight before the pilot came out: he was very welcome, however, when he did come, informing us that the Capt. Klenkoffstrein would be alongside in a quarter of an hour with a steamer to tow us in. Accordingly, at two o'clock on the twenty-second day, we were safely moored in the harbour of New Archangel.

"This is formed by a cluster of small islands immediately in front of the settlement, and has two entrances, one North, the other South. Once inside, it is as smooth as a pond, no port can be safer; and it is impregnable, owing chiefly to its natural defences, though the Russians have taken good care to erect batteries, which command it at every point. The town or settlement is built on a flat strip of land, jutting out here, as if on purpose, from the high belt of mountains which form the extremity of the sound. The governor's house is perched on a rock, about 80 feet in height, and is 140 feet long by 70 wide; of two good stories, roofed with sheet iron, painted red, and capped by a lighthouse, which can be distinguished by vessels at sea. The whole is defended by a battery, which commands every point of the harbour, and encircles one-half of the house to the S.E. The N.W. end is approached by a flight of steps. Halfway up sentinels are placed day and night, and here also are posted brass guns on light field carriages. The upper story is divided into one grand saloon in the centre, flanked by a drawing-room and billiard-room at one end, and a drawing-room at the other; all are well proportioned, painted, not papered, and the walls adorned with good engravings of British victories by sea and land. In the saloon is a magnificent full-length painting of Nicholas. The lower story contains (so to speak) a dining-room, drawing-room, study, and the domestic establishment of the governor and his lady. The dining-room is hung with prints of English towns, principally on the seaboard: the drawing-room with views in Switzerland

* See Kotzebue's New Voyage Round the World, vol. ii. pp. 138—142.

and Germany. The whole is plainly and substantially furnished, and heated with stoves of the continental custom. This, as well as all other houses, is built of wood, immense logs, dovetailed into each other, squared and painted.

"The arsenal is the next object which arrests the attention of a stranger, from the number of men employed either building new or repairing old vessels. At this moment they are building a new steamer, destined, I think, for Mr. Leidesdorff, of California. The workmanship appears good and solid; everything for her is made on the spot, for which purpose they have casting-houses, boiler-makers, coopers, turners, and all the other *ers* requisite for such an undertaking. The boiler is almost completed, and is of *copper*. They have also their tool-makers, workers in tin and brass, chart engravers, sawyers, and saw-mills, for all which occupations suitable establishments have been erected.

"The climate is moist; out of fourteen days there were only two upon which nautical observations could be made; these two were as fine and as warm as are experienced in any country."*

It was visited by Sir George Simpson on his overland journey round the world, in 1842.

"Sitka, (the late) governor Etholine's residence, consisted of a suite of apartments communicating, according to the Russian fashion, with each other, all the public rooms being handsomely furnished and richly decorated. It commanded a view of the whole establishment, which was, in fact, a little village; while about halfway down the rock two batteries on terraces frowned respectively over the land and the water. Behind the bay, which forms the harbour, rise stupendous hills of conical mountains, with summits of everlasting snow. To seaward, Mount Edgcumbe also, in the form of a cone, rears its truncated peak, still remembered as the source of smoke and flame, of lava and ashes; but now known—so various are the emergencies of nature—to be the repository of the accumulated snows of an age.

"We sat down to a good dinner in the French style, the party, in addition to our host and hostess (the latter a pretty and ladylike woman from Helsingförs, in Finland) and ourselves, comprising twelve of the company's officers. We afterwards visited the schools, in which there were twenty boys and as many girls, principally half-breeds; such of the children as were orphans were supported by the company, and the others by their parents. The scholars appeared to be clean and healthy. The boys, on attaining the proper age, would be drafted into the service, more particularly into the nautical branch of the same; and the girls would, in due time, become their wives, or the wives of others.

"Nor did religion seem to be neglected at Sitka, any more than education. The Greek church had its bishop, with fifteen priests, deacons, and followers; and the Lutherans had their clergyman. Here, as in other parts of the empire, the ecclesiastics were all maintained by the imperial government, without any expense, or at least without any direct expense, to the Russian-American Company.

"The good folks of New Archangel appear to live well. The surrounding country abounds in the cheveril, the finest meat I ever ate, with the single exception of the moose; while halibut, cod, herrings, flounders, and many other

* Nautical Magazine, June, 1849, pp. 310-1.

sorts of fish, are always to be had for the taking in unlimited quantities. In a little stream, which is within a mile of the port, salmon are so plentiful at the proper season, that, when ascending the river, they have literally been known to embarrass the movements of a canoe. About 100,000 of the last-mentioned fish, equivalent to 1,500 barrels, are annually salted for the use of the establishment; they are so inferior, however, in richness and flavour, to such as are caught farther to the southward, that they are not adapted for exportation." *

Point Wodehouse, as previously mentioned, is the S.W. point of Sitka Sound. An extensive group of islets and rocks extend N.N.W. from it for 3 or 4 miles from the shore, which, from that point, with little variation, takes a course of S. 36° E. This part of the coast is much broken into small openings, with islets and detached rocks lying off it.

PORT BANKS, of Capt. Dixon, is in this interval. It is in lat. 56° 35′, and the following are Dixon's remarks on it:—" The prospect at Port Banks, though rather confined, yet has something more pleasing and romantic than any we had seen on the coast. The land to the northward and southward rises sufficiently to an elevation to convey every idea of winter; and though its sides are perpetually covered with snow, yet the numerous pines, which ever and anon pop out their lusty heads, divest it of that dreary and horrific cast with the barren mountains to the N.W. of Cook's River (Inlet). To the eastward the land is considerably lower, and the pines appear to grow in the most regular and exact order; these, together with the brushwood and shrubs on the surrounding beaches, form a most beautiful contrast to the higher land, and render the appearance of the whole truly pleasing and delightful." †

CAPE OMMANEY, the southern extremity of the Sitka Archipelago, lies 45 miles from Point Wodehouse, and has been previously described, page 454.

Thus the entire circuit of this archipelago has been imperfectly noticed.

We now return to the northward, taking up the description at the point where Cross Sound terminates.

CAPE SPENCER, the point above named, has been noticed on page 458.

From Cape Spencer the coast takes a direction of N.W. It is steep and entire, well wooded, and, with the exception of one small opening between it and Cape Fairweather, appears not likely to afford shelter for shipping. The coast is completely bounded at a little distance by steep, compact mountains, which are a continuation of the same undivided range stretching from the eastward.

CAPE FAIRWEATHER is placed by Vancouver in lat. 58° 50½′, lon. 137° 50′. This cape cannot be considered as a very conspicuous promontory; it is most distinguished when seen from the southward, as the land to the West of it retires a few miles back to the North, and there forms a bend in the coast, and is the most conspicuous point eastward of Cape Phipps, at Behring's Bay, to the northward.

MOUNT FAIRWEATHER is one of the most remarkable mountains on the N.W. coast of America. It is placed in lat. 58° 54′, lon. 137° 38′, and 9 miles from the nearest shore. Capt. Cook says: " This mountain is the highest of a

* Sir George Simpson, vol. i. pp. 219—221; 227.

† Dixon, p. 195. (June, 1787.)

chain, or rather ridge of mountains, that rise at the N.W. entrance of Cross Sound, and extend in a N.W. direction, parallel with the coast. These mountains were wholly covered with snow (in May, 1778) from the highest summit down to the sea-coast (which was 12 leagues distant), some few places excepted, when we could perceive trees rising, as it were, out of the sea; and which, therefore, we supposed, grew on low land, or on islands bordering on the shore of the continent.

Capt. Portlock says of it:—"This mountain, or rather ridge of mountains, as it forms into several, is by far the highest land on this part of the coast, much loftier than Mount Edgcumbe, and, I think, nearly the height of Mount St. Elias."*

From Cape Fairweather to *Cape Phipps*, at the entrance of Behring's Bay, the distance is 73 miles; the intermediate coast is a low border extending from the base of the mountains, well wooded, and in some parts appears to be much inundated, the waters finding their way to the sea in shallow rivulets, through two or three breaks in the beach.

It is between these points that Cook, who sailed at a considerable distance from the shore, considered that there was an appearance of an extensive bay, and a wooded island, off the South point of it, in lat. 59° 18′, and where he supposed that Commodore Behring anchored.† A close examination of it by Vancouver, however, has dispelled these views:—"But in this neighbourhood no such bay or island exists, and Capt. Cook must have been led into the mistake by the great distance at which he saw the coast, in consequence of which he was prevented from noticing the extensive border of low land that stretches from the foot of the vast range of lofty mountains, and forms the sea-shore. The irregularity of the base of these mountains, which retire in some places to a considerable distance, and especially in the part now alluded to, would, on a more remote view than we have taken, lead the most cautious observer to consider the appearance on the coast as indicating deep bays or openings, likely to afford tolerable, and even good shelter; and had it not been for the information we had previously received from Mr. Brown, who had been close in with these shores, we should have still supposed, until thus far advanced, that we had Behring's Bay in view, with an island lying near its south-eastern part. This deception is occasioned by a ramification of the mountains stretching towards the ocean, and terminating in a perpendicular cliff, as if at the sea-side, having a more elevated part of the low border covered with wood, lying to the S.W. of it; the former, at a distance, appears to form the East point of an extensive bay, and the latter an island lying off from it; but both these are at the distance of some miles from the sea-shore, which hence takes a direction of N.W. by W. ¾ W., and is chiefly composed of a very low tract of land, terminating in sandy beaches; over which, from the mast-head, were seen considerable pools or lagoons of water, communicating with the ocean by shallow breaks in the beach, across all of which the sea broke with much violence. Where this low country was not intersected by the

* Portlock, p. 256.
† Cook's Third Voyage, vol. II. p. 347; and Muller's Voyages et Découvertes des Russes, pp. 248—254.

inland waters, it was tolerably well wooded; but as we advanced to the eastward, this border became less extensive, was more elevated, much less covered with wood, and for a few miles totally destitute of either wood or verdure; and composed of naked fragments of rocks of various magnitudes, lying, as it were, in the front of Mount Fairweather."—(Vancouver, vol. iii. pp. 208-9).

Behind this supposed bay, or rather to the South of it, the chain of mountains is interrupted by a plain of a few leagues in extent; beyond which the sight was unlimited. Vancouver says:—"The interruption on the summit of these very elevated mountains was conspicuously evident to us as we sailed along the coast this day, and looked like a plain composed of a solid mass of ice, or frozen snow, inclining gradually towards the low border; which, from the smoothness, uniformity, and clean appearance of its surface, conveyed the idea of extensive waters having existed beyond the then limits of our view, which had passed over this depressed part of the mountains, until their progress had been stopped by the severity of the climate, and that by the accumulation of succeeding snow, freezing on this body of ice, a barrier had become formed, that had prevented such waters from flowing into the sea. This is not the only place where we had noticed the like appearances. Since passing the Icy Bay, to the E.N.E. of Behring's Bay, other valleys had been seen, strongly resembling this, but none were so extensive, nor was the surface of any of them so clean, most of them appearing very dirty."—(Vancouver, vol. iii. p. 209.)

BEHRING'S BAY, the true situation and character of which was first elicited by Vancouver, runs inland, to the N.E., between Cape Phipps and Point Manby. Capt. Cook, supposing, as before stated, that a bay existed to the S.E., conceived it to be the bay that Chetrow, the master of Behring's fleet, reconnoitred. This mistake was also followed by Capt. Dixon, who gave the name of Admiralty Bay to that in question; but as Behring certainly was the discoverer of a bay in this locality, and there being but one, the name of that navigator has supplanted that applied by Dixon.

CAPE PHIPPS, the south-easternmost point of Behring's Bay, is in lat. 59° 33′, lon. 139° 47′. About 2 miles within it, the coast taking a S.E. direction, there is a small opening in the low land, accessible only for boats, near which was found an Indian village. Capt. Sir Edward Belcher states that he was driven much to the *westward* by the current near Cape Phipps (vol. i. p. 82).

Point Turner, which is a low narrow strip of land, forming the S.E. point of the island that protects Port Mulgrave from the ocean, is E. ¾ S. 2¾ miles from the inner or North point of Cape Phipps. About a league E. ¼ N. from Point Turner is a point on the main land which is the East end of a rounding bay, about 4 miles across to Cape Phipps. It is necessary to give a good berth to Cape Phipps, in order to avoid a small reef that stretches from it into the sea. Cape Turner, on the contrary, is bold, and must be kept close on board, for the purpose of avoiding the shoals that lie a little distance to the eastward of it; between these shoals and the point good anchorage is found, in 8 to 14 fathoms, clear good holding ground.

The rise and fall of tide here are about 9 feet; and it is high water about 30′ after the moon passes the meridian.

PORT MULGRAVE lies to the N.E. of Point Turner, and is protected from the ocean, as before stated, by an island lying in a N.E. and S.W. direction.

It was possibly first discovered by Capt. Dixon, June, 1787, who named it after that nobleman. It contains a number of small low islands, which, in common with the rest of the coast, are entirely covered with pines, intermixed with brushwood. To the North and West are high mountains covered with snow, 10 leagues distant.*

It was visited by Sir Edward Belcher, in H.M.S. *Sulphur*, who stayed a short time here. Fish, halibut, and salmon of two kinds, were abundant and moderate, of which the crews purchased and cured great quantities. Game very scarce. The remains of the Russian establishments were observed; a blockhouse pitched on a cliff, on the East side; and on the low point, where the astronomical observations were taken, the remains of another; also a staff, with a vane and cross, over a grave. Sir Edward Belcher says that a good leading mark for the entrance to the harbour is with Mount Fairweather over Cape Turner (or N. 88° E.).

An off-shore shoal, 7 fathoms, sand, was probably crossed by the *Sulphur* in coming out of Port Mulgrave, when no land could be seen within 3 miles.—(Belcher, vol. i. p. 89.)

The island or islands before alluded to, which form the outer face of Port Mulgrave, and of which Point Turner is the S.W. extremity, extend for 8 miles in length. They are almost joined to the continental shore by a spit incommoded with many rocks and huge stones, but leaving a very narrow channel, by which Vancouver's vessel, the *Chatham*, passed from one part of the inlet to the other, from the northward. The channel leading along the continent was found on examination to be not more than 50 yards wide, though nearly at high water; for a small space the depth was only 15 feet, but it quickly increased to 10 fathoms, and then to 17 fathoms. This passage is about 600 yards long, lying from the (northern) entrance of the channel S. 60° W. 2 miles, in which space the continental shore forms a small bay, and to the southward of the narrow part it takes a more southerly direction; along it are some islets and rocks, and the western side of the channel is much broken. About a league from the narrow part the depth continued to be from 17 to 12 fathoms, until a shoal was reached that lies right across the passage, on which the *Chatham* grounded, but was got off into deep water, without any damage. About the conclusion of the ebb a boat was sent to examine two places on the bar that were not dry; one of these, though narrow, was found to be sufficiently deep and free from danger, provided its line of direction was marked with buoys; this was done, and at half flood the *Chatham* passed through it, having not less than 3 fathoms water, which soon after deepened to 15 fathoms.....Thus, by persevering, Mr. Puget made his way through a channel which, though he found practicable, he does not recommend to be followed, especially as the communication between Port Mulgrave and the ocean is easy and commodious by the passage to the South and westward of Point Turner.—(Vancouver, vol. iii. pp. 228-9.)

* Dixon, p. 170.

KNIGHT'S ISLAND is 5 miles N.N.E of the northern entrance to the channel just described. It admits of a navigable passage all round it, but there are some rocks that lie about half a mile from its West point; and there is an islet situated between it and the main land, on its N.E. side.

ELEANOR'S COVE.—From the North entrance to Port Mulgrave the continental coast takes a N. 30° E. direction, 6 miles, to this cove, which is the eastern extremity of Behring's Bay. It is protected from the westward by Knight's Island, which is about 2 miles long in a N.E. and S.W. direction, and about a mile broad, lying at the distance of a mile from the main land.

The shore here is low, and trends about N. 14° W. 6 miles to *Point Latouche*, the S.E. limit of *Digges's Sound* or *Bay* :—"This was the only place in the bay that presented the least prospect of any interior navigation, and this was necessarily limited, by the closely connected range of lofty snowy mountains that stretched along the coast at no great distance from the sea-side. It was in reality found to be closed from side to side by a firm and compact body of ice; beyond which, at the back of the ice, a small inlet appeared to extend to N.E. by E., about a league. The depth of water at the entrance of the opening is great, and on its N.E. side is a bay, which afforded good anchorage, but had a most dreary aspect, from its vicinity to the ice, notwithstanding which, vegetation was in an advanced state of forwardness."—(Vancouver: July, 1794.)

The two points forming the entrance to Digges's Sound, thus named by Vancouver, are bluff, lying nearly East and West of each other, half a league asunder, the easternmost of them being Point Latouche, as above mentioned. The shores are composed of a continuation of the low border, extending from the foot of the mountain to the sea-side, and are bounded by frozen ice or snow, especially in the sound.

The continental coast, forming the North side of Behring's Bay, runs to the southward of West, and is nearly straight and compact. At 8 miles from the opening the land falls back, forming a small bay, with a low island about 2 miles long to the N.N.E. of it. The coast here trends S. 63° W. 8 miles, and then S. 85° W. 2 leagues, to Point Manby.

POINT MANBY forms the N.W. point of Behring's Bay. It is in lat. 59° 42′, lon. 140° 13′. To the eastward of it the country is well wooded, and proceeding northward it loses its verdant and more fertile appearance; the coast still continues to be a low compact border of plain land.

"On the evening of June 26th, 1794, the *Chatham* (Vancouver's ship) arrived off Point Manby; the water was found to be much discoloured at the distance of 4 miles from the shore, where bottom could not be gained with the hand-line, nor were any ripplings, or other indications of shallow water, or hidden dangers, noticed. The same appearances had been observed by us on board the *Discovery*, in several instances, to the eastward of Cape Suckling (presently described), which I concluded were occasioned by the vast quantity of fresh water produced by the dissolving ice and snow on the sides of the mountains at this season of the year; this, draining through the low border of land, becomes impregnated with the soil, and being specifically lighter than the sea-water, on which it floats, produces the effect noticed by Mr. Puget." This remarkable effect of the great changes in the temperature of the climate produces most extensive alterations in the form of the coast, as will be noticed presently.

POINT RIOU, though no longer existing, was a tolerably well-marked promontory at the period of Vancouver's survey, and to whom it owes its name. He describes it as being low, well wooded, with a small islet detached at a little to the westward of it. The coast is still composed of a spacious margin of low land, rising, with a gradual and uniform ascent, to the foot of the still connected chain of lofty mountains, whose summits are but the base from whence Mount St. Elias towers majestically conspicuous in regions of perpetual frost. Vancouver's charts, from the extensive changes continually going on, present but little to recognise in this part at the present period.

ICY BAY lies to the N.W. of what was Point Riou. It is terminated by steep cliffs, from whence the ice descends to the sea. At the eastern side of the bay the coast is formed of low, or rather moderately elevated, land. Its West point is a high, abrupt, cliffy point, bounded by a solid body of ice or frozen snow.

This portion of the coast was visited by H.M.S. *Sulphur*, in her voyage round the world, in 1837, and the following are Sir Edward Belcher's remarks, made during that visit:—

"Icy Bay is very aptly so named, as Vancouver's Point Riou must have dissolved, as well as the small island also mentioned, and on which I had long set my heart, as one of my principal positions. At noon we tacked, in 10 fathoms, mud, having passed through a quantity of small ice, all of a soft nature. The whole of this bay, and the valley above it, was now found to be composed of (apparently) snow-ice, about 30 feet in height at the water cliff, and probably based on a low, muddy beach; the water for some distance in contact not even showing a ripple; which, it occurred to me, arose from being charged with floating vegetable matter, probably pine-bark, &c.

"The small bergs, or reft masses of ice, forming the cliffy outlines of the bay, were veined and variegated by mud streaks, like marble, and, where they had been exposed to the sea, were excavated into arches, similar to some of our chalk formations. The *base* of the point named by Vancouver Point Riou probably remains; but being free, for some distance, of the greater bergs, it presented only a low sand or muddy spit, with ragged, dirty-coloured ice, grounded. No island could be traced, and our interest was too deeply excited in seeking for it, to overlook such a desirable object.

"The *current* was found to set 1½ miles per hour, *West*, varying but slightly in force, and not at all in direction. At this position we anchored in 50 fathoms, mud, near Mount St. Elias; not a single drift tree was noticed. We were within the white water about 2 miles, which I am now satisfied flows from the ice, but why it preserves its uniformity of strength and direction, is yet a problem to be solved." *

MOUNT ST. ELIAS is one of the most remarkable features of North-West America. It is a noble conical mountain, rising far into the clouds, and although in a climate far from temperate, and of such an elevation as to lead to the

* Voyage of the *Sulphur*, vol. i. pp. 78—80.

conclusion that it rises far into the limits of perpetual snow, yet Sir Edward Belcher says:—"Its edges, to the very summit, present a few black wrinkles, and the depth of snow does not, even in the drifts, appear to be very deep. It stands, as it were, as before mentioned, upon the summit of the lofty range which runs parallel with the sea-coast." Its elevation, according to angular measurement, is 14,987 feet above the sea, and, even when visible at 150 miles distant, appears to be a majestic mountain. Its discoverer was the celebrated Behring, who made the coast here on the 20th of August, 1741, the name being applied from the saint to whom that day is dedicated. Its latitude is 60° 18′ N., lon. 140° 52′ W.

PAMPLONA ROCK, &c.—According to some information given to Vancouver's party by the Russian officers he met here, there is a very dangerous rocky shoal, about 15 miles in length, lying by compass in a direction S. by W., 63 miles from a place called by them *Leda Unala*. This Mr. Puget conceived to be near the point called Point Riou. The Russian officer, Portoff, himself had been on the shoal, taking sea-otters, and stated that the first discovery of it was owing to a Russian galliot having had the misfortune, some years before, to be wrecked upon it; two of the crew were drowned, but the rest escaped in their boats. After that period an annual visit had been made to it, for the purpose of killing sea-otters, which were there met with.

From the Spaniards, also, Vancouver learnt that a very dangerous rock existed in this neighbourhood, the situation of which they had taken great pains to ascertain, and had found it to lie S. 41° E. from Cape Suckling, at the distance of 26 leagues, and which was called by them Rock Pamplona. By this bearing it appears to lie E.S.E., 8 miles distant from the rocky shoal described by the Russians above; here it may be inferred that Portoff and the Spaniards intended the same shoal, though it is not stated by the latter to be so extensive as by the former.

It is without doubt dangerously situated for the navigation of this coast, and it may possibly have proved fatal to Mr. Meares's consort, Mr. Tipping, who, with his vessel, was never heard of after leaving Prince William's Sound, in 1786.

The *coast*, from Icy Bay, extends nearly East and West, without anything remarkable for 40 miles, where there is a small river, called by the Russians *Riko Bolshe Unala*. It has a bar, and but little depth of water.

A few leagues farther to the westward is another small river, emptying itself into a shallow bay. Its entrance is obstructed by a bar, on which, with easterly winds, the sea breaks with great violence, and in the finest weather is only navigable for boats; but within the bar the depth increases for a little distance, and then it stretches towards the mountains.

The coast between this and Cape Suckling shoots out into small projecting points, with alternate low, cliffy, or white sandy beaches, being the termination of a border of low woodland country, extending some distance within, until it joins the foot of a closely united chain of lofty frozen mountains, which is connected with the same range that extends to the north-westward around Prince William's Sound and Cook's Inlet. From these low projecting points some shoals stretch into the ocean. Vancouver passed one of these at the distance of about 4 miles, sounding in 35 fathoms; it extends in a southerly direction, 2 miles from a low point of land that forms the West point of a bay, apparently very shoal, and, from the quantity of white muddy water that flowed from it into the sea, it was concluded to be the outlet of the floods formed on the low land by the dissolving ice and snow on the sides of the neighbouring mountains, which, at that season of the year (midsummer) must be copious, as the temperature was generally between 50° and 65°, and the elevated parts of the coast were still covered with snow, as low down as where the pine trees began to grow. From the West point of this bay, in lat. 60° 3′ 30″, lon. 142° 54′ W., the shore towards Cape Suckling makes a small bend to the north-westward, but the general direction of the coast is nearly East and West, and appears to be firm and compact.

CAPE SUCKLING, so named by Cook in his third voyage, is conspicuous. Vancouver, differing much from Cook, places it in lat. 60° 1′, lon. 143° 41′, but is determined by Lieut. Raper as lon. 143° 54′.* The point of the cape is low, but within it is a tolerably high hill, which is disjoined from the mountains by low land, so that at a distance the cape looks like an island.

When near Cape Suckling, Capt. Sir Edward Belcher says:—"Our attention was suddenly attracted by the peculiar outline of the ridge in profile, which one of our draughtsmen was sketching, apparently toothed. On examining it closely with a telescope, I found that although the surface presented to the naked eye a comparatively even outline, it was actually one mass of small, four-sided, truncated pyramids, resembling salt-water mud which has been exposed several days to the rays of a tropical sun (as in tropical salt marshes), or an immense collection of huts!

"For some time we were lost in conjecture, probably from the dark ash colour; but our attention being drawn to nearer objects, and the sun lending his aid, we found the whole slope, from ridge to base, similarly composed; and as the rays played on those near the beach, the brilliant illumination distinctly showed them to be ice. We were divided between admiration and astonishment. What could produce these special forms? If one could fancy himself perched on an eminence, about 500 feet above a city of snow-white pyramidal houses, with smoke-coloured flat roofs, covering many square miles of surface, and rising ridge above ridge in steps, he might form some faint idea of this beautiful freak of nature." †

In one direction from the southward, Cape Suckling exhibits on its lower profile the brow, nose, and lips of a man. It is a low neck, stretching out from a mountainous, isolated ridge, which terminates about 3 miles from it easterly, where the flats of the ice pyramids just alluded to terminate. Apparently the river or opening near Cape Suckling flows round its base. There is little doubt but that we may attribute the current to this outlet, arising, probably, from the melting of the snow. We had less strength of current after passing this position. Immense piles of drift wood were noticed on each side of the opening, but *none*

* It may be premised that the charts drawn up by Vancouver were found by Sir Edward Belcher to be plainly erroneous about this region. All his transit bearings and other observations indicated this. A river appears to flow near Cape Suckling, which has not been noticed.—Voyage of the *Sulphur*, vol. i. p. 175.

† Voyage of the *Sulphur*, vol. i. pp. 75-6.

elsewhere. Floating trees of considerable magnitude were numerous, and one was probably 200 feet in length. The current was northerly; and the water within 3 miles of the land was whitish, showing a distinct division, doubtless snow water and mud.—(Sir Edward Belcher.)

KAYE'S ISLAND, to the West and S.W. of the cape, is long and narrow. Its South point, named by Vancouver *Cape Hamond*, is very remarkable, being a naked rock, elevated considerably above the land within it. There is also an elevated rock lying off it, which, from some points of view, appears like a ruined castle. Towards the sea the island terminates in a kind of bare, sloping cliffs, with a narrow, stony beach at their foot, and interrupted with some gullies, in each of which is a rivulet or torrent, and the whole surmounted with a growth of smallish pine trees.

"Kaye's Island, viewed from the eastward, presents the appearance of two islands. The southern is a high table-rock, free from trees or vegetation, and of a whitish hue; the other is moderately high land for this region, with three bare peaks, its lower region being well wooded.

"At dawn the snowy ranges of mountains, from the termination of Montagu Island as far as Cape Suckling, or in the direction of Kaye's Island, were entirely free from clouds or vapours, a sight not common in these regions, and generally a warning for bad weather."—(Sir Edward Belcher, vol. i. p. 70.)

WINGHAM ISLAND.—Off the N.W. point of Kaye's Island is Wingham Island, and off its N.E. point, *Point Mesurier*, are some elevated rocks. Within these, and to the N.W. of Cape Suckling, is *Comptroller's Bay*, which is shoal, and extends 20 miles north-westward to Point Hey.

Wingham Island, which can be seen to nearly its whole length between Cape Suckling and Point Le Mesurier (the North part of Kaye's Island), is moderately elevated, rising in three hummocks, which are bare on their summits. The southern at a distance, owing to the lowness of the neck, appears separated. The whole is well clothed with trees.—(Sir Edward Belcher.)

PRINCE WILLIAM'S SOUND was first explored by Capt. Cook in his last voyage.* Although this extensive inlet was before known to the Russians, the coast took him ten days to traverse, 11th to 21st of May, 1778, a week of which was spent in the inlet; but, from the subsequent survey of Capt. Vancouver, it was found that no portion of his celebrated predecessor's labours were so defectively described and delineated as this, which leads to the supposition that some important authority has been omitted in the drawing up of the narrative, which would not have occurred had the unfortunate circumnavigator survived to superintend its publication.† From the minute examination which was made of it by Vancouver, it proved to be a branch of the ocean that requires the greatest circumspection to navigate; and although it diverges into many extensive arms, yet none of them can be considered as commodious harbours, on account of the rocks and shoals that obstruct the approaches to them, or of the very great depth

* Prince William's Sound, and particularly its N.E. part, was visited by Senr. Fidalgo, in 1790, for the purpose of inquiring into the nature and extent of the Russian establishments in these regions.

† See Cook's Third Voyage, vol. ii. pp. 353—366; and Vancouver, vol. iii. pp. 193-4.

of water about their entrances. Of the former innumerable were discovered, which led to the supposition that many others existed.

The N.E. point of the coast, where the sound commences, is *Cape Witshed*, which is 43 miles from Point Hey, last described, the coast between being fronted by a very extensive sand-flat. The outward coast of the sound is formed by Hinchinbrook and Montagu Islands; between and to the West of which are the entrances to it.

HINCHINBROOK ISLAND is the north-easternmost of those before Prince William's Sound. Its N.E. point, named *Point Bentinck*, is opposite to Point Witshed, a league asunder, the space between occupied by a low, barren, uninterrupted sand at low water, being a continuation of the sand-bank extending from Comptroller's Bay, and also along the coast to the N.E. of Point Witshed. It is dry at low water, but at high water it was stated that there is a boat channel, though Vancouver's party found the whole space occupied by a most tremendous surf, rendering any passage at that time impracticable. Cape Hinchinbrook, the S.W. point of the island, is 20 miles S.W. of Point Bentinck, and is placed by Vancouver in lat. 60° 16½′, lon. (corrected) 146° 27′. In a direction S.W. ½ S. 7 miles from the cape, is a barren, flat, rocky islet, with several rocks lying at a small distance from it. This lies, therefore, off the entrance to the sound between Hinchinbrook and the N.E. end of Montagu Islands.

Between Montagu and Hinchinbrook Islands Capt. Sir Edward Belcher found shoal water, contrary to Vancouver's idea, rendering it necessary to anchor in 17 fathoms: the tide running at 3 knots.*

PORT ETCHES† is on the eastern side of the entrance into the sound, consequently on the eastern end of Hinchinbrook Island. The depth off the North point of entrance is very great; no bottom with 100 fathoms could be found within a quarter of a mile of the shore. Off this point are some rocky islets, and there are some within the entrance (the *Porpoise Rocks*), and until these are past, there is no depth for anchorage. On the North side of the port is a lagoon (*Constantine Harbour*), within which Vancouver found a Russian establishment, on a situation commanding the low narrow peninsula, and formed in 1793, when some ship-building was being carried on.‡

Port Etches was visited by H.M.S. *Sulphur* in August, 1837. In the account of the voyage is the following:—"This establishment of the Imperial Russian Fur Company consists of the official resident, eight Russians, and fifty Aleutian and other allies. The houses are included in a substantial wooden quadrangle, furnished at its sea angles with two octagonal turrets, capped in the old English style, and pierced with loop-holes and ports; the summits of the lines are armed with spikes of wood. It is calculated to stand a tolerable siege, under determined

* Sir E. Belcher, vol. i. p. 70.

† Port Etches derives its name from Richard Cadman Etches, a merchant, who, with others, entered into a trading partnership (May, 1785), under the title of the King George's Sound Company, for carrying on the fur trade on this coast, having procured a licence for this purpose from the South Sea Company. The voyages of Capts. Portlock and Dixon, in the *King George* and *Queen Charlotte*, in and subsequent to 1785 were undertaken for this company.

‡ Upon Garden Island Capt. Belcher found a line that was marked by Portlock, July 22, 1787, and was very nearly destroying it.—(Belcher, vol. i. p. 73). At present the island is covered with pine trees, without many traces of the garden.

hands. The sleeping apartments, or ' 'tween decks,' as we should term them, are desperately filthy. The whole range is warmed by Dutch ovens, and the sides, being 18 inches in thickness, are well calculated to withstand the cold, as well as to defy musketry.

"The native allies, who live in huts outside, are filthier than the Esquimaux; arising, doubtless, from their life of inactivity, resulting from doubtful dependence. I was taken through the fish and oil establishment, which was inches deep in hardened filth and seal oil, and thence to the room containing peltry. I was much disappointed at the quality of the furs. They comprised sea-otter, sable, rat, squirrel, fox, bear, wolf, seal, and beaver, very large and heavy. The only desirable skins were those of the sea-otter and sable, and they were not first rate. As it is strictly forbidden to sell anything, and our visit bound us in honour not to permit anything of the sort, I felt little inclination to remain in this valuable repository." *

Port Etches might furnish a most complete harbour, if vessels frequented these regions, or a station should ever be required in so high a latitude. The currents, however, between it and Montagu Island render it difficult of approach; and the Russian commandant stated that many sunken rocks (but perhaps of 10 or 15 fathoms) lie off Cape Hinchinbrook. Capt. Belcher's observations make Vancouver nearly as much in error in longitude as he ascribes to Cook.†

Phipps Point, the N.E. or opposite extremity of the peninsula, on which is the Russian establishment, is placed by Sir Edward Belcher in lat. 60° 21′ 12″ N., lon. 146° 50′ 15″ W. High water, full and change, 1ʰ 15′; rise, 9½ feet. Variation, 31° 38′ E.

HAWKINS' ISLAND is to the N.E. of Hinchinbrook Island, and is about 20 miles long, N.E. and S.W. On its southern side is the channel before mentioned, which is contracted by the sand-bank on the S.E. shore to a narrow channel against the southern shore of the island. Off the entrance of the passage between the West end of Hawkins' Island and Hinchinbrook Island is a shoal, and in the passage are some islets and rocks.

On the North side of Hawkins' Island is *Port Cordova*, an arm extending from its N.W. point about 13 miles in an easterly direction. Within these limits are a bay and a small branch, in which are several rocks and rocky islets. These shores are in general low, ending in pebbly beaches, where shoal water extends some distance, and renders landing at low tide very unpleasant. To the northward is *Port Gravina*. The S.E. point of its entrance is placed by Vancouver in lat. 60° 41′, lon. 146° 11½′. To the S.W. of it are an islet and some rocks. Its northern shore extends from its West point East 5 miles, and then E. by S. ½ S. 12 or 13 miles, affording some small coves, with rocks and islets lying off it, and then turns to N.N.E., for 4½ miles, to its termination. Its East and South shores are encumbered with islets and rocks.

Snug Corner Bay is on the N.W. end of the peninsula, separating Ports Gravina and Fidalgo. Its West point is in lat. 60° 45′, lon. 146° 35′. "And a very snug place it is: I went, accompanied by some of the officers, to view the

* Sir Edw. Belcher, vol. i. p. 73. † *Ibid.* vol. i. p. 74.

head of it, and we found that it was sheltered from all winds, with a depth of water from 7 to 3 fathoms, over a muddy bottom. The land near the shore is low, part clear and part wooded."—(Capt. Cook, vol. ii. p. 361.)

PORT FIDALGO is so named after the Spanish commander who visited it in 1790. It extends in a winding direction to lat. 60° 55′, lon. 145° 48′; its width being about 2 miles, and its length 28 miles. Towards its upper end are some islands, and its shores are in general low, bounded by a pebbly beach, and pleasingly diversified by trees.

A small inlet runs in, 2 miles in a N.N.E. direction, at the N.W. point of Port Fidalgo, and S.S.W. ½ W. 4½ miles from its West point is the South end of Bligh's Island; between this island and the main land are some islets and rocks. The shores are also rocky. Bligh's Island is 7 miles long N.N.E. and S.S.W., and some islands off its North end form the southern side of the entrance to Puerto de Valdes.

PUERTO DE VALDES was so named by Senr. Fidalgo, and extends N.E. by N. for 12 miles, where a small brook, supplied by the dissolving snow and ice, flows into the arm, and from thence extends 5 miles in an East direction to its termination, in shallow water. Its eastern shores are indented with small bays, and lined with rocks and islets. The port is from half a league to a league in breadth. Its West point is called *Point Freemantle*, and is in lat. 60° 57′, lon. 146° 49′. This inlet was thought by some of the party under Capt. Cook to form the entrance to an inlet extending indefinitely to the N.E., but which opinion was controverted (justly) by others.

Southward of Port Freemantle is an island 7 miles long, in a S.W. by W. direction, and a league broad; within it is a passage half a league wide; and on its continental shore are two bays; that immediately North of Point Freemantle is the larger of the two, with an island in the N.E. corner. "It is a circumstance not unworthy of remark in these bays, so near as they are to each other, that the eastern one presents a southern, and the other a south-eastern aspect; and that the westernmost should be nearly free from ice, while the easternmost, with a full South exposure, should be terminated by a solid body of compact elevated ice; both being equally bounded at no great distance by a continuation of the high ridge of snowy mountains."—(Vancouver, vol. iii. p. 185: June, 1794). In passing the easternmost of the bays, the thundering noise of the falling (or calving) of large masses of ice was heard.

Westward of the island before mentioned is an arm extending about 4 leagues to the North, and terminating at the foot of a continuation of the range of lofty mountains. Its upper parts were much encumbered with ice (June, 1794), as were both sides of it with innumerable rocks and some islets. It is, in general, about a league wide, and its western coast terminates to the South, or *Point Pellew*, and from this the coast takes an irregular direction, about W.S.W. 10 miles, toward the East point of a passage leading northward. The shores which compose this extent of coast are formed by a low border of land, extending from the base of the mountains, much indented by small bays, and at high tide greatly intersected by water. It produces a few dwarf trees and other insignificant vegetable productions, and, like other parts of the continent bordering upon the sound, is bounded by small islands, islets, and rocks, extending as far as the eye can discern, rendering the progress of the boats tedious and intricate.

From the point before mentioned a channel extends about 3 leagues in length, to the N.W. ½ N. This in some places is a mile, and in others not a quarter of a mile broad; its West side formed by *Esther Island.* Four miles North from its further end is *Point Pakenham,* which is the S.W. point of Port Wells. This extends in a N.N.E. direction, and terminated in a firm and compact body of ice, but, at the time of the visit, was so encumbered with floating masses of ice, that it was highly dangerous to proceed in it. Here the party witnessed the falling of three tremendous bodies of ice from the cliffs, the shock of one of which was sensibly felt, though 2 leagues distant. To the West of the point is another but unimportant bay. Hence the coast pursues a southerly direction, 5 leagues to *Point Pigot.* The continent is here composed of a stupendous range of snowy mountains, from whose base low projecting land extends, jutting out into points, and forming the shores, which are thinly wooded with dwarf pines and stunted alders.

Point Pigot and *Point Cochrane,* opposite to it, form the entrance to *Passage Canal.* The principal branch extends from Point Pigot West 13 miles, and then S.W. by S. 4 miles farther, terminating in lat. 60° 48′. Here the head of the inlet reaches within 12 miles of Turnagain Arm, at the head of Cook's Inlet, hereafter described. On either side of the isthmus the country appeared to be composed of lofty, barren, impassable mountains, enveloped with perpetual snow. The isthmus itself is a valley of some breadth, which, though containing elevated land, was very free from snow (in June), and appeared to be perfectly easy of access. By it the Russians, and Indians also, communicated with either of these extensive sounds. The other branch extends 2½ leagues W.S.W. from Point Cochrane, which is 1½ miles South from Point Pigot. Eight miles E. by S. from Point Cochrane is *Point Culross;* and immediately East of the former is a large bay, about 3 miles deep, terminating in a boundary of frozen snow and ice, reaching from a compact body of lofty frozen mountains to the water's edge. Here it was singular that the shores between these icy bays are mostly composed of a border of very low land, well wooded with trees of the pine and alder tribes, stretching from the base of stupendous mountains into the sea.

Off Point Culross is an island about a league from the shore, and about 4 miles long; and following the coast southward for 6 miles, we arrive at an opening about 2 miles wide, leading to three small branches; two of them taking a southerly, and the third a north-westerly course. From the South point of the entrance, 11 miles along a shore broken into small bays, lined by innumerable rocks, and exposed to the whole range of the N.E. swell from the sound, brings you to *Point Nowell,* in lat. 60° 27′.

Between this coast and the ocean are a considerable number of large islands, which lie generally in a N.N.E. and S.S.W. direction. The coast of the continent runs, but in a very irregular manner, in the same direction, to *Cape Puget,* in lat. 59° 55′, lon. 148° 3′, this being the point where the shores of the main land form the seaward face. The continental coast will be first described; then the islands before it.

From Point Nowell the main coast turns to S.W. ½ S. for about 11 miles, to a point where an arm extends first N.W., and then terminates to the South, in a circular basin full of rocks; but about midway between the point and this are two arms extending parallel to each other, and 4 miles long, in a N.W. direction. Before this coast is an island, following its direction at 2 miles distant, and 3½ leagues long, forming a passage, but so full of rocks, that it is only navigable for boats or canoes. Five and a half miles S.E. from the South point of the arm first mentioned, is *Point Countess*, in lat. 60° 13′. Immediately West of it is an opening leading southward, and terminating in two small arms, a league within the entrance. Still farther West is a bay about 4½ miles deep, terminating in a compact body of ice that descended from high perpendicular cliffs to the water-side. The coast southward of Point Countess forms the N.W. side of a narrow channel, 11 miles in length, in a S.W. ½ W. direction. It is in most parts less than a mile broad, and there are several sunken rocks in it. Its shores are composed of steep rocky mountains. The North point of the southern end of this strait* is Point Waters; it has some rocks and breakers before it. This point is on the eastern side of Port Bainbridge, an inlet from the ocean, extending 18 miles in length from its entrance, in a North direction. Its termination is in lat. 60° 13½′, in a small tract of low land, before which are some rocks. From its being directly open to the ocean, although at this distance from it, the wind, when it sets up or down the channel, sends such a violent sea upon it, that landing is dangerous. *Point Pyke*, on the western side, is 6 miles from Point Waters; it is remarkable for its sugar-loaf form. Between them are two bays, surrounded on all sides by lofty, abrupt, snowy mountains. S.S.E. 5 miles from Point Pyke is *Point Elrington*, the south-eastern point of Port Bainbridge, and the south-westernmost part of a high, rugged cluster of islands. It is a high, steep, barren promontory of small extent, connected to the island near it by a narrow isthmus, which was covered with various kinds of sea-fowl. Between these points are some bays, and a large opening leading N.E., with many rocks about the shores, just above water. Opposite to Point Elrington is Cape Puget, before mentioned, on the main land. The western shore of the port, northward of Cape Puget, is compact, although somewhat indented with small bays and coves.

The island of which Points Elrington and Pyke form a portion, is high and rugged, and about 6 leagues in length, in a general N.N.E. direction. *Latouche Island* lies off its eastern side, separated by a channel half a league broad. Its northern point is named *Point Grace*. *Knight's Island* lies to the northward of these, and is upwards of 9 leagues in length in the same direction; and beyond this, again, are some others of less dimensions.

Between this and Montagu Island is *Green Island*, so named by Cook, in May, 1778, from its being entirely free from snow, and covered with wood and verdure. The islands near the open sea are, as before stated, elevated and rocky;

* It was in this strait that Vancouver's party encountered a violent storm, June, 1794; a very heavy gust of wind brought down from a considerable height on the mountain side an immense mass of earth, trees, and frozen snow, which fell at a distance not exceeding a hundred yards from the assembled party. They observed in other places the effects of similar storms, which will serve as a warning to any one on these shores.

those within are low ones. Off the North point of Green Island, a league or a league and a half North, are some ledges of rocks, some above and others under water, making it very unsafe plying in this neighbourhood.

MONTAGU ISLAND is the largest and principal island of Prince William's Sound, it being, according to Vancouver's survey, 46 miles in length, from S.W. by S. to N.E. by N.; its average breadth is about 2 leagues. Its South point is in lat. 59° 46', lon. 147° 30'. The passage on the inside, or to N.W. of the island, forms an entrance into Prince William's Sound, between it and Latouche and Green Islands, of course varying in breadth.

At 16 miles from the South end of Montagu Island is *Point Bazil*, in lat. 60° 1'. To the North of this are tolerably good soundings, on the island side of the channel, while to the southward of it no bottom could be reached with 60 or 80 fathoms, within a mile of the shore.

The two bays, one named by Portlock *Hemming's Bay*, and the other M'Leod's Harbour, are stated by Mr. Whidbey to be very exposed anchorages, and nothing more than stopping places in navigating this channel.

M'Leod's Harbour is thus described by Capt. Portlock:—"It may not be amiss to observe, that all ships coming into this harbour ought to keep the shore of Montagu Island on board as close as they can; for if they get off into the the channel, and over toward the West shore, they will soon bring 60, 70, and 80 fathoms water, and that depth too close in shore for anchoring."

M'Leod's Harbour is 5 or 6 leagues within the S.W. point of Montagu Island. Its outer points, Point Bryant on the South, and Point Woodcock on the North, are about 2 miles apart, and joined by a bank of 7 and 8 fathoms, black sand and mud, within which is a depth of 21 to 12 fathoms. Within it takes a turn to the North, round a point which is quite bold-to, and may be passed close. A ship can lie in 4½ or 5 fathoms water, with the South point of the bay just shut in with this point, at about a cable's length from the shore."*

PORT CHALMERS, on the West side, and toward the North end of Montagu Island, is in lat. 60° 16' N., lon. 146° 50'; variation, in June, 1794, 28° 30' E. Vancouver says:—"The place of our anchoring in Port Chalmers can only be considered as a small cove, on a rugged rocky coast, so very difficult of access or egress, that our utmost vigilance in sounding was unequal to warn us of a rock (off the harbour's mouth), on which the ship grounded." *Stockdale's Harbour*, too, is only a bay full of rocks, and of course not worthy of particular attention. The shores about Port Chalmers are in general low, and very swampy in many places, on which the sea appeared to be making rapid encroachments, the remains of the forests being seen below high-water mark. The trees around are not very luxuriant, but make rather a dwarfish forest. The only fish obtained were a few indifferent crabs from the shores; a little wild celery, and excellent spruce beer, may be procured from the land. Off the entrance to the harbour are several lurking rocks, which make its approach very dangerous, as before stated. In front of the entrance to the harbour is a small woody islet, lying about a mile from the point forming its entrance.

* Portlock, pp. 206-7.

The *South Passage Rock* lies from the North point of the harbour West something less than a mile distant, and from the small woody islet, N. by W. ½ W. about three-quarters of a mile. To the North of this is the North Passage Rock, lying from the North point of the harbour N.N.W. ¼ W. 2½ miles distant, and W.S.W. three-quarters of a mile from *Stockdale's Harbour*. These rocks must be carefully avoided, as they are not always visible, but are covered at high tide with scarcely any weeds or other indications.

The approaches to this harbour, from the South, are also much incommoded by two shoals in mid-channel, between the South point of the harbour and a rugged rock that lies about a mile from the East side of the largest of Green Island. The southernmost of these shoals is from 19 to 6 fathoms, without any weeds or other signs; and the other, three-quarters of a mile N.E. of it, shoals equally quick, but has a small patch of weeds in 3 fathoms, with 5 and 7 fathoms close around it. They are both very small, and on each side of them is a clear channel.

It is high water at Port Chalmers about one hour after the moon passes the meridian: the current sets southward, and there is no draught into the harbour. Springs rise 13 and 14½ feet, the night tides rising above a foot more than those in the day.

The strait between Montagu and Green Islands, to the northward of Port Chalmers, is embarrassed by a line of sunken rocks, which are very steep-to, affording no indication of their proximity by the lead. About half a league from the N.W. point of Montagu Island the depth is 65 fathoms: from this point a ledge of rocks extends half a mile. Off the North point of Green Island, 3 miles distant, is a ledge of rocks, and to the northward of these again is another. The N.E. end of Montagu Island is divided into bays or sounds, two of which appear capacious, but from their points of entrance (as well as within them) rocks extend a considerable distance. As this side of the island is greatly exposed to the prevailing winds, great caution ought to be observed in navigating near its shores.

From the N.E. point of Montagu Island its shores run compactly to the S.W. for 31 miles, to a low projecting point covered with wood. Off it lies a cluster of six rocky islets, chiefly composed of steep cliffs, nearly level on their tops, which may serve as a direction in thick or gloomy weather to the South point of Montagu Island, lying from them S.W. by W. ½ W. distant 17 miles. They are tolerably well wooded, and are not liable to be mistaken, particularly for the Chiswell Isles (21 leagues to the West), because those appear to be entirely barren.

The South point of Montagu Island, Point Elrington, and Cape Puget, between which are the S.W. entrances to Prince William's Sound, have been before described.

The Chiswell Isles are a group of naked rugged rocks, seemingly destitute of soil and any kind of vegetation. The centre of the southernmost group is in lat. 59° 31′, lon. 149° 2′. From this, the easternmost, which is a single detached rock, lies N.E. ¾ E. about a league distant; and the northernmost, which has several less islets and rocks about it, lies N. by E. ½ E. 5 miles distant.

Blying Sound of the Russians, called by Portlock *Port Andrews*, lies within the Chiswell Isles.

To the south-westward of the Chiswell Isles the coast presents a broken appearance as far as *Pie's Islands*, the South extreme of the southernmost of which lies in lat. 59° 19′, lon. 149° 51′. This island, in several points of view, forms a very conspicuous peak, and although not remarkable for its great height, yet from its singular appearance it is not likely to be mistaken in this neighbourhood, as it descends with great regularity from its summit to the water's edge. A group of rocks lying W. by S. ½ S. 4 miles from it, must be very dangerous in thick weather, as it is probably covered at high water, spring tides.

Between Pie's Islands and *Point Gore*, a distance of 18 miles, the coast is in most parts very mountainous, and descends rather quickly into the ocean, excepting in those places where it is broken into valleys, some of which are extensive. In the interval are two openings, and several low, detached parcels of rocks lie at a greater distance from the land than usual on this part of the coast.

Point Gore is placed by Vancouver in lat. 59° 11′, lon. (corrected) 150° 22′. Towards the sea this projecting promontory terminates in an abrupt cliff, moderately elevated, and is connected to the main land by a low peninsula covered with trees. To the westward of the point is *Port Dick*, described by Portlock.

CAPE ELIZABETH is the S.E. point of the mouth of Cook's Inlet. It is placed by Vancouver in lat. 59° 9′, lon. (corrected) 151° 18′. The coast here is composed of high land, before which lie three small islands and some rocks; the cape is itself the largest of these, and the westernmost of them. They appear to afford a navigable channel between them and the land, nearly in an East and West direction; but this is somewhat doubtful, for between the cape and the middle island, some low, lurking rocks were discerned, which had the appearance of being connected with a cluster of rocks above the surface, lying S. E. ½ E. 3 or 4 miles from the cape. To the S.W. of the middle isle is another cluster of rocks, both above and below the water's surface.

PORT CHATHAM, so named by Vancouver from his tender, is situated behind the island which forms Cape Elizabeth, and from that promontory extends to a point in a N.E. direction 5½ miles, and from thence it terminates in an excellent harbour, about 2 miles long from West to East, and 1 broad, North and South, affording secure and convenient anchorage. The passage into it, passing to the N.W. of Cape Elizabeth, is free from all obstructions but such as are sufficiently conspicuous or easily avoided. These consist principally of shoals that extend a little distance from each point of the cove on the North side of the entrance, and an islet about which are some rocks that lie to the S.W. of the S.E. point of entrance into the harbour. A narrow channel exists between the rocks and the main land, from 7 to 12 fathoms deep. The soundings in Port Chatham are tolerably regular, from 5 to 25 fathoms, the bottom a stiff clay; the shores in most places are a low border, very well wooded with pine trees and some shrubs. This border occupies a small space between the water-side and the part of the mountains that compose the neighbouring country, up which, to a certain height, trees and other vegetables were produced; but their more elevated parts appeared to be barren, and their summits were covered with snow, in all probability perpetual. The Chatham anchorage, off an excellent run of water, was found to be in lat. 59° 14′, lon. 150° 56′; var. (May, 1794), 24° 0′ E. The rise and fall of the tide, near the change of the moon, were 14 feet, but during neap tides not more than 10 or 11 feet. High water about an hour after the moon had passed the meridian; but this and other circumstances relative to the tides were found to be greatly influenced by the form and direction of the winds. The situation of the harbour in respect to its vicinity to the ocean, its free access and egress, and convenient communication with the shore, was considered by Mr. Puget to be equal, if not superior, to the generality of the ports visited in these regions. The Russian establishment, *Fort Alexandroffsk*, is in a bay to the westward of Port Chatham.

COOK'S INLET.

POINT BEDE, so named by Cook, May 26, 1788, is a lofty promontory, and from this the coast trends N.E. by E., with a chain of mountains inland extending in the same direction. The land on the coast is woody, and there seemed to be no deficiency of harbours.

GRAHAM'S HARBOUR is 7 miles from Point Bede:—"Graham's Harbour," says Capt. Portlock, "I found a most excellent one indeed, with great plenty of wood everywhere, and several fine runs of water. For a considerable distance it runs up nearly E.S.E., and then trends rather to the southward, with 14 fathoms water over a bottom of muddy sand. The East side affords plenty of black birch and other kinds of wood, which grew close to a beach where the boats could have easy access."

The entrance, according to Portlock's sketch, is between Russian Point on the South, off which a rocky shoal dries at half ebb nearly 1½ miles out, and Coal Bay on the North, 4 or 5 miles apart. In the entrance is Passage Island, on either side of which is an open channel. From this it runs up about 9 miles to the E.S.E., and terminates in a fresh-water river. There are several projecting points on each side of the harbour, that form very good and snug bays, where a ship might, if necessary, be hauled on shore in the greatest safety.

The Russian establishment here was considered only temporary, and situated on a pleasant piece of flat land, about 3 miles long and 200 yards wide, bounded by a good sandy beach on one side, and a fresh-water lake on the other. The Russians were twenty-five in number.

Coal Bay on the North side, to the East of the North point, is a pretty good one, carrying soundings in 14, 12, and 8 fathoms, fine black sand.

Capts. Portlock and Dixon landed on the West side of the bay, and in walking round discovered two veins of cannel *coal*, situated near some hills just by the beach, about the middle of the bay, and with very little trouble several large pieces were got out of the bank.*

The best time to run into this harbour is as near low water as possible. Whatever danger there is may then be seen, either from beds of kelp, or the rocks showing themselves above water.

TSCHOUGATSCHOUK BAY lies to the N.E., and its N.W. extremity is

* Portlock, p. 108; Dixon, pp. 60-1.

Anchor Point, in lat. 59° 39′; and hence, according to Vancouver's chart, the coast pursues a nearly straight direction 60 miles to the Russian establishment, 8 miles to the S.E. of the East Foreland.

The S.W. limit of Cook's Inlet may be placed at CAPE DOUGLAS, in lat. 58° 52′, lon. 152° 51′. The coast hereabout is composed of a low tract of country, stretching into the sea from the base of very lofty mountains, wrapped in snow, which also covered the surface of the land (May, 1794) quite down to the water's edge. This was likewise the case with that which appeared to be the cape; off which, a few miles to the northward, lies a very low flat island, *Shaw's Island*, off the N.E. point of which a ledge of rocks stretches, the extent of which increases materially at low water. To the northward of the mountains that form the promontory of Cape Douglas, is a lofty, rugged ridge, that at a distance appears to be detached, and to give an appearance of many openings in the coast, but a nearer approach shows it to be firmly connected by land less elevated, and forming a deep bay between the cape and the lower borders of Ouchouganat Island, or Mount St. Augustin. The shores of this bay, *Bourdieu's Bay*, in most directions seem compact, but encumbered with large rocks and stones; the depth of water across it North and South is from 9 to 12 fathoms. The bottom of the bay is formed by an extensive low country lying between the base of the rugged range of mountains and the water-side.

OUCHOUGANAT ISLAND, or MOUNT ST. AUGUSTIN, is a very remarkable island, rising with a uniform ascent from the shores to its lofty summit, which is nearly perpendicular, to the centre of the island, inclining somewhat to its eastern side, and being in lat. 59° 22′, lon. 153° 0′. It is about 9 leagues in circuit; towards the sea-side it is very low, from whence it rises, though regular, with rather a steep ascent, and forms a lofty, uniform, conical mountain, presenting nearly the same appearance from every point of view, and clothed with snow and ice down to the water's edge, through which neither tree nor shrub were seen to protrude. Landing on it is difficult, from the shore being bounded at the distance of a quarter of a mile by innumerable large detached rocks, which, however, extend farthest on the North side of the island. The width of the passage between it and the main land is about 6 miles.

Advancing northward along the shores of the main land, it will appear indented and broken into small coves and bays, that appear likely to afford secure anchorage. The points of these bays are in general steep and rocky, and behind the coast rises a continuation of the lofty range, extending from Cape Douglas, clad in perpetual snow. In lat. 59° 42′ are three islets, against the shore, behind which there is appearance of anchorage and shelter. There is nothing remarkable on the coast* until we come to the northward of lat. 60°, where there are two openings, the northern of which is the principal. It runs to the West, and

* The weather now (April 18, 1794), though extremely cold (the mercury standing at 25°), was very cheerful, and afforded us an excellent view of the surrounding region, composed, at a little distance from the river, of stupendous mountains, whose rugged and romantic forms, clothed in a perpetual sheet of ice and snow, presented a prospect, though magnificently grand, yet dreary, cold, and inhospitable. In the midst of these appeared the volcano, near the summit of which, from two distinct craters on its south-eastern side, were emitted large columns of whitish smoke, unless, as was supposed by some on board, it was vapour arising from hot springs in that

then S.W. towards the foot of a conspicuous volcano, which lies in lat. 60° 6′, lon. 152° 36′. The S.W. part of the sound is a small shallow opening, formed by two low points, covered with wood, and quite unimportant.

From the mouth of this opening to the West and East Foreland, where the breadth of Cook's Inlet is considerably contracted, the distance is 43 miles; the distance between its shores at this part being about 30 miles. In the intermediate space lies an island, named by the Russians *Coulgiack Island*, which divides the inlet here into two channels, the N.W. of which is much encumbered by dangerous and extensive shoals. The island itself is about 13 miles long, nearly N.E. and S.W., and is narrow. It is covered, in most parts, with small pine and alder trees. Vancouver landed on it, on the South point of a shallow bay on its N.W. side, towards its S.W. extremity. The snow, which was lying very deep on the ground (April 17, 1794), confined their walk to the beach, on which was lodged some small drift-wood, and on it they found some pieces of COAL, resembling cannel coal. The more important part of this island to the navigator is a dangerous shoal, which extends, in its direction from the S.W. end, for the distance of at least 2 leagues. Vancouver crossed it in 4 fathoms at a league from its South extremity. From the great variety of soundings on passing over it, it appears to be very uneven, as in several instances the ship struck violently, when the rise and fall of the waves were by no means equal to the depth shown by the lead. It is not improbable but that it may have some of the innumerable large fragments of rock lodged on it, which are to be found on the shores of the island. If so, it is infinitely more dangerous than a mere spit of sand to contend with. This shoal continues all along the S.E. shore of the island to 2 miles' distance off it.

Abreast of the S.W. point, on the West shore, is *Point Harriet*, which is a moderately high steep cliff. The shore on either side of it is a low beach, particularly to the northward, where the margin of low land is of greater extent than to the southward. Off the point a shoal extends a league off, on the outer edge of which is only 3 fathoms. To the N.E. of this again, the channel between the island and the western shore has some extensive shoals. On one of these Vancouver grounded. It lies off the middle of the island, and stretches to the northward. It is between 6 and 7 miles from the main land, and is near a league from the West side of the island, where a flat extends some distance into the river.

Beyond this the shores of the river are comparatively low, or only moderately elevated, jutting out into three remarkable steep cliffy points, named the East, West, and North Forelands; the two former forming the Narrows. Between the N.E. end of Coulgiack Island and the centre of the Narrows is a shoal, observed by Messrs. Portlock and Dixon. It is of very small extent, and bears from the point N.E. ½ E. 6 miles distant.

The *West Foreland* is in lat. 60° 42′, lon. 151° 12′, and is about 8½ miles nearly due West from the East Foreland. At the distance of about a mile off

---

neighbourhood; but how far this conjecture was consistent with the severity of the climate at the top of that lofty mountain, is not within the limits of my judgment to determine.—*Vancouver*, vol. iii. p. 100.

the former, the soundings are from 7 to 12 fathoms. A rock, that is visible only at half tide, lies about the fourth of a mile from the extremity of the point.

Between the West Foreland and the North Foreland, both of which are on the western shore, the coast forms a spacious open bay, called by Portlock *Trading Bay*. Shallow water extends from the former to within about 5 leagues of the latter point, from whence a depth of 5 fathoms will be found close to the main land.

The south-east shore eastward of the East Foreland forms a shallow bay, between it and a point 6 miles N.E. by N. above it, with soundings of 7 and 8 fathoms within a convenient distance of the shore, sheltered from the East, South, and S.W. winds, and not much exposed to those which blow from the opposite quarters. Beyond this point, and between it and a point lying 7 miles W.S.W. from Point Possession, a distance of 21 miles, the outer bank forms a perfect labyrinth of conical rocks, detached from each other on a bank of sand and small stones, extending a league and a league and a half from the shore. These rocks are of different elevations, and as few of them are of sufficient height to appear above high-water level, the navigation of these shores with such rapid tides requires the utmost circumspection in boats—with any large vessel it would be madness. These dangerous pyramidal rocks rise perpendicularly from a base at the depth of 4 to 9 fathoms, and are perfectly steep on every side within the distance of a boat's length. This very extraordinary rugged region appears to join to the southern side of the shoal on which Cook's ship, the *Resolution*, grounded in 1778 (vol. ii. p. 399); hence it must be considered as a fortunate circumstance that neither he nor Vancouver attempted to pass to the South of it. This shoal extends half way over the strait, and its outer end is about 9 miles nearly North from the East Foreland.

The *North Foreland* is in lat. 61° 4′, lon. 150° 35′, and on it Vancouver found the Russian factory, which consisted of one large house, the residence of nineteen Russians, who had established it in 1790. For 2 leagues to the North of this, along the western shore, tolerable anchorage is found, and commodious communication with the shore, abounding with wood close to the water-side, and affording several streams of excellent water. But this space is greatly exposed to the East and S.E. winds, the prevalent and most violent in this country; for it was remarked that all the trees that had fallen were lying with their heads toward the West and N.W., and all other perennial vegetables bent in the same direction. From this extent the shoals gradually stretch to the distance of 5 miles from the shore, until they join on to Point Mackenzie; the land between forming a low and perfectly compact shore, without the smallest discernible opening.

Turnagain Island lies at the head of the more extensive part of Cook's Inlet. Its West end is in lat. 61° 8′. It is about 3½ miles long, E.N.E. and W.N.W., and half a league broad. From the West point a shoal stretches half a league in a N.W. direction, which narrows the channel between it and the North bank to scarcely a mile in width. The island lies in the entrance of a branch diverging from the main inlet in a N.E. direction. Its entrance lies between Point Mackenzie on the North, and Point Woronzow, S.W. by S. 2 miles from it. Cook's vessel penetrated this inlet a short distance, but left its termination undiscovered. Vancouver anchored 5 miles above its entrance, and found that all above him become, at low water, a succession of dry sand-banks, occupying the whole of the space up to its head, 18 miles farther on, in lat. 61° 29′, lon. 148° 55′. At high tide it becomes an extensive sheet of water.

TURNAGAIN ARM.—The southernmost branch is the principal. It was called by Cook *Turnagain River*, and by Vancouver *Turnagain Arm*, he having decided its real character. Its entrance lies between *Point Campbell*, which is S. by W. ¾ W. 4 miles from Point Woronzow, and *Point Possession* on the southern shore. At 14 or 16 miles above these the shores converge again, up to which points they are 3 or 4 leagues asunder, each side forming a bay at high water, but their shores cannot be approached on account of the shallow flat that extends from the North side from 3 to 5 miles, and from the opposite shore about half the distance, between which is a channel 1½ leagues wide; but this is interrupted by a shoal, which dries in many places, 1½ leagues long, N.E. and S.W., leaving a channel only half a league broad at its South end.

The country bordering upon the bays between the outer and inner points of the arm is low, well wooded, and rises with a gradual ascent, until, at the inner point of entrance, when the shores suddenly rise to lofty eminences, in nearly perpendicular cliffs, and compose stupendous mountains, that are broken into chasms and deep gullies. Down these rush immense torrents of water, rendering the naked sides of these precipices awfully grand; on their tops grow a few stunted pine trees, but are nearly destitute of every other vegetable production. The tide here rises 13 feet perpendicularly, so that at low water the remaining portion of the arm is dry, or nearly so. It extends 22 miles above these points, and thus approaches to within 4 leagues of the head of Passage Canal, in the N.W. part of Prince William's Sound, described on page 476. Across this isthmus the Russians and Indians communicate with these two extensive inland waters, as there mentioned.

Vancouver found the time of high water, at his northernmost point in the N.E. arm, to be about six hours after the moon passes the meridian, and the rise and fall at springs he roughly estimated at about 27 feet. Like many other extensive inlets, which are closed at their upper ends (as, for example, the Bay of Fundy, and the Bristol Channel), the great range of the tide at its head is to be accounted for by the converging nature of the inlet, which forces a much larger body of water into its upper portion; consequently the tidal current rushes with great velocity, and Vancouver found the ebb running 5 and 6 knots, the flood not much less, above the Forelands. Lower down their velocity is proportionably less.

We have thus described the shores of this very extensive arm of the ocean. When Cook explored it in 1778, he supposed that it might be navigated much higher than he penetrated into Turnagain Arm. He formed this opinion from the water, though very considerably fresher, still retaining much of its saltness, and therefore assumed that a very extensive inland communication was connected with it. "If the discovery of this great river,* which promises to vie with some of the most considerable ones already known to be capable of extensive inland naviga-

* Capt. Cook having here left a blank, which he had not filled up with any particular name, Lord Sandwich directed that it should be called Cook's River.

tion, should prove of use either to the present or to any future age, the time we spent in it ought the less to be regretted. But to us, who had a much greater object in view, the delay thus occasioned was an essential loss."* Had the great navigator but penetrated a few miles farther, his views would have been of a different character, and it was reserved for Vancouver to determine its real nature, which, as it does not partake of that of a river, he has properly called Cook's Inlet.

CHAPTER XVI.

THE KODIACK ARCHIPELAGO, ALIASKA, AND THE ALEUTIAN ARCHIPELAGO.

The KODIACK ARCHIPELAGO was first seen by Behring, on his voyage of discovery, on returning from the American coast, in 1741. They were seen in 1763 by the Russian merchant Glotoff. In 1768 Chelighoff took possession of them in the name of a company of merchants for the trade in furs, of which he was the chief; and, in 1799, they were granted in full possession to a society of merchants formed from that of Chelighoff, and some other similar associations, under the title of the Russian American Company. This company is, to this day, in possession of the islands, as well as all those lying between Kamtschatka and the N.W. coast of America.

The Kodiack Archipelago is composed of two principal isles, Kodiack, and Afognack, and several smaller islets in their neighbourhood.

Kodiack (or Cadiack, as it is called by Lisiansky) is very mountainous, and surrounded by deep bays, into which a number of small rivers fall. On the shores of these many settlements might be formed; but the country elsewhere is in general too elevated, and is, besides, for the greater part of the year, covered with snow. The materials of which the island is composed are chiefly slate and common grey-stone. The climate, from the account given of it by the inhabitants, and from what Lisiansky experienced, is by no means agreeable; the air is seldom clear, and, even in summer, there are few days which may be called warm; the weather, indeed, depends entirely on the winds; so long as they continue to blow from the North, the West, or the South quarter, it is fine; when from other points of the compass, fogs, damps, and rain, are sure to prevail. The winters very much resemble what is felt in Russia in a bad autumn, which is, however, not without exceptions.

Poplar, alder, and birch grow on the island, though in no great quantity; and pine is only to be found in the vicinity of the Harbour of St. Paul, and farther to the northward of it. Some culinary plants, as cabbages, turnips, potatoes, &c., have been cultivated since the Russians have been here, but not generally throughout the islands. The dark and rainy weather is unfavourable to horticulture.

* Cook's Third Voyage, vol. ii. p. 396.

The native animals are few, consisting of bears, foxes, ermines, &c. Birds are much more numerous, both in numbers and variety. Kodiack also abounds in fish, which are halibut, cod, flounders, &c., and salmon, which last come into the rivers, from May to October, in such abundance, that hundreds may be caught in a short time with the hands only. The marine animals were formerly much more numerous, but from the indiscriminate slaughter they have been much thinned. This, however, from the better system pursued, is less manifest than formerly. Fur seals were formerly one of the staple products of the group.

The population is small, compared with the size of the island. We have no recent accounts, but they were estimated at a total of 4,000 by Lisiansky, in 1805. It was stated that, previous to the arrival of the Russians (who are accused of very great cruelties and oppressions), it was more than double this. Chelighoff stated that he subjected 50,000 men to the crown of Russia. They resemble, in many points, the Indian natives described in other portions of the American coasts, possessing much the same features and many of the habits. One atrocious custom is, that of men, called schoopans, living with men as women, to which they are educated from infancy. The inhabitants are almost entirely occupied in the chace of the wild and fur-bearing animals, in the service of the Russian American Company.*

KODIACK, as before stated, is high, hilly, and very much intersected. Its greatest diameter is about 30 leagues in a N.E. and S.W. direction, and its breadth may be assumed as 15 leagues. Although we have not an exact acquaintance with the whole of the island, its eastern portion is sufficiently well known, because the Russian American Company send, nearly every year, some ships which are always commanded by able officers, who sometimes make a long stay here. Its western coast is nearly altogether unknown. The shore, on all sides of the island, is indented with a great number of large and deep bays, which contain excellent harbours. That of Tschiniatskoy is the largest, and at the same time the most important; for it is in the bottom of this bay that the establishment of the Russian American Company, formerly the principal in the Pacific Ocean, lies. This is the town and harbour of St. Paul. It is, therefore, the only port frequented by strangers, and we shall be more particular in its description.

TSCHINIATSKOY BAY is formed by the cape of that name on the South, and Long or Barren (Sterile) Island on the North, an opening of 8 miles in a N. ¼ W. and S. ¼ E. direction. It is 6 miles deep, and its S.W. portion is filled with rocks.

The frequent and lasting fogs which occur here would render the approach to this bay very difficult, if it were not for the island named Ougack, lying 15 miles South of Cape Tschiniatskoy, and 2½ miles from the land. This being the only island on the East coast to the South of the bay, it becomes an infallible point of recognizance on approaching it.

The HARBOUR of ST. PAUL is excellent in every respect; the depth 5, 6, and

* Many particulars of the group, in addition to those contained in the accounts of the Russian discovery by Dr. Coxe, and Pallas, will be found in Lisiansky's Voyage, chap. x. p. 190, *et seq.*; Billing's Voyage, by Martin Sauer; Langsdorff's Travels; Cook's Third Voyage, vol. iii.; and Vancouver's Voyage, vol. iii. These will give a good idea of the condition and resources of this inhospitable country.

7 fathoms, good holding ground. Properly speaking, it is a narrow channel, formed by Proche Island, which will hold but few vessels at a time; the outer road is equally well sheltered and secure. We give Admiral Krusenstern's translations of Capt. Golownin's directions:—

"The Port of St. Paul has two entrances; one from the South, by the Bay of Tschiniatskoy; the other, from the North, passes through the outer road. Neither are dangerous if the wind be favourable and the weather sufficiently clear to distinguish the shores around the port; but it should not be approached during the night or in fog, for there are no lights, and the currents may carry you easily on to the shoals and rocks, which are abundant on all sides.

"As soon as you have cleared Cape Tschiniatskoy, you find before you a rock called *Gorbun* by Capt. Lisiansky; steer *N.W.* $\frac{1}{2}$ *W.*, or *N.W.* $\frac{3}{4}$ *W.*, *true*, and you will soon sea ahead a small high island, *Toporkowa*, upon which you must be careful to direct your course. This island will show you the direction of the current; steer right upon it, leaving to starboard *Barren* (or Sterile) *Island*, and then the channel which separates it from another called *Woody* (Boisée) *Island*. When abreast of the South point of the latter, which may be readily known by the rocks surrounding it, bear to the North, ranging along the western shore of Woody Isle as near as possible, paying attention to the soundings, which diminish regularly on either side up to the entrance of the port. Following these directions, if the wind be not contrary, and carrying short sail, you may pass, without a pilot, the shoals on the western side near two isles (marked A and B on Krusenstern's chart) and reach the entrance of the port. Although buoys are marked at the entrance, they must not be expected, but there is no doubt but the agent of the company will constantly attend to these matters.

If, after passing along the West coast of Woody Isle, the wind or other obstacles prevent an advance, you may anchor in perfect security until the weather becomes more favourable. In case a vessel may have entered the bay, and the wind will not allow her to follow the foregoing route, and it is absolutely necessary that she should reach the port, she will find a good shelter very near *Cape Escarpé* (steep), on the western side of the bay. In this case, after nearing the Gorbun Rock, run directly for this cape, or to W. $\frac{3}{4}$ N., until a remarkable jutting point bears W.N.W.; it is readily distinguished on this low coast by its elevation, and its peaked form. As soon as Cape Escarpé is passed, change the course towards the starboard, to anchor under Toporkowa Island, from whence you may readily reach the harbour, either under sail or by towing.

To enter the Harbour of St. Paul by the northern passage, steer for Cape Pine; then, being near to this cape, which ought to bear to N.W., distant a mile or half a mile, run into the middle of the channel between Kodiack and Woody Isle, carefully observing not to go into less than 18 or 20 fathoms depth, steering directly for the islands (A and B) before mentioned, to the West of Woody Isle, until the town of St. Paul is seen; you may then enter the port itself, or rather anchor in the outer road. The best anchorage is under Woody Isle, in 13, 14, or 15 fathoms, sand. Nearer the port the bottom is of mud, but here you are not so well sheltered as under Woody Isle.

If you wish to enter the port under sail, you must take care of the contrary current, or have good cables; the breadth of the harbour not allowing you to bear up, you must drop anchor when under way. In the summer it would be better to anchor in the road, mooring in the direction of the tides; the flood running to N.E., and the ebb to S.W. The starboard anchor ought to be laid towards S.W., and the larboard N.E., having an open hawse for N.W. and West winds, which blow strongly and in gusts. The tides change regularly every six hours.

To the South of Cape Greville, or Tolstoy (great, or large), is Cape Tonkoy (fine, or small), of the Russian charts. It forms the N.E. point of *Igatskoy* or *Ihack Bay*, a deep inlet of 16 miles to the West, but only 2½ miles in breadth. There are some good harbours in it, and Capt. Lisiansky, above all, praises one lying in the S.W. part of it. In the bay the Russian American Company have an establishment. In entering the bay, keep close to the South shore, the North is bestrewed with rocks.

Twelve miles to the South from Igatskoy Bay is *Kiloudenskoy* or *Kiluden Bay*, where the company also have an establishment. According to Lisiansky, it much resembles the former bay, only that it is not so deep. It is properly composed of two bays, either of which afford shelter.

In the S.E. part of Kodiack, to the South of the last-named bay, lies *Saltchidack* or *Siachladack Island*, which is nearly 20 miles broad S.W. and N.E. Neither Cook nor Vancouver, who passed it,* could see that it was an island, because the strait that separates it from Kodiack is not above 250 yards wide in parts, though some miles at the openings. *Cape Barnabas* of Cook is the N.E. point of this island.

Two-headed Point of Capt. Cook is, as he supposed, on a small island, the position of which was determined by Vancouver, at 8 miles S.W. by W. ½ W. from the S.W. point of Saltchidack Island. It is probably the same island that Admiral Sarytscheff calls *Nasikok*, and according to him is distinguished by a high mountain, and is the northernmost of four isles that must be doubled before entering the *Port of the Epiphany*. This accords with what Vancouver says, that it forms the S.W. point of the great road or the channel between Kodiack and Saltchidack. The opening of this is 7 miles wide; there are two harbours in it on the Kodiack shore; the first, called Kiack, is opposite Cape Bay (Myssoff of the Russian charts), and 10 miles North of the former is Nayoumlack Bay.

At a few miles North of this last is *Epiphany Bay*; it is small, only a mile in circumference, and 60 yards in the opening. The depth is 10, 8, 7, and 4½ fathoms, muddy bottom. The inhabitants call it Manikoks, and it was here that Chelighoff first landed, and Billings remained some days, in 1790. The charts hereabout are very defective and require revision.

The southern point of Kodiack was called by Cook *Cape Trinity*. It was also placed by Vancouver. At 11½ miles South of Cape Trinity lie two isles, named by Cook *Trinity Isles*; they are so close together that they might almost be considered as one island; together they are 12 leagues in length, East and West, and 2 or 3 leagues from the coast. On the Russian charts the eastern is called *Sitchunak*; the western, *Tugidack*.

* Cook's Third Voyage, vol. iii.; and Vancouver, vol. iii.

The western shore of Kodiack, although the Russian Company have had it for many years, is, as before stated, but little known. To the North of Cape Trinity is *Alitock Bay*, where the company have an establishment; the westernmost point of Kodiack, *Cape Ykolik*, in 57° 14′ North, lies N.N.W. ½ W. 38 miles from Cape Trinity; and at 18 miles N.E. of it is the company's establishment, named *Carluck*. It is from here that the baïdares destined for the opposite shore depart, the Strait of Chelighoff being narrowest here.

At 12 miles to the North of Carluck is *Oujack* or *Oohiack Bay*. It is a deep indentation, extending 27 miles in a S.S.E. ¾ E. direction, the distance between its head and that of Kiludenskoy Bay, on the opposite side of Kodiack, being only 8 miles. In the opening of the bay is an island, which forms a channel on either side.

The northern point of the bay is formed by a very projecting cape of the same name. To the right of it is a second bay, which is but 10 miles deep, to the N.E. of which is Ouganick Isle.

The N.W. point of Kodiack, according to Lisiansky, is in lat. 57° 28′, and 2 miles from this point lies the extreme of North Island, which extends 15 miles N.N.W. and E.N.E. This is separated by a narrow channel from Afognack Isle, an eastern point of which was named by Cook *Cape Whit-Sunday*. A channel about half a mile broad separates it from *Chonjack Island*, which is 14 miles from South to North. It is this island that Cook calls *Cape Banks*.

The northern coast of Kodiack, North Island, and the South part of Afognack, form a channel 20 miles long and 2 wide, in which 10 to 20 fathoms water is found.

At 4 miles from Cape Whit-Sunday is an island 8 miles in length, called by the Russians *Evratschey*, the name of a small animal very abundant there, from the skins of which the inhabitants make their *parques*, or fur shirts. This island is what Behring supposed to be the N.E. point of Kodiack, and called it Cape St. Hermogenes. Cook found it to be an island, and preserved Behring's appellation. The Trinity Bay of Cook lies between it and the N.E. part of Kodiack; but a better knowledge shows that the term is not applicable, though Cook did not doubt the existence of the North channel separating the islands. To the North of St. Hermogenes are some rocks discovered by Cook.*

CHELEKHOFF STRAIT.†—Cook called the North entrance of Chelighoff Strait Smoky Bay. It separates Kodiack from the continent North of the Peninsula of Aliaska, and derives its name from the Russian commander who first brought the inhabitants of the adjoining countries under subjection.

Up to a recent period the western side of the strait in question was comparatively unknown; but in 1832 it was examined and surveyed by Mr. Wassilieff, an officer of the Russian navy, in the service of the Russian American Company, in baïdares, and it is from the manuscript chart then drawn up that the charts have been corrected. They show that the strait is narrower than was at first supposed; it does not exceed 25 or 30 miles.

CAPE DOUGLAS, which has been before described, page 486, is the N.W.

* Cook's Third Voyage, vol. iii. p. 384.

† *Cheligheff*, Krusenstern.

limit of the strait, and from this point the survey extended to a bay in lat. 56° 36′, opposite to which are the Evdokeeff Islands. A great number of bays were examined by Capt. Wassilieff within this space, which appears to offer good shelter. Of these it is only necessary to particularize one, that called *Poualo*, in lat. 47° 46′, lon. 155° 0′ West, which is 5 miles distant from the great lake *Nanouantoughat*, from which the *River Ougagouk* flows. This river has been adopted by Krusenstern as the northern limit of the Peninsula of Aliaska. Here is the portage for the merchandise which the agents of the Russian Company transmit to their establishment situated on the shores of Bristol Bay.

The PENINSULA of ALIASKA is a remarkable tongue of land extending from the River Ougagouk, above mentioned, to the Strait of Isanotzky, separating it from Ounimack, the easternmost of the Aleutian Archipelago, an extent of 330 miles; its breadth diminishing from 90 miles in the North to 25 miles in the southern parts.

From its configuration it may be regarded as a continuation of the Aleutian Islands. In early times the knowledge of this land was most vague. Prior to Capt. Cook's visit to Behring's Sea, the only geographical authority was a curious map, evidently derived from oral information, on which the numerous islands and lands were distributed without any regard to their actual relative size and position, but which carries with it some authenticity, because many of the names in it are now more definitely applied to known points. This map is prefixed to an account of these regions by Von Stæhlin, in 1774.* It is also appended to Müller's Voyages from Asia, 1761.† In these works the land in question figures as the large Island of Alaschka, filling up a large portion of the space now known to be occupied by Behring's Sea.

The first authentic notice of its shores was that given in the account of the third and disastrous voyage of Capt. Cook, who examined, though but very slightly, both sides of the peninsula at different points. His observations are but necessarily slight, from his having but imperfectly seen the land. The southern side remained in the same imperfect state until the examination, previously noticed, by Capt. Wassilieff in 1832. Its northern side, of which we shall speak hereafter, is somewhat better known.

From the *Bay of Poualo*, the north-eastern limit of the peninsula noticed above, Capt. Wassilieff's examination extended to a large bay in lat. 56° 40′, and abreast of the Evdokeeff Islands. This bay has been named *Wassilieff Bay*. The space between Poualo Bay and this point contains a great number of bays, and all along the coast are numerous islands, all of which are named on Capt. Wassilieff's chart, but of which we have no especial description.‡

The EVDOKEEFF ISLANDS were discovered by Behring on August 4th, 1741, and named by him in honour of the saint of the day. They form a group

* An Account of the New Northern Archipelago, by J. Von Stæhlin, 8vo, London, 1774.

† Voyages from Asia, &c., by S. Müller, 4to, London, 1761.

‡ Capt. Lütke states that M. Wassilieff's journal was placed in his hands, but that, instead of elucidating his chart, which seems to merit confidence, he absolutely found it to differ very greatly from his delineations. Not giving any particulars as to the nature, appearance, or productions, of the coast, but little could be gathered from it.—*Voyage du Séniavine*, Part. Naut. p. 274.

of seven islands, the three largest of which are called *Simidin*, *Alexinoy*, and *Ageach*. According to Admiral Sarytscheff, who passed between these islands, they are very close to each other, very high, and surrounded with rocks, some under water, others uncovered. It was generally supposed that the largest of these islands, Simidin, was the same as that which Cook named Foggy Island; but this is controverted by Krusenstern, who thinks that this is to be found between this group and the land. Capt. Golownin did not see the large Island of Simidin, but determined the position of the southernmost island of the group as lat. 56° 0′ North, lon. 156° 22′ West.

ST. STEPHEN'S ISLAND.—The following is Admiral Krusenstern's remarks on the Islands of St. Stephen and Tschirikoff:—"In the memoir accompanying the chart of the Kodiack Islands I have already said that Vancouver considered that the island named by him Tschirikoff was identical with the Foggy Island* of Behring; but as Cook and he have taken different islands for Foggy Island, I have attentively examined Behring's Journal as to the position of this island, and this is what it says.

"During the night of the 1st August, 1741, Behring suddenly saw land, and as a strong current bearing to the South, accompanied by a thick fog and a calm, was carrying the ship directly on to the land, he was obliged to anchor in 16 fathoms. On the morrow, at 8ʰ A.M., he saw that the land, from which he was about 4 miles distant, was an island which, in an East and West direction, was about 3 miles in length; a reef projected for 3 miles off the East point, which bore E.S.E. from him. At 8ʰ P.M. he weighed anchor in a thick fog, which constantly hung over him, and on the following day, at 7ʰ A.M., he made an island to the South at the distance of 7 German leagues. Behring places this island in 55° 32′, and gives it the name of St. Stephen, a name which has not been preserved; that of *Foggy*, by which it is now known, was probably applied to it by the officers of his ship, on account of the fogs which occurred at the time of its discovery.

"The 4th of August he found himself near the Evdokeeff Islands, the southernmost of which bore to the S.S.W. ¾ W. at 20 miles' distance; the latitude on this day is marked as 55° 45′. But we know that the Evdokeeff Islands lie in 56° 10′; it follows that Behring's latitude is nearly half a degree too far to the South; if this error is applied to St. Stephen, we get lat. 56° 0′. Then if the latitude and extent of this island are compared with the latitude and size of Tschirikoff Island, which is 10 leagues in circuit, it must be clearly seen that they cannot be identical, as has hitherto been believed; and it is in consequence of this supposed identity that the Island of St. Stephen has been entirely omitted from the charts. I will now prove that Cook, Vancouver, and Admiral Sarytscheff, have all seen an island where the Island of St. Stephen ought to be placed. Cook makes no mention in his journal of this island, but it is found on his chart; and a passage in Vancouver, which I will presently quote, demonstrates that he saw it. Admiral Sarytscheff being, June 25th, 1721, in the midst of the Evdokeeff Islands, observed in lat. 56° 10′. The day following he found himself in

* *Tumannoi-ostrow*, c'est-à-dire, L'Isle Nebuleuse.—*Müller*, p. 261.

lat. 56° 20′, and half a degree more to the East. It is stated in his journal for this day:—'At one o'clock, P.M., we saw a low island to the S. 56° E., at the distance of 26 miles, which bears the name of Oukamock; the latitude of this island ought then to be 56° 6′.' On the original draft of his voyage it is even placed in 56° 14′. The mean of these is 56° 10′, which does not differ much from that of the St. Stephen Island of Behring, which was made to bear E. ½ S., when the Evdokeeff Islands bore from S.S.W. ¾ W. to W.S.W. ½ W.

"Vancouver observed, April 4th, 1794, in lat. 55° 48′, and lon. 154° 56′. Having run from noon to six, P.M., 40 miles to N. 65° E., the latitude at this time would be 56° 5′, and the longitude, 153° 50′. Trinity Island then bore N. 10° E., and another island from W. ½ N. to W. by S. To judge by the latitude and the direction of the wind, this other island could be no other than the St. Stephen Island of Behring. On this occasion Vancouver remarks:—'The latter I took to be that which was laid down in Capt. Cook's chart to the S.W. of Trinity Island.' This land, though not noticed in Capt. Cook's journal, was seen and passed on its southern side by the *Discovery* in that voyage, which proves that the *Resolution* and *Discovery* could not have gone far to the North of Tschirikoff's Island, which was obscured at that time by thick foggy weather.*

"For these reasons I do not hesitate to assign a place in my chart for the discovery of Behring. I have placed it in 56° 10′, and 155° 22′, to the N. 20° W. 7 leagues from the northern point of Tschirikoff's Island, and I have preserved the name of St. Stephen, not only on account of its being the name given to it by Behring, but also because Foggy Island is given by Cook to another island."†

TSCHIRIKOFF'S ISLAND was, therefore, discovered by Vancouver, April 4th, 1794, and named by him after the companion of Behring. He states the circumstances thus:—"The N.E. point of the island bore by compass N. 55° W., distant about 2 leagues; its eastern extremity, which is a low, rocky point, and was our nearest shore, S. 66° W. 2 miles; and its South point, S. 30° W., about 2 leagues distant. In the point of view in which we saw the south-western, southern, and eastern sides of this island, it appeared to form a somewhat irregular, four-sided figure, about 10 leagues in circuit, having from its western part, which is low and flat, and which had the appearance of being insular, a remarkably high, flat, square rock, lying in a direction S. 66° W., at the distance of 2 miles, between which and the island is a ledge of smaller rocks. The season of the year greatly contributed to increase the dreary and inhospitable aspect of the country; in addition to which it seemed to be entirely destitute of trees or shrubs, or they were hidden beneath its winter garment of snow, which appeared to be very deep about its S.E. parts, consisting of high, steep cliffs; but on its western side, which was considerably lower, this appearance was not so general. About its shores were some small whales, the first we had noticed during this passage to the North."‡

A *rock* is marked on the charts to the S.W. of the Island of Simidin, in lat. 55° 50′; evidently a different position to those recorded by Cook as having

* Vancouver, vol. iii. p. 88. † Krusenstern, part ii. pp. 105-6.
‡ Vancouver, vol. iii. pp. 86-7.

been seen June 16, 1778, a cluster of small islets, or rocks, lying about 9 leagues from the coast, which would be in about lat. 56° 3′, and lon. 158° 0′ W.

Of this part of the coast of the peninsula, as before stated, our knowledge is very scanty. Capt. Cook, who is almost the only navigator who tells us anything about it, says:—"For some distance to the S.W. (of Foggy Cape) this country is more broken or rugged than any part we had yet seen, both with respect to the hills themselves, and to the coast, which seemed full of creeks, or small inlets, none of which appeared to be of any great depth. Perhaps, upon a closer examination, some of the projecting points between these inlets will be found to be islands. Every part had a very barren aspect, and was covered with snow from the summits of the hills down to a very small distance from the sea coast."

SCHUMAGIN ISLANDS.—This group, which is the next considerable collection West of the Evdokeeff group, according to a notice inserted in the Memoir of Capt. Lütke, is composed of fifteen islands, and seven smaller islets. They received the name of Schumagin (Choumaguine) from Behring, in memory of one of his sailors, who was buried here.* Admiral Sarytscheff, in his journal, names the two largest islands of the group *Ounga* and *Nagay*. The first, according to him, extends 12 leagues from North to South, with a breadth of 7 leagues; he places its northern extremity in lat. 55° 42′. The Island Nagay, with a similar direction, is 8 leagues in length. Besides the Islands of Ounga and Nagay, Sarytscheff names those of *Kagai*, *Sajouliucktusigh*, *Nuinack*, *Tagh-Kiniagh*, and *Kiuniutanany*; all these, and several others not named, lie very close together. Kagay Island, according to Sarytscheff, ought to be placed in lat. 55° 5′ North, and lon. 160° 33′ West. Capt. Golownin saw none of these islands except Tagh-Kiniagh, which he places in 54° 56′ North, and 159° 40′ West. The Island Nuinack lies 5 leagues to the S.W. of this. Cook took the largest of the group to be Kodiack.

The state of our knowledge respecting this group may be summed up in few words,—it is very imperfect and unsatisfactory. There is no apparent analogy between the remarks of any two observers. In Capt. Lütke's work he gives some details respecting them from the observations of *John Veniaminoff*, a priest who has visited them, and who has also given a sketch map of all this part, but, as it differs so much from all others, and those necessarily imperfect, no decision can be arrived at as to the comparative merits. Under these circumstances we shall confine our extracts to that of the Island Ounga, on account of the fact *of coal* existing on it.

OUNGA is the largest of all, and the westernmost of the group. According to the observations of *Stépanoff*, of the Russian Company, its North extreme is in lat. 55° 37′, that of its South part 55° 11′, and its length about 26 miles. Its breadth is about half its length.—(Memoir of M. Tébenkoff.) This island is mountainous and cliffy, particularly on its South coast, but the N.W. side extends in a plain, which terminates in the low cape called Tonkoï. The island has three bays: the largest, *Zakharovskaïa*, is on the N.E. side; it is open to the N.E., but the anchorage may be kept; here the vessels of the company formerly

* Müller's Découvertes des Russes, pp. 262—277.

wintered. There are some islets in its opening. The second, on the East side, penetrates considerably into the land, but it has but very little water. On this bay stands a village, called by the Russians *Delarovskoï*, and by the Aleutes, *Ougnagak*. The third is on the South coast. There are few lakes on the island, but there are as many as ten small rivers affording fish. On the shores there is a great deal of drift-wood to be found, though but few whales. The rocks are generally of a silicious character.

On the West side of *Zakharovskaïa Bay*, there are, in two places, *some beds of coal*, arranged in perfectly horizontal strata, at 100 yards above the level of the sea. They have commenced working them for use. On the North side much petrified wood is met with.

Upon the island are foxes and reindeer, and in the sea, cod, turbot, and navaga. The land produces bushes of alder and willow, and three or four species of bay. Turnips and potatoes grow well in the gardens of the inhabitants, who also raise pigs and fowls.*

Cook makes the channel which separates Ounga from the islands which line the coast of Aliaska at this parallel only 5 or 6 miles broad. When in the middle of this channel he observed the latitude to be 55° 18′, and it is according to this observation that Krusenstern placed upon his chart the islands which Cook states to be near the coast, but which are too distant from the Schumagins to be considered as belonging to them.

Capt. Cook says :—" I believe these islands to be the same that Behring calls Schumagin Islands, or those islands which he called by that name to be a part of them, for this group is pretty extensive. We saw islands as far to the southward as an island could be seen : they commence in the longitude of 200° 15′ East, and extend a degree and a half, or two degrees, to the westward. I cannot be particular, as we could not distinguish all the islands from the coast of the continent. Most of these islands are of a good height, very barren and rugged, abounding with rocks and deep cliffs, and exhibiting other romantic appearances. There are several snug bays and coves about them; streams of fresh water run from their elevated parts; some drift-wood was floating around, but not a tree nor a bush was to be seen growing on the land. A good deal of the snow still lay on many of them ; and the parts of the continent which showed themselves between the innermost islands were quite covered with it." †

Between the Schumagin Islands and the western extremity of Aliaska, the coast is bordered with a large number of small islands. Admiral Sarytscheff, who passed here, says in his journal that eight of them, of which he gives the names, are larger than the rest.

Nanimack Island, nearly the westernmost, is 4 leagues to the North of Sannagh (presently described). To the S.E. of it lie a quantity of small islets and rocks above water.

Animack, or *Reindeer Island*, lies 6 miles to the North of Nanimack. To the

* Lütke, Voyage du *Séniavine*, Part. Naut., p. 267-8.
† Cook's Third Voyage, vol. iii. pp. 412-13.

S.E. and East of this island there is a group of rocks and islets similar to those projecting to the S.E. from Nanimack Island.

Lialiuskigh lies to the N.E. of Animack, at the distance of 14 miles.

Two islands, without names, lie at the distance of 3 miles from this; one to the North, the other to the N.E.

Kuegdogh lies to the East, 2 miles off from the fifth island.

Kitagotagh lies to the E.S.E., at the distance of 3 miles from the last-named island.

Ounatchogh, 2 miles to the N.E. of the preceding; between these two last there is a high and pointed rock.

Cook passed these islands June 20th, 1778, and estimated their distance from the coast at 7 leagues. At noon on this day, being in lat. 54° 44′, Halibut Island bearing S. 65° W., he made some land to the northward, which he named *Rock Point*. It is apparently one of the islands lying near the coast, one of which in reality is in lat. 54° 59′, a latitude which corresponds with that of Rock Point.

Opposite to *Ounatchogh Island*, on the coast of Aliaska, is a very lofty *volcano*, the summit of which fell, in 1786, during an eruption. It is perhaps the same mountain that Cook saw emitting smoke:—"The rocks and breakers before mentioned forced us so far from the continent that we had but a distant view of the coast between Rock Point and Halibut Island; over this and the adjoining islands we could see the main land covered with snow, but particularly some hills, whose elevated tops were seen towering above the clouds to a most stupendous height. The most south-westerly of these hills was discovered to have a *volcano*, which continually threw up vast columns of black smoke. It stands not far from the coast, and in the lat. of 54° 48′, and the lon. of 195° 45′. It is also remarkable from its figure, which is a complete cone, and the volcano is at the very summit. We seldom saw this (or indeed any other of these mountains) wholly clear of clouds. At times both base and summit would be clear, when a narrow cloud, sometimes two or three, one above another, would embrace the middle like a girdle; which, with the column of smoke, rising perpendicularly to a great height out of its top, and spreading before the wind into a tail of vast length, made a very picturesque appearance. It may be worth remarking, that the wind, at the height to which the smoke of this volcano reached, moved sometimes in a direction contrary to what it did at sea, even when it blew a fresh gale. In the afternoon, having three hours' calm, our people caught upwards of a hundred halibuts, some of which weighed a hundred pounds, and none less than twenty pounds." This was in 35 fathoms water, and 3 or 4 miles from the shore. —(Cook's Voyage, vol. iii. pp. 416-17.)

SANNAGH, or HALIBUT ISLAND, which is the westernmost of those on the coast of Aliaska, received its second name from Cook, on account of the circumstance last quoted. It is separated from the coast by a channel 4 leagues in breadth. "This island is 7 or 8 leagues in circuit, and, except the head, the land of it is low and very barren. There are several small islands near it, all of the same appearance; but there seemed to be a passage between them and the main, 2 or 3 leagues broad."—(Cook.) *Halibut Head*, is a round hill in the centre,* which he made when 13 leagues off to the E.N.E.

Since Admiral Sarytscheff, who passed in sight of this island, June 20th, 1791, none of the Russian navigators have remarked it. It is bestrewed all around its circumference with naked rocks and islets. Admiral Sarytscheff makes particular mention of three of these last, which lie at the eastern part of the island, and says farther, that from the western extremity of this island a line of rocks above water extends towards the East (?) for a distance of 2 leagues. Cook did not see them.

The island is full of lakes, from whence rivers flow, chiefly to the South side, and are very abundant in fish. There are foxes, sea-calves, and a great quantity of birds of all species. There are but few spots for landing. The natives say, that at 30 versts (20 miles) to the S.E. or S.S.E. of Sannagh there is a bank with from 2½ to 9 fathoms of water on it; between the island and the bank the depth is 18 to 27 fathoms; there are also six smaller banks visible at high water. Sea-weed grows at times on them. No vessel sailing here has seen them, so their existence is doubtful.†

The STRAIT of ISANOTSKOY,‡ separating Aliaska from the Aleutian Islands, was known to exist prior to 1768; for subsequent to that year the Russian charts have shown it, though upon some English ones, that of Arrowsmith in particular, it was omitted until many years afterwards.

Isanotskoy Strait not only separates Aliaska from Ounimack, but it divides the latter from *Ikatun Island*, lying 3½ miles South of the S.W. point of Aliaska. The upper or northern part of the strait extends for 12 miles N. ¼ W., and S. ¼ E.; its breadth does not anywhere exceed 4 miles. At its northern extremity, that is, between Aliaska and the N.E. point of Ounimack (behind which lies Krenitzin Bay), the strait is only 2 miles broad; it is even narrower according to the account given by Krenitzin, and moreover is obstructed by a large number of banks. This is what he says:—The N.W. entrance of this strait is extremely difficult, on account of the sand-banks and currents which are felt during the ebb and flood tides. That to the S.E. is very much easier, and the soundings do not give less than 4½ fathoms.§ A shoal of considerable extent is marked on Krenitzin's chart in the middle of the strait, which almost fills up its entire breadth. On the chart by Khoudiakoff (which, with the former, are the only representations of any pretensions) this shoal is not found; but the N.W. point of Aliaska is here surrounded by rocks; the breadth of the entrance is, notwithstanding, not less than 2 miles; and as on the chart soundings are marked throughout its extent, it must be inferred that M. Khoudiakoff examined the strait in detail. Which of the two charts merits the preference cannot be decided until a fresh examination is made.

* Admiral Sarytscheff places Halibut Head in the N.W. part of the island.

† Capt. Lütke, p. 274.

‡ Capt. Lütke says that it is not Issanotsky or Isanotskoy. The name of the strait is the same as the Island *Sannakh* (Sannagh or *Issannakh*).—*Memoire*, p. 295.

§ Nouvelles Découvertes des Russes, by Coxe, p. 254.

The lower part of the strait, that is, the portion between Ounimack and Ikatun Island, is 8 miles long by 4 broad. This breadth, however, is contracted by one half by *Kitenamagan Island*, lying half a mile from Ikatun.

From the North end strait of Isanotskoy, the coast of Aliaska runs to the north-eastward, in nearly the same direction as the southern coast of that peninsula. This will be described in the next chapter, in connexion with the remainder of the coasts of the Sea of Behring. The Aleutian Archipelago, forming, as it were, a broken continuation of the peninsula, will follow.

ALEUTIAN ARCHIPELAGO.

It is to the celebrated Behring, as we have mentioned regarding the Kodiack Islands, that Russia owes the discovery of the Aleutian Islands. It was during his return from the coast of America in 1741 that he discovered several of them, now known under the names of Semitsch, Kiska, and Amtschitka. In 1745 an enterprising merchant, named Basoff, made a voyage hither in search of the sea-otters which he had heard were abundant on the shores. After this period they were more frequently visited, and they daily became better known.* The geographical positions of the group we owe principally to the Russian Vice-Admiral Sarytscheff, who accompanied Capt. Billings in his expedition in 1791-92. He determined, by astronomic observations, the positions of the greater part of the islands, and constructed charts of many of them, among others, a detailed survey of his own of the Island of Ounalashka, which is up to the present the only one we possess. To Capt. Cook, too, we owe some observations on this island and some others near it. Capts. Golownin and Kotzebue, in the years 1817 and 1818, determined the position of some of their points. Capt. Lütke has given a long article upon this archipelago, drawn from the journals of several Russian navigators, who have, at different periods, visited these islands, and particularly from the observations of Lieut. Tébenkoff, and M. Inghestrōm, an officer in the service of the Russian-American Company.† Many of these remarks are considered by Admiral Krusenstern not to be of sufficient authority to carry much weight; still, in our general ignorance of this archipelago, they become valuable. Capt. Beechey has also added slightly to our knowledge of them; and several other navigators, whose names will be alluded to, have added something to the general stock. The islands of the group that are nearest to Kamtschatka are the least known, and the archipelago, as mentioned relative to the Kodiack Islands, are in the possession of the Russian-American Company, who have some establishments upon them.

The Aleutian Islands form a chain, which extends nearly East and West from the Isle of Attou, in lon. 172° 45′ E., to the peninsula of Aliaska, comprising an extent of 23° of longitude, and between 51° and 55° of North latitude.

They have been divided into several groups. The western or Blignie group,

* See Coxe's Russian Discoveries; Pallas, in his Northern Memoirs, and Journal de St. Petersbourg, 1781-82.

† Voyage du *Séniavine*, Part. Naut., pp. 279—330.

is composed of four islands: Attou, Agattou, Semitsch, and Bouldyr. Another group is named Rat Islands; a third, the Andréanoff Isles; and the eastern group, the Fox Islands, because these animals are only found on the islands composing that particular group. Krusenstern considers these subdivisions unnecessary; he comprehends them all under the one title of Aleutian Islands, as more simple and more convenient. Capt. Lütke remarks that these distinctions, though not absolutely necessary, assist the memory, and have some advantages. He therefore follows the divisions formerly adopted.

On all these islands traces of volcanic action are evident. On many of them there are volcanoes in activity, and some, as for example, Ounimack, are subject to continual volcanic eruptions and shocks. The Fox Islands exceed all the others in height; the farther we advance to the West the lower they become. The direction in which almost all the islands of the Fox group lie, lengthwise, is S.W. to N.E. They are low and narrow to the S.W., and increase in breadth and elevation to the N.E. But beyond the Island of Amkhitka, where the general direction of the chain runs to the N.W., this law alters, and the S.E. extremities of the island are lower and narrower, and their N.W. extremities higher and broader.

We owe the subsequent descriptions to Admiral Krusenstern's Memoires Hydrographiques de l'Atlas de l'Ocean Pacifique, 1827, part ii. pp. 75, *et seq.*, and Supplement, 1835; Capt. Lütke's Voyage du Séniavine, 1836; and other authors as quoted.

The following description commences with the easternmost of the Archipelago, and proceeds westward in succession.

FOX ISLANDS.

This group, extending from Ounimack to Amoukta, is the most important of the Archipelago, commercially, on account of the produce of the chase which is annually drawn from them; and, geographically, from their central situation, and the ports they contain. They are better known than the others, but much less so than would be supposed, from the number of seamen who have visited them. But the one object of these visits has merged all other considerations, and the world has benefited but little from the acquaintance.

OUNIMACK is the easternmost of the group, and is separated from the Peninsula of Aliaska by the Strait of Isanotskoy. In a harbour in this strait Krenitzin wintered in 1768. Krusenstern's delineation of this island is chiefly drawn from the operations of M. Khoudiakoff in 1792, and is probably very erroneous. In 1826 Capt. Beechey passed through the strait separating it from the islands to the southward. He calls its S.W. point *Wedge-shaped Cape*, before which lies a rock, and the narrowest part of the strait is formed by the Isle Kougalga; Beechey making the distance 9½ miles, while Kotzebue makes it 16 miles, and 25 miles between Ounimack and Akoun Islands. Beechey's observations have been, in some degree, confirmed by others. These and other discrepancies render it very desirable that a more complete knowledge of these islands should be obtained; because, although this strait may not be the best for passing through the chain to the northward, yet, in going from Behring's Strait or Ounalashka to the southward, it is to be preferred. Krusenstern has called it *Rurick Strait*, after Kotzebue's vessel.

The southernmost point of Ounimack is called in Krusenstern's chart *Cape Hitsou;* by Capt. Lütke, *Cape Kithouck* or *Khitkhoukh;* and, from its shape, by Capt. Beechey, *Wedge-shaped Cape.* From this southern cape the coast runs to the N.E. to *Cape Lütke*, beyond which to the eastern strait the coast has not been examined. The S.W. point of the island is *Cape Sarytscheff*, so named by Capt. Stanikowitch, who found before it a large rock similarly situated with respect to it as that of the southern cape. From this cape Capt. Stanikowitch followed the coast to another cape in lat. 54° 51′ North, which he called *Mordvinoff*, beyond which the coast still remains unknown.*

There is some doubt as to the true size of the island; Krusenstern makes it 50 miles; Capt. Stanikowitch, in the *Moller*, makes it 65 miles in a N. 52° E. and S. 52° W. direction, and its greatest breadth about 25 miles. It is, so to speak, but the cover to a furnace, continually burning: on the summit of this a high mountain chain extends throughout the island, having several spiracles, by which the pent-up fires find vent, and eject, without cessation, their burning products. Notwithstanding the number of craters, this subterranean fire causes frequent earthquakes. The highest of these summits, the *Chichaldinskoï Volcano*, was measured by Capt. Lütke as 8,935 English feet high. It is a regular cone; and to the East of it is another, with a double summit. It stands nearly in the centre of this island, in about 54° 45′ and 163° 59′. At 6 miles from the S.W. side is another equally conical volcano, called *Pogrommoï*, or *Nosovskoï*, which Kotzebue says is 5,525 English feet in height. Cook mentions this as being entirely covered with snow in July. The *Issannakh Chain* has also two high peaks towards the N.E. extremity of the island. The whole of the mountain chains are nearer the South than the North side.†

A broad bed of gravel forms the N.E. extremity of the island, and a low coast extends as far as the village of Chichaldinskoï without any shelter. *Chichaldinskoï Village* is two-thirds the distance from the N.E. extremity to a cape 3½ miles East from Cape Mordvinoff. The land is low and level here, and a river yielding abundance of fish flows past the village. Some vegetables are cultivated by the inhabitants. Beyond this the coast is higher. The N.W. extremity, *Cape Mordvinoff* (*Cape Noïsak*), is in lat. 54° 51′, lon. 164° 29′. From *Cape Chichkoff*, which is bluff, and very remarkable, because the land on each side of it is very low, to the West extremity of the island, *Cape Sarytscheff*, the coast forms the base of the *Pogrommoï Volcano.* The latter cape is bluff, and of a moderate height. At 7 or 8 miles from it, on the summit of the coast, is the village of Pogrommoï; off it there is a boat-landing. Much drift-wood, sea-weed, and animals, are thrown on the shore here. At 4 or 5 miles S.E. of Cape Sarytscheff is the small village of *Nosovskoï*, where there is easy landing. At about 6 miles to the S.E. of this is the high steep, *Cape Khitkhoukh*, which is to be known by a high pile of stones before it, and which is called *Ounga.* This cape is mentioned above, and to the N.E. of it, which is the direction of the

* Krusenstern, p. 95.

† Lütke, Voyage, &c., Part. Naut., pp. 292-3.

South coast, there is considerable difficulty in reconciling the different authorities.* We will therefore let it pass.

RURICK or OUNIMACK STRAIT, through which Kotzebue passed in 1817, separates Ounimack from the Krenitzin Isles to the southward. Its narrowest part, as before mentioned, is about 9½ miles broad. The currents are very violent in it. Capt. Beechey found them to be at the rate of 3 miles an hour to the S. 33° W., and mentions an American who experienced one of 6 miles an hour.

Although Rurick Strait is the widest and safest for traversing the Aleutian chain from North to South, and *vice versâ*, and also the most convenient for passing into the northern part of the Sea of Behring, it is not so advantageous for ships which, coming from the Pacific, are destined for Ounalashka. They are then obliged to make a circuit of nearly 20 leagues; and after having cleared the strait, they must run at least 7 or 8 leagues to the N.W. before they can bear up for the northernmost cape of Ounalashka. For this reason Capt. Wrangel recommends the Strait of Akoutan, farther westward, for this route.

KRENITZIN ISLANDS.—The islands next in succession to Ounimack, Tébenkoff calls the Krenitzin Islands, from the navigator who first saw them. They are five in number, and were first correctly placed by Kotzebue, though there is a detailed description of them by a Russian navigator, Solovieff, in the relation of his voyage in 1770 to 1775.† The N.E. of them, called by Krenitzin *Kougalga*, in reality consists of two islands, *Ougamock* and *Ouektock*. They are very close to each other, and the separation was not observed by either Cook or Beechey. There is a peak on the N.E. extreme of Ougamock.

TIGALGA, or KIGALGA, or TIGALDA, is the next to the S.W., and is about 4 leagues long, East and West. Its centre is in lat. 54° 5′ N., and lon. 165° 0′ W. A small island, connected by a chain of rocks to Tigalda, lies off its northern extremity.

Tigalda is mountainous, and intersected by three isthmuses. Sea-lions and sea-calves frequent the island, and a large quantity of birds' eggs are collected. The sea throws a great quantity of drift-wood on its coast, and COAL is found on the shore of Derbinskoï Strait. There is one village on the N.E. side of the island.

Derbinskoï Strait separates it from Abatanock, and is remarkable among all the others for the extraordinary rapidity of the current and its strong tide races.

ABATANOCK lies West of Tigalda, and is about the same size, and lies in the same direction. At 2 miles West of the western point of Abatanock is the small island of *Aektock*, or *Goly* (bare), which is about a league in circumference, and lies 2 miles South of the South point of Akoun. Beside these five islands thus described by Kotzebue, there is a sixth, mentioned by Solovieff, called *Nangarnan*, which ought to lie to the S.E. of Tigalda, but is not so placed on Kotzebue's chart; therefore it must be presumed that he did not see it, and it must be sought for elsewhere.

The ISLAND of AKOUN forms the S.W. portion of Rurick Strait, and lies to the West of the Krenitzin Islands. It is about 14 miles long, in a N.E. and

* Lütke, Voyage, &c., Part. Naut., p. 293—295. † Journal de St. Petersbourg, 1782.

S.W. direction; its breadth is unequal. There are two small bays on it, one in the N.E. part, the other in the N.W. part of the island.

The island is mountainous, and particularly cliffy on its N.E. and North sides. On the South side of the island is a sort of column, which, seen from the East or West, resembles a tower, or steeple surrounded by houses. On its N.W. side is a smoking volcano, and near the village on the side of Akounskoï Strait, are some hot springs. There are only three villages in the island, and the Russian Company have attempted to raise cattle on it.

Akounskoï Strait, separating Akoun from Akoutan, is not more than 2 miles wide, is throughout bestrewed with rocks, and subject to strong currents and tide races.*

AKOUTAN lies half a league to the West of Akoun. Kotzebue sent Lieut. Chramtschenko around it in baïdars, and we therefore possess a tolerable knowledge of its coasts. It is large, mountainous, of a round form, and having a diameter of 12 or 13 miles. It has no good harbour; there are some coves on the northern coast, but they cannot be serviceable to any but very small vessels.

With the exception of Ounalashka, it is higher than the neighbouring islands. Nearly in its centre is an active volcano, measured by Capt. Lütke as 3,332 feet. The coasts are steep, particularly on the South side: on the North they slope more gradually and evenly. Volcanic evidences are everywhere abundant. Between its S.W. end and the opposite shore of Ounalashka is the Island of *Ounalga*, the same which Cook calls *Oonella*. It is immediately before the Bay of Samganooda, 2 miles from the N.E. point of Ounalashka.

The ISLAND of OUNALASHKA, which is the largest and the best known of the Aleutian Archipelago, follows. It extends 70 miles from N.E. to S.W. The S.W. extreme is in lat. 53° 13′, and lon. 167° 47′ W.; and the N.E. part in lat. 54° 1′ N., and lon. 166° 22′ W.

The name here given as generally known to Europeans is a contraction of *Nagounalaska*, the correct name. It is the most important of the group, because it is the residence of the chief of the section of the Russian Company's hunting operations.†

There are many deep bays on the coasts of Ounalashka, which have nearly all been examined and surveyed by our navigators. The northern shore has the greatest number, as for example:—Captain's Bay, the Bay of Otters, Illuluck Bay, Kaleghta Bay, and Samganooda Bay, visited by Cook.

CAPTAIN'S BAY was so named, because Captain Levacheff was obliged to winter here in 1768-69. There is a plan of this bay, on a large scale, in the atlas accompanying Kotzebue's voyage. It is formed by Capes *Kaleghta* and *Wessiloffsky*, which lie in an E.N.E. and W.S.W. direction, 9 miles from each other, and is about 13 miles to its southern part. The upper part of the bay contains three distinct smaller bays, the eastern, northern, and western bays. Capt. Levacheff wintered in the southern, to which Krusenstern has given his name, the better to distinguish it. It has not much to recommend it; its entrance, scarcely more than a quarter of a mile broad, is formed on the West by a projecting point of the coast, and to the East by the southern point of the Island

* Lütke, p. 290.

† *Ibid.* p. 280.

of Oumacknagh. A small island, a mile in length, named Ouknadagh, lies to the North of the entrance of Port Levacheff.

The eastern bay bears the name of PORT ILLULUCK, from the village of that name, where the company has an establishment. Admiral Sarytscheff and Capt. Kotzebue, who visited it, have given a detailed description of it. The latter says that it would be the best harbour in the universe if the entrance to it was not so difficult; but the great depth of water in the outer or Captain's Bay presents great difficulties for a vessel entering Port Illuluck; if it should fall calm, she would remain exposed to the violent currents and squalls which often occur here. *Oumacknagh Island* forms the western side of the port, which, like the island itself, has a N.E. and S.W. direction; its depth is 3¾ miles, and its opening formed by the northern part of Oumacknagh, and the coast opposite is 2 miles wide. It lies 7 miles South from Cape Kaleghta, and communicates with Port Levacheff by a narrow channel 100 yards in breadth. The soundings in the centre of Port Illuluck are from 7 to 14 fathoms.

The mean of the observations made by Kotzebue places the port in lat. 53° 52′ 25″ N., and lon. 166° 32′ 0″ W. Variation of the compass, 19° 24′ E. The establishment of the port, 7h 30′; the highest tide observed, 7 ft. 6 in.

KALEGHTA BAY, which lies next, to the eastward of Illuluck, is open and deep, and only merits attention on account of a village of the same name at its head, at the mouth of a small river flowing from a lake, in which is an abundance of fish of a species of salmon.

The BAY of OTTERS, or BOBROVAIA, adjoins Kaleghta Bay on the East, and is the largest of those which intersect Ounalashka, being 18 miles deep in a N.E. and S.W. direction. Its breadth, as well at the mouth as the rest of the bay, excepting the North part, is about 4 miles. It was minutely examined by Admiral Sarytscheff in baïdars. Both shores of the bay present a large number of small coves, of 2 or 3 miles deep, which contain good anchorages; the rivulets which fall from the mountains afford good water. The western part of the Bay of Otters is formed by a peninsula, which is the same land which forms the eastern side of Captain's Bay. At the extremity of this peninsula is Samganooda Bay, where Cook anchored twice. Cook found the variation to be 19° 59′ 15″ E., or half a degree more than Kotzebue found it forty years later. It is high water, full and change, at 6h 30′, rise and fall, 3½ and 4 feet.

Judging from the description given by Cook, this port is preferable to Port Illuluck. It is 4 miles long in a S. ½ W. direction, and affords safe anchorage throughout its extent; the mouth of the bay is a mile in width, and it narrows toward the bottom to a quarter of a mile, where you may anchor in from 4 to 7 fathoms, on a bottom of sand and mud, being entirely land-locked. This port has also the advantage of being nearer the open ocean, and has not therefore the inconvenience complained of by Capt. Kotzebue when speaking of Illuluck, as mentioned before. The principal merit of this bay consists in the fact that here there is an establishment of the Russian American Company.

The bay which Cook afterwards entered, but which he was obliged to quit directly, on account of the great depth of water, is called *Kaleghta*, and lies some miles to the West of Samganooda. It is mentioned above.

The *Island of Ounalga* (Cook's *Oonella*) lies before Samganooda Bay, as mentioned before. Between it and the Island of Akoutan, to the N.E. of it, is the *Strait of Akoutan*. Capt. Wrangel recommends this strait to be preferred for passing between the islands, because it leads directly to all the ports lying on the N.E. side of Ounalashka. This strait is 2½ miles broad; but it is somewhat narrowed by a chain of islets lying half a mile off the N.E. part of Ounalga. These islets are called, in Capt. Lütke's Memoir, *Egg Islets*.

Spirkin Island, which forms the eastern point of the Bay of Otters, is 10 miles long in a N. ½ E. and S. ½ W. direction. The *Oudagagh Channel*, which separates it from Ounalashka, is about a mile wide, and 3½ miles long, in a N. by W. ½ W. direction. The depth in it is 45 fathoms.

At the N.E. end of Spirkin Island lies the small isle *Ougalgan*, being separated by a clear channel about a mile long. It was traversed by Cook, Sarytscheff, and Kotzebue.

There are two rocks near the eastern coast of Spirkin Island, between which Cook passed in the night of June 26th, 1778. According to Cook the first rock lies 4 miles S. ½ E. from Ougalgan Isle; the other at 5 miles directly to the South of this isle; their distance apart is 1½ leagues, and the relative bearing N.E. and S.W.*

Up to the present time the eastern shores of Ounalashka have remained unexplored, with the exception of the small bay of *Killiliack*, which was examined by Admiral Sarytscheff, and of which he has given a plan in the account of this voyage. It is easily known by a remarkable cape, named *Amtschitka*, standing a little to the North of the entrance to the bay. Besides this there is another mark which will point out the situation of the bay in approaching the land on its parallel. Ounalashka here presents the appearance of being divided into two parts. The Bay of Killiliack is perfectly sheltered from all winds. Its mouth is half a league wide; farther inside it narrows to 250 yards. The depth is about 10 fathoms. Admiral Sarytscheff is the only navigator who has visited it.

The western side of Ounalashka has been examined in detail as well by Admiral Sarytscheff as by Kotzebue, in baïdars; we are therefore intimate with all the indentations, which are here in great number on the coast.

Makouchinskoy Bay lies about the middle of the island. It is 2¾ miles wide at its opening, and 11 or 12 miles long in an easterly direction. In it there are many coves, which may prove to be good anchorages. The head of this bay approaches that of the Bay of Otters within 3 miles, and within 7 miles of Captain's Bay, in such a manner that this portion of Ounalashka forms a peninsula of 40 miles in circumference, composed of high mountains, among which a very high volcano is to be distinguished. There are still several bays on the western shore, which we have not mentioned, that are given in the charts of Admiral Sarytscheff and Capt. Kotzebue.

OUMNACK ISLAND lies next West to Ounalashka, and, next to that island, is the largest of the Archipelago. The strait which separates them is 4 miles wide

* Cook's Third Voyage, vol. ii. p. 240.

in its southern part; but this is diminished to one-half by Tinginack Island, which lies in mid-channel; this renders the passage difficult for large vessels. Outside the strait, at 6 miles to the South of Tinginack, is a reef, which covers at high water. This was first shown on Kotzebue's chart, and must increase the danger of this navigation.

Oumnack is nearly 20 leagues in length, in a S.W. by S. and N.E. by N. direction. Its height increases in the same direction; and its North end is composed of very high mountains, among which may be distinguished a very high volcano, covered with eternal snow. Cook saw this island October 29th, 1778, some days after quitting Ounalashka, and calls it *Amoughta.*

Upon the island are two active volcanoes, the first, *Vcevidovskoï*, is nearly in the centre of the island, and is its highest point; the other, *Toulikskoï*, is 10 miles from the N.E. side. The S.W. extremity of the island, *Cape Sigak*, lies, according to Capt. Kotzebue's chart, in lat. 52° 50′, and lon. 168° 42′. A short distance from this the S.E. coast forms some small open bays, one of which is called the *Old Port*, which is somewhat sheltered from the South from seaward by a bank. Beyond this is the *Black Cape*, projecting considerably into the sea, and forming the open cove called *Drovianaïa* (wood), on account of the great quantity of drift-wood thrown on to it. Beyond this the coast runs nearly straight, and not very high, to *Vcevidovskaïa Cove*, open to the South, before which lie the Vcevidovskaïa Islands, mentioned presently. Here the coast is low and sloping, and thus extends to *Gloubokaïa* (deep) *Cove*, into which a river discharges itself, abounding with fish. Farther to the N.E., beyond a mass of rocks 90 feet in height, inclining to the N.E., is the village *Egorkovskoï*, in a small creek midway between Cape Sigak and Oumnack Strait. The neighbourhood of the village affords great resources; the grass flourishes, and potatoes and turnips are cultivated. A rude, sandy, and straight coast surrounds this as far as the village *Toulikskoi*, lying in front of the islet *Tanghinakh*, in the Strait of Oumnack. Near the S.E. coast there are many reefs and banks.

The eastern face of the island facing Ounalashka is steep and rocky in some places, but is not high. The North part is high, sandy, and even, frequently intersected with ravines, but without a single remarkable inlet. The West coast is mountainous, but not steep. On this side, at 8 miles from the S.W. extremity, is the largest village of the island, *Retchechnoï*, standing on a small hill between some lakes inland and the sea-shore. Before it is a small and safe harbour for small vessels. On each side of the narrow entrance is a rocky islet, one of which is called *Anangouliak*. Fish is abundant here.

Nearly in the middle of the West side of the island is the large but open bay called *Ingakoadak*.

On the S.E. side of the island, and in front of the Vcevidovskoï volcano and the bay of the same name, are situated the small islands called also *Vcevidovskïes.* They are six in number, and are 2 miles off the coast, the interval being full of banks. On Kotzebue's chart, there are but two, the largest named *Ouegakh.*

Oumnack, like the rest of the islands, is deficient of wood, some willow and other bushes only growing on it. It is, next to Ounimack, the most subject to volcanic eruptions. One feature is an evidence of this; it is the abundance of hot springs, one of which resembles the Geysers of Iceland. Capt. Lütke gives an account of many of these phenomena, and similar may be found in Kotzebue's and in Langsdorff's travels.*

To the northward of Oumnack is a long reef, stretching for 26 miles in a nearly North (*true*) direction, at the outer point of which is the *Ship Rock*. It was so named by Cook, and is in the form of a tower.

At 200 fathoms within the Ship Rock is the small island of JOANN BOGOSLOFF. It is of volcanic origin, and did not appear until 1796, after an earthquake. The length of this small island, from N.W. by N. to S.E. by S., is 1⅜ miles. Its breadth is about half its length. A chain of rocks projects 2 miles beyond its N.W. extremity, and another a mile from its N.E. point. According to the observations of Capt. Wassilieff, the peak in the centre of the island is 2240 feet high.† This island, as before stated, is connected with Oumnack by a reef of rocks, which doubtless owe their origin to a similar cause; for in 1778 Cook, and thirty years later Sarytscheff, sailed between the Ship Rock and the Island of Oumnack.

To the westward of this island is a group of four—or, according to Lütke, five—volcanic islands, which bear the name of the *Isles of the Four Mountains*; they are all very near to each other. Kotzebue saw but three. The name of the S.W. isle is *Ouliaga*; of the N.W., *Tano*; the N.E. is called *Tschiganock*; and that to the S.E., *Chagamil*.‡ The first and the last are the largest, being 5 or 6 leagues in circumference. There are also two islets to the North of the group, but they have no name.

YOUNASKA.—To the S.W. of these islands is Younaska, which is about 5 leagues from N.E. to S.W. According to Capt. Kotzebue, there is a high mountain in the centre of the island. *Tschegoula*, or *Tchougoul*, a small island, lies West of Younaska, and near the N.E. point of Amoughta. Kotzebue says it is of a circular form, and 3 miles in diameter.

It seems as if formed of fragments of rock ready to fall down, and has no landing place. At about a mile from it, in the direction of Amoughta, is a small isolated rock.

AMOUGHTA, or AMOUKHTOU, is the westernmost of the chain of the Fox Islands. It is nearly round, and about 6 miles in diameter. Its centre is mountainous,

* Lütke, Voyage, &c., Part. Naut., pp. 298—302.

† Admiral Krusenstern adds the account of this phenomenon from a report, dated June 10th, 1817, from Mr. Baranoff, chief of the American Company's establishment:—"In 1806, a new volcano appeared on one of the Aleutian Islands, and on May 1st, in that year, a violent tempest from the North occurred, and during its force a rumbling noise, and distant explosions, similar to thunder-claps, were heard at Ounalashka. At the commencement of the third day the tempest abated, and the sky became clear. They then observed, between Ounalashka and Oumnack, to the North of the latter, a flame jetting out of the sea, and soon after, smoke, which continued for ten consecutive days. After this, a white body, of a round form, was observed to rise out of the water, and increase rapidly in size. At the end of a month the flame ceased, but the smoke increased considerably, and the island kept on increasing. On June 1st, 1814, they sent a baidar to examine it, but they could scarcely land, on account of the violent currents and the pointed rocks. The island was formed by precipices, covered with small stones, which were being continually ejected from the crater. In 1815, a second expedition found the island very much lower than in the previous year, and its appearance entirely changed. The precipices had fallen, and were continually crumbling away."

‡ Krusenstern, Memoire, &c., p. 86. Lütke names them differently: *Kigalga*, *Kigamiliakh*, *Tanakh-Angounakh*, *Oulliayhin*, and *Tchegoulak*; this is according to the priest John. Tébenkoff and Sarytscheff each give different designations to them.—See Lütke, p. 297.

and its summit irregular. The coasts are low, but steep. A short distance from its South end a high column of rock rises above the water. There is neither bay nor river on it, and although formerly in volcanic activity, its fires are extinct.

All the islands of the Fox chain, to the West of Oumnack, are now uninhabited, but formerly were so. They are all more or less frequented by otters, sea-horses, seals, and birds.*

The CHANNELS between this group are those generally used by ships either going or returning from the Sea of Behring. The Russian Company's vessels generally prefer that of the Strait of Ounimack; but Capt. Wrangel prefers the Strait of Akoutan, between that island and Akoun, as being much shorter. In returning by this strait, with the prevalent S.W. winds, you may run to the S.E. without being cramped by the coasts. By the Ounimack or Rurick Channel, the length of coast is much greater, and it is also embarrassed by the dangerous Sannakh Island.

The *Strait of Ounalga* should not be used except in case of necessity, on account of its narrowness, its strong currents, and its terrible tide races, which, when wind and waves oppose each other, are sufficient to dismast a vessel.

The *Strait of Akoutan* is considered by Lütke the best to quit by. It is 2 miles broad between Akoutan and the five *Tchaitchi* (sea-mew) *Islands*, which are of an oblong form, distant half a mile from the N.E. point of Ounalga, and hitherto have not been shown on the chart. Care should be taken of these islands in coming from the southward, as a mistake might be fatal.

Vessels do not use any other than the three channels here mentioned, but they may be attempted. With respect to the rapidity of the currents in the different straits, they may be placed in the following order: that of *Derbinskoi*, the most rapid, and subject to tide races, the current frequently running at 8 miles an hour; then comes those of *Ounalga*, *Akoun*, and *Abatanock*; and lastly that of *Ounimack* or *Rurick*, the most tranquil, and never with any overfalls, the current not exceeding 4 knots.†

ANDRÉANOFF ISLANDS.‡

This group extends from Segouam, or Sigouam, to *Goreli*, or the "Burnt Island," as Lütke also calls the first-named island.

There is a larger interval between the foregoing islands and the next to the westward, than between themselves, the distance from Amoughta and Segouam being 1° 9′ West in longitude. This distance (55 miles) is thought by Lütke to be too great. From the imperfect knowledge that Admiral Krusenstern could collect concerning them, this may be true; but as no data was given for correcting it, they have been left by him as before.

SEGOUAM, or *Goreli*, according to the promychlenniks or hunters, is the easternmost of the group. It is nearly double the size of Amoughta, though on the charts it is made smaller. It is intersected by a chain of mountains, divided into three masses, smoke issuing at times from the central mass. In the N.E.

* Lütke, p. 304. † *Ibid.* p. 306. ‡ Or Andréanoffsky, according to Lütke.

part they rise perpendicularly from the water. There are neither rocks, shoals, nor dangers around it. Birds are abundant. There are many hot springs and vapour apertures, and on the West side there are some convenient landing places for baïdars.

The *Amoughta Channel*, to the East of it, is clear, safe, and has the same currents as the rest. The channel to the West, between it and Amlia, is far from being as convenient. From the extent of the latter island, and also that of Segouam, a barrier of nearly 100 miles is opposed to the periodic current passing between them, and consequently the water rushes violently from either side through the strait, causing terrible and dangerous races.*

AMLIA, which succeeds Segouam, to the westward, is long and narrow, and extends nearly East and West, *true*. Its length, according to Lütke, is not exactly determined; from the charts, it is 40 miles. The centre of the island is occupied with a chain of mountains, the greater part of a conical form, but, compared with the others, of a moderate height. There is no active volcano on it, and its shores are in general clear. The West cape projects to N.N.W., and, according to observation made on land, is in lat. 52° 6½′. On the South coast, about a mile from the West end, is a large open bay, and an Aleutian village. At this part the island is not more than a verst (two-thirds of an English mile) in breadth, but it is broader in the middle. The South side forms several bays, but all are open save one, *Svetchnikoff Harbour*, described by the pilot Tchernoff, in 1832. According to the Memoir of M. Tébenkoff, this port is 16 miles from the East extremity of the island, and penetrates 1½ miles to the N.N.W., and is about half a mile broad. It is sheltered from seaward by a small, narrow island, formed of rocks, about three-quarters of a mile in length, North and South, half a mile South from the East cape of the port. This space is full of rocks, so that, to enter, this island must be rounded to West. Inside there are 14 fathoms water, and farther inside, 6 to 4½ fathoms, sandy bottom. A high rock, lying South 82° East, at 9½ miles from the entrance, and 2½ miles from the nearest coast, will point out the situation of this harbour.†

The greater part of Amlia is covered with high grass, among which are found some edible roots; some potatoes, &c., are also raised.

At 5 miles N.N.E. from the East point of Amlia, M. Klotschkoff, an officer of the Russian marine, commanding a small vessel, the *Tschirikoff*, for the American Company, discovered a high rock, of 20 fathoms in diameter, on which he found an immense number of sea-calves. This rock was named by Krusenstern after its discoverer. But in Lütke's Memoir, p. 308, quoting M. Inghetröm, it states that this rock does not exist in that position. It must therefore be sought for elsewhere. Nevertheless, until it is refound, Admiral Krusenstern retains the position assigned by M. Klotschkoff.‡

The *strait* between Amlia and Atkha is not more than 1½ miles broad, is still farther narrowed by some rocks, and cannot be passed by a sailing vessel, on account of the furious currents.

ATKHA, or ATCHA, the largest and principal of this group, Lütke says is

* Lütke, pp. 307-8. † *Ibid.* pp. 308-9. ‡ Krusenstern, Memoire, p. 85.

the worst represented on the charts, being placed 10′ too far North. Admiral Sarytscheff passed along its South coast, and Capt. Golownin on the northern.*

The length of the island, from the S.W. extremity to the farthest point to the N.E., is more than 50 nautic miles. And here we find the same feature so general in the eastern group of a narrow and low S.W. extremity, enlarging and increasing in height to the N.E. Like the Peninsula of Makouchin, on Ounalashka, the North part of Atkha also forms a peninsula, crowned with high mountains, the northernmost of which is the active and smoking volcano, called *Korovinskoï*, lying on the North coast, and 4,852 English feet in height. Seen from the N.E., it presents two summits, and on the North is very steep, and the shore inaccessible. Four miles to the South rises the volcano of *Klutchevskoï*, and not far from the N.E. extremity is another. The base of the Korovinskoï volcano advances to the North into the sea, forming a rocky escarpment, which is the North extremity of the island. To the East of this cape the coast trends to the S.E. On the other side it runs nearly on the parallel without curvature to the N.W. extremity, called *Cape Potaïnikoff*, and forms a low, even, flat point, and dropping perpendicularly, which M. Inghestrõm compares to an artificial mole.

Cape Potaïnikoff† is thus called on account of a reef extending thence 2 miles to the W.N.W. On the cape is a steep, conical volcano. From this the high and cliffy coast runs to the South, to the low and narrow isthmus of *Sergheïeff*, which connects a smaller peninsula, also called Sergheïeff, to the larger one. Its West extremity, *Cape Korovinskoï*, rises out of the sea in a steep cliff, with many slips. From most of the headlands here rocks extend to 4 cables' length off.

Korovinskaïa Bay, which opens to the West, is formed by the large peninsula and the connecting isthmus, which is called *Amlinskoï*, because it faces Amlia. Two coves, which penetrate on the North side of this bay, and are covered by the Sergheieff Peninsula, form the *harbour* of Korovinskoï, which is perfectly sheltered, and only has the grave objection of an extremely difficult entrance. It is formed by two very low spits of gravel, which run over to a great distance, leaving only a very tortuous and narrow channel between, which must be well buoyed to be passed.

The inconvenience of the place would not allow M. Inghestrõm to make regular tidal observations. But they were found to be very irregular and much influenced by the wind. It is high water on full and change between one and two o'clock. The range of the tide was about 4 feet; in spring and autumn 5 or 6 feet at most. Sometimes there was but one tide in the twenty-four hours.

The outer Bay of Korovinskaïa is 6¼ miles broad in its opening between Cape Korovinskoï and *Cape Iaïtchnoï*, bearing S. 16° E., *true*, from it. There is only anchorage on the North side, in 14 fathoms, before the entrance to the harbour. It is without danger in the fine season, but in autumn terrible gales passing from N.W. to N.E. occur, when a ship could not remain here; and at such times the entrance is covered with breakers for several weeks together.

There are two coves on the South side of Korovinskaïa Bay; the nearest to the

* Voyage de Capt. Golownin, tome i. p. 172, orig.

† Rocks beneath the surface of the water, which show by *breakers* only at intervals, such as at low water, or at spring tides, or at equinoctial tides, are called *Potaïniki* in this country.

isthmus is *Pechtchunaïa* (sandy) *Cove*, the West point, which bears S. 67° E. by compass from the S.W. extremity of the Sergheïevsky Peninsula. This cove, in which there is very little water, and is open to the North, merits no attention whatever in a maritime view; but on its East side, near the entrance, lies a rock, an enormous mass of (fossil) wood, the bark and branches of which may be very clearly distinguished. This wood is of a grayish colour, and does not burn in the fire like coal; it consumes slowly, and therefore is of no use; but it leads to the supposition that a careful search would lead to the discovery of true *coal*, which would be of very great importance.

The other cove, *Sarannaïa*, is to the S.W. of the former, and is the only place where a ship can anchor conveniently and also may procure water.

At 6 miles S. 52½° W. from Cape Iaïtchnoi, 2 or 3 miles off shore, is the solitary Island *Soleny* (salt): it is small, and not high, and between it and the cape is the long *Staritchkoff Reef*. Beyond Cape Iaïtchnoi the North coast of Atkha curves to the S.W., and forms several bays; the first is *Iaïtchnaïa Bay*, lying beyond the above reef, and connected with it is *Obetavannaïa Bay*. They are both open and unimportant. To E. by S. of Soleny Island, and to the South of a cape E. by N. from that island, is the deep Bay of *Gloubokaïa*, which, it is said, affords excellent shelter. To the S.E., or S.E. ¼ S. from Soleny Island, is a land-locked bay, in which are some excellent harbours. Inside it separates into two, one named *Bannerskaïa*, after M. Banner, mentioned by Langsdorff, and the other like it called *Pestsovaïa*.

Cape Tolstoï projects considerably into the sea. On its East side is an open bay; on its West side is *Koourovskaïa Bay*, extending first S.E. and then E. by S. It is 24 or 25 miles from Korovinskoï Harbour, and, in fine weather, is serviceable, but in bad or foggy weather its entrance is difficult. To reach this bay, M. Inghestrŏm recommends making the Island of Koniougi, 10 or 11 miles to N. 59° W., by compass, from the entrance, and then bear away for it; on its North side is the high cape, Tolstoï, and on the S.W. side is a conical peak, which rises high and isolated near the coast; and between them, at some miles in the offing, the islets which shelter the bay. Then steer for the largest of them, and round it carefully by the lead, and, when once it is doubled, the entrance is not difficult. Near the cape lying near the conical peak, called for this reason *Podsopotchnoï* (under the peak), is a sunken rock, which only breaks at times: to avoid this pass between the islands.

From this last cape the coast trends 4 miles to the West to *Cape Betchevinskoï*, from which a reef extends for 1½ miles to the N.W.; and in this direction, at 8 or 9 miles from the coast, is the small island *Koniouge* (Kanugy? Krusenstern), which is an enormous rock, perpendicular to the North, and a low point to the S.W. Its surface is constantly changing from volcanic effects, and the Aleutes say that it regularly and slowly keeps rising out of the water: around and on it are an innumerable number of birds, called Koniougi, attracted, it is thought, by the warmth of the island, to which their name is applied.

Beyond Cape Betchevinskoï the coast turns to the South, and forms a bay which penetrates 2 or 3 miles inland, the bottom being separated from the South coast of Atkha by a narrow, marshy isthmus, not more than 300 yards broad.

The bay is shallow. Beyond this the coast trends 2 miles to W.S.W., not far from the mountains, and then turns again to the South, forming two small bays. Beyond this again the coast extends N. to N.W. by W. and West, and then forms an obtuse bluff cape; then at a mile farther, another peaked cape; after which the island narrows so as to be not more than one-third of a mile broad, and forms a low, sandy isthmus. To the S.W. of this isthmus a mountain, the last in the island, forms the S.W. extremity of Atkha. This S.W. extreme is 37 or 38 miles from the village in a direct line, and about 18 miles from Cape Tolstoï.

The South coast of Atkha was not visited by M. Ingheström. All the information is therefore derived from an intelligent Toyon (native chief), who has given a sketch of it, which accompanies Capt. Lütke's Memoir. As this can scarcely be of service in a nautical view, we omit the short details of it.

There is but one village upon Atkha, called *Nikolskoï*, on the South side of the inner harbour of *Korovinskoï*. It consists of a few houses for the employés of the Russian Company, a church, &c. It is in a low and damp situation, and has many disadvantages. One great inconvenience of Atkha is the extreme scarcity of provisions. But few fish are taken on the coast or in the rivers. Supplies are brought from the Commander Islands to the West, but all are insufficient to avert severe famine in the winter.

It may be said that there is no summer in Atkha; for during those months fogs and rain are particularly prevalent. In winter, on the contrary, the weather is generally clear. The island has abundant evidence everywhere of its volcanic fires. We have before spoken of some of the volcanoes. Mineral and hot springs are frequently met with; and one remarkable feature is very common: these are volcanic spiracles, or blowing-holes, emitting, at intervals, burning clay, or else the same in a state of fusion; others only send forth hot and sulphureous vapours; and these, in the S.W. side, form almost the entire surface, and have considerable effect on the climate.

Kassatotchy Island, which bears N. 54° W. 9 or 10 miles from the mountain on the S.W. extremity of Atkha, is one of these volcanoes. It is a mountain rising at once out of the sea; the crater on its summit, it is stated by the inhabitants, being full of water.

The TCHASTIE ISLANDS, a group of small but high islands to the West of Atkha, are thus called (*tchastie*, crowded) from their arrangement; they have been but little known hitherto. There are thirteen of different sizes, and six large isolated rocks. M. Ingheström saw them several times, and supposes that the danger of approaching them, from the terrible tide races and currents through them, and the want of shelter, has prevented any knowledge being gained.

The island nearest to Atkha is *Oglodak*, 3½ miles distant. It is high, steep, without landing, and is the only one that is tenanted, and that by foxes. A little to the West of this is the high Islet of *Nerpitchy* (sea-calf), and to the N.W., quite close also, a high rock called *Sivoutchy* (sea-lion's). Beyond this, in the direction of Adakh, is a range of rocks, then the Island of *Togalak*; after this, *Tchougoul*, or *Tchougoulak*, which is some miles in extent, and was formerly inhabited; beyond this is an isolated rock; and lastly, we reach three conspicuous islands close together, of which the last, *Kagalaska*, is not far from the coast of Adakh. From these islands, in the direction of Sitkhin, there are six more islets, one of which is large, the rest small.

SITKHIN ISLAND, which, to distinguish it from another of the name to the West, is sometimes called East Sitkhin, is in lat. 52° 4′ or 5′, and, according to the observations made by Capt. Stanikowitch, in the corvette *Moller*, in lon. 176° 2′ (centre). It is about 25 miles in circumference, and in its centre is a volcano covered with perpetual snow, which was ascertained by M. Ingheström to be 5,033 English feet in elevation.

ADAKH is a large mountainous island, but lower than Sitkhin; it is covered with perpetual snow in some parts. Its North end is in about lat. 52° 4′ 6″. The bay on the North side is open, and there are others on the N.E., South, and West sides; that to the South offers the best shelter. It is separated by a small isthmus from the bay on the West coast. The position and details of the island are very imperfect, as are also those of the next island.*

KANAGA, or KONNIAGA, is the island next West to Adakh, to which it is similar in size, being 7 or 8 leagues in length, by half that breadth. The northern part of this island is remarkable by a high smoking volcano, one of the most lofty in the chain; the rest of the island is not very high. Near to its western part is a small island bearing the name of the Isle of Otters; these islands, however, are but incorrectly delineated and placed on the charts.

TANAGA is separated from Kanaga by a channel 2 leagues broad, extending 8 leagues in a N.E. direction. According to the observations of Admiral Sarytscheff, in 1791, this land is about 11 leagues in extent, from East to West, and 4 in breadth. It is easily distinguished by an elevated volcano, which stands at its S.W. point. Near the point is a bay, which Sarytscheff visited, and made a plan of in his voyage. The entrance of this bay may be about 4 miles broad, and is about 8 miles deep. At this distance, in the North part of the bay, the vessel in which he penetrated anchored on a bottom of fine black sand, in front of the entrance of two rivers, whose sources were in the mountains, and here entered the bay. Watering is very easily performed in this bay, the boats ascending the rivers without any obstacles.

At 6 leagues to the West of Tanaga is *Goreloy* or *Burnt* Island. It has a very high volcano, whose summit is covered with perpetual snow. From this volcano it derives its name, Goreli or Goreloy. Sarytscheff gives it a circumference of 6 leagues; and M. Inghström considers this volcano, and those on Kanaga and Tanaga, as the highest in the Aleutian chain.

South of this are two small islands, with those South of it, forming the westernmost of the group of the Andréanoffsky Islands. They are mentioned by Sarytscheff as being 16 miles S.W. of the S.W. point of Tanaga. He found the passage between quite clear. As they had no name, Krusenstern gave them the name of *Délaroff*, one of the first founders of the Russian American Company.

Lütke says that they are in reality seven in number, without reckoning detached rocks. Of the two Délaroff Islands, the West is called *Ogloga*, the East *Skakhoul*.

* Lütke, pp. 320—322.

At some distance East of the last are two rocks. At about 15 versts from these rocks, or 40 versts S.S.W. from Goreloy, is a third island, *Kakhvalga*; and 10 or 12 versts to the West of the latter is *Ounalga*, a low island. *Ulak*, also low, and the largest of all, is to the South of Tanaga.*

AMATIGNACK and *Illack* are also two small isles to the South of these again. The southernmost, Amatignack (" a chip" in Aleute), is the larger of the two, and the highest of all.

KRYCI OR RAT ISLANDS.

The islands next West of the foregoing are included by Lütke and others under the above denomination. Krusenstern says that it is more properly confined to a small island West of Amtschitka.

SEMISOPOCHNOI.—At 16 or 17 leagues West of Goreloy or Burnt Island, is the *Isle of the Seven Mountains*. The name (*Semisopochnoi*) is significant of its character. It is of a circular form, and is 10 leagues in circumference.

According to the observations of M. Inghesträm, its lat. is 51° 59′, and its lon. 180° 14′ 3″ West, as ascertained in the corvette *Moller*. Its position is the best fixed of all the neighbouring islands. The mountains do not exceed 3,000 feet in height, and in summer the snow lies on it in bands. One of the mountains, in the North part, sends forth smoke. The North and East sides have a wild and desolate aspect; on the South and West there are several green spots. The shores are in general clear.

The *strait* between Semisopochnoi and Goreloy is the best for crossing the Aleutian chain. It is safe throughout, is not less than 45 miles broad, and is not subject to tide races.†

AMTSCHITKA is a large island; it is not very hilly, and extends about 11 leagues N. 72° W. and S. 72° E. Near its West point are two small islands, the larger of which is called *Rat Island*, a name, as above mentioned, which has been extended to the whole group. Behring saw this island on October 25th, and gave it the name of St. Marcian (Markiana). He says it was moderately high, and covered with snow. Although his latitudes are erroneous, yet it must be considered that this island is intended, and not Amatignac, as has been supposed.

The island is about 35 miles in length, E.S.E. ½ E. and W.N.W. ½ W. by compass. It is low, and is not more than 4 miles wide at the S.E. part, but is broader and higher in the N.W. Its S.E. extremity forms a peninsula, on which a hillock rises, and from it a reef extends for 2 miles. The South coast runs nearly in a straight line, without any bays or coves, and in general with a less depth than the North side, and lined with a larger number of rocks and reefs. At about a third the length of the island a chain of moderately high mountains rises, and falls again toward the N.W. extremity, where it forms a low but steep cape, called by the Aleutes *Satanna*, or Bird's Cape.

On the North side of the island, at 10 miles from its East end, is *Kirilovskaïa Bay*, the only place in the island where you can stay at anchor. This anchorage

* Lütke, pp. 322-3.
† *Ibid.* pp. 325-6. This terminates the remarks of M. Inghesträm, as here quoted.

is somewhat sheltered from seaward by a reef off its middle, which requires caution in entering; there are also some reefs on either side of it. It is dangerous to remain here in Autumn or winter, when northerly gales are prevalent. The lat. of the bay is 51° 27′ 1″, lon. 180° 40′ 6″ W. Magnetic variation, 14° 5′ E. in 1830. And high water occurs about 10 o'clock at full and change.

The inhabitants say that there is no other bay than this here described.

Many persons here believe in the existence of a bank 10 or 15 miles to the South of Amtschitka; but M. Inghestrõm, who could not discover any appearance of such at 8 miles off, is inclined to doubt it. He equally disbelieves in the existence of some land said to be in the same direction, in about lat. 50¾°.

To the N.W. of Amtschitka are the *Tschegoula Isles*, a group of four small hilly islands, which extend about 6 leagues East and West. In his journal Sarytscheff calls the westernmost Tschegoula, a name applied by Admiral Krusenstern to the group. One of these is called by Lütke *Little* or *West Sitkhin*, another is named *Dawydoff* on the chart.

KRYCI, Rat, or *Ayougadagh Island*, before mentioned, is 7 miles long, mountainous, and in lat. 51° 45′, lon. 180° 40′ W.

KISKA, a hilly island, with the exception of its eastern part, which is low, lies to the West of the preceding. Its length, North and South, is 25 miles. A rock in the form of a column lies 3 miles N.W. of the island. The island is laid down in Krusenstern's chart, from the Russian Admiralty chart of Sarytscheff, which differs in position somewhat from that given in his Memoir. Behring saw it October 28th, 1741. He says it is very hilly, and named St. Stephan; he also speaks of the small islands to the East of Kiska, but instead of four he only mentions three. His latitude would appear to be half a degree too far South. In the atlas attached to Capt. Lütke's Memoir (No. xiii.) is a small chart of this island, executed by M. Inghestrõm, whom we have before mentioned. He places it 10° still farther North; but Admiral Krusenstern considers that neither the configuration nor the position are more correct than from the previous observations. According to M. Inghestrõm, there exists to the East of Kiska, at 1½ miles distant, a small isle, which he calls *Little Kiska;* and 3 miles from this, in the same direction, the small Island *Tannadagh*, and a rock. He also mentions some rocks, which do not uncover, between the islands Bouldyr and Kiska, at the distance of 5 leagues from the former. Capt. Lütke, however, considers this position as doubtful.

BOULDYR lies to the W.N.W. of Kiska. It is a hilly island, surrounded by peaked rocks, which extend to half a league beyond the S.W. point of the island. It is about 4 leagues in circumference, and two large rocks exist at the western part of the island.

BLIJNI GROUP.

This group, composed of two islands and a collection of separate rocks, is called *Blijni* (the nearest), because it is the nearest to Kamtschatka, of the Aleutian chain. It was also the first visited by the hardy hunters. The discovery belongs to Behring, as presently mentioned.

Semitsch forms a portion of this westernmost group of the Aleutian Islands.

It consists of two small islands half a league apart, and extending E.N.E. and W.S.W. 6 or 7 miles. The low land which Behring saw October, 1741, to which he gave the name of *Abraham Isle*, is certainly the Semitsch Isles, which he took for a single island. On the charts of Sarytscheff and the Russian Admiralty a group of rocks was inserted, as lying 16 leagues to the East of the Isle of Attou. These were placed from the verbal reports of the fur traders, and were considered as very doubtful. Their existence, nevertheless, was confirmed by a letter received by Admiral Krusenstern from M. Inghesträm, which informed him that the latter had distinctly seen these rocks, and that he had approached very near to them. He estimated their distance from the N.E. extremity of the Semitsch Islands at 5 leagues in a S. 79° E. direction; but the latitude and longitude given by Inghesträm, 52° 37′ and 175°, does not at all accord with the observations of Admiral Sarytscheff. Unless, therefore, an error of 25′ in the latitude of Semitsch can be admitted, we must suppose an error in these figures.

AGATTOU, to the South of the Semitsch Islands, according to Sarytscheff has a circumference of 34 miles. It is separated from Attou by a strait of 15 miles wide, according to the Russian chart.

ATTOU is one of the largest of the Aleutian Islands. Its eastern extreme lies 6 leagues W.S.W. from the Semitsch Islands. According to Capt. Golownin's observations, in 1808, it is 27¾ miles in length East and West; but the Russian chart makes it 48 miles long. As Golownin was not sure that the point he saw was really the western point or not, the latter dimensions by Admiral Sarytscheff may be correct. In the S.E. part of the island is a bay, in which a vessel belonging to the Russian American Company anchored. In the plan of this, by Lieut. Dawydoff, it is called *Massacre Bay*. It is about 3 miles wide, and midway between the two outer capes is a group of small islets and rocks, within which there also appears to be a low reef, and another lies outside them, half a mile to the S.W. Lieut. Etolin discovered on the North coast of Attou an excellent bay, which he called *Tschitschagoff Bay*, where the Russian American Company have now an establishment. It is in lat. 52° 56′, and 9 miles from the eastern point of the island.

This is the western extremity of the Aleutian chain, which forms the south-eastern limit of the Sea of Behring. The two islands lying near the coast of Kamtschatka, Copper and Behring's Islands, though lying 180 miles to the N.W., might almost be considered as a continuation of the chain, but, as they are more connected with the western coast, they will be described with it hereafter.

CHAPTER XVII.

THE SEA OF BEHRING, BEHRING'S STRAIT, ETC.

The extensive and inclement sea, whose shores and islands are described in the present chapter, derives its name from the celebrated Russian navigator who first sailed in it, and made known to Europe the real character of the extremities of the New and Old Worlds.

The appellation of the Sea of Behring was first and justly applied to it by Capt. Golownin. This replaced those by which it had been improperly designated, such as the Sea of Otters (Bobrovoïé); but it has for many years lost the exclusive claim to this title. There is no more reason why it should be termed the Sea of Kamtschatka, sometimes given to it, than that of Aliaska or the Aleutian. The name of Behring is therefore most apt.

After the revival of learning in the 15th and 16th centuries, the desire of gaining increased knowledge of everything that could be acquired of the actual nature of the world, then known but to a very limited extent, led to the numerous expeditions of discovery whose results and operations we have alluded to in the introductory remarks to the preceding chapters. For the most part these expeditions were despatched from Europe to the westward, and thus the progress of acquaintance with the shores of the Great Ocean was towards the North, and it will be seen that territorial acquisitions proceeded, chronologically, in this way. In this part, however, the farthest extremity of the world, a new system of operations was brought into action. Here the progress of discovery was in an opposite direction. The Russian Czar, Peter the Great, to whom the half of a great continent still seemed not sufficient, drew up with his own hand, shortly before his death, the instructions for a voyage whose object was to ascertain whether Asia was separated from America by a strait, and then to extend the Russian dominion beyond such a limit, should nature have thus marked it.

But in the distant regions of this vast empire there were no means by which such an exploratory voyage could be organized. They were therefore sent from Russia: Vitus Behring and Alexoi Tschirikoff were chosen by the Empress Catherine to execute this part of the will of her great partner and predecessor in power. The two ships destined for this expedition were constructed at Kamtschatka, the first of their kind that had been seen in this the extremity of a land then scarcely known. They did not set sail from the mouth of the Kamtschatka River until July 20th, 1728. Behring shaped his course to the N.E., never losing sight of the Asiatic coast. On August 15th he reached lat. 67° 18′ N., at a point (now Cape Serdze Kamen) where the coast turned to the westward, from which he returned to the port whence he sailed, without having seen the American coast. He had thus entered the Icy Ocean without knowing it: he had solved the great problem, and posterity has imposed the name of Behring upon this strait; the reality supplying the place of the fabulous Strait of Anian.

Behring and Tschirikoff made a second voyage in 1729, but it yielded no new information.

A third time did the same commanders set sail from Kamtschatka, June 4th, 1741; this time with the intention of making the American continent to the eastward, the result of which voyage is alluded on to page 436. Behring then sailed through the chain of islands skirting the great Peninsula of Aliaska. On August 29th he anchored at the Schumagin Islands. After this he struggled against contrary winds till September 24th, when he made the S.W. extremity of Aliaska. In the course of the next month he discovered part of the Aleutian Islands. By this time the commander was ill and decaying. The greatest part of the crew, too, were attacked by that frightful disease, scurvy.

Behring's ships and crew were reduced to the greatest distress by the weather and sufferings they encountered. Worn out with these, they made for the island now bearing his name; but the ship was stranded. Many of the crew died as they were removed into the air from the terrible scourge of scurvy that had attacked them. On November 9th the captain himself was brought on shore on a hand-barrow, and died on December 8th, 1741. "He was a Dane by birth, and had in his youth made voyages to the East and West Indies, when the glorious example of the immortal Emperor Peter the Great for the marine tempted him to seek his fortune in Russia. I have found it somewhere that in the year 1707 he was a lieutenant, and in 1710 captain-lieutenant, in the Russian fleet. When he was made a captain I cannot exactly determine. Having thus served in the Cronstadt fleet from its beginning, and been in all the expeditions by sea in the war with the Swedes, he joined to the capacity requisite for his office a long experience, which made him particularly worthy of such exploits as were the discoveries wherewith he had been twice entrusted. It is a pity that it was his fate to end his life in such an unfortunate manner. He may be said to have been buried half alive, for the sand rolling down occasionally from the side of the ditch in which he lay, and covering his feet, he at last would not suffer it to be removed, and said that he felt some warmth from it, which otherwise he should want in the remaining parts of his body, and thus the sand increased to his belly; so that, after his decease, they were obliged to scrape him out of the ground, in order to inter him in a proper manner."—(Müller, p. 55.)

The ship, the *St. Peter*, was subsequently wrecked at her anchors, and the survivors constructed a small vessel from her timber, by means of which they reached Port St. Peter and St. Paul on the 27th August ensuing, having existed on the seals' and whales' flesh thrown on the shores of the island. Among those who reached Europe safely were Müller, the historian of the expedition, and the celebrated philosopher, Gmelin. Steller, the naturalist, died on his journey.

Tschirikoff's voyage was less unfortunate than his commander's; but he scarcely had fewer hardships. He returned to Russia, after having explored part of the American coast.

Such was the success of the first Russian expeditions. They were followed at intervals by many others, among which the names of Glotoff, Synd, Krenitzin, and Levacheff, stand most conspicuous. The details of these voyages, mostly in the Aleutian chain, are given in the interesting work of the Rev. William Coxe,* whose influence at St. Petersburg led to the advancement of scientific research in this part.

Our great navigator, James Cook, entered the Icy Ocean, and ascertained the exact nature of the continental separation first traversed by Behring in August, 1778; and his second in command, Capt. Clerke, after his sad catastrophe at Hawaii, again penetrated to the icy barrier in July, 1779. The abortive Russian expedition under Commodore Joseph Billings, an Englishman, made by order of the Empress Catherine II., in 1785 to 1794, did not penetrate the strait, one great object of the expedition, after having traversed the Sea of Behring. For the account of this voyage, which cost so much time and expense in preparation, we are indebted to the secretary, Martin Sauer, who published an edition in English, and to Capt. Sarytscheff, the second in command, who published another account in Russian. Otherwise we should have received no information whatever from this expedition.

In 1817 Capt. Otto von Kotzebue examined and named most of the points on the American coast of Behring's Straits; and in 1820 his lieutenant, Schischmareff, again visited its shores.

The Asiatic coast, from East Cape to Kamtschatka, is amply described from the admirable survey and examinations of Capt. (since Rear-Admiral) Frederic Lütke. In 1826 the Russian corvettes, the *Moller*, Capt. Stanikowitch, and the *Séniavine*, under Capt. Lütke, were despatched to these seas. The *Séniavine* examined almost the whole of the western shores, and to the account of this voyage we owe our descriptions.

To our own country we are indebted for the accurate knowledge we possess of the American shores. This is from the memorable voyage of H.M.S. *Blossom*, under Capt. F. W. Beechey, who minutely surveyed the eastern shores of the strait. This ship was sent from England to co-operate with the Arctic Expedition, despatched to examine the northern coast of America. It was anticipated that one branch under Dr. (afterwards Sir John) Franklin, would have reached the Pacific, but, after two attempts made by the *Blossom* in the summers of 1826 and 1827, the object was relinquished.

To the intrepid voyager, Sir John Franklin, we are indebted for the last of these expeditions. The intense anxiety of the whole world to aid in the succour of the well-known Arctic Expedition, commenced in 1846, caused our Admiralty to send H.M. ships *Herald*, Capt. Kellett, and *Plover*, Commander Moore, through the straits in 1849. In company with these the schooner yacht, *Nancy Dawson*, under her owner, Robert Shedder, Esq., penetrated to the extreme North and East. They made several interesting discoveries, hereafter alluded to, and proceeded farther North than any before, though, as is but too well known, without succeeding in the object of the expedition.

At the present time (1850) an expedition, consisting of the well-tried ships, *Enterprise* and *Investigator*, under Capt. Collinson, R.N., well known as the talented surveying officer on the China coast, is endeavouring to relieve the

* Russian Discoveries, by W. Coxe, 4to, London, 1780.

public anxiety by gaining some information of the missing expedition, the results of which will doubtless furnish much additional knowledge.

The *islands* in the open sea, South of the strait, are occupied by the Russian Company, who have collected on them large quantities of fur-seals; as this is a very important branch of their trade, we have given in the Appendix some account of their mode of capture. The nature of these islands is described in the ensuing pages.

Whales have been found to be numerous in the Icy Ocean. Capt. Beechey notices this:—"Off Icy Cape we saw a great many black whales, more than I ever remember to have seen in Baffin's Bay."—(Voyage of the *Blossom*, vol. i. p. 276.) The pursuit may be followed during July and August; and in consequence of the high latitude, there is light enough throughout the night for the purpose,—no small advantage. Another point, too, is, that the depth not being great, seldom exceeding 20 or 30 fathoms, there are no floating icebergs of formidable dimensions, as compared with other icy regions, and there is good anchorage everywhere. On the other hand, thick fogs, strong currents, and imperfect charts, render the operations somewhat precarious.

The ports and their capabilities, on the American side, are amply described. Of the American inhabitants we will say a few words:—

The Western Esquimaux seem to be very nearly connected with the other tribes inhabiting the northern and north-eastern shores of America, in their features, language, and habits. They also resemble the Tchuktchi, from whom they are probably descended. Those occupying the N.W. coast of America, between 64° 34′, and 71° 24′ N., are a nation of fishermen, dwelling near the sea-shore, from which they draw almost their entire subsistence. They construct their yourts, or winter residences, partly excavated in the earth, and partly covered with moss laid upon poles of drift-wood, upon convenient positions for their occupation. They form themselves into communities, seldom exceeding 100 persons. Beechey reckoned nineteen of these villages within the above limits, the total number not exceeding 2,500 persons. They are taller than the Eastern Esquimaux, averaging 5 ft. 7½ in.; they are also better looking than those of Baffin's Bay, &c. They differ widely, too, from the inhabitants of Greenland, being more continent, industrious, and provident, and also more warlike, irascible, and uncourteous, and nearly resembling in these respects the Tchuktchi. Hospitality forms one distinguishing feature of their disposition; smoking is their favourite habit, to which they are passionately devoted. The lip ornament, or labret, before mentioned, is here peculiar to the *males* of the Western Esquimaux, and are only in use from Norton Sound to Mackenzie River, on the North coast. It is an ancient custom, as it is described by Deschnew in 1648. It is singular that it is here confined to the men only for a short distance, while southward the custom prevails among the females on a great extent of coast. Their language so nearly resembles that of the eastern tribes, that they are mutually understood; South of Norton Sound, however, this remark does not hold good.

When Capt. Beechey was on the coast, in 1826-27, he found them somewhat annoying and treacherous at times. It is evident, however, that this must have arisen from their imperfect conception of the object of his visit. Since that period they have greatly improved in this respect; and during the progress of the expedition in 1849, Capts. Kellett, Moore, and Shedden found them pleasant, docile, and very tractable. This was in part owing to their having a proper interpreter, who explained what their object was. The Russian settlements, too, have been very instrumental in causing this alteration in their conduct. Many of them were found with shirts, gaudy handkerchiefs, cottons printed with walrus, reindeer, and all other animals they are in the habit of catching, and representations in ivory of knives and kettles, all procured from the Russians. They were at last very anxious to obtain muskets, and evinced no fear in discharging them.

The PENINSULA of ALIASKA,* whose south-eastern coast has been previously described, forms, with the Aleutian Archipelago, the southern limits of Behring's Sea, or, as it has been sometimes called, the Sea of Kamtschatka. The description is therefore resumed at the point where the former left off, namely, at the Strait of Isanotzkoy.

There appears to be considerable uncertainty as to the correctness of the charts of the coast, immediately to the N.E. of the Strait of Isanotzkoy. It was partially explored in baïdars from the corvette *Moller*.

Point Krenitzin appears to be the first prominent point to the northward. It is the extremity of a low bed of gravel, and is the N.W. point of an open bay, on the East side of which is the village *Morjovskoï* (Morses). This village stands on low, marshy land, intersected with numerous lakes. Wood or trees are very rare here; but the natives live well, and cultivate potatoes and turnips in their gardens, and raise poultry.

Izenbek Bay (or, according to Krusenstern, Comte Heiden Bay†) is the next port in proceeding northward. Its S.W. point, *Cape Glazenap*, or *Mitkoff* of Krusenstern, according to the observations made by Lütke, is in lat. 55° 14′·8, and lon. 162° 50′·7. This cape is very remarkable by a considerable elevation, and by its form; at a distance it is like an island separated from the low land to the South of it. The mouth of the bay is filled up by a long and very low island, which at high water is divided into two or three. Its South point is about a mile distant from Cape Glazenap, and its North point above 2 miles from Cape Moffet. This extensive gulf has not been fully described. The depth in the South entrance was found to be 4½ and 5 fathoms, bottom of fine black sand. From the top of Cape Glazenap the depth appeared greater to the southward; and on the East side of the island there was a large quantity of whalebone, which shows that there is depth enough for these animals. From these circumstances it was considered that Izenbek Bay would offer anchorage, and therefore merits a more minute examination. The eastern coast of the bay is surrounded by high mountains, covered with snow.

Amak or Aamak Island lies off this part of the coast. It is an extinct volcano, covered with calcined stones and lava. It is rather less than 4 miles in

* Capt. Lütke says, that in spelling this *Aliasha* he follows the orthography generally adopted in the colonies. It was sometimes previously spelt *Aliaksa*. The natives pronounce it *Aliakskha*; so that both modes are right or wrong.

† It will be seen hereafter that Stanikowitch gives this name to a bay to the N.E.

length, in a N. 17° W. and S. 17° E. direction. Its South extremity is in lat. 55° 25′ and lon. 163° 1′ 30″. At 2 miles N.W. by compass from its North end is a rock called the *Sivoutchy* (or Sea-Lions') *Rock*. Between the island and the main the depth is from 9 to 14 fathoms. The soundings in the offing are not deep. Capt. Cook found 19 fathoms at 3 or 4 leagues off the South part of the island. They also caught here a good number of fine cod-fish.* Capt. Stanikowitch found the magnetic variation to be 21½° E. in 1826.

The coast to the northward presents nothing very remarkable. It trends first N.E. and then E. For a distance of 50 miles there are only two capes, somewhat higher than the rest; to these the names of the lieutenants of the corvette *Moller*, *Leontovitch* and *Leskoff*, were applied. In this space there is much drift-wood, and the nature of the coast is generally a damp turf, covered with moss, frequently interrupted by lakes of fresh water. *Cape Leontovitch* is the most conspicuous, as beyond it the coast trends more to the East. It is low, and the land extends in a mossy and marshy plain to the foot of the mountains, parallel with the coast, at 15 miles distant.

Cape Rojnoff, or *Roshnoff*, which is at the extremity of the extent of coast above alluded to, is very low; and at 1½ miles from it is the western point of Kritskoï Island, also very low; the two form the entrance to a shallow bay, which runs in first W. by S., and then S. by W., by compass, to a low isthmus, not more than 5 versts in breadth, separating it from Pavlovskaia Bay, on the South coast of Aliaska. When the tide is in this bay appears of vast extent; but at low water it is nearly dry throughout, leaving only a narrow and sinuous channel through it. The chain of mountains which extends through the peninsula is interrupted in this part. The rise of the tide is 15 feet, and the (approximate) time of high water is 7^h 30′.

MOLLER BAY is a large indentation of the land, which narrows the peninsula to 12 miles in breadth at this part. A large portion of this bay is filled with shoals, which uncover at low water; but in its S.E. angle is a small cove, where there is sufficient water for every description of vessels, and which is covered from seaward by a bed of gravel which advances to the South. This is the only port which exists on all the North coast of Aliaska. It is about a mile in length and breadth, and the lead gives from 4½ to 8 fathoms, muddy bottom. In entering it, you must range close to the bed of gravel, the edges of which are clear and deep, to avoid a shoal, which is less than a mile from it; and as soon as you have made out this bed of gravel, steer for a point on the East side of the port, which is particularly remarkable by its blackish colour, and which lies about 1½ miles N.E. by E. by compass, from the extremity of the gravel bed. The mountains, which are of a moderate height, advance here to the coast, leaving no room for the mossy plains which generally occupy the space between the foot of the mountains and the sea-shore. This part, the most remarkable on the N.W. shore of Aliaska, deserves a more particular examination, as it was only superficially explored in the boats of the corvette *Moller*.

The bay is about 6 miles wide in its opening. Its entrance, on the West

* Cook's Third Voyage, vol. ii.

side, is formed by the East point of *Kritskoï Island*, which was determined to be in 56° 0'·7, and lon. 160° 41' W. The extremity of the bed of gravel which forms the port is 7½ miles to the S.E. by E. (by compass) from this. Kritskoï Island extends 9 miles along the coast in a W. by S. and E. by N. direction. It is so low in several parts, that at high water it must be overflowed. The strait which separates it from the continent is 1½ miles broad, and has very little water; at low water, spring tides, it must be nearly dry. A great number of morses frequent the eastern end of the island, the coast being covered by thousands of them at times. Opposite its West end is Cape Rojnoff, before mentioned.

From Moller Bay the coast trends to the N. by E., and at 20 miles distant is *Cape Koutouzoff*. This cape is high and abrupt, and 13 miles N. E. from it is *Cape Séniavine*, also high and steep, the space between being low, either marshy or covered with grass and furrowed by numerous fresh-water rivulets. About midway between are two hillocks on the coast. The peaks of some of the mountains in the background are so high as to be covered with snow throughout the summer. Cape Séniavine is in lat. 56° 23' 45", and lon. 160° 2' 45" W.

The coast beyond this offers nothing remarkable for a considerable distance. It consists of a low, level, mossy plain. At about 9 leagues from Cape Séniavine, a low bed of gravel commences, which extends in front of the coast for 10 miles, separated from it by a shallow canal, which at low water becomes a lake, discharging itself by two small streams at either end. Beyond this the low level coast extends 10 miles farther to *Cape Strogonoff*. This is extremely low, and projects 2 miles to the North. In the interval on the shore is a great quantity of drift-wood, some large trees being amongst it.

COMTE HEIDEN BAY (or, according to Krusenstern, Houdobin Bay*) is formed by the low Cape Strogonoff to the westward. Before this cape is an islet, equally low, called on the chart *Chestakoff Islet*, and between this islet, or rather between the shoal off its North end and the coast of the continent, is the entrance of the bay, which extends 6 miles to the S.E., and the opening is 2½ miles broad. It is not known whether there be any anchorage in it.

From this the coast, equally low, uniform, and covered with moss, trends to the N.N.E. In lat. 57° 5' are two moderately high capes, terminating to seaward in sandy downs.

Cape Menchikoff (named after the Prince) is in lat. 57° 30'·4, and lon. 157° 58'·5. It consists of a mound of sand, from the flattened summit of which may be seen the marshy land surrounding it on all sides, which at high water may be covered and form the cape into an island.

At 8 miles N.E. from this is the mouth of the *River Ougatchik*, called the *Soulima* in the journal of the *Moller* and by Krusenstern, which is 2 miles broad: the South side is a marsh, the North side higher. The depth in its entrance is from 10 to 18 feet at low water. The current in it is extremely rapid, and the natives say that it separates into two branches in the interior, one to N.E., the other to S.E.

* See note on page 524.

The North point of its mouth is named *Cape Greig*, after the Admiral. It is high and steep: its lat. is 57° 43′, and lon. 157° 47′·2, and is surrounded by sand-banks, dry at low water. From this the coast trends, low, level, and straight, in a general N. 17° E. direction to the mouth of the Agougack River.

Capt. Cook, in passing along the coast, July 9th, 1778, says:—"At noon we were in lat. 57° 49′, and about 2 leagues from the land, which extended from S. by E. to E.N.E., being all a low coast, with points shooting out in some places, which from the deck appeared like islands; but from the masthead low land was seen to connect them. In this situation the depth of water was 15 fathoms, the bottom a fine black sand."*

The River Agougack, or Ugagouk, Krusenstern considers to be the northern limit of the Peninsula of Aliaska, as it really separates Aliaska from the continent of America. It here begins to extend in breadth, and the river traverses it from East to West, and rises in a lake called *Nanouantoughat*, which is only separated from the Strait of Chelighoff by a neck of land 5 miles in breadth, at the back of the Bay of Pascalo. Over this space is a portage for the merchandise collected by the Russian American Fur Company at the establishments in Bristol Bay, which is thus transported, by means of the river, to the head-quarters at Sitka.

The breadth of the river is about 2 miles at its mouth, and at about 7 miles N.E. by compass from it is a hill, which, with Cape Tschitchagoff, serves to point out the position of the entrance. Its shores are very low, particularly the South side. The current is very rapid, the tide rising 20 feet.

Cape Tschitchagoff is bluff without being high, and surrounded for 2 miles distant by shoals, which, being left dry at low water, form a series of sand islets.

From this cape a level marshy coast succeeds. Its aspect is most extraordinarily monotonous, and for 30 miles in a N.E. by N. direction, which it takes, there is only one spot, in about lat. 58° 35′, where there is a hill, which rises about 200 yards from the sea, forming an elevation like a wall composed of sand and clay.

Cape Souvoroff, which is at the distance above named, was the last point reached by Stanikowitch in the corvette *Moller*. It forms the North point of the mouth of the *River Nanek* or *Nackneck*.

This river, flowing from a lake of the same name, traverses the Peninsula of Aliaska. At its mouth it is about two-thirds of a mile in breadth, and there are 2 fathoms water in it. Its steep sides are formed of an adhesive clay. On each side of the mouth are villages. That to the right is *Koutchougoumut*, that to the left *Paougvigumut*. The position of the latter was ascertained to be in 58° 42′ 5″ N., and 157° 0′ 30″ W. The population of the two villages was about 500. The River Nanek abounds with an extraordinary quantity of fish. This, with reindeer and birds, forms the supplies of the natives. From the evidences seen on its borders, the tide rises in the river to 30 feet.†

"As we had advanced to the N.E. we had found the depth of water gradually decreasing, and the coast trending more and more northerly. But the ridge of

* Third Voyage, vol. ii. p. 428.

† Lutke, Voyage, &c., Part. Naut., p. 257, *et seq.*

mountains behind it continued to lie in the same direction as those more westerly; so that the extent of low land between the foot of the mountains and the sea-coast insensibly increased. Both high and low grounds were perfectly destitute of wood, but seemed to be covered with a green turf, except the mountains, which were covered with snow."—(Capt. Cook.)

BRISTOL BAY.—The coast we have been describing forms the southern portion of the coast of the bay, that named by Capt. Cook after the Admiral, the Earl of Bristol.

This great indentation was sailed around by Capt. Cook, but the shoals which obstruct it prevented his making a detailed examination of its shores. This work was undertaken in 1818 by Oustugoff, at the desire of the late Count Romanzoff, who also bore a part of the expenses; but, as this officer did not possess sufficient scientific knowledge to carry it completely into effect, Lieut. Chramtschenko, an officer of merit, was sent in 1821 to examine it in detail.*

Cape Newenham forms the northern point of this bay, and, according to Admiral Krusenstern, Cape Ounimack, the western point of the island of that name off Aliaska, may be taken as its southern limits, and which lies 82 leagues to the S.S.W. of Cape Newenham, so that the N.W. coast of Aliaska forms the southern part of Bristol Bay.

This bay contains another inner bay, which is formed by a promontory, named *Cape Constantine* on the Russian charts, and within which three rivers debouch. The northernmost of these is the *Bristol River*. The Ouglaghmoutes (as the inhabitants of the shores of the bay are called) give it the name of *Quitshak*. It takes its rise in a lake named on the Russian charts *Lake Cheleghoff*, which is again connected with another lake, *Ilimen*, by a second river. The latter lake is in lat. 59° 50′.

Capt. Cook's account of his making the mouth of this river is thus:—"Behind this point is a river, the entrance of which seemed to be a mile broad, but I can say nothing as to its depth. The water appeared discoloured, as upon shoals, but a calm would have given it the same aspect. It seemed to have a winding direction, through the great flat that lies between the chain of mountains to the S.E. and the hills to the N.W. It must abound with salmon, as we saw many leaping in the sea before the entrance; and some were found in the maws of the cod which we had caught."—(Vol. ii. p. 429.)

Cook's determination of its position, lat. 58° 27′, lon. 201° 55′, may thus not be so exact as usual. A short distance East, and ten miles to the North of the mouth of this river, as it is marked by Cook, is that of the *Naneck*, which has an East and West direction, and near which is a village of Ouglaghmoutes.

The third river which enters this inner bay, to the West of the Bristol River, is called the *Nouchagack*, or *Noushagak*. Its course is in a N.N.W. direction. The opening formed on the North by the coast, and by Cape Constantine on the South, may be taken as its mouth, which is 20 miles broad, and preserves

* Subsequently to 1821, Lieut. Chramtschenko (who accompanied Kotzebue in the *Rurick*), together with M. Etoline, a naval officer in the Russian American Company's service, were annually despatched, in two vessels, to explore the American coasts. To these officers we owe the more accurate knowledge of all the coast from Cape Newenham to Norton Sound, which Capt. Cook could not approach on account of the shoal water.

this breadth to the distance of 30 miles; it then turns rather more to the eastward, and its breadth begins to diminish. In lat. 58° 57′ the river is still 3 miles broad, and it is here, on the left bank, that the company have founded an establishment, which is called *Alexandroffsk*. A bank 15 miles in length lies before this establishment; and it is between this bank and the main land that there is a passage to the Road of Alexandroffsk. Baron Wrangel, who explored the Nouchagak in 1832, discovered a passage between these banks, with which the river is filled during the ebb. The tides are very strong in the river; they rise 23 feet in the summer months, and 47 feet in the autumn.* The ebb lasts an hour longer than the flood tide, and its force is from 4 to 5 miles an hour.

The extremity of Cape Constantine is in lat. 58° 29′ N., and 158° 45′ W. It is the same land that Cook saw on July 10th, 1778, but could not approach for the shoal water, as the ships were obliged to stand off in 4 fathoms water. The examination of Lieut. Chramtschenko proves that Cape Constantine is surrounded by sand-banks to the distance of 4 leagues to the southward; in general, all the coast between the mouths of the Nouchagak and Bristol Rivers is bounded by similar sand-banks, which extend so far out to sea that the space comprised between Cape Constantine and the coast, considered as the outer entrance to the Nouchagak River, is narrowed to 7 miles. The eastern point of this bay, forming this entrance, is named *Cape Etoline*. The bay itself is called *Chramtschenko Bay*, after the Russian surveyor.

From Cape Constantine the coast takes a westerly direction, and forms, with Cape Newenham, an extensive bay of 35 leagues in extent, into which the two rivers, *Kululack* and *Tujugiak*, discharge themselves. There are several islands in this bay. Cook only saw the easternmost, which he named *Round Island*. It is an elevated hill, about 7 miles from the continent, and obtained this name from its figure. To the N.W. of it are two or three hillocks, that appeared like islands, but, as the ships passed at too great a distance, this was not determined. At the distance of 20 miles to the West of Round Island is a larger one, about 50 miles in circumference, which has been named *Hagemeister Island*, after the captain of that name, who was for some time commander of the Russian Company's colonies here. Cook saw its southern point without suspecting it to be part of an island:—"It is an elevated point, which obtained the name of *Calm Point*, from our having calm weather when off it."—(Cook, vol. ii. p. 431.) Between this point and Round Island there are four other islands in a N.E. direction. Lieut. Chramtschenko passed between these islands, and between the main land and Hagemeister Island. This channel is about 8 miles in length, and it may be passed through either from the North or South; the anchorage in it is safe throughout. It is called the *Hagemeister Channel*. At 8 leagues to the West from Calm Point is a point which is without any name on Cook's chart, and behind which, according to Chramtschenko, there is a small bay, which affords an anchorage.

* In this great rise of tide it resembles the channel of the same name in the West of England, and all other similarly constituted bays.

CAPE NEWENHAM is 5 leagues to the West of this bay. The following is Capt. Cook's account of it:—"At five in the morning of the 16th July, 1778, the fog having cleared up, we found ourselves nearer the land than we expected. Calm Point bore N. 72° E., and a point 8 leagues from it, in the direction of West, bore N. 3° E. 3 miles distant. Between these two points the coast forms a bay, in some parts of which the land was hardly visible from the masthead. There is also a bay on the N.W. side of this last point, between it and an elevated promontory, which at this time bore N. 36° W. 16 miles distant. At nine I sent Lieut. Williamson to this promontory, with orders to land, and see what direction the land took beyond it, and what the country produced; for from the ships it had but a barren appearance. We found here the flood tide setting strongly to the N.W. along the coast. At noon it was high water, and we anchored in 24 fathoms, 4 leagues distant from the shore. At five in the afternoon, the tide making in our favour, we weighed, and drove with it, for there was no wind.

"Soon after Mr. Williamson returned, and reported that he had landed on the point, and having climbed the highest hill, found that the farthest part of the coast in sight bore nearly North. He took possession of the country in his Majesty's name, and left on the hill a bottle, on which was inscribed, on a piece of paper, the names of the ships, and the date of the discovery. The promontory to which he gave the name of *Cape Newenham*, is a rocky point of tolerable height, situated in lat. 58° 42′, and in lon. 197° 36′. Over, or within it, are two elevated hills, rising one behind the other. The innermost, or easternmost, is the highest. The country, as far as Mr. Williamson could see, produces neither tree nor shrub. The hills are naked; but on the lower grounds grew grass, and other plants, very few of which were in flower. He saw no other animal but a doe and her fawn, and a dead sea-horse, or cow, on the beach. Of these animals we had lately seen a great many." *

Off the westernmost point of the cape there is a small island, according to the Russian charts, named *Sea-Lions' Island*. From Cape Newenham the coast trends to the northward, as before stated, and here commence the sand-banks and shoals lying before the mouth of the great river *Kuskowime*, or *Kouskoquim*, which discharges itself into the sea in lat. 59° 50′, and which has been explored by Lieuts. Chramtschenko and Etoline.

To the N.E. of Cape Newenham is *Tchagvan Bay*, 4½ miles long and 3 broad. It is 2 cables' length in width at the entrance, the sides of which are lined with shoals.

The BAY OF GOOD NEWS is 16 miles to the North of Tchagvan Bay. It was reached by the land expedition of Oustugoff and Korsakoff † in 1818-19. If it received its name from the reports gathered of some white-bearded men on the banks of the Kvikhpak River, it would be more properly called the Bay of False Alarms. It was examined in 1821 by M. Etoline. From his observations, the extremity of the gravel bed, which bounds the opening of the bay to the

* Cook's Third Voyage, vol. ii. pp. 431-2. Capt. Lütke says, that the sailors in this part call it the *Black Cape*, from its gloomy appearance.—*Memoire*, p. 256.

† See Hist. Chronol. de Berg., part ii. p. 21.

North, is in lat. 59° 3′ 9″, and lon. 161° 53′. Cape Newenham bears S. 24° W. 24 miles from it. It penetrates 8½ miles in an E.N.E. direction, and is 5½ miles broad. The opening is narrowed by beds of gravel to half a mile wide. Its shores are surrounded with shoals, so that there is only good anchorage in the middle, and that not too far in. In entering keep on the North side. The establishment of the port is near 6¼ hours, the greatest rise 13 to 14 feet. Magnetic variation, in 1821, 22° 17′ E.

Capt. Cook endeavoured to proceed northward along this coast, and penetrated to lat. 69° 37½′, but was prevented from getting farther from the shallowness and intricacy of the channel, which, though he thought might have a northern outlet, yet the attempting it would lead to great risk and loss of time. He then attempted to proceed to the westward, but was prevented by the extensive shoals lying to the West of the channel he had entered, and which dried partly at low water. It was not until he had returned nearly to the parallel of Cape Newenham, by the same route, that he could get to the West, clear of the banks of sand and stones which confine it. To the northward of the cape the coast is composed of hills and low land, and appeared to form several bays. The northernmost part of the coast that he could see he judged to be in lat. 60°, and seemed to form a low point, which obtained the name of *Shoal Ness.*

The tide of flood sets to the North, and the ebb to the South. It rises and falls, upon a perpendicular, 5 or 6 feet; and it was reckoned that it was high water, on the full and change days, at eight o'clock. Variation, 22° 56′ 51″ E.

NUNIWACK ISLAND* was discovered by Capt. Wassilieff in 1821. It is to the N.W., and 40 leagues distant from Cape Newenham. It is 70 miles in extent from East to West, and about half that in breadth. Its N.E. extremity is in lat. 60° 32′ N., lon. 165° 30′ W.; and its S.E. point is 60° N., and 165° 3′ W. A channel, 20 miles in breadth, separates it from the continent, which here forms a large cape in lat. 60° 44′ N., lon. 165° W., discovered by M. Etoline, to which he gave the name of *Cape Vancouver*, and that of Cook to the strait; but Admiral Krusenstern says, that as there is another strait bearing the name of the famous circumnavigator, it will be better to name it *Etoline Strait*, after the officer who first passed it.

According to M. Etoline, its N.E. extremity lies 18 or 19 miles to S. 82° W. (*true*) from Cape Vancouver. Its West extreme, which forms a moderately high, steep cape, was determined by M. Chramtschenko (who discovered the island in the company's vessels nearly at the same time as M. Etoline) to be in lat. 60° 13′. From this the coast on one side trends to E.N.E., and on the other to S.E. by compass. In approaching it from the West, the island presents a level coast, not high, and terminating to seaward in reddish cliffs. In the centre of the island some mountains of moderate height rise in a gentle slope. There is neither tree nor bush; but some villages were seen. There are several places where anchorage may be found; but the best place is in the strait on the continental side, where the depth is 6 to 8 fathoms, gravel bottom.

* *Nuniwack*, or, as it is otherwise spelt, *Nounivak*, or *Nounivok* (Lutke), was so named by the company's officers, Etoline and Chramtschenko, who, simultaneously with Wassilieff, discovered it in 1821, after the native appellation. The latter called it, from his ship, *Discovery Island*, but the first name is far the best.

Cape Avinoff, the S.E. limit of this strait, is in lat. 59° 42′. It is not high, but at a distance resembles an island. It is surrounded by shoals to the distance of 7 or 8 miles, so that it cannot be approached even in a boat within this distance. This bank lines the coast as far as Cape Vancouver itself.

Capt. Cook, who preceded all other navigators in examining the coast, says:—"No land was seen to the southward of Point Shallow Water,* which I judge to lie in the latitude of 63°; so that between this latitude and Shoal Ness, in lat. 60°, the coast is entirely unexplored. Probably it is only accessible to boats or very small vessels; or, at least, if there be channels for large vessels, it would require some time to find them; and I am of opinion that they must be looked for near the coast. From the masthead the sea within us appeared to be chequered with shoals; the water was very much discoloured and muddy, and considerably fresher than that of any of the places where we had lately anchored. From this I inferred that a considerable river runs into the sea in this unknown part."†

CAPE ROMANZOFF, or ROUMIANTSOFF, was thus named after the great statesman by MM. Chramtschenko and Etoline, although Capt. Schischmareff had already seen it two months previously, that is, in June, 1821. It forms the western extremity of that vast and marshy appendage to the American continent, bounded on the North by the Kvikhpak River, and on the South by Cook's River. M. Etoline determined its latitude as 61° 53′; M. Chramtschenko as 61° 50′ 5″; and its longitude was deduced as 166° 28′. It is thus about 85 miles N.W. from Cape Vancouver. It is high and bluff, and in the middle of August was still partly covered with snow, which well distinguishes it from the low and sandy shores to the North and South of it. It is entirely free from wood, like the adjacent coasts. Seen from a great distance, it shows like islands, and all navigators believed it to be so until its character was determined. On either side of it the coast trends N.E. and S.E.

Between this cape and Cape Shallow Water the coast has never been seen by any navigator, and if it is apparently well marked on the charts, it is rather from conjecture and from native report. All that is certainly known is that it is low, marshy, with here and there capes and small hills. At 20 miles from the coast a chain of low hills rises. There is no wood to be seen on any part between this and Golovnine Bay. The whole coast is intersected by great and small rivers, some of which, as the *Kvikhpak* and the *Kvikhluak*, form immense gulfs on reaching the sea. The first of these rivers has a long course, and its banks are well peopled. One of the company's vessels attempted to ascend it in 1831, but without success. Almost all these rivers communicate with each other.

The shoals which line the space we have been speaking of, according to what the inhabitants say, do not reach to the coast itself, but leave between a channel deep enough for ships; it is only off the mouths of the rivers that it is broken by other banks, between which are the passes. This formation is attributed to the ices brought down by the streams. Thus the waters from the rivers Kvikhpak and *Postoliuk*, running to the North, and those of the Kvikhluak and the Kouïmla, to the South, form channels on either side. M. Tébenkoff supposes that great

* Called by the natives *Asiatchak*.

† Cook's Third Voyage, vol. ii. pp. 480-90.

depths will be found in many places between the banks. The bottom throughout the extent they occupy is a vast collection of sand.

STUART'S ISLAND lies in lat. 63° 35′. It is 6 or 7 leagues in circuit. Some parts of it are of a middling height; but in general it is low, with some rocks lying off the western part. The coast of the continent is for the most part low land, but high land was seen up the country. It forms a point opposite the island, which was named by Cook *Cape Stephens*, and lies in lat. 63° 33′, and lon. 197° 41′. Some drift-wood was seen upon the shores both of the island and of the continent, but not a tree was perceived growing upon either. One might anchor, upon occasion, between the N.E. side of this island and the continent, in a depth of 5 fathoms, sheltered from westerly, southerly, and easterly winds. But this station would be wholly exposed to the northerly winds, the land in that direction being at too great a distance to afford any security.*

To the northward of Stuart's Island, Capt. Cook passed two small islands against the coast, and as he ranged it several people appeared upon the shore, and, by signs, seemed to invite them to approach.

Besborough Island was seen at 15 leagues off by Cook, and though it lies 6 or 7 miles from the continent, has no channel inside it for ships, on account of the shallowness of the water; and to the N.E. of it is *Chaktolimout Bay*. Capt. Cook anchored in it September 7th to 16th, 1778. He says it is but an indifferent station, being exposed to the South and S.W. winds. But he was so fortunate as to have the wind from the North and N.E. all the time, with remarkably fine weather. This gave them the opportunity of making seventy sets of lunar observations. The mean result of these made the longitude of the anchoring place on the West side of the sound to be 197° 13′, lat. 64° 31′. Variation, 25° 45′ E. Dip of the needle, 76° 25′.

Of the tides it was observed, that the night flood rose about 2 or 3 feet, and that the day flood was hardly perceivable.†

CHAKTOLIMOUT BAY, called by Cook *Chacktoole*, is an open bay between Capes Denbigh and Stephens. It is surrounded by a low shore, where the water is so shoal that, as before mentioned, there is no passage for ships between Besborough Island and the main, though it is 6 or 7 miles off it. The southern part of the bay was examined in 1831 by Lieut. Tébenkoff, whose description is given at length by Lütke. The following is an extract:—

"TEBENKOFF COVE.—On the North side of Cape Stephens, at 11 miles to the East of the North extremity of Stuart Island, is a cove discovered by me in 1831. It penetrates 1½ miles to the S.S.W., and is closed on the South by a small low island, on each side of which is a strait. The bay is not more than a mile broad. At its West extremity are two islets, very close to the coast. As far as the middle of the bay there are 21 to 24 feet water. The anchorage under the West side, opposite the village, is only exposed to N.N.W. and N.E.; but even with these winds there is no heavy sea.

"On the West side, two-thirds of a mile to the South of the village, the Russian Company have recently erected a small fort.

* Cook's Third Voyage, vol. ii. p. 488.

† *Ibid.* p. 485.

"The entrance into the cove is not at all difficult. After reaching Stuart Island you can run parallel with the coast at the distance of a mile in not less than 4 fathoms water; then you can range very close to the two islets on the West side of the cove. Coming from the North you must make either Besborough or Egg Islands; the first lies N. 5° W. from the cove, and the second at 9 miles N.N.E. by compass. Egg Island is smaller and lower than Besborough. The latitude of the anchorage is 63° 28′ 30″, the longitude 161° 52′ West. Magnetic variation, 30° East."*

CAPE DENBIGH is 17 leagues from Besborough Island, in a direction N. 27° E. It is the extremity of a peninsula, united to the continent by a low neck of land, on each side of which the coast forms a bay, that to the southward being the Chaktolimout Bay just mentioned.

"The berries to be got here were wild currant berries, hurtle berries, partridge berries, and heath berries. I also went ashore myself, and walked over part of the peninsula. In several places there was very good grass; and I hardly saw a spot on which some vegetable was not growing. The low land, which connects this peninsula with the continent, is full of narrow creeks, and abounds with ponds of water, some of which were already frozen over. There were a great many geese and bustards, but so shy, that it was not possible to get within musket-shot of them. We also met with some snipes; and on the high ground were partridges, of two sorts. When there was any wood, mosquitoes were in plenty. Some of the officers, who travelled farther than I did, met with a few of the natives of both sexes, who treated them with civility.

"It appeared to me that this peninsula must have been an island in remote times, for there were marks of the sea having flowed over the isthmus; and, even now, it appeared to be kept out by a bank of stones, sand, and wood, thrown up by the waves. By this bank it was evident that the land was here encroaching upon the sea, and it was easy to trace its gradual formation." — (Capt. Cook.)†

The whole of the beach around the bay seemed to be covered with drift-wood; but on account of the shoals, which extend quite around, to the distance of 2 or 3 miles from the shore, it is impossible to get it off.

The *head of Norton Sound* was partially explored by Mr. King, one of Cook's officers. From the heights, on the West side of the inlet, the two coasts were seen to join, and the inlet to terminate in a small river or creek, before which were banks of sand or mud, and everywhere shoal water. The land, too, was low and swampy for some distance to the northward; then it swelled into hills; and the complete junction of those on each side of the inlet was easily traced.

From the elevated spot on which Mr. King surveyed the sound, he could distinguish many extensive valleys, with rivers running through them, well wooded, and bounded by hills of a gentle ascent and moderate height. One of these rivers, to the N.W., appeared to be considerable; and, from its direction, he was inclined to think that it emptied itself into the sea at the head of the bay.

* Voyage du *Séniavine*, Part. Naut., pp. 249-50.

† Vol. ii. pp. 483-4.

Some of his people, who penetrated beyond this into the country, found the trees larger the farther they advanced.*

Bald Head forms the north-western limit of this inner sound, and is 20 miles to the northward of Cape Denbigh. On the West side of Bald Head the shore forms a bay, in the bottom of which is a low beach, where Capt. Cook saw a number of huts, or habitations of the natives. At about 20 miles to the W.S.W. of this point the coast projects out into a bluff head, composed of perpendicular strata of a rock of a dark-blue colour, mixed with quartz and glimmer. There joins to the beach a narrow border of land, covered, when Cook landed (September 10th, 1778), with long grass and some angelica. Beyond this the ground rises abruptly. At the top of this elevation they found a heath, abounding with a variety of berries; and farther on the country was level, and thinly covered with small spruce trees, and birch and willows, no bigger than broom stuff. They observed the tracks of deer and foxes on the beach; on which also lay a great quantity of drift-wood; and there was no want of fresh water.

The soundings off this shore are very shoal, not being more than 6 fathoms at a league off, and decreasing to 3 and under to the eastward. To the S.W. of this point the country is covered with wood, an agreeable sight, compared with that to the North. The coast continues in a S.W. direction as far as Cape Darby, where it turns to the North and West.

CAPE DARBY is in lat. 64° 21′, lon. 197°. Capt. Cook anchored off it in a quarter less 5 fathoms, half a league from the coast, the South point of which bore S. 26° W.; Bald Head, N. 60° E. 26 miles distant; and Besborough Island, S. 52° E. 15 leagues distant:—"As this was a very open road, and consequently not a safe station, I resolved not to wait to complete water, as that would require some time; but only to supply the ships with wood, and then go to a more convenient place in search of the other article. We took off the drift-wood that lay upon the beach. In the afternoon I went ashore, and walked a little way into the country, which, when there was no wood, was covered with heath and other plants, some of which produce berries in abundance. All the berries were ripe; the hurtle berries too much so; and hardly a single plant was in flower. The undermost, such as birch, willows, and alders, rendered it very troublesome walking amongst the trees, which were all spruce, and none of them above 6 or 8 inches in diameter. But we found some lying upon the beach more than twice this size. All the drift-wood in these northern parts was fir. I saw not a *stick* of any other sort."†

GOLOVNINE BAY.—On the West side of Cape Darby is Golovnine Bay, discovered in 1821 by Capt. Chramtschenko. The natives here term it *Tatchik*. Its opening is limited on the East by Cape Darby, and on the West by *Cape Kamennoï*, or Rocky, lying 7 miles W.N.W. of the first. The latter cape was so called from a flat and bare rock close to it. These two capes are high and steep, but Cape Darby is the highest. The bay extends first to North, and then to N.W. At 3 miles to the North of Cape Kamennoï, there is a bed of gravel across

* On the recent maps of this part of America a river is represented as falling into this sound, up which is a trading post.

† Cook, vol. ii. p. 479.

is, running off to the East from the West side. At 2 miles from the East side of the bay, which is opposite to it, it is terminated by a reef of uncovered rocks, which, at a distance, is extremely like an artificial pier or mole, whence it is called the *Stone Mole*, or *Kamennaia Pristan*: at the extremity of this bed of gravel is a temporary or summer village. On the left or N.W. side, in a valley, is a forest of pine trees; and beyond the bed of gravel the bay extends 8 miles farther to the N.W., but is so shallow as to be impassable for boats in some places.

There is anchorage in all parts of the outer bay as far as the Stone Mole: bring it to bear from W. to W.S.W.; within this the depth rapidly decreases. M. Tebenkoff anchored in 3½ fathoms, with the Stone Mole bearing N.W. or N.W. ½ N.; Cape Darby, S. 72° E.; and Cape Kamennoi, S. 2° W. by compass. In this situation there is the double advantage of collecting more readily the drift-wood thrown on to its West side, and to water in the small river which discharges itself into the West side of the gulf. This point is in lat. 64° 26′ 42″. The longitude was assumed to be 163° 8′. The bay is perfectly clear throughout; but, as it is open to winds from the South, the anchorage is not without some danger from this cause.

The time of high water, the establishment, is 6h 23′. At full moon it rises 3 feet 8 inches.

The inner bay communicates with the outer by a narrow gullet. At its head a large river discharges by five mouths, which makes the water of the bay fresh. This great river, it is probable, communicates with an opening in the mountains, which was seen to N. 40° W. from the head of the bay, and, according to some native information acquired by M. Chramtschenko, there is river communication all the way to the Lake Imou-rouk, in Grantley Harbour, Port Clarence, so that this part of the coast is in reality an island. It took them five days to traverse this inland navigation in their baidars.*

Aziak, *Azjiak*, or *Ayak Island*, called by Cook *Sledge Island*, on account of a sledge being found by him on it, is 10 or 12 miles S.E. ½ E. of Cape Rodney. Its latitude is there given as 64° 30′. Its longitude will be about 166°. According to Cook it is about 12 miles in circumference. M. Tebenkoff does not think it to be more than a mile. He describes it as a rock rising 642 feet above the water. A low point projects on its North side; and, on the East, a village stands on the slope of the rugged coast. The island may be approached on all sides. The anchorage is bad to the East; the bottom is of large stones. It is better to the North, near the point, although the current runs here 3 knots, but the bottom is much better. There is neither tree nor shrub on the island, but among the moss a number of edible plants may be found. The variation is 34° E.

Oukivok, or King's Island of Capt. Cook, is a rock 756 feet in height, not more than a mile in circuit, and cliffs on all sides. There is a village, the houses excavated in the rocks, on a rugged slope, at 150 feet above the sea. The inhabitants are doubtlessly attracted by the number of morses which come round the island.

Capt. Cook's account of it is thus:—" The surface of the ground is composed

* Lütke, Voyage, &c., Part. Naut., pp. 246-7.

chiefly of large loose stones that are in many places covered with moss and other vegetables, of which there were above twenty or thirty different sorts, and most of them in flower; but I saw neither shrub nor tree, either upon the island or on the continent. On a small low spot, near the beach where we landed, was a good deal of wild purslain, peas, longwort, &c., some of which we took on board for the pot. We saw one fox, a few plovers, and some other small birds; and we met with some decayed huts that were partly built below ground. People had lately been on the island, and it is pretty clear that they frequently visit it for some purpose or other, as there was a beaten path from the one end to the other. We found, a little way from the shore where we landed, a sledge, which occasioned this name being given to the island." *

POINT RODNEY is a low point to the N.W. ½ W, of Sledge Island. "The island before us, which we supposed to be the continent of America, appeared low next the sea, but inland it swelled into hills, which rise one behind another, to a considerable height. It had a greenish hue, but seemed destitute of wood, and free from snow. While we lay at anchor we found that the flood tide came from the East, and set to the West, till between 10 and 11 o'clock; the stream set to the eastward, and the water fell 3 feet. The flood ran both stronger and longer than the ebb, from which I concluded that, besides the tide, there was a westerly current."†

Point Rodney being low, and the water shallow, it is difficult to land. From the beach to the foot of the mountain there is a plain, about 2 miles wide, covered with lichens and grass, upon which Capt. Beechey observed several herds of reindeer feeding; but the communication is in places interrupted by narrow lakes, which extend several miles along the coast. Upon the beach is a greater abundance of drift-wood than is found on other parts of the coast. Several indications of the presence of inhabitants were found. About 2 miles from the coast the country becomes mountainous, and far inland rises to peaked hills of great height, covered with perennial snow.

PORT CLARENCE, which was explored and named by Capt. Beechey, August, 1827,‡ is 5 leagues to the northward of Point Rodney. It was passed unnoticed by Cook in his passage through the strait, but this is not surprising, from the character of the land forming it.

POINT SPENCER, the North extremity of a low spit of land, projecting about 10 miles from the coast, forms the southern protection of this spacious harbour. It here forms a right angle, having a channel about 2 miles wide between its extremity and the northern shore. This southern side of Port Clarence is a low diluvial formation, covered with grass, and intersected by narrow channels and lakes. It projects from a range of cliffs which appear to have been once upon the coast, and sweeping round, terminates in the low shingly point before named, Point Spencer. In one place this point is so narrow and low, that in a heavy gale of wind the sea must almost inundate it; to the northward, however, it becomes wider and higher; and by the remains of some yourts upon it has at one

* Cook, vol. ii. p. 441. † *Ibid.* vol. ii. p. 440.

‡ Port Clarence was for a long time previously known to the Russians as *Kaviayak Bay*, but they did not know that it contained its excellent port.—*Lütke.*

time been the residence of Esquimaux. Like the land just described, it is intersected with lakes, some of which rise and fall with the tide, and is covered, though scantily, with a coarse grass. Near Point Spencer the beach has been forced up, by some extraordinary pressure, into ridges, of which the outer one, 10 or 12 feet above the sea, is the highest. Upon and about these ridges there was a great quantity of drift timber, but more on the inner side of the point than on the outer.

The northern and eastern shores of Port Clarence slope from the mountains to the sea, and are occasionally terminated by cliffs. The soil is covered with a thick coating of moss, among which are a few plants. The valleys and hollows are all filled with dwarf willow or birch.

GRANTLEY HARBOUR forms an inner harbour to the extensive and excellent bay just described. The channel into it from the outer harbour is extremely narrow, the entrance being contracted by two sandy spits; but the water is deep, and in one part there is not less than 12 fathoms. At the upper end of the harbour is a second strait, about 300 yards in width, formed between steep cliffs, but this channel, too, is contracted by sandy points. This strait probably communicates with a large inland lake (*Imou Rouk*), as described by the Esquimaux, as a large body of nearly fresh water made its exit through it. At the entrance of the strait, called *Toshook* by the natives, Beechey found an Esquimaux village, and upon the northern and eastern parts of the harbour there were two others.

There are very few natives in the outer harbour. On the northern side there is a village of yourts, to which the inhabitants apparently only resorted in the winter. Some Esquimaux fishermen, upon the low point of the inner harbour, had been very successful.

"These two ports," says Capt. Beechey, "situated so near Behring's Strait, may, at some future time, be of great importance to navigation, as they will be found particularly useful by vessels which may not wish to pass the strait in bad weather. The outer harbour, which for convenience and security surpasses any other near Behring's Strait with which we are acquainted, I attached the name of Port Clarence, in honour of his most gracious Majesty, then Duke of Clarence. To the inner, which is well adapted to the purposes of repair, and is sufficiently deep to receive a frigate, provided she lands her guns, which can be conveniently done upon the sandy spit at the entrance, I gave the name of Grantley Harbour, in compliment to Lord Grantley."*

Point Spencer is in lat. 65° 16′ 40″, lon. 166° 47′ 50″ W.; variation, 26° 36′ E. High water, full and change, in the port, 4h 25′.

Point Jackson, named, like the last, from a distinguished naval officer, forms the North side of the entrance to Port Clarence. Off it the water is more shallow than usual.

CAPE YORK, named after the Duke of York, is a bold promontory, and near it there is probably a river, called *Youp-nut* by the natives. From hence to Cape Prince of Wales the coast is of quite a different character to that to the north-

* Voyage of the *Blossom*, p. 543.

ward of the latter, being bounded by steep, rocky cliffs, and broken by deep valleys, while the other is low, swampy ground.

CAPE PRINCE OF WALES is the westernmost extreme of America. This celebrated promontory is the western termination of a peaked mountain, which, being connected with the main by low ground at a distance, has the appearance of being isolated. The promontory is bold, and remarkable by a number of ragged points and large fragments of rock lying upon the ridge, which connects the cape with the peak. About a mile to the northward of the cape some low land begins to project from the foot of the mountain, taking first a northerly, and then a north-easterly direction, to Schischmareff Inlet. Off this point is a dangerous shoal, upon which the sea breaks heavily. The natives have a village upon the low land near the cape, called *Eidannoo*, and another island called *King-a-ghe*, which appears to be a considerable mart; and as they generally select the mouths of rivers for their residence, it is not improbable that a stream may here empty itself into the sea, which, meeting the current through the strait, may occasion the shoal. About 14 miles inland from Eidannoo there is a remarkable conical hill, often visible when the mountain tops are covered, which, being well fixed, will be found useful at such times by ships passing through the strait. Twelve miles farther inland the country becomes mountainous, and is remarkable for its sharp ridges. The altitude of one of the peaks, which is nearly the highest in the range, is 2,596 feet. These mountains, being covered with snow when the *Blossom* was here (August, 1827), gave the country a very wintry aspect.

Off the cape is a very dangerous shoal, stretching to the N.E. from it. It takes the direction of the current, and is extremely dangerous, in consequence of the water shoaling so suddenly, and having deep water within it, by which a ship coming from the northward may be led down between the shoal and the land, without any suspicion of her danger. Though the *Blossom* found nothing less than twenty-seven feet water, as near as the soundings could be ascertained in so high a sea, yet, from the appearance of the breakers outside the place where she crossed, the depth is probably less. It is remarkable that this spit of sand, extending so far as it does from the land, should have escaped the observation of the Russians, as well as of our countrymen. Cook, in his chart, marks 5 fathoms close off the cape, and Kotzebue 3 fathoms, but this spit appeared to extend 6 or 7 miles from it. The spit may be extending rapidly.

THE DIOMEDE ISLANDS are three small islands occupying a conspicuous geographical position, as they lie between the nearest points of the two great continents of Asia and America, being thus in the very narrowest portion of Behring's Strait.

They have been the subject of some slight dispute as to their real number. Our celebrated Capt. Cook places *three* islands here in the middle of the strait. Kotzebue imagined that he saw a *fourth*, which must either have been overlooked by our navigators, Cook and Clerke, or that it had been subsequently raised by an earthquake; but this statement was received with some doubt, even by the Russians themselves. The subject was set at rest by Capt. Beechey, in the *Blossom*, in 1826. They are *three* in number.

The south-eastern of the three islands is a high square rock, named by Capt.

Beechey the *Fairway Rock*, and by the native who drew a chart of this region, Oo-ghe-e-ak. It is an excellent guide to the eastern channel, which is the widest and best.

The central island was named, after the Admiral, *Krusenstern Island*, and by the above-mentioned authority, Igna-look. It is an island with perpendicular cliffs and a flat surface.

The third, or north-western one, which is the largest, was named by Capt. Beechey, after Kotzebue's supposed discovery, *Ratmanoff Island*, and Noo-nar-book by the native. It is 3 miles long, high to southward, and terminates, in the opposite direction, in low rocky cliffs, with small rocky points off them.

East Cape, in almost every direction, is so like an island, that Capt. Beechey had no doubt it was the occasion of the mistake committed by the Russian navigator.*

From Cape Prince of Wales the coast trends to the northward, the water being shallow just to the North of it, and from it there extends a shoal for 8 or 10 miles, a tongue of sand and stones, perhaps the effect of current, on which the *Blossom* was nearly lost in 1827. Outside it the bottom is mud, but within the spit the water deepens and again changes to mud, and having 10½ fathoms within 2½ miles of the coast.

The coast itself is low, with a ridge of sand extending along it, on which are or were some Esquimaux habitations. The land behind is marshy in the summer, and extends without anything remarkable for 35 miles to the entrance of Schischmareff Inlet.

SCHISCHMAREFF INLET was so named by Kotzebue. The width between the opening, *Cape Lowenstern* being the N.E. point, is 10 miles, and the inlet itself extends to the E.S.E. as far as the eye can reach. It has not been explored, but it has been surmised that it may be part of the mouth of a very large river, the other branches of which debouch in the Bay of Good Hope, or Kotzebue Sound. But this is mere conjecture, and the Esquimaux give no indication of such a fact.

Sarytscheff Island lies immediately before its entrance. This island is low and sandy, and is apparently joined under water to the southward to the strip of sand which fronts the coast extending towards Cape Prince of Wales. Capt. Beechey noticed upon it a considerable village of yourts, the largest that he had then met with, coming from the southward. They appeared to prefer having their dwellings upon this sandy foundation to the main land, probably on account of the latter being swampy, which is the case everywhere in the vicinity of this inlet and Kotzebue Sound. Mountains were seen at the back of the Schischmareff Inlet, but the coast was not visible—probably it is low.

The channel into this inlet, on the northern side of Sarytscheff Island, is extremely narrow and intricate, and the space is strewed with shoals.†

The coast to the N.W. is low and swampy, with small lakes inland. The ridge of sand continues along the coast to Cape Espenburg, and then terminates.

KOTZEBUE SOUND.—The land on the South side of the Sound, as far as the Bay of Good Hope, is higher, more rocky, and of a bolder character than

* Voyage of the *Blossom*, vol. i. p. 246.

† Kotzebue.

the opposite shore, though it still resembles it in its swampy, superficial covering, and in the occurrence of lakes wherever the land is flat. Under water, also, it has a bolder character than the northern side, and has soundings of 4 and 5 fathoms, quite close to the promontories. There are two or three places under these headlands which, in case of necessity, will afford shelter to boats, but each with a particular wind only; and in resorting thither the direction of the wind, and the side of the promontory, must be taken into consideration.

CAPE ESPENBURG is about 30 miles from Schischmareff Inlet. It is formed by the sea to the southward of it into a narrow strip of land, upon which are some high sand-hills. Capt. Beechey found a great many poles placed erect on it, and the traces of Esquimaux, though then quite deserted. The beach was composed principally of dark-coloured volcanic sand, and strewed with dead shells; the sand-hills partly covered with grass and shrubs producing berries. Upon the peninsula, also, are some lakes, frequented by wild fowl.

From Cape Espenburg the shores of Kotzebue Sound trend to the northward, but are fronted by an extensive bar of sand, extending nearly 2 miles off the land; this continues for 10 miles, to the entrance of an inlet, which enters the sound between low mud capes.

The whole of the western side of Kotzebue Sound is formed by a vast plain of low ground, extending to Cape Espenburg, on which are several lakes and creeks, which may communicate with the inlet above alluded to, or they may, altogether, with Schischmareff Inlet, form the mouths of some large river. But as this cannot serve any useful purpose for navigation, it has not been explored.

The inlet in question was considered by Kotzebue to have evidences that it would lead to some great inland discovery; with this view Capt. Beechey attempted its exploration, but in beating up for it in the *Blossom* he was obliged to desist, from the shallowness of the water. He therefore dispatched his gig, but from the shoal she could not get within a cable's length of the breakers. At the top of the tide, when the water is smooth, perhaps small boats may enter, but under any other circumstances a ducking must be the consequence, as was the case with Kotzebue on repassing it.

The BAY of GOOD HOPE forms the S.W. angle of Kotzebue Sound. In its southern part a river enters, which the Esquimaux say extends inland five days' journey for their baïdars, but its entrance is too shallow for boats. To the eastward of this river the coast is low, swampy, and intersected by lakes and rivers. Here a range of rounded hills, which bound the southern part of the Sound, branch off inland to the West, and a distant range of a totally different character rises over the vast plain that we have been describing, as extending to Cape Espenburg.

To the eastward of this low coast commences a series of low points, with small bays between them: and these points are singular as affording volcanic rocks and lava. Upon them, as well as upon the shores of the bay, large blocks of vesicular and other lava are accumulated. The country here slopes gradually from some hills to the beach, and is so well overgrown, that Capt. Beechey could not examine its substrata, but they did not, in outward formation, exhibit indication of volcanic agency.

Gull Head, a narrow, rocky peninsula, stretching a mile into the sea, is 4 miles East of these points. It is principally composed of slaty limestone, containing particles of talc, in larger or smaller quantities, as it is elevated above or on a level with the sea, but without any visible stratification. This part of the coast is interesting in a geological point of view, as being the only part of the Sound where volcanic rocks occur.

Cape Deceit, a bold promontory, with a conspicuous rock off it, so named by Kotzebue, is 3 miles beyond Gull Head. It is also of compact limestone, devoid of any visible stratification. To the eastward of it, in the first and second bays, are the mouths of two rivers, at which Capt. Beechey found several spars and logs of drift-wood placed erect, showing that the natives had occupied these stations in the summer, for the purpose of catching fish, but they were then deserted. Both the rivers had bars across their entrances, preventing access for a boat. This part of the Sound appears to have but few temptations to the Esquimaux, as Capt. Beechey only saw two parties on it, and they were on an excursion; and at two places were some deserted yourts, unworthy of notice.

Spafarief Bay, at the S.E. angle of Kotzebue Sound, commences about 9 miles East of Cape Deceit. It terminates to the South in a small creek, navigable a very short distance, and that by boats only. Its whole extent inland is about 3 miles, when it separates into a number of small branches communicating with several lakes, which in spring no doubt discharge a large quantity of fresh water into the Sound; though at the dry season of the year (September) they were of inconsiderable size. A little to the northward of the creek there is a pointed hill, just 640 feet high by measurement, from whence the surrounding country is visible; it is all covered, except the summit of this hill, with a deep swampy moss.

The beach was strewed with a great quantity of drift-wood, some in a very perfect state; they were all pine trees except one, apparently a silver birch. From the mouth of this river the coast trends nearly North, by compass, about 8 miles, when it turns to the eastward, forming the southern side of Eschscholtz Bay.

ESCHSCHOLTZ BAY is a deep indentation of the S.E. shore of Kotzebue Sound, extending for 10 miles to the eastward within its entrance, which is formed by Chamisso Island and Choris Peninsula, extending from the North shore.

The land about this part of the Sound is generally characterized by rounded hills from about 600 feet to 1,000 feet above the sea, with small lakes and rivers; its surface is rent into deep furrows, which, until a very late period in the summer, were filled with water, and were covered with a thick swampy moss, and in some places with long grass or bushes; they are extremely edious to traverse on foot. Early in the summer myriads of mosquitoes infest this swampy shore, and almost preclude the possibility of continuing any pursuit; but in August they begin to die off, and soon afterwards entirely disappear.

Almost the extreme area of Eschscholtz Bay within its entrance is very shallow, so that a ship cannot advance far beyond this point, and is the North extreme part of the bay. The shore is of difficult access, on account of long muddy flats extending into the bay, and at low water drying in some places a quarter of a mile from the beach. From the destruction of the earthy cliffs surrounding it, by the summer thawing of the ice, the bay is most probably fast filling up, and at no very distant period will be left scarcely navigable for boats.

It is in the southern part of the bay that a point very interesting to naturalists occurs, which received much attention from European savans at the time of its exploration by Kotzebue, and therefore was more closely examined by Capt. Beechey, whose own words will afford the best description of the part in question.

"While the duties of the ship were being forwarded under my first lieutenant, Mr. Peard, I took the opportunity to visit the extraordinary ice formations in Eschscholtz Bay, mentioned by Kotzebue as 'being covered with a soil half a foot thick, producing the most luxuriant grass,' and containing abundance of mammoth bones. We sailed up the bay, which was extremely shallow, and landed at a deserted village on a low sandy point, where Kotzebue bivouacked when he visited the place, and to which I afterwards gave the name of Elephant Point, from the bones of that animal being found near it.

"The cliffs in which this singular formation was discovered begin near this point, and extend westward nearly in a straight line to a rocky cliff of primitive formation at the entrance of the bay, whence the coast takes an abrupt turn to the southward. The cliffs are from 20 to 80 feet in height, and rise inland to a rounded range of hills between 400 and 500 feet above the sea. In some places they present a perpendicular front to the northward; in others a slightly inclined surface; and are occasionally intersected by valleys and watercourses, generally overgrown by low bushes. Opposite each of these valleys there is a projecting flat piece of ground, consisting of the materials that have been washed down the ravine, where the only good landing for boats is afforded. The soil of the cliffs is a bluish-coloured mud, for the most part covered with mud and long grass, full of deep furrows, generally filled with water or frozen snow. Mud, in a frozen state, forms the surface of the cliffs in some parts; in others, the rock appears with the mud above it, or sometimes with a bank halfway up it, as if the superstratum had gradually slid down and accumulated against the cliff. By the large rents near the edges of the mud cliffs they appear to be breaking away, and contributing daily to diminish the depth of water in the bay."*

The appearance noticed by Kotzebue, so nearly resembling an iceberg capped with earth and grass, was found to be occasioned either by the water from the thawing ice or snow trickling down the surface of the earthy cliff from above, or by the snow being banked up against the cliff, or collected in hollows in the winter, and afterwards converted into ice by the partial thawings and freezings; the upper soil, becoming loosened by the thaw, is itself ultimately projected over the edge of the cliff, and thus gave rise to the deceptive appearance which misled the Russian officers. This fact was fully established by Capt. Beechey.

Elephant Point, before alluded to, is a low sandy projection from the interesting cliffs above described, and was named thus from the fossil remains discovered in abundance near it, and that of the animal from which it is called among the number. Some of these remains have been deposited in the British

* Voyage of the *Blossom*, vol. i. pp. 257-8.

Museum, adding one item to the wonders of geology. They are described in the Appendix to the Voyage of the *Blossom*.

BUCKLAND RIVER, so named in compliment to the learned and excellent professor, enters Eschscholtz Bay about 3 miles East of Elephant Point. The width of it here is about 1½ miles; but this space is broken into narrow and intricate channels by banks, some dry, others partly so. The stream at the time of Capt. Beechey's visit (September, 1826) passed rapidly between them, and at an early period of the season a considerable body of water must be poured into the Sound, though from the comparative width of the channels the current in the latter is not much felt. The shore around is flat, broken by several lakes, in which there were a great many wild fowl.

The country on the northern side of Eschscholtz Bay was found to be almost impassable from swamp in September, 1826, notwithstanding that the season was so far advanced. It seemed as if the peaty covering obstructed the drainage, which was kept on the surface by the frozen subsoil. The power of the sun was great, the thermometer rising to 62° in the shade. On the side of the hill that sloped to the southward, the willow and birch grew to the height of 18 feet, and formed so dense a wood that it could not be penetrated. The trees bordering on the beach were quite dead, apparently in consequence of their bark having been rubbed through by the ice, which had been formed about 9 feet above high-water mark, and had left a steep ridge of sand and shingles. The berries (in September) were in great perfection and abundance, and proved a most agreeable addition to the salt diet of the seamen.

The cliffs to the East of Choris Peninsula are composed of a green-coloured mica slate, in which the mica predominates, and containing garnets, &c.

CHORIS PENINSULA forms the western side of the bay, and extends a little to the East of South, *true*, for about 3 miles, and is separated into two portions by a sandy neck. Off its South end shoal water extends for three-quarters of a mile, and this shallow depth continues eastward, and fills the bay. Its S.W. extremity was named *Point Garnet*, and at its S.E. portion are some lakes on the low shore, where fresh water may be procured, towards the end of the season, when the Chamisso springs are exhausted.

The watering place at Chamisso Island was on the N.E. side, and also at the East end. In July it was abundant, but at the end of August, upon its being revisited for this supply, it was found that there was not a drop to be had, in consequence of the streams at which they had formerly filled the casks being derived from beds of thawing ice and snow, which were then entirely dissolved.

The eastern side of Choris Peninsula was examined, in 1849, by Capt. Kellett, R.N., for a wintering station for H.M.S. *Plover*. He found shoal water in all of them, shoaling gradually northerly towards the sandy peninsula. He was of opinion that if a vessel did winter there, she would be greatly exposed, and probably, on the breaking up of the ice, be either carried into the straits, or shoved up on to the beach.

Upon a sandy bay on the western side of the peninsula a few Esquimaux were seen. In this bay they caught enough salmon and other fish to give an acceptable meal to the whole ship's company; subsequent trials, however, were unsuccessful.

CHAMISSO ISLAND, named after the naturalist who accompanied Kotzebue, is 2 miles South of the extremity of Choris Peninsula, and in the space between them is the anchorage. The island is about a mile and a quarter in length, N.W. and S.E., by half that breadth. The highest part of it, which is 231 feet above the sea, is steep, except to the eastward, where it ends in a low, sandy point, upon which were the remains of some Esquimaux habitations.* It has the same swampy covering as the land previously described, from which, until late in the summer, several streams descend, and are very convenient for procuring water. Detached from Chamisso there is a steep rock, named, by way of distinction, *Puffin Island*, composed of mouldering granite, which has broke away in such a manner that the remaining part assumes the form of a tower. During the period of incubation of the aquatic birds every hole and projecting crag on the sides of this rock is occupied by them. Its shores resound with the chorus of thousands of the feathery tribe, and its surface presents a curiously mottled carpet of brown, black, and white.† The island is accessible in almost every quarter, and its centre is in lat. 66° 13′ 11″ North, lon. 161° 46′ 0″ West. Variation, 31° 10′ East; high water, full and change, at the anchorage, 4^h 42′.

The coast northward from Choris Peninsula extends nearly N.W. for 13 miles to *Cape Blossom*, named after Capt. Beechey's ship. This cape is an ice formation of a similar nature to that noticed in Eschscholtz Bay, only more extensive, and having a contrary aspect. The ice here was found, on examining it from above, to be detached from the cliff at the back of it, additional evidence as to its formation.

Upon the beach under the cliffs there was an abundance of drift-wood, beech and pine; one of the latter was a tree of three feet in diameter, which was fresh, and washed up between July and September, 1826. They here met with some natives laying out their nets for seals, in which they were unsuccessful. The party procured about two bushels of whortle berries that they had gathered.

The bottom of this side of the Sound is very even at 6 or 7 miles from the land to the southward of Cape Blossom, but to the northward of it a shoal extends 8 miles off the land, and is very dangerous, as the soundings give very short warning of its proximity. The distance from the shore, could it be judged of under ordinary circumstances, would on some occasions be a most treacherous guide, as the mirage in fine weather plays about it, and gives the land a very different appearance at one moment from that which it assumes at another. There are no good landmarks for it. The bearings from its extremity, in 2¼ fathoms of water, are Cape Blossom, *S.* 66° 40′ *E. true;* Western High Mount, *N.* 17° 30′ *W. true;* and the West extreme, a bluff cape near Cape Krusenstern,

* Capt. Kellett's party found (July, 1849) the cask of flour left by Beechey twenty-three years before to be quite perfect. The sand around it was frozen so hard that it required enormous exertion to get it out.

† Some notice of the visit of Kotzebue in July, 1816, and also of his lieutenant, then Capt. von Schischmareff, who paid a second visit in 1820, was found by Capt. Beechey, in 1826, upon the island.

N. 37° *W. true.* But the best way to avoid it is to go about directly the soundings decrease to 6 fathoms; as after that depth they shoal so rapidly to 2½ fathoms, that there is scarcely room to put the ship round.

HOTHAM INLET, off the entrance of which the foregoing shoal lies, escaped the notice of Kotzebue, and was named by Capt. Beechey after Sir Henry Hotham, K.C.B. It is a broad sheet of water, and extends to the S.E. 30 or 40 miles, and, at some distance up, it was fresh. This was ascertained by landing in the sound to the southward of Cape Blossom, at which place it was found that the inlet approached the sea within 1½ miles. It is only navigable for small boats, and, being fresh, it cannot lead to any sea beyond. It has but one small entrance, so very narrow and intricate, that the boats grounded repeatedly in pursuing it.

To the North and West of the opening is a sandy point, at the back of which is another inlet, which is about 5 miles long N.W. and S.E., but its entrance is so shallow, that the gulls waded across it. On this sandy neck there was a native burying ground; some of the bodies were placed on platforms of drift-wood, covered with a double tent of drift-wood, forming a conical pile.

CAPE KRUSENSTERN, about 12 miles from this point, is a low tongue of land, intersected by lakes, lying at the foot of a high cluster of hills, not in any way remarkable. The land slopes down from thence to several rocky cliffs, which, until the low point is seen at the foot of them, appear to be the entrance of the Sound, but they are nearly a mile inland from it. The coast here takes an abrupt turn to the northward, and the current sets strong against the bend; which is probably the reason of there being deep water close to the beach, as also the occasion of a shoal in a north-westerly direction from the point, which appears to have been thrown up by the eddy water.

To the northward of this point there is a shingly beach, sufficiently steep to afford very good landing when the water is smooth. Behind it there is a plain about a mile wide, extending from the hills to the sea, composed of elastic bog earth, intersected by small streams, on the edges of which the buttercup, poppy, bluebell, &c., thrive very well. In other parts, however, the vegetation was stinted.

The coast hence takes a N.W. direction, and, at 15 miles from Cape Krusenstern, there is a range of hills terminating about 4 miles from the sea, which must be the Cape Mulgrave of Capt. Cook, who navigated this part of the coast at too great a distance to see the land in front of the hills, which is extremely low, and, after passing the *Mulgrave Range*, forms an extensive plain, intersected by lakes, near the beach. These lakes are situated so close together, that, by transporting a small boat from one to the other, a very good inland navigation, if necessary, might be performed. They are supplied by the draining of the land and the melting snow, and discharge their waters through small openings in the shingly beach, too shallow to be entered by anything larger than a baïdar; one of them excepted, through which the current ran too strong for soundings to be taken.

Cape Seppings is 35 miles from the Mulgrave Hills, and 9½ miles from Cape Thompson, to the N.W. This latter, named after one of the navy commission, is a bold promontory, 450 feet in height, and marked with differently coloured strata. Capt. Beechey met here, as he had done at most parts where he landed, some Esquimaux, who were friendly and extremely good-natured and honest. He visited their village of skin tents in the adjoining valley, upon a fine stream of fresh water. Here the party were offered the utmost hospitality; several dishes were placed before them, their two choicest being the entrails of a fine seal, and a bowl of coagulated blood. These not tempting, another was offered, the raw flesh of the narwhal, cut into lumps, with equal distribution of the black and white fat. Near this cape is one that has been named Cape Ricord by the Russians.

POINT HOPE, which is 11 miles from Cape Thompson, is the western extremity of a point of land which projects almost 16 miles from the general line of coast; it is intersected by several lakes and small creeks, the entrances to which are on the North side. There is also a great abundance of drift-wood. Upon the extremity of the point there was a village of Esquimaux, the inhabitants of which were very wretched and forbidding, but gave the party a very welcome reception. The opening to the lakes is on the North side, and across this opening is a bar, consisting of pebbles and mud, which has every indication of being on the increase; but when the water is smooth a boat may enter, and she will find excellent security within from all winds. It is remarkable that both Cook and Clerke, who passed within a very short distance of this point, mistook the projection for ice that had been driven against the land, and omitted to mark it in their charts.

Capt. Beechey, in passing along the coast between this point and Kotzebue Sound, in July and August, found a *current* setting along the coast in a northerly direction, so strong, at times, that it carried the ship to leeward, notwithstanding she beat the whole day with every sail set. It varied in velocity from 1½ to 3½ miles per hour, and was strongest in-shore. It was very constant, and the water was much fresher than the ordinary sea-water. Capt. Beechey makes the following observations on it:—" It is necessary here to give some farther particulars of this current, in order that it may not be supposed that the whole body of water between the two continents was setting into the Polar Sea at so considerable a rate. By sinking the patent log first 5 fathoms, and then 3 fathoms, and allowing it to remain in the first instance six hours, and in the latter twelve hours, it was clearly ascertained that there was no current at either of those depths; but, at the distance of 9 feet from the surface, the motion of the water was nearly equal to that at the top. Hence we must conclude that the current was superficial, and confined to a depth of between 9 and 12 feet.

" By the freshness of the water alongside, there is every reason to believe that the current was occasioned by the many rivers which, at this time of the year, empty themselves into the sea, at different parts of the coast, beginning with Schischmareff Inlet. The specific gravity of the sea, off that place, was 1·02502, from which it gradually decreased, and at our station, off Point Hope, was 1·0173, the temperature at each being 58°. On the other hand the strength of the stream had gradually increased from half a mile an hour to 3 miles, which was its greatest rapidity. So far there is nothing extraordinary in the fact; but why this body of water should continually press to the northward, in preference to taking any other direction, or gradually expending itself in the sea, is a question of considerable interest."—(Part i. pp. 265-6.)

From the entrance to the lakes previously mentioned the coast runs nearly North, *true*, for 7 and 8 miles, to *Capes Dyer* and *Lewis*, respectively.

CAPE LISBURNE, of Capt. Cook, is 5½ miles North of Cape Lewis. It is a mountain which, at the point, is 849 feet above the sea, and at so short a distance from it on one side, that it was fearful to look down upon the beach below. The basis of the mountain was flint of the purest kind, and limestone, abounding in fossil shells, enchinites, and marine animals.

The mountain was ascended by a valley, which collected the tributary streams of the mountain, and poured them in a cascade from the beach. There was very little soil in the valley; the stones were covered with a thick, wet, swampy moss, which was traversed with difficulty. Vegetation, however, was as luxurious as in Kotzebue Sound, more than 100 miles to the southward, or, what is of more consequence, more than that distance farther from the great barrier of ice. Several reindeer were feeding on this luxuriant pasture; the cliffs were covered with birds, and the swamps generated myriads of mosquitoes, which were more persevering, if possible, than those at Chamisso Island.

It was at this point that the *Blossom's* barge, which had been despatched to examine the coast to the northward, met on her return with such tempestuous weather, that she doubled this windy cape with great difficulty. "At 2^h p.m., September 6th, we got within the influence of the variable winds, occasioned by the steep and high land of the cape. The bubble and agitation of the sea exceeded any idea of the kind we had formed, and broke over the boat in every direction; we had no means of extricating her. The gusts of wind that came from every quarter, lasting but a moment, left us no prospect of getting clear. We were at this time about 2 miles from the land. The wind in-shore of us blew with astonishing violence; the eddies from the hills making whirlwinds, which carried up the spray equal in height to the mountain."—(Part ii. p. 320.)

At Cape Lisburne the coast turns to the eastward at nearly right angles, and the coast being lower, some lakes are formed within the beach, which were open to the sea when the *Blossom* passed them in August, but which entrances were entirely obliterated by the waves in the following month. At about 5 miles from the cape is one of these lakes, larger than the others. Continuing eastward along the coast, at 32 miles from Cape Lisburne, we reach *Cape Sabine*, which is low, and projects but slightly from the general line of coast, which here begins to trend to the north-eastward.

CAPE BEAUFORT is situated in the depth of a great bay formed between Cape Lisburne and Icy Cape, and is the last point where the hills come close down by the sea, by reason of the coast line curving to the northward, while the range of hills continues its former direction. From the rugged mountains of limestone and flint at Cape Lisburne, there is a uniform descent to the rounded hills of sandstone at Cape Beaufort. The range, however, is broken by extensive valleys, intersected by lakes and rivers. Some of these lakes border upon the sea, and in the summer months are accessible to baïdars, or even large boats; but as soon as the current from the beds of thawing snow inland ceases, the sea throws up a bar across the mouths of them, and they cannot be entered. The beach, at the places where Capt. Beechey's party landed, was shingle and mud, the country mossy and swampy, and infested with mosquitoes. They noticed recent tracks of wolves, and of some cloven-footed animals, and saw several ptarmigans, ortolans, and a lark. Very little drift-wood had found its way upon this part of the coast.

Cape Beaufort, as before mentioned, is composed of sandstone, enclosing bits of petrified wood and rushes, and is traversed by narrow *veins of coal*, lying in an E.N.E. and W.S.W. direction. That at the surface was dry, but some pieces, which had been thrown up by some burrowing animal, probably the ermine, burned very well.

The coast from Cape Beaufort trends more to the northward, and at 17 leagues distant from it is *Point Lay*. About this point the outer coast is formed of a strip of shingle or sand, 150 yards in width, and about 6 feet above the level of the sea. Within this outer line is an extensive lake, which extends, with some interruptions, for above 120 miles, and from 2 to 6 miles broad. This lake is connected with the sea at high tide, and the water is consequently salt; yet, by digging at the distance of less than a yard from its margin, some water sufficiently fresh to drink was obtained; a resource of which the natives appeared to be well aware. An abundance of drift-wood was heaped up on the upper part of the shingle. The trees were torn up by the roots, and some were worm-eaten; but the greater part appeared to have been only a short time at sea, and all of it that was examined was pine.

At the time of the *Blossom's* visit, at the end of August, 1826, immense flocks of ducks, consisting entirely of young ones and females, were seen migrating to the southward. The young ones could not fly. From the desolate appearance of the coast, at the point on which they landed, they scarcely expected to find a human being; but they had no sooner set their foot on shore than a baïdar full of Esquimaux landed just by. The ebb set to S.S.W. at the rate of half a knot: the flood ran to N. ½ E. at the same rate, an evidence that it comes from the southward.

Drift-wood was everywhere abundant, though least so on such parts of the coast as had a western aspect, but without any apparent reason for this difference. The soundings off the coast are everywhere regular, and should the ice be drifted against this outer strip of land, by transporting a small boat over the narrow necks separating the lakes, a good inland navigation might be performed. The habitations of the Esquimaux are invariably upon the low strips of sand bordering on the lakes.

ICY CAPE is 38 miles northward of Point Lay, and is the farthest point reached by Capt. Cook, who applied the name to it from the quantity of ice which surrounded it; but none was visible when the *Blossom* was here in August, 1826.

The cape is very low, and has the large lake previously alluded to at the back of it, which here receives the waters of a considerable river, and communicating with the sea through a narrow channel, much encumbered with shoals. The main land on both sides of Icy Cape, from Wainwright Inlet on one side to Cape

Beaufort on the other, is flat, and covered with swampy moss. It presents a line of low mud cliffs, between which and the shingly beach there is the succession of narrow lakes capable of being navigated by baïdars or small boats.

Off here a great many black whales were seen, more than was remembered even in Baffin's Bay. There are several winter habitations of the Esquimaux upon the cape.

The Blossom Shoals lie immediately off Icy Cape. They consist of several successive banks, about three-quarters of a mile apart, lying parallel with the coast line. Upon the outer ones there were only 3½ and 4 fathoms; and upon the inner bank so little water that the sea broke continually over it. Between the shoals there were 9 and 10 fathoms, with very irregular casts. These shoals are probably the effect of the large river which here empties itself into the sea; though they may be occasioned by heavy ice grounding off the point and being fixed to the bottom, as upon endeavouring to weigh the *Blossom's* anchor, the chain broke after a very heavy strain. It may here be remarked that these shoals might afford shelter from the pressure of ice, should it be necessary for a vessel to get within them.

From Icy Cape the outer coast continues in one low unbroken line to the eastward, and is covered with a thick peat, which retains the water, making it very swampy and almost impassable. Upon the beach is *abundance of coal* and driftwood. The former is interspersed with the shingle and sand; and the latter, in some parts, is very abundant, and is forced high upon the beach, probably by the pressure of the ice when driven against the coast.

Wainwright Inlet is 40 miles eastward of Icy Cape, and appears to be a spacious opening or lake, the estuary of a considerable river. Its South entrance point was named *Point Marsh;* its northern, *Cape Collie;* between them is the mouth, having a shoal across connected with the land on the northern side, but with a channel for boats in the opposite direction. The country around is low, covered with a brown moss, and intersected with watercourses. To the northward of the entrance of the lake the coast becomes higher, and presents an extensive range of mud cliffs, terminating in a cape, which, however, is some little distance inland.

The natives, taking advantage of this elevated ground, had constructed their winter residences in it; they were very numerous, and extended some distance along the coast. Here, too, they form their stores of pemmican, &c.

During the progress of the expedition in search of Sir John Franklin, in 1849, this inlet was examined by Capt. Kellett, for the purpose of finding a wintering station for the *Plover.* The report returned was :—" That the channel was very narrow and winding: that 9 feet was the most water that could with certainty be carried in (10 feet was found after), and that even to ensure that depth the channel would require close buoying; that a fair wind or a calm, so that a vessel might either sail or be towed in, was necessary, the channel being too narrow and intricate to warp through. Once in, there was a sufficiency of water, and a convenient spot for the *Plover* to winter, alongside a bank, well sheltered. The natives told them that a considerable river runs into it; at least one that they can in their baïdars navigate for many days; and that it ran to the S.E.; that on its banks, and in the neighbourhood of the inlet, the reindeer collected in great numbers, in their progress northerly and on their return South."*

To the northward of Cape Collie the coast preserves the same N.E. direction, and at 16 miles from it is *Point Belcher*. The outer cast, midway between, is low, but more dry than that in the vicinity of Wainwright Inlet, with a beach of sand and gravel mixed, upon which there was also an abundance of *coal* and drift-wood. Several native yourts were seen here, and the coast was more populous than anywhere to the southward.

POINT FRANKLIN, which is 18 miles E.N.E., *true*, from Point Belcher, is the outermost point of a chain of sandy islands, lying at some distance from the main land, and extending towards the latter point. They were named the *Seahorse Islands*. The surface of the beach at Cape Franklin was a fine sand; but by digging a few inches down it was mixed with *coal*. Here, also, as in the preceding shores, there is a great quantity of drift-wood.

From the cape the coast, still consisting of a chain of islands lying off the main land, turns to the S.E., and unites with the land, forming a bay, which was named *Peard Bay*, after the first lieutenant of the *Blossom*.

From the bottom of Peard Bay the coast assumes a different aspect, and a few miles beyond consists of clay cliffs, about 60 feet high, and presented an ice formation resembling that which has been described in Eschscholtz Bay. The interior of the country is flat, and was only partially covered with snow in the middle of August. At about 13 miles from the commencement of the higher coast a river discharges itself into a lake within this shingly beach, the water of which being perfectly fresh will afford supplies.

Refuge Inlet is about 15 miles N.E. of this river, into which the current was setting very fast, and carrying the ice with it, when the *Blossom's* barge was returning at the end of August, occasioning great difficulty in tracking her against the wind, which was dead against them.

Cape Smyth, so named after the officer who accompanied the boat expedition, is 10 miles from Refuge Inlet. It is about 45 feet in height. Upon it was a village, as there was also upon the N.E. point of Refuge Inlet.

POINT BARROW, which is 16 miles N.E. of Cape Smyth, is the north-westernmost extreme of the American continent. It is 126 miles to the N.E. of Icy Cape and only 146 from the extreme of Capt. (afterwards Sir John) Franklin's exploration westward from the mouth of the Mackenzie River; so that the two expeditions designed to co-operate with each other were only separated by this short interval, afterwards filled up by the overland expeditions of Messrs. Dease and Simpson, and subsequently explored in 1849 by the *Nancy Dawson*.

The cape was named after Sir John Barrow, of the British Admiralty, the chief promoter of the arctic expedition.

Point Barrow is the northern extremity of a spit of land, which juts out 6 or 8 miles from the more regular coast line. From Cape Smyth the coast slopes

* On approaching Wainwright Inlet, July 25th, 1849, Capt. Kellett says that the vast number of walruses that surrounded them kept up a continual bellowing or grunting. The barking of the innumerable seals, the small whales, and the immense flocks of ducks continually rising from the water as they neared them, warned them of the approach to the ice, although the temperature was still high. They made the land about 10 miles to the North of Wainwright Inlet.

regularly to the northward, and this peninsula is joined to the main land by a neck, which does not exceed 1½ miles in breadth, and appears in some places less. The extremity was broader than any other part, and had several small lakes of water on it, which were frozen over at the end of August, 1829. On the eastern side of this peninsula was a large Esquimaux village, consisting entirely of yourts, the natives of which were inclined to be troublesome to the boat's party, leaving no doubt as to what would be the fate of a crew falling into their power.

The bay on the eastern side of the cape was named *Elson Bay*, in compliment to the officer of the barge.

The position of Point Barrow was ascertained to be lat. 71° 23′ 31″ N., lon. 156° 21′ 30″ W. The azimuth sights made the variation 41° E.; that observed on thin ice was 42° 15′ E.

The *current* about Point Barrow was found to be setting at the rate of 3½ miles an hour, N.E. *true*, and the ice was all drifting to that quarter.

The western coast of America terminates at Point Barrow. Beyond this it extends to the eastward, and hitherto has been untraversed by any sailing vessel. Even Point Barrow itself cannot always be approached by boats during the open season. The coast to the eastward has only been explored through the indomitable courage and submission to privation exercised by the officers of the arctic expeditions despatched at different intervals. In a nautical view, it is therefore utterly unimportant; in a scientific light, it is most interesting. It may very fairly be presumed, that, with the exploring expeditions now in operation in search of Sir John Franklin's party, that our knowledge will be completed, and all farther curiosity as to the nature and phenomena of these remote regions will remain unsatisfied for many years, perhaps for ever.

We have thus brought the reader from one extremity of America to the other; from the stormy and inclement, but comparatively well known country of Tierra del Fuego and Patagonia, through all the varieties of temperate and tropical climates, to the El Dorado of the present day, and by the fertile and productive countries as yet only frequented by the British or Russian fur-hunters, to the icy regions last described. In all this wonderful extent of coast line, unparalleled for the uniformity of its direction, the reader will have seen the vast variety of aspects under which nature presents itself. But up to the present time the whole of the capabilities of the countries within it have remained comparatively dormant. In some parts the climate utterly precludes all prospect of advancement. In others political dissensions, and the consequent insecurity of property, are the barriers to all commerce. In the more remote North, possessing every requisite for the comfort and support of a vast population in many parts, the great distance from all other parts of the civilized world has acted as a barrier to colonization. All these causes, which diminish the importance of the countries we have described, appear, however, to be on the wane; and it is more than probable that a very few years will see the relative value of the commerce of the eastern Pacific on a very different basis to what it is when these pages are issued.

ICY OR POLAR SEA.

It was, as has been before alluded to, long a problem of very great interest to the world, what was the termination to the North of the Great Ocean? which was solved, though not satisfactorily, by Behring in 1741. The true nature of this limit to the world of waters was first more exactly ascertained by Capt. Cook's last expedition. The *Resolution* and *Discovery* reached this icy barrier in the summer of 1778; and again, under the orders of Capt. Clerke, in 1779. Since that time, more than one navigator has visited this remote region, though it must be acknowledged that scientific interests rather than commercial importance have led to these explorations. But as of late years these waters have been much frequented for the pursuit of the whale fishery, more is known concerning them Yet, from the analogy of other tracts of the ocean, it must be presumed that this region in its turn will become abandoned, ere long, to its original solitudes.

The BARRIER OF ICE extending from the shores of America to those of Asia possesses all the peculiarities incident to the similar natural features in other regions. It varies in its extent, or rather its encroachments, on the open sea to the southward of it in different seasons, and therefore no absolute description can be given of its limits. The few following extracts will be sufficient to indicate its character.

Icy Cape, on the American shore, would appear to be a frequent limit to its southern side. This cape was much encumbered when Cook discovered it in August, 1778; hence its name. When Capt. Beechey visited it in the *Blossom*, in the same month of 1826, there was none visible; but, on sending his barge to explore the coast to the northward, it was found that strong northerly currents had drifted it in that direction, and it was not until they approached Cape Smyth and Point Barrow, that the pack was found to close with the coast. But this advance was not unattended with danger, for, before they could return, the body of ice had closed, or nearly closed, with the coast as far as Point Franklin, in lat. 71°. The packed ice to the northward was from 6 to 14 feet high. At 100 miles westward of this part the main pack was found to reach to 71° 10′.

In the following year, or in August, 1827, a new arrangement was found. The main body of the ice was many miles to the southward of its former limits, and reached within a few miles of Icy Cape; and from this its edge extended in a West and W.S.W. direction, to below 70°, in lon. 168° W. Thus in one year it had increased nearly a degree beyond its first limits.

When Capt. Cook came here in 1778, he came on the main pack on August 17th, in lat. 70° 41′, that is, a little to the North of Icy Cape. It was here quite impenetrable, and extended W. by N. and E. by S. as far as the eye could reach. It was compact as a wall, and 10 feet high, at least; its surface extremely rugged, and farther North it appeared much higher. The vicinity of the edge of the main body was indicated by a prodigious number of sea-horses (morses), whose roaring or braying gave them notice of its proximity. The whole of it was a moveable mass, for within two or three days it had drifted down to 69° 40′. When approaching the Asiatic coast, which he had done by coasting, more or less, along the barrier, on August 27, he says:—"Having but little wind,

I went with the boats to examine the state of the ice; I found it consisting of loose pieces, of various extent, and so close together that I could hardly enter the outer edge with a boat; and it was as impossible for ships to enter it as if it had been so many rocks. I took particular notice, that it was all pure, transparent ice,* except the upper surface, which was a little porous. It appeared to be entirely composed of frozen snow, and to have been all formed at sea; for, setting aside the improbability, or rather impossibility, of such large masses floating out of rivers, in which there is hardly water for a boat, none of the productions of the land were found incorporated or fixed in it, which must have unavoidably been the case had it been formed in rivers, either great or small. The pieces of ice which formed the outer edge of the field were from 40 to 50 yards in extent to 4 or 5; and I judged that the larger pieces reached 30 feet, or more, under the surface of the water. It also appeared to me very improbable that this ice could have been the production of the preceding winter alone; I should suppose it rather to have been the production of a great many winters. Nor was it less improbable, in my judgment, that the little that remained of the summer could destroy the tenth part of what now subsisted of this mass; for the sun had already exerted upon it the full influence of his rays. Indeed, I am of opinion, that the sun contributes very little toward reducing these great masses; for, although that luminary is a considerable while above the horizon, it seldom shines out for more than a few hours at a time, and often is not seen for several days in succession. It is the wind, or rather the waves raised by the wind, that brings down the bulk of these enormous masses, by grinding one piece against another, and by undermining and washing away those parts that lie exposed to the surge of the sea. This was evident, from our observing that the upper surface of many pieces had been partly washed away, while the base or under part remained firm for several fathoms round that which appeared above water, exactly like a shoal round an elevated rock. We measured the depth of water upon one, and found it to be fifteen feet, so that the ships might have sailed over it. If I had not measured this depth I would not have believed that there was sufficient weight of ice above the surface to have sunk the other so much below it. Thus it may happen, that more ice is destroyed in one stormy season than is formed in several winters, and an endless accumulation is prevented. But that there is always a remaining store, every one who has been upon the spot will conclude, and none but closet studying philosophers will dispute."† This was about latitude 69° 20′. The following day he sailed for and discovered the Asiatic coast at Cape North, which may thus be taken as the western limit of this barrier.

In the ensuing year, 1779, after the unfortunate death of the commodore, Capt. Clerke, the second in command, took the two ships up to the northward again; but in this season it was a month earlier. It will be needless to follow them in their progress; but so far as their experience went, it appeared that the

* Clear and transparent ice has recently been shown by Dr. Faraday and others to be one of the purest substances in nature, and therefore eminently adapted for supplies of water. Some remarks on this will be given in the Appendix.

† Vol. ii. pp. 462-3.

sea to the northward of Behring's Strait is clearer of ice in August than in July. The following is a summary of the conclusions arrived at by the two attempts:—

"It may be observed, that in the year 1778, we did not meet with the ice till we advanced to the latitude of 70°, and on August 17th, and then we found it in compact bodies, extending as far as the eye could reach, and of which a part or the whole was moveable, since, by its drifting down upon us, we narrowly escaped being hemmed in between it and the land. After experiencing both how fruitless and dangerous it would be to attempt to penetrate farther North, between the ice and the land, we stood over to the Asiatic side, between the latitude of 69° and 70°, frequently encountering, in this tract, large and extensive fields of ice; and though by reason of the fogs, and thickness of the weather, we were not able absolutely and entirely to trace a connected line of it across, yet we were sure to meet with it before we reached the latitude of 70°, whenever we attempted to stand to the northward. On the 26th of August, in lat. 69¾°, and lon. 184° E., we were obstructed by it in such quantities as made it impossible for us to pass either to the North or West, and obliged us to run along the edge of it to the S.S.W. till we saw land, which we afterwards found to be the coast of Asia. With the season thus far advanced, the weather setting in with snow and sleet, and other signs of approaching winter, we abandoned our enterprise for that time.

"In this second attempt we could do little more than confirm the observations we had made in the first; for we were never able to approach the continent of Asia higher than the latitude of 67°, nor that of America on any parts, excepting a few leagues between the latitude of 68° and 68° 20′, that were not seen last year. We were now obstructed by ice 3° lower, and our endeavours to push to the northward were principally confined to the mid-space between the two coasts. We penetrated near 3° farther on the American side than on the Asiatic, meeting with the ice both years sooner, and in greater quantities, on the latter coast. As we advanced to the North we still found the ice more compact and solid; yet as, in our different traverses from side to side, we passed over spaces which had before been covered with it, we conjectured that most of what we saw was moveable. Its height, on a medium, we took to be from 8 to 10 feet, and that of the highest from 16 to 18. We again tried the currents there, and found them unequal, but never to exceed a mile an hour; by comparing the reckoning with the observations, we also found the current to set in different ways, yet more from the S.W. than any other quarter; but whatever their direction might be, their effect was so trifling, that no conclusions respecting the existence of any passage to the northward could be drawn from them. We found the month of July to be infinitely colder than that of August. The thermometer in July was once at 28°, and very commonly at 30°; whereas the last year in August it was rare to have it as low as the freezing point. In both seasons we had some high winds, all of which came from the S.W. We were subject to fogs, whenever the wind was moderate, from whatever quarter; but they attended southerly winds more constantly than contrary ones."*

* Capt. Clerke, in Cook's Third Voyage, vol. iii. pp. 276-7.

These extracts will give a sufficient idea of the character of the ice at the respective periods. They have been visited more recently by the expeditions under Capts. Kellett and Moore, in H.M.S. *Herald* and *Plover*, in search of the parties under Sir John Franklin. Without entering into any irrelevant detail, we will insert the account of the progress along the ice as given in the newspapers at the end of January, 1850. Capt. Kellett's letter has the following :—

"July 26th, 1849.—At 4^h a.m. the ice could be seen in heavy masses, extending from the shore near the Sea-horse Islands, S.E. from Point Barrow. At 6^h we were obliged to heave-to, in consequence of a dense fog : this cleared off at 11^h 30′. The *Plover* was close to, but neither the boats nor the yacht were in sight.

"We both made sail, steering true North, and were at 1^h p.m. in lat. 71° 5′, where we made the heavily-packed ice, extending as nearly as far as the eye could reach from N.W. by W. to N.E. At this time we had soundings in 40 fathoms mud, the deepest water we have had since leaving the Island of St. Lawrence. We continued running along the pack until 8^h p.m., when, a thick fog coming on, we ran 2 or 3 miles South, and hove to, the wind blowing from N.N.E., and directly off the ice. We had run along it 30 miles.

"The pack was composed of a dirty-coloured ice, not more than 5 or 6 feet high, except some pinnacles deeply seated in the pack, which had no doubt been thrown up by the floes coming in contact. Every few miles the ice streamed off from the pack, through which the *Plover* sailed.

"July 29th.—At 1^h 30′ the fog cleared off; the pack from N.N.W. to N.N.E., distant about 6 miles. Made sail during the forenoon, running through streams of loose ice. At 10^h passed some large and heavy floes. Commander Moore, considering them sufficiently heavy and extensive to obtain a suite of magnetical observations, dropped the *Plover* through between them, and made fast with the ice anchors under the lee of the largest in a most seamanlike manner.

"I landed on the floe, with Lieut. Trollope. The latitude, time, and variation were obtained on it (lat. 71° 30′ N., lon. 162° 5′ W.); but the other observations were vitiated by its motion in azimuth; and, by its constant breaking away, the level would not stand. We had 28 fathoms, mud, alongside it, and no current.

"I found the ice driving slowly to the southward, with the N.N.E. wind blowing fresh. Very few walruses and but a single diver seen. The general height of this floe was 5 feet, and about a mile in extent; on it were found pebbles and mud, which led Commander Moore to suppose that it had been in contact with the land.

"At 3^h p.m. the *Plover* slipped from the ice, and both ships, with a N.E. wind, made sail westerly, until 6^h p.m., when we hauled up true North, having no ice in sight in that direction, and only from the masthead on the weather beam. A fine clear night, running along 6 and 7 knots; temperature of the water, 40°; depth, 21 fathoms (increasing).

"At midnight the latitude was obtained by the inferior passage of the sun. At 5^h a.m. the temperature of the water had fallen to 36°, and almost at the same instant the ice was reported from the masthead. Between this time and 7^h a.m. (when we hove-to within half a mile of the pack) we ran 10·5, so that I consider 11 miles to be about the distance that pack ice can be seen in clear weather from a ship's masthead.

"The pack was of dirty-coloured ice, showing an outline without a break in it 5 or 6 feet high, with columns and pinnacles much higher some distance in. Although the wind was off the pack, there was not a particle of loose or drift ice from it; our soundings had gradually increased to 35 fathoms, soft blue mud. The only living things seen were a pair of small divers, black, with a white ball in the back, and two remarkable birds, very much like the female of the tropical man-of-war bird, a dingy black colour with excessively long wings, and the same flight when soaring. We could not succeed in shooting any of either species. We remained hove-to off the pack for an hour. In the dredge we got muscles, and a few bivalves common to these seas.

"This was our most northern position, lat. 72 °51′ N., lon. 168° W. The ice, as far as it could be seen from the masthead, trended away W.S.W. (by compass), Commander Moore and the ice-master reporting a water sky to the North of the pack, and a strong iceblink to the S.W.

"It was impossible to gain this reported open water, as the pack was impenetrable. The pack we had just traced for 40 leagues made in a series of steps westerly and northerly; the westerly being about 10 or 12 miles, and the northerly 20 miles. We made sail at 9ʰ a.m., steering for the coast a little to the westward of our track up; wind N.E., gradually decreasing as we got southerly.

"Five o'clock, a.m.—Fell a dead calm, the sea glassy smooth, and so transparent that a white plate was distinctly seen at a depth of 80 feet. As we approached the coast, we again met *numbers of whales*, walruses, seals, and flights of ducks and sea-birds. July 30, 8ʰ a.m.—Packed in-shore in 8 fathoms, close to the northward of the Blossom Shoals off Icy Cape."

Subsequently the *Herald* proceeded to the westward. Capt. Kellett's journal continues:—

"August 9th.—In the morning passed the carcass of a dead whale, and another in the afternoon. I sent a boat to this one, stuck a flag in it, and buried a bottle, containing a current paper, a notice of my whereabouts, and of my intention to go westerly, for the information of the *Plover*, should she fall in with it. Many reports of land from the masthead; a land-bird seen.

"Having this favourable wind for examining the pack by the westward, I continued to steer as high as the wind would permit on the starboard tack. The wind continued to lighten until the morning of the 10th, when it fell a dead calm.

"The sea was literally covered in streams with particles of a pink colour, like wood ashes, or coarse sawdust from cedar, a tenth of an inch long, and 0·5 in diameter, and round. On placing it under the microscope, no appearance of circulation could be detected. Mr. Goodridge, the surgeon of this ship, supposes it to have proceeded from the carcasses of the whales he saw yesterday, the oil having been forced through the pores by the pressure of the water; giving the uniform size and shape in which we found it. I endeavoured to dry some in blotting paper, but it was absorbed by the paper, and nothing left but an oily stain. Tried the current, and found it running to the westward, one-third of a mile an hour. Walruses grunting around in groups of eight and ten together; quantities of small pieces of drift-wood, all pine, which appeared to have been washed from some beach. The temperature of the water at the surface in 29 fathoms was 45°, and at the bottom 43°. The dredge produced (in soft blue mud) a good many mussels, star-fish (found in all parts of this sea), a few bivalves (got before), and some very small shrimps.

"A light southerly wind sprang up, gradually increasing, and veering to the eastward. At 10h 30′, after standing to the S.W. for 15 miles, the loom of the land in the neighbourhood of North Cape could be seen. I tacked to the N.E., with the wind fresh from E.S.E.; not wishing to run the risk of being caught with a south-easter between the land and the ice-floe, which I considered could not be far off, from the smoothness of the water, the numbers of walruses, and particularly a little black and white diver, which we never saw except in its vicinity.

"August 11th.—Steering until this day in thick and bad weather to the N.E., at which time we were in lat. 70° 1′, lon. 173° 53′. Bore up North to endeavour to fall in with the pack. By 6h p.m., a dense fog came on; we hauled to the wind on the port tack under reduced sails, ship heading S.E., with a short jumping sea.

"August 12th, a.m.—The wind shifted suddenly to N.N.E., and afterwards N.N.W., blowing hard; reduced to treble topsails and reefed foresail, our soundings having decreased to 17 fathoms, mud. No observations. Our reckoning placed us in lat. 70° 20′, lon. 171° 23′, in 18 fathoms, sand.

"SHOAL.—Shortly after noon our depth decreased to 16 fathoms, the colour of the water becoming lighter, with a breaking sea all round. Our soundings decreased a fathom each cast until 1h 30′ p.m., when we wore in 11 fathoms shingle, getting in wearing 9 fathoms, then 12; and when trimmed to go back, as we went on, had several casts of 8, and one of 7 fathoms; then suddenly got into 14, which gradually increased. The sun came out, verifying our noon position. Until midnight it blew a strong gale.

"August 13th, a.m.—Fine; wore to stand back to the shoal. Shoaled our water 13 fathoms, and at 10h I imagined I saw breakers on the lee bow. Ship refused stays, wore, but had no less water at midnight; passed over the tail of the bank in 8 fathoms, 5 miles N.W. of our former position. Continued to stand to the eastward until I could weather the South end of the shoal; then tacked, passing, in 16 fathoms, 3 miles South of our first position. When I bore up North to fix its western edge, a slight easterly current took me rather farther in that direction than I intended. I have, however, confined it within a radius of 5 miles.

"The weather would not allow of our anchoring so as to make a closer examination of the shoal with our boats, and the sea was too hollow and heavy to attempt taking the ship herself into less water. In approaching the shoal, the bottom changes from mud to fine sand, and when in the least water coarse gravel and stones. We found nothing less than 7 fathoms; but I am of opinion that a bank exists which would bring a ship up.

"August 14th.—We experienced very strong, variable, and S.E. breezes, with rain, until midnight of the 14th, when the wind changed to the westward, and brought with it fine weather. Continued to stand to northward and westward until noon on the 15th, being in lat. 71° 12′, and lon. 170° 10′; bore up W. ½ S., passing several pieces of drift-wood. Our soundings increased as we left the bank (westerly) to 25 fathoms, mud.

"August 16th.—Wind very variable, and direction S.S.W. to S.E. Large flocks of phalaropes, divers and gulls numerous. At midnight wind very fresh from S.S.E., steering W.S.W.; depth increasing to 10 fathoms. At 3h a.m. on the 17th, the temperature of the sea fell from 40° to 36°; the wind light and cold. Shortened sail, supposing that I was very near the ice; frequent snow showers.

"At 5h a.m. wind shifted suddenly from the N.W. in a sharp squall, with heavy snow. Shortly after 8h, when one of these snow storms cleared off, the packed ice was seen from the masthead from S.S.W. to N.N.W., 5 miles distant. The weather was so bad that I bore up for the rendezvous. The weather, however, as suddenly cleared up. I hauled my wind for the north-western extreme of the ice that had been seen. At 9h 40′ the exciting report of 'Land ho!' was made from the masthead.

"HERALD and PLOVER ISLANDS.—In running a course along the pack towards our first discovery, a small group of islands was reported on our port beam, a considerable distance within the outer margin of the ice. Lanes of water could be seen reaching almost up to the group, but too narrow to enter unless the ship had been sufficiently fortified to force a hole for herself.

"These small islands at intervals were very distinct, and were not considered at the time very distant. Still more distant than this group (from the deck) a very extensive and high land was reported, which I had been watching for some time, and anxiously awaited a report from some one else. There was a fine clear atmosphere (such a one as can only be seen in this climate), except in the direction of this extended land, where the clouds rolled in numerous immense masses, occasionally leaving the very lofty peaks uncapped, where could be distinctly seen columns and pillars, very broken, which is characteristic of the higher headlands in this sea—East Cape and Cape Lisburne, for example.

"With the exception of the N.E. and S.E. extremes, none of the lower land could be seen, unless, indeed, what I took at first for a small group of islands, within the pack edge, was a point of this great island.

"This island, or point, was distant 25 miles from the ship's track, higher parts of the land seen not less, I consider, than 60. When we hove-to off the first land seen, the northern extreme of the great land showed out to the eastward for a moment, and so clear as to cause some who had doubts before to cry out, 'There, sir, is the land, quite plain.'

"From the time land was reported until we hove-to under it, we ran 25 miles directly for it. At first we could not see that the pack joined it, but as we approached the island we found the pack to rest on the island, and to extend from it as far as the eye could reach to the E.S.E.

"The weather, which had been fine all day, now changed suddenly to dense clouds and snow showers, blowing from the South, with so much sea that I did not anchor as I intended. I left the ship with two boats; the senior lieutenant, Mr. Maguire; Mr. Seemann, naturalist; and Mr. Collinson, mate, in one: Mr. Goodridge, surgeon; Mr. Pakenham, midshipman; and myself, in the other, almost despairing of being able to reach the island.

"The ship kept off and on outside the thickest part of the loose ice, through which the boats were obliged to be very careful in picking their way, on the S.E. side, where I thought I might have ascended. We reached the island, and found running on it a very heavy sea; the first lieutenant, however, landed, having backed his boat in until he could get foot-hold (without swimming), and then jumped overboard. I followed his example; the others were anxious to do the same, but the sea was so high that I could not permit them.

"We hoisted the jack, and took possession of the island, with the usual ceremonies, in the name of her most gracious Majesty Queen Victoria.

"The extent we had to walk over was not more than 30 feet. From this space and a short distance that we scrambled up, we collected eight species of plants; specimens of the rock were also brought away.

"With the time we could spare and our materials, the island was perfectly inaccessible to us. This was a great disappointment, as from its summit, which is elevated above the sea 1,400 feet, much could have been seen, and all doubt set aside, more particularly as I knew the moment I got on board I should be obliged to carry sail to get off the pack and out of the bight of it we were in; nor could I expect that at this period of the season, the weather would improve.

"The island on which I landed is 4½ miles in extent East and West, and about 2½ North and South, in the shape of a triangle, the western end being its apex. It is almost inaccessible on all sides, and a solid mass of granite. Innumerable black and white divers (common to this sea) here found a safe place to deposit their eggs and bring up their young; not a walrus or seal was seen on its shore, or on the ice in its vicinity. We observed here none of the small land-birds that were so numerous about us before making the land.

"It becomes a nervous thing to report a discovery of land in these regions, without actually landing on it, after the unfortunate mistake to the southward; but, as far as a man can be certain, who has 130 pair of eyes to assist him, and all agreeing, am certain we have discovered an extensive land. I think, also, it is more than probable that these peaks we saw are a continuation of the range of mountains seen by the natives off Cape Jakan (coast of Asia), mentioned by Baron Wrangel, in his Polar Voyages.* I returned to the ship, and reluctantly made all the sail we could carry from this interesting neighbourhood to the S.E.

"August 20th.—Sighted Cape Lisburne in a thick fog; hauled off to await clear weather; passed several carcasses of whales."

There is one feature of this sea which might appear somewhat remarkable, and that is its shallowness. There is anchorage almost all over it. The depth varies from 20 to 30 fathoms, seldom exceeding the latter; the bottom composed of sand, mud, and stones. This, therefore, simplifies the navigation during the few weeks that it may be said to be open to navigation. It has been traversed in almost every portion; and, with the exception of the shoal discovered by Capt. Kellett, no permanent danger appears to exist in the open space between the shores of the two continents.

* There is an account of this land extant, in which it is stated that Andreef reached it in baïdars in 1762; that it is called *Tikigen*, and inhabited by a race called *Khrahay*.

THE COAST OF ASIA.

The merit of discovery of this coast is due to Behring, as we have repeatedly remarked previously. It had been slightly and cursorily examined by few subsequent to that great navigator's first voyage until Capt. Cook saw it, and first declared its true character. Capts. Clerke and King passed along it in the following year. Capt. Kotzebue in the *Rurick*, Capts. Billings, Sarytscheff, and Wrangel, also added slightly to our previous acquaintance. But all these authorities collectively gave a very vague and imperfect notion of the whole, and but little serviceable for navigation could be gleaned from their works.

All this, however, was obviated by the surveying expedition under Capt. (afterwards Rear-Admiral) Lütke, whose excellent and ample work leaves little to desire. This expedition, which left St. Petersburg in August, 1826, consisted of two corvettes, the *Moller*, under Capt. Stanikowitch, and the *Séniavine*, under Capt. Lütke. The operations of the latter are our present object. After making many excellent observations in the North Pacific, he proceeded to Awatska Bay, and thence surveyed the greater portion of the coasts of Kamtschatka and Eastern Asia to the northward, as far as the East Cape of Behring Strait. The account of this voyage has furnished us with most of the subsequent particulars.*

The TCHUKTCHIS, the inhabitants of Eastern Asia, may demand a short notice here. Of all the Asiatic races inhabiting Siberia these are the only ones that have not submitted to the tribute of peltries demanded by the Russians.

The Tchuktchis inhabit the north-eastern part of Asia, extending from Tchaun Bay to Behring Straits in one direction, and in the other from the Anadyr, and the upper coasts of the Aniui, to the Polar Sea. To the South are the Koriaks, and to the West the Tchuwanzes and Jukahirs of the Aniui. They formerly occupied a more extensive territory, before the Cossacks. The Tchuktchis, though still in a great measure a nomade race, have less of the characteristics which usually accompany such a mode of life than the wandering Tunguses; they are more covetous and more saving than belongs to the character of the genuine nomade races. They lay up stores for the future, and in general do not remove their dwellings without an object, but only when it is necessary to seek fresh pasture for their reindeer. They are disgraced by the most shameless licentiousness.†

Capt. Lütke has given a detailed account of these people in his work. He states that under this name are designated two distinct races, distinct in mode of life, in language, and in appearance: the one nomade or wandering, which he terms the Reindeer Tchuktchis; the other dwelling in fixed habitations on the sea-side, called the Sedentary Tchuktchis. The first are the same people as the Kariaks or Koriaks, to the southward, only with some slight differences. The first call themselves *Tchaoukthous*, the second *Namollos*. They have been found to be more friendly than earlier writers have given them credit for, and were serviceable to Capt. Moore when he anchored here in 1848-49. A good notion

* Voyage Autour du Monde sur la Corvette *Le Séniavine*, 1826-27-28-29, par Frederic Lütke, Capt. de Vass., &c., translated (into French) from the original Russian, by Cons. J. Boyé. 3 vols., Paris, 1835; and Partie Nautique, St. Petersburg, 1836.

† See Wrangel's Expeditions in Northern Siberia, translated by Mrs. Sabine, p. 357.

may be formed of their habits and appearance from the works of Cook, Billings, and Lütke.

CAPE NORTH of Capt. Cook, or *Ir-Kaipie*, according to Admiral von Wrangel, is the most northern point of the Asiatic coast that was made by Capt. Cook in coming westward from the opposite shores of America, August, 1778; hence its name. The land here is, in every respect, says Cook, like that of the opposite one of America; that is, low land next the sea, with elevated land farther back. It was perfectly destitute of wood and even snow; but was, probably, covered with a mossy substance, that gave it a brownish cast. In the low ground, lying between the high land and the sea, was a lake extending to the S.E., farther than could be seen. The land, on first making it, showed itself in two hills like islands, but afterwards the whole appeared connected, the western extreme, terminating in a bluff point, being one of these hills. As he stood off again, the westernmost of the two hills came open off the bluff point in the direction of N.W. It had the appearance of being an island; but it might be joined to the other by low land, though it was not seen. If so, there is a twofold point, with a bay between them. This point, which is steep and rocky, was named Cape North.

Cape North, or Ir-Kaipie, is in lat. 68° 55′ 16″. The longitude, dependent on that of C. Jakan, is 179° 57′ E.; variation, 21° 40′ E. This was ascertained by Admiral von Wrangel, in his fourth Siberian expedition, April 22, 1823. Here terminated the expedition, which returned to the westward, toward St. Petersburg.*

The coast to the S.E. of this is represented from the Russian charts; the names and details there given will be sufficient for the guidance of those who may chance to visit this inhospitable country. The most graphic descriptions are those of Capt. Cook, with which we have commenced. His subsequent remarks will fill up all deficiency.

He made the land again to the S.E. of that previously described, a very low point or spit bearing S.S.W. 2 or 3 miles distant, to the East of which there appeared to be a narrow channel, leading into some water that was seen over the point. Probably the lake before mentioned communicates here with the sea.—(Vol. ii. pp. 466-7.)

According to the chart, it would appear to be the various branches of the mouths of the river *Ekehtagh*, the outer points of which are named *Points Emua-en* and *Tenkourguin*, that are here referred to.

Still farther eastward some parts appeared higher than others; but in general it was very low, with high land up the country (p. 467). In this part the *Rivers Kentel*, *Amguina*, and *Vankarema*, debouch.

Burney or Koliutchin Island (South point) is in lat. 67° 27′, lon. 184° 24′ E. Its first name was given by Cook. It is about 4 or 5 miles in circuit, of a middling height, with a steep rocky coast, situated about 3 leagues from the main. The inland country, hereabout, is full of hills, some of which are of considerable height.

CAPE SERDZE KAMEN, in lat. 67° 12′, lon. 188° 20′ E., is the extent to

* Wrangel, trans. by Mrs. Sabine, p. 355.

which Behring reached, August 15, 1728. Here he thought that it was time to think of his return, "as it was not advisable to winter in these parts, since the well-known want of wood in all the northerly regions towards the frozen sea, the savages of the country not yet reduced to the obedience of the Russian government, and the steep rocks everywhere found along the shore, between which there was not anchorage nor harbour, rendered it too dangerous."* By this voyage, however, he established the fact of the separation of the American and Asiatic continents.

Capt. Cook says:—"The coast seemed to form several rocky points, connected by a low shore, without the least appearance of a harbour. At some distance from the sea the low land appeared to swell into a number of hills. The highest of these were covered with snow, and in other respects the whole country seemed naked. At seven in the evening two points of land, at some distance beyond the eastern head, opened off in the direction of S. 37° E. I was now well assured of what I believed before, that this was the country of the Tchuktchi, or the N.E. coast of Asia, and that thus far Behring proceeded in 1728; that is, to this head, which Müller says is called Serdze Kamen, on account of a rock upon it, shaped like a heart. But I conceive that M. Müller's knowledge of the geography of these parts to be very imperfect. There are many elevated rocks upon this cape, and possibly some one or other of them may have the shape of a heart. It is a pretty lofty promontory, with a steep rocky cliff facing the sea. To the eastward of it the coast is high and bold, but to the westward it is low, and trends N.N.W. and N.W. by W., which is nearly the direction all the way to Cape North. The soundings are everywhere the same at an equal distance from the shore, which is also the case on the opposite shore of America. The greatest depth we found in ranging along it was 23 fathoms; and in the night, or in foggy weather, the soundings are no bad guide in sailing along either of these shores."†

EAST CAPE, the extremity of Asia, has been mentioned before as forming, with Cape Prince of Wales, the westernmost point of America, the narrowest part of Behring Strait. It is a peninsula of considerable height, joined to the continent by a very low, and, to appearance, narrow neck of land. It shows a steep, rocky cliff against the sea; and off the extreme point are some rocks like spires. It is in lat. 66° 3′ N., lon. 190° 16′ E. From its general appearance it might be taken for an island, and this doubtless occasioned an error in the number of the St. Diomede Islands lying off it.

ST. LAWRENCE BAY lies to the S.W. of East Cape, and was so named by Cook, he having anchored in it on St. Lawrence's day, August 10, 1778. It is remarkable that Behring sailed past it just half a century before, that is, August 10, 1728, on which account the neighbouring island was called St. Lawrence Island. "This Bay of St. Lawrence is at least 5 leagues broad at the entrance, and 4 leagues deep, narrowing towards the bottom, where it appeared to be tolerably well sheltered from the sea winds, provided there be a sufficient depth of water for ships. I did not wait to examine it, although

* Müller, Voyages et Découvertes des Russes, p. 4.
† Cook's Third Voyage, vol. ii. pp. 468-9.

I was very desirous of finding a harbour in those parts, to which I might resort next spring. But I wanted one where wood might be got, and I knew that none was to be found here." *

The bay was minutely surveyed by Capt. Lütke in July, 1828, and here commence the sailing directions given by that navigator. *Cape Nouniagmo* is the N.E. extremity of St. Lawrence Bay. It is distinguished by a remarkable hill, not from its elevation but from its rounded summit. *Cape Krleougoun*, which forms the S.W. extremity, is 11½ miles S. 52° W. from it. The western slope of this mountain declines very gradually to form a large opening, through which runs a rapid but shallow river, on which is a village of Stationary Tchuktchis, named *Nouniagmo*. It is 2 miles from the cape of the name. *Cape Pnaougoun*, beyond which commences the interior bay, is 3½ miles W. by N. from this village. Between these the shores are level and low, terminating abruptly at the sea-shore. From these escarpments, entirely covered with snow in July, avalanches were constantly falling with great noise. Not a single shrub breaks the monotony of the interior plains.

The opposite shores are similar to this, but rather higher. Cape Krleougoun is high and very steep; beyond it the coast turns rapidly to the N.W., towards Mitchigmensk Bay. Upon this cape there is a mountain, very remarkable from some sharp peaks. It is a very well determined position on the chart. The cape is in lat. 65° 29′ 40″ N., and lon. 171° 0′ W.† At half a mile from its extremity is a large village.

From this cape the coast extends, rounding to the N.N.E. and N., for 7 miles; where a bed of gravel projects, forming a tolerably large lake. It is 3½ miles S. 58° W. from Cape Pnaougoun, and may be taken as the other point of the inner bay. Above it is a large village.

The depth in the centre of the bay is 27 fathoms. Against the North shore, at the distance of 1 to 2 miles, there are 5, 6, and 9 fathoms, sand and gravel. Farther off, and nearer the centre, it rapidly increases. At 1 or 1½ miles from the South shore there are from 7 to 12 fathoms, muddy bottom; on approaching the inner bay the depth increases, and opposite the bed of gravel there are 23 fathoms. No indication of reefs or dangers was perceived.

These gravel deposits will be found to be so frequent in occurrence, that they certainly form a moiety of the entire coast between East Cape and the South extremity of Lopatka. A summary description of them may therefore be here given. What is meant by a *bed of gravel* is a formation or collection of shingle, rising from a few inches to 6 or 7 feet above the surface of the water. They are generally covered with a turfy moss and plants similar to those on the land. They generally extend in a straight line, or gradually and slightly curve. They sometimes form distinct islands, and sometimes join on to the continent, forming the coast itself, or else points projecting from it. Their breadth varies; some are almost washed over by the sea, and none exceed a mile. There is generally a great depth on their edges, and frequently at 10 or 12 yards off there are

* Cook, vol. ii. p. 472.

† The longitudes by Capt. Lütke, between East Cape and Cape Tchapline, have been placed 10′ farther East, to accord with the charts.

4 or 5 fathoms. At 2 or 3 miles off the depth gradually increases, the bottom frequently muddy; so that, wherever one of these gravel beds are met with on the coast, so sure are you to find anchorage. Nevertheless it sometimes occurs that detached and similarly deep banks lie before these. In digging holes in these banks water is found at the level of the sea, but always among the shingle.

Such shingle banks are met with in other regions, but nowhere so frequently as in the Seas of Behring and Okhotsk. They are seen at every step, and a glance is sufficient to demonstrate that they are formed by the sea, but in what manner is not so evident.

The inner bay extends W.N.W. and N.W. for 19 miles, and throughout maintains nearly an equal breadth of 3½ miles. The opening is rather narrowed by a low gravel bed, at half a mile West of Cape Pnaougoun. Its highest part is covered with a turfy and humid moss, in the centre of which is a lake of fresh water—a strange circumstance. Its distance from the South Cape is 2¼ miles. Here is the chief entrance; there are more than 27 fathoms water, and no danger. In the East passage there are not more than 11 feet water.

Cape Pnaougoun and the coast, for a mile distant, are formed by a bed of gravel; farther off, though low, it is perpendicular, and covered with snow. Extending from Cape Pnaougoun to the N.N.E. and N.W. it forms a cove, 1½ miles long and wide, in which is secure anchorage. Capt. Kotzebue says:—"At 12 miles from the entrance, the bay takes a N.W. direction. At 3 miles farther are two high and steep islands; the easternmost was named *Chramtschenko;* the other and smaller one, *Petrof Isle*, from the first and second pilots of the *Rurick*. The bay terminates in a circular and shallow cove, in which two small rivers throw their cataracts of the purest water."

At the extremity of St. Lawrence Bay the termination of a chain of high and peaked mountains abuts, which has every appearance of being a branch of the chain traversing the Tchuktchis country from East to West, and joining the Stanovoi chain.

But few birds or fish, for provision, were seen here; a few salmon were all that were procured. But these privations were amply compensated by the abundance of reindeer which may be procured from the wandering Tchuktchis, always near the coast in the autumn, for iron articles, &c., or, above all, for tobacco.

A large quantity of sea-calves and morses are brought hither in the winter upon the ice. Fresh water is to be had, and of very excellent quality, but not everywhere readily procurable. Capt. Lütke took his from a brook one mile from Cape Pnaougoun; it may be taken from the beach by means of a hose; on the other hand, not a morsel of wood can be got. It is worthy of remark that, although the opposite or American coast abounds with it, both growing and drift, not a single piece is brought here by the sea. The tides are very insignificant; the greatest difference observed was 15 inches; and were usually very irregular. As near as could be ascertained, the establishment of the port was 4^{h} 20′. It is said that it never rises more than 4 feet in the autumn, when strong gales occur. The currents are strong, but apparently as irregular as the tides. Winds generally light; those between South and East bring the fog, and those from the North or N.W. dissipate it.

The observations made upon the bed of gravel make its West point lat. 65° 37′ 30″ N., and lon. 170° 53′ 30″ W.; the latter differing half a degree from Kotzebue's position. Variation, 24° 4′ E.; dip, 76° 42′.

METCHIGMENSK BAY.—From Cape Krleougoun the coast turns rapidly in the N.W. part, and, curving in an open bay, extends for 20 miles to the West to a moderately elevated but very steep cape, on which is the large village of *Lugren*. Up to this the coast is covered with moss, and rises insensibly to uniform hills. The coast appeared clear, without any danger. A bed of gravel, separating Metchigmensk Bay from the sea, extends from Cape Lugren for 20 miles to the West, curving to the South.

The entrance to the bay is very difficult, from its narrowness and the lowness of the points forming it. Before making them out, the people on them will be seen, as if walking on the water. Besides this, the entrance is so placed, that its opening cannot be made out until it is brought to bear N.W. by compass, and consequently when near the western side. It must thus be sought from the mast-head, like the opening to a coral reef, which, in foggy weather, is impracticable. The village of *Igouan* (called Agutkino on Billings' plan, which was found correctly delineated) would be a good guide if it were permanent, but it is only there in summer.

The village of *Metchigm*, on the West side, at 2 miles from the point of the gravel bed, is a sure mark. The winter yourts show themselves by a thick verdure on and around them. The best mark to find the entrance is a cape on the North of the bay, which, on the continental side, projects to the South. It is tolerably high, even, and ends in a low point to the S.W., appearing, at a distance, like two or three islands. The northernmost and longest of these apparent islands, which is distinguished by a cliff, lies W.N.W. by compass from the entrance. Bringing it on this bearing, and steering for it, you go right for the entrance. The Bay of Metchigmensk penetrates the land for a great distance, but it was not explored. The remarks as to supplies at St. Lawrence Bay equally apply to this. The entrance of the bay was assumed to be in lat. 65° 30′ 30″ N., and lon. 172° 0′ W.

The bed of gravel which forms the West side of the entrance of Metchigmensk Bay extends 5 or 6 miles to the N.W.: about halfway it is cut by a rivulet in the S.W. angle of the bay, through which the Tchuktchis say their baïdars alone can pass.

At 15 miles from Metchigmensk Bay *Cape Khaluetkin* projects, very remarkable for a round-topped mountain. To the South of the cape is *Héliaghyn Bay*, surrounded by a very low coast, apparently terminating in an inner bay, but of no importance.

Thence the coast turns to S.E. to *Cape Nygtchygan*, which, from North and N.E., at 15 miles distant, appears to be an island, on account of the lowness of the land between it and Héliaghyn Bay. It is steep; to the N.W. of it a bed of gravel extends 3 or 4 miles, which unites at its other end to the coast, forming a lake or bay. Beyond Cape Nygtchygan is the opening of the extensive Strait of Séniavine.

STRAIT of SE'NIAVINE.—The existence of this remarkable strait was not suspected until the voyage of Capt. Lütke, who applied the name of his vessel to it. The *entrance* was noticed by Behring as a gulf. Cook took it to be a shallow bay.* Capt. Sarytscheff also saw it through the fog.

Séniavine Strait is formed by two large islands, *Arakamtchetchen* and *Ittygran.* It runs first towards the S.W., then South, and to the East, nearly 30 miles, and from 6 miles to half a mile in breadth. Its entrance is between *Capes Neegtchan* and *Kougouan,* bearing S. ½ E. and N. ½ W., 5 miles apart. Each of them is distinguished by tolerably high mountains. Neegtchan lies some distance from the coast; but Kougouan falls perpendicularly into the sea, and was distinguished by Cook. Cape Neegtchan, in lat. 64° 55′ 30″ N., and lon. 172° 17′ 30″ W., is the northern limit of the strait; as *Cape Mertens,* in lat. 64° 33′ 15″, and lon. 172° 20′, is its southern extremity.

At 2 miles from Cape Neegtchan is the small river *Maritch.* Its mouth is a good harbour for small vessels, as they can moor against the land. The current runs strongly out of it, but is not deep far up it. Near the mouth, to the N.W., is the Tchuktchi village, *Yaniakinon. Penkegneï Bay* extends beyond the entrance, first N.W. ½ W. 5 miles, then as far to S.W. ½ S., then 2 miles to West. It is surrounded by high mountains, advancing to the coast itself. It is deep and safe, but, from being so far from the sea, is inconvenient.

The continental coast from this bay runs 6 miles to South and S.W. to *Abolecheff Bay,* partly steep, partly sloping, but mountainous throughout. Its opening is in front of the South point of Arakamtchetchen Island, and it extends 6 miles to the West and W.S.W. Its breadth is 1 to 1½ miles. Its North shore consists of a gravel bed, behind which, at a short distance, high mountains rise, among which *Tagleokou* is remarkable for its perfectly conical summit. The upper part of the bay is surrounded by a very low and sandy shore. There is good anchorage throughout above the *second* cape; but to be perfectly sheltered you must double the third cape, and lie in 17 to 19 fathoms, sticky mud. Fresh water abounds everywhere, but no wood to be obtained.

From the *first,* or S.E. cape of Abolecheff Bay, the coast runs 5 miles to S.W. ½ S., and forms a bay open to the N.N.E. The surrounding mountains will not permit a ray of the sun to penetrate into it: it is therefore cold, sombre, and frozen. From this icy bay the coast runs 3 miles to the East, and approaches the West extremity of Ittygran Island. A bay on the latter corresponds to a gravel bed running to the N.E., and is made remarkable by the high pyramidal mountain, *Elpynghyn;* the two together form a sheltered harbour, with 9 to 20 fathoms.

From the mountain Elpynghyn the coast trends evenly to East and E.S.E. for 6 miles; then with steep, reddish cliffs, 2 miles farther to Cape Mertens, the South termination of Séniavine Strait.

CAPE MERTENS is high, steep, and is distinguished by a mountain with three summits. Between it and the Elpynghyn Mountain there is no shelter.

ARAKAMTCHETCHEN ISLAND, the largest of the islands forming the Strait of Séniavine, is 16 miles long from S.W. to N.E., and 8½ miles in its

* Cook's Third Voyage, vol. ii. p. 472.

greatest breadth. From the S.W. point to nearly one-half its length it is traversed by a chain of hills, moderately high, with flattened summits; the highest of which, *Mount Athos*, has two separate granitic rocks crowning its summit, a short distance apart. This was one of the most important points in Capt. Lütke's examination; from its top they had a superb view, extending all over the strait, from Cape Nouneagino in the North, to St. Lawrence Island in the South. *Cape Kyghynin*, the East point of the island, and the easternmost point of the land forming the Strait of Séniavine, is in lat. 64° 46′ N., and lon. 172° 7′ W., and 28 miles due East of the bottom of Penkegneï Bay, its western extremity.

Cape Kougouan, the North point of the island, forms, with Cape Neegtchan, the North entrance of the strait, to which the natives give the name of *Tchiarloun*.

PORT RATMANOFF, at 2 miles S.W. from Cape Kougouan, is the port. It is small but good, and is preferable to all others on account of its proximity to the sea. It is easy to make out by Cape Kougouan and another cape equally steep but lower, at 3½ miles W.S.W. from it: Port Ratmanoff is midway between them. The port is formed by a gravel bed, extending 1,000 yards W.S.W. from the coast it joins. A portion of the space behind it, 1½ cables' length in diameter, has 24 to 36 feet water, muddy bottom. Vessels can moor to the gravel bed, where neither wind nor swell can incommode them. At 2½ cables' length from the South point of the bed of gravel is an isolated sand-bank, with 8 feet water over it; this point must therefore be kept not more than 150 or 200 yards off in rounding it. Water may be got from a rivulet at the South point.

Cape Paghelian, the S.W. extremity of the island, is 8 miles from this port, the coast between being nearly straight. There is good anchorage in this slight bay formed by it, and tolerable shelter. The whole of the coast is low and strong. Cape Paghelian is the extremity of a bed of sandy gravel, extending 500 yards South, 74° West from the S.W. extremity of the island itself. It is scarcely above the surface of the water. From its commencement the rocky shores run 1½ miles to the East, rising quickly to form the *Meinghyngaï Mountain*, conspicuous from its rounded top. Then begins a gravel bed, which, trending in a curve to S.E. and S.W., forms the excellent road of *Glasenapp*. The extremity of this gravel bed, called *Yerghin*, is 1¾ miles to South, 65½° East from Cape Paghelian. There is good anchorage in the bay thus formed in from 10 to 16 fathoms, mud. You may even moor to the bed of gravel. Fresh water is scarcer here than in other parts, but in its N.E. angle are some good ponds. On Cape Yerghin is a village of two or three yourts.

From this cape the gravel bed runs 2 miles to the N.E. to a pointed and steep cape; then the shore gradually trends to *Cape Ryghynin*.

ITTYGRAN ISLAND, 2 miles to the South of the previous island, is 6 miles long East and West, and 2 or 3 miles broad. Its N.W. extremity is distinguished by a blackish and perpendicular rock. From thence the North coast of the island runs directly East, and then turns to S.E., to South, and S.W., to *Cape Postels*, the S.E. extremity of the island, lying 3 miles N.W. ½ N. from Cape Mertens, and 2½ miles from the nearest part of the continent. This forms the breadth of the South entrance to the Strait of Séniavine, called by the Tchuktchis

Tchetchekouïoum. *Cape Postels* is distinguished by a moderately high hill, with a perfectly round top.

Kynkaï Island, which is not more than three-quarters of a mile in circuit, lies 1¾ miles to S. 62° W. from Cape Paghelian, and 1¼ miles to N.W. of the N.W. extreme of Ittygran Island. It is moderately high, rocky, slopes rapidly, and has a bare and flattish top.

Nouneangan, a small rocky islet, is outside the strait, lying 4½ miles N.E. ¼ N. from Cape Mertens. It is cliffy all round, about 80 feet high, and covered with a pleasing verdure.

One remarkable feature of the Strait of Séniavine, which also occurs at the Bay of St. Lawrence, is, that in these straits, enclosed by coasts, that the depth is greater than in the middle of the adjacent sea, which does not exceed, except in some parts, 24 fathoms. On the American coast the depth is not great; but it is still more singular that this depth is separated from the shallower open sea by a bank with still less water over it; so that the soundings first decrease on approaching the coast, and then increase when on it. In the middle of Behring's Strait the depths diminish equally on either side.

The *tides* were almost imperceptible two or three days after the new moon, but a strong North wind raised the level, temporarily, 2 or 3 feet.

The *wind* naturally affects the atmosphere; with those between North and West it is clear; the South brings clouds, and S.E. moisture.

The variation of the compass, at the River Maritch, was 23° 5′ East.

Vessels coming here to trade with the Tchuktchis generally visit the Bay of St. Lawrence, but this doubtless will be, or has been, abandoned for the Strait of Séniavine. It is 60 miles farther South, and more sheltered from ice and North winds. Its superior ports, too, are a great advantage. For a short stay, and to procure water, *Glasenapp Bay* will be found excellent. If a longer stay is to be made, Abolecheff Bay will answer. Should it be necessary to entirely discharge a vessel and heave her down, Ratmanoff Bay offers every facility. The Strait of Séniavine also offers more trade, because the Reindeer Tchuktchis assemble here in greater numbers, from the superiority of the pasturage, even over those of the Bay of Anadyr.

CAPE TCHAPLIN.—The coast from Cape Mertens runs to the South. The mountains recede into the interior, and from the coast a bed of gravel projects, which, trending in a curve to S.E. and East, forms the long point called Cape Tchaplin, in lat. 64° 24′ 30″, and lon. 172° 14′ West. The Tchuktchis have a large summer habitation on the cape.

At the commencement of this bed of gravel high mountains, with pointed summits, advance to the shore. They are the last branch on this side, and are not seen again until we approach the head of the Gulf of St. Croix. To the West of this steep coast is an open cove, extending 2 miles to the North; from this cove the coast runs to the South, but the mountains forming it are lower. They, for the most part, jut out in steep and high escarpments, with large gaps between, in which, in some instances, are gravel beds extending, on which the Tchuktchis have some temporary habitations.

CAPE TCHOUKOTSKOÏ.—Farther on the coast trends towards the S.W.

and West, and gradually towards the N.W. quarter. Capt. Lütke considered the southernmost point of this extent of coast as Cape Tchoukotskoï. Geographers had for a long period denominated the most projecting part of this coast by that title, and it seemed only reasonable that this usage should be followed. It is a bluff headland, declining in a narrow crest, from which rise some high rocks, terminating in points. It abuts on the sea with a small roundish hill, on which some similar isolated rocks rise. These may all be very readily distinguished in coming East or West, but, viewed from the South, it is all confounded with the coast.

The Cape lies in lat. 64° 16′ N., and lon. 173° 10′ W. Cook assigned its position as 64° 13′, and 173° 24′, but he saw it from a great distance, and near its parallel; this discrepancy of longitude is therefore not remarkable. According to Behring's journal its latitude is 64° 20′. Beyond this commences the Gulf of Anadyr.

The GULF of ANADYR.—The S.W. limit of this gulf may be placed at Cape St. Thaddeus, lying 200 miles S. 65° W. from Cape Tchoukotskoï. With this breadth in its opening, the gulf is 420 miles in circuit, without reckoning the smaller sinuosities and the Gulf of St. Croix, which is 180 miles in circuit.

Up to the time of the visit of Capt. Lütke, Behring had been the only navigator who had sailed in it. Behring went around it, anchored on the West coast in lat. 63° 47′, discovered the Gulf of St. Croix, took in water at the small open Bay of Transfiguration, after which he followed the coast at a short distance, as far as the strait now bearing his name. Behring had not the means at command for making observations with the accuracy required in modern times; but the direction of the coasts, traced simply from his route, had more resemblance to their real bearings than all the details that were to be found on the charts.

From *Cape Tchoukotskoï* the coast extends to N.W. At 12 miles to N. 70° W. from this cape we reach *Cape Stolétié* (of the century), which much resembles the former, of a blackish colour, and having, in a similar manner, isolated rocks on its crest. The coast between is steep, and without any remarkable sinuosities.

At 7½ miles from Cape Stolétié, *Cape Ouliakhpen* projects in a steep declivity, and is high. The rocks of this cape, and also of those farther to the N.W., are not so black as those which extend towards Cape Tchoukotskoï; and the isolated and pointed rocks on their crests are not seen here. On the East side of this cape is an open bay, into which the small river *Vouten* falls.

PORT PROVIDENCE.—It is presumed that the open bay just alluded to by Capt. Lütke is the same with that which has recently afforded winter shelter for H.M.S. *Plover*, in 1848-49, which was despatched in search of the missing expedition of Sir John Franklin. From the brief notice that has as yet appeared, it is extensive, with safe anchorage, protected from the sea by a long low spit. A supply of water could be conveniently obtained from the anchorage first selected. Commander Moore's subsequent proceedings he describes thus:—

"On the 20th October, 1848, finding the direction and force of the wind to continue, the temperature of the air to fall as low as 23½ Fahrenheit, and the sea-water to 28½, I deemed it prudent to take the opinion of the officers as to whether an endeavour to proceed to the northward should be made. These opinions were strictly in accordance with my own sentiments, viz., that it would be better to remain in this secure harbour for the winter, than make a useless attempt to proceed northward, with a probability of being unable to regain my advantageous position (from which I could send out overland expeditions), and on account of the advanced season to lose the chance of wintering even in Petropaulovski. I therefore determined that, should no favourable change take place before the 26th, to select a convenient spot in which to place the ship for the winter.

"On the 23rd a still further reduction of temperature took place; the upper part of the harbour was reported freezing over, and large masses of ice forming during the night about the ship; in consequence of which, after a personal examination of an inner harbour possessing many advantages, I removed thither on the 24th, anchoring at 3^h p.m. in 7 fathoms.

"EMMA HARBOUR, to which I had now removed, communicated with the larger one by an opening a mile wide, forming a basin 4 miles long, and 1½ in breadth, surrounded on every side by lofty mountains, except to the southward, where it was separated from the sea by a tract of low land and an extensive lagoon, and having deep water at the entrance and middle, with good anchorage on each side close to the shore. On the low land to the South was a native settlement of seven huts, to which belonged a large herd of reindeer, from which I hoped from time to time to obtain supplies of fresh meat. Considering it, however, safer for the ship, on account of the force and prevalence of the N.E. winds, as well as the probability of the ice drifting, and on the whole better to be at a little distance from a people whose friendly disposition was not yet established, I removed to the North side of the harbour on the 25th, and there secured the ship for the winter on the 28th of October.

"From the 29th the people were employed in dismantling the ship, leaving nothing but the lower rigging over the mastheads, building a house of stone for the convenience of working the forge, drying clothes, &c., and housing the ship in, all which was completed by the 8th of November.

"During this time ice was continually forming round, and frequently broken up by squalls and strong N.E. winds, so that the ship was not finally frozen in until the 18th, when the natives were first enabled to visit us alongside the ship in their sledges, drawn by dogs.

"The different tribes of natives near my winter quarters at first appeared to hesitate about coming on board, but on making a few presents, and allowing some traffic to be carried on with them, they gained confidence, at least so far as to enter the ship readily when invited to do so, we being careful on all occasions to guard against treachery on their part, on account of the warlike and relentless character attached to the people of these coasts by some authors.

"During the months of November and December the ship was daily visited, not only by those in the vicinity, but also by others from a distance along the coast and inland, by intercourse with whom I was enabled to satisfy myself that they were not only peaceful, but disposed to be actively friendly towards myself and the officers and men under my command."

The position of the harbour, ascertained during this stay, is lat. 64° 25′ 55″ N., lon. 173° 7′ 15″ W.

The officers of the *Plover* made several excursions into the neighbourhood, and in February one of these parties reached within sight of the East Cape. They were all, however, conducted under great hardships.

CAPE YAKKOUN, like Cape Ouliakhpen, is very high and very steep. It is conspicuous, from a pyramidal rock rising from its summit.

CAPE TCHING-AN falls from a great height, almost perpendicularly, into the sea. It is very remarkable by a red band which intersects the cape from its summit to its base. Between it and Cape Yakkoun is an open bay, around which the coast slopes down. A village was seen in it.

From Cape Tching-an the coast, consisting chiefly of perpendicular rocks, trends to N.W. and W.N.W. as far as *Cape Spanberg*. It is high, and in lat. 64° 42⅓′ N., and lon. 174° 42′ W. On the South side of the cape is a high steep rock, with a rounded top; and on the West side is a hill equally rounded, the flanks of which gradually slope on either side. Between this cape and *Cape Halgan*, 9 miles distant, to the N. 71° W., a bay penetrates into the land, which circumstances did not permit Capt. Lütke to examine in detail.

Cape Halgan is high and very steep. Seen from the South side, a small head slopes to the right; but on the West side is a serrated and pointed ridge. In front of it is a large detached rock.

Cape Ninirlioun is as high and as bluff as the preceding, and in general the intervening coast is equally so. This cape is very remarkable by its flat top, but more so from its entirely different appearance from that which follows it, *Cape Attcheun*, in lat. 64° 46′, lon. 175° 28′. This latter cape, moderately elevated, steep to seaward, is separated from Cape Ninirlioun by a bay surrounded by a low coast, from which a long chain of mountains extends to the N.W., a country covered with moss. That which forms the S.E. side of Cape Attcheun, at the distance of 20 miles or more, seems to be detached.

TRANSFIGURATION BAY.—A coast with a similar appearance extends in a winding manner 4 miles to the N.W. to a small open bay, which Lütke recognised as Behring's Bay of Transfiguration, or Preobrayenia. It is surrounded by a low shore, and towards its extremity it receives the *River Ledianaya* (frozen), which the Tchuktchis call Kouïvaem.

From this bay the coast is high, nearly perpendicular, and like a wall; it extends 9 miles to *Cape Enmelian*. Cape Behring is equally high and perpendicular; but between them there is a small space, where the even, uniform coast curves into the form of an open bay.

CAPE BEHRING is situated in lat. 65° 0′ 30″ N., lon. 175° 57′ W. It is particularly noticeable, because here suddenly terminate the steep rocks which, with small exceptions, form the entire extent of coast as far as Cape Tchoukotskoï, and farther North the coast becomes still lower. The mountains in this space are similar to those at Cape Tchoukotskoï; of a mean height, level at the summit, sloping, and even flat, which particularly characterizes the mountains about Cape Ninirlioun. The high and peaked mountains, like those in the Bay of St. Lawrence, will no longer be seen, even in the distance. From Cape Behring the coast turns abruptly to the N.E., then to North, sloping gradually, and terminating perpendicularly in some parts, as far as *Cape Tchirikoff*, which is steep, and forms an open bay. In many places, on the mossy land, traces of rivulets or small streams are seen, and on their borders Tchuktchi villages. Cape Tchirikoff has a tolerably high hill, with a peaked top on it.

Beyond this only a single bluff and high cape can be distinguished, lying 4 or 5 miles to the N.W. of Cape Tchirikoff. The coast thence trends towards the mouth of a large river, from whence it takes a westerly direction. Capt. Lütke, not being able to get any communications with the inhabitants near this part, could not ascertain the name of this river. The direction of its course is generally North and South, but it falls into the sea to S.W. It flows through low lands, so that, at the distance of 15 or 20 miles, only a large opening is seen, beyond which nothing is visible. It is doubtful whether sufficient depth of water for ships would be found, because, at 10 miles from its mouth, the lead only gave 6 to 7 fathoms. It most probably takes its rise near the same place where the rivers that flow into the Bays of Koulutchinskoï and Metchigmenskoï do. Lands covered with moss and marshes apparently occupy all the space comprised between the head of these bays and the river. Towards the South, as far as Cape Behring, the hills of moderate height are alone covered with moss, with the exception of the two capes that have been mentioned.

All the eastern shore of the Gulf of Anadyr is destitute of wood.

The soundings from Cape Tchoukotskoï up to Cape Attcheun show from 27 to 31 fathoms at the distance of 6 to 10 miles from the coast, the bottom mud or gravel. Off Cape Behring there are 18 to 22 fathoms at 8 miles distant; off Cape Tchirikoff there are not more than 10 or 11 fathoms at the same distance; and still farther there are not more than 7 or 8 fathoms. Here the bottom is generally gravelly. No sort of danger was perceived from the *Séniavine*. It must therefore be considered that it is clear throughout, because Behring, who kept close to the land all the way beyond the cape now bearing his name, does not either make mention of any shoal or reef whatever.

The variation of the compass in this part is from 19° to 20° E.

To the West of the river above described the coast is low for 4 miles, and then commences to become hilly. The mountains, higher than those on the East coast, are peaked or flat at the summit, but are all dispersed without any order, with valleys at intervals very deeply cut in, through which rivulets and small rivers flow. The coast in this form extends 15 miles W.N.W. ½ W. and W.N.W., forming a small open bay, into which a small river falls, and terminated on the South by a high bluff cape. The bottom of the bay is in lat. 64° 36½′, and lon. 176° 48′, and is properly the northern extremity of the Gulf of Anadyr. On the West side of the cape just mentioned is another equally open bay, into which a tolerably large river flows from the North, and winding in large valleys between the mountains.

At 3 or 4 miles from this last river the most remarkable bed of gravel that had been seen commences. It extends without interruption to S.W. and W. for 45 nautic miles, as far as Cape Meetchken, in the Gulf of St. Croix, and consequently forming the largest portion of the North coast of the Gulf of Anadyr. Some portions of this were hidden by the fog, but that part which was seen from the *Séniavine* was of a wearying uniformity, the same waste and sterility; it was throughout nothing but a heap of bare shingle, with the exception of a very few spots, where there had been, or still was, a habitation; there was also some grass, and other herbs that accompany man.

A narrow and shallow canal separates this gravel bed from the continental coast, which runs parallel to it, and bounds the sea with low reddish cliffs. Farther off a mossy plain extends to the first of the low and even mountains, 7 or 8 miles from the sea.

Off Cape Meetchken the North coast of the gulf is broken, to form the entrance to the great Gulf of St. Croix.

The eastern angles of the Gulf of Anadyr are the portions which have the least depth. There the number of fathoms equal the number of miles (Italian) that you are off the coast. At 12 miles you have 12 fathoms; at 8 miles, 8 fathoms, or rather somewhat less than this. Within 3 or 4 miles of the coast there are, however, still 5 or 6 fathoms. Along the gravel bed there is similar depth, but it increases to the westward. Off Cape Meetchken, at only 2 miles distant, you will find 27 fathoms; but at 1 mile or 1½ miles from thence is a bank on which the *Séniavine* ran great risk of grounding. In the North part of the gulf the bottom is gravelly throughout, but after passing the bed of gravel it begins to be muddy.

THE GULF OF ST. CROIX occupies a space of 54 miles of latitude, and 35 miles from East to West. It reaches within 10 miles of the arctic circle. Its shores, to the distance of 35 miles from its entrance, run nearly parallel to each other, to N.N.W., and 20 miles apart. Farther on they approach each other, and narrow the gulf to less than 4 miles.

CAPE MEETCHKEN, the western extremity of the bed of gravel previously described, forms the East point of the entrance; it is in lat. 65° 28′ 40″, and lon. 178° 47′. The shortest distance to the opposite shore to the West is 13½ miles.

There is good anchorage on the North side of Cape Meetchken, open, however, to N.W. and W.N.W.; the coast in this direction, being 40 miles distant, affords not much protection, though Capt. Lütke considers that, with proper precaution, a vessel might hold out in bad weather. The depth is 5 to 9 fathoms, and the best place is to bring Cape Meetchken to bear S.W. by compass. Care must be taken, in entering, of the rocky bank before alluded to, which lies 1½ miles S.W. of the cape. On the bed of gravel is a permanent establishment of the Stationary Tchuktchis, called *Meetchken*, composed of seven or eight summer yourts. They were the most peaceable and the poorest that were met with; though they were not in want, a ship could not procure anything from them.

The eastern side of the gulf, the nearest part of which is 8 miles from Cape Meetchken, has but very little depth. At 3½ or 4 miles there are but 3 fathoms. Farther to the N.W. a shoal extends 1½ and 2 miles from the coast, which is a low cliff of 12, 20, and in some parts 60 feet in height. There are no mountains whatever along the coast. It is a mossy plain, which, in some parts, gently rises into small hills. Only near the entrance a branch of the mountains advances, of which the nearest to the gulf is called, by the Tchuktchis, *Linglingaï*, a word in their language meaning "heart rock," in Russian *Serdze Kamen*. It lies in lat. 65° 36½′ N., and lon. 178° 17′, and its height is 1,462 feet above the level of the sea. It is one of the best determined points in the gulf.

At 26 miles from Cape Meetchken a tolerably large and high bed of gravel advances from the coast to the N.W. It is covered with dry moss, and forms a cove 2 miles in circuit, exposed to the N.W., in which there is safer anchorage than in that at Cape Meetchken. There is no running water here, nor is there on all the coast, but ponds and small lakes are frequent, and afford good snow or ice water.

At about 8 miles from this point a long and low point projects, forming the South limit of the *Bay of Kanghynin*, which is nearly 6 miles wide at its opening, and not less than 40 miles in circumference; but, on account of its shallow depth, it does not merit any attention.

The northern side of the gulf presents an entire contrast to those of the East and western sides. High mountains here advance in three abrupt capes, of a sombre appearance, the extremities of which offer a resemblance to the letter M. The superior angles of this letter are, Egvekinot and Etelkouïum Bays, and the base of its eastern line, the North point of Kanghynin Bay.

Egvekinot Bay penetrates 7 miles due North, with a breadth of 1 or 1½ miles. The high mountains which surround it leave all round a narrow band of low shore, and at the bottom a large valley covered with moss. There is no part of the bay worthy of the name of a harbour.

Etelkouïum Bay lies by the side of the former. The entrance only was examined, because as early as September 3rd, 1827, it was already covered, for the distance of more than 5 miles, with new ice. The depth in the entrance was 13 to 18 fathoms, muddy bottom. It had every appearance of being a good port. At the entrance of the bay, on its North side, a bed of gravel forms *Krusenstern Cove*, the opening of which is about a fourth of a mile wide, with 7 to 12 fathoms water. Quiet anchorage may be had within it.

Beyond these two bays the North shore of the gulf trends 10 miles to the West, sometimes perpendicularly, at others sloping, but mountainous throughout, and then turns suddenly to the South. At the angle of this curve is *Engaoughin Bay*, a round cove of 9 miles in circuit, sheltered from the South by a low point projecting 2 miles to the West, and by a gravel bed standing alone in front of the point. This forms an excellent harbour, the only one worthy of the character in the Gulf of St. Croix.

In the N.W. angle of this port there is a rivulet of fresh water, but it is only serviceable at high water. There are several ponds and lakes of fresh water.

From this bay the western coast of the gulf runs S.S.W., and then curves gradually to the S.E., without forming a single remarkable bay or cove. At the distance of 10 miles from the port the mountains advance very near to the sea, and reach it in places with high cliffs; but farther on the coast assumes an appearance exactly like that on the opposite or eastern side. This side of the bay is distinguished by its superior depth. There is not a single shoal throughout its extent. In lat. 65° 38′ it forms a pointed cape, which curves rapidly to the South, and preserves this direction as far as the limits of the gulf. In the centre the depths are from 22 to 40 fathoms, muddy bottom throughout.

The most remarkable mountain about it is that of *Matatchingaï*, at the bottom of Etelkouïum Bay. It is distinguished from all others as well by its elevation as by its sombre and rugged flanks. From the entrance of the gulf, at the distance of 60 miles, it appears at the sea level not to be more than 20 or 30 miles. Its height was calculated at 9,180 feet.

On the West side, up as far as the Port of Engaoughiu, a large quantity of drift-wood is found, even long and large trunks of trees; on the East and North coasts, on the contrary, not a single piece is met with. This circumstance is worthy of note; it proves that the current from the River Anadyr, from which it comes in entering the Gulf of St. Croix, bears chiefly to the West, although from the bearing of its shores the contrary would have been anticipated. On no part of the shores of the gulf is the smallest trace of growing wood to be met with.

The tides were carefully observed. The establishment of the port appears to be 8^h 50′. The greatest rise was 7 feet, but usually it was 4½ to 5½ feet; some former traces showed a rise of 9 feet.

The variation of the compass, on Cape Meetchken, was 21° 45′ East. The dip, 75° 30′.

The RIVER ANADYR, which gives its name to the gulf which receives its waters, is the most considerable which falls into the Sea of Behring. Capt. Lütke was prevented by circumstances from examining this portion of the gulf, which must therefore be left to the imperfect delineations of the chart for all further ideas.

CAPE ST. THADDEUS is the S.W. cape of the Gulf of Anadyr. Behring gave this name to a cape on August 21, o.s., being in lat. 62° 42′, and from his data the term has been defined to apply to the high bluff cape situated in lat. 62° 42′, and lon. 179° 38′ East. It has been considered that the cape to the southward (Cape Navarin) was the headland in question, but which is 26 miles more to the South.

Cape St. Thaddeus is the point which projects farthest to the East in this portion of the coast, while beyond the cape turns to the N.W. and S.W., so that it forms a sort of natural limit to the Gulf of Anadyr. At 15 miles to the S.W. ¼ S. is another high cape, to which Capt. Lütke applied the name of *King* (most probably the same that that navigator took for Cape St. Thaddeus).

The Bay of ARCHANGEL GABRIEL.—From Cape King the coast turns suddenly to the N.W., forming a bay, which penetrates the land to a depth not less than 15 miles, with a breadth of 6 miles. To this bay Capt. Lütke gave the name of Behring's vessel.

CAPE NAVARIN.—From Archangel Gabriel Bay the coast runs South to this cape, in lat. 62° 16′, and lon. 179° 4½′ East, beyond which it turns abruptly to the N.W., to form an open gulf, into which it is likely the River Khatyrka falls. In addition to its conspicuous situation, Cape Navarin is remarkable for a high mountain on its point, 2,512 feet in height, the flanks of which descend nearly perpendicularly into the sea. This has been taken as the real Cape St. Thaddeus of Behring, but it does not accord with his journal.

Cape Navarin is the South extremity of the peninsula which bounds the Bay of Archangel Gabriel on the South. A chain of high mountains extends through it. *Mount Heiden* surpasses the rest in elevation (2,230 feet), and is distinguished by its conical form. In the middle of September (1827) it was entirely covered with snow.

All this part of the coast, according to the examination by Capt. Lütke, differs very greatly from all previous delineations. After Behring, Capt. King, or rather Clerke, was the only visitor previously, but this was at such a distance that the greatest discrepancies exist, especially in longitudes. From these discordances it is almost impossible to fix exactly the points intended by each navigator. On all old charts this portion of the coast is shown nearly half a degree too far North.

The variation of the compass on Cape Navarin was 13° 35′ E.

From the cape we have a long interval of coast, upwards of 350 miles in extent, of which we know nothing. Capt. Clerke passed it at a great distance, and Capt. Lütke, both in his progress to the North and on his return, was prevented by bad and foggy weather from making any observation on it. It is the country inhabited by the Koriaks (Kariaks), who have been previously noticed.

CAPE OLUTORSKOI is the first point described by Capt. Lütke. It is in lat. 59° 58′, and lon. 170° 28′ East. It is remarkable by a high mountain with three summits (2,537 feet), with a steep ascent from the sea. At a short distance to the North there is a hillock of a conical form. From this cape the coast extends on one side to W.N.W., towards the *Gulf of Olutorskoï;* and on the other, first 4 miles to E.N.E., then 30 miles to the North, rather inclining to the East. In all this extent it is mountainous, and falls into the sea in cliffy headlands. In the latter half of September there was much snow on the ground, and not a plant could be seen.

The *Gulf of Olutorskoï* was not examined by Capt. Lütke on account of the fog and its distance. Its western termination is a cape, which was supposed to be *Cape Govenskoï.* According to an imperfect observation, it is in lat. 59° 50′, and lon. 166° 18′. It is high, bluff, and cliffy, and over it are some high mountains which were covered with snow.

CAPE ILPINSKOI.—From this cape the last-named coast trends nearly West to Cape Ilpinskoï, where the coast suddenly becomes lower. This cape is in lat. 69° 48½′ and lon. 165° 57′. Projecting from mountains of a moderate height, it advances to the S.W. in an even point, not very high, and falling perpendicularly into the sea. According to Krachenninikoff, it is joined to the continent by an isthmus so low and narrow that the sea washes over it.

VERKHOTOURSKY or Little KARAGHINSKY ISLAND lies directly before Cape Ilpinskoi. Its lat. is 59° 37½′, and its lon. 165° 43′. It is of a round form, and 3 or 4 miles in circumference. On all sides except the N.W. it falls perpendicularly into the sea. To the N.W. it projects a short distance in a low point and a bed of gravel, on which were some habitations inhabited by the Koriaks, who come hither to hunt black foxes.

The strait between Verkhotoursky Island and Cape Ilpinskoi is 12 miles broad. Nearly in the middle of it is a dangerous reef, awash, extending 1½ or 2 miles East and West. In the centre is a small but high rock. A little to the North is a stony islet, and it may be presumed that there are other dangers.

From Cape Ilpinskoi the coast curves to the West and S.W., forming a large gulf, which is bounded to the South by the large Island of Karaghinsky.

As this bay forms one of the narrowest and the lowest portion of the Peninsula of Kamtschatka, it is usually taken as the northern limit of that country; the Bay of Penjinsk, in the Sea of Okhotsk, forming the opposite coast.

Although a small portion of the coast to the southward may be included in the shores of the Sea of Behring, we shall for the present quit them, leaving them to be described in connexion with the peninsula in the next chapter.

Having thus described the shores and the adjacent islands of the Sea of Behring, it remains, to complete this chapter, to describe the detached islands which are found in it. In this we have derived much information from the voyages of Capts. Lütke, Cook, Billings, Kotzebue, Beechey, and other navigators, whose works are quoted in the respective places.

ST. LAWRENCE ISLAND.

This island is the northernmost of those which lie in the open sea. It was discovered by Behring on St. Lawrence's day, August 10th, 1728. He stated that he passed by it without observing anything particular on it except the cottages of some fishermen.

Capt. Cook gave it the name of *Clerke Island*, but Capt. King, who composed the third volume of the account of Cook's Third Voyage, does not use this designation. It was seen by Capt. Kotzebue, who examined the East and S.E. sides, but did not observe the union of the East and West portions. In the early days of our knowledge of it it was supposed to consist of several islands. This arises from the nature of the land, the high extremities being connected by a tract of very low land, which consequently cannot be seen at a great distance.

From this cause, beyond doubt, the islands Macarius, St. Stephen, St. Theodore, and St. Abraham of Lieut. Syndt, are only the higher hills, which are all that are seen of St. Lawrence at a distance. Cook thus named a part of its extreme Anderson Island, after his deceased and much-respected surgeon.

In 1828 Capt. Schischmareff made a detailed examination of its shores, with the exception of that part examined by his former commander, Capt. Kotzebue, in 1817.* On the S.W. side is a small open bay, where the officers of the *Rurick* landed; this spot is readily recognised by the small rocky island in its vicinity.

From these examinations it appears that the island is above 29 leagues in extent from East to West. The N.W. point, to which Admiral Krusenstern has given the name of the Russian surveyor, *Schischmareff Point*, is in lat. 63° 46′ N., lon. 188° 19′ E. This exactly accords with the position determined by Capt. Clerke, "who took it for an island." This island, if its boundaries were at this time within our view, is about 3 leagues in circuit. The North part may be seen at the distance of 10 or 12 leagues; but as it falls in low land to the S.E., the extent of which we could not see, some of us conjectured that it might be joined to the land to the eastward of it (which is now known to be the case), but we were prevented by the haziness of the weather from ascertaining this. They were covered with snow, and presented a most dreary picture.†

* Kotzebue's Voyage, p. 195.

† Cook's Third Voyage, vol. iii. p. 243.

A very projecting point on the North side of the island is in lat. 63° 12′ N., lon. 159° 50′ W. Capt. Kotzebue places the eastern point of the island in lat. 63° 18′, and lon. 168° 48′, from which Schischmareff differs slightly, as he does, too, in the configuration of the coast.

The island which Cook saw near this point, in lat. 63° 10′, and lon. 159° 50′, is composed, according to Kotzebue, of two islands; Schischmareff says there are three. The inhabitants call the eastern part of the coast *Kaegalack*, and the western *Chibocko*.

The eastern point of the island is named Cape Anderson, and here an historic doubt existed.

A shoal of 11 fathoms was found by the *Blossom* precisely in the situation assigned to a small island named by Cook after his respected surgeon, Mr. Anderson. This island had never been seen after, and the veracity of the great navigator had been in consequence impeached. Capt. Beechey, however, rectifies this error, having found that it was intended for the East end of St. Lawrence Island; the compilers of his chart appear to have overlooked certain data collected here, which would not have been omitted had Cook's life been spared.*

We have no detailed description of the shores, or capabilities of the island. The foregoing are the principal facts relating to it, scattered through the works quoted.

ST. MATTHEW ISLAND.

This island was discovered by Lieut. Syndt, in August, 1766. Capt. Cook, ignorant of this circumstance, considered it as a new discovery in 1778, and called it *Gore's Island*. He only saw the S.E. part from a distance, and probably only made out the small island lying separately to the North, which the Russian promychlenniks call *Morjovi* or Morses' Island, as it is here only that these animals visit. Since Cook's time, it has been seen by several Russian navigators. Sarytscheff anchored here; Schischmareff passed close to it; but none have given a detailed description of it. On the Russian charts it has always borne its original name, *Matvoi*, or St. Matthew; but to preserve the name by Cook, Lütke has called the West extremity of the island *Cape Gore*.

St. Matthew Island lies N.W. and S.E., and in a direct line is 27 miles long, and 3½ to 4½ in breadth. Its shores consist partly of high rocks, partly of low land. The S.E. extremity of the island, most justly called by Cook *Cape Upright*, rises out of the water like a wall to the height of 1,400 feet. This is the highest point of the island. It falls suddenly to the N.W., forming a very low and very narrow isthmus; not being seen beyond 4 or 5 miles, causes Cape Upright, even at this distance, to appear as a separate island. Beyond this isthmus, the island increases in breadth and elevation, and then again contracts, forming another isthmus, similar to the first, at 9 miles from it, then a third, from which formation, St. Matthew at a distance appears like several islands. The S.E. or outer point of Cape Upright is in lat. 60° 18′, and lon. 172° 4.′

At 12 miles W. 6° N. from this cape is *Sugar-loaf Cape*, thus named from an

* Beechey, Voyage of the *Blossom*, p. 563.

extremely remarkable mountain which surmounts it. This mountain is 1,438 feet in height, and on every side appears as an irregular cone. Its pointed summit, the only one on the island, could be seen over every lower portion of the island, and Capt. Lütke says, was of the greatest utility as a mark to connect these observations. Between Cape Sugar-loaf and Cape Upright are two bays, entirely unprotected, surrounded by low shores. On the North side of the Sugar-loaf is a similar bay, and an isthmus similar to that connecting Cape Upright. From this towards the N.W. to the West extreme, Cape Gore, are almost perpendicular rocks, intersected in many parts by ravines.

Cape Gore terminates to seaward in a low cliff. Off it are some rocky islets. At 9 miles North from the cape, and 3 from the North end of the island, on the coast quite by itself, is a remarkable rock of a rhomboidal form.

The North point of the island, named by Capt. Lütke after Capt. Sarytscheff's vessel, is in lat. 60° 38′, and lon. 172° 41′. It is steep, but much lower than Cape Upright. The eastern shore of the island much resembles the opposite one. There are corresponding bays on either side, which form the narrow *isthmuses*.

Morjovi Island is steep on every part except the S.W. Its North extreme, in lat. 60° 44′, and lon. 172° 52′, equals Cape Upright in elevation, and much resembles it. The South end extends in a low point to the S.E., where the officers of the *Slava Rossii*, Billings' ship, landed.

Pinnacle Island, justly so named by Cook, lies 16 miles W.S.W. from Cape Upright. Two sides, nearly perpendicular, unite at the elevation of 990 feet in a pointed crest, with a number of pointed rocks on it; at the steep S.W. extremity are some isolated rocks; and the N.E. point terminates in an entire range of connected and extraordinary pointed rocks.

The shores of St. Matthew are clear, and the depth very great. There might not be great difficulty in landing in fine weather in the bays. The *island is not* inhabited, and is scarcely capable of being so. The Russian Company attempted to fix a small colony here in 1809, but one-half perished from scurvy, and it was then abandoned. The formation of the island is volcanic. The variation was found to be 19° 5′ E.*

PRIBUILOFF ISLANDS.

These are a group of three small rocky islands, two of which were discovered by M. Pribuiloff, in 1768; this officer was under Capt. Billings' expedition, in 1790. At first they were called *Novy* (new); then *Lebedevski*, from the name of the owner of the vessel which discovered them: M. Chelekoff called them *Zouboff*; more recently they have been called *Kotovy* (sea-bears), and *Severny* (north), from the immense quantity of the animals found there, and their position relative to Ounalashka. Admiral Sarytscheff has placed them on his chart under the name of the officer who discovered them, as here repeated. They are most commonly called in the colonies here *Ostrovki*, the little islands.†

* Lütke, Voyage du *Séniavine*, Part. Naut., pp. 341—343.
† The navigators who have described more or less fully this small group (besides that which stands first, in the Voyage of the *Séniavine*, Part. Naut., pp. 336—340), are—a rough account in

ST. GEORGE'S ISLAND is the southernmost. The southern and western parts are surrounded by rocks; but the North is easy of approach, and affords good anchorage in a commodious bay for small vessels, not drawing above 8 or 9 feet water. The whole island is volcanic, destitute of inhabitants, and only produces the bulbs, plants, and berries which are to be met with in all the Aleutian Islands. Pribuiloff found the low lands and surrounding rocks covered with sea animals, particularly the ursine seal (*kotic*) and sea-lion (*sivoutchi*); and with the skins of these animals they nearly loaded their vessels.

On the Island of St. George they passed the winter, and found the inland parts overrun with foxes, which afforded them a profitable chase. It also abounded with the tusks of the walrus, which they picked up on the shores. It is about 3 miles wide, and extending E. by N. ½ E. 19 miles; or, according to Lütke, 13½ miles in length. Drift-wood was at first abundant, but that, with the fur-bearing animals, soon became scarce.

Capt. Lütke makes the following remarks on it:—"Its East extremity was determined by us to be in longitude 169° 10′. Its latitude, according to Capt. Tchistiakoff, is 56° 38′. The aspect of the S.E. coast is very monotonous; on its level surface there is but one point rising above the rest, and this is 1,083 feet, English, above the sea. The two extremities of the island terminate in very steep rocks. The North coast, which we examined, consists entirely of rocks, of 300 feet in height, the greater part rising perpendicularly out of the water. In one position, at 5 miles from the N.E. point, the coast slopes inward, and is covered with a thick herbage. Here is the company's establishment. A small cove between the rocks serves to shelter the baïdars; you may even anchor there in South and S.E. winds: at a mile off there are 17 fathoms water, black sandy bottom. This anchorage is slightly sheltered from the East by a low point between the village and the East point of the island. The surface of the N.W. part is perfectly flat and horizontal, and is covered with grass. The coasts in general are clear, but at 13 or 15 miles to the East there was a bank seen, in 1824, by Capt. Chramtschenko."*

ST. PAUL'S ISLAND, the second discovered by Pribuiloff, is much smaller than that of St. George; this, as well as the former, was the retreat of immense herds of seals.

St. Paul is 44 miles to the North of St. George, which is 190 miles N. 39° W., *true*, from the North point of Ounalashka.†

The Russian hunters have stated that they saw from the summit of the highest mountain of St. Paul some land to the S.W. Capt. Kotzebue did not find it, though he ran above 25 leagues in this direction. A vessel was also despatched by the Russian Company, in 1831, 1832, and 1833, with the same result. The land in question may possibly be farther off; but Capt. Lütke concludes, from all the remarks, that banks of clouds have caused the deception.

Chamisso's Geological Memoir of Kotzebue's New Voyage Round the World; in *Martin Sauer's* Account of Billings' Expedition, pp. 311, 233; another in *Langsdorff's* Travels; in *Lisiansky's* Voyage; and in *Capt. Beechey's* Voyage of the *Blossom*.

* Voyage du *Séniavine*, 1827, Par. Naut., pp. 336-7.

† *Billings'* Voyage.

St. Paul was not examined by Capt. Lütke, and generally speaking it has not been hitherto described in detail. It is placed on the chart from an imperfect sketch by one of the colonial officers. The island extends to the South by a low bed of gravel, on which stands the village. At half a mile to the S.W. is an islet called *Sivoutchi* or Sea-Lion Island, which, according to Capt. Tchistiakoff, is in lat. 57° 5′ and 41° W. of the East end of St. George, and consequently in 169° 51′. Between the bed of gravel and the West end of the island, 7 or 8 miles distant, to the N.W. or N.W. ½ W., the coast curves into a bay, and forms some small coves, in one of which is a tolerably good shelter for small vessels. The eastern and northern parts of the island are low, and the coasts sloping and sandy; but the West side is mountainous, and terminates to seaward on a high steep cape, which is distinguished by a remarkable height surrounding it. There is on the East side of the island another mountain equally remarkable. These are of a moderate elevation; their summits appear to be broken, and the volcanic stones found here prove that they have been in a state of volcanic activity.

At 5 miles W.S.W. from the Sivoutchi Rock, and nearly due South (*true*) from the West end, is a small high island, 7 miles in circuit, called *Bobrovi* or *Sea-Otter Island;* a reef extends from this island for half a mile to the S.W., and between this island and St. Paul are some hidden dangers. At 4 miles S. 75° E., by compass, from the East extremity is another low and rocky island, called *Morjovi* or *Morses' Island.* The relative bearing of Bobrovi and Morjovi is N. 43° E. and S. 43° W. (*true*), and the distance 14½ miles. There are some reefs to the East and North of the island, and also at the West extremity. On the sketch is shown, at 12 miles to the East of its N.E. end, a bank which uncovers at low water; but this is all we know of it. Perhaps it may be the same as that seen to the East of St. George's Island.

The vessels which usually come in June and July to St. Paul for the chase, stay on the S.E. side of the bed of gravel spoken of above, in front of the village, at three-quarters of a mile from the coast, in 9 to 13 fathoms water; but there is no security. There is sufficient fresh water in the lakes and rivulets of the two islands. There is no species of wood growing on the islands, and but very little drift-wood on the beaches.

On St. Paul the principal chase is sea-bears; on St. George, sea-lions. Foxes are found on both.

The climate of these islands is as humid and disagreeable as possible. Verdure does not show itself until the end of April or May. Dense fogs prevail in summer, the atmosphere is rarely clear, and the sun is still more rarely to be seen. Snow falls in October. In December, North winds bring the ice, which remains here frequently until May.

It is sometimes difficult to find these "small islands" in the condensed fogs which prevail here. Lieut. Tébenkoff has given some instructions to follow in this case; but they are perhaps unnecessary here. It may only be stated that, at times, the land may be seen from the masthead when below it is very thick; and it may sometimes be found by the roaring of the sea-lions on the beach.

Capt. Beechey, on his first return from his exploration North of Behring's Strait, passed these island; we transcribe his remarks:—

"On the 21st October, 1826, we came in sight of the Island of St. Paul, the northern island of a small group which, though long known to English geographers, has been omitted in some of our most esteemed modern charts. The group consists of three islands, named St. George, St. Paul, and Sea-Otter. We saw only the two latter in this passage; but in the following year passed near to the other, and on the opposite side of St. Paul to that on which our course was directed at this time. The Islands of St. Paul and St. George are both high, with bold shores, and without any port, though there is said to be anchoring ground off both, and soundings in the offing at moderate depths. At a distance of 25 miles from Sea-Otter Island, in the direction of N. 37° W., *true*, and in lat. 59° 22′ N., we had 52 fathoms, hard ground; after this, proceeding southward, the water deepens. St. Paul is distinguished by three small peaks, which, one of them in particular, have the appearance of craters; St. George consists of two hills, united by moderately high ground, and is higher than St. Paul; both were covered with a brown vegetation. Sea-Otter Island is very small, and little better than a rock. The Russians have long had settlements upon both the large islands, subordinate to the establishments at Sitka, and annually send thither for peltry, consisting principally of the skins of amphibious animals, which, from their fine furry nature, are highly valued by the Chinese and Tartar nations."*

COMMANDER ISLANDS.

These two islands, Behring and Medny or Copper Islands, do not in reality form a portion of the Aleutian Archipelago, but must be considered as a part of the chain connecting the volcanoes of America with those of Kamtschatka.

The first Russian navigators gave them their present name, in memory of one of the most tragic events in the annals of navigation—the death of Behring (known in these countries under his title of *Commander*), on the westernmost island, which now bears his name.

Like all the rest of the islands in this sea, there has hitherto—that is, prior to Capt. Lütke's work—been no connected description of them. Capt. Billings saw the North coast of Behring Island, and Capt. Golownin determined the position of the South extremities of the two islands, but did not describe the whole of the coasts. Capts. Kotzebue and Beechey saw the N.W. part of the same, and Capt. Lütke saw the N. and N.E. coasts; and, lastly, the Russian American Company's officers have furnished some notices of the bays. Capt. Lütke's Voyage of the *Séniavine* has furnished us with the accounts of them.

BEHRING ISLAND is nearly 50 miles long from N.W. to S.E. Its greatest breadth at the North end is 16 or 17 miles; to the S.E. it narrows, and forms a pointed cape, in lat. 54° 41′ 5″, lon. 123° 17′.† A chain of mountains, of 2,200 feet in height, extend throughout the island; in its centre are some peaks. They are in general higher in the South, and lower and more even in the North. The South cape, called *Cape Manati* by Behring's companions, is conspicuous by some high-peaked rocks terminating it. From this the East coast

* Beechey, Voyage, part i. pp. 339-40. † Golownin, from Petropaulovski, is 201° 17′.

trends North in steep cliffs to *Cape Khitroff*, in lat. 54° 56′, lon. 193° 17′ W. From this to the N.E. point, *Cape Waxell*, the coast trends generally N.W. ½ N. and S.E. ½ S., forming some insignificant curves, frequently intersected by ravines and cavities. The N.E. extreme is an obtuse, low head, projecting 3 miles into the sea. Reefs project from its North and East angles to a mile or more, and it seems that all this coast is bestrewed with rocks. In the curve formed by the East coast is a small bay with a sandy beach, on which is a large quantity of drift-wood.

Cape Youchin, the low N.W. extremity of the island, is in lat. 55° 25′, and lon. 194° 2′. From this point a dangerous covered reef extends 2 miles to the North, on which, at 4 cables' length from the shore, is a large uncovered rock. Between Capes Waxell and Youchin the coast forms an open bay, and about midway between them, in a ravine covered with verdure, with a small river, is a temporary establishment of the Russian American Company, consisting of a few yourts and huts, used by the promychlenniks, who hunt the polar foxes. *These* men say the whole of the bay is bestrewed with rocks.

From Cape Youchin the coast trends to S.W. to the West extremity of the island, in lat. 55° 17′, and lon. 194° 10′ 3″, according to Capt. Beechey; and thence to the S.E., in which direction, at 10 miles farther on, is the company's factory, on the shore of a small bay open to N.W., where, in summer, is tolerably good anchorage in 4 or 5 fathoms, sand, at half a verst from shore. This bay is called here the *port;* but it must be by contrast to the other unapproachable points. Two islets abreast of the village, due west by compass, are good marks to make the port; the one, *Toporkoff*, is 2 miles, and the other, the *Arii Rock** (the Alcas Rock), at nearly 6 miles. The first is a mile in circumference, and is not very high; the other is a high rock, inhabited by a multitude of alcas. Between the two, rather nearest to Toporkoff, is a sunken rock, that only uncovers at low water, called *Polovintchaty*. To the N.E. of this again are some indications of sunken reefs, so that the North sides of these islands should be avoided.

The S.W. coast of the island, from the port to the South extreme, is entirely unknown.

The spot where Behring died, as related in a former page, is on the East side of the island, at three-quarters of a mile W.N.W. from Cape Khitroff.

The water is very deep around the island. At from 4 to 6 miles off the N.E. and North shores the depth was found to be 58 to 67 fathoms, muddy bottom on the North side; further to the East, stony bottom.

The magnetic variation was found by Lütke to be 5° 50′ E. (1827), or 3° less than Kotzebue in 1821, so that there is some error in one or both. Tébenkoff makes it 10° E. in 1831 at Medny Island.

MEDNY or COPPER ISLAND is remarkable for its long and narrow figure. The only island it resembles hereabout is Amlia. It is about 30 miles in length, and its greatest breadth towards the middle is not more than 5 miles; it frequently

* On Lütke's chart this is called *Sivoutchy;* on that of Kotzebue, which has furnished the extent and figure of the island, it is named *Novy* (new).

does not exceed 2 miles. It seems to be the crest of a mountain rising out of the sea in a S.E. and N.W. direction. Medny Island is scarcely lower than its neighbour, Behring Island; seen from the *Behring Cross*, as the spot where the commander perished is termed, it appears to consist of three islands. Its shores are very steep, clear in most parts, and the depth around very great. There are some reefs at its N.W. and S.W. extremities, and at some other points, but they do not extend far off. The island is entirely without anchorage for large ships; but on its N.E. side, at 10 miles from its N.W. extreme, is a small port, where small vessels may ride. The coast here forms a small bay to the West and South, two-thirds of a mile in circumference, having in its entrance 6 and 7 fathoms, which, farther, is diminished to 2, bottom of sand and stones.

The outer coasts of the bay are high; the S.W. side is clear, but on the East side is a multitude of isolated rocks and stones, which shelter the port a little from the North. These rocks, and a high conical mountain on the S.E. side of the bay, serve as marks for entering. The rocks must be left to starboard, and then steer direct for the village, and as soon as you reach as high as a stone column on the West side of the little harbour, you must cast anchor instantly, and at the same time moor the poop to the shore, for there is no room for her to swing. You will have 2 and 2½ fathoms at half a cable's length from the shore. The harbour is badly sheltered from the North, and to guard against North winds always keep a sufficient scope of cable.

The company's establishment is on the South side of the harbour. Its latitude is 54° 47′. According to the observations of Capt. Golownin, the S.E. extremity of the island is in lat. 54° 32′ 24″, and lon. 168° 9′ E. The latitude of the N.W. extremity is 54° 52′ 25″, and its longitude 167° 31′ E.

Medny (Mednoï or Copper, as the Russian name signifies) was thus named on account of the native copper found here, and which was attempted to be worked in the middle of the last century. This fact is not generally known, or has been forgotten. It was said that the copper was a portion of a Japanese vessel wrecked here; a party of miners was sent here in 1755,* but the poverty of the mine led to its abandonment.

The CLIMATE of these islands is not very rigorous. There are no very intense frosts in winter, but they have at times very heavy snow storms. In January and February the N.W. and West winds bring the ice on the coasts in large quantities. The weather is clear with N.E. and East winds; it is overcast with those from *East* and S.E. Fogs and cold prevail throughout the spring; the snow does not disappear entirely before June; the best part of the year is the month of August.

There are no active volcanoes on either of the islands, but earthquakes are frequent, the shocks of which are sometimes felt for a long time, as, for example, in June, 1827, the oscillations lasted for four minutes by the watch, without interruption. Sometimes, during an earthquake, the sea rapidly rises 10 feet or more, and falls again with the same rapidity.†

The TIDES at both islands rise generally about 6 or 7 feet at the new and full moon. No particular current has been noticed between the islands.

* See Courrier de Sibérie, 1822, tome xviii.
† See the effect of these waves from earthquakes, as noticed on pages 112, 113.

After violent and long-continued winds, a large quantity of drift-wood is thrown on to the shores, principally of those species that grow at Kamtschatka, but sometimes the cypress that grows on the American coast, and even the wood which only grows at Japan. Sometimes, also, lacquered vessels of wood, of Japanese manufacture, have been found, which goes to prove that in this part of the ocean the currents trend to North or N.E.

There are no trees on either island, but in the ravines and along the small rivers there are some bushes of willow and the service tree. Of plants fit for food, angelica, nettles, sorrel, and parsley are found; the yellow raspberry and whortle berries are in great quantities. The only land animals are polar foxes, particularly blue ones; of amphibious ones, the bear, sea-lion, and sea-calf are found. Whales are seldom taken. Partridges are met with, and so are swans; of sea-birds there is an abundance.

Behring Island abounds in small rivers rich in fish that come here to spawn.

The Aleutes say that in some parts of the strait, between the islands, there are sunken rocks, but as they have not yet been seen, this may be doubted.

We have adverted generally to the climate of this dominion of fogs, and a sufficient notion of its peculiarity may be gleaned from what has been said. Of the *currents*, we will reserve for the present the observations that have been made; they will be found in connexion with the general subject of currents, in the section hereafter devoted to that purpose.

CHAPTER XVIII.

KAMTSCHATKA, OKHOTSK, AND THE KURILE ARCHIPELAGO.

The first country described in this chapter is the great Peninsula of Kamtschatka, a part of the Russian government of Irkutsk. It lies between the parallels of 62° and 51° North latitude, and is consequently about 800 miles in length.

The honour of the first discovery of Kamtschatka is attributed to Feodot Alexeieff, a merchant, who sailed from the Kovyma, on the North coast of Siberia, round the Peninsula of the Tschutki, in company with seven other vessels, about the year 1648. Tradition says that he was driven from the rest, and wintered in Kamtschatka, and afterwards sailed round to Okhotsk, but was killed on his road to the Anadirsk. This account is also in part corroborated by Simeon Deshneff, who commanded one of the vessels. But, as he did not live to make any report of his discovery, Volodimir Atlassoff, a Cossack, who invaded and conquered the peninsula in 1699, and returned to his fort at Irkutsk, has the claim of first making it known. The conquest was completed in 1706, and it has ever since paid tribute, in furs, to the governor of Irkutsk.

The natural limit of the peninsula would seem to be, as we have before stated, at the bay to the West of Cape Ilpinsk. The civil division extends beyond this to the River Olioutor. It is divided into four districts, each of which is governed by a *toïon* or lieutenant. The commander of the troops resides at Petropaulovski, which has been for some years the chief town.

The natives are of two races, the Kamtschadales and the Kariaks or Koriaks, whose territories are divided at Cape Oukinskoï. The latter have been before described. The Kamtschadales differ from them more in mode of life than physical conformation. They seem to partake of the Mongolian type. The result of European intercourse has had no favourable effect on them. The diseases and vicious habits introduced have sadly thinned their numbers, which do not now exceed between 4,000 and 5,000.

Of the geography of the peninsula a few words may be said. Of the eastern coast, with the exception of the few points imperfectly seen or observed by Cook and other navigators, the only delineation that existed up to a recent period was that furnished by Behring. Of the general deficiency of this little need be said. Capt. Lütke was despatched from St. Petersburg to minutely survey this coast in the *Séniavine*, in 1827-28. Delays and contrarieties prevented this being done to the extent intended, and only some of the more prominent features received the great attention which that commander was capable of exercising in this exploration. The results of this survey, which included the elimination of the large island of Karaghinskoï, among other important points, is given from that officer's work, published in 1835-36. Of its principal port, Awatska Bay, we have numerous accounts, so that the difficulty is rather one of selection from this *embarras des richesses*. Professor Adolph Erman also employed some time in the examination of various points on land. The results of these are given in his Reise um die Erde. From the combination of these observations the dimensions of the country have been reduced to two-fifths of the extent falsely assigned to it by some; so that this Russian province might have well been considered as a *terra incognita*. "I might add," says the professor, "in support of this remark, that the latest Russian maps had, so to speak, doubled the existence of a whole series of localities; for we saw on them the names of many villages situated between Port St. Peter and St. Paul, and the sources of the River Kamtschatka, noted a second time along the direct line between the said town and the Port of Bolcheresk. The truth is, all these villages twice mentioned really exist only once, and that they were not aware, at St. Petersburg, that steep mountains prevented travellers from passing direct from St. Peter and St. Paul to Bolcheresk, but, on the contrary, obliged them to go first towards the North, by the same road which leads to the sources of the Kamtschatka."

Of the western coast of the peninsula but little new has been acquired for many years. This is perhaps of less importance commercially, though a more intimate acquaintance would be desirable, inasmuch as the Sea of Okhotsk, which it bounds on the East, has been for some years the resort of a portion of those hardy and intrepid whalemen who have drawn out more of the resources of the great ocean we are describing than any other commercial body.

Kamtschatka is pre-eminently a country of volcanoes. Some of the highest peaks in the world surmount its mountain ranges. These mountains, which cover about two-thirds of the entire surface, form an irregular chain in a S.S.W. direction. Many of their summits are in a high state of volcanic action; and, considered as a whole, it may be supposed that they form a portion of the great volcanic belt which extends through Aliaska and the Aleutian Islands, and is continued on through the Kuriles, Japan, and Formosa, to the Asiatic Archipelago.

In the principal range running North from Cape Lopatka, its South extremity, thirteen summits, with craters and hot springs, have been observed, one other height being isolated, and lying West of the main range. The most active of these are Assatchinskoï (8,340 feet), Avatcha (8,760 feet), and Klutchevskoï (16,512 feet). During an eruption of the first, in 1828, the scoria and ashes were carried as far as Petropaulovski, 120 versts (80 English miles) distant. In 1827 Avatcha was in violent eruption, and ejected a large quantity of water, besides lava and stones. Klutchevskoï broke out during Dr. Erman's visit, in 1829, after having been dormant for forty years. Much more might be extracted here on this interesting subject, but their general appearances, of most interest to the mariner, will be found in the subsequent descriptions.* Among other of their products, Erman found fossil amber, reminding him strongly of the shores of the Baltic.

There are no large rivers in Kamtschatka: the configuration and formation of the peninsula preclude this. The largest is the Kamtschatka River, which, however, is said to be capable of admitting vessels of 100 tons about 150 miles up the stream.

The severity of the climate has been exaggerated, though it is severe. In some of the sheltered valleys, which possess great natural beauty, the temperature is not very inclement. Perhaps a similar train of remarks would hold good both for Japan and Kamtschatka, that there is great difference between the East and West faces of the country; the former differing from the piercing West winds passing over the ice and snow of the continent of Asia. Of course agriculture has been but little pursued. Its slender population know but few wants, and these are supplied from the produce of the chase, as bears, lynxes, otters, reindeer, foxes, &c.; the skins of these form the principal export, and but few supplies can be calculated on by vessels touching here.†

KARAGHINSKY ISLAND.

This island, and the adjacent coast, until the time of Lütke's exploration in 1828, had not been seen by any known navigator, except Syndt, since the time that Behring had seen one or two of its points through the fog. It is scarcely necessary to add that the charts drawn up from this had not a shadow

* There is an excellent and lengthened account of these *Sopki*, as the volcanoes are termed, in the third volume of the Voyage of the *Séniavine* (Capt. Lütke), by M. Alex. Postels, pp. 63, *et seq.*

† For accounts of the country, its inhabitants, and the productions, besides the works which may be subsequently quoted, much, indeed all available, information will be found in Coxe's Russian Discoveries; Steller's Account; Billings' Voyage of Martin Sauer, chap. xxi. pp. 289—319; Cochrane's Travels in Siberia, vol. ii. pp. 27—56; Cook's Third Voyage, vol. iii., by Capt. King, chaps. vi. and vii. pp. 324—376; Dobell's Kamtschatka, vol. i. pp. 1—188; Erman, Reise um die Erde, vol. i. p. 415—420. An Account of Kamtschatka, by Krachenninikoff, in the original Russian, has not been translated; Capts. Beechey, Lütke, Du Petit Thouars, and others, have related something concerning it, as will be seen hereafter.

of resemblance to the reality. From some information, which, however, proved to be false, of an extensive inland harbour on its West side, Capt. Lütke, was induced to spend a longer time in its examination than other parts; to him, therefore, we owe all we know of it.

It is 55 miles in length, and an uninterrupted chain of mountains traverses its length, declining towards the S.W., and rising again at the South end, forming a mountain about 700 feet in elevation. The western coast is of an insignificant height. The sandy beaches are about 100 or 200 feet broad, and then gently rising; bare rocks are seldom seen. All the shore of the North and East sides of the island is higher and steeper. This causes a great difference in the appearance of the opposite coast, the steep ascents, the rugged or rounded summits, frequently rising to 1,250 feet above the sea, and an Alpine vegetation give it a mountainous character.*

Its N.E. extremity, *Cape Golenichtcheff*, in lat. 59° 13½′, and lon. 164° 40′ E., is 25 miles nearly direct South from the Island of Verkhotoursky. Its South end, *Cape Krachenninikoff*, in lat. 58° 28′, and lon. 163° 32′, is at the distance of 40 miles from Cape Oukinskoï, on the coast of Kamtschatka. From the N.E. extremity to the S.E., in the centre of the island, a chain of steep mountains extends, of 2,000 feet in height, on the two flanks of which are chains of less elevation. On the S.E. side they reach to the sea-coast, forming generally high abrupt capes. Throughout the mountains are deep ravines and gorges. The coast curves in slight, but entirely open, bays. On the south side of Cape Golenichtcheff is a cove rather larger than the rest, but altogether exposed; in its bottom are some indications of a large river.

At 13 or 14 miles from the S.W. end of the island, the mountains decline considerably in height, becoming more even and sloping, and at 6 miles from this extremity give place to a low isthmus of 1½ miles broad, which, beyond 20 miles off, gives the appearance of a separation of the higher hills to the southward.

Beyond the S.W. end, the coast trending to the East forms a bay open to N.N.E. and N. by W., but where there is very convenient anchorage. With the exception of a small shoal at its S.W. end, the depth throughout is very great along the coast, but North winds may send in a heavy sea.

At 31 miles from the S.W. end, and at 27 miles from Cape Golenichtcheff, a bed of gravel running off the coast extends 7 miles to the West and S.W. Its point, *Cape Sémenoff*, is 13½ miles S. 62° E. from Cape Kouzmichtcheff on Kamtschatka: it is low, and is from half a mile to 300 yards in breadth. With the coast it forms a bay, 5 miles wide in the opening, which Lütke called the "Bay of False Intelligence," because the Kariaks gave them reason to expect a good port here. It is open to all the S.W. quarter; the coast lies too far distant to afford any shelter. Notwithstanding all this it is an excellent roadstead.

At the point where the bed of gravel joins on to the land, a creek opens on to this bed of gravel and towards the bay above described: it is three-quarters of a mile long, S.E. and N.W., and 160 or 200 yards broad. It has 10 feet at low

* See Voyage du *Séniavine*, tome ii. p. 177, by Capt. Lütke, and tome iii., by M. Alex. Postels, pp. 91-4. The subsequent descriptions are from Lütke, in vol. iv.

water, muddy bottom. The tide generally rises 3 or 4 feet. This creek forms an excellent harbour for small vessels.

The depth in the strait separating Karaghinsky from the continent is from 23 to 27 fathoms, most commonly a muddy bottom. It seemed as if there was some bank at 4 or 5 miles to the S.E. of Cape Krachenninikoff, as a change in the colour of the water was observed, and the soundings rapidly decreased to 12 fathoms.

The island was formerly inhabited by Kariaks, but now is only temporarily so by the Olutores and Kamtschadales, who come here to hunt. The remains of the old habitations are seen in many places. There is no wood on it; a few bushes may serve as fuel, and there is none drifted on to it. There is a great abundance of the vegetable tribe, and for this reason the bears abound, which, with red foxes, were the only animals seen. Large quantities of geese and ducks frequent the lakes, but they are very wild. Fish abounds in almost all the small rivers, and fresh water is abundant.

Cape Ilpinskoï, which has been before alluded to, is the North point of the large gulf which washes the eastern shores of the narrowest part of the Kamtschatka Peninsula. It is in lat. 59° 48½′, and lon. 165° 57′. It is joined to the continent by a low and narrow isthmus, over which the sea washes.

Cape Kouzmichtcheff, the position of which is well determined as lat. 59° 5′, and lon. 163° 19′, is steep, and is conspicuous by the direction of the coast on either side, as to the northward it trends S.E. towards it, and to the southward it runs W.N.W.

Karaghinskaïa Bay is formed to the northward by Cape Kouzmichtcheff. It penetrates the land for 9 miles in a N.W. direction, its breadth being from 4 to 8 miles. At the head of the bay the *River Karaga* discharges itself; its mouth is in lat. 59° 8′, and lon. 162° 59′. The shores of the river itself are low, but mountains covered with wood rise at no great distance. On the left side of its mouth, a bed of gravel, covered with verdure, extends 8 miles to the E S.E.; parallel with it on the coast, a high cliff extends in the same direction, and abuts on to Cape Kouzmichtcheff. A bed of gravel also extends for 4 miles to the South of the mouth of the Karaga, and then the coast, but little elevated, level but cliffed, turns gradually to the S.E., and forming a cape in lat. 58° 55′, and lon. 163° 2′ E., which is the southern limit of Karaghinskaïa Bay.

The River Karaga is not deep; it is even doubtful whether it can be penetrated far beyond the bay, inasmuch at the distance of above 6 miles the lead did not give more than 6½ fathoms.

From the river towards the interior, a chain of very high mountains extend: they are partly covered with snow, and present a strange appearance of serrated peaks, the summits formed into spiracles.

Cape Oukinskoï, which was passed at a great distance by Capt. Lütke, and was consequently not determined with very great accuracy, forms the southern limit of a very extensive gulf, 60 miles in extent from North to South, of which Karaghinskaïa Bay may be placed on the North. The coast is almost entirely unknown, and only the points here mentioned could be determined. In the southern part of this bay the *River Ouka* discharges itself, into which, according to the journal of Lieut. Syndt, vessels not drawing more than 15 feet could formerly enter. Of its present condition we know nothing. Cape Oukinskoï, or Natchikinskoï, was considered by Capt. Lütke to be in lat. 57° 58′, lon. 162° 47′ East. It is low and level; the position of the high and remarkable mountain upon it is probably determined with greater precision; its lat. is 57° 54′, and its lon. 162° 52′. This cape is the boundary between the Kamtschadales and the (Sedentary) Kariaks; the first dwelling to the South, the second to the North of it.

The coast extends for 60 miles to the S.E., to a cape marked on the charts as *Cape Ozernoï;* this was called so because in the old charts such a name is marked in the vicinity, though it was not ascertained whether the inhabitants knew it by such. It is in lat. 57° 18′, and lon. 163° 14′ East. It is distinguished by a mountain slightly peaked.

The coast beyond this is formed of high and sloping mountains, and nothing remarkable occurs until the *River Stolbovskaïa* is reached. The mouth of this river is very distinct, and was determined to be in 56° 40½′, and 162° 39′. Before reaching it the coast trends S.E. and then S.S.W., forming a bay which falls back 45 or 50 miles. At 10 miles from its mouth the coast begins to be mountainous, and continues so for 15 miles; farther on to the northward it becomes level and lower for the extent of 25 or 30 miles, when the high and mountainous coast, before alluded to, begins.

Cape Stolbovoï is a high cliff, in lat. 56° 40½′, lon. 163° 21′ East. Before it are three detached rocks, one very large, the two others smaller. The coast from the cape towards the River Stolbovskaïa turns abruptly to the N.W., and soon afterwards to West and W.S.W. The mountains decline in height, a low cliff then runs along the seaboard, and afterwards a low sloping coast extends to the mouth of the river.

At 12 miles South from Cape Stolbovoï, in lat. 56° 27′, the chain of mountains is interrupted to give place to a very low valley, through which, to the W.S.W. from seaward, there was no elevation visible between the Klutchevskoï Volcano; for which reason it was supposed that in this part a mouth or large lake extended to a great distance.

The coast in question trends nearly upon a meridian for 35 miles South from Cape Stolbovoï, with some few inflexions. Throughout this extent, with the exception just alluded to, the coast is high and mountainous, terminating on the sea-coast often in slopes, but with cliffs in some parts. From the above distance, which will be in lat. 56° 9′, and lon. 163° 24′, the coast runs 15 miles to S.E. to Cape Kamtschatskoï.

CAPE KAMTSCHATSKOÏ.—There was some doubt as to which was the actual cape bearing this name. The coast here forms a sloping and slightly elevated cape, which, seen from the S.E. at a great distance, would have the appearance of a remarkably prominent point; this renders it very difficult to determine which was precisely the point in this space to which the name Kamtschatskoï should belong. If, as it is on the ancient charts, it is the southernmost, its longitude would then be about 162° 57′. Capt. King determined it to be in 163° 20′; but it is beyond doubt that he, being 20 miles, at least, distant from the nearest land, could not see the South extreme itself on account of its low elevation. Under this view his determination accords with that of Capt. Lütke.

From this cape the coast turns gradually to the S.W. and West, then runs N.W. and West to the mouth of the *River Kamtschatka*. This part of the coast was not seen by Lütke.

The KLUTCHEVSKOÏ VOLCANO.—The great mountain of Kamtschatka lies at the back of the bay to the West and South of Cape Kamtschatskoï. Its latitude is 56° 8′ N., lon. 160° 45′ E. Of this determination the longitude accords with that given by Dr. Adolph Erman, but the latitude varies. Capt. Lütke says this point, as well as the greater portion of the principal points of the peninsula, had never been determined by any one, and consequently it was placed half a degree farther South than the reality.

This volcano, called also *Kamtschatskoï*, and surnamed *Klutchevskoï* from the name of the village *Klutchi* (springs), lying at its foot on the South or right bank of the River Kamtschatka, is of a truncated, but very steep, conical form. On its S.W. and N.E. sides are two other, but lower mountains, the first with a serrated summit, called by the Kamtschadales the *Needle*, the second even. The volcano bears S. 76° W. from the mouth of the River Kamtschatka, and the extremity of Cape Kamtschatskoï is 20 miles S. 63° E. from it.

From an angular measurement, which, however, could not be repeated, Capt. Lütke calculates its height as 16,502 English feet. Professor Erman measured it as 15,766 English feet. He terms it the *Peak of Kliuchevsk.**

We quote Capt. Lütke's description of it, when he saw it to the W.S.W. through the mountains to the South of Cape Stolbovoï before alluded to:—

"As soon as we had passed beyond the mountains lying to the South of this break, the majestic Klutchevskoï Volcano and his satellites showed themselves; the serrated mountain to his left, and to his right two flat mountains. All these are incomparably lower than the volcano itself; they were nevertheless entirely covered with snow; there was also a great quantity on the volcano itself, but it only covered it in patches. The morning sun, reflected from their sides, showed them clearly against the azure of the sky, notwithstanding their immense distance (the volcano was then just 104 miles off). Unfortunately a cloud continually hid its summit, and prevented the possibility of re-measuring its height. To make up for this, exact bearings furnished the means of determining its position with exactness. It presented the same form here, of a cone rather truncated, whose axis is to its diameter at the base as 2½ to 2. The lower country between the surrounding mountains, over which we were looking, was doubtless the eastern side of the *Great Seal Lake*, the outlet of which joins to the mouth of the Kamtschatka River."†

Dr. Erman states, that he saw it in a picturesque and sublime acclivity, and approached the burning lava, which poured forth a continuous stream, till he reached the height of 8,000 feet.

* See Reise um die Erde, and the Journal of the Royal Geographical Society, vol. ix. 1839, p. 509.

† Voyage du *Séniavine*, tome ii. pp. 173–4

CAPE KRONOTSKOI is in lat. 54° 54′, and lon. 162° 13′. The intervening coast between it and the river trends S.W., South, and then 30 or 40 miles S.E. by S. From what could be seen of it from the *Séniavine*, it is low and level on the shore. In the distance is seen the chain of high snowy mountains, extending to the Klutchevskoï Volcano. According to Krachenninikoff, in the space of 100 versts (66⅓ British miles) there is not a river, and the coast is throughout mountainous. There is a large detached rock off Cape Kronotskoï. There was in the middle of June much snow on the cape, except on its southern face, where none was visible. The side also appeared to be well wooded.

The northern shores of the Gulf of Kronotskoï, which extend inward to the South and West of the cape, were not examined by Lütke, and all that he could say of it was, that it did not seem to fall back so much as is represented on the charts. The North shore extends to the W.N.W., and is low on the sea. At 10 or 12 miles from the coast about the cape steep mountains rise; they are irregular and much broken. The coast itself is not very high; although abrupt and broken in nearly every part, it forms several small capes, between which are some open coves. The appearance of the coast is very uniform. In all this space no trace whatever of any river could be discerned, and from the configuration of the coast none would be suspected there.

The KRONOTSKOI VOLCANO, 10,610 English feet in height, stands on the North side of the bay in question. Its lat. is 54° 45′, and lon. 160° 37′ E. It appears to be entirely isolated, and may be seen at 120 miles distant. Capt. Lütke, in the narrative of his voyage, says:—"On the morning of the 28th June, we saw at the same time the volcanoes of Avatchinsky, Koriatsky, Joupanoff, and Kronotskoï; the two first at 82 miles, the last at 68 miles off. The Kronotskoï Volcano, similar to that of Villeuchinski, has the form of a regular cone, but it seems to be less steep than the latter. We saw to the left of it a mountain, the summit of which was flattened, and close to it a peaked hill, probably the same that was overturned during the passage of the Chevelutch Mountain from its ancient to its present site.* The snow covered the Peak of Kronotskoï, but there were some places on its flanks which were uncovered. Its elevation, carefully taken as above, shows it to be rather higher than Etna. We saw it from Cape Shipounskoï at nightfall, through a not very clear atmosphere, at 95 miles distant, and on our return to Kamtschatka, at 120 miles off very distinctly."†

Between it and the Joupanoff Volcano to the southward, many high mountain summits were seen, but not forming a continuous chain. In the southern part of the bay, at 30 miles to the North of Cape Shipounskoï, a cape projects, in the neighbourhood of which the mountains recede into the interior of the country, leaving only a low coast. There seemed to be the mouth of a river here, which might be the *Chopkhod* or *Joupanova*. From this cape to Cape Shipounskoï, the direction of the coast is generally North and South for an extent of 25 or 30 miles. The coast itself is formed by an extension of the mountain chain to the southward. The mountains are not very high, but they are steep, irregular, and intersected in many parts by perpendicular ravines.

* See Description of Kamtschatka by Krachenninikoff, second edition (Russian), vol. i. p. 40.
† Voyage du *Séniavine*, vol. ii. p. 171.

CAPE SHIPOUNSKOI is in lat. 53° 6′, and is 1° 11′ E. of Petropaulovski. It is the extremity of some level land, which advances 3 miles from the chain extending to the Joupanoff Volcano, and terminates on the sea-coast throughout in rocky cliffs 200 feet high. Seen from the S.W. or N.E., it has the same aspect as that of a projecting and even cape, but on the S.E. the level appearance is confounded with the other mountains, because their flanks to the N.E. and S.W. reach to the sea. Beyond the extreme point are some detached rocks, which seem to be united by a reef.

Capt. Lütke observed a strong current off the cape which produced some overfalls.

Cape Nalatcheff is 22 miles W.N.W. ½ W. from *Cape Shipounskoï*. It is a high, steep mountain, the summit irregularly rounded; it projects in a point to the South. Besides its elevation, it is distinguished from the neighbouring coast by its black colour. Seen from S.W. it seems to be detached. The coast to the eastward of it is low and sandy near the sea, and rises towards a chain of moderately high mountains, but which are steep, and terminate in peaks; these extend to Cape Shipounskoï. Viewed from the S.W., this chain seems interrupted in a part where *Betchevinskaïa Bay* opens. This would be an excellent harbour if vessels could anchor, but there are only 4 feet water in its opening. Merchant vessels used formerly to visit the little *River Vakhilskaia*, which debouches 5 miles N.W. of this bay. The coast westward of Cape Nalatcheff is low and level, and rises gradually on all sides towards the summit of the Koletskoï Volcano. The little *River Kalakhtyrka*, which enters the sea at 7 miles from the lighthouse cape of Awatska Bay, is pointed out by a rock of moderate height, whitened by the dung of the sea-birds, lying 2 versts (1⅓ English miles) to the South of it. This river abounds with fish; and many of the inhabitants of Petropaulovski have fisheries here. There is very little water in its entrance.

The coast between this river and Awatska Bay is lofty, and terminates on the coast, in many parts, in high cliffs. This space is intersected in one part by a low isthmus, between the bottom of Rakovya Bay and the sea, across which the hunters who go to take birds from Toporkoff Island transport their canoes.

AWATSKA BAY.*

This bay, the principal port of the Peninsula of Kamtschatka, derives its chief interest from its containing the Port of St. Peter and St. Paul, Petropaulovski, as much from its intrinsic superiority. It is so extensive and excellent, that it would allow all the navies in the world to anchor in perfect security in its capacious basin. Yet the navigator in entering it will at first see no sign of human habitation or commerce on its shores, unless, perchance, some vessel may be approaching or quitting its only port, the little town above mentioned. As it is the principal point of interest in this remote region, we have been the more diffuse in its description, more so than its own importance would warrant.

It was visited by Capt. Beechey in H.M.S. *Blossom*, who made an accurate

* It is thus written by Capt. Beechey. Capt. Du Petit Thouars and others write *Avatcha*; Müller says *Awatscha*; or properly, according to Kamtschadalian pronunciation, *Suaatscha*.—*Voyages et Découv.* p. 36.

and ample survey of it; and has also furnished part of the subsequent directions. In the narrative of the voyage of the French frigate *La Venus*, under Capt. Du Petit Thouars, is a lengthened account of it commercially and nautically; and from this source, too, much abridged, we have drawn up the following account of it.*

The Road of Avatscha, or Awatska, on the eastern coast of Kamtschatka, lies at the bottom of the bay of the same name; it is reached through a narrow channel, which is 4 miles long and about 1 mile broad. This strait, although thus narrow, is not dangerous, because there is anchorage throughout its whole extent; in it, as in nearly all close channels, the winds are almost always either directly in or out of it; that is, they are either contrary or favourable for passing it. The immense Bay of Awatska, which leads to this channel, is formed by the retreat of the coast line between *Capes Gavareah* and *Shipounskoï*, or *Cheponskoï*; these two capes are the best landfalls for making the Port of Petropaulovski. In fact, whether Cape Gavareah or Cape Shipounskoï is closed with, if the vessel should be overtaken in either of these positions by thick fogs or strong winds from East or S.E., it is always possible to keep at sea; should the endeavour be to make the channel at once on its parallel, not only will the making the coast be retarded without any advantage being gained, but should she then be surprised by any contrariety, there is no means of making an advantageous tack in order to keep off, and the situation of the ship will be troublesome, there being no soundings on the coast, and neither do they offer any anchorage which could be taken in such circumstances. By proceeding in this way several vessels destined for Petropaulovski have been lost, and that which brought the government provisions in May, 1836, ran ashore on the reefs off Toporkoff Island.

In coming from the East or South, Cape Gavareah, then, should be first made; this point may be approached very nearly, and is nearer the port than Shipounskoï; but if coming from the North, it is more natural and convenient to make for Shipounskoï. In the first case, if the weather be clear, before arriving close to Cape Gavareah, the Peak of the Kouatskoi or Avatscha Mountain † will be seen to the N. by W.; it is nearly as high as that of Teneriffe, and may be seen in clear weather 30 or 40 leagues off. To the right of this peak, and close to it on the East, is the Koselskoï Volcano: ‡ this mountain, not quite so high as the first, is equally remarkable. If, with the prospect of these mountains, you should see, further to the West, the Peak of Villeuchinski,§ which is to the S.S.W. of the volcano and the town of Petropaulovski, there cannot be any doubt as to the position of the ship, more particularly should you also make out, to the South of Villeuchinski, Mount Gavareachinski,‖ standing above Cape Gavareah, and at 16 miles to the W.N.W., *true*, from that cape. The view of these four mountains will alone be sufficient to make sure of the port; and if, as occurs very frequently, the coast should be enveloped in fog, and these mountains only are seen, by directing the course either on to the Villeuchinski Peak, or that of Avatscha, the track will lead very near to the entrance, the points of which are readily made out.

* See also Capt. King, in Cook's Third Voyage; Capt. Rikord, in Golownin's account of Japan: Capt. Lütke, Voyage du *Séniavine*, Part. Naut., &c.

† 11,554 feet, according to Capt. Beechey. ‡ 9,058 feet, *ibid.* § 7,372 feet, *ibid.*

‖ Flat Mountain, 7,932 feet, *ibid.*

Having reached to 4 or 5 miles to the East of Cape Gavareah, and running N. by W., *true*, for 33 miles, you will reach the lighthouse at the entrance, in the middle of the channel; if the vessel is to the East of, and the same distance from, Cape Shipounskoï, the route will be W.S.W. and W.S.W. ½ W., *true*, for about 50 miles; in either case account must be taken of the currents occasioned by the tide.

Lastly, being the latitude, and approaching Cape Gavareah from seaward, after making Mount Gavareachinski, you will see to the right, towards the N.W., that of Villeuchinski, and nearly at the same time, more to the northward, the peaks of Avatscha and the Koselskoï Volcano. In approaching the coast will be observed beneath the two first mountains, and it will be seen stretching away to the South, and on the other side trending and decreasing in height towards the North; but the land in the bottom of the bay, being too distant, will not be seen. The coast, for this reason, appears to terminate here; but as it shows itself again more to the North, beneath the peaks of Avatscha and the volcano, the part of the coast that is not seen will have the appearance of an entrance or strait; and *it is* in this part that the entrance lies. In nearing it the coast successively shows itself. After that will be seen, under the volcano peak, a point that is higher and blacker than the neighbouring coasts, and which successively extends by degrees towards the West; on this blackish point a small hut will be seen like a small white speck. This hut is the dwelling of the signal men in charge of the fire which is lighted on this point as soon as night approaches; but only when a vessel is expected or in sight. Whatever may be the position of the vessel, as soon as the entrance light is seen she may steer for it, rather to windward of it; but it must not be approached at less than 2 miles off to the East of a line due North and South, *true*, on account of a ledge of rocks which is attached to the point, and extends 1½ miles to the S.E. by E., *true*, from the point and lighthouse. *It is* necessary then, in beating in, to pay attention to this bearing of the light, or, which is preferable, if it is daylight, not to shut in with the lighthouse point three high rocks lying about a mile to N.W., *true*, from the point; these basaltic rocks, called the *Three Brothers*, are remarkable for their needle-like form, and are nearly steep-to; outside of them they may be approached within a cable's length.

The point which forms the South entrance of the channel was named *Point Venus*. Close to it, to the East, there is a breaker; but it is so near the shore, that it is not dangerous by keeping a cable's length off. It is on this side of the passage that the depth is greatest. After doubling Point Venus, a small bay will be seen on the same side; and beyond it, to N.N.W., *true*, from the point, a second, the extremity of which is very perpendicular, and is surrounded by high rocks in the form of pyramids and needles. Two rocky islets lie outside to S.E. by E. of this last point. From these islets to Point Staniski, which follows to N. by E., all the bay is bestrewed with shoals and rocks, the easternmost of which forms the West limit of the channel. On this shoal is a rock, which points out its place; this serves as a beacon, and does not cover but seldom, and then it shows by breakers. This shoal, named *Staniski*, very much narrows the space for beating through; it lies $1\frac{1}{10}$ miles to the West, *true*, from the Three Brothers Shoal, and will be cleared by keeping the eastern part of the Babouschka Rock (on the West side) on the eastern point of the *South Signal Post*, which is beyond it on the same side. If it is night, it will be sufficient to steer so as to keep in sight the light on the South Signal Post to the right of Babouschka, which is easily done ; for, should you get to the West of it, the light will be hidden by Babouschka.

Having reached Staniski Point, which ought not to be approached within 2 cables' length, on account of a sunken rock, which does not always show, and on account of the flood current setting on to this part, the ebb setting on to the Staniski Bank, steer for *Ismenaï Islet*, then you will come up more West after doubling the Babouschka Rock, so as to keep at an equal distance from either coast, or rather closer to the windward side.

If it is necessary to beat through, on running the eastern board, Pinnacle Point must not be approached, for it is surrounded by rocks and shoals. A good bearing for clearing these dangers is not to open the Brothers to the right of the Lighthouse Point, or better not to shut in the North Signal Post by Point Ismenaï; but this last is not always visible from fog. The South Signal Point is clear, and may be passed close-to. Point Ismenaï is not so ; it projects 2 cables' length to the West. This may be cleared by keeping Point Venus hidden by Point Staniski, or by not uncovering it.

After passing to the North of Ismenaï Point, the eastern side may be ranged close to, on the West side of the entrance; and to the North of the South Signal Post there is a sand-bank, extending for 2½ miles to N.N.W., *true*, from this point; but on this side of the passage the depth diminishes gradually, and there is sufficient time to about-ship when less than 5 fathoms is found ; it will be enough, on the eastern tack, not to hide the Babouschka Rock by the South Signal Point.

There is another bank in the Road of Petropaulovski ; it lies to the North of the North Signal Post, and about 1¼ miles from this point. It is called the *Rakovya Bank*, and separates the road of Rakovya from that of Petropaulovski. It is a rocky shoal, with 4½ feet on it; it is sometimes marked by a buoy, with a small flagstaff on it ; it is dangerous during the ebb which sets over it, but is easily avoided both when it is marked by the buoy, or by not shutting the Brothers in with Point Ismenaï. At night, by not hiding the entrance light by Point Ismenaï, you will also have nothing to fear. With the exception of these banks, there is deep water to the shore throughout the bay, and the bottom is excellent anchorage.

There are no cross marks sufficiently good to know when you have doubled to the North of the Rakovya Bank, but you may judge by the distance of Point Shakoff, at the entrance of the port, or by the bearing and distance of the buoy on it.

If you arrive by night off the entrance to the bay, and the wind should be contrary for entering, it will be dangerous to attempt to enter the port without the assistance of a pilot, or unless well acquainted with it.

With contrary winds, with wind too light to steer, or during calm, the currents and narrowness of the entrance render the navigation difficult ; but the possibility of anchoring throughout diminishes the danger ; with a leading wind, the entering or leaving Awatska Bay offers much difficulty, no danger.

If the currents affect the steering, which, in a light breeze, frequently happens in the entrance, it is well to anchor in Ismenaï Bay, or, if necessary, in any part of the channel.

The lights at the entrance are very judiciously placed. The outer light may be approached without any risk, by means of the lead and anchor close to it, should the wind be contrary; but if the wind is favourable for entering, from the middle of the channel steer on to the South Signal Light, and keep it on the East targent of Babouschka; this will bring the ship abreast of Point Staniski, and from this point, steering North or N. $\frac{1}{8}$ E., will bring you to the middle of the coast, between Point Ismenaï and the North Signal Post, taking care to bear North, or even N.W. $\frac{1}{8}$ N., as soon as you are to the North of the North Signal Post, and steering thus you will reach safely the anchorage of Petropaulovski. Care must be taken, in this course, not to shut in the entrance light by the land of Point Ismenaï, so as to clear the Rakovya Bank.

The lights shown from the Lighthouse Point, from the South Signal Post and the North Signal Post, are not done so regularly; they are only so when a vessel is seen or expected at Petropaulovski. This circumstance renders prudence necessary in approaching. The lights are not shown from lighthouses. One ought to be erected on the entrance point, but at present they are wood fires, kept up by the signal-men on the different points.

There is no particular precaution necessary for safety in anchoring in Awatska Bay: the sea is never so heavy as to occasion any trouble; but as the bay is surrounded by high mountains, violent gusts are sometimes felt, so that, for greater security and quietude, it is better to have a long hawser out.

The tidal currents are very irregular, both in form and duration; they were never found more than at 2 miles in the entrance, or $1\frac{4}{10}$ miles in the road.

There was no sand found throughout the bay, except on the bank off the South Signal and off some other point: but on the shore pure sand was never found; it was always mixed with earth, rocks, or pebbles.

The HARBOUR of PETROPAULOVSKI, on the eastern side of this bay, is small, deep, and well shut it. It is defended by three small raking batteries, mounted with guns of small calibre. A vessel, of whatever size, can enter it, and undertake any description of repairs.

TAREINSKI HARBOUR, lying in the S.W. part of the bay, is immense and excellent, but as there is neither population nor commerce in it, it has, up to the present time, been of no utility.

RAKOVYA HARBOUR also forms, to the South of Petropaulovski, an equally excellent port, but it is of less easy access than the foregoing, on account of the Rakovya Bank, lying in the middle of the channel leading to it.

In fine weather the morning breeze is from the North to N.N.W., lasting until eight or ten o'clock, and sometimes even until eleven o'clock; then, shifting to the West and South, it sinks altogether: in the afternoon, about one or two o'clock, the breeze from the offing sets in, varying from South towards East.*

* Voyage sur la Frégate *La Venus*, tome ii. pp. 55—68.

DIRECTIONS BY CAPT. BEECHEY.*—It is desirable to make the coast well to the southward of Cape Gavareah, and to round it as closely as possible, as the wind will, in all probability, veer to the northward on passing it. If the weather be clear, two mountains will be seen to the West and N.W. of the cape, and one far off to the northward and eastward. The eastern one of the two former, called Villeuchinski, is 7,372 feet high, and peaked like a sugar-loaf, and is in lat. 52° 39′ 43″ N., and lon. 49′ 46″ W. of Petropaulovski, (158° 22′ E.) The highest and most northern of the three latter is the Mountain of Awatska, in lat. 53° 20′ 1″ N., and 3′ 47″ E. of the before-mentioned town. Its height is 11,500 feet, and in clear weather it may be seen a very considerable distance. The centre hill of the three is the volcano, but it emits very little smoke. These peaks are the best guide to Awatska Bay, until near enough to distinguish the entrance, which will then appear to lie between high perpendicular cliffs. Upon the eastern one of these, the *lighthouse bluff*, there are a hut and signal staff, and when any vessel is expected a light is sometimes shown. If the harbour be open, a large rock, called the *Babouschka*, will be seen on the western side of the channel, and three others, named the Brothers, on the eastern side, off the lighthouse. The channel lies in a N. by W. direction, *true*; and when the wind is fair it may be sailed through by keeping mid-channel; but it frequently happens that vessels have to beat in, and as the narrowness of the channel renders it necessary to stand as close to the dangers as possible, in order to lessen the number of tacks, it is requisite to attend strictly to the leading marks.

The outer dangers are a reef of rocks lying S.E., about 2 miles from the lighthouse, and a reef lying off a bank which connects the two capes opposite, *i. e.*, *Staniski Point*, with the cape to the southward. To avoid the lighthouse reef, do not shut in the land to the northward of the lighthouse bluff, unless certain of being at least 2½ miles off shore, and when within three-quarters of a mile only, tack when the lighthouse bluff bears North, or N. ½ E. The Brothers Rock, in one with the lighthouse, is close upon the edge of the reef. The first western danger has a rock above water upon it, and may be avoided by not opening the Babouschka with the cape beyond, with a flagstaff upon it, or by keeping Staniski Point well open with the said signal bluff. In standing towards this rock, take care that the ebb tide in particular does not set you upon it. A good working mark for all this western shore is the Babouschka open with *Direction Bluff*, the *last* cape or hill on the *left upon the low land*, at the head of Awatska Bay. The bay South of Staniski Point is filled with rocks and foul ground. The lighthouse reef is connected with the Brothers, and the cape must not be approached in any part within half a mile, nor the Brothers within a full cable's length. There are no good marks for the exact limit of this reef off the Brothers, and consequently ships must estimate that short distance. They must also here, and once for all, in beating through this channel, allow for shooting in stays, and for the tides, which, ebb and flood, sweep over toward these rocks, running S.E. and N.E. They should also keep good way on the vessel, as the eddy currents may otherwise prevent her coming about.

* Voyage of the *Blossom*, part ii. Appendix, pp. 649-50.

To the northward of the Brothers, two-thirds of the way between them and a ragged cape, at the South extreme of a large sandy bay (Ismenaï Bay), there are some rocks nearly awash; and off the rugged cape called *Pinnacle Point* (N.N.W. 1⅝ miles from the lighthouse), there is a small reef, one of the outer rocks of which dries at half-tide. These dangers can almost always be seen; their outer edges lie nearly in a line, and they may be approached within a cable's length. If they are not seen, do not shut in the Rakovya signal bluff. Off Pinnacle Point the lead finds deeper water than mid-channel, and very irregular soundings.

To the northward of Staniski Point the Babouschka may be opened to the eastward a little with the *signal* staff bluff, but be careful of a shoal which extends about 3 cables' length South of the Babouschka. Babouschka has no danger to the eastward, at a greater distance than a cable's length; and when it is passed, there is nothing to fear on the western shore, until N.N.W. of the signal staff, off which there is a long shoal, with only 2 and 2½ fathoms. The water shoals gradually toward it, and the helm may safely be put down in 4⅝ fathoms; but a certain guide is not to open the western tangent of Babouschka with Staniski Point South of it. There is no other danger on this side of the entrance.

When a cable's length North of Pinnacle Reef, you may stretch into Ismenaï Bay, guided by the soundings, which are regular, taking care of a 3-fathom knoll which lies half-way between Pinnacle Point and the cape North of it. This bay affords good anchorage, and it may be convenient to anchor there for a tide. There is no other danger than the above-mentioned knoll. The large square rock at the northern part of this bay (Ismenaï Rock) may be passed at a cable distance. This rock is connected with the land to the northward by a reef, and in standing back towards it, the *Pinnacle* Point must be kept *open* with the *lighthouse*. When *in one*, there are but 3½ fathoms. Rakovya signal staff to the northward, in one with the bluff South of it (which has a large green bush overhanging its brow), will place you in 5 fathoms, close to the rocks.

Off the North bluff of Ismenaï Bay there extends a small reef to a full cable's length from the shore; until this is passed do not shut in Pinnacle Point with the lighthouse. But to the northward of it you may tack within a cable's length of the bluffs, extending that distance a little off the signal staff bluff, in consequence of some rocks which lie off them.

Northward of Rakovya signal staff, the only danger is the Rakovya Shoal, upon the West part of which there is a buoy in summer, and to clear this keep the Brothers *in sight*.

There is no good mark for determining when you are to the northward of this shoal, and as the tides in their course up Rakovya Harbour are apt to set you towards it, it is better to keep the Brothers open until you are certain, by your distance, of having passed it (its northern edge is seven-eighths of a mile from Rakovya bluff), particularly as you may now stretch to the westward as far as you please, and as there is nothing to obstruct your beat up to the anchorage. The ground is everywhere good, and a person may select his own berth.

Rakovya Harbour, on the eastern side of Awatska Bay, will afford good security to a vessel running in from sea with a southerly gale, at which time she might find difficulty in bringing up at the usual anchorage. In this case, the Rakovya Shoal must be rounded, and left to the northward; 5 and 5½ fathoms will be close upon the edge of it, but the water should not be shoaled under 9 fathoms.

The little Harbour of Petropaulovski is a convenient place for a refit of any kind. In entering it is only necessary to guard against a near approach to the signal staff on the peninsula on the West. The sandy point may be passed within a few yards' distance.

Weighing from the anchorage, off the peninsular flagstaff, with light winds, and with the beginning of the ebb, it is necessary to guard against being swept down upon the Rakovya Shoal, and when past it, upon the signal bluff on the same side. There are strong eddies all over this bay, and when the winds are light, ships often become unmanageable. It is better to weigh with the last drain of the flood.

Tareinski Harbour, at the S.W. angle of Awatska Bay, is an excellent port, but it is not frequented. It has no dangers, and may be safely entered by a stranger.

It is high water at Petropaulovski at 3^h 30′ full and change; the tide rises 6 feet 7 inches spring tides, and 2 feet 2 inches neap tides.

The church at Petropaulovski is in lat. 53° 1′ 0″ N., lon. 158° 43′ 30″ E.

The TOWN of PETROPAULOVSKI, which is now at the head of the harbour, stands in an amphitheatre on the slopes of two hills, which form the valley, and is simply composed of a group of small wooden houses, covered with reeds or dry grass, and surrounded by courts and gardens, with palisades. At the lower part of the town, in the bottom of the valley, is the church; it is remarkable for its fantastic construction, and for its roof, which, painted green, seems to add considerably to the effect of the picture, surrounded as it is by lofty mountains.

In approaching *Point Shakoff*, as the extremity of the peninsula forming the harbour was named, and in which is a battery, a white buoy will be seen, marking the extremity of a bank, extending S.S.E. (*true*) from it; this may be passed close to it, leaving it to the left, and thence steer to the end of a low point of land which projects at an angle of about 45° from the direction of the coast, and nearly closes the bottom of the bay, making it into an excellent natural harbour, the best that can be desired. This tongue of land, like an artificial causeway, is but little above the surface of the water, and is now covered with *balagans*, huts raised on piles above the ground, serving to dry fish. In the early days of the Russian occupation it was the site of the colony. Arrived at the bottom of the port you land on a plank, which holds the place of a mole, and pass directly before a guardhouse, near which is a small battery. Turning to the left down a good street, broad and macadamized, after passing the government workshops in the centre of Petropaulovski, turning to the right after passing them, and crossing a wooden bridge, you pass the church on the right hand, and then reach the government offices. These two streets are all that merit the name. The greater part of the houses outside of them are placed without any arrangement, and without any attempt at convenience or comfort. The general aspect of Petropaulovski greatly resembles the French establishments at Newfoundland. The appearance of the fish-dryers' houses, and the strong smell of fish, give a greater degree to the similitude. The towns of St. Pierre and Miguelon, however, are larger, and are much more important commercially.

The houses are generally alike, and are called *isbas*, log-houses, the windows sometimes of glass, but more generally with talc, from Okhotsk. When La Pérouse visited it the inhabitants generally lived in *balagans*, now there is not a single one so used. There is not a monument in Petropaulovski, except one, a simple column surmounted by a globe, surrounded by a railing, which bears an inscription, "To Captain Vitus Behring," in Russian. No edifice demands particular attention. In the church the rites of the Greek church are conducted with great richness and solemnity. There is, besides, an hospital and a school.

The population at the time of the visit of the *Venus* amounted to 385 men and 221 women; the greater part employed by the government.*

CLIMATE.—Capt. Du Petit Thouars says, from the best information he could procure on the climate of Kamtschatka, it is evident that, up to the 15th of October, the weather is frequently fine at Petropaulovski; but after this period it becomes very wet, and the land begins to be covered with snow, which becomes permanent, and does not disappear until May or June in the ensuing year. In the months of November, December, and January, violent storms are experienced.

During winter the cold is severe; the snow falls in an abundance far from common, and frequently rises as high as the houses, which thus become buried until the return of spring. The inhabitants are then obliged to open galleries to communicate from one house to another and to go to the church; in the meantime, whatever may be the intensity of the cold, it is very rare that the roadstead is entirely frozen over; the ice does not generally extend more than a cable's length off shore; and further, after bad weather occasioned by winds on shore, as well as those from West and North, the ice becomes detached from the shore, and is carried out of the road. One of the most severe winters remembered at Petropaulovski was that of 1814. In that year the road was almost entirely blocked up; and there was only a small space clear immediately in front of the entrance between the northern and the southern signal. In ordinary winters, the coves, the bays, and the rivers, are only covered, and the ice is not always too thick to hinder a passage by breaking or cutting.

The resources of the port and road of Awatska are almost nothing; there is no certainty or reckoning on wood or water; still less to procure any refitments for the ship. A vessel in need of repair will only find here a safe anchorage; besides this, she must depend on her own resources, both for provisions and workmen. It is, however, possible to obtain, in urgent cases, some slight aid from the government stores, and some workmen of the port; but these assistances, besides being very limited, are very precarious.

The rearing of cattle has made good progress, and bullocks have sufficiently

* Voyage Autour du Monde sur la Frégate *La Venus*, 1836—39, tome ii. pp. 37—40.

increased to assure a supply of refreshments of that sort to a ship requiring it. There is also sometimes fresh butter to be procured; it is made by the Kamtschadales. It is very difficult to get poultry or eggs; these are objects of luxury too rare in this country. There are no sheep nor pigs; the dogs, it is said, prevent their being reared. A few legumes are at rare times to be found.

Fish is very abundant in the bay, and the pursuit is successful in the good season; it begins with the cod and herrings, and is followed by the salmon and salmon-trout. These fish, on being taken, are salted for the winter provisions of the inhabitants and their dogs. It might be made as productive as the fisheries of Newfoundland were it followed with any commercial spirit. At present home consumption is alone attended to.

In general, there are but few species of shell-fish in the bay. Neither oysters nor other crustacea were found by the French, with the exception of crabs.

In winter the communication with the interior being easy, the Kamtschadales bring into the market at Petropaulovski, reindeer, argali sheep, bears, and also hares and partridges, which, like other Arctic animals, turn white in winter.*

The eastern coast of Kamtschatka, between Cape Gavareah and Cape Lopatka, trends to the S.W. South of *Achachinskoï* the land is not so high and broken as between that bay and the mouth of Awatska Bay, being of only a moderate elevation toward the sea, with hills gradually rising farther back in the country. The coast is steep and bold, and full of white chalky patches.

About 7 leagues S. by W. of Cape Gavareah is a high headland, and between them are two narrow but deep inlets, which may unite, it was thought by Capt. King, behind what appeared to be a high island. The coast of these inlets is steep and cliffy. The hills break abruptly, and form chasms and deep valleys, which are well wooded.

Achachinskoï Bay, in lat. 51° 54′, is formed to the northward by a point, and penetrates deeply into the land, in the distant bottom of which Capt. King supposed a large river might empty itself, the land behind being unusually low. South of this bay, the land is not so rugged as the country to the northward.†

It was on this part of the coast that the singular occurrence of the wreck of a Japanese vessel occurred in July, 1729. In a former page we have spoken of such an event having occurred on the coast of Oregon, another will be mentioned at the Sandwich Islands. All these facts, which doubtless might be multiplied, would tend to prove that the winds and currents in the western portion of the North Pacific have a great analogy to those of the North Atlantic; the same progress of the cyclones, or revolving storms, and the same drift of the N.E. currents, like the great gulf stream. All these will be alluded to. The vessel in question was from Satsma, in Japan, bound for another Japanese port called Azaka (Ohosaka?), laden with rice, cotton, and silks. She was driven from her course by a violent storm to sea, where they remained for six months, and at last reached this coast and cast anchor. The crew, seventeen in number, landed and encamped; and were by chance seen twenty-three days after by a Cossack chief, Andrew Tschinnikov, and some Kamtschadales. The unfortunate Japanese received them with the utmost joy, and loaded them with presents, but the

* Voyage de *La Venus*, tome ii. pp. 68—70.

† Cook's Third Voyage, vol. iii. p. 381.

treacherous Cossack abandoned them directly. They then took to their boat, which Tschinnikov seeing, ordered them all to be shot but two; one a boy of eleven, the other the supercargo, Sofa, a middle-aged man. The Cossack soon received a halter for his barbarity, and the two strangers were conducted to St. Petersburg, where they served as instructors to several pupils in their language. They survived five and six years. Their portraits are in the imperial cabinet at St. Petersburg.*

CAPE LOPATKA is the South part of Kamtschatka, and is in lat. 51° 2′, and lon. 156° 50′. It is a very low, flat cape, sloping gradually from the high level land to the North, and to the N.W. of it is a remarkably high mountain. Its name, Lopatka, signifies the bladebone of a man, or a shovel, and is expressive of its form. It extends from the South end of the peninsula 10 or 15 miles, and is about half a mile broad.

The passage between this cape and the N.W. Kurile Island is about 3 miles broad, and very dangerous, on account of the strong currents and the sunken rocks off the cape. It may be here stated that this portion of the coast, from Awatska Bay southward, is delineated from the survey of Capt. Krusenstern. Lütke commenced at that port proceeding northward.

KURILE ISLANDS.

This extensive chain of islands extends nearly in a uniform N.E. and S.W. line from the South extremity of Kamtschatka to the North point of the Island of Jesso, a distance of 650 miles.

The *Boussole Channel* separates the chain into two portions; that to the northward belonging to Russia, the southern islands forming a portion of the Japanese possessions.

The northern or Russian portion is all apparently of volcanic origin; indeed the whole chain may be looked upon as a series of submerged mountains, a continuation of the mountain chain traversing Kamtschatka through its whole length. Of these northern islands but few of them were inhabited at the time Krusenstern drew up his great work, and altogether there were but very few people on them, and there was no Russian establishment.

Of the Japanese portion the most considerable islands are Ouroup, Itouroup, Kounashire, and Tschikotan or Spanberg Island. On these there are military posts for defence, and establishments for facilitating the commerce with the Aïnos, the native inhabitants.

According to Malté Brun, the Japanese term the entire chain of islands Kouroumissa, or "the road of sea-weeds;" and, according to Capt. Golownin, they denominate the portion belonging to them Toi-sima, or "the distant islands." The name Kurile is derived from the Kamtschadale word for "smoke," the volcanic islands having been seen from Lopatka.

Our acquaintance with the configuration of the Kurile Islands may be considered as tolerably complete. This result is one of the most difficult problems in hydrography. The fog in which the group is constantly enveloped; the violent currents experienced in all the passages or straits separating them; the steepness of their coasts, and the impossibility to anchor near the land, are such formidable

* See Krachenninikoff, vol. ii. p. 4; and Müller, p. 8.

obstacles, that it tries to the utmost the patience and perseverance of the mariner to acquire any knowledge respecting them. La Pérouse and Capt. Broughton, who certainly must be considered as most intrepid and excellent navigators, have found these to be great obstacles, and, as the vast extent of their voyages and the lateness of the season did not allow them to remain a great while on these coasts, they must not be the less eulogized, because Capt. Golownin (or Golovnine), whose sole object was the examination of the archipelago, has done so more completely.

We are indebted, then, to this Russian commander, who, in 1811, was charged with the survey in the *Diana;* this was completed with the exception of the North sides of Kounashire and Itouroup. Besides this, we have the observations of Admiral Krusenstern in the *Nadiêjeda*, in 1805; of La Pérouse and Broughton, before alluded to; of Langman, Spanberg, and by other Russian officers; and also a chart by the surveyor Gilaeff, made in 1790.

ALAID is the northernmost of the Kurile Islands. It is small, and, according to the observations of Admiral Krusenstern, is in lat. 50° 54′ N., lon. 155° 32′ E. It lies rather within or to the West of the general line of the archipelago.

SOUMSHOU ISLAND would, therefore, be reckoned as the first island in reckoning from Kamtschatka, Alaid not properly forming part of this group.

Its southern extreme terminates in a tongue of low land; the North end is the same, and is distant 10 miles from Cape Lopatka, in a S.W. ¼ W. direction. This last cape is equally low, and perhaps was formerly united with Soumshou. It is said that the channel at present separating them is filled with shoals. The island extends about 10 miles in a North and South direction, and its centre lies in lat. 50° 46′, and lon. 156° 26′ E.

POROMOUSHIR ISLAND is among the largest of the archipelago, being 20 leagues in length from N.E. to S.W. Its southern part is very mountainous, the S.W. portion less so. There is also a high mountain in lat. 50° 15′, and lon. 155° 24′ 15″. Krusenstern says, "We could not approach the N.E. extremity, which was hidden by the Island of Soumshou, separated from that of Poromoushir by a channel of a mile at most in breadth. We nevertheless saw the South point of this last island over the low land of the extremity of Soumshou, and on a line with Alaid Island, which then bore N. 66° W. It is not impossible, therefore, but that the North point of Poromoushir may be more to the North than 50° 50′ N. According to the survey by Gilaeff, it reaches to lat. 51°; but we ought not to be surprised at a discrepancy of 10′, when it is known that Russian observers frequently have made errors of 20′ and 30′ of latitude in these parts."

SHIRINKY ISLAND, which lies off the S.W. extremity of Poromoushir, according to Krusenstern's chart, and on which point is marked a peak, probably a volcano, is small, being not more than 2 miles in diameter. It lies in lat. 50° 10′ N., lon. 154° 58′ E. On August 26, 1805, at noon, the *Nadiêjeda* was within 4 miles of it. It then bore from N. 2° E. to N. by E.

MONKONRUSHY ISLAND is rather larger than the last, and is nearly of the same form; it lies in lat. 49° 51′ N., and lon. 154° 32′ E.

AVOS ROCK.—Lieut. Khwostoff, commanding the Russian American Company's vessel *Juno*, discovered, in June, 1806, a rock lying 8 miles to the S.W. of Monkonrushy Island, to which he gave the name of Avos, because, in first seeing it, he thought it was his consort, which was so named. This rock is surrounded by a dangerous reef, formed of rocks even with the water's edge. Krusenstern says that he did not see it, and that he had not found it on any chart except that published in 1802 by the Russian government, where it was named *Magetshadack* (or Sea-Lion Island). It is there made too large; but its position to the South of Monkonrushy is correct, and the name of Sea-Lion applied to it, leaves no doubt as to its identity.

ONNEKOTAN ISLAND is 28 miles in extent from N.E. ¼ N. to S.W. ¼ S. Admiral Krusenstern having sailed along its western shores at a short distance, it was distinctly seen, which was not the case with its eastern face, which was passed a long way off. The N.E. point was seen but very indistinctly through a thick fog. The S.W. point, then named *Cape Krenitsin*, lies, according to observations made in the *Nadiéjeda*, in lat. 49° 19′ N., and lon. 154° 44′ E.; and its S.E. extreme at about 2 miles still farther to the South.

The channel or strait which separates this island from that of Poromoushir is 19 miles broad. It is very safe; all ships going from Okhotsk to Kamtschatka, or to the American coast and returning, use this channel in preference.

KHARAMOUKOTAN ISLAND lies S.W. ¼ S. 8 miles from Onnekotan. Although the channel separating these two islands is safe, the currents in it are very violent, and Admiral Krusenstern is of opinion, that with light winds, or if overtaken by a calm, the passage would become dangerous. It is true that he traversed it without any accident, but then the wind was very fresh.

The island is of a round form; its diameter is 7 miles; a peak rising in its centre lies, according to the observations, in lat. 49° 8′ N., and lon. 154° 39′ E.

SHIASHKOTAN ISLAND lies 8 miles S.W. ½ W. from Kharamoukotan, and is 12 miles long in a N.N.E. and S.S.W. direction. Its centre is in lat. 48° 52′ N., and lon. 154° 8′ E. *Ekarma Island* is placed on Krusenstern's first chart to the South of Shiashkotan; it is situated to the North of it, not more than a mile off, according to the chart of Gilaeff. This is probably correct, because subsequently an old Russian chart was found in which it is also thus placed. Krusenstern had but two charts that showed them, one just published by the Russian government, the other attached to the voyage of Admiral Sarytscheff; both of them indicated a small island to the South of Shiashkotan; and, as on May 31, 1805, Krusenstern discovered land in this direction, without being able to distinguish through the fog whether it was a small island, or rather the extremity of a larger land, he thought he saw Ekarma. Taking this to be the case, that the northern part of the island Shiashkotan on his chart is Ekarma, this would then be in lat. 49° N., and lon. 154° E., and thus it is shown on the new chart. Nevertheless, an error of two or three minutes may exist in both of the islands, seeing that the land was very distant and the weather hazy.

TSHIRINKOTAN ISLAND is small, and scarcely more than 7 miles in circuit. Krusenstern saw it twice, May 30th through the fog, and July 11th in clear weather. It lies 8 leagues to the West of the South end of Shiashkotan, in lat. 48° 44′ N., lon. 153° 24′ E.

The SNARES.—On August 30, 1805, Krusenstern discovered four small islets, or rather rocks, one of which is awash. He passed them within 9 miles, and at noon they bore West. The current was so violent that, notwithstanding he stood off under all sail in a very fresh breeze, he ran great hazard of being drifted on to them. He named them the Snares, on account of the danger he risked so unexpectedly. The strong currents around these rocks will always cause great embarrassment to every ship that passes near them. They lie S.E. ¼ E. from Tshirinkotan, in lat. 48° 35′ N., and lon. 153° 44′ E. The hazy weather he met with prevented the exact determination of their position, and it is doubted whether any other navigator has seen them either before or since that time.

RAUKOKO ISLAND is small but hilly. It has a high peak, lying, according to Krusenstern's observations, in lat. 48° 16′ 20″ N., and lon. 153° 15′ E.

MATAUA ISLAND lies directly to the South of Raukoko. The strait which separates these two islands, according to Krusenstern's measurements, is 8 miles, according to those of Capt. Golownin, 10 miles in breadth, and was named by the latter *Golownin Strait*. Mataua Island is 6 miles in length from North to South. *Sarytscheff Peak*, standing in its centre, is situated, according to a large number of excellent observations made on board the *Nadiéjeda*, in lat. 48° 6′ N., and lon. 153° 12′ 30″ E.

RASHAU ISLAND, in lat. 47° 47′ N., and lon. 152° 55′ E., is about 5 leagues in circumference; it lies to the South of Mataua Island. Krusenstern named the strait which separates these two islands *Nadiéjeda Strait*, because his vessel was the first which passed it, in 1805. This channel is 16 miles broad; it is very safe, but the currents in it are violent.

At this part terminates the observations of Admiral Krusenstern, and those of Capt. Golownin commence.

USHISHIR ISLAND succeeds to Rashau. It is composed of two islands, connected by a reef of 400 yards in length; each of these two small islands is about half a league in length, N.N.E. and S.S.W. A reef of rocks extends from the northernmost of these islands towards Rashau; these rocks are terminated by a small islet, named *Trednoy*, after the name of one of the officers of the *Diana*, Golownin's ship. It lies 10 miles to the S.W. from the southern extremity of Rashau, and N.N.E. 3¾ miles from the northernmost of the Ushishir Island; its extent, East and West, is 1 mile. Capt. Golownin passed through the channel which separates Ushishir from Ketoy Island, and found it very safe. The southern point of Ushishir is in lat. 47° 32′ 40″ N., and lon. 152° 38′ 30″ E. It ought to be remarked that, in the Russian government chart, published in 1802, the Island of Ushishir is also shown to be composed of two islands.

KETOY ISLAND lies 12½ miles to the S.W. of Ushishir; it is high and mountainous, and about 8 miles in circumference. Its South extremity lies in lat. 47° 17′ 30″ N., and lon. 152° 24′ E. Some rocks and islets extend for a considerable distance off its N.E. and East sides.

SIMUSIR ISLAND is 27 miles in extent N.E. and S.W., and 5 miles in breadth. Although it has a circumference of 25 leagues, there is notwithstanding no good anchorage to be found on its shores. In the northern part of it there is a bay, which was also seen by Capt. Broughton, after whom it is named; although it is very spacious, this port is only navigable for small vessels, on account of a reef lying in the middle of its entrance. At high water there are only 12 to 15 feet, and at low water 6 feet depth in it. The peak named by La Pérouse *Prevost*

Peak, is situated about 10 miles to the S. 44° W. from the N.E. point of the island, and lies, according to Capt. Golownin, in lat. 47° 2′ 50″ N., and lon. 151° 52′ 50″ E. The longitude of the peak, according to the observations of La Pérouse, as corrected by Dagelet, differs only three minutes from the determination of Golownin. At the southern extremity of the island, named by La Pérouse *Cape Rollin*, there is a high mountain, in lat. 46° 51′ N., and lon. 151° 37′ E. Capt. Golownin was the first person who passed through the strait separating the Islands of Ketoy and Simusir; he gave it the name of *Diana Strait*, after that of his vessel.

Between the Islands of Simusir and Ouroup are the islands which are named *Les Quatre Frères* (the Four Brothers) on La Pérouse's chart, but there are only three of them in reality. The northernmost is the *Round Island* of Capt. Broughton; but on the chart by Admiral Krusenstern, and also on that by Golownin, it is called *Broughton Island.* The two others are named by the Kuriles *Tschirpoy*, or *Torpoy;* they lie N.N.E. and S.S.W. from each other, half a league apart; the northernmost of the two is distinguished by three pointed paps. The southern island is an extinct volcano. According to Capt. Golownin's observations, the northern Torpoy is in lat. 46° 32′ 45″ N., and lon. 150° 37′ 10″ E.; the southernmost in lat. 46° 29′ 15″ N., and lon. 150° 33½′ E. From the North point of the northern island, a reef extends toward the East, which is half a mile long, and very much resembles an artificial breakwater; at its extremity there is a high rock.

Broughton Island is a naked but very high rock. The channel between these three islands and that of Simusir was named by La Pérouse the *Boussole Channel.* According to Capt. Golownin, Broughton Island is in lat. 46° 42′ 30″ N., and lon. 150° 28′ 30″ E.; and according to Capt. Broughton, it is in lat. 49° 58′ E. As was stated in the introductory remarks, the islands South of this belong to Japan, those to the North of the Boussole Channel being part of the Russian possessions.

OUROUP ISLAND is the nearest land to the S.W. It was named by the Dutch, *Staaten Island.* The N.W. point is hilly, as is all the island, which is covered with mountains, many of which are very high. It is 18 leagues in extent from N.E. to S.W., and its greatest breadth is about 5 leagues. A chain of rocks runs off from its N.E. point for a distance of 5 miles, in an E.N.E. direction, and at the distance of a mile from the shore there is a large rock of a pyramidal form, with two others smaller; the first is sufficiently high to be seen in clear weather at 7 leagues' distance. The northernmost point of the island, named by La Pérouse *Cape Castricum*, lies, according to Capt. Golownin's chart, in lat. 46° 16′ N., and lon. 150° 22′ E., and the southernmost point, named by the Dutch *Cape Van der Lind*, in lat. 45° 39′ N., and lon. 149° 34′ E. At half a mile off the South point of the island, bearing S.W., is a rock of a circular form; but the reef marked on the Dutch charts does not exist.

ITOUROUP ISLAND is separated from Ouroup by the *Strait of De Vries*, discovered in 1643. The N.E. point of the former is high and perpendicular, and is also remarkable by three paps. According to Capt. Golownin, it lies in lat. 45° 38′ 30″ N., and lon. 149° 14′ E., the breadth of the strait is consequently

13½ miles in an East and West direction. The South point of Itouroup, which Krusenstern names *Cape Rikord*, lies, according to the observations of Golownin and Rikord, in 44° 29′ N., and 126° 34′ E. The island, from these calculations, will therefore be 47 leagues in extent from N.E. ¼ E. to S.W. ¼ W. Its greatest breadth is about 6 leagues. Capt. Golownin only examined the southern part of the island; the only part of the North coast that he saw was the N.W. cape, named by La Pérouse *Cape de Vries*; according to his observations it lies in lat. 45° 37′ N., and lon. 149° 1′ E. Capt. Broughton examined the S.W. part of the island more closely than La Pérouse did, that is, from Cape *Trou* to the S.W. extreme; but we know nothing more of the N.W. portion, comprised between Cape Trou and Cape Vries. On La Pérouse's chart Cape Trou is placed in 45° 35′, and in 45° 0′ in that of Broughton; the first latitude is certainly too far North. The Japanese have two establishments, *Sana* and *Urbitsh*, on the S.W. part of Itouroup.

TSCHIKOTAN or SPANBERG ISLAND.—To the South of the S.W. point of Itouroup is the Island Tschikotan, which is called by Capt. Broughton Spanberg Island. It is called in Cook's voyage, *Nadeegsda*. Capt. Broughton's appellation is derived from the fact that Spanberg watered here.*

The centre of the island, according to Golownin, is in lat. 43° 53′ N., and lon. 146° 43′ 30″ E., and its S.W. point is in lat. 43° 51′, and lon. 146° 45′ 30″. The island is only 5 miles long in one direction, East and West, and about the same North and South. In the centre of the island there rises a mount, even and uniform to the summit. It is said that a good *harbour* will be found in the S.W. part of the island.

At 9 or 10 miles' distance from Tschikotan (or *Chikotan*), bearing S.W. ¼ W., there are several rocky islets, also seen by Capt. Broughton, October 5th, 1796. Capt. Golownin passed between these islets, and very close to them: he particularly noticed one of them, which was about a mile in length; it is flat, narrow, and covered with verdure. He compared its appearance to that of the Saltholm, in the Cattegat. The Island of Tschikotan was seen from on board the *Castricum*, but, supposing that it was the eastern extremity of the Island of Jesso, it was named Channel Cape. The space comprised between Jesso and Tschikotan, is the *Walvis Bay* of De Vries, in the same way that the small islets we have just described, between Jesso and Tschikotan, were named Walvis Islands.

KOUNASHIRE ISLAND is the last of the Kuriles. It is separated from Itouroup by the *Pico Channel*. It was first traversed in 1643, by Capt. Vries: next Capt. Loffzoff, in the Russian ship *St. Catherine*, in 1793; Capt. Broughton, in 1797; and Capt. Golownin, in 1813, have successively passed through this strait. Krusenstern named the N.E. point of Kounashire *Cape Loffzoff*; it is in lat. 44° 29′ 15″ N., lon. 146° 8′ W.; the latitude is precisely the same as that of the South point of Itouroup, so that the channel between is 16 miles in width, East and West.

St. Antony's Peak, called by the natives Tschatschanoboury, stands near the N.E. point, and, according to several observations made by Capts. Golownin and

* See Broughton's Voyage, p. 126.

Rikord, it is in lat. 44° 31′ N., and lon. 145° 46′ E. To the N.E. of St. Antony's Peak there is another, less elevated. The Dutch also mention this, under the name of *Mount Maria*.

It may be remarked, that the eastern part of the Strait of Pico, that is, the S.W. coast of Itouroup Island, is differently drawn on Golownin's chart to that of Capt. Broughton's; but as the latter coasted very near to this part of the island, from Cape Trou to the southern extremity of Itouroup Island, he saw it more distinctly than could have been done on board the *Diana*.

The S.W. part of Kounashire forms a bay, named by the officers of the *Diana* the *Bay of Traitors*, because it was here that the Japanese, after inviting Capt. Golownin to land, seized him and made him prisoner. The two points forming the bay lie in a N. 60° W. and S. 66° E. direction, 11¼ miles one from the other. The depth of this bay is not equal on both sides; the S.E. side, of which the coast is low and sandy, forms a tongue of land nearly 7 miles in length; the opposite side, also forming a tongue of land, is only 5 miles. The flood tide, which hardly rises beyond 4½ feet, comes from the East, and directs itself along the coast, and turns around the S.W. point towards the strait which separates Kounashire from Jesso, and called, on Krusenstern's chart, the *Strait of Jesso*. The Japanese establishment on the Bay of Traitors is in lat. 43° 44′ N., and lon. 144° 59′ 30″ E.; and the N.E. point of the Kounashire Island being situated, as we have seen above, in lat. 44° 29′, and lon. 146° 8′, its total extent will be 70 miles in a N.E. and S.W. direction. Capt. Golownin, though not sufficiently near to minutely examine the S.E. coast of the island, nor to the N.E., yet was able to make out sufficient to furnish a tolerably accurate notion of what it is.

From the reports of the Japanese (Kachi,* who had several times coasted the Island of Kounashire, in his voyages from Itouroup to Matzmay), it is surrounded with rocks and dangers. As for the northern shore of the island, we know nothing of it, so that Krusenstern was obliged to complete his chart from a Japanese drawing of Jesso, on which Kounashire was also shown.

Of the prevalent winds in the vicinity of the Kuriles, in the chapter devoted to that subject Admiral Krusenstern's remarks will be found hereafter.

SEA OF OKHOTSK.

The Sea of Okhotsk, surrounded as it is on all its northern and western sides by the continent, and to the S.E. by the range of the Kurile Archipelago, may be considered as completely land-locked. In this respect, as well as in size and general situation, it is not unlike Hudson's Bay.

A large portion of its shores are comparatively unknown, at least in a nautical sense, for with the exception of its single important port, from which it derives its name, we have no accurate or recent description of its details.

The western coast of the Kamtschatka Peninsula is partly known: the great gulfs in its N.E. part are almost *terra incognita*, having only been visited, and that superficially, by some land travellers. Okhotsk, as has been before mentioned, has been more recently described; the coast to the S.W. of it has been surveyed,

* See Voyage de Capt. Rikord.

but not verbally described, by Sarytscheff and Tomine; and of the Shantar Islands, and the great Peninsula of Saghalin, we have the details given by Broughton and Krusenstern.

We have no particulars respecting the open sea, its currents or its soundings; but it is known to resemble the Sea of Behring in its shallowness, so that, at its centre, it is never above 200 fathoms, nor more than 50 fathoms at 50 miles off the land. The shores are closely surrounded by mountains, which occasions all the rivers, with one exception, to be insignificant in their magnitude.

The only river falling into this vast basin is the Amour, if indeed the Amour can fairly be said to do so, terminating as it does in a bay, which, being bounded in front by the Island or Peninsula of Saghalin, opens by one strait into the Sea of Okhotsk, and perhaps by another into the Sea of Japan.

One feature of interest in this remote expanse of waters is the field it offers for the whale fishery. These have at times been found to be very numerous, and of late years the hardy American whalemen have here pursued successfully their gigantic game.

CAPE LOPATKA, the South extremity of Kamtschatka, has been described on page 604.

The western coast of Kamtschatka is uniformly low and sandy, to the distance of about 25 to 30 miles inland, when the mountains commence. It produces only willow, alder, and mountain ash, with some scattered patches of stunted birch trees. The runs of water into the sea from the mountains do not, with the exception of the Bolchoireka, deserve the name of rivers, though they are all well stocked with fish from the sea in the season, as trout, and different species of salmon. They are generally at the distance of 15 to 20 miles from each other. The *Itsha* and the *Tigil* are the most considerable; and neither of them have a course, with all the windings, of more than 100 miles.

BOLCHERETSKOÏ is the place that is best known to Europeans on the West coast, though its present insignificance scarcely deserves notice. It was the seat of the government of Kamtschatka previous to its removal to Petropaulovski. Of course this abstraction has diminished its little importance, and it is therefore seldom or never visited now by commercial vessels. The expeditions under Capts. Cook and Clerke visited it overland from Awatska Bay. Capt. King's description of it is as follows:—" Bolcheretskoï is situated in a low swampy plain, that extends to the Sea of Okhotsk, being about 40 miles long, and of a considerable breadth. It lies on the North side of the *Bolchoireka* (or great river), between the mouths of the *Gottsofka* and the *Bistraia*, which here empty themselves into this river; and the peninsula on which it stands has been separated from the continent by a large canal, the work of the present commander, which has not only added much to its strength as a fortress, but has made it much less liable than it was before to inundations. Below the town the river is from 6 to 8 feet deep, and about a quarter of a mile broad. It empties itself into the Sea of Okhotsk, at the distance of 22 miles; where, according to Krachenninikoff, it is capable of admitting vessels of a considerable size. There is no corn of any species cultivated in this part of the country; and Major Behm informed me that his was the only garden that had yet been planted. The ground was for the most part covered with snow (this was in May, 1779); that which was free from it appeared full of small hillocks, of a black, turfy nature.

The houses in Bolcheretskoï are all of one fashion, being built of logs, and thatched. That of the commander was much larger than the rest, consisting of three rooms, which might have been considered handsome, if the tall windows had not given them a poor and disagreeable appearance. The town consists of several rows of low buildings, each consisting of five or six dwellings. Besides these are barracks for the Russian soldiers and Cossacks; a well-looking church and a court room; and, at the end of the town, a great number of *balagans* (log houses, on piles), belonging to the Kamtschadales. The inhabitants, taken altogether, amount to between 500 and 600.*

It is in lat. 52° 54½′ N., lon. 158° 22′ E.

Off the western coast of Kamtschatka the sea is shallow to a considerable distance; and the commanders of transport vessels, who never lose sight of the exposed coast if they can help it, judge of their distance from the land, in foggy weather, by the soundings, allowing a fathom for a mile; nor is there at the entrance into any of the rivers more than 6 feet at low water, with a considerable surf breaking on the sandy beach.

The villages on this coast, beyond Bolcheretskoï, are Itshinsk and Tigilsk, situated on the *Tigil* and *Itsha Rivers*. *Itshinsk* contains a church, and about ten houses with fifty inhabitants.

TIGILSK, in lat. 58° 1′ N., lon. 158° 15′ E., is the principal place. Sauer says it contained forty-five wooden houses and a church. It is called by the Russians a fortified town, is surrounded by wooden palisades, and was built in 1752. The number of inhabitants was 338, including women and children.†

Besides these, there are eight inconsiderable villages, containing each three or four houses, on the West coast.

The GULFS of IGIGHINSKOÏ, or Jieghinsky, and PENJINSKOÏ, which form the N.E. portion of the Sea of Okhotsk, are but very little known, perhaps altogether unknown in a nautical view. They run to the N.E., between *Cape Outholotskoï* on Kamtschatka, lat. 57° 28′ N., lon. 155° 45′ E., and *Cape Bligan*, in lat. 59° 20′, and lon. 152° 50′, and extending as far North as 62° 25′. They are separated by a promontory, terminating in *Cape Tainotskoï*.

In 1787, after La Pérouse touched at Awatska Bay, the naturalist, M. de Lesseps, started thence on an overland journey to Europe, through Siberia. He crossed over the Isthmus of Kamtschatka from Karaghinskaïa Bay, and reached the eastern coast of Penjinskoï Gulf at Poustaresk. We extract the following from his narrative:—

POUSTARESK, lat. 61° 0′ N., lon. 162° 30′ E., is a small village on the side of a hill, the foot of which is bathed by the sea. The river cannot be called such; it is merely a narrow inlet of the sea, reaching to the foot of the above mountain. The hamlet was only composed of two yourts, and when M. Lesseps arrived here, in March, 1788, it was deserted, and he was reduced to the greatest strait for

* Cook's Third Voyage, vol. iii. pp. 215-6.
† Billings' Voyage, by Martin Sauer, pp. 202—204.

provisions. The inhabitants, it would appear, chiefly depended on fish for subsistence, but it was evidently not abundant.

After a terrible journey over the ice and frozen land, M. de Lesseps reached the *Talofka River*; its shores are clothed with wood, some of it of large size; and then after passing large plains covered with broom, he reached the mouth of the *Penginsk River*. "Its breadth was imposing, and the ices which covered it, and which were heaped up to a prodigious height, would have appeared still more picturesque, if we could have taken any other road than across them."

KAMINOI, at the head of *Penjinskoï* Gulf, is an *ostrog*, or village, about 300 versts distant from Poustaresk. It is on an elevation nearly on the sea-shore, and at the mouth of the River Pengina, in lat. 62° 0′ N., lon. 162° 50′ E. It encloses a large number of balagans and a dozen yourts, all very large. The palisade encloses a collection of houses, which, though close together, cover a considerable space, and was ornamented with bows, arrows, guns, and lances, under the shelter of which miserable defence the Koriaks believe themselves impregnable. The population of Kaminoi did not then exceed 300, men, women, and children.

At 20 versts from Kaminoi the chain of mountains again reaches the sea, and between the River *Chestokova* and Kaminoi, the Rivers *Oklana* and *Egatcha* fall into the head of the gulf. M. de Lesseps was here overtaken by a terrific hurricane of snow, March 26th, 1788. He then proceeded towards Ingiga.

INGIGA, or Fort *Jiejiginsk*, lat. 61° 40′, lon. 160° E., at the head of the gulf to which it gives its name, stands on the river of the same name, at 30 versts from its mouth, and from without appears as a square enclosure defended by a palisade, of a height and thickness that astonished M. de Lesseps. At its four angles there are four bastions armed with cannons, guarded continually by sentinels, as well as the three gates of the town, one of which only is opened. Before the house of the commander there is a small open space, a small guardhouse on one side preventing access to it. The houses are all of wood, very low, and nearly all of a uniform elevation. The population was 400 or 500, all members or attached to the Russian service; the last formed the larger part, and composed the garrison. The commerce is chiefly in furs, and that principally reindeer skins.*

From this fort M. de Lesseps proceeded to the S.W. nearly along the shores of the gulf, but does not give any description of then passing through *Jamsk*, lat. 59° 29′, lon. 153° 0′, which lies at the S.W. angle of Igighinskoï Gulf, and then bearing to the westward passed *Taouinsk*, lat. 59° 56′, lon. 148° 30′, which gives its name to a large gulf, called *Taouinsk Bay*, apparently open to the South, and having several arms branching from it, but of which we have no particulars. The extent of coast between this and Okhotsk we have not found any description of.

OKHOTSK is the principal seaport, if it deserve such a name, of the sea to which it gives the appellation. It stands on the N.W. side of the sea, in lat. 59° 20′ N., lon. 143° 14′ E.

The shallowness of the water a long way off from the entrance of the harbour, and the violence and cross set of the tides at the harbour's mouth, preclude the

* Lesseps, vol. ii. pp. 70-1.

possibility of Okhotsk being an easily accessible port, except for a small vessel. Necessity alone, resulting from the loss of the *Amour*, can induce the Russian government to keep it at such an expense, and under its present circumstances.*

Okhotsk was visited by Sir George Simpson in his overland journey. He arrived here from Sitka, June 27th, 1842. "Okhotsk (Ochotsk), now that we had reached it, appeared to have little to recommend it to our favour, standing on a shingly beach so low and flat as not to be distinguished at our distance from the adjacent waters. We saw nothing but a number of wretched buildings, which seemed to be in the sea, just as much as ourselves; while, from their irregularity, they looked as if actually afloat; and even of this miserable prospect one of the characteristic fogs of this part of the world begrudged us fully one-half."

The Russian Company's post stands near the end of a tongue of land, about three-quarters of a mile in length, and one-quarter of a mile in width, so little elevated above the level of the sea, that, when the southerly wind blows hard or continues long, the whole is almost sure to be inundated. The town lies about half a mile distant, situated on the left bank of the Kuchtin. It has stood on this site only for a few years, having formerly occupied a low point between the sea and the Ochota, and it appears to have been removed just in time, for the river has, since then, formed the tip of the point into an island, sending the main body of its waters through this new channel of its own cutting. Even now the town is not secure, being subject, as well as the company's post, to inundations in southerly gales.

The population of Okhotsk is about 800 souls; though, forty years ago, it amounted, according to Langsdorff's estimate, to about 2,000. This arises, probably, from the town having been supplanted as a penal colony by the mining, a change which cannot be regretted on either side, for the convicts were of the worst class, nor can they look back with any regret to Okhotsk from any place whatever.

A more dreary scene can scarcely be conceived. Not a tree, and hardly even a green blade, is to be seen within miles of the town, and a stagnant marsh in the midst of it must be, except when it is frozen, a nursery for all sorts of malaria. The soil is on a par with the climate. Summer consists of three months of damp and chilly weather, succeeded by nine months of dreary winter, as raw as it is intense. The principal food of the inhabitants is fish. The Sea of Okhotsk yields as many as fourteen varieties of the salmon alone, one of them, the nerker, being the finest thing of the kind ever tasted. Fish is also the staple food of cattle and poultry. All other supplies for the table are ruinously extravagant, as much of the stores is burdened with a land-carriage of 7,000 miles. On such fare, and in such a climate, no people could be healthy. Scurvy, in particular, rages here every winter.

Okhotsk possesses a ship-builder's yard, in consequence of its connexion with Kamtschatka; and, owing to the frequent losses of the transports, they must be kept in full employ. The carpenters do their duty well.

Bad as the Harbour of Okhotsk is, it is believed to be the best in the Sea of Okhotsk. Capt. Kadnikoff (who, however, was lost in the autumn of 1842)

* Cochrane's Pedestrian Journey, vol. i. chaps. viii., ix. See also Langsdorff's Travels.

intended that year to survey what is called Jan Harbour, lying some distance to the S.W. of Okhotsk, and if his report had been favourable, the Russian American Company would have removed their establishment thither on account of its collateral advantages.*

The coast between Okhotsk and *Fort Oudskoi* has been surveyed and described by the Russian Vice-Admirals, Sarytscheff and Tomine, so we may suppose them to be well represented on the charts, but as their accounts are in the untranslated Russian, we have no description to give here.

JONAS ISLAND, a collection of naked rocks, discovered by Capt. Billings in the Russian corvette *Slava Rossii*, lies off this portion of the coast. Its position was ascertained by Admiral Krusenstern, in 1805, as lat. 56° 25′ 30″, and lon. 143° 16′ East. It is about 2 miles in circumference, and 1,200 feet high. Off its West side this islet is entirely surrounded by detached rocks, against which the waves beat with great violence, and which extend a considerable distance below the surface of the water.

Fort Oudskoï (Oudskoï Ostrog), a considerable establishment, was determined by Lieut. Kosmin, in 1829-30, to be in lat. 54° 29′ N., lon. 134° 58′ E. This portion of the coast, until the survey made by that officer in the years above mentioned, was quite unknown. Fort Oudskoï, or Ouda, lies on the left bank of the River Ouda, at above 20 miles from its mouth, which is in lat. 54° 44′ N., lon. 134° 25′ E.

The SHANTAR ISLANDS were also surveyed by Lieut. Kosmin, who accompanied Capt. Wrangel in his Siberian expeditions. From that survey it appears that the principal of the group is *Great Shantar Island*, which is 35 miles long, in an East and West direction, and of an equal breadth from North to South. Notwithstanding this extent of coast, it does not appear to afford any port, but the S.W. point of the island projects to the S.W., so as to form an open bay to the eastward of it. Between this point and *Cape Nikla* and *Dougangea*, the nearest point of the continent, 14 miles distant, are two islets, *Barrier* and *Duck Islands*. These are surrounded by rocks and reefs; one of these rocks is 30 feet high. The North point of Great Shantar is in lat. 55° 11′ N., lon. 137° 40′ E.; its South point is in lat. 54° 56′. To the South of it are some small islands, which were not examined. To the East of it are two islands, one in lat. 55° 2′ N., and lon. 138° 22′, is called *Prokofieff*, the other, in lat. 54° 43′, and lon. 138° 12′, *Koassoff*. At the distance of 6 miles to the West of the Shantar Islands is *Feklistoff Island*. It is 20 miles in extent, from N.W. to S.E.; its breadth being about one-half. This island has also no port nor shelter, and, as far as is known, was the only one inhabited in 1806.

The RIVER and BAY of TOUGOURA are to the southward of the Shantar Islands. They were examined, in 1806, by an officer of the Russian navy, Borissoff; but he had not sufficient means at his command to perfect his work.

From the mouth of the River Ouda, before described, the coast, for an extent of 50 miles, runs nearly in an East direction to a point behind which is a small bay, named the *Bay of Swans*. The western point of this bay is *Cape Dougandsha*,

* Overland Journey Round the World, vol. ii. pp. 244—257.

and from this cape to the mouth of the *River Tougoura* the coast runs South; it forms, with a promontory lying 25 miles more to the South than Cape Dougandsha, the *Bay of Tougoura*, the opening of which is 12 miles, and the depth, as above, 25 miles. This promontory is called *Cape Linekinskoy*, and on the latest charts is in lat. 54° 14′, lon. 137° 24. The River Tougoura falls into the head of the bay in lat. 53° 40′.

The coast between this and *Cape Khabaroff*, in lat. 53° 40′, lon. 141° 22′, remains almost entirely unknown.

CAPE ROMBERG is the easternmost projection of the Asiatic continent here, being in lon. 141° 45′ E. It forms, with Cape Golovatcheff on the Peninsula of Saghalin, the North entrance to the great River Amour, which debouches into the extensive bay formed by that peninsula and the coast.

The RIVER AMOUR is very imperfectly known. Its navigation was ceded to the Chinese before modern science and enterprise had been brought to bear on it; consequently it may rest in its present darkness for a lengthened period.

In almost every point of view the Amour is the most valuable stream in northern Asia. Of all the large rivers of that boundless region it is the only one that empties itself into a navigable part of the universal ocean. It is, in fact, the only highway of nature that directly connects the central steppes of Asia with the rest of the world. But the political arrangements of man have decreed otherwise; and at this moment the Amour is infinitely less useful as a channel of traffic than almost any one of the land-locked rivers of Siberia. In one word, it belongs, not to Russia, but to China.

It was in 1689 that a treaty was entered into by China with Peter the Great, by which the latter gave up a portion of his conquest on this great river for the advantage of a regular land trade between the two empires. But it was soon found that a fair at Kiakhta, or a factory at Pekin, was a poor compensation for the loss of this valuable artery to Central Asia, and by which cession the possession of Kamtschatka, and the islands beyond, are reduced to half their value. Had the navigation of the river been open to the commerce of the world, it would be difficult to tell what might have been the consequences of the immense commercial and civilizing advantages it would have conferred.*

PENINSULA OF SAGHALIN.

The name applied to this great island is derived from the Mantchous, who thus call it after their great river, Saghalin Ula, the Amour, which falls into the sea opposite the N.W. coast of this island. Besides the term Saghalin, geographers and navigators have given it the names of *Oku-Jesso*, of *Karafta*, of *Tschoka*, and of *Sandan*.

The first of these appellations is apparently due to a geographic error of the Japanese, who, having heard of the great unknown land to the northward of Jesso, called it Oku-Jesso, that is, Great Jesso, in contradistinction to Little Jesso, with which they were acquainted.

* See Sir George Simpson's Journey, vol. ii. pp. 239-40.

According to Malté Brun,* Oku Jesso means Upper or North Jesso, in either of these cases signifying a distant and unknown country. Karafto, it is said, is the name given to it by the Aïnos, or natives of the country; but the Japanese informed Krusenstern, that under the term Karafto, or Karafuto, was only comprehended that portion of Saghalin occupied by themselves. Sandan was supposed to be a land to the North, separated from an island, Karafuto, by a strait.† The name of Tschoka does not appear to be the correct name for the whole, as it was not then known by the Japanese; so that it would appear that Saghalin is the more appropriate name, because the others, at least, only apply to portions of it.

Saghalin is at present in the occupation of the Japanese and Chinese; the first occupying the southern part, and the last the northern portion. Capt. Golownin‡ says, that before the period of La Pérouse's visit there was no Japanese establishment either on Saghalin or on the southern Kuriles. But the appearance of this navigator led both the Japanese and Chinese to fear that the country would be taken by the foreigners; and under these apprehensions, both governments had instant possession of either end. Capt. Krusenstern visited Aniwa Bay in the South, and the recent construction of the buildings there was a convincing proof of the freshness of the colonization. But it hardly seemed probable to him that the Chinese government had taken any part of the colonization of the North. A colony of Mantchoos, come without doubt from the opposite coast, seemed to them to have been established there for a very long period, but they saw no Chinese; in the Aniwa Bay establishment, the administration was conducted by Japanese military officers. It would appear that the Russians had some notion of taking possession of the southern end of the peninsula, at least it was so hinted at in 1805.

The natives would appear to be very few in number, as scarcely any signs whatever were seen on all the eastern side by Capt. Krusenstern, who closely examined this coast, and from the second volume of whose voyage the subsequent particulars are chiefly gleaned. They were principally found in the service of the Japanese in Aniwa Bay. They are called *Aïnos*, as are the natives of the adjacent Island of Jesso, and are certainly the same people that, since Spanberg's time, have been called Hairy Kuriles. The Aïnos are rather below the middle stature; of a dark, nearly black, complexion, with a thick bushy beard, and black, rough, straight hair; except in the beard they resemble the Kamtschadales. The women are ugly, but modest in the highest degree. Their characteristic quality is goodness of heart, as is expressed in their countenance. Their dress consists chiefly of the skins of seals and dogs, of which latter they keep great abundance. They appeared to live in the most happy way in their domestic life. Their numbers must be very inconsiderable, as only about 300 were seen in Aniwa Bay, and very few elsewhere; it may be presumed that there are none inland, as their food is chiefly fish.

Capt. Krusenstern is inclined to doubt the ancient accounts of its inhabitants

* Précis de la Geographie Universelle, vol. iii. p. 466.
† Voyage de *La Nadiéjeda* et *La Neva*, vol. ii. p. 59.
‡ Recollections of Japan.

being covered with hair. This originated with the Chinese, who stated that the country is filled with a wild people, whose whole body is covered with hair, and with such enormous beards, that they are obliged to raise them to drink. These accounts were repeated by the Dutch captain, De Vries, in 1643; and again by the Russians, under Spanberg, in 1739: but Krusenstern could not find, after close questions and examination, anything of the sort, and believes, therefore, that some exaggeration or falsification exists.*

It is not absolutely determined whether Saghalin be an island or a peninsula; but as all evidence certainly tends towards the latter opinion, that appellation has been retained. The following facts will tell what is known respecting it. La Pérouse, expecting to find a channel through this place to the Sea of Okhotsk from the southward, penetrated as far to the northward as his vessel allowed, but the depth at last decreased at the rate of a fathom a mile. He then sent two boats to sound, but they only advanced 3 miles into 6 fathoms. He also was told that a sand-bank, overgrown with sea-weed, connected the two coasts. There was no current observed, and therefore La Pérouse concluded, from these facts, that if there was a channel, it must be very narrow and shallow. To the North Krusenstern found the water to be nearly fresh, and of a dirty yellow colour; as soon as he had doubled Cape Maria, the North extremity, and in the channel separating Saghalin from the continent, the water, undoubtedly from the Amour, was quite fresh.

Capt. Broughton advanced 9 miles farther than La Pérouse from the southward, and found the channel closed on all sides by low sand-hills, without the smallest appearance of a passage. It must therefore be considered that Saghalin is joined to the continent by a flat sandy neck of land, over which, it is possible, the sea may wash when the strong southerly gales which occur here drive the waters to a higher level; and that this isthmus may be of comparatively recent date, and still on the increase from the deposits from the Amour, so that the older Chinese charts may be correct.

The Peninsula of Saghalin extends in a North and South direction about 170 leagues. Its breadth, in general, does not exceed 25 leagues; but in its northern part it is not more than 3 leagues. The southern part, as well as the northern, terminates in a bay: the Dutch gave the name of Aniwa Bay to the southernmost; Capt. Krusenstern called the other North Bay.

CAPE ELIZABETH is the North point of the peninsula; it is in lat. 54° 24′ 30″ N., and lon. 142° 47′ E., and is a high mass of rock, forming the extremity of a continuous chain of mountains. It is very remarkable from a number of high pointed hills, or rather naked rocks, upon which neither tree nor verdure is visible. It descends gradually toward the sea, and at the brink of the precipice is a pinnacle or small peak. Seen from the West it is exceedingly like Cape Lopatka, except that it is higher. On the West side of the cape a point projects, and between them is a small open bay.

CAPE MARIA is in lat. 54° 17′ 30″, and lon. 142° 17′ 45″ E. It is lower than Cape Elizabeth, and consists of a chain of hills all nearly of the same elevation.

* Voyage of the *Nadiejeda*, vol. ii. p. 72.

It slopes gently down to the sea, and terminates in a steep precipice, from whence a dangerous reef runs to the N.E.

NORTH BAY lies between these two capes, which lie N. 65° E. and S. 65° W. 18 miles from each other. The bay lies very open, but appeared to be safe, especially in summer, when North winds are rare.

In a valley quite in the bottom of the bay, in lat. 54° 15′ 45″, and lon. 142° 37′, or 9 miles South of Cape Elizabeth, Capt. Krusenstern found a colony of Tartars. At some distance before the houses appear in sight, this spot is remarkable from having the appearance of two islands, between which any one would expect to find secure anchorage. The neighbourhood is remarkably delightful. It is high water about Cape Maria at new and full moon, at two o'clock; the tide does not rise to any considerable height.

NADIE'JEDA BAY (or Nadeshda Bay), to the S.E. of Cape Maria, on the West coast, is rather open, and consequently not safe for anchorage, as the ground everywhere is rocky. It lies in lat. 54° 10′ 15″ N., lon. 142° 27′ 34″ E. A plentiful supply of wood and water may easily be procured here; but the situation is such as will preclude it ever being much visited by navigators. When Capt. Krusenstern anchored here, in August, 1805, he found another Tartar settlement, but did not see a single Aïno or aborigine of Saghalin. The southern point of the bay was named *Cape Horner*, after Dr. Horner, the naturalist, on board the *Nadiéjeda.*

The N.W. coast of Saghalin is infinitely preferable to the S.W.; between the mountains, which are entirely overgrown with the thickest forests, are valleys which appear very capable of cultivation. The shores are broken, and almost everywhere of a yellow colour, which gives the coast the appearance of being hemmed in by an artificial wall. The confines of the high and low lands are precisely in the same parallel as on the opposite side; and beyond the limits, to the S.S.W., as far as the eye could reach, nothing could be seen but the low sandy shore, with here and there a few insulated but picturesque sand-hills. On the extreme point of land visible from the N.N.E. is a high hill, which is remarkable in this ocean of sand; at no great distance from it is a large rock in the form of a pyramid.

CAPE GOLOVATCHEFF, in lat. 53° 30′ 15″ N., and lon. 141° 55′ E., forms with Cape Romberg, on the coast of Tartary, the entrance into the lake of the River Amour, before mentioned. The depth found midway between the points was 4 fathoms, gradually decreasing to 3½. The water was found to be perfectly fresh, and a strong current ran from S.S.E. and S.

Having before alluded to the Amour River and this part of the coast, we return again to the North point, and proceed to describe the eastern coast of the peninsula.

CAPE LÖWENSTERN is in lat. 54° 3′ 15″ N., and 143° 12′ 30″ E. It was named after Krusenstern's third lieutenant. The appearance of the coast between this cape and Cape Elizabeth is very dreary; no traces of vegetation are apparent, and the whole coast is iron-bound, consisting of one mass of black granite rock, with here and there a white spot; the depth at 3 miles off shore was 30 fathoms, rocky bottom. There are four capes between these two headlands. In front of Cape Löwenstern there is a large rock.

Southward of it the shore is everywhere steep, and in several places consists of rocks of a chalk-like appearance. The land is high and mountainous, with narrow spaces between the hills.

Cape Klokatcheff is in lat. 53° 46′, and lon. 143° 7′ E., and near it appeared to be the mouth of a considerable river, as the land appeared to be unconnected. The coast about it consists of flattish land, gradually increasing in height, the shores being perfectly flat and sandy.

Cape Würst is in lat. 52° 57′ 30″, and lon. 143° 17′ 30″. A long way inland there are several considerable high lands, the coast being, as far as the eye can reach, composed of flat sand.

Shoal Point, in lat. 52° 32′ 30″, and lon. 143° 14′ 30″, may easily be known by a hill of tolerable height, which on this flat coast almost merits the name of a mountain, and forms a very remarkable object. At this point the coast recedes to the westward, and here a *dangerous shoal* lies. It is in lat. 52° 30′ N., and stretches probably for some miles North and South at a distance of 10 miles from the shore. This is the only one met with off the coast, and has 4½ *to 8 fathoms* on its outer edge.

Downs Point is in lat. 51° 53′ N., and lon. 144° 13′ 30″ E. It is remarkable for a round hill. To the northward of it is a chain of five hills, of a billowy form, having the appearance of islands in this extended plain. The whole coast here, like that to the southward, is scarcely raised above the water's edge; it is entirely of sand, and a little way inland is covered with a seemingly impenetrable thicket of low shrubs.

CAPE DELISLE, named after the astronomer Delisle de la Croyère, is in lat. 51° 0′ 30″, lon. 143° 43′, and forms the boundary of the mountainous part of Saghalin, for to the northward of it there is neither high land nor a single mountain, the shore everywhere consisting of sand, of a most dangerous *uniformity*.

Cape Ratmanoff is in lat. 50° 48′ N., and lon. 143° 53′ 15″ W. It terminates in a flat neck of land, stretching a considerable distance into the sea. The coast hereabouts is invariably craggy, and of a yellow colour. Capes Ratmanoff and Delisle are connected by a flat sandy beach, with mountains in the background between them.

Cape Rimnik is in lat. 50° 12′ 30″, lon. 144° 5′ E. At the back of it, some miles inland, is *Mount Tiara*, so named by Krusenstern from its form, a tolerably high flat hill, remarkable for having three points on its summit; it is in lat. 50° 3′ N., lon. 216° 23′ W. From its parallel the coast trends S. 30° E.

Cape Bellingshausen is in lat. 49° 35′ N., and lon. 144° 25′ 45″ E. Seven miles S.S.W. of it is a point which was thought to offer a good harbour. The shore is very abrupt and entirely white. Between two hills that project considerably, the southernmost apparently insulated, is this apparent harbour, and perhaps a small river. It was, however, unexplored. The country about it is very regular in appearance.

Flat Bay, in lat. 49° 5′, is surrounded on all sides by a country very low. It is a deep opening, in which, even from the masthead, no land could be descried: from this circumstance it was thought by Capt. Krusenstern that it was the mouth of a large river.

CAPE PATIENCE is the most prominent and the easternmost cape of Saghalin. It is in lat. 48° 52′ N., and lon. 144° 46′ 15″ E., and is a very low promontory, formed by a double hill, terminating abruptly. From this a flat tongue of land projects pretty far to the South, and on the North side of the cape the land is likewise very low, the flat hill near Flat Bay being the first high land in that direction. By this hill Cape Patience, which, owing to its little elevation, is not easily perceived, may soon be recognised.

PATIENCE BAY is extensive, and limited to the East by the cape of the same name, and to the West by Cape Soimonoff. Cape Patience is surrounded by a rocky shoal, extending a considerable distance.

ROBBEN ISLAND, surrounded by a very dangerous reef, lies off Cape Patience. Capt. Krusenstern saw and examined the extent of this reef. The N.E. front he places in lat. 48° 36′, and lon. 144° 33′, and that part which may be considered as the S.W. extremity is in lat. 48° 28′, and lon. 144° 10′ E., so that its whole circumference is about 35 miles. The waves broke violently over it, and to the northward there appeared, as far as the eye could reach, a large field of ice, under which, in all probability, the reef continued. The middle of Robben Island is in lat. 48° 32′ 15″, and lon. 144° 23′. With respect to the channel between the cape and the reef, it will seldom or never be required, so it was not examined. The ship *Castricom*, under the Dutch commander De Vries, anchored here in 1643, and gave the names to the bay, &c. In the N.E. angle of Patience Bay is the mouth of a river, which Lieut. Ratmanoff ascended. The mouth is only 15 fathoms wide and 7 feet deep; he proceeded up about 5 miles, and found it abounding with fish, and wood and game near the banks. No houses were seen, but marks of fire in several places. Three Aïnos, in their seal-skin dresses, were met, but they could not be induced to approach the party.

The North coast of the bay is mountainous, and the beach craggy. Far inland are lofty snow-topped mountains, except in one part, where an even country stretches away to the northward as far as the eye can reach.

In the N.E. corner of the bay is the mouth of a tolerably large river, named the *Neva*. Its entrance, in lat. 49° 14′ 40″, and lon. 216° 58′, is about half a mile wide. Off its mouth the water was fresher, and branches of trees were among the clayey soundings. A smaller river debouches to the southward.

CAPE SOIMONOFF, in lat. 48° 53′ 20″, and lon. 143° 2′, is the western limit of Patience Bay; it is a high promontory, projecting very much to the eastward, and was taken for an island when it bore to the northward.

CAPE DALRYMPLE, named after the English hydrographer, is in lat. 48° 21′ N., and lon. 142° 50′. It is formed by a high mountain, lying close upon the beach in a North and South direction, and is the more easily known from being altogether isolated, except that to the northward, 12 or 15 miles, is another group, very unlike this, apparently consisting of four separate mountains. The coast between is, with the exception of a peak of moderate height, quite low.

The coast, from the former cape, trends S. by W., consisting of lofty mountains, divided by deep valleys, the shore being steep and rocky. In several places are inlets between the rocks, which might afford anchorage, one in lat. 48° 10′, looked more promising than the rest. The whole country is more agreeable in prospect than farther South, and its pleasing and fertile appearance gave it a decided advantage over middle and northern Saghalin. *Cape Muloffsky*, a projecting point of land, is in lat. 47° 57′ 45″, and lon. 142° 44′.

BERNIZET PEAK of La Pérouse is probably the same as *Mount Spanberg* of the Dutch. It is a lofty, rounded mountain, in lat. 47° 33′, and lon. 142° 20′. It is near the N.W. end of a lofty chain of mountains, running through the valley from N.W. to N.E.

Cape Séniavine is a high point of land, in lat. 47° 16′ 30″, lon. 142° 59′ 30″. To the northward of it the coast is low, and falls suddenly off to the westward; to the southward are lofty mountains, covered with snow in May, 1805.

MORDWINOFF BAY lies to the southward of this, and is limited to the East by Cape Tonin. Plenty of water was found in it in many places, and abundance of firewood. On the shores of the bay several dwelling-houses were found, but most of them were empty. A few people, however, were seen; these Aïnos appeared to be superior to the others in the South. They trade with the Japanese with train-oil and furs; thus all their utensils and furniture were of Japanese manufacture.

CAPE TONIN is in lat. 46° 50′, lon. 143° 33′. It is of moderate height, and entirely overgrown with fir trees. A chain of rocks stretches to the northward from it; southward of it the bottom is rocky, with small stones; to the northward it is entirely of clay.

CAPE LÖWENORN is in lat. 46° 23′ 10″, lon. 143° 40′. It is a steep projecting rock, easily to be distinguished from the rest of this coast by its yellow colour. North of it the coast assumes rather a westerly direction, and consists of a chain of large lofty mountains, covered with snow in May. A number of whales and seals sported around the *Nadiéjeda*, and a boat rowed off the shore, but returned before reaching the ship.

CAPE ANIWA, the south-easternmost projection of Saghalin, is in every respect a remarkable promontory, the more so from a chain of high mountains near it, stretching away to the northward, between which and the cape is a hollow that gives it the appearance of a saddle. The headland itself is a steep, abrupt mass of rocks, perfectly barren, and having a deep inlet at its point. The position was very carefully observed by Capt. Krusenstern; it is in lat. 46° 2′ 20″ N., lon. 143° 30′ 20″ E.

ANIWA BAY occupies the southern end of Saghalin, and in some respects is the most important locality, as it is here that the Japanese establishments are formed. Its opening is between Cape Aniwa on the East and Cape Crillon on the West, 64 miles apart; its head is at Salmon Bay, the site of the Japanese establishment, and 50 miles within the line of opening.

From Cape Aniwa it runs first in a northerly direction, then inclining a little to the West to a headland, which projects also to the West, and from this, as far as the head of the bay, it runs North and South. In this part of the coast, which was not examined by Krusenstern, is a rock called the *Pyramid* on the charts.

Tamary Aniwa is apparently the name given by the Dutch to the projecting point on the East side of the bay above alluded to. Here was a Japanese establishment, perhaps more considerable than that at Salmon Cove; probably the chief is Aniwa Bay. The harbour is somewhat sheltered against the South wind, but too small for a ship of considerable size to be there. Capt. Krusenstern's officers found the houses of the Japanese in a beautiful vale, through which a stream of clear water ran. They received the Russians in the handsomest manner, without any apprehension. There were about one hundred dwellings of the Aïnos.

Lachsforellen or Salmon-Trout Bay is entirely exposed to the South, which are here said to be the prevailing winds, and consequently the road is by no means safe. The great surf is also an obstacle to landing. Tamary Aniwa is its S.E. point. The Japanese have here also a large establishment. *Salm* or *Salmon Bay* is at the head of Aniwa Bay. Krusenstern's anchorage off it was in lat. 46° 41′ 15″, and lon. 142° 32′. The Japanese factory, at the mouth of the small river, bore N. 49° W. 2½ miles. The establishment is on both sides of this river, and consists of a few dwelling-houses, and eight or nine new warehouses, quite filled with fish, salt, and rice. The officers were afraid at their visit, but soon became friendly.

Aniwa Bay would afford many refreshments. The shores at this part are covered with crabs and oysters. Fish is also most abundant, and very readily taken. A larger quantity of whales was perhaps nowhere to be found. The woods would afford admirable timber for shipbuilding. All these advantages led the Russian officers to covet possession of it for their country.*

The West side of Aniwa Bay is throughout very mountainous; and even in May was covered in parts by snow. A flat and rather projecting mountain, in the direction of the coast, which trends S.S.W., is alone distinguished for its greater height. The shore is throughout lined with steep rocks, between some of which there appeared to be entrances, though not deserving the name of bays.

CAPE CRILLON is the S.W. limit of Aniwa Bay, and the South extremity of Saghalin. It is in lat. 45° 54′ 15″ N., lon. 141° 57′ 56″. At a short distance from it is a small round rock, and another small rock is at its extremity.

La Dangereuse Rock lies off it. This, with the cape, were thus named by La Pérouse. The rock is correctly termed; it is nearly even with the surface, and is 10 miles S. 48° E. from Cape Crillon, and is in lat. 45° 47′ 15″, and lon. 142° 8′ 45″. When Krusenstern passed it a number of sea-horses lay on it, and made so horrible a noise that it was heard from the ship 2½ miles distant. It lies in the middle of the strait of La Pérouse, which separates Saghalin from the Island of Jesso.

The variation of the compass along the whole coast was found by Capt. Krusenstern in his survey, in 1805, not to exceed 1°.

* See on this subject the remarks in Krusenstern's Voyage, &c., vol. ii. pp. 67—70.

CHAPTER XIX.

THE JAPANESE ARCHIPELAGO.

This very extensive and important country has been almost a *terra incognita* to Europeans; its commerce, its resources, its people, and even its physical character, remain still enveloped in much mystery. The celebrated and long mysterious traveller, Marco Polo, was the first to announce to the western world the existence of the rich and powerful island of *Xipangu*, now known to be Japan. In 1542 a Portuguese, Mendez Pinto, was cast by a storm on its shores, and a Portuguese settlement from Malacca was soon afterwards made on Kiusiu.

Like China it has been a comparatively sealed country to Europeans, yet it has by no means remained unvisited and unexplored by them, and this from very early times, for commercial enterprise. In June, 1588, some citizens of Rotterdam fitted out a small fleet of five ships to trade in the Indian Archipelago, and injure, as much as possible, the commerce and power of Spain. Among several Englishmen in this fleet, were William Adams, of Gillingham, near Rochester, and Timothy Shotter, who had accompanied the famous Cavendish in his circumnavigation. The venture was pre-eminently unfortunate. Only one ship, and that the smallest, the *Joyous Message*, commanded by Siebold de Weert, returned to Holland, and this, too, after reaching the further end of the Strait of Magalhaens. Two of the others were destroyed, and the fourth, in which were these two Englishmen, reached Japan a mere wreck. They were taken prisoners, and every means that the Spaniards and Portuguese, then allowed to be there, could invent or exert, were raised to cause the crucifixion of these heretical pirates. This did not succeed, and, after some confinement, Adams was taken into the confidence of the emperor; the rest departed. He was raised to great honours; became of first importance in the political and commercial affairs of the empire; but did not succeed to the extent of his intentions, having gained privileges only for the Dutch, who have studiously avoided mention of his part in their establishment. Some extensive privileges were also granted, at his instigation, to the English East India Company, to establish a factory at Firando.*

The Japanese have hitherto resisted any visits from foreigners to their country with far greater pertinacity than the Chinese. The recent embassies of the Americans and French give but little hope of any concessions being granted in this respect: whether it is to remain a sealed country for many years longer, or, like its celestial brother, is to be thrown open by might to the commerce of the world, must be left for time to show. It would be out of place to dwell upon this here. Siebold's work, in connexion with those of Kämpfer, Thunberg, Titsingh, and other older authors, will give a good insight into their domestic policy. As a

* The first English who visited it were with Capt. Saris, who came to the relief of Adams, from England, in 1611, arriving at Firando June 9th, 1613.

more available source to many sailors, a general outline of a portion of Siebold's work is given in the Nautical Magazine for 1842-43.* Some few ships have touched at different ports, but have been repulsed with the utmost strictness and firmness. Thus an American ship, the *Manhattan*, Capt. Cooper, picked up some shipwrecked Japanese, and carried them into Jedo; but neither he nor his crew were suffered to set foot on shore on pain of death. The Japanese were neither rude nor uncourteous, but every advance was firmly resisted. This incident will serve as an example of Japanese intercourse.†

As is well known, the only port now allowed to be open to foreigners, and this permission is limited to the Dutch and Chinese, is the Port of Nagasaki, or rather for the Dutch, the Island of Dezima, lying before it. We have but few European accounts of the country. After Kæmpfer, Thunberg, the physician to the Dutch embassy to the court of Jedo, in 1775-76, was the only one who had told us about the interior. Isaac Titsingh, president of the Dutch commerce, 1780-84, brought an immense collection of Japanese objects, but their scientific value was lost by his premature death in 1812. Admiral Krusenstern and his officers, Tilesius and Langsdorff, have added to our knowledge of the country. Golownin could learn but little from his condition as a prisoner. All these deficiencies have been, however, obviated by the collection from the notices of the Dutch presidents, by Dr. Ph. Fr. von Siebold, who visited Japan in the period between 1823-30. This magnificent work, which is worthy of any nation, will afford as much as can be desired as to the internal economy of the nation, but is not so explicit in a nautical sense. To the work of Admiral Krusenstern we therefore have recourse for the descriptions, correcting, however, various points from the later, and, as must be considered, more perfect, work of Von Siebold.

CLIMATE.—The following observations upon this subject, connected with Japan, are made by Siebold:—

"In speaking of ice, frost, and snow, within 32° of the equator, we should consider the geographic position of the Japanese Islands, and cite an observation which has been more than once made, and at last confirmed by Alex. de Humboldt.‡ The eastern part of Europe, and the immense continent of Asia, are vastly more cold, under the same latitude, than western Europe, making allowance for the greater or less elevation above the sea level. The climate of islands being much milder than that of continents, it can scarcely be comprehended that the temperature should be lower in Japan than those European countries under the same latitude. But the cause of this contradiction is found in the low temperature of Asia, which, surrounding the Japanese and Kurile Islands on the West and North, has a very decided influence on their climate. From the proximity of the continent, and the winds blowing off that coast during a portion of the year, the cold arises which prevails in Japan, particularly in the North and N.W. Thus in lat. 32° N. the thermometer descends on the coast

* Notices of Japan,—Nautical Magazine, 1842-43.

† An account of the Dutch trade and their enormous profits at Japan will be found in Sir Stamford Raffles' interesting account of Java, vol. i. pp. 242, *et seq.*; also in Charlevoix, Histoire et Description du Japon, vol. ii. p. 465. See also the Introduction to Golownin's Recollections of Japan.

‡ Fragmens de Geologie et de Climatologie Asiatiques, par A. de Humboldt, Paris, 1831.

to 30°, and 29° Fahr. It freezes to several lines in thickness, and snow falls that remains on the ground for several days. In lat. 36° the lakes, as those of Suwa on the Sinano, are covered with a bed of ice, which, between 38° and 40°, becomes thick enough for the river to be crossed on foot. In the Island of Tsusima (lat. 34° 12′ N., lon. 126° 55′ E.) rice will not grow; near Matsmaë, in the Island of Jedo, wheat returns but a very poor harvest; and on Cape Soja (lat. 45° 21′ N., lon. 140° 29′ E.) the wild Aïnos, a vigorous race, are obliged to retire into caverns, to preserve themselves from the intolerable rigour of winter. On the other hand, the S.E. and eastern sides, protected from the freezing winds of Asia by high chains of mountains, which traverse these great islands of Kiusiu, Sikok, and Nippon, in a direction parallel to the continent, have a more fertile and more temperate climate. In those parts of the country between lat. 31° and 34°, the palm, the banana, myrtle, and other plants of the torrid zone, are found. In some parts the sugar-cane is successfully cultivated, and they gather two rice harvests each year. The environs of Sendai, a city in lat. 38° 16′ N., and lon. 138° 36′ E., produce this grain in such abundance, that, notwithstanding their northern position, they are in reality, as they are called, the granaries of Jedo, the most populous city of the universe. But it is more particularly in the rigorous season, which lasts from the commencement of January to the end of February, that this difference between the western and eastern shores of Japan becomes most remarkable. At Dezima, for example, in lat. 32° 45′ N., lon. 127° 31′ E., the thermometer marks 45° Fahr.; while at Jedo, in lat. 35° 41′, lon. 137° 22′ E., it rises to 56°; so that the position of the capital, more easterly by 9° 51′ than the factory, raises its temperature higher by 11°, although it is only 3° nearer the pole. Thus in the two months of winter in which these observations were made, the coasts facing the Asiatic continent were exposed for thirty-seven days consecutively to the freezing winds from N.W. and North. This circumstance explains, besides, why the white mountain (*Siro jama*), which is on the western coast of Nippon, in lat. 36°, is covered with perpetual snow at 8,200 feet above the sea; and why *Fusi jama*, at the eastern extremity of the island, with its summit at 12,450 feet, remains without snow for months together.

"During the hot weather in July and August, when the winds blow from South and S.E., this disproportion in the temperature disappears, and the mean height of the thermometer for this season is 79° at Dezima, and 76° at Jedo. On the South and S.E. coasts, then refreshed by these winds, it hardly exceeds 85°; nevertheless in the South and S.W. parts of Kiusiu, and chiefly in the bays sheltered from the breezes, it often rises to 90° and 98°, and sometimes even to 100°."*

As the shores of this extensive empire to be in any degree completely elucidated would require a separate volume, we shall confine our notices to the most brief details.

The empire of Japan is composed of three great principal islands, *Nippon* (Nipon, Niphon), *Kiusiu*, and *Sikok* (Sikokf). These are surrounded by a prodigious number of others; those of Kiusiu and Sikok so much so, that the approach to them becomes very dangerous. The shores of the principal islands

* Siebold, vol. i. pp. 230—232.

are also much intersected and irregular; from this reason Kæmpfer compared the Japanese Islands to Great Britain. To these may be added their dependencies, the Island of Jesso and the southern portion of the Peninsula of Saghalin. Of the latter we have spoken in the last chapter. We commence, therefore, with the northernmost, proceeding southwards.

ISLAND OF JESSO.

This island, in its time, has been the object of much geographical discussion and criticism. The Dutch commander *De Vries* was the first to give a distinct notion of its existence and general character to the world. He sailed along the eastern side of it; but his voyage was not sufficient to afford any accurate ideas as to its position. More lately, the detailed researches made in the years 1787 and 1797, by La Pérouse and Capt. Broughton, and then the voyage of the *Nadiéjeda*, by Capt. Krusenstern, in 1805, have cleared up all doubts on the subject, so that, with the exception of the North coast, its geography is now known.

Malté Brun, in his Précis de Géographie, tome iii., mentions two Japanese works on Jesso, brought to Europe by Titsingh. Each of these bears the title of *Jesso-ki*, or a description of Jesso, one of them written in 1720, and the other in 1752. From these works it appears that the Japanese are acquainted with this island under the name of Jesso, that is, *the coast*, and they called the inhabitants *Mosin*, signifying " bodies covered with beard." The Mosins, it is stated, had already made some conquests in Japan, and had seized on the northern coast as far as the mountain Ojama; but they were soon driven back to their island, and not being able to defend themselves against the Japanese, they were subjugated by them, so that they could only preserve their independence in the southern part of Saghalin. Capt. Golownin, who had, during his stay in Jesso, several occasions to communicate with the literary Japanese, whose learning and acquirements he highly praises, has given some very curious particulars relating to this island and its inhabitants; but which differ much from the authorities quoted by Malté Brun.

Capt. Golownin says, that 400 years ago a Japanese prince purchased of the natives of Jesso a portion of the S.W. coast of that island, and which was known at the time by the name of the Japanese land. The prince having in addition to his title that of sovereign of Matsoumay, transferred this denomination to his new dominion, and soon the whole island acquired this name, *Matsoumay*. Fish being of the chief articles of consumption among the Japanese, it is probable that the great quantities of this article procurable on the shores of Jesso induced the Japanese government to form several establishments, which, up to the present time, have extended up to Saghalin and the southern part of the Kuriles.

At the time of the first Japanese establishment in the island, the inhabitants, according to Golownin, called themselves *Einso*, from which word the names *Jesso*, *Aïno*, and *Insu* are derived. Broughton is the only writer who has made use of the latter term, the nearest approach to the original name, Einso, which is preserved in different parts of the island, and particularly in the environs of Volcano Bay, which was visited by Broughton, who remained there some time. The name of Jesso has been adopted by most recent authors, because it is that by which it is most generally known by Europeans, although the name Aïno is, beyond doubt, that given to the original inhabitants of the island. The Japanese themselves also make use of it; thus they call the portions of Jesso on which they have no establishment, by the name *Ainokfuki*, the land of the Aïno, probably the interior of the island.

Our present knowledge of the island is drawn from the notices of its original discovery in the Dutch ships *Castricum* and *Breskes*, under Capt. De Vries, in 1643, from that of Laxman in 1792, from Capt. Broughton in 1796, and from Capt. Rikord, of the Russian navy, in the corvette *Diana*. The northern coast has not been visited, but is copied from the Japanese representations. All the western coast, with the gulfs, bays, and islands in its neighbourhood, are exhibited from the observations of Capt. Krusenstern in 1805.

The Island of Jesso is of a triangular form, the sides of which are 100, 85, and 73 leagues in length. The three extremities of this triangle are Cape Soya, the North point of the island, in lat. 45° 31′ 15″ N., and lon. 141° 51′ E.; Cape Nadiéjeda, the South point, in lat. 41° 25′ 10″ N., and lon. 140° 9′ 30″ E.; and Cape Broughton, its eastern extreme, in lat. 43° 38′ 30″ N., and lon. 146° 7′ 30″ E.

Jesso is separated from the Kurile Islands on the S.E. by a strait, named by Krusenstern the *Strait of Jesso*, which is about 8 miles broad in the narrowest part, that is, opposite the S.W. point of the Island of Kounashire, according to Capt. Golownin's chart. This part of the coast has not been yet explored, and is not likely to be for some time. All that is known of it is that it forms a deep bay, and that the extremity of Kounashire advances very far into this bay, so that the eastern cape of Jesso entirely hides the strait which separates the two islands, from which cause those navigators who have sailed along here have taken Kounashire and Jesso to be but one island, an error which had been followed by nearly all geographers.

CAPE BROUGHTON,* the eastern cape of Jesso, according to the observations of the commander whose name it now bears, lies S. 65° W. 10 leagues distant from the Island of Tschikotan; from this, its proper position will be lat. 43° 38½′N., and lon. 146° 7⅓′ E.

The eastern point of the island, according to a minute chart of the S.E. part of Jesso by Capt. Laxman, forms the extremity of a tongue of land of 10 miles in length, surrounded by rocks and islets.

PORT NEMORO is on the inner side of the tongue of land, at 6 miles to the W.S.W. from its extreme. Laxman anchored here in the *Elizabeth*, and found a Japanese establishment. The islands named on his chart, *Sinshi*, *Moshmari*, and *Mashiri*, which are connected to the eastern point by a line of rocks, lie 2 miles directly to the East of that point, and are 4 miles in extent from North to South. Two other islands, *Fakarero* and *Erou*, lie 3 miles farther East. These

* Admiral Krusenstern says:—"In honour of the English navigator who has, with a slight exception, made the circuit of the coasts of Jesso, and first determined the geographical position of its eastern extremity, I have named it Cape Broughton."—(*Memoir*, vol. ii. p. 205.)

are probably the same islands that Capt. Broughton saw, October 5th, 1796, and which he says are barren, and surrounded by rocks.

The strait separating Jesso from Tschikotan is 10 leagues in length, but it is much contracted by rocks and breakers, which occupy an extent of 20 miles, according to Golownin's chart, so that, to pass through this strait, the only safe channel is between Walvis and Tschikotan Islands; both Golownin and Broughton passed through it. It may be stated, however, that some charts give a different extent to this strait.

CAPE SPANBERG, a name also applied by Admiral Krusenstern in honour of the first Russian navigator who visited these parts, lies in lat. 44° 35′, and lon. 145° 0′. This position is assumed only from the representations given in the Japanese charts before alluded to, and therefore is but a very wide approximation, but, for want of better data, it may be taken as near the truth. Had it not been for the unfortunate adventure that befel Capt. Golownin at Kounashire, we should have been made as well acquainted with this coast as we are now with that on the West. It will be remembered that this officer was charged with the examination of all the northern side as far as the Strait of La Pérouse.

Cape Broughton, and another cape lying 13 miles N. 60° W. from it, form the two extremities of a bay which Krusenstern has named *Laxman Bay*, in which is found *Port Fureck*, 7 miles to the S.W. of Nemoro. Near the N.W. point of this bay there is also another port, named *Notsky*, the position of which is somewhere about lat. 43° 45′ N., and lon. 145° 52′ E.

Laxman's chart shows another harbour, *Port Atkis*, where he anchored during his passage from Nemoro to the Strait of Sangar. Its position, assumed from that of Cape Broughton, from which it lies S.W. ¼ W. 38 miles, according to the chart, will be in lat. 43° 20′ N., and lon. 145° 30′ E. The name *Atkis* so strongly resembles that of the Dutch *Acqueis*, that it leads to the conviction that they indicate the same place, the more so that the description given by the Dutch is very conformable to Laxman's Atkis Bay. According to the Dutch, Acqueis lies in the bottom of a large bay, which is 8 miles deep and 2 in breadth. On the other chart it is shown as 6 miles deep and 1½ broad. The latitudes given, however, of these two ports differ nearly 2°, the Dutch placing it in 45° 10′; but, as they do not say in their description of Jesso whether it is on the eastern or the northern side on which this port lies, and as the ship *Castricum* ranged along the eastern coast, Admiral Krusenstern considers that 45° 10′ is a typographical error for 43° 10′:* Burney thought that Port Acqueis would be found to the W.S.W. of Cape Aniwa.† It is probable that the Dutch discovered this harbour in returning from the North, and immediately before arriving at the Bay of Good Hope.

It has been before mentioned that the bay named by Capt. Vries *Walvis Boght*, is the space between Spanberg Island and the eastern point of Jesso, and that his Walvis Islands are those lying on this strait, at 10 miles W.S.W. from Spanberg Islands. At the distance of 40 miles from this Walvis Boght, in a S.W. direction, a bay is marked on Jansen's chart, and in the opening an island, with the name Island of the Three Kings, and to the North of it several others. This

* Krusenstern, Memoir, &c., vol. ii. pp. 206-7.
† Burney, Chronological History of Voyages, &c., vol. iii. p. 159.

bay cannot be the same as Atkis Bay. Somewhere on this coast, too, is a mountain, named *Tamary Peak* on the Dutch charts.

Between Atkis and Cape Broughton, Laxman places several islands near the land, one of which is before the entrance to Port Atkis, and two others joined by a reef called *Rikimushiri* and *Chigab*, at 7 miles N.E. of this entrance. The two *Eroro Islands* are 12 miles farther in the same direction. At 4 miles S.W. from Cape Broughton are two others, named *Imoshiri* and *Somoshiri*.

The BAY of GOOD HOPE, according to Jansen's chart, lies to the S.W. of Atkis. It is deep, but its mouth is narrowed by breakers, extending off both points of the entrance. De Vries entered it, and anchored to the South of a small island near its bottom. The soundings throughout vary from 5 to 16 and 17 fathoms. A projecting point, called *Cape Swars*, is on the right side of the bay, and forms the entrance of the inner bay. The eastern point of the bay is called *Cape Matsuyker*, in lat. 43° 0′ N. Capt. Broughton passed it without entering it; but he saw a hill in it which he named *Peaked Hill*, placing it in lat. 43° 0′ N., lon. 144° 12′ E.

CAPE EROEN or EVOSN, the S.E. extremity of Jesso, is very imperfectly known and placed on the charts. Capt. Rikord's position is perhaps the best, lat. 41° 59′ N., and lon. 142° 55′ E.

VOLCANO BAY is to the West of Cape Eroen. It was visited and named by Broughton in September, 1796. He thus speaks of it:—"I have seen few lands that bear a finer aspect than the northern side of Volcano Bay. It presents an agreeable diversity of rising grounds, and a most pleasing variety of deciduous trees shedding at this time their summer foliage.

"The entrance into this extensive bay is formed by the land making the harbour, which the natives call Endermo, and the South point, which they call *Esarmi*. They bear from each other N. 17° W. and S. 17° E. 11 leagues. There are no less than three volcanoes in the bay, which induced me to call it by that name. There are 50 fathoms of water in the centre, and the soundings decrease on the approach to either shore. During our stay at the period of the equinoxes, we experienced generally very fine weather, with gentle land and sea winds from the N.E. and S.E., and no swell to prevent a ship riding in safety, even in the bay, and the harbour of Endermo is perfectly sheltered from all bad weather.

"Endermo Harbour, as before said, affords good shelter from all winds, bringing the bluff on the extreme part of the isthmus, which forms the starboard point in coming in to bear N.W. In this situation we found 4 or 5 fathoms; and the larboard entry point on the North shore was on with the bluff. In running for the harbour, the island must be kept open with the starboard entry point till within half a mile of a small islet (which is only so at half tide), and then you must steer in to the S.W., when the water will be shoaled, and any berth taken you may prefer. The soundings gradually decrease from 10 to 2 fathoms, soft bottom. A few houses were scattered on the South side of the harbour; and towards the head the shores are low and flat, so much so as to prevent boats landing within 100 yards. In all other parts wood and water are procured with the utmost convenience. The small island was named *Hans*

Olason Island, from one of Broughton's seamen who was buried there. The harbour is formed by the apparent island, which is an extensive peninsula, of a circular figure. Lat. of the entrance, 42° 19′ 29″ N., lon. 141° 7′ 36″ E. High water, full and change, 5h 30′; rise and fall, 6 feet."*

The astronomer made the following observations on shore, opposite the anchorage:—

Mean of observed latitudes, 42° 33′ 11″ N., longitudes deduced from magnetic observations, 140° 50′ 32″ E. Variation on shore by three compasses, 0° 16′ 30″ W.; ditto, on board by all the compasses, 1° 27′ 20″ E. High water, full and change, 4h 30′; rise and fall between 4 and 5 feet. We experienced no tide at anchor.

The STRAIT of SANGAR separates Jesso from Nippon. Cape Esarmi forms its eastern entrance. Its breadth is 9 leagues. Laxman was the first European who entered it, but Broughton was the first who passed through it. On the North side of the strait is *Khakodade Bay*, or, as Broughton writes it, Agodaddy Bay, visited in 1792 by Laxman, and in 1813 by Capt. Rikord. It is of considerable extent, and is formed to the South by a peninsula 3½ miles in length, on which is built the town of *Khakodade*, which, next to Matsoumay, is the largest town on the island. The latitude of this point is about 41° 43′ 30″, and its lon. 141° 58′.

CAPE NADIEJEDA, the S.W. point of Jesso, was observed with great precision by Capt. Krusenstern to be in lat. 41° 25′ 10″ N., lon. 140° 9′ 30″ E. It therefore forms the N.W. point of the entrance to the Strait of Sangar; the opposite point, on Nippon, being Cape Sangar, which is in lat. 41° 16′ 30″, and lon. 140° 14′ E., so that the strait is here 9 miles broad.

The two islands, *O-sima* and *Ko-sima*, are only black, rocky mountains, of volcanic origin. In coming from the North, there would not be a better guide for entering the strait than by bringing Ko-sima exactly before its entrance. O-sima is of a round form, in lat. 41° 31′ 30″, and lon. 139° 19′ 15″, 6 miles in circumference. The other is long, and 10 miles in circuit, lying in lat. 41° 21′ 30″, and lon. 139° 46′. A high rock lies some distance to the North of the latter island. The passage between the two islands is 10 miles broad, and perfectly safe.

MATSOUMAY (or Matsmaï), the capital of the Island of Jesso, lies in a bay of the same name to the N.W. of Cape Nadiéjeda. The two capes forming this bay lie N. 70° W. and S. 70° E. from each other 4 leagues apart. The northernmost, *Cape Matsoumay* is in lat. 41° 30′, and lon. 139° 57′. The city of Matsoumay is directly to the East of this cape, 7 miles distant, that is, in lat. 41° 30′, and lon. 140° 4′. According to Broughton, a small island, apparently joined to the land by reefs, lies near its N.W. point. On it he perceived a small building, perhaps a guardhouse.

CAPE SINEKO is to the N.W. of Cape Matsoumay, in lat. 41° 39′ 30″, and lon. 139° 54′ 15″, according to Capt. Krusenstern's observations in 1805. At 40 miles N. 8° W. from it is *Cape Oote Nizavou* of the Japanese charts; this also was determined by Capt. Krusenstern; lat. 42° 18′ 10″, lon. 139° 46′. To the S.E. of it is the *Island of Okosir*, distant 12 miles from the coast. It is 11 miles long, in a N.N.E. ¾ E. and S.S.W. ¾ W. direction.

* Broughton's Voyage, pp. 102—104.

Cape Koutousoff is in lat. 42° 38′, and lon. 139° 46′. It lies to the northward of Cape Oote Nizavou, and is remarkable for a high mountain, the position of which is lat. 42° 38′ N. and 140° 1′ E. Between these two capes is *Koutousoff Bay*.

Sukhtelen Bay is formed to the South by Cape Koutousoff, and to the North by *Cape Novosilzov*, in lat. 43° 14′ 30″, lon. 140° 25′ 30″. It projects into the sea more than 20 miles in a North and South direction. The bay is only separated from Volcano Bay on the South side by an isthmus 20 miles in breadth.

Strogonoff Bay lies to the northward of Cape Novosilzov. It was thus named by Capt. Krusenstern in 1805, and is 12 leagues in depth from N.W. to S.E., by 14 leagues from Cape Novosilzov to *Cape Malespina*, in lat. 43° 42′ 51″, and lon. 141° 18′ 30″.

To the N.E. of this latter cape is *Mount Peak* or *Pallas*, in lat. 44° 0′ N., and lon. 141° 54′ E. This will point out the position of a bay with low shores, which lies between Capes Malespina and *Schischkoff*. This is lat. 44° 20′, and lon. 141° 37′.

Off this latter cape are two small islets, *Teurire* and *Yanikessery*, lying 12 miles N.W. of the cape.

CAPE ROMANZOFF is the N.W. extremity of Jesso; it was thus named by Krusenstern, and is placed by him in lat. 45° 25′ 50″, and lon. 141° 34′ 20″. A narrow and low tongue of land extends nearly a mile to the N.W. from this point. There is a large bay between Cape Romanzoff and another lying N. 62° E. 14 miles distant, called *Soya* by the inabitants. The *Na diejeda*, Capt. Krusenstern's ship, anchored in this bay, which also received the name of Romanzoff, at the entrance of a small bay in the southern part of the greater bay, at 2 miles from the nearest shore, in 9 fathoms, an excellent bottom of fine sand and mud.

Near to Cape Romanzoff are two islands, *Refunshery* and *Rioshery*. The first is the Cape Guibert, and the second the Pic de Langle, of La Pérouse, who thought that both formed part of Jesso. The Pic de Langle, according to Capt. Krusenstern's observations, is in lat. 45° 11′ N., lon. 141° 12′ 15″ E., and is probably the mountain which the Dutch called Blyde Berg. Cape Guibert, that is, the N.E. point of Refunshery, is in lat. 45° 27′ 45″ N., lon. 141° 4′ E. It is high in the centre, and extends 12 miles in a N. by E. and S. by W. direction. The Strait of La Pérouse, which separates Jesso from Saghalin, has been before noticed.*

The large island just described is usually considered as an appendage to Japan rather than forming an integral portion of the empire. This consists of the large Island of Nippon, of Kiusiu, and Sikok, as before mentioned. For the representation of a large portion of these we owe our knowledge to the Japanese charts, which, however, do not come up to the requirement of Europeans. Their features in general are closely represented, but are naturally distorted. The coasts will be slightly alluded to in the subsequent pages, in consequence of the paucity and uncertainty of our information.

The shores of these islands offer numerous harbours, of which the best for large

* The chief portion of this has been taken from Krusenstern's Memoirs.

ships are, Nagasaki, Simonoséki, Fiôgo, Jedo, Isinomaki, and Awomori. The two last are on the North of the Island of Nippon. The seaport most frequented is that of Ohosáka, although its shallowness will only allow small vessels to enter. All these towns have offices (*tofi ja*) where business is transacted, and where the anchorage dues and other imposts are received, and also where the captains procure their papers and cargoes. In the large commercial cities there are customs' officers and maritime intendants, who severely inspect the entry and export of merchandise.—(Siebold, vol. i. p. 219.)

ISLAND OF NIPPON.

The Japanese do nôt confine this name of Nippon to this island alone, but, in general, extend to the whole of the Japanese empire the title Nippon (Nipon, Niphon, or Nifon), signifying the foundation or origin of the sun.* It is more than 700 miles in length in a N.E. and S.W. direction; and its breadth varies from 50 to 150 miles. The country is high, and covered with mountains, some of which, known on the charts as *King*, *Fusi*, *Tilesius*, &c., are of considerable height.

The STRAIT of TSUGAR, or SANGAR, separates its North extremity from Jesso. Its southern side is but little known, and that little is from the relations of Capt. Broughton. In the North extremity of Nippon is an extensive bay, 7 miles wide at the mouth, and 35 miles from East to West, when within. This bay forms a peninsula to the East, the northern point of which, *Toriwi-saki*, is in lat. 41° 33′ N., lon. 141° 18′ E.; according to Capt. Broughton it runs out into a low point.

The eastern entrance of the Strait of Sangar is 9 leagues broad between Cape Esarne, or *Jesan*, or Jesso, and *Sirija-saki*, the N.E. point of Nippon. This point is in lat. 41° 25′ N., lon. 141° 45′ E. Off this N.E. point there are several islands marked, the largest of which is called *Dodo-sima*; but by Siebold's chart they do not reach far off.

The particulars of the East coast, proceeding southward, would appear to be a series of surmises, inasmuch as the observations of Broughton, King, and others, do not coincide with each other, nor at all with the Japanese charts.

The first point which, in the *present* state of Japan, would interest Europeans, is the capital city, Jedo.

JEDO lies in lat. 35° 41′ N., lon. 139° 44′ 30″ E. It lies at the bottom of an extensive bay, which is, according to Kæmpfer and Thunberg, full of mud at the head, so that vessels of light draught cannot reach the city, but discharge their freights a league or two below it. Several rivers discharge themselves into the bay. The entrance to it lies on the eastern part of a bay, named on Krusenstern's chart *Odawara Bay* (*Wodawara*, Siebold). This bay is formed, to the East, by a promontory on either side. That to the East is named *Awa*, or Ava, the S.E. point of which is *Cape King*, or *Firutatsi*, in lat. 54° 45′ N., lon. 139° 48′ E. The Bay of Jedo consists of two bays, separated by a tongue of land, leaving a narrower channel.

* Kæmpfer, vol. i. p. 93, French edition.

Off the Bay of Odawara is a group of islands, seen by De Vries and by Capt. Broughton.

The northernmost of these is OHO-SIMA, or OSIMA, the centre of which is in lat. 34° 43′ N., lon. 139° 23′ E. On one of the charts extant a volcano is marked on it, and was seen by Broughton July 31, 1797. To distinguish this from the Volcano Island, to the southward, Krusenstern applied the name of *Vries's Island* to it. On Siebold's chart it is called *Barneveld's Island.* Smoke was seen to issue from the western part of the summit of the high mountain in the centre.

The BROKEN ISLANDS lie to the southward of Oho-sima, extending 15 miles North and South. Capt. Broughton passed between them, and he says:—"We found the passage 5 or 6 leagues wide betwixt these islands, and no dangers. The North point of Broken Island is rather high, with perpendicular whitish cliffs. Off the S.W. part is a large detached rock, with several small ones about it. Its greatest extent in a N.E. and S.W. direction is 4 or 5 miles. To the N.E. of it are two more islands. The first is low and flat, but the northernmost is more extended, of moderate height, and connected in parts by low land, which makes it appear at a distance like separate islands, with a conspicuous white mark on the southern one, and a rock lying off it to the westward. Directly North of this island, at 4 or 5 miles' distance, are the two hummocks before mentioned; the northern one is the largest. In the afternoon we had a fine view of the famous *Mount Fusi* towering above the high land, and covered with snow." *

From the nature of the charts, we cannot identify these with the Japanese authorities, but on Siebold's chart two of them are named *Tosi-sima*, or *Nü-sima*, and *Nikine-sima.*

VOLCANO ISLAND, or *Mitake*, is in lat. 34° 6′, lon. 139° 32′. It was named Brandten Eyland, or Burning Island, by De Vries, who places a group of rocks to the S.W. of it. Broughton says that there are, in addition, some black rocks at the distance of 2 or 3 miles from the East point of the island. According to him Volcano Island is large, well cultivated, and covered with verdure to the summit of a very high mountain which stands on it, and presents a very agreeable prospect.

PRINCE ISLAND (of De Vries) is 9 miles S.S.E. from Volcano Island, according to Siebold. It is probably the same called *Outer Island* by Broughton, but without reason. It is in lat. 34° 8′, lon. 139° 36′. It is named *Mikura*, on the chart (Siebold's), and is there marked as De Vries's Ongelukkig Eyland (Unlucky Island).

FATSISIO, or *Fatsizioo.*—This is a group of several islands, the largest of which is named as above. At each end of it, East and West, are *Awo-sima* and *Ko-sima*, and several smaller islands lie to the North of it. The latitude of the group is 33° 6′ 30″, its longitude 3° 50′ 30″ E. of Miako (or 135° 40′ E., Green).

Ko-sima, according to Broughton, is only a high peaked hill, not more than a league in circuit, and, if it had not been inhabited, would have been thought inaccessible.

Fatsisio, from the same authority, appeared to extend N.W. and S.E. 3 or 4

* Broughton's Voyages, pp. 142-3.

leagues, and presented a very fertile appearance. Siebold marks an anchorage, *Mitsune*, on its North side, and there is also another bay, *Funetsuki*, to the East of the former.

Between Fatsisio and Mikura a current to the eastward is marked, called the *Kuro Siwo Stream*. This is shown on all the Japanese charts. Krusenstern calls it *Kourose-gawa*, that is, *Current of the Black Gulf*, with the following remark:—"This current is 20 matsi (5-9ths of a Japanese *ri*, that is, about three-quarters of a mile) broad. For 10 matsi it has a very rapid course. In winter and spring it is very difficult to navigate, but in summer and autumn vessels can pass it." It is probable that the difficulty of this current has been exaggerated in this note, and that there is only a very rapid current, stronger in winter and spring than in summer and autumn. On one of the charts an island named *Oye* is marked at the West end of this current; but, as it was not seen by Colnett nor Broughton, it probably does not exist.

South Island, or *Onango-sima*, was discovered by the Dutch, but was subsequently seen by Colnett.

MOUNT FUSI stands on the promontory of Idsu, or Izou, before mentioned as being the West side of Odowara Bay. This mountain stands on the southern part of the peninsula, and is the loftiest in Japan. It is estimated to be 10,000 or 12,000 feet high, is always crowned with snow, and was formerly an active and peculiarly dreaded volcano, but for upwards of a century has been dormant.*

On the West side of Idsu is another spacious bay, entering the land nearly as far as that of Odowara. It is called *Tootomi* (*Tohodomi*, Siebold) *Bay* by Krusenstern. Another bay to the West of this, but still more extensive, and advancing farther inland, is called on the charts *Owari Bay*, from the name of the province at its head. It is called *Iseno Umi* or *Lake* by Siebold. This bay is very large, and contains two others, separated by a promontory, which extends in a North and South direction, nearly 8 leagues, the South point of which is called *Moro Saki*.

Cape Sima is the South point of this bay, and is in lat. 34° 20′, and lon. 6° 12′ E. of Nagasaki. Beyond this the coast takes a S.W. direction to the South point of Nippon. The island *Oko-sima* lies off this cape. Capt. Broughton passed it at a short distance, and determined its position to be in lat. 33° 25′, and lon. 135° 47′ E. The cape being low, and the island lying off it, gives it the appearance of being a peninsula. Broughton took it for such, and says, the insulated appearance of the peninsula will always make it known (p. 256).

From this point the coast takes a N.W. direction, and forms, with the S.E. point of Sikok, an extensive bay, at the bottom of which is a channel, the entrance of which is 10 miles broad. It leads between Sikok and Nippon to the Bay of Ohosáka.

OHOSA'KA is one of the busiest seaports of Japan, being, in fact, the port of Miako, one of the chief cities. Ohosáka stands near the mouth of the largest of the Japanese rivers. It rises in the great lake *Buva-no-oumi*, and falls into the Bay of Ohosáka, after a course of 60 or 80 miles. Miako stands between the

* Parker's Journal of a Voyage to Japan.

lake and the bay. On account of its draining a large area, the head of the bay is filled with shoals, so that only small ships come quite up to Ohosáka.

To the West the bay is formed by the island *Avad-sima*, which is nearly 30 miles in length from North to South. From Ohosáka Bay the South coast of Nippon runs rather to the South of West for 4° of longitude to its West extremity. The Island of Sikok lies before a great part of this extent, and, according to the Japanese charts, our only authority, there was an immense number of islands distributed along the space. As there is nothing known of the numerous ports and towns on it but their names as given on these charts, they need not here be noticed.

SIMONOSAKI, the seaport where the Dutch travellers usually cross from Kiusiu to Nippon, in their periodical journeys to Jedo, lies at the S.W. extremity of Jedo, which is here separated by a very narrow channel from Kiusiu; the port on which, opposite to Simonosaki, is Kokura.

From this point the coast rounds to the West, North, and N.E., and is lined with a great number of small islands. Numerous bays and towns *exist* on the coast, but they are unknown to Europeans.

CAPE ITSOUMO, which was seen and named by Capt. Krusenstern April 22, 1805, is in lat. 35° 44′ N., and lon. 132° 34′ E., and is probably the N.W. extremity, or some other point of a peninsula which extends 40 miles in an East and West direction, from the Japanese charts. Admiral Krusenstern found considerable discrepancies between the positions observed and those marked on the charts, so that they cannot be taken as very accurate, though apparently full of detail.

Directly to the South of Itsoumo Peninsula is a lake called *Mitzou-Oumi*, communicating with the sea by a channel, the North entrance of which is at the eastern point of the peninsula.

The OKI ISLANDS lie to the North of Itsoumo, according to the Japanese charts, 50 miles North; but, from the observations of Admiral Krusenstern, he supposes them not to be more than 20 miles. The group is composed of four large islands, and several smaller ones, which extend about 16 leagues in a S.S.W. and N.N.E. direction. In the southern part of the largest is a tolerably large bay, on which stands the chief town of the island, *Yematso*. The North extremity of the large isle is placed by Admiral Krusenstern in 36° 30′, but in the Japanese charts it reaches to 37° N.

From the Peninsula of Itsoumo the coast runs nearly East to a cape named *Tanga* by Krusenstern, in lat. 32° 53′, and lon. 4° 27′ E. of Nagasaki. From this cape the coast running to the South forms a large bay, which is nearly 15 leagues wide in the opening, and 10 deep. This has been named *Wakasa Bay*. From the eastern part of this bay the coast runs in about a N.N.E. direction to *Cape Noto*. This cape was determined by La Pérouse to be in lat. 37° 36′ N., and lon. 137° 20′ E. (corrected). On the Japanese charts it is called *Sousnomissaki*, and is the North extreme of a promontory which extends 50 miles in a North and South direction.

On the Japanese charts the western part of the Noto Promontory is lined with small islands, one of which, at 2 leagues from the coast, W.S.W. from the cape, is named *Boukoura ;* but the island that La Pérouse saw at 5 leagues off from the cape, in lat. 37° 57′, and lon. 136° 56′, is not marked on them; he called it *Jootsima*, and described it as small, flat, wooded, and only 2 leagues in circumference. La Pérouse marks another rocky islet, in lat. 37° 36′, and lon. 136° 50′, West of Cape Noto; it is too distant from the land to be thought identical with Boukoura, so it has been placed on the charts as *Astrolabe Island.*

The extremity of a promontory, seen by Krusenstern, and by him named Russians' Cape, is the third point determined, astronomically, on the North coast of Nippon. It lies 200 miles N.E. from Cape Noto. The intervening coast has not been seen by any European navigator, so that it has been entirely delineated from the Japanese charts. The first point to be noticed is *Cape Yetsiou*, which lies in about lat. 37° 17′, and lon. 6° 22′ E. of Nagasaki. It projects to the northward, and, with the Noto Promontory, forms a deep bay, which receives a great number of rivers, at the mouth of one of which is the town of *Togama*, a name by which Krusenstern has distinguished the bay.

From Cape Yetsiou the coast extends for 10 leagues to a point called *Yatsou-saki*, from whence it begins to trend towards the N.E. by N., and then due North. At 2 miles from the coast in this part is an island called *Awa-sima*, the only one off this extent of coast.

RUSSIANS' CAPE, named *Oga-sima* and *Nankaba* on the Japanese charts, is a large promontory, which projects 35 miles to the S.W. It is very soon recognised by a mountain with a rounded summit, which rises in its centre. Towards the North it joins the land by a narrow tongue, which is shown on the charts as not more than a mile broad, for which reason it was taken by Krusenstern for a long time to be an island, until a nearer approach showed the connexion. The point is in lat. 39° 20′, and lon. 139° 44′ E.; its northern point is in lat. 40° 0′, and its southern in lat. 39° 40′. The whole peninsula is generally mountainous, and consists of a series of jutting joints; its coasts are steep and rocky.

SADO ISLAND lies between Cape Noto and Russians' Cape. No European navigator has seen it. By the charts it is about 15 leagues long and 7 or 8 broad. Near the South point is the town of *Siro*, apparently the capital of the island. The narrowest part of the channel separating the island from the main land is about 10 leagues broad. It is said that gold works exist on Sado: Kæmpfer also speaks of the gold mines found on this island, which he calls Sador.

There are several islands near the coast of Sado; one of the largest of which is called *Tabou-sima*, lying 10 leagues N.E. by E. of Cape Wakissaki, the North point of Sado.

At about 8 leagues to the North of Russians' Cape, Krusenstern saw, near the mouth of a river, a large town; several vessels were at anchor off it, and four boats, well armed, came off, probably to attack the *Nadiéjeda ;* but having observed the ship, they hurried back again. This would appear to be the River *Nosiri-guwa* and the town of *Nosiri*, in lat. 40° 15′ N., lon. 140° 5′ E.

CAPE GAMALLY, in lat. 40° 37′ N., and lon. 139° 49′ E., is named on some of the Japanese charts *Nangasaky*, and *Tilesius Peak*, on the same parallel, *Tsougar Fouzi.* The cape is very remarkable, because here the coast assumes quite a different direction. Beyond, it trends N.E., E.N.E., and E., so that the coast here forms a large bay, of which *Cape Greig*, a very projecting point, is the northern limit. This latter was well determined as in lat. 41° 9′ 16″ N., lon. 140° 8′ E.; and at only 10 miles to the N.E. is Cape Sangar, or Tsugar, the N.W. extreme of Nippon.

ISLAND OF SIKOK.

The name of *Sikok*, as Siebold writes it, or *Sikokf* according to Krusenstern, is not found in the Japanese charts. It is the smallest of the three principal islands of the Japanese group. In a N.E. and S.W. direction it is 140 miles in extent, by 60 in breadth. It lies directly to the East of Kiusiu, from which it is separated by the *Strait of Boungo*, which, throughout its extent, is limited by the western coast of Sikok, and in its narrowest part is not more than 10 miles in breadth. The straits to the North and East, which divide Sikok from Nippon, are scarcely broader; besides this, they are so encumbered with such an infinity of small islands, that the passage is frequently not more than a mile in breadth. Kæmpfer and Thunberg, whose route from Kokura towards Ohosáka (Osacca) led them along these coasts, attest to the existence of these thousands of islands.

The Island *Avadsi-sima*, which has been before mentioned, lies off the N.E. end of Sikok, and before the Bay of Ohosáka; Kæmpfer probably took the North point of it for a separate island. The town of *Smoto* appears to be the only one on it; in lat. 34° 20′, and lon. 4° 26′ E. of Nagasaki.

The N.W. extremity of Sikok is named by Krusenstern *Cape Yemafar;* it forms, at the same time, the western point of a large bay, also called *Yemafar.*

Cape Toubakino-misaky, in lat. 33° 52′, and lon. 4° 8′ E. of Nagasaki, is the easternmost point of Sikok. A channel, 10 miles in breadth, which on the Japanese chart bears the name of *Foukai-Yerioumi*, separates Sikok from Nippon. Krusenstern names it the *Kino Channel*, from the province on its eastern side.

The southern point of Sikok does not bear any name on the Japanese chart; it is named by Krusenstern *Cape Tosa.* Arrowsmith calls it by the Russian name of Oblakoff, or "cloudy." Capt. Broughton saw this cape in 1796 and 1796, and Krusenstern saw it September 30, 1805, but at a great distance. It seemed to be high land; but it might have been a mountain behind the cape. The following is the description made at the time:—"Beyond the southern point the coast turns abruptly to the North, and seemed to form a small bay, the North and West shores of which we could distinguish. From this bay the coast runs to W.N.W., and seemed to form another small bay. The land is mountainous near the cape, and gradually lowers to this bay, and then rises again, so that the bottom of the bay showed us a large valley, bounded on the East by a chain of steep mountains, which renders this part of the coast very easy to be made out." On the Japanese charts a small bay, called *Semitsououra*, is marked at 8 miles to the N.W. of the southern cape; to the North of this a second, and then a third at 8 leagues to the North of Cape Tosa. This is called the *Gulf of Tentsi*, and has in its opening an island called *Oki*. This all accords very well with what was seen from the *Nadiéjeda*. Behind this bay a mountain is marked, named *Mount Sasayama*, which is probably the same as that seen by Krusenstern, and what Broughton called Saddle Point.

From the Gulf of Tentsi to a very projecting point, called *Naka-oura*, the direction of the coast is N. 60° W. for 15 leagues. From this point the coast runs direct to the North, and continues in this direction as far as *Cape Misaky*, which is the westernmost point of Sikok. Cape Misaky is made in lat. 33° 18′ N., and lon. 2° 5′ E. of Nagasaki, and forms the N.W. point of a large bay, which is 4 leagues in the opening and 7 leagues in depth. The distance between Cape Misaky and Cape Yemafar is 24 leagues. The direction of the coast is N.E. Several islands are strewn along it, and on one of them a mountain named *Jono-kosoutsy* is marked.

From the South point of Sikok to its eastern extreme the coast runs in a N.E. direction. It is less intersected than the West and North sides, and on it are nine small islands. The only remarkable point is *Cape Mourodonosaky*, in lat. 33° 8′, and lon. 3° 45′ E. of Nagasaki. From this cape to that of Awa the distance is 18 leagues.

ISLAND OF KIUSIU.

The word Kiusiu, according to Kæmpfer, signifies "country of nine," in reference to its division into nine larger provinces. It is 65 leagues in length from North to South, and 40 leagues in breadth. It is separated from Nippon by a channel, which, in one part, is narrowed to 2 miles in width. The strait which separates it from Sikok also is contracted to 10 miles broad at one point. The Strait of Van Diemen forms its southern boundary, and the Strait of Corea, or Koraï, is on the western side. It contains the only port now open to foreign commerce, Nagasaki, which is on its western side.*

The North coast of Kiusiu is very much intersected by bays and inlets, and fronted by a great number of islands. It has not been examined by any European, so that we are entirely indebted to the Japanese charts for what knowledge we have of it. Near the N.W. point is marked a deep bay, to which Krusenstern applied the name of *Fisen*, from the province it lies in. To the East of this is a second, designated by the name *Foukouaka*, the same as a town lying in the South part of the bay.

The North point of Kiusiu, which is surrounded by several small islands, is named *Cape Kanero Misaky*, and is placed in lat. 35° 5′, and lon. 50° E. of Nagasaki. Off the N.W. point of Kiusiu is the Island *Iki*, or *Yki*, in lat. 33° 48′ N., and lon. 129° 48′ E.

The strait formed by the N.E. part of Kiusiu and the opposite coast of Nippon is 14 miles wide at its entrance; but farther to the East it contracts to not more than a mile. It continues of this breadth for 2 or 3 miles, and then commences to get wider. It is not in the narrowest part of this channel that the passage from

* *Coal* exists in Kiusiu. At Koyanosi Siebold saw a coal fire, which was very acceptable in the winter season, during which his journey to Jedo was made. He visited a coal mine at Wukumoto, and though not allowed to descend the shaft more than halfway, or about sixty steps, he saw enough to satisfy him that the mine was well and judiciously worked. The upper strata which he saw was only a few inches thick, but he was told the lower beds were of many feet, and he says the blocks of coal drawn up confirmed the statement. The coal, being bituminous in its nature, appears to be made into coke for use, as perhaps being more agreeable to people who generally use charcoal.

Kiusiu to Nippon is made; the traverse is made to the S.W. of this channel, when they embark at the town of *Kokura*, in the province of *Bousen*, to pass over to the celebrated Port of *Simonoséki*, in the province of Nangata, at the West extremity of Nippon. Between Kokura and Simonoséki there is an island named by Kæmpfer *Kikiusima*, or *Firosima*, or, on one chart, *Fiki*. Kæmpfer and Siebold passed through Kokura on their road from Nagasaki to Jedo.

Of the eastern coast of Kiusiu no European voyager has visited it, with the exception of those portions seen by Broughton and Krusenstern. To the East of the narrow part of the channel, near Kokura, the sea is called *Swoo-nada*, a name which ought to be extended to the West of that strait. The N.E. point, Kiusiu, Krusenstern places in lat. 33° 8′, and lon. 131° 54′. To the East of it is the small Island *Fima-sima*.

The strait separating the Islands of Kiusiu and Nippon is named the *Strait of Boungo*, from the province on its western side. It is filled with islets and rocks. The narrowest part is between a point named on the chart Cape Boungo on Kiusiu, and Cape Misaky on Sikok, their distance asunder being 10 *miles*. Cape Boungo is the southern point of a bay in which are three large towns. To the South of it the coast is very much broken; and, in lat. 32° 14′, lon. 131° 42′, is *Cape Tschirikoff*, thus named by Krusenstern. Cape Cochrane, from his observations, lies in lat. 31° 51′ N., and lon. 131° 27′ E. In its parallel there is a high mountain, named on the chart *Kirisimasan*. The cape itself is easy to recognise, by a high mountain, of a conical form, behind the point.

To the South of Cape Cochrane another headland, *Cape Danville* (or, on the Japanese charts, Daynomisaky, or Tachy-saky), was seen; it is in lat. 31° 28′ N., and lon. 131° 27′, and, with *Cape Nagaeff*, in lat. 31° 15′, and lon. 131° 11′, it forms the two extremities of a bay, named by Krusenstern after the province it lies in, *Ousoumi*.

The southern point of Kiusiu is *Cape Tchitschagoff*, in lat. 30° 56′ 45″, and lon. 130° 36′ 30″. It forms, with the point on which *Peak Horner* stands, in lat. 31° 9′ 30″, and lon. 130° 36′, the entrance of a very deep bay, running in a N.N.E. direction.

Capt. Krusenstern tells, that they had scarcely weathered Cape Tchitschagoff, "when we perceived a lofty mountain, of a conical form, the base of which was quite at the water's edge; I called it, after our astronomer, Peak Horner. The situation of this remarkable mountain was ascertained, with the greatest precision, by Dr. Horner; and this, with Volcano Island, form two unerring marks of the Straits of Van Diemen. In the N.E. a large bay opened upon us, extending far to the northward, and having, apparently, a passage in that direction, though it probably is bounded there. This bay, of which Cape Tchitschagoff forms the S.E. point, and Peak Horner the N.W., had a very picturesque appearance, a number of small islands lying in irregular shapes on the N.W. side of it, two of which, forming a large bow in appearance, were very remarkable. The whole bay, excepting to the North, is surrounded by high mountains, whose summits were covered with the most beautiful verdure. Peak Horner stands on a point of land, and seemed to rise out of the sea, adding very much to the picturesque appearance of the country.

"I now steered N.W. ½ W. towards a point of land forming, on the other side of the above-mentioned peak, another very beautiful bay, and divided into two parts by the land to the North projecting very far forward. The western bay, where there was a small town, was surrounded by a charming valley, divided into large fields and regular plantations of large trees. A high pointed needle-rock stands at a short distance from the shore, and forms the entrance to this small bay, in which some vessels were lying at anchor. Behind the valley, and far inland, was a mountain, of a regular and unbroken appearance, from the middle of which arose a lofty peak. Our latitude, at noon, was, by observation, 31° 9′ 17″ N., and agreed very nearly with that of our reckoning; a proof that the currents, which are so strong as to render the ship, when there is but little wind, quite ungovernable, proceed from a regular change of ebb and flood. By our observations it is high water in the Straits of Van Diemen at nine o'clock at the new and full moon; the flood setting from the S.W., the ebb from the N.E."

At the head of the bay, between Cape Tschitschagoff and Peak Horner, is the large town Kago-sima, and in its upper part is the Island Sakoura, with a high mountain, Mitake, in lat. 31° 30′, and lon. 130° 42′.

Off the South end of Kiusiu are some islands, two of which are tolerably large. One is named Tanega-sima, and the other Jakŭno-sima (Yakouno-sima, Kr.)

Dr. Siebold says:—"There is still great confusion in the geography of the islands to the S.S.E. and S.W. of Japan. In those which are met with in the track near the South and S.W. coast it may be incidentally remarked, that the Pinnacle Islands, which were seen by Broughton and Colnett, are the same group as Linschoten describes under the name of *De Zeven Zusters* (Dutch), or *As seze Yrmas** (The Seven Sisters), and, on the most recent Japanese charts, that of *Nana-sima*, which has the same meaning. These, and some others lying more to the South and West, Siebold names under the title of *Linschoten Archipelago*."

Tanega-sima is level and covered with trees, which gives it a pleasant appearance. It lies nearly North and South, aud is about 18 miles in length. The North point, according to Krusenstern's observations, is in lat. 30° 42′ 30″, lon. 131° 0′.

Jakŭno-sima, the other island, is hilly, and on it a mountain, Mitake, is marked.

The five small islands in the Strait of Van Diemen, which were seen from the *Nadiéjeda*, and named by Krusenstern, are all marked and named on the Japanese charts. A bare rock, about a mile in diameter, lies 4 leagues West of the North end of Tanega-sima. It was named *Seriphos*, and is called *Make-sima*, or *Oumawo* (horse's tail), on the charts. To the West of it, 6 leagues off, is a tolerably high island, named *Apollos* by Krusenstern, the *Take-sima* of the charts. At 2 leagues from this island, in lat. 30° 43′, and lon. 130° 17′, is an island with a volcano, called *Yewo-sima*, or *Sulphur Island*. An island, 5 leagues to the South of Volcano Island, is not on the Japanese charts. It is called *Julia Island* by Krusenstern, and lies off the N.W. end of Jakŭno-sima. A fifth island, to the West of Volcano Island, to which the old name of *St. Clair* has been preserved in Krusenstern's work, lies, from his observations, in lat. 30° 45′ 15″ N., and

* These last Portuguese words have some resemblance in sound to "*asses' ears*": is there any connexion between them? Such derivations are much more frequent than are generally imagined.

lon. 129° 54′ 15″ E. On the Japanese charts it bears the name of *Kuros-sima*, or Black Island.

The strait to the southward of Jakŭno-sima is called *Colnett Strait* on Siebold's chart; and on its South side two islands are marked: the northern one, in lat. 30° 0′, lon. 130° 0′, is called *Jerabou*; and another, in lat. 29° 30′, which bears the name of *Kikiay*. It appears that there is some doubt as to the existence or position of these islands, though Capt. Broughton and Capt. Colnett, it is supposed, had both some knowledge o their existence. The Pinnacle Island, marked S.W. of Jerabou, as has been before mentioned, does not exist there.

The western point of the S.W. peninsula of Kiusiu is named *Cape Tchesmé*, or, on the Japanese charts, *Nomano-misaky*, in lat. 31° 24′ N., and lon. 130° 2′ E. With *Cape Kagul*, or *Fasimisaky*, it forms the opening of a bay named Satzouma Bay.

The channel between Kiusiu and the islands to the South of it is called the Strait of Van Diemen. Capt. Krusenstern passed through it in the *Nadiéjeda* in very favourable weather, so that he observed the relative positions of many of its more important points.

The KOSIKI GROUP lie off this part of Kiusiu. They are called the Meac-sima Islands in Krusenstern's work. They lie on the meridian of 129° 42′ E., lat. 31° 40′, and are surrounded by rocks on all sides. At 2 leagues to the S.W. from the South Island are two islets, that Krusenstern named the *Symplegades*, the *Tsukurase* of Siebold's chart. To the West of these is another group of rocks, black and pointed, lying N. 39° W. 7 miles from the South point of the South Island. Krusenstern estimated the breadth of the channel separating the Kosiki group from Kiusiu to be 16 miles.

To the N.E. of the Kosiki Islands lies the Island *Naka-sima*, which is close to the Kiusiu, and was supposed by Krusenstern, when he passed it, to be a portion of it. It is the southernmost of a chain which fills all the bay on the West side of Kiusiu to the southward of the *Peninsula of Simabara*. It is upon this peninsula, which is connected to the island by a low isthmus to the N.W., that the famous volcano Wunzendake stands.

WUNZENDAKE (" peak of the hot-water springs ") forms one of the line of volcanic spiracles, the outlets of that immense subterranean river of fire which, from the Molucca and Philippine Islands, extends through Liukiu and the Japanese Archipelago, along the Kurile islands, through Kamtschatka, and expires amid the eternal ice of the North. Wunzendake is 4,110 feet in height, and occupies nearly the centre of the Peninsula of Sima-bara, which forms the district of Takaku, the East part of the province of Fizen. A very low isthmus attaches this peninsula, which is also low, and lies between lats. 32° 33′ and 32° 51′ N., and lons. 127° 52′ and 128° 10′, being there 2¼ German miles in length, by 1½ miles broad. Beyond the isthmus the land gradually increases in elevation, and is commanded by several heights, in the middle of which Wunzendake rises in the form of a truncated pyramid. There was a terrible eruption from it in 1792, since which it has been the terror of the country around it; and it is still in action, threatening a new catastrophe.

Within the peninsula is the extensive Gulf of Sima-bara, running to the N.W.

This is formed to the westward by an irregular peninsula, constituting the *Province of Fizen*, and it is at the S.W. part of it that the famous city of Nagasaki stands.

The BAY OF KIUSIU is formed to the eastward by the foregoing Peninsula of Fizen, to the westward by the Goto group; and to the northward is the once more important Island of Fira-to or Firando.

TSUS-SIMA, a group of islands, lies to the northward of Fira-to, midway between that island and the S.E. extremity of the Korean Peninsula.

The positions of the points of this isle or isles have been determined by Krusenstern; from his description it appears that, when the East end bore exactly West, a smaller island (Colnett's) bore due East. The latitude, at noon, was 34° 35′ 55″ N., and the longitude, by three chronometers, which agreed with each other, on an average, within 30″, was 129° 43′ 15″ E. The northern extremity of Tsus, at that time, bore W. by N.; and a high, flat mountain, not far from this point, S. 85° W.

"Tsus lies in an almost North and South direction, in which its greatest length is 35 miles. We could form no accurate idea of its width, but I conceive it to be not less than 10 or 12 miles, and perhaps more; for we saw high mountains at some distance inland. From the southern end the island takes an almost N.E. direction as far as a point of land that runs very much out to the eastward, and, behind this, it appears to divide into two; at least, here the shore forms a deep bay, in the background of which the land probably closes again. From this cape (Fida-Buengono) the island takes rather a westerly direction. The North point of Tsus, according to Dr. Horner's observations, is situated in lat. 34° 40′ 30″ N., and lon. 129° 29′ 30″; and the above-mentioned flat mountain, not far from this point, in lat. 34° 32′ N. The northern and eastern parts of the island are represented as much more hilly than the southern; yet even here some of a tolerable height were seen, with white spots upon them, which were probably chalk cliffs. The whole island consists of a chain of pretty high hills, divided by deep valleys.

Colnett's Island is a great naked rock, of a circular form, in lat. 34° 16′ 30″ N., and lon. 129° 56′ E., 6 or 7 miles in circumference, and situated about 23 miles to the eastward of Cape Fida-Buengono, on Tsus-sima."—See, further, Krusenstern's Voyage (English), vol. ii. p. 13.

At the S.E. point of Koraï, or Korea, is the harbour described by Broughton, *Tchosan*, or *Chosan*, which would appear to afford good shelter.*

The GOTO ISLANDS lie to the S.S.W. of the Tsus-sima Islands, and form

* This harbour was visited by Capt. Broughton in October, 1797, by whom a particular plan of it has been given. Capt. Broughton says, "It is situated on the S.E. part of the coast of Corea, in lat. 35° 2′ N., and lon. 120° 7′ E., and bears N.N.W. from the North part of the Island Tzima (Tsus-sima), at 10 leagues' distance. It has a safe entrance, and no dangers to be apprehended on either shore. Two miles to the West of the black rocks on the North side of the entrance is an abrupt high headland, which I named Magnetic Head, from its affecting our compass-needles. North of this head is a fine sandy bay, with good anchorage; where we remained during our stay, having the sea open for two points of the compass, in which angle we saw distinctly the Island Tzima (Tsus-sima)."

Capt. Broughton proceeds in noticing the little opportunity they had of making any remarks upon the customs and manners of the people, from their avoiding all intercourse. Wood and water were obtained from them, but not without difficulty; and these are not to be expected, unless by an armed ship, possessing the means of enforcing its demands.

Some useful remarks, on passing from the N.E. end of Tsus-sima to this harbour, are given in the West side of the Bay of Kiusiu, West of Nagasaki. The name is said to signify the five islands. They in reality consist of that number, but there are a great many smaller ones around them. That to the S.W. is the largest, and has a town marked on it as *Fukajé*; the other islands bear the names of *Fisago-sima*, *Naru-sima*, *Nisi-sima*, *Figasi-sima*. The northernmost is Uki-sima, and this is only 10 or 12 miles from Fira-to. CAPE GOTO, or *Ohoseno saki*, is in lat. 32° 34′, lon. 128° 44′. Krusenstern saw near the cape a large rock, which appeared to be divided into three parts.

CAPE NOMO, in lat. 32° 35′ N., and lon. 129° 43′ E., forms the S.E. point of the Bay of Kiusiu.

Cape Nomo is a point of great importance in making the Bay of Nagasaki. This cape, *Nomosaki*, forms, with a tongue of land which faces it on the North, and which bears the name of Ohosáki, the bay, full of shoals, rocks, and islands, at the eastern extremity of which stands the city of Nagasaki.

Nomosaki, in lat. 32° 35′ N., lon. 127° 23′ E., is 1,600 feet in height, on a narrow and mountainous tongue of land, extending toward the S.W., and recognisable by its steep, rocky walls, its rounded summit, and a deep cavity in the form of a saddle, by which it is attached to the hills of the cape, of which it forms the point. The small Island of Kawasima, near this promontory, and to the N.W. the rocks called by the Japanese *Mitsu-se*, by the Dutch, *De Hen met de Kuikens* (the hen with the chickens), point it out in a still more certain manner. Among the numerous mountains and hills on this tongue of land, the highest summit is that of Kawara-jama (mountain of tiles), which rises to 632 metres above the sea.

The shores of the western part of the Islands Kiusiu form, with the neighbouring heights of the Goto Group, a semicircle, which embraces all the horizon from West and North to East.

MESIMA ISLANDS, LINSCHOTEN ARCHIPELAGO, or the ASSES' EARS. These islands were first described by Jan Huigen Van Linschoten. His description is :—" A high land, steep, but not large; on the highest part are two paps. On a nearer approach you will see another more oblong, flat, and even in height; then several small rocks, among which are two larger, which resemble organ pipes."*

In later times these two rocks received the name of the *Asses' Ears*, from the English navigators. They will serve to point out the entrance to Nagasaki.

The group consists of four islands; the largest, which is to the North, is *Taka*, the southernmost is *Kusakaki*, and the two smaller ones are called *Osima* (Men's Island), and *Mesima* (Women's Island). The name Meaksima may be derived from the last.

Capt. Broughton's Voyage, pp. 326—329. " Many villages appeared scattered about the harbour; and in the N.W. part was seen a large town, encircled with stone walls, and battlements upon them. Several junks were lying in a basin near it, protected by a pier. Another mole, or basin, appeared to the S.W. of the other, near some white houses, of a superior construction, enclosed by a thick wood. The villages seemed to abound with people; and the harbour was full of boats, sailing about on their different avocations."—*Broughton*, p. 331.

* Linschoten, Reysgeschrift (Journal du Voyage, p. 80).

Herklots and Broughton say they saw five islands; Admiral Krusenstern has taken the Kosiki group for that of Mesima group, an error which has given rise to many mistakes.*

FIRA-TO, the first European factory, lies on the North of the Bay of Kiusiu. It is in lat. 33° 21′, lon. 121° 31′ E., near the North end of an island lying in a N.E. and S.W. direction.

The first factory (Fira-to, Firando, or Fyrando, as it is variously spelt) founded by the Dutch in Japan was in the Island of Fira-to; this was in 1609-11. Fira-to is an oblong island to the N.E. of Kiusiu, that is, at the head of the great Bay of Kiusiu. This was destroyed and abandoned November 9, 1640. Dezima, the island in front or opposite to Nagasaki, had been granted in 1635 and 1636 as a residence to the Portuguese, but they were banished with much ignominy at the end of 1639. The Dutch then established themselves there instead of Fira-to.†

NAGASAKI, as Siebold writes it, or *Nangasaky* of former authors, is, as has been frequently mentioned, the only port of Japan open to foreigners. It stands at the head of a bay, running in a N.E. direction on a peninsula at the western extremity of Kiusiu. From its being the only point that has been frequently visited, we know more of it, and therefore the description and directions are more perfect. For the notices of the city itself and of Dezima, the residence of the Dutch, we have availed ourselves of the work of Dr. Von Siebold; for the nautical directions for the port, the Voyage of Krusenstern has been the authority.‡

The following is Siebold's description of Nagasaki:—

Nagasaki, one of the five imperial cities, situated on the western coast of the Island of Kiusiu, is in lat. 32° 45′ N., and lon. 127° 31′ 30″ E. In 1826 it contained 29,127 inhabitants, independent of the military, the employés of the *Sjögun* and the princes, the priests and monks, which together form an additional amount of about 6,000 souls.

In the town and its dependencies there are 92 streets, 11,457 houses, 62 temples and Buddhist cloisters, a large edifice for religious worship, and five small chapels for the Kami sect. It is the only port in Japan open to foreign vessels, Dutch and Chinese. Besides this, it is the residence of a governor, whose colleague, representing the city of Jedo, relieves every three years; of the superintendent of the domains of the Sjögun, of a commandant, two mayors, a mint or bank belonging to the foreign commerce, a college of interpreters for the Dutch, Chinese, and Corean. There are two government palaces, those of the princes of Fizen and Tsikuzen, who furnish alternately the garrison of the port; the offices of the chargés d'affaires of the princes of Satsuma, of Tsusima, and of some other provinces of Kiusiu; a Dutch factory in the artificial Island of Dezima, and a Chinese factory, named *Tö-zin Jasiki*, forming the southern

* Siebold, vol. i. pp. 88-9. † *Idem.*

‡ The first plan of the bay and town of Nagasaki was given by Kæmpfer from a Japanese chart, which was subsequently corrected at Nagasaki in 1802-21. An excellent chart appeared in London in 1794, from a manuscript of the Dutch East India Company. It is to be regretted that Admiral Krusenstern did not know of these charts at the time of his visit to Japan; the plan which this illustrious navigator has given us would then have been much more perfect.—*Siebold.*

suburb; a prison, a house for aliens, public magazines, an arsenal, a *funa-kuru*, or covered yard to shelter the smaller vessels of war; a botanic garden, a place for executions; then several theatres, a great number of tea-houses, and other public places, frequented by crowds of dancing girls and musicians.

There exists great industry and an animated commerce; a manufacture of porcelain, breweries of *Sáké* and *Soja*, medicine shops, grocery stores, drapers' and silk shops, Chinese magazines, and many trifles. This city, the single point of union between Japan and the rest of the world, has a continual fresh accession of merchants, learned men, and idlers, from all parts of the empire, and its harbour is always filled with national ships.*

DEZIMA, to the South of the city of Nagasaki, near the coast extending to the N.E. of this bay, has the shape of the Japanese *fan*. When it was asked of the Sjôgun what was to be the figure of the small island that was to be raised from the bottom of the sea, it is said that the autocrat gave his fan as the model for the prison for the Portuguese.

This artificial island was made from the earth of a neighbouring hill, which was levelled for this purpose. A wall of basaltic stones protects it from the water, and at high water the island is 6 feet above its level. Its length to the South is 624 Rhenish feet, the North, 516 feet, and its breadth, 216 feet. To the South and West it is bounded by the sea; on the North and East it faces the city of Nagasaki, from which it is separated by a narrow canal, but communicates with it by a stone bridge and a gate (Landpoort), which is constantly guarded. Another gate (Waterpoort), to the West, is open to the commerce of vessels anchored before the city. It is on this narrow space of land that the wooden houses of the Dutch, their magazines, and some other serviceable buildings are constructed. These buildings, crowded together, leave scarcely room enough for a street, which, with an open space for the flagstaff, the botanic garden, and that of the office, is the only promenade of the foreigners shut up and guarded in the island.

The house of the chief has several tolerably large rooms of European construction, which, in 1823, were decorated and furnished in perfect taste, at the expense of the Dutch government. On the open space mentioned the spar rises, surmounted with the Dutch flag, which indicates to the ships of that nation, on their arrival, the solitary residence of their compatriots. These cherished colours regularly announce to the inhabitants of Nagasaki the anniversary of the national festivals. The flagstaff is in lat. 32° 45′ N., and lon. 127° 31′ E.

Among the government establishments which belong to the domain of natural science the botanic garden may be noticed. This was formed at Dezima in 1823 and 1824. In 1829 it already contained nearly a thousand rare plants of the Japanese flora. There is here a monument to the honour of the naturalists Kæmpfer and Thunberg.

DIRECTIONS.—The entrance of the Harbour of Nagasaki lies in lat. 32° 43′ 45″ N., and lon. 230° 15′ 0″ W., in the middle of the Bay of Kiusiu, which is formed by Cape Nomo to the South, and Cape Seurote to the North. From Cape Goto, in lat. 32° 34′ 50″, and lon. 231° 16′, the entrance to the harbour bears

* Siebold, Voyage au Japan, tome i. pp. 210-11.

E. by N. 51 miles. The distance from the easternmost of the Goto Islands is only 33 miles, and perhaps still less, from a chain of small rocky islands, which stretch to the N.E. from the Gotos, and probably join to Cape Seurote, and seem, at this point at least, to render a passage impracticable; and which, according to the report of the Japanese, is only navigable for boats. Having correctly ascertained the entrance, no doubt can exist as to the course to be steered; but should the want of an observation occasion any uncertainty, the mountainous nature of this part of the coast renders Nagasaki very remarkable. The land at Cape Nomo and Cape Seurote is not particularly high; but Nagasaki, on the contrary, is surrounded by very lofty mountains, among which is a chain higher than the rest, at the southern extremity, which lies rather E. by S. of the entrance. It is best to keep as much as possible in the middle between the Goto Islands and Kiusiu, and to steer a N.E. course until the parallel of the entrance, and then due East. In this direction the hill behind Nagasaki soon becomes visible, and is a certain mark, even at a very considerable distance. When within about 9 or 10 miles of the entrance, a large tree is seen on the Island of Iwo-sima, on the South side of it; and this tree, which is visible at a greater distance than 10 miles, being brought to bear S. 85° E., is then in a line with the point of the above-mentioned hill. With these two very particular marks it is impossible to miss the course to be pursued: but if on making the land of Kiusiu you steer for Cape Nomo, as we did, believing the entrance of Nagasaki to be 12 miles more South than we found, and then along the coast, you are not only in danger (either in a calm, or by the tides, which at the time of the full and new moon are very strong) of being driven too near the rocks, but might very easily mistake an entrance in lat. 32° 40′ for the true one, and which, though it really leads to Nagasaki, might prove dangerous, never having been explored.

Cape Nomo, the southern point of the Bay of Nagasaki, lies in lat. 32° 35′ 10″, and lon. 230° 17′ 30″. This promontory consists of a hill, with a split or double summit, and, at a little distance, has the appearance of an island; and when near it is very remarkable by a large rock which lies in its front. Between Cape Nomo and the entrance into the harbour are a number of rocks and small rocky islands, one of which is of considerable height; and others, like the Papenberg in the Bay of Nagasaki, are remarkable from being planted with trees from the base entirely up to the summit. Behind the islands and rocks is a bay, the South side of which is bounded mostly by a flat and very well cultivated country; farther inland it is more mountainous, the hills stretching in a N.W. direction as far as Nagasaki, in large ranges adjoining each other, and planted with avenues and groups of trees. Behind Cape Nomo the coast assumes a S.E. direction; and here there appears a large bay, which in the Japanese charts is called Arima, but which we were unable to examine. The last point seen by us is in lat. 30° 32′, and lon. 230° 11′.

Cape Seurote bears N. by W. of Cape Nomo, 25 miles, and from the entrance, N. 30° W. 17½ miles, and is in lat. 32° 58′ 30″, and lon. 230° 25′. The cape itself is not of an extraordinary height, and may be known by a hollow to the S.E., from which the land rises to the North, and is, on the whole, more mountainous than Nomo. Southward of Seurote are several islands, of which the largest and nearest to the cape is called Natsima; and the most to the South, Kitsima: but these, as well as the cape itself, we only saw on the 8th of October, the day of our arrival, and on quitting our first anchorage on the 9th.

The Harbour of Nagasaki may be divided into three parts; for it contains three different roads, which are all perfectly safe. The first without, to the westward of the Island Papenberg; the second, in the middle, to the eastward of that island; and the third, at the bottom of the harbour, forms the inner road in front of the city. As we lay for a considerable time in all of them, I am enabled to describe them very circumstantially. The entrance is formed to the southward by the North end of the Island of Iwo-sima, and to the northward by Cape Facunda;* which two points lie N.E. and S.W. 40°, distant about 2⅓ miles from each other. In the middle between them the depth is 33 fathoms, and with this water we anchored over a bottom of fine gray sand. In the direction of E.S.E., E.S.E. ½ E., and East (the course of the outer road), it gradually decreases, till you anchor in 22 or 25 fathoms, over a bottom of thick green oaze, with fine sand. This outer road, to the West of Papenberg, is completely sheltered from every wind except the N.W. and W.N.W.; but as this wind blows but seldom during the N.E. monsoon, and never very strong, it is perfectly safe at this time of year. The anchorage is excellent, and we had considerable trouble in weighing our anchor after it had lain in the ground during eight days, in which time it had not blown at all fresh; nor was it weighed the second time without trouble, although we passed but one night here: so that, unless a vessel intends to remain here any time, it will be found sufficient to cast out a kedge instead of a second sheet-anchor. Ours lay to the North in a depth of 18 fathoms. This road is formed by the following islands: to the West and S.W. is the lofty Island of Iwo-sima, which lies nearly North and South, and is 1½ miles in length; the hill which forms it is divided in the middle by a low valley, whence there are some houses, and upon the top of the northern half of the island a large tree, standing in an insulated situation, and visible at a considerable distance, marks the entrance to the harbour, and was of particular service to us in combining the plan of the harbour with the sea-marks. In an almost N.E. direction of the hill from the tree is a valley with a considerable village, surrounded by a very fine wood; and in the same direction, about a quarter of a mile from the shore, is a rock, which I believe is covered with high water. E.S.E. of the Island of Iwo-sima is another, called Taka-sima, divided by a channel not half a mile wide, but most likely free from rocks. There is probably no passage between Kagack-sima and Taka-sima. To the North of Kagack-sima lie some rocks, called Kanda-sima; and farther to the N.E., half a mile off, the small Island of Amcabier, 1½ miles in circumference. On its N.E. point is a Japanese fort. These last islands surround the outer road from the S.W. to the S.E.; to the East about 2 miles from the main land, to the N.E. Papenberg, and the Island of Kamino-sima to the North. From these another chain of rocks stretches to the West, between which there does not appear a passage for the smallest craft.

* The northern cape at the entrance I have called, for want of another name, after the town of Facunda, which lies in an open bay not far from there.

The middle road, or that to the East of Papenberg, is surrounded on all sides by land, and is equally safe with the inner one, to which I should prefer it, as its anchorage is better. To the West lies Papenberg: a small island, though not half a mile in circumference, is the highest in the harbour, and particularly remarkable from its being planted on both sides with a row of trees from its base to its summit. Its name, Papenberg, is derived from the tradition that, during the extirpation of the Christians from Japan, the Catholic priests were thrown from the top of this mountain. To the S.W. lie the islands Amiahur, Rajack-sima, and Taka-sima; and, in a rather more southerly direction, the broad channel, open to the sea, but in which, during the S.W. storms, the waves are broken by small islands lying as well without it as within it; and on this account it is necessary to anchor rather nearer to Papenberg, in order to be perfectly secured. To the southward and eastward is the right bank of the channel leading to the city; to the N.E., Nagasaki; to the N. and N.W., a part of the left bank of the channel of Nagasaki, and the Island Kamino-sima. From the outer road to the centre the depth decreases gradually from 25 to 17 fathoms. In this passage the only thing to be observed is, to keep closer to the Papenberg than to the opposite shore, and the former may be approached within a cable's length, as, even at this distance, there is a depth of 18 or 20 fathoms.

North-east of Papenberg lies, about three-quarters of a mile distant, a small flat island, entirely overgrown with wood, and bearing the name of *Nosumi-sima* (Rat Island); it is about the same size as Papenberg; and 130 fathoms farther, in the same direction, is the small Bay of Kibatsch, in which there are from 6 to 10 fathoms water. This, in all the harbour of Nagasaki, is the best place to refit a ship, for in the inner one the shore is everywhere so muddy that no ship can approach it. It was on the left side of this little bay that we were allowed a small space, scarcely longer than the ship itself, surrounded with bamboos, as a walk.

I would recommend ships coming for the first time to Nagasaki not to suffer themselves to be detained by any Japanese boats, which come out several miles to meet them, but to sail straight for the outer road. They may even run at once into the middle road, without the least danger, particularly during the S.W. monsoon. The assistance of the Japanese in this passage is perfectly unnecessary, and by rejecting it they will avoid the unpleasant predicament of being kept two days in the middle of the entrance, when, if anything of a storm were to spring up, they would be exposed to the greatest danger. Unless my advice be adopted, they must hire a hundred boats to tow them to Papenberg, when they will experience the additional mortification of losing 100 fathoms of towing line, which the Japanese will cut off, the moment they have carried them in.

From the middle to the inner road, or to the city of Nagasaki, the course lies N. 40° E.; the distance is about $2\frac{1}{2}$ miles, and the depth decreases gradually from 18 to 5 fathoms. Nearly halfway, where the channel is not more than 400 fathoms wide, are situated the imperial batteries, or the emperor's guard. These consist of a number of buildings, but without a single cannon, similar batteries being erected along both shores; and indeed, as the breadth of the channel does not exceed 500, and in some places not more than 300 fathoms, it would be impossible to conquer the city of Nagasaki if the Japanese knew how to fortify it, though in its present state it is not more formidable than the most miserable fishing town in Europe. A single frigate, with a few fire ships, would destroy the whole of Nagasaki in a few hours, notwithstanding its population, who could not possibly make any resistance. In the vicinity of the emperor's guard, on the right bank, there is a bay, which was always full of small ships, and where there is no doubt plenty of water for large ships, and on both sides of the channel there are several similar bays. This one, owing to its romantic appearance, was very striking, and seemed to be the largest; but we were not allowed to examine any one of them.

The anchorage near Nagasaki is not so good as either in the middle or outer road, as the bottom is a very thin clay; besides that, as the south-west channel is here quite open to the sea, there is less shelter than when lying close under the Papenberg.

The mean of a great number of observations, taken during the stay of the *Nadiĕjeda*, made the latitude of the flagstaff at Dezima to be 32° 44′ 18″, and of Nagasaki 32° 43′ 40″. The longitude of the centre of the town of Nagasaki was calculated from 1,028 lunar distance by Dr. Horner and Capt. Krusenstern as 230° 7′ 53″, or, in round numbers, 230° 8′ W. The mean of the observations for the variation made it 1° 45′ 36″ W. The mean time of high water, full and change, was 7^h 52′ 41″. The greatest range, April 2, 1805, was 11 feet 5 inches; the lowest, March 25, 1 foot 2 inches.—(Voyage round the World, by Capt. A. J. Von Krusenstern, in 1803-6, translated by Richard Belgrave Hoppner, Esq., vol. i. pp. 296—305.)

In thus concluding these brief and very imperfect notices of this important country, it is necessary to remark, once more, that for the most part they are drawn up from the very vague data afforded by the Japanese charts. These singular and interesting examples of geography differ so much between themselves and from other observations, that their indications can afford but little assistance to the science of navigation. The two maps drawn up by Admiral Krusenstern, and that by Dr. Von Siebold, differ so widely in many points, that their features would scarcely be recognised; yet in their main points they are identical, and they would certainly be of far more importance to the geographer than to the seaman. The few points determined by European observers are certainly too vague to draw up any satisfactory conclusion from their combined results. A far more complete series of observations are necessary to form a just conception of the Japanese Archipelago.

INDEX TO PART I.

Aamak or Amak Island (Aliaska Peninsula), 524
Abatanock Island (Aleutian Archipelago), 505
Abolecheff Bay (Siberia), 567
Abraham Island (Aleutian Islands), 519
Abreojos Point (Lower California), 309
Abtao, Island of (Chiloe), 98
Acajutla Port, or Sonsonate Roads (Central America), 247
Acapulco (Mexico), 269
Acari (Peru), 157
Acari, Morro (Peru), 157
Achachinskoï Bay (Kamtschatka), 603
Achilles Bank (Chile), 103
Aconcagua Volcano (Andes), 102, 115
Adakh Island (Aleutian Islands), 516
Adams, Point (Oregon), 359
Adamson, Cape (Russian America), 448
Addington, Cape (Russian America), 447
Adelaide's Queen, Archipelago, (Patagonia), 70
Adèle Bank (Guayaquil River), 191
Admiralty Inlet (Oregon), 379
Admiralty Island (Russian America), 455
Admiralty Sound (Strait of Magalhaens), 14
Adventure Bay (Patagonia), 86
Adventure Bridge (Strait of Magalhaens), 10
Adventure Cove (Tierra del Fuego), 60
Aektock or Goly Island (Aleutian Archipelago), 505
Affleck's Canal (Russian America), 447
Afuera Bank (Panamá Bay), 217
Afuera and Afuerita Island (Central America), 220
Agattou Island (Aleutian Islands), 519
Ageach Island (Aliaska Peninsula), 496
Agnes Islands (Tierra del Fuego), 65
Agougack, River (Aliaska Peninsula), 527
Agûa, Volcan de (Central America), 250
Aguatulco (Mexico), 261
Agnea or Darwin Channel (Patagonia), 85
Aguirre Bay (Tierra del Fuego), 51
Aguja or Needle Point (Peru), 180
Aguy Point (Chiloe), 90
Ahoni Point (Chiloe), 96
Aborcados Islets (Guayaquil River), 199
Aid Basin (Patagonia), 72
Akoun, Island of (Aleutian Archipelago), 505
Akounskoï Strait (Aleutian Archipelago), 506
Akoutan Island (Aleutian Archipelago), 506, 508
Akoutan, Strait of (Aleutian Islands), 511
Alacran, Island of (Peru), 147
Alaid Island (Kurile Islands), 605
Alan, Point (Oregon), 384
Alausi, Valley of (Andes), 187
Alava, Point (Russian America), 438
Albert Head (Vancouver Island) 416
Alcalde Point (Chile), 128
Alcatraces Island (San Francisco), 335
Alcatras Rock (Mexico), 269
Alcatrasses Island (Upper California), 340
Aldarien River (Concepcion Bay), 109
Aldunate Inlet (Patagonia), 78
Alexander, Point (Russian America), 444
Alexinoy Island (Aliaska Peninsula), 496
Aleutian Archipelago (North Pacific), 502
Algarroba Point (Chile), 114
Algodon Bay (Bolivia), 144
Aliaska, Peninsula of (Russian America), 495, 524
Aliau Island (Chiloe), 95
Alijos Rocks (Lower California), 309
Alikhoolip, Cape (Tierra del Fuego), 63
Alitock Bay (Kodiack Island), 494
Almejas Bay (Lower California), 306
Almendral, The (Valparaiso), 115
Aloe Shoal (Concepcion Bay), 110
Alquilqua Bay (Strait of Magalhaens), 36
Althorp, Port (Russian America), 461
Altunchagua Peak (Andes), 170
Amak or Aamak Island (Aliaska Peninsula), 524
Amapala Mountain (Central America), 242
Amapala or Conchagua Gulf (Central America), 242
Amargos Point (Valdivia), 104
Amatape, Mountain of (Peru), 182
Amatignack Island (Aleutian Islands), 517
Amatitlan Lake (Central America), 249
Ambato, Valley of (Andes), 187
Ameca, Bay of (Mexico), 276
Amelia, Point (Sitka Islands), 463
Amelius, Point (Russian America), 446
American Fork (Upper California), 337
Amguina River (Siberia), 562
Amilpas Range (Mexico), 258
Amlia Island (Aleutian Islands), 512
Amlinskoï Peninsula (Atkha Island), 513
Amortajada or Santa Clara Island (Guayaquil River), 190
Amoughta or Amoukhtou Island (Aleutian Islands), 510
Amoughta Channel (Aleutian Islands), 512
Amour River (Siberia), 616
Amtschitka Island (Aleutian Islands), 517
Amtschitka Cape (Ounalashka Island), 508
Anacapes or Enneeapah Island (Upper California), 326

Anacachi Rock (Chile), 132
Anaclache, Nevado de (Andes), 149
Anadyr, River (Siberia), 576
Analao Islet (Patagonia), 85
Anangouliak Islet (Oumnack Island), 509
Anchor Point (Cook's Inlet), 486
Ancon Sin Salida (Patagonia), 81
Ancon, Bay of (Peru), 170
Ancud, Gulf of (Chiloe), 89
Ancud or S. Carlos (Chiloe), 90
Anda-nivel, The (Central America), 250
Andes, The (Chile) 101
Andes, The (Peru), 170
Andes, The (Colombia), 187
Andes, The (New Granada), 188
Andréanoff or Andreanoffsky Islands (Aleutian Islands), 511
Andrews Bay (Strait of Magalhaens), 24
Andrews, Port (Russian America), 484
Anegadiza Point (Chile), 106
Angel Island (Upper California), 341
Angele de la Guardia Island (Gulf of California), 300
Angeles, Port (Mexico), 278
Angelos Point (De Fuca Strait), 376
Angosto, Puerto (Strait of Magalhaens), 34
Animack or Reindeer Island (Aliaska Peninsula), 499
Aniwa Cape and Bay (Saghalin), 622
Anmer, Point (Russian America), 452
Anna Pink Bay (Patagonia), 84
Año Nuevo Point (Monterey), 328, 329
Antuco, Peak of (Andes), 102
Anvil Island (Howe's Sound), 395
Anxious Point (Magdalen Sound), 18
Apabon Point (Chiloe), 95
Apache Indians (Mexico), 296
Apian Island (Chiloe), 95
Apisa River (Mexico), 274
Apollos Island, or Take-sima (Japan), 641
Apostle Rocks (Tierra del Fuego), 38, 68
Apple Tree Coves (Admiralty Inlet), 380
April Peak (Patagonia), 70
Arakamtchetchen Island (Siberia), 576
Aranta Cove (Peru), 154
Araucania (Chile), 104
Arauco (Chile), 108
Arcana Hill (Peru), 179
Arce Bay (Strait of Magalhaens), 30
Archangel Gabriel, Bay of (Siberia), 576
Archangel, New (Sitka Islands), 466
Arden, Point (Russian America), 452
Arena, Punta de (Concepcion), 112
Arena Point (Bolivia), 144
Arena, Point Baira de (Upper California), 345
Arena Point (Guayaquil River), 191
Arenas Point (Tierra del Fuego), 44
Arenas, Punta de, or Nicoya (Central America), 226, 228
Arequipa (Peru), 152
Arguello, Point (Upper California), 326
Arica (Peru), 147
Arii Rock (Sea of Behring), 584
Arispe (Mexico), 296
Armstrong, Port (Russian America), 459
Arro or Haro Strait (N.W. America), 419
As seze Yrmas Islands (Japan), 641
Ascension, Rio de la (Gulf of California), 299
Asia Islands and Peak (Peru), 162
Asseradores Island (Realejo), 235, 237
Asses' Ears (Patagonia), 5, 6
Asses' Ears, or Mesima Islands (Japan), 645
Assuay, Department of (Ecuador), 186
Astley, Point (Russian America), 452
Astoria (Columbia River), 361
Astrolabe Island (Japan), 637
Atacamà, Desert of (Bolivia), 137
Atacames or Tacames (Columbia), 204
Atahuanqui (Peru), 171
Atequipa Valley (Peru), 156
Athos, Mount (Siberia), 568
Atico Point and Valley (Peru), 156
Atkha or Atcha Island (Aleutian Islands), 512
Atkinson, Point (New Georgia), 394
Atkis, Port (Jesso Island), 629
Atrato River (Colombia), 208
Attcheun, Cape (Siberia), 572
Attou Island (Aleutian Islands), 519
Audiencia, Punta de la (Mexico), 274
Augusta Island (Patagonia), 70
Augusta, Point (Russian America), 459
Austro Cape (Realejo) 237
Avadsi-sima Island (Japan), 638
Avad-sima (Japan), 636
Avatcha or Awatska Bay (Kamtschatka), 594
Avos Rock (Kurile Islands), 605
Awa Cape (Japanese Islands), 633
Awatska Bay (Kamtschatka), 594
Awo-sima Island (Japanese Islands), 634
Ayak, or Aziak, or Sledge Island (Behring's Strait), 536
Ayanqui Point (Guayaquil River), 199
Ayautau Islands (Patagonia), 76
Aymond, Mount (Patagonia), 5
Ayougadagh Island (Aleutian Islands), 518
Aytay Point (Chiloe), 96
Ayuta, Morro and River of (Mexico), 261
Ayuta, Morro de (Mexico), 260
Azada, Island of (Mexico), 291
Aziak, or Ayak, or Sledge Island (Behring's Strait), 536
Azua Point (Peru), 158
Azucar, Pan de (Chile), 109

Babouschka Rock (Awatska Bay), 597, 599
Bad Bay (Patagonia), 79
Bagona, Rio (Mexico), 287
Bahia Honda (Central America), 221
Bahia de la Campaña, or Bell Bay (Strait of Magalhaens), 26
Bainbridge, Port (Prince William's Sound), 481
Baja de Afuera (Panamá Bay), 217
Baker Point (Russian America), 445
Baker, Mount (Oregon), 379

Baker's Bay (Columbia River), 360
Balao River, 191
Bald Head (Russian America), 535
Ballena Point (Chile), 136
Ballenita Port (Chile), 136
Ballista Islands (Peru), 160
Bamba, Bay of (Mexico), 260
Bamba, Playa (Mexico), 261
Banderas, Valle de (Mexico), 276
Banks's Island (N.W. America), 428
Banks' Hill (Tierra del Fuego), 50
Banks, Port (Sitka Islands), 468
Banks, Cape (Kodiack Island), 494
Bannerskaia Bay (Atkha Island), 514
Baracura Heights (Chiloe), 90
Baranoff Island (Sitka Islands), 457
Barbara Channel (Tierra del Fuego), 20
Barielo Bay (Strait of Magalhaens), 30
Barlow's Cove (Russian America), 454
Barnabas, Cape (Kodiack Islands), 493
Barneveld's Island (Japanese Islands), 634
Barnett, Cape (Russian America), 448
Barnevelt's Isles (Tierra del Fuego), 53
Barra Falsa (Realejo), 237
Barra de Arena, Point (Upper California), 345
Barranca Point (Strait of Magalhaens) 7
Barranca (Peru), 171
Barranquilla de Copiapo (Chili), 131
Barren or Sterile Island (Kodiack Islands), 491
Barrie, Point (Russian America), 445, 446
Barrister Bay (Tierra del Fuego), 68
Barros de Zuniga (Upper California), 317
Barrow, Point (Russian America), 551
Bartolom, Cape San (Russian America), 448
Basil Hall, Port (Staten Island), 47
Basket Island (Tierra del Fuego), 63
Bat or Murciellagos Islands (Central America), 230
Batchelor's River (Strait of Magalhaens), 25
Baxa, Point (Strait of Magalhaens), 7
Baylio Bucareli, Puerto del (Russian America), 447
Bazil, Point (Prince William's Sound), 482
Beaubasin, Port (Strait of Magalhaens), 25
Beauclerc, Port (Russian America), 446
Beaufort, Cape (Russian America), 548
Beaufort Bay (Patagonia), 89
Beaufoy, Mount (Tierra del Fuego), 58
Beagle Island (Patagonia), 70
Beagle Channel (Tierra del Fuego), 52
Beagle Mountains (Peru), 171
Beaver Harbour (Vancouver Island), 408
Bede, Point (Cook's Inlet), 485
Bedford Bay (Barbara Channel), 21
Beechey Head (Vancouver Island), 416
Becher Bay (Vancouver Island), 416
Begueta, Bay of (Peru), 171
Behm's Canal (Russian America), 438
Behring, Cape (Siberia), 572
Behring Island (Sea of Behring), 583
Behring's Bay (Russian America), 470
Behring's Strait, 539
Behring's Sea (North Pacific), 520
Belcher, Point (Russian America), 551
Belen Bank (Concepcion Bay), 109, 111
Bell Bay, or Bahia de la Campaña (Strait of Magalhaens), 26
Bellingham Bay (Oregon), 387
Bellingshausen, Cape (Saghalin), 620
Bell of Quillota (Valparaiso), 115, 118
Bell's Island (Russian America), 440
Benicia (Upper California), 336
Bentinck, Point (Prince William's Sound), 477
Bentinck Island (Vancouver Island), 416
Bentinck's Arms (N.W. America), 424
Bereford, Islands de (Vancouver Island), 409
Berkeley or Nitinat Sound (Vancouver Island), 414
Berner's Bay (Russian America), 456
Bernizet Peak, or Mount Spanberg (Saghalin), 622
Besborough Island (Russian America), 533
Bestraïa River (Sea of Okhotsk), 611
Betchevinskaïa Bay (Kamtschatka), 594
Betchevinskoï Cape (Atkha Island), 514
Betton's Island (Russian America), 441
Beware Point (Peru), 157
Beware Harbour (N.W. America), 397
Bingham, Point (Russian America), 461
Bio Bio River (Chile), 109
Bio Bio, Paps of, 110
Birch Bay (Oregon), 388
Bird's or Satanna Cape (Aleutian Islands), 517
Biruquete Province (Colombia), 207
Black Cape (Oumnack Island), 509
Black Creek (Oregon), 383
Black Point (Lower California), 310
Black Rock, Christmas Sound (Tierra del Fuego), 59
Blanca, Ysla (Chile), 136
Blanca Island (Peru), 160
Blanco, Cape (Central America), 227
Blanco, Cape (Peru), 182
Blanco or Orford, Cape (Oregon), 354
Blanco or Lobo Point (Peru), 145
Blaquiere, Point (Russian America), 443
Bligan, Cape (Sea of Okhotsk), 612
Bligh's Island (Prince William's Sound), 479
Blijni Islands (Aleutian Islands), 518
Blossom Rock (Upper California), 340
Blossom Rock (San Francisco), 335
Blossom, Cape (Russian America), 545
Blossom's Shoals (Russian America), 550
Blying Sound (Russian America), 484
Bobrovaia or Otters, Bay of (Ounalashka Island), 507
Bobrovi or Sea Otter Island (Sea of Behring), 582
Boca Chica, Acapulco (Mexico), 270
Boca del Infierno, 413
Boca de Quadra (Russian America), 438
Bocca Barra (Mexico), 259
Bodega Cove, La (Chile), 114
Bodega, Port (Upper California), 344

Bogota, Santa Fé de (New Granada), 186
Boisée or Woody Island (Kodiack Islands), 492
Bolbones, Sierras (Upper California), 337
Bolcheretskoï (Sea of Okhotsk), 611
Bolivia, Republic of, 137
Boneta Point (San Francisco), 335
Bonifacio Head (Chile), 105
Bonilla or Smith Island (Oregon), 385
Boqueron of Pisco (Peru), 159
Boqueron, Mount (Magdalen Sound), 19
Boqueron Channel (Callao), 168
Boqueron, Cape (Strait of Magalhaens), 10
Boquita Point (Chile), 113
Borja Bay (Strait of Magalhaens), 28, 29
Boruca River (Central America), 222
Bouchage or Cantin Bay (Strait of Magalhaens), 16
Boukouru Island (Japan), 637
Bouldyr Island (Aleutian Islands), 518
Boungo, Strait of (Japan), 640
Bourdieu's Bay (Cook's Inlet), 486
Bourchier Bay (Tierra del Fuego), 58
Bourgo, Strait of (Japan), 638
Bousen, Province of (Japan), 640
Boussole Channel (Kurile Islands), 604
Boussole Channel (Kurile Islands), 608
Boyaca, Province of (New Granada), 186
Boyle's Point (Vancouver Island), 408
Brackenridge Bluff (Oregon), 371
Bradfield Canal (Russian America), 443
Bradley Cove (Strait of Magalhaens), 26
Brava or San Lorenzo Point (Panamá Bay), 210
Brazo Ancho or Wide Channel (Patagonia), 80
Brazo de Balda and Baldinat (New Hanover), 406
Brazo de Cardenas, or Port Neville (N.W. America), 402
Brazo de Toba (N.W. America), 398
Brazo de Retamal, or Call's Canal (N.W. America), 403
Brazo de Mazarredo, or Jervis Canal (Howe's Sound), 396
Brazo de Carmelo, or Howe's Sound (New Georgia), 394
Brazo Ancho, Point (Patagonia), 83
Brazo de Vernacci or Knight's Canal (New Hanover), 406
Brazo de Malaspina (N.W. America), 398
Breaker, Point (Vancouver Island), 409, 411
Breaker Bay (Tierra del Fuego), 66
Break-pot or Quebra Olla Rock (Concepcion Bay), 109
Breaksea Island (Patagonia), 75
Brecknock Passage (Tierra del Fuego), 63
Brenton Cape (Patagonia), 73
Bridget, Point (Russian America), 456
Bristol Bay (Russian America), 528
Brito (Central America), 232, 233
Broderip Bay (Barbara Channel), 22
Broken Islands (Japanese Islands), 634
Brooks, Port (Vancouver Island), 409
Brotchy Ledge (Vancouver Island), 417
Broughton's Archipelago (N.W. America), 403–5
Broughton Island (Kurile Islands), 608
Broughton, Cape (Jesso Island), 628
Brown Point (Oregon), 370
Brown's Passage (N.W. America), 430
Bruja Point (Mexico), 269
Brunswick Peninsula (Strait of Magalhaens), 28
Bryant, Point (Prince William's Sound), 482
Bucalemo Head (Chile), 114
Bucareli, Port (Russian America), 447
Buck, Point (Queen Charlotte Island), 434
Buckland River (Russian America), 544
Buckland, Mount (Strait of Magalhaens), 14, 15
Buchon, Monte de (Upper California), 327
Budd's Inlet (Puget Sound), 381
Budd's Harbour (De Fuca Strait), 376
Buen Tiempo, or Fairweather Cape (Patagonia), 3
Buena Point (Port Culebra), 229
Buenaventura, San (Upper California), 322
Buenaventura Bay (Colombia), 206
Bueno, River (Chile), 103
Buey Rock (Concepcion Bay), 111
Bufadero Rock (Mexico), 266, 268
Bufadero Cliff (Peru), 172
Bulkeley's Channel, or Gulf of Xaultegua (Strait of Magalhaens), 33
Burgess Island (Strait of Magalhaens), 26
Burica Point (Central America), 222
Burke's Canal (N.W. America), 424
Burney or Koliutchin Island (Siberia), 562
Burney, Mount (Patagonia), 81
Burnt Island (Tierra del Fuego), 60
Burnt or Goreloy Islands (Aleutian Islands), 516
Burrard's Canal (New Georgia), 394
Burrough's Bay (Russian America), 439
Bushy Island (Russian America), 445
Bute's Canal or Inlet (N.W. America), 399
Buva-no-oumi Lake (Japan), 635
Byers's Strait (Patagonia), 71
Bynoe Island (Tierra del Fuego), 20
Bynoe Point (Patagonia) 75
Byron Island (Patagonia), 75, 76

Caamano, Cape (Russian America), 441
Caamano Island (Oregon), 384
Caballos Road (Peru), 158
Cabeza de Vaca (Chile), 134
Cabra, Las Tetas de (Mexico), 297
Cachos Point (Chile), 131
Cacique Point (Port Culebra), 229
Cadiack or Kodiack Island (Russian America), 490
Caduhuapi Rocks (Chiloe), 89
Cahoos River (Oregon), 355
Cahuache Island (Chiloe), 95
Calandare or Palu, Point (Mexico), 291
Calavaros, Rio (Upper California) 340
Calbuco or El Fuerte (Chiloe), 98

Calder, Mount (Russian America), 445
Caldera, Port (Chile), 134
Caldera Bay (Central America), 227
California, Gulf of (Mexico), 287
California, Upper, 295
California, Lower, 291
Calla-calla River (Valdivia), 104
Callan Bay (De Fuca Strait), 376
Callao (Peru), 165
Callo Point and Island (Guayaquil River), 200
Call's Canal or Creek (N.W. America), 403
Calm Point (Russian America), 529
Calvario Point (Peru), 174
Calver, Cape, 416
Calvert Island (N.W. America), 423
Camana, Monte and Valley of (Peru), 155
Cambridge Island (Patagonia), 70
Camden, Port (Russian America), 449
Camden Islands (Tierra del Fuego), 63
Cammusan or Camosack Harbour (Vancouver Island), 417
Camosack or Cammusan Harbour (Vancouver Island), 417
Campaña Island (Patagonia), 73
Campana de Quillotta (Valparaiso), 115, 118
Campana, Mount (Peru), 178
Campania, Isle de la (N.W. America), 428
Campbell, Point (Cook's Inlet), 489
Camuta, Rio (Mexico), 274
Canal of San Juan (Central America), 231
Canal, Siriano River (Central America), 242
Canal de Ballenas (Gulf of California), 309
Canal Peligroso (Gulf of California), 299
Canal of the Mountains (Patagonia), 82
Canales, or Afuera Is. (Central America), 220
Canals, Atlantic and Pacific (Atrato), 208
Canaveral Cove (Patagonia), 84
Canaveral de Puerto (N.W. America), 429
Candadillo, Point (Central America), 244-5
Candalaria, Point (Patagonia), 72
Canel River (De Fuca Strait), 376
Cañete or Canyete (Peru), 161
Canning Isles (Patagonia), 83
Caño Island (Central America), 226
Canoas Rocks (Strait of Magalhaens), 27
Canoitad Rocks (Chiloe), 89
Cantarras, or Afuerita Island (Central America), 220
Cantin or Bouchage Bay (Strait of Magalhaens), 16
Capa Point (Peru), 186
Capalita River (Mexico), 266
Cape or Myssoff Bay (Kodiack Islands), 493
Capistrano, San Juan de (Upper California), 320
Capitanes Point (Chile), 103
Capstan Rocks (Tierra del Fuego), 63
Captain's Bay (Ounalashka Island), 506
Caracas Bay (Colombia), 202
Carampangue River (Chile), 108
Carbon, Morro de (Mexico), 260
Cardero Canal (N.W. America), 402
Cardon Island (Realejo), 235, 239
Cardon Head and Channel (Realejo), 237
Carelmapu, or Doña Sebastiana Islets (Chiloe), 90
Carelmapu Head (Chiloe), 91
Carelmapu Islet (Chile), 103
Carluck, Fort (Kodiack Islands), 494
Carlos III. Island (Strait of Magalhaens), 27
Carmelo, Bay of (Upper California), 328
Carmen, Isle del (Gulf of California), 302
Carnero Bay and Head (Chile), 107
Carquin, Bay and Head of (Peru), 171
Carranza Point (Chile), 113
Carrasco Point (Vancouver Island), 414
Carrasco, Mount (Peru), 145
Carretas Head (Peru), 158
Carrisal Point (Chile), 127
Carrisal Cove (Chile), 130
Carrisal, Herradura de (Chile), 129
Carr's Inlet (Puget Island), 381
Cartagena Beach (Chile), 114
Carter's Bay (N.W. America), 426
Cartwright's Sound (Queen Charlotte's Island), 434
Carra Island (Chiloe), 98
Casalla (Panama Bay), 212
Casares (Central America), 233
Cascade Canal (N.W. America), 425
Cascade Harbour (Strait of Magalhaens), 25
Cascajal Bay and River (Colombia), 207
Case's Inlet (Puget Sound), 381
Case's Inlet (Oregon), 383
Casma (Peru), 174
Castañon Bluff (Realejo), 237
Castañon Island (Realejo), 235
Castlereagh, Cape (Tierra del Fuego), 63
Castricum Cape (Kurile Islands), 608
Castro (Chiloe), 94
Casualidad Rock (Chiloe), 122
Catala Island (Vancouver Island), 410
Catalana, Island of (Gulf of California), 302
Catalina or Gorda Point (Port Culebra), 229
Catalina Bay (Strait of Magalhaens), 11
Catalon Island (Gulf of California), 301
Catharine Point (Patagonia), 3
Cathedral Mount (Patagonia), 73
Catherine Point (Tierra del Fuego), 43
Cauca, Province of (New Granada), 186
Caucahuapi Head (Chiloe), 89
Caucahue Island (Chiloe), 92
Caucahue Strait (Chiloe), 92
Cauten River and Head (Chile), 105
Caution, Cape (N.W. America), 422
Caxa Chica Rock (Chile), 131
Caxa Grande Rock (Chile), 132
Cayambe Urcu Mountain (Andes), 188
Cayetano Island (Barbara Channel), 22
Cayetano Island (Strait of Magalhaens), 27
Caylin (Chiloe), 97
Cayuela Entrance (Vancouver Island), 414
Cebaco Islands (Central America), 219
Cedros or Cerros Is. (Lower California), 310
Cenizas or S. Hilario Island (Lower California), 311-2
Centinela Point (Chiloe), 98

Centinela Point (Chiloe), 98
Centinela, Point (Guayaquil River), 194
Central America or Guatemala, 222
Cerrabo Island (Gulf of California), 303
Cerro de las Animas, 191
Cerro de la Giganta, El (Gulf of California), 302
Cerro Azul (Peru), 161
Cerro del Guasco (Chile), 129
Cerros or Cedros Island (Lower California), 310
Chacalu (Mexico), 277
Chacansi (Peru), 144
Chacao Narrows (Chiloe), 91
Chacon, Cape de (Russian America) 442
Chagamil Island (Aleutian Islands), 510
Chaktolimout Bay (Russian America), 533
Chala Point (Peru), 156
Challenger, wreck of H.M.S. (Chile), 106
Chalmers, Port (Prince William's Sound), 482
Chametla or Rosario, Rio del (Mexico), 287
Chametla, Hillocks of (Mexico), 287
Chamisso Island (Russian America), 544-5
Chancay, Bay of (Peru), 170
Chancery Point (Tierra del Fuego), 68
Chanchan Cove (Chile), 105
Chañeral, Port of (Chile), 127
Changarni Islet (Panama Bay), 217
Channel's Mouth (Patagonia), 76
Chanticleer Island (Tierra del Fuego), 55
Chao Islands (Peru), 176
Chapline or Tchaplin, Cape (Siberia), 569
Charles Islands (Strait of Magalhaens), 27
Charles Island (Tierra del Fuego), 66
Charles's Islands (Strait of Magalhaens), 25
Chasco, Cove of (Chile), 131
Chatham Island (Patagonia), 82
Chatham Strait (Russian America), 454
Chatham, Port (Russian America), 484
Chatham, Point (N.W. America), 401
Chatham's Sound (N.W. America), 430
Chaugues Islands (Chiloe), 92-3
Chaulin Islands (Chiloe), 97
Chaulinec Island (Chiloe), 95
Chavini Point (Peru), 156
Chayhuao Point (Chiloe), 97
Chaylaime Point (Patagonia), 86
Cheap Channel (Patagonia), 77
Cheapa, Province of (Mexico), 257
Cheapo River (Panamá Bay), 211
Checo, Mines of (Andes), 102
Chelegboff Lake (Russian America), 528
Chelighoff Strait (Aliaska Peninsula), 494
Chelin Island (Chiloe), 95
Chenoke or Chenook Point and Village (Columbia River), 360
Chepillo or Chepelio Island (Panamá Bay), 211
Chequetan (Mexico), 273
Chestakoff Islet (Aliaska Peninsula), 526
Chestokova River (Sea of Okhotsk), 613
Cheuranatta, Bay of (Peru), 145
Chicapa River (Mexico), 260
Chicarene Point (Central America), 241
Chicarene, La Playeta de (Central America), 244
Chichagoff Island (Sitka Islands), 458
Chichaldinskoï Volcano (Ounimack Is.), 504
Chichkoff, Cape (Ounimack Island), 504
Chickeeles or Chicaylis River (Oregon), 370-1
Chidhuapi Island (Chiloe), 98
Chigab Island (Jesso Island), 639
Chigua Loco Cove (Chile), 123
Chihuahua, Province of (Mexico), 288
Chikotan or Tschikotan Island (Kurile Islands), 609
Chilca Point and Port (Peru), 162
Chilé, Republic of, 100
Chilen Bluff (Chiloe), 92
Chileno Point (Peru), 144
Chilhat Village (Russian America), 456
Chillan, Peak of (Andes), 102
Chiloe, Island of (Chile), 87
Chimborazo, Mountain (Andes), 187
Chimbote (Peru), 175
Chimu, Valley of (Peru), 176
Chincha Islands (Peru), 160
Chinook or Chenoke Point and Village (Columbia River), 360
Chinquiquira River (Colombia), 207
Chipana, Bay of (Peru), 144
Chipicani, Nevado of (Andes), 149
Chipounsky, see Shipounskoi (Kamtschatka), 594
Chiquirin or Chicarene Point (Central America), 242
Chiriqui, Mountain of (Central America), 226
Chisinche, Alto de (Andes), 187
Chiswell Isles (Russian America), 483
Chiut Island (Chiloe), 96
Chivilingo River (Chile), 108
Chocó, District of (Colombia), 207
Chocó Mountains (Andes), 188
Chocoy Head (Chiloe), 91
Chogon Point (Chiloe), 92
Choiseul Bay (Strait of Magalhaens), 27
Cholmondeley Sound (Russian America), 442
Chomache (Peru), 145
Chonjack Island (Kodiack Islands), 494
Chonga, Cerro del (Mexico), 265
Chonos Archipelago (Patagonia), 83—86
Chontales (Central America), 234
Chopkod River (Kamtschatka), 593
Chorillo (Peru), 163
Choris Peninsula (Russian America), 544
Choros Point (Chile), 127
Choros Bank (Concepcion Bay), 109
Chosan or Tchosan Harbour (Japan), 643
Choumaguine or Schumagin Sound (Aliaska Peninsula), 498
Chramtschenko Bay (Russian America), 529
Chramtschenko Island (Siberia), 565
Christian's Sound (Russian America), 449
Christmas Sound (Tierra del Fuego), 59, 62
Christopher Point (Vancouver Island), 416
Chuapa, River (Chile), 123

Chulin Island (Chiloe), 96
Chuloteca River (Central America), 242
Chungara Mount (Andes), 149
Chungunga Island (Chile), 126
Chupador Inlet (Guayaquil River), 190
Chupador, Little Point (Guayaquil River), 195
Church Rock, The (Chile), 113
Church, Cape (Vancouver Island), 416
Churruca, Point (Strait of Magalhaens), 36
Ciervo, Island of (Mexico), 290
Cinaloa or Sinaloa River (Mexico), 297
Cirujano Island (Patagonia), 78
Citera (Colombia), 208
Clalam Indians (De Fuca Strait), 378
Clapperton Inlet (Patagonia), 80
Clarence, Port (Behring's Strait), 537
Clarence Islands (Strait of Magalhaens), 25
Clarence Strait, Duke of (Russian America), 441
Classet, Cape (Oregon), 374
Classet Village (Oregon), 374
Clatsop Village (Colombia River), 360
Clayoquot Sound (Vancouver Island), 413
Clearbottom Bay (Tierra del Fuego), 58
Clements Island (Patagonia), 85
Clerke or St. Lawrence Island (Behring's Sea), 578
Clerke, Port (Tierra del Fuego), 61
Cliff Cove (Patagonia), 84
Climate, Sea of Behring, 582, 585
Climate, Chile, 101
Climate, Upper California 317; Lower California, 296
Climate, Guatemala, 225
Climate, Japan, 625
Climate, Awatska Bay and Kamtschatka, &c., 602
Climate, Mexico, West coast, 256
Climate, Panamá, 209
Climate, Patagonia, West coast, 69
Climate, Peru, 140
Climate, San Blas (Mexico), 285
Climate, Staten Island, 46
Climate, Vancouver Island (N.W. America), 392
Cloak Bay (Queen Charlotte Island), 435
Clonard Bay (Queen Charlotte Island), 435
Coaguanaja River (Mexico), 274
Coal, Aliaska Peninsula (Russian America), 499
Coal, Cape Beaufort (Russian America), 549
Coal, Chiloe, 88
Coal, Columbia River (Oregon), 367
Coal, Concepcion (Chile), 109
Coal, Cook's Inlet, 485, 487
Coal, Icy Cape (Russian America), 559
Coal, Island of Kiusiu (Japan), 940
Coal, Krenitzin Island (Aleutian Archipelago), 505
Coal, Vancouver's Island, 408
Coal Bay (Cook's Islet), 485
Coalcaman, Rio (Mexico), 274
Cobija or La Mar (Bolivia), 143
Cocale Head (Chile), 105
Cochinos Islet (Chiloe), 90
Cochrane, Point (Prince William's Sound), 480
Cockburn Channel (Tierra del Fuego), 19
Cocos Bay (Port Culebra), 229
Cocotue Head (Chiloe), 89
Coffin, Mount (Oregon), 367
Coffin Rock (Columbia River), 368
Cogimies Shoals (Colombia), 203
Coke, Point (Russian America), 452
Colan (Peru), 181
Colanche River (Guayaquil River), 198
Colcura (Chile), 108
Coles Point (Peru), 150
Colima, Volcan de (Mexico), 274-5
Colima, City of (Mexico), 275
Colita Island (Chiloe), 97-8
Coliumo, Bay of (Chile), 113
College Rocks (Tierra del Fuego), 66
Collie, Cape (Russian America), 550
Colnett, Cape (Lower California), 312
Colnett, Cape (Staten Island), 48
Colnett Strait (Japan), 642
Colnett's Island (Japan), 643
Colocao Heights (Chile), 108
Colorada, Punta (Gulf of California), 302
Colorado, Rio (Gulf of California), 299
Colombia, Republic of, 185
Colpoys, Point (Russian America), 445
Colseed Inlet (Oregon), 384
Columbia Islands (Panama Bay), 212
Columbia River, The (Oregon), 357
Colworth, Cape (Patagonia), 80
Commander Islands (Sea of Behring), 583
Commencement Bay (Admiralty Inlet), 381
Company Point (Vancouver Island), 416
Comptroller's Bay (Prince William's Sound), 477
Comptroller's Bay (Russian America), 476
Compu Inlet (Chiloe), 97
Comte Heiden or Izenbek Bay (Aliaska Peninsula), 524-5
Concepcion Strait (Patagonia), 82
Concepcion Bay (Chile), 109
Concepcion Strait (Patagonia), 80
Conception, Point (Upper California), 324
Concha or Harwood's Island (N.W. America), 398
Conchagua or San Carlos (Central America), 241
Conchagua or Fonseca Gulf (Central America), 240
Conchali Bay (Chile), 123
Conclusion Island (Russian America), 445
Conclusion, Port (Russian America), 456
Concon Point and Rocks (Chile), 121
Condado Bank (Panamá Bay), 217
Condesa Bay (Strait of Magalhaens), 30
Cone Creek (Patagonia), 83
Conejo Islet (Central America), 242
Congo River (Panamá Bay), 210
Constantine Harbour (Prince William's Sound), 477

Constantine, Cape (Russian America), 528
Constitucion (Chile), 114
Constitucion Road (Bolivia), 142
Convents Hill (Patagonia), 3
Conversion Point (Upper California), 322
Cook Bay (Tierra del Fuego), 63
Cook, Port (Staten Island), 47
Cook's Inlet (Russian America), 485
Copiapó (Chile), 131
Copper or Medny Island (Sea of Behring), 584
Copper Cove (Bolivia), 144
Coquils River (Oregon), 355
Coquimbo or La Serena (Chile), 124
Coquimbo, Herradura de (Chile), 124
Corcobado Mountain (Andes), 87
Corcovado Gulf (Chiloe), 89
Cordes Bay (Strait of Magalhaens), 24
Cordonazo de San Francisco (Mexico), see Winds
Cordon Island (Vancouver Island), 409
Cordova, Port (Prince William's Sound), 478
Cordova Islet (Strait of Magalhaens), 27
Cornejo Point (Peru), 153
Cornejos Islet (Peru), 173
Cornwallis, Point (Russian America), 449
Corona Head (Chiloe), 89
Coronados Islets (Upper California), 317
Coronados, Is. of (Gulf of California), 302
Coronation Island (Russian America), 447
Coronel Point (Chile), 108
Coronel Point (Chiloe), 91
Coronilla Cove (Strait of Magalhaens), 28
Coroumilla Point (Valparaiso), 115
Corral Fort (Valdivia), 104
Corrientes, Cape (Colombia), 207
Corrientes, Cape (Mexico), 276
Corso, Cape or Mount (Patagonia), 73
Corso, Cape (Magdalena Bay), 308
Cortado, Cape (Strait of Magalhaens), 37
Cortes Island (N.W. America), 400
Cortes, Sea of, or Gulf of California (Mexico), 296
Corvetana, La (Mexico), 276
Coseguina Volcano (Central America), 240
Cotamyts River (Oregon), 355
Cotocache Mountain (Andes), 188
Cotopaxi Volcano (Andes), 187
Coulgiack Island (Cook's Inlet), 487
Countess, Point (Prince William's Sound), 481
Couverden, Point (Russian America), 459
Coventry Cape (Strait of Magalhaens), 24
Cowlitz River (Oregon), 367
Cox Island (Vancouver Island), 409
Cox, Port (Vancouver Island), 414
Coyuca, Beaches of (Mexico), 273
Coyuca, Paps of (Mexico), 270, 273
Craig, Point (Russian America), 444
Crescent Bay (De Fuca Strait), 376
Cresciente Island (Magdalena Bay), 306
Creston Island (Mexico), 287, 289
Crillon, Mount (Russian America), 462
Crillon, Cape (Saghalin), 623
Croker Peninsula (Strait of Magalhaens), 33
Crooked Reach (Strait of Magalhaens), 28
Crooze's Island (Sitka Islands), 457
Cross, Cape (Russian America), 462
Cross Sound (Russian America), 461
Cruz de la Ballena, Point (Chile), 122
Cucao, District of (Chiloe), 89
Cucao Bay and Heights (Chiloe), 89
Cuello Point (Chiloe), 97
Cuenza, Valley of (Andes), 187
Cuevas, Cape (Strait of Magalhaens), 37
Culebra, Port (Central America), 229
Culebra Island (Panamá Bay), 217
Culebras Point (Peru), 173
Culiacan (Mexico), 296
Culiacan, River (Mexico), 197
Cullin, Island of (Chiloe), 99
Culross, Point (Prince William's Sound), 480
Cumba River (Peru), 149
Cumming, Point (N.W. America), 427
Cundinamarca, Province of (New Granada), 186
Cupica or Tupica (Colombia), 207
Curauma Head (Valparaiso), 115
Curaumilla Point (Valparaiso), 115
Cutler Channel (Patagonia), 81
Cyprees Point (Upper California), 329
Cyprees Island (Oregon), 387

Dagua River (Colombia), 207
Dahap Inlet (Oregon), 384
Dalcahue (Chiloe), 93
Dallas Point (Chile), 131
Dalrymple, Cape (Saghalin), 621
Damas Bay (Quibo Island), 219
Dame, Point (Upper California), 321
Dangereuse Rock (Saghalin), 623
Danville, Cape (Japan), 640
Danzantes, Isle de los (Gulf of California), 302
Darby Cove (Strait of Magalhaens), 36
Darby, Cape (Russian America), 535
Darca Point (Concepcion Bay), 112
Dardo Head (Peru), 158
Dark Hill (Patagonia), 84
Darwin or Aguea Channel (Patagonia), 85
David Point (Quibo Island), 219
Davison, Point (Russian America), 432—442
Dawson Island (Strait of Magalhaens), 14
Dawydoff Island (Aleutian Islands), 518
Day, Point (N.W. America), 425
Dead Tree Island (Patagonia), 78
Dean Harbour (Barbara Channel), 22
Dean's Canal (N.W. America), 425
Deceit Island (Tierra del Fuego), 33
Deceit Cape (Russian America), 542
Deceit or Mistaken Cape (Tierra del Fuego), 54
Deception Passage (Oregon), 385
Decision, Cape (Russian America), 447
Deepwater Sound (Tierra del Fuego), 66
Deep Harbour (Patagonia), 80
Deep-sea Bluff (New Hanover), 406
Defiançe, Point (Admiralty Inlet), 381

De Fuca Strait (N.W. America), 373, 414
Delaroff Islands (Aleutian Islands), 516
Delarovskoï (Aliaska Peninsula), 499
Delgada Point (Magdalena Bay), 306
Delgada, Point (Strait of Magalhaens), 7
Delgado, Point (Upper California), 345
Delicada Point (Chile), 108
Delisle, Cape (Saghalin), 620
Del Mar, Hacienda (Mexico), 287
Del Palmito, Hacienda (Mexico), 287
Denbigh, Cape (Russian America), 534
Derbinskoï Strait (Aleutian Archipelago), 505
Descarte Point (Central America), 230
Deseado, Cape (Tierra del Fuego), 68
Desertores Islands (Chiloe), 95
Desolada, Cape (Central America), 233-4
Desolation Sound (N.W. America), 398
Desolation, Cape (Tierra del Fuego), 63
Despensa Island (Central America), 230
Destruction Island (Oregon), 372
Detif Point (Chiloe), 95
Devil's Basin (Tierra del Fuego), 60
Devil's Peak (Peru), 162
De Vries, Strait of (Kurile Islands), 608
Dezima, Island of (Japan), 647
Diamante Point (Mexico), 269
Diana Islands (Patagonia), 81
Diana Strait (Kurile Islands), 608
Diana Peak (Patagonia), 70
Diavolo, Sierras (Upper California), 337
Dick, Port (Russian America), 484
Diego Ramirez Islands (Tierra del Fuego), 57
Digges's Sound or Bay (Russian America), 472
Dighton Cove (Barbara Channel), 22
Dinero, Mount (Patagonia), 5
Dinner Cove (Barbara Channel), 22
Diomede Islands (Behring's Strait), 539
Direction Bluff (Awatska Bay), 601
Direction Bluff (Peru), 157
Direction Islands or Evangelists (Strait of Magalhaens), 38
Direction Hills (Patagonia), 6
Disappointment, Cape (Oregon), 359
Discovery or Nuniwack Island (Russian America), 531
Discovery, Port (De Fuca Strait), 377
Discovery Passage (N.W. America), 400
Dislocation Harbour (Tierra del Fuego), 68
Division, Mount (Peru), 175
Dixon's Channel (N.W. America), 435
Dodo-sima Island (Japanese Creek), 633
Doña Sebastiana or Carelmapu Islets (Chiloe), 90
Doña Maria, Table of (Peru), 158
Doña Paula, Estero (Realejo), 235
Doña Sebastiana Islet (Chile), 103
Don Diego, Port (Mexico), 260
Don Martin Island (Peru), 171
Doris Cove (Tierra del Fuego), 63
Dormido Rock (Chile), 108
Dos Hermanos Islands (Strait of Magalhaens), 25
Dongandsha, Cape (Sea of Okhotsk), 615
Dongangea Island (Sea of Okhotsk), 615
Douglas or San Carlos Island (Russian America), 448
Douglas, Cape (Aliaska Peninsula), 495
Douglas Island (Russian America), 454
Douglas, Cape (Cook's Inlet) 486
Downs Point (Saghalin), 620
Dragon Rocks (Oregon) 348
Drake Harbour (Upper California), 343
Drovianaia Cove (Oumnack Island), 509
Duck Harbour (Patagonia), 70
Duff Point (Vancouver Island), 405
Duke of Clarence Strait (Russian America), 441
Dulce, Gulf of (Central America), 226
Duncan Harbour (Patagonia), 70
Duncan's Canal (Russian America), 444
Duncan Rock (Patagonia), 70
Duncan Rock (Oregon), 374
Dundas Island (N.W. America), 430
Dundee Rock (Patagonia), 74
Dungeness Point (Patagonia), 3, 5
Dungeness, New (De Fuca Strait), 376
Duntze Head (Vancouver Island), 417
Du Petit Thouars Channel (Magdalena Bay), 305
Durango, Province of (Mexico), 288
Dyer, Cape (Russian America), 548
Dyer, Cape (Patagonia), 74
Dyneley Sound (Cockburn Channel), 20
Dyneley Bay (Patagonia) 73-4
Dynevor Castle (Strait of Magalhaens), 28

Eagle Bay (Strait of Magalhaens), 16
Earle Cove (Barbara Channel), 22
East Cape of Asia (Behring's Strait), 563
East Foreland (Cook's Inlet), 487
East Furies (Tierra del Fuego), 20
Earthquakes, Peru, 141
Earthquake, Concepcion (Chile), 112
Echenique, Point (Strait of Magalhaens), 63
Ecuador, Republic of (Colombia), 185
Edgecumbe, Cape and Mount (Sitka Islands), 464
Edgecumbe, Cape and Mount (Sitka Islands) 466
Edgeworth Cape (Barbara Channel), 23
Ediz Hook (De Fuca Strait), 376
Edmund, Point (N.W. America), 424
Edward, Cape (Sitka Islands), 462
Edward, Point (N.W. America), 425
Egatcha River (Sea of Okhotsk), 613
Egg Island (Russian America), 534
Egg Islets (Ounalashka Island), 508
Egorkvoskoï (Oumnack Island), 509
Egvekinot Bay (Siberia), 575
Eidannoo Cape (Behring's Strait), 539
Ekarma Island (Kurile Islands), 606
Ekeptagh River (Siberia), 562
Eld's Island (Oregon), 370
Eld's Inlet (Puget Sound), 381
Eleanor's Cove (Russian America), 472

Elena, Point Santa, and Bay (Guayaquil River), 198
Elephant Point (Russian America), 543
El Frayle Rock (Chile), 108
Eliza Bay (Cockburn Channel), 20
Elizabeth Bay (Strait of Magalhaens), 27
Elizabeth, Cape (Saghalin), 618
Elizabeth, Cape (Russian America), 484
Elizabeth Island (Strait of Magalhaens), 9
Ellice, Point (Columbia River), 361
Ellis, Point (Russian America), 449
El Morion, or St. David's Head (Strait of Magalhaens), 29
El Platina Mountains (Central America), 232
Elpynghyn Mountain (Siberia), 567
Elrington, Point (Prince William's Sound), 481
Elson Bay (Russian America), 552
El Viejo Volcano (Central America), 238
Elvira, Point (Strait of Magalhaens), 26
Elwha, River (De Fuca Strait), 376
Emma Harbour (Siberia), 571
Emua-en Point (Siberia), 562
Endermo Harbour (Jesso Island), 630
Engano, Cabo del (Sitka Islands), 466
Enganno or Deceit, Cape (Tierra del Fuego), 54
Engaoughin, Port of (Siberia), 576
Englefield Bay (Queen Charlotte Island), 434
English Cove (Peru), 160
English Creek (Chile), 106
English Narrows (Patagonia), 83
Enmelian, Cape (Siberia), 572
Enneeapah or Anacapes Island (Upper California), 326
Entrada Point (Magdalena Bay), 306
Entrade de Heceta (Oregon), 364
Epiphany, Port of the (Kodiack Islands), 493
Equipalito Rock (Gulf of California), 301
Ernest's Sound, Prince (Russian America), 443
Eroen or Evosn, Cape (Jesso Island), 630
Eroro Island (Jesso Island), 630
Erou Island (Japanese Archipelago), 628
Esarmi Point (Jesso Island), 630
Esarne or Jesan, Cape (Nippon), 633
Escape Point (Russian America), 441
Escarpé, Cape (Kodiack Islands), 492
Escheran or Zipegua Island (Mexico), 261
Eschscholtz Bay (Russian America), 542
Esclavos, Rio de los (Central America), 249
Escuintla, Province of (Central America), 249
Esmeraldo River and Town (Colombia), 205
Espagnole Riviére, or Rio Colorado (Gulf of California), 299
Española Point, 191
Espenburg, Cape (Russian America), 541
Esperanza Inlet (Vancouver Island), 409
Esperanza Island (Patagonia), 82
Espinosa Arm (Vancouver Island), 410
Espinosa Ridge (Concepcion Bay), 111
Espiritu Santo, Cape (Patagonia), 3
Espiritu Santo, Cape (Tierra del Fuego), 43
Esposicion Island (Central America), 242
Esquimalt or Squimalt Harbour (Vancouver Island), 417
Essington, Port (N.W. America), 428, 430
Estaguillas Point (Chile), 103
Estapa (Mexico), 274
Estero Doña Paula (Realejo), 235
Estero Basin (N.W. America), 402
Estero del Arsenal de San Blas (W. co. Mexico), 284
Estero Real (Central America), 242-3
Esteros Bay and Point (Upper California), 327
Estevan Channel (Patagonia), 80
Esther Island (Prince William's Sound), 480
Estrada Rock (Mexico), 274
Estrada, Port (Queen Charlotte Island), 435
Estrecho de San Pablo (Upper California), 341
Estrete Island (Mexico), 260
Etches, Port (Prince William's Sound), 477
Etelkoulum Bay (Siberia), 575
Eten Point (Peru), 179
Etoline Strait (Russian America), 531
Etoline, Cape (Russian America), 529
Euston Bay (Tierra del Fuego), 65
Euston Opening (Strait of Magalhaens), 28
Evangelists, The, or Isles of Direction (Strait of Magalhaens), 38
Evdokeeff Islands (Aliaska Peninsula), 495
Evouts Isles (Tierra del Fuego), 53
Evratschey Island (Kodiack Islands), 494

Facunda Cape (Japan), 648
Fairway Isles (Patagonia), 80
Fairway Rock, or Oo-ghe-e-ak (Behring's Strait), 540
Fairweather, Cape (Patagonia), 3
Fairweather, Mount (Russian America), 468
Fakarero Island (Japanese Archipelago), 628
Fallos Channel (Patagonia), 72, 75
False Point (Chile), 103
False Cape (Tierra del Fuego), 57
False Cape Horn (Tierra del Fuego), 56
False Point (Bolivia), 143
Falso, Puerto (Upper California), 317
Famisaky or Kagul Cape (Japan), 642
Fanshaw, Point (Russian America), 450
Farallones, The (Upper California), 338, 343
Farallones Alijos Rocks (Lower California), 309
Farallon Rock (Panamá Bay), 216
Fatsisio or Fatsizioo Island (Japanese Islands), 634
Favida or Texada Island, 397
Feklistoff Island (Sea of Okhotsk), 615
Felipe de Jesus, San (Gulf of California), 300
Felipe, Point (Upper California), 323
Felix, Point (Strait of Magalhaens), 36
Fermin, Point (Upper California), 321
Fernando de Magalhaens, History of his Voyage, 1
Ferrer Cove (Vancouver Island), 412

Ferrol, Bay of (Peru), 175
Fida Buengono Cape (Japan), 643
Fidalgo Island (Oregon), 385
Fidalgo, Port (Prince William's Sound), 478-9
Field's Bay (Barbara Channel), 21
Fife's Passage or Sound (New Hanover), 406
Figasi-sima Island (Japan), 644
Fima-sima Island (Japan), 640
Fincham Islands (Tierra del Fuego), 66
Fira-to, Firando or Fyrando (Japan), 645
First Narrow (Strait of Magalhaens), 6
Fisago-sima Island (Japan), 644
Fisen, Bay (Japan), 640
Fisgard Island (Vancouver Island), 417
Fish Cove (Strait of Magalhaens), 9
Fisher's Canal (N.W. America), 425
Fisherman's Cove (N.W. America), 427
Flamenco Bay (Chile), 135
Flat Bay (Saghalin), 620
Flattery, Cape and Rocks (Oregon), 373
Flinn Sound (Patagonia), 75
FitzGibbon Point (Russian America), 437
Fitzhugh's Sound (N.W. America), 423
FitzRoy Island (Cockburn Channel), 19
FitzRoy Channel (Strait of Magalhaens), 28
Five Hummocks Point (Lower California), 311
Fizen, Province of (Japan), 643
Flores Bay (Strait of Magalhaens), 30
Florida Blanca, Cape (Queen Charlotte Island), 435
Foca Point (Peru), 181
Foggy Island (Aliaska Peninsula), 496
Foggy Cape (Russian America), 438
Fogs on Tierra del Fuego (Strait of Magalhaens), 40
Fonseca or Conchagua Gulf (Central America), 240
Forelius Peninsula (Patagonia), 78
Fore-top, The (Valparaiso), 115
Forrester's or San Carlos Island (Russian America), 448
Forsyth Island (Bolivia), 142
Fort George or Astoria (Columbia River), 361
Fort Bluff (Mexico), 279
Fortescue Bay (Strait of Magalhaens), 24-5
Fortune Bay (Patagonia), 81
Foukai-Yerioumi Channel (Japan), 638
Foukouaka Bay (Japan), 640
Foulweather Bluff (Admiralty Inlet), 379
Foulweather, Cape (Oregon), 356
Four Brothers or Quatre Frères Islands (Kurile Islands), 608
Four Mountains, Islands of the (Aleutian Islands), 510
Fox, Cape (Russian America), 438
Fox, Cape (N.W. America), 432
Fox Islands (Aleutian Archipelago), 503
Fox Bay (Chile), 113
Français, Baie des (Russian America), 462
Francis Point (Oregon), 387
Franklin, Point (Russian America), 551
Fraser River (N.W. America), 393
Frazer Island (Vancouver Island), 416
Frederick, Point (Queen Charlotte Island), 435
Frederick, Port (Russian America), 461
Frederick's, Prince, Sound (Russian America), 449
Freemantle, Port (Prince William's Sound), 479
Freshwater Bay (De Fuca Strait), 376
Freshwater Bay (Strait of Magalhaens), 11
Friar's Hill (Patagonia), 3
Friendly Cove (Nootka Sound), 412
Fronton Reef (Concepcion Bay), 111
Fronton Island (Callao), 168
Froward, Cape (Strait of Magalhaens), 17, 23
Fuca, Strait of Juan de (N.W. America), 414
Fuegians, Description of, 2
Fuego, Volcan de (Central America), 252
Fuerte, Rio del (Gulf of California), 296-7
Fuerte, El, or Calbuco (Chiloe), 98
Fuerto Viejo Cove (Chile), 108
Fukajé (Japan), 644
Funetsuki Bay (Japanese Islands), 635
Fureck, Port (Jesso Island), 629
Furies, East and West (Tierra del Fuego), 20
Furies Rocks, East and West (Tierra del Fuego), 64
Fury Island and Harbour (Tierra del Fuego), 64
Fury Harbour (Tierra del Fuego), 20
Fury Peaks (Tierra del Fuego), 64
Fusi, Mount (Japanese Islands), 634-5

Gabriel Channel (Strait of Magalhaens), 14
Galao, Boca del (Guayaquil River), 193
Galapagos Islands (Gulf of California), 300
Galera Island (Panamá Bay), 212
Galera, Punta de la (Chile), 103
Galera Point (Callao), 168
Galera Point (Colombia), 204
Galiano and Valdes Islands (Vancouver Island), 408
Gallant, Port (Strait of Magalhaens), 25
Gallegos, Cape (Patagonia), 83
Gallegos River (Patagonia), 3
Gallo Point (Chile), 114
Gallo Island (Colombia), 205
Gamaley, Cape (Japan), 637
Gambier, Point (Russian America), 452
Gamble, Port (Oregon), 383
Gamboa River (Chiloe), 94
Gap Peak (Strait of Magalhaens), 7
Garachina, Point (Panamá Bay), 210
Garden Island (Prince William's Sound), 477
Gardner, Point (Russian America), 449-50
Gardner, Port (Oregon), 384
Gardner's Canal (N.W. America), 427
Garita, Morro de (Peru), 176
Garnet, Point (Russian America), 544
Garrido Island (Patagonia), 85
Garrobo Island (Central America), 242
Garzos Point (Concepcion Bay), 110

Gaston Bay (Oregon), 387
Gavareah, Cape (Awatska Bay), 595
Gente Grande Point (Strait of Magalhaens), 10
George's, King, Archipelago (Russian America), 461
George's Island, Tres Marias (Mexico), 277
George, Cape (Patagonia), 70
Gil, Isle de (N.W. America), 427-8
Gila, Rio (Gulf of California), 299
Gilbert Island (Tierra del Fuego), 63
Giquilisco Port, or Triunfo de los Libres (Central America), 244
Glacier Sound (Strait of Magalhaens), 35
Glascott Point (Strait of Magalhaens), 17
Glasenapp Harbour (Siberia), 568
Glazenap or Mitkoff, Cape (Aliaska Peninsula), 524
Gloubokaïa Bay (Atkha Island), 514
Gloucester, Cape (Tierra del Fuego), 65-6
Gloutokaïa Cove (Ounmack Island), 509
Gobernador Hill, or Cerro Verde (Chile), 120, 122
Godoy Point (Chile), 103
Gold-dust Isle (Tierra del Fuego), 59
Golenichtcheff, Cape (Karaghinsky Island), 589
Goletas Channel (Vancouver Island), 408
Golovatcheff, Cape (Saghalin), 619
Golownin Bay (Russian America), 535
Golownin Strait (Kurile Islands), 607
Gomez, or Azade, Island of (Mexico), 291
Gonzales Head (Valdivia), 104
Gonzalez Narrows (Strait of Magalhaens), 27
Gonzalo, Point (Vancouver Island), 418
Good Harbour (Patagonia), 76
Good's Bay (Patagonia), 80
Good Hope, Bay of (Jesso Island), 630
Good Hope, Bay of (Russian America), 541
Good Success Bay and Cape (Tierra del Fuego), 49, 51
Good News, Bay of (Russian America), 530
Good Luck Bay (Strait of Magalhaens), 30
Gorbun Rock (Kodiack Islands), 492
Gorda Point (Central America), 229
Gorda Point (Colombia), 205
Gordon, Point (New Hanover), 406
Gordon River (Vancouver Island), 416
Gore, Cape (St. Matthew Island), 580
Gore, Point (Russian America), 484
Goree Road (Tierra del Fuego), 52
Gore's or St. Matthew Island (Sea of Behring), 579
Goreli or Segouam Island (Aleutian Islands), 511
Goreloy or Burnt Island (Aleutian Islands), 516
Gorgona Island (Colombia), 205
Goto, Cape, or Ohoseno-saki (Japan), 644
Goto Islands (Japan), 543
Gottsofka River (Sea of Okhotsk), 611
Goulding's Harbour (Sitka Islands), 463
Govenskoï, Cape (Siberia), 578
Grace, Point (Prince William's Sound), 481
Gracia Cape (Strait of Magalhaens), 9
Grafton Islands (Tierra del Fuego), 65-6
Graham's Harbour (Cook's Inlet), 485
Grajero Point (Lower California), 312
Gramadel Bay (Peru), 172
Grande, Isla (Chile), 131
Grantley Harbour (Russian America), 538
Graves Island (Tierra del Fuego), 67
Graves, Mount (Strait of Magalhaens), 14
Gravina, Island (Russian America), 438
Gravina, Port (Prince William's Sound), 478
Gray's Harbour (Oregon), 370
Gray's Bay (Columbia River), 361
Green Island (Prince William's Sound), 481
Green Island (Guayaquil River), 194
Greenough Peninsula (Strait of Magalhaens), 25
Gregory Range (Strait of Magalhaens), 7
Gregory, Cape (Oregon), 355
Greig, Cape (Aliaska Peninsula), 527
Greig, Cape (Japan), 638
Grenada, Town of (Central America), 234
Grenville's Canal (N.W. America), 428
Grenville, Point (Oregon), 372
Greville or Tolstoy, Cape (Kodiack Islands), 493
Grey, Point (New Georgia), 394
Grifo or Roqueta Island (Mexico), 269
Grindall, Point (Russian America), 442
Grueso Point (Peru), 145
Guabun Head (Chiloe), 89
Guacalat, River (Central America), 249
Guadaloupe, Mission of (Upper California), 327
Guaianeco Island (Patagonia), 75-6
Gualatieri or Schama Peak (Andes), 149
Gualillas Pass (Andes), 149
Guambacho or Samanco Bay (Peru), 174-5
Guanaco Point (Tierra del Fuego), 53
Guañape, Hill and Islands of (Peru), 176
Guarmey (Peru), 172
Guasco or Huasco (Chile), 128-9
Guano, 139, 143, 144, 149, 150, 160, 169
Guascoma River (Central America), 243
Guatemala, Description of, 222
Guatemala, Volcan de (Central America), 246
Guatlan (Mexico), 276
Guatulco, Port (Mexico), 266
Guayaquil (Ecuador), 189
Guayaquil City of, 192
Guaymas (Mexico), 297
Guaytecas Islands (Patagonia), 86
Guemas Island (Oregon), 387
Guia Narrows (Patagonia), 82
Guibert or Refunshery Island (Jesso), 632
Guiranas, Punta de (Upper California), 317
Guirior Bay (Strait of Magalhaens), 30
Gull Head (Russian America), 542
Gulf of California (Mexico), 287
Gun Bay (Strait of Magalhaens), 16

Hagemeister Island (Russian America), 529
Hahamish Harbour (Oregon), 384

Half Port Bay (Strait of Magalhaens), 32
Halgan, Cape (Siberia), 572
Halibut or Sannagh Island (Aliaska Peninsula), 500
Hammersley's Inlet (Paget Sound), 381
Hamond, Cape (Russian America), 476
Hamper Bay (Patagonia), 81
Hanover Island (Patagonia), 80, 82
Hans Olason Island (Japanese Archipelago), 631
Hanson, Point (Oregon), 370
Hapana Reach, or Tlupana Arm (Vancouver Island), 413
Harbour Islet (Peru), 173
Hardwicke's Island (N.W. America), 402
Hardy Peninsula (Tierra del Fuego), 53, 56
Hardy Bay (Vancouver Island), 408
Harmless Point (Peru), 157
Haro Strait (Vancouver Island), 419
Harriet, Point (Cook's Inlet), 487
Harrington, Point (Russian America), 445
Harris, Port (Russian America), 449
Harvey Bay (Patagonia), 76
Harwood's Island (N.W. America), 398
Hawkesbury's Island (N.W. America), 427
Hawkins' Island (Prince William's Sound), 478
Hawkin's Bay (Strait of Magalhaens), 25
Hazel, Point (Oregon), 384
Hazy Island (Russian America), 454
Hector Rock (Chile), 107
Heiden, Mount (Siberia), 576
Héliaghya Bay (Siberia), 566
Hellyer Rocks (Patagonia), 84
Hemming's Bay (Prince William's Sound), 482
Henderson Island (Tierra del Fuego), 58
Henderson's Inlet (Puget Sound), 382
Henry, Port (Patagonia), 71
Henry, Cape (Queen Charlotte's Island), 434
Herald Island (Arctic Sea), 559
Hermite Isles (Tierra del Fuego), 53
Hermosillo (Mexico), 296
Hermoso River (Central America), 242
Hermoso, Morro (Lower California), 309
Herradura de Carrisal, Bay of (Chile), 129
Herradura de Mexillones, Bay of (Bolivia), 142
Herradura de Coquimbo (Chile), 124
Herradura or Pichidanque Bay (Chile), 122
Hewett Bay (Tierra del Fuego), 20
Hewett Bay (Barbara Channel), 21
Hey, Point (Prince William's Sound), 477
Heywood's Passage (Patagonia), 81
Hicarita Island (Central America), 219
Hicaron or Quicara Island (Central America), 219
Hidden Harbour (Strait of Magalhaens), 26
Higgins, Point (Russian America), 441
Highfield, Point (Russian America), 443
Hill's Islands (Barbara Channel), 22
Hill Rock (Quibo Island), 219
Hinchinbrook Island (Prince William's Sound), 477
Hippah Island (Queen Charlotte Island), 434
Hitson, Cape (Ounimack Island), 504
Hobart, Point (Russian America), 452
Holkham Bay (Russian America), 452
Holland, Cape (Strait of Magalhaens), 24
Holloway Sound (Patagonia), 79
Hood's Canal (Oregon), 383
Hood's Bay (Russian America), 455
Hope, Point (Russian America), 547
Hope Harbour (Magdalen Sound), 18
Hope Harbour (Tierra del Fuego), 65
Hope Island (Tierra del Fuego), 59
Hope Bay (Vancouver Island), 409
Hopkins, Point (N.W. America), 427
Hoppner, Port (Staten Island), 48
Hoppner Sound (Patagonia), 79
Horace Peaks (Tierra del Fuego), 63
Horadada Island (Callao), 168
Horca Hill (Peru), 172
Horcon Head (Chile), 121
Hormigas, The (Peru), 169
Horn, Cape (Tierra del Fuego), 54
Horn, False Cape (Tierra del Fuego), 56
Horner, Cape (Saghalin), 619
Horner, Peak (Japan), 640
Hose Harbour (Patagonia), 81
Hotham Inlet (Russian America), 546
Houdobin or Comte Heiden Bay (Aliaska Peninsula), 526
Houghton, Port (Russian America), 451
Howe's Sound (New Georgia), 394
Howe, Point (Russian America), 444
Huacaneo Island (Patagonia), 86
Huacas Point (Peru), 159
Huacho, Bay of (Peru), 171
Huafo or No-man's Island (Patagonia), 85, 87
Huamblin or Socorro Island (Patagonia), 85, 86
Huamblin or San Pedro Island (Chiloe), 89
Huamilulu (Mexico), 261
Huanaquero Point and Hill (Chile), 124
Huanchaco (Peru), 176
Huanchaco Peak (Peru), 178
Huano Creek (Peru), 150
Huapacho Shoal (Chiloe), 89, 90
Huapilacuy (Chiloe), 90
Huapilinao Head (Chiloe), 92
Huar Island (Chiloe), 99
Huara Islands (Peru), 170
Huasco or Guasco (Chile), 128-9
Huaytecas or Guaytecas Islands (Patagonia), 86
Huechucucuy Head (Chiloe), 89
Hueñauca or Osorno Volcano (Chile), 99
Hugh, Point (Russian America), 452
Huildad, Islet of (Chiloe), 97
Huilotepec Hills (Mexico), 260
Humos Cape (Chile), 113
Hunter, Point (Queen Charlotte Island), 434
Hunt, Point (N.W. America), 429
Hurtado Point (N.W. America), 398
Hyde, Mount (Tierra del Fuego), 54
Iaïtchnoï, Cape (Atkha Island), 513

Ibbertson's Sound (Queen Charlotte Island), 434
Ibbetson, Cape (N.W. America), 429
Icy Sound (Barbara Channel), 22
Icy Sound (Strait of Magalhaens), 35
Icy Bay (Russian America), 473
Icy Cape (Russian America), 549
Icy (Arctic) or Polar Sea, 553
Icy (Arctic) Barrier, 553
Igatskoy or Ihack Bay (Kodiack Islands), 493
Igighinskoï or Jieghinsky, Gulf of (Sea of Okhotsk), 612
Ignacio Bay (Patagonia), 77
Ignacio, Point and Rock (Gulf of California), 297
Igouan (Siberia), 566
Ihack or Igatskoy Bay (Kodiack Islands), 493
Ikatun Island (Aliaska Peninsula), 501
Iki or Yki Island (Japan), 640
Ilay (Peru), 152
Ilay Point (Peru), 151, 153
Ildefonso Islands, St. (Tierra del Fuego), 58
Ildefonso or Upright Cape (Strait of Magalhaens), 35
Ileñao Island (Panamá Bay), 217
Ilimen, Cape (Russian America), 528
Illack Islands (Aleutian Islands), 517
Ilpinskoï, Cape (Siberia), 578
Ilpinskoï, Cape (Kamtschatka), 590
Imeldeb Island (Chiloe), 95
Imoshiri Island (Jesso Island), 630
Imou Rouk Lake (Russian America), 538
Inchin or San Fernando Islands (Patagonia), 85
Indian Reach (Patagonia), 83
Indian Cove (Tierra del Fuego), 58
Infantes, Isle de los (Strait of Magalhaens), 27
Ingiga or Fort Jiejiginsk (Sea of Okhotsk), 613
Inglefield, Cape (Strait of Magalhaens), 26
Inlet Bay (Patagonia), 81
Inman Bay (Strait of Magalhaens), 25
Inman, Cape (Tierra del Fuego), 67
Inskip Islands (Vancouver Island), 417
Ipswich Island (Tierra del Fuego), 65
Iquique (Peru), 145
Ir-Kaipie or North Cape (Siberia), 562
Isabel, Cape (Patagonia), 70
Isabella Island (Mexico), 287, 289
Isalco, Volcano of (Central America), 249
Isanotskoy, Strait of (Aliaska), 501
Iscuandé River (Colombia), 207
Iseno Umi Lake (Japan), 635
Island or Borja Bay (Strait of Magalhaens), 28
Islands, Bay of (Sitka Islands), 463
Islas del Rey (Panamá Bay), 212
Ismenaï Islet (Awatska Bay), 597
Isquiliac Island and Mount (Patagonia), 85
Issannakh or Sannagh Island (Aliaska Peninsula), 501
Isthmus of Panamá, 214
Istapa, Port of (Central America), 249
Isthmus Bay (Patagonia), 81
Istmo, Province of (New Granada), 186
Itouroup Island (Kurile Islands), 608
Itsha River (Sea of Okhotsk), 611
Itsoumo, Cape (Japan), 636
Ittygran Island (Siberia), 568
Izenbek Bay (Aliaska Peninsula), 524

Jack's Harbour (Strait of Magalhaens), 17
Jackson, Point (Russian America), 536
Jacobi Island (Sitka Islands), 456
Jaguey Point (Peru), 172
Jakuno-sima Island (Japan), 641
Jalisco, Province of (Mexico), 286
Jalisco, Province of (Mexico), 275
Jambeli Creek (Guayaquil River), 190
James Islands (Strait of Magalhaens), 27
Jampa Point (Guayaquil River), 199
Jamsk (Sea of Okhotsk), 613
Jan Harbour (Sea of Okhotsk), 614
Japanese Archipelago, 624
Japanese Junk, Wreck of (*Oregon*), 372
Japanese Wreck (Kamtschatka), 603
Jara Head (Bolivia), 142
Jaseur Reef (Strait of Le Marie), 50
Jedo (Japanese Archipelago), 633
Jémenoff Cape (Karghinsky Island), 589
Jequepa or Tequepa, Point (Mexico), 273
Jerabou Island (Japan), 642
Jerdan's Peak (Tierra del Fuego), 65
Jerome Channel (Strait of Magalhaens), 28
Jervis's Canal (Howe's Sound), 396
Jesso, Island of (Japanese Archipelago), 627
Jesuit Sound (Patagonia), 77
Jesus, City of (Strait of Magalhaens), 12
Jieghinsky or Igighinskoï, Gulf of (Sea of Okhotsk), 612
Jiquilisco, San Salvador de (Central America), 244-5
Jirausa or Siriano River (Central America), 242
Joann Bogosloff Island (Aleutian Islands), 510
John Begg Reef (Upper California), 395-6
John Port (N.W. America), 425
John Renwick Rock (Chile), 106
Johnstone's Straits (Vancouver Island), 403
Jonas Island (Sea of Okhotsk), 615
Jono-kosoutsy Mountain (Japan), 639
Jootsima Island (Japan), 637
Jordan Island (Tierra del Fuego), 56
Jordan River (Vancouver Island), 416
Jorgino, Mount (Bolivia), 142
Jorullo, Volcan de (Mexico), 274
Josef Bay (Vancouver Island), 409
Joupanoff Volcano (Kamtschatka), 594
Joupanova River (Kamtschatka), 593
Juan de Fuca, Strait of (Oregon), 373, 414
Juan Gomez, Mesas or Tables of (Lower California), 313
Juanilla Island (Central America), 230
Juchitan River (Mexico), 260
Judas, Cape (Magdalena Bay), 306

Judges Rocks (Tierra del Fuego), 68
Julia Island (Japan), 641
Juluapan Point (Mexico), 274

Kaegalack Coast (St. Lawrence Island), 579
Kaeghdagh Island (Aliaska Peninsula), 506
Kagai Island (Aliaska Peninsula), 498
Kagalaska Island (Aleutian Islands), 515
Kago-sima (Japan), 641
Kagul Cape, or Fasimisaky (Japan), 642
Kakhvalga Island (Aleutian Islands), 517
Kalakhtyrka, River (Kamtschatka), 594
Kaleghta Bay (Ounalashka Island), 507
Kalegbta, Cape (Ounalashka Island), 506
Kamennoï Cape (Russian America), 535
Kamennaia Pristan, or Stone Mole (Russian America), 536
Kaminoë (Sea of Okhotsk), 613
Kamtschatkoï, Cape (Kamtschatka), 591
Kamtschatka, Peninsula of, 587
Kamtschatka River (Kamtschatka), 592
Kanaga or Konniaga Island (Aleutian Islands), 516
Kanero Misaky, Cape (Japan), 640
Kanghynin, Bay of (Siberia), 575
Karafta or Saghalin, 616
Karaga, River (Kamtschatka), 590
Karaghinsky Island (Kamtschatka), 588
Karaghinsky Bay (Kamtschatka), 590
Karaghinsky, Little, or Verkhotoursky Island (Siberia), 577
Karquines, Straits of (San Francisco), 336
Kassatotchy Island (Aleutian Islands), 515
Katalamet Range (Oregon), 361
Kater's Peak (Tierra del Fuego), 54
Kaviayak Bay, or Port Clarence (Russian America), 537
Kawa-sima, Island (Japan), 644
Kaye's Island (Russian America), 476
Keats's Sound (Magdalen Sound), 19
Kellim or Kellmso Pond (Oregon), 383
Kelly Harbour (Patagonia), 77-8
Kempe Peaks (Tierra del Fuego), 64
Kempe Harbour (Strait of Magalhaens), 25
Kentel, River (Siberia), 562
Ketoy Island (Kurile Islands), 607
Ketsoth Village (De Fuca Strait), 376
Khabaroff, Cape (Sea of Okhotsk), 616
Khakodade Bay (Japanese Archipelago), 631
Khaluetkin, Cape (Siberia), 566
Kharamoukotan Island (Kurile Islands), 606
Khitzoff, Cape (Sea of Behring), 584
Kiack Harbour (Kodiack Islands), 493
Kigalga Island (Aleutian Islands), 505, 510
Kigamiliakh Island (Aleutian Islands), 510
Kikiay Island (Japan), 642
Kikiusima or Firosima Island (Japanese Archipelago), 640
Killiliack Bay (Ounalashka Island), 508
Kiloudenskoy Bay (Kodiack Islands), 493
Kiluden Bay (Kodiack Islands), 493
King Island (Cockburn Channel), 19
King, Cape (Japanese Islands), 633
King-a-ghee Island (Behring's Strait), 539
King George III.'s Archipelago (Russian America), 456
King George's or Nootka Sound (Vancouver Island), 410
King's Island (N.W. America), 425
King's Island (Panamá Bay), 212
King's or Oukivok Island (Behring's Strait), 536
Kingsmill Point (Russian America), 449
Kinnaird, Point (Tierra del Fuego), 51
Kino Channel (Japan), 638
Kirke's Rocks (Cockburn Channel), 19
Kirilovskaia Bay (Aleutian Islands), 517
Kirimasan Mountain (Japan), 640
Kiska Island (Aleutian Islands), 518
Kitagotagh Island (Aliaska Peninsula), 500
Kitenomagan Island (Aliaska Peninsula), 502
Kithouck or Khitkhoukh Cape (Ounimack Island), 504
Kitson Island (Oregon), 383
Kiuniutanany Island (Aliaska Peninsula), 498
Kiusiu, Island of (Japan), 639
Kiusiu, Bay of (Japan), 643
Klaholoh Rock (De Fuca Strait), 376
Klamet or Too-too-tut-na River (Oregon), 348
Klatsop Village (Colombia River), 360
Kliutchevsk or Klutchevskoï Volcano (Kamtschatka), 592
Klokatcheff, Cape (Saghalin), 620
Klutchevskoï Volcano (Aleutian Islands), 513
Klutchevskoï or Kamtschatkoï (Kamtschatka), 592
Knight's Island (Prince William's Sound), 481
Knight's Island (Russian America), 472
Knight's Canal or Inlet (New Hanover), 406
Kodiack Archipelago (Russian America), 490
Kokura (Japan), 640
Koliutchin or Burney Island (Siberia), 562
Koniouge Island (Atkha Island), 514
Konniaga or Kanaga Island (Aleutian Islands), 516
Korovinskaia, Cape and Bay (Atkha Island), 513
Korovinskoi Volcano and Bay (Aleutian Islands), 513
Kosiki Group, or Meac-sima (Japan), 642
Ko-sima Island (Japanese Islands), 634
Ko-sima Island (Jesso), 631
Kotzebue Sound (Russian America), 540
Kouatskoï, Peak of (Kamtschatka), 595
Kougalga Island (Aleutian Archipelago), 505
Kougouan, Cape (Siberia), 567
Kounashire Island (Japanese Archipelago), 628
Kounashire Island (Kurile Islands), 609
Kourose-gawa Stream (Japanese Islands), 635
Kourovskaia Bay (Atkha Island), 514
Kouskoquim River (Russian America), 530
Kousoff Island (Sea of Okhotsk), 615
Koutchougoumut (Russian America), 527

Koutousoff, Cape and Bay (Jesso), 632
Koutouzoff, Cape (Aliaska Peninsula), 526
Kouzmichtcheff, Cape (Kamtschatka), 590
Koyanosi (Japan), 640
Krachenninikoff, Cape (Karaghinsky Island), 589
Krenitzin Point (Aliaska Peninsula), 524
Krenitzin Cape (Kurile Islands), 606
Krenitsin Islands (Aleutian Archipelago), 505
Kritskoi Island (Aliaska Peninsula), 526
Krleougoun, Cape (Siberia), 564
Kronotskoi, Cape and Volcano (Kamtschatka), 593
Krusenstern, Cape (Russian America), 546
Krusenstern Island, or Igna-look (Behring's Strait), 540
Kryoi or Rat Island (Aleutian Islands), 517-8
Kululack River (Russian America), 529
Kurile Islands (North Pacific), 604
Kuro Siwo Stream (Japanese Islands), 635
Kuskowime River (Russian America), 530
Kvikhluak River (Russian America), 532
Kvikhpak River (Russian America), 532
Kydaka Point (De Fuca Strait), 376
Kyghynin, Cape (Siberia), 568
Kynkai Island (Siberia), 569

La Asuncion Island (Lower California), 309
Labyrinth Islands (Magdalen Sound), 19
Lachira Bay (Peru), 163
Lachaforellen Bay (Saghalin), 623
Ladrone or Zedrones Islands (Central America), 229
La Guata Cove (Peru), 154
Laguna, Morro de la (Mexico), 261
La Mar, or Cobija (Bolivia), 143
Lambayeque Road (Peru), 179
Lambert, Point (N.W. America), 428
Lami Bank (Chiloe), 98
Landfall Island (Tierra del Fuego), 66-7
Land of Desolation (South America), 20
Langara Island (Queen Charlotte Island), 435
Langara Bay (Strait of Magalhaens), 30
Langara, Port (Strait of Magalhaens), 27
Langley, Fort (N.W. America), 393
Lanz Islands (Vancouver Island), 409
La Paz (Gulf of California), 302
La Pérouse, Strait of (Japanese Archipelago), 632
La Playa, San Blas (Mexico), 281
Laraquete Beach (Chile), 108
Laredo Bay (Strait of Magalhaens), 10
Las Animas Island (Gulf of California), 300
Las Animas (Chile), 135
Las Bajas Islet (Peru), 170
La Serena or Coquimbo (Chile), 125
La Soledad Valley (Upper California), 328
Lasuen, Point (Upper California), 320
Last Harbour (Strait of Magalhaens), 27
Last Hope Inlet (Patagonia), 82
Latitude Bay (Tierra del Fuego), 67
Latouche, Point (Russian America), 472
Latouche Island (Prince William's Sound), 481
Launches Rocks (Tierra del Fuego), 68
La Union, Port (Central America), 241
Laura Basin (Tierra del Fuego), 66
Lavapie Point (Chile), 107
Lavata Bay (Chile), 136
La Visitacion (Gulf of California), 300
Lavinia, Point (Russian America), 461
Lawrence, Port (De Fuca Strait), 379
Laxa Rock, Manzanilla (Mexico), 290
Laxman Bay (Jesso Island), 629
Lay, Point (Russian America), 549
Laytec Island (Chiloe), 97
Lazaro, Cape San (Lower California), 308-9
Leading Bluff (Bolivia), 142
Leading Island (Tierra del Fuego), 65
Leading-in Cliff (Columbia River), 364
Ledo Unala or Pamplona Rock (Russian America), 474
Ledianaya River (Siberia), 572
Lees, Point (Russian America), 440
Legarto Head (Peru), 173
Leigan or Roses Harbour (Queen Charlotte Island), 436
Lelbun Point (Chiloe), 96
Le Marie, Strait of (Tierra del Fuego), 48
Le Mesurier, Point (Russian America), 442
Lempa, River (Central America), 245
Lemuy (Chiloe), 93
Lemuy Island (Chiloe), 95
Lengua de Vaca Point (Chile), 123
Lengua Point (Magdalena Bay), 306
Lennox Harbour (Tierra del Fuego), 52
Lennox Island (Tierra del Fuego), 52
Leon, City of (Central America), 234
Leones Point (Chile), 128
Leontovitch, Cape (Aliaska Peninsula), 525
Leskoff, Cape (Aliaska Peninsula), 525
L'Etoile Cape (Strait of Magalhaens), 32
Leübu River (Chile), 107
Levacheff, Port (Ounalashka Island), 507
Lewis, Cape (Russian America), 548
Lialinskigh Island (Aliaska Peninsula), 500
Libertad, Port (Central America), 245-6
Lighthouse of Valparaiso, 117
Lighthouse, Guayaquil River, 190
Ligua River (Chile), 122
Liles Point (Chile), 121
Lima (Peru), 165, 167
Limari River (Chile), 123
Lime Rock (Upper California), 341
Linekinskoi, Cape (Sea of Okhotsk), 616
Linlin, Island of (Chiloe), 93
Liulingaï Cape (Siberia), 575
Linschoten Archipelago (Japan), 641
Linschoten Archipelago, or Asses' Ears (Japan), 645
Lintinao Island (Chiloe), 93
Linna, Island of (Chiloe), 93
Lion Cove (Strait of Magalhaens), 30
Lirquen Point (Concepcion Bay), 109
Lisburne, Cape (Russian America), 548
Lliuco (Chiloe), 92

Loa, Gully and River of (Bolivia), 144
Lobo Point (Chile), 129
Lobo or Blanca Point (Peru), 145
Lobo Point (Chile), 109
Lobos Point (San Francisco), 335
Lobos Head (Chiloe), 92
Lobos Marinos Island (Gulf of California), 297
Lobos de Afuera Islands (Peru), 180
Lobos de Tierra Islands (Peru), 180
Locos Island (Chile), 122
Loffzoff, Cape (Kurile Islands), 609
Logan Rock (Patagonia), 79
Loma, Point de la (Upper California), 317
Lomas Bay (Strait of Magalhaens), 14
Lomas Bay (Tierra del Fuego), 43
Lomas Bank (Strait of Magalhaens), 5
Lomas Valley and Point (Peru), 156
London Island (Tierra del Fuego), 63
Long Reach (Strait of Magalhaens), 28
Long Island (Kodiack Islands), 491
Long Island (Patagonia), 81
Look-out, Cape (Oregon), 356
Lopatka, Cape (Kamtschatka), 604
Lopez Island (Oregon), 386
Lopez Island (Vancouver Island), 419
Lora Point (Chile), 114
Lord Nelson Strait (Patagonia), 70, 80
Los Custodios, Cape (Mexico), 277
Los Frayles Rocks (Mexico), 274
Lotillo Cove (Chili), 108
Loughborough's Canal (N.W. America), 402
Low, Port (Patagonia), 86-7
Low Peak (Tierra del Fuego), 68
Löwenorn, Cape (Saghalin), 622
Löwenstern, Cape (Behring's Strait), 540
Löwenstern, Cape (Saghalin), 619
Lucan, Point (Russian America), 461
Luco Bay (Chile), 107
Ludlow, Port (Oregon), 383
Lugren Cape (Siberia), 566
Luri River (Peru), 163
Lütke, Cape (Ounimack Island), 504

Macabi Island (Peru), 178
Macartney, Point (Russian America), 450
Machado, Cape (Patagonia), 76
Mackenzie Point (Cook's Inlet), 488
Macnamara, Point (Russian America), 445
Madan, Point (Russian America), 443
Madera Island (Lake Nicaragua), 233
Madison, Port (Admiralty Inlet), 380
Madre de Dios Archipelago (Patagonia), 70, 72
Madre de Dios Islands (Patagonia), 80, 82
Madre, Sierra (Central America), 249
Magalhaens, Strait of, 1
Magdalen Sound (Tierra del Fuego), 18
Magdalena, Gulf of (Lower California), 305
Magdalena Bay (Lower California), 306
Magdalena, Province of (New Granada), 186
Magetshadack Island (Kurile Islands), 606
Magill Islands (Tierra del Fuego), 64
Magill's Islands (Tierra del Fuego), 20
Maglares, Point (Gulf of California), 302
Magnetic Head (Japan), 643
Maguinana Island (Guayaquil River), 195
Main-top, The (Valparaiso), 115
Makouchinskoy Bay (Ounalashka Island), 508
Mala Point (Central America), 209
Mala, Point (Panamá Bay), 219
Mala Hill, 191
Mala, Baja de, 191
Malabrigo Road (Peru), 178
Malaspina Strait (N.W. America), 397-8
Malaspina Cape (Jesso), 632
Malmesbury, Port (Russian America), 449
Malpelo Point (Peru), 182
Malpelo Island (Colombia), 207
Mamilla, Gully of (Bolivia), 144
Managua, Lake of (Central America), 233-4
Managua Volcano (Central America), 226
Manati, Cape (Sea of Behring), 583
Manby, Point (Russian America), 473
Mancora, Valley of (Peru), 182
Mandinga Punta, or Puna Bluff, 191
Mangrove Island (Magdalena Bay), 306
Manguera or Miauguera Island (Central America), 242
Mangles Point (Colombia), 203
Manta, Port of (Colombia), 201
Mantos, Port (Central America), 226
Manzanilla, or Salagua (Mexico), 274
Manzano Bank (Concepcion Bay), 109
Manzano Cove (Chile), 103
Manzera Shoal and Islet (Valdivia), 104
March Harbour (Tierra del Fuego), 62
Maria, Cape (Saghalin), 618
Maria, Mount (Kounashire Island), 610
Mariato, Cape (Central America), 219
Marian's Cove (Strait of Magalhaens), 32
Marietas, Las (Mexico), 276
Marine Islands (Patagonia), 79
Maritch River (Siberia), 567
Marques, Port (Mexico) 269, 271
Mar Rojo, or Gulf of California (Mexico), 295
Marrowstone Point (De Fuca Strait), 379
Marsden, Port (Russian America), 459
Marsden, Point (Russian America), 455
Marsh, Point (Russian America), 550
Marshall, Point (N.W. America), 397
Marvinos Bay (Vancouver Island), 413
Mary Island and Point (N.W. America), 400
Mary, Port (Sitka Islands), 464
Mas-al-Oeste, Punta de (Patagonia), 81
Masaredo, Port (Queen Charlotte Island), 435
Maskelyne, Point (N.W. America), 430
Massacre Bay (Attow Island), 519
Mastiri Island (Japanese Archipelago), 628
Matalqui, Cape (Chiloe), 89
Matana Island (Kurile Islands), 607
Matanchel (Mexico), 277
Matatchingaï Mountain (Siberia), 575
Matorillos, Island of (Guayaquil River), 195
Matsumay or Matsmaï (Jesso), 361

Matsuyker, Cape (Jesso Island), 630
Maule River (Chile), 113
Maullin Inlet (Chile), 103
Maxwell, Port (Tierra del Fuego), 56
Mayo, Rio (Mexico), 297
Maytensillo Cove (Chile), 123
Mazatlan (Mexico), 287
Mazorque Island (Peru), 170
Mazzaredo Bay (Strait of Magalhaens), 25
M'Intyre's Bay (Queen Charlotte Island), 435
M'Intyre, Cape (Patagonia), 74
M'Leod's Harbour (Prince William's Sound), 482
M'Loughlin Point (Vancouver Island), 417
M'Neil, Port (Vancouver Island), 408
Meac-sima or Kosiki Group (Japan), 642
Meares, Port (Russian America), 448
Medal Bay, or Puerto de la Medalla (Strait of Magalhaens), 33
Medio Point (Chile), 133
Medny or Copper Island (Sea of Behring), 584
Mee-na Point (De Fuca Strait), 375
Meetchken, Cape (Siberia), 574
Mehuin River (Chile), 105
Meinghyngaï Mountain (Siberia), 568
Meli-moyu, Mount (Andes), 87
Melizzos or Twin Peaks (Andes), 149
Melville Sound (Cockburn Channel), 20
Menchikoff, Cape (Aliaska Peninsula), 526
Menchuan Island (Patagonia), 85
Mendocino, Cape (Upper California), 346
Menzies Bay (N.W. America), 401
Mercenarios (Gulf of California), 302
Mercy, Harbour of (Strait of Magalhaens), 37
Mertens, Cape (Siberia), 567
Mesas (or Tables) de Juan Gomez (Lower California), 313
Mesas (or Tables) de Narvaez (Lower California), 305
Mesier Channel (Patagonia), 76, 80, 83
Mesima Islands, or Asses' Ears (Japan), 645
Mesurier, Point (Russian America), 476
Metchigmensk Bay (Siberia,) 566
Meulin Island (Chiloe), 95
Mexicana Point (Vancouver Island), 409
Mexico, Republic of, 253
Mexico Point (Peru), 151
Mexillones, Bay of (Bolivia), 143
Mexillones, Herradura de (Bolivia), 142
Mianguera Island (Central America), 242
Michatoyat River (Central America), 249
Michoacan, Province of (Mexico), 275
Middle Cape (Staten Island), 50
Middle Island (Patagonia), 73
Middle Bay (Chile), 131
Mier, Port (Vancouver Island), 409
Miga Point (Port Culebra), 229
Mikura Island (Japanese Islands), 634
Milagro Cove (Chile), 103
Milbank Sound (N.W. America), 425
Minchen-madom Mountain (Andes), 87
Milky Way, The (Tierra del Fuego), 64
Millar's Cove (Strait of Magalhaens), 26
Millon Point (Chile), 107
Mill Point, or Punta del Molino (Valdivia), 104
Misaky, Cape (Japan), 639
Misericordia, Puerto de la (Strait of Magalhaens), 37
Misione de Caborca (Gulf of California), 299
Mistaken or Deceit Cape (Tierra del Fuego), 54
Mita, Point (Mexico), 276
Mitake Mountain (Japan), 641
Mitchell, Point (Russian America), 444
Mitkoff or Glazenap, Cape (Aliaska Peninsula), 524
Mitzou-Oumi (Japan), 636
Mitsu-se, or Hen and Chickens (Japan), 644
Mizen Top, The (Valparaiso), 116
Mocha Island (Chile), 105
Moché, 177
Mocuina Basin (Vancouver Island), 410
Mogote (Central America), 233
Moguegua (Peru), 151
Moira Sound (Russian America), 442
Molate Island (Upper California), 340
Moleje, Bay of (Gulf of California), 300
Moleje (Gulf of California), 300
Molguilla Point (Chile), 106
Molino, Punta del, or Mill Point (Valdivia), 104
Mollendo, Port of (Peru), 151
Molendito Cove (Peru), 154
Moller, Cape and Bay (Aliaska Peninsula), 525
Momotombo Volcano (Central America), 226
Mondrayon, Island of (Guayaquil River), 194
Mongon, Cerro or Mount (Peru), 174
Mongoncilla Point (Peru), 173
Monkonrushy Island (Kurile Islands), 605
Monmouth Islands (Strait of Magalhaens), 27
Monmouth, Cape (Strait of Magalhaens), 10
Monserrate, Island of (Gulf of California), 302
Montague, Cape (Patagonia), 73
Montague Bay (Patagonia), 81
Montagu Island (Prince William's Sound), 482
Montanita Point (Guayaquil River), 199
Montezuma, City of (Upper California), 336
Monte Christo, Mount (Columbia), 201
Monte de Buchon (Upper California), 327
Monterey, Bay and Town of (Upper California), 328, 333
Montijo Bay (Central America), 219
Montrose or Xavier Island (Patagonia), 77
Montuosa Island (Central America), 221
Mordwinoff, Bay (Saghalin), 622
Mordvinoff, Cape (Ounimack Island), 504
Moreno, Morro, or Monte Jorge (Bolivia), 142
Moreno, Bay of (Bolivia), 142
Morion, El, or St. David's Head (Strait of Magalhaens), 29

Morjovi or Morse Island (Sea of Behring), 582
Morjovi or St. Matthew's Island (Sea of Behring), 579
Morjovskoï (Aliaska Peninsula), 524
Moro Saki Point (Japan), 635
Morro of Santa Agueda (Strait of Magalhaens), 18
Morro de Puercos (Central America), 219
Morro Hermoso (Lower California), 309
Morro Solar (Peru), 163
Morro, Port de la (Patagonia), 72
Morton Island (Tierra del Fuego), 58
Mount Corso Island (Patagonia), 73
Mount, St. Augustin (Cook's Inlet), 486
Mountain of Guatimala (Central America), 250
Mourodonosaky, Cape (Japan), 639
Mudge, Point, 400
Muelles Point (Chile), 122
Muerto or Amortajada Island (Guayaquil River), 190
Mulatus Point (Peru), 170
Mulgrave, Port (Russian America), 471
Mulgrave Range (Russian America), 546
Muloffsky, Cape (Saghalin), 622
Multonomah or Wappatoo Island (Columbia River), 368
Murciellagos, or Bat Islands (Central America), 230
Murray Cove (Strait of Magalhaens), 26
Murray Narrow (Tierra del Fuego), 53
Muscle Bay (Strait of Magalhaens), 27
Muscle Island (Strait of Magalhaens), 24
Mussel Canal (N.W. America), 426
Mutico Point (Chiloe), 90
Myssoff or Cape Bay (Kodiack Islands), 493

Nacascolo, Naguiscolo Port, or Playa Hermosa (Central America), 232
Nacoame River (Central America), 242
Nadiéjeda or Nadeshda Bay (Saghalin), 619
Nadiéjeda, Cape (Jesso), 631
Nadiéjeda Strait (Kurile Islands), 607
Nagaeff, Cape (Japan), 637
Nagasaki (Japan), 645
Nagay Island (Aliaska Peninsula), 498
Naguiscolo Port (Central America), 242
Naguiscolo Port, or Playa Hermosa (Central America), 232
Nahuelhuapi Lake (Chile), 99
Naka-oura Point (Japan), 639
Naka-sima Island (Japan), 642
Nalatcheff, Cape (Kamtschatka), 594
Namollos Tribe (Siberia), 561
Nana-sima Islands (Japan), 641
Naneck or Nackneck River (Russian America), 527
Nangarnan Island (Aleutian Archipelago), 505
Nangasaki or Nagasaki (Japan), 645
Nanimack Island (Aliaska Peninsula), 499
Nanouantoughat Lake (Russian America), 495, 527
Narborough, Sir John, Islands (Strait of Magalhaens), 37
Narborough or Ypun Island (Patagonia), 85-6
Nargasaky (Japan), 637
Naro, Cape (Guaymas Harbour), 298
Narrows, Admiralty Inlet (Admiralty Inlet), 381
Narrow Bank (Strait of Magalhaens), 5
Narrow Creek (Patagonia), 81
Naru-sima Island (Japan), 644
Narvaez, Tables or Mesas of (Lower California), 305
Nasaya Volcano (Central America), 226
Nasca Point (Peru), 157
Nash Harbour (Strait of Magalhaens), 27
Nasikok Island (Kodiack Islands), 493
Nass Bay (N.W. America), 431
Nassau Island (Strait of Magalhaens), 17
Nata, Town of (Central America), 213
Natan, Cape (Central America), 230
Natchikinskoi or Oukinskoi Cape (Kamtschatka), 591
Natividad, Island of (Lower California), 309
Nativity, Point (Tierra del Fuego), 59
Navarin Island (Tierra del Fuego), 53
Navidad (Mexico), 276
Nayahue Island (Chiloe), 96
Nayoumlack Bay (Kodiack Islands), 493
Neeah Bay, or Scarborough Harbour (De Fuca Strait), 375
Neegtchan Cape (Siberia), 567
Neesham Bay (Patagonia), 73
Negada Point (Quibo Island), 219
Negro Cape (Strait of Magalhaens), 10
Nelson, Point (Russian America), 438
Nemoro, Port (Jesso Island), 628
Nepean, Point (Russian America), 456
Nepean Sound (N.W. America), 428
Nerpitchy Islet (Aleutian Islands), 515
Nesbitt, Point (Russian America), 445
Neuke, Mount (Chile), 113
Neuman Inlet (Patagonia), 79
Neva River (Saghalin), 621
Nevadas, Sierra (Upper California), 337
Nevado de Guaylillas (Andes), 170
Nevados (Andes), 149
Neville, Port (N.W. America), 402
New Albion, or California (Lower California), 313
New Archangel (Sitka Islands), 466
New Cornwall (N.W. America), 420
New Dungeness (De Fuca Strait), 376
New Eddystone Rock (Russian America), 439
Newenham, Cape (Russian America), 530
New Granada, Republic of (Colombia), 186
New Hanover (N.W. America), 420
New Helvetia (Upper California), 337
New Island (Tierra del Fuego), 52
New, Point (Oregon), 370
New Year Sound (Tierra del Fuego), 58
New Year, or Año Nuevo Point (Monterey), 328-9

New Year Islands and Harbour (Staten Island), 46
Niapippi River (Colombia), 208
Nicaragua, Lake of (Central America), 233
Nicholson Rocks (Tierra del Fuego), 63
Nicoya, Gulf of, Punta de Arenas (Central America), 226
Niebla Castle (Valdivia), 104
Nihuel Island (Chiloe), 96
Nikla, Cape (Sea of Okhotsk), 615
Nikolskoï (Atkha Island), 515
Nimpkish or Nimkish River (Vancouver Island), 408
Nindiri, Volcano (Central America), 226
Ninirlioun (Siberia), 572
Nipple Hill (Lower California), 310
Nippon, Nipon, or Niphon, Island of (Japanese Archipelago), 633
Nisqually, Fort (Oregon), 382
Nisi-Sima Islands (Japan), 644
Nitinat Sound (Vancouver Island), 414
Nixon, Cape (Patagonia), 74
Nodales Peak (Strait of Magalhaens), 17
Nodales Channel (N.W. America), 402
Noir Island (Tierra del Fuego), 64
Noir, Cape (Tierra del Fuego), 65
Noïsak or Mordvinoff, Cape (Ounimack Island), 504
Nomano-misaky or Tchesmé Cape (Japan), 642
No-Man's Island, or Huafo (Patagonia), 85
Nombre Head (Tierra del Fuego), 43
Nomo, Cape (Japan), 644
Nonura Point (Peru), 180
Nootka Sound (Vancouver Island), 410
Noratos Cove (Peru), 154
Norfolk or Sitka Sound (Sitka Islands), 465
North Hill (Patagonia), 3
North Bay (Saghalin), 619
North Foreland (Cook's Inlet), 488
North, Cape, or Ir-Kaipie (Siberia), 562
North Anchorage (Patagonia), 80
North Anchorage (Barbara Channel), 21
Northumberland, Cape (Russian America), 441
Norton Sound (Russian America), 534
Nodrigawa River (Japan), 637
Nosovskoï Volcano (Ounimack Island), 504
Nosumi-sima, or Rat Island (Japan), 649
Notch, Cape (Strait of Magalhaens), 30-1
Noto, Cape (Japan), 636
Notsky, Port (Jesso Island), 629
Nouchagack River (Russian America), 528
Nouneangan Island (Siberia), 569
Nouneagino, Cape (Siberia), 568
Nouniagmo (Siberia), 564
Nouniagmo, Cape (Siberia), 564
Novita (Colombia), 208
Novosilzov, Cape (Jesso), 632
Nowell, Point (Prince William's Sound), 480
Nuinack Island (Aliaska Peninsula), 498
Nuniwack Island (Russian America), 531
Nutland Bay (Barbara Channel), 21
Nygtchygan, Cape (Siberia), 566
Oak Point (Columbia River), 367
Oak Cove, or Port Lawrence (De Fuca Strait), 379
Oake Bay (Patagonia), 81
Oazy Harbour (Strait of Magalhaens), 9
Obetavannaïa Bay (Atkha Island), 514
Obispito Cove (Chile), 134
Obispo Rock (Mexico), 271
Observatory Rocks (Vancouver Island), 416
Observatory Point (De Fuca Strait), 376
Observatory Inlet (N.W. America), 431
Obstruction Sound (Patagonia), 82
Ocoña, Valley of (Peru), 156
Odawara or Wodawara Bay (Japanese Archipelago), 633
Oeste, Punta del (Patagonia), 81
Ogden Point (Vancouver Island), 417
Oglodak Island (Aleutian Islands), 515
Ogloga Islands (Aleutian Islands), 516
Ohosaka (Japan), 635
Ohoseno-saki, or Cape Goto (Japan), 644
Oho-sima or O-sima Islands (Japanese Islands), 634
Okanagan (N.W. America), 393
Okho River (De Fuca Strait), 376
Okhotsk (Siberia), 613
Okhotsk, Sea of (Siberia), 610
Oki Islands (Japan), 636
Oklana River (Sea of Okhotsk), 613
Okosir, Island of (Japanese Archipelago), 631
Oku Jesso, or Saghalin, 616
Old Port (Oumnack Island), 509
Olleta Point (Chiloe), 89
Olutorskoi Cape and Gulf (Siberia), 577
Ommaney, Cape (Russian America), 454
Ommaney, Cape (Sitka Islands), 468
Omotepe Island (Central America), 233
Omotepeque Volcano (Central America), 234
Onango-sima or South Island (Japanese Islands), 635
One Mile Rock (Upper California), 339
Onnekotan Island (Kurile Islands), 606
Onslow, Point (Russian America), 442
Oojack or Oohiack Bay (Kodiack Islands), 494
Oonella or Ounalga Island (Aleutian Archipelago), 506
Oote Nizavou Cape (Jesso), 631
Orange Bay (Tierra del Fuego), 56-7
Orange, Cape (Tierra del Fuego), 6
Orchard, Port (Admiralty Inlet), 380
Orcas Island (Vancouver Island), 419
Oregon Territory (Oregon), 348, 354
Orford or Blanco, Cape (Oregon), 354
Orozco, Table of (Tierra del Fuego), 45
Oscuro Cove (Chiloe), 92
Osima Island (Japan), 645
O-sima Island (Jesso), 631
Osorno Bay (Strait of Magalhaens), 30
Osorno or Purraraque Volcano (Chile), 99
Otoque Island (Panamá Bay), 219
Otter Islands (Patagonia), 81
Otters or Bobrovaia, Bay of (Ounalashka Island), 507
Otway Bay (Tierra del Fuego), 66-7

Otway, Port (Patagonia), 79
Otway Water (Strait of Magalhaens), 28
Ouchougnat Island, or Mount St. Augustin (Cook's Inlet), 486
Oudagagh Channel (Ounalashka Island), 508
Oudskoi, Fort (Sea of Okhotsk), 615
Ouegakh Island (Aleutian Islands), 509
Ouektock Island (Aleutian Archipelago), 505
Ougagouk Lake (Aliaska Peninsula), 496
Ougalgan Island (Ounalashka Island), 508
Ougamock Island (Aleutian Archipelago), 505
Ouganeck Island (Kodiack Islands), 494
Ougatchik, River (Aliaska Peninsula), 526
Oughlaghmoutes Indians (Russian America), 528
Ougnagak (Aliaska Peninsula), 499
Ouka River (Kamtschatka), 590
Oukamock Island (Aliaska Peninsula), 497
Oukinskoï, Cape (Kamtschatka), 590
Oukivok or King's Island (Behring's Strait), 536
Ouknadagh Island (Ounalashka Island), 507
Ouliaga Island (Aleutian Islands), 510
Ouliaklipen, Cape (Siberia), 570
Oulliaghin Island (Aleutian Islands), 510
Oumacknagh, Island (Ounalashka Island), 507
Oumnack Island (Aleutian Archipelago), 508
Ounakhagh Island (Aliaska Peninsula), 500
Ounalashka, Island of (Aleutian Archipelago), 506
Ounalga, Strait of (Aleutian Islands), 511
Ounalga or Oonella Island (Aleutian Archipelago), 506, 508
Ounga Island (Aliaska Peninsula), 498
Ounga Cape (Ounimack Island), 504
Ounimack, Island of (Aleutian Archipelago), 503
Ouroup Island (Kurile Islands), 608
Ousoumi Bay (Japan), 640
Outholotskoï, Cape (Sea of Okhotsk), 612
Owari Bay (Japanese Islands), 635
Owen Point and Island (Vancouver Island), 416
Oyarvide, Heights of (Peru), 145
Ozernoï, Cape (Kamtschatka), 591

Pabellon of Pica (Peru), 145
Pacasmayo Road (Peru), 178
Pacayo, Volcan de (Central America), 252
Pachacamac Islands (Peru), 162
Pachea, Pacheque, or Pacheca Island (Panamá Bay), 212
Pacocha (Peru), 150
Pacora River (Panamá Bay), 215
Padilla Bay (Oregon), 387
Paghelian, Cape (Siberia), 568
Pagoobnoy or Pernicious Strait (Sitka Islands), 457
Painter's Muller (Staten Island), 48
Pajaros Ninos (Concepcion Bay), 110
Pajaros, Island (Mexico), 297
Pajaros, Islas de los (Mexico), 287, 290
Pajaros Islets (Chile), 126
Pajaros Niños Islets (Chile), 124
Pajonal, Cove of (Chile), 130
Pajuras Point (Chile), 128
Pakenham Point (Prince William's Sound), 480
Pala, Point (Mexico), 291
Pallas Mount or Peak (Jesso), 632
Palmo, Cape (Lower California), 305
Palominos Rocks (Callao), 168
Palos Colorados (Upper California), 340
Pamplona Rock (Russian America), 474
Panaloya or Tipitapa River (Central America), 234
Panamá, Bay of (Central America), 209
Panamá, City of (Central America), 212
Pancha Point (Peru), 170
Paougvigumut (Russian America), 527
Papagayo, Bight of (Central America), 229
Papagayos Winds (Central America), 229
Papenberg Island (Japan), 648
Paposa (Bolivia), 141
Papudo (Chile), 120, 122
Paquiqui Cape (Bolivia), 144
Para Reef (Concepcion Bay), 110
Paracas, Peninsula (Peru), 159
Parallel Peak (Patagonia), 73
Parallel Roads of Coquimbo (Chile), 126
Parga Cuesta Mountain (Chile), 103
Parina Point (Peru), 182
Parinacota Mount (Andes), 149
Park Bay (Cockburn Channel), 19
Parker, Point (Russian America), 455
Parker, Cape (Strait of Magalhaens), 37
Parredon River (Central America), 222
Parry Head (Vancouver Island), 416
Parry, Port (Staten Island), 48
Partridge Point (Oregon), 385
Pasquiel (Central America), 234
Passage Canal (Prince William's Sound), 480
Passage Point (Strait of Magalhaens), 27
Passage Island (Howe's Sound), 395
Passage Island (Cook's Inlet), 485
Passado or Passao, Cape (Colombia), 203
Patache, Point (Peru), 145
Patagonia, West coast of, 69
Patch Cove (Patagonia), 84
Patience, Cape and Bay (Saghalin), 621
Patillo Point (Peru), 172
Patos, Isle de los (Gulf of California), 299
Payaros, Island (Gulf of California), 299
Payana Shoals (Guayaquil River), 190
Paysan (Peru), 178
Payta (Peru), 181
Paz and Liebre Islands (Patagonia), 86
Peacock Rock (Peru), 160
Peaked Hill (Jesso Island), 630
Pearce, Point (N.W. America), 429
Peard Bay (Russian America), 551
Pearl Rocks (N.W. America), 423
Pearl Islands (Panamá Bay), 215
Pearl Banks (Gulf of California), 302
Pearl Fishery (Panamá Bay), 217

Pecherais Tribe (Tierra del Fuego), 2
Pecheura Point (Chiloe), 91
Pechtchanaia Cove (Atkha Island), 514
Pechucura Shoal (Chiloe), 90
Peckett's Harbour (Strait of Magalhaens), 9
Pedder Bay (Vancouver Island), 416
Pedernales Point (Colombia), 203
Pedro Gonzales Islands (Panamá Bay), 212
Peel Inlet (Patagonia), 82
Pelado Island (Peru), 170
Pelado, Islet of (Guayaquil River), 199
Pelican Bay (Oregon), 348
Pelicanos Rocks (Chile), 124
Pellew, Point (Prince William's Sound), 479
Pelly's or Stikine River (Russian America), 444
Peñas, Gulf of (Patagonia), 77, 83
Peñas, Cape (Tierra del Fuego), 44
Penco Shoal (Concepcion Bay), 110
Penguin or Magdalena Island (Strait of Magalhaens), 10
Penitente Point (Chile), 122
Penkegneï Bay (Siberia), 567
Penjinak River (Sea of Okhotsk), 613
Penjinskoi, Gulf of (Sea of Okhotsk), 612
Penn's Cove (Oregon), 385
Pequena Bay (Magdalena Bay), 306
Percy, Point (Russian America), 441
Periagua Islet (Strait of Magalhaens), 25
Periagua Rocks (Chiloe), 91
Perico Island (Panamá Bay), 217
Pernicious or Pagoobnoy Strait (Sitka Islands), 457
Perpetua, Cape (Oregon), 356
Peru, Republic of, 137
Pescado Blanco Bay (Lower California), 310
Pescador Islands (Peru), 170
Pescadores Point (Peru), 156
Pestsovaia Bay (Atkha Island), 514
Petatlan, Morro de (Mexico), 273
Petrof Island (Siberia), 565
Petropaulovski Harbour (Awatska Bay), 598, 601
Petucura Rock (Chiloe), 91
Peulla River (Chile), 99
Philip, Point (New Hanover), 407
Philip, Cape (Strait of Magalhaens), 35
Philip, Cape (Patagonia), 80
Philip Rocks (Tierra del Fuego), 63
Phillip, Cape (Strait of Magalhaens), 37
Phipps, Cape (Russian America), 469, 470
Phipps Point (Prince William's Sound), 478
Piastla River (Mexico), 297
Piazza Island (Patagonia), 81
Pic de Langle (Jesso), 632
Pichalo Point (Peru), 146
Pichidanque or Herradura Bay (Chile), 122
Pichincha Mountain (Andes), 188
Pichinqua Bay (Gulf of California), 302
Pickersgill Cove (Tierra del Fuego), 61
Pico Channel (Kurile Islands), 609
Picos Point (Peru), 182
Picton Opening (Patagonia), 73
Piedra, Point (Guayaquil River), 405
Piedra Blanca Reef (Mexico), 267
Piedras Point (Peru), 146
Piedro de Tierra (Mexico), 279
Piedro de Mer (Mexico), 278
Pie's Islands (Russian America), 484
Pigot, Point (Prince William's Sound), 480
Pillar Cape (Tierra del Fuego), 68
Pillar Rock (Columbia River), 361
Pillar, Cape (Strait of Magalhaens), 38
Pillar Point (De Fuca Strait), 376
Pilot's Cove (Admiralty Inlet), 380
Pinas, Port (Colombia), 209
Pine, Cape (Kodiack Islands), 492
Pinero Rock (Peru), 159
Pinnacle Island (St. Matthew Island), 580
Pinnacle Point (Awatska Bay), 601
Pinos, Point (Monterey), 328-9
Piojo Point (Valdivia), 104
Pirulil Head (Chiloe), 89
Pisagua (Peru), 146
Pisco (Peru), 159
Pisura Point and River (Peru), 180
Pitt Island, or Crooze Island (Sitka Islands), 457
Pitt's Archipelago (N.W. America), 428-9
Piura, San Miguel de (Peru), 181
Plata Isle (Guayaquil River), 200
Plata Point (Bolivia), 142
Playa Maria Bay (Lower California), 310
Playa Parda Cove (Strait of Magalhaens), 32
Playa Brava, La (Bolivia), 142
Playeta de Chicarene, La (Central America), 244
Plover Island (Arctic Sea), 559
Pnaougoun, Cape (Siberia), 564
Podsopotchnoï Cape (Atkha Island), 514
Pogrommoï or Nosovskoï Volcano (Ounimack Island), 504
Point Palmer (Patagonia), 81
Point Victoria, Sacramento River (Upper California), 337
Poison Cove (N.W. America), 426
Pole, Cape (Russian America), 447
Policarpo Cove (Tierra del Fuego), 45
Polillao, Cove of (Chile), 127
Polovintchaty Rock (Sea of Behring), 584
Pond, Bay and Mount (Strait of Magalhaens), 26
Ponente Point (Realejo), 237
Poqueldon (Chiloe), 94
Poromoushir Island (Kurile Islands), 605
Porpesse Point (Strait of Magalhaens), 10
Porpoise Rocks (Prince William's Sound), 477
Port Famine (Strait of Magalhaens), 12
Portland Canal (Russian America), 438
Portland's Canal (N.W. America), 431
Portlock's Harbour (Sitka Islands), 463
Posadas Bay (Strait of Magalhaens), 30
Possession, Cape (Patagonia), 5
Possession Bay (Strait of Magalhaens), 5
Possession Sound (Oregon), 384
Possession, Point (Cook's Inlet), 489
Postels, Cape (Siberia), 568
Postotiak River (Russian America), 532

Potainikoff, Cape (Atkha Island), 513
Poualo Bay (Aliaska Peninsula), 495
Poustaresk (Sea of Okhotsk), 612
Pratt Passage (Tierra del Fuego), 64
Prevost Peak (Kurile Islands), 607
Pribuiloff Islands (Sea of Behring), 581
Primero, Cape (Patagonia), 71, 73
Prince Island (Japanese Islands), 634
Prince of Wales, Cape (Behring's Strait), 539
Prince Ernest's Sound (Russian America), 443
Prince Frederick's Sound (Russian America), 449
Prince George's Island (Mexico), 277-8
Prince William's Sound (Russian America), 476
Princess Royal Islands (N.W. America), 425
Principe, Canal de (N.W. America), 429
Pringle Point (Patagonia), 83
Proche Island (Kodiack Islands), 492
Prokofieff Island (Sea of Okhotsk), 615
Protection, Port (Russian America), 445
Protection Island (De Fuca Strait), 378
Providence Cape (Strait of Magalhaens), 34
Providence, Port (Siberia), 570
Puchachailgua Harbour (Strait of Magalhaens), 36
Pucari Shoal (Chiloe), 99
Puebla Nueva, or Santiago (Central America), 221
Puercos, Morro de (Central America), 219
Puerto de la Medalla or Medal Bay (Strait of Magalhaens), 33
Puerto, Escondido (Gulf of California), 299
Puerto del Sororro (Oregon), 387
Puerto de los Angelos (De Fuca Strait), 376
Puerto Angosto (Strait of Magalhaens), 34
Puerto Bueno (Patagonia), 82
Puffin Island (Russian America), 546
Puget Island (Columbia River), 367
Puget Sound (Oregon), 381
Puget, Cape (Russian America), 480
Pulberia Reefs (Panamá Bay), 277
Pulmun Bank (Chiloe), 93
Pulpito de San Juan (Gulf of California), 302
Puluqui Island (Chiloe), 98
Puna, Island of (Guayaquil River), 190
Puna Vieja (Guayaquil River), 193
Punoun Point (Chiloe), 91
Punta de Arenas, or Nicoya (Central America), 226
Punta de Arena (Guayaquil River), 193
Punta Arenas, La Union (Central America), 242
Purcell Island (Patagonia), 78
Purraque or Osorno Volcano (Chile), 99
Puyo Island (Patagonia), 85
Puyallup River (Admiralty Inlet), 381
Pybus, Point (Russian America), 450
Pyke, Point (Prince William's Sound), 481
Pyramid Hill (Magdalen Sound), 19
Pyramid Rock (Gulf of California), 300
Quadra and Vancouver Island (N.W. America), 389
Quadra, Boca de (Russian America), 438
Quatre Frères Islands (Kurile Islands), 608
Quatre Fils d'Aymon Hills (Patagonia), 5-6
Quebra Olla, or Breakpot Rock (Concepcion Bay), 109
Quebrada and Canal de la Raspadura (Colombia), 208
Quebrada Grande (Central America), 232
Quedal, Cape (Chile), 103
Queen's or South Channel (Columbia River), 359
Queen Charlotte Sound (N.W. America), 422
Queen Charlotte Island (N.W. America), 432
Queen Adelaide's Archipelago (Patagonia), 70, 80
Quehuy Island (Chiloe), 95
Quelan Point (Chiloe), 96
Quemado Point (Peru), 158
Quemao Point (Chiloe), 92
Quenu Point and Island (Chiloe), 98
Quenac Island (Chiloe), 95
Quibo Island (Central America), 219
Quicara or Hicaron Island (Central America), 219
Quicavi, Laguna of (Chiloe), 93
Quilan, Cape (Chiloe), 89
Quilca (Peru), 154
Quillahua Point (Chile), 103
Quintero Bay and Rocks (Chile), 121
Quilimari (Chile), 122
Quillota Campana or Bell (Valparaiso), 115, 118
Quinched (Chiloe), 93
Quinten Cove (Chile), 115
Quintergen Point (Chiloe), 92
Quiriquina Island (Concepcion Bay), 109
Quisiguina or Coseguina Volcano (Central America), 240
Quito (Ecuador), 188
Quitshak River (Russian America), 528
Quod, Cape (Strait of Magalhaens), 29-31
Quoin Hill (Strait of Magalhaens), 8

Rainier, Mount (Oregon), 379
Rakovya Bank (Awatska Bay), 597
Rakovya Harbour (Awatska Bay), 598, 600
Raleon River (Chile), 99
Ramsden, Point (N.W. America), 431
Rancheria Island (Central America), 220
Rapel Shoal (Chile), 114
Raper, Cape (Patagonia), 79
Rare Cove (Chile), 113
Rasa Isle, or Wolf Rock (Russian America), 448
Rashan Island (Kurile Islands), 607
Raspadura, Quebrada and Canal de la (Columbia), 208
Raspberry Island (N.W. America), 429
Rasposo Province (Colombia), 207
Rat or Ayougadagh Island (Aleutian Islands), 518
Rat Island, or Nosumi-sima (Japan), 649

Rat or Krysi Islands (Aleutian Islands), 517
Ratmanoff, Cape (Saghalin), 620
Ratmanoff Island, or Noo-nar-book (Behring's Strait), 540
Ratmanoff, Port (Siberia), 568
Raukoko Island (Kurile Islands), 607
Real de Loreto, Mission of (Gulf of California), 302
Realejo (Central America), 235, 240
Redonda Island (N.W. America), 400
Redondo, Cape (Magdalena Bay), 306
Refuge Inlet (Russian America), 551
Refuge, Port (Patagonia), 84
Refunshery Island (Jesso), 632
Rehusa Channel (Magdalena Bay), 306
Reindeer or Animaok Island (Aliaska Peninsula), 499
Relan (Chiloe), 93
Relief Harbour (Patagonia), 82
Reloncavi Sound (Chilee), 99
Remedios, Point (Central America), 247-8
Rennell's Sound (Queen Charlotte Island), 434
Renwick, John, Rock (Chile), 108
Rescue Point (Patagonia), 84
Resolution Cove (Vancouver Island), 413
Restoration Cove (N.W. America), 424
Restoration, Point (Admiralty Inlet), 380
Retchechnoi (Oumnack Island), 509
Retreat Bay (Patagonia), 81
Retreat, Point (Russian America), 454
Revilla Gigedo, Canal of (Russian America), 438, 441
Revilla Gigedo, Island of (Russian America), 439
Rey or King's Islands (Panamá Bay), 212
Reyes, Punta de los (Upper California), 338, 343
Rice Trevor Islands (Tierra del Fuego), 67
Richardson, Mount (Staten Island), 47
Ricord, Cape (Kurile Islands), 609
Rikimushiri Island (Jesso Island), 630
Rimac River (Lima), 167
Rimnik Cape (Saghalin), 620
Rincon de Tanque (Chile), 124
Rincon Point (Bolivia), 141
Rio de San Juan (Strait of Magalhaens), 15
Rioshery Bay (Jesso), 632
Riou, Point (Russian America), 473
Rivers Canal (N.W. America), 423
Robben Island (Saghalin), 621
Roberts, Point (Oregon), 388
Rock of Dundee (Patagonia), 74
Rock Point (Aliaska Peninsula), 500
Rocky Point (Strait of Magalhaens), 11
Rocky Cove (Patagonia), 81
Rocky Island (Vancouver Island), 416
Rocky Point (Upper California), 346
Rodd Point (Vancouver Island), 417
Rodney, Point (Behring's Strait), 537
Roguan Flat (Concepcion Bay), 110
Rojnoff, Cape (Aliaska Peninsula), 525
Roldan's Bell (Strait of Magalhaens), 14
Rollin, Cape (Simusir Island), 608
Romanzoff, Cape (Jesso), 632
Romanzoff, Cape (Russian America), 532
Romberg Cape (Sea of Okhotsk), 616
Roqueta or Grifo Island (Mexico), 269
Rosario (Mexico), 296
Rosario, Port (Patagonia), 72
Rosario Shoal (Chiloe), 99
Rosario Strait (Vancouver Island), 419
Rosario Strait (N.W. America), 397
Rosario Strait (Oregon), 386
Rosario, Bay of (Mexico), 261
Rosario or S. Francis de Agnatulco (Mexico), 261
Rose or Ymbisible Point (Queen Charlotte Island), 435
Rose's Harbour, or Bay de Lujan (Queen Charlotte Island), 435
Roshnoff, Cape (Aliaska Peninsula), 525
Ross, Port (Upper California), 344
Rothsay, Point (Russian America), 443
Roumiantzoff or Romanzoff, Cape (Russian America), 532
Round Island (Kurile Islands), 608
Round Island (Strait of Magalhaens), 34
Round Island (Russian America), 529
Rous Sound (Tierra del Fuego), 59
Royal Bay (Vancouver Island), 417
Royal Road (Strait of Magalhaens), 9
Rumena, Cape (Chile), 107
Rundle Pass (Patagonia), 75, 76
Rupert Island (Strait of Magalhaens), 27
Rurick Strait (Ounimack Island), 505
Russian America (N.W. America), 438
Russian American Company, 438
Russian Point (Cook's Inlet), 485
Russians' Cape (Japan), 637
Ryghyain, Cape (Siberia), 568

Sabine, Cape (Russian America), 548
Sacate Island (Central America), 242
Sacatula or Zacatulu (Mexico), 274
Sachalin or Saghalin Peninsula, 616
Sacraficios Island (Mexico), 268
Sacramento City (Upper California), 337
Saddle Island (Tierra del Fuego), 56
Sado Island (Japan), 637
Safety Cove or Port (N.W. America), 423-4
Saghalin, Peninsula of (Eastern Asia), 616
Sagouliuoktasigh Island (Aliaska Peninsula), 498
Sakoura Island (Japan), 641
Sal Point (Peru), 182
Sal, Point (Upper California), 327
Sal si Puedes, Island of (Gulf of California), 300
Salado, Bay of (Chile), 131
Salado Point (Chile), 131
Salagna or Manzanilla (Mexico), 274
Salinas, Bay of (Peru), 170-1
Salinas Bay (Central America), 230
Salinas Island (Central America), 230
Salinas Port (Mexico), 260
Salinas, Point (Guayaquil River), 190
Salinas de Rosario, Morro de los (Mexico), 261

Salisbury, Point (Russian America), 453
Salm or Salmon Bay (Saghalin), 623
Salmon River (N.W. America), 427
Salmon Cove (N.W. America), 431
Saltchidack or Siachladack Island (Kodiack Islands), 493
Sama, Point and Morro of (Peru), 149
Samanco Head (Peru), 175
Samanco or Guambacho Bay (Peru), 174-5
Sambo or Sambu River (Panamá Bay), 210
Samganooda Bay (Ounalashka Island), 507
Samuel, Point (Russian America), 455
San Andres Bay (Patagonia), 83
San Andres, Hacienda (Mexico), 287
San Andres (Central America), 232
San Antonio, Port (Strait of Magalhaens), 14
San Antonio Point (Chile), 103
San Barnabé Islet (Gulf of California), 300
San Bartholomew or Turtle Bay (Lower California), 309
San Bartolom, Cape (Russian America), 448
San Benito Island (Lower California), 310
San Blas Channel (Patagonia), 70
San Bruno Hills (Upper California), 334, 335, 338
San Bruno Cove (Gulf of California), 302
San Buenaventura, Cape (Gulf of California), 300
San Buenaventura (Upper California), 322
San Buenaventura (Colombia), 206
San Carlos (Mexico), 271
San Carlos Fort (Valdivia), 104
San Carlos Island (Russian America), 448
San Carlos, Port (Chiloe), 89
San Carlos, or Conchagua (Central America), 241
San Carlos de Chiloe, Town of (Chiloe), 90
San Carlos de Monterey (Upper California), 328—333
San Clemente Island (Upper California), 325
San Diego, Cape (Tierra del Fuego), 45
San Diego, Port (Upper California), 317
San Estanislao (Gulf of California), 300
San Estevan Port (Patagonia), 84
San Estevan, Cape, 413
San Estevan Sound (N.W. America), 428
San Eugenio Point (Lower California), 309
San Felipe, Port Famine (Strait of Magalhaens), 12
San Fermen (Gulf of California), 300
San Fernando or Inchin Islands (Patagonia), 85
San Francisco (Upper California), 334
San Francisco Island, Pachacamac (Peru), 162
San Francisco, Cape (Colombia), 203
San Francisco Cape (Bolivia), 144
San Francisco de Burja (Gulf of California), 300
San Francisco or San Quentin Bay (Lower California), 310
San Gabriel, or Pueblo de los Angeles (Upper California), 320
San Gallan Point (Chiloe), 91
San Gallan, Island of (Peru), 159
San Geraldo or St. Balardo River (Upper California), 326
San Geronimo Island (Lower California), 310
San Hilario, or Cenizas Island (Lower California), 312
San Isidro, Cape (Strait of Magalhaens), 16
San João, Fort (Concepcion Bay), 111
San Jose Bank (Panamá Bay), 212
San Jose Island (Panamá Bay), 212
San Jose del Cabo, Bay of (Lower California), 304
S. José de los Ures, Mission of (Mexico), 296
San Jose Shoal (Chiloe), 99
San Juan River (Colombia), 208
San Juan Island (Upper California), 325
San Juan, Port (Peru), 157
San Juan, Port (Vancouver Island), 416
San Juan Capistrano (Upper California), 320
San Juan Nepomuceno (Gulf of California), 302
San Juan del Sur, Port (Central America), 230
San Juan y San Pablo (Gulf of California), 300
San Lorenzo, Cape (Colombia), 201
San Lorenzo or Brava Point (Panamá Bay), 210
San Lorenzo Island (Gulf of California), 300
San Lorenzo Island (Callao), 165, 168
San Lucas Bay (Lower California), 304
San Lucas Cape (Lower California), 305
San Luis Obispo, Mission of (Upper California), 327
San Luiz de Gonzaga (Gulf of California), 300
San Marcos Island and Cape (Gulf of California), 300
San Miguel, Gulf of (Colombia), 209
San Miguel Point (Lower California), 312
San Miguel de Piura (Peru), 181
San Miguel Volcano (Central America), 226
San Miguel, Port (Strait of Magalhaens), 24
San Miguel, Volcano (Central America), 245
San Nicolas Island (Upper California), 325
San Nicolas, Port (Peru), 157
San Pablo, Bay of (San Francisco), 336
San Pablo, Cape (Tierra del Fuego), 45
San Pedrito Rock (Mexico), 274
San Pedro or Huamblin Island (Chiloe), 89
San Pedro (Upper California), 320
San Pedro Island (Chiloe), 89
San Pedro Passage (Chiloe), 97
San Pedro Sound (Strait of Magalhaens), 26
San Pedro Point (Bolivia), 141
San Pedro Point (Chile), 136
San Pedro, Isle de (Gulf of California), 299
San Pedro de Yoco (Peru), 179
San Policarpo, Port (Tierra del Fuego), 45
San Quentin, Port (Lower California), 310
San Quentin or Virgenes, Cape (Lower California), 310
San Ramon or De los Virgenes, Bay (Lower California), 312

San Roman, Cape (Patagonia), 76
San Roque Island (Lower California), 309
San Salvador Volcano (Central America), 226
San Salvador, City of (Central America), 246
San Sebastian Bay (Tierra del Fuego), 43
San Sebastian, Cape (Tierra del Fuego), 44
San Sebastian Vizcaino Bay (Lower California), 309
San Tadeo River (Patagonia), 78
San Vicente, Port (Chile), 109
San Vincente Volcano (Central America), 226
San Vincente, Cape (Tierra del Fuego), 45
Santa Agueda Island (Gulf of California), 300
Santa Anna Islet (Gulf of California), 300
Santa Anna Point (Strait of Magalhaens), 9, 12
Santa Barbara Island (Upper California), 325
Santa Barbara Channel (Upper California), 325
Santa Barbara (Upper California), 322
Santa Barbara, Port (Patagonia), 74
Santa Brigida, Point (Strait of Magalhaens), 17
Santa Casilda, Cape (Strait of Magalhaens), 36
Santa Catalina Island (Upper California), 325
Santa Clara or Woody Point (Vancouver Island), 409
Santa Clara, or Amortajada Island (Guayaquil River), 190
Santa Cruz Island (Upper California), 326
Santa Cruz, Hacienda (Mexico), 287
Santa Elena or Tomas Bay (Central America), 230
Santa Inez Medio, Cape (Tierra del Fuego), 45
Santa Island and Head (Peru), 175
Santa Lucia Hills (Upper California), 327
Santa Lucia, Cape (Patagonia), 70
Santa Margarita (Lower California), 305
Santa Margarita, Cape (Queen Charlotte Island), 435
Santa Margarita Island (Magdalena Bay), 306
Santa Maria Island (Chile), 107
Santa Maria de Aome, Rio (Gulf of California), 297
Santa Maria Point (Lower California), 310
Santa Maria Point (Peru), 158
Santa Maria River (Panamá Bay), 210
Santa Marina Point (Magdalena Bay), 306
Santa Monica, Port (Strait of Magalhaens), 36
Santa Rosa Island (Upper California), 326
Santa Teresa, Barra de (Mexico), 259
Santa Teresa Point (Gulf of California), 302
Santa Teresa Point (Chiloe), 91
Santa Ynez Hill (Chiloe), 122
Santa Ysabel (Gulf of California), 300
Sana (Kurile Islands), 609
Sandan or Saghalin, 616
Sanderson Island (Tierra del Fuego), 58
Sandy Bay (Patagonia), 81
Sandy Point (Strait of Magalhaens), 11
Sangar or Tsugar, Strait of (Japanese Archipelago), 631, 633
Sannagh or Halibut Island (Aliaska Peninsula), 500
Santano Cove (Peru), 154
Santiago, Cape (Patagonia), 70-1
Santiago Bay, Manzanilla (Mexico), 274
Santiago de Ystapa, Morro de (Mexico), 261
Santo, Barra del Espiritu, River Lempa (Central America), 245
Santo Domingo, Point (Lower California), 309
Sapa, Valley of (Peru), 147
Sarah, Point (N.W. America), 498
Sarannaïa Cove (Atkha Island), 514
Sarco, Bay of (Chile), 128
Sarmiento Bank (Strait of Magalhaens), 4, 5
Sarmiento Channel (Patagonia), 80, 82
Sarmiento, Mount (Strait of Magalhaens), 14
Sarytscheff, Cape (Ounimack Island), 504
Sarytscheff Island (Russian America), 540
Sarytscheff Peak (Kurile Islands), 607
Sasayama, Mount (Japan), 638
Satanora or Bird's Cape (Aleutian Islands), 517
Saumarez Island (Patagonia), 83
Sausalito Bay, or Whaler's Harbour (San Francisco), 335
Savary's Island (N.W. America), 398
Sawyer Bank (Realejo), 237
Scarborough Harbour, or Neeah Bay (De Fuca Strait), 375
Schapenham Bay (Tierra del Fuego), 57
Schetky, Cape (Tierra del Fuego), 67
Schischkoff, Cape (Jesso), 632
Schischmareff Point (St. Lawrence Island), 578
Schischmareff, Inlet (Behring's Strait), 540
Schomberg Cape (Tierra del Fuego), 20
Schumagin Island (Aliaska Peninsula), 496
Scotch Fir Point (Gulf of Georgia), 397
Scotchwell Harbour (Patagonia), 86
Scott Islands (Vancouver Island), 409
Scott, Cape (Vancouver Island), 409
Scuchadero (Panamá Bay), 211
Seabeck Island (Oregon), 384
Seahorse Islands (Russian America), 551
Sea Lion's Island (Russian America), 530
Sea Otter or Bobrovi Island (Sea of Behring), 582
Sea Otter Harbour (Vancouver Island), 409
Sea Otter Sound (Russian America), 447
Seal Cove (Strait of Magalhaens), 28
Seal Rocks (Patagonia), 73
Seasons, Tierra del Fuego, 42
Sechura, Bay of (Peru), 181
Second Narrow (Strait of Magalhaens), 8
Secretary Island (Vancouver Island), 416
Secretary Wren's Island (Strait of Magalhaens), 27
Sedger River (Strait of Magalhaens), 13, 15

Seduction Point (Russian America), 456
Segouam or Goreli Island (Aleutian Islands), 511
Sehama or Gualatieri Peak (Andes), 149
Sekan Point (De Fuca Strait), 376
Selango Island (Guayaquil River), 199
Seluian Rock (Chiloe), 91
Semisopochnoï Island (Aleutian Islands), 517
Semitsch Island (Aleutian Islands), 518
Semitsououra Bay (Japan), 638
Séniavine, Strait of (Siberia), 566
Séniavine, Cape (Saghalin), 622
Séniavine, Cape (Aliaska Peninsula), 526
Seno Lauretaneo, or Gulf of California (Mexico), 295
Separation or Mercy Harbour (Strait of Magalhaens), 37
Seppings, Cape (Russian America), 546
Sequalchin River (Oregon), 355
Serdze Kamen, Cape (Siberia), 562
Sergheïeff Isthmus (Atkha Island), 513
Seriphos Island, Make-sima or Oumawo (Japan), 641
Seris Indians (Mexico), 299
Serrate Entrance (Peru), 158
Sesambre Isles (Tierra del Fuego), 53
Sesga Point (Port Culebra), 229
Seurote, Cape (Japan), 647
Seven Mountains, Island of the (Aleutian Islands), 517
Seymour's Canal (Russian America), 452
Shag Island (Tierra del Fuego), 61
Shag Narrows (Barbara Channel), 22
Shag Narrows (Strait of Magalhaens), 27
Shakoff Point (Awatska Bay), 601
Shantar Islands (Sea of Okhotsk), 615
Shaste, Mount (Upper California), 347
Shaw's Island (Cook's Inlet), 486
Shelter Island (Strait of Magalhaens), 32
Shiashkotan Island (Kurile Islands), 606
Ship Rock (Aleutian Islands), 510
Shipounskoi, Cape (Kamtschatka), 594
Shirinky Island (Kurile Islands), 605
Shoal Ness (Russian America), 531
Shoal Point (Saghalin), 620
Shoalwater Bay (Oregon), 370
Sholl Bay (Magdalen Sound), 19
Sholl Harbour (Strait of Magalhaens), 25
Sholl's Bay (Strait of Magalhaens), 35
Shoulder Peak (Tierra del Fuego), 68
Shucartie, Port (Vancouver Island), 409
Siachladack Island (Kodiack Islands), 493
Sielata Point (Mexico), 269
Sigak, Cape (Oumnack Island), 509
Signal Hill (Coquimbo), 125
Sigouam or Goreli Island (Aleutian Islands), 511
Siguantanejo or Sihuantanejo Port (Mexico), 273
Sikok, Island of (Japan), 638
Silla Hill, La (Chile), 120
Silla or Saddle of Payta (Peru), 181
Silvester Point (Strait of Magalhaens), 10
Sima, Cape (Japan), 635
Simabara, Peninsula of (Japan), 642
Simala River (Central America), 253
Simidin Island (Aliaska Peninsula), 496
Simonosaki (Japan), 636
Simonoséki (Japan), 640
Simpson River (N.W. America), 431
Simpson, Fort (N.W. America), 430
Simusir Island (Kurile Islands), 607
Sinahomis River (Oregon), 384
Sinaloa, Province of (Mexico), 288, 296
Sineko, Cape (Jesso), 631
Sinshi Island (Jesso Island), 628
Sintalapa or Tilapa River (Mexico), 252, 257
Sirano or Siriano River (Central America), 242
Sir Francis Drake, Port (Upper California), 343
Sir George Eyre Sound (Patagonia), 83
Siriano River, or Estero Jirausa (Central America), 242
Sir John Narborough's Islands (Strait of Magalhaens), 37
Siro (Japan), 637
Sitchunak Island (Kodiack Islands), 494
Sitka or Norfolk Sound (Sitka Islands), 465
Sitka or King George the Third's Archipelago (Russian America), 456
Sitkhin Islet (Aleutian Islands), 516, 518
Sivoutchi Islet (Sea of Behring), 582
Sivoutchy Islet (Aleutian Islands), 515
Sivoutchy Rock (Aliaska Peninsula), 525
Skakhoul Islands (Aleutian Islands), 516
Skidegats (Queen Charlotte Island), 435
Skyring, Mount (Tierra del Fuego), 20, 64
Skyring Harbour (Strait of Magalhaens), 37
Skyring Water (Strait of Magalhaens), 28
Skyring, Island (Patagonia), 85
Slate Islet (Russian America), 438
Sledge or Ayak Island (Behring's Strait), 536
Slip Point (De Fuca Strait), 376
Smith or Bonilla Island (Oregon), 385
Smith's River (Upper California), 347
Smith's Inlet (N W. America), 423
Smoto (Japan), 638
Smyth Channel (Patagonia), 80, 81
Smyth Harbour (Barbara Channel), 22
Smyth Head (Vancouver Island), 416
Smyth, Cape (Russian America), 551
Smyth's Channel (Strait of Magalhaens), 35
Snares Islets, The (Kurile Islands), 606
Snettisham, Point (Russian America), 452
Snowy Channel (Strait of Magalhaens), 31
Snowy Sound (Strait of Magalhaens), 30
Snug Bay (Strait of Magalhaens), 23
Snug Corner Bay (Prince William's Sound), 478
Soconusco, District of (Mexico), 257
Socorro or Huamblin Island (Patagonia), 85, 86
Soimonoff, Cape (Saghalin), 621
Solar, Morro (Peru), 163
Solentiname Island (Nicaragua Lake), 233
Soleny Island (Atkha Island), 514
Sombrerito Hill, 300

Sombrio River (Vancouver Island), 416
Somoshiri Island (Jesso Island), 630
Sona, Island of (Guayaquil River), 195
Sonate Island (Nicaraqua Lake), 233
Sonora, Province of (Mexico), 288, 296
Sonsonate Roads, or Port Acajutla (Central America), 247
Sooke, Inlet (Vancouver Island), 416
Sophia, Point (Russian America), 460
Sopki or Volcanoes of Kamtschatka, 588
Soulima or Ougatchik River (Aliaska Peninsula), 526
Soumshou Island (Kurile Islands), 605
South Channel (Columbia River), 366
South Island, or Onango-sima (Japanese Islands), 635
South Passage Rock (Russian America), 483
Souvoroff, Cape (Russian America), 527
Soya Bay (Jesso), 632
Spafarief Bay (Russian America), 542
Spanberg or Bernizet Peak (Saghalin), 622
Spanberg or Tschikotan Island (Kurile Islands), 609
Spanberg, Cape (Siberia), 572
Spanberg, Cape (Jesso Island), 629
Spartan Passage (Patagonia), 73
Speedwell Bay (Patagonia), 76
Spencer, Cape (Tierra del Fuego), 56
Spencer Cape (Russian America), 468
Spencer, Cape (Russian America), 460
Spencer, Point (Russian America), 537
Spirkin Island (Ounalashka Island), 508
Squally Point (Magdalen Sound), 19
Squimalt or Esquimalt Harbour (Vancouver Island), 417
Srogg Rocks (Vancouver Island), 417
St. Albans, Point (Russian America), 446
St. Andrew's Sound (Patagonia), 82
St. Ann's Island (Strait of Magalhaens), 83
St. Anthony, Cape (Staten Island), 50
St. Anthony's Peak (Kurile Islands), 609
St. Augustin, Mount (Cook's Inlet), 486
St. Augustin, Fort (Concepcion Bay), 111
St. Balardo or San Geraldo River (Upper California), 326
St. Bartholomew, Cape (Staten Island), 50
St. Clair Island, or Kuros-sima (Japan), 641
St. Croix, Gulf of (Siberia), 574
St. David's Head (Strait of Magalhaens), 29
St. David's Sound (Strait of Magalhaens), 27
St. Elena Point (Central America), 230
St. Elias, Point (Russian America), 473
St. Estevan Gulf (Patagonia), 77, 78
St. Francis Bay (Tierra del Fuego), 54
St. George, Point (Upper California), 348
St. George's Island (Sea of Behring), 581
St. Helen's, Mount (Oregon), 367, 383
St. Hermogenes, Cape (Kodiack Islands), 494
St. Ildefonso Islands (Tierra del Fuego), 58
St. Isidro Point (Strait of Magalhaens), 9
St. Jago River (Colombia), 205
St. James, Cape (Queen Charlotte Island), 434
St. Joachim's Cove (Tierra del Fuego), 56
St. John, Cape (Staten Island), 47
St. John's Harbour (Staten Island), 47
St. Juan, Mount, San Blas (Mexico), 278
St. Lawrence or Clerke Island (Behring's Sea), 578
St. Lawrence Bay (Siberia), 563
St. Marcien or Markiana Island (Aleutian Islands), 517
St. Martin's Island (Patagonia), 71
St. Martin's or Wigwam Cove (Tierra del Fuego), 55
St. Mary, Point (Strait of Magalhaens), 9, 11
St. Mary's Point (Russian America), 456
St. Matthew or Gore's Island (Sea of Behring), 579
St. Nicholas Bay (Strait of Magalhaens), 17
St. Paul, Harbour of (Kodiack Islands), 492
St. Paul's Dome (Mount) (Tierra del Fuego), 64
St. Paul's Dome (Patagonia), 78
St. Paul's Island (Sea of Behring), 581
St. Pedro Nolasco, Island of (Gulf of California), 296
St. Peter and St. Paul, Port (Kurile Islands), 598, 601
St. Philip, Mount, Port Famine (Strait of Magalhaens), 13
St. Quentin's Sound (Patagonia), 78
St. Simon's Bay (Strait of Magalhaens), 26
St. Simon's Head (Strait of Magalhaens), 8
St. Stephen's Island (Aliaska Peninsula), 496
St. Thaddeus, Cape (Siberia), 576
St. Vincent, Cape, or Sweepstakes Foreland (Strait of Magalhaens), 8
Staines Peninsula (Patagonia), 82
Stanhope, Point (Russian America), 445
Staniforth, Point (N.W. America), 427
Staniski, Point (Awatska Bay), 597, 599
Staples Inlet (Strait of Magalhaens), 25
Staritchkoff Reef (Atkha Island), 514
Staten Island (Tierra del Fuego), 45
Station Peak (Lower California), 310
Stearn's Bluff (Oregon), 371
Stephens, Mount (New Hanover), 407
Stephens, Port (N.W. America), 430
Stephens, Cape (Russian America), 533
Stephen's Passage (Russian America), 457
Stephen's Island (N.W. America), 429
Sterile or Barren Island (Kodiack Islands), 491
Stewart Islands (Tierra del Fuego), 63
Stewart, Port (Russian America), 440
Stewart Bay (Patagonia), 84
Stewart Harbour (Tierra del Fuego), 63
Stikine, River and Fort (Russian America), 443
Stockdale's Harbour (Prince William's Sound), 482
Stokes's Inlet (Magdalen Sound), 19
Stokes's Bay (Tierra del Fuego), 65
Stolbovoï, Cape (Kamtschatka), 591
Stolbovskaïa, River (Kamtschatka), 591
Stolétié, Cape (Siberia), 570

Stone Mole, or Kamennaïa Pristan (Russian America), 536
Stormy Bay (Cockburn Channel), 19
Stormy or Tempestad Channel (Strait of Magalhaens), 36
Strawberry Bay (Oregon), 387
Strogonoff Bay (Jesso), 632
Strogonoff, Cape (Aliaska Peninsula), 526
Stuart Bay (Strait of Magalhaens), 31
Stuart's Lake (N.W. America), 393
Stuart's Island (N.W. America), 400
Stuart's Island (Russian America), 533
Sturgeon Bank (New Georgia), 394
Styleman, Point (Russian America), 452
Success Bay (Tierra del Fuego), 49
Suchiltepeques, Province (Central America), 253
Suckling, Cape (Russian America), 475
Sugar-loaf Cape (St. Matthew Island), 580
Sugar-loaf Island (Chile), 135
Sugar-loaf Island (Strait of Magalhaens), 38
Sugar-loaf Mountain (Patagonia), 79
Suisun Bay (San Francisco), 336
Sukteteleu Bay (Jesso), 632
Sullivan, Point (Russian America), 449
Sulphur Island, or Yewo-sima (Japan), 641
Summer Isles (Patagonia), 81
Sunday, Cape (Tierra del Fuego), 44, 67
Supé (Peru), 171
Suquamish Harbour (Oregon), 383
Susan, Port (Oregon), 384
Susannah Cove (Strait of Magalhaens), 8
Sutersville (Upper California), 337
Svetchnikoff Harbour (Aleutian Islands), 512
Swaine, Cape (N.W. America), 425
Swallow Harbour (Strait of Magalhaens), 30
Swans, Bay of (Sea of Okhotsk), 615
Swans, Cape (Jesso Island), 630
Sweepstakes Foreland (Strait of Magalhaens), 8
Swoo-nada Channel (Japan), 640
Syclopish Rock (Oregon), 384
Sykes, Point (Russian America), 438
Symplegades Islets, or Tsurukase (Japan), 642

Tabla Point (Chile), 122
Table of Orozco (Tierra del Fuego), 45
Table Hill (Upper California), 338
Tables or Mesas of Narvaez (Lower California), 305
Taboga Island (Panamá Bay), 216
Taboguilla Island (Panamá Bay), 216
Tabon, Island of (Chiloe), 98
Tabou-sima (Japan), 637
Tacames or Atacames (Colombia), 204
Taco, H. B. Company's Establishment (Russian America), 453
Tagh-Kiniagh Island (Aliaska Peninsula), 498
Tagidack Island (Kodiack Islands), 494
Taglaokou Mountain (Siberia), 567
Tainotskoï, Cape (Sea of Okhotsk), 612
Tajumulco, Volcan de (Central America), 252
Talara Point (Peru), 132
Talcahuana Peninsula (Concepcion Bay), 110
Talcan Island (Chiloe), 95
Talinay, Mount (Chile), 123
Talofka River (Sea of Okhotsk), 613
Taltal Point (Bolivia), 141
Taltal Point (Chile), 136
Taluaptea, or Pillar Rock (Columbia River), 361
Tamar Cape and Harbour (Strait of Magalhaens), 34
Tamar Island (Strait of Magalhaens), 35
Tamarindo (Central America), 233
Tamary Peak (Jesso Island), 630
Tamatlan (Mexico), 276
Tamazula, River (Mexico), 297
Tambo, Valley of (Peru), 151
Tanaga Island (Aleutian Islands), 516
Tanagh-AngounakhIsland(AleutianIslands), 510
Tanega-sima Island (Japan), 641
Tanga, Cape (Japan), 636
Tanghinakh (Ounimack Island), 509
Tangolatangola, Island of (Mexico), 265
Tannedagh Island (Aleutian Islands), 518
Tano Island (Aleutian Islands), 510
Tanque, Playa de (Chile), 123
Taouinsk (Sea of Okhotsk), 613
Tareinski Harbour (Awatska Bay), 598, 601
Tarn Bay (Patagonia), 76, 83
Tarn, Mount (Strait of Magalhaens), 16
Tase's Canal (Vancouver Island), 409
Tatchik, or Golownin Bay (Russian America), 535
Tate, Cape (Tierra del Fuego), 66-7
Tatouche Islands (Oregon), 374
Taylor, Point (Strait of Magalhaens), 26
Taytaohaohuon or Taytao, Cape (Patagonia), 84
Tchagran Bay (Russian America), 530
Tchaitchi Islands (Aleutian Islands), 511
Tchaoukthous Tribe (Siberia), 561
Tchaplin, Cape (Siberia), 569
Tchastie Islands (Aleutian Islands), 515
Tchesmé, Cape, or Nomano-misaky (Japan), 642
Tchetchekouïoum Strait (Siberia), 569
Tchiarloun Strait (Siberia), 568
Tching-an, Cape (Siberia), 572
Tchinkitânay Bay, or Norfolk Sound (Sitka Islands), 466
Tchirikoff, Cape (Siberia), 572
Tchitschagoff, Cape (Japan), 640
Tchosan or Chosan Harbour (Corea), 643
TchougoulIsland (Aleutian Islands), 510, 515
Tchoukotskoï, Cape (Siberia), 569
Tchuktchi Nation (Siberia), 561
Teacapan, Boca de (Mexico), 287
Teatinos Point (Chile), 126
Tebenkoff Cove (Russian America), 533
Tecusitan Point (Mexico), 277
Tehuantepec, City of (Mexico), 259
Tehuantepec, Gulf of (Mexico), 258

Tehuantepec Road (Mexico), 260
Tejupan, Paps of (Mexico), 274, 284
Tekeenica Tribe (Tierra del Fuego), 2
Temblador Cove (Chile), 126
Temblenque Point (Guayaquil River), 190
Tempestad, or Stormy Channel (Strait of Magalhaens), 36
Tenkourguin Point (Siberia), 562
Tenola, Barra de (Mexico), 259
Tenoun Point (Chiloe), 93
Tenquehuen Island (Patagonia), 85
Tenquelil Island (Chiloe), 95
Tentsi, Gulf of (Japan), 638
Tenuy Point (Chiloe), 89
Tepic (Mexico), 282
Tequepa or Jequepa Point (Mexico), 273
Terron Point (Vancouver Island), 414
Terupa Rock (Panamá Bay), 216
Tessan Channel (Magdalena Bay), 306
Tetas de Cabra, Las (Mexico), 297
Teurire Island (Jesso), 632
Texada or Favida Island (Gulf of Georgia), 397
Thelupan or Tejupan, Paps of (Mexico), 274
Thieves Sound (Cockburn Channel), 19
Third Bay (Strait of Magalhaens), 36
Thomas Point (Vancouver Island), 408
Thomas, Point (Peru), 171
Three Brothers Rocks (Awatska Bay), 596
Three Brothers Hills (Tierra del Fuego), 45
Three Finger Island (Patagonia), 85
Three Island Bay (Strait of Magalhaens), 28
Three Points or Tres Puntas, Cape (Patagonia), 71
Thurlow Island (N.W. America), 402
Tiara, Mount (Saghalin), 620
Tiburon, Isle de (Gulf of California), 299
Tides, Chiloe, 99
Tides, Cockburn Channel, 19
Tides, Cape Virgins, 4
Tides, First Narrow (Strait of Magalhaens), 7
Tides, Second Narrow, &c. (Strait of Magalhaens), 9, 11
Tides, Strait of Le Maire, 50
Tides, Peru, 173
Tides, Port Famine, &c. (Strait of Magalhaens), 12
Tierra del Fuego, Observations on (Strait of Magalhaens), 39
Tierra del Fuego, Inhabitants of, 2
Tigalga or Kigalga Island (Aleutian Archipelago), 505
Tiger Island (Central America), 242
Tigil River (Sea of Okhotsk), 611
Tigilsk (Sea of Okhotsk), 612
Tilapa or Sintilapa River (Central America), 252, 257
Tilema Marsh (Mexico), 260
Tilesius Peak (Japan), 637
Tilgo Island (Chile), 126
Tingimack Island (Aleutian Archipelago), 509
Tipitapa or Panaloya River (Central America), 234
Tiquia Reef (Chiloe), 95
Tirua Cape (Chile), 106
Tisingal Gold Mines (Central America), 223
Todos Santos Lake (Chile), 99
Todos los Santos, Mission of (Lower California), 305
Todos los Santos Bay (Lower California), 312
Togalak Island (Aleutian Islands), 515
Togama (Japan), 637
Tohodomi Bay (Japanese Islands), 635
Tokutki Point (Oregon), 384
Toldo de la Nieve Peak (Andes), 170
Tolega Volcano (Central America), 226
Tolstoï Cape (Atkha Island), 514
Tolten Island (Chile), 105
Tomaco or Tumaco Port (Colombia), 205
Tomas or Santa Elena Bay (Central America), 230
Tom's Narrows (Strait of Magalhaens), 27
Touala, Plains of (Mexico), 258
Tongog Bay and Peninsula (Chile), 123
Tongue Point and Channel (Columbia River), 363
Tonin, Cape (Saghalin), 622
Tonkoy, Cape (Kodiack Island), 493
Tootomi or Tohodomi Bay (Japanese Islands), 635
Too-too-tut-na or Klamet River (Oregon), 354
Topacalma, Point and Shoal (Chile), 114
Toporkoff Islet (Sea of Behring), 584
Toporkowa Island (Kodiack Islands), 492
Toriwi-saki Point (Nippon), 633
Toro Point (Chile), 114
Toro Reef (Chile), 127
Torpoy or Tschirpoy Island (Kurile Islands), 608
Tortola and Tortolita Islands (Panamá Bay), 217
Tortolas Islets, 136
Tortoral de Lengua de Vaca (Chile), 123
Tortoralillo Bay (Chile), 134
Tortoralillo (Chile), 126
Tortuga, Isle de la (Gulf of California), 299
Tortuga Island (Gulf of California), 300
Tortuga Island (Peru), 174
Tortuga, Mount (Peru), 175
Tortuguitas Island (Gulf of California), 300
Tosa, Cape (Japan), 638
Tosco, Cape (Magdalena Bay), 306
Toshook Strait (Russian America), 538
Totoral, Point of (Chile), 130
Totten's Inlet (Puget Sound), 381
Toubakino-misaky, Cape (Japan), 638
Tougoura River and Bay (Sea of Okhotsk), 615
Toulikskoï (Oumnack Island), 509
Toulikskoï Volcano (Aleutian Archipelago), 509
Tower Bay (Mexico), 270
Tower Rock (Patagonia), 71
Tower Rocks (Tierra del Fuego), 65
Townsend, Port (De Fuca Strait), 379
Townshend Harbour (Tierra del Fuego), 63

Trading Bay (Cook's Inlet), 488
Traitors, Bay of (Kounashire Island), 610
Traitor's Cove (Russian America), 441
Transfiguration Bay (Siberia), 572
Tranque Island (Chiloe), 96
Transition Bay (Magdalen Sound), 19
Treble Island (Tierra del Fuego), 63
Trednoy Islet (Kurile Islands), 607
Trefusis Bay (Tierra del Fuego), 59
Tres Cruces Point (Chile), 114
Tres Marias Islands (Mexico), 277
Tres Montes, Cape (Patagonia), 79, 80
Tres Ojitos, Los (Gulf of California), 299
Tres Puntas or Three Points, Cape (Patagonia), 71
Triangulo Island (Vancouver Island), 409
Trinidad Cape (Gulf of California), 300
Trinidad, Gulf of (Patagonia), 72, 83
Trinidad Bay (Upper California), 346
Trinity, Cape and Isles (Kodiack Islands), 493
Triton Bank (Strait of Magalhaens), 7
Triunfo de los Libres, or Giquilisco Port (Central America), 244
Trollope, Point (Russian America), 439
Trollope's River (Queen Charlotte Island), 435
Trujillana Entrance (Peru), 158
Truxillo Bay (Strait of Magalhaens), 37
Truxillo (Peru), 176-7
Tschegoula Isles (Aleutian Islands), 518
Tschegoula Island (Aleutian Islands), 510
Tschiganock Island (Aleutian Islands), 510
Tschikotan or Spanberg Island (Kurile Islands), 609
Tschiniotskoy Bay (Kodiack Islands), 491
Tschirikoff's Island (Aliaska Peninsula), 497
Tschirikoff, Cape (Japan), 640
Tshirinkotan Island (Kurile Islands), 606
Tschirpoy or Torpoy Island (Kurile Islands), 608
Tschitchagoff, Cape (Russian America), 527
Tschitschagoff Bay (Alton Island), 519
Tschoka or Saghalin, 616
Tschougatschouk Bay (Cook's Inlet), 485
Tsougar Fouzi (Japan), 637
Tsugar or Sangar, Strait of (Japanese Archipelago), 633
Tsus-sima Islands (Japan), 643
Tubul River (Chile), 107
Tucapel Head (Chile), 107
Tucapel Point (Chile), 106
Tuesday Bay (Strait of Magalhaens), 37
Tujugiak River (Russian America), 529
Tumaco, Port of (Colombia), 205
Tumbes, Heights of (Chile), 109
Tumbes Bay (Guayaquil River), 190
Tumbes, Rio (Peru), 183
Tungo (Peru), 158
Tunquen Bight (Chile), 115
Tupica (Colombia), 207
Tupongati or Tupungato Peak (Andes), 102
Turn Point (Tierra del Fuego), 58
Turnagain Island (Cook's Inlet), 488
Turnagain River or Arm (Cook's Inlet), 489
Turner, Point (Russian America), 470
Turtle or San Bartholomew Bay (Lower California), 309
Tussac Rocks (Tierra del Fuego), 20
Two-headed Point (Kodiack Islands), 493
Tyke, Point (Peru), 149

Ugagouk or Agougack River (Aliaska Peninsula), 527
Uki-sima Island (Japan), 644
Ulak Island (Aleutian Islands), 517
Ulloa Peninsula (Strait of Magalhaens), 27
Umaya River (Mexico), 297
Umpqua River and Fort (Oregon), 356
Union, Port La (Central America), 241
Union Sound (Patagonia), 81
Upper California (Lower California), 313
Upright, Cape (St. Matthew Island), 579
Upright, Cape (Strait of Magalhaens), 34
Upright or Ildefonso Cape (Strait of Magalhaens), 35
Upwood, Point (N.W. America), 397
Urava Islands (Panamá Bay), 216
Urbitsh (Kurile Islands), 609
Uriarte, Port (Strait of Magalhaens), 36
Usborne, Mount (Peru), 172
Useless Cove (Patagonia), 83
Useless Bay (Strait of Magalhaens), 14
Ushishir Island (Kurile Islands), 607

Vaca, Cabeza de (Chile), 134
Vagares Rocks (Central America), 230
Vakhilskaïa, River (Kamtschatka), 594
Valcarcel or Eagle Bay (Strait of Magalhaens), 16
Váldes Island (N.W. America), 460
Valdez, Puerto de (Prince William's Sound), 479
Valdez, Port (Strait of Magalhaens), 14
Valdez, Port (Vancouver Island), 409
Valdivia (Chile), 104
Valdivia Hills (Chile), 103
Valentine Harbour (Strait of Magalhaens), 26, 35
Valentyn, Cape (Strait of Magalhaens), 14
Valentyn Bay (Tierra del Fuego), 51
Vallenar, Point (Russian America), 441
Vallenar Islands (Patagonia), 85
Valparaiso, 115
Van Isles (Patagonia), 73
Vancouver Island (N.W. America), 389
Vancouver Island (Patagonia), 82
Vancouver, Port (Staten Island), 47
Vancouver, Fort (Oregon), 368
Vancouver, Cape (Russian America), 531
Van der Lind, Cape (Kurile Islands), 608
Vandeput, Point (Russian America), 450
Vankarema River (Siberia), 562
Vascuñan Cape (Chile), 128
Vaspon's Island (Admiralty Inlet), 381
Vcevidovskoï Volcano and Islands (Aleutian Archipelago), 509
Venado (Central America), 228

Venados, De los, Island (Mexico), 287, 290
Ventanilla Point (Chile), 121
Ventosa or Tehuantepec Road (Mexico), 260
Venus, Point, Awatska Bay (Kamtschatka), 596
Venus Channel (Magdalena Bay), 306
Verkhotoursky or Little Karaghinsky Island (Siberia), 578
Vermejo, Mar, or Gulf of California (Mexico), 295
Vermilion Sea, or Gulf of California (Mexico), 295
Vernal Mountain, The (Magdalen Sound), 18, 25
Verugate (Central America), 228
Vicente, Point (Upper California), 321
Victor, Quebrador of (Peru), 147
Victoria, and Harbour (Vancouver Island), 417
Victory, Cape (Patagonia), 70
Victory Passage (Patagonia), 81
Vieja Island (Peru), 159
Villa de los Castillos, or Mazatlan (Mexico), 288
Villarica Mountain (Andes), 102
Villena Cove (Strait of Magalhaens), 30
Villeuchinski, Peak of (Kamtschatka), 595
Villiers, Point (Strait of Magalhaens), 28
Vinda Island (Peru), 174
Violin Island (Central America), 242
Viradores, North and South (Central America), 229
Virgenes or San Ramon Bay (Lower California), 312
Virgenes or San Quentin Cape (Lower California), 310
Virgenes, Cape de los (Gulf of California), 300
Virgin and Pearl Rocks (N.W. America), 423
Virgins, Cape (Patagonia), 3
Vogelborg Rock (Chile), 108
Volcan Nevado (Strait of Magalhaens), 14
Volcan Viejo (Peru), 151
Volcano Bay (Jesso Island), 630
Volcano or Mitake Island (Japanese Islands), 634
Vouten, Cape (Siberia), 570
Vrie's Island (Japanese Islands), 634

Wager Island (Patagonia) 75
Wager, Wreck of the (Patagonia), 75-6
Wainwright Inlet (Russian America), 550
Wakasa Bay (Japan), 636
Wakash Nation (Vancouver Island), 392
Wakefield Passage (Tierra del Fuego), 65
Wales, Point (N.W. America), 430
Walker Shoal (Strait of Magalhaens), 10
Walker's Island (Columbia River), 367
Walker's Cove (Russian America), 439
Wallis Shoal (Strait of Magalhaens), 5
Wallis's Mark (Strait of Magalhaens), 27
Walpole, Point (Russian America), 451
Walvis Boght (Jesso Island), 629
Walvis Bay (Kurile Islands), 609
Wapautoo or Wappatoo or Multonomah Island (Columbia River), 368
Warde, Point (Russian America), 443
Warp Bay (Cockburn Channel), 19
Warren's Island (Russian America), 447
Warrior Point and Branch (Columbia River), 368
Washington or Vancouver Island (N.W. America), 389
Washington or Queen Charlotte Island (N.W. America), 432
Wassilieff Bay (Aliaska Peninsula), 495
Waterfall, Port (Strait of Magalhaens), 15
Water Cove, Orange Bay (Tierra del Fuego), 57
Waterman Island (Tierra del Fuego), 63
Waxell, Cape (Sea of Behring), 584
Weather Point (Patagonia), 87
Webster (Upper California), 337
Wedge Island (Russian America), 442
Wedge-shaped Cape (Ounimack Island), 503
Week Islands (Tierra del Fuego), 66, 67
Welcome Bay (Patagonia), 81
Wellington Island (Patagonia), 72, 76, 80
Well's Passage (Vancouver Island), 408
Wessiloffsky, Cape (Ounalashka Island), 506
West Furies (Tierra del Fuego), 20
West Channel (Patagonia), 71
West Point (Strait of Magalhaens), 24
West Foreland (Cook's Inlet), 487
Westminster Hall Islands (Strait of Magalhaens), 37
Wet Island (Barbara Channel), 22
Whale Point and Sound (Strait of Magalhaens), 27
Whale's Back Sand (Callao), 168
Whaler's Harbour, or Sausalito Bay (San Francisco), 335
Whaley Point (Russian America), 39
Whidbey Island (Oregon), 385
Whit Sunday, Cape (Kodiack Islands), 494
White Horse Rock (Patagonia), 70
White Rock Point (Chile), 114
Wicananish Harbour (Vancouver Island), 414
Wide Channel (Patagonia), 73, 80
Wigwam Point (Strait of Magalhaens), 25
Wigwam or St. Martin's Cove (Tierra del Fuego), 55
William Point (Oregon), 387
William Head (Vancouver Island), 416
Willumette River (Oregon), 368
Wimbledon, Point (Russian America), 461
Wind, Cockburn and Barbara Channels, 23
Windham, Point (Russian America), 452
Winds, Cordonazos (Mexico), 257
Winds, Cordonazo de San Francisco (Mexico), 286
Winds, Gulf of California (Mexico), 287
Winds (Tierra del Fuego), 41
Winds, Tierra del Fuego (Strait of Magalhaens), 40
Winds, Papagayos, 229

Windward Bay (Patagonia), 73
Wingham Island (Russian America), 476
Witshed, Cape (Prince William's Sound), 477
Wodawara Bay (Japanese Archipelago), 633
Wodehouse, Point (Sitka Islands), 466, 468
Wolf Island (Vancouver Island), 416
Wolfe Rock, or Rosa Isle (Russian America), 448
Wollaston Island (Tierra del Fuego), 53
Wood Island (Tierra del Fuego), 59
Wood Cove (Strait of Magalhaens), 28
Wood's Bay (Strait of Magalhaens), 24
Woodcock, Point (Prince William's Sound), 482
Woodcock, Mount (Barbara Channel), 22
Wooden's Rock (Russian America), 454
Woody or Boisée Island (Kodiack Islands), 492
Woody Point (Vancouver Island), 409
Works Canal (N.W. America), 430
Woronzow, Point (Cook's Inlet), 488
Worsley Bay and Sound (Patagonia), 82
Wreck Point (Patagonia), 75
Wreck, Japanese (Kamtschatka), 603
Wukumoto (Japan), 640
Wunzendake Peak (Japan), 642
Würst, Cape (Saghalin), 620
Wyadda or Neeah Island (De Fuca Strait), 375

Xaultegua, Gulf of (Strait of Magalhaens), 32, 33
Xavier Island and Port (Patagonia), 77
Xicalapa River (Central America), 253
Xipixapa (Colombia), 206

Yacana-kunny Tribe (Tierra del Fuego), 2
Yakkoun, Cape (Siberia), 572
Yakouno-sima Island (Japan), 641
Yal, Point and Bay of (Chiloe), 95
Yalad Cove (Chiloe), 97
Yaniakinon Village (Siberia), 567
Yanikessery Island (Jesso), 632
Yamnas Cove (Chile), 107
Yaqui, Rio (Gulf of California), 297
Yatsou-saki (Japan), 637
Yca (Peru), 158
Yca River (Peru), 158
Yemafar, Cape (Japan), 638
Yematso (Japan), 636
Yerba Buena or San Francisco (Upper California), 336
Yerba Buena Island (San Francisco), 335
Yerghin Point (Siberia), 568
Yetsiou Cape (Japan), 637
Yewo-sima or Sulphur Island (Japan), 641
Ykolik, Cape (Kodiack Islands), 494
Ylas del Rey (Panamá Bay), 212
Yliniza Peak (Andes), 187
Ylo (Peru), 150
Ymbisible, Point (Queen Charlotte Island), 435
Ymerquiña Island, 96
Ynche-mo Island (Patagonia), 85
Yndependencia Bay (Peru), 158
Ynfiernillo Rock (Peru), 158
Yngles, Port (Chile), 134
Yngles Bank (Chiloe), 90
Yoco, San Pedro de (Peru), 179
York, Cape (Russian America), 538
York Minster (Tierra del Fuego), 62
Youchin, Cape (Sea of Behring), 584
Younaska Island (Aleutian Islands), 510
Young, Point (Russian America), 454
Young's Bay and River (Columbia River), 361
Youp-nut River (Russian America), 538
Ypun or Narborough Island (Patagonia), 85-6
Ystapa, Morro de (Mexico), 260

Zacatula or Sacatula (Mexico), 274
Zach Peninsula (Patagonia), 81
Zadan, Cerro de (Mexico), 265, 268
Zakharovskaïa Bay (Aliaska Peninsula), 498
Zapallar Point (Chile), 122
Zapote Island (Nicaragua Lake), 233
Zapotera Island (Nicaragua Lake), 233
Zebadea Mountains (Central America), 232
Zeballos Arm (Vancouver Island), 410
Zeven Zusters, The (Japan), 641
Zedzones or Ladrone Islands (Central America), 222
Zipegua, Punta de (Mexico), 260
Zipegua or Machaguista Rocks (Mexico), 261
Zonzonate or Acajutla (Central America), 247
Zuniga, Point (Lower California), 310
Zuniga, Barros de (Upper California), 317

London: J. & W. Rider, Printers, 14, Bartholomew Close

www.ingramcontent.com/pod-product-compliance
Lightning Source LLC
LaVergne TN
LVHW012209040326
832903LV00003B/204
9781016341288